T0348481

Handbook of GAME THEORY

VOLUME 4

Handbook of GAME THEORY

VOLUME 4

Edited by

H. Peyton Young
Department of Economics,
University of Oxford, Oxford, UK

Shmuel Zamir
Department of Economics,
University of Exeter Business School, UK;
The Center for the Study of Rationality,
The Hebrew University of Jerusalem, Israel

Amsterdam • Boston • Heidelberg • London • New York • Oxford
Paris • San Diego • San Francisco • Singapore • Sydney • Tokyo

North-Holland is an imprint of Elsevier
Radarweg 29, PO Box 211, 1000 AE Amsterdam, The Netherlands
The Boulevard, Langford Lane, Kidlington, Oxford OX5 1GB, UK

British Library Cataloguing-in-Publication Data
A catalogue record for this book is available from the British Library

Library of Congress Cataloging-in-Publication Data
A catalog record for this book is available from the Library of Congress

ISBN: 978-0-444-53766-9

For information on all North-Holland publications
visit our website at store.elsevier.com

Typeset by SPi Global, India

CONTENTS

CONTRIBUTORS

Ken Binmore
Philosophy Department, Bristol University, Bristol, UK

Colin F. Camerer
California Institute of Technology, Pasadena, CA, USA

Eddie Dekel
Eitan Berglas School of Economics, Tel Aviv University, Tel Aviv, Israel
Economics Department, Northwestern University, Evanston, IL, USA

Aviezri S. Fraenkel
Department of Computer Science and Applied Mathematics, Weizmann Institute of Science, Rehovot, Israel

Peter Hammerstein
Institute for Theoretical Biology, Humboldt University, Berlin, Germany

Robert A. Hearn
H3 Labs LLC, Palo Alto, CA, USA

Teck-Hua Ho
University of California, Berkeley, Berkeley, CA, USA

Matthew O. Jackson
Department of Economics, Stanford University, Stanford, CA, USA
The Santa Fe Institute, Santa Fe, NM, USA
CIFAR, Toronto, Ontario, Canada

Todd R. Kaplan
Department of Economics, University of Exeter Business School, Exeter, UK
Department of Economics, University of Haifa, Haifa, Israel

Edi Karni
Department of Economics, Johns Hopkins University, Baltimore, MD, USA
The Warwick Business School, Warwick University, Coventry, UK

Rida Laraki
CNRS in LAMSADE (Université Paris-Dauphine), France
Economics Department, Ecole Polytechnique, France

Olof Leimar
Department of Zoology, Stockholm University, Stockholm, Sweden

Fabio Maccheroni
Department of Decision Sciences and IGIER, Università Bocconi, Milano, Italy

George J. Mailath
Department of Economics, University of Pennsylvania, Philadelphia, PA, USA

Jason R. Marden
Department of Electrical, Computer and Energy Engineering, Boulder, CO, USA

Massimo Marinacci
Department of Decision Sciences and IGIER, Università Bocconi, Milano, Italy

Noam Nisan
Microsoft Research, Hebrew University of Jerusalam, Jerusalam, Israel

Wojciech Olszewski
Department of Economics, Northwestern University, Evanston, IL, USA

Christos Papadimitriou
University of California, Berkeley, CA, USA

Debraj Ray
Department of Economics, New York University, New York, NY, USA

Larry Samuelson
Department of Economics, Yale University, New Haven, CT, USA

William H. Sandholm
Department of Economics, University of Wisconsin, Madison, WI, USA

Jeff S. Shamma
School of Electrical and Computer Engineering, Georgia Institute of Technology, Atlanta, GA, USA

Aaron N. Siegel
Twitter, San Francisco, CA, USA

Marciano Siniscalchi
Economics Department, Northwestern University, Evanston, IL, USA

Sylvain Sorin
Mathematics, CNRS, IMJ-PRG, UMR 7586, Sorbonne Universités, UPMC Univ Paris 06, Univ Paris Diderot, Sorbonne Paris Cité, Paris, France

Rajiv Vohra
Department of Economics, Brown University, Providence, RI, USA

Rakesh V. Vohra
Department of Economics, University of Pennsylvania, Philadelphia, PA, USA
Department of Electrical and Systems Engineering, University of Pennsylvania, Philadelphia, PA, USA

Chris Wallace
Department of Economics, University of Leicester, Leicester, UK

H. Peyton Young
Department of Economics, University of Oxford, Oxford, UK

Shmuel Zamir
Department of Economics, University of Exeter Business School, Exeter, UK
The Center for the Study of Rationality, The Hebrew University of Jerusalem, Israel

Yves Zenou
Department of Economics, Stockholm University, Stockholm, Sweden
Research Institute of Industrial Economics (IFN), Stockholm, Sweden

PREFACE

Over a decade has passed since the publication of the previous volume in this series. In the interim game theory has progressed at a remarkable rate. Topics that were merely in their infancy have become well-established; others that had reached maturity have expanded so much that they warrant a fresh treatment. The former category includes games on networks (chapter 3), evolutionary game theory (chapters 6 and 13), nonstandard utility theory (chapter 17), combinatorial games (chapter 15), and combinatorial auctions (chapter 8). None of these subjects was covered in the earlier volumes. The latter category includes advances in the theory of zero-sum games (chapter 2), reputations in repeated games (chapter 4), cooperative games and coalition formation (chapter 5), auctions (chapter 7), experiments (chapter 10), and epistemic game theory (chapter 12). The central role that rationality plays in game theory is considered anew in chapter 1.

There have been several other important developments. One is the blurring of lines between cooperative and noncooperative game theory. For example, games played on networks, the dynamics of coalition formation, and applications of game theory to computer science use a combination of noncooperative and cooperative techniques. Another important development is the rapid expansion of game theory into fields outside of economics. These include computer science (chapters 9 and 14), biology (chapter 11), distributed control (chapter 16), and expert testing in statistics (chapter 18). Game theory has also found an increasing number of applications, such as the design of matching markets, auctions, utility markets, internet protocols, and many others. For this reason the present volume bears the title *The Handbook of Game Theory*, in contrast to the earlier three volumes, which were called *Handbook of Game Theory with Economic Applications.*

The chapters are intended to provide authoritative introductions to the current literature at a level accessible to students and nonspecialists. Contributors were urged to provide their own perspective as opposed to an encyclopedic review of the literature. We also urged them to draw attention to unsolved problems and alternative perspectives. The result is a set of chapters that are more diverse in style than is typical for review articles. We hope that they succeed in drawing readers into the subject and encourage them to extend the frontiers of this extraordinarily vibrant field of research.

ACKNOWLEDGMENTS

We wish to thank Scott Bentley, Joslyn Chaiprasert-Paguio, Pauline Wilkinson, and the production team at Elsevier for bringing this volume to fruition. We are particularly indebted to Romina Goldman, who provided us with invaluable assistance throughout the process and helped keep the trains running on time.

H. Peyton Young
Shmuel Zamir

INTRODUCTION TO THE SERIES

The aim of the Handbooks in Economics series is to produce handbooks for various branches of economics, each of which is a definitive source, reference, and teaching supplement for use by professional researchers and advanced graduate students. Each handbook provides self-contained surveys of the current state of a branch of economics in the form of chapters prepared by leading specialists on various aspects of this branch of economics. These surveys summarize not only received results but also newer developments, from recent journal articles and discussion papers. Some original material is also included, but the main goal is to provide comprehensive and accessible surveys. The handbooks are intended to provide not only useful reference volumes for professional collections but also possible supplementary readings for advanced courses for graduate students in economics.

Kenneth J. Arrow
Michael D. Intriligator

CHAPTER 1

Rationality

Ken Binmore*
*Philosophy Department, Bristol University, Bristol BS8 1TB, UK

Contents

Handbook of game theory
http://dx.doi.org/10.1016/B978-0-12-420056-2.00001-X

Abstract

This chapter is a very compressed review of the neoclassical orthodoxy on the nature of rationality on economic theory. It defends the orthodoxy both against the behavioral criticism that it assumes too much and the revisionist view that it assumes too little. In places, especially on the subject of Bayesianism, the paper criticizes current practice on the grounds that it has gone astray in departing from the principles of its founding fathers. Elsewhere, especially on the modeling of knowledge, some new proposals are made. In particular, it is argued that interpreting counterfactuals is not part of the function of a definition of rationality.

Keywords: Substantive rationality, Neoclassical economics, Revealed preference, Risk, Uncertainty, Bayesianism, Small worlds, Common prior, Common knowledge, Nash equilibrium, Equilibrium selection, Refinement theory, Nash program

JEL Codes: B41, C70

1.1. NEOCLASSICAL RATIONALITY

The *Oxford Handbook of Rationality* (Mele and Rawling, 2003) contains 22 essays with at least 22 opinions on how rationality can or should be construed. None of these views come anywhere near the strict neoclassical orthodoxy to which this article is devoted—even when the authors imagine they are discussing rationality as understood in economics.

Would the authors have done better to abandon their various rationality models in favor of the neoclasssical orthodoxy? I think it would be a bad mistake to do so in the metaphysical belief that the neoclasssical orthodoxy is the one and only true model of rationality. The best that one can reasonably hope for in a model is that it should work well in a properly defined context. On this criterion, the neoclasssical orthodoxy continues to be hugely successful when applied in the kind of economic context for which it was created. However, this article argues that the economics profession is suffering from all kinds of indigestion as a result of being too ready to swallow various attempts to extend the orthodoxy to contexts for which it is not suited. For such purposes, we need new models of rationality—not to replace the current orthodoxy—but to take over in contexts for which the current orthodoxy is ill-adapted. In brief, one may not like the 22 rationality models of the Oxford handbook any more than I do, but there is no reason to think that their authors are wasting their time in contemplating alternatives to the neoclassical orthodoxy.

1.1.1 Substantive or procedural?

One limitation of the neoclassical orthodoxy is that it is a theory of substantive rationality rather than procedural or algorithmic rationality (Simon, 1976). That is to say, the orthodoxy is concerned only with *what* decisions are made and not with *how* they are made. But what hope is there of predicting what errors economic agents are likely to make in practice without some theory of how their minds work?

For this reason, some behavioral economists advocate abandoning the neoclassical orthodoxy in favor of empirically based theories of bounded rationality. Nobody denies that scientific theories of bounded or algorithmic rationality are much to be desired, but even if such theories became more firmly established than at present, the example of evolutionary biology explains why it does not follow that it would be wise to abandon the neoclassical orthodoxy.

1.1.1.1 Evolution

Evolutionary biology is perhaps the application in which game theory has had its greatest successes, sometimes predicting the behavior of animals in simple situations with considerable accuracy. But none of these successes have been achieved by tracking what goes on inside the heads of the players of biological games. Biologists speak of *proximate* causes when they explain behavior in terms of enzymes catalyzing chemical reactions or electrical impulses propagating through neurons. However, such proximate chains of causation are usually leapfrogged in favor of *ultimate* causes, which assume that evolution will somehow create internal processes that lead the players to act *as though* seeking to maximize their expected Darwinian fitness. The end-products of impossibly complicated evolutionary processes are thereby predicted by assuming that they will result in animals playing their games of life in a rational way.

Education, imitation, and individual learning can play the same role in economics as evolution plays in biology (Alchian, 1950, and many others). In fact, when one passes to continuous time, the simplest models of imitation and individual learning lead to the same replicator dynamics that biologists use to model biological evolution. When any monotonic dynamics like the replicator dynamics converge, they necessarily converge on a Nash equilibrium of whatever game is being played (Samuelson and Zhang, 1992). Even if the boundedly rational decision procedures of bankers or stockbrokers remain forever cloaked in mystery, substantive theories of rationality will therefore nevertheless be useful in predicting their behavior in situations to which their professions have had time to adapt.

Advocates of theories of procedural rationality are right to complain that substantive theories are often applied to agents who have had no chance to adapt to the situations in which they find themselves, but substantive theories like the neoclassical orthodoxy will always be worth retaining if only because they provide a simple means of describing the limiting outcome of convergent economic processes. However, seeing substantive rationality in such a light imposes restrictions on the extent to which it is useful to elaborate the theory. Philosophers will always be interested in how hyper-rational supermen might behave, but economists who want to use the theory to predict the behavior of real agents need to bear in mind that it will usually only be useful to the extent that it is capable of approximating actual behavior that has been refined by experience.

1.1.2 Rationality as consistency

Hobbes (1986) said that one can characterize a man in terms of his strength of body, his passions, his experience, and his reason. In modern economics, an agent's strength of body becomes his feasible set—the set of actions it is possible for him to take. His passions become his preferences. His experience is summarized by his beliefs. His reason becomes the set of rationality principles that guide his choice of an optimal action from his feasible set, given his preferences over the possible consequences and his beliefs about matters over which he has no control.

Hume (1739) was the first to tie down the implications for rationality within such a formulation: "Reason is, and ought only to be, the slave of the passions." In this quote, Hume denies that reason can tell us what we ought to want. Contrary to Immanuel Kant and numerous philosophers before and after, Hume argues that rationality should be about means rather than ends. As he extravagantly put it: "Tis not contrary to reason to prefer the destruction of the whole world to the scratching of my finger . . . Tis as little contrary to reason to prefer even my own acknowledge'd lesser good to my greater."

Following Samuelson (1947) and others, the neoclassical orthodoxy develops Hume's doctrine into a theory that reduces rationality to *nothing more* than consistency. In particular, it imposes no restriction at all on an agent's utility function.

The orthodoxy is therefore totally at odds with the popular misconception of economic theory, according to which agents are supposedly always modeled as being totally selfish. Some behavioral economists (Henrich et al., 2004) encourage this prejudice by inventing a "selfishness axiom" to which neoclassical economists are said to subscribe. But when the same behavioral economists fit unselfish utility functions to laboratory data, they are not stepping outside the neoclassical paradigm. On the contrary, in assuming that human behavior can be successfully modeled in terms of maximizing utility functions, they take for granted that their subjects make choices in the consistent fashion that is the neoclassical organizing principle.

To challenge the neoclassical paradigm in the laboratory, it is necessary to produce evidence of *inconsistent* decision behavior—which isn't hard to do in certain contexts, even with experienced and well-paid subjects. We are much in need of experimental work that pursues this inconsistency issue in a systematic way, rather than continually "refuting" the neoclassical orthodoxy by giving examples in which inexperienced subjects fail to maximize monetary rewards in the laboratory. We need instead to work harder to find the boundaries that separate those contexts in which the orthodoxy works reasonably well from those in which it doesn't.

1.1.3 Positive or normative?

Is the neoclassical orthodoxy a descriptive or a prescriptive theory? Most economists would probably reply that it is a prescriptive theory that succeeds in being descriptive to the extent that agents make their decisions well, but I think this characterization is too quick. Everybody would doubtlessly agree that the decisions made by an ideally rational agent should be consistent with each other, but are we entitled to assume that the ideal of consistency on which the neoclassical orthodoxy is based should take precedence over everything else?

Physicists laugh at the idea that they should value consistency between their models above the accuracy that each model has in its own domain of application[1]—and why should matters be any different for us? Appealing to a substantive theory based on consistency assumptions without being able to explain how the necessary consistency requirements came to be satisfied is rather like taking for granted that we already have the equivalent of the theory-of-everything that physicists regard as their Holy Grail. However, more on this subject will have to await a discussion of the foundations of Bayesian decision theory (Section 4.2).

1.2. REVEALED PREFERENCE

Although it is perhaps more honored in the breach than the observance, the official orthodoxy of neoclassical economics is Paul Samuelson's theory of revealed preference.[2] Leaving any speculation about what goes on in people's heads to psychologists, the theory assumes that we already know what an agent chose (or would choose) in some situations, and uses this data to deduce what they will choose in other situations. For this purpose, we need to assume that the agent's choices in these different situations are consistent with each other.

Samuelson's (1947) own consistency assumptions are set in a market context, but the orthodoxy follows Arrow (1959), Sen (1971), and many others in working within an abstract setting. In the absence of information about prices and budgets, the theory then has much less bite, but is correspondingly simpler. The next section is a brief reminder of how the theory goes in a simplified case.[3]

[1] Quantum theory and relativity are the leading examples in physics of theories that are inconsistent where they overlap.

[2] I would prefer to say "attributed preference" so as not to give the impression that the theory necessarily relies on agents consciously maximizing utility functions in their heads. For example, Samuelson's Weak Axiom of Revealed Preference apparently works better for pigeons than for people (Kagel et al., 1995).

[3] Kreps (1988) provides a full but unfussy account.

1.2.1 Independence of irrelevant alternatives

A decision problem for an agent to be called Pandora can be characterized as a function $D: A \times B \to C$, where A is her set of feasible actions, B is a set of possible states of the world over which she has no control, and C is the set of possible consequences or outcomes. Beliefs about the current state of the world are trivialized in this section by assuming that B contains only one state. Nothing much is then lost in identifying actions and outcomes.

Suppose it is known or assumed that Pandora will always choose a single outcome $\gamma(S)$ from any subset S of C.[4] The orthodoxy deems these choices to be consistent if the Independence of Irrelevant Alternatives is satisfied:

If B is ever chosen when A is feasible,
then A is never chosen when B is feasible.

1.2.1.1 Aesop

Many of Aesop's fables illustrate irrationalities disbarred by the neoclassical orthodoxy. The story of the fox and the grapes illustrates a case in which it does not make sense to allow what is feasible in the set A to influence one's assessment of the value of the outcomes in the set C (or one's beliefs about the states in the set B). This obviously matters a lot to the neoclassical orthodoxy because the final theory firmly separates what goes on in the three sets A, B, and C. But this cannot be taken for granted. It is sometimes not understood that it is often necessary to *reformulate* a decision problem so that A, B, and C are properly separated when applying the neoclassical orthodoxy to a decision problem.

The example favored by textbooks is that a person is likely to feel differently about umbrellas and ice-creams in sunny or rainy weather. The standard response is that umbrellas and ice-creams should not be treated as raw consequences in the set C. One should instead substitute the four states of mind associated with having an umbrella or an ice-cream in sunny or rainy weather.

Sen (1993) offers another example that is directly relevant to the Independence of Irrelevant Alternatives. A respectable lady is inclined to accept an invitation to take tea but changes her mind when told that she will also have the opportunity to snort cocaine. She would never choose to snort cocaine, but it is not an irrelevant alternative because its appearance in her feasible set changes her opinion about the kind of tea party to which she has been invited. Sen apparently thinks that such examples undermine the neoclassical orthodoxy, but my own view is that one might as well seek to refute arithmetic by pointing out that $1 + 1 = 1$ for raindrops. The neoclassical orthodoxy can only be expected to apply to a particular decision problem when the problem has been properly formulated.

[4] If $\gamma(S)$ is allowed to be a set with more than one element so that indifference becomes possible, a slightly more elaborate theory is required (Binmore, 2009, p. 11).

Some critics complain that the need to choose carefully among all the many ways of formulating a decision problem before applying the neoclassical orthodoxy reduces the theory to the status of a tautology. It is true that the consistency requirement can be rendered empty by including enough parameters in a formulation, but this is a criticism of bad modeling technique rather than of the theory itself.

1.2.1.2 Utility

It follows from the Independence of Irrelevant Alternatives that the revealed preference relation $\prec$ defined by $A \prec B$ if and only if $B = \gamma\{A, B\}$ is transitive. If C is a finite set, it follows in turn that we can construct an infinity of utility functions $u : C \to \mathbb{R}$ that represent Pandora's choice behavior in the sense that

$$x = \gamma(S) \text{ if and only if } u(x) = \max_{s \in S} u(s)\,. \qquad [1.1]$$

The function u is said to be an ordinal utility function because any strictly increasing function of u would serve equally well to describe Pandora's choices.

1.2.1.3 Causal utility fallacy

Most philosophers think that economists still hold fast to the utilitarian doctrine that human choice behavior is caused by an internal need or compulsion to maximize some correlate of happiness. Such a proximate theory of human psychology provides one possible explanation of why human choice behavior might be consistent, but the neoclassical orthodoxy does not endorse the reverse implication. To claim otherwise is sometimes called the causal utility fallacy.

For example, the success of evolutionary game theory in biology is based on the assumption that animals consistently behave as though seeking to maximize their expected fitness in environments to which they are adapted, but nobody thinks that animals are wired up with meters that register how many extra children each available action is likely to generate. The same goes for human behavior to which utility functions can be fitted by choosing the available parameters carefully. It then follows that the behavior is consistent in the neoclassical sense, but not that it is *caused* by subjects—consciously or unconsciously—maximizing the fitted utility function inside their heads.

1.2.2 Revealed preference in game theory

Von Neumann and Morgenstern (1944) saw the object of game theory as being to deduce how people will behave in games from how they behave in one-person decision problems. To make this approach work in the Prisoners' Dilemma, one can ask Alice two questions. Would you cooperate if Bob is sure to cooperate? Would you cooperate if Bob is sure to defect? If the answer is always *no* in both cases, then the neoclassical orthodoxy tells us to write smaller utilities in the row of Alice's payoff matrix that corresponds to cooperating than in the row that corresponds to defecting. Defecting then strictly dominates cooperating. If the answer is sometimes *yes*, then the game being played is not an example of the Prisoners' Dilemma.

This trivial argument generates much heat. Some critics complain that it makes the claim that it is rational to defect in the Prisoners' Dilemma into a tautology. Others argue that how Alice would reply to various choices Bob might make is only hypothetical. To both of which I say: so what? More to the point are critics who want the hypothesized data to be derived only from the play of games like the game to be studied, to which I say: why tie our hands behind our backs in this way?

1.3. DECISIONS UNDER RISK

When the set B in a decision problem $D: A \times B \to C$ contains more than one state, we need a theory of revealed or attributed belief to stand alongside the theory of revealed preference. The beginnings of such a theory actually predate the theory of revealed preference, going back at least as far as Ramsey (1931).

When objective probabilities or frequencies can be assigned to the states of world in B, Pandora is said to make her decisions under risk. When no objective probabilities are on the table, she is said to choose under uncertainty.

1.3.1 VN&M utility functions

In risky situations, the set C of consequences can be expanded to the set $\text{lott}(C)$ of lotteries with prizes in C. Each such lottery specifies a probability $p(x)$ with which each consequence x occurs when the lottery is implemented. As in the case without risk, we suppose that Pandora's choices from each subset of $\text{lott}(C)$ are given. If the Independence of Irrelevant Alternatives is satisfied, Pandora chooses as though she were seeking to optimize relative to a transitive preference relation $\prec$ defined on $\text{lott}(C)$.

Von Neumann and Morgenstern (VN&M) famously proposed a further set of consistency assumptions for decisions made in risky situations that are too well known to be repeated here. They imply the existence of a function $u: C \to \mathbb{R}$, whose expected value $\mathcal{E}u$ is a utility function for $\prec$ (Binmore, 2009, p. 41). Equation (1) is then satisfied with C replaced by $\text{lott}(C)$ and u by $\mathcal{E}u$.

The utility function $\mathcal{E}u$ is ordinal on $\text{lott}(C)$, but the VN&M utility function u is cardinal on C, which means that any other VN&M utility function that represents $\prec$ has the form $Au + B$, where $A > 0$ and B are real constants.

1.3.1.1 Attitudes to risk

If Pandora's VN&M utility function on some commodity space is concave, she is said to be risk averse. If it is convex, she is risk loving. All that needs to be said here is that it is a mistake to deduce that risk-loving people like gambling and risk-averse people do not. If people actually liked or disliked the act of gambling itself, then VN&M's compound lottery postulate would not apply (Binmore, 2009, p. 54). After all, it wouldn't be much fun to go to Las Vegas and slap all your money down at once on a single number at the Roulette table.

1.3.1.2 Unbounded utility?

All kinds of paradoxes become possible if VN&M utility functions are allowed to be unbounded (Binmore, 2009, p. 55). For example, suppose the devil offers Pandora a choice between two identical sealed envelopes, one of which is known to contain twice as much money as the other. Pandora chooses one of the envelopes and finds that it contains $\$ 2n$. So the other envelope contains either $\$ n$ or $\$ 4n$. Pandora computes its expected dollar value to be

$$\tfrac{1}{2} \times n + \tfrac{1}{2} \times 4n = 5n/2 > 2n .$$

If she is risk neutral, Pandora will therefore always want to swap whatever envelope she chose for the other envelope.

Some philosophers believe that such "paradoxes of the infinite" have metaphysical significance, but the consensus in economics is to eliminate them from consideration by disbarring unbounded VN&M utility functions where it matters. But this proviso doesn't prevent our assuming, for example, that Pandora's VN&M utility is linear in money as long as we only consider bounded sums of money.

1.3.1.3 Utilitarianism

Luce and Raiffa's (1957) famous *Games and Decisions* lists various fallacies associated with VN&M utilities, of which two remain relevant in social choice theory because of a continuing controversy over Harsanyi's (1977) use of VN&M utilities in defending utilitarianism (Binmore, 2009; Weymark, 1991).

Luce and Raiffa's first fallacy is the claim that VN&M's consistency postulates imply that the intensity with which Pandora prefers x to y is measured by the difference $u(x) - u(y)$ (Binmore, 2009, p. 66). It is true that nothing in the VN&M postulates allows any deductions to be made about intensities of preference, because no assumptions about such intensities are built into the postulates. However, nothing says that we can't add additional assumptions about intensities and follow up the implications.

In fact, VN&M Neumann and Morgenstern (1944, p.18) themselves offer a simple definition of intensity under which differences of VN&M utilities can reasonably be said to measure the intensity of Pandora's preferences. If Pandora reveals the preferences $a \prec b$ and $c \prec d$, their definition deems her to hold the first preference more intensely than the second if and only if she would always be willing to swap a lottery in which the prizes a and d each occur with probability $\frac{1}{2}$ for a lottery in which the prizes b and c each occur with probability $\frac{1}{2}$.

Luce and Raiffa's second fallacy is that one can compare the VN&M utilities of different people. Denying that interpersonal comparison is possible has a long history in economics (Robbins, 1938), but VN&M Neumann and Morgenstern (1944) clearly did not share this view, since the part of their book devoted to cooperative game theory takes for granted that utilities can be compared across individuals. However, as in the case of intensity of preference, extra assumptions need to be added to the VN&M postulates for this purpose. My own view is that Harsanyi's (1977) approach to the problem works perfectly well, but the immediate point is that there is nothing intrinsically nonsensical about comparing VN&M utilities (Binmore, 2009, p. 67).[5]

1.4. BAYESIAN DECISION THEORY

Philosophers distinguish various types of probability other than the frequencies considered in the previous section (Gillies, 2000). In particular, they distinguish subjective or betting probabilities from logical or epistemic probabilities (also called credences or degrees of belief). If an adequate theory of logical probabilities existed, it would quantify the extent to which the available evidence supports any given proposition. Bayesian decision theory is sometimes taken to be such a theory, but its probabilities are avowedly subjective in character. Its roots are to be found in the work of Ramsey (1931) and de Finetti (1937), but it was brought to fruition by Savage (1954) in his famous *Foundations of Statistics.*

1.4.1 Savage's theory

Savage replaces the set lott(C) of lotteries in the VN&M theory by a set gamble(C) of gambles. Instead of assigning objective probabilities to the outcomes in the set C, a gamble assigns an event $E(x)$ from a partition of the belief space B to each x in C. If the Independence of Irrelevant Alternatives is satisfied, Pandora then chooses as though she were seeking to optimize relative to a transitive preference relation $\prec$ defined on gamble(C).

Savage then proposes yet more consistency assumptions[6] for decisions made under uncertainty that are again too well known to be repeated here. It is common to summarize the assumptions by saying that they ensure that Pandora is not vulnerable to a Dutch book.[7]

[5] Although I do not think a similar justification of VN&M's notion of transferable utility is possible except in the case when the players' utility functions are linear in money.

[6] Anscombe and Aumann (1963) greatly simplify Savage's treatment by regarding his theory as an expansion of the VN&M theory. One might say that they replace gamble(C) by gamble(lott(C)).

[7] A Dutch book is a system of bets which ensure that you must lose if forced to take one side or the other in each bet.

Savage's assumptions imply the existence of a VN&M utility function $u : C \to \mathbb{R}$ and a subjective probability function $p : B \to [0, 1]$ with the property that the expected value $\mathcal{E}u$ taken relative to p is a utility function for $\prec$ (Binmore, 2009, p. 117). Equation (1) is then satisfied with C replaced by gamble (C) and u by $\mathcal{E}u$.

1.4.1.1 *Bayes' rule*

Savage's theory seems to have become known as Bayesian decision theory because classical statisticians like Ronald Fisher sought to make fun of it by saying that it was all about Bayes' rule and nothing else. Bayes' rule is inescapable when working with objective probabilities or frequencies, but Savage needed a new argument to justify the rule for the subjective probabilities of his theory. A suitable argument is not very difficult (Binmore, 2005, p. 126), but it is important to register the fact that an argument is necessary because it is commonly taken for granted that anything called a probability automatically satisfies Bayes' rule without the need for any justification at all. Things then go badly wrong when Savage's subjective probabilities are reinterpreted as logical probabilities, and his theory taken to be a solution to the deep and difficult problem of scientific induction—as in the kind of naive Bayesianism to be considered below.

1.4.2 Small worlds

When is it rational to be consistent in the sense required to justify Bayesian decision theory? Here I advocate diverging from what has become the orthodoxy among modern neoclassical economists and returning to Savage's original ideas.

It certainly cannot always be rational to be consistent, otherwise scientists would not inconsistently abandon old theories in favor of new theories when the data refutes a premise they have been taking for granted. Savage therefore restricted the application of his theory to what he called *small worlds* in which it makes sense to insist on consistency. He describes the idea that one can sensibly use his theory in any world whatever as "utterly ridiculous" and "preposterous" (Savage, 1954, p.16).

According to Savage, a small world is one within which it is always possible to "look before you leap." Pandora can then take account *in advance* of the impact that all conceivable future pieces of information might have on the underlying model that determines her subjective beliefs. Any mistakes built into her original model that might be revealed in the future will then *already* have been corrected, so that no possibility remains of any unpleasant surprises.

In a large world—like the worlds of macroeconomics or finance—the possibility of an unpleasant surprise that reveals some consideration overlooked in Pandora's original model cannot be discounted. As Savage puts it, in a large world, Pandora can only "cross certain bridges when they are reached." Knee-jerk consistency is then no virtue.

Someone who insists on acting consistently come what may is just someone who obstinately refuses to admit the possibility of error.

1.4.2.1 Bayesianism?

Savage's subjective probabilities are widely misinterpreted as logical probabilities. Bayesianism develops this mistake into the creed that the problem of scientific induction is solved. Rationality is said somehow to endow us with prior probabilities. One then simply uses Bayes' rule to update these prior probabilities as new information becomes available.

What would Savage's own reaction have been to this development? His response when confronted with Allais' paradox illustrates his attitude. When Allais pointed out an inconsistency in his choices, Savage recognized that he had behaved irrationally, and modified his choices accordingly. Luce and Raiffa (1957, p. 302) summarize his general approach as follows:

> *Once confronted with inconsistencies, one should, so the argument goes, modify one's initial decisions so as to be consistent. Let us assume that this jockeying—making snap judgments, checking up on their consistency, modifying them, again checking on consistency etc—leads ultimately to a bona fide, prior distribution.*

1.4.2.2 Where do Savage's priors come from?

To enlarge on Luce and Raiffa's story, imagine that Pandora contemplates a number of possible future histories. Bayesianism would have Pandora look within herself for a prior before receiving any information, but Savage would argue that Pandora would do better to consult her gut feelings with more information rather than less. So she should imagine what subjective probabilities would seem reasonable *after* experiencing each possible future history. The resulting posteriors are unlikely to be consistent. She should then use her confidence in the various judgments built into her posteriors to massage them until they become consistent. But consistent posteriors can be deduced by Bayes' rule from a prior. Instead of deducing her posterior probabilities from prior probabilities chosen on *a priori* grounds, Pandora therefore constructs her prior by massaging the unmassaged posterior probabilities with which her analysis begins—thereby reversing the procedure that naive Bayesianism takes for granted.[8]

Savage thought that, without going through such a process of reflective introspection, there is no particular virtue in being consistent at all. But when the world in which we are making decisions is large and complex, there is no way that such a process could be

[8] One might say that this story explains how Savage thought that his theory of subjective probability could be used as part of a theory of logical probability, but the story leaves the details of the process of how consistency is achieved completely unmodeled. However, it is in these unmodeled details that nearly all the action is to be found.

carried through successfully. For this reason, Savage restricted the sensible application of his theory to small worlds.[9]

1.4.2.3 When are the worlds of game theory small?

The games normally studied by game theorists are small worlds almost by definition. The standard use of Bayesian decision theory in such games is therefore beyond reproach. However, there are at least three cases in which doubts arise.

The first occurs when games are constructed that attempt to model large-scale economic phenomena. Bayesian methods may sensibly be employed in analyzing such models, but not in modeling the decision processes of the agents who appear in the model. From the point of view of the people caught up in them, the worlds of finance or macroeconomics are large, and it would therefore be irrational for them to use Bayesian decision theory. Nor do I know of any experimental evidence which suggests that real people come close to updating using Bayes' rule, except in a broad qualitative sense. On the contrary, human beings seem to find anything involving conditional probabilities very hard to handle.

How then should we analyze such models? I don't think anyone knows the answer. But it doesn't follow that it is therefore OK to go on attributing Bayesian rationality to agents who are not only not Bayesian rational in fact—but shouldn't be Bayesian rational in theory. Saying nothing at all on a subject about which one knows nothing will often be a better option.

The second case arises if players are allowed to mix their strategies with devices constructed using the principles of algorithmic randomization, that cannot adequately be described using Bayesian methods (Binmore, 2009). But space does not permit any exploration of this recondite possibility.

The third case applies when foundational game theorists attempt to subsume the workings of the minds of the players within a Bayesian model. The modeler then claims to be able to construct a model that is more complex than his own mind. I am among those who have adapted standard Gödelian arguments to obtain contradictions from this assumption (Binmore, 2009), but the following quote from Lyall Watson says all that really matters: "If the brain were so simple we could understand it, we would be so simple that we couldn't."

I think the third case applies to various attempts to provide Bayesian foundations for game theory, notably rationalizability and subjective correlated equilibrium. The iterated deletion of dominated strategies is a useful tool, but the claims of rationalizers that *only* the end-products of some refinement of this tool are truly rational require

[9] However, it does not follow that modern Bayesian statistics is worthless—only that it can claim no philosophical superiority over classical statistics when used in a large world. In fact, Bayesian statisticians seem quite happy to accept that both their approach and that of classical statistics have only empirical foundations.

applying Bayesian methods where they do not apply (Bernheim, 1984; Pearce, 1984). The same criticism applies to Auman's (1987) alternative idea that Nash equilibria should be replaced as the basic concept in game theory by correlated equilibria (Section 7.2). If my critique of Bayesianism applies anywhere, it is surely to Aumann's modeling of the entire universe (including all its thinking beings) as a small world.

1.4.2.4 Common priors?

Are we entitled to assume that all players will share the same prior in games that can reasonably be treated as small worlds? Harsanyi argued in favor of this assumption on the grounds that all ideally rational folk in the same state of ignorance will necessarily select the same prior, but this story is too naive even for stout defenders of Bayesianism like Aumann, for whom the common prior assumption is a step too far. Nevertheless, the literature often takes the common prior assumption for granted without any discussion at all. It is therefore perhaps worthwhile to ask when the common prior assumption can be justified using Savage's story about how a rational person should construct a prior.

When Savage's massaging process is applied in a game, Pandora must not only massage her own tentative subjective beliefs in an attempt to achieve consistency; she must also seek to mimic the massaging attempts of the other players. If the final result on which her considerations converge is invulnerable to a Dutch book, her massaging efforts must generate a *common* prior from which the posterior beliefs she attributes to each player can be deduced by conditioning on their information (Nau and McCardle, 1990). But why should the common prior to which Pandora is led be the *same* common prior as those obtained by other players going through the same process?

In complicated games, one can expect the massaging process to converge on the same outcome for all players only if the gut feelings with which they begin the process are similar. But we can only expect the players to have similar gut feelings if they all share a common culture, and so have a similar history of experience. Or to say the same thing another way, only when the players of a game are members of a reasonably close-knit community can they be expected to avoid leaving themselves open to a Dutch book being made against their group *as a whole.*

1.5. KNOWLEDGE

Numerous attempts have been made to expand the neoclassical orthodoxy by inventing refinements of the rationality concept that introduce implicit assumptions about what a player would know or believe if supposedly impossible events were to occur. But do such assumptions belong in a definition of rationality?

1.5.1 Knowledge as commitment

The philosophical tradition that knowledge is justified true belief is now commonly denied but without any firm new consensus emerging. Game theorists usually treat knowledge simply as belief with probability one. My unorthodox opinion is that neither attempt at a definition is adequate (Binmore, 2009, 2011).

We make a commitment to a model in game theory when we specify the rules of the game we plan to study. The states of the world we plan to investigate are then normally the possible plays of the game. However the game is played, the rules of the game remain the same. Philosophers would say that the rules are then the same in all possible worlds. I suggest that we use the same criterion for knowledge in game theory, so that our current conception of knowledge-as-belief is replaced by knowledge-as-commitment, even though we then end up with a notion in which Pandora may know something that is neither true nor justified.

In the spirit of the theory of revealed preference, Pandora can then be said to know something if she acts *as though* it were true in all possible worlds that might arise within whatever model she is using. For example, when physicists are asked to explain their ontology, they do not usually imagine that they are being asked a metaphysical question about the fundamental nature of reality. They simply reply by stating the propositions that are taken to be axiomatic in their models. Their faith in these axioms may only be provisional, but commitment to a model does not require the kind of faith that martyrs die for.

With this definition of knowledge, nothing can happen to shake Pandora's commitment to what she knows. This accords with the convention in real life according to which we say that Pandora was mistaken in saying that she knew something if she later changes her mind about it.[10]

I have some sympathy with those who think that economists should leave philosophizing about the nature of knowledge to philosophers, but the status of backward induction in game theory is at least one topic on which my proposal would sweep away much confusion (Section 6.3).

1.5.1.1 Contradicting knowledge?

What if a possible world occurs that embodies an in-your-face contradiction of something that Bob knows. Knowledge-as-commitment requires ignoring the contradiction and continuing to uphold what was previously held to be known. An analogy is to be found when mathematicians are led to a contradiction after adjoining a new proposition to an axiom system. One could deal with the problem by throwing away one or more axioms, but it is actually the new proposition that is said to be refuted.

It seems to me that we often similarly dismiss data that does not match our preconceptions. Such behavior need not be intellectually dishonest, because few models (if any) succeed in taking account of everything that might matter. Off-model explanations

[10] What then of certitude or belief-with-probability-one? I agree with those Bayesians who disbar zero subjective probabilities, so that one never needs to condition on a zero probability event. Models in which non-trivial events are accorded a subjective probability of one may nevertheless be useful, but only as limiting cases of models in which extreme probabilities do not appear.

of apparent contradictions are therefore sometimes a practical necessity. However, it is important that such off-model explanations should not have any implications for the future. Otherwise, the knowledge that is being insulated from a current challenge would be laid open to future challenges and hence not properly counted as knowledge. For this reason, the error distributions in refinements of Nash equilibrium like Selten's (1975) trembling-hand equilibrium are taken to be independent of past error distributions. Philosophers similarly urge interpreting counterfactuals so as to minimize the extent to which the basic model needs to be modified to accommodate the counterfactual phenomena (Lewis, 1973).

1.5.2 Common knowledge

An event is mutual knowledge if everybody knows it. If everybody knows that everybody knows it, and everybody knows that everybody knows that everybody knows it, and so on, then the event is common knowledge. In a brilliant paper, Aumann (1976) showed that the infinite regress in this formulation can be eliminated, thereby providing a characterization of common knowledge that makes it amenable to serious analysis.[11]

Since that time, game theorists (including me) have habitually guarded their flanks by asserting that everything in sight is to be taken to be common knowledge, but I think we need to think harder about this issue lest we part company with the idea that our models should be capable of approximating actual behavior that has been refined by experience (Section 1.1).

Milgrom (1981) has shown that an event is common knowledge if and only if it is implied by a public event—which is an event that cannot occur without becoming common knowledge. But this definition of a public event is very hard to satisfy. The usual requirement is said to be that everybody needs to observe everybody else observing it, but then one looks in vain for a public event that implies that French is the language of France. So it is not common knowledge that French is the language of France, let alone that the probability distributions from which bidders' valuations are drawn in auction models are common knowledge.[12] Nevertheless, such assumptions are so standard that they are almost never discussed outside the literature on epistemic game theory.

Nor is it always so hard to do without common knowledge assumptions (Mertens and Zamir, 1985). For example, consider a Cournot duopoly with demand equation $p + q =$

[11] The definition of common knowledge is often attributed to the philosopher Lewis (1969). He was doubtlessly the first to speak of common knowledge in this context, but his account is such a mess that it is not even clear that he was talking about the same concept as Aumann. In any case, it turns out that the priority dispute has been settled by Brian Skyrms' recent discovery that both were scooped by the sociologist Friedell (1969).

[12] My own experience in consulting on telecom auctions is that bidders only have a very rough idea of their *own* valuations let alone those of other potential bidders. Nor is it easy to persuade them to spend the necessary time and money valuing the assets they plan to bid for.

K and incomplete information about unit costs, but without common knowledge of any underlying probability distribution. Then it is fairly easy (Binmore, 2007b, p. 444) to show that the equilibrium outputs are:

$$\alpha(A) = \tfrac{1}{2}(K - A) - \tfrac{1}{4}(K - \overline{B}) + \tfrac{1}{8}(K - \overline{\overline{A}}) - \tfrac{1}{16}(K - \overline{\overline{\overline{B}}}) + \cdots,$$
$$\beta(B) = \tfrac{1}{2}(K - B) - \tfrac{1}{4}(K - \overline{A}) + \tfrac{1}{8}(K - \overline{\overline{B}}) - \tfrac{1}{16}(K - \overline{\overline{\overline{A}}}) + \cdots,$$

where A and B are the players' unit costs, and the bars indicate expectations (so that $\overline{\overline{A}}$ is player I's expectation of player II's expectation of player I's unit cost A).

1.5.3 Common knowledge of rationality?

Common knowledge of rationality is often mistakenly said to be essential for game theory arguments. Let us hope not, because we never come anywhere near such a state of bliss when applying game theory in practice. What does it mean anyway? Here I shall proceed on the orthodox assumption that knowledge is to be interpreted as certitude—belief-with-probability-one.

Suppose that all the players believe that they are all Bayesian rational at the outset of the game. According to orthodox Bayesian theory, the implication is that there is a probability space S whose states specify (among other things) whether or not the players are rational. Some states must therefore describe situations in which one or more of the players are irrational. But there are many possible ways of being irrational. To specify the state space fully, we would therefore need to itemize all the different ways in which a player might be irrational. To model the players' beliefs about the rationality of their opponents, we then need to elaborate the game by imagining an opening chance move that selects a state from S. The players then believe with probability one that the selected state will have the property that all the players are rational.

1.5.3.1 Counterfactuals

Given this set-up, how are the counterfactuals that necessarily arise in extensive games to be interpreted? What beliefs should be attributed to players in the event that they find themselves at an information set to which they have assigned a zero probability? Selten and Leopold (1982) argue that the way to deal with counterfactuals in game theory is to use the context in which a game arises to construct an elaboration of the game so that a counterfactual in the original game ceases to be counterfactual in the elaborated game. That is to say, they argue in favor of formalizing the kind of off-model stories used to "explain away" events that contradict a treasured model. They then follow Kolmogorov in suggesting that the newly invented parameters of the elaborated game be allowed to converge to zero, so that the elaborated game converges on the original game. One then interprets a counterfactual in terms of the limiting form of whatever factual statement in the elaborated game corresponds to the counterfactual in the original game.

Selten's (1975) trembling-hand story is one way of constructing such an elaborated game, but nothing says that rational players must necessarily interpret counterfactuals in this particular way. Personally, I think it a mistake to build mechanisms for interpreting counterfactuals into the definition of a rational player. Where it matters, modelers need to include the *context* in which a game arises as part of its specification to make it clear how any relevant counterfactuals should be interpreted.

In practice, the leading explanation of irrational play will usually be that the player who made it is irrational—as in the story that ought to accompany any assertion of common knowledge of rationality. Unlike Selten's trembling-hand story, a player's revealed irrationality then has implications for his or her future play. Selten (1978) himself uses his Chain-Store Paradox to make this point. In particular, the example shows that trembling-hand equilibria do not always make sense. More generally, Fudenberg et al. (1988) have shown that there is always some way to interpret counterfactuals so as to justify any Nash equilibrium whatever (Section 6.3).

1.6. NASH EQUILIBRIUM

Without claiming that there is only one viable approach to modeling rational players, this paper has defended the traditional orthodoxy on rationality, as perfected by Savage (1954) in his *Foundations of Statistics*. This section similarly defends the traditional approach to game theory taken by Von Neumann and Morgenstern (1944) and Nash (1951), again without claiming that other approaches may not have their virtues in some contexts.

It is sometimes argued that the idea of a Nash equilibrium was a major departure from Von Neumann's approach to game theory. The reason that formulating the idea of an equilibrium was left to Nash (1951) was presumably because Von Neumann's aim was to identify unambiguous rational solutions of games. He apparently thought it a mistake to make the best-reply criterion fundamental in defining the rational solution of a noncooperative game. He probably would have said, on the contrary, that the best-reply criterion should follow from an independent definition of a rational solution, as it does in the case of two-person, zero-sum games. However, Von Neumann would certainly not have denied that, where a rational solution can be identified, it would need to be a Nash equilibrium. Nor does Nash give the impression that he thought he was doing more than generalizing Von Neumann's ideas.[13] In any case, this section offers a case for not gilding the lily by seeking to replace the notion of a Nash equilibrium as currently used in the economics literature by more elaborate notions.

[13] Except in his PhD thesis, where he offers what we would now call an evolutionary defense of Nash equilibria—which the editor of *Annals of Mathematics* removed from his published paper on the grounds that it was of no interest!

1.6.1 Evolutionary game theory

One can regard game theory as a branch of mathematical philosophy, but if one hopes to use it to predict actual behavior, we need to avoid theories of rationality that depart too far from the outcomes to be expected from the evolutionary processes that govern how real people play most real games (Section 1.1).

Smith and Price (1972) define an evolutionarily stable strategy (ESS) to be a strategy for a symmetric game that leads to Nash equilibrium if used by all the players. There is further small print that biologists seem largely to ignore, which is as well since the replicator dynamics sometimes converges on Nash equilibria in 3×3 symmetric games that do not satisfy the ESS smallprint (Hofbauer and Sigmund, 1998). The same goes for various refinements popular in the economics literature, notably the deletion of weakly dominated strategies. Even in a game as simple as the Ultimatum Game, there are trajectories that converge on weakly dominated Nash equilibria (Binmore et al., 1995).

One can find dynamic systems that converge on Nash equilibria with more satisfactory properties, but so what? We don't get to choose the adjustment processes used by real people. In brief, to move away from the primitive notion of a Nash equilibrium would seem only to drive an unnecessary wedge between rational and evolutionary game theory.

1.6.2 Knowledge requirements

I have already argued against the indiscriminate claim that everything needs to be common knowledge for game theory to work. In particular, there is no need to invoke common knowledge of rationality in defending Nash equilibria. For example, it is enough if each player knows what strategies the other players will use. A rational player will then necessarily make a best reply to the strategies of the other players, and the result will be that a Nash equilibrium is played (Aumann and Brandenburger, 1995).

How might it come about that a player knows (or believes with a sufficiently high probability) what strategies the other players are going to use? The simplest possibility is that the player has access to social statistics that record how the game has been played in the past under similar circumstances. Or one may follow Von Neumann in imagining a great book of game theory that everyone knows all players consult before choosing a strategy. A player encodes what he or she knows or believes about the game to be played (including what is known or believed about the other potential players), and the book recommends a strategy. Of course, the game theorists who wrote the book will be aware that rational players will not follow the book's recommendations unless it advises the play of a best reply to the recommendations it makes to other potential players of the game who will also be consulting the book. So again a Nash equilibrium will be played. Note, in both cases, the close connection between the story being told and evolutionary game theory. Evolution—whether genetic or cultural—does not write game theory books or compile social statistics, but the end-product is *as if* it had.

But how do game theorists know what to write in the great book of game theory? It is at this stage that heavy equipment from epistemic game theory may be necessary, but none of this equipment needs to be built into the minds of the players themselves for game theory to make sense—which is just as well, because game theory would otherwise never predict anything. In brief, we need to separate what goes on in the minds of game theorists from what goes on in the minds of people who actually play games in real life.

1.6.3 Equilibrium selection problem

My own view is that the equilibrium selection problem is the leading difficulty holding up new advances in game theory. Games that arise in real life typically have many more equilibria than the games that economists usually study. For example, the folk theorem tells us that repeated games generically have an infinity of Nash equilibria.

Sometimes attempts are made to solve the multiplicity problem by going beyond the methodological individualism of the neoclassical orthodoxy by appealing to notions of "collective rationality." I think that such an approach is misguided—even when all that is involved is deleting weakly dominated equilibria or equilibria that are not Pareto efficient.

The alternative is to accept that rationality is not the place to look for certain kinds of answers. This is entirely obvious when the issue is whether we should all drive on the left or on the right in the Driving Game. As with counterfactuals, writing such arbitrary coordinating conventions into the great book of game theory necessarily takes us outside the province of rationality. After all, we do not expect mathematicians to tell us which of the two solutions of a quadratic equation is correct or to express their preferences over the solutions of a differential equation. Such matters are resolved by examining the *context* in which equations with multiple solutions arise.

If not to rationality, where do we look for answers? I am with Schelling (1960), Young (2001), and many others in seeing social norms (or conventions or focal points) as evolution's answer to the equilibrium selection problem. One may respond that this takes us into territory in which economists enjoy no comparative advantage, but that is not my own experience (Binmore, 2005).

1.6.3.1 Refinements of Nash equilibrium

One approach to the equilibrium selection problem in extensive-form games is embodied in the refinement literature. Attempts are made to build particular interpretations of counterfactuals into new definitions of rationality in the manner argued against in Section 5.3. The result is often to cut down very substantially on the number of Nash equilibria between which a choice needs to be made. Sometimes only one equilibrium remains.

The leading refinement is the notion of a subgame-perfect equilibrium, which is obtained by employing backward induction in the case of finite games of perfect information (Selten, 1975). For example, all splits of the money in the famous Ultimatum Game are Nash equilibrium outcomes, but only the equilibrium in which the proposer gets everything is subgame perfect. However, we seldom observe anything close to the subgame-perfect outcome in the laboratory; nor is it necessarily the limit of the replicator dynamics (Binmore et al., 1995; Guth et al., 1982).

Behavioral economists often say that such results refute the neoclassical orthodoxy, in which claim they are assisted by a theorem of Aumann (1995) which asserts that common knowledge of rationality implies backward-induction play in finite games of perfect information. However, it seldom seems to be understood that the kind of rationality Aumann is talking about is not Bayesian rationality as normally understood, but an invention of his own called "substantive rationality" (which has no connection with Herbert Simon's substantive rationality). Aumann's definition of substantive rationality is constructed so that his underlying hypothesis would be maintained even if contradicted by evidence that becomes available as the game is played. He therefore builds a naive view on how counterfactuals should be interpreted into his definition of rationality. As observed earlier, I think that clarity demands that assumptions about what the players would believe if things that are not going to happen were to happen do not belong in a definition of rationality. As for Aumann's theorem, why not just say that it is obvious that common knowledge of Bayesian rationality implies backward induction in finite games of perfect information with knowledge interpreted as knowledge-as-commitment but not with knowledge-as-belief? (Binmore, 2011).

To summarize, I think the refinement literature is misconceived as a contribution to the notion of rationality. But it doesn't follow that all refinements of Nash equilibrium should be thrown away. When the context in which a game is to be played validates the interpretation of counterfactuals built into a refinement, then it can be very useful indeed. For example, the fact that the folk theorem of repeated games applies with only minor changes when Nash equilibria are replaced by subgame-perfect equilbria is very important in certain applications because it has the potential to answer the age-old question of who guards the guardians.

1.7. BLACK BOXES

It is sometimes very useful to pack away some of the complications that arise when the strict neoclassical orthodoxy is applied into one or more black boxes. Economists sometimes refer to the result as a reduced-form model. It is difficult to see how progress would be made in some situations without the use of reduced-form models, but it is important never to forget that the stuff packed away in a black box is still there and may need to be unpacked if the model is to be applied in a context for which it was not originally designed. Two examples are offered in this section.

1.7.1 Nash program

What of cooperative solution concepts like the Shapley value or the Nash bargaining solution? Sometimes, it is said that such notions need no more justification than their axiomatic characterizations provide, but the plausibility of such axioms is often based on implicit appeals to collective rationality principles. Fortunately, John Nash offered us a different way of thinking about cooperative solution concepts that is entirely consistent with the neoclassical orthodoxy.

VN&M's (Neumann and Morgenstern, 1944) *Theory of Games and Economic Behavior* splits game theory into cooperative game theory and noncooperative game theory. Critics sometimes ask why economists insist on appealing to noncooperative theory, when everybody knows that human beings are social animals and hence naturally cooperative. Such critics are usually under the misapprehension that noncooperative game theory is the study of conflict, which is perhaps not so surprising since the only noncooperative games that VN&M analyze are zero sum. But noncooperative game theory does not assume that rational players will compete unrelentlessly. It is better described as the study of games in which any cooperation is fully explained by the choice of strategies the players make.

Cooperative game theory differs from noncooperative game theory in abandoning any pretension at explaining *how* cooperation is sustained. From Nash's viewpoint, it postulates instead that the players have access to an unmodeled black box whose contents somehow resolve all problems of commitment and trust. In management jargon, cooperative theory assumes that the problem of how cooperation is sustained is solved 'off-model' rather than 'on-model' as in noncooperative theory.

As a bare minimum, the cooperative black box needs to contain a preplay negotiation period. During this negotiation period, the players are free to sign whatever agreements they choose about the strategies to be used in the game they are about to play. These preplay agreements are understood as *binding.* Once the players have signed an agreement, cooperative game theory assumes that there is no way they can wriggle out of their contractual obligations should they later prove to be inconvenient. In economic applications, one can sometimes justify this strong assumption by arguing that the black box contains all the apparatus of the legal system. The players then honor their contracts for fear of being sued if they don't. In social applications, the black box may contain the reasons why the players care about the effect that behaving dishonestly in the present may have on their reputation for trustworthy behavior in the future. One can even argue that the black box contains the results of our childhood conditioning, or an inborn aversion to immoral behavior.

The *Nash program* invites us to open the cooperative black box to see whether the mechanism inside really does work in the way the axioms characterizing a cooperative solution concept assume. John Nash observed that any negotiation is itself a species of game, in which the moves are everything the players may say or do while bargaining. If we model any bargaining that precedes a game $\mathcal{G}$ in this way, the result is an enlarged game $\mathcal{N}$. A strategy for this negotiation game first tells a player how to conduct the preplay negotiations, and then how to play $\mathcal{G}$ depending on the outcome of the negotiations.

Negotiation games must be studied *without* presupposing any preplay bargaining, all preplay activity having already been built into their rules. Analyzing them is therefore a task for noncooperative game theory. We then have a way of checking up on cooperative game theory. If a cooperative solution concept says that the result of a rational agreement on how to play $\mathcal{G}$ will be s, then we should also get s from solving $\mathcal{N}$ using noncooperative game theory. The leading example is the fact that the asymmetric Nash bargaining solution emerges as the unique subgame-perfect equilibrium outcome of Rubinstein's (1982) bargaining model.

Sometimes, of course, the class of negotiation games that support a particular cooperative solution concept may be very small or involve structural features incompatible with the context in which the concept is commonly applied. I fear that this is true of the very popular Shapley value, but the Shapley value also has a second interpretation as a fair arbitration scheme.

1.7.2 Other preplay activity

The Nash program requires that the black box of cooperative theory be unpacked so that its contents can be built into the structure of a noncooperative game. The same reasoning ought properly to be applied to all other black boxes too—like those refinements of Nash equilibria we have already considered that conceal assumptions about the interpretation of counterfactuals.

The black boxes to be considered in this section are those that implicitly postulate some preplay interaction between the players. Harsanyi's (1967) theory of incomplete information is a case in point. The theory converts a situation with incomplete (or asymmetric) information into a game of imperfect information by inventing an opening chance move that assigns each player a type that specifies the player's preferences and beliefs. The players are then assumed to use Bayes' rule to update their beliefs given the type they have been assigned. The game of imperfect information they are playing is therefore traditionally called a Bayesian game. Following Harsanyi, this Bayesian game is analyzed by computing its Nash equilibria.[14] What are commonly called

[14] The methodology is actually no different to that used by Von Neumann and Morgenstern (1944) in analyzing his Poker models (Binmore, 2007a, p. 435).

Bayesian equilibria are therefore the Nash equilibria of a Bayesian game. However, it is commonplace to write down formulas which are said to represent "Bayesian equilibria of games of asymmetric information" that conceal not only the structure of the chance move that assigns the players' types, but the knowledge implicitly attributed to the players about its nature (Section 5.2).

The same point applies when it is argued that correlated equilibria should replace Nash equilibria as the basic concept of game theory. A correlated equilibrium of a game $\mathcal{G}$ is just a Nash equilibrium of an expanded game $\mathcal{C}$ created by prefixing $\mathcal{G}$ by a chance move that may send correlated signals to the players. Section 4.2 argues that the Bayesian arguments offered in favor of replacing Nash equilibria by correlated equilibria are unsound,[15] but the reason urged here is that one ought not to put anything at all into a black box when attempting a noncooperative analysis, let alone anything so important as the mechanism by means of which players may succeed in correlating their strategic behavior.

1.8. CONCLUSION

There is no ideal model of rationality waiting to de discovered in some Platonic limbo. Any model has to be judged by how well it works in practice. To the extent that the neoclassical orthodoxy does work well in practice, I argue that it is because it allows game theorists to short-circuit the detailed study of the evolutionary processes that are the genuine reason that real people sometimes end up playing games as game theory predicts.

This paper reviews some attempts that have been made to drive a wedge between rational game theory and evolutionary game theory by abandoning the foundations created by Savage, Von Neumann, Nash, and other pioneers. My conservative conclusion is that we should leave these proposals to bubble away on a back burner, and carry on doing game theory in the traditional style that has proved so successful in the past.

ACKNOWLEDGMENTS

I gratefuly acknowledge support from the European Research Council Seventh Framework Program (FP7/2007-2013), ERC Grant agreement no. 295449.

[15] An additional reason sometimes given is that it is much easier to compute the set of all correlated equilibria than the set of all Nash equilibria, but nobody similarly argues for replacing the inverse-square law of gravity by something more mathematically convenient. In the other direction, one has the problem that duplicating a row in a payoff matrix ought to be strategically neutral but actually changes the set of correlated equilibria (Aumann and Dreze, 2008). Moreover, one needs a common prior assumption to recover Nash equilibria from correlated equilibria.

REFERENCES

Alchian, A., 1950. Uncertainty, evolution and economic theory. J. Polit. Econ. 58, 221–221.

Anscombe, F., Aumann, R., 1963. A definition of subjective probability. Ann. Math. Statist. 34, 199–205.

Arrow, K., 1959. Rational choice functions and ordering. Economica 26, 121–127.

Aumann, R., 1976. Agreeing to disagree. Ann. Statist. 4, 1236–1239.

Aumann, R., 1987. Correlated equilibrium as an expression of Bayesian rationality. Econometrica 55, 1–18.

Aumann, R., 1995. Backward induction and common knowledge of rationality. Games Econ. Behav. 8, 6–19.

Aumann, R., Brandenburger, A., 1995. Epistemic conditions for Nash equilibrium. Econometrica 63, 1161–1180.

Aumann, R., Dreze, J., 2008. Rational expectations in games. Am. Econ. Rev. 98, 72–86.

Bernheim, D., 1984. Rationalizable strategic behavior. Econometrica 52, 1007–1028.

Binmore, K., 2005. Natural Justice. Oxford University Press, New York.

Binmore, K., 2007a. The origins of fair play. Proc. Brit. Acad. 151, 151–193.

Binmore, K., 2007b. Playing for Real. Oxford University Press, New York.

Binmore, K., 2009. Rational Decisions. Princeton University Press, Princeton.

Binmore, K., 2011. Interpreting knowledge in the backward induction problem. Episteme 8, 248–261.

Binmore, K., Gale, J., Samuelson, L., 1995. Learning to be imperfect: the Ultimatum Game. Games Econ. Behav. 8, 56–90.

de Finetti, B., 1937. La prevision: Ses lois logique, ses sources subjectives. Ann. Inst. Henri Poincare, 7, 1–68.

Friedell, M., 1969. On the structure of shared awareness. Behav. Sci. 14, 28–39.

Fudenberg, D., Kreps, D., Levine, D., 1988. On the robustness of equilibrium refinements. J. Econ. Theory 44, 354–380.

Gillies, D., 2000. Philosophical Theories of Probability. Routledge, London.

Guth, W., Schmittberger, R., Schwarze, B., 1982. An experimental analysis of ultimatum bargaining. J. Behav. Org. 3, 367–388.

Harsanyi, J., 1967. Games with incomplete information played by "Bayesian" players, I–III. Manage. Sci. 14, 159–182.

Harsanyi, J., 1977. Rational Behavior and Bargaining Equilibrium in Games and Social Situations. Cambridge University Press, Cambridge.

Henrich, J., Boyd, R., Bowles, S., Camerer, C., Fehr, E., Gintis, H., 2004. Foundations of Human Sociality: Economic Experiments and Ethnographic Evidence from Fifteen Small-Scale Societies, Oxford University Press, New York.

Hobbes, T., 1986. Leviathan. Penguin Classics, London. (Edited by C.B. Macpherson. First published 1651).

Hofbauer, J., Sigmund, K., 1998. Evolutionary Games and Population Dynamics. Cambridge University Press, Cambridge.

Hume, D., 1739. A Treatise of Human Nature, Second ed. Clarendon Press, Oxford. (Edited by L.A. Selby-Bigge. Revised by P. Nidditch. First published 1739).

Kagel, J., Battalio, R., Green, L., 1995. Economic Choice Theory: An Experimental Analysis of Animal Behavior. Cambridge University Press, Cambridge.

Kreps, D., 1988. Notes on the Theory of Choice. Westview Press, Boulder, CO.

Lewis, D., 1969. Conventions: A Philosophical Study. Harvard University Press, Cambridge, MA.

Lewis, D., 1973. Counterfactuals. Blackwell, Oxford.

Luce, R., Raiffa, H., 1957. Games and Decisions. Wiley, New York.

Maynard Smith, J., Price, G., 1972. The logic of animal conflict. Nature 246, 15–18.

Mele, A., Rawling, P., 2003. Oxford Handbook of Rationality. Oxford University Press, Oxford.

Mertens, J.F., Zamir, S., 1985. Formulation of Bayesian analysis for games with incomplete information. Int. J. Game Theory 14, 1–29.

Milgrom, P., 1981. An axiomatic characterization of common knowledge. Econometrica 49, 219–222.

Nash, J., 1951. Non-cooperative games. Ann. Math. 54, 286–295.

Nau, R., McCardle, K., 1990. Coherent behavior in noncooperative games. J. Econ. Theory 50, 424–444.

Pearce, D., 1984. Rationalizable strategic behavior and the problem of perfection. Econometrica 52, 1029–1050.
Ramsey, F., 1931. Truth and probability. In: Ramsey, F. (Ed.), Foundations of Mathematics and other Logical Essays. Harcourt, New York.
Robbins, L., 1938. Inter-personal comparisons of utility. Econ. J. 48, 635–641.
Rubinstein, A., 1982. Perfect equilibrium in a bargaining model. Econometrica 50, 97–109.
Samuelson, P., 1947. Foundations of Economic Analysis. Harvard University Press, Cambridge, MA.
Samuelson, L., Zhang, J., 1992. Evolutionary stability in asymmetric games. J. Econ. Theory 57, 364–391.
Savage, L., 1954. The Foundations of Statistics. Wiley, New York.
Schelling, T., 1960. The Strategy of Conflict. Harvard University Press, Cambridge, MA.
Selten, R., 1975. Reexamination of the perfectness concept for equilibrium points in extensive-games. Int. J. Game Theory 4, 25–55.
Selten, R., 1978. The chain-store paradox. Theory Decis. 9, 127–159.
Selten, R., Leopold, U., 1982. Subjunctive conditionals in decision theory and game theory. In: Stegmuller, W., Balzer, W., and Spohn, W. (Eds.), Studies in Economics. Springer-Verlag, Berlin. Philosophy of Economics, vol. 2.
Sen, A., 1971. Choice functions and revealed preference. Rev. Econ. Stud. 38, 307–317.
Sen, A., 1993. Internal consistency of choice. Econometrica 61, 495–522.
Simon, H., 1976. From substantive to procedural rationality. In Method and Appraisal in Economics. Cambridge University Press, Cambridge.
Von Neumann, J., Morgenstern, O., 1944. The Theory of Games and Economic Behavior. Princeton University Press, Princeton.
Weymark, J., 1991. A reconsideration of the Harsanyi-Sen debate on utilitarianism. In: Elster, J., Roemer, J. (Eds.), Interpersonal Comparisons of Well-Being. Cambridge University Press, Cambridge.
Young, P., 2001. Individual Strategy and Social Structure: An Evolutionary Theory of Institutions. Princeton University Press, Princeton.

CHAPTER 2

Advances in Zero-Sum Dynamic Games

Rida Laraki[*,†]**, Sylvain Sorin**[‡]
[*]CNRS in LAMSADE (Université Paris-Dauphine), France
[†]Economics Department at Ecole Polytechnique, France
[‡]Mathematics, CNRS, IMJ-PRG, UMR 7586, Sorbonne Universités, UPMC Univ Paris 06, Univ Paris Diderot, Sorbonne Paris Cité, Paris, France

Contents

Handbook of game theory
http://dx.doi.org/10.1016/B978-0-12-420056-2.00002-1

Abstract

The survey presents recent results in the theory of two-person zero-sum repeated games and their connections with differential and continuous-time games. The emphasis is made on the following points:

(1) A general model allows to deal simultaneously with stochastic and informational aspects.
(2) All evaluations of the stage payoffs can be covered in the same framework (and not only the usual Cesàro and Abel means).
(3) The model in discrete time can be seen and analyzed as a discretization of a continuous time game. Moreover, tools and ideas from repeated games are very fruitful for continuous time games and vice versa.
(4) Numerous important conjectures have been answered (some in the negative).
(5) New tools and original models have been proposed. As a consequence, the field (discrete versus continuous time, stochastic versus incomplete information models) has a much more unified structure, and research is extremely active.

Keywords: repeated, stochastic and differential games, discrete and continuous time, Shapley operator, incomplete information, imperfect monitoring, asymptotic and uniform value, dual game, weak and strong approachability.

JEL Codes: C73. C61, C62

AMS Codes: 91A5, 91A23, 91A25

2.1. INTRODUCTION

The theory of repeated games focuses on situations involving multistage interactions, where, at each period, the Players are facing a stage game in which their actions have two effects: they induce a stage payoff, and they affect the future of the game (note the difference with other multimove games like pursuit or stopping games where there is no stage payoff). If the stage game is a fixed zero-sum game G, repetition adds nothing: the value and optimal strategies are the same as in G. The situation, however, is drastically different for nonzero-sum games leading to a family of so-called Folk theorems: the use of plans and threats generates new equilibria (Sorin's (1992) chapter 4 in Handbook of Game Theory (HGT1)).

In this survey, we will concentrate on the zero-sum case and consider the framework where the stage game belongs to a family G^m, $m \in M$, of two-person zero-sum games played on action sets $I \times J$. Two basic classes of repeated games that have been studied and analyzed extensively in previous volumes of HGT are *stochastic games* (the subject of Mertens's (2002) chapter 47 and Vieille's (2002) chapter 48 in HGT3) and *incomplete information games* (the subject of Zamir's (1992) chapter 5 in HGT1). The reader is referred to these chapters for a general introduction to the topic and a presentation of the fundamental results.

In stochastic games, the parameter m, which determines which game is being played, is a publicly known variable, controlled by the Players. It evolves over time and its value m_{n+1} at stage $n+1$ (called the state) is a random stationary function of the triple (i_n, j_n, m_n) which are the moves, respectively the state, at stage n. At each period, both Players share the same information and, in particular, know the current state. On the other hand, the state is changing and the issue for the Players at stage n is to control both the current payoff g_n (induced by (i_n, j_n, m_n)) and the next state m_{n+1}.

In incomplete information games, the parameter m is chosen once and for all by nature and kept fixed during play, but at least one Player does not have full information about it. In this situation, the issue is the trade-off between using the information (which increases the set of strategies in the stage game) and revealing it (this decreases the potential advantage for the future).

We will see that these two apparently very different models—evolving known state versus unknown fixed state—are particular incarnations of a more general model and share many common properties.

2.1.1 General model of repeated games (RG)

The general presentation of this section follows Mertens et al. (1994). To make it more accessible, we will assume that all sets (of actions, states, and signals) are finite; in the general case, measurable and/or topological hypotheses are in order, but we will not treat such issues here. Some theorems will be stated with compact action spaces. In that case, payoff and transition functions are assumed to be continuous.

Let M be a parameter space and g a function from $I \times J \times M$ to $\mathbb{R}$. For every $m \in M$, this defines a two Player zero-sum game with action sets I and J for Player 1 (the maximizer) and Player 2, respectively, and with a payoff function $g(\cdot, m)$. The initial parameter m_1 is chosen at random and the Players receive some initial information about it, say a_1 (resp. b_1) for Player 1 (resp. Player 2). This choice of nature is performed according to some initial probability distribution π on $A \times B \times M$, where A and B are the signal sets of each Player. The game is then played in discrete time.

At each stage $n = 1, 2, \ldots$, Player 1 (resp. Player 2) chooses an action $i_n \in I$ (resp. $j_n \in J$). This determines a stage payoff $g_n = g(i_n, j_n, m_n)$, where m_n is the current value of the state parameter. Then, a new value m_{n+1} of the parameter is selected and the Players get some information about it. This is generated by a map Q from $I \times J \times M$ to the set of probability distributions on $A \times B \times M$. More precisely, at stage $n+1$, a triple $(a_{n+1}, b_{n+1}, m_{n+1})$ is chosen according to the distribution $Q(i_n, j_n, m_n)$ and a_{n+1} (resp. b_{n+1}) is transmitted to Player 1 (resp. Player 2).

Note that each signal may reveal some information about the previous choice of actions (i_n, j_n) and/or past and current values (m_n and m_{n+1}) of the parameter:

Stochastic games (with standard signaling: perfect monitoring) (Mertens, 2002) correspond to *public signals* including the parameter: $a_{n+1} = b_{n+1} = \{i_n, j_n, m_{n+1}\}$.

Incomplete information repeated games (with standard signaling) (Zamir, 1992) correspond to an absorbing transition on the parameter ($m_n = m_1$ for every n) and no further information (after the initial one) on the parameter, but previous actions are observed: $a_{n+1} = b_{n+1} = \{i_n, j_n\}$.

A *play* of the game induces a sequence $m_1, a_1, b_1, i_1, j_1, m_2, a_2, b_2, i_2, j_2, \ldots$, while the information of Player 1 before his move at stage n is a private history of him of the form $(a_1, i_1, a_2, i_2, \ldots, a_n)$ and similarly for Player 2. The corresponding sequence of payoffs is $g_1, g_2, \ldots$ and it is not known to the Players (unless it can be deduced from the signals).

A (behavioral) strategy σ for Player 1 is a map from Player 1 private histories to $\Delta(I)$, the space of probability distributions on the set of actions I: in this way, σ defines the probability distribution of the current stage action as a function of the past events known to Player 1: a behavioral strategy τ for Player 2 is defined similarly. Together with the components of the game, π and Q, a pair (σ, τ) of behavioral strategies induces a probability distribution on plays, and hence on the sequence of payoffs. $\mathsf{E}_{(\sigma,\tau)}$ stands for the corresponding expectation.

2.1.2 Compact evaluations

Once the description of the repeated game is specified, strategy sets are well defined as well as the play (or the distribution on plays) that they induce. In turn, a play determines a flow of stage payoffs $g = \{g_n; n \geq 1\}$, indexed by the positive integers in $\mathbb{N}^* = \mathbb{N}\backslash\{0\}$. Several procedures have been introduced to evaluate this sequence.

Compact evaluations associate to every probability distribution $\mu = \{\mu_n; n \geq 1\}$ on $\mathbb{N}^*$ a game G_μ with evaluation function $\langle \mu, g \rangle = \sum_n g_n \mu_n$. μ_n is interpreted as the (normalized) weight of stage n. Under standard assumptions on the data of the game, the strategy sets are convex-compact and the payoff function is bilinear and continuous in the product topology. Consequently, Sion's minmax theorem implies that the game has a value, denoted by v_μ.

2.1.3 Asymptotic analysis

The *asymptotic analysis* focuses on the problems of existence and the characterization of the asymptotic value (or limit value) $v = \lim v_{\mu^r}$ along a sequence of distributions μ^r with maximum weight (interpreted as the mesh in the continuous time discretization)

$\|\mu^r\| = \sup_n \mu_n^r \to 0$, and the dependence of this limit on a particular chosen sequence (see also Section 9.1.1). The connection with continuous time games is as follows. The RG is considered as the discretization in time of some continuous-time game (to be defined) played between time 0 and 1 and such that the duration of stage n is μ_n.

Two standard and well studied RG evaluations are:

(i) The *finitely repeated n-stage game* G_n, $n \geq 1$, with the Cesàro average of the stream of payoffs $\overline{g}_n = \frac{1}{n}\sum_{r=1}^{n} g_r$ and value v_n,

(ii) The *λ-discounted repeated game* G_λ, $\lambda \in]0,1]$, with the Abel average of the stream of payoffs $\overline{g}_\lambda = \sum_{r=1}^{\infty} \lambda(1-\lambda)^{r-1} g_r$ and value v_λ.

More generally, instead of deterministic weights, we can also consider stochastic evaluations. This has been introduced by Neyman and Sorin (2010) under the name "random duration process." In this framework, μ_n is a random variable, the law of which depends upon the previous path of the process (see Section 2.3).

2.1.4 Uniform analysis

A drawback of the previous approach is that even if the asymptotic value exists, the optimal behavior of a Player in the RG may depend heavily upon the exact evaluation (n for finitely repeated games, λ for discounted games) (Zamir, 1973). In other words, the value of the game with many stages is well defined, but one may need to know the exact duration to play well. The uniform approach considers this issue by looking for strategies that are almost optimal in any sufficiently long RG (n large enough, λ small enough). More precisely:

Definition 2.1. *We will say that $\underline{v}$ is the uniform maxmin if the following two conditions are satisfied:*

(i) Player 1 can guarantee *$\underline{v}$: for any $\varepsilon > 0$, there exists a strategy σ of Player 1 and an integer N such that for all $n \geq N$ and every strategy τ of Player 2:*

$$\mathsf{E}_{(\sigma,\tau)}(\overline{g}_n) \geq \underline{v} - \varepsilon,$$

(ii) Player 2 can defend *$\underline{v}$: for all $\varepsilon > 0$ and any strategy σ of Player 1, there exist an integer N and a strategy τ of Player 2 such that for all $n \geq N$:*

$$\mathsf{E}_{(\sigma,\tau)}(\overline{g}_n) \leq \underline{v} + \varepsilon.$$

Note the strong requirement of uniformity with respect to both n and τ in (*i*) and with respect to n in (*ii*). In particular existence has to be proved. A dual definition holds for the *uniform minmax* $\overline{v}$.

Whenever $\underline{v} = \overline{v}$, the game is said to have a *uniform value*, denoted by v_∞.

The existence of the uniform value does not always hold in general RG (for example, for games with incomplete information on both sides or stochastic games with signals on the moves (imperfect monitoring)). However, its existence implies the existence of the asymptotic value for decreasing evaluation processes (μ_n^r decreasing in n). In particular: $v_\infty = \lim_{n\to\infty} v_n = \lim_{\lambda\to 0} v_\lambda$.

The next sections will focus on the following points:

- in the compact case, corresponding to the asymptotic approach, a unified analysis is provided through the extended Shapley operator that allows us to treat general repeated games and arbitrary evaluation functions;
- the links between the asymptotic approach and the uniform approach, initiated by the tools in the Theorem of Mertens and Neyman (1981);
- the connection with differential games: asymptotic approach versus games of fixed duration, uniform approach versus qualitative differential games.

2.2. RECURSIVE STRUCTURE

2.2.1 Discounted stochastic games

The first and simplest recursive formula for repeated games was established by Shapley (1953), who characterizes the λ-discounted value of a finite stochastic game with state space Ω as the only solution of the equation (recall that g is the payoff and Q the transition):

$$v_\lambda(\omega) = \mathtt{val}_{X\times Y}\big\{\lambda g(x,y,\omega) + (1-\lambda)\sum_{\omega'} Q(x,y,\omega)[\omega']v_\lambda(\omega')\big\}. \qquad [2.1]$$

where $X = \Delta(I)$ and $Y = \Delta(J)$ are the spaces of mixed moves, $\mathtt{val}_{X\times Y} = \sup_{x\in X}\inf_{y\in Y} = \inf_{y\in Y}\sup_{x\in X}$ is the value operator (whenever it exists, which will be the case in almost all the chapter where moreover, max and min are achieved). Also, for a function $h : I\times J \to \mathbb{R}$, $h(x,y)$ denotes $\mathsf{E}_{x,y}h = \sum_{i,j} x(i)y(j)h(i,j)$ which is the bilinear extension to $X\times Y$.

This formula expresses the value of the game as a function of the current payoff and the value from the next stage and onwards. Since the Players know the initial state ω and learn at each stage the new state ω', they can perform the analysis for each state separately and can use the recursive formula to compute an optimal strategy for each ω. In particular, they have an optimal stationary strategy and the "state" ω of the stochastic game is the natural "state variable" to compute the value and optimal strategies.

The *Shapley operator* $\mathbf{T}$ associates to a function f from Ω to $\mathbb{R}$ the function $\mathbf{T}(f)$ defined as:

$$\mathbf{T}(f)(\omega) = \mathtt{val}_{X\times Y}\{g(x,y,\omega) + \sum_{\omega'} Q(x,y,\omega)[\omega']f(\omega')\} \qquad [2.2]$$

Thus, v_λ is the only fixed point of the mapping $f \mapsto \lambda\mathbf{T}\left(\frac{(1-\lambda)f}{\lambda}\right)$.

2.2.2 General discounted repeated games

2.2.2.1 Recursive structure

The result of the previous section can be extended to any general repeated game, following Mertens (1987) and Mertens et al. (1994). Let us give a brief description.

The recursive structure relies on the construction of the universal belief space, Mertens and Zamir (1985), which represents the infinite hierarchy of beliefs of the Players: $\Theta = M \times \Theta^1 \times \Theta^2$, where Θ^i, homeomorphic to $\Delta(M \times \Theta^{-i})$, is the universal space of types of Player i (where $i = 1, 2, -i = 3 - i$).

A consistent probability ρ on Θ is such that the conditional probability induced by ρ at θ^i coincides with θ^i itself, both as elements of $\Delta(M \times \Theta^{-i})$. The set of consistent probabilities is denoted by $\mathbb{P} \subset \Delta(\Theta)$. The signaling structure in the game, just before the actions at stage n, describes an information scheme (basically a probability on $M \times \hat{A} \times \hat{B}$ where $\hat{A}$ is a general signal space to Player 1 and $\hat{B}$ for Player 2) that induces a consistent probability $\mathcal{P}_n \in \mathbb{P}$ (see Mertens et al., 1994, Sections III.1, III.2, IV.3). This is referred to as the "entrance law." Taking into account the existence of a value for any finite repeated game with compact evaluation, one can assume that the strategies used by the Players are announced to both. The entrance law $\mathcal{P}_n$ and the (behavioral) strategies at stage n (say α_n and β_n), which can be represented as measurable maps from type sets to mixed actions sets, determine the current payoff and the new entrance law $\mathcal{P}_{n+1} = H(\mathcal{P}_n, \alpha_n, \beta_n)$. Thus the initial game is value-equivalent to a game where at each stage n, before the moves of the Players, a new triple of parameter and signals to the Players is generated according to $\mathcal{P}_n$, and $\mathcal{P}_{n+1}$ is determined given the stage behavioral strategies. This updating rule is the basis of the *recursive structure* for which $\mathbb{P}$ is the "state space." The stationary aspect of the repeated game is expressed by the fact that H does not depend on the stage n.

The (generalized) Shapley operator $\mathbf{T}$ is defined on the set of real-bounded functions on $\mathbb{P}$ as:

$$\mathbf{T}(f)(\mathcal{P}) = \sup_{\alpha} \inf_{\beta} \left\{ g(\mathcal{P}, \alpha, \beta) + f(H(\mathcal{P}, \alpha, \beta)) \right\}. \qquad [2.3]$$

Then the usual recursive equations hold (see Mertens et al. (1994), Section IV.3). For the discounted game, one has:

$$v_\lambda(\mathcal{P}) = \mathtt{val}_{\alpha \times \beta} \left\{ \lambda g(\mathcal{P}, \alpha, \beta) + (1 - \lambda) v_\lambda(H(\mathcal{P}, \alpha, \beta)) \right\} \qquad [2.4]$$

where $\mathtt{val}_{\alpha \times \beta} = \sup_\alpha \inf_\beta = \inf_\beta \sup_\alpha$ is the value operator for the "one stage game at $\mathcal{P}$." This representation corresponds to a "deterministic" stochastic game on the state space $\mathbb{P} \subset \Delta(\Theta)$.

Hence to each compact repeated game G, one can associate an *auxiliary game* Γ having the same compact values on $\mathbb{P}$. The discounted values satisfy the recursive equation [2.4]. However, the play and strategies in the two games differ, since, in the auxiliary game, an additional signal corresponding to the stage behavioral strategies is given to the payers.

2.2.2.2 Specific classes of repeated games

In the framework of a standard stochastic game with state space Ω, the universal belief space representation of the previous section would correspond to the level of probabilities on the state space, thus $\mathbb{P} = \Delta(\Omega)$. One recovers the initial Shapley formula [2.1] by letting $\mathcal{P}$ be the Dirac mass at ω, in which case (α, β) reduce to (x, y) (i.e., only the ω-component of $(\Delta(I) \times \Delta(J))^{\Omega}$ is relevant), $H(\mathcal{P}, \alpha, \beta)$ corresponds to $Q(\omega, x, y)$, and finally $v_\lambda(H(\mathcal{P}, \alpha, \beta)) = \mathsf{E}_{Q(\omega,x,y)} v_\lambda(\cdot)$.

Let us describe explicitly the recursive structure in the framework of repeated games with incomplete information (independent case with standard signaling). M is a product space $K \times L$, π is a product probability $\pi(k, l) = p^k \times q^l$ with $p \in \Delta(K)$, $q \in \Delta(L)$, and the first signals to the Players are given by: $a_1 = k$ and $b_1 = \ell$. Given the parameter $m = (k, \ell)$, each Player knows his own component and holds a prior on the other Player's component. From stage 1 on, the parameter m is fixed and the information of the Players after stage n is $a_{n+1} = b_{n+1} = \{i_n, j_n\}$.

The auxiliary stochastic game Γ corresponding to the recursive structure is as follows: the "state space" Ω is $\Delta(K) \times \Delta(L)$ and is interpreted as the space of beliefs on the realized value of the parameter. $\mathbf{X} = \Delta(I)^K$ and $\mathbf{Y} = \Delta(J)^L$ are the type-dependent mixed action sets of the Players; g is extended on $\Omega \times \mathbf{X} \times \mathbf{Y}$ as $g(p, q, x, y) = \sum_{k,\ell} p^k q^\ell g(k, \ell, x^k, y^\ell)$, where $g(k, \ell, x^k, y^\ell)$ denotes the expected payoff at $m = (k, \ell)$ where Player 1 (resp. 2) plays according to $x^k \in \Delta(I)$ (resp. $y^\ell \in \Delta(J)$). Given (p, q, x, y), let $\bar{x}(i) = \sum_k x^k(i) p^k$ be the total probability of action i by Player 1 and $p(i)$ be the conditional probability on K given the action i, explicitly $p^k(i) = \frac{p^k x^k(i)}{\bar{x}(i)}$ (and similarly for y and q). Since actions are announced in the original game, and stage strategies are known in the auxiliary game, these posterior probabilities are known by the Players, so one can work with $\Theta = \Omega$ and take as $\mathbb{P} = \Delta(\Omega)$. Finally, the transition Q (from Ω to $\Delta(\Omega)$) is defined by the following formula:

$$Q(p, q, x, y)(p', q') = \sum_{i,j;(p(i),q(j))=(p',q')} \bar{x}(i)\bar{y}(j).$$

The probability to move from (p, q) to (p', q') under (x, y) is the probability of playing the actions that will generate these posteriors. The resulting form of the Shapley operator is:

$$\mathbf{T}(f)(p, q) = \sup_{x \in \mathbf{X}} \inf_{y \in \mathbf{Y}} \left\{ \sum_{k,\ell} p^k q^\ell g(k, \ell, x^k, y^\ell) + \sum_{i,j} \bar{x}(i)\bar{y}(j) f(p(i), q(j)) \right\} \quad [2.5]$$

where with the previous notations:

$$\sum_{i,j} \bar{x}(i)\bar{y}(j)f(p(i), q(j)) = \mathsf{E}_{Q(p,q,x,y)}\left[f(p', q')\right] = f(H(p, q, x, y))$$

and again:

$$v_\lambda = \lambda \mathbf{T}\left[\frac{(1-\lambda)}{\lambda} v_\lambda\right].$$

The corresponding equations for v_n and v_λ are due to Aumann and Maschler (1966–67) and are reproduced in Aumann and Maschler (1995), and Mertens and Zamir (1971).

Recall that the auxiliary game Γ is "equivalent" to the original one in terms of values but uses different strategy spaces. In the true game, the strategy of the opponent is unknown, hence the computation of the posterior distribution is not feasible.

Most of the results extend to the *dependent case*, introduced by Mertens and Zamir (1971). In addition to the space M endowed with the probability π, there are two signaling maps from M to A for Player 1 (resp. B for Player 2) that correspond to the initial (correlated) information of the Players on the unknown parameter. $\pi(.|a)$ then denotes the belief of Player 1 on M given his signal a (and similarly for Player 2).

2.2.3 Compact evaluations and continuous time extension

The recursive formula expressing the discounted value through the Shapley operator can be extended for values of games with the same plays but alternative evaluations of the stream of payoffs. Introduce, for $\varepsilon \in [0, 1]$ the operator Φ given by:

$$\Phi(\varepsilon, f) = \varepsilon \mathbf{T}\left(\frac{(1-\varepsilon)f}{\varepsilon}\right).$$

Then v_λ is a fixed point of $\Phi(\lambda, .)$:

$$v_\lambda = \Phi(\lambda, v_\lambda) \qquad [2.6]$$

and v_n (the value of the n-stage game) satisfies the recursive formula:

$$v_n = \Phi\left(\frac{1}{n}, v_{n-1}\right), \qquad [2.7]$$

with $v_0 = 0$. Note that [2.7] is equivalent to:

$$n v_n = \mathbf{T}((n-1)v_{n-1}) = \mathbf{T}^n(0), \qquad [2.8]$$

More generally, any probability μ on the integers induces a partition $\Pi = \{t_n; n \geq 0\}$ of $[0, 1]$ with $t_0 = 0, t_n = \sum_{m=1}^{n} \mu_m$. Consequently, the repeated game is naturally represented as a discretization of a continuous time game played between times 0 and 1, where the actions are constant on each subinterval (t_{n-1}, t_n) with length μ_n, which is the weight of stage n in the original game. Let v_Π (or equivalently v_μ) denote its value. The recursive equation can then be written as:

$$v_\Pi = \Phi(t_1, v_{\Pi_{t_1}}) \qquad [2.9]$$

where Π_{t_1} is the renormalization on $[0,1]$ of the restriction of Π to the interval $[t_1, 1]$.

The difficulty with the two recursive formulas [2.7] and [2.9], expressing v_n and v_Π, is the lack of stationarity compared to [2.6]. One way to deal with this issue is to add the time variable to the state space and to define $V_\Pi(t_n)$ as the value of the RG starting at time t_n, i.e., with evaluation μ_{n+m} for the payoff g_m at stage m. The total weight (length) of this game is no longer 1 but $1 - t_n$. With this new time variable, one obtains the equivalent recursive formula:

$$V_\Pi(t_n) = (t_{n+1} - t_n)\mathbf{T}\left(\frac{V_\Pi(t_{n+1})}{t_{n+1} - t_n}\right) \qquad [2.10]$$

which is a functional equation for V_Π. Observe that the stationarity properties of the game induce time homogeneity:

$$V_\Pi(t_n) = (1 - t_n)V_{\Pi_{t_n}}(0). \qquad [2.11]$$

By taking the linear extension of $V_\Pi(t_n)$, one can now define, for every partition Π, a function $V_\Pi(t)$ on $[0,1]$. A key lemma is the following (see Cardaliaguet et al. 2012):

Lemma 2.1. *Assume that the sequence μ_n is decreasing. Then V_Π is $2C$-Lipschitz in t, where $C = \sup_{i,j,m} |g(i,j,m)|$.*

Similarly, one can show that RG with random duration satisfies a recursive formula such as the one described above (Neyman and Sorin, 2010). Explicitly, an *uncertain duration process* $\Theta = \langle (A, \mathcal{B}, \mu), (s_n)_{n\geq 0}, \theta \rangle$ is a triple where θ is an integer-valued random variable defined on a probability space $(A, \mathcal{B}, \mu)$ with finite expectation $E(\theta)$, and each signal s_n (sent to the Players after their moves at stage n) is a measurable function defined on the probability space $(A, \mathcal{B}, \mu)$ with finite range S. An equivalent representation is through a random tree with finite expected length where the nodes at distance n correspond to the information sets at stage n. Given such a node ζ_n, known to the Players, its successor at stage $n+1$ is chosen at random according to the subtree defined by Θ at ζ_n. One can define the random iterate $\mathbf{T}^\Theta$ of the nonexpansive map $\mathbf{T}$ (see Neyman, 2003). Then, a recursive formula analogous to [2.8] holds for the value v_Θ of the game with uncertain duration Θ (see Theorem 3 in Neyman and Sorin, 2010):

$$E(\theta)v_\Theta = \mathbf{T}^\Theta(0). \qquad [2.12]$$

Note that the extension to continuous time has not yet been done for RG with random duration.

2.3. ASYMPTOTIC ANALYSIS

The asymptotic analysis aims at finding conditions for: (1) the existence of the asymptotic values $\lim_{\lambda\to 0} v_\lambda$, $\lim_{n\to\infty} v_n$, and more generally $\lim_{r\to\infty} v_{\mu^r}$, (2) equality of those limits, (3) a characterization of the asymptotic value by a formula or variational inequalities expressed from the basic data of the game. Most of the results presented in this section belong to the class of finite stochastic games with incomplete information which will be our benchmark model. While the existence of the asymptotic value is still an open problem in this framework, it has been solved in some particular important subclasses.

2.3.1 Benchmark model

A zero-sum stochastic game with incomplete information is played in discrete time as follows. Let I, J, K, L, and Ω be finite sets. At stage $n = 0$, nature chooses independently $k \in K$ and $l \in L$ according to some probability distributions $p \in \Delta(K)$ and $q \in \Delta(L)$. Player 1 privately learns his type k, and Player 2 learns l. An initial state $\omega_1 = \omega \in \Omega$ is given and known to the Players. Inductively, at stage $n = 1, 2, \ldots$ knowing the past history of moves and states $h_n = (\omega_1, i_1, j_1, \ldots, i_{n-1}, j_{n-1}, \omega_n)$, Player 1 chooses $i_n \in I$ and Player 2 chooses $j_n \in J$. The payoff at stage n is $g_n = g(k, l, i_n, j_n, \omega_n)$. The new state $\omega_{n+1} \in \Omega$ is drawn according to the probability distribution $Q(i_n, j_n, \omega_n)(\cdot)$ and (i_n, j_n, ω_{n+1}) is publicly announced. Note that this model encompasses both stochastic games and repeated games with incomplete information (with standard signaling).

Let $\mathcal{F}$ denote the set of real-valued functions f on $\Delta(K) \times \Delta(L) \times \Omega$ bounded by C, concave in p, convex in q, and $2C$-Lipschitz in (p, q) for the L_1 norm, so that for every $(p_1, q_1, p_2, q_2, \omega)$ one has:

$$|f(p_1, q_1, \omega) - f(p_2, q_2, \omega)| \leq 2C(\|p_1 - p_2\|_1 + \|q_1 - q_2\|_1),$$

where $\|p_1 - p_2\|_1 = \sum_{k\in K} |p_2^k - p_2^k|$ and similarly fo q.

In this framework, *the Shapley operator* $\mathbf{T}$ associates to a function f in $\mathcal{F}$ the function:

$$\mathbf{T}(f)(p, q, \omega) = \mathtt{val}_{x\in\Delta(I)^K \times\ y\in\Delta(J)^L} \left[\begin{array}{c} g(p, q, x, y, \omega) \\ \sum_{i,j,\tilde{\omega}} \bar{x}(i)\bar{y}(j) Q(i, j, \omega)(\tilde{\omega}) f(p(i), q(j), \tilde{\omega}) \end{array} \right] \quad [2.13]$$

where $g(p, q, x, y, \omega) = \sum_{i,j,k,l} p^k q^l x^k(i) y^l(j) g(k, l, i, j, \omega)$ is the expected stage payoff.

$\mathbf{T}$ maps $\mathcal{F}$ to itself (see Laraki, 2001a, 2004). The associated *projective operator* (corresponding to the game where only the future matters) is:

$$\mathbf{R}(f)(p, q, \omega) = \mathtt{val}_{x\in\Delta(I)^K \times\ y\in\Delta(J)^L} \left[\sum_{i,j,\tilde{\omega}} \bar{x}(i)\bar{y}(j) Q(i, j, \omega)(\tilde{\omega}) f(p(i), q(j), \tilde{\omega}) \right]. \quad [2.14]$$

Any accumulation point (for the uniform norm on $\mathcal{F}$) of the equi-Lipschitz family $\{v_\lambda\}$, as λ goes to zero or of $\{v_n\}$, as n goes to infinity, is a fixed point of the projective operator. Observe, however, that any function in $\mathcal{F}$ which is independent of ω is a fixed point of $\mathbf{R}$.

2.3.2 Basic results

2.3.2.1 Incomplete information

When Ω is a singleton, the game is a repeated game with incomplete information (Aumann and Maschler, 1995). Moreover, when information is incomplete on one side (L is a singleton), Aumann and Maschler proved the existence of the asymptotic value:

$$v = \lim_{n\to\infty} v_n = \lim_{\lambda\to 0} v_\lambda$$

and provided the following famous explicit formula:

$$v(p) = \mathtt{Cav}_{\Delta(K)}(u)(p)$$

where:

$$u(p) = \mathtt{val}_{\Delta(I)\times\Delta(J)} \sum_k p^k g(k, x, y)$$

is the value of the *nonrevealing game* and $\mathtt{Cav}_C$ is the concavification operator: given ϕ, a real bounded function defined on a convex set C, $\mathtt{Cav}_C(\phi)$ is the smallest concave function on C greater than ϕ.

The Aumann-Maschler proof works as follow. A splitting lemma (Zamir (1992), Proposition 3.2.) shows that if Player 1 can guarantee a function $f(p)$, he can guarantee $\mathtt{Cav}_{\Delta(K)}(f)(p)$. Since Player 1 can always guarantee the value of the nonrevealing game $u(p)$ (by ignoring his information), he can also guarantee $\mathtt{Cav}_{\Delta(K)}(u)(p)$. As for Player 2, by playing a best response in the nonrevealing game stage by stage given his updated belief, he can prevent Player 1 to obtain more than $\mathtt{Cav}_{\Delta(K)}(u)(p)$ up to some error term which is at most, using martingale arguments, vanishing on average.

For RG of incomplete information on both sides, Mertens and Zamir (1971) proved the existence of $v = \lim_{n\to\infty} v_n = \lim_{\lambda\to 0} v_\lambda$. They identified v as the unique solution of the system of functional equations with unknown real function ϕ on $\Delta(I) \times \Delta(J)$:

$$\phi(p, q) = \mathtt{Cav}_{p\in\Delta(K)} \min\{\phi, u\}(p, q), \quad \phi(p, q) = \mathtt{Vex}_{q\in\Delta(L)} \max\{\phi, u\}(p, q). \quad [2.15]$$

u is again the value of the nonrevealing game with $u(p, q) = \mathtt{val}_{\Delta(I)\times\Delta(J)} \sum_{k,\ell} p^k q^\ell g(k, \ell, x, y)$. The operator $u \mapsto \phi$ given by [2.15] will be called the Mertens-Zamir system and denoted by **MZ**. It associates to any continuous function w on $\Delta(K) \times \Delta(L)$, a unique concave-convex continuous function $\mathbf{MZ}(w)$ (see Laraki, 2001b).

One of Mertens and Zamir's proofs is as follows. Using sophisticated reply strategies, one shows that $h = \liminf v_n$ satisfies:

$$h(p, q) \geq \mathtt{Cav}_{p \in \Delta(K)} \mathtt{Vex}_{q \in \Delta(L)} \max\{h, u\}(p, q). \qquad [2.16]$$

Define inductively dual sequences of functions $\{c_n\}$ and $\{d_n\}$ on $\Delta(K) \times \Delta(L)$ by $c_0 \equiv -\infty$ and

$$c_{n+1}(p, q) = \mathtt{Cav}_{p \in \Delta(K)} \mathtt{Vex}_{q \in \Delta(L)} \max\{c_n, u\}(p, q)$$

and similarly for $\{d_n\}$. Then they converge respectively to c and d satisfying:

$$c(p, q) = \mathtt{Cav}_{p \in \Delta(K)} \mathtt{Vex}_{q \in \Delta(L)} \max\{c, u\}(p, q),$$

$$d(p, q) = \mathtt{Vex}_{q \in \Delta(L)} \mathtt{Cav}_{p \in \Delta(K)} \min\{d, u\}(p, q) \qquad [2.17]$$

A comparison principle is then used to deduce that $c \geq d$. In fact, consider an extreme point (p_0, q_0) of the convex hull of the set where $d - c$ is maximal. Then one shows that the $\mathtt{Vex}$ and $\mathtt{Cav}$ operators in the above formula [2.15] at (p_0, q_0) are trivial (there is no use of information) which implies $c(p_0, q_0) \geq u(p_0, q_0) \geq d(p_0, q_0)$. Finally $h \geq c$ implies by symmetry that $\lim_{n \to \infty} v_n$ exists.

2.3.2.2 *Stochastic games*

For stochastic games (K and L are reduced to a singleton), the existence of $\lim_{\lambda \to 0} v_\lambda$ in the finite case (Ω, I, J finite) was first proved by Bewley and Kohlberg (1976a) using algebraic arguments: the optimality equations for strategies and values in the Shapley operator can be written as a finite set of polynomial equalities and inequalities and thus define a semi-algebraic set in some euclidean space $\mathbb{R}^N$. By projection, v_λ has an expansion in Puiseux series hence has a limit as λ goes to 0.

An alternative, more elementary approach has been recently obtained by Oliu-Barton (2013).

The existence of $\lim_{n \to \infty} v_n$ may be deduced from $\lim_{\lambda \to 0} v_\lambda$ by comparison arguments, see Bewley and Kohlberg (1976b) or, more generally, Theorem 2.1.

2.3.3 Operator approach

The operator approach corresponds to the study of the asymptotic value trough the Shapley operator. It was first introduced by Kohlberg (1974) in the analysis of finite absorbing games (stochastic games with a single non-absorbing state). The author uses the additional information obtained from the derivative of the Shapley operator (which is defined from $\mathbf{R}$ to itself in this case) at $\lambda = 0$ to deduce the existence of $v = \lim_{\lambda \to 0} v_\lambda = \lim_{n \to \infty} v_n$ and a characterization of v through variational inequalities.

Rosenberg and Sorin (2001) extended this approach to general RG. This tool provides sufficient conditions under which v exists and exhibits a variational characterization.

2.3.3.1 Nonexpansive monotone maps

We introduce here general properties of operators that will be applied to repeated games through the Shapley operator **T**. We study iterates of an operator T mapping $\mathcal{G}$ to itself, where $\mathcal{G}$ is a subspace of the space of real bounded functions on some set E.

Assume:

1) $\mathcal{G}$ is a convex cone, containing the constants and closed for the uniform norm, denoted by $\|\cdot\|$.

2) T is monotonic: $f \geq g$ implies $Tf \geq Tg$, and translates the constants $T(f+c) = T(f) + c$.

The second assumption implies in particular that T is non-expansive.

In our benchmark model, $E = \Delta(K) \times \Delta(L) \times \Omega$ and $\mathcal{G}=\mathcal{F}$. In general RG, one may think of E as the set of consistent probabilities on the universal belief space.

Define the iterates and fixed points of T as:

$$V_n = T^n[0], \quad V_\lambda = T[(1-\lambda)V_\lambda]$$

and by normalizing $v_n = V_n$, $v_\lambda = \lambda V_\lambda$. Introducing the family of operators:

$$\Phi(\varepsilon, f) = \varepsilon T\left[\frac{1-\varepsilon}{\varepsilon} f\right], \qquad [2.18]$$

one obtains:

$$v_n = \Phi\left(\frac{1}{n}, v_{n-1}\right), \qquad v_\lambda = \Phi(\lambda, v_\lambda) \qquad [2.19]$$

and we consider the asymptotic behavior of these families of functions (analogous to the n stage and discounted values) which thus relies on the properties of $\Phi(\varepsilon, \cdot)$, as ε goes to 0.

A basic result giving a sufficient condition for the existence and equality of the limits is due to Neyman (2003):

Theorem 2.1. *If v_λ is of bounded variation in the sense that:*

$$\sum_i \|v_{\lambda_{i+1}} - v_{\lambda_i}\| < \infty, \qquad [2.20]$$

for any sequence λ_i decreasing to 0. Then $\lim_{n\to\infty} v_n = \lim_{\lambda\to 0} v_\lambda$.

The result extends to random duration processes ($\lim_{\mathsf{E}(\theta)\to+\infty} v_\Theta = \lim_{\lambda\to 0} v_\lambda$) when the expected duration of the current stage decreases along the play (Neyman and Sorin, 2010).

Following Rosenberg and Sorin (2001), Sorin (2004) defines spaces of functions that will correspond to upper and lower bounds on the families of values.

(i) *Uniform domination.* Let $\mathcal{L}^+$ be the space of functions $f \in \mathcal{G}$ that satisfy: there exists $M_0 \geq 0$ such that $M \geq M_0$ implies $T(Mf) \leq (M+1)f$. ($\mathcal{L}^-$ is defined similarly.)

(ii) *Pointwise domination.* When the set E is not finite, one can introduce the larger class $\mathcal{S}^+$ of functions satisfying $\limsup_{M\to\infty}\{T(Mf)(e) - (M+1)f(e)\} \leq 0$, $\forall e \in E$. ($\mathcal{S}^-$ is defined in a dual way).

The comparison criteria for uniform domination is expressed by the following result:

Theorem 2.2. *Rosenberg and Sorin (2001)*
If $f \in \mathcal{L}^+$, $\limsup_{n\to\infty} v_n$ and $\limsup_{\lambda\to 0} v_\lambda$ are less than f.
Consequently, if the intersection of the closure of $\mathcal{L}^+$ and $\mathcal{L}^-$ is not empty, then both $\lim_{n\to\infty} v_n$ and $\lim_{\lambda\to 0} v_\lambda$ exist and coincide.

And for pointwise domination by the next property:

Theorem 2.3. *Rosenberg and Sorin (2001)*
Assume E compact. Let $\mathcal{S}_0^+$ (resp. $\mathcal{S}_0^-$) be the space of continuous functions in $\mathcal{S}^+$ (resp. $\mathcal{S}^-$). Then, for any $f^+ \in \mathcal{S}_0^+$ and $f^- \in \mathcal{S}_0^-$, $f^+ \geq f^-$.
Consequently, the intersection of the closures of $\mathcal{S}_0^+$ and $\mathcal{S}_0^-$ contains at most one point.

These two results provide sufficient conditions for the uniqueness of a solution satisfying the properties. The next one gives a sufficient condition for the existence of a solution.

T has the *derivative property* if for every $f \in \mathcal{G}$ and $e \in E$:

$$\varphi^*(f)(e) = \lim_{\varepsilon\to 0^+} \frac{\Phi(\varepsilon, f)(e) - f(e)}{\varepsilon}$$

exists in $\overline{\mathbb{R}}$. If such a derivative exists, $\mathcal{S}^+$ is the set of functions $f \in \mathcal{G}$ that satisfy $\varphi^*(f)(e) \leq 0$, for all $e \in E$ and similarly for $\mathcal{S}^-$, $\varphi^*(f)(e) \geq 0$.

Theorem 2.4. *Rosenberg and Sorin (2001)*
Assume that T has the derivative property and E is compact. Let f be such that φ^* "changes sign" at f, meaning that there exist two sequences $\{f_n^-\}$ and $\{f_n^+\}$ of continuous functions converging to f such that $\varphi^*(f_n^-) \leq 0 \leq \varphi^*(f_n^+)$. Then, f belongs to the closures of $\mathcal{S}_0^+$ and $\mathcal{S}_0^-$.

Definition 2.2. *T has the recession property if $\lim_{\varepsilon\to 0} \Phi(\varepsilon, f)(\theta) = \lim_{\varepsilon\to 0} \varepsilon T(\frac{f}{\varepsilon})(\theta)$, written $R(f)(\theta)$, exists.*

Theorem 2.5. *Vigeral (2010b)*

Assume that T has the recession property and is convex. Then v_n (resp. v_λ) has at most one accumulation point.

The proof uses the inequality: $R(f+g) \leq T(f) + R(g)$ and relies on properties of the family of operators T_m defined by:

$$T_m(f) = \frac{1}{m} T^m(mf). \qquad [2.21]$$

2.3.3.2 Applications to RG

The Shapley operator **T** satisfies the derivative and recession properties, so the results of the previous section can be applied.

Absorbing games are stochastic games where the state can change at most once on a play Kohlberg (1974).

Theorem 2.6. *Rosenberg and Sorin (2001)*

$\lim_{n\to\infty} v_n$ *and* $\lim_{\lambda\to 0} v_\lambda$ *exist and are equal in absorbing games with compact action spaces.*

Recursive games are stochastic games where the payoff is 0 until the state becomes absorbing (Everett, 1957).

Theorem 2.7. *Sorin (2003), Vigeral (2010c)*

$\lim_{n\to\infty} v_n$ *and* $\lim_{\lambda\to 0} v_\lambda$ *exist and are equal in recursive games with finite state space and compact action spaces.*

Notice that the algebraic approach (for stochastic games) cannot be used when action or state spaces are not finite. However, one cannot expect to get a proof for general stochastic games with finite state space, see Vigeral's counterexample in Section 9.

Pointwise domination is used to prove existence and equality of $\lim_{n\to\infty} v_n$ and $\lim_{\lambda\to 0} v_\lambda$ through the derived game φ^* and the recession operator **R** in the following cases.

Theorem 2.8. *Rosenberg and Sorin (2001)*

$\lim_{n\to\infty} v_n$ *and* $\lim_{\lambda\to 0} v_\lambda$ *exist and are equal in repeated games with incomplete information.*

This provides an alternative proof of the result of Mertens and Zamir (1971). One shows that any accumulation point w of the family $\{v_\lambda\}$ (resp. $\{v_n\}$) as $\lambda \to 0$ (resp. $n \to \infty$) belongs to the closure of $\mathcal{S}_0^+$, hence, by symmetry, the existence of a limit follows using Theorem 2.3. More precisely, Theorem 2.4 gives the following characterization of the asymptotic value: given a real function f on a linear set X, denote by $\mathcal{E}f$ the projection on X of the extreme points of its epigraph. Then: $v = \lim_{n\to\infty} v_n = \lim_{\lambda\to 0} v_\lambda$ exists and is the unique saddle continuous function satisfying both inequalities:

$$\mathbf{Q1}: p \in \mathcal{E}v(\cdot, q) \Rightarrow v(p,q) \le u(p,q), \mathbf{Q2}: q \in \mathcal{E}v(p,\cdot) \Rightarrow v(p,q) \ge u(p,q), \quad [2.22]$$

where u is the value of the nonrevealing game. One then shows that this is equivalent to the characterization of Mertens and Zamir [2.15] (see also Laraki (2001a) and Section 3.4).

All the results above also hold for (decreasing) random duration processes (Neyman and Sorin, 2010).

Theorem 2.9. *Rosenberg (2000)*

$\lim_{n\to\infty} v_n$ and $\lim_{\lambda\to 0} v_\lambda$ exist and are equal in finite absorbing games with incomplete information on one side.

This is the first general subclass where both stochastic and information aspects are present and in which the asymptotic value, $\lim_{\lambda\to 0} v_\lambda = \lim_{n\to\infty} v_n$ exists.

Theorem 2.10. *Vigeral (2010b)*

$\lim_{n\to\infty} v_n$ (resp. $\lim_{\lambda\to 0} v_\lambda$) exists in repeated games where one Player controls the transition and the family $\{v_n\}$ (resp. $\{v_\lambda\}$) is relatively compact.

This follows from the convexity of $\mathbf{T}$ in that case. It applies in particular to dynamic programming or games with incomplete information on one side (see also Sections 5.3 and 5.4).

Vigeral (2010b) provides a simple stochastic game in which the sets $\mathcal{L}^+$ and $\mathcal{L}^-$ associated to $\mathbf{T}$ do not intersect, but the sets $\mathcal{L}_m^+$ and $\mathcal{L}_m^-$ associated to the operator $\mathbf{T}_m$ do intersect for some m large enough. This suggests that, for games in the benchmark model, the operator approach should be extended to iterations of the Shapley operator.

2.3.4 Variational approach

Inspired by the tools used to prove existence of the value in differential games (see Section 6), Laraki (2001a,b, 2010) introduced the variational approach to obtain existence of $\lim v_\lambda$ in a RG and to provide a characterization of the limit via variational inequalities. Rather than going to the limit between time t and $t+h$ in the dynamic programming equation, one uses the fact that v_λ is a fixed point of the Shapley operator [2.6] (as in the operator approach). Given an optimal stationary strategy of the maximizer (resp. minimizer) in the λ-discounted game, one deduces an inequality (involving both value and strategy) that must be satisfied as λ goes to zero. Finally, a comparison principle is used to deduce that only one function satisfies the two inequalities (which have to be sharp enough to specify a single element).

The extension of the variational approach to $\lim v_n$ or $\lim v_\mu$ was solved recently by Cardaliaguet et al. (2012) by increasing the state space to $\Omega \times [0,1]$, that is, by introducing time as a new variable, and viewing each evaluation as a particular discretization of the time interval $[0,1]$ (Section 2.3). From [2.10], one shows that accumulation points are viscosity solutions of a related differential equation and finally, comparison tools give uniqueness.

To understand the approach (and the exact role of the time variable), we first describe it for discounted games and then for general evaluations in three classes: RG with incomplete information, absorbing games, and splitting games.

2.3.4.1 Discounted values and variational inequalities

(A) RG with incomplete information. We follow Laraki (2001a) (see also Cardaliaguet et al., 2012). To prove (uniform) convergence, it is enough to show that $\mathcal{V}_0$, the set of accumulation points of the family $\{v_\lambda\}$, is a singleton. Let $\mathcal{V}$ be the set of fixed points of the projective operator $\mathbf{R}$ [2.14] and observe that $\mathcal{V}_0 \subset \mathcal{V}$.

Given $w \in \mathcal{V}_0$, denote by $\mathbf{X}(p,q,w) \subseteq \mathbf{X} = \Delta(I)^K$ the set of optimal strategies for Player 1 (resp. $\mathbf{Y}(p,q,w) \subseteq \mathbf{Y} = \Delta(J)^L$ for Player 2) in the projective operator $\mathbf{R}$ for w at (p,q). A strategy $x \in \mathbf{X}$ of Player 1 is called nonrevealing at p, $x \in NR_{\mathbf{X}}(p)$ if $p(i) = p$ for all $i \in I$ with $x(i) > 0$ and similarly for $y \in \mathbf{Y}$. The value of the nonrevealing game is given by:

$$u(p,q) = \mathtt{val}_{NR_{\mathbf{X}}(p) \times NR_{\mathbf{Y}}(q)} g(p,q,x,y)\ . \qquad [2.23]$$

Lemma 2.2. *Any $w \in \mathcal{V}_0$ satisfies:*

$$\textbf{\textit{P1}}: \quad \textit{If } \mathbf{X}(p,q,w) \subset NR_{\mathbf{X}}(p) \textit{ then } w(p,q) \le u(p,q) \qquad [2.24]$$

$$\textbf{\textit{P2}}: \quad \textit{If } \mathbf{Y}(p,q,w) \subset NR_{\mathbf{Y}}(q) \textit{ then } w(p,q) \ge u(p,q). \qquad [2.25]$$

The interpretation is straightforward. If by playing optimally a Player should not use asymptotically his information, he cannot get more that the value of the nonrevealing game, because the other Player has always the option of playing nonrevealing. What is remarkable is that the two above properties, plus geometry (concavity in p, convexity in q) and smoothness (continuity) are enough to characterize the asymptotic value.

The proof is simple. Let $v_{\lambda_n} \to w$ and x_n optimal for the maximizer in the Shapley equation $v_{\lambda_n}(p,q) = \Phi(\lambda_n, v_{\lambda_n})(p,q)$. Then, for every nonrevealing pure strategy j of Player 2, one has:

$$v_{\lambda_n}(p,q) \le \lambda_n g(x_n, j, p, q) + (1-\lambda_n) \sum_i \bar{x}_n(i) v_{\lambda_n}(p_n(i), q).$$

Jensen's inequality implies $\sum_i \bar{x}_n(i) v_{\lambda_n}(p_n(i), q) \le v_{\lambda_n}(p,q)$ hence $v_{\lambda_n}(p,q) \le g(x_n, j, p, q)$. Thus, if $\bar{x} \in \mathbf{X}(p,q,w)$ is an accumulation point of $\{x_n\}$, one obtains $w(p,q) \le g(\bar{x}, j, p, q)$, $\forall j \in J$. Since by assumption, $\mathbf{X}(p,q,w) \subset NR_{\mathbf{X}}(p)$, one gets $w(p,q) \le u(p,q)$.

For uniqueness, the following comparison principle is established:

Lemma 2.3. *Let w_1 and w_2 be in $\mathcal{V}$ satisfying P_1 and P_2 respectively, then $w_1 \le w_2$.*

The proof follows an idea by Mertens and Zamir (see Section 3.2.1.). Let (p_0, q_0) be an extreme point of the compact set where the difference $(w_1 - w_2)(p,q)$ is maximal. Then, one has that $\mathbf{X}(p_0, q_0, w_1) \subset NR_{\mathbf{X}}(p_0)$, and $\mathbf{Y}(p_0, q_0, w_2) \subset NR_{\mathbf{Y}}(q_0)$. In fact, both functions being concave, w_1 has to be strictly concave at p_0 (if it is an interior point) and the result follows (see the relation with [2.22] **Q1**, **Q2**). Thus, $w_1(p_0, q_0) \le u(p_0, q_0) \le w_2(p_0, q_0)$.

Consequently one obtains:

Theorem 2.11. $\lim_{\lambda \to 0} v_\lambda$ *exists and is the unique function in $\mathcal{V}$ that satisfies* ***P1*** *and* ***P2***.

This characterization is equivalent to the Mertens-Zamir system (Laraki, 2001a; Rosenberg and Sorin, 2001) and to the two properties **Q1** and **Q2** established above [2.22].

(B) Absorbing games. We follow Laraki (2010). Recall that only one state is nonabsorbing, hence in the other states one can assume without loss of generality that the payoff is constant (and equal to the value). The game is thus defined by the following elements: two finite sets I and J, two (payoff) functions f, g from $I \times J$ to $\mathbb{R}$ and a function π from $I \times J$ to $[0,1]$: $f(i,j)$ is the current payoff, $\pi(i,j)$ is the probability of nonabsorption, and $g(i,j)$ is the absorbing payoff. Define $\pi^*(i,j) = 1 - \pi(i,j)$, $f^*(i,j) = \pi^*(i,j) \times g(i,j)$, and extend bilinearly any $\varphi : I \times J \to \mathbf{R}$ to $\mathbf{R}^I \times \mathbf{R}^J$ as usual: $\varphi(\alpha, \beta) = \sum_{i \in I, j \in J} \alpha^i \beta^j \varphi(i,j)$.

The variational approach proves the following new characterization of $\lim_{\lambda \to 0} v_\lambda$ as the value of a strategic one-shot game that can be interpreted as a "limit" game (see Section 8).

Theorem 2.12. *As $\lambda \to 0$, v_λ converges to v given by:*

$$v = \mathtt{val}_{((x,\alpha),(y,\beta)) \in (\Delta(I) \times \mathbf{R}_+^I) \times (\Delta(J) \times \mathbf{R}_+^J)} \frac{f(x,y) + f^*(\alpha, y) + f^*(x, \beta)}{1 + \pi^*(\alpha, y) + \pi^*(x, \beta)}. \qquad [2.26]$$

Actually, if x_λ is an optimal stationary strategy for Player 1 in the λ-discounted game, then: $v_\lambda \leq \lambda f(x_\lambda, j) + (1-\lambda)(\pi(x_\lambda, j) v_\lambda + f^*(x_\lambda, j), \forall j \in J$. Accordingly:

$$v_\lambda \leq \frac{f(x_\lambda, j) + f^*(\frac{(1-\lambda)x_\lambda}{\lambda}, j)}{1 + \pi^*(\frac{(1-\lambda)x_\lambda}{\lambda}, j)}, \quad \forall j \in J. \qquad [2.27]$$

Consequently, given an accumulation point w of $\{v_\lambda\}$, there exists $\overline{x} \in \Delta(I)$ an accumulation point of $\{x_\lambda\}$ such that for all $\varepsilon > 0$, there exists $\overline{\alpha} \in \mathbf{R}_+^I$ $\left(\text{of the form } \frac{(1-\overline{\lambda})x_{\overline{\lambda}}}{\overline{\lambda}}\right)$ satisfying:

$$w \leq \frac{f(\overline{x}, j) + f^*(\overline{\alpha}, j)}{1 + \pi^*(\overline{\alpha}, j)} + \varepsilon, \quad \forall j \in J. \qquad [2.28]$$

By linearity, the last inequality extends to any $y \in \Delta(J)$. On the other hand, w is a fixed point of the projective operator and $\overline{x}$ is optimal there, hence:

$$w \leq \pi(\overline{x}, y)\, w + f^*(\overline{x}, y), \qquad \forall y \in \Delta(J). \qquad [2.29]$$

Inequality [2.29] is linear thus extends to $\beta \in \mathbf{R}_+^J$ and combining with [2.28] one obtains that w satisfies the inequality:

$$w \leq \sup_{(x,\alpha)\in\Delta(I)\times\mathbf{R}_+^I} \inf_{(y,\beta)\in\Delta(J)\times\mathbf{R}_+^J} \frac{f(x,y) + f^*(\alpha, y) + f^*(x, \beta)}{1 + \pi^*(\alpha, y) + \pi^*(x, \beta)}. \qquad [2.30]$$

Following the optimal strategy of Player 2 yields the reverse inequality. Uniqueness and characterization of $\lim v_\lambda$ is deduced from the fact that $\sup\inf \leq \inf\sup$.

2.3.4.2 General RG and viscosity tools

We follow Cardaliaguet et al. (2012).

(A) RG with incomplete information. Consider an arbitrarily evaluation probability μ on $\mathbb{N}^*$ inducing a partition Π. Let $V_\Pi(t_k, p, q)$ be the value of the game starting at time t_k. One has $V_\Pi(1, p, q) := 0$ and:

$$V_\Pi(t_n, p, q) = \mathtt{val}[\mu_{n+1} g(x, y, p, q) + \sum_{i,j} x(i) y(j) V_\Pi(t_{n+1}, p(i), q(j))]. \qquad [2.31]$$

Moreover V_Π belongs to $\mathcal{F}$. Given a sequence $\{\mu^m\}$ of decreasing evaluations ($\mu_n^m \geq \mu_{n+1}^m$), Lemma 2.1 implies that the family of $V_{\Pi(m)}$ associated to partitions $\Pi(m)$ is equi-Lipschitz.

Let $\mathcal{T}_0$ be the nonempty set of accumulation points (for the uniform convergence) as $\mu_1^m \to 0$. Let $\mathcal{T}$ be the set of real continuous functions W on $[0,1] \times \Delta(K) \times \Delta(L)$ such that for all $t \in [0,1]$, $W(t, ., .) \in \mathcal{V}$. Recall that $\mathcal{T}_0 \subset \mathcal{T}$. $\mathbf{X}(t, p, q, W)$ is the set of optimal strategies for Player 1 in $\mathbf{R}(W(t, ., .))$ and $\mathbf{Y}(t, p, q, W)$ is defined accordingly. Define two properties for a function $W \in \mathcal{T}$ and a $\mathcal{C}^1$ test function $\phi : [0,1] \to \mathbb{R}$ as follows:

P1′ : If $t \in [0,1)$ is such that $\mathbf{X}(t,p,q,W)$ is nonrevealing and $W(\cdot,p,q) - \phi(\cdot)$ has a global maximum at t, then $u(p,q) + \phi'(t) \geq 0$.

P2′ : If $t \in [0,1)$ is such that $\mathbf{Y}(t,p,q,W)$ is nonrevealing and $W(\cdot,p,q) - \phi(\cdot)$ has a global minimum at t then $u(p,q) + \phi'(t) \leq 0$.

The variational counterpart of Lemma 2.2 is obtained by using [2.31].

Lemma 2.4. *Any $W \in \mathcal{T}_0$ satisfies **P1′** and **P2′**.*

The proof is as follows. Let (t,p,q) as is **P1′**. Adding the function $s \mapsto (s-t)^2$ to ϕ if necessary, we can assume that this global maximum is strict. Let $W_{\mu^k} \to W$. Let Π^k be the partition associated to μ^k and $t^k_{n(k)}$ be a global maximum of $W_{\mu^k}(\cdot,p,q) - \phi(\cdot)$ on Π^k. Since t is a strict maximum, one has $t^k_{n(k)} \to t$, as $k \to \infty$.

Proceeding as in Section 2.3.4.1. one obtains with $x_k \in \mathbf{X}$ being optimal for W_{μ^k}:

$$0 \leq g(x_k,j,p,q)\left(t^k_{n(k)+1} - t^k_{n(k)}\right) + \left[W_{\mu^k}\left(t^k_{n(k)+1},p,q\right) - W_{\mu^k}\left(t^k_{n(k)},p,q\right)\right].$$

Since $t^k_{n(k)}$ is a global maximum of $W_m(\cdot,p,q) - \phi(\cdot)$ on Π^k one deduces:

$$0 \leq g(x_k,j,p,q)\left(t^k_{n(k)+1} - t^k_{n(k)}\right) + \left[\phi\left(t^k_{n(k)+1},p,q\right) - \phi\left(t^k_{n(k)},p,q\right)\right].$$

so $0 \leq g(\bar{x},j,p,q) + \phi'(t)$ for any accumulation point $\bar{x}$ of $\{x_k\}$. The result follows because $\bar{x}$ is nonrevealing by **P1′**. Uniqueness is proved by the following comparison principle:

Lemma 2.5. *Let W_1 and W_2 in $\mathcal{T}$ satisfy* **P1′**, **P2′**, *and also the terminal condition:*

$$W_1(1,p,q) \leq W_2(1,p,q), \quad \forall (p,q) \in \Delta(K) \times \Delta(L).$$

Then $W_1 \leq W_2$ on $[0,1] \times \Delta(K) \times \Delta(L)$.

Consequently, we obtain:

Theorem 2.13. *V_π converges uniformly to the unique point $W \in \mathcal{T}$ that satisfies the variational inequalities* **P1′** *and* **P2′** *and the terminal condition $W(1,p,q) = 0$. In particular, $v_\mu(p,q)$ converges uniformly to $v(p,q) = W(0,p,q)$ and $W(t,p,q) = (1-t)v(p,q)$, where $v = \mathbf{MZ}(u)$.*

To summarize the idea of the proof, (A) one uses viscosity solutions (smooth majorant or minorant) to obtain first-order conditions on the accumulation points of the sequence of values and on the corresponding optimal strategies. (B) One then considers two accumulation points (functions) and establishes property of an extreme point of the set of states where their difference is maximal. At this point, the optimal strategies have specific aspects (C) that imply that (A) gives uniqueness.

On the other hand, the variational approach proof is much stronger than needed. Continuous time is used as a tool to show eventually, that the asymptotic value $W(t) = (1-t)v$ is linear in t. However, one must first keep in mind that linearity is a valid conclusion only if the existence of the limit is known (which is the statement that needs to be shown). Second, if g in equation [2.31] is time-dependent in the game on $[0, 1]$, the same proof and characterization still work (see Section 8.3.3. for continuous-time games with incomplete information).

(B) Absorbing games. Consider a decreasing evaluation $\mu = \{\mu_n\}$. Denote by v_μ the value of the associated absorbing game. Let $W_\mu(t_m)$ be the value of the game starting at time t_m defined recursively by $W_\mu(1) = 0$ and:

$$W_\mu(t_m) = \mathtt{val}_{(x,y)\in\Delta(I)\times\Delta(J)}\left[\mu_{m+1}f(x,y) + \pi(x,y)W_\mu(t_{m+1}) + (1-t_{m+1})f^*(x,y)\right]. \quad [2.32]$$

Under monotonicity of μ, the linear interpolation of W_μ is a $2C$-Lipschitz continuous in $[0, 1]$. Set for any $(t, a, b, x, \alpha, y, \beta) \in [0, 1] \times \mathbb{R} \times \mathbb{R} \times \Delta(I) \times \mathbb{R}_+^I \times \Delta(J) \times \mathbb{R}_+^J$,

$$h(t,a,b,x,\alpha,y,\beta) = \frac{f(x,y) + (1-t)[f^*(\alpha,y) + f^*(x,\beta)] - \left[\pi^*(\alpha,y) + \pi^*(x,\beta)\right]a + b}{1 + \pi^*(\alpha,y) + \pi^*(x,\beta)}.$$

Define the Hamiltonian of the game as:

$$H(t,a,b) = \mathtt{val}_{((x,\alpha),(y,\beta))\in(\Delta(I)\times\mathbf{R}_+^I)\times(\Delta(J)\times\mathbf{R}_+^J)}h(t,a,b,x,\alpha,y,\beta).$$

Theorem 2.12 implies that H is well defined (the value exists).

Define two variational inequalities for a continuous function U on [0,1] as follows: for all $t \in [0, 1)$ and any $\mathcal{C}^1$ function $\phi : [0, 1] \to \mathbb{R}$:

R1: If $U(\cdot) - \phi(\cdot)$ admits a global maximum at $t \in [0, 1)$ then $H^-(t, U(t), \phi'(t)) \geq 0$.
R2: If $U(\cdot) - \phi(\cdot)$ admits a global minimum at $t \in [0, 1)$ then $H^+(t, U(t), \phi'(t)) \leq 0$.

Lemma 2.6. *Any accumulation point $U(\cdot)$ of $W_\mu(\cdot)$ satisfies **R1** and **R2**.*

That is, U is a viscosity solution of the HJB equation $H(t, U(t), \nabla U(t)) = 0$. The comparison principle is as one expects:

Lemma 2.7. *Let U_1 and U_2 be two continuous functions satisfying **R1-R2** and $U_1(1) \leq U_2(1)$. Then $U_1 \leq U_2$ on $[0, 1]$.*

Consequently, we get:

Theorem 2.14. *W_μ converges to the unique C Lipschitz solution to* ***R1–R2*** *with $U(1) = 0$. Consequently, $U(t) = (1 - t)v$ and v_μ converges to v given by [2.26].*

(C) Splitting games. In RG with incomplete information on the one side, the use of information has an impact on the current payoff and on the state variable. However, the difference between the payoff of the mean strategy and the expected payoff is proportional to the L^1 variation of the martingale of posteriors at that stage hence on the average vanishes (Zamir, 1992, Propositions 3.8, 3.13, and 3.14). One can assume thus that the informed Player is playing optimally in the non-revealing game and the remaining aspect is the strategic control of the state variable: the martingale of posteriors. This representation, introduced in Sorin (2002) p. 50, leads to a continuous time martingale maximization problem $\max_{p_t} \int_0^1 u(p_t)dt$ (where the max is over all càdlàg martingales $\{p_t\}$ on $\Delta(K)$ with $p_0 = p$). The value is clearly $Cav_{\Delta(K)}(u)(p)$. Moreover, the optimal strategy in the RG that does the splitting at the beginning is maximal in the continuous time problem and inversely, any maximal martingale is ϵ-optimal in the RG for a sufficiently fine time partition.

Since then, this representation has been extended to nonautonomous continuous time games with incomplete information on the one side (see Section 9.4).

The search for an analog representation for the asymptotic value of RG with incomplete information on *both* sides naturally leads to the splitting game (Sorin, 2002, p. 78). It is a stochastic game where each Player controls a martingale. A 1-Lipschitz continuous stage payoff function U from $\Delta(K) \times \Delta(L)$ to $\mathbb{R}$ is given and the RG is played as follows. At stage m, knowing the state variable (p_m, q_m) in $\Delta(K) \times \Delta(L)$, Player 1 chooses p_{m+1} according to some $\theta \in M^K_{p_m}$ and Player 2 chooses q_{m+1} according to some $\nu \in M^L_{q_m}$ where M^K_p stands for the set of probabilities on $\Delta(K)$ with expectation p (and similarly for M^L_q). The current payoff is $U(p_{m+1}, q_{m+1})$ and the new state (p_{m+1}, q_{m+1}) is announced to both Players.

Existence of $\lim v_\lambda$ and its identification as $\mathbf{MZ}(U)$ was proved in that case by Laraki (2001a). One shows that the associated Shapley operator (called the *splitting operator*):

$$\mathbf{T}(f)(p, q) = \mathtt{val}_{\theta \in M^K_p \times \nu \in M^L_q} \int_{\Delta(K)\times\Delta(L)} [U(p', q') + f(p', q')]\theta(dp')\nu(dq') \qquad [2.33]$$

preserves 1-Lipschitz continuity (for the L^1 norm on $\Delta(K) \times \Delta(L)$): note the difficulty is due to the fact that the strategy sets depend on the state. This result is a consequence of a splitting lemma proved in Laraki (2004), and guarantees that the discounted values form a family of equi-Lipschitz functions, so one can mimic the variational approach proof as in RG with incomplete information on both sides. For the extension of preservation of equi-Lipschitz continuity in MDP see Laraki and Sudderth (2004).

The extension to $\lim v_\mu$ for general evaluation μ appears in Cardaliaguet et al. (2012) using the same time extension trick as in the previous paragraph (A) starting from the Shapley equation:

$$V_\Pi(t_n,p,q) = \texttt{val}_{\theta\in M_p^K\times\nu\in M_q^L}\int_{\Delta(K)\times\Delta(L)}[\mu_{n+1}U(p',q') + V_\Pi(t_{n+1},p',q')]\theta(dp')\nu(dq'). \quad [2.34]$$

Laraki (2001b) extends the definition of the splitting game from $\Delta(K)$ and $\Delta(L)$ to any convex compact subsets C and D of $\mathbf{R}^n$ and uses it to show existence and uniqueness of a concave-convex continuous function ϕ on $C\times D$ that satisfies the Mertens-Zamir system: $\phi = \texttt{Cav}_C\min(\phi,U) = \texttt{Vex}_D\max(\phi,U)$ under regularity assumptions on C and D. Namely, they need to be face-closed (FC): the limiting set of any converging sequence of faces of C is also a face of C.

The FC condition on a set C is necessary and sufficient to guarantee that $\texttt{Cav}_C\psi$ is continuous for every continuous function ψ (Laraki, 2004), hence for the splitting operator to preserve continuity (Laraki, 2001b). The proof of convergence of $\lim v_\lambda$ is more technical than on $\Delta(K)\times\Delta(L)$ and uses epi-convergence technics because the family $\{v_\lambda\}$ is not equi-Lipschitz.

The problem of existence and uniqueness of a solution of the Mertens-Zamir system without the FC condition or continuity of U remains open.

The above analysis extends to the non-autonomous case where U depends on t (Cardaliaguet et al., 2012). Any partition Π of $[0,1]$ induces a value function $V_\Pi(t,p,q)$ that converges uniformly to the unique function satisfying:

P1$''$: If $t\in[0,1)$ is such that $\mathbf{X}(t,p,q,W)$ is nonrevealing and $W(\cdot,p,q)-\phi(\cdot)$ has a global maximum at t, then $U(t,p,q)+\phi'(t)\geq 0$.

P2$''$: If $t\in[0,1)$ is such that $\mathbf{Y}(t,p,q,W)$ is nonrevealing and $W(\cdot,p,q)-\phi(\cdot)$ has a global minimum at t, then $U(t,p,q)+\phi'(t)\leq 0$.

2.3.4.3 *Compact discounted games and comparison criteria*

An approach related to the existence of the asymptotic value for discounted games has been proposed in Sorin and Vigeral (2012). The main tools used in the proofs are:

- the fact that the discounted value v_λ satisfies the Shapley equation [2.26],
- properties of accumulation points of the discounted values, and of the corresponding optimal strategies,
- comparison of two accumulation points leading to uniqueness and characterization.

In particular, this allows to cover the case of absorbing and recursive games with compact action spaces and provides an alternative formula for the asymptotic value of absorbing games, namely:

$$v = \mathtt{val}_{X \times Y} W(x, y)$$

with:

$$W(x, y) = med\{f(x, y), \sup_{x', \pi^*(x', y) > 0} \bar{f}^*(x', y), \inf_{y', \pi^*(x, y') > 0} \bar{f}^*(x, y')\}$$

where $\bar{f}^*(x, y)$ is the expected absorbing payoff: $\pi^*(x, y)\bar{f}^*(x, y) = f^*(x, y)$.

2.4. THE DUAL GAME

In this section, we focus on repeated games with incomplete information. The dual game has been introduced by De Meyer (1996a,b) and leads to many applications.

2.4.1 Definition and basic results

Consider a two-person zero-sum game with incomplete information on one side defined by two sets of actions S and T, a finite parameter space K, a probability distribution $p \in \Delta(K)$, and for each $k \in K$ a real-valued payoff function G^k on $S \times T$. Assume S and T convex and for each k, G^k bounded and bilinear on $S \times T$.

The game is played as follows: $k \in K$ is selected according to p and revealed to Player 1 (the maximizer) while Player 2 only knows p. In the normal form, Player 1 chooses $\mathbf{s} = \{s^k\}$ in S^K, Player 2 chooses t in T and the payoff is $G^p(\mathbf{s}, t) = \sum_k p^k G^k(s^k, t)$. Let $\underline{v}(p) = \sup_{S^K} \inf_T G^p(\mathbf{s}, t)$ and $\overline{v}(p) = \inf_T \sup_{S^K} G^p(\mathbf{s}, t)$. Then both value functions are concave in p, the first thanks to the splitting procedure (see, e.g., Zamir, 1992, p. 118) and the second as an infimum of linear functions.

Following De Meyer (1996a,b), one defines for each $z \in \mathbb{R}^k$, the "dual game" $G^*(z)$, where Player 1 chooses $k \in K$ and $s \in S$, while Player 2 chooses $t \in T$ and the payoff is:

$$h[z](k, s; t) = G^k(s, t) - z^k.$$

Define by $\underline{w}(z)$ and $\overline{w}(z)$ the corresponding maxmin and minmax.

Theorem 2.15. *De Meyer (1996a,b), Sorin (2002)*

The following duality relations hold:

$$\underline{w}(z) = \max_{p \in \Delta(K)} \{\underline{v}(p) - \langle p, z \rangle\}, \qquad \underline{v}(p) = \inf_{z \in \mathbb{R}^K} \{\underline{w}(z) + \langle p, z \rangle\}. \qquad [2.35]$$

$$\overline{w}(z) = \max_{p \in \Delta(K)} \{\overline{v}(p) - \langle p, z \rangle\}, \qquad \overline{v}(p) = \inf_{z \in \mathbb{R}^K} \{\overline{w}(z) + \langle p, z \rangle\}. \qquad [2.36]$$

In terms of strategies, one obtains the following correspondences:

Corollary 2.1.

1) *Given z (and $\varepsilon \geq 0$), let p attain the maximum in (2.35) and let* $\mathbf{s}$ *be ε-optimal in G^p, then $(p, \mathbf{s})$ is ε-optimal in $G^*(z)$.*
2) *Given p (and $\varepsilon \geq 0$), let z attain the infimum up to ε in (2.36) and let t be ε-optimal in $G^*(z)$, then t is also 2ε-optimal in G^p.*

To see the link with approachability (Section 7 and Sorin, 2002, Chapter 2) define

$$B = \{z \in \mathbb{R}^k : \overline{v}(p) \leq \langle p, z \rangle, \forall p \in \Delta(K)\}.$$

Then, B is the set of "reachable" vectors for Player 2 in the sense that for any $z \in B$, and any $\varepsilon > 0$, there exists a $t \in T$ such that $\sup_s G^k(s,t) \leq z^k + \varepsilon$, $\forall k \in K$. In particular $\overline{w}(z) \leq 0$ if and only if $z \in B$, which means that, from his point of view, the uninformed Player 2 is playing in a game with "vector payoffs."

2.4.2 Recursive structure and optimal strategies of the noninformed player

Consider now a RG with incomplete information on one side and recall the basic recursive formula for G_n:

$$(n+1)\ v_{n+1}(p) = \max_{x \in X^K} \min_{y \in Y} \left\{ \sum_k p^k g(k, x^k, y) + n \sum_i \bar{x}(i)\ v_n(p(i)) \right\} \qquad [2.37]$$

Let us consider the dual game G_n^* and its value w_n which satisfies:

$$w_n(z) = \max_{p \in \Delta(K)} \{v_n(p) - \langle p, z \rangle\}.$$

This leads to the dual recursive equation (De Meyer, 1996b):

$$(n+1) w_{n+1}(z) = \min_{y \in Y} \max_{i \in I} n w_n \left(\frac{n+1}{n} z - \frac{1}{n} g(i, y) \right). \qquad [2.38]$$

In particular, Player 2 has an optimal strategy in $G_{n+1}^*(z)$ that depends only on z (and on the length of the game). At stage 1, he plays y optimal in the dual recursive equation and from stage 2 on, given the move i_1 of Player 1 at stage 1, plays optimally in $G_n^*(\frac{n+1}{n} z - \frac{1}{n} g(i_1, y))$. A similar result holds for any evaluation μ of the stream of payoffs. Thus z is the natural state variable for Player 2.

Recall that the recursive formula for the primal game (2.37) allows the informed Player to construct inductively an optimal strategy since he knows $p(i)$. This is not the case for Player 2, who cannot compute $p(i)$, hence the first interest of the dual game is to obtain an explicit algorithm for optimal strategies of the uninformed Player via Corollary 2.1.

These properties extend to RG with incomplete information on both sides. In such a game, Player 2 must consider all possible realizations of $k \in K$ and so plays in a game with vector payoffs in $\mathbb{R}^K$. On the other hand, he reveals information and so generates a martingale $\tilde{q}$ on $\Delta(L)$.

There are thus two dual games: the Fenchel conjugate with respect to p (resp. q) allows to compute an optimal strategy for Player 2 (resp. 1). In the first dual, from $w_n(z, q) = \max_{p \in \Delta(K)}\{v_n(p, q) - \langle p, z\rangle\}$, De Meyer and Marino (2005) deduce that

$$(n+1)w_{n+1}(z, q) = \min_{y,\{z_{i,j}\}} \max_{i \in I} \sum_j \overline{y}(j) n w_n(z_{i,j}, q(j)) \qquad [2.39]$$

where the minimum is taken over all $y \in \Delta(J)^L$ and $z_{i,j} \in \mathbb{R}^K$ such that $\sum_j \overline{y}(j) z_{i,j} = \frac{n+1}{n} z - \frac{1}{n} g(q, i, y)$. Hence, Player 2 has an optimal strategy, which is Markovian with respect to (z, q).

A similar conclusion holds for stochastic games with incomplete information on both sides where Player 2 has a optimal strategy Markovian with respect to (z, q, ω) (Rosenberg, 1998).

To summarize, to a game with incomplete information on both sides are associated three games having a recursive structure:

- The usual auxiliary game related to the Shapley operator [2.6] with state parameter (p, q).
- For Player 2, a "dual game" where the value satisfies [2.39] and the state variable, known by Player 2, is (z, q).
- Similarly for Player 1.

2.4.3 The dual differential game

Consider a RG with incomplete information on one side. The advantage of dealing with the dual recursive formula [2.38] rather than with [2.37] is that the state variable evolves smoothly from z to $z + \frac{1}{n}(z - g(i, y))$ while the martingale $p(i)$ may have jumps. De Meyer and Rosenberg (1999) use the dual formula to provide a new proof of Aumann and Maschler's result via the study of approximate fixed points and derive a heuristic partial differential equation for the limit. This leads them to anticipate a link with differential game theory. This is made precise in Laraki (2002) where it is proved that w_n satisfying [2.38] is the value of the time discretization with mesh $\frac{1}{n}$ of a differential game on $[0, 1]$ with dynamics $\zeta(t) \in \mathbb{R}^K$ given by:

$$\frac{d\zeta}{dt} = g(x_t, y_t), \quad \zeta(0) = -z$$

$x_t \in X$, $y_t \in Y$, and terminal payoff $\max_k \zeta^k(1)$. Basic results of differential games of fixed duration (see Section 6) (Souganidis, 1985) show that the game starting at time t from state ζ has a value $\varphi(t, \zeta)$, which is the only viscosity solution of the following Hamilton-Jacobi equation on $[0, 1]$ with terminal condition:

$$\frac{\partial \varphi}{\partial t} + u(D\varphi) = 0, \quad \varphi(1, \zeta) = \max_k \zeta^k. \qquad [2.40]$$

Hence $\varphi(0, -z) = \lim_{n\to\infty} w_n(z) = w(z)$. Using Hopf's representation formula, one obtains:

$$\varphi(1-t, \zeta) = \sup_{a\in\mathbb{R}^K} \inf_{b\in\mathbb{R}^K} \{\max_k b^k + \langle a, \zeta - b\rangle + tu(a)\}$$

and finally $w(z) = \sup_{p\in\Delta(K)}\{u(p) - \langle p, z\rangle\}$. Hence $\lim_{\lambda\to 0} v_\lambda = \lim_{n\to\infty} v_n = \texttt{Cav}_{\Delta(K)} u$, by taking the Fenchel conjugate. Moreover, this is true for any compact evaluation of payoffs.

An alternative identification of the limit can be obtained through variational inequalities by translating in the primal game, the viscosity properties in the dual expressed in terms of local sub/super-differentials. This leads exactly to the properties **P1** and **P2** in the variational approach (Section 3.4). The approach has been extended recently by Gensbittel (2012) to games with infinite action spaces.

Interestingly, the dynamics of this differential game is exactly the one introduced by Vieille (1992), to show, in the context of Blackwell approachability, that any set is either weakly approachable or weakly excludable (see Section 7).

2.4.4 Error term, control of martingales, and applications to price dynamics

The initial objective of De Meyer (1996a,b) when he introduced the dual game was to study the error term in the Aumann and Maschler's RG model. The proof of $Cav(u)$ theorem shows that $e_n(p) := v_n(p) - \lim_{n\to\infty} v_n(p) = O\left(n^{-\frac{1}{2}}\right)$. The precise asymptotic analysis of $e_n(p)$ was first studied by Mertens and Zamir (1976b, 1995) when $I = J = K$ have cardinality 2 (see also Heuer, 1992b). They show that the speed of convergence can be improved to $O\left(n^{-\frac{2}{3}}\right)$ except in a particular class of "fair games" where $u(p) = 0$ for every p (without information, no Player has an advantage). In this class, the limit $\Psi(p)$ of $\sqrt{n}e_n(p)$ is shown to be related to the normal density function using a differential equation obtained by passing to the limit in the recursive primal formula. Moreover, $\Psi(p)$ appears as the limit of the maximal normalized L^1 variation of a martingale. In fact, an optimal strategy of the informed Player in G_n induces a martingale $\{p^n_m; 1 \leq m \leq n\}$ on $\Delta(K)$ starting at p. This martingale has a n-stage L^1 variation $D^1_n(p^n) := \mathbb{E}[\sum_{m=1}^n \|p^n_m - p^n_{m-1}\|_1]$ and asymptotically $D^1_n(p^n)/\sqrt{n}$ is maximal:

$$\lim_n \frac{1}{\sqrt{n}} |\max_{q^n} D^1_n(q^n) - D^1_n(p^n)| \to 0,$$

where the maximum is taken over all martingales on $\Delta(K)$ with length n starting at p.

For games where $I = J$ are finite, the error term is analyzed in depth by De Meyer in a series of papers (De Meyer, 1996a,b, 1999) where the dual game and the central limit theorem play a crucial role. For this purpose, De Meyer introduces a heuristic limit game with incomplete information and its dual, which is a stochastic differential game played on the time interval $[0, 1]$. He proves that it has a value V which satisfies a dynamic programming principle that leads to a second-order PDE. De Meyer proves then that if V is smooth, then $\sqrt{n}e_n$ converges uniformly to the Fenchel conjugate of V.

The main application of this work is achieved in De Meyer and Moussa-Saley (2002). They show that when two asymmetrically informed risk neutral agents repeatedly exchange a risky asset for a numéraire, they are playing a "fair RG" with incomplete information. The model may be seen as a particular RG à la Aumann-Maschler where action sets are infinite. Their main result is a characterization of the limit of the martingales of beliefs induced by an optimal strategy of the informed Player. Those discrete-time martingales are mapped to the time interval $[0, 1]$ and are considered as piecewise constant stochastic processes. The limit process for the weak topology is shown to be a Brownian motion.

De Meyer (2010) extends the result to general spaces I, J, and $K = \mathbf{R}$ and shows that the limit diffusion process does not depend on the specific "natural" trading mechanism, but only on the initial belief $p \in \Delta(K)$: it belongs to the CMMV class (continuous time martingales of maximal variation). It contains the dynamics of Black and Scholes and Bachelier as special cases.

The main step of De Meyer's results is the introduction of a discrete-time stochastic control problem whose value is equal to v_n (the value of the n-stage game) and whose maximizers coincide with a posteriori martingales at equilibrium. This generalizes the above maximization of the L^1-variation of a martingale. The first idea is to measure the variation at stage m by the value of the one-stage game where the transmission of information by Player 1 corresponds to the step from p_m to p_{m+1}. The next extension is, starting from W a real valued function defined on the set of probabilities over $\mathbf{R}^d$, to define the W-variation of a martingale $p = \{p_0, p_1, \ldots, p_n\}$ by:

$$V_n^W(p^n) := \mathbb{E}\left[\sum_{m=1}^{n} W(p_m - p_{m-1}|p_0, \ldots, p_{m-1})\right].$$

De Meyer (2010) solved the problem for $d = 1$, corresponding to a financial exchange model with one risky asset, and Gensbittel (2013) extended the result to higher dimensions (corresponding to a portfolio of $d \geq 1$ risky assets). Under quite general conditions on W, he obtains:

Theorem 2.16. *De Meyer (2010) for $d = 1$, Gensbittel (2013) for $d \geq 1$.*

$$\lim_n \sqrt{n} V_n(t,p) = \max_{\{p_s\}} \mathbb{E}[\int_t^1 \phi\left(\frac{\mathrm{d} < p_s >}{\mathrm{d}s}\right) \mathrm{d}s],$$

where the maximum is over càdlàg martingales p_t on $\Delta(K)$ that start at p and

$$\phi(A) = sup_{\mu \in \Delta(\mathbf{R}^d):cov(\mu)=A} W(\mu).$$

Here, $< p_s >$ denotes the quadratic variation and $cov(\mu)$ the covariance. The maximizing process above is interpreted by De Meyer as the dynamics of the equilibrium price of the risky assets. When $d = 1$ or under some conditions on the correlation between the risky assets, this process belongs to the CMMV class.

2.5. UNIFORM ANALYSIS

We turn now to the uniform approach and first recall basic results described in the previous chapters of HGT by Zamir (1992), Mertens (2002), and Vieille (2002).

2.5.1 Basic results

2.5.1.1 Incomplete information

Concerning games with lack of information on one side, Aumann and Maschler (1966) show the existence of a uniform value, (see Aumann and Maschler, 1995) and the famous formula:

$$v_\infty(p) = \mathtt{Cav}_{p \in \Delta(K)} u(p), \tag{2.41}$$

first for games with *standard signaling* (or perfect monitoring, i.e., where the moves are announced), then for general signals on the moves. u is as usual the value of the nonrevealing game and the construction of an optimal strategy for the uninformed Player is due to Kohlberg (1975).

For games with lack of information on both sides and standard signaling, Aumann and Maschler (1967) show that the maxmin and minmax exist (see Aumann and Maschler, 1995). Moreover, they give explicit formulas:

$$\underline{v}(p,q) = \mathtt{Cav}_{p \in \Delta(K)} \mathtt{Vex}_{q \in \Delta(L)} u(p,q), \;\; \overline{v}(p,q) = \mathtt{Vex}_{q \in \Delta(L)} \mathtt{Cav}_{p \in \Delta(K)} u(p,q). \tag{2.42}$$

They also construct games without a value. For several extensions to the dependent case and state independent signaling structure, mainly due to Mertens and Zamir (see Mertens et al., 1994).

2.5.1.2 Stochastic games

In the framework of stochastic games with standard signaling, the first proof of existence of a uniform value was obtained for the "Big Match" by Blackwell and Ferguson (1968), and then for absorbing games by Kohlberg (1974). The main result is due to Mertens and Neyman (1981):

$$v_\infty \textit{ exists for finite stochastic games.} \qquad [2.43]$$

The proof uses two ingredients:

(i) properties of the family $\{v_\lambda\}$ obtained by Bewley and Kohlberg (1976a) through their algebraic characterization,

(ii) the knowledge of the realized payoff at each stage n, to build an ε-optimal strategy as follows. One constructs a map $\bar{\lambda}$ and a sufficient statistics L_n of the past history at stage n such that σ is, at that stage, an optimal strategy in the game with discount parameter $\bar{\lambda}(L_n)$.

2.5.1.3 Symmetric case

A first connection between incomplete information games and stochastic games is obtained in the so-called symmetric case. This corresponds to games where the state in M is constant and may not be known by the Players, but their information during the play is symmetric (hence includes their actions). The natural state space is the set of probabilities on M and the analysis reduces to a stochastic game on $\Delta(M)$, which is no longer finite but on which the state process is regular (martingale), see Kohlberg and Zamir (1974), Forges (1982), and for alternative tools that extend to the non-zero-sum case, Neyman and Sorin (1998).

2.5.2 From asymptotic value to uniform value

Recall that the existence of v_∞ implies that it is also the limit of any sequence v_μ (with μ decreasing) or more generally v_θ (random duration with $\mathsf{E}(\theta) \longrightarrow \infty$).

On the other hand, the proof in Mertens and Neyman (1981) shows that in a stochastic game with standard signaling the following holds:

Theorem 2.17. *Assume that $w :]0,1] \to \mathbb{R}^\Omega$ satisfies:*

1) $\|w(\lambda) - w(\lambda')\| \le \int_\lambda^{\lambda'} f(x)\mathrm{d}x$, *for* $0 < \lambda < \lambda' < 1$, *with* $f \in L^1(]0,1])$,

2) $\Phi(\lambda, w(\lambda)) \ge w(\lambda)$, *for every* $\lambda > 0$ *small enough.*

Then Player 1 can guarantee $w(\lambda)$.

In the initial framework of finite stochastic games, one can take $w(\lambda) = v_\lambda$, hence (2) follows and one deduces property (1) from the fact that v_λ is semi-algebraic in λ.

More generally, this approach allows to prove the existence of v_∞ for continuous games with compact action spaces that are either absorbing (Mertens et al., 2009), recursive (Vigeral, 2010c) using the operator approach of Rosenberg and Sorin (2001), or definable (Bolte et al., 2013) (see Section 9.3.4).

2.5.3 Dynamic programming and MDP

Stronger results are available in the framework of general dynamic programming: this corresponds to a one person stochastic game with a state space Ω, a correspondence C from Ω to itself (with non empty values), and a real-valued, bounded payoff g on Ω. A play is a sequence $\{\omega_n\}$ satisfying $\omega_{n+1} \in C(\omega_n)$.

Lehrer and Sorin (1992) give an example where $\lim_{n\to\infty} v_n$ and $\lim_{\lambda\to 0} v_\lambda$ both exist and differ. They also prove that uniform convergence (on Ω) of v_n is equivalent to uniform convergence of v_λ and then the limits are the same. For a recent extension to the continuous time framework, see Oliu-Barton and Vigeral (2013).

However, this condition does not imply existence of the uniform value v_∞ (see Lehrer and Monderer, 1994; Monderer and Sorin, 1993).

Recent advances have been obtained by Renault (2011) introducing new notions like the values $v_{m,n}$ (resp. $\nu_{m,n}$) of the game where the payoff is the average between stage $m+1$ and $m+n$ (resp. the minimum of all averages between stage $m+1$ and $m+\ell$ for $\ell \leq n$).

Theorem 2.18. *Assume that the state space Ω is a compact metric space.*

(1) *If the family of functions v_n is uniformly equicontinuous, then $\lim_{n\to\infty} v_n = v$ exists, the convergence is uniform and:*

$$v(\omega) = \inf_{n\geq 1} \sup_{m\geq 0} v_{m,n}(\omega) = \sup_{m\geq 0} \inf_{n\geq 1} v_{m,n}(\omega).$$

(2) *If the family of functions $\nu_{m,n}$ is uniformly equicontinuous, then the uniform value v_∞ exits and*

$$v_\infty(\omega) = \inf_{n\geq 1} \sup_{m\geq 0} \nu_{m,n}(\omega) = \sup_{m\geq 0} \inf_{n\geq 1} \nu_{m,n}(\omega) = v(\omega)$$

For part (1), no assumption on Ω is needed and $\{v_n\}$ totally bounded suffices.

For (2), the construction of an ε-optimal strategy is by concatenation of strategies defined on large blocks, giving good payoffs while keeping the "level" of the state. Condition (2) plays the role of (i) in Mertens and Neyman's proof. It holds, for example, if g is continuous and C is non-expansive.

In particular for Markov Decision Process (finite state space K, move space I, and transition probability from $K \times I$ to K), the natural state space is $X = \Delta(K)$. In the case of partial observation (signal space A and transition probability from $K \times I$ to $K \times A$), the natural state space is $\Delta_f(X)$ on which C is non-expansive and the previous result implies:

Theorem 2.19. *Renault (2011)*
MDP processes with finite state space and partial observation have a uniform value.

This extends previous tools and results by Rosenberg et al. (2002).

Further developments to the continuous time setup lead to the study of asymptotic and uniform value in control problems defined as follows: the differential equation $\dot{x}_s = f(x_s, u_s)$ describes the control by $\mathbf{u} \in \mathbf{U}$ (measurable functions from $[0, +\infty)$ to U) of the state $x \in \mathbb{R}^n$ and one defines the value function $V_{m,t}$, $m, t \in \mathbb{R}^+$ by:

$$V_{m,t}(x) = \sup_{\mathbf{u} \in \mathbf{U}} \frac{1}{t} \int_m^{m+t} g(x_s, u_s) \mathrm{d}s, \qquad x_0 = x.$$

Quincampoix and Renault (2011) prove that if g is continuous, the (feasible) state space X is bounded and the nonexpansiveness condition

$$\forall x, y \in X, \qquad \sup_{u \in U} \inf_{v \in U} \langle x - y, f(x, u) - f(y, v) \rangle \leq 0$$

holds, then the uniform value exists, the convergence $V_t (= V_{0,t}) \to V_\infty$ is uniform and:

$$V_\infty(x) = \inf_{t \geq 1} \sup_{m \geq 0} V_{m,t}(x) = \sup_{m \geq 0} \inf_{t \geq 1} V_{m,t}(x).$$

For similar results in the framework of differential games, see Bardi (2009).

2.5.4 Games with transition controlled by one player

Consider now a game where Player 1 controls the transition on the parameter: basic examples are stochastic games where the transition is independent of Player's 2 moves, or games with incomplete information on one side (with no signals); but this class also covers the case where the parameter is random, its evolution independent of Player 2's moves, and Player 1 knows more than Player 2.

Basically, the state space will be the beliefs of Player 2 on the parameter, which are variables controlled and known by Player 1. The analysis in Renault (2012) first constructs an auxiliary stochastic game on this space, and then reduces the analysis of the game to a dynamic programming problem by looking at stage by stage best reply of Player 2 (whose moves do not affect the future of the process). The finiteness assumption on the basic data implies that one can apply Theorem 2.18 part 2 to obtain:

Theorem 2.20. *In the finite case, games with transition controlled by one Player have a uniform value.*

The result extends previous work of Rosenberg et al. (2004) and also the model of Markov games with lack of information on one side introduced by Renault (2006) (see also Krausz and Rieder, 1997): here the parameter follows a Markov chain and is known at each stage by Player 1 while Player 2 knows only the initial law. The moves are observed. Neyman (2008) extends the analysis to the case with signals and constructs an optimal strategy for the informed Player.

This class is very interesting but an explicit formula for the value is not yet available.

Here is an example: there are two states $k = 1, 2$. At each stage, the state changes with probability ρ and the initial distribution is $(1/2, 1/2)$. The payoff is given by:

$k = 1$	L	R
T	1	0
B	0	0

$k = 2$	L	R
T	0	0
B	0	1

If $\rho = 0$ or 1, the game reduces to a standard game with incomplete information on one side with value $1/4$. By symmetry, it is enough to consider the interval $[1/2, 1]$; for $\rho \in [1/2, 2/3]$ the value is $\rho/(4\rho - 1)$, and still unknown otherwise (Marino, 2005; Hörner et al., 2010).

2.5.5 Stochastic games with signals on actions

Consider a stochastic game and assume that the signal to each Player reveals the current state but not necessarily the previous action of the opponent. By the recursive formula for v_λ and v_n, or more generally v_Θ, these quantities are the same as in the standard signaling case since the state variable is not affected by the change in the information structure. However, for example, in the Big Match, when Player 1 has no information on Player 2's action the max min is 0 (Kohlberg, 1974) and the uniform value does not exist anymore.

It follows that the existence of a uniform value for stochastic games depends on the signaling structure on actions. However, one has the following property:

Theorem 2.21. *Maxmin and minmax exist in finite stochastic games with signals on actions.*

This result, due to Coulomb (2003), and Rosenberg et al. (2003a) is extremely involved and relies on the construction of two auxiliary games, one for each Player.

Consider the maxmin and some discount factor λ. Introduce an equivalence relation among the mixed actions y and y' of Player 2 facing the mixed action x of Player 1 by $y \sim y'$ if they induce the same transition on the signals of Player 1 for each action i having significant weight ($\geq L\lambda$) under x. Define now the maxmin value of a discounted game where the payoff is the minimum with respect to an equivalence class of Player 2. This quantity will satisfy a fixed point equation defined by a semialgebraic set and will play the role of $w(\lambda)$ in Theorem 2.17. It remains to show, first for Player 1, that this auxiliary payoff indeed can be achieved in the real game. Then for Player 2, he will first follow a strategy realizing a best reply to σ of Player 1 up to a stage where the equivalence relation will allow for an indistinguishable switch in action. He will then change his strategy to obtain a good payoff from then on, without being detected. Obviously, a dual game is defined for the minmax (involving the structure of signals for Player 2).

An illuminating example, due to Coulomb (2003), is as follows:

	α	β	γ
a	1^*	0^*	L
b	0	1	L

Payoffs ($L \geq 1$)

	α	β	γ
a	?	?	?
b	A	B	A

Signals to Player 1

Given a strategy σ of Player 1, Player 2 will start by playing $(0, \varepsilon, 1-\varepsilon)$ and switch to $(1-\varepsilon, \varepsilon, 0)$ when the probability under σ of playing a in the future, given the distribution $(1-\varepsilon, \varepsilon)$ on the signals (A, B), is small enough. Hence the maxmin is 0.

For a nice overview, see Rosenberg et al. (2003b), Coulomb (1992, 1996, 1999, 2001, 2003), and Neyman and Sorin (2003).

2.5.6 Further results

Different examples include Sorin (1984), Sorin (1985a), and Chapter 6 in Sorin (2002) where several stochastic games with incomplete information are analyzed. Among the new tools are approachability strategies for games with vector payoffs and absorbing states and the use of a time change induced by an optimal strategy in the asymptotic game, to play well in the uniform game.

Rosenberg and Vieille (2000) consider a recursive game with lack of information on one side. The initial state is chosen in a finite set K according to some $p \in \Delta(K)$. After each stage, the moves are announced and Player 1 knows the state. If one leaves K, the payoff is absorbing and denoted by a. Denote by $\pi(p, x, y)$ the probability to stay in K and by $\tilde{p}$ the random conditional probability on K. The Shapley operator is:

$$\mathbf{T}(f)(p) = \mathtt{val}_{X^K \times Y}\{\pi(p, x, y)\mathsf{E}f(\tilde{p}) + (1 - \pi(p, x, y))\mathsf{E}(a)\}. \qquad [2.44]$$

Consider $w(p)$ an accumulation point of $v_\lambda(p)$. To prove that Player 1 can guarantee w, one alternates optimal strategies in the projective game if the current state satisfies $w(p_n) \leq \varepsilon$ and in G_λ (with $\|v_\lambda - w\| \leq \varepsilon^2$) otherwise. See Solan and Vieille (2002) for another extension.

In all these games with standard signaling, whenever Player 1 is fully informed, one has: $\lim_{n\to\infty} v_n = \lim_{\lambda\to 0} v_\lambda = \max\min$. A counter example in the general case is given in Section 9.3.2.

2.6. DIFFERENTIAL GAMES

2.6.1 A short presentation of differential games (DG)

Differential games (DG) are played in continuous time. A state space Z and control sets U for Player 1 and V for Player 2 are given. At each time t, the game is in some state z_t and each Player chooses a control ($u_t \in U, v_t \in V$). This induces a current payoff $\gamma_t = \gamma(z_t, t, u_t, v_t)$ and defines the dynamics $\dot{z}_t = f(z_t, t, u_t, v_t)$ followed by the state (see Friedman (1994) Chapter 22 in HGT2). Notice that in the autonomous case, if the Players use piece-wise constant controls on intervals of size δ, the induced process is like a RG.

There are many ways of defining strategies in differential games. For simplicity of the presentation, only nonanticipative strategies with delay are presented here. The main reasons are (1) they allow to put the game in normal form and (2) they are the most natural since they suppose that a Player always needs a delay (that may be chosen strategically) before reacting to a change in the behavior of the other Player.

Let $\mathcal{U}$ (resp. $\mathcal{V}$) denote the set of measurable control maps from $\mathbb{R}^+$ to U (resp. V). $\alpha \in \mathcal{A}$ (resp. $\beta \in \mathcal{B}$) is a nonanticipative strategy with delay (NAD) if α maps $\mathbf{v} \in \mathcal{V}$ to $\alpha(\mathbf{v}) = \mathbf{u} \in \mathcal{U}$ and there is $\delta > 0$ such that if $\mathbf{v}_s = \mathbf{v}'_s$ on $[0, t]$ then $\alpha(\mathbf{v}) = \alpha(\mathbf{v}')$ on $[0, t+\delta]$, for all $t \in \mathbb{R}^+$. A pair (α, β) defines a unique pair $(\mathbf{u}, \mathbf{v}) \in \mathcal{U} \times \mathcal{V}$ with $\alpha(\mathbf{v}) = \mathbf{u}$ and $\beta(\mathbf{u}) = \mathbf{v}$, thus the solution $\mathbf{z}$ is well defined. The map $t \in [0, +\infty[\mapsto (\mathbf{z}_t, \mathbf{u}_t, \mathbf{v}_t)$ specifies the trajectory $(\mathbf{z}, \mathbf{u}, \mathbf{v})(\alpha, \beta)$ (see, e.g., Cardaliaguet, 2007; Cardaliaguet and Quincampoix, 2008).

As in RG, there are many ways to evaluate payoffs.

Compact evaluations, or quantitative DG. This concerns the class of DG with total evaluation of the form:

$$\Gamma(\alpha, \beta)(z_0) = \int_0^T \gamma_t \, \mathrm{d}t + \overline{\gamma}(z_T) \tag{2.45}$$

where $\overline{\gamma}$ is some terminal payoff function, or:

$$\Gamma(\alpha, \beta)(z_0) = \int_0^\infty \gamma_t \mu(\mathrm{d}t)$$

where μ is a probability on $[0, +\infty]$ like $\frac{1}{T}\mathbf{1}_{[0,T]}\mathrm{d}t$ or $\lambda \exp(-\lambda t)\mathrm{d}t$.

The game is now well defined in normal form and the issues are the existence of a value, its characterization and properties of optimal strategies.

Uniform criteria, or qualitative DG. The aim is to control the asymptotic properties of the trajectories like: the state $\mathbf{z}_t$ should stay in some set C for all $t \in \mathbb{R}^+$ or from some time $T \geq 0$ on. Basic references include Krasovskii and Subbotin (1988), Cardaliaguet (1996), and Cardaliaguet et al. (2007).

2.6.2 Quantitative differential games

We describe very briefly the main tools in the proof of existence of a value, due to Evans and Souganidis (1984), but using NAD strategies, compare with Friedman (1994).

Consider the case defined by (2.45) under the following assumptions:

1) U and V are compact sets in $\mathbb{R}^K$,

2) $Z = \mathbb{R}^N$,

3) All functions f (dynamics), γ (running payoff), and $\bar{\gamma}$ (terminal payoff) are bounded, jointly continuous and uniformly Lipschitz in z,

4) Define the Hamiltonians $H^+(p, z, t) = \inf_v \sup_u\{\langle f(z, t, u, v), p\rangle + \gamma(z, t, u, v)\}$ and $H^-(p, z, t) = \sup_u \inf_v\{\langle f(z, t, u, v), p\rangle + \gamma(z, t, u, v)\}$ and assume that Isaacs's condition holds: $H^+(p, z, t) = H^-(p, z, t) = H(p, z, t)$, for all $(p, z, t) \in \mathbb{R}^N \times \mathbb{R}^N \times [0, T]$.

For $T \geq t \geq 0$ and $z \in Z$, consider the game on $[t, T]$ starting at time t from state z and let $\overline{v}[z, t]$ and $\underline{v}[z, t]$ denote the corresponding minmax and maxmin. Explicitly:

$$\overline{v}[z, t] = \inf_{\beta} \sup_{\alpha}[\int_t^T \gamma_s \mathrm{d}s + \bar{\gamma}(Z_T)]$$

where $\gamma_s = \gamma(\mathbf{z}_s, s, \mathbf{u}_s, \mathbf{v}_s)$ is the payoff at time s and $(\mathbf{z}, \mathbf{u}, \mathbf{v})$ is the trajectory induced by (α, β) and f on $[t, T]$ with $\mathbf{z}_t = z$. Hence $\mathbf{u}_s = \mathbf{u}_s(\alpha, \beta)$, $\mathbf{v}_s = \mathbf{v}_s(\alpha, \beta)$, and $\mathbf{z}_s = \mathbf{z}_s(\alpha, \beta, z, t)$.

The first property is the following dynamic programming inequality:

Theorem 2.22. *For $0 \leq t \leq t + \delta \leq T$, $\overline{v}$ satisfies:*

$$\overline{v}[z, t] \leq \inf_{\beta} \sup_{\alpha} \left\{\int_t^{t+\delta} \gamma(\mathbf{z}_s(\alpha, \beta, z, t), s, \mathbf{u}_s(\alpha, \beta), \mathbf{v}_s(\alpha, \beta))\mathrm{d}s + \overline{v}[\mathbf{z}_{t+\delta}(\alpha, \beta, z, t), t + \delta]\right\}. \quad [2.46]$$

In addition $\overline{v}$ is uniformly Lipschitz in z and t.

Property [2.46] implies in particular that for any C^1 function Φ on $[0, T] \times Z$ with $\Phi[t, z] = \overline{v}[t, z]$ and $\Phi \geq \overline{v}$ in a neighborhood of (t, z) one has, for all $\delta > 0$ small enough:

$$\inf_{\beta} \sup_{\alpha} \left\{\frac{1}{\delta}\int_t^{t+\delta} \gamma(\mathbf{z}_s(\alpha, \beta, z, t), s, \mathbf{u}_s(\alpha, \beta), \mathbf{v}_s(\alpha, \beta))\mathrm{d}s + \frac{\Phi[\mathbf{z}_{t+\delta}(\alpha, \beta, z, t), t + \delta] - \Phi[z, t]}{\delta}\right\} \geq 0. \quad [2.47]$$

Letting δ going to 0 implies that Φ satisfies the following property:

$$\inf_{v} \sup_{u} \{\gamma(z, t, u, v) + \partial_t\Phi[z, t] + \langle D\Phi[z, t], f((z, t, u, v)\rangle\} \geq 0$$

which gives the differential inequality:

$$\partial_t \Phi[z, t] + H^+(D\Phi[z, t], z, t) \geq 0. \qquad [2.48]$$

The fact that any smooth local majorant of $\overline{v}$ satisfies [2.48] can be expressed as:

Proposition 2.1. *$\overline{v}$ is a viscosity subsolution of the equation $\partial_t W[z, t] + H^+ (DW[z, t], z, t) = 0$.*

Obviously a dual property holds. One use then Assumption (3) and the next comparison principle:

Theorem 2.23. *Let W_1 be a viscosity subsolution and W_2 be a viscosity supersolution of*

$$\partial_t W[z, t] + H(DW[z, t], z, t) = 0$$

then $W_1[T, .] \leq W_2[T, .]$ implies $W_1[t, z] \leq W_2[z, t], \forall z \in Z, \forall t \in [0, T]$.

One obtains finally:

Theorem 2.24. *The differential game has a value:*

$$\overline{v}[z, t] = \underline{v}[z, t].$$

In fact, the previous Theorem 2.23 implies $\overline{v}[z, t] \leq \underline{v}[z, t]$.

Note that the comparison Theorem 2.23 is much more general and applies to W_1 u.s.c., W_2 l.s.c., H uniformly Lipschitz in p and satisfying: $|H(p, z_1, t_1) - H(p, z_2, t_2)| \leq C(1 + \|p\|)\|(z_1, t_1) - (z_2, t_2)\|$. Also $\overline{v}$ is in fact, even without Isaacs's condition, a viscosity solution of $\partial_t W[z, t] + H^+(DW[z, t], z, t) = 0$.

For complements, see, e.g., Souganidis (1999), Bardi and Capuzzo Dolcetta (1996) and for viscosity solutions Crandall et al. (1992).

The analysis has been extended by Cardaliaguet and Quincampoix (2008) to the symmetric case where the initial value of the state $z \in \mathbb{R}^N$ is random and only its law μ is known. Along the play, the Players observe the controls but not the state. Assuming $\mu \in M$, the set of measures with finite second moment, the analysis is done on M endowed with the L^2-Wasserstein distance by extending the previous tools and results to this infinite-dimensional setting.

Another extension involving mixed strategies when Isaacs' condition is not assumed is developed in Buckdahn et al. (2013).

2.6.3 Quantitative differential games with incomplete information

An approach similar to the one for RG has been introduced by Cardaliaguet (2007) and developed by Cardaliaguet and Rainer (2009a) to study differential games of fixed duration with incomplete information. Stochastic differential games with incomplete information have been analyzed by Cardaliaguet and Rainer (2009a) (see also Buckdahn et al., 2010).

The model works as follows. Let K and L be two finite sets. For each (k, ℓ), a differential game $\Gamma^{k\ell}$ on $[0, T]$ with control sets U and V is given. The initial position of the system is $z_0 = \{z_0^{k\ell}\} \in Z^{K\times L}$, the dynamics is $f^{k\ell}(z^{k\ell}, t, u, v)$, the running payoff is $\gamma^{k\ell}(z^{k\ell}, t, u, v)$, and the terminal payoff is $\bar{\gamma}^{k\ell}(z^{k\ell})$. $k \in K$ is chosen according to a probability distribution $p \in \Delta(K)$, similarly $\ell \in L$ is chosen according to $q \in \Delta(L)$. Both Players know p and q and in addition Player 1 learns k and Player 2 learns l. Then, the game $\Gamma^{k\ell}$ is played starting from $z_0^{k\ell}$. The corresponding game is $\Gamma(p, q)[z_0, 0]$. The game $\Gamma(p, q)[z, t]$ starting from $z = \{z^{k\ell}\}$ at time t is defined similarly. One main difference with the previous section is that even if Isaacs' condition holds, the Players have to use randomization to choose their controls in order to hide their private information. $\alpha \in \bar{\mathcal{A}}$ is the choice at random of an element in $\mathcal{A}$. Hence a strategy for Player 1 is described by a profile $\hat{\alpha} = \{\alpha^k\} \in \bar{\mathcal{A}}^K$ (α^k is used if the signal is k). The payoff induced by a pair of profiles $(\hat{\alpha}, \hat{\beta})$ in $\Gamma(p, q)[z, t]$ is $G^{p,q}[z, t](\hat{\alpha}, \hat{\beta}) = \sum_{k,\ell} p^k q^\ell G^{k\ell}[z, t](\alpha^k, \beta^\ell)$ where $G^{k\ell}[z, t](\alpha^k, \beta^\ell)$ is the payoff in the game $\Gamma^{k\ell}$ induced by the (random) strategies (α^k, β^ℓ).

Notice that $\Gamma(p, q)[z, t]$ can be considered as a game with incomplete information on one side where Player 1 knows which of the games $\Gamma(k, q)[z, t]$ will be played, where k has distribution p and Player 2 is uninformed. Consider the minmax in $\Gamma(p, q)[z, t]$:

$$\overline{V}(p, q)[z, t] = \inf_{\hat{\beta}} \sup_{\hat{\alpha}} G^{p,q}[z, t](\hat{\alpha}, \hat{\beta}) = \inf_{\{\beta^\ell\}} \sup_{\{\alpha^k\}} \sum_k p^k \left\{ \sum_\ell q^\ell G^{k\ell}[z, t](\alpha^k, \beta^\ell) \right\}$$

The dual game with respect to p and with parameter $\theta \in \mathbb{R}^K$ has a minmax given by:

$$\overline{W}(\theta, q)[z, t] = \inf_{\hat{\beta}} \sup_{\alpha \in \bar{\mathcal{A}}} \max_k \left\{ \sum_\ell q^\ell G^{k\ell}[z, t](\alpha, \beta^\ell) - \theta^k \right\}$$

and using (2.36):

$$\overline{W}(\theta, q)[z, t] = \max_{p \in \Delta(K)} \{\overline{V}(p, q)[z, t] - \langle p, \theta \rangle\}.$$

Note that $\overline{V}(p, q)[z, t]$ does not obey a dynamic programming equation: the Players observe the controls not the strategy profiles, and the current state is unknown but $\overline{W}(\theta, q)[z, t]$ will satisfy a subdynamical programming equation. First the max can be taken on $\mathcal{A}$, then if Player 2 ignores his information, one obtains:

Proposition 2.2.

$$\overline{W}(\theta, q)[z, t] \leq \inf_{\beta \in \mathcal{B}} \sup_{\alpha \in \mathcal{A}} \overline{W}(\theta(t+\delta), q)[\mathbf{z}_{t+\delta}, t+\delta] \qquad [2.49]$$

where $\mathbf{z}_{t+\delta} = \mathbf{z}_{t+\delta}(\alpha, \beta, z, t)$ *and* $\theta^k(t+\delta) = \theta^k - \sum_\ell q^\ell \int_t^{t+\delta} \gamma^{k\ell}(\mathbf{z}_s^{k\ell}, s, \mathbf{u}_s, \mathbf{v}_s)ds.$

Assume that the following Hamiltonian H satisfies Isaacs's condition:

$$\begin{aligned} H(z, t, \xi, p, q) &= \inf_v \sup_u \{\langle f(z, t, u, v), \xi\rangle + \sum_{k,\ell} p^k q^\ell \gamma^{k\ell}(z^{k\ell}, t, u, v)\} \\ &= \sup_u \inf_v \{\langle f(z, t, u, v), \xi\rangle + \sum_{k,\ell} p^k q^\ell \gamma^{k\ell}(z^{k\ell}, t, u, v)\}. \end{aligned}$$

Here $f(z, ., ., .)$ stands for $\{f^{k\ell}(z^{k\ell}, ., ., .)\}$ and $\xi = \{\xi^{k\ell}\}$.

Given $\Phi \in \mathcal{C}^2(Z \times [0, T] \times \mathbb{R}^K)$, let $\overline{L}\Phi(z, t, \bar{p}) = \max\{\langle D^2_{pp}\Phi(z, t, \bar{p})\rho, \rho\rangle; \rho \in T_{\bar{p}}\Delta(K)\}$ where $T_{\bar{p}}\Delta(K)$ is the tangent cone to $\Delta(K)$ at $\bar{p}$.

The crucial idea is to use (2.36) to deduce from (2.49) the following property on $\overline{V}$:

Proposition 2.3. $\overline{V}$ *is a viscosity subsolution for* H *in the sense that: for any given* $\bar{q} \in \Delta(L)$ *and any test function* $\Phi \in \mathcal{C}^2(Z \times [0, T] \times \mathbb{R}^K)$ *such that the map* $(z, t, p) \mapsto \overline{V}(z, t, p, \bar{q}) - \Phi(z, t, p)$ *has a local maximum on* $Z \times [0, T] \times \Delta(K))$ *at* $(\bar{z}, \bar{t}, \bar{p})$ *then*

$$\max\left\{\overline{L}\Phi(\bar{z}, \bar{t}, \bar{p}); \partial_t\Phi(\bar{z}, \bar{t}, \bar{p}) + H(\bar{z}, \bar{t}, D_z\Phi(\bar{z}, \bar{t}, \bar{p}), \bar{p}, \bar{q})\right\} \geq 0. \qquad [2.50]$$

A similar dual definition, with $\underline{L}$, holds for a viscosity supersolution.

Finally a comparison principle extending Theorem 2.23 proves the existence of a value $v = \overline{V} = \underline{V}$.

Theorem 2.25. *Let* F_1 *and* F_2: $Z \times [0, T] \times \Delta(K) \times \Delta(L) \mapsto \mathbb{R}$ *be Lipschitz and saddle (concave in* p *and convex in* q*). Assume that* F_1 *is a subsolution and* F_2 *a supersolution with* $F_1(., T, ., .) \leq F_2(., T, ., .)$*, then* $F_1 \leq F_2$ *on* $Z \times [0, T] \times \Delta(K) \times \Delta(L)$.

Basically the idea of the proof is to mimick the complete information case. However, the value operator is on the pair of profiles $(\hat{\alpha}, \hat{\beta})$. Hence, the infinitesimal version involves vectors in $U^K \times V^L$ while only the realized controls in U or V are observed; consequently, the dynamic programming property does not apply in the same space. The use of the dual games allows, for example, for Player 2 to work with a variable that depends only on the realized trajectory of his opponent. The geometrical properties (convexity) of the minmax imply that it is enough to characterize the extreme points and then Player 2 can play non-revealing. As a consequence, the dynamic programming inequality on the dual of the minmax involving a pair (α, β) induces an inequality with the infinitesimal value operator on $U \times V$ for the test function. The situation being symmetrical for Player 1, a comparison theorem can be obtained.

Using the characterization above, Souquière (2010) shows that in the case where f and γ are independent of z and the terminal payoff is linear, $v = \mathbf{MZ}(u)$, where u is the value of the corresponding non-revealing game, and thus recovers Mertens-Zamir's result through differential games. This formula does not hold in general (see examples in Cardaliaguet, 2008). However, one has the following approximation procedure. Given a finite partition Π of $[0, 1]$ define inductively V_Π by:

$$V_\Pi(z, t_m, p, q) = \mathbf{MZ}\left[\sup_u \inf_v \{V_\Pi(z + \delta_{m+1} f(z, t_m, u, v), t_{m+1}, p, q) + \delta_{m+1} \sum_{k\ell} p^k q^\ell \gamma^{k\ell}(z^{k\ell}, t_m, u, v)\}\right]$$

where $\delta_{m+1} = t_{m+1} - t_m$. Then using results of Laraki (2001b, 2004), Souquière (2010) proves that V_Π converges uniformly to v, as the mesh of Π goes to 0. This extends a similar construction for games with lack of information on one side in Cardaliaguet (2009), where moreover an algorithm for constructing ε-optimal strategies is provided. Hence, the **MZ** operator (which is constant in the framework of repeated games: this is the time homogeneity property) appears as the true infinitesimal operator in a nonautonomous framework.

Cardaliaguet and Souquière (2012) study the case where the initial state is random (with law μ) and known by Player 1 who also observes the control of Player 2. Player 2 knows μ but is blind: he has no further information during the play.

2.7. APPROACHABILITY

This section describes the exciting and productive interaction between RG and DG in a specific area: approachability theory, introduced and studied by Blackwell (1956).

2.7.1 Definition

Given an $I \times J$ matrix A with coefficients in $\mathbb{R}^k$, a two-person infinitely repeated game form G is defined as follows. At each stage $n = 1, 2, \ldots$, each Player chooses an element in his set of actions: $i_n \in I$ for Player 1 (resp. $j_n \in J$ for Player 2), the corresponding vector outcome is $g_n = A_{i_n j_n} \in \mathbb{R}^k$ and the couple of actions (i_n, j_n) is announced to both Players. $\bar{g}_n = \frac{1}{n}\sum_{m=1}^n g_m$ is the average vector outcome up to stage n. The aim of Player 1 is that $\bar{g}_n$ approaches a target set $C \subset \mathbb{R}^k$. Approachability theory is thus a generalization of max-min level in a (one shot) game with real payoff where C is of the form $[v, +\infty)$.

The asymptotic approach corresponds to the following notion:

Definition 2.3. *A nonempty closed set C in $\mathbb{R}^k$ is* **weakly approachable** *by Player 1 in G if, for every $\varepsilon > 0$, there exists $N \in \mathbb{N}$ such that for any $n \geq N$ there is a strategy $\sigma = \sigma(n, \varepsilon)$ of Player 1 such that, for any strategy τ of Player 2:*

$$\mathsf{E}_{\sigma,\tau}(d_C(\overline{g}_n)) \leq \varepsilon.$$

where d_C stands for the distance to C.

If v_n is the value of the n-stage game with payoff $-\mathsf{E}(d_C(\bar{g}_n))$, weak-approachability means $v_n \to 0$. The uniform approach is expressed by the next definition:

Definition 2.4. *A nonempty closed set C in $\mathbb{R}^k$ is* **approachable** *by Player 1 in G if, for every $\varepsilon > 0$, there exists a strategy $\sigma = \sigma(\varepsilon)$ of Player 1 and $N \in \mathbb{N}$ such that, for any strategy τ of Player 2 and any $n \geq N$:*

$$\mathsf{E}_{\sigma,\tau}(d_C(\overline{g}_n)) \leq \varepsilon.$$

In this case, asymptotically the average outcome remains close in expectation to the target C, uniformly with respect to the opponent's behavior. The dual concept is excludability.

The "expected deterministic" repeated game form $G^\star$ is an alternative two-person infinitely repeated game associated, as the previous one, to the matrix A but where at each stage $n = 1, 2, \ldots$, Player 1 (resp. Player 2) chooses $u_n \in U = \Delta(I)$ (resp. $v_n \in V = \Delta(J)$), the outcome is $g_n^\star = u_n A v_n$ and (u_n, v_n) is announced. Hence $G^\star$ is the game played in "mixed actions" or in expectation. Weak $\star$approachability, $v_n^\star$, and $\star$approachability are defined similarly.

2.7.2 Weak approachability and quantitative differential games

The next result is due to Vieille (1992). Recall that the aim is to obtain an average outcome at stage n close to C.

First consider the game $G^\star$. Use as state variable the accumulated payoff $z_t = \int_0^t \gamma_s \mathrm{d}s$, $\gamma_s = u_s A v_s$ being the payoff at time s and consider the differential game Λ of fixed duration played on $[0, 1]$ starting from $z_0 = 0 \in \mathbb{R}^k$ with dynamics:

$$\dot{z}_t = u_t A v_t = f(u_t, v_t) \qquad [2.51]$$

and terminal payoff $-d_C(\mathbf{z}(1))$. Note that Isaacs's condition holds: $\max_u \min_v \langle f(z, u, v), \xi\rangle = \min_v \max_u \langle f(z, u, v), \xi\rangle = \mathtt{val}_{U\times V}\langle f(z, u, v), \xi\rangle$ for all $\xi \in \mathbb{R}^k$. The n stage game $G_n^\star$ appears then as a discrete time approximation of Λ and $v_n^\star = V_n(0, 0)$ where V_n satisfies, for $k = 0, \ldots, n-1$ and $z \in \mathbb{R}^k$:

$$V_n\left(\frac{k}{n}, z\right) = \mathtt{val}_{U\times V} V_n\left(\frac{k+1}{n}, z + \frac{1}{n}uAv\right) \qquad [2.52]$$

with terminal condition $V(1, z) = -d_C(z)$. Let $\Phi(t, z)$ be the value of the game played on $[t, 1]$ starting from z (i.e., with total outcome $z + \int_t^1 \gamma_s ds$). Then basic results from DG imply (see Section 6):

Theorem 2.26.
(1) $\Phi(z, t)$ *is the unique viscosity solution on* $[0, 1] \times \mathbb{R}^k$ *of:*

$$\partial_t \Phi(z, t) + \mathtt{val}_{U\times V}\langle D\Phi(z, t), uAv\rangle = 0$$

with $\Phi(z, 1) = -d_C(z)$.

(2)

$$\lim_{n\to\infty} v_n^\star = \Phi(0, 0).$$

The last step is to relate the values in $G_n^\star$ and in G_n.

Theorem 2.27.

$$\lim v_n^\star = \lim v_n$$

The idea of the proof is to play by blocks in G_{Ln} and to mimic an optimal behavior in $G_n^\star$. Inductively at the mth block of L stages in G_{Ln} Player 1 will play i.i.d. a mixed action optimal at stage m in $G_n^\star$ (given the past history) and $y_m^\star$ is defined as the empirical distribution of actions of Player 2 during this block. Then the (average) outcome in G_{Ln} will be close to the one in $G_n^\star$ for large L, hence the result.

Corollary 2.2.
Every set is weakly approachable or weakly excludable.

2.7.3 Approachability and B-sets

The main notion was introduced by Blackwell (1956). $\pi_C(a)$ denotes the set of closest points to a in C.

Definition 2.5. *A closed set* C *in* $\mathbb{R}^k$ *is a* **B**-*set for Player 1 (for a given matrix* A*), if for any* $a \notin C$*, there exists* $b \in \pi_C(a)$ *and a mixed action* $u = \hat{u}(a)$ *in* $U = \Delta(I)$ *such that the hyperplane through* b *orthogonal to the segment* $[ab]$ *separates* a *from* uAV:

$$\langle uAv - b, a - b\rangle \le 0, \qquad \forall v \in V$$

The basic result of Blackwell (1956) is:

Theorem 2.28. *Let C be a* **B***-set for Player 1. Then it is approachable in G and ⋆approachable in $G^\star$ by that Player. An approachability strategy is given by $\sigma(h_n) = \hat{u}(\bar{g}_n)$ (resp. $\sigma^\star(h_n^\star) = \hat{u}(\bar{g}_n^\star)$), where h_n (resp. $h_n^\star$) denotes the history at stage n.*

An important consequence of Theorem 2.28 is the next result due to Blackwell (1956):

Theorem 2.29. *A convex set C is either approachable or excludable.*

A further result due to Spinat (2002) characterizes minimal approachable sets:

Theorem 2.30. *A set C is approachable iff it contains a* **B***-set.*

2.7.4 Approachability and qualitative differential games

To study ⋆approachability, we consider an alternative qualitative differential game Γ where both the dynamics and the objective differ from the previous quantitative differential game Λ. We follow here Assoulamani et al. (2009). The aim is to control asymptotically the average payoff that will be the state variable and the discrete dynamics is of the form:

$$\bar{g}_{n+1} - \bar{g}_n = \frac{1}{n+1}(g_{n+1} - \bar{g}_n).$$

The continuous counterpart is $\bar{\gamma}_t = \frac{1}{t}\int_0^t u_s A v_s ds$. A change of variable $z_t = \bar{\gamma}_{e^t}$ leads to:

$$\dot{z}_t = u_t A v_t - z_t. \qquad [2.53]$$

which is the dynamics of an autonomous differential game Γ with $f(z, u, v) = uAv - z$, that still satisfies Isaacs' condition. In addition, the aim of Player 1 in Γ is to stay in a certain set C.

Definition 2.6. *A nonempty closed set C in $\mathbb{R}^k$ is a* **discriminating domain** *for Player 1, given f if:*

$$\forall a \in C, \ \ \forall p \in NP_C(a), \qquad \sup_{v \in V} \inf_{u \in U} \langle f(a, u, v), p \rangle \leq 0, \qquad [2.54]$$

where $NP_C(a) = \{p \in \mathbb{R}^K; d_C(a + p) = \|p\|\}$ is the set of proximal normals to C at a.

The interpretation is that, at any boundary point $x \in C$, Player 1 can react to any control of Player 2 in order to keep the trajectory in the half space facing a proximal normal p.

The following theorem, due to Cardaliaguet (1996), states that Player 1 can ensure remaining in a discriminating domain.

Theorem 2.31. *Assume that f satisfies Isaacs' condition, that $f(x, U, v)$ is convex for all x, v, and that C is a closed subset of $\mathbb{R}^k$. Then C is a discriminating domain if and only if for every z belonging to C, there exists a nonanticipative strategy $\alpha \in \mathcal{A}'$, such that for any $\mathbf{v} \in \mathbf{V}$, the trajectory $\mathbf{z}[\alpha(\mathbf{v}), \mathbf{v}, z](t)$ remains in C for every $t \geq 0$.*

The link with approachability is through the following result:

Theorem 2.32. *Let $f(z, u, v) = uAv - z$. A closed set $C \subset \mathbb{R}^k$ is a discriminating domain for Player 1, if and only if C is a $\mathbf{B}$-set for Player 1.*

The main result is then:

Theorem 2.33. *A closed set C is $\star$approachable in $G^\star$ if and only if it contains a $\mathbf{B}$-set.*

The direct part follows from Blackwell's proof. For the converse, first one defines a map Ψ from strategies of Player 1 in $G^\star$ to nonanticipative strategies in Γ. Next, given $\varepsilon > 0$ and a strategy σ_ε that ε-approaches C in $G^\star$, one shows that the trajectories in the differential game Γ that are compatible with $\alpha_\varepsilon = \Psi(\sigma_\varepsilon)$ approach C up to ε. Finally, one proves that the ω-limit set of any trajectory compatible with some α is a discriminating domain.

In particular, approachability and $\star$approachability coincide.

In a similar way, one can explicitly construct an approachability strategy in the repeated game G starting from a preserving strategy in Γ. The proof is inspired by the "extremal aiming" method of Krasovskii and Subbotin (1988) which is in the spirit of proximal normals and approachability.

2.7.5 Remarks and extensions

1. In both cases, the main ideas to represent a RG as a DG is first to take as state variable either the total payoff or the average payoff, but in both cases the corresponding dynamics is (asymptotically) smooth; the second aspect is to work with expectation so that the trajectory is deterministic.

2. For recent extensions of approachability condition for games on more general spaces, see Lehrer (2002) and Milman (2006). For games with signals on the outcomes, see Lehrer and Solan (2007) and Perchet (2009, 2011a,b) which provides a characterization for convex sets. Perchet and Quincampoix (2012) present a general perspective by working on the space of distribution on signals that can be generated during the play. Approachability is then analyzed in the space of probability distributions on $\mathbb{R}^n$ with the

L^2-Wasserstein distance using tools from As Soulaimani (2008) and Cardaliaguet and Quincampoix (2008).

2.8. ALTERNATIVE TOOLS AND TOPICS

2.8.1 Alternative approaches

2.8.1.1 A different use of the recursive structure

The use of the operator approach does not allow to deal easily with games with signals: the natural state space on which the value is defined is large. In their original approach, Mertens and Zamir (1971) and Mertens (1972) introduce thus majorant and minorant games having both simple recursive structure, i.e., small level in the hierarchy of beliefs in the auxiliary game.

Similarly, a sophisticated use of the recursive structure allows to obtain exact speed of convergence for games with incomplete information on one side (Mertens, 1998).

2.8.1.2 No signals

For a class of games with no signals, Mertens and Zamir (1976a) introduced a kind of normal form representation of the infinite game for the maxmin (and for the minmax): this is a collection of strategies (and corresponding payoffs) that they prove to be an exhaustive representation of optimal plays. In any large game, Player 1 can guarantee the value of this auxiliary game and Player 2 defend it.

A similar approach is used in Sorin (1989) for the asymptotic value. One uses a two-level scale: blocks of stages are used to identify the regular moves and sequence of blocks are needed for the exceptional moves.

2.8.1.3 State dependent signals

For RG with incomplete information, the analysis in Sorin (1985b) shows that the introduction of state dependent signals generates absorbing states in the space of beliefs, hence the natural study of absorbing games with incomplete information. In Sorin (1984,1985a), two classes are studied and the minmax, maxmin, and asymptotic values are identified.

2.8.1.4 Incomplete information on the duration

These are games where the duration is a random variable on which the Players have private information. For RG with incomplete information, these games have maxmin and minmax that may differ (Neyman, 2012a).

2.8.1.5 Games with information lag

There is a large literature on games with information lag starting with Scarf and Shapley (1957). A recent study in the framework of stochastic games is due to Levy (2002).

2.8.2 The "Limit Game"

2.8.2.1 Presentation

In addition to the convergence of the values $\{v_\mu\}$, one looks for a game $\mathcal{G}$ on $[0, 1]$ with strategy sets $\mathbf{U}$ and $\mathbf{V}$ and value w such that:

(1) the play at time t in $\mathcal{G}$ would be similar to the play at stage $m = [tn]$ in G_n (or at the fraction t of the total weight of the game for general evaluation μ),

(2) ε-optimal strategies in $\mathcal{G}$ would induce 2ε-optimal strategies in G_n, for large n. More precisely, the history for Player 1 up to stage m in G_n defines a state variable that is used to define with a strategy U in $\mathcal{G}$ at time $t = m/n$ a move in G_n.

Obviously then, the asymptotic value exists and is w.

2.8.2.2 Examples

One example was explicitly described (strategies and payoff) for the Big Match with incomplete information on one side in Sorin (1984). $\mathbf{V}$ is the set of measurable maps f from $[0, 1]$ to $\Delta(J)$. Hence Player 2 plays $f(t)$ at time t and the associated strategy in G_n is a piecewiese constant approximation. $\mathbf{U}$ is the set of vectors of stopping times $\{\rho^k\}$, $k \in K$, i.e., increasing maps from $[0, 1]$ to $[0, 1]$ and $\rho^k(t)$ is the probability to stop the game before time t if the private information is k.

The auxiliary differential game introduced by Vieille (1992) to study weak approachability, Section 7.2 is also an example of a limit game.

A recent example deals with absorbing games (Laraki, 2010). Recall the auxiliary game Γ corresponding to (2.26). Then one shows that given a strategy (x, α) ε-optimal in Γ, its image $x + \lambda\alpha$ (normalized) is 2ε-optimal in G_λ.

For games with incomplete information on one side, the asymptotic value $v(p)$ is the value of the splitting game (Section 3.4.2. C) with payoff $\int_0^1 u(p_t)dt$, where u is the value of the nonrevealing game and p_t is a martingale in $\Delta(K)$ starting from p at $t = 0$.

For the uninformed Player, an asymptotically optimal strategy has been defined by Heuer (1992a) and extends to the case of lack of information on both sides. His approach has the additional advantage to show that, assuming that the Mertens-Zamir system has a solution v, then the asymptotic value exists and is v.

2.8.2.3 Specific Properties

This representation allows also to look for further properties, like stationarity of the expected payoff at time t along the plays induced optimal strategies, see Sorin et al. (2010) or robustness of optimal strategies: to play at stage n optimally in the discounted game with $\lambda_n = \frac{\mu_n}{\sum_{m \geq n} \mu_m}$ should be asymptotically optimal for the evaluation μ.

2.8.3 Repeated games and differential equations

2.8.3.1 RG and PDE

As already remarked by De Meyer (1996a, 1999), the fact that the asymptotic value of game satisfies a limit dynamic principle hence a PDE can be used in the other direction. To prove that a certain PDE has a solution, one constructs a family of simple finite RG (corresponding to time and space discretizations) and one then shows that the limit of the values exists.

For similar results in this direction see, e.g., Kohn and Serfaty, (2006), and Peres et al. (2009).

2.8.3.2 RG and evolution equations

We follow Vigeral (2010a). Consider again a nonexpansive mapping $\mathbf{T}$ from a Banach space X to itself. The nonnormalized n stage values satisfy $V_n = \mathbf{T}^n(0)$ hence:

$$V_n - V_{n-1} = -(Id - \mathbf{T})(V_{n-1})$$

which can be considered as a discretization of the differential equation:

$$\dot{x} = -Ax \qquad [2.55]$$

where the maximal monotone operator A is $Id - \mathbf{T}$.

The comparison between the iterates of $\mathbf{T}$ and the solution U of (2.55) is given by Chernoff's formula (see Brézis, 1973, p. 16):

$$\|U(t) - \mathbf{T}^n(U(0))\| \leq \|U'(0)\|\sqrt{t + (n-t)^2}.$$

In particular with $U(0) = 0$ and $t = n$, one obtains:

$$\|\frac{U(n)}{n} - v_n\| \leq \frac{\|\mathbf{T}(0)\|}{\sqrt{n}}.$$

It is thus natural to consider $u(t) = \frac{U(t)}{t}$ which satisfies an equation of the form:

$$\dot{x}(t) = \Phi(\varepsilon(t), x(t)) - x(t) \qquad [2.56]$$

where as usual $\Phi(\varepsilon, x) = \varepsilon\mathbf{T}\left(\frac{1-\varepsilon}{\varepsilon}x\right)$ and notice that (2.56) is no longer autonomous.

Define the condition (L) by:

$$\|\Phi(\lambda, x) - \Phi(\mu, x)\| \leq |\lambda - \mu|(C + \|x\|).$$

Theorem 2.34. *Let $u(t)$ be the solution of [2.56], associated to $\varepsilon(t)$.*
(a) *If $\varepsilon(t) = \lambda$, then $\|u(t) - v_\lambda\| \to 0$*
(b) *If $\varepsilon(t) \sim \frac{1}{t}$, then $\|u(n) - v_n\| \to 0$*
Assume condition (L).
(c) *If $\frac{\varepsilon'(t)}{\varepsilon^2(t)} \to 0$ then $\|u(t) - v_{\varepsilon(t)}\| \to 0$*

Hence $\lim v_n$ and $\lim v_\lambda$ mimic solutions of similar perturbed evolution equations and in addition one has the following robustness result:

Theorem 2.35. *Let u (resp. $\bar{u}$) be a solution of [2.56] associated to ε (resp. $\bar{\varepsilon}$).*
Then $\|u(t) - \bar{u}(t)\| \to 0$ as soon as
(i) $\varepsilon(t) \sim \bar{\varepsilon}(t)$ *as $t \to \infty$ or*
(ii) $|\varepsilon - \bar{\varepsilon}| \in L^1$.

2.8.4 Multimove games

We describe here very briefly some other areas that are connected to RG.

2.8.4.1 Alternative evaluations

There are games similar to those of Section 1 where the sequence of payoffs $\{g_n\}$ is evaluated through a single functional like limsup, or liminf.

Some of the results and of the tools are quite similar to those of RG, like "recursive structure," "operator approach," or construction by iteration of optimal strategies. A basic reference is Maitra and Sudderth (1996).

2.8.4.2 Evaluation on plays

More generally, the evaluation is defined here directly on the set of plays (equipped with the product topology) extending the games of Gale and Stewart (1953) and Blackwell (1969). The basic results, first in the perfect information case then in the general framework, are due to Martin (1975, 1998) and can be expressed as:

Theorem 2.36. *Borel games are determined.*

For the extension to stochastic games and related topics, see Maitra and Sudderth (1992, 1993, 1998, 2003). For a recent result in the framework of games with delayed information, see Shmaya (2011).

2.8.4.3 Stopping games

Stopping games (with symmetric information) have been introduced by Dynkin (1969). They are played on a probability space $(\Omega, \mathcal{A}, P)$ endowed with a filtration $\mathcal{F} = (\mathcal{F}_t)$ describing the common information of the Players as time go on. Each Player i chooses a measurable time θ_i when he stops and the game ends at $\theta = \min\{\theta_1, \theta_2\}$. The payoff is $g_i(\theta, \omega)$ if Player $i = 1, 2$ stops first the game and is $f(\theta, \omega)$ if they stop it simultaneously. Payoff functions are supposed to be uniformly integrable.

Neveu (1975) in discrete time and Lepeltier and Maingueneau (1984) in continuous time proved the existence of the value in *pure strategies* under the "standard" condition $g_1 \leq f \leq g_2$ (at each moment, each Player prefers the other to stop rather than himself).

Without this condition, mixed strategies are necessary to have a value. Rosenberg et al. (2001) proved the existence of the value in discrete time. As in Mertens and Neyman (1981), they let the Players play an optimal discounted strategy, where the discount factor may change from time to time, depending on their information. Shmaya and Solan (2004) provided a very elegant proof of the result based on a stochastic variation of Ramsey's theorem. Finally, Laraki and Solan (2005) proved the existence of the value in continuous time.

For stopping games with incomplete information on one side, the value may not exist even under the standard assumption (Laraki, 2000). This is due to the fact that Player 2 prefers to wait until Player 1 uses his information, while Player 1 prefers to use his information only if he knows that Player 2 will never stop.

For finite horizon stopping games with incomplete information on one side, the value exists in discrete time using Sion's minmax theorem. In continuous time, the value exists under the standard assumption and may be explicitly characterized by using viscosity solutions combined with BSDE technics (Grün, 2013). Without the standard condition, the value may not exist (Bich and Laraki, 2013). One example is as follows. One type of Player 1 has a dominant strategy: to stop at time zero. The other type prefers to stop just after 0, but before Player 2. However, Player 2 also prefers to stop just after zero but before type 2 of Player 1.

For a survey on the topic, see Solan and Vieille (2004) and for the related class of duels see Radzik and Raghavan (1994).

2.9. RECENT ADVANCES

We cover here very recent and important advances and in particular counter examples to conjectures concerning the asymptotic value and new approaches to multistage interactions.

2.9.1 Dynamic programming and games with an informed controller

2.9.1.1 General evaluation and total variation

Recall that an evaluation μ is a probability on $\mathbb{N}^*$ which corresponds to a discrete time process and the associated length is related to its expectation. However, some regularity has to be satisfied to express that the duration goes to ∞: $\|\mu\| \to 0$ is not sufficient (take for payoff an alternating sequence of 0 and 1 and for evaluation the sequences $\mu^n = (0, 1/n, 0, 1/n, \ldots)$ and $\nu^n = (1/n, 0, 1/n, 0, \ldots)$).

One measure of regularity is the *total variation*, $TV(\mu) = \sum_n |\mu_n - \mu_{n+1}|$, and a related ergodic theorem holds for a Markov chain P: there exists Q such that $\sum_n \mu_n P^n$ converges to Q as $TV(\mu) \to 0$ (Sorin, 2002, p. 105). Obviously for decreasing evaluations, one has $\mu_1 = \|\mu\| = TV(\mu)$.

One can define stronger notions of convergence associated to this criteria, following Renault (2013), Renault and Venel (2012):

v is a TV-asymptotic value if for each $\varepsilon > 0$, there exists $\delta > 0$ such that $TV(\mu) \leq \delta$ implies $\|v_\mu - v\| \leq \varepsilon$. (In particular then $v = \lim_{n\to\infty} v_n = \lim_{\lambda\to 0} v_\lambda$.)

Similarly v_∞ is a TV-uniform value if for each $\varepsilon > 0$, there exists $\delta > 0$ and σ^* strategy of Player 1 such that $TV(\mu) \leq \delta$ implies $E_{\sigma^*,\tau}\langle \mu, g\rangle \geq v_\infty - \varepsilon$, for all τ (and a dual statement for Player 2).

Obviously a TV-uniform value is a TV-asymptotic value.

2.9.1.2 Dynamic programming and TV-asymptotic value

In the framework of dynamic programming (see Section 5.3), Renault (2013) proves that given a sequence of evaluations $\{\mu^k\}$ with $TV(\mu^k) \to 0$, the corresponding sequences of values v_{μ^k} converge (uniformly) iff the family $\{v_{\mu^k}\}$ is totally bounded (pre-compact). In this case the limit is v^* with:

$$v^*(\omega) = \inf_{\mu\in\mathcal{M}} \sup_m v_{m\circ\mu}(\omega) = \inf_k \sup_m v_{m\circ\mu^k}(\omega)$$

where $\mathcal{M}$ is the set of evaluations and $m \circ \mu$ is the m-translation of the evaluation μ $(m \circ \mu_n = \mu_{m+n})$. In particular a TV-asymptotic value exists if Ω is a totally bounded metric space and the family $\{v_\mu, \mu \in \mathcal{M}\}$ is uniformly equicontinuous.

2.9.1.3 Dynamic programming and TV-uniform value

Renault and Venel (2012) provide condition for an MDP to have a TV-uniform value and give a characterization of it, in the spirit of gambling, namely: excessive property and invariance. They also introduce a new metric on $\Delta_f(X)$ where $X = \Delta(K)$ with K finite, that allows them to prove the existence of a TV-uniform value in MPD with finite state space and signals.

2.9.1.4 Games with a more informed controller

Renault and Venel (2012) prove that game with an informed controller has a TV-uniform value extending the result in Renault (2012).

An extension in another direction is due to Gensbittel et al. (2013). Define a game with a more informed controller as a situation where:

(i) the belief $\zeta_n^1(1) \in M^1(1) = \Delta(M)$ of Player 1 on the state m_n at stage n (i.e., the conditional probability given his information h_n^1) is more precise than the belief

$\zeta_n^2(1) \in M^2(1) = \Delta(M)$ of Player 2 (adding the information h_n^2 to h_n^1 would not affect $\zeta_n^1(1)$).

(ii) Player 1 knows the belief $\zeta_n^2(2) \in M^2(2) = \Delta(M^1(1))$ of Player 2 on his own beliefs.

(iii) Player 1 control the process $\{\zeta_n^2(2)\}$.

Then the game has a uniform value.

The analysis is done trough an auxiliary game with state space $M^2(2)$ and where the result of Renault (2012) applies. The difficulty is to prove that properties of optimal strategies in the auxiliary game can be preserved in the original one.

2.9.1.5 Comments

Note that the analysis in Cardaliaguet et al. (2012) shows in fact that a TV-asymptotic value exists in incomplete information, absorbing and splitting games. Even more, v_μ is close to v as soon as $\|\mu\|$ is small enough. However, these results are very specific to these classes.

Recently, Ziliotto (2014) has given examples of a finite stochastic game with no TV-asymptotic value and of an absorbing game with no TV-uniform value (actually the Big Match).

In particular, the above results in Section 9.1 do not extend from the one-Player to the two-Player case.

2.9.2 Markov games with incomplete information on both sides

Gensbittel and Renault (2012) proved recently the existence of $\lim_{n\to\infty} v_n$ in repeated games with incomplete information in which the types of Players 1 and 2 evolve according to two independent Markov chains $\mathbf{K}$ and $\mathbf{L}$ on two finite sets, respectively K and L. The parameter $\{k_m\}$ follows $\mathbf{K}$, starting from some distribution p at stage one and k_m is observed by Player 1 at stage m. The distribution of the state k_m is thus given by $p\mathbf{K}^m$. A similar situation, with $\mathbf{L}$ and q, holds for Player 2.

Moves are revealed along the play and the stage m payoff is given by $g_m = g(k_m, l_m, i_m, j_m)$.

The authors show that $v = \lim_{n\to\infty} v_n$ exists and is the unique continuous solution of a "Mertens-Zamir"-like system of equations with respect to a function $u^*(p, q)$. In addition to be concave/convex, the function v has to be invariant through the Markov chain: $v(p, q) = v(p\mathbf{K}, q\mathbf{L})$. As for u^*, it is the uniform limit $u_n^*(p, q)$ of the sequence of values of a n-stage repeated "non-revealing game" $\Gamma_n^*(p, q)$. This game is similar to the one introduced by Renault (2006) to solve the one side incomplete information case (see Section 5.4). In $\Gamma_n^*(p, q)$, Player 1 plays as in the original game but is restricted to use strategies that keep the induced beliefs of Player 2 constant on the partition

$\tilde{K}$ of K into recurrence classes defined by the Markov chain $\mathbf{K}$ (and similarly for Player 2).

The proof of uniform convergences of u_n^* is difficult and uses, in particular, a splitting lemma in Laraki (2004) to show equi-continuity of the family of functions $\{u_n^*\}$.

2.9.3 Counter examples for the asymptotic approach

2.9.3.1 Counter example for finite state stochastic games with compact action spaces

Vigeral (2013) gives an example of a continuous stochastic game on $[0, 1]^2$ with four states (two of them being absorbing) where the asymptotic value does not exists: the family $\{v_\lambda\}$ oscillates as $\lambda \to 0$. The payoff is independent of the moves and the transition are continuous. However, under optimal strategies, the probability to stay in each state goes to zero and the induced occupation measure on the two states oscillate as $\lambda \to 0$. An analog property holds for v_n.

This example answers negatively a long standing conjecture for stochastic games with finite state space. In fact, the operator and variational approaches were initially developed to prove the existence of $\lim_{\lambda\to 0} v_\lambda$ and $\lim_{n\to\infty} v_n$. They work in the class of irreversible games (in which, once one leaves a state, it cannot be reached again, such as absorbing, recursive, and incomplete information games). Outside this class, the asymptotic value may not exist.

2.9.3.2 Counter examples for games with finite parameter sets

A phenomena similar to the previous one is obtained by Ziliotto (2013) for a game, say G_4 played with countable action set and finite state space.

More interestingly, this game is obtained by a series of transformations preserving the value and starting from a game G_1 which is a finite stochastic game where the Players have symmetric information, hence know the moves, but do not observe the state. In that case, the game is equivalent to a stochastic game with standard signaling and a countable state space $\Omega' \subset \Delta(\Omega)$.

Explicitly, Player 1 has three moves: Stay1, Stay2, and Quit, and Player 2 has two moves: Left and Right. The payoff is -1 and the transition are as follows:

A	**Left**	**Right**
Stay1	A	$(\frac{1}{2}A + \frac{1}{2}B)$
Stay2	$(\frac{1}{2}A + \frac{1}{2}B)$	A
Quit	A^*	A^*

B	**Left**	**Right**
Stay1	A	B
Stay2	B	A
Quit	B^+	B^+

The initial state is A and A^* is an absorbing state with payoff -1. During the play, the Players will generate common beliefs on the states $\{A, B\}$. Once B^+ is reached, a dual game with a similar structure and payoff 1 is played. Now, Player 2 has three moves: Player 1 two moves: the transition is similar but different from the one above.

In an equivalent game G_2, only one Player plays at each stage and he receives a random signal on the transition (like Player 1 facing $(1/2, 1/2)$ in the above game).

Finally, G_3 is a stochastic game with known state and a countable state space corresponding to the beliefs of the Player on $\{A, B\}$, which are of the form $\{1/2^n, 1 - 1/2^n\}$.

In all cases, the families $\{v_\lambda\}$ and $\{v_n\}$ oscillate.

The main point is that the discrete time in the repeated interaction generates a discrete countable state space while the evolution of the parameter λ is continuous.

This spectacular result answers negatively two famous conjectures of Mertens (1987): existence of the asymptotic value in games with finite parameter spaces and its equality with the maxmin whenever Player 1 is more informed than Player 2. In fact, the example shows more, the asymptotic value may not exist even in the case of symmetric information!

2.9.3.3 Oscillations

To understand the nature and links between the previous examples, Sorin and Vigeral (2013) construct a family of configurations which are zero-sum repeated games in discrete time where the purpose is to control the law of a stopping time of exit. For a given discount factor $\lambda \in]0, 1]$, optimal stationary strategies define an inertia rate Q_λ ($1 - Q_\lambda$ is the normalized probability of exit during the game).

When two such configurations (1 and 2) are coupled, this induces a stochastic game where the state will move from one to the other in a way depending on the previous rates Q_λ^i, $i = 1, 2$ and the discounted value will satisfy:

$$v_\lambda^i = a^i Q_\lambda^i + (1 - Q_\lambda^i) v_\lambda^{-i}, \quad i = 1, 2$$

where a^i is the payoff in configuration i. The important observation is that the discounted value is a function of the ratio $\frac{Q_\lambda^1}{Q_\lambda^2}$. It can oscillate as $\lambda \to 0$, when both inertia rates converge to 0.

The above analysis shows that oscillations in the inertia rate and reversibility allow for nonconvergence of the discounted values.

2.9.3.4 Regularity and o-minimal structures

An alternative way to keep regularity in the asymptotic approach is to avoid oscillations. This is the case when the state space is finite (due to the algebraic property of v_λ). Recently, Bolte et al. (2013) extend the algebraic approach to a larger class of stochastic games with compact action sets.

The concept of o-minimal structure was introduced recently as an extension of semi-algebraic geometry through an axiomatization of its most important properties (see van den Dries, 1998). It consists in a collection of subsets of $\mathbf{R}^n$, for each $n \in \mathbb{N}$, called *definable* sets. Among other natural requirements, the collection need to be stable by linear projection, its "one-dimensional" sets must consist on the set of finite unions of intervals, and must at least contain all semi-algebraic sets. It shares most of the nice properties of semi-algebraic sets, in particular, finiteness of the number of connected components.

Theorem 2.37. *If (the graph of) the Shapley operator is definable, v_λ is of bounded variations hence the uniform value exists.*

This raises the question: under which condition is the Shapley operator definable?

A function f on $\Omega \times I \times J$ is separable if there are finite sets S and T, and functions a_s, b_t, and $c_{s,t}$ such that $f(i,j,\omega) = \sum_{s\in S}\sum_{t\in T} c_{s,t}(\omega)a_s(i,\omega)b_t(j,\omega)$. This is the case for polynomial games or games where one of the payers has finitely many actions. Then one proves that if transition and payoff functions are separable and definable in some o-minimal structure, then the Shapley operator is definable in the *same* structure.

The Stone-Weierstrass theorem is then used to drop the separability assumption on payoffs, leading to the following extension of fundamental result of Mertens and Nemann.

Theorem 2.38. *Any stochastic games with finitely many states, compact action sets, continuous payoff functions, and definable separable transition functions has a uniform value.*

An example shows that with semi-algebraic data, the Shapley operator may not be semi-algebraic, but belongs to a higher o-minimal structure. Hence, the open question: do a definable stochastic game has a definable Shapley operator in a larger o-minimal structure?

2.9.4 Control problem, martingales, and PDE

Cardaliaguet and Rainer (2009b) consider a continuous time game $\Gamma(p)$ on $[0, T]$ with incomplete information on one side and payoff function $\gamma^k(t, u, v)$, $k \in K$. Player 1 knows k and both Players know its distribution $p \in \Delta(K)$. The value of the nonrevealing local game is $U(t,p) = \mathtt{val}_{U\times V} \sum_k p^k \gamma^k(t, u, v)$. This is a particular differential game with incomplete information where the state variable is not controlled. Hence, all previous primal and dual PDE characterizations of the value $V(t,p)$ style hold (Section 6.3.). Moreover, there is a martingale maximization formulation (compare with the splitting game, Section 3.4.2 C). Explicitly:

Theorem 2.39. *$V(\cdot\,,\cdot)$ is:*

(a) the smallest function, concave in p and continuous viscosity solution of $\frac{\partial f}{\partial t}(t,p) + U(t,p) \le 0$, with boundary condition $f(1,p) = 0$.

(b) the value of the control problem $\max_{p_t} \mathbb{E}[\int_0^1 U(s,p_t)dt]$ where the maximum is over all continuous time càdlàg martingales $\{p_t\}$ on $\Delta(K)$ that starts at p.

Any maximizer in the above problem induces an optimal strategy of the informed Player in the game. Cardaliaguet and Rainer (2009b) provide more explicit computations and show in particular that, unlike in Aumann-Maschler's model where only one splitting is made at the beginning and no further information is revealed, in the nonautonomous case information may be revealed gradually or in the middle of the game, depending on how U varies with t. Such phenomena cannot be observed in RG: they are time-independent.

Grün (2012) extends these characterizations (PDE in the primal and dual games, martingale maximization) to stochastic continuous time games with incomplete information on one side.

Cardaliaguet and Rainer (2012) generalized the result to the case where $K = \mathbb{R}^d$ so that the PDE formulation is on $\Delta(\mathbb{R}^d)$, like in Cardaliaguet and Quincampoix (2008).

The martingale maximization problem appears also when the state evolves along the play according to a Markov chain (Cardaliaguet et al., 2013) (see Section 9.5.5.).

Very recently, Gensbittel (2013) generalizes the characterizations to continuous time games where Players are gradually informed, but Player 1 is more informed. Formally, information is modeled by a stochastic process $Z_t = (X_t, Y_t)$, where X_t is a private information for Player 1 and Y_t is a public information. This is a first step toward understanding De Meyer's financial model when Players get information gradually and not only at once.

2.9.5 New links between discrete and continuous time games

2.9.5.1 Multistage approach

The game is played in discrete time, stage after stage. However, stage n represents more the nth interaction between the Players rather than a specific time event. In addition stage n has no intrinsic length: given an evaluation μ, its duration in the normalized game on $[0,1]$ is μ_n. Finally, the variation of the state variable at each stage is independent of the evaluation.

2.9.5.2 Discretization of a continuous time game

An alternative framework for the asymptotic approach is to consider an increasing frequency of the interactions between the Players. The underlying model is a game in continuous time played on $[0, +\infty]$ with a state variable Z_t which law depends upon the actions on the Players.

The usual repeated game G^1 corresponds to the version where the Players play at integer times $1, 2, \ldots$ (or that their moves are constant on the interval of time $[n, n+1]$). Then the law of Z_{n+1} is a function of Z_n, i_n, j_n.

In the game G^δ, the timing of moves is still discrete and the play is like above; however, stage n corresponds to the time interval $[n\delta, (n+1)\delta)$ and the transition from Z_n to Z_{n+1} will be a function of Z_n, i_n, j_n and of the length δ (see Neyman, 2012b).

In particular, as $\delta \to 0$ the play should be like in continuous time but with "smooth" transitions.

In addition, some evaluation criteria has to be given to integrate the process of payoffs on $[0, +\infty]$.

2.9.5.3 Stochastic games with short stage duration

A typical such model is a stochastic game with finite state space Ω where the state variable follows a continuous Markov chain controlled by the Players trough a generator $A(i,j), i \in I, j \in J$ on $\Omega \times \Omega$ (see the initial version of Zachrisson, 1964).

Consider the discounted case where the evaluation on $[0, +\infty]$ is given by re^{-rt}. We follow Neyman (2013). Given a duration $h \geq 0$ let $P_h = \exp(hA)$ be the transition kernel associated to A and stage duration h.

Then the value of the corresponding $G^{\delta,r}$ game satisfies:

$$v^{\delta,r}(\omega) = \mathtt{val}_{X\times Y}\{(1 - e^{-r\delta})g(x, y, \omega) + e^{-r\delta} \sum_{\omega'} P_\delta(x, y, \omega)[\omega']v^{\delta,r}(\omega')\} \quad [2.57]$$

and converges, as $\delta \to 0$, to v^r solution of:

$$v^r(\omega) = \mathtt{val}_{X\times Y}\{g(x, y, \omega) + \sum_{\omega'} A(x, y, \omega)[\omega']v^r(\omega')/r\} \quad [2.58]$$

which is also the value of the $r\delta$-discounted repeated game with transition $Id + \delta A/(1 - r\,\delta)$, for any δ small enough.

Optimal strategies associated to (2.58) induce stationary strategies that are approximatively optimal in the game $G^{\delta,r}$ for δ small enough. The convergence of the values to v^r holds for all games with short duration stages, not necessarily uniform (compare with the limit game 8.3).

Replacing the discounted evaluation by a decreasing evaluation θ on $[0, +\infty]$ gives similar results with Markov optimal strategies for the limit game G^θ of the family $G^{\delta,\theta}$.

In the finite case (I and J finite), the above system defining v^r is semi algebraic and $\lim_{r\to 0} v^r = v^0$ exists. Using an argument of Solan (2003), one shows that the convergence of $v^{\delta,r}$ to v^r is uniform in r.

In addition, Neyman (2013), considers a family of repeated games with vanishing stage duration and adapted generators and gives condition for the family of values to converge.

Finally, the existence of a uniform value (with respect to the duration t for all δ small enough) is studied, using Theorem 2.17.

2.9.5.4 Stochastic games in continuous time

Neyman (2012b) also introduced directly a game Γ played in continuous time. Basically, one requires the strategy of each Player to be independent of his previous short past behavior. In addition, Players use and observe mixed strategies. This implies that facing "simple strategies" a strategy is inducing a single history (on short time interval) and the play is well defined.

Optimal strategies associated to the game [2.58] induce optimal stationary strategies in the discounted game Γ^r, which value is v^r (see also Guo and Hernandez-Lerma, 2003 and 2005).

Consider the operator $\mathbf{N}$ on $\mathbb{R}^\Omega$ defined by :

$$\mathbf{N}[v](\omega) = \mathtt{val}_{X\times Y}\{g(x,y,\omega) + \sum_{\omega'} A(x,y,\omega)[\omega']v(\omega')\}. \qquad [2.59]$$

$\mathbf{N}$ defines a semigroup of operators with $S_0 = Id$ and $\lim_{h\to 0} \frac{S_h - Id}{h} = \mathbf{N}$. Then the value of the continuous game of length t, Γ_t is $S_t(0)/t$ (compare with [2.8] and [2.10]). Optimal strategies are obtained by playing optimally at time s in the game associated to $\mathbf{N}(S_{t-s}(0))(\omega_s)$.

Finally, the game has a uniform value.

2.9.5.5 Incomplete information games with short stage duration

Cardaliaguet et al. (2013) consider a model similar to 9.4.2 with a given discount factor r. Only Player 1 is informed upon the state which follows a continuous time Markov chain with generator A. The initial distribution is p and known by both Players. This is the "continuous time" analog of the model of Renault (2006); however, there are two main differences: the asymptotics is not the same (large number of stages versus short stage duration) and in particular the total variation of the sate variable is unbounded in the first model (take a sequence of i.i.d random variables on K) while bounded in the second. Note that in the initial Aumann-Maschler framework where the state is fixed during the play, the two asymptotics are the same.

Cardaliaguet et al. (2013) considers the more general case where the Markov chain is controlled by both Players (thus A is a family $A(i,j)$) and proves that the limit (as $\delta \to 0$) of the values $v^{\delta,r}$ of $G^{\delta,r}$, the r-discounted game with stage duration δ, exists and is the unique viscosity solution of an Hamilton-Jacobi equation with a barrier (compare with [2.50]):

$$\min\left\{rv(p) - \mathtt{val}_{X\times Y}[rg(x,y,p) + \langle \nabla v(p), pA(x,y)\rangle], -\bar{L}v(p)\right\} = 0. \qquad [2.60]$$

A similar result holds for the case with lack of information on both sides where each Player controls his privately known state variable that evolves through a continuous time Markov chain.

Note that the underlying nonrevealing game corresponds to a game where the state follows a continuous time Markov chain, but the Players are not informed, which is in the spirit of Cardaliaguet and Quincampoix (2008). Its value satisfies:

$$u^{\delta,r}(p) = \mathtt{val}_{X\times Y}\left\{(1-e^{r\delta})g(x,y,p) + e^{r\delta}\mathsf{E}_{x,y}[v^{\delta,r}(pP_\delta(i,j))]\right\}. \qquad [2.61]$$

hence at the limit

$$u^r(p) = \mathtt{val}_{X\times Y}\left\{g(x,y,p) + \langle \nabla u^r(p)/r, pA(x,y)\rangle\right\}. \qquad [2.62]$$

2.9.6 Final comments

Recent developments in the field of two-Player zero-sum repeated games are promising and challenging. On the one hand, one sees the emergence of many new models, deep techniques, and a unifying theory of dynamic games, including both discrete- and continuous-time considerations and dealing with incomplete information, stochastic, and signaling aspects at the same time. On the other hand, we know now that v_λ and v_n may not converge even in finite RG and the characterization of the class of regular games (where both $\lim_{n\to\infty} v_n$ and $\lim_{\lambda\to 0} v_\lambda$ exist and are equal) is an important challenge.

In any case, a reflexion is necessary on the modeling aspect, and especially on the relation between discrete and continuous time formulations. The natural way to take the limit in a long repeated interaction is by considering the *relative* length of a stage compared to the total length, obtaining thus a continuous-time representation on the compact interval $[0,1]$. However, in the presence of an exogenous duration process on $\mathbb{R}^+$ (like the law of a state variable), this normalization on $[0,1]$ is no longer possible, and one would like the actual length of a stage to go to zero, leading to a continuous time game on $\mathbb{R}^+$. Also, some continuous-time games do not have faithful discretization (such as stopping games), but this is not only due to the variable "time," it also occurs with discrete versus continuum strategy spaces.

Finally, we should stress here that all these recent advances are expected to have a fundamental impact on the study of *nonzero-sum* games as well, such as nonexistence of limiting equilibrium payoffs in discounted games, folk theorems in continuous-time stochastic games, modeling imperfect monitoring in continuous-time games, nonautonomous transition and payoff functions, etc.

ACKNOWLEDGMENTS

The authors thank N. Vieille for precise and useful comments.

This research was supported by grant ANR-10-BLAN 0112 (Laraki) and by the European Union under the 7th Framework Program "FP7-PEOPLE-2010-ITN", grant agreement number 264735-SADCO (Sorin).

REFERENCES

As Soulaimani, S. 2008. Viability with probabilistic knowledge of initial condition, application to optimal control, Set-Valued Anal. 16, 1037–1060.

Assoulamani, S., Quincampoix, M., Sorin, S., 2009. Repeated games and qualitative differential games: approachability and comparison of strategies, SIAM J. Control Optim. 48, 2461–2479.

Aumann, R.J., Maschler, M., 1995. Repeated Games with Incomplete Information. M.I.T. Press (with the collaboration of R. Stearns), Cambridge MA, USA.

Bardi, M., 2009. On differential games with long-time-average cost. In: Advances in Dynamic Games and their Applications, Bernhard P., Gaitsgory, V., Pourtalier, O., (Eds.), Annals of ISDG, vol. 10, Birkhauser, pp. 3–18.

Bardi, M., Capuzzo Dolcetta, I., 1996. Optimal Control and Viscosity Solutions of Hamilton-Jacobi-Bellman Equations. Birkhauser.

Bewley, T., Kohlberg, E., 1976a. The asymptotic theory of stochastic games. Math Operat. Res. 1, 197–208.

Bewley, T., Kohlberg, E., 1976b. The asymptotic solution of a recursion equation occurring in stochastic games. Math. Operat. Res. 1, 321–336.

Bich, P., Laraki, R., 2013. On the existence of approximated and sharing rule equilibria in discontinuous games. Preprint hal-00846143.

Blackwell, D., 1956. An analog of the minmax theorem for vector payoffs. Paci. J. Math. 6, 1–8.

Blackwell, D., 1969. Infinite G_δ games with imperfect information, Applicationes Mathematicae X, 99–101.

Blackwell, D., Ferguson, T., 1968. The Big Match. Ann. Math. Stat. 39, 159–163.

Bolte, J., Gaubert, S., Vigeral, G., 2013. Definable zero-sum stochastic games. Preprint. arXiv:1301.1967v1.

Brézis, H., 1973. Opérateurs Maximaux Monotones et Semi-Groupes de Contractions dans les Espaces de Hilbert, North Holland, Amsterdam.

Buckdahn, R., Li, J., Quincampoix, M., 2013. Value function of differential games without Isaacs conditions: An approach with nonanticipative mixed strategies. Int. J. Game Theory 42, 989–1020.

Buckdahn, R., Cardaliaguet, P., Quincampoix, M., 2010. Some recent aspects of differential game theory. Dyn. Games Appl. 1, 74–114. ellman-Isaacs equations. SIAM J. Control Optim. 47, 444–475.

Cardaliaguet, P., 1996. A differential game with two Players and one target. SIAM J. Control Optim. 34, 1441–1460.

Cardaliaguet, P., 2007. Differential games with asymmetric information. SIAM J. Control Optim. 46, 816–838.

Cardaliaguet, P., 2008. Representation formulas for differential games with asymmetric information. J. Opimi. Theory Appl. 138, 1–16.

Cardaliaguet, P., 2009. Numerical approximation and optimal strategies for dfferential games with lack of information on one side. Advances in Dynamic Games and their Applications. In: Bernhard P., Gaitsgory, V., Pourtalier, O., (Eds.), Annals of ISDG, vol. 10. Birkhauser, pp. 159–176.

Cardaliaguet, P., Quincampoix, M., 2008. Deterministic differential games under probability knowledge of initial condition. Int. Game Theory Rev. 10, 1–16.

Cardaliaguet, P., Rainer, C., 2009a. Stochastic differential games with asymmetric information, Appl. Math. Optim., 59, 1–36.
Cardaliaguet, P., Rainer, C., 2009b. On a continuous time game with incomplete information. Math. Operat. Res. 34, 769–794.
Cardaliaguet, P., Rainer, C., 2012. Games with incomplete information in continuous time and for continuous types. Dyn. Games Appl. 2, 206–227.
Cardaliaguet, P., Souquière, A., 2012. A differential game with a blind Player. SIAM J. Control and Optimization 50, 2090–2116.
Cardaliaguet, P., Laraki, R., Sorin, S., 2012. A continuous time approach for the asymptotic value in two-person zero-sum repeated games. SIAM J. Control Optim. 50, 1573–1596.
Cardaliaguet, P., Quincampoix, M., Saint-Pierre, P., 2007. Differential games through viability theory: Old and recent results. Advances in Dynamic Game Theory. In: Jorgensen S., Quincampoix, M., Vincent, T., (Eds.), Annals of ISDG, vol. 9. Birkhauser, pp. 3–36.
Cardaliaguet, P., Rainer, C., Rosenberg, D., Vieille, N., 2013. Markov games with frequent actions and incomplete information. Preprint hal 00843475.
Coulomb, J.-M., 1992. Repeated games with absorbing states and no signals. Int. J. Game Theory 21, 161–174.
Coulomb, J.-M., 1996. A note on 'Big Match'. ESAIM.: Probab. Stat. 1, 89–93, http://www.edpsciences.com/ps/.
Coulomb, J.-M., 1999. Generalized Big Match. Math. Operat. Res. 24, 795–816.
Coulomb, J.-M., 2001. Repeated games with absorbing states and signalling structure. Math. Operat. Res. 26, 286–303.
Coulomb, J.M., 2003a. Stochastic games without perfect monitoring. Int. J. Game Theory 32, 73–96.
Coulomb, J.M., 2003b. Games with a recursive structure. Stochastic Games and Applications. In: Neyman A. and Sorin, S., (eds), vol. C 570. NATO Science Series, Kluwer Academic Publishers, pp. 427–442.
Crandall, M.G., Ishii, H., Lions, P.-L., 1992. User's guide to viscosity solutions of second order partial differential equations. Bull. Amer. Soc. 27, 1–67.
De Meyer, B., 1996a. Repeated games and partial differential equations. Math. Operat. Res. 21, 209–236.
De Meyer, B., 1996b. Repeated games, duality and the Central Limit theorem. Math. Operat. Res. 21, 237–251.
De Meyer, B., 1999. From repeated games to Brownian games. Ann. Inst. Henri Poinc. Prob. Statist. 35, 1–48.
De Meyer, B., 2010. Price dynamics on a stock market with asymmetric information. Games Econ. Behav. 69, 42–71.
De Meyer, B., Marino, A., 2005. Duality and optimal strategies in the finitely repeated zero-sum games with incomplete information on both sides. Cahier de la MSE 2005–27.
De Meyer, B., Moussa-Saley, H., 2002. On the strategic origin of Brownian motion in finance. Int. J. Game Theory 31, 285–319.
De Meyer, B., Rosenberg, D., 1999. "Cav u" and the dual game. Math. Operat. Res. 24, 619–626.
Dynkin, E.B., 1969. Game variant of a problem on optimal stopping. Soviet Math. Dokl. 10, 270–274.
Evans, L.C., Souganidis, P.E., 1984. Differential games and representation formulas for solutions of Hamilton-Jacobi equations. Indiana Univ. Math. J. 33, 773–797.
Everett, H., 1957. Recursive games. Dresher M., Tucker, A.W., Wolfe, P., (Eds.), Contributions to the Theory of Games, III, Annals of Mathematical Studies. vol. 39. Princeton University Press, pp. 47–78.
Forges, F., 1982. Infinitely repeated games of incomplete information: Symmetric case with random signals. Int. J. Game Theory 11, 203–213.
Friedman, A., 1994. Differential games. Aumann, R.J., Hart, S., (Eds.), Handbook of Game Theory. vol. 2. Elsevier, pp. 781–799.
Gale, D., Stewart, F.M., 1953. Infinite games with perfect information. Kuhn H., Tucker, A.W., (Eds.), Contributions to the Theory of Games, II, Annals of Mathematical Studies. vol. 28. Princeton University Press, pp. 245–266.

Gensbittel, F., 2012. Extensions of the Cav(u) theorem for repeated games with one sided information. Math. Operat. Res. Preprint, hal-00745575.
Gensbittel, F., 2013. Covariance control problems of martingales arising from game theory. SIAM J. Control Optim. 51(2), 1152–1185.
Gensbittel, F., 2013. Continuous-time limit of dynamic games with incomplete information and a more informed Player. Preprint hal 00910970.
Gensbittel, F., Renault, R., 2012. The value of Markov chain games with incomplete information on both sides. Preprint arXiv:1210.7221.
Gensbittel, F., Oliu-Barton, M., Venel, X., 2013. Existence of the uniform value in repeated games with a more informed controller. Preprint hal-00772043.
Grün, C., 2012. A BSDE approach to stochastic differential games with incomplete information. Stoch. Proc. Appl. 122(4), 1917–1946.
Grün, C., 2013. On Dynkin games with incomplete information. Preprint arXiv:1207.2320v1.
Guo, X., Hernandez-Lerma, O., 2003. Zero-sum games for continuous-time Markov chains with unbounded transition and average payoff rates. J. Appl. Prob. 40, 327–345.
Guo, X., Hernandez-Lerma, O., 2005. Zero-sum continuous-time Markov games with unbounded transition and discounted payoff rates. Bernoulli 11, 1009–1029.
Heuer, M., 1992a. Asymptotically optimal strategies in repeated games with incomplete information. Int. J. Game Theory 20, 377–392.
Heuer, M., 1992b. Optimal strategies for the uninformed Player. Int. J. Game Theory 20, 33–51.
Hörner, J., Rosenberg, D., Solan, E., Vieille, N., 2010. On a Markov game with one-sided incomplete information. Operat. Res. 58, 1107–1115.
Kohlberg, E., 1974. Repeated games with absorbing states. Ann Stat. 2, 724–738.
Kohlberg, E., 1975. Optimal strategies in repeated games with incomplete information. Int. J. Game Theory 4, 7–24.
Kohlberg, E., Zamir, S., 1974. Repeated games of incomplete information: the symmetric case. Ann. Stat. 2, 1040.
Kohn, R.V., Serfaty, S., 2006. A deterministic-control-based approach to motion by curvature. Communi. Pure Appl. Math. 59, 344–407.
Krasovskii, N.N., Subbotin, A.I., 1988. Game-Theoretical Control Problems. Springer.
Krausz, A., Rieder, U., 1997. Markov games with incomplete information. Math. Methods Operat. Res. 46, 263–279.
Laraki, R., 2000. Repeated games with incomplete information: a variational approach. PhD thesis, UPMC, Paris, France.
Laraki, R., 2001a. Variational inequalities, systems of functional equations and incomplete information repeated games. SIAM J. Control Optim. 40, 516–524.
Laraki, R., 2001b. The splitting game and applications. Int. J. Game Theory 30, 359–376.
Laraki, R., 2002. Repeated games with lack of information on one side: the dual differential approach. Math. Operat. Res. 27, 419–440.
Laraki, R., 2004. On the regularity of the convexification operator on a compact set, J. Convex Anal. 11, 209–234.
Laraki, R., 2010. Explicit formulas for repeated games with absorbing states. Int. J. Game Theory 39, 53–69.
Laraki, R., Solan, E., 2005. The value of zero-sum stopping games in continuous time. SIAM J. Control Optimi. 43, 1913–1922.
Laraki, R., Sudderth, W.D., 2004. The preservation of continuity and Lipschitz continuity by optimal reward operators. Mathe. Operat. Res. 29, 672–685.
Lehrer, E., 2002. Approachability in infinite dimensional spaces. Int. J. Game Theory 31, 253–268.
Lehrer, E., Monderer, D., 1994. Discounting versus averaging in dynamic programming. Games and Econ. Behav. 6, 97–113.
Lehrer, E., Solan, E., 2007. Learning to play partially specified equilibrium. Preprint.
Lehrer, E., Sorin, S., 1992. A uniform Tauberian theorem in dynamic programming. Math. Operat. Res. 17, 303–307.
Lepeltier, J.P., Maingueneau, M.A., 1984. Le jeu de Dynkin en théorie générale sans l' hypothèse de Mokobodski. Stochastics 13, 25–44.

Levy, Y., 2002. Stochastic games with information lag. Games Econ. Behav. 74, 243–256.
Maitra, A.P., Sudderth, W.D., 1992. An operator solution of stochastic games. Israel J. Math. 78, 33–49.
Maitra, A.P., Sudderth, W.D., 1993. Borel stochastic games with lim sup payoff. Ann. Probab. 21, 861–885.
Maitra, A.P., Sudderth, W.D., 1996. Discrete Gambling and Stochastic Games. Springer.
Maitra, A.P., Sudderth, W.D., 1998. Finitely additive stochastic games with Borel measurable payoffs, Int. J. Game Theory 27, 257–267.
Maitra, A.P., Sudderth, W.D., 2003. Borel stay-in-a-set games. Int. J. Game Theory 32, 97–108.
Marino, A., 2005. The value and optimal strategies of a particular Markov chain game. Preprint.
Martin, D.A., 1975. Borel determinacy. Ann. Math. 102, 363–371.
Martin, D.A., 1998. The determinacy of Blackwell games. J. Symbolic Logic 63, 1565–1581.
Mertens, J.-F., 1972. The value of two-person zero-sum repeated games: the extensive case. Int. J. Game Theory 1, 217–227.
Mertens, J.-F., 1987. Repeated games. Proceedings of the International Congress of Mathematicians, Berkeley 1986, American Mathematical Society, pp. 1528–1577.
Mertens, J.-F., 1998. The speed of convergence in repeated games with incomplete information on one side. Int. J. Game Theory 27, 343–359.
Mertens, J.-F., 2002. Stochastic games. In: Aumann, R.J., Hart, S., (Eds.), Handbook of Game Theory, vol. 3. Elsevier, pp. 1809–1832.
Mertens, J.-F., Neyman, A., 1981. Stochastic games. Int. J. Game Theory 10, 53–66.
Mertens, J.-F., Zamir, S., 1971. The value of two-person zero-sum repeated games with lack of information on both sides. Int. J. Game Theory 1, 39–64.
Mertens, J.-F., Zamir, S., 1976a. On a repeated game without a recursive structure. Int. J. Game Theory 5, 173–182.
Mertens, J.-F., Zamir, S., 1976b. The normal distribution and repeated games. Int. J. Game Theory 5, 187–197.
Mertens, J.-F., Zamir, S., 1985. Formulation of Bayesian analysis for games with incomplete information. Int. J. Game Theory 14, 1–29.
Mertens, J.-F., Zamir, S., 1995. Incomplete information games and the normal distribution. CORE D.P. 9520.
Mertens, J.-F., Neyman, A., Rosenberg, D., 2009. Absorbing games with compact action spaces. Math. Operat. Res. 34, 257–262.
Mertens, J.-F., Sorin, S., Zamir, S., 1994. Repeated Games, CORE D.P. 9420-21-22, to appear Cambridge University Press 2014.
Milman, E., 2006. Approachable sets of vector payoffs in stochastic games. Games Econ. Behav. 56, 135–147.
Monderer, D., Sorin, S., 1993. Asymptotic properties in dynamic programming. Int. J. Game Theory 22, 1–11.
Neveu, J., 1975. Discrete-Parameter Martingales. North-Holland, Amsterdam.
Neyman, A., 2003. Stochastic games and nonexpansive maps. In: Neyman, A., Sorin, S., (Eds.), Stochastic Games and Applications, NATO Science Series C 570, Kluwer Academic Publishers, pp. 397–415.
Neyman, A., 2008. Existence of optimal strategies in Markov games with incomplete information. Int. J. Game Theory 37, 581–596.
Neyman, A., 2012a. The value of two-person zero-sum repeated games with incomplete information and uncertain duration. Int. J. Game Theory 41, 195–207.
Neyman, A., 2012b. Continuous-time stochastic games. DP 616, CSR Jerusalem.
Neyman, A., 2013. Stochastic games with short-stage duration. Dyn. Games Appl. 3, 236–278.
Neyman, A., Sorin, S., 1998. Equilibria in repeated games of incomplete information: the general symmetric case. Int. J. Game Theory 27, 201–210.

Neyman, A., Sorin, S., (Eds.) 2003. Stochastic Games and Applications, NATO Science Series, vol. C 570. Kluwer Academic Publishers.

Neyman, A., Sorin, S., 2010. Repeated games with public uncertain duration process. Int. J. Game Theory 39, 29–52.

Oliu-Barton, M., 2013. The asymptotic value in finite stochastic games. Math. Operat. Res. to appear.

Oliu-Barton, M., Vigeral, G., 2013. A uniform Tauberian theorem in optimal control. In: Cardaliaguet, P., Cressman, R., (Eds.), Advances in Dynamic Games, Annals of the International Society of Dynamic Games, vol. 12. pp. 199–215.

Perchet, V., 2009. Calibration and internal no-regret with random signals. In: Gavalda, R., Lugosi, G., Zegmann, T., Zilles, S., (Eds.), Algorithmic Learning Theory, vol. LNAI 5809, Springer, pp. 68–82.

Perchet, V., 2011a. Approachability of convex sets in games with partial monitoring. J. Optim. Theory Appl. 149, 665–677.

Perchet, V., 2011b. Internal regret with partial monitoring calibration-based optimal algorithm. J. Mach. Learn. Res. 12, 1893–1921.

Perchet, V., Quincampoix, M., 2012. On an unified framework for approachability in games with or without signals. Preprint hal- 00776676.

Peres, Y., Schramm, O., Sheffield, S., Wilson, D., 2009. Tug of war and the infinity Laplacian. J. Am. Math. Soc. 22, 167–210.

Quincampoix, M., Renault, J., 2011. On the existence of a limit value in some non expansive optimal control problems. SIAM J. Control Optim. 49, 2118–2132.

Radzik, T., Raghavan, T.E.S., 1994. Duels. In: Aumann, R.J., Hart, S., (Eds.), Handbook of Game Theory, vol. 2, pp. 761–768.

Renault, J., 2006. The value of Markov chain games with lack of information on one side. Math. Operat. Res. 31, 490–512.

Renault, J., 2011. Uniform value in dynamic programming. J. Eur. Math. Soc. 13, 309–330.

Renault, J., 2012. The value of repeated games with an informed controller. Math. Operat. Res. 37, 154–179.

Renault, J., 2013. General limit value in dynamic programming. Preprint hal 00769763.

Renault, J., Venel, X., 2012. A distance for probability spaces, and long-term values in Markov decision processes and repeated games. Preprint hal-00674998.

Rosenberg, D., 1998. Duality and Markovian strategies. Int. J. Game Theory 27, 577–597.

Rosenberg, D., 2000. Zero-sum absorbing games with incomplete information on one side: asymptotic analysis. SIAM J. Control Optim. 39, 208–225.

Rosenberg, D., Sorin, S., 2001. An operator approach to zero-sum repeated games. Israël J. Math. 121, 221–246.

Rosenberg, D., Vieille, N., 2000. The maxmin of recursive games with incomplete information on one side. Math. Operat. Res. 25, 23–35.

Rosenberg, D., Solan, E., Vieille, N., 2001. Stopping games with randomized strategies. Prob. Theory Rel. Fields 119, 433–451.

Rosenberg, D., Solan, E., Vieille, N., 2002. Blackwell optimality in Markov decision processes with partial observation. Ann. Stat. 30, 1178–1193.

Rosenberg, D., Solan, E., Vieille, N., 2003a. The maxmin value of stochastic games with imperfect monitoring. Int. J. Game Theory 32, 133–150.

Rosenberg, D., Solan, E., Vieille, N., 2003b. Stochastic games with imperfect monitoring. In: Haurie, A., Muto, S., Petrosjan, L.A., Raghavan, T.E.S., (Eds.), Advances in Dynamic Games: Applications to Economics, Management Science, Engineering, and Environmental Management. Annals of the ISDG, vol. 8, pp. 3–22.

Rosenberg, D., Solan, E., Vieille, N., 2004. Stochastic games with a single controller and incomplete information. SIAM J. Control Optim. 43, 86–110.

Scarf, H.E., Shapley, L.S., 1957. Games with partial information. In: Dresher, M., Tucker, A.W., Wolfe, P., (Eds.), Contributions to the Theory of Games, III, Annals of Mathematical Studies, vol. 39. Princeton University Press, pp. 213–229.

Shapley, L.S., 1953. Stochastic games. Proc. Nat. Acad. Sci. U.S.A 39, 1095–1100.

Shmaya, E., 2011. The determinacy of infinite games with eventual perfect monitoring. Proc. AMS 139, 3665–3678.
Shmaya, E., Solan, E., 2004. Zero-sum dynamic games and a stochastic variation of Ramsey's theorem. Stoch. Process. Appl. 112, 319–329.
Solan, E., 2003. Continuity of the value of competitive Markov decision processes. J. Theoret. Probab. 16, 831–845.
Solan, E., Vieille, N., 2002. Uniform value in recursive games. Ann. Appl. Probab. 12, 1185–1201.
Solan, E., Vieille, N., 2004. Stopping games -recent results. In: Nowak, A.S., Szajowski, K., (Eds.), Advances in Dynamic Games, Annals of the ISDG, vol. 7. Birkhauser, pp. 235–245.
Sorin, S., 1984. "Big Match" with lack of information on one side (Part I). Int. J. Game Theory 13, 201–255.
Sorin, S., 1985a. "Big Match" with lack of information on one side, Part II. Int. J. Game Theory 14, 173–204.
Sorin, S., 1985b. On repeated games with state dependent matrices. Int. J. Game Theory 14, 249–272.
Sorin, S., 1989. On repeated games without a recursive structure: existence of $\lim v_n$. Int. J. Game Theory 18, 45–55.
Sorin, S., 1992. Repeated games with complete information. In: Aumann, R.J., Hart, S., (Eds.), Handbook of Game Theory, vol. 1. Elsevier, pp. 71–107.
Sorin, S., 2002. A First Course on Zero-Sum Repeated Games. Springer.
Sorin, S., 2003. The operator approach to zero-sum stochastic games. In: Neyman, A., Sorin, S., (Eds.), Stochastic Games and Applications, NATO Science Series C 570, Kluwer Academic Publishers, pp. 417–426.
Sorin, S., 2004. Asymptotic properties of monotonic nonexpansive mappings. Discrete Events Dyn. Syst. 14, 109–122.
Sorin, S., Vigeral, G., 2012. Existence of the limit value of two person zero-sum discounted repeated games via comparison theorems. J. Opim. Theory Appl. 157, 564–576.
Sorin, S., Vigeral, G., 2013. Reversibility and oscillations in zero-sum discounted stochastic games. Preprint hal 00869656.
Sorin, S., Venel, X., Vigeral, G., 2010. Asymptotic properties of optimal trajectories in dynamic programming. Sankhya 72A, 237–245.
Souganidis, P.E., 1985. Approximation schemes for viscosity solutions of Hamilton-Jacobi equations. J. Differ. Equat. 17, 781–791.
Souganidis, P.E., 1999. Two Player zero sum differential games and viscosity solutions. In: Bardi, M., Raghavan, T.E.S., Parthasarathy, T., (Eds.), Stochastic and Differential Games, Annals of the ISDG, vol. 4. Birkhauser, pp. 70–104.
Souquière, A., 2010. Approximation and representation of the value for some differential games with asymmetric information. Int. J. Game Theory 39, 699–722.
Spinat, X., 2002. A necessary and sufficient condition for approachability. Math. Operat. Res. 27, 31–44.
van den Dries, L., 1998. Tame Topology and o-Minimal Structures. London Mathematical Society Lecture Notes Series. vol. 248. Cambridge University Press, Cambridge.
Vieille, N., 1992. Weak approachability. Math. Operat. Res. 17, 781–791.
Vieille, N., 2002. Stochastic games: recent results. In: Aumann, R.J., Hart, S., (Eds.), Handbook of Game Theory, vol. 3. Elsevier, pp. 1834–1850.
Vigeral, G., 2010a. Evolution equations in discrete and continuous time for non expansive operators in Banach spaces. ESAIM COCV 16, 809–832.
Vigeral, G., 2010b. Iterated monotonic nonexpansive operators and asymptotic properties of zero-sum stochastic games. Preprint.
Vigeral, G., 2010c. Zero-sum recursive games with compact action spaces. Preprint.
Vigeral, G., 2013. A zero-sum stochastic game with compact action sets and no asymptotic value. Dyn. Games Appl. 3, 172–186.

Zachrisson, L.E., 1964. Markov Games. In: Dresher, M., Shapley, L.S., Tucker, A.W., (Eds.), Advances in Game Theory, Annals of Mathematical Studies, vol. 52. Princeton University Press, pp. 210–253.
Zamir, S., 1973. On the notion of value for games with infinitely many stages. Ann. Stat. 1, 791–796.
Zamir, S., 1992. Repeated games of incomplete information: zero-sum. In: Aumann, R.J., Hart, S., (Eds.), Handbook of Game Theory, vol. 1. Elsevier, pp. 109–154.
Ziliotto, B., 2013. Zero-sum repeated games: counterexamples to the existence of the asymptotic value and the conjecture $maxmin = lim v_n$. Preprint hal-00824039.
Ziliotto, B., 2014. General limit value in stochastic games. Preprint.

CHAPTER 3

Games on Networks

Matthew O. Jackson[*,†,‡], **Yves Zenou**[‡,¶]
[*]Department of Economics, Stanford University, Stanford, CA, USA
[†]The Santa Fe Institute, Santa Fe, NM, USA
[‡]CIFAR, Toronto, Ontario, Canada
[§]Department of Economics, Stockholm University, Stockholm, Sweden
[¶]Research Institute of Industrial Economics (IFN), Stockholm, Sweden

Contents

Prepared for the Handbook of Game Theory Vol. 4, edited by Peyton Young and Shmuel Zamir, to be published by Elsevier Science in 2014.

Handbook of game theory
http://dx.doi.org/10.1016/B978-0-12-420056-2.00003-3

Abstract

We provide an overview and synthesis of the literatures analyzing games in which players are connected via a network structure. We discuss, in particular, the impact of the structure of the network on individuals' behaviors. We focus on game theoretic modeling, but also include some discussion of analyses of peer effects, as well as applications to diffusion, employment, crime, industrial organization, and education.

Keywords: Network games, Social networks, Games on networks, Graphical games, Games with incomplete information, Peer effects

JEL Codes: A14, C72, D85

3.1. INTRODUCTION AND OVERVIEW

Social networks are important in many facets of our lives. Most decisions that people make, from which products to buy to whom to vote for, are influenced by the choices of their friends and acquaintances. For example, the decision of an individual of whether to adopt a new technology, attend a meeting, commit a crime, find a job is often influenced by the choices of his or her friends and acquaintances (be they social or professional). The emerging empirical evidence on these issues motivates the theoretical study of network effects.

Here, we provide an overview of literatures on the analysis of the interaction of individuals who are connected via a network and whose behaviors are influenced by those around them.[1] Such interactions are natural ones to model using game theory, as the payoffs that an individual receives from various choices depends on the behaviors of his or her neighbors.

This particular view of games on networks, where an agent chooses an action and then the payoffs of each player is determined by those of his or her neighbors is a special perspective, but one that applies to many different contexts including the peer effects mentioned above. There are also other applications that involve strategic decision making and networks of relationships, such as exchange or trade on networks (networked markets). These sorts of analyses tend to be much more particular in their structure (e.g., following a specific bargaining protocol or timing on interactions) whereas there are many things that can be said about games on networks in the more basic context. We very briefly discuss other settings, but our focus is on the canonical case.

Of course, one can view these settings as special cases of game theory more generally, and so some results from the general literature directly apply: for example, existence of various forms of equilibria can be deduced from standard results. The interest, therefore, is instead in whether there is anything we can deduce that holds systematically regarding how play in a game depends on the network structure of interactions. For example, if individuals only wish to buy a new product if a sufficient fraction of their friends do, can we say something about how segregation patterns in the network of friendships affects the purchase of the product? Can we say anything about who is the most influential individual in a network where people look to their peers in choosing an effort level in education? Thus, our discussion of the literature focuses on investigations relating network characteristics to behavior.

The main challenge that is faced in studying strategic interaction in social settings is the inherent complexity of networks. Without focusing in on specific structures in terms of the games, it is hard to draw any conclusions. The literature has primarily taken three approaches to this challenge, and these form the basis for our discussion. One involves looking at games of strategic complements and strategic substitutes, where the interaction

[1] As game theoretic models of network formation have been surveyed extensively elsewhere (see, e.g., De Martí Beltran and Zenou, 2011; Goyal, 2007; Jackson, 2003, 2004, 2005b, 2008a, 2011), we concentrate here on games on networks. We also do not survey the literature on communication structures in cooperative games. This is another large literature, following Myerson (1977) that has been surveyed elsewhere (see Slikker and van den Nouweland, 2001, for the graph-based literature and Jackson, 2005a, 2008a, for the allocation rule literature and allocation rules). There is also a large literature on agent-based models including networked interactions (e.g., see Tesfatsion, 2006, for some references) that is already surveyed elsewhere, and not included here. Exploiting the growing capabilities of computers, agent-based methods have been used to analyze dynamic systems of interacting agents in cases where the network is fixed (see Wilhite, 2006) as well as where the relationships are endogenous (Vriend, 2006). Finally, the games on networks literature has informed the recent empirical literature incorporating network analyses into studies of peer effects and behavior (see Jackson, Rogers and Zenou, 2014, and Jackson, 2014, for surveys).

in payoffs between players satisfies some natural and useful monotonicity properties. With strategic complementarities, a player's incentives to take an action (or a "higher" action) are increasing in the number of his or her friends who take the (higher) action; and with strategic substitutes the opposite incentives are in place. We show that the monotonicity properties of these structured interactions have allowed the literature to deduce a number of results related to equilibrium behavior as well as dynamics, and how those depend on network structure. A second approach relies on looking at a quite tractable "linear-quadratic" setting where agents choose a continuous level of activity. That simple parametric specification permits an explicit solution for equilibrium behavior as a function of a network, and thus leads to interesting comparative statics and other results that are useful in empirical work. A third approach considers settings with an uncertain pattern of interactions, where players make choices (such as learning a language) without being certain about with whom they will interact. The uncertainty actually simplifies the problem since behavior depends on anticipated rates of interaction, rather than complex realizations of interactions. Together all of these various approaches and models make a number of predictions about behavior, relating levels of actions to network density, relating players' behaviors to their position in the network, and relating behavior to things like the degree distribution and cost of taking given actions. The theory thus makes predictions both about how a player's behavior relates to his/her position in a network, as well as what overall behavior patterns to expect as a function of the network structure.

3.2. BACKGROUND DEFINITIONS

We begin with a class of canonical and widely applicable games; specifically, games where there is a fixed and given network of interactions. Links indicate which players' strategies affect which others' payoffs. In particular, a given player's payoff depends only on the play of his or her neighbors. Of course, this results in indirect network effects since there may be chains of influence.

We provide some basic definitions of games on networks.[2]

3.2.1 Players and networks

We consider a finite set of players $N = \{1, \ldots, n\}$ who are connected in a network.

A *network* (or *graph*) is a pair $(N, \mathbf{g})$, where $\mathbf{g}$ is a network on the set of nodes N. These represent the interaction structure in the game, indicating the other players whose actions impact a given player's payoff.

[2] We provide terse definitions here. For a reader new to these definitions, Chapter 2 in Jackson (2008a) provides more discussion and background.

We abuse notation and let $\mathbf{g}$ denote the two standard ways in which networks are represented: by their adjacency matrices as well as by listing the pairs of nodes that are connected. Thus, $\mathbf{g}$ will sometimes be an $n \times n$ adjacency matrix, with entry g_{ij} in $\{0, 1\}$ denoting whether i is linked to j and can also include the intensity of that relationship (in which case $\mathbf{g}$ takes on more general real values). At other times $\mathbf{g}$ denotes the set of all relationships that are present, and so we use notation $ij \in \mathbf{g}$ to indicate that i is linked to j.

A network is *undirected* if $\mathbf{g}$ is required to be symmetric so that relationships are necessarily reciprocal and $g_{ij} = g_{ji}$ for all i and j, and is *directed* if relationships can be unidirectional.

A relationship between two nodes i and j, represented by $ij \in \mathbf{g}$, is referred to as a link. Links are also referred to as edges or ties in various parts of the literature; and sometimes also directed links, directed edges, or arcs in the specific case of a directed network. Shorthand notations for the network obtained by adding or deleting a link ij to or from an existing network $\mathbf{g}$ are $\mathbf{g} + ij$ and $\mathbf{g} - ij$, respectively.

A *walk* in a network $(N, \mathbf{g})$ refers to a sequence of nodes, $i_1, i_2, i_3, \ldots, i_{K-1}, i_K$ such that $i_k i_{k+1} \in \mathbf{g}$ for each k from 1 to $K - 1$.[3] The *length* of the walk is the number of links in it, or $K - 1$.

A *path* in a network $(N, \mathbf{g})$ is a walk in $(N, \mathbf{g})$, $i_1, i_2, i_3, \ldots, i_{K-1}, i_K$, such that all the nodes are distinct.

A *cycle* in a network $(N, \mathbf{g})$ is a walk in $(N, \mathbf{g})$, $i_1, i_2, i_3, \ldots, i_{K-1}, i_K$, such that $i_1 = i_K$.

A network $(N, \mathbf{g})$ is *connected* if there is a path in $(N, \mathbf{g})$ between every pair of nodes i and j.[4]

A *component* of a network $(N, \mathbf{g})$ is a subnetwork $(N', \mathbf{g}')$ (so $N' \subset N$ and $\mathbf{g}' \subset \mathbf{g}$) such that

- there is a path in $\mathbf{g}'$ from every node $i \in N'$ to every other node $j \in N', j \neq i$,
- $i \in N'$ and $ij \in g$ implies $j \in N'$ and $ij \in g'$.

Thus, a component of a network is a maximal connected subnetwork with all adjacent links, so that and there is no way of expanding the set of nodes in the subnetwork and still having it be connected.

The *distance* between two nodes in the same component of a network is the length of a shortest path (also known as a *geodesic*) between them.

[3] Standard definitions of walks, paths, and cycles specify them as sets of nodes together with sets of links. The definitions here simplify notation, and for the purposes of this chapter the difference is inconsequential.

[4] Each of these definitions has an analog for directed networks, simply viewing the pairs as directed links and then having the name directed walk, directed path, and directed cycle. In defining connectedness for a directed network one often uses a strong definition requiring a directed path from each node to every other node.

The *neighbors* of a node i in a network $(N, \mathbf{g})$ are denoted by $N_i(\mathbf{g})$. As we predominantly discuss settings where N is fixed, we omit dependence on the set of nodes N, and so write $N_i(\mathbf{g})$ rather than $N_i(N, \mathbf{g})$. Thus,

$$N_i(\mathbf{g}) = \{j | ij \in \mathbf{g}\}$$

The *degree* of a node i in a network $(N, \mathbf{g})$ is the number of neighbors that i has in the network, so that $d_i(\mathbf{g}) = |N_i(\mathbf{g})|$.[5]

The kth power $\mathbf{g}^k = \mathbf{g} \times \overset{(k\ times)}{\cdots} \mathbf{g}$ of the adjacency matrix $\mathbf{g}$ keeps track of indirect connections in $\mathbf{g}$. More precisely, the coefficient $g_{ij}^{[k]}$ in the (i,j) cell of $\mathbf{g}^k$ gives the number of walks of length k in $\mathbf{g}$ between i and j.

An *independent set* relative to a network $(N, \mathbf{g})$ is a subset of nodes $A \subset N$ for which no two nodes are adjacent (i.e., linked). A is a maximal independent set if there does not exist another independent, set $A' \neq A$, such that $A \subset A' \subset N$. A *dominating set* relative to a network $(N, \mathbf{g})$ is a subset of nodes $A \subset N$ such that every node not in A is linked to at least one member of A. For example, the central node in a star forms a dominating set and also a maximal independent set, while each peripheral node is an independent set and the set of all peripheral nodes is a maximal independent set. Any set including the central node and some peripheral nodes is a dominating set, but not an independent set.

Let $G(N)$ be the set of networks on the nodes N under consideration, which will often be the set of simple[6] networks unless otherwise stated.

3.2.2 Games on networks

Players in N have action spaces A_i. Let $A = A_1 \times \cdots A_n$. In most of our discussion, the action spaces are finite sets or subsets of a Euclidean space.

Player i's payoff function is denoted $u_i : A \times G(N) \to \mathbb{R}$.

Unless otherwise indicated *equilibrium* refers to a pure strategy Nash equilibrium[7]: a profile of actions $\mathbf{a} \in A = A_1 \times \cdots A_n$, such that

$$u_i(a_i, \mathbf{a}_{-i}, \mathbf{g}) \geq u_i(a_i', \mathbf{a}_{-i}, \mathbf{g})$$

for all i and $a_i' \in A_i$.

[5] Unless otherwise stated, let us suppose that $g_{ii} = 0$, so that nodes are not linked to themselves.

[6] Simple networks are undirected, unweighted and with at most one link between any pair of nodes.

[7] Mixed strategy equilibria also exist in such settings, but in many cases might be less applicable. While often modeled as a simultaneous game, many applications of games on networks are ones in which players are able to adjust their actions over time (e.g., changing technologies). In many (but not all) such games mixed strategy equilibria are unstable and so would be less likely to apply well. In addition, mixed strategy equilibria in such settings can be very difficult to characterize. For instance in games of strategic complements, there can exist many mixed strategy equilibria that are computationally infeasible to find and catalog, whereas extremal equilibria, which are pure strategy equilibria, are very easy to find.

A given player's payoff depends on other players' actions, but only on those to whom the player is linked in the network. In fact, without loss of generality the network can be taken to indicate the payoff interactions in the society. More formally, i's payoff depends only on a_i and $\{a_j\}_{j \in N_i(\mathbf{g})}$, so that for any i, a_i, and $\mathbf{g}$:

$$u_i(a_i, \mathbf{a}_{-i}, \mathbf{g}) = u_i(a_i, \mathbf{a}'_{-i}, \mathbf{g}),$$

whenever $\mathbf{a}_j = \mathbf{a}'_j$ for all $j \in N_i(\mathbf{g})$.

To fix ideas, let us consider a couple of examples.

Example 3.1. *The Majority Game.*[8]

Players' action spaces are $A_i = \{0, 1\}$. This covers applications where a player can choose to either do something or not to, for instance, buying a product, attending a party, and so forth. In this particular game, if more than one half of i's neighbors choose action 1, then it is best for player i to choose 1, and if fewer than one half of i's neighbors choose action 1 then it is best for player i to choose action 0.

Specifically, the payoff to a player from taking action 1 compared to action 0 depends on the fraction of neighbors who choose action 1, such that

$$u_i(1, a_{N_i(\mathbf{g})}) > u_i(0, a_{N_i(\mathbf{g})}) \quad \text{if} \quad \frac{\sum_{j \in N_i(\mathbf{g})} a_j}{|N_i(\mathbf{g})|} > \frac{1}{2}$$

and

$$u_i(1, a_{N_i(\mathbf{g})}) < u_i(0, a_{N_i(\mathbf{g})}) \quad \text{if} \quad \frac{\sum_{j \in N_i(\mathbf{g})} a_j}{|N_i(\mathbf{g})|} < \frac{1}{2}.$$

There are clearly multiple equilibria in this game. For example, all players taking action 0 (or 1) is an equilibrium. Figure 3.1 displays another equilibrium of the majority game.

Example 3.2. *"Best-Shot" Public Goods Games*

Another canonical example of a game on a network is based on what are known as "best-shot" public goods games (see Hirshleifer, 1983). For instance, the action might be learning how to do something, where that information is easily communicated; or buying a book or other product that is easily lent from one player to another. Taking the action 1 is costly and if any of a player's neighbors takes the action then the player is better off not taking the action; but, taking the action and paying the cost is better than having nobody in a player's neighborhood take the action.

[8] This game has been studied extensively in physics (see, e.g., Conway's "game of life") and in various agent-based models that followed, such as the "voter model" (see, e.g., Clifford and Sudbury, 1973; Holley and Liggett, 1975).

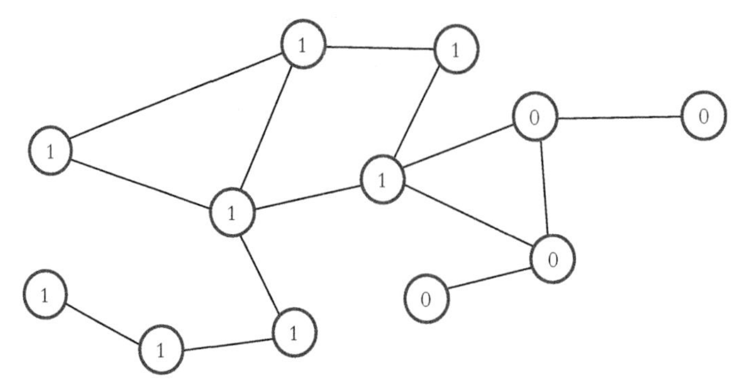

Figure 3.1 An equilibrium in the majority game.

$$u_i(\mathbf{a}, \mathbf{g}) = \begin{cases} 1 - c & \text{if } a_i = 1, \\ 1 & \text{if } a_i = 0, \; a_j = 1 \text{ for some } j \in N_i(\mathbf{g}) \\ 0 & \text{if } a_i = 0, \; a_j = 0 \text{ for all } j \in N_i(\mathbf{g}), \end{cases}$$

where $1 > c > 0$. So, a player would prefer that a neighbor take the action than having to do it himself or herself, but will take the action if no neighbors do.

There are many possible equilibria in the best-shot public goods game and Figure 3.2 displays one of them. Interestingly, the equilibria in this game correspond exactly to having the set of players who choose action 1 form a maximal independent set of nodes in the network (as noted by Bramoullé and Kranton (2007a)); that is, a maximal set of nodes that have no links to each other in the network. In Figure 3.2, it is clear that nobody wants to deviate from their Nash equilibrium actions. Take, for example, the central player who chooses action 1. His/her utility is $1 - c$. Since all his/her neighbors choose action 0, deviating by choosing action 0 would give him/her a utility of $0 < 1 - c$. Similarly, for each player who chooses action 0, his/her utility is 1 since at least one of his/her neighbors choose action 1. Choosing action 1 would give him/her $1 - c < 1$.

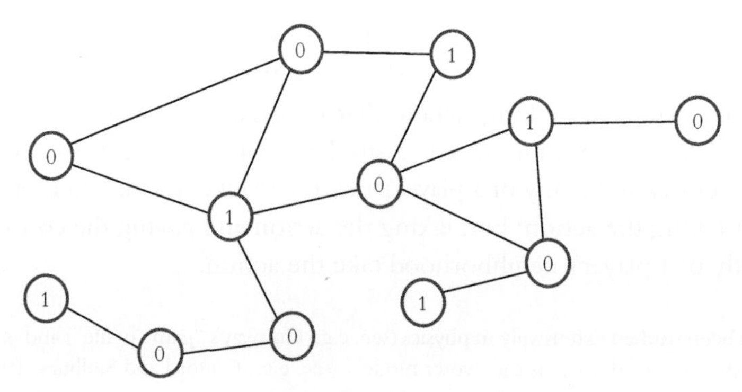

Figure 3.2 An equilibrium in the best-shot public good game, and a maximal independent set.

3.3. STRATEGIC COMPLEMENTS AND STRATEGIC SUBSTITUTES

Although there are many forms that games on networks can take, there are two prominent and broadly encompassing classes of games. In fact, the two previous examples are typical members of these two classes of games. The distinction between these types of games relates to whether a given player's relative payoff to taking an action versus not is increasing or decreasing in the set of neighbors who take the action. To see the nature of the distinction, let us take the actions in the games to be well-ordered, such as a subset of the real line (or more generally a lattice, as detailed shortly). The first class of examples, of which coordination games are the canonical example, are games of *strategic complements*. In games of strategic complements, an increase in the actions of other players leads a given player's higher actions to have relatively higher payoffs compared to that player's lower actions. Examples of such games include things like the adoption of a technology, human capital decisions, criminal efforts, smoking behaviors, etc. Games of *strategic substitutes* are such that the opposite is true: an increase in other players' actions leads to relatively lower payoffs to higher actions of a given player. Applications of strategic substitutes include, for example, local public good provision and information gathering.

3.3.1 Defining strategic complements and substitutes

Let us take A_i (the action space) to be a complete lattice with an associated partial order $\geq_i$, for each i.[9] Then it is easy to see that A is also complete lattice, if we define $\mathbf{a} \geq \mathbf{a}'$ if and only if $a_i \geq a_i'$ for every i, and where for any $S \subset A$ we define $\inf(S) = (\inf_i\{a_i : \mathbf{a} \in S\})_i$ and $\sup(S) = (\sup_i\{a_i : \mathbf{a} \in S\})_i$.

A game exhibits *strategic complements* if it exhibits *increasing differences*; that is, for all i, $a_i \geq_i a_i'$ and $a_{-i} \geq_{-i} a_{-i}'$:

$$u_i(a_i, a_{-i}, \mathbf{g}) - u_i(a_i', a_{-i}, \mathbf{g}) \geq u_i(a_i, a_{-i}', \mathbf{g}) - u_i(a_i', a_{-i}', \mathbf{g}).$$

A game exhibits *strategic substitutes* if it exhibits *decreasing differences*; that is, for all i, j, with $i \neq j$, $a_i \geq_i a_i'$ and $a_{-i} \geq_{-i} a_{-i}'$:

$$u_i(a_i, a_{-i}, \mathbf{g}) - u_i(a_i', a_{-i}, \mathbf{g}) \leq u_i(a_i, a_{-i}', \mathbf{g}) - u_i(a_i', a_{-i}', \mathbf{g}).$$

These notions are said to apply strictly if the inequalities above are strict whenever $a_i >_i a_i'$ and $a_{-i} \geq_{-i} a_{-i}'$ with $a_j >_j a_j'$ for some $j \in N_i(\mathbf{g})$.

The majority game (Example 3.1) is one of strategic complements while the best-shot public goods game (Example 3.2) is one of strategic substitutes.

[9] Let $\geq_i$ be a partial order on a (nonempty) set A_i (so $\geq_i$ is reflexive, transitive and antisymmetric). $(A_i, \geq_i)$ is a lattice if any two elements a_i and a_i' have a least upper bound (supremum for i, $\sup_i$, such that $\sup_i(a_i, a_i') \geq a_i$ and $\sup_i(a_i, a_i') \geq a_i'$), and a greatest lower bound (infimum for i, such that $\inf_i(a_i, a_i') \leq a_i$ and $\inf_i(a_i, a_i') \leq a_i'$), in the set. A *lattice* $(A_i, \geq_i)$ is *complete* if every nonempty subset of A_i has a supremum and an infimum in A_i.

3.3.2 Existence of equilibrium

3.3.2.1 Games of strategic complements

Beyond capturing many applications, games of strategic complements are well-behaved in a variety of ways. Not only do equilibria generally exist, but they form a lattice so that they are well-ordered and there are easy algorithms for finding the maximal and minimal equilibria.

Theorem 3.1. *Consider a game of strategic complements such that:*

- *for every player i, and specification of strategies of the other players, $a_{-i} \in A_{-i}$, player i has a nonempty set of best responses $BR_i(a_{-i})$ that is a closed sublattice of the complete lattice A_i,*[10] *and*
- *for every player i, if $a'_{-i} \geq a_{-i}$, then $\sup_i BR_i(a'_{-i}) \geq_i \sup_i BR_i(a_{-i})$ and $\inf_i BR_i(a'_{-i}) \geq_i \inf_i BR_i(a_{-i})$.*

An equilibrium exists and the set of equilibria form a (nonempty) complete lattice.[11]

Variations on this theorem can be found in Topkis (1979) and Zhou (1994), and for arbitrary sets of players in Acemoglu and Jackson (2011).

In games of strategic complements such that the set of actions is finite, or compact and payoffs are continuous, the conditions of the theorem apply and there exists an equilibrium. Note that the equilibria exist in pure strategies, directly in terms of the actions A without requiring any additional randomizations. The same is not true for games of strategic substitutes.

Finding maximal and minimal equilibria in a game of strategic complements is then quite easy. Let us describe an algorithm for the case where A is finite. Begin with all players playing the maximal action $a^0 = \overline{a}$. Let $a_i^1 = \sup_i(BR_i(a_{-i}^0))$ for each i and, iteratively, let $a_i^k = \sup_i\left(BR_i\left(a_{-i}^{k-1}\right)\right)$. It follows that a point such that $a^k = a^{k-1}$ is the maximal equilibrium point, and given the finite set of strategies this must occur in a finite number of iterations. Analogously, starting from the minimal action and iterating upward, one can find the minimal equilibrium point.[12]

This also means that dynamics that iterate on best response dynamics will generally converge to equilibrium points in such games (e.g., see Milgrom and Roberts, 1990).

[10] Closure requires that $\sup_i(BR_i(a_{-i})) \in BR_i(a_{-i})$ and $\inf_i(BR_i(a_{-i})) \in BR_i(a_{-i})$.

[11] The set of equilibria is not necessarily a sublattice of A (see Topkis, 1979; Zhou, 1994). That is, the sup in A of a set of equilibria may not be an equilibrium, and so sup and inf have to be restricted to the set of equilibria to ensure that the set is a complete lattice, but the same partial order can be used.

[12] Calvó-Armengol and Jackson (2004, 2007) use this technique to calculate the maximal equilibrium in their dynamic game with strategic complementarities. They propose an application of this game by looking at labor-market networks and showing that investing in human capital depends on having access to job information.

3.3.2.2 Games of strategic substitutes and other games on networks

Moving beyond games of strategic complements, existence of equilibria and the structure of the set are no longer so nicely behaved.

Existence of equilibria can be guaranteed along standard lines: for instance equilibria exist if A_i is a nonempty, compact, and convex subset of a Euclidean space and u_i is continuous and quasi-concave for every i. This covers the canonical case where A_i are the mixed strategies associated with an underlying finite set of pure actions and u_i is the expected payoff and hence quasi-concave. Nonetheless, this means that pure strategy equilibria may not exist unless the game has some specific structure (and we discuss some such cases below).

In addition, with the lack of lattice structure, best responses are no longer so nicely ordered and equilibria in many network games can be more difficult to find.[13]

Nonetheless, some games of strategic substitutes on networks still have many important applications and are tractable in some cases. For example, consider the best-shot public goods game discussed above (Example 3.2). As we showed above, best-shot public goods games on a network always have pure strategy equilibria, and in fact those equilibria are the situations where the players who take action 1 form a maximal independent set.

Finding all of the maximal independent sets is computationally intensive, but finding one such set is easy. Here is an algorithm that finds an equilibrium.[14] At a given step k, the algorithm lists a set of the providers of the public good (the independent set of nodes), P_k, and a set of non-providers of the public good (who will not be in the eventual maximal independent set of nodes), NP_k, where the eventual maximal independent set will be the final P_k. In terms of finding an equilibrium to the best-shot game, the final P_k is the list of players who take action 1, and the final NP_k is the set of players who take action 0.

Step 1: Pick some node i and let $P_1 = \{i\}$ and $NP_1 = N_i(\mathbf{g})$.

Step k: Iterate by picking one of the players j who is not yet assigned to sets P_{k-1} or NP_{k-1}. Let $P_k = P_{k-1} \cup \{j\}$ and $NP_k = NP_{k-1} \cup N_j(\mathbf{g})$.[15]

End: Stop when $P_k \cup NP_k = N$.

More generally, one might ask the question of whether it is possible to find the "best" equilibrium in the best-shot game. Given that in every equilibrium all players get a payoff of either 1 or $1 - c$, minimizing the number of players who pay the cost c would be one

[13] For results on the complexity of finding equilibria in games on networks beyond the strategic complements and strategic substitutes cases see, for example, Kearns et al. (2001), Kakade et al. (2004), Daskalakis et al. (2009), and Papadimitriou and Roughgarden (2008).

[14] This is from Jackson (2008a, pp. 304-306), based on an obvious algorithm for finding a maximal independent set.

[15] Note that this is well-defined, since no neighbors of j can be in P_{k-1} as otherwise j would have been in NP_{k-1}.

metric via which to rank equilibria. As discussed by Dall'Asta et al. (2011), finding such equilibria can be difficult but finding them (approximately) through an intuitive class of mechanisms that tradeoff accuracy against speed is possible.[16]

There are other games of strategic substitutes where at least some equilibria are also easy to find.

Example 3.3. *A "Weakest-Link" Public Goods Game*

Another example of a local-public goods game on a network is based on what are known as "weakest-link" public goods games (see Hirshleifer, 1983).[17]

Here each player chooses some level of public good contribution (so $A_i = \mathbb{R}_+$) and the payoff to a player is the *minimum* action taken by any player in his or her neighborhood (in contrast to the maximum, as in the best-shot game). In particular,

$$u_i(a_i, a_{N_i(\mathbf{g})}) = \min_{j \in N_i(\mathbf{g}) \cup \{i\}} \{a_j\} - c(a_i)$$

where c is an increasing, convex and differentiable cost function.

If there is a smallest a^* such that $c'(a^*) \geq 1$, and each player has at least one neighbor in the network g, then any profile of actions where every player chooses the same contribution $a_i = a^*$ is an equilibrium of this game. Note that in a network in which every player has at least one neighbor, everyone playing $a_i = 0$ is also an equilibrium (or any common $a \leq a^*$), and so the game will have multiple equilibria when it is nondegenerate.

3.3.2.3 Games with strategic substitutes, continuous action spaces and linear best-replies

We have seen that equilibria in games with strategic substitutes are difficult to characterize and multiple equilibria rather than unique equilibrium are the rule. If, however, the best-reply functions are linear, then some further results can be obtained, both in terms of characterization of equilibria and comparative statics. Bramoullé and Kranton (2007a) and Bramoullé et al. (2014) study these type of games. In their models, players experiment to obtain new information and benefit from their neighbors' experimentation. Each player i selects an action $a_i \geq 0$, and obtains a payoff $u_i(\mathbf{a}, \mathbf{g})$ that depends on the action profile $\mathbf{a}$ and on the underlying network $\mathbf{g}$, in the following way:

[16] See also Dall'Asta et al. (2009) and Boncinelli and Pin (2012).

[17] Another example is that of anticoordination games as in Bramoullé (2007).

$$u_i(\mathbf{a}, \mathbf{g}) = v\left(a_i + \phi \sum_{j=1}^{n} g_{ij} a_j\right) - c\, a_i \qquad [3.1]$$

where $v(.)$ is an increasing, differentiable and strictly concave function on $\mathbb{R}_+$ and $c > 0$ is the constant marginal cost of own action such that $v'(0) > c > v'(x)$ for some $x > 0$. As in Bramoullé and Kranton (2007a), consider the case when $\phi = 1$. This is clearly a game with (strict) strategic substitutes since

$$\frac{\partial u_i(\mathbf{a}, \mathbf{g})}{\partial a_i \partial a_j} = v''\left(a_i + \sum_{j=1}^{n} g_{ij} a_j\right) < 0$$

Denote by a^* the action level of a player who experiments by him/herself, i.e. $a^* = v'^{-1}(c)$. Then, the best-reply function, for each individual i, i's best response to $\mathbf{a}_{-i}$ is linear and given by:

$$a_i^* = \begin{cases} a^* - \sum_{j=1}^{n} g_{ij} a_j & \text{if } a^* > \sum_{j=1}^{n} g_{ij} a_j \\ 0 & \text{if } a^* \leq \sum_{j=1}^{n} g_{ij} a_j \end{cases}$$

We can distinguish between two types of equilibria. An action profile $\mathbf{a}$ is *specialized* if players actions are such that $a_i = 0$ or $a_i = a^*$ for every i. A player for which $a_i = a^*$ is a "specialist." An action profile $\mathbf{a}$ is *distributed* when all players choose a positive action less than the individually optimal action level: $0 < a_i < a^*$, $\forall i \in N$. *Hybrid* equilibria are other than these extremes.

Because actions are strategic substitutes, maximal independent sets are a natural notion in this model (see Section 3.3.2.2). Indeed, in equilibrium, no two specialists can be linked. Hence, specialized equilibria are characterized by this structural property of a network, i.e. the specialists are equal to a maximal independent set of the network. A result of Bramoullé and Kranton (2007a) can be stated as follows:

Proposition 3.1. *A specialized profile is a Nash equilibrium of the above game if and only if its set of specialists is a maximal independent set of the structure* $\mathbf{g}$. *Since for every* $\mathbf{g}$ *there exists a maximal independent set, there always exists a specialized Nash equilibrium.*

Figure 3.3 illustrates this proposition for a star-shaped network with four players. Observe that, in a star network, there are two maximal independent sets: the one that only includes the central player and the one that includes all peripheral players. As a result, using Proposition 3.1, there are two specialized equilibria: (i) the center is a specialist and provides action a^* while peripheral players choose actions of 0 (Figure 3.3, left panel); (ii) the center chooses action 0 while peripheral players are specialists and exert effort a^* each (Figure 3.3, right panel).

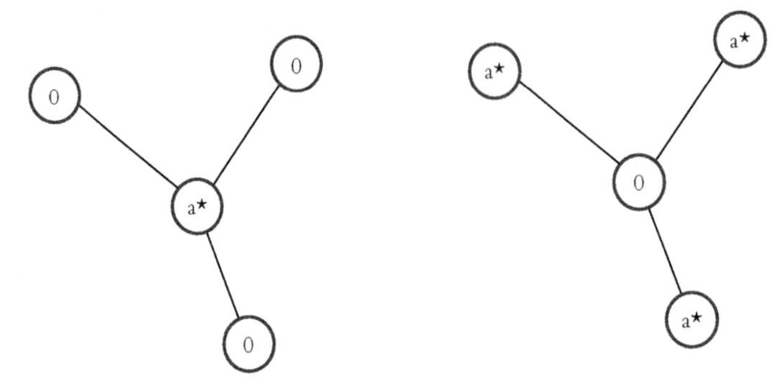

Figure 3.3 Equilibria in a local public good game.

Bramoullé et al. (2014) provide more results on these types of games. Denote by $\mu_0(\mathbf{g})$, the lowest eigenvalue of the adjacency matrix **g**. They show that if $\phi < -1/\mu_0(\mathbf{g})$, there exists a unique Nash equilibrium of this game while, when $\phi > -1/\mu_0(\mathbf{g})$, there always exists a corner equilibrium, i.e. for some player i, $a_i^* = 0$. In terms of comparative statics, they also show that any increase in ϕ or any additional link to the network leads to an equilibrium with lower total actions. Thus, while some players may increase their actions, the decreases dominate.

3.3.3 Two-action games on networks

The class of (complete information) games on networks where players can choose one of two actions, so that $A_i = \{0, 1\}$ for all i, is an important class of games in terms of its applications, one that has been widely studied, and one that allows us to see some general insights. It includes coordination games, and generally all sort of games where players choose whether to do something (adopt a new technology, participate in something, provide a public good effort) or not. These were called *graphical games* by Kearns et al. (2001) and Kakade et al. (2004) who studied the complexity of finding equilibria in such games.

In particular, let us concentrate on a class of such games that are referred to as "semi-anonymous" by Jackson (2008a). These games are not fully anonymous because players are connected in a network and only care about their neighbors' actions, but players care about their neighbors equally. That is, they care only about how many of their neighbors take action 1 and how many take action 0, but not which particular neighbors choose 1 versus 0. Thus, such games are anonymous except to the extent of the interaction patterns governed by the network. In addition, in keeping with the semi-anonymity, we can consider payoff functions that do not depend on a player's identity, but only on how many neighbors that he or she has.

Such games then have some nice properties and best responses are easily described. Letting d be a player's degree, in the case of strategic complements there is a threshold

$t(d)$ such that if more than $t(d)$ neighbors choose action 1 then the player prefers to choose action 1, while if fewer than $t(d)$ neighbors choose 1 then the player prefers to choose 0. It is possible to have situations where an individual is exactly indifferent at the threshold, although that would not occur for generic specifications of payoff functions. Analogously, for the case of strategic substitutes, there is also a threshold, but the best response of the player is reversed, so that he or she prefers to take action 0 if more than $t(d)$ neighbors take action 1, and prefers action 1 if fewer than $t(d)$ neighbors take action 1.

The majority game (Example 3.1) is a game of strategic complements where the threshold is $d/2$, whereas the best-shot public good game (Example 3.2) is a game of strategic substitutes where the threshold is anything between 0 and 1.

3.3.3.1 Changes in behaviors as the network varies

The threshold expression of the two-action semi-anonymous games on networks allows us to easily deduce a few useful comparative statics.

For example, as discussed by Galeotti et al. (2010),[18] it is easy to see that in games of complements where the threshold is nonincreasing in degree, adding links will lead to (weakly) higher actions as players will have higher numbers of neighbors taking action 1.

Proposition 3.2. *Consider a semi-anonymous two-action game of strategic complements on a network* $(N, \mathbf{g})$ *and such that the threshold for taking action 1 is nonincreasing as a function of degree (so that* $t(d+1) \leq t(d)$ *for each* d*). If* $\mathbf{g}'$ *is obtained by adding links to the network* $\mathbf{g}$ *(so that* $\mathbf{g} \subset \mathbf{g}'$*), then for any equilibrium* $\mathbf{a}$ *under* $\mathbf{g}$*, there exists an equilibrium* $\mathbf{a}'$ *under* $\mathbf{g}'$ *such that* $\mathbf{a}' \geq \mathbf{a}$*, so that all players play at least as high an action under* $\mathbf{a}'$ *as under* $\mathbf{a}$.

The case of strategic substitutes does not lead to the same clear-cut conclusions since adding links can change the structure of payoffs in unpredictable ways. Decreasing actions for some players can lead to increasing actions for others in the case of strategic substitutes, and so changing network structure leads to more complex changes in behavior (see Figure 2 in Galeotti et al., 2010, for an illustration of this, and Jackson, 2008a, for additional examples and results).

The comparative statics in equilibrium behavior become clearer in games with incomplete information, as detailed in Section 3.5.

3.3.3.2 Coordination games

Perhaps one of the most important and extensively studied classes of games of strategic complements on networks are coordination games. It is an important class of games because of its many applications: including the choice of a language, a technology, whether to buy a new product, adopt a particular behavior, and so forth; when there are complementarities in actions between friends or acquaintances.

[18] See the working paper version, or Jackson (2008a) for details.

The majority game is an example of this class, but more generally the threshold for wanting to match one's neighbors may differ from 50%, depending both on the payoff to matching versus failing to match one's neighbors. A standard representation of such a game would be as follows (normalizing the payoff to action 0 to zero and then keeping track of the difference in payoffs):

	1	0
1	(b, b)	$(-c, 0)$
0	$(0, -c)$	$(0, 0)$

where b and c are both strictly positive. Thus, coordinating on action 1 is overall better for society, but involves some risk from miscoordination $-c$. A player has to choose either 0 or 1 and then matches against each neighbor. Here there is a threshold fraction $q = \frac{c}{c+b}$ such that action 1 is a best response for a given player if and only if at least a fraction q of the player's neighbors choose 1. This fraction is the same for all players independent of their degrees. Thus, this is a game of strategic complements where the threshold in terms of numbers of neighbors of degree d is simply qd.

Here the Pareto efficient equilibrium (and sometimes referred to as the payoff dominant equilibrium) is for all players to play action 1. However, reaching this play will depend on what players expect their neighbors to play. In setting where $q > 1/2$ (so $c > b$), then the equilibrium in which players both play action 0 is said to be the "risk dominant" equilibrium (as named by Harsanyi and Selten, 1988), in the sense that if a player had a uniform prior over other players' plays—so an equal chance of each other player playing 0 or 1, then action 0 would be the best expected payoff maximizing choice for the player. In a case where $q < 1/2$, then the equilibrium in which both players play action 1 is both risk dominant and payoff dominant (Pareto efficient).

We know from Theorem 3.1 that the set of equilibria form a lattice, and here the maximum equilibrium is where all players choose action 1 while the minimum equilibrium is where all players choose action 0.

What else can we deduce about the set of equilibria? Do there exist equilibria where some players choose 1 and others choose 0? What happens if we start with some small initial seed of players choosing 1 and others choosing 0, and then iterate on best replies? Morris (2000) provides some answers to these questions.[19]

If we let S be the set of players who play action 1 in an equilibrium, then it is clear that each player in S must have at least a fraction q of his/her neighbors in the set S, and also each player outside of S must a fraction of no more than q of his/her neighbors in S, or equivalently has a fraction of at least $1 - q$ of his/her neighbors outside of S.

[19] The analysis here follows Jackson's (2008a) adaptation of Morris's results to a finite population setting. See also Ellison (1993) and Blume (1993) for some related analyses.

To capture this, given $1 \geq r \geq 0$, Morris (2000) defined the set of nodes S to be *r-cohesive* with respect to a network $(N, \mathbf{g})$ if each node in S has at least a fraction r of its neighbors in S. That is, S is r-cohesive relative to $\mathbf{g}$ if

$$\min_{i \in S} \frac{|N_i(\mathbf{g}) \cap S|}{d_i(\mathbf{g})} \geq r, \qquad [3.2]$$

where $0/0$ is set to 1.

The *cohesiveness* of a given set S relative to a network $(N, \mathbf{g})$ is then the maximum r such that S is r-cohesive. The following proposition of Morris (2000) follows directly:

Proposition 3.3. *Consider a network $(N, \mathbf{g})$ and a coordination game as described above. There exists an equilibrium where action 1 is played by $S \subset N$ and action 0 by $N \backslash S$ if and only if S is q-cohesive and such that its complement $N \backslash S$ is $(1 - q)$-cohesive.*

Cohesiveness provides enough of a "separation" in a network for different behaviors to exist on different parts of a network.

With this proposition as background, it is then easy to see how behavior might spread in a network. In particular, start from a network $(N, \mathbf{g})$ with all players choosing action 0. Next, "infect" a set of players by switching them to play action 1 (and they can never switch back). Next, let players (other than the initially infected) best respond to the current actions of their neighbors, switching players to action 1 if their payoffs are at least as good with action 1 as with action 0 against the actions of the other players. Repeat this process starting from the new actions, and stop at a stage where no new players change to action 1. If there is some set of players S whose initial infection leads to all players taking action 1 under the best response dynamics, then we say that there is a *contagion from S*.

Define a set S to be *uniformly no more than* r-cohesive if there is no nonempty subset of S that is more than r-cohesive. We then can deduce the following proposition.

Proposition 3.4. *Consider a network $(N, \mathbf{g})$ and a coordination game as described above. Contagion from the set S occurs if and only if its complement is uniformly no more than $(1 - q)$-cohesive.*

The proof of this proposition is straightforward: If the complement of S has a subset S' that is more than $(1 - q)$-cohesive, then S' will all play 0 under the process above, at every step. Thus, it is necessary for the complement to be uniformly no more than $(1 - q)$-cohesive in order to have contagion to all remaining players outside of S. To see that this condition is also sufficient, note that if the complement is uniformly no more than $(1 - q)$-cohesive, then the complement is no more than $(1 - q)$-cohesive. This means that there must be at least one player in the complement who has at least a fraction of q of his or her neighbors in S. So, at the first step, at least one player changes strategies. Subsequently, at each step, the set of players who have not yet changed strategies is no more than $(1 - q)$-cohesive, and so some player must have at least q neighbors who are playing 1 and will change. Thus, as long as some players have not yet changed, the process will have new players changing, and so every player must eventually change.

The cohesion conditions used in the above results can be difficult to check in a network, but the concept is still quite useful in terms of outlining what is necessary to maintain different behaviors in a society: one needs a sufficient schism between (at least) two groups. This is closely related to homophily in the network, whereby a group's members tend to be relatively more connected to each other than with outsiders (see e.g., Bramoullé et al., 2012; Currarini et al., 2009, 2010; Golub and Jackson, 2010, 2012a,b,c; Jackson, 2008b; Jackson and Lopez-Pintado, 2013; McPherson et al., 2001).[20]

3.3.3.3 Stochastically stable play in coordination games on networks

While analyses of equilibria in complete information games on networks provide some insights and intuitions, they often lack the randomness (and heterogeneity) that are pervasive in many applications. In fact, in some cases adding randomness can substantially refine the multiplicity of equilibria that exist in the complete information noiseless setting. To see how this works, let us visit some of the literature on stochastic stability in coordination games.

Let us consider the following variation on settings considered by Kandori et al. (1993) and Young (1993).[21] Start with $(N, \mathbf{g})$ being the complete network, so that each player plays with all others and chooses strategy either 1 or 0 in the coordination game from Section 3.3.3.2. Players play the game repeatedly over time. However, they are myopic: in each period a player chooses a best reply to the play of the others in the previous period. So far, this is similar to our analysis of the previous section except for the complete network. Indeed, here there are two obvious trajectories of society: all playing 1 and all playing 0 in every period. In fact, it is clear that if the starting proportion is sufficiently

[20] Chwe (1999, 2000) studies related classes of games, but instead where players care not only about the behavior of their own neighborhood, but of the collective, with an application to collective actions like participating in a large protest or revolution. His motivation is similar, in looking at the network structure and investigating where it is possible to sustain a collective action when players can only be sure of their neighbors' play and worry about a failed collective action: so they will only participate if they are sure that some quota is exceeded. Granovetter (1978) and Schelling (1978) also provide dynamic models to analyze such issues, but avoiding a clear linkage of outcomes to the social network structure in society. In particular, a snowball effect might be generated only if there are initially enough activists who are willing to participate, independently of others decisions. Once these activists start the process.

[21] For more background on the origins of stochastic stability arguments in game theory, see Wallace and Young (2014).

different from q (at least $1/(n-1)$ above or below suffices), then one of these two states is reached after the first period, and even away from that, if q is not too close to $1/2$ (at least $1/n$ above or below suffices), then one of the states is reached after at most two periods.[22]

However, to refine the predictions of the system, let us add slight perturbations to the choice of actions: in each period, a player's choice is reversed (1 to 0 or 0 to 1) with a slight probability $\varepsilon > 0$, and this is done independently across players. Players still best reply to what was played in the previous period, but sometimes their chosen action is changed by some exogenous perturbation. This dynamic process is now a Markov chain that is irreducible and aperiodic. This process thus has very nice ergodic properties that are easy to deduce from standard results in Markov theory. In fact, without even relying on much of the mathematics, it is easy to see some basic properties of this process:

- Any configuration of play is possible in any period since there is at least an ε probability of each player playing either action in every period, given that the perturbations are independent across players.
- If ε is sufficiently small then once the population has sufficiently more or fewer than a fraction of q playing 1, then the next period with a high probability, all players play 1 or all players play 0.
- Thus, the process will continue to exhibit all possible plays infinitely often over time, but, as ε tends to 0, it will spend most of its time with all players choosing the same action.

The important insight that emerged from Kandori et al. (1993) and Young (1993) is that if q is not too close to $1/2$ (more than $1/(n-1)$ away), then, as ε tends to 0:

- if $q > 1/2 + 1/(n-1)$, then the fraction of periods where all players play action 0 tends to one, and
- if $q < 1/2 - 1/(n-1)$, then the fraction of periods where all players play action 1 tends to one.

Thus, if q is not too close to $1/2$, then stochastic stability picks out the risk dominant action: the action that is better for a player when others play uniformly at random.

The intuition behind this result is relatively straightforward. Considering that most of the time all players choose the same action, the important determinant of the system is how long it takes to transition from an extreme where all play 0 to the other extreme where all play 1, and vice versa. Here q enters. If we start with all playing 0, then we need roughly qn "perturbations" of the best replies within a single period before the system transitions to all playing 1, and, in the reverse direction, at least $(1-q)n$ perturbations

[22] Some starting configurations that have a fraction of players sufficiently close to q playing 1, when q is sufficiently close to $1/2$, can cycle endlessly, where two approximately equal-sized groups continue to switch back and forth between 1 and 0, each miscoordinating with the other group over time.

are needed. For small ε, the probabilities of this happening are effectively on the order of ε^{qn} and $\varepsilon^{(1-q)n}$, respectively. As ε tends to 0, if $q < 1/2$ then the first becomes infinitely more likely than the second, and if $q > 1/2$ then the reverse is true.

Thus, introducing very slight trembles to a coordination game can end up making very stark predictions in terms of the ergodic play of a society. What are some issues with relying on this approach to make predictions about behavior? One is that as ε tends to 0, it could take an enormous amount of time to reach a transition. For example, suppose that $q = 1/3$ and we start with all players choosing 0. Even though we know that in the "long run," for small ε the process spends most of its time with all players choosing 1, it could take a very long time before we leave the starting state of everybody playing 0. In a sense, the "short run" could last a very long time in a large society, as many very unlikely simultaneous trembles could be needed to transition.[23]

The more interesting case in terms of many applications is when one turns to a setting of less than complete networks. A first work in this direction was Ellison (1993) who pointed out that this could have important implications for the speed of transitions from one state to another. In particular, as shown by Ellison (1993), for example, if players are connected in a "circle" so that each player has two adjacent neighbors (so, i's neighbors in the network are $i-1$ and $i+1$, with the wrap-around convention that n is connected to 1), then the transition can happen much more quickly. Instead of waiting for $n/3$ perturbations when $q = 1/3$, if two adjacent players are perturbed, then that is enough to spread behavior 1 to the whole society; something which is much more likely to happen. Thus, network structures significantly affect the times to transition in the network.

The network structure can also have profound effects on the long-run distribution of play, as shown by Jackson and Watts (2002a,b).

Jackson and Watts (2002b) propose a model in which the interaction pattern is an arbitrary network $\mathbf{g}$. In each period t, a player i chooses an action $a_i^t \in \{1, 0\}$ and then receives a payoff equals to

$$u_i(\mathbf{a}^t, \mathbf{g}) = \sum_{j \neq i} g_{ij} v_i(a_i^t, a_j^t)$$

where $v_i(a_i^t, a_j^t)$ is a payoff that depends on the actions chosen. The dynamic process is described as follows. In each period one player is chosen at random (with equal probability across players) to update his/her strategy. A player updates his/her strategy myopically, best responding to what the other players with whom he/she interacts did

[23] Another issue is raised by Bergin and Lipman (1996) who point out that this depends on the error structure being the same ε regardless of circumstances. If the probability of trembles is enriched arbitrarily, then the relative probability of various transitions can be made arbitrary and the selection is lost.

in the previous period. There is also a probability $1 > \varepsilon > 0$ that a player trembles, and chooses a strategy that he/she did not intend to. Thus, with probability $1 - \varepsilon$, the strategy chosen is $a_i^t = \arg\max_{a_i} u_i(a_i, a_{-i}^{t-1}, \mathbf{g})$ and with probability ε the strategy is $a_i^t \neq \arg\max_{a_i} u_i(a_i, a_{-i}^{t-1}, \mathbf{g})$. The probabilities of trembles are identical and independent across players, strategies, and periods. Again, this is a well-defined Markov chain where the state is the vector of actions $\mathbf{a}^t$, that are played in period t. Since this process is aperiodic and irreducible, the Markov chain has a unique stationary distribution, denoted $\mu^\varepsilon(\mathbf{a})$.

If we consider the *complete network* where all players are connected to each other so that each player has $n - 1$ links, then the analysis is as above from the models of Kandori et al. (1993) and Young (1993)[24] and the unique stochastically stable state is the risk-dominant equilibrium (provided that q is not too close to 1/2).

However, instead let us consider a *star network* where player 1 is the at the center of the star, connected to every other player but where other players are only connected to player 1. Jackson and Watts (2002b) show that, in this case, there are two stochastically stable states: one where all players play 1 and the other where all players play 0; regardless of q. Note that now peripheral players $i > n$ care only about what player 1 is playing, and they update to play whatever 1 played last period when called on to update. Player 1, in contrast, cares about what all the players are doing. Thus one tremble by player 1 can lead from a network where all play 1 to one where all play 0. Thus starting from either equilibrium of all play 1 or all play 0, we need only one tremble to have updating lead naturally to the other equilibrium. As the relative number of trembles is the important factor in determining the set of stochastically stable states, both of these states are stochastically stable.

The important difference from the complete network setting is that regardless of the starting condition, if the central player changes actions, that can change the subsequent play of all other players. Thus, all playing 1 and all playing 0 are both more "easily" destabilized, and the long-run distribution does not place all weight on just one equilibrium.[25]

[24] The Jackson and Watts process involves changes of players' actions one at a time, which is easier to handle when the network is endogenized (see Section 3.6.1), but differs from the Kandori et al. (1993) and Young (1993) where decisions and trembles are simultaneous. That difference is largely inconsequential when applied to a fixed network.

[25] The results on this do depend on the perturbation structure. As shown by Blume (1993) and Young (1998), if one uses a structure in which errors are exponentially proportional to the payoff loss of making the error, then the center makes infinitely fewer errors in moving away from 1 than away from 0 when $q < 1/2$ (and vice versa when $q > 1/2$), which can then restore the original conclusion that the risk-dominant play is obtained. However, that then relies on some errors by individual players becoming infinitely more likely than others, which seems implausible if errors can result from things like errors in calculation or some unmodeled preference shock, and so forth. For more discussion of this error structure see Wallace and Young (2014).

This shows that network structure can influence the play of a society, even in terms of selecting among (strict) Nash equilibria. Exactly what emerges depends on a number of details and how the long-run play depends on the structure of the network can be quite complex. Nonetheless, there are some clean results that emerge when one further enriches the structure to allow players also to choose the network along with their play, as discussed in Section 3.6.

3.4. A MODEL WITH CONTINUOUS ACTIONS, QUADRATIC PAYOFFS, AND STRATEGIC COMPLEMENTARITIES

Although models with finite numbers of actions capture many applications, there are others that are well-approximated by continuous actions; for instance in which players have choices of how much time or effort to exert in an activity like education or crime. In this section, we analyze a simple model of a game with strategic complements where the utility function is linear-quadratic. An advantage of this formulation is that it allows for an easy characterization of equilibrium as a function of network structure.[26]

3.4.1 The benchmark quadratic model

Consider a game in which players decide how much time or effort to exert in some activity, denoted $a_i \in \mathbb{R}_+$. The payoff to player i as a function of the action profile and network, $u_i(\mathbf{a}, \mathbf{g})$, is given by

$$u_i(\mathbf{a}, \mathbf{g}) = \alpha\, a_i - \frac{1}{2} a_i^2 + \phi \sum_{j=1}^{n} g_{ij} a_i a_j, \qquad [3.3]$$

where $\alpha, \phi > 0$. In this formulation, players are ex ante homogeneous in terms of observable characteristics (i.e., they all have the same α, ϕ) and their heterogeneity stems entirely from their position in the network. The first two terms of [3.3], $\alpha\, a_i - \frac{1}{2} a_i^2$, give the benefits and costs of providing the action level a_i. The last term of this utility function, $\phi \sum_{j=1}^{n} g_{ij} a_i a_j$, reflects the strategic complementarity between friends' and acquaintances' actions and own action. This peer effect depends on the different locations of players in the network $\mathbf{g}$. When i and j are directly linked, i.e. $g_{ij} = 1$, the cross derivative is $\phi > 0$ and reflects *strategic complementarity* in efforts. When i and j are not direct friends, i.e. $g_{ij} = 0$, this cross derivative is zero.

Ballester et al. (2006) have analyzed this game, determining its unique Nash equilibrium in pure strategies (when it exists, provided that ϕ is not too large). The

[26] The definitions of strategic complements and substitutes apply to such games exactly as before. We focus in this section mainly on games of strategic complements due to space constraints. For more discussion of the case of strategic substitutes, see Bramoullé et al. (2014).

first-order necessary condition for each player i's choice of action to maximize his or her payoff is:

$$\frac{\partial u_i(\mathbf{a},\mathbf{g})}{\partial a_i} = \alpha - a_i + \phi \sum_{j=1}^{n} g_{ij} a_j = 0,$$

which leads to:

$$a_i^* = \alpha + \phi \sum_{j=1}^{n} g_{ij} a_j^*. \qquad [3.4]$$

In matrix form:

$$\mathbf{a}^* = \alpha \mathbf{1} + \phi \mathbf{g} \mathbf{a}^*$$

where $\mathbf{1}$ is the column vector of 1 and $\mathbf{g}$ is the adjacency matrix. Solving this leads to:

$$\mathbf{a}^* = \alpha \left(\mathbf{I} - \phi \mathbf{g}\right)^{-1} \mathbf{1} \qquad [3.5]$$

where $\mathbf{I}$ is the identity matrix (and provide the inverse of $\left(\mathbf{I} - \phi \mathbf{g}\right)$ is well-defined).

3.4.1.1 Katz-Bonacich network centrality and strategic behavior

The equilibrium action profile of this quadratic model is related to a centrality measure. Let

$$\mathbf{M}\left(\mathbf{g},\phi\right) = \left(\mathbf{I} - \phi \mathbf{g}\right)^{-1} = \sum_{p=0}^{+\infty} \phi^p \mathbf{g}^p$$

As defined by Katz (1953) and Bonacich (1987),[27] given $\mathbf{w} \in \mathbb{R}_+^n$ and $\phi \geq 0$, the vector of *weighted Katz-Bonacich centralities* relative to a network $\mathbf{g}$ is:

$$\mathbf{b}_{\mathbf{w}}\left(\mathbf{g},\phi\right) = \mathbf{M}\left(\mathbf{g},\phi\right)\mathbf{w} = \left(\mathbf{I} - \phi \mathbf{g}\right)^{-1}\mathbf{w} = \sum_{p=0}^{+\infty} \phi^p \mathbf{g}^p \mathbf{w}. \qquad [3.6]$$

In particular, when $\mathbf{w} = \mathbf{1}$, the *unweighted* Katz-Bonacich centrality of node i is $b_{1,i}(\mathbf{g},\phi) = \sum_{j=1}^{n} M_{ij}(\mathbf{g},\phi)$, and counts the *total* number of walks in $\mathbf{g}$ starting from i, discounted exponentially by ϕ. It is the sum of all loops $M_{ii}(\mathbf{g},\phi)$ from i to i itself, and of all the outer walks $\sum_{j \neq i} M_{ij}(\mathbf{g},\phi)$ from i to every other player $j \neq i$, that is:

$$b_{1,i}(\mathbf{g},\phi) = M_{ii}(\mathbf{g},\phi) + \sum_{j \neq i} M_{ij}(\mathbf{g},\phi).$$

By definition, $M_{ii}(\mathbf{g},\phi) \geq 1$, and thus $b_i(\mathbf{g},\phi) \geq 1$, with equality when $\phi = 0$.

[27] For more background and discussion of this and related definitions of centrality, see Jackson (2008a).

3.4.1.2 Nash equilibrium

We now characterize the Nash equilibrium of the game where players choose their effort level $a_i \geq 0$ simultaneously. Let $\mu_1(\mathbf{g})$ denote the spectral radius of $\mathbf{g}$.

Proposition 3.5. *If $\phi\mu_1(\mathbf{g}) < 1$, the game with payoffs [3.3] has a unique (and interior) Nash equilibrium in pure strategies given by:*

$$\mathbf{a}^* = \alpha\, \mathbf{b}_1\left(\mathbf{g}, \phi\right). \quad [3.7]$$

This results shows that Katz-Bonacich centrality embodies the feedback calculations that underlie equilibrium behavior when utility functions are linear-quadratic. In [3.3], the local payoff interdependence is restricted to neighbors. In equilibrium, however, this local payoff interdependence spreads indirectly through the network. The condition $\phi\mu_1(\mathbf{g}) < 1$ stipulates that local complementarities must be small enough compared to own concavity, which prevents excessive feedback which can lead to the absence of a finite equilibrium solution.

There are different ways of proving Proposition 3.5. Ballester et al. (2006) show that the condition $\phi\mu_1(\mathbf{g}) < 1$ ensures that the matrix $\mathbf{I} - \phi\mathbf{g}$ is invertible by Theorem III of Debreu and Herstein (1953, p. 601). Then, using [3.5], it is straightforward to see that the optimal efforts are equal given by Katz-Bonacich centrality. Finally, they rule out corner solutions to establish uniqueness. Since $\frac{\partial u_i(\mathbf{a},\mathbf{g})}{\partial a_i} |_{\mathbf{a}=\mathbf{0}} = \alpha > 0$, then $\mathbf{a} = \mathbf{0}$ cannot be a Nash equilibrium. It is also clear from the first-order condition that a deviation to positive effort is profitable if only some subgroup $S \subset N$ of players provides zero effort.[28] As a result, the Nash equilibrium is unique and each a_i^* is interior.

Another simple way of proving Proposition 3.5 is, as noted by Bramoullé et al. (2014) and König (2013), is to observe that this game is a potential game (as defined by Monderer and Shapley, 1996)[29] with potential function[30]:

$$P(\mathbf{a}, \mathbf{g}, \phi) = \sum_{i=1}^{n} u_i(\mathbf{a}, \mathbf{g}) - \frac{\phi}{2} \sum_{i=1}^{n} \sum_{j=1}^{n} g_{ij} a_i a_j$$

[28] From Theorem I of Debreu and Herstein (1953, p. 600), $\phi\mu_1(\mathbf{g}) < 1$ also guarantees that $\mathbf{I} - \phi\mathbf{g}_S$ is invertible for $\mathbf{g}_S \subset \mathbf{g}$, where $\mathbf{g}_S$ is the adjacency matrix for the subgroup $S \subset N$ of players.

[29] A game is a potential game if there is a function $P : X \to \mathbb{R}$ such that, for each $i \in N$, for each $x_{-i} \in X_{-i}$, and for each $x_i, z_i \in X_i$,

$$u_i(x_i, x_{-i}) - u_i(z_i, x_{-i}) = P(x_i, x_{-i}) - P(z_i, x_{-i}).$$

[30] Here the potential $P(\mathbf{x}, \mathbf{g}, \phi)$ is constructed by taking the sum of all utilities, a sum that is corrected by a term which takes into account the network externalities exerted by each player i.

$$= \sum_{i=1}^{n} \alpha\, a_i - \frac{1}{2}\sum_{i=1}^{n} a_i^2 + \frac{\phi}{2}\sum_{i=1}^{n}\sum_{j=1}^{n} g_{ij} a_i a_j,$$

or in matrix form:

$$P(\mathbf{a},\mathbf{g},\phi) = \alpha \mathbf{a}^\top \mathbf{1} - \frac{1}{2}\mathbf{a}^\top \mathbf{a} + \mathbf{a}^\top \frac{\phi}{2}\mathbf{g}\mathbf{a}$$

$$= \alpha \mathbf{a}^\top \mathbf{1} - \frac{1}{2}\mathbf{a}^\top \left(\mathbf{I} - \phi \mathbf{g}\right)\mathbf{a}.$$

It is well-known (see, e.g., Monderer and Shapley, 1996) that solutions of the program $\max_{\mathbf{a}} P(\mathbf{a},\mathbf{g},\phi)$ are a subset of the set of Nash equilibria.[31] This program has a unique interior solution if the potential function $P(\mathbf{a},\mathbf{g},\phi)$ is strictly concave on the relevant domain. The Hessian matrix of $P(\mathbf{a},\mathbf{g},\phi)$ is easily computed to be $-\left(\mathbf{I}-\phi\mathbf{g}\right)$. The matrix $\mathbf{I}-\phi\mathbf{g}$ is positive definite if for all non-zero $\mathbf{a}$:

$$\mathbf{a}^\top \left(\mathbf{I}-\phi\mathbf{g}\right)\mathbf{a} > 0 \Leftrightarrow \phi < \left(\frac{\mathbf{a}^\top \mathbf{g}\mathbf{a}}{\mathbf{a}^\top \mathbf{a}}\right)^{-1}.$$

By the Rayleigh-Ritz theorem, we have $\mu_1(\mathbf{g}) = \sup_{\mathbf{a}\neq \mathbf{0}}\left(\frac{\mathbf{a}^\top \mathbf{g}\mathbf{a}}{\mathbf{a}^\top \mathbf{a}}\right)$. Thus, a necessary and sufficient condition for having a strict concave potential is that $\phi\mu_1(\mathbf{g}) < 1$, as stated in Proposition 3.5.

To illustrate the results and to describe a *social multiplier*, let us consider the case of a dyad, i.e. $n = 2$. In that case, the utility [3.3] can be written as:

$$u_i(\mathbf{a},\mathbf{g}) = \alpha\, x_i - \frac{1}{2}a_i^2 + \phi a_i a_j.$$

If there were no social interactions, the unique Nash equilibrium would be:

$$a_1^* = a_2^* = \alpha,$$

while, for a dyad where the two individuals are linked to each other (i.e., $g_{12} = g_{21} = 1$), the unique Nash equilibrium is given by (if $\phi < 1$)[32]:

$$a_1^* = a_2^* = \frac{\alpha}{1-\phi}.$$

[31] To establish uniqueness of the equilibrium, one has to show a one-to-one correspondence between the set of Nash equilibria and the solutions to the first-order conditions of the maximization problem (see Bramoullé et al., 2014 for details).

[32] It is straightforward to check that:

$$\mathbf{b}_1(\mathbf{g},\phi) = \left(\mathbf{I}-\phi\mathbf{g}\right)^{-1}\mathbf{1} = \begin{pmatrix} 1/(1-\phi) \\ 1/(1-\phi) \end{pmatrix}.$$

In the dyad, complementarities lead to an effort level above the equilibrium value for an isolated player ($a_1^* = a_2^* = \alpha$). The factor $1/(1-\phi) > 1$ is often referred to as a *social multiplier*.[33,34]

In terms of comparative statics, it is clear that, by standard strategic-complementarity arguments, increasing the number of links of any player raises his/her effort, and for this specification of utility functions also increases his/her payoff.

3.4.1.3 Welfare

It is obvious that the Nash equilibrium in such a game is inefficient, as there are positive externalities in efforts. There is *too little effort* at the Nash equilibrium as compared to the social optimum outcome, because each individual ignores the positive impact of her effort on others. As a result, there are benefits from subsidizing effort, as we now detail.

To analyze welfare, we first relate the equilibrium utility of player i to Katz-Bonacich centrality:

$$u_i(\mathbf{a}^*, \mathbf{g}) = \alpha\, a_i^* - \frac{1}{2} a_i^{*2} + \phi \sum_{j=1}^{n} g_{ij} a_i^* a_j^*$$

$$= a_i^* \left(\alpha + \phi \sum_{j=1}^{n} g_{ij} a_j^* \right) - \frac{1}{2} a_i^{*2}.$$

By [3.4] where $a_i^* = \alpha + \phi \sum_{j=1}^{n} g_{ij} a_j^*$:

$$u_i(\mathbf{a}^*, \mathbf{g}) = a_i^{*2} - \frac{1}{2} a_i^{*2} = \frac{1}{2} a_i^{*2} = \frac{1}{2} \left[b_{1,i}\left(\mathbf{g}, \phi\right)\right]^2. \quad [3.8]$$

The total equilibrium welfare (i.e., the sum of all equilibrium utilities) is then:

$$\mathcal{W}(\mathbf{a}^*, \mathbf{g}) = \mathbf{u}(\mathbf{a}^*, \mathbf{g}) \cdot \mathbf{1} = \frac{1}{2} \mathbf{b}_1^\top\left(\mathbf{g}, \phi\right) \mathbf{b}_1\left(\mathbf{g}, \phi\right). \quad [3.9]$$

[33] See, for instance, Glaeser et al. (2003), and references therein.

[34] Belhaj and Deroïan (2011) have extended the previous model by considering two substitute activities, so that player i chooses both $a_{1,i}$ and $a_{2,i}$ such that $a_{1,i} + a_{2,i} = 1$. The payoff is:

$$u_i(\mathbf{a}_1, \mathbf{a}_2, \mathbf{g}) = \alpha_1\, a_{1,i} - \frac{1}{2} a_{1,i}^2 + \alpha_2 a_{2,i} - \frac{1}{2} a_{2,i}^2 + \sum_{j=1}^{n} g_{ij} \left(\phi_1 a_{1,i} a_{1,j} + \phi_2 a_{2,i} a_{2,j} \right).$$

The model incorporates local synergies and both lower and upper bounds on efforts, which facilitates the analysis of equilibrium. Belhaj and Deroïan (2014) study both interior and corner solutions, and provide comparative statics with respect to activity cost.

Following Helsley and Zenou (2014), let us show that the Nash equilibrium [3.5] is not efficient. For that, the planner chooses $a_1, \ldots, a_n$ that maximize total welfare, that is:

$$\max_{\mathbf{a}} \mathcal{W}(\mathbf{a}, \mathbf{g}) = \max_{a_1,\ldots,a_n} \sum_{i=1}^{i=n} u_i(\mathbf{a}, \mathbf{g}) = \max_{a_1,\ldots,a_n} \left\{ \sum_{i=1}^{i=n} \left[\alpha\, a_i - \frac{1}{2} a_i^2 \right] + \phi \sum_{i=1}^{i=n} \sum_{j=1}^{n} g_{ij} a_i a_j \right\}.$$

The first-order conditions are that for each $i = 1, \ldots, n$:

$$\alpha - a_i + \phi \sum_j g_{ij} a_j + \phi \sum_j g_{ji} a_j = 0,$$

which implies that (since $g_{ij} = g_{ji}$):

$$a_i^O = \alpha + 2\phi \sum_j g_{ij} a_j^O, \qquad [3.10]$$

where the superscript O refers to the "social optimum." In matrix form,

$$\mathbf{a}^O = \alpha \left(\mathbf{I} - 2\phi\, \mathbf{g} \right)^{-1} \mathbf{1} = \alpha \mathbf{b}_1 \left(\mathbf{g}, 2\phi \right). \qquad [3.11]$$

Examination of [3.7] and [3.11] shows that the two solutions differ and that the Nash equilibrium effort of each individual i is inefficiently low.

Analogously to the derivation of [3.9], we see that

$$\mathcal{W}(\mathbf{a}^O, \mathbf{g}) = \frac{1}{2} \mathbf{b}_1^\top \left(\mathbf{g}, 2\phi \right) \mathbf{b}_1 \left(\mathbf{g}, 2\phi \right). \qquad [3.12]$$

Here, a Pigouvian *subsidy* can lead individuals to choose the socially optimal effort levels. Let us derive this optimal subsidy that can restore the first-best outcome [3.10].

The timing is as follows. First, the government announces the per-effort subsidy $s_i(\mathbf{a}) \geq 0$ for each individual $i = 1, \ldots, n$. Then each individual i chooses effort a_i to maximize his or her payoff accounting for the subsidy. Because of the planner's subsidy, individual i's utility is now given by:

$$u_i(\mathbf{a}, \mathbf{g}; s_i) = [\alpha + s_i]\, a_i - \frac{1}{2} a_i^2 + \phi \sum_{j=1}^{n} g_{ij} a_i a_j. \qquad [3.13]$$

Helsley and Zenou (2014) show the following result:

Proposition 3.6. *Assume that $2\phi\mu_1(\mathbf{g}) < 1$. If subsidies that satisfy $s_i = \phi \sum_j g_{ij} a_j^O$ are in place, then the first-best efforts form a Nash equilibrium. In equilibrium, this per-effort subsidy is equal to: $s_i = \frac{1}{2}\left[b_{1,i}\left(\mathbf{g}, 2\phi\right) - \alpha \right]$. Thus, the planner gives a higher per-effort subsidy to more central players in the network.*

This proposition does not address the financing of the subsidy. Here since the anticipated total cost of the subsidies is calculable ex ante (knowing the network and players' preferences), the planner can raise the value of the subsidies by any tax scheme that is independent of $\mathbf{a}$.

3.4.2 The model with global congestion

In some applications, in addition to local complementarities, players might experience global competitive effects. When global competition or congestion matters (e.g., see our discussion of applications of this model to crime, Cournot competition, etc.), we can modify the utility function [3.3] to allow for global competitive effects. One such formulation is:

$$u_i(\mathbf{a},\mathbf{g}) = \alpha\, a_i - \frac{1}{2}a_i^2 + \phi a_i \sum_{j=1}^{n} g_{ij}a_j - \gamma\, a_i \sum_{j=1}^{n} a_j \qquad [3.14]$$

where global congestion is captured by the factor $-\gamma \sum_{j=1}^{n} a_j$ that multiplies player i's action. Compared to the previous case without global substitutes where no player had interest in providing zero effort (see [3.4]), it is now possible that corner solutions arise at the Nash equilibrium. Ballester et al. (2006) show that, if $\phi\mu_1(\mathbf{g}) < 1+\gamma$, then there exists a unique interior Nash equilibrium given by:

$$\mathbf{a}^* = \frac{\alpha}{1+\gamma\left[1+b_1(\mathbf{g},\phi/(1+\gamma))\right]}\mathbf{b}_1\left(\mathbf{g},\phi/(1+\gamma)\right), \qquad [3.15]$$

where $b_1(\mathbf{g},\phi/(1+\gamma)) = \mathbf{1}^{\top}\mathbf{M}\left(\mathbf{g},\phi,\gamma\right)\mathbf{1}$ (where $\mathbf{M}\left(\mathbf{g},\phi,\gamma\right) = \left[(1+\gamma)\,\mathbf{I}{-}\phi\mathbf{g}\right]^{-1}$) is the sum of unweighted Bonacich centralities of all players, i.e.,

$$b_1(\mathbf{g},\phi/(1+\gamma)) = b_{1,1}(\mathbf{g},\phi/(1+\gamma)) + \cdots + b_{n,1}(\mathbf{g},\phi/(1+\gamma)).$$

Thus, the model is easily adapted to include such effects, provided they also fit into a linear-quadratic form.

In terms of comparative statics, the standard complementarity argument for the benchmark model, which implies that equilibrium efforts increase (on a player by player basis) in any component with in which links are added, does not apply here because of the competition effect. However, the following result regarding total aggregate effort can be shown:

Proposition 3.7. *Let* $\mathbf{g}$ *and* $\mathbf{g}'$ *be symmetric and such that* $\mathbf{g}' \geq \mathbf{g}$ *and* $\mathbf{g}' \neq \mathbf{g}$. *If* $\phi\mu_1(\mathbf{g}) < 1$ *and* $\phi'\mu_1(\mathbf{g}') < 1$, *then* $\sum_i a^*(\mathbf{g}')_i > \sum_i a^*(\mathbf{g})_i$.

Proposition 3.7 shows that an increase in network relationships (in a partial order sense) increases aggregate activity. This result is due to the fact that neighbors are the source of local complementarities. As a result, players obtain more local complementarities in $\mathbf{g}'$ than in $\mathbf{g}$, and equilibrium aggregate activity is thus higher in $\mathbf{g}'$ than in $\mathbf{g}$. Symmetry of the adjacency is not really needed here: using Farkas' lemma, Belhaj and Deroïan (2013) have shown that, even with asymmetric $\mathbf{g}$ and $\mathbf{g}'$, Proposition 3.7 holds. They show that if the transposed system admits a positive solution, then any perturbation of the linear system that enhances complementarities leads to an increase of average effort.

3.4.3 The model with ex ante heterogeneity

So far, players were only heterogeneous due to their positions in the network. Let us now extend the model to allow for players who are also heterogeneous in terms of characteristics; e.g., age, race, gender, education, etc., which is important when one would like to bring the model to the data and to test it. One way to introduce some aspects of heterogeneity while still maintaining tractability is to allow the utility function of player i to be given by:

$$u_i(\mathbf{a}, \mathbf{g}) = \alpha_i\, a_i - \frac{1}{2}a_i^2 - \gamma \sum_{j=1}^{n} a_i a_j + \phi \sum_{j=1}^{n} g_{ij} a_i a_j, \qquad [3.16]$$

where α_i depends on the characteristics of player i. In many empirical applications (see, e.g., Calvó-Armengol et al., 2009, for education, or Patacchini and Zenou, 2012, for crime), α_i could also includes the average characteristics of individual i's neighbors, i.e. the average level of parental education of i's friends, etc., which are referred to as contextual effects. In particular, assume that

$$\alpha_i = \sum_{h=1}^{H} \beta_h x_i^h + \frac{1}{d_i(\mathbf{g})} \sum_{h=1}^{H} \sum_{j=1}^{n} \gamma_h g_{ij}\, x_j^h \qquad [3.17]$$

where $(x_i^h)_h$ is a set of H variables accounting for individual characteristics and β_h, γ_h are parameters.

Let $b_{\boldsymbol{\alpha}}\left(\mathbf{g}, \phi/(1+\gamma)\right) = \boldsymbol{\alpha}^{\top}\mathbf{M}\left(\mathbf{g}, \phi, \gamma\right)\boldsymbol{\alpha}$ be the weighted sum of weighted Bonacich centralities of all players, i.e. $b_{\boldsymbol{\alpha}} = \alpha_1 b_{\boldsymbol{\alpha},1} + \cdots + \alpha_n b_{\boldsymbol{\alpha},n}$. Calvó-Armengol et al. (2009) show the following.

Proposition 3.8.

(i) If $\boldsymbol{\alpha} = \alpha\mathbf{1}$, then the game has a unique Nash equilibrium in pure strategies if and only if $\phi\mu_1(\mathbf{g}) < 1 + \gamma$. This equilibrium $\mathbf{a}^*$ is interior and described by [3.15].

(ii) If $\boldsymbol{\alpha} \neq \alpha\mathbf{1}$, then let $\overline{\alpha} = \max\{\alpha_i \mid i \in N\} > \underline{\alpha} = \min\{\alpha_i \mid i \in N\} > 0$. In this case, if $\phi\mu_1(\mathbf{g}) + n\gamma\left(\overline{\alpha}/\underline{\alpha} - 1\right) < 1 + \gamma$, then this game has a unique Nash equilibrium in pure strategies and it is interior and described by:

$$\mathbf{a}^* = \frac{1}{1+\gamma}\left[\mathbf{b}_\alpha\left(\mathbf{g}, \phi/(1+\gamma)\right)) - \frac{\gamma b_\alpha\left(\mathbf{g}, \phi/(1+\gamma)\right)}{1+\gamma+\gamma b_1\left(\mathbf{g}, \phi/(1+\gamma)\right)}\right.$$
$$\left.\times\, \mathbf{b}_1\left(\mathbf{g}, \phi/(1+\gamma)\right)\right]. \qquad [3.18]$$

3.4.4 Some applications of the quadratic model

Part of the usefulness of the quadratic model is that it can provide explicit relationships between network structure and behavior, and thus can make sharp predictions in context. Let us examine some specific contexts where it generates further results.

3.4.4.1 Crime

Criminal activity is, to some extent, a group phenomenon, and the crime and delinquency are related to positions in social networks (see, e.g., Sarnecki, 2001; Sutherland, 1947; Warr, 2002). Indeed, delinquents often have friends who have committed offenses, and social ties are a means of influence to commit crimes. In fact, the *structure* of social networks matters in explaining an individual's delinquent behavior: in adolescents' friendship networks, Haynie (2001), Patacchini and Zenou (2008), Calvó-Armengol et al. (2005) find that individual Katz-Bonacich centrality together with the density of friendship links affect the delinquency-peer association. This suggests that the properties of friendship networks should be taken into account to better understand peer influence on delinquent behavior and to craft delinquency-reducing policies.

Glaeser et al. (1996) were among the first to model criminal social interactions. Their model clearly and intuitively shows how criminal interconnections act as a social multiplier on aggregate crime. They consider, however, only a specific network structure where criminals are located on a circle. Following Calvó-Armengol and Zenou (2004) and Ballester et al. (2010), we examine a model that can encompass any social network. Let us reinterpret the local-complementarity model with global congestion described in Section 3.4 in terms of criminal activities. Let a_i be the criminal effort level of delinquent i. Following Becker (1968), assume that delinquents trade off the costs and benefits of delinquent activities. The expected delinquency gains to delinquent i are:

$$u_i(\mathbf{a}, \mathbf{g}) = \underbrace{y_i(\mathbf{a})}_{\text{benefits}} - \underbrace{p_i(\mathbf{a}, \mathbf{g})}_{\text{prob.caught}} \underbrace{f}_{\text{fine}}, \qquad [3.19]$$

where

$$\begin{cases} y_i(\mathbf{a}) = \alpha'_i\, a_i - \frac{1}{2}a_i^2 - \gamma \sum_{j=i}^n a_i a_j \\ p_i(\mathbf{a}, \mathbf{g}) = p_0 a_i \max\left\{1 - \phi' \sum_{j=1}^n g_{ij} a_j, 0\right\} \end{cases}$$

The proceeds $y_i(\mathbf{a})$ include the *global* payoff interdependence. The expected cost of criminal activity, $p_i(\mathbf{a}, \mathbf{g})f$, is positively related to a_i as the apprehension probability increases with one's involvement in delinquency. The crucial assumption is that delinquents' activity has complementarities with their friends' criminal activity, but a criminal also faces global competition as well as increased expected costs as he or she increases activity.

By direct substitution:

$$u_i(\mathbf{a}, \mathbf{g}) = \left(\alpha_i' - p_0 f\right) a_i - \frac{1}{2} a_i^2 - \gamma \sum_{j=i}^{n} a_i a_j + p_0 f \phi' \sum_{j=1}^{n} g_{ij} a_i a_j. \quad [3.20]$$

It should be clear that these utilities, [3.20] and [3.14], are equivalent if $\alpha_i = \alpha_i' - p_0 f > 0$ and $\phi = p_0 f \phi'$. Thus, we can apply Proposition 3.8 (ii): if $p_0 f \phi' \mu_1(\mathbf{g}) + n\gamma\left(\overline{\alpha}/\underline{\alpha} - 1\right) < 1 + \gamma$, then there exists a unique Nash equilibrium:

$$\mathbf{a}^* = \frac{1}{1+\gamma}\left[\mathbf{b}_{\boldsymbol{\alpha}}\left(\mathbf{g}, p_0 f \phi'/(1+\gamma)\right) - \frac{\gamma b_{\boldsymbol{\alpha}}\left(\mathbf{g}, p_0 f \phi'/(1+\gamma)\right)}{1+\gamma+\gamma b_{\mathbf{1}}\left(\mathbf{g}, p_0 f \phi'/(1+\gamma)\right)} \times \mathbf{b}_{\mathbf{1}}\left(\mathbf{g}, p_0 f \phi'/(1+\gamma)\right)\right].$$

An interesting aspect of a network analysis of criminal activity is that it allows us to derive a *key player* policy: If we could choose to push one player's activity to 0 and remove all his/her existing links, which player's removal would lead to the highest overall criminal activity reduction when examining the resulting impact on others' behaviors?[35] Given that delinquent removal has both a direct and an indirect effect on the group outcome, the choice of the key player results from considering both effects. In particular, the key player need not necessarily be the one with the highest criminal activity (the one with the highest Bonacich centrality measure). Formally, the planner's problem is:

$$\max_i \{a^*(\mathbf{g}) - a_{-i}^*(\mathbf{g}_{-i})\},$$

where $a^* = \mathbf{a}^{*\top}\mathbf{1}$. The program above is equivalent to:

$$\min_i \{a_{-i}^*(\mathbf{g}_{-i})\} \quad [3.21]$$

Following Ballester et al. (2006, 2010), define a network centrality measure $d_{\boldsymbol{\alpha}}(\mathbf{g}, \phi)$ that corresponds to this program. Recall that $b_{\boldsymbol{\alpha}}(\mathbf{g}, \phi) = \sum_{i=1}^{n} b_{\boldsymbol{\alpha}}(\mathbf{g}, \phi)_i$.

(i) If $\boldsymbol{\alpha} = \alpha\mathbf{1}$, then the *intercentrality measure* of player i is:

$$d_{\mathbf{1}}(\mathbf{g}, \phi)_i = b_{\mathbf{1}}(\mathbf{g}, \phi) - b_{\mathbf{1}}(\mathbf{g}_{-i}, \phi) = \frac{\left[b_{\mathbf{1}}(\mathbf{g}, \phi)_i\right]^2}{M_{ii}(\mathbf{g}, \phi)}. \quad [3.22]$$

[35] See Dell (2011) for related analyses of spillover effects and reactions of criminals to enforcement with respect to Mexican drug cartels.

(ii) If $\boldsymbol{\alpha} \neq \alpha\mathbf{1}$, then the *intercentrality measure* of player i is:

$$d_{\boldsymbol{\alpha}}(g, \phi)_i = b_{\alpha}(\mathbf{g}, \phi) - b_{\alpha}(\mathbf{g}_{-i}, \phi) = \frac{b_{\boldsymbol{\alpha}}(\mathbf{g}, \phi)_i \sum_{j=1}^{n} M_{ji}(\mathbf{g}, \phi)}{M_{ii}(\mathbf{g}, \phi)}. \qquad [3.23]$$

The intercentrality measure $d_{\boldsymbol{\alpha}}(\mathbf{g}, \phi)_i$ of player i is the sum of i's centrality measures in $\mathbf{g}$, and i's contribution to the centrality measure of every other player $j \neq i$ also in $\mathbf{g}$. The following result establishes that intercentrality captures the two dimensions of the removal of a player from a network: the direct effect on criminal activity and the indirect effect on others' criminal activities.

Proposition 3.9. *A player i^* is the key player who solves [3.21] if and only if i^* has the highest intercentrality in g: $d_{\boldsymbol{\alpha}}(\mathbf{g}, \phi)_{i^*} \geq d_{\boldsymbol{\alpha}}(\mathbf{g}, \phi)_i$, for all $i = 1, \ldots, n$.*

The key player policy is such the planner perturbs the network by removing a delinquent and all other delinquents are allowed to change their effort after the removal but the network is not "rewired," and so is a *short-term policy*.[36]

The individual Nash equilibrium efforts are proportional to the Katz-Bonacich centrality network measures, while the key player is the delinquent with the highest intercentrality measure. This difference is not surprising, as the equilibrium index derives from individual considerations, while the intercentrality measure solves the planner's optimality collective concerns. The measure $d_{\boldsymbol{\alpha}}(\mathbf{g}, \phi)$ goes beyond the measure $\mathbf{b}_{\boldsymbol{\alpha}}(\mathbf{g}, \phi)$ by keeping track of the cross-contributions that arise.

3.4.4.2 Education

The influence of peers on education outcomes has been studied extensively (see, e.g., Evans et al., 1992; Sacerdote, 2001; Zimmerman, 2003; for a survey, see Sacerdote, 2011).[37] Following Calvó-Armengol et al. (2009), let us reinterpret the benchmark model of Section 3.4 in terms of education: a_i is an educational effort. Applying Proposition 3.5, if $\phi\mu_1(\mathbf{g}) < 1$, the equilibrium effort levels are $a_i^* = b_1(\mathbf{g}, \phi)_i$.

In practice, we are not interested in the effort per se, but in the outcome of effort; i.e. the educational achievement obtained by each student. Let the educational achievement of student i be denoted by β_i^* and described by

[36] Liu et al. (2012) develop a dynamic network formation model where, once a delinquent is removed from the network, the remaining delinquents can form new links. They do not have a formula like the intercentrality measure here, but test the model empirically. The invariant and dynamic models often lead to the same key player in the AddHealth data.

[37] For recent models of human capital investment, social mobility and networks, see Calvó-Armengol and Jackson (2009), Jackson (2007), and Bervoets et al. (2012). The papers study the intergenerational relationship between parents' and offsprings' educational outcomes. They show that a positive correlation emerges between their education status, *without any direct interaction*, because of the overlap in the surroundings that influence their education decisions.

$$\beta_i^* = \alpha_i + a_i^*$$

where α_i is again a parameter that depends on the characteristics of student i. Thus,

$$\beta_i^* = \alpha_i + b_1(\mathbf{g}, \phi)_i.$$

Calvó-Armengol et al. (2009) estimate this last equation using the AddHealth data. Estimating such an equation is useful because it allows the authors to disentangle between the effects of own characteristics α_i from a student's location in the network on the educational outcome (i.e., the grades of the student).[38] In terms of magnitude, they find that a standard deviation increase in the Katz-Bonacich centrality of a student increases his or her performance in school by about 7% of one standard deviation.

Again, one can ask what an optimal subsidy would be in this setting. Ballester et al. (2011) determine the optimal per-effort subsidy for each student in order to maximize total output (i.e., the sum of all students' efforts). They take the planner's objective to be to choose $\mathbf{s}$ to maximize

$$a^* = \sum_i a_i^* = \sum_i b_{\boldsymbol{\alpha}+\mathbf{s}}(\mathbf{g}, \phi)_i = \mathbf{1}^\top \mathbf{M}(\boldsymbol{\alpha} + \mathbf{s})$$

subject to a budget constraint:

$$\sum_i s_i a_i^* \leq B \text{ and } s_i \geq 0, \forall i,$$

where $B > 0$ is the budget that the government has for this policy.

Assume that $\alpha_1 < \ldots < \alpha_n$. Ballester et al. (2011) demonstrate the following result.

Proposition 3.10. *Assume* $\phi\mu_1(\mathbf{g}) < 1$ *and* $\alpha_n < \sqrt{\frac{4B+b_{w\boldsymbol{\alpha}}}{b_1}}$. *There exists a unique solution and it is interior and described by*

$$\mathbf{s}^* = \frac{1}{2}\left[\sqrt{\frac{4B + b_{w\alpha}}{b_1}}\mathbf{1} - \boldsymbol{\alpha}\right]. \quad [3.24]$$

The optimal subsidy $\mathbf{s}^*$ for each student depends on the heterogeneity α_i and on overall network characteristics (i.e., b_1 and $b_{w\alpha}$), but not on the individual position in the network. Ballester et al. (2011) show that

[38] To evaluate the effect of Katz-Bonacich centrality, they first estimate $a_i^* = \alpha_i + \phi \sum_{j=1}^n g_{ij} a_j^*$. This leads to an estimated value of ϕ, denoted by $\widehat{\phi}$, for each network of friends in the AddHealth data. Remember that $\widehat{\phi}$ not only captures the intensity of complementarities in the utility function [3.3] but also the decay factor in the Katz-Bonacich centrality, i.e., how much weight to put on distant walks in the network.

$$\frac{\partial s_i^*}{\partial \alpha_i} \gtreqless 0 \Leftrightarrow b_1 \lesseqgtr \frac{\left(b_{\alpha,i}\right)^2}{4\left(4B+b_{w\alpha}\right)},$$

implying that the planner does not always give higher per-effort subsidies to either the more or less advantaged students in terms of background characteristics.

3.4.4.3 Industrial organization

Another application of the model is to collaboration between firms. Collaboration takes a variety of forms which includes creation and sharing of knowledge about markets and technologies (via joint research activities), setting market standards and sharing facilities (such as distribution channels). Recent studies (see for instance Hagedoorn, 2002) suggest that joint ventures seem to have become less popular while non-equity contractual forms of R&D partnerships, such as joint R&D pacts and joint development agreements, have become more prevalent modes of inter-firm collaborations. Little is known, however, about how the structure of these collaboration networks affects the profits of firms. The linear-quadratic model of Section 3.4 can be applied to this question.[39]

Following, König et al. (2014b), we study competition in quantities a la Cournot between n firms with homogeneous products and price, given the following linear inverse market demand for each firm i:

$$p_i = \theta_i - \sum_{j \in N} q_j,$$

where $\theta_i > 0$ and q_j is the quantity produced by firm j. Assume throughout that θ_i is assumed to be large enough for all $i = 1, \ldots, n$ so that price and quantities are always strictly positive. The marginal cost of each firm i is:

$$c_i(\mathbf{g}) = \phi_0 - \phi\left[\sum_{j=1}^{n} g_{ij} q_j\right], \qquad [3.25]$$

where $\phi_0 > 0$ represents the firm's marginal cost when it has no links and $\phi > 0$ influences the cost reduction induced by each link formed by a firm. The marginal cost of each firm i is a decreasing function of the quantities produced by all firms $j \neq i$ that have a direct link with firm i. Assume that ϕ_0 is large enough so that $c_i(g) \geq 0$, $\forall i \in N$, $\forall \mathbf{g}$. The profit function of firm i in a network $\mathbf{g}$ is thus:

[39] There is a growing literature on industrial organization and networks, much of which has focused on network formation. See, e.g., Manshadi and Johari (2009) for Bertrand competition; Goyal and Moraga (2001), Goyal and Joshi (2003), Deroian and Gannon (2006), Goyal et al. (2008), and Westbrock (2010) for Cournot competition and R&D networks; Candogan et al. (2012) and Bloch and Quérou (2013) for monopoly pricing.

$$\pi_i(\mathbf{q}, \mathbf{g}) = pq_i - c_i(\mathbf{g})q_i = \alpha_i q_i - q_i^2 - \sum_{j \neq i} q_i q_j + \phi \sum_{j=1}^{n} g_{ij} q_i q_j \qquad [3.26]$$

where $\alpha_i \equiv \theta_i - \phi_0$. We can apply Proposition 3.8 (ii) to obtain the following result. Let $\overline{\alpha} > \underline{\alpha} > 0$. If $\phi\mu_1(\mathbf{g}) + n\left(\overline{\alpha}/\underline{\alpha} - 1\right)/3 < 1$, then this game has a unique Nash equilibrium in pure strategies $\mathbf{q}^*$, which is interior and given by:

$$\mathbf{q}^* = \mathbf{b_a}\left(\mathbf{g}, \phi\right) - \frac{b_{\mathbf{a}}\left(\mathbf{g}, \phi\right)}{1 + b_{\mathbf{1}}\left(\mathbf{g}, \phi\right)} \mathbf{b_1}\left(\mathbf{g}, \phi\right). \qquad [3.27]$$

This characterizes the equilibrium quantities produced by firms as a function of their position in the network (again, as measured by their Katz-Bonacich centrality).

Proposition 3.7 then provides comparative statics for the total activity in the industry. Overall industry output increases when the network of collaboration links expands, irrespective of the network geometry and the number of additional links.[40]

3.4.4.4 Cities

Helsley and Zenou (2014) use the quadratic model to investigate the interaction between the social space (i.e., the network) and the geographical space (i.e., the city). For that, they consider a city which consists of two locations, a *center*, where all social interactions occur, and a *periphery*. Players are located in either the center or the periphery. The distance between the center and the periphery is normalized to one.

Let $l_i \in \{0, 1\}$ represent the location of player i, defined as his/her distance from the interaction center. Players derive utility from a numeraire good z_i and interactions with others according to the utility function:

$$u_i(\mathbf{a}, \mathbf{g}) = z_i + \alpha\, a_i - \frac{1}{2} a_i^2 + \phi \sum_{j=1}^{n} g_{ij} a_i a_j \qquad [3.28]$$

where a_i (effort) is the number of visits that player i makes to the center. Thus, utility depends on the visit choice of player i, the visit choices of other players and on player i's position in the social network $\mathbf{g}$. Players located in the periphery must travel to the center to interact with others. Letting I represent income and τ represent marginal transport

[40] For the case of a linear inverse demand curve, this generalizes the findings in Goyal and Moraga (2001) and Goyal and Joshi (2003), where monotonicity of industry output is established for the case of regular collaboration networks, where each firm forms the same number of bilateral agreements. For such regular networks, links are added as multiples of n, as all firms' connections are increased simultaneously.

cost, budget balance implies that expenditure on the numeraire is $z_i = I - \tau\, l_i\, a_i$. Using this expression to substitute for z_i in [3.28], we obtain:

$$u_i(\mathbf{a}, \mathbf{g}) = I + \alpha_i a_i - \frac{1}{2}a_i^2 + \phi \sum_{j=1}^{n} g_{ij} a_i a_j, \qquad [3.29]$$

where $\alpha_i = \alpha - \tau\, l_i > 0$. Consider the case where location l_i of each player i is fixed. Then, using Proposition 3.8 (ii) for $\gamma = 0$, if $\phi\mu_1(\mathbf{g}) < 1$, there is a unique (and interior) Nash equilibrium given by: $\mathbf{a}^* = \mathbf{b}_{\boldsymbol{\alpha}}(\mathbf{g}, \phi)$. The Nash equilibrium number of visits a_i^* thus depends on position in the social network and geographic location. This implies that a player who is more central in the social network makes more visits to the interaction center in equilibrium, as he or she has more to gain from interacting with others and so exerts higher interaction effort for any vector of geographic locations.

Helsley and Zenou (2014) extend this model to allow players to choose between locating in the center and the periphery. They assume that because the center has more economic activity, there is an exogenous cost differential $C > 0$ associated with locating at the center. This cost differential might arise from congestion effects or reflect a difference in location rent from competition among other activities for center locations. They show that, in equilibrium, players who are most central in the social network locate at the interaction center, while players who are less central in the social network locate in the periphery. This expresses the salient relationship between position in the social network and geographic location. If interaction involves costly transportation, then players who occupy more central positions in the social network have the most to gain from locating at the interaction center.

Applying the logic of Proposition 3.6, it is again clear that the Nash equilibrium outcome is inefficient and there will be too little social interaction. The first-best outcome can be restored if the planner subsidizes i's activity with the optimal subsidy $\phi \sum_j g_{ij} a_j$. In this case, it is optimal for the planner to give higher subsidies to more central players in the social network.

3.4.4.5 Conformity and conspicuous effects

Let us consider one last application of the quadratic model, with some modifications. In that model, it is the *sum* of friends' activities that impact a player's utility from increasing his or her action [3.3]. This is clearly not always the right model, as it might be some other function of friends' activities that matters. Patacchini and Zenou (2012) and Liu et al. (2014) alter the model so that it is the *average effort level* of friends that affects a player's marginal utility of own action.

Let $\widehat{g}_{ij} = g_{ij}/d_i(\mathbf{g})$, and set $\widehat{g}_{ii} = 0$. By construction, $0 \leq \widehat{g}_{ij} \leq 1$.
Let $\overline{a}_i$ the average effort of individual i's friends: given by:

$$\overline{a}_i(\mathbf{g}) = \frac{1}{d_i(\mathbf{g})} \sum_{j=1}^{n} g_{ij} a_j = \sum_{j=1}^{n} \widehat{g}_{ij} a_j. \quad [3.30]$$

Player i's payoff is described by:

$$u_i(\mathbf{a}, \mathbf{g}) = \alpha\, a_i - \frac{1}{2} a_i^2 - \frac{\delta}{2} (a_i - \overline{a}_i(\mathbf{g}))^2 \quad [3.31]$$

with $\delta \geq 0$. The last term $\delta\,(a_i - \overline{a}_i)^2$ is a standard manner of capturing conformity.[41] Each player wants to minimize the distance between his or her action and the average action of his or her reference group, where δ is a parameter describing the taste for conformity.

First-order conditions imply that:

$$a_i^* = \frac{\alpha}{1+\delta} + \frac{\delta}{(1+\delta)} \sum_{j=1}^{n} \widehat{g}_{ij} a_j^*$$

or in matrix form

$$\mathbf{a}^* = \frac{\alpha}{1+\delta} \left(\mathbf{I} - \frac{\delta}{(1+\delta)} \widehat{\mathbf{g}} \right)^{-1} \mathbf{1} = \frac{\alpha}{1+\delta} \mathbf{b}_\mathbf{1}(\widehat{\mathbf{g}}, \delta/(1+\delta)).$$

Using the AddHealth data, Liu et al. (2014) test which model (local average or local aggregate) does a better job of matching behaviors with regards to effort in education, sport, and screen activities (e.g., video games) for adolescents in the United States. They find that peer effects are not significant for screen activities. For sports activities, they find that students are mostly influenced by the aggregate activity of their friends; while for education they show that both the aggregate performance of friends as well as conformity matter, although the magnitude of the effect is higher for the latter.[42]

[41] See, for instance, Kandel and Lazear (1992), where peer pressure arises when individuals deviate from a well-established group norm, e.g., individuals are penalized for working less than the group norm; Berman (2000), where praying is more satisfying the more average participants there are; and Akerlof (1980, 1997), Bernheim (1994), where deviations from an average action imply a loss of reputation and status. Calvó-Armengol and Jackson (2010) model explicit peer pressure (as a strategy in the game).

[42] Ghiglino and Goyal (2010) propose an alternative model with *conspicuous effects* so that individuals are happier the higher is their effort compared to that of their peers (direct links) and derive negative utility if they effort is below that of their peers. They also compare the local aggregate and the local average model in the context of a pure exchange economy where individuals trade in markets and are influenced by their neighbors. They found that with *aggregate* comparisons, networks matter even if all people have same wealth. With *average* comparisons networks are irrelevant when individuals have the same wealth. The two models are, however, similar if there is heterogeneity in wealth.

Finally, we can provide a more general function for how own action interacts with the average of a player's friends actions. For that, we can adapt the models proposed by Clark and Oswald (1998) to include a network. Consider the following utility function:

$$u_i(\mathbf{a}, \mathbf{g}) = \alpha\, a_i - \frac{1}{2}a_i^2 + \phi\, v(a_i - \overline{a}_i(\mathbf{g})),$$

where $\overline{a}_i$ is defined by [3.30], $v(.)$ is an increasing function of $a_i - \overline{a}_i$ and $v'' < 0$. First-order conditions imply that if there is an interior equilibrium then it must satisfy:

$$a_i = \alpha + \phi\, v'(a_i - \overline{a}_i(\mathbf{g}))$$

and also

$$\frac{\partial a_i}{\partial \overline{a}_i} = \frac{v''(a_i - \overline{a}_i(\mathbf{g}))}{-1 + \phi\, v''(a_i - \overline{a}_i(\mathbf{g}))}.$$

If $v'' < 0$, then a rise in others' efforts leads to an increase in own effort. This is due to the fact that if other individuals set a high $\overline{a}_i(\mathbf{g})$, that reduces $a_i - \overline{a}_i(\mathbf{g})$, and this increases the marginal benefit from effort a_i for those with such a comparison-concave utility. If $v'' > 0$, then a rise in others' efforts leads to a decrease in own effort.

To illustrate this model, consider an example in which:

$$u_i(\mathbf{a}, \mathbf{g}) = \alpha\, a_i - \frac{1}{2}a_i^2 + \phi\, \frac{a_i}{\overline{a}_i(\mathbf{g})}, \qquad [3.32]$$

and where ϕ can be positive or negative. If $\phi > 0$, then there is a conspicuous effect since individuals increase their utility by having higher effort than their peers. If $\phi < 0$, then it becomes costly for individuals to differentiate themselves from their peers. First-order conditions imply that if there is an interior equilibrium then:

$$a_i^* = \alpha + \frac{\phi}{\overline{a}_i^*(\mathbf{g})}.$$

An advantage of [3.32] is that the characterization of the Nash equilibrium is easy.

3.5. NETWORK GAMES WITH INCOMPLETE INFORMATION

While settings with a fixed and known network are widely applicable, there are also many applications where players choose actions without fully knowing with whom they will interact. For example, learning a language, investing in education, investing in a software program, and so forth. These can be better modeled using the machinery of incomplete information games.

It is also important to point out that this class of games also helps in tractability relative to complete information games. The results on games on networks that we have outlined so far primarily rely on either games of strategic complements or in cases with a continuum of actions and quadratic payoff functions. The analysis of other games, and in particular of games of strategic substitutes, even with just two actions, is difficult in the context of complete information, but becomes much more tractable in the context of incomplete information, as we detail below.

3.5.1 Incomplete information and contagion effects

The following discussion builds on the models of Jackson and Yariv (2005, 2007, 2011), Jackson and Rogers (2007), Sundararajan (2007), and Galeotti et al. (2010), primarily incorporating aspects of Jackson and Yariv (2007) and Galeotti et al. (2010).

3.5.1.1 A model of network games with incomplete information

Instead of a known network of interactions, players are now unsure about the network that will be in place in the future, but have some idea of the number of interactions that they will have. To fix ideas, think of choosing whether to adopt a new software program that is only useful in interactions with other players who adopt the software as well, but without being sure of with whom one will interact in the future.

In particular, the set of players N is fixed, but the network $(N, \mathbf{g})$ is unknown when players choose their actions. A player i knows his or her own degree d_i, when choosing an action, but does not yet know the realized network.

Players choose actions in $\{0, 1\}$, and we normalize the payoff to choosing 0 to be 0, and so effectively consider the difference in payoffs between choosing action 0 and 1. Player i has a cost of choosing action 1, denoted c_i. Player i's payoff from action 1 when i has d_i neighbors and expects them each independently to choose 1 with a probability x is

$$v(d_i, x) - c_i,$$

and so action 1 is a best response for player i if and only if $c_i \leq v(d_i, x)$.

It is easy to see how this incorporates some of the games we considered earlier. For instance, in the case of a best-shot public goods game of Example 3.2[43]:

$$v(d_i, x) = (1 - x)^{d_i}.$$

In the case of our coordination game from Section 3.3.3.2, the payoff is

$$v(d_i, x) = \sum_{m=0}^{d_i} B_{d_i}(m, x)\left[mb - (d_i - m)c\right],$$

where $B_{d_i}(m, x)$ is the binomial probability of having exactly m neighbors out of d_i play action 1 when they independently choose action 1 with probability x.

[43] We have normalized the payoff to action 0 to 0, so this is the difference between action 1 and action 0. If no other player chooses action 1, then the difference in overall payoff is $1 - c_i$ and otherwise it is $-c_i$.

The uncertainty about the network affects a player's beliefs about the play of his or her neighbors.

In particular, let us consider a case where a player's probability distribution over the types of neighbors that he or she faces is governed by beliefs about the distribution of other players types. To keep things simple, let us examine a case where c_i is described by an atomless distribution F, independently across players, and independently of a players' degree.

Let a player's beliefs about the degree of each of his or her neighbors be described by a degree distribution $\widetilde{P}(d)$, independently across neighbors.[44] One example, is one in which players are matched uniformly at random, and players have degrees distributed according to some P, in which case the probability of matching to another player of degree d is $P(d)d/E_P[d] = \widetilde{P}(d)$. So for instance, a player is twice as likely to be matched with someone who has two interactions as someone who has one. Galeotti et al. (2010) discuss the more general case where the distribution of neighbors' degrees can depend on own degree.

With this structure, we can then consider *(symmetric) Bayesian equilibria*: players choose an action based on their type d_i, c_i, and so let the strategy be denoted by $\sigma(d_i, c_i) \in \{0, 1\}$. Given the atomless distribution of costs, it is enough to consider pure strategy equilibria, as at most one (probability zero) cost type is ever indifferent for any given d_i.

A simple equation is then sufficient to characterize equilibria. Again, letting x be the probability that a randomly chosen neighbor chooses action 1, a player of type d, c plays action 1 if and only if (up to ties)

$$v(d, x) \geq c.$$

Thus, $F(v(d, x))$ is the probability that a random (best-responding) neighbor of degree d chooses the action 1. A characterization of equilibrium is then that

$$x = \Phi(x) = \sum_d \widetilde{P}(d)F(v(d, x)). \qquad [3.33]$$

In cases where F and v are continuous, existence of equilibrium follows directly from the fact that the right hand side is a continuous function mapping from $[0, 1]$ into $[0, 1]$.

It is easy to keep track of equilibria directly in terms of x since that ties down behaviors of all types of players (up to sets of measure 0).

[44] When fixing a size of a society, only certain configurations of degrees are possible, and so in general there are some interdependencies in the degrees of any player's neighbors. This is thus a limiting case.

3.5.1.2 Monotonicity of equilibria

The nice aspect of the equilibria in the incomplete information setting, is that behavior can now be nicely ordered, depending on the properties of v. In many applications, v is either increasing in x (strict strategic complements) or decreasing in x (strict strategic substitutes). It also tends to be either increasing or decreasing in d, although there are other cases of interest (e.g., as in a game where the player cares precisely about the average payoff from play with neighbors).

The strategic substitute or complement nature of a game imply that strategies of players can be represented (up to indifferences) by threshold functions $\tau(d_i, c_i)$, so that i plays 1 if and only if x exceeds τ in the case of complements or is below it in the case of substitutes.

The monotonicity in degree affects how the thresholds behave as a function of d, as seen in the following proposition, of which variations appear in Jackson and Yariv (2005, 2007) and Galeotti et al. (2010).

Proposition 3.11. *If v is increasing in d for every x, then there exists a symmetric pure strategy Bayesian equilibrium such that equilibrium strategies are increasing in the sense that higher degree players have higher (or at least as high) probabilities of choosing action 1 compared to lower degree players, and correspondingly higher degree players have lower thresholds given the same c_i. Similarly, if v is decreasing in d for every x, then the reverse conclusion holds.*

These observations are useful in understanding dynamics of equilibria and comparative statics of equilibria in response to various changes in the primitives of the environment.

3.5.1.3 A dynamic best reply process

Let us consider a more general version of the contagion/diffusion process that we discussed in Section 3.3.3.2. At time $t = 0$, a fraction x^0 of the population is exogenously and randomly assigned the action 1, and the rest of the population is assigned the action 0. At each time $t > 0$, each player[45] best responds to the distribution of players choosing the action 1 in period $t - 1$ under the distributions described above. We work with expectations, or a mean-field approximation of the system.[46]

[45] In contrast to our earlier discussion, in this process every player adjusts, including those assigned to action 1 at the outset.

[46] A mean-field version of a model is a deterministic approximation of the statistical system where interactions take place at their expected rates. See Vega-Redondo (2007) or Jackson (2008a) for some discussion of these techniques in network analysis.

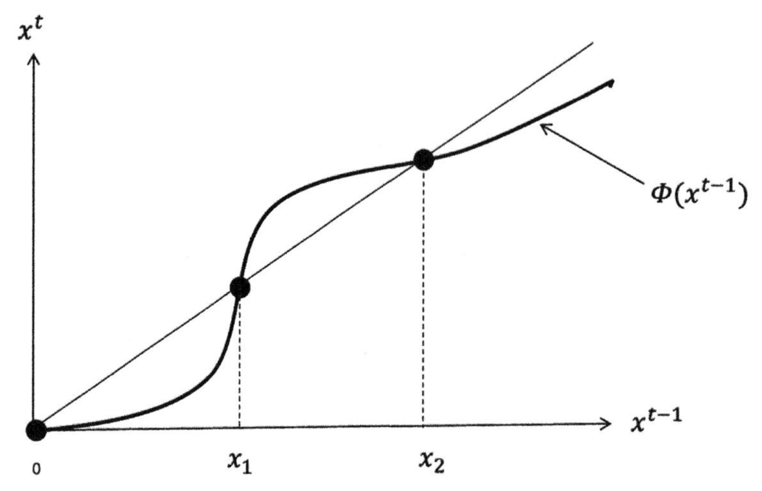

Figure 3.4 Best response curve and equilibria.

In particular, let x^t denote the expected probability that a player's neighbor will choose action 1 at time t as a function of x^{t-1}:

$$x^t = \Phi(x^{t-1}) \equiv \sum_d \widetilde{P}(d)F(v(d, x^{t-1})).$$

Equilibria are again fixed points, but now we view this as a dynamic. In a game of strategic complements, convergence of behavior from any starting point to some equilibrium is monotone, either upwards or downwards, and any player switches behaviors at most once. In a game of strategic substitutes, convergence is not ensured from some starting points in some cases, but there are still things that we can deduce about the equilibria.

We can see that there can exist multiple equilibria, as pictured in Figure 3.4.

Equilibria are points where the curve Φ intersects the 45° line. In games of strategic substitutes, Φ is a decreasing function, and so the equilibrium is unique. In contrast, in games of strategic complements, Φ is increasing and can have multiple equilibria.[47]

Galeotti et al. (2010) and Jackson and Yariv (2005, 2007) provide comparative statics in equilibria, as a function of various changes in networks and payoffs. In particular, it is easy to see that as one shifts a continuous Φ upwards or downwards, equilibria shift in predictable ways. In particular, as Φ is shifted upwards (so $\overline{\Phi}(x) > \Phi(x)$ for all x, the

[47] As pointed out in Jackson and Yariv (2007), some of these equilibria are robust to small perturbations and others are not, depending on the shape of Φ and how it intersects the 45° line. In short, stable equilibria are ones where the curve Φ cuts through the 45° line from the left. In cases where the intersection is a single point, such equilibria are stable in that the dynamics from a slight perturbation return naturally to the original equilibrium. In contrast, if Φ cuts from the right or below the 45° line, then the equilibrium is unstable. For example, in Figure 3.4, the equilibrium corresponding to x_1 is unstable while those corresponding to 0 and x_2 are stable.

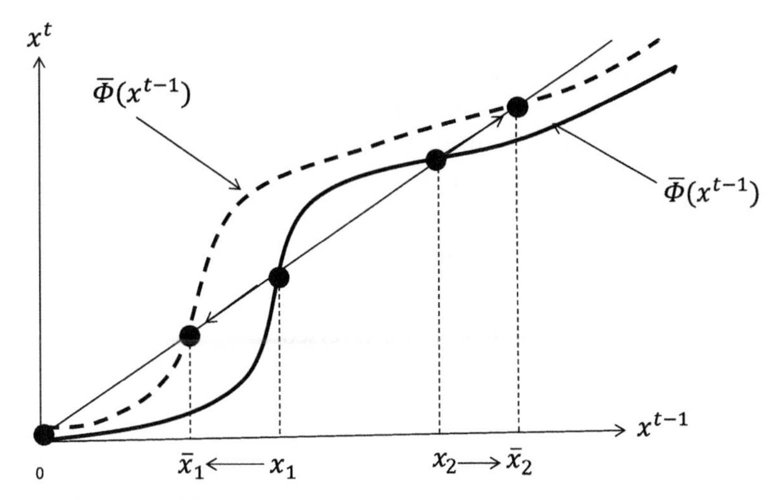

Figure 3.5 The effects of shifting Φ.

greatest equilibrium under $\overline{\Phi}$ moves shifts upward (unless it was already 1), as pictured in Figure 3.5[48]: To understand what leads to such shifts, let us examine Φ as it is defined in [3.33].

First, let us consider changes in the relative costs of the actions; for instance, an increase in the cost of adopting action 1. This can be captured by a strict first-order stochastic dominance (FOSD) shift of the cost distribution from F to $\overline{F}$ (so $\overline{F}(y) \leq F(y)$ for each y with strict inequalities for some y's in the relevant range). It then follows that

$$\overline{\Phi}(x) = \sum_d \widetilde{P}(d)\overline{F}(v(d,x)) \leq \sum_d \widetilde{P}(d)F(v(d,x)) = \Phi(x)$$

for every x. Thus, an increase in costs shifts the curve downwards, and so the greatest equilibrium shifts downward (as does any stable equilibrium for small changes).[49] This has a simple explanation: increasing the cost of choosing action 1 raises the thresholds for adopting that action and lowers the adoption.

One can also deduce how changes in network structure, here captured via the degree distribution, affect equilibrium, based on arguments developed by Jackson and Rogers (2007) and Jackson and Yariv (2005, 2007), and also studied extensively in Galeotti et al. (2010).

For example, consider an increase in the anticipated degree of each random neighbor that a player has in the sense of FOSD. That is, suppose $\widetilde{P}'$ FOSD $\widetilde{P}$, so that

[48] One can deduce more about the shifts in equilibria: the stable equilibria (weakly) decrease from local shifts, and the unstable ones weakly increase from such local shifts of Φ. See Jackson and Yariv (2005, 2007) for details.

[49] One can also deduce similar statements for the least and other equilibria.

$\sum_{d \leq d'} \widetilde{P}'(d) \leq \sum_{d \leq d'} \widetilde{P}(d)$ for each d'. Couple this with $v(d,x)$ being non-decreasing in d, so that individuals who have more interactions are at least as likely to adopt action 1 at the same level of behavior per neighbor. For instance, this holds if it is the number of neighbors adopting action 1 that matters, or even if it is just the fraction that matters. Then, it follows from the definition of FOSD that

$$\Phi'(x) = \sum_d \widetilde{P}'(d)F(v(d,x)) \geq \sum_d \widetilde{P}(d)F(v(d,x)) = \Phi(x),$$

and, so under P', the greatest equilibrium increases. If we reverse things so that $v(d,x)$ is non-increasing in d, then the conclusion is reversed.

Again, this has an intuition related to the best replies. If $v(d,x)$ is increasing in d, then higher degree individuals are more likely to adopt a behavior for any given anticipated level of activity of neighbors. Thus, starting from the maximal equilibrium, as we increase the anticipated degrees of each player's neighbors, we increase the activity generated for any given x. Regardless of how players respond to this (strategic substitutes or complements), this shifts the whole curve upwards and the new equilibrium higher.[50]

This sort of analysis can also be extended to other changes in anticipated network structure, such as mean preserving spreads (MPSs). For example, Jackson and Yariv (2007) show that if $F(v(d,x))$ is non-decreasing and convex in d, then MPSs in the degree distributions lead to nicely ordered Φs. For example, power, Poisson, and regular degree distributions with identical means can be ordered in terms of MPSs and then lead to respective Φ^{power}, Φ^{Poisson}, and Φ^{regular} such that

$$\Phi^{\text{power}}(x) \geq \Phi^{\text{Poisson}}(x) \geq \Phi^{\text{regular}}(x)$$

for all x, which leads to a corresponding ranking of maximal equilibria.[51]

Finally, using this sort of approach, one can study the time series of adoption, since $x^t - x^{t-1} = \Phi(x^{t-1}) - x^{t-1}$ is described by $\Phi(x) - x$. Thus by studying the shape of this curve as a function of x, one can deduce things about classic diffusion patterns (e.g., "S-Shaped" rates of adoption), as shown by Jackson and Yariv (2007) as follows:

Proposition 3.12. *Let $F(v(d,x))$ be twice continuously differentiable and increasing in x for all d. If $F(v(d,x))$ is strictly concave (convex) in x for all d, then there exists $x^* \in [0,1]$ such that $\Phi(x) - x$ is increasing (decreasing) up to x^* and then decreasing (increasing) past x^* (whenever $\Phi(x) \neq 0, 1$).*

[50] In the case of strategic complements, all players end up with (weakly) higher activity, while in the case of strategic substitutes, some players' actions may increase and others decrease, but the overall effect from the shift in ϕ has to be upwards.

[51] The welfare effects of changes in equilibria depend on changes in the overall population utilities, which require a bit more structure to discern. See Galeotti et al. (2010) and Jackson and Yariv (2005, 2007) for discussion.

The idea is that as x increases, the concavity of $F(v(d,x))$ initially leads high feedback effects of changes in x leading to greater changes in $\Phi(x)$, but then it eventually slows down.

3.5.2 Incomplete information about payoffs

The previous analysis was centered on uncertainty about the interaction patterns. Instead, one might also consider situations where the network of interactions is fixed and known, but there is some uncertainty about payoffs.

Consider the payoff function from [3.3]. De Martí Beltran and Zenou (2012) analyze the quadratic game of Section 3.4.1 when the parameter α is common to but unknown by all players. Players know, however, the value of the synergy parameter ϕ.[52] Let α take two possible values: $\alpha_H > \alpha_L > 0$. All individuals share a common prior, which is $\mathbb{P}(\alpha = \alpha_H) = p \in (0,1)$. Each player receives a private signal, $s_i \in \{h, l\}$, such that $\mathbb{P}(s_i = h|\alpha_H) = \mathbb{P}(s_i = l|\alpha_L) = q \geq 1/2$.

There is no communication between players. Player i chooses an action $a_i(s_i) \geq 0$ as a function of the signal $s_i \in \{h, l\}$. Let $\underline{a}_i = a_i(l)$ and $\overline{a}_i = a_i(h)$. The expected utility of player i is thus equal to:

$$E_i\left[u_i(\mathbf{a},\mathbf{g}) \mid s_i\right] = E_i\left[\alpha \mid s_i\right] a_i - \frac{1}{2}a_i^2 + \phi \sum_{j=1}^{n} g_{ij} a_i E_i\left[a_j \mid s_i\right].$$

Let $\widehat{\alpha}_L = E_i[\alpha \mid s_i = l]$ and $\widehat{\alpha}_H = E_i[\alpha \mid s_i = h]$. The best-reply functions are thus:

$$\underline{a}_i^* = \widehat{\alpha}_L + \phi \sum_{j=1}^{n} g_{ij}\left[\mathbb{P}\left(s_j = l|s_i = l\right)\underline{a}_j^* + \mathbb{P}\left(s_j = h|s_i = l\right)\overline{a}_j^*\right]$$

$$\overline{a}_i^* = \widehat{\alpha}_H + \phi \sum_{j=1}^{n} g_{ij}\left[\mathbb{P}\left(s_j = l|s_i = h\right)\underline{a}_j^* + \mathbb{P}\left(s_j = h|s_i = h\right)\overline{a}_j^*\right]$$

De Martí Beltran and Zenou (2012) show that if $\phi\mu_1(\mathbf{g}) < 1$, then there exists a unique Bayesian-Nash equilibrium of this game where equilibrium efforts are given by:

$$\underline{\mathbf{a}}^* = \widehat{\alpha}\,\mathbf{b}_1\left(\mathbf{g},\phi\right) - \left(\tfrac{1-\gamma_H}{2-\gamma_H-\gamma_L}\right)\left(\widehat{\alpha}_H - \widehat{\alpha}_L\right)\mathbf{b}_1\left(\mathbf{g},\left(\gamma_H+\gamma_L-1\right)\phi\right)$$

$$\overline{\mathbf{a}}^* = \widehat{\alpha}\,\mathbf{b}_1\left(\mathbf{g},\phi\right) + \left(\tfrac{1-\gamma_H}{2-\gamma_H-\gamma_L}\right)\left(\widehat{\alpha}_H - \widehat{\alpha}_L\right)\mathbf{b}_1\left(\mathbf{g},\left(\gamma_H+\gamma_L-1\right)\phi\right)$$

[52] We can also analyze a game where ϕ is unknown. The results are similar.

where

$$\widehat{\alpha} = \frac{(1-\gamma_H)\,\widehat{\alpha}_L + (1-\gamma_L)\,\widehat{\alpha}_H}{2-\gamma_H-\gamma_L}.$$

Here, the equilibrium efforts become a combination of two Katz-Bonacich centralities, $\mathbf{b_1}(\mathbf{g},\phi)$ and $\mathbf{b_1}(\mathbf{g},(\gamma_H+\gamma_L-1)\,\phi)$, reflecting the uncertainty between the two values that the parameter could take. Thus, the model can be adapted to incorporate some uncertainty, and that ends up mitigating the effort levels.

3.5.3 Incomplete information with communication in networks

When players desire to coordinate their behaviors, and they face some uncertainty, it is natural for them to communicate. This has interesting implications for who talks to whom in a network setting.

For example, Hagenbach and Koessler (2011) develop an interesting model in which players have to take actions and care about how those actions relate to three things: (i) some private parameter, (ii) a state of nature common to all players, and (iii) other players' actions. The private information that players' have about the state of nature motivates communication. Players would like to convey information to others, but their private biases may mean that revealing information to others could move those players' actions in the wrong directions. The tension is due to the fact that a player wishes to have other players' choose actions similar to his or her action, which involves matching the state of nature; but those players have different desires in terms of how close they wish to be to the state of nature, based on their private parameters.[53]

This is a challenging setting to analyze, as even settings with two players are complex (e.g., Crawford and Sobel, 1982). Nonetheless, Hagenbach and Koessler (2011) find some very interesting features of equilibria. They find that there are multiple equilibria, in terms of which players reveal their information to which others, and that players can only talk to people whose private biases are similar enough to their own. However, communication patterns between any two players depend not only on the private biases of the two players, but also on the relative private biases of the other players with whom each of them communicates, as that affects how information translates into actions. This leads to the multiplicity of equilibria.

There are a variety of other related models, including Calvó-Armengol and de Martí Beltran (2009), Galeotti et al. (2013), Acemoglu et al. (2011), Calvó-Armengol et al. (2014). Also related is a literature on updating and learning originating from DeGroot

[53] This is also related to Chwe (2000), who analyzes a game in which players want to coordinate their binary decisions and guess the action of others based on the local information that players communicate to their neighbors in the network about their preferences.

(1974), with some earlier roots. That literature examines whether behavior converges over time, how quickly it converges, whether a consensus opinion or behavior is reached, and what the limit is (e.g., Bala and Goyal, 1998; DeMarzo et al., 2003; Golub and Jackson, 2010, 2012a,b,c; see Jackson 2008a, 2011 for more background). The DeGroot model can be interpreted as one where players wish to behavior in ways similar to their neighbors, and they myopically best respond.

3.6. CHOOSING BOTH ACTIONS AND LINKS

An important aspect of strategic behavior in network settings that has been looked at but is far from being completely understood is the coevolution of networks and behavior. Much of the literature that we have discussed to this point focused primarily on the influence of a network on behavior. On the other side, there is also a large literature that we have not discussed here that analyzes network formation taking the payoffs from forming links as a given.[54] It is clear, however, that there is feedback: People adjust their behaviors based on that of their friends and they choose their friends based on behaviors. Kandel (1978) provides interesting evidence suggesting that both effects are present in friendship networks, so that over time players adjust actions to match that of their friends and are more likely to maintain and form new friendships with other individuals who act similarly to themselves. Bringing the formation and behavior literatures together is thus important. In this section, we therefore study models where players choose both actions and links. In particular, we would like to see (i) how this changes the conclusions one gets from analyzing the choices separately, and (ii) whether this produces some interesting patterns/dynamics.

3.6.1 Coordination games

Let us return to consider the widely applicable class of coordination games, such as those examined in Sections 3.3.3.2 and 3.3.3.3.

We slightly re-normalize the game structure so that payoffs from any pairwise interaction are given by

	1	0
1	(b, b)	(z, y)
0	(y, z)	(x, x)

[54] For example, see Jackson and Wolinsky (1996), Dutta and Mutuswami (1997), Bala and Goyal (2000), Dutta and Jackson (2000), and the overview in Jackson (2008a). Those models can allow for the equilibrium of a game on a network to generate payoffs as a function of the network, but do not explicitly include such an analysis. For an analysis that takes endogenous network formation into account in peer effects, see Goldsmith-Pinkham and Imbens (2013) and the accompanying papers in the same issue.

where $x > z$ and $b > y$. This allows for a coordination game with arbitrary payoffs.[55]

The threshold fraction $q = \frac{x-z}{x-z+b-y}$ is such that action 1 is a best response for a given player if and only if at least a fraction q of the player's neighbors choose 1. Here, the Pareto efficient equilibrium (payoff dominant equilibrium) depends on whether x or b is larger. The risk dominant equilibrium is governed by the size of q. If $q > 1/2$ (so $x - z > b - y$), then action 0 is risk dominant, and if $q < 1/2$, then action 1 is both risk dominant and payoff dominant (Pareto efficient).

As we saw in Section 3.3.3.3, in an evenly matched society, the stochastically stable equilibrium was the risk dominant profile of actions, as it takes more trembles to move away from the risk dominant profile than to it. We also saw that the network structure of interactions can matter: for instance, in a star network all playing 0 and all playing 1 are both stochastically stable.

An important question emerges as to how this all depends on choice of partners. A basic intuition arises that it is better to choose partners who are playing an action that leads to a higher payoff. However, this is complicated by history: if one has many friends choosing a low payoff action, how easy is it to change partners and get to better play?

Ely (2002)[56] provides an analysis of this by allowing players to choose both actions and neighborhoods (i.e., with whom they want to interact by choosing a location). In other words, when a revision opportunity arises, a player simultaneously chooses a strategy and location in order to maximize his/her expected payoff. In the model of Ely (2002), if some player randomly moves to an unoccupied location and plays the efficient strategy, then other players would like to move to that location and play the efficient strategy rather than staying at a location where they play the inefficient strategy. As a result, Ely shows that the limit distribution (with small trembles) places probability one on the efficient outcome, so that risk-dominance ceases to play a role in determining long-run play.

This result depends on the locational aspect of the interaction patterns, and caring about average play rather than numbers of interactions. In particular, in changing locations, players can sever all old ties, form new ties, and switch technologies simultaneously, and the number of new versus old ties is not important to the players.

Jackson and Watts (2002b) instead propose a model which is not a location one, but rather one where players choose their interaction patterns on an individual-by-individual basis. In other words, they model the interaction pattern as a network where individuals periodically have the discretion to add or sever links to other players.

In each period t, a player i chooses an action $a_i^t \in \{1, 0\}$ and then receives a payoff of

$$u_i(\mathbf{a}^t, \mathbf{g}^t) = \sum_{j \neq i} g_{ij}^t \left[v_i(a_i^t, a_j^t) - k(d_i(\mathbf{g}^t)) \right]$$

[55] In the previous analysis, it was fine to normalize payoffs of one of the actions to 0. Here, since there are also costs of links to consider, absolute payoffs of actions matter (not just relative differences), and so we keep track of the value all possible combinations of actions.

[56] See also Mailath et al. (2001).

where $v_i(a_i^t, a_j^t)$ is a payoff that depends on the actions chosen, the network $\mathbf{g}^t$, and a cost $k(d_i(\mathbf{g}^t))$ of having $d_i(\mathbf{g}^t)$ links.

The full version of the Jackson and Watts (2002b) dynamic process is as follows.

- Period t begins with network $\mathbf{g}^{t-1}$ in place.
- One link ij_t is chosen at random with probability p_{ij}. If the link is not in $\mathbf{g}^{t-1}$ then it is added if both players weakly benefit from adding it and at least one strictly benefits, under the myopic assumption that the rest of the network stays fixed and actions will be $\mathbf{a}^{t-1}$. If the link is in $\mathbf{g}^{t-1}$ then it is deleted if either player strictly benefits from deleting it, under the myopic assumption that the rest of the network stays fixed and actions will be $\mathbf{a}^{t-1}$. With a probability γ, $1 > \gamma > 0$, the decision to add or delete the link is reversed. This results in a new network $\mathbf{g}^t$.
- Next, one player i is randomly chosen with some probability q_i. That player chooses a_i^t to maximize $u_i(a_i^t, \mathbf{a}_{-i}^{t-1}, \mathbf{g}^t)$. With a probability ε, $1 > \varepsilon > 0$, the action of player i is reversed. This results in a new action profile $\mathbf{a}^t$.[57]

The probabilities of trembles are identical and independent across players, strategies, and periods. This is well-defined Markov chain where the state is the vector of actions $\mathbf{a}^t$, that are played in period t. Since this process is aperiodic and irreducible, the Markov chain has a unique stationary distribution, denoted $\mu^{\gamma,\varepsilon}(\mathbf{g}, \mathbf{a})$.

Jackson and Watts (2002b) analyze a variety of different cost structures, but let us just consider one of those and refer the reader to the paper for other details.

Consider a cost of link structure $k(d)$ that is equal to kd for $d \leq D$ and infinite if $d > D$. So, players have a constant marginal cost of links up to some level degree D, and then the cost of maintaining additional costs is arbitrarily large, so that they effectively have a capacity constraint on friendships.

Stochastic stability now involves looking at the limit of the process $\mu^{\gamma,\varepsilon}(\mathbf{g}, \mathbf{a})$ as γ and ε both go to zero (at similar rates, so take $\gamma = f\varepsilon$ for some constant f). Jackson and Watts (2002b) show[58]:

Proposition 3.13. *Let D be even and such that $n > D > 1$. Also, let $1 - 2/D > q > 2/D$ and $q \neq 1/2$.*

(i) *If $x - k > 0$ and $b - k < 0$, then the set of stochastically stable states involve all players playing 0; and, analogously, if $x - k < 0$ and $b - k > 0$, then the set of stochastically stable states involve all players playing 1. There can be a variety of network configurations in stochastically stable states.*

[57] Disconnected players choose a best response to the last period distribution of play of the others.

[58] Jackson and Watts (2002b) provide more analysis of this model, and Goyal and Vega-Redondo (2005) find similar results for a model with unilateral link formation. There is also an analysis by Droste et al. (2000) for the case of geographic link costs.

(ii) *If $y - k > 0$ and $z - k > 0$, then in any stochastically stable state each player has D links and plays the risk dominant action.*

(iii) *If $y - k < 0$ and/or $z - k < 0$, and $x - k > 0$ and $b - k > 0$, then in stochastically stable states involve all players playing 0 as well as all players playing 1. There can be a variety of network configurations in stochastically stable states.*

Although the proof of the proposition is quite involved, some of the intuitions are relatively easy to see. In case (i), there is only one of the actions that can lead to a positive payoff and so links can only exist in the long run between players playing the action leading to a positive payoff, and indeed that turns out to be stochastically stable as if players randomly start playing the action that leads to a positive payoff then they end up linking to each other, and so only a couple of trembles can start growing a network of people playing the action that leads to positive payoffs. In case (ii), both actions lead to positive payoffs (up to the capacity constraint) regardless of what ones neighbors do. Here, standard risk dominance arguments take over as players form the full number of D links, and then trembles in actions play the leading role in determining the outcome. Case (iii) is perhaps the most subtle and interesting one. It addresses the situation where at least one of the actions leads to sufficiently poor payoffs from miscoordination, that it is better to sever a link to a player with whom one is not coordinating when playing that action, than to maintain that link. In this situation, trembles in actions can lead to changes in network structure, which then aids in changes in behavior. For example, suppose that all players are initially playing the risk dominant action. A tremble can lead a player who changes action to be disconnected from the network. With two such trembles, two disconnected players can then form a component playing the other action, which continues to grow as trembles accumulate. This process moves symmetrically between all playing the risk dominant action, and the other case (and each intermediate equilibrium with some mixture of play is destabilized by a single tremble).

There are also questions as to the relative rates at which actions change compared to network structure, and that can affect the overall convergence to equilibrium, as one sees in Ehrhardt et al. (2006), as well as Holme and Newman (2006) and Gross and Blasius (2008). There are also analyses of equilibrium play in games when players are matched in heterogeneous roles (e.g., firms and workers, husbands and wives, etc.) as in Jackson and Watts (2010).

While the details can be complicated, the main message is that endogenizing the interaction structure between players can have a profound impact on the way in which play evolves in a society. Thus, the endogeneity of relationships cannot be ignored when trying to make robust predictions about behaviors in a society, despite the complexity that endogenous relationships introduce.

3.6.2 Network formation in quadratic games

In addition to the models discussed above, there are other models that have combined action and link choices. For example, Bloch and Dutta (2009) proposed a model with both link intensities and communication.

We expose two models, one static (Cabrales et al., 2011) and one dynamic (König et al. 2010, 2014a), which both use the quadratic model from Section 3.4.

Consider a simultaneous move game of network formation (or social interactions) and investment. $T = \{1, \ldots, t\}$ is a finite set of types for the players. We let n be a multiple of t: $n = mt$ for some integer $m \geq 1$, and there is the same number of players of each type. The case where $n = t$ is referred to as *the baseline game* and the case where $n = mt$ as *the m-replica* of this baseline game. Let $\tau(i) \in T$ be player i's type.

Let $c > 0$. Player i's payoff is described by:

$$u_i(\mathbf{a}, \mathbf{s}) = \alpha_{\tau(i)} a_i + \phi \sum_{j=1, j \neq i}^{n} g_{ij}(\mathbf{s}) a_j a_i - \frac{1}{2} c a_i^2 - \frac{1}{2} s_i^2 \qquad [3.34]$$

where $a_i \geq 0$ is the action (productive investment) taken by player i while $s_i \geq 0$ is the socialization effort of player i, with $\mathbf{s} = (s_1, \ldots, s_n)$. The returns to the investment are the sum of a private component and a synergistic component. The private returns are heterogeneous across players and depend on their type.

The network of interactions $g_{ij}(\mathbf{s})$ between players i and j are determined by

$$g_{ij}(\mathbf{s}) = \rho(\mathbf{s}) s_i s_j \qquad [3.35]$$

where

$$\rho(\mathbf{s}) = \begin{cases} 1/\sum_{j=1}^{n} s_j, & \text{if } \mathbf{s} \neq \mathbf{0} \\ 0, & \text{if } \mathbf{s} = \mathbf{0} \end{cases} \qquad [3.36]$$

so that $g_i(\mathbf{s}) = s_i$. That is, players decide upon their total interaction intensity.

This sort of approach to network formation appears in the models of Curarrini et al. (2009, 2010) and has also been used by Golub and Livne (2011), among others. An important aspect is that the synergistic effort $\mathbf{s}$ is *generic* within a community—a scalar decision. Network formation is not the result of an earmarked socialization process. Using [3.35] and [3.36], one can see that the probability of forming a link, $g_{ij}(\mathbf{s})$, is equal to $s_i s_j / \sum_{j=1}^{n} s_j$. This means that the more time two players i and j spend socializing, the more likely they form a link together.

Let

$$\phi(\boldsymbol{\alpha}) = \phi \frac{\sum_{\tau=1}^{t} \alpha_\tau^2}{\sum_{\tau=1}^{t} \alpha_\tau}. \qquad [3.37]$$

Cabrales et al. (2011) demonstrate the following result:

Proposition 3.14. *Suppose that* $2\,(c/3)^{3/2} > \phi(\boldsymbol{\alpha}) > 0$. *Then, there exists an* m^* *such that for all m-replicas with* $m \geq m^*$, *there are two (stable) interior pure strategy Nash equilibria. These pure strategy Nash equilibria are such that for all players i of type* τ, *the strategies* (s_i, a_i) *converge to* $(s^*_{\tau(i)}, a^*_{\tau(i)})$ *as m goes to infinity, where* $s^*_{\tau(i)} = \alpha_{\tau(i)}s$, $a^*_{\tau(i)} = \alpha_{\tau(i)}a$, *and* (s, a) *are positive solutions to*

$$\begin{cases} s = \phi(\boldsymbol{\alpha})a^2 \\ a\left[c - \phi(\boldsymbol{\alpha})s\right] = 1 \end{cases} \qquad [3.38]$$

When $\phi(\boldsymbol{\alpha})$ is small enough compared to the infra-marginal cost for a productive investment, the system of two equations [3.38] with two unknowns has exactly two positive solutions. As m gets large, each such solution gets arbitrarily close to a pure strategy Nash equilibrium of the corresponding m-replica game. The multiplicity reflects complementarities in socialization and in actions.

The approximated equilibria characterized in Proposition 3.14 display important features. First, the level of socialization per unit of productive investment is the same for all players, that is, $s_i^*/a_i^* = s_j^*/a_j^*$, for all i, j. This is equivalent to having a marginal rate of substitution of socialization versus action uniform across all players. Second, differences in actions reflect differences in idiosyncratic traits. More precisely, $a_i^*/a_j^* = \alpha_{\tau(i)}/\alpha_{\tau(j)}$, for all i, j. Third, in the presence of synergies, actions are all scaled up (compared to the case without synergies) by a multiplier. This multiplier, which is homogeneous across all players, is a decreasing function of the cost c, and an increasing function of the second-order average type $\phi(\boldsymbol{\alpha})$.

Figure 3.6 plots equations [3.38].

From this figure, it is clear that the system [3.38] need not always have a positive solution. The upper bound on $\phi(\boldsymbol{\alpha})$ in Proposition 3.14 is a necessary and sufficient condition for the two graphs to cross in the positive orthant of the space (s, a). When $\phi(\boldsymbol{\alpha})$ is too large, the synergistic multiplier operates too intensively and there is no intersection: there is a feedback where increases in socialization lead to increase actions and vice versa.

The equilibrium actions can be ranked component-wise, and the equilibrium payoffs can be Pareto-ranked accordingly. There is a Pareto-superior approximate equilibrium ($(\mathbf{s}^*, \mathbf{a}^*)$ in Figure 3.6) and a Pareto-inferior approximate equilibrium ($(\mathbf{s}^{**}, \mathbf{a}^{**})$ in Figure 3.6) while the socially efficient outcome lies in between the two equilibria.[59] Formally, $(\mathbf{s}^*, \mathbf{a}^*) \geq (\mathbf{s}^O, \mathbf{a}^O) \geq (\mathbf{s}^{**}, \mathbf{a}^{**})$ and $\mathbf{u}(\mathbf{s}^O, \mathbf{a}^O) \geq \mathbf{u}(\mathbf{s}^*, \mathbf{a}^*) \geq \mathbf{u}(\mathbf{s}^{**}, \mathbf{a}^{**})$.

Another model based on the quadratic model where players choose with whom they want to interact is that of König et al. (2014a). At each period of time, players play a two-stage game: in the first stage, players play their equilibrium actions in the quadratic

[59] The superscript O refers to the "social optimum" outcome.

game, while in the second stage a randomly chosen player can update his/her linking strategy by creating a new link as a best response to the current network. Links do not last forever but have an exponentially distributed life time. A critical assumption is that the most valuable links (i.e., the ones with the highest Bonacich centrality) decay at a lower rate than those that are less valuable. As in the literature on evolutionary models described in Section 3.6.1, the authors introduce some noise to this model. Indeed, there is a possibility of error, captured by the stochastic term in the utility function. The authors then analyze the limit of the invariant distribution, the stochastically stable networks, as the noise vanishes.

The network generated by this dynamic process is a nested split graph when the noise tends to zero. The authors also show that the stochastically stable network is a nested split graph. These graphs, known from the applied mathematics literature (see, in particular, Mahadev and Peled, 1995), have a simple structure that make them tractable. In order to define nested split graphs, we first have to define the degree partition of a graph.

Let $(N, \mathbf{g})$ be a graph whose distinct positive degrees are $d_{(1)} < d_{(2)} < \ldots < d_{(k)}$, and let $d_0 = 0$ (even if no player with degree 0 exists in $\mathbf{g}$). Further, define $\mathcal{D}_i = \{j \in N : d_j = d_{(i)}\}$ for $i = 0, \ldots, k$. Then the set-valued vector $\mathcal{D} = (\mathcal{D}_0, \mathcal{D}_1, \ldots, \mathcal{D}_k)$ is the *degree partition* of $\mathbf{g}$.

A network $(N, \mathbf{g})$ is *a nested split graph* if it satisfies the following. Let $\mathcal{D} = (\mathcal{D}_0, \mathcal{D}_1, \ldots, \mathcal{D}_k)$ be its degree partition. Then the nodes N can be partitioned into

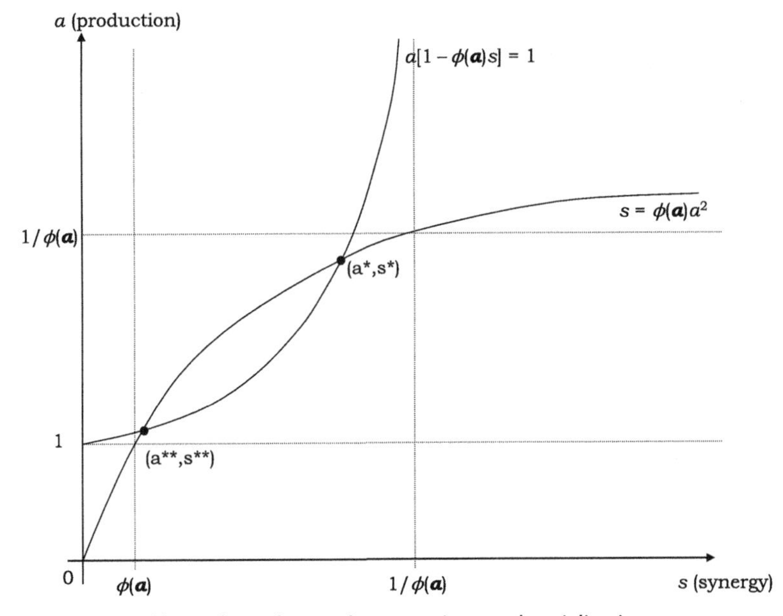

Figure 3.6 Multiple equilibria when players choose actions and socialization.

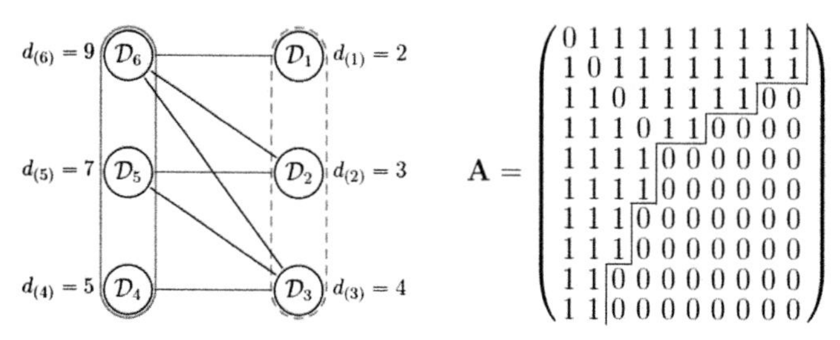

Figure 3.7 Representation of a connected nested split graph (left) and the associated adjacency matrix (right) with $n = 10$ agents and $k = 6$ distinct positive degrees. A line between $\mathcal{D}_i$ and $\mathcal{D}_j$ indicates that every node in $\mathcal{D}_i$ is linked to every node in $\mathcal{D}_j$. The solid frame indicates the dominating set and the nodes in the independent sets are included in the dashed frame. Next to the set $\mathcal{D}_i$ the degree of the nodes in the set is indicated. The neighborhoods are nested such that the degrees are given by $d_{(i+1)} = d_{(i)} + |\mathcal{D}_{k-i+1}|$ for $i \neq \left\lfloor \frac{k}{2} \right\rfloor$ and $d_{(i+1)} = d_{(i)} + |\mathcal{D}_{k-i+1}| - 1$ for $i = \left\lfloor \frac{k}{2} \right\rfloor$. In the corresponding adjacency matrix $\mathbf{A}$ to the right the zero-entries are separated from the one-entries by a stepfunction.

independent sets $\mathcal{D}_i$, $i = 1, \ldots, \left\lfloor \frac{k}{2} \right\rfloor$ and a "dominating" set $\bigcup_{i=\left\lfloor \frac{k}{2} \right\rfloor+1}^{k} \mathcal{D}_i$ in the network $(N \backslash \mathcal{D}_0, \mathbf{g}')$, where g' is the corresponding subnetwork.[60] Moreover, the neighborhoods of the nodes are nested: for each node $j \in \mathcal{D}_i$, $i = 1, \ldots, k$,

$$N_j(\mathbf{g}) = \begin{cases} \bigcup_{j=1}^{i} \mathcal{D}_{k+1-j} & \text{if } i = 1, \ldots, \left\lfloor \frac{k}{2} \right\rfloor, \\ \bigcup_{j=1}^{i} \mathcal{D}_{k+1-j} \setminus \{j\} & \text{if } i = \left\lfloor \frac{k}{2} \right\rfloor + 1, \ldots, k. \end{cases} \quad [3.39]$$

Figure 3.7 illustrates a (path-connected) nested split graph. From the stepwise property of the adjacency matrix it follows that a connected nested split graph contains at least one spanning star, that is, there is at least one player that is connected to all other players.

Let us give some more intuition of the result that the stochastically stable network is a nested split graph. In this model, because of complementarities, players always want to link to others who are more central since this leads to higher actions (as actions are proportional to centrality) and higher actions raise payoffs. Similarly, links decay to those with lower centrality as these players have lower actions and hence lower payoff effects. Notice moreover that, once a player loses a link, he/she becomes less central and this makes it more likely that another link decays. Thus link gains and losses are self reinforcing. This intuition suggests that if the probability of adding links is large then the process should approximate complete network while if it is small then the process should approximate the star network. The key insight of this model is that, for intermediate values of this probability, the network is a nested split graph.

[60] $\lceil x \rceil$ denotes the smallest integer larger or equal than x (the ceiling of x). Similarly, $\lfloor x \rfloor$ denotes the largest integer smaller or equal than x (the floor of x).

König et al. (2014a) also explicitly characterize the degree distribution $P(d)$ in the stochastically stable networks. Instead of relying on a mean-field approximation of the degree distribution and related measures as some dynamic network formation models do, because of the nature of nested split graphs, the authors are able to derive explicit solutions for network statistics of the stochastically stable networks (by computing the adjacency matrix). They find that, as rates at which linking opportunities arrive and links decay, a sharp transition takes place in the network density. This transition entails a crossover from highly centralized networks when the linking opportunities are rare and the link decay is high to highly decentralized networks when many linking opportunities arrive and only few links are removed.

Interestingly, some aspect of nestedness is seen empirically. For example, the organization of the New York garment industry (Uzzi, 1996), the trade relationships between countries (De Benedictis and Tajoli, 2011) and of the Fedwire bank network (Soramaki et al., 2007) show some nested features in the sense that their organization is strongly hierarchical. If we consider, for example, the Fedwire network, it is characterized by a relatively small number of strong flows (many transfers) between banks with the vast majority of linkages being weak to non-existing (few to no interbank payment flows). In other words, most banks have only a few connections while a small number of interlinked-"hub nodes" have thousands.

3.6.3 Games with strategic substitutes

We now turn to discuss network formation in the context of games with strategic substitutes. Let us extend the model of Bramoullé and Kranton (2007a) exposed in Section 3.3.2.3 to introduce link formation. Remember that in a game with no link formation, there were specialized equilibria in which the set of players exerting full effort forms a maximal independent and vice versa.

Galeotti and Goyal (2010) consider a variation in which the utility function of player i is given by:

$$u_i(\mathbf{a}, \mathbf{g}) = v\left(a_i + \phi \sum_{j=1}^{n} g_{ij} a_j\right) - c\, a_i - k\, d_i(\mathbf{g}), \qquad [3.40]$$

where $v(.)$ is a strictly increasing and strictly concave function on $\mathbb{R}_+$ and $c > 0$ is the constant marginal cost of own action such that $v'(0) > c > v'(x)$ for some x, and $k > 0$ is the cost of linking with an other player.

Consider *directed networks*, in which a player can unilaterally form a link with another player without his/her consent and pays the cost of forming this link. As in Section 3.3.2.3, let a^{**} be the effort level of an individual who experiments by him/herself, i.e. $v'(a^{**}) = c$. Assume $k < c\, a^{**}$ so that it is cheaper to link to another player who exerts effort than to exert effort directly. This ensures that some players form links with others. Every (strict) equilibrium of this game has a core-periphery architecture so that the players in the core acquire information personally, while the peripheral players acquire no information personally but form links and get all their information from the core players. Galeotti and Goyal (2010) also show that, under some conditions, the socially optimal outcome is a star network in which the hub chooses some positive level of effort while all other players choose 0.[61]

López-Pintado (2008) presents an interesting dynamic network formation model with strategic substituabilities. She assumes that players make a binary decision, whether or not to exert effort, rather than exerting a continuous effort. In each period, a player, uniformly selected at random from the population, updates his/her strategy and chooses a myopic best response. In other words, taking as given the behavior of others, the player chooses the action that maximizes his/her current benefits. López-Pintado (2008) shows that this dynamic process converges to a Nash equilibrium of the static model. She then studies a mean-field version of the myopic best response dynamics and considers the asymptotic properties of the equilibria in (general) random networks when the network size becomes large. In this model, López-Pintado shows that the dynamics converge to a unique, globally stable fraction of free-riders. She also demonstrates that the higher is the degree of players in a homogeneous network, the higher is the proportion of players free-riding and the proportion of links pointing to a free-rider. Under additional conditions on the degree distribution, she shows that the proportion of links pointing to a free-rider increases under a FOSD shift of the degree distribution and decreases under a MPS of the degree distribution. These results suggest that there tends to be more free-riding in denser or more equal networks.

3.7. REPEATED GAMES AND NETWORK STRUCTURE

Several papers have provided a theoretical analysis of a repeated games in network settings. This includes Raub and Weesie (1990), Vega-Redondo (2006), Ali and Miller (2009, 2012), Lippert and Spagnolo (2011), Mihm et al. (2009), Jackson et al. (2012), Fainmesser (2012), and Dall'Asta et al. (2012).[62]

There are several ideas that emerge from this literature. One relates to network structure and information travel. The basic idea is that to enforce individual cooperation in prisoner's dilemma sorts of games, other players have to be able to react to a

[61] The equilibrium and socially optimal networks are similar to those obtained by König et al. (2014a) in the context of a dynamic network formation model with strategic complementarities (see Section 3.6.2) since nested-split graphs have clearly a core-periphery structure.

[62] There is also a related literature in evolutionary game theory where this some structure to interactions, such as Nowak and May (1992) and Eshel et al. (1998).

given player's deviations. Raub and Weesie (1990) and Ali and Miller (2009) show how completely connected networks shorten the travel time of contagion of bad behavior which can quicken punishment for deviations. Players only observe their own interactions, and so punishments travel through the network only through contagious behavior, and the challenge to enforcing individual cooperation is how long it takes for someone's bad behavior to come to reach their neighbors through contagion.[63]

A second idea relates to heterogeneity. Haag and Lagunoff (2006) show how heterogeneity can also favor cliques. They allow for preference differences that can preclude cooperative behavior within a repeated game, and so partitioning society into homogeneous subgroups can enable cooperative behavior that might not be feasible otherwise.

A third idea relates to robustness of networks, which favor networks that look like "quilts" of small cliques pasted together. Jackson et al. (2012) examine self-enforcing favor exchange, where players can periodically do favors for their friends but at a cost. In order to ensure that a player performs a favor when called upon, the player fears losing several friendships and thus many future favors. There are many such networks that can enforce favor exchange, by threatening to sever links to players who fail to perform favors when they are asked to. However, as Jackson et al. (2012) show, only very special networks are "robust." Robustness requires two things: a form of renegotiation so that the punishments are really credible, and immunity to large contagions—the only people who end up losing links are friends of the player failing to do a favor. These networks consist of (many) small cliques of players, where if a player fails to perform a favor then a particular clique disintegrates, but that does not lead to further contagion. Other networks experience further contagions where society loses more of its links.

The many settings of repeated interactions between friends makes understanding repeated games on networks essential. This literature is still at its beginning and many open questions remain.

3.8. CONCLUDING REMARKS AND FURTHER AREAS OF RESEARCH

The settings where social network structure has profound implications for human behavior are quite numerous. Thus, it is not surprising that the literature that relates to this subject is enormous. We have focused mainly on the central theoretical underpinnings of games played on networks.

[63] See Lippert and Spagnolo (2011) for more discussion on optimal strategies with word-of-mouth communication. Contagion strategies are costly, and can deter truthful communication as players may not wish to have the contagion occur. Other strategies that only result in punishment of the initial deviator can improve in some environments. More generally, the issue of incentives for communication is a complex one in such settings, as also discussed in Ali and Miller (2012).

Networks are inherently complex, and so much of the progress that has been made required some focus on specific game structures. There are three main ways in which progress has been made. One involves looking at games of strategic complements and strategic substitutes, where the interaction in payoffs between players satisfies some natural and useful monotonicity properties. That monotonicity provides a basis for understanding how patterns of behaviors relate to network structure. A second approach relies on looking at a simple parametric specification of a game in terms of a quadratic form that permits an explicit solution for equilibrium behavior as a function of a network. A third approach considers settings where the specific pattern of interactions is uncertain, in which case equilibrium can be expressed nicely as a function of the number of interactions that players expect to have.

These models make a number of predictions about behavior, relating levels of actions to network density, relating players' behaviors to their position in the network, and relating behavior to things like the degree distribution and cost of taking given actions. Thus, we end up both with predictions about how a specific player's behavior relates to his/her position in a network, as well as what overall behavior patterns to expect as a function of the network structure. While much progress has been made, the innumerable applications and importance of the subject cry out for additional study.

Let us close with a brief discussion of various areas of research that are closely related, but we have not covered due to space constraints.

3.8.1 Bargaining and exchange on networks

An important area in terms of applications to economics concerns bargaining and exchange on networks. Many business relationships are not centralized, but take place through networks of interactions. A standard modeling technique in the literature that has modeled this is to have only pairs of players connected in the network can trade. The key questions addressed by this literature are: How does the network structure influence market outcomes? How does a player's position in the network determine his/her bargaining power and the local prices he/she faces? Who trades with whom and on what terms? Are trades efficient? If players can choose their links, do the efficient networks form? How does this depend on the bargaining protocol and the costs of links?

The literature on this subject includes the early experimental investigations of Cook and Emerson (1978), and what followed in the literature on "exchange theory."[64] That literature documents how positions in a network influence terms of trade, testing theories of power and brokerage.

[64] Early writings on exchanges in networks include Homans (1958, 1961), Blau (1964), and eventually included explicit consideration of social network structure as in Emerson (1962, 1976).

The theoretical analyses of this subject later emerged using models of trade based on tools including auction theory, noncooperative analyses of bargaining games, and use of the core. Much of that literature assumes that buyers have unit demands and sellers have unit supply, and that these are pre-identified. For example, Kranton and Minehart (2001) considered a setting where prices are determined by an ascending-bid auction mechanism. They showed that the unique equilibrium in weakly undominated strategies leads to an efficient allocation of the goods. They found that the network formation was also efficient in a setting where buyers pay the cost of connection. Jackson (2003) shows that this result does not generalize, but is special to buyers bearing the entire connection cost, and that cost being born before buyers know their valuations for the good. He shows that both under-connection (in cases where some players see insufficient shares of the gains from trade) and over-connection (as players try to improve their bargaining position) are possible sources of inefficiency.

The literature that followed explored a variety of different exchange mechanisms and settings in terms of valuations. Calvó-Armengol (2003) explores the power of a position in a network, in a context of pairwise sequential bargaining with neighbor players. Polanski (2007) assumes that a maximum number of pairs of linked players are selected to bargain every period. Corominas-Bosch (2004) considered an alternating offer bargaining game. One of the most general analyses is by Elliott (2011) who allows for general valuations among pairs of players and then characterizes the core allocations. Using that as a basis, he then documents the inefficiencies that arise in network formation, showing how under-connection and over-connection depend on how costs of links are allocated. Elliott also documents the size of the potential inefficiencies, and shows that small changes in one part of a network can have cascading effects. Manea (2011) and Abreu and Manea (2012) provide a non-cooperative models of decentralized bargaining in networks when players might not be ex ante designated as buyers or sellers, but where any pair may transact. Condorelli and Galeotti (2012) provide a look at a setting where there is incomplete information about potential valuations and possibilities of resale.

Related models include those studying oligopoly games that are played by networked firms as in Nava (2009), who analyzes when it is that competitive outcomes are reached, and Lever (2011) who examines network formation. Blume et al. (2009) examine the role of middlemen in determining profits and efficient trade, Fainmesser (2011) examines a repeated game with reputations concerning past transactions, and Campbell (2013) examines the role of selling products in the presence of network word-of-mouth effects.

There are also studies, such as Kakade et al. (2004, 2005), that examine exchange on random networks. One finding is that if the network is sufficiently symmetric so that buyers have similar numbers of connections, as well as sellers, then there are low levels of price dispersion, while sufficient asymmetry allows for more price dispersion.

The literature to date provides some foundations for understanding how networks influence terms of trade. Still largely missing are empirical tests of some of the theory with field data. Some experimental evidence (e.g., Charness et al., 2005) validates some of the bargaining predictions in lab settings.

3.8.2 Risk-sharing networks

Another important application where networks play a prominent role in shaping behavior is in terms of risk-sharing. In much of the developing world, people face severe income fluctuations due to weather shocks, diseases affecting crops and livestock, and other factors. These fluctuations are costly because households are poor and lack access to formal insurance markets. Informal risk-sharing arrangements, which help cope with this risk through transfers and gifts, are therefore widespread.[65,66]

Again, this is an area where simple game theoretic models can add substantial insight. Some studies to date include Bramoullé and Kranton (2007b), Bloch et al. (2008), Karlan et al. (2009), Belhaj and Deroïan (2012), Jackson et al. (2012), and Ambrus et al. (2014). This is an area where rich repeated game studies should help deepen our understanding of the relationship between networks of interactions and the (in)-efficiency of risk-sharing.

3.8.3 Dynamic games and network structure

Most of our discussion has focused on static games or else simple variations on dynamics with perturbations. Beyond those, and the repeated games mentioned above, another important area where networks can have profound implications is in terms of the dynamics of patterns that emerge. Just as an example, Calvó-Armengol and Jackson (2004) study the dynamics of labor markets: the evolution over time of employment statuses of n workers connected by a network of relationships where individuals exchange job information only between direct neighbors.[67] Accounting for the network of interactions can help explain things like the duration of unemployment of each worker. A worker whose friends are unemployed has more difficulty in hearing about job openings and this leads to increased spells of unemployment, as well as decreased incentives to invest in human capital. The complementarities

[65] Rosenzweig (1988) and Udry (1994) document that the majority of transfers take place between neighbors and relatives (see also Townsend, 1994). Other empirical work confirms this finding with more detailed social network data (Dercon and De Weerdt, 2006; Fafchamps and Gubert, 2007; Fafchamps and Lund, 2003).

[66] Although this is clearly vital in the developing world, substantial sharing of risks is also prevalent in the developed world.

[67] See also the model of Calvó-Armengol et al. (2007) who study the importance of weak ties in crime and employment.

in that setting, coupled with the network structure, helps provide an understanding of various empirical observations of employment. Calvó-Armengol and Jackson (2004) also show that education subsidies and other labor market regulation policies display local increasing returns due to the network structure. Subsidies and programs are more tightly clustered with network structure in mind can then be more effective.

That is just an example of an area where the coupling of game theoretic analysis and network structure can have important implications for dynamics and for policy. This is an important and still under-studied area.

3.8.4 More empirical applications based on theory

Part of the reason that accounting for networks in studying behavior is so essential, is that understanding the relationship can shape optimal policy. For example, in Section 3.4.4.1, we discussed the idea of key player in crime, i.e. the criminal who once removed generates the highest reduction in crime. Liu et al. (2012) examined this idea using data on adolescent youths in the US (AddHealth) and show that, indeed, a key player policy reduces crime more than a random-targeted policy. In other words, targeting nodes or players in a network can have snow-ball effects because of the interactions between players in the network. This is the idea of the social multiplier developed in Section 3.4.1.

Other examples of areas where models of games on networks help inform policy and learn from field work include financial networks, diffusion of innovations, and political interactions. For example, financial markets can be considered as a network where links are transactions of dependencies between firms or other organizations (Leitner, 2005; Cohen-Cole etal., 2011; Elliott et al., 2014). Network analyses in financial settings can enhance our understanding of the interactions and optimal regulation and policy. It is also clear that networks influence adoption of technologies. There is indeed empirical evidence of social learning (e.g., Conley and Udry, 2010). Theory (e.g., Section 3.5.1.3) tells us that the adoption of a new technology is related to network structure. In a recent paper, Banerjee et al. (2013) study the adoption of a microfinance program in villages in rural India. They find that participation to the program is higher in villages when the first set of people to be informed are more important in a network sense in that they have higher diffusion centrality (a new centrality measure). As a last example, there is evidence that personal connections amongst politicians have a significant impact on the voting behavior of U.S. politicians (Cohen and Malloy, 2010). Here there is a need for both theory and additional empirical work, that helps us understand how the network of interactions between politicians shapes legislation and a host of other governmental outcomes.

As the theory of games on networks continues to develop, the interaction with field work will continue to become richer and more valuable.

3.8.5 Lab and field experiments

While our knowledge of peer effects is growing, the complexities involved mean that there is still much to be learned. Given the enormous difficulty of identifying social effects in the field, essential tools in this area of research are laboratory and field experiments, where one can control and directly measure how players' behaviors relate to network structure.

Experiments have been used to study strategic network formation (e.g., Callander and Plott, 2005; Charness and Jackson, 2007; Falk and Kosfeld, 2003; Goeree et al., 2009; Pantz and Zeigelmeyer, 2003), learning in network settings, (e.g., Celen et al., 2010; Chandrasekhar et al., 2011; Choi et al., 2012; Möbius et al., 2010) as well as games played on networks (see Jackson and Yariv, 2011; Kosfeld, 2004, for additional background).[68]

For instance, Goeree et al. (2010) and Leider et al. (2009) find that play in games is related to social distance in a network, with players being more cooperative or generous to those who are direct friends or close in a social distance sense compared to those who are more distant in the network. Kearns et al. (2009) find that a society's ability to reach a consensus action in a game of complementarity depends on network structure.

Moreover, experiments can be very useful in directly testing some of the theoretical predictions given above. For example, Charness et al. (2014) test games of networks by looking at two important factors: (i) whether actions are either strategic substitutes or strategic complements, and (ii) whether subjects have complete or incomplete information about the structure of a random network. They find that subjects conform to the theoretical predictions of the Galeotti et al. (2010) model that we exposed in Section 3.5.1. They also find some interesting selections of equilibria that suggest that additional theory would be useful.

ACKNOWLEDGMENTS

We gratefully acknowledge financial support from the NSF under grants SES-0961481 and SES-1155302, grant FA9550-12-1-0411 from the AFOSR and DARPA, and ARO MURI award No. W911NF-12-1-0509. We thank Yann Bramoullé, Montasser Ghachem, Michael König, Rachel Kranton, Paolo Pin, Brian Rogers, Yasuhiro Sato, Giancarlo Spagnola, Yiqing Xing, and a reviewer for the Handbook for comments on earlier drafts.

[68] There are also various field experiments that effectively involve games on networks, such as Centola (2010).

REFERENCES

Abreu, D., Manea, M., 2012. Bargaining and efficiency in networks. J. Econ. Theory 147, 43-70.

Acemoglu, D., Dahleh, M.A., Lobel, I., Ozdaglar, A., 2011. Bayesian learning in social networks. Rev. Econ. Stud. 78, 1201–1236.

Acemoglu, D., Jackson, M.O., 2011. History, expectations, and leadership in the evolution of cooperation. NBER Working Paper No. 17066.

Akerlof, G.A., 1980. A theory of social custom of which unemployment may be one consequence. Q. J. Econ. 94, 749–775.

Akerlof, G.A., 1997. Social distance and social decisions. Econometrica 65, 1005–1027.

Ali, S.N., Miller, D.A., 2009. Enforcing cooperation in networked societies. Unpublished manuscript, University of California at San Diego.

Ali, S.N., Miller, D.A., 2012. Ostracism. Unpublished manuscript, University of California at San Diego.

Ambrus, A., Möbius, M., Szeidl, A., 2014. Consumption risk-sharing in social networks. Am. Econ. Rev. 104, 149–182.

Bala, V., Goyal, S., 1998. Learning from neighbors. Rev. Econ. Stud. 65, 595–621.

Bala, V., Goyal, S., 2000. A non-cooperative model of network formation. Econometrica 68, 1181–1231.

Ballester, C., Calvó-Armengol, A., Zenou, Y., 2006. Who's who in networks: wanted the key player. Econometrica 74:5, 1403–1417.

Ballester, C., Calvó-Armengol, A., Zenou, Y., 2010. Delinquent networks. J. Eur. Econ. Assoc. 8, 34–61.

Ballester, C., Calvó-Armengol, A., Zenou, Y., 2014. Education policies when networks matter. Berkeley Electronic Journal of Economic Analysis and Policy, forthcoming.

Banerjee, A., Chandrasekhar, A.G., Duflo, E., Jackson, M.O., 2013. The diffusion of microfinance. Science 341(6144). DOI: 10.1126/science.1236498.

Becker, G., 1968. Crime and punishment: an economic approach. J. Polit. Econ. 76, 169–217.

Belhaj, M., Deroïan, F., 2014. Competing activities in social networks. Berkeley Electronic Journal of Economic Analysis and Policy, forthcoming.

Belhaj, M., Deroïan, F., 2012. Risk taking under heterogenous revenue sharing. J. Dev. Econ. 98, 192–202.

Belhaj, M., Deroïan, F., 2013. Strategic interaction and aggregate incentives. J. Math. Econ. 49, 183–188.

Bergin, J., Lipman, B.L., 1996. Evolution with state-dependent mutations. Econometrica 64, 943–956.

Berman, E., 2000. Sect, subsidy, and sacrifice: an economist's view of ultra-orthodox Jews. Q. J. Econ. 115, 905–953.

Bernheim, B.D., 1994. A theory of conformity. J. Polit. Econ. 102, 841–877.

Bervoets, S., Calvó-Armengol, A., Zenou, Y., 2012. The role of social networks and peer effects in education transmission. CEPR Discussion Paper No. 8932.

Blau, P., 1964. Exchange and Power. John Wiley and Sons, New York.

Bloch, F., Dutta, B., 2009. Communication networks with endogenous link strength. Games Econ. Behav. 66, 39–56.

Bloch, F., Quérou, N., 2013. Pricing in social networks. Games Econ. Behav. 80, 243–261.

Bloch, F., Genicot, G., Ray, D., 2008. Informal insurance in social networks. J. Econ. Theory 143, 36–58.

Blume, L.E., 1993. The statistical mechanics of strategic interaction. Games Econ. Behav. 5, 387–424.

Blume, L.E., Easley, D., Kleinberg, J., Tardos, É., 2009. Trading networks with price-setting agents. Games Econ. Behav. 67, 36–50.

Bonacich, P., 1987. Power and centrality: a family of measures. Am. J. Sociol. 92, 1170–1182.

Boncinelli, L., Pin, P., 2012. Stochastic stability in the best shot network game. Games Econ. Behav. 75, 538–554.

Bramoullé, Y., 2007. Anti-coordination and social interactions. Games Econ. Behav. 58, 30–49.

Bramoullé, Y., Kranton, R.E., 2007a. Public goods in networks. J. Econ. Theory 135, 478–494.

Bramoullé, Y., Kranton, R.E., 2007b. Risk-sharing networks. J. Econ. Behav. Organ. 64, 275–294.

Bramoullé, Y., Currarini, S., Jackson, M.O., Pin, P., Rogers, B., 2012. Homophily and long-run integration in social networks. J. Econ. Theory 147, 1754–1786.

Bramoullé, Y., Kranton, R., D'Amours, M., 2014. Strategic interaction and networks. Am. Econ. Rev. 104, 898–930.

Cabrales, A., Calvó-Armengol, A., Zenou, Y., 2011. Social interactions and spillovers. Games Econ. Behav. 72, 339–360.
Callander, S., Plott, C.R., 2005. Principles of network development and evolution: an experimental study. J. Public Econ. 89, 1469–1495.
Calvó-Armengol, A., 2003. A decentralized market with trading links. Math. Soc. Sci. 45, 83–103.
Calvó-Armengol, A., Zenou, Y., 2004. Social networks and crime decisions: the role of social structure in facilitating delinquent behavior. Int. Econ. Rev. 45, 935–954.
Calvó-Armengol, A., Jackson, M.O., 2004. The effects of social networks on employment and inequality. Am. Econ. Rev. 94, 426–454.
Calvó-Armengol, A., Jackson, M.O., 2007. Networks in labor markets: wage and employment dynamics and inequality. J. Econ. Theory 132, 27–46.
Calvó-Armengol, A., de Martí Beltran, J., 2009. Information gathering in organizations: equilibrium, welfare and optimal network structure. J. Eur. Econ. Assoc. 7, 116–161.
Calvó-Armengol, A., Jackson, M.O., 2009. Like father, like son: labor market networks and social mobility. Am. Econ. J. Microecon. 1, 124–150.
Calvó-Armengol, A., Jackson, M.O., 2010. Peer pressure. J. Eur. Econ. Assoc. 8, 62–89.
Calvó-Armengol, A., Patacchini, E., Zenou, Y., 2005. Peer effects and social networks in education and crime. CEPR Discussion Paper No. 5244.
Calvó-Armengol, A., Verdier, T., Zenou, Y., 2007. Strong ties and weak ties in employment and crime. J. Public Econ. 91, 203–233.
Calvó-Armengol, A., Patacchini, E., Zenou, Y., 2009. Peer effects and social networks in education. Rev. Econ. Stud. 76, 1239–1267.
Calvó-Armengol, A., de Martí Beltran, J., Prat, A., 2014. Communication and influence. Theoretical Economics, forthcoming.
Campbell, A., 2013. Word-of-mouth communication and percolation in social networks. Am. Econ. Rev. 103, 2466–2498.
Candogan, O., Bimpikis, K., Ozdaglar, A., 2012. Optimal pricing in networks with externalities. Oper. Res. 60, 883–905.
Celen, B., Kariv, S., Schotter, A., 2010. An experimental test of advice and social learning. Management Science 56, 1678–1701.
Centola, D., 2010. The spread of behavior in an online social network experiment. Science, 329, 1194–1197.
Chandrasekhar, A.G., Larreguy, H., Xandri, J.P., 2011. Testing models of social learning on networks: evidence from a lab experiment in the field. Consortium on Financial Systems and Poverty Working Paper.
Charness, G., Jackson, M.O., 2007. Group play in games and the role of consent in network formation. J. Econ. Theory 136, 417–445.
Charness, G., Corominas-Bosch, M., Frechette, G.R., 2005. Bargaining on networks: an experiment. J. Econ. Theory 136, 28–65.
Charness, G., Feri, F., Meléndez-Jiménez, M.A., Sutter, M., 2014. Experimental Games on Networks: Underpinnings of Behavior and Equilibrium Selection. Econometrica, forthcoming.
Choi, S., Gale, D., Kariv, S., 2012. Social learning in networks: a quantal response equilibrium analysis of experimental data. Rev. Econ. Des. 16, 93–118.
Chwe, M., 1999. Structure and strategy in collective action. Am. J. Sociol. 105, 128–156.
Chwe, M., 2000. Communication and coordination in social networks. Rev. Econ. Stud. 67, 1–16.
Clark, A.E., Oswald, A.J., 1998. Comparison-concave utility and following behaviour in social and economic settings. J. Public Econ. 70, 133–155.
Clifford, P., Sudbury, A., 1973. A model for spatial conflict. Biometrika 60, 581–588.
Cohen-Cole, E., Patacchini, E., Zenou, Y., 2011. Systemic risk and network formation in the interbank market. CEPR Discussion Paper No. 8332.
Cohen, L., Malloy, C., 2010. Friends in high places. NBER Working Paper No. 16437.
Conley, T.J., Udry, C.R., 2010. Learning about a new technology: pineapple in Ghana. Am. Econ. Rev. 100, 35–69.
Condorelli, D., Galeotti, A., 2012. Bilateral trading in networks. Unpublished manuscript, University of Essex.

Cook, K.S., Emerson, R.M., 1978. Power, equity and commitment in exchange networks. Am. Sociol. Rev. 43, 721–739.
Corominas-Bosch, M., 2004. On two-sided network markets. J. Econ. Theory 115, 35–77.
Crawford, V.P., Sobel, J., 1982. Strategic information transmission. Econometrica 50, 1431–1451.
Currarini, S., Jackson, M.O., Pin, P., 2009. An economic model of friendship: homophily, minorities, and segregation. Econometrica 77, 1003–1045.
Currarini, S., Jackson, M.O., Pin, P., 2010. Identifying the roles of race-based choice and chance in high school friendship network formation. Proc. Nat. Acad. Sci. U.S.A. 107(11), 4857–4861.
Dall'Asta, L., Pin, P., Ramezanpour, A., 2009. Statistical mechanics of maximal independent sets. Phys. Rev. E, 80, 061136.
Dall'Asta, L., Pin, P., Ramezanpour, A., 2011. Optimal equilibria of the best shot game. J. Public Econ. Theory 13, 885–901.
Dall'Asta, L., Pin, P., Marsili, M., 2012. Collaboration in social networks. Proc. Nat. Acad. Sci. 109, 4395–4400.
Daskalakis, C., Goldberg, P.W., Papadimitriou, C.H., 2009. The complexity of computing a Nash equilibrium. SIAM J. Comput. 39, 195–259.
Debreu, G., Herstein, I.N., 1953. Nonnegative square matrices. Econometrica 21, 597–607.
De Benedictis, L., Tajoli, L., 2011. The world trade network. World Econ. 34, 1417–1454.
DeGroot, M.H., 1974. Reaching a consensus. J. Am. Stat. Assoc. 69, 118–121.
Dell, M., 2011. Trafficking networks and the Mexican drug war. Unpublished manuscript, MIT.
De Martí Beltran, J., Zenou, Y., 2011. Social networks. In: Jarvie, I., Zamora-Bonilla, J. (Eds.), Handbook of Philosophy of Social Science. SAGE Publications, London, pp. 339–361.
De Martí Beltran, J., Zenou, Y., 2012. Network games with incomplete information. Unpublished manuscript, Stockholm University.
DeMarzo, P., Vayanos, D., Zwiebel, J., 2003. Persuasion bias, social influence, and unidimensional opinions. Q. J. Econ. 1183, 909–968.
Dercon, S., De Weerdt, J., 2006. Risk sharing networks and insurance against illness. J. Dev. Econ. 81, 337–356.
Deroian, F., Gannon, F., 2006. Quality-improving alliances in differentiated oligopoly. Int. J. Ind. Organ. 24, 629–637.
Droste, E., Gilles, R.P., Johnson, C., 2000. Evolution of conventions in endogenous social networks. Unpublished manuscript, Queen's University Belfast.
Dutta, B., Mutuswami, S., 1997. Stable networks. J. Econ. Theory 76, 322–344.
Dutta, B., Jackson, M.O., 2000. The stability and efficiency of directed communication networks. Rev. Econ. Des. 5, 251–272.
Elliott, M., 2011. Inefficiencies in networked markets. Unpublished manuscript, California Institute of Technology.
Elliott, M., Golub, B., Jackson, M.O., 2014. Financial networks and contagions. American Economic Review, inpress.
Ellison, G., 1993. Learning, local interaction, and coordination. Econometrica 61, 1047–1071.
Ely, J.C., 2002. Local conventions. Adv. Theor. Econ. 2, Article 1.
Ehrhardt, G., Marsili, M., Vega-Redondo, F., 2006. Diffusion and growth in an evolving network. Int. J. Game Theory 34, 383–397.
Emerson, R.M., 1962. Power-dependence relations. Am. Soc. Rev. 27, 31–41.
Emerson, R.M., 1976. Social exchange theory. Annu. Rev. Sociol. 2, 335–362.
Eshel, I., Samuelson, L., Shaked, A., 1998. Altruists, egoists, and hooligans in a local interaction model. Am. Econ. Rev. 88, 157–179.
Evans, W.N., Oates, W.E., Schwab, R.M., 1992. Measuring peer group effects: a study of teenage behavior. J. Polit. Econ. 100, 966–991.
Fafchamps, M., Lund, S., 2003. Risk-sharing networks in rural Philippines. J. Dev. Econ. 71, 261–287.
Fafchamps, M., Gubert, F., 2007. The formation of risk sharing networks. J. Dev. Econ. 83, 326–350.
Fainmesser, I., 2011. Bilateral and Community Enforcement in a Networked Market with Simple Strategies. *SSRN research paper*, Brown University.
Fainmesser, I., 2012. Community structure and market outcomes: a repeated games in networks approach. Am. Econ. J. Microecon. 4, 32–69.

Falk, A., Kosfeld, M., 2003. It's all about connections: evidence on network formation. IZA Discussion Paper No. 777.
Galeotti, A., Goyal, S., 2010. The law of the few. Am. Econ. Rev. 100, 1468–1492.
Galeotti, A., Goyal, S., Jackson, M.O., Vega-Redondo, F., Yariv, L., 2010. Network games. Rev. Econ. Stud. 77, 218–244.
Galeotti, A., Ghiglino, C., Squintani, F., 2013. Strategic information transmission in networks. J. Econ. Theory 148, 1751–1769.
Ghiglino, C., Goyal, S., 2010. Keeping up with the neighbors: social interaction in a market economy. J. Eur. Econ. Assoc. 8, 90–119.
Glaeser, E.L., Sacerdote, B.I., Scheinkman, J.A., 1996. Crime and social interactions. Q. J. Econ. 111, 508–548.
Glaeser, E.L., Sacerdote, B.I., Scheinkman, J.A., 2003. The social multiplier. J. Eur. Econ. Assoc. 1, 345–353.
Goeree, J.K., McConnell, M.A., Mitchell, T., Tromp, T., Yariv, L., 2010. The 1/d law of giving, American Economic Journal: Microeconomics 2, 183–203.
Goeree, J.K., Riedl, A., Ule, A., 2009. In search of stars: network formation among heterogeneous agents. Games Econ. Behav. 67, 445–466.
Goldsmith-Pinkham, P., Imbens, G., 2013. Social networks and the identification of peer effects. J. Busin. Econ. Stat. 31(3), 253–264.
Golub, B., Jackson, M.O., 2010. Naive learning and influence in social networks: convergence and wise crowds. Am. Econ. J. Microecon. 2, 112–149.
Golub, B., Livne, Y., 2011. Strategic random networks and tipping points in network formation. Unpublished manuscript, MIT.
Golub, B., Jackson, M.O., 2012a. How homophily affects the speed of learning and best response dynamics. Q. J. Econ. 127, 1287–1338.
Golub, B., Jackson, M.O., 2012b. Does homophily predict consensus times? Testing a model of network structure via a dynamic process. Rev. Net. Econ. 11, 1–28.
Golub, B., Jackson, M.O., 2012c. Network structure and the speed of learning: measuring homophily based on its consequences. Ann. Econ. Stat. 107/108.
Goyal, S., 2007. Connections: An Introduction to the Economics of Networks. Princeton University Press.
Goyal, S., Moraga, J-L., 2001. R&D networks. Rand J. Econ. 32, 686–707.
Goyal, S., Joshi, S., 2003. Networks of collaboration in oligopoly. Games Econ. Behav. 43, 57–85.
Goyal, S., Vega-Redondo, F., 2005. Network formation and social coordination. Games Econ. Behav. 50, 178–207.
Goyal, S., Konovalov, Moraga-Gonzalez, J., 2008. Hybrid R&D. J. Eur. Econ. Assoc. 6, 1309–1338.
Granovetter, M.S., 1978. Threshold models of collective behavior. Am. J. Sociol. 83, 1420–1443.
Gross, T., Blasius, B., 2008. Adaptive coevolutionary networks: a review. J. R. Soc. Interface 5, 259–271.
Haag, M., Lagunoff, R., 2006. Social norms, local interaction, and neighborhood planning. Int. Econ. Rev. 47, 265–296.
Hagedoorn, J., 2002. Inter-firm R&D partnerships: an overview of major trends and patterns since 1960. Res. Policy 31, 477–492.
Hagenbach, J., Koessler, F., 2011. Strategic communication networks. Rev. Econ. Stud. 77, 1072–1099.
Harsanyi, J.C., Selten, R., 1988. A General Theory of Equilibrium Selection in Games. MIT Press, Cambridge, MA.
Haynie, D.L., 2001. Delinquent peers revisited: does network structure matter? Am. J. Sociol. 106, 1013–1057.
Helsley, R., Zenou, Y., 2014. Social networks and interactions in cities. J. Econ. Theory, 150, 426–466.
Hirshleifer, J., 1983. From weakest-link to best-shot: the voluntary provision of public goods. Public Choice 41, 371–386.
Holley, R.A., Liggett, T.M., 1975. Ergodic theorems for weakly interacting infinite systems and the voter model. Ann. Probab. 3, 643–663.
Holme, P., Newman, M.E.J., 2006. Nonequilibrium phase transition in the coevolution of networks and opinions. Phys. Rev. E 74, 056108.

Homans, G.C., 1958. Social behavior as exchange. Am. J. Sociol. 62, 596–606.
Homans, G.C., 1961. Social Behavior: Its Elementary Forms. Harcourt Brace and World, New York.
Jackson, M.O., 2003. The stability and efficiency of economic and social networks. In: Koray, S., Sertel, M. (Eds.), *Advances in Economic Design*. Springer-Verlag, Heidelberg.
Jackson, M.O., 2004. A survey of models of network formation: stability and efficiency. In: Demange, G., Wooders, M. (Eds.), *Group Formation in Economics. Networks, Clubs and Coalitions*. Cambridge University Press, Cambridge, UK, pp. 11–57.
Jackson, M.O., 2005a. Allocation rules for network games. Games Econ. Behav. 51, 128–154.
Jackson, M.O., 2005b. The economics of social networks. In: Blundell, R., Newey, W., Persson, T. (Eds.), *Proceedings of the 9th World Congress of the Econometric Society*. Cambridge University Press, Cambridge, UK.
Jackson, M.O., 2007. Social structure, segregation, and economic behavior. Nancy Schwartz Memorial Lecture, given in April 2007 at Northwestern University, printed version: http://www.stanford.edu/~jacksonm/schwartzlecture.pdf.
Jackson, M.O., 2008a. Social and Economic Networks. Princeton University Press, Princeton, NJ.
Jackson, M.O., 2008b. Average distance, diameter, and clustering in social networks with homophily. In: Papadimitriou, C., Zhang, S. (Eds.) *Proceedings of the Workshop in Internet and Network Economics (WINE)*, Lecture Notes in Computer Science. Springer Verlag, Berlin.
Jackson, M.O., 2011. An overview of social networks and economic applications. In: Benhabib, J., Bisin, A., Jackson, M.O. (Eds.), *Handbook of Social Economics Volume 1A*. Elsevier Science, Amsterdam, pp. 511–579.
Jackson, M.O., 2014. Networks and the identification of economic behaviors. SSRN WP 2404632.
Jackson, M.O., Wolinsky, A., 1996. A strategic model of social and economic networks. J. Econ. Theory 71, 44–74.
Jackson, M.O., Watts, A., 2002a. The evolution of social and economic networks. J. Econ. Theory 106, 265–295.
Jackson, M.O., Watts, A., 2002b. On the formation of interaction networks in social coordination games. Games Econ. Behav. 41, 265–291.
Jackson, M.O., Yariv, L., 2005. Diffusion on social networks. Économie Publique 16, 3–16.
Jackson, M.O., Rogers, B.W., 2007. Relating network structure to diffusion properties through stochastic dominance. B.E. J. Theor. Econ. 7, 1–13.
Jackson, M.O., Yariv, L., 2007. The diffusion of behavior and equilibrium structure properties on social networks. Am. Econ. Rev. Pap. Proc. 97, 92–98.
Jackson, M.O., Watts, A., 2010. Social games: matching and the play of finitely repeated games. Games Econ. Behav. 70, 170–191.
Jackson, M.O., Yariv, L., 2011. Diffusion, strategic interaction, and social structure. In: Benhabib, J., Bisin, A., Jackson, M.O. (Eds.), *Handbook of Social Economics Volume 1A*. Elsevier Science, Amsterdam, pp. 645–678.
Jackson, M.O., Lopez-Pintado, D., 2013. Diffusion in networks with heterogeneous agents and homophily. Netw. Sci. 1, 49–67.
Jackson, M.O., Rodriguez-Barraquer, T., Tan, X., 2012. Social capital and social quilts: network patterns of favor exchange. Am. Econ. Rev. 102, 1857–1897.
Jackson, M.O., Rogers, B.W., Zenou, Y., 2014. The impact of social networks on economic behavior. SSRN 2467812.
Kakade, S.M., Kearns, M., Ortiz, L.E., 2004. Graphical economics. Proc. Annu. Conf. Learn. Theory (COLT) 23, 17–32.
Kakade, S.M., Kearns, M., Ortiz, L.E., Pemantle, R., Suri, S., 2005. Economic properties of social networks. Adv. Neural Infor. Proc. Syst. (NIPS) 17.
Kandel, D.B., 1978. Homophily, selection, and socialization in adolescent friendships. Am. J. Sociol. 14, 427–436.
Kandel, E., Lazear, E.P., 1992. Peer pressure and partnerships. J. Polit. Econ. 100, 801–817.
Kandori, M., Mailath, G., Rob, R., 1993. Learning, mutation, and long-run equilibria in games. Econometrica 61, 29–56.

Karlan, D., Mobius, M., Rosenblat, T., Szeidl, A., 2009. Trust and social collateral. Q. J. Econ. 124:3, 1307–1361.

Katz, L., 1953. A new status index derived from sociometric analysis. Psychometrica 18, 39–43.

Kearns, M.J., Littman, M., Singh, S., 2001. Graphical models for game theory. In: Breese, J.S., Koller, D. (Eds.), *Proceedings of the 17th Conference on Uncertainty in Artificial Intelligence*. Morgan Kaufmann, SanFrancisco, pp. 253–260.

Kearns, M.J., Judd, S., Tan, J., Wortman, J., 2009. Behavioral experiments on biased voting in networks. Proc. Nat. Acad. Sci. U.S.A. 106, 1347–1352.

König, M.D., 2013. Dynamic R&D networks. Working Paper Series/Department of Economics No. 109, University of Zurich.

König, M.D., Tessone, C., Zenou, Y., 2010. From assortative to dissortative networks: the role of capacity constraints. Adv. Complex Syst. 13, 483–499.

König, M.D., Tessone, C., Zenou, Y., 2014a. Nestedness in networks: a theoretical model and some applications. Theor. Econ., forthcoming.

König, M.D., Liu, X., Zenou, Y., 2014b. R&D Networks: Theory, Empirics and Policy Implications. Unpublished manuscript, Stockholm University.

Kosfeld, M., 2004. Economic networks in the laboratory: a survey. Rev. Netw. Econ. 30, 20–42.

Kranton, R., Minehart, D., 2001. A theory of buyer-seller networks. Am. Econ. Rev. 91, 485–508.

Leider, S., Möbius, M., Rosenblat, T., Do, Q.-A., 2009. Directed altruism and enforced reciprocity in social networks. Q. J. Econ. 124, 1815–1851.

Leitner, Y., 2005. Financial networks: contagion, commitment, and private sector bailouts. J. Finance 6, 2925–2953.

Lever, C., 2011. Price competition on a network. Banco de México Working Paper 2011-04.

Lippert, S., Spagnolo, G., 2011. Networks of relations and word-of-mouth communication. Games Econ. Behav. 72, 202–217.

Liu, X., Patacchini, E., Zenou, Y., Lee, L-F., 2012. Criminal networks: who is the key player? CEPR Discussion Paper No. 8772.

Liu, X., Patacchini, E., Zenou, Y., 2014. Endogenous peer effects: Local aggregate or local average? Journal of Economic Behavior and Organization 103, 39–59.

López-Pintado, D., 2008. The spread of free-riding behavior in a social network. Eastern Econ. J. 34, 464–479.

Mahadev, N., Peled, U., 1995. Threshold Graphs and Related Topics. Amsterdam, North Holland.

Mailath, G.J., Samuelson, L., Shaked, A., 2001. Endogenous interactions. In: Nicita, A., Pagano, U. (Eds.), *The Evolution of Economic Diversity*. New York, Routledge, pp. 300–324.

Manea, M., 2011. Bargaining on networks. Am. Econ. Rev. 101, 2042–2080.

Manshadi, V., Johari, R., 2009. Supermodular network games. Annual Allerton Conference on Communication, Control, and Computing.

McPherson, M., Smith-Lovin, L., Cook, J.M., 2001. Birds of a feather: homophily in social networks. Annu. Rev. Sociol. 27, 415–444.

Mihm, M., Toth, R., Lang, C., 2009. What goes around comes around: a theory of strategic indirect reciprocity in networks. CAE Working Paper No. 09-07.

Milgrom, P., Roberts, J., 1990. Rationalizability, learning and equilibrium in games with strategic complementarities. Econometrica 58, 1255–1278.

Möbius, M., Phan, T., Szeidl, A., 2010. Treasure hunt: social learning in the field. Unpublished manuscript, Duke University.

Monderer, D., Shapley, L.S., 1996. Potential games. Games Econ. Behav. 14, 124–143.

Morris, S., 2000. Contagion. Rev. Econ. Stud. 67, 57–78.

Myerson, R.B., 1977. Graphs and cooperation in games. Math. Oper. Res. 2, 225–229.

Nava, F., 2009. Quantity competition in networked markets outflow and inflow competition. STICERD Research Paper No. TE542, London School of Economics.

Nowak, M., May, R., 1992. Evolutionary games and spatial chaos. Nature 359, 826–829.

Pantz, K., Ziegelmeyer, A., 2003. An experimental study of network formation. Unpublished manuscript, Max Planck Institute.

Papadimitriou, C., Roughgarden, T., 2008. Computing correlated equilibria in multi-player games. J. ACM 55, 14:1–14:29.
Patacchini, E., Zenou, Y., 2008. The strength of weak ties in crime. Eur. Econ. Rev. 52, 209–236.
Patacchini, E., Zenou, Y., 2012. Juvenile delinquency and conformism. J. Law Econ. Organ. 28, 1–31.
Polanski, A., 2007. Bilateral bargaining in networks. J. Econ. Theory 134, 557–565.
Raub, W., Weesie, J., 1990. Reputation and efficiency in social interactions: an example of network effects. Am. J. Sociol. 96, 626–654.
Rosenzweig, M., 1988. Risk, implicit contracts and the family in rural areas of low-income countries. Econ. J. 98, 1148–1170.
Sacerdote, B., 2001. Peer effects with random assignment: results from Dartmouth roomates. Q. J. Econ. 116, 681–704.
Sacerdote, B., 2011. Peer effects in education: how might they work, how big are they and how much do we know thus far? In: Hanushek, E.A., Machin, S., Woessmann, L. (Eds.), *Handbook of Economics of Education*, vol. 3. Elevier Science, Amsterdam, pp. 249–277.
Schelling, T.C., 1978. Micromotives and Macrobehavior. Norton, New York.
Slikker, M., van den Nouweland, A., 2001. Social and Economic Networks in Cooperative Game Theory. Kluwer Academic Publisher, Norwell, MA.
Sarnecki, J., 2001. Delinquent Networks: Youth Co-Offending in Stockholm. Cambridge University Press, Cambridge.
Soramaki, K., Bech, M., Arnold, J., Glass, R., Beyeler, W., 2007. The topology of interbank payment flows. Phys. A: Stat. Mech. Appl. 379, 317–333.
Sundararajan, A., 2007. Local network effects and complex network structure. B.E. J. Theor. Econ. 7.
Sutherland, E.H., 1947. Principles of Criminology, fourth ed. J.B. Lippincott, Chicago.
Tesfatsion, L., 2006. Agent-based computational economics: a constructive approach to economic theory. In: Tesfatsion, L., Judd, K.L. (Eds.), *Handbook of Computational Economics*, vol. 2. Elsevier Science, Amsterdam, pp. 831–880.
Topkis, D., 1979. Equilibrium points in nonzero-sum *n* person submodular games. SIAM J. Control Optim. 17, 773–787.
Townsend, R., 1994. Risk and insurance in village India. Econometrica 62, 539–591.
Udry, C., 1994. Risk and insurance in a rural credit market: an empirical investigation in Northern Nigeria. Rev. Econ. Stud. 61, 495–526.
Uzzi, B., 1996. The sources and consequences of embeddedness for the economic performance of organizations: the network effect. Am. Sociol. Rev. 61, 674–698.
Vega-Redondo, F., 2006. Building up social capital in a changing world. J. Econ. Dyn. Control 30, 2305–2338.
Vega-Redondo, F., 2007. Complex Social Networks. Cambridge University Press, Cambridge.
Vriend, N.J., 2006. ACE models of endogenous interactions. In: Tesfatsion, L., Judd, K.L. (Eds.), *Handbook of Computational Economics*, vol. 2. Elsevier, Amsterdam, pp. 1047–1079.
Wallace, C., Young, H.P., 2014. Stochastic Evolutionary Game Dynamics. In: Young, P., Zamir, S. (Eds.), *Handbook of Game Theory*, vol. 4. Elsevier Science.
Warr, M., 2002. Companions in Crime: The Social Aspects of Criminal Conduct. Cambridge University Press, Cambridge.
Westbrock, B., 2010. Natural concentration in industrial research collaboration. RAND J. Econ. 41, 351–371.
Wilhite, A., 2006. Economic activity on fixed networks. In: Tesfatsion, L., Judd, K.L. (Eds.), Handbook of Computational Economics, vol. 2. Elsevier, Amsterdam, pp. 1013–1045.
Young, H.P., 1993. The evolution of conventions. Econometrica 61, 57–84.
Young, H.P., 1998. Individual Strategy and Social Structure. Princeton University Press, Princeton.
Zhou, L., 1994. The set of stable equilibria of a supermodular game is a complete lattice. Games Econ. Behav. 7, 295–3000.
Zimmerman, D., 2003. Peer effects in academic outcomes: evidence from a natural experiment. Rev. Econ. Statis. 85, 9–23.

CHAPTER 4

Reputations in Repeated Games

George J. Mailath*, **Larry Samuelson**†

*Department of Economics, University of Pennsylvania, Philadelphia, PA 19104, USA
†Department of Economics, Yale University, New Haven, CT 06520, USA

Contents

Handbook of game theory
http://dx.doi.org/10.1016/B978-0-12-420056-2.00004-5

Abstract

This paper surveys work on reputations in repeated games of incomplete information. We first develop the adverse-selection approach to reputations in the context of a long-lived player, who may be a "normal" type or one of a number of "commitment" types, and who faces a succession of short-lived players. We use entropy-based arguments both to establish a lower bound on the equilibrium payoff of the long-lived player (demonstrating ex ante reputation effects) and to show that this lower bound is asymptotically irrelevant under imperfect monitoring (demonstrating the impermanence of reputation effects). The chapter continues by examining the (necessarily weaker) reputation results that can be established for the case of two long-lived players, and by examining variations in the model under which reputation effects can persist indefinitely. The chapter closes with brief remarks on alternative approaches.

Keywords: Repeated games, Entropy, Games of incomplete information, Commitment, Stackelberg types

JEL Codes: C73, D82

4.1. INTRODUCTION

4.1.1 Reputations

The word "reputation" appears throughout discussions of everyday interactions. Firms are said to have reputations for providing good service, professionals for working hard, people for being honest, newspapers for being unbiased, governments for being free from corruption, and so on. Reputations establish links between past behavior and expectations of future behavior—one expects good service because good service has been provided in the past, or expects fair treatment because one has been treated fairly in the past. These reputation effects are so familiar as to be taken for granted. One is instinctively skeptical of a watch offered for sale by a stranger on a subway platform, but more confident of a special deal on a watch from an established jeweler. Firms proudly advertise that they are fixtures in their communities, while few customers would be attracted by a slogan of "here today, gone tomorrow."

Repeated games allow for a clean description of both the myopic incentives that agents have to behave opportunistically and, via appropriate specifications of future behavior (and so rewards and punishments), the incentives that deter opportunistic behavior. As a consequence, strategic interactions within long-run relationships have often been studied using repeated games. For the same reason, the study of reputations has been particularly fruitful in the context of repeated games, the topic of this chapter. We do not provide a comprehensive guide to the literature, since a complete list of the relevant repeated-games papers, at the hurried rate of one paragraph per paper, would leave us no room to discuss the substantive issues. Instead, we identify the key points of entry into the literature, confident that those who are interested will easily find their way past these.[1]

4.1.2 The interpretive approach to reputations

There are two approaches to reputations in the repeated-games literature. In the first, an equilibrium of the repeated game is selected whose actions along the equilibrium path are not Nash equilibria of the stage game. Incentives to choose these actions are created by attaching less favorable continuation paths to deviations. For perhaps the most familiar example, there is an equilibrium of the repeated prisoners' dilemma (if the players are sufficiently patient) in which the players cooperate in every period, with any deviation from such behavior prompting relentless mutual defection.

The players who choose the equilibrium actions in such a case are often interpreted as maintaining a reputation for doing so, with a punishment-triggering deviation interpreted as the loss of one's reputation. For example, players in the repeated prisoners' dilemma are interpreted as maintaining a reputation for being cooperative, while the first instance of defection destroys that reputation.

In this approach, the link between past behavior and expectations of future behavior is an equilibrium phenomenon, holding in some equilibria but not in others. The notion of reputation is used to interpret an equilibrium strategy profile, but otherwise involves no modification of the basic repeated game and adds nothing to the formal analysis.

4.1.3 The adverse selection approach to reputations

The adverse selection approach to reputations considers games of incomplete information. The motivation typically stems from a game of complete information in which the players are "normal," and the game of incomplete information is viewed as a perturbation of the complete information game. In keeping with this motivation, attention is typically focused on games of "nearly" complete information, in the sense that a player whose type is unknown is very likely (but not quite certain) to be a normal type. For example, a player in a repeated game might be almost certain to have stage-game payoffs given by the prisoners' dilemma, but may with some small possibility have no other option than

[1] It should come as no surprise that we recommend Mailath and Samuelson (2006) for further reading on most topics in this chapter.

to play tit-for-tat.[2] Again, consistent with the perturbation motivation, it is desirable that the set of alternative types be not unduly constrained.

The idea that a player has an incentive to build, maintain, or milk his reputation is captured by the incentive that player has to manipulate the beliefs of other players about his type. The updating of these beliefs establishes links between past behavior and expectations of future behavior. We say "reputations effects" arise if these links give rise to restrictions on equilibrium payoffs or behavior that do not arise in the underlying game of complete information.

We concentrate throughout on the adverse selection approach to reputations. The basic results identify circumstances in which reputation effects necessarily arise, imposing bounds on equilibrium payoffs that are in many cases quite striking.

4.2. REPUTATIONS WITH SHORT-LIVED PLAYERS

4.2.1 An example

We begin with the example of the "product-choice" game shown in Figure 4.1. Think of the long-lived player 1 ("he") as a firm choosing to provide either high (H) or low (L) effort. Player 2 ("she") represents a succession of customers, with a new customer in each period, choosing between a customized (c) or standardized (s) product. The payoffs reveal that high effort is costly for player 1, since L is a strictly dominant strategy in the stage game. Player 1 would like player 2 to choose the customized product c, but 2 is willing to do so only if she is sufficiently confident that 1 will choose H.

The stage game has a unique Nash equilibrium in which the firm provides low effort and the customer buys the standardized product. In the discrete-time infinite horizon game, the firm maximizes the average discounted sum of his payoffs. In the infinite horizon game of complete information, every payoff in the interval $[1, 2]$ is a subgame perfect equilibrium payoff if the firm is sufficiently patient.

Equilibria in which the firm's payoff is close to 2 have an intuitive feel to them. In these equilibria, the firm frequently exerts high effort H, so that the customer will play her best response of purchasing the customized product c. Indeed, the firm should be able to develop a "reputation" for playing H by persistently doing so. This may be initially costly for the firm, because customers may not be immediately convinced that

	c	s
H	2, 3	0, 2
L	3, 0	1, 1

Figure 4.1 The product-choice game.

[2] This was the case in one of the seminal reputation papers, Kreps et al. (1982).

the firm will play H and hence customers may play s for some time, but the subsequent payoff could make this investment worthwhile for a sufficiently patient firm.

Nothing in the structure of the repeated game captures this intuition. Repeated games have a recursive structure: the continuation game following any history is identical to the original game. No matter how many times the firm has previously played H, the standard theory provides no reason for customers to believe that the firm is more likely to play H now than at the beginning of the game.

Now suppose that in addition to the normal type of firm, i.e., the type whose payoffs are given by Figure 4.1, there is a commitment, or behavioral, type $\xi(H)$. This type invariably and necessarily plays H (and hence the description "commitment" or "behavioral" type). We also refer to this type as the Stackelberg type, since H is the action the firm would take in the unique subgame perfect equilibrium of a sequential-move game of perfect information in which the firm chooses publicly before the customer. Even if the prior probability of this behavioral type is very small, it has a dramatic effect on the set of equilibrium payoffs when the firm is sufficiently patient. It is an implication of Proposition 4.1 that there is no Nash equilibrium of the incomplete information game with a payoff to the normal firm near 1.

The flavor of the analysis is conveyed by asking what must be true in an equilibrium in which the normal type of player 1 receives a payoff close to 1, when he has discount factor close to 1. For this to occur, there must be many periods in which the customer chooses the standardized product, which in turn requires that in those periods the customer expects L with high probability. But if player 1 plays H in every period, then the customer is repeatedly surprised. And, since the H-commitment type has positive prior probability, this is impossible: each time the customer is surprised the posterior probability on the H-commitment type jumps a discrete amount (leading eventually to a posterior above the level at which customers choose the customized product).

Reputation effects do much more than eliminate unintuitive outcomes; they also rescue intuitive outcomes as equilibrium outcomes. For example, the *finitely* repeated complete information product-choice game has a unique subgame perfect equilibrium, and in this equilibrium, the static Nash profile Ls is played in every period. Nonetheless, our intuition again suggests that if the game has a sufficiently long horizon, the firm should here as well be able to develop a "reputation" for playing H by persistently doing so. It is an immediate implication of the logic underlying Proposition 4.1 that the firm can indeed develop such a reputation in the game with incomplete information.[3]

Similarly, if customers can only imperfectly monitor the firm's effort choice, the upper bound on the firm's complete information equilibrium payoff will be less than 2.

[3] Kreps et al. (1982) first made this point in the context of the finitely repeated prisoners' dilemma with the commitment type playing tit-for-tat. Kreps and Wilson (1982) and Milgrom and Roberts (1982) further explore this insight in the finite horizon chain store game.

Moreover, for some monitoring distributions, the complete information game has, for *all* discount factors, a unique sequential equilibrium, and in this equilibrium, the static Nash profile Ls is played in every period.[4] Proposition 4.1 nonetheless implies that the firm can develop a reputation for playing H in the game with incomplete information.

4.2.2 The benchmark complete information game

Player i has a finite action space A_i. Player 1's actions are monitored via a public signal: the signal y, drawn from a finite set Y, is realized with probability $\rho(y|a_1)$ when the action $a_1 \in A_1$ is chosen. We make the analysis more convenient here by assuming that the public signal depends only on player 1's action, returning to this assumption in Section 4.2.5. Since player 2 is short-lived, we further simplify notation by assuming that the period t player 2's action choice is not observed by subsequent player 2's (though it is observed by player 1). The arguments are identical (with more notation) if player 2's actions are public.

Player i's ex post stage game payoff is a function $u_i^* : A_1 \times A_2 \times Y \to \mathbb{R}$ and i's ex ante payoff is $u_i : A_1 \times A_2 \to \mathbb{R}$ is given by

$$u_i(a) := \sum_y u_i^*(a_1, a_2, y)\rho(y|a_1).$$

We typically begin the analysis with the ex ante payoffs, as in the product-choice game of Figure 4.1, and the monitoring structure ρ, leaving the ex post payoff functions u_i^* to be defined implicitly.

Player 2 observes the public signals, but does not observe player 1's actions. Player 2 might draw inferences about player 1's actions from her payoffs, but since player 2 is short-lived, these inferences are irrelevant (as long as the period t player 2 does not communicate any such inference to subsequent short-lived players). Following the literature on public monitoring repeated games, such inferences can also be precluded by assuming player 2's ex post payoff does not depend on a_1, depending only on (a_2, y).

The benchmark game includes perfect monitoring games as a special case. In the perfect monitoring product-choice game for example, $A_1 = \{H, L\}$, $A_2 = \{c, s\}$, $Y = A_1$, and $\rho(y \mid a_1) = 1$ if $y = a_1$ and 0 otherwise. An imperfect public monitoring version of the product-choice game is analyzed in Section 4.2.4.3 (Example 4.2).

Player 1 is long-lived (and discounts flow payoffs by a discount factor δ), while player 2 is short-lived (living for one period). The set of private histories for player 1 is $H_1 := \cup_{t=0}^{\infty}(A_1 \times A_2 \times Y)^t$, and a behavior strategy for player 1 is a function $\sigma_1 : H_1 \to \Delta(A_1)$. The set of histories for the short-lived players is $H_2 := \cup_{t=0}^{\infty} Y^t$, and a behavior

[4] We return to this in Example 4.2.

strategy for the short-lived players is a function $\sigma_2 : H_2 \to \Delta(A_2)$. Note that the period t short-lived player does not know the action choices of past short-lived players.[5]

4.2.3 The incomplete information game and commitment types

The type of player 1 is unknown to player 2. A possible type of player 1 is denoted by $\xi \in \Xi$, where Ξ is a finite or countable set of types. Player 2's prior belief about 1's type is given by the distribution μ, with support Ξ.

The set of types is partitioned into payoff types, Ξ_1, and commitment types, $\Xi_2 := \Xi \setminus \Xi_1$. Payoff types maximize the average discounted value of payoffs, which depend on their type. We accordingly expand the definition of the ex post payoff function u_1^* to incorporate types, $u_1^* : A_1 \times A_2 \times Y \times \Xi_1 \to \mathbb{R}$. The ex ante payoff function $u_1 : A_1 \times A_2 \times \Xi_1 \to \mathbb{R}$ is now given by

$$u_1(a_1, a_2, \xi) := \sum_y u_1^*(a_1, a_2, y, \xi)\rho(y|a_1).$$

It is common to identify one payoff type as the normal type, denoted here by ξ_0. When doing so, it is also common to drop the type argument from the payoff function. It is also common to think of the prior probability of the normal type $\mu(\xi_0)$ as being relatively large, so the games of incomplete information are a seemingly small departure from the underlying game of complete information, though there is no requirement that this be the case.

Commitment types do not have payoffs, and simply play a specified repeated game strategy. While a commitment type of player 1 can be committed to any strategy in the repeated game, much of the literature focuses on simple commitment or action types: such types play the same (pure or mixed) stage-game action in every period, regardless of history.[6] For example, one simple commitment type in the product-choice game always exerts high effort, while another always plays high effort with probability $\frac{2}{3}$. We denote by $\xi(\alpha_1)$ the (simple commitment) type that plays the action $\alpha_1 \in \Delta(A_1)$ in every period.

Remark 4.1. (Payoff or commitment types) The distinction between payoff and commitment types is not clear cut. For example, pure simple commitment types are easily modeled as payoff types. We need only represent the type $\xi(a_1)$ as receiving the

[5] Under our assumption that the public signal's distribution is a function of only the long-lived players' action, the analysis to be presented proceeds unchanged if the period t short-lived player knows the actions choices of past short-lived players (i.e., the short-lived player actions are public). In more general settings (where the signal distribution depends on the complete action profile), the analysis must be adjusted in obvious ways (as we describe below).

[6] This focus is due to the observation that with short-lived uninformed players, more complicated commitment types do not lead to higher reputation bounds.

stage-game payoff 1 if he plays action a_1 (regardless of what signal appears or what player 2 chooses) and zero otherwise. Note that this makes the consistent play of a_1 a strictly dominant strategy in the repeated game, and that it is not enough to simply have the action be dominant in the stage game.

A behavior strategy for player 1 in the incomplete information game is given by

$$\sigma_1 : H_1 \times \Xi \to \Delta(A_1),$$

such that, for all simple commitment types $\xi(\alpha_1) \in \Xi_2$,[7]

$$\sigma_1(h_1^t, \xi(\alpha_1)) = \alpha_1, \qquad \forall h_1^t \in H_1.$$

A behavior strategy for player 2 is (as in the complete information game) a map $\sigma_2 : H_2 \to \Delta(A_2)$.

The space of outcomes is given by $\Omega := \Xi \times (A_1 \times A_2 \times Y)^\infty$, with an outcome $\omega = (\xi, a_1^0 a_2^0 y^0, a_1^1 a_2^1 y^1, a_1^2 a_2^2 y^2, \ldots) \in \Omega$, specifying the type of player 1, the actions chosen and the realized signal in each period.

A profile of strategies (σ_1, σ_2), along with the prior probability over types μ (with support Ξ), induces a probability measure on the set of outcomes Ω, denoted by $\mathbf{P} \in \Delta(\Omega)$. For a fixed commitment type $\hat{\xi} = \xi(\hat{\alpha}_1)$, the probability measure on the set of outcomes Ω conditioning on $\hat{\xi}$ (and so induced by $(\hat{\sigma}_1, \sigma_2)$, where $\hat{\sigma}_1$ is the simple strategy specifying $\hat{\alpha}_1$ in every period irrespective of history), is denoted $\widehat{\mathbf{P}} \in \Delta(\Omega)$. Denoting by $\widetilde{\mathbf{P}}$ the measure induced by (σ_1, σ_2) and conditioning on $\xi \neq \hat{\xi}$, we have

$$\mathbf{P} = \mu(\hat{\xi})\widehat{\mathbf{P}} + (1 - \mu(\hat{\xi}))\widetilde{\mathbf{P}}. \qquad [4.1]$$

Given a strategy profile σ, $U_1(\sigma, \xi)$ denotes the type-ξ long-lived player's payoff in the repeated game,

$$U_1(\sigma, \xi) := E_{\mathbf{P}}\left[(1-\delta) \sum_{t=0}^{\infty} \delta^t u_1(a^t, y^t, \xi) \,\middle|\, \xi \right].$$

Denote by $\Gamma(\mu, \delta)$ the game of incomplete information.

As usual, a Nash equilibrium is a collection of mutual best responses:

Definition 4.1. *A strategy profile (σ_1', σ_2') is a* Nash equilibrium *of the game $\Gamma(\mu, \delta)$ if, for all $\xi \in \Xi_1$, σ_1' maximizes $U_1(\sigma_1, \sigma_2', \xi)$ over player 1's repeated game strategies, and if for all t and all $h_2^t \in \mathcal{H}_2$ that have positive probability under (σ_1', σ_2') and μ (i.e., $\mathbf{P}(h_2^t) > 0$),*

$$E_{\mathbf{P}}\left[u_2(\sigma_1'(h_1^t, \xi), \sigma_2'(h_2^t)) \mid h_2^t\right] = \max_{a_2 \in A_2} E_{\mathbf{P}}\left[u_2(\sigma_1'(h_1^t, \xi), a_2) \mid h_2^t\right].$$

[7] For convenience, we have omitted the analogous requirement that σ_1 is similarly restricted for nonsimple commitment types.

4.2.4 Reputation bounds

The link from simple commitment types to reputation effects arises from a basic property of updating. Suppose an action α_1' is statistically identified (i.e., there is no α_1'' giving rise to the same distribution of signals) under the signal distribution ρ, and suppose player 2 assigns positive probability to the simple type $\xi(\alpha_1')$. Then, if the normal player 1 persistently plays α_1', player 2 must eventually place high probability on that action being played (and so will best respond to that action). Intuitively, since α_1' is statistically identified, in any period in which player 2 places low probability on α_1' (and so on $\xi(\alpha_1')$), the signals will typically lead player 2 to increase the posterior probability on $\xi(\alpha_1')$, and so eventually on α_1'. Consequently, there cannot be too many periods in which player 2 places low probability on α_1'.

When the action chosen is perfectly monitored by player 2 (which requires the benchmark game have perfect monitoring and α_1' be a pure action), this intuition has a direct formalization (the route followed in the original argument of Fudenberg and Levine, 1989). However, when the action is not perfectly monitored, the path from the intuition to the proof is less clear. The original argument for this case (Fudenberg and Levine, 1992) uses sophisticated martingale techniques, subsequently simplified by Sorin (1999) (an exposition can be found in Mailath and Samuelson, 2006, Section 15.4.2). Here we present a recent simple unified argument, due to Gossner (2011), based on relative entropy (see Cover and Thomas, 2006, for an introduction).

4.2.4.1 Relative entropy

Let X be a finite set of outcomes. The relative entropy or Kullback-Leibler distance between probability distributions p and q over X is

$$d(p\|q) := \sum_{x\in X} p(x) \log \frac{p(x)}{q(x)}.$$

By convention, $0 \log \frac{0}{q} = 0$ for all $q \in [0, 1]$ and $p \log \frac{p}{0} = \infty$ for all $p \in (0, 1]$.[8] In our applications of relative entropy, the support of q will always contain the support of p. Since relative entropy is not symmetric, we often distinguish the roles of p and q by saying that $d(p\|q)$ is the relative entropy of q with respect to p. Relative entropy is always nonnegative and only equals 0 when $p = q$ (Cover and Thomas, 2006, Theorem 2.6.3).[9]

The relative entropy of q with respect to p measures an observer's expected error in predicting $x \in X$ using the distribution q when the true distribution is p. The probability of a sample of n draws from X identically and independently distributed according to p

[8] These equalities, for $p, q > 0$, are justified by continuity arguments. The remaining case, $0 \log \frac{0}{0} = 0$, is made to simplify statements and eliminate nuisance cases.
The logs may have any base. Both base 2 and base e are used in information theory. We use base e.

[9] Apply Jensen's inequality to $-d(p\|q) = \sum_{x\in \operatorname{supp} p} p(x) \log[q(x)/p(x)]$.

is $\prod_x p(x)^{n_x}$, where n_x is the number of realizations of $x \in X$ in the sample. An observer who believes the data is distributed according to q assigns to the same sample probability $\prod_x q(x)^{n_x}$. The log likelihood ratio of the sample is

$$\mathcal{L}(x_1, \ldots, x_n) = \sum_x n_x \log \frac{p(x)}{q(x)}.$$

As the sample size n grows large, the average log likelihood $\mathcal{L}(x_1, \ldots, x_n)/n$ converges almost surely to $d(p\|q)$ (and so the log likelihood becomes arbitrarily large for any $q \neq p$, since $d(p\|q) > 0$ for any such q).

While not a metric (since it is asymmetric and does not satisfy the triangle inequality), relative entropy is usefully viewed as a notion of distance. For example, Pinsker's inequality[10] bounds the relative entropy of q with respect to p from below by a function of their L^1 distance:

$$\left\|p - q\right\| \leq \sqrt{2d(p\|q)}, \tag{4.2}$$

where

$$\left\|p - q\right\| := \sum_x \left|p(x) - q(x)\right| = 2\sum_{\{x:p(x)\geq q(x)\}} \left|p(x) - q(x)\right|.$$

Thus, $\left\|p_n - q\right\| \to 0$ if $d(p_n\|q) \to 0$ as $n \to \infty$. While the reverse implication does not hold in general,[11] it does for full support q.

The usefulness of relative entropy arises from a chain rule. Let P and Q be two distributions over a finite product set $X \times Y$, with marginals P_X and Q_X on X, and conditional probabilities $P_Y(\cdot|x)$ and $Q_Y(\cdot|x)$ on Y given $x \in X$. The chain rule for relative entropy is

$$\begin{aligned} d(P\|Q) &= d(P_X\|Q_X) + \sum_x P_X(x) d\left(P_Y(\cdot|x)\|Q_Y(\cdot|x)\right) \\ &= d(P_X\|Q_X) + E_{P_X} d\left(P_Y(\cdot|x)\|Q_Y(\cdot|x)\right). \end{aligned} \tag{4.3}$$

The chain rule is a straightforward calculation (Cover and Thomas, 2006, Theorem 2.5.3). The error in predicting the pair xy can be decomposed into the error predicting x, and conditional on x, the error in predicting y.

The key to the reputation bound is bounding the error in the one-step ahead predictions of the uninformed players when the long-lived player plays identically to a commitment type. The presence of the commitment type ensures that there is a "grain of truth" in player 2's beliefs which, together with the chain rule, yields a useful bound on relative entropy. The basic technical tool is the following lemma.

[10] See Cesa-Bianchi and Lugosi (2006, page 371) or Cover and Thomas (2006, Lemma 11.6.1); note that Cover and Thomas (2006) define relative entropy using log base 2.

[11] Suppose $X = \{0, 1\}$, $p_n(1) = 1 - \frac{1}{n}$, and $q(1) = 1$. Then, $\left\|p_n - q\right\| \to 0$, while $d(p_n\|q) = \infty$ for all n.

Lemma 4.1. *Let X be a finite set of outcomes. Suppose $q = \varepsilon p + (1-\varepsilon)p'$ for some $\varepsilon > 0$ and $p, p' \in \Delta(X)$. Then,*

$$d(p\|q) \leq -\log \varepsilon.$$

Proof. Since $q(x)/p(x) \geq \varepsilon$, we have

$$-d(p\|q) = \sum\nolimits_x p(x) \log \frac{q(x)}{p(x)} \geq \sum\nolimits_x p(x) \log \varepsilon = \log \varepsilon.$$

4.2.4.2 Bounding the one-step ahead prediction errors

Fix a (possibly mixed) action $\hat{\alpha}_1 \in \Delta(A_1)$ and suppose the commitment type $\hat{\xi} = \xi(\hat{\alpha}_1)$ has positive prior probability.

At the history h_2^t, player 2 chooses an action $\sigma_2(h_2^t)$ that is a best response to $\alpha_1(h_2^t) := E_{\mathbf{P}}[\sigma_1(h_1^t, \xi) \mid h_2^t]$, that is, a_2 has positive probability under $\sigma_2(h_2^t)$ only if it maximizes[12]

$$\sum_{a_1} \sum_{y} u_2^*(a_1, a_2, y) \rho(y|a_1) \alpha_1(a_1|h_2^t).$$

At the history h_2^t, player 2's predicted distribution of the period t signals is $p(h_2^t) := \rho(\cdot|\alpha_1(h_2^t)) = \sum_{a_1} \rho(\cdot|a_1)\alpha_1(a_1|h_2^t)$, while the true distribution when player 1 plays $\hat{\alpha}_1$ is $\hat{p} := \rho(\cdot|\hat{\alpha}_1) = \sum_{a_1} \rho(\cdot|a_1)\hat{\alpha}_1(a_1)$. Hence, if player 1 is playing $\hat{\alpha}_1$, then in general player 2 is not best responding to the true distribution of signals, and his one-step ahead prediction error is $d\left(\hat{p}\|p(h_2^t)\right)$. However, player 2 is best responding to an action profile $\alpha_1(h_2^t)$ that is $d\left(\hat{p}\|p(h_2^t)\right)$-close to $\hat{\alpha}_1$ (as measured by the relative entropy of the induced signals). To bound player 1's payoff, it suffices to bound the number of periods in which $d\left(\hat{p}\|p(h_2^t)\right)$ is large.

Since player 2's beliefs assign positive probability to the event that player 1 is always playing $\hat{\alpha}_1$, then as player 1 does so, player 2's one-step prediction error must disappear asymptotically. A bound on player 1's payoff then arises from noting that *if* player 1 relentlessly plays $\hat{\alpha}_1$, then player 2 must eventually just as persistently play a best response to $\hat{\alpha}_1$. "Eventually" may be a long time, but this delay is inconsequential to a sufficiently patient player 1.

For any period t, denote the marginal of the unconditional distribution $\mathbf{P}$ on H_2^t, the space of t period histories of public signals, by $\mathbf{P}_2^t$. Similarly, the marginal on H_2^t of $\widehat{\mathbf{P}}$ (the distribution conditional on $\hat{\xi}$) is denoted $\widehat{\mathbf{P}}_2^t$. Recalling [4.1] and applying Lemma 4.1 to these marginal distributions (which have finite supports) yields

$$d(\widehat{\mathbf{P}}_2^t \| \mathbf{P}_2^t) \leq -\log \mu(\hat{\xi}). \qquad [4.4]$$

[12] The probability assigned to $a_1 \in A_1$ by the distribution $\alpha_1(h_2^t)$ is denoted by $\alpha_1(a_1|h_2^t)$.

It is worth emphasizing that this inequality holds for all t, and across all equilibria. But notice also that the bounding term is unbounded as $\mu(\hat{\xi}) \to 0$ and our interest is in the case where $\mu(\hat{\xi})$ is small.

Suppose $\mu(\hat{\xi})$ is indeed small. Then, in those periods when player 2's prediction under $\mathbf{P}$, $p(h_2^t)$, has a large relative entropy with respect to $\hat{p}$, she will (with high probability) be surprised,[13] and so will significantly increase her posterior probability that player 1 is $\hat{\xi}$. This effectively increases the size of the grain of truth from the perspective of period $t+1$, reducing the maximum relative entropy $p(h_2^{t+1})$ can have with respect to $\hat{p}$. Intuitively, for fixed $\mu(\hat{\xi})$, there cannot be too many periods in which player 2 can be surprised with high probability.

To make this intuition precise, we consider the one-step ahead prediction errors, $d(\hat{p}\|p(h_2^\tau))$. The chain rule implies

$$d(\widehat{\mathbf{P}}_2^t\|\mathbf{P}_2^t) = \sum_{\tau=0}^{t-1} E_{\widehat{\mathbf{P}}} d(\hat{p}\|p(h_2^\tau)),$$

that is, the prediction error over t periods is the total of the t *expected* one-step ahead prediction errors. Since [4.4] holds for all t,

$$\sum_{\tau=0}^{\infty} E_{\widehat{\mathbf{P}}} d\left(\hat{p}\|p(h_2^\tau)\right) \le -\log \mu(\hat{\xi}). \qquad [4.5]$$

That is, all but a finite number of expected one-step ahead prediction errors must be small.

4.2.4.3 From prediction bounds to payoffs

It remains to connect the bound on prediction errors [4.5] with a bound on player 1's payoffs.

Definition 4.2. *An action $\alpha_2 \in \Delta(A_2)$ is an ε-*entropy confirming best response *to $\alpha_1 \in \Delta(A_1)$ if there exists $\alpha_1' \in \Delta(A_1)$ such that*

1. *α_2 is a best response to α_1'; and*

2. *$d(\rho(\cdot|\alpha_1)\|\rho(\cdot|\alpha_1')) \le \varepsilon$.*

The set of ε-entropy confirming best responses to α_1 is denoted $B_\varepsilon^d(\alpha_1)$.

Recall that in a Nash equilibrium, at any on-the-equilibrium-path history h_2^t, player 2's action is a $d(\hat{p}\|p(h_2^t))$-entropy confirming best response to $\hat{\alpha}_1$. Suppose player 1 always plays $\hat{\alpha}_1$. We have just seen that the expected number of periods in which $d(\hat{p}\|p(h_2^t))$

[13] Or, more precisely, the period $(t+1)$ version of player 2 will be surprised.

is large is bounded, independently of δ. Then for δ close to 1, player 1's equilibrium payoffs will be effectively determined by player 2's ε-entropy confirming best responses for ε small.

Define, for all payoff types $\xi \in \Xi_1$,

$$\underline{v}^{\xi}_{\alpha_1}(\varepsilon) := \min_{\alpha_2 \in B^d_\varepsilon(\alpha_1)} u_1(\alpha_1, \alpha_2, \xi),$$

and denote by $\underline{w}^{\xi}_{\alpha_1}$ the largest convex function below $\underline{v}^{\xi}_{\alpha_1}$. The function $\underline{w}^{\xi}_{\alpha_1}$ is nonincreasing in ε because $\underline{v}^{\xi}_{\alpha_1}$ is. The function $\underline{w}^{\xi}_{\alpha_1}$ allows us to translate a bound on the total discounted expected one-step ahead prediction errors into a bound on the total discounted expected payoffs of player 1.

Proposition 4.1. *Suppose the action type $\hat{\xi} = \xi(\hat{\alpha}_1)$ has positive prior probability, $\mu(\hat{\xi}) > 0$, for some potentially mixed action $\hat{\alpha}_1 \in \Delta(A_1)$. Then, player 1 type ξ's payoff in any Nash equilibrium of the game $\Gamma(\mu, \delta)$ is greater than or equal to $\underline{w}^{\xi}_{\hat{\alpha}_1}(\hat{\varepsilon})$, where $\hat{\varepsilon} := -(1-\delta)\log \mu(\hat{\xi})$.*

It is worth emphasizing that the only aspect of the set of types and the prior that plays a role in the proposition is the probability assigned to $\hat{\xi}$. The set of types may be very large, and other quite crazy types may receive significant probability under the prior μ.

Proof. Since in any Nash equilibrium (σ'_1, σ'_2), each payoff type ξ has the option of playing $\hat{\alpha}_1$ in every period, we have

$$\begin{aligned} U_1(\sigma', \xi) &= (1-\delta)\sum\nolimits_{t=0}^{\infty} \delta^t E_{\mathbf{P}}[u_1(\sigma'_1(h^t_1), \sigma'_2(h^t_2), \xi) \mid \xi] \\ &\geq (1-\delta)\sum\nolimits_{t=0}^{\infty} \delta^t E_{\widehat{\mathbf{P}}} u_1(\hat{\alpha}_1, \sigma'_2(h^t_2), \xi) \\ &\geq (1-\delta)\sum\nolimits_{t=0}^{\infty} \delta^t E_{\widehat{\mathbf{P}}} \underline{v}^{\xi}_{\hat{\alpha}_1}(d(\hat{p} \| p(h^t_2))) \\ &\geq (1-\delta)\sum\nolimits_{t=0}^{\infty} \delta^t E_{\widehat{\mathbf{P}}} \underline{w}^{\xi}_{\hat{\alpha}_1}(d(\hat{p} \| p(h^t_2))) \end{aligned}$$

(and so, by an application of Jensen's inequality)

$$\geq \underline{w}^{\xi}_{\hat{\alpha}_1}\left((1-\delta)\sum\nolimits_{t=0}^{\infty} \delta^t E_{\widehat{\mathbf{P}}} d(\hat{p} \| p(h^t_2))\right)$$

(and so, by an application of [4.5])

$$\geq \underline{w}^{\xi}_{\hat{\alpha}_1}\left(-(1-\delta)\log \mu(\hat{\xi})\right).$$

Corollary 4.1. *Suppose the action type $\hat{\xi} = \xi(\hat{\alpha}_1)$ has positive prior probability, $\mu(\hat{\xi}) > 0$, for some potentially mixed action $\hat{\alpha}_1 \in \Delta(A_1)$. Then, for all $\xi \in \Xi_1$ and $\eta > 0$, there exists a $\bar{\delta} < 1$ such that, for all $\delta \in (\bar{\delta}, 1)$, player 1 type ξ's payoff in any Nash equilibrium of the game $\Gamma(\mu, \delta)$ is greater than or equal to*

$$\underline{v}^{\xi}_{\hat{\alpha}_1}(0) - \eta.$$

Proof. Since the distribution of signals is independent of player 2's actions, if a mixture is an ε-entropy confirming best response to $\hat{\alpha}_1$, then so is every action in its support.[14] This implies that $B^d_0(\hat{\alpha}_1) = B^d_\varepsilon(\hat{\alpha}_1)$ for ε sufficiently small, and so $\underline{v}^{\xi}_{\hat{\alpha}_1}(0) = \underline{v}^{\xi}_{\hat{\alpha}_1}(\varepsilon)$ for ε sufficiently small. Hence,

$$\underline{v}^{\xi}_{\hat{\alpha}_1}(0) = \underline{w}^{\xi}_{\hat{\alpha}_1}(0) = \lim_{\varepsilon \searrow 0} \underline{w}^{\xi}_{\hat{\alpha}_1}(\varepsilon).$$

The Corollary is now immediate from Proposition 4.1.

Example 4.1. To illustrate Proposition 4.1 and its corollary, consider first the perfect monitoring product-choice game of Figure 4.1. Let ξ denote the payoff type with the player 1 payoffs specified. Recall that in the perfect monitoring game, the set of signals coincides with the set of firm actions. The action c is the unique best response of a customer to any action α_1 satisfying $\alpha_1(H) > \frac{1}{2}$, while s is also a best response when $\alpha_1(H) = \frac{1}{2}$. Thus, $B^d_\varepsilon(H) = \{c\}$ for all $\varepsilon < \log 2$ (since $d(H\|\alpha_1) = -\log \alpha_1(H)$), while $B^d_\varepsilon(H) = \{c, s\}$ for all $\varepsilon \geq \log 2 \approx 0.69$. That is, c is the unique ε-entropy confirming best response to H, for ε smaller than the relative entropy of the equal randomization on H and L with respect to the pure action H. This implies

$$\underline{v}^{\xi}_{H}(\varepsilon) = \begin{cases} 2, & \text{if } \varepsilon < \log 2, \\ 0, & \text{if } \varepsilon \geq \log 2. \end{cases}$$

The functions $\underline{v}^{\xi}_{H}$ and $\underline{w}^{\xi}_{H}$ are graphed in Figure 4.2.

Consider now the mixed action $\hat{\alpha}_1$ which plays the action H with probability $\frac{2}{3}$ (and the action L with probability $\frac{1}{3}$). The relative entropy of the equal randomization on H and L with respect to the mixed action $\hat{\alpha}_1$ is $\frac{5}{3}\log 2 - \log 3 =: \bar{\varepsilon} \approx 0.06$, and so any action with smaller relative entropy with respect to $\hat{\alpha}_1$ has c as the unique best response. This implies

$$\underline{v}^{\xi}_{\hat{\alpha}_1}(\varepsilon) = \begin{cases} 2\frac{1}{3}, & \text{if } \varepsilon < \bar{\varepsilon}, \\ \frac{1}{3}, & \text{if } \varepsilon \geq \bar{\varepsilon}. \end{cases}$$

Suppose the customer puts positive probability on firm type $\xi_H := \xi(H)$. If $\mu(\xi_H)$ is large, then its log is close to zero, and we have a trivial and unsurprising reputation bound

[14] This is not true in general, see Mailath and Samuelson (2006, fn. 10, p. 480).

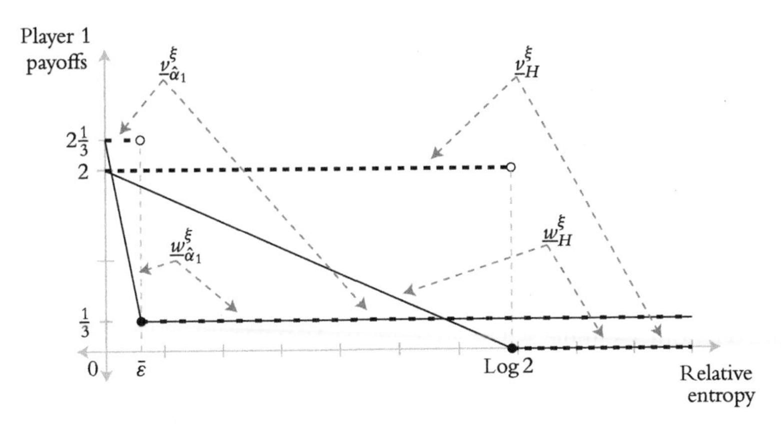

Figure 4.2 The functions $\underline{v}^{\xi}_{H}$, $\underline{w}^{\xi}_{H}$, $\underline{v}^{\xi}_{\hat{\alpha}_1}$, and $\underline{w}^{\xi}_{\hat{\alpha}_1}$ for the perfect monitoring version of the product-choice game in Figure 4.1 (the payoff type ξ has the specified player 1 payoffs). The relative entropy of $\frac{1}{2} \circ H + \frac{1}{2} \circ L$ with respect to H is $\log 2 \approx 0.69$. For higher relative entropies, s is an ε-entropy confirming best response to H. The relative entropy of $\frac{1}{2} \circ H + \frac{1}{2} \circ L$ with respect to $\hat{\alpha}_1$ is $\bar{\varepsilon} \approx 0.06$. For higher relative entropies, s is an ε-entropy confirming best response to $\hat{\alpha}_1$.

(since the customer will best respond to H from the beginning). Suppose, though, that the probability is close to zero. Then it may take many periods of imitating the action type ξ_H before the customer predicts H in every period (reflected in a large magnitude log). But since the number of periods is independent of the discount factor, for δ sufficiently close to 1, the lower bound is still close to 2: the term $(1-\delta)\log \mu(\xi)$ is close to zero.

This bound is independent of customer beliefs about the possible presence of the type $\hat{\xi} := \xi(\hat{\alpha}_1)$. A possibility of the type $\hat{\xi}$ can improve the lower bound on payoffs. However, since this type's behavior (signal distribution) is closer to the critical distribution $\frac{1}{2} \circ H + \frac{1}{2} \circ L$, the critical value of the relative entropy is significantly lower ($\bar{\varepsilon} << \log 2$; see Figure 4.2), and hence more periods must pass before the type ξ firm can be assured that the customer will play a best response. If the prior assigned equal probability to both $\hat{\xi}$ and ξ_H, then the bound on δ required to bound the type ξ firm's payoff by $\underline{v}^{\xi}_{\hat{\alpha}_1}(0) - \eta$ is significantly tighter than that required to bound the type ξ firm's payoff by $\underline{v}^{\xi}_{H}(0) - \eta$.

Example 4.2. One would expect reputation-based payoff bounds to be weaker in the presence of imperfect monitoring. While there is a sense in which this is true, there is another sense in which this is false.

To illustrate, we consider an imperfect monitoring version of the product-choice game. The actions of the firm are private, and the public signal is drawn from $Y := \{\underline{y}, \bar{y}\}$ according to the distribution $\rho(\bar{y} \mid a_1) = p$ if $a_1 = H$ and $q < p$ if $a_1 = L$. The ex ante payoffs for the type ξ firm and customer are again given by Figure 4.1.

While the best responses of the customer are unchanged from Example 4.1 , the ε-entropy confirming best responses are changed: the relative entropy of the mixture α_1 with respect to H is now

$$d(H\|\alpha_1) = p\log\frac{p}{\alpha_1(H)p+\alpha_1(L)q} + (1-p)\log\frac{1-p}{\alpha_1(H)(1-p)+\alpha_1(L)(1-q)}.$$

For the parameterization $p=1-q=\frac{2}{3}$, $d(H\|\frac{1}{2}\circ H+\frac{1}{2}\circ L)=\bar{\varepsilon}\approx 0.06$, the critical value for $\hat{\alpha}_1$ from Example 4.1, and so $\underline{v}_H^{\xi}(\varepsilon) = 2$ for $\varepsilon < \bar{\varepsilon}$, and 0 for $\varepsilon \geq \bar{\varepsilon}$. Moreover, $\underline{w}_H^{\xi}(0) = 2$.

Finally, the relative entropy of the critical mixture $\frac{1}{2}\circ H+\frac{1}{2}\circ L$ with respect to the mixture $\hat{\alpha}_1$ from Example 4.1 is approximately 0.006. Since the firm's actions are statistically identified by the signals, we also have $\underline{v}_{\hat{\alpha}_1}^{\xi}(0) = \underline{w}_{\hat{\alpha}_1}^{\xi}(0) = 2\frac{1}{3}$.

Recall that in the perfect monitoring product-choice game, the bound on the relative entropy to get a strictly positive lower bound on payoffs from $\underline{w}_H^{\xi}(\varepsilon)$ is $\log 2 \approx 0.69$ rather than $\bar{\varepsilon}\approx 0.06$. This implies that the required lower bound on the discount factor to get the same lower bound on payoffs is larger for the imperfect monitoring game, or equivalently, for the same discount factor, the reputation lower bound is lower under imperfect monitoring.

It is worth recalling at this point our earlier observation from Section 4.2.1 that reputation effects can rescue intuitive outcomes as equilibrium outcomes. Assume the actions of the customers are public; as we noted earlier, this does not affect the arguments or the reputation bounds calculated (beyond complicating notation). For the parameterization $p = 1-q = \frac{2}{3}$, the complete information repeated game has a unique sequential equilibrium outcome, in which L is played in every period (Mailath and Samuelson, 2006, Section 7.6.2 and Proposition 10.1.1). The firm's equilibrium payoff in the complete information game is 1. In particular, while the reputation payoff bound is weaker in an absolute sense in the presence of imperfect monitoring, the relative bound (bound less the maximal payoff in any equilibrium of the complete information game) is stronger.

4.2.4.4 The Stackelberg bound

The reputation literature has tended to focus on Stackelberg bounds. Player 1's pure-action Stackelberg payoff (for type ξ) is defined as

$$\bar{v}_1^{\xi} := \sup_{a_1\in A_1}\ \min_{\alpha_2\in B(a_1)} u_1(a_1,\alpha_2,\xi),$$

where $B(a_1)$ is the set of player 2 myopic best replies to a_1. Since A_1 is finite, the supremum is attained by some action a_1^* and any such action is an associated "Stackelberg action,"

$$a_1^{\xi} \in \arg\max_{a_1\in A_1}\ \min_{\alpha_2\in B(a_1)} u_1(a_1,\alpha_2,\xi).$$

This is a pure action to which the type ξ player 1 would commit, if player 1 had the chance to do so (and hence the name Stackelberg action), given that such a commitment induces a best response from player 2.

The mixed-action Stackelberg payoff is defined as

$$\bar{\bar{v}}_1^{\xi} := \sup_{\alpha_1 \in \Delta(A_1)} \min_{\alpha_2 \in B(\alpha_1)} u_1(\alpha_1, \alpha_2, \xi).$$

Typically, the supremum is not achieved by any mixed action, and so there is no mixed-action Stackelberg type. There are, of course, mixed commitment types that, if player 2 is convinced she is facing such a type, will yield payoffs arbitrarily close to the mixed-action Stackelberg payoff.

If the signals are informative about the actions of player 1, then the set of zero-entropy confirming best replies coincides with the set of best replies and so we have the following corollary:

Corollary 4.2. *Suppose the actions of player* 1 *are statistically identified, i.e.,* $\rho(\cdot|a_1) \neq \rho(\cdot|a_1')$ *for all* $a_1 \neq a_1' \in A_1$. *Suppose the action type* $\xi(a_1^{\xi})$ *has positive prior probability for some Stackelberg action* a_1^{ξ} *for the payoff type* ξ. *Then, for all* $\eta > 0$, *there exists* $\bar{\delta} \in (0, 1)$ *such that for all* $\delta \in (\bar{\delta}, 1)$, *the set of player* 1 *type* ξ*'s Nash equilibrium payoffs of the game* $\Gamma(\mu, \delta)$ *is bounded below by* $\bar{v}_1^{\xi} - \eta$.

Suppose the mixed actions of player 1 *are statistically identified, i.e.,* $\rho(\cdot|\alpha_1) \neq \rho(\cdot|\alpha_1')$ *for all* $\alpha_1 \neq \alpha_1' \in \Delta(A_1)$. *Suppose the support of* μ *includes a set of action types* $\{\xi(\alpha_1) : \alpha_1 \in \Delta^*\}$, *where* Δ^* *is a countable dense subset of* $\Delta(A_1)$. *Then, for all* $\eta > 0$, *there exists* $\bar{\delta} \in (0, 1)$ *such that for all* $\delta \in (\bar{\delta}, 1)$, *the set of player* 1 *type* ξ*'s Nash equilibrium payoffs of the game* $\Gamma(\mu, \delta)$ *is bounded below by* $\bar{\bar{v}}_1^{\xi} - \eta$.

4.2.5 More general monitoring structures

While the analysis in Section 4.2.4 is presented for the case in which the distribution of signals is a function of the actions of player 1 only, the arguments apply more generally. It is worth first noting that nothing in the argument depended on the signals being public, and so the argument applies immediately to the case of private monitoring, with signals that depend only on player 1's actions.

Suppose now that the distribution over signals depends on the actions of *both* players. The definition of ε-confirming best responses in Definition 4.2 is still valid, once condition 2 is adjusted to reflect the dependence of ρ on both players' actions:

$$d(\rho(\cdot|(\alpha_1, \alpha_2)) \| \rho(\cdot|(\alpha_1', \alpha_2))) < \varepsilon.$$

Proposition 4.1 and Corollary 4.1 are true in this more general setting as written. While the proof of Proposition 4.1 is unchanged (as a few moments of reflection will reveal), the proof of the Corollary is not. In particular, while $B_{\varepsilon}^{d}(\alpha_1)$ is still upper hemicontinuous in ε, it is not locally constant at 0 (see Gossner (2011) for details).

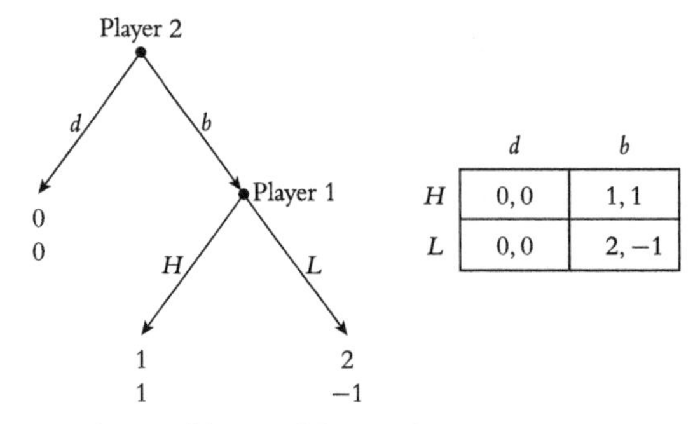

	d	b
H	0, 0	1, 1
L	0, 0	2, −1

Figure 4.3 The extensive and normal forms of the purchase game.

The obtained reputation bounds, however, may be significantly weakened. One obvious way for signals to depend on the actions of both players is that the stage game has a nontrivial extensive form, with the public signal in each period consisting of the terminal node reached in that period. In this case, the Stackelberg action of player 1 may not be statistically identified, limiting the effects of reputations. Consider the purchase game illustrated in Figure 4.3. The short-lived customer first decides between "buy" (b) and "don't buy" (d), and then after b, the long-lived firm decides on the level of effort, high (H) or low (L). The extensive and normal forms are given in Figure 4.3. The game has three public signals, corresponding to the three terminal nodes. The distribution over the public signal is affected by the behavior of both players, and as a consequence the zero-entropy confirming best replies to H (the Stackelberg action) consist of b (the best response) and d (which is not a best response to H). There is no useful reputation bound in this case: even if the short-lived customers assign positive probability to the possibility that the firm is the Stackelberg type, there is a sequential equilibrium in which the firm's payoff is 0 (Fudenberg and Levine, 1989, Example 4).

4.2.6 Temporary reputations under imperfect monitoring

As the analysis of Section 4.2.4 reveals, the lack of perfect monitoring of actions does not pose a difficulty for the formation of reputations. It does however pose a difficulty for its maintenance. Cripps et al. (2004, 2007) show that under imperfect monitoring, reputations in repeated games are temporary. We present a new simpler entropy-based proof due to Gossner (private communications).

Earlier indications that reputations are temporary can be found in Benabou and Laroque (1992), Kalai and Lehrer (1995), and Mailath and Samuelson (2001). Benabou and Laroque (1992) show that the long-lived player eventually reveals her type in any Markov perfect equilibrium of a particular repeated game of strategic information transmission. Kalai and Lehrer (1995) use merging arguments to show that, under weaker conditions than we impose here, play in repeated games of incomplete information must converge to a subjective correlated equilibrium of the complete information continuation game.[15] We describe Mailath and Samuelson (2001) in Section 4.4.3.3.

For simplicity, we restrict attention to the case in which there are only two types of player 1, the normal type ξ_0 and the action type $\hat{\xi} = \xi(\hat{\alpha}_1)$.[16] We assume the public signals have full support (Assumption 4.1) under $\hat{\alpha}_1$.[17] Reputations are temporary under private monitoring as well, with an identical proof (though the notation becomes a little messier). We also assume that with sufficiently many observations, either player can correctly identify, from the frequencies of the signals, any fixed stage-game action of their opponent (Assumptions 4.2 and 4.3). We now allow the signal distribution to depend upon both players' actions (since doing so does not result in any complications to the arguments, while clarifying their nature).

Assumption 4.1. (Full Support) *$\rho(y|\hat{\alpha}_1, a_2) > 0$ for all $a_2 \in A_2$ and $y \in Y$.*

Assumption 4.2. (Identification of 1) *For all $\alpha_2 \in \Delta(A_2)$, the $|A_1|$ columns in the matrix $[\rho(y|a_1\alpha_2)]_{y\in Y, a_1 \in A_1}$ are linearly independent.*

Assumption 4.3. (Identification of 2) *For all $a_1 \in A_1$, the $|A_2|$ columns in the matrix $[\rho(y|a_1 a_2)]_{y\in Y, a_2 \in A_2}$ are linearly independent.*

We assume the actions of the short-lived player are not observed by player 1 (but see footnote 17), and so player 1's period t private history consists of the public signals and his own past actions, denoted by $h_1^t \in H_1^t := (A_1 \times Y)^t$. We continue to assume that the short-lived players do not observe past short-lived player actions, and so the private history for player 2 is denoted $h_2^t \in H_2^t := Y^t$.[18]

Given a strategy profile (σ_1, σ_2) of the incomplete information game, the short-lived player's belief in period t that player 1 is type $\hat{\xi}$ is

$$\mu^t(h_2^t) := \mathbf{P}(\hat{\xi}|h_2^t),$$

and so μ^0 is the period 0, or prior, probability assigned to $\hat{\xi}$.

[15] This result is immediate in our context, since we examine a Nash equilibrium of the incomplete information game.

[16] The case of countably many simple action types is also covered, using the modifications described in Cripps et al. (2004, Section 6.1).

[17] This is stronger than necessary. For example, we can easily accommodate public player 2 actions by setting $Y = Y_1 \times A_2$, and assuming the signal in Y_1 has full support.

[18] As for the earlier analysis on reputation bounds, the disappearing reputation results also hold when short-lived players observe past short-lived player actions (a natural assumption under private monitoring).

Proposition 4.2. *Suppose Assumptions 4.1–4.3 are satisfied. Suppose player 2 has a unique best response $\hat{a}_2$ to $\hat{\alpha}_1$ and that $(\hat{\alpha}_1, \hat{a}_2)$ is not a Nash equilibrium of the stage game. If (σ_1, σ_2) is a Nash equilibrium of the game $\Gamma(\mu, \delta)$, then*

$$\mu^t \to 0, \qquad \widetilde{\mathbf{P}}\text{-a.s.}$$

Hence, conditional on player 1 being the normal type, player 2's posterior belief that 1 is normal almost-surely converges to one.

By Assumption 4.1, Bayes' rule determines μ^t after all histories. At any Nash equilibrium of this game, the belief μ^t is a bounded martingale and so converges $\mathbf{P}$-almost surely (and hence $\widetilde{\mathbf{P}}$- and $\widehat{\mathbf{P}}$-almost surely) to a random variable μ^∞. The idea of the proof is:

1. En route to a contradiction, assume that there is a positive $\widetilde{\mathbf{P}}$-probability event on which μ^∞ is strictly positive.
2. On this event, player 2 believes that both types of player 1 are eventually choosing the same distribution over actions $\hat{\alpha}_1$ (because otherwise player 2 could distinguish them).
3. Consequently, on a positive $\widetilde{\mathbf{P}}$-probability set of histories, eventually, player 2 will always play a best response to $\hat{\alpha}_1$.
4. Assumption 4.3 then implies that there is a positive $\widetilde{\mathbf{P}}$-probability set of histories on which player 1 infers that player 2 is for many periods best responding to $\hat{\alpha}_1$, irrespective of the observed signals.
5. This yields the contradiction, since player 1 has a strict incentive to play differently than $\hat{\alpha}_1$.

Since $\hat{\alpha}_1$ is not a best reply to $\hat{a}_2$, there is a $\gamma > 0$, an action $\hat{a}_1 \in A_1$ receiving positive probability under $\hat{\alpha}_1$, and an $\varepsilon_2 > 0$ such that

$$\gamma < \min_{\alpha_2(\hat{a}_2) \geq 1-\varepsilon_2} \left(\max_{a_1} u_1(a_1, \alpha_2) - u_1(\hat{a}_1, \alpha_2) \right). \qquad [4.6]$$

Define $\hat{p}(h_2^t) := \rho(\cdot|(\hat{\alpha}_1, \sigma_2(h_2^t))$ and redefine $p(h_2^t) := \rho(\cdot|\alpha_1(h_2^t), \sigma_2(h_2^t))$. Recall that $\|\cdot\|$ denotes L^1 distance. With this notation, Assumption 4.2 implies that there exists $\varepsilon_3 > 0$ such that, if $\left\| p(h_2^t) - \hat{p}(h_2^t) \right\| < \varepsilon_3$ (so that player 2 assigns sufficiently high probability to $\hat{\alpha}_1$), then player 2's unique best response is $\hat{a}_2$. Assumptions 4.1 and 4.2 imply that there exists $\varepsilon_1 > 0$, with

$$\varepsilon_1 < \min \left\{ \varepsilon_3, \min_{(\alpha_1, \alpha_2): \alpha_1(\hat{a}_1)=0} \|\rho(\cdot|(\alpha_1, \alpha_2)) - \rho(\cdot|(\hat{\alpha}_1, \alpha_2))\| \right\},$$

such that

$$\underline{\rho} := \min_{\alpha_1, a_2} \left\{ \rho(y|\alpha_1, a_2) : \left\| \rho(\cdot|\alpha_1, a_2) - \rho(\cdot|\hat{\alpha}_1, a_2) \right\| \leq \varepsilon_1 \right\} > 0. \qquad [4.7]$$

On the event

$$X^t := \left\{ \left\| p(h_2^t) - \hat{p}(h_2^t) \right\| < \varepsilon_1 \right\},$$

player 2's beliefs lead to her best responding to $\hat{\alpha}_1$, i.e., $\sigma_2(h_2^t) = \hat{a}_2$. The first lemma bounds the extent of player 2's $\widetilde{\mathbf{P}}$-expected surprises (i.e., periods in which player 2 both assigns a nontrivial probability to player 1 being $\hat{\xi}$ and believes $p(h_2^t)$ is far from $\hat{p}(h_2^t)$):

Lemma 4.2. ***(Player 2 either learns the type is normal or doesn't believe it matters)***

$$\sum_{t=0}^{\infty} E_{\widetilde{\mathbf{P}}} \left[(\mu^t)^2 (1 - \mathbb{1}_{X^t}) \right] \leq -\frac{2\log(1-\mu^0)}{\varepsilon_1^2}$$

where $\mathbb{1}_{X^t}$ is the indicator function for the event X^t.

Proof. Since (where $\tilde{p}(h_2^t)$ is the predicted signal distribution conditional on ξ_0)

$$p(h_2^t) = \mu^t \hat{p}(h_2^t) + (1 - \mu^t)\tilde{p}(h_2^t),$$

we have

$$\mu^t (p(h_2^t) - \hat{p}(h_2^t)) = (1 - \mu^t)(\tilde{p}(h_2^t) - p(h_2^t)),$$

and so

$$\mu^t \left\| p(h_2^t) - \hat{p}(h_2^t) \right\| \leq \left\| p(h_2^t) - \tilde{p}(h_2^t) \right\|.$$

Then,

$$\begin{aligned}
\frac{\varepsilon_1^2}{2} \sum_{t=0}^{\infty} E_{\widetilde{\mathbf{P}}} \left[(\mu^t)^2 (1 - \mathbb{1}_{X^t}) \right] &\leq \frac{1}{2} \sum_{t=0}^{\infty} E_{\widetilde{\mathbf{P}}} \left[(\mu^t)^2 \left\| p(h_2^t) - \hat{p}(h_2^t) \right\|^2 \right] \\
&\leq \frac{1}{2} \sum_{t=0}^{\infty} E_{\widetilde{\mathbf{P}}} \left[\left\| p(h_2^t) - \tilde{p}(h_2^t) \right\|^2 \right] \\
&\leq \sum_{t=0}^{\infty} E_{\widetilde{\mathbf{P}}} \left[d(\tilde{p}(h_2^t) \| p(h_2^t)) \right] \\
&\leq -\log(1 - \mu^0),
\end{aligned}$$

where the penultimate inequality is Pinsker's inequality (4.2) and the final inequality is (4.5) (with the normal type in the role of $\hat{\xi}$).

4.2.6.1 The implications of reputations not disappearing

Since posterior beliefs are a bounded martingale under $\mathbf{P}$, they converge $\mathbf{P}$ (and so $\widetilde{\mathbf{P}}$) almost surely, with limit μ^{∞}. If reputations do not disappear almost surely under $\widetilde{\mathbf{P}}$, then

$$\widetilde{\mathbf{P}}(\mu^{\infty} = 0) < 1,$$

and so there exists a $\lambda > 0$ and T_0 such that

$$\widetilde{\mathbf{P}}(\mu^{t} \geq \lambda, \forall t \geq T_0) > 0.$$

Define

$$F := \{\mu^{t} \geq \lambda, \forall t \geq T_0\}.$$

We first show that if reputations do not disappear almost surely under $\widetilde{\mathbf{P}}$, then eventually, with $\widetilde{\mathbf{P}}$-positive probability, player 2 must believe ξ_0 almost plays $\hat{\alpha}_1$ in every future period.

Lemma 4.3. ***(On F, eventually player** 2 **believes** ξ_0 **plays** $\hat{\alpha}_1$**)*** *Suppose* $\mu^t \not\to 0$ $\widetilde{\mathbf{P}}$*-a.s. There exists* T_1 *such that for*

$$B := \bigcap\nolimits_{t \geq T_1} X^t,$$

we have

$$\widetilde{\mathbf{P}}(B) \geq \widetilde{\mathbf{P}}(F \cap B) > 0.$$

Proof. We begin with the following calculation:

$$\sum_{t=0}^{\infty} E_{\widetilde{\mathbf{P}}}\left[(\mu^t)^2(1 - \mathbb{1}_{X^t})\right] \geq \widetilde{\mathbf{P}}(F) \sum_{t=0}^{\infty} E_{\widetilde{\mathbf{P}}}\left[(\mu^t)^2(1 - \mathbb{1}_{X^t}) \mid F\right] \tag{4.8}$$

$$\geq \widetilde{\mathbf{P}}(F)\lambda^2 \sum_{t=T_0}^{\infty} E_{\widetilde{\mathbf{P}}}\left[1 - \mathbb{1}_{X^t} \mid F\right].$$

Lemma 4.2 implies the left side of (4.8) is finite, and so there exists $T_1 \geq T_0$ such that

$$\sum_{t \geq T_1} E_{\widetilde{\mathbf{P}}}\left[1 - \mathbb{1}_{X^t} \mid F\right] < 1.$$

Then,

$$\begin{aligned}
\widetilde{\mathbf{P}}(F \cap B) &= \widetilde{\mathbf{P}}(F) - \widetilde{\mathbf{P}}(F \setminus B) \\
&= \widetilde{\mathbf{P}}(F) - \widetilde{\mathbf{P}}(F \cap (\Omega \setminus B)) \\
&= \widetilde{\mathbf{P}}(F)\left(1 - \widetilde{\mathbf{P}}(\Omega \setminus B \mid F)\right) \\
&\geq \widetilde{\mathbf{P}}(F)\left(1 - \sum\nolimits_{t \geq T_1} \widetilde{\mathbf{P}}(\Omega \setminus X^t \mid F)\right) > 0.
\end{aligned}$$

The next lemma effectively asserts that when player 2 is eventually (under $\widetilde{\mathbf{P}}$) always playing $\hat{a}_2$, the best response to $\hat{\alpha}_1$, then the normal player 1 sees histories that lead him to be confident that for many periods, player 2 is indeed playing $\hat{a}_2$. Note that on the set B, player 2 is playing $\hat{a}_2$ in every period after T_1. Denote the filtration describing player i's information by $(\mathcal{H}_i^t)_t$.

Lemma 4.4. (*On B, eventually player 1 figures out that player 2 is best responding to* $\hat{\alpha}_1$) *Suppose $\mu^t \not\to 0$ $\widetilde{\mathbf{P}}$-a.s. For the event B from Lemma 4.3, for all τ, there is a subsequence (t_n) such that as $n \to \infty$,*

$$\sum_{k=1}^{\tau} \left\{1 - E_{\widetilde{\mathbf{P}}}\left[\sigma_2(h_2^{t_n+k})(\hat{a}_2) \middle| \mathcal{H}_1^{t_n}\right]\right\} \mathbb{1}_B \to 0 \quad \widetilde{\mathbf{P}}\text{-a.s.} \tag{4.9}$$

The convergence in [4.9] *holds also when $\widetilde{\mathbf{P}}$ is replaced by* $\mathbf{P}$.

Proof. We prove [4.9]; obvious modifications to the argument proves it for $\mathbf{P}$. Recall that $h_1^{t+1} = (h_1^t, a_1^t y^t)$, and, for $k \geq 1$, denote player 1's k-step ahead prediction of his action and signal by $\beta_t^k(h_1^t) \in \Delta(A_1 \times Y)$, so that $\beta_t^k(h_1^t)(a_1^{t+k-1}y^{t+k-1}) = \widetilde{\mathbf{P}}(a_1^{t+k-1}y^{t+k-1} \mid h_1^t)$. Similarly, denote player 1's k-step prediction conditional on B by $\beta_{t,B}^k(h_1^t)$. The chain rule implies that, for all t and k,

$$\begin{aligned}
d(\beta_{t,B}^k(h_1^t) \| \beta_t^k(h_1^t)) \\
&\leq d\left(\widetilde{\mathbf{P}}(a_1^t y^t, \ldots, a_1^{t+k-1}y^{t+k-1} \mid h_1^t, B) \middle\| \widetilde{\mathbf{P}}(a_1^t y^t, \ldots, a_1^{t+k-1}y^{t+k-1} \mid h_1^t)\right) \\
&= E_{\widetilde{\mathbf{P}}(\cdot|h_1^t,B)} \sum_{k'=1}^{k} d\left(\widetilde{\mathbf{P}}(a_1^{t+k'-1}y^{t+k'-1} \mid h_1^{t+k'-1}, B) \middle\| \widetilde{\mathbf{P}}(a_1^{t+k'-1}y^{t+k'-1} \mid h_1^{t+k'-1})\right) \\
&= E_{\widetilde{\mathbf{P}}(\cdot|h_1^t,B)} \sum_{k'=1}^{k} d\left(\beta_{t+k'-1,B}^1(h_1^{t+k'-1}) \| \beta_{t+k'-1}^1(h_1^{t+k'-1})\right).
\end{aligned}$$

Consequently, for all $h_1^{T_1}$ satisfying $\widetilde{\mathbf{P}}(h_1^{T_1}, B) > 0$, and for all k,

$$\begin{aligned}
\sum_{t \geq T_1} E_{\widetilde{\mathbf{P}}}\left[d(\beta_{t,B}^k(h_1^t) \| \beta_t^k(h_1^t)) \middle| h_1^{T_1}, B\right] \\
&\leq \sum_{t \geq T_1} E_{\widetilde{\mathbf{P}}}\left[\sum_{k'=1}^{k} d\left(\beta_{t+k'-1,B}^1(h_1^{t+k'-1}) \| \beta_{t+k'-1}^1(h_1^{t+k'-1})\right) \middle| h_1^{T_1}, B\right] \\
&\leq k \sum_{t \geq T_1} E_{\widetilde{\mathbf{P}}}\left[d\left(\beta_{t,B}^1(h_1^t) \| \beta_t^1(h_1^t)\right) \middle| h_1^{T_1}, B\right] \\
&\leq -k \log \widetilde{\mathbf{P}}(B \mid h_1^{T_1}).
\end{aligned}$$

The last inequality follows from Lemma 4.1 applied to the equality

$$\widetilde{\mathbf{P}}\left(y^{T_1},\ldots,y^{T_1+\ell-1}\,\middle|\,h_1^{T_1}\right)=\widetilde{\mathbf{P}}(B\mid h_1^{T_1})\widetilde{\mathbf{P}}\left(y^{T_1},\ldots,y^{T_1+\ell-1}\,\middle|\,h_1^{T_1},B\right)$$
$$+(1-\widetilde{\mathbf{P}}(B\mid h_1^{T_1}))\widetilde{\mathbf{P}}\left(y^{T_1},\ldots,y^{T_1+\ell-1}\,\middle|\,h_1^{T_1},\Omega\setminus B\right)$$

and the chain rule (via an argument similar to that leading to [4.5]).

Thus, for all k,

$$E_{\widetilde{\mathbf{P}}}\left[d(\beta^k_{t,B}(h_1^t)\|\beta^k_t(h_1^t))\,\middle|\,h_1^{T_1},B\right]\to 0 \text{ as } t\to\infty,$$

and so (applying Pinsker's inequality)

$$E_{\widetilde{\mathbf{P}}}\left[\|\beta^k_{t,B}(h_1^t)-\beta^k_t(h_1^t)\|\,\middle|\,h_1^{T_1},B\right]\to 0 \text{ as } t\to\infty,$$

yielding, for all τ,

$$\sum_{k=1}^{\tau}E_{\widetilde{\mathbf{P}}}\left[\|\beta^k_{t,B}(h_1^t)-\beta^k_t(h_1^t)\|\,\middle|\,h_1^{T_1},B\right]\to 0 \text{ as } t\to\infty. \qquad [4.10]$$

Since

$$\|\beta^k_{t,B}(h_1^t)-\beta^k_t(h_1^t)\|\geq\kappa\left(1-E_{\widetilde{\mathbf{P}}}[\sigma_2(h_2^{t+k})(\hat{a}_2)|h_1^t]\right),$$

where

$$\kappa:=\min_{\substack{a_1\in A_1,\\ \alpha_2\in\Delta(A_2),\alpha_2(\hat{a}_2)=0}}\left\|\rho(\cdot|a_1,\hat{a}_2)-\rho(\cdot|a_1,\alpha_2)\right\|$$

is strictly positive by Assumption 4.3, [4.10] implies

$$\sum_{k=1}^{\tau}E_{\widetilde{\mathbf{P}}}\left[1-E_{\widetilde{\mathbf{P}}}[\sigma_2(h_2^{t+k})(\hat{a}_2)|h_1^t]\,\middle|\,h_1^{T_1},B\right]\to 0 \text{ as } t\to\infty.$$

This implies [4.9], since convergence in probability implies subsequence a.e. convergence (Chung, 1974, Theorem 4.2.3).

Using Lemma 4.4, we next argue that on $B\cap F$, player 1 believes that player 2 is eventually ignoring her history while best responding to $\hat{\alpha}_1$. In the following lemma, ε_2 is from [4.6]. The set $A_t(\tau)$ is the set of player 2 t-period histories such that player 2 ignores the next τ signals (the positive probability condition only eliminates 2's actions inconsistent with σ_2, since under Assumption 4.1, every signal realization has positive probability under $\hat{\alpha}_1$ and so under $\mathbf{P}$).

Lemma 4.5. ***(Eventually player 1 figures out that player 2 is best responding to $\hat{\alpha}_1$, independently of signals)*** *Suppose $\mu^t \not\to 0$ $\widetilde{\mathbf{P}}$-a.s. For all τ, there is a subsequence (t_m) such that as $m \to \infty$,*

$$\widetilde{\mathbf{P}}(A_{t_m}(\tau)|\mathcal{H}_1^{t_m})\mathbb{1}_{B\cap F} \to \mathbb{1}_{B\cap F} \quad \widetilde{\mathbf{P}}\text{-a.s.},$$

where

$$A_t(\tau) := \{h_2^t : \sigma_2(h_2^{t+k})(\hat{a}_2) > 1 - \varepsilon_2/2,\ \forall h_2^{t+k} \text{ s.t. } \mathbf{P}(h_2^{t+k}|h_2^t) > 0,$$
$$\forall k = 1, \ldots, \tau, \}.$$

Proof. Define

$$A_t^\tau := \{h_2^{t+\tau} : \sigma_2(h_2^{t+k})(\hat{a}_2) > 1 - \varepsilon_2/2,\ k = 1, \ldots, \tau\}.$$

The set A_t^τ is the set of $(t+\tau)$-period histories for player 2 at which she essentially best responds to $\hat{\alpha}_1$ for the *last* τ periods. Note that, viewed as subsets of Ω, $A_t(\tau) \subset A_t^\tau$.[19]

We then have

$$\sum_{k=1}^{\tau} E_{\widetilde{\mathbf{P}}}\left\{1 - E_{\widetilde{\mathbf{P}}}\left[\sigma_2(h_2^{t+k})(\hat{a}_2)\middle| h_1^t\right]\middle| B\cap F\right\}$$
$$= E_{\widetilde{\mathbf{P}}}\left\{E_{\widetilde{\mathbf{P}}}\left[\sum\nolimits_{k=1}^{\tau}\left(1 - \sigma_2(h_2^{t+k})(\hat{a}_2)\right)\middle| h_1^t\right]\middle| B\cap F\right\}$$
$$= E_{\widetilde{\mathbf{P}}}\left\{E_{\widetilde{\mathbf{P}}}\left[\sum\nolimits_{k=1}^{\tau}\left(1 - \sigma_2(h_2^{t+k})(\hat{a}_2)\right)\middle| h_1^t, A_t^\tau\right]\widetilde{\mathbf{P}}(A_t^\tau|h_1^t)\middle| B\cap F\right\}$$
$$+ E_{\widetilde{\mathbf{P}}}\left\{E_{\widetilde{\mathbf{P}}}\left[\sum\nolimits_{k=1}^{\tau}\left(1 - \sigma_2(h_2^{t+k})(\hat{a}_2)\right)\middle| h_1^t, \Omega\setminus A_t^\tau\right](1 - \widetilde{\mathbf{P}}(A_t^\tau|h_1^t))\middle| B\cap F\right\}.$$

Dropping the first term and using the implied lower bound from $\Omega \setminus A_t^\tau$ on $\sum_{k=1}^{\tau}(1 - \sigma_2(h_2^{t+k})(\hat{a}_2))$ yields

$$\sum_{k=1}^{\tau} E_{\widetilde{\mathbf{P}}}\left\{1 - E_{\widetilde{\mathbf{P}}}\left[\sigma_2(h_2^{t+k})(\hat{a}_2)\middle| h_1^t\right]\middle| B\cap F\right\}$$
$$\geq \frac{\varepsilon_2}{2}\left(1 - E_{\widetilde{\mathbf{P}}}\left\{\widetilde{\mathbf{P}}(A_t^\tau|h_1^t)\middle| B\cap F\right\}\right).$$

Lemma 4.4 then implies

$$\lim_{n\to\infty} E_{\widetilde{\mathbf{P}}}\left\{\widetilde{\mathbf{P}}(A_{t_n}^\tau|\mathcal{H}_1^{t_n})\middle| B\cap F\right\} = 1.$$

[19] More precisely, if $h_2^t(\omega)$ is the t-period player 2 history under $\omega \in \Omega$, then

$$\{\omega : h_2^t(\omega) \in A_t(\tau)\} \subset \{\omega : h_2^{t+\tau}(\omega) \in A_t^\tau\}.$$

As before, this then implies that on a subsequence (t_ℓ) of (t_n), we have, as $\ell \to \infty$,

$$\widetilde{\mathbf{P}}(A^\tau_{t_\ell}|\mathcal{H}^{t_\ell}_1)\mathbb{1}_{B\cap F} \to \mathbb{1}_{B\cap F} \quad \widetilde{\mathbf{P}}\text{-a.s.} \qquad [4.11]$$

Thus, the normal player 1 eventually (on $B \cap F$) assigns probability 1 to $(t_\ell + \tau)$-period histories for player 2 at which player 2 essentially best responds to $\hat{\alpha}_1$ for the *last* τ periods. It remains to argue that this convergence holds when $A_{t_\ell}(\tau)$ replaces $A^\tau_{t_\ell}$.

A similar argument to that proving [4.11] shows that as $m \to \infty$,

$$\mathbf{P}(A^\tau_{t_m}|\mathcal{H}^{t_m}_1)\mathbb{1}_{B\cap F} \to \mathbb{1}_{B\cap F} \quad \mathbf{P}\text{-a.s.}$$

(where the subsequence (t_m) can be chosen so that [4.11] still holds).[20]

Since $B \cap F \in \mathcal{H}^\infty_2$ and $\mathcal{H}^t_2 \subset \mathcal{H}^t_1$, this implies

$$\mathbf{P}(A^\tau_{t_m}|\mathcal{H}^{t_m}_2)\mathbb{1}_{B\cap F} \to \mathbb{1}_{B\cap F} \quad \mathbf{P}\text{-a.s.}$$

We claim that for all $\omega \in B \cap F$, for sufficiently large m if $h^{t_m+\tau}_2(\omega) \in A^\tau_{t_m}$, then $h^{t_m}_2(\omega) \in A_t(\tau)$. This then implies the desired result.

Suppose not. Then, for infinitely many t_m,

$$h^{t_m}_2(\omega) \notin A_{t_m}(\tau) \quad \text{and} \quad h^{t_m+\tau}_2(\omega) \in A^\tau_{t_m}.$$

At any such t_m, since there is at least one τ period continuation of the history $h^{t_m}_2(\omega)$ that is not in $A^\tau_{t_m}$, we have (from Assumption 4.1) $\widehat{\mathbf{P}}(A^\tau_{t_m} \mid \mathcal{H}^{t_m}_2)(\omega) \le 1 - \underline{\rho}^\tau$, where $\underline{\rho} > 0$ is defined in [4.7]. Moreover, on F, $\mu^{t_m} \ge \lambda$ for $t_m \ge T_0$. But this yields a contradiction, since these two imply that $\mathbf{P}(A^\tau_{t_m} \mid \mathcal{H}^{t_m}_2)(\omega)$ is bounded away from 1 infinitely often:

$$\begin{aligned}\mathbf{P}(A^\tau_{t_m} \mid \mathcal{H}^{t_m}_2)(\omega) &\le (1 - \mu^{t_m}) + \mu^{t_m}\widehat{\mathbf{P}}(A^\tau_{t_m} \mid \mathcal{H}^{t_m}_2)(\omega)\\ &\le 1 - \mu^{t_m} + \mu^{t_m}(1 - \underline{\rho}^\tau) = 1 - \lambda\underline{\rho}^\tau.\end{aligned}$$

Indeed, player 2 "knows" that player 1 believes that player 2 is eventually ignoring her history:

Lemma 4.6. ***(Eventually player 2 figures out that player 1 figures out...)*** *Suppose $\mu^t \not\to 0$ $\widetilde{\mathbf{P}}$-a.s. For all τ, there is a subsequence (t_m) such that as $m \to \infty$,*

$$E_{\widetilde{\mathbf{P}}}\left[\widetilde{\mathbf{P}}(A_{t_m}(\tau)|\mathcal{H}^{t_m}_1)\middle|\, \mathcal{H}^{t_m}_2\right]\mathbb{1}_{B\cap F} \to \mathbb{1}_{B\cap F}, \qquad \widetilde{\mathbf{P}}\text{-a.s.}$$

[20] Assumption 4.1 is weaker than full support imperfect monitoring, requiring only that all signals have positive probability under $\widehat{\mathbf{P}}$. Under full support imperfect monitoring ($\rho(y|a) > 0$ for all a), the following argument is valid with $\widetilde{\mathbf{P}}$ replacing $\mathbf{P}$, without the need to introduce a new subsequence.

Proof. Let (t_m) be the subsequence identified in Lemma 4.5. Conditioning on $\mathcal{H}_2^{t_m}$, Lemma 4.5 implies

$$E_{\widetilde{\mathbf{P}}}\left[\widetilde{\mathbf{P}}(A_{t_m}(\tau)|\mathcal{H}_1^{t_m})\mathbb{1}_{B\cap F}\middle|\,\mathcal{H}_2^{t_m}\right] - E_{\widetilde{\mathbf{P}}}\left[\mathbb{1}_{B\cap F}\middle|\,\mathcal{H}_2^{t_m}\right] \to 0, \qquad \widetilde{\mathbf{P}}\text{-a.s.}$$

The observation that $E_{\widetilde{\mathbf{P}}}[\mathbb{1}_{B\cap F} \mid \mathcal{H}_2^{t_m}]$ converges $\widetilde{\mathbf{P}}$-almost surely to $\mathbb{1}_{B\cap F}$ (since $B \cap F$ is in $\mathcal{H}_2^{\infty}$) yields

$$E_{\widetilde{\mathbf{P}}}\left[\widetilde{\mathbf{P}}(A_{t_m}(\tau)|\mathcal{H}_1^{t_m})\mathbb{1}_{B\cap F}\middle|\,\mathcal{H}_2^{t_m}\right] \to \mathbb{1}_{B\cap F}, \qquad \widetilde{\mathbf{P}}\text{-a.s.},$$

implying the lemma.[21]

4.2.6.2 *The contradiction and conclusion of the proof*

For fixed δ, there is a τ such that

$$(1-\delta)\gamma > 2\delta^{\tau} \max_a |u_1(a)|,$$

that is, the loss of γ (defined in [4.6]) in one period exceeds any possible potential gain deferred τ periods.

For this value of τ, for $\widetilde{\mathbf{P}}(A_{t_m}(\tau)|h_1^{t_m})$ close enough to 1, optimal play by player 1 requires $\sigma_1(h_1^{t_m})(\hat{a}_1) = 0$. Lemma 4.6 then implies that, on $B \cap F$, eventually player 2 predicts $\sigma_1(h_1^{t_m})(\hat{a}_1) = 0$, which implies

$$\lim_{m\to\infty} \widetilde{\mathbf{P}}(X^{t_m}|B \cap F) = 0,$$

which is a contradiction, since $\widetilde{\mathbf{P}}(X^t|B) = 1$ for all $t \geq T_1$.

4.2.7 Interpretation

There is a tension between Propositions 4.1 and 4.2. By Proposition 4.1 and its Corollaries 4.1 and 4.2, the normal player 1's ability to masquerade as a commitment type places a lower bound on his payoff. And yet, according to Proposition 4.2, when the player 1's actions are imperfectly monitored, if player 1 is indeed normal, then player 2 must learn that this is the case. Indeed, more can be said if there is a little more structure on the game.[22] If in addition to public imperfect monitoring of player 1's actions, the

[21] Define $g_t := E_{\widetilde{\mathbf{P}}}\left[\mathbb{1}_{B\cap F}\middle|\,\mathcal{H}_2^t\right]$. Then, since g_t is $\mathcal{H}_2^t$-measurable,

$$\begin{aligned}E_{\widetilde{\mathbf{P}}}&\left[\widetilde{\mathbf{P}}(A_t(\tau)|\mathcal{H}_1^t)\mathbb{1}_{B\cap F}\middle|\,\mathcal{H}_2^t\right] - E_{\widetilde{\mathbf{P}}}\left[\widetilde{\mathbf{P}}(A_t(\tau)|\mathcal{H}_1^t)\middle|\,\mathcal{H}_2^t\right]\mathbb{1}_{B\cap F}\\ &= E_{\widetilde{\mathbf{P}}}\left[\widetilde{\mathbf{P}}(A_t(\tau)|\mathcal{H}_1^t)(\mathbb{1}_{B\cap F} - g_t)\middle|\,\mathcal{H}_2^t\right] + E_{\widetilde{\mathbf{P}}}\left[\widetilde{\mathbf{P}}(A_t(\tau)|\mathcal{H}_1^t)\middle|\,\mathcal{H}_2^t\right](g_t - \mathbb{1}_{B\cap F}),\end{aligned}$$

which converges to 0 $\widetilde{\mathbf{P}}$-almost surely, since $g_t \to \mathbb{1}_{B\cap F}$ $\widetilde{\mathbf{P}}$-almost surely.

[22] One might seek refuge in the observation that one can readily find equilibria in perfect monitoring games in which reputation effects persist indefinitely. The initial finite horizon reputation models of Kreps and Wilson (1982) and Milgrom and Roberts (1982) exhibited equilibria in which player 1 chooses the

actions of player 2 are public, then not only does player 2 learn the type of player 1, but the continuation play of any Nash equilibrium converges to a Nash equilibrium of the complete information game (Cripps et al., 2004, Theorem 2).

The tension is only apparent, however, since the reputation bounds are only effective for high discount factors and only concern ex ante payoffs. In contrast, the disappearing reputation results fix the discount factor, and consider the asymptotic evolution of beliefs (and behavior).

Suppose player 2's actions are public and player 1 is either normal or the Stackelberg type. Fix $\eta > 0$ and a discount factor δ', and suppose player 1 is the normal type with probability $1 - \mu^0$. Suppose moreover, δ' is strictly larger than the bound $\bar{\delta}$ from Corollary 4.2, so that the normal player 1's expected payoff in any Nash equilibrium is at least $v_1^* - \eta$, where v_1^* is 1's Stackelberg payoff. By Proposition 4.2, over time the posterior will tend to fall. As illustrated in Figure 4.4, since the discount factor is fixed at δ', the posterior will eventually fall below μ', and so Corollary 4.2 no longer applies. However, since all signals have positive probability, there is positive probability that even after the belief has fallen below μ', some history of signals will again push the

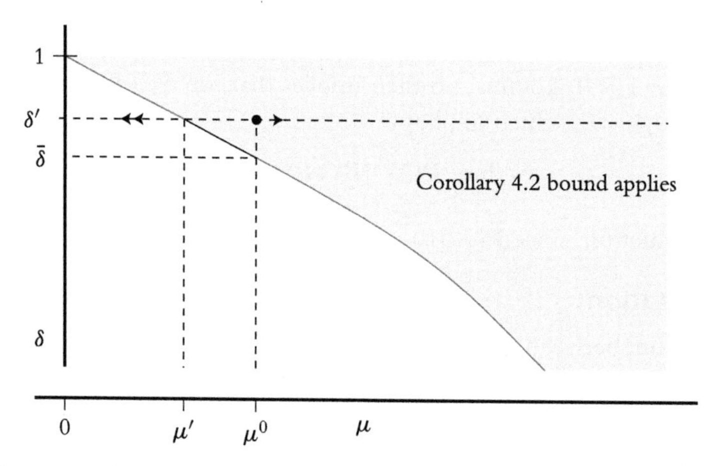

Figure 4.4 Illustration of reputation effects under imperfect monitoring. Suppose $\hat{\xi}$ is a Stackelberg type. For fixed $\eta > 0, \bar{\delta}$ is the lower bound from Corollary 4.2 for prior probability $\mu(\hat{\xi}) = \mu^0$. For players 1's discount factor $\delta' > \bar{\delta}$, Corollary 4.2 applies in period 0. The arrows indicate possible belief updates of player 2 about the likelihood of player 1 being the Stackelberg type $\hat{\xi}$. When player 1 is normal, almost surely the vertical axis is reached.

Stackelberg action in all but a relatively small number of terminal periods (see also Conlon, 2003). The analogous equilibria in infinitely repeated games feature the Stackelberg action in every period, enforced by the realization that any misstep reveals player 1 to be normal. However, it is somewhat discomforting to rely on properties of equilibria in perfect monitoring games that have no counterparts in games of imperfect monitoring.

posterior above μ'. Since continuation play forms a Nash equilibrium of the incomplete information game, Corollary 4.2 again applies. The probability of such histories becomes vanishingly small as the histories become long.

As an example, recall the imperfect monitoring product-choice game (with public player 2 actions) with the pure Stackelberg type $\hat{\xi} = \xi(H)$ considered in Example 4.2. For the parameterization in that example, the complete information repeated game has a unique sequential equilibrium outcome, in which L is played in every period. Nonetheless, by Corollary 4.2, given any prior probability μ^0 on $\xi(H)$, for sufficiently high δ, player 1's ex ante equilibrium payoff is at least $1\frac{3}{4}$, higher than the complete information equilibrium payoff of 1. Fix such a δ. We then know that in *any* equilibrium, for many periods, the short-lived player must be playing c (which requires that the short-lived player in those periods assign sufficient probability to H). However, the equilibrium incentive player 1 has to cheat in each period (if only with low probability), arising from the imperfection in the monitoring, implies that (with high probability) the posterior will eventually fall sufficiently that the lack of complete information intertemporal incentives forces player 1 to play L (and player 2 to respond with s). Indeed, from Cripps et al. (2004, Theorem 3), for high δ and for any $\varepsilon > 0$ there is an equilibrium of the incomplete information game that has ex ante player 1 payoffs over $1\frac{3}{4}$, and yet the $\widetilde{\mathbf{P}}$ probability of the event that eventually Ls is played in *every* period is at least $1 - \varepsilon$.

This discussion raises an important question: why do we care about beliefs after histories that occur so far into the future as to be effectively discounted into irrelevance when calculating ex ante payoffs?

While the short run properties of equilibria are interesting, we believe that the long run equilibrium properties are relevant in many situations. For example, an analyst may not know the age of the relationship to which the model is to be applied. We do sometimes observe strategic interactions from a well-defined beginning, but we also often encounter on-going interactions whose beginnings are difficult to identify. Long run equilibrium properties may be an important guide to behavior in the latter cases. Alternatively, one might take the view of a social planner who is concerned with the continuation payoffs of the long-lived player and with the fate of all short-lived players, even those in the distant future. Finally, interest may center on the steady states of models with incomplete information, again directing attention to long run properties.

4.2.8 Exogenously informative signals

Section 4.2.6 established conditions under which reputations eventually disappear. Section 4.2.7 explained why this result must be interpreted carefully, with an emphasis on keeping track of which limits are taken in which order. This section, considering the impact of exogenously informative signals, provides a further illustration of the subtleties that can arise in taking limits. This section is based on Hu (2014), which provides a general analysis; Wiseman's (2009) original analysis considered the chain store example.

We consider the product-choice game of Figure 4.1 with the set of types $\Xi = \{\xi_0, \hat{\xi}\}$, where ξ_0 is the normal type and $\hat{\xi} = \xi(H)$ is the Stackelberg type. The stage game is a game of perfect monitoring. *Imperfect* monitoring played an important role in the disappearing-reputation result of Section 4.2.6. In contrast, there is no difficulty in constructing equilibria of the perfect monitoring product-choice game in which both the normal and Stackelberg types play H in every period, with customers never learning the type of player 1. It thus seems as if a game of perfect monitoring is not particularly fertile ground for studying temporary reputations.

However, suppose that at the end of every period, the players observe a public signal independent of the players' actions. In each period t, the signal z^t is an identically and independently distributed draw from $\{z_0, \hat{z}\}$. The signal is informative about player 1's type: $0 < \pi(\hat{z} \mid \xi_0) := \mathbb{P}(\hat{z} \mid \xi_0) < \mathbb{P}(\hat{z} \mid \hat{\xi}) =: \pi(\hat{z} \mid \hat{\xi}) < 1$. A sufficiently long sequence of such signals suffices for player 2 to almost surely learn player 1's type.

It now appears as if reputation arguments must lose their force. Because the public signals are unrelated to the players' actions, there is a lower bound on the rate at which player 2 learns about player 1. If player 1 is very patient, the vast bulk of his discounted expected payoff will come from periods in which player 2 is virtually certain that 1 is the normal type (assuming 1 is indeed normal). It then seems as if there is no chance to establish a reputation-based lower bound on 1's payoff. Nonetheless, for all $\varepsilon > 0$, there exists $\bar{\delta} < 1$ such that if $\delta \in (\bar{\delta}, 1)$, then player 1's equilibrium payoff is at least

$$\underline{w}_H(\ell) - \varepsilon, \qquad [4.12]$$

where $\underline{w}_H$ is the function illustrated in Figure 4.2, and $\ell := d(\pi(\cdot \mid \xi_0) \| \pi(\cdot \mid \hat{\xi}))$ is the relative entropy of $\pi(\cdot \mid \hat{\xi})$ with respect to $\pi(\cdot \mid \xi_0)$.

Suppose $\pi(\hat{z} \mid \xi_0) = 1 - \pi(\hat{z} \mid \hat{\xi}) = \alpha < \frac{1}{2}$. Then, $\ell = (1 - 2\alpha)\log[(1 - \alpha)/\alpha]$. For α near zero, individual signals are very informative, and we might expect that the reputation arguments would be ineffective. This is what we find: for $\alpha < 0.22$, we have $\ell > \log 2$, and so only the conclusion that player 1's equilibrium payoff is bounded above 0, which is *less* than player 1's minmax payoff in the stage game. Hence, sufficiently precise signals can indeed preclude the construction of a reputation bound on payoffs. On the other hand, as α approaches $1/2$, so that signals become arbitrarily uninformative, ℓ approaches 0 and so the bound approaches 2, the reputation bound in the game without exogenous signals. For intermediate values of α that are not too large, reputations have some force, though not as much as if the public signals were completely uninformative.

Reputations still have force with exogenous signals because the signals have full support. Suppose customers have seen a long history of exogenous signals suggesting that the firm is normal (which is likely when the firm is indeed normal). If they do not expect the normal type to exert high effort after some such history, high effort in that period results in a dramatic increase in the posterior that the firm is the Stackelberg type and hence will exert high effort in the future. While this can happen infinitely often, it can't happen too frequently (because otherwise the resulting increases in posterior overwhelm the exogenous signals, leading to a similar contradiction as for the canonical reputations argument), resulting in [4.12].[23]

We conclude this subsection with the proof that [4.12] is a lower bound on equilibrium payoffs for player 1. The space of uncertainty is now $\Omega := \{\xi_0, \hat{\xi}\} \times (A_1 \times A_2 \times Z)^\infty$. The set of endogenous signals is A_1, while the the set of exogenous signals is given by Z. Fix an equilibrium $\sigma = (\sigma_1, \sigma_2)$ of the incomplete information game. As before, σ induces the unconditional probability measure $\mathbf{P}$ on Ω, while $\hat{\xi}$ (with σ_2) induces the measure $\widehat{\mathbf{P}}$.

While it is no longer the case that there is a uniform bound on the number of large expected one-step ahead prediction errors of the form [4.5], there is a useful nonuniform bound.

Let $\widehat{\mathbf{Q}}$ be the measure induced on Ω by the normal type ξ_0 playing H in every period. Then, for any history $h^t \in (A_1 \times A_2 \times Z)^t$, since the exogenous signals are independent of actions,

$$\widehat{\mathbf{Q}}^t(h^t) = \widehat{\mathbf{P}}^t(h^t) \prod_{\tau=0}^{t-1} \frac{\pi(z^\tau(h^t) \mid \xi_0)}{\pi(z^\tau(h^t) \mid \hat{\xi})},$$

where, as usual, we denote the marginals on t-period histories by a superscript t. Then, since $\widehat{\mathbf{P}}^t(h^t)/\mathbf{P}^t(h^t) \le 1/\mu(\hat{\xi})$,

$$\begin{aligned} d\left(\widehat{\mathbf{Q}}^t \,\|\, \mathbf{P}^t\right) &= \sum_{h^t} \widehat{\mathbf{Q}}^t(h^t) \log \frac{\widehat{\mathbf{P}}^t(h^t)}{\mathbf{P}^t(h^t)} + \sum_{h^t} \widehat{\mathbf{Q}}^t(h^t) \sum_{\tau=0}^{t-1} \log \frac{\pi(z^\tau(h^t) \mid \xi_0)}{\pi(z^\tau(h^t) \mid \hat{\xi})} \\ &= \sum_{h^t} \widehat{\mathbf{Q}}^t(h^t) \log \frac{\widehat{\mathbf{P}}^t(h^t)}{\mathbf{P}^t(h^t)} + \sum_{\tau=0}^{t-1} \sum_{z^\tau \in Z} \pi(z^\tau \mid \xi_0) \log \frac{\pi(z^\tau \mid \xi_0)}{\pi(z^\tau \mid \hat{\xi})} \\ &\le -\log \mu(\hat{\xi}) + t\ell. \end{aligned} \qquad [4.13]$$

It remains to bound the total discounted number of one-step ahead prediction errors. In the notation of Section 4.2.4.2, $\hat{p}$ is the degenerate distribution assigning probability 1 to H, while $p(h^t)$ is the probability that 2 assigns to player 1 choosing H in period t, given the history h^t.

[23] In the model with perfect monitoring with no exogenous signals, customers can only be surprised a finite number of times under the considered deviation (if not, the increases in the posterior after each surprise eventually result in a posterior precluding further surprises, since posteriors never decrease). In contrast, with exogenous signals, there is the possibility of an infinite number of surprises, since the expected posterior decreases when there are no surprises. Nonetheless, if this happens too frequently, the increasing updates from the surprises dominate the decreasing updates from the exogenous signals and again the increases in the posterior after surprises eventually preclude further surprises.

Then, from the chain rule [4.3], where the last term is the expected relative entropy of the period t z-signal predictions, we have

$$d\left(\widehat{\mathbf{Q}}^{t+1} \,\|\, \mathbf{P}^{t+1}\right) = d\left(\widehat{\mathbf{Q}}^{t} \,\|\, \mathbf{P}^{t}\right) + E_{\widehat{\mathbf{Q}}^t} d(\hat{p} \| p(h^t)) + E_{\widehat{\mathbf{Q}}^t, \hat{p}} d(\widehat{\mathbf{Q}}_Z^{t+1} \| \mathbf{P}_Z^{t+1})$$
$$\geq d\left(\widehat{\mathbf{Q}}^{t} \,\|\, \mathbf{P}^{t}\right) + E_{\widehat{\mathbf{Q}}^t} d(\hat{p} \| p(h^t)).$$

Thus, where we normalize $d(\widehat{\mathbf{Q}}^0 \| \mathbf{P}^0) = 0$,

$$(1-\delta) \sum_{t=0}^{\infty} \delta^t E_{\widehat{\mathbf{Q}}^t} d(\hat{p} \| p(h^t)) \leq (1-\delta) \sum_{t=0}^{\infty} \delta^t \left[d\left(\widehat{\mathbf{Q}}^{t+1} \,\|\, \mathbf{P}^{t+1}\right) - d\left(\widehat{\mathbf{Q}}^{t} \,\|\, \mathbf{P}^{t}\right) \right]$$
$$= (1-\delta)^2 \sum_{t=1}^{\infty} \delta^{t-1} d\left(\widehat{\mathbf{Q}}^{t} \,\|\, \mathbf{P}^{t}\right)$$
$$\leq -(1-\delta) \log \mu(\hat{\xi}) + \ell,$$

where the second inequality follows from [4.13] and some algebraic manipulation. The bound now follows from an argument similar to the proof of Proposition 4.1.

4.3. REPUTATIONS WITH TWO LONG-LIVED PLAYERS

Section 4.2 studied reputations in the most common context, that of a long-lived player facing a succession of short-lived players. This section examines the case in which player 2 is also a long-lived player.

Reputation results for the case of two long-lived players are not as strong as those for the long-lived/short-lived case, and a basic theme of the work presented in this section is the trade-off between the specificity of the model and the strength of the results. To get strong results, one must either restrict attention to seemingly quite special games, or must rely on seemingly quite special commitment types.

The standard models with two long-lived players fix a discount factor for player 2 and then examine the limit as player 1's discount factor approaches one, making player 1 arbitrarily relatively patient. There is a smaller literature that examines reputations in the case of two equally patient long-lived players, with even weaker results. As we make the players progressively more symmetric by moving from the case of a short-lived player 2, to the case of a long-lived player 2 but arbitrarily more patient player 1, to the case of two equally patient long-lived players, the results become successively weaker. This is unsurprising. Reputation results require some asymmetry. A reputation result imposes a lower bound on equilibrium payoffs, and it is typically impossible to guarantee such a payoff to both players, For example, it is typically impossible for both players to receive their Stackelberg payoffs. Some asymmetry must then lie behind a result that guarantees such a payoff to one player, and the weaker this asymmetry, the weaker the reputation result.

4.3.1 Types vs. actions

Suppose we simply apply the logic of Section 4.2, hoping to obtain a Stackelberg reputation bound when both players are long-lived and player 1's characteristics are unknown. To keep things simple, suppose there is perfect monitoring. If the normal player 1 persistently plays the Stackelberg action and player 2 assigns positive prior probability to a type committed to that action, then player 2 must eventually attach high probability to the event that the Stackelberg action is played in the future. This argument depends only upon the properties of Bayesian belief revision, independently of whether the person holding the beliefs is long-lived or short-lived.

If this belief suffices for player 2 to play a best response to the Stackelberg action, as is the case when player 2 is short lived, then the remainder of the argument is straightforward. The normal player 1 must eventually receive very nearly the Stackelberg payoff in each period of the repeated game. By making player 1 sufficiently patient, we can ensure that this consideration dominates player 1's payoffs, putting a lower bound on the latter.

The key step when working with two long-lived players is thus to establish conditions under which, as player 2 becomes increasingly convinced that the Stackelberg action will appear, she must eventually play a best response to that action. This initially seems obvious. If player 2 is "very" convinced that the Stackelberg action will be played not only now but for sufficiently many periods to come, there appears to be nothing better she can do than play a stage-game best response.

This intuition misses the following possibility. Player 2 may be choosing something other than a best response to the Stackelberg action out of fear that a current best response may trigger a disastrous future punishment. This punishment would not appear if player 2 faced the Stackelberg type, but player 2 can be made confident only that she faces the Stackelberg action, not the Stackelberg type. The fact that the punishment lies off the equilibrium path makes it difficult to assuage player 2's fear of such punishments.

The short-lived players of Section 4.2 find themselves in the same situation: convinced that their long-lived opponent will play the Stackelberg action, but uncertain as to what affect their own best response to this Stackelberg action will have on future behavior. However, because they are short-lived, this uncertainty does not affect their behavior. The difference between expecting the Stackelberg action and expecting the Stackelberg type (or more generally between expecting any action and expecting the corresponding type committed to that action) is irrelevant in the case of short-lived opponents, but crucial when facing long-lived opponents.

4.3.2 An example: The failure of reputation effects

This section presents a simple example, adapted from Schmidt (1993), illustrating the new issues that can arise when building a reputation against long-lived opponents.

	L	C	R
T	$10,10$	$0,0$	$-z,9$
B	$0,0$	$1,1$	$1,0$

Figure 4.5 A modified coordination game, $z \in [0,8)$.

Consider the game in Figure 4.5. Not only is the profile TL a Nash equilibrium of the stage game in Figure 4.5, it gives the player 1 his largest feasible payoff. At the same time, BC is another Nash equilibrium of the stage game, so the complete information repeated game with perfect monitoring has the infinite repetition of BC as another equilibrium outcome. It is immediate that if player 2 only assigns positive probability to the normal type (the payoff type of player 1 with payoffs shown in Figure 4.5) and the simple Stackelberg-action type (which always plays T), then this particular form of incomplete information again generates reputation effects.

In the example, we add another commitment type, an enforcement type, whose behavior depends on history (so this type is not simple). The idea is to use this type to induce player 2 to not always statically best respond to T. Instead, on the candidate equilibrium path, player 2 will play L in even periods, and R in odd periods. The enforcement commitment type plays T initially, and continues with T unless player 2 stops playing L in even periods and R in odd periods, at which point the enforcement type plays B thereafter.

It is player 2's fear of triggering the out-of-equilibrium behavior of the enforcement type that will prevent player 1 from building an effective reputation. The prior distribution puts probability 0.8 on the normal type and probability 0.1 on each of the other types. At the cost of a somewhat more complicated exposition, we could replace each of these commitment types with a payoff type (as in Schmidt, 1993), and derive the corresponding behavior as part of the equilibrium.

We describe a strategy profile with an outcome path that alternates between TL and TR, beginning with TL, and then verify it is an equilibrium for patient players. An automaton representation of the profile is given in Figure 4.6.

- *Normal type of player* 1: Play T after any history except one in which player 1 has at least once played B, in which case play B.
- *Player* 2:
  - After any history h^t featuring the play of TL in even periods and TR in odd periods, play L if t is even and R if t is odd. After any history h^t featuring the play of TL in even periods preceding $t-1$ and TR in odd periods preceding $t-1$, but in which player 1 plays B in period $t-1$, play C in period t and for every continuation history (since player 1 has revealed himself to be normal).
  - After any history h^t in which h^{t-1} features the play of TL in even periods and TR in odd periods, and player 2 does not play L if $t-1$ is even or R if $t-1$ is odd,

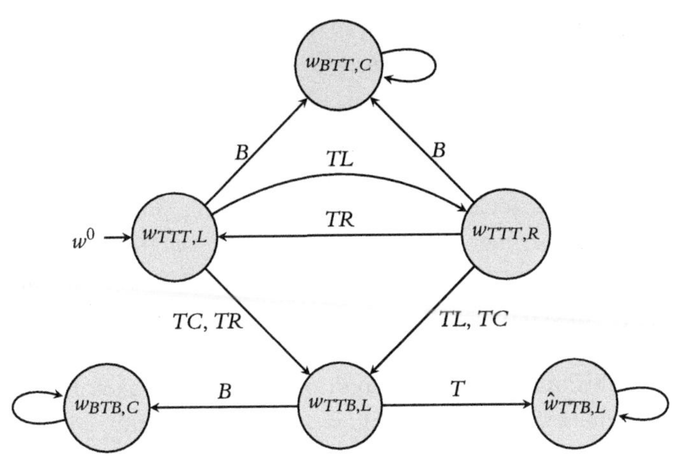

Figure 4.6 An automaton representation of the strategy profile illustrating the failure of reputation effects in Section 4.3.2. In state $w_{a_1^n a_1^s a_1^e, a_2}$ the normal type plays a_1^n, the Stackelberg type plays a_1^s (which is always equal to T), the enforcement type plays a_1^e, and player 2 plays a_2. Play begins in state $w^0 = w_{TTT,L}$. Transitions are determined by the observed actions $a_1 a_2$, and states $w_{BTB,C}$, $w_{BTT,C}$, and $\hat{w}_{TTB,L}$ are absorbing.

play L in period t. If 1 plays T in period t, play L in period $t+1$ and for every continuation history (interpreted as the result of attaching probability zero to the enforcement type and making no further belief revisions). If 1 plays B in period t, play C in period $t+1$ and for every continuation history (interpreted as the result of attaching probability one to the enforcement type and making no further belief revisions).

The normal player 1's behavior depends only on his previous actions, featuring the constant play of T in equilibrium and with any switch to B triggering the subsequent constant play of B. The first item in the description of player 2's behavior describes her actions after histories in which 2 has made no deviations from equilibrium play, and the last item describes how player 2 behaves after she has deviated from equilibrium play.

We now argue that these strategies constitute an equilibrium for $z < 8$ and sufficiently large δ. First consider player 1. Along the equilibrium path, the normal type of player 1 earns a payoff that, for large δ is very close to $(10 - z)/2$ respectively. A deviation leads to a continuation payoff of 1. Hence, for any $z < 8$, there is a sufficiently large $\overline{\delta} < 1$ such that for $\delta \geq \overline{\delta}$, it is optimal for the normal player 1 to play T after every equilibrium history.

Should play ever leave the equilibrium path as a result of player 1's having chosen B, subsequent play after any continuation history constitutes a stage-game Nash equilibrium

(BC) for player 2 and the normal type of player 1, and hence is optimal for player 1. Should play leave the equilibrium path because player 2 has deviated, then following the equilibrium strategies earns the player 1 a payoff of 10 in every subsequent period, and player 1 is thus playing optimally.

Now consider player 2. Along the equilibrium path, player 2 learns nothing about player 1. If player 2 deviates from the equilibrium, she has a chance to screen the types of player 1, earning a continuation payoff of 10 against the normal or Stackelberg type and a continuation payoff of at most 1 against the enforcement type. The resulting expected payoff is at most 9.1, falling short of the equilibrium payoff of almost 9.5 (for a patient player 2). This completes the argument that these strategies are an equilibrium under the conditions on discount factors that appear in the reputation result, namely that we fix δ_2 (allowed to be sufficiently large) and then let δ_1 approach 1.

By increasing the absolute value of z in Figure 4.5, while perhaps requiring more patience for player 1, we obtain an equilibrium in which the normal player 1's payoff is arbitrarily close to his pure minmax payoff of 1. It is thus apparent that reputation considerations can be quite ineffective.

In equilibrium, player 2 *is* convinced that she will face the Stackelberg action in every period. However, she dares not play a best response out of fear that doing so has adverse future consequences, a fear made real by the possibility of the enforcement type.[24]

Reputation arguments with two long-lived players thus cannot be simple extensions of long-lived/short-lived player results. Player 1 has the option of leading a long-lived player 2 to expect commitment behavior on the path of play, but this no longer suffices to ensure a best response from player 2, no matter how firm the belief. Despite a flurry of activity in the literature, the results are necessarily weaker and more specialized. In particular, the results for two long-lived players exploit some structure of the game or the setting to argue that player 2 will play a best response to a particular action, once convinced that player 1 is likely to play that action.

The remainder of this section illustrates the arguments and results that have appeared for the case of two long-lived players. As has the literature, we concentrate on the case in which there is uncertainty only about player 1's type, and in which player 2's discount factor is fixed while player 1's discount factor is allowed to become arbitrarily large. Hence, while we are making the players more symmetric in the sense of making both long-lived, we are still exploiting asymmetries between the players. There has been some work, such as Cripps et al. (2005) and Atakan and Ekmekci (2012,

[24] Celentani et al. (1996, Section 5) describe an example with similar features, but involving only a normal and Stackelberg type of player 1, using the future play of the normal player 1 to punish player 2 for choosing a best response to the Stackelberg action.

2013), which we will not discuss here, on games with two long-lived players who are equally patient, leaving only the one-sided incomplete information as the source of asymmetry.

As in the case of short-lived player 2's, Cripps et al. (2004, 2007) establish conditions under which a long-lived player 1 facing a long-lived opponent in a game of imperfect monitoring can only maintain a temporary reputation. If player 1 is indeed normal, then with probability 1 player 2 must eventually attach arbitrarily high probability to 1's being normal. Once again, we are reminded that the ex ante payoff calculations that dominate the reputations literature may not be useful in characterizing long run behavior.

4.3.3 Minmax-action reputations

This section presents a slight extension of Schmidt (1993), illustrating a reputation effect in games with two long-lived players.

We consider a perfect monitoring repeated game with two long-lived players, 1 and 2, with finite action sets. Player 1's type is determined by a probability distribution μ with a finite or countable support. The support of μ contains the normal type of player 1, ξ_0, and a collection of commitment types.

4.3.3.1 Minmax-action types

Let $\underline{v}_1(\xi_0, \delta_1, \delta_2)$ be the infimum, over the set of Nash equilibria of the repeated game, of the normal player 1's payoffs. For the action $\alpha_1 \in \Delta(A_1)$, let

$$v_1^*(\alpha_1) := \min_{a_2 \in B(\alpha_1)} u_1(\alpha_1, a_2),$$

where $B(\alpha_1)$ is the set of player 2 stage-game best responses to α_1.

The basic result is that if there exists a simple commitment type committed to an action $\hat{\alpha}_1$ minmaxing player 2, then the normal player 1, if sufficiently patient, is assured an equilibrium payoff close to $v_1^*(\hat{\alpha}_1)$:[25]

Proposition 4.3. *Suppose $\mu(\xi(\hat{\alpha}_1)) > 0$ for some action $\hat{\alpha}_1$ that minmaxes player 2. For any $\eta > 0$ and $\delta_2 < 1$, there exists a $\underline{\delta}_1 \in (0, 1)$ such that for all $\delta_1 \in (\underline{\delta}_1, 1)$,*

$$\underline{v}_1(\xi_0, \delta_1, \delta_2) > v_1^*(\hat{\alpha}_1) - \eta.$$

Fix an action $\hat{\alpha}_1$ satisfying the criteria of the proposition. The key to establishing Proposition 4.3 is to characterize the behavior of player 2, on histories likely to arise

[25] An action $\hat{\alpha}_1$ minmaxes player 2 if

$$\hat{\alpha}_1 \in \arg\min_{\alpha_1 \in \Delta A_1} \max_{a_2 \in A_2} u_2(\alpha_1, a_2).$$

Schmidt (1993) considered the case in which the action $\hat{\alpha}_1$ is a pure action, i.e., that there is a pure action that mixed-action minmaxes player 2.

when player 1 repeatedly plays $\hat{\alpha}_1$. Let $\hat{\Omega}$ be the event that player 1 always plays $\hat{\alpha}_1$. For any history h^t that arises with positive probability given $\hat{\Omega}$ (i.e., $\mathbf{P}\{\omega \in \hat{\Omega} : h^t(\omega) = h^t\} > 0$), let $E_{\mathbf{P}}[U_2(\sigma|_{h^t}) \mid \hat{\Omega}]$ be 2's expected continuation payoff, conditional on the history h^t and $\hat{\Omega}$. Let $\underline{v}_2$ be player 2's minmax payoff.

Lemma 4.7. *Fix $\delta_2 \in (0,1)$ and $\varepsilon > 0$. There exists $L \in \mathbb{N}$ and $\kappa > 0$ such that, for all Nash equilibria σ, pure strategies $\tilde{\sigma}_2$ satisfying $\tilde{\sigma}_2(\bar{h}^{t'}) \in \operatorname{supp}\sigma_2(\bar{h}^{t'})$ for all $\bar{h}^{t'} \in H^{t'}$, and histories $h^t \in H^t$ with positive probability under $\hat{\Omega}$, if*

$$E_{\mathbf{P}}[U_2((\sigma_1,\tilde{\sigma}_2)|_{h^t}) \mid \hat{\Omega}] \leq \underline{v}_2 - \varepsilon, \qquad [4.14]$$

then there is a period $t+\tau$, $0 \leq \tau \leq L$, and a continuation history $h^{t+\tau}$ that has positive probability when $\hat{\alpha}_1$ is played in periods $t, t+1, \ldots, t+\tau-1$, and 2 plays $\tilde{\sigma}_2$, such that at $h^{t+\tau}$, player 2's predicted distribution of player 1's action α_1 in period $t+\tau$ satisfies $d(\hat{\alpha}_1\|\alpha_1) > \kappa$.

Intuitively, condition [4.14] indicates that player 2's equilibrium strategy gives player 2 a payoff below her minmax payoff, conditional on $\hat{\alpha}_1$ always being played. But no equilibrium strategy can ever give player 2 an expected payoff lower than her minmax payoff, and hence it must be that player 2 does *not* expect $\hat{\alpha}_1$ to always be played.

Proof. Suppose, to the contrary, that for the sequence $(L_n, \kappa_n)_n$, where $L_n = n$ and $\kappa_n = n^{-1}$, there is a sequence of equilibria σ^n violating the result. Fix n, and let h^t be a history that occurs with positive probability under $\hat{\Omega}$ for which [4.14] holds, and such that in each of the next $L_n + 1$ periods, if player 1 has always played $\hat{\alpha}_1$ and player 2 follows $\tilde{\sigma}_2$, then player 2's predicted distribution of player 1's action α_1 satisfies $d(\hat{\alpha}_1\|\alpha_1) \leq \kappa_n$ after every positive probability history. We will derive a contradiction for sufficiently large n.

Given such posterior beliefs for player 2, an upper bound on player 2's period t expected continuation payoff is given by (where $u_2^{t+\tau}(\kappa_n)$ is player 2's expected payoff under the strategy profile $\tilde{\sigma}^n \equiv (\sigma_1^n, \tilde{\sigma}_2^n)$ in period $t+\tau$ when player 1 plays an action α_1 with $d(\hat{\alpha}_1\|\alpha_1) \leq \kappa$ in period $t+\tau$, and M is an upper bound on the magnitude of player 2's stage-game payoff),[26]

$$\begin{aligned} E_{\mathbf{P}}[U_2(\tilde{\sigma}^n|_{h^t}) \mid h^t] &\leq (1-\delta_2)u_2^t(\kappa_n) + (1-\delta_2)\delta_2 u_2^{t+1}(\kappa_n) + (1-\delta_2)\delta_2^2 u_2^{t+2}(\kappa_n) \\ &\quad + \cdots + (1-\delta_2)\delta_2^{L_n} u_2^{t+L_n}(\kappa_n) + \delta_2^{L_n+1}M \\ &= (1-\delta_2)\sum_{\tau=0}^{L_n} \delta_2^{\tau} u_2^{t+\tau}(\kappa_n) + \delta_2^{L_n+1}M. \end{aligned}$$

[26] The history h^t appears twice in the notation $E_{\mathbf{P}}[U_2(\tilde{\sigma}|_{h^t}) \mid h^t]$ for 2's expected continuation value given h^t. The subscript h^t reflects the role of h^t in determining player 2's continuation strategy. The second h^t reflects the role of the history h^t in determining the beliefs involved in calculating the expected payoff given that history.

For L_n sufficiently large and κ_n sufficiently small (i.e., n sufficiently large), the upper bound is within $\varepsilon/2$ of $E_{\mathbf{P}}[U_2(\tilde{\sigma}^n|_{h^t}) \mid \hat{\Omega}]$, i.e., it is close to player 2's expected continuation payoff conditioning on the event $\hat{\Omega}$. Hence, using [4.14], for large n

$$E_{\mathbf{P}}[U_2(\tilde{\sigma}^n|_{h^t}) \mid h^t] \leq E_{\mathbf{P}}[U_2(\tilde{\sigma}^n|_{h^t}) \mid \hat{\Omega}] + \frac{\varepsilon}{2} < \underline{v}_2.$$

But then player 2's continuation value at history h^t, $E_{\mathbf{P}}[U_2(\tilde{\sigma}^n|_{h^t}) \mid h^t]$, is strictly less than her minmax payoff, a contradiction.

Proof of Proposition 4.3. Fix a Nash equilibrium with player 1 payoff v_1. Fix δ_2 and η, and denote by $B_{\varepsilon'}(\hat{\alpha}_1)$ the ε'-neighborhood of $B(\hat{\alpha}_1)$. There exists $\varepsilon' > 0$ such that if $\sigma_2(h^t) \in B_{\varepsilon'}(\hat{\alpha}_1)$, then

$$u_1(\hat{\alpha}_1, \sigma_2(h^t)) > v_1^*(\hat{\alpha}_1) - \frac{\eta}{2}.$$

Let $\varepsilon := \underline{v}_2 - \max_{\alpha_2 \notin B_{\varepsilon'}(\hat{\alpha}_1)} u_2(\hat{\alpha}_1, \alpha_2) > 0$, and let L and κ be the corresponding values from Lemma 4.7. Lemma 4.7 implies that if the outcome of the game is contained in $\hat{\Omega}$, and player 2 fails to play ε'-close to a best response to $\hat{\alpha}_1$ in some period t, then there must be a period $t + \tau$, with $0 \leq \tau \leq L$ and a continuation history $h^{t+\tau}$ that has positive probability when $\hat{\alpha}_1$ is played in periods $t, t+1, \ldots, t+\tau-1$, and 2 plays $\tilde{\sigma}_2$, such that at $h^{t+\tau}$, player 2's one-step ahead prediction error when 1 plays $\hat{\alpha}_1$ is at least κ, i.e.,

$$d(\hat{p} \| p^{t+\tau}(h^{t+\tau})) \geq \kappa.$$

Define $\underline{\hat{\alpha}}_1 := \min\{\hat{\alpha}_1(a_1) : \hat{\alpha}_1(a_1) > 0\}$ and let $2M$ be the difference between the largest and smallest stage-game payoffs for player 1. We can bound the amount by which player 1's equilibrium payoff v_1 falls short of $v_1^*(\hat{\alpha}_1)$ as follows:

$$\begin{aligned}
v_1^*(\hat{\alpha}_1) - v_1 &\leq 2M(1-\delta_1) \sum_{t=0}^{\infty} \delta_1^t E_{\hat{\mathbf{P}}}[\mathbb{1}_{\{\tilde{\sigma}_2(h^t) \notin B_{\varepsilon'}(\hat{\alpha}_1)\}}] + \frac{\eta}{2} \\
&\leq 2M(1-\delta_1) \frac{L+1}{\underline{\hat{\alpha}}_1^{L+1} \delta_1^L} \sum_{t=0}^{\infty} \delta_1^t E_{\hat{\mathbf{P}}}[\mathbb{1}_{\{d(\hat{p} \| p^t(h^t)) > \kappa\}}] + \frac{\eta}{2} \\
&\leq 2M(1-\delta_1) \frac{L+1}{\underline{\hat{\alpha}}_1^{L+1} \delta_1^L} \sum_{t=0}^{\infty} \delta_1^t \frac{1}{\kappa} E_{\hat{\mathbf{P}}}[d(\hat{p} \| p(h^t))] + \frac{\eta}{2} \\
&\leq 2M(1-\delta_1) \frac{L+1}{\underline{\hat{\alpha}}_1^{L+1} \delta_1^L} \frac{1}{\kappa} (-\log \mu(\xi(\hat{\alpha}_1))) + \frac{\eta}{2},
\end{aligned}$$

where

1. $\tilde{\sigma}_2$ in the first line is some pure strategy for 2 in the support of 2's equilibrium strategy,
2. the second inequality comes from the observation that player 2's failure to best respond to $\hat{\alpha}_1$ in different periods may be due to a belief that 1 will not play $\hat{\alpha}_1$

in the same future period after the same positive probability history (there can be at most $L+1$ such periods for each belief, and the probability of such a history is at least $\underline{\hat{\alpha}}_1^{L+1}$), and

3. the last inequality follows from [4.5].

Since the last term in the chain of inequalities can be made less than η for δ_1 sufficiently close to 1, the proposition is proved.

If there are multiple actions that minmax player 2, the relevant payoff bound corresponds to the maximum value of $v_1^*(\hat{\alpha}_1)$ over the set of such actions whose corresponding simple commitment types are assigned positive probability by player 2's prior.

4.3.3.2 Conflicting interests

The strength of the bound on player 1's equilibrium payoffs depends on the nature of player 1's actions that minmax player 2. The highest reputation bound is obtained when there exists an action α_1^* that mixed action minmaxes player 2 and is also close to player 1's Stackelberg action, since the reputation bound can then be arbitrarily close to player 1's Stackelberg payoff. Restricting attention to pure Stackelberg actions and payoffs, Schmidt (1993) refers to such games as games of conflicting interests:

Definition 4.3. *The stage game has* conflicting interests *if a pure Stackelberg action a_1^* mixed-action minmaxes player* 2.

The payoff bound derived in Proposition 4.3 is not helpful in the product-choice game (Figure 4.1). The pure Stackelberg action H prompts a best response of c that earns player 2 a payoff of 3, above her minmax payoff of 1. The mixtures that allow player 1 to approach his mixed Stackelberg payoff of $5/2$, involving a probability of H exceeding but close to $1/2$, similarly prompt player 2 to choose c and earn a payoff larger than 1. The normal player 1 and player 2 thus both fare better when 1 chooses either the pure Stackelberg action or an analogous mixed action (and 2 best responds) than in the stage-game Nash equilibrium (which minmaxes player 2). This coincidence of interests precludes using the reasoning behind Proposition 4.3.

The prisoners' dilemma (Figure 4.7) is a game of conflicting interests. Player 1's pure Stackelberg action is D. Player 2's unique best response of D yields a payoff of $(0, 0)$, giving player 2 her minmax level. Proposition 4.3 then establishes conditions, including

	C	D
C	$2, 2$	$-1, 3$
D	$3, -1$	$0, 0$

Figure 4.7 Prisoners' dilemma.

	In	Out
A	$2, 2$	$5, 0$
F	$-1, -1$	$5, 0$

Figure 4.8 A simultaneous-move version of the chain-store game.

the presence of a player 1 type committed to D and sufficient patience on the part of player 1, under which the normal player 1 must earn nearly his Stackelberg payoff. In the prisoners' dilemma, however, this bound is not significant, being no improvement on the observation that player 1's payoff must be weakly individually rational.

In the normal form version of the chain store game (Figure 4.8), player 1 achieves his mixed Stackelberg payoff of 5 by playing the (pure) action F, prompting player 2 to choose *Out* and hence to receive her mixed minmax payoff of 0. We thus have a game of conflicting interests, in which (unlike the prisoners' dilemma) the reputation result has some impact. The lower bound on the normal player 1's payoff is then close to his Stackelberg payoff of 5, the highest player 1 payoff in the game.

Consider the game shown in Figure 4.9. The Nash equilibria of the stage game are TL, BC, and a mixed equilibrium $\left(\frac{1}{2} \circ T + \frac{1}{2} \circ B, \frac{2}{5} \circ L + \frac{3}{5} \circ C\right)$. Player 1 minmaxes player 2 by playing B, for a minmax value for player 2 of 0. The pure and mixed Stackelberg action for player 1 is T, against which player 2's best response is L, for payoffs of $(3, 2)$. Proposition 4.3 accordingly provides no reason to think it would be helpful for player 1 to commit to T. However, if the set of possible player 1 types includes a type committed to B (perhaps as well as a type committed to T), then Proposition 4.3 implies that (up to some $\eta > 0$) the normal player 1 must earn a payoff no less than 2. The presence of a type committed to B thus gives rise to a nontrivial reputation bound on player 1's payoff, though this bound falls short of his Stackelberg payoff.

4.3.4 Discussion

Our point of departure for studying reputations with two long-lived players was the observation that a long-lived player 2 might not play a best response to the Stackelberg action, even when convinced she will face the latter, for fear of triggering a punishment. If we are to achieve a reputation result for player 1, then there must be something in the particular application that assuages player 2's fear of punishment. Proposition 4.3 considers cases in which player 2 can do no better when facing the Stackelberg action

	L	C	R
T	$3, 2$	$0, 1$	$0, 1$
B	$0, -1$	$2, 0$	$0, -1$

Figure 4.9 A game without conflicting interests.

than achieve her minmax payoff. No punishment can be worse for player 2 than being minmaxed, and hence no punishment type can coerce player 2 into choosing something other than a best response.

What if a commitment type does not minmax player 2? The following subsections sketch some of the alternative approaches to reputations with two long-lived players.

4.3.4.1 Weaker payoff bounds for more general actions

For any action a_1', if player 2 puts positive prior probability on the commitment type $\xi(a_1')$, then when facing a steady stream of a_1', player 2 must eventually come to expect a_1'. If the action a_1' does not minmax player 2, we can no longer bound the number of periods in which player 2 is not best responding. We can, however, bound the number of times player 2 chooses an action that gives her a payoff less than her minmax value. This allows us to construct an argument analogous to that of Section 4.3.3, concluding that a lower bound on player 1's payoff is given by

$$v_1^{\dagger}(a_1') \equiv \min_{\alpha_2 \in \mathcal{D}(a_1')} u_1(a_1', \alpha_2),$$

where

$$\mathcal{D}(a_1') = \{\alpha_2 \in \Delta(A_2) \mid u_2(a_1', \alpha_2) \geq \underline{v}_2\}$$

is the set of player 2 actions that, in response to a_1', ensure that 2 receives at least her mixed minmax utility $\underline{v}_2$. In particular, Cripps et al. (1996) show that if there exists $a_1' \in A_1$ with $\mu(\xi(a_1')) > 0$, then for any fixed $\delta_2 < 1$ and $\eta > 0$, then there exists a $\underline{\delta}_1 < 1$ such that for all $\delta_1 \in (\underline{\delta}_1, 1)$, $\underline{v}_1(\xi_0, \delta_1, \delta_2) \geq v_1^{\dagger}(a_1) - \eta$. This gives us a payoff bound that is applicable to any pure action for which there is a corresponding simple commitment type, though this bound will typically be lower than the Stackelberg payoff.

To illustrate this result, consider the battle of the sexes game in Figure 4.10. The minmax utility for player 2 is 3/4. Player 1's Stackelberg action is T, which does not minmax player 2, and hence this is not a game of conflicting interests.

The set of responses to B in which player 2 receives at least her minmax payoff is the set of actions that place at least probability 1/4 on L. Hence, if 2 assigns positive probability to a simple player 1 type committed to B, then we have a lower bound on 1's payoff of 1/4. This bound falls short of player 1's minmax payoff, and hence is not very informative. The set of responses to T in which player 2 receives at least her minmax payoff is the set of actions that place at least probability 3/4 on R. Hence, if 2 assigns

	L	R
T	0, 0	3, 1
B	1, 3	0, 0

Figure 4.10 Battle of the sexes game.

positive probability to a simple player 1 type committed to T, then we have a lower bound on 1's payoff of 9/4. This bound falls short of player 1's Stackelberg payoff, but nonetheless gives us a higher bound that would appear in the corresponding game of complete information.

4.3.4.2 Imperfect monitoring

The difficulty in Section 4.3.2 is that player 2 frequently plays an action that is not her best response to 1's Stackelberg action, in fear that playing the best response will push the game off the equilibrium path into a continuation phase where she is punished. The normal and Stackelberg types of player 1 would not impose such a punishment, but there is another punishment type who would. Along the equilibrium path player 2 has no opportunity to discern whether she is facing the normal type or the punishment type.

Celentani et al. (1996) observe that in games of imperfect public monitoring, the sharp distinction between being on and off the equilibrium path disappears. Player 2 may then have ample opportunity to become well acquainted with player 1's behavior, including any punishment possibilities.

Consider a game with two long-lived players and finite action sets A_1 and A_2. Suppose the public monitoring distribution ρ has full support, i.e., there is a set of public signals Y with the property that for all $y \in Y$ and $a \in A_1 \times A_2$, there is positive probability $\rho(y \mid a) > 0$ of observing signal y when action profile a is played. Suppose also that Assumption 4.2 of Section 4.2.6 holds. Recall that this is an identification condition. Conditional on player 2's mixed action, different mixed actions on the part of player 1 generate different signal distributions. Given arbitrarily large amounts of data, player 2 could then distinguish player 1's actions.

It is important that player 2's actions be imperfectly monitored by player 1, so that a sufficiently wide range of player 1 behavior occurs in equilibrium. It is also important that player 2 be able to update her beliefs about the type of player 1, in response to the behavior she observes. Full-support public monitoring delivers the first condition, while the identification condition (Assumption 4.2) ensures the second.

We now allow player 1 to be committed to a strategy that is not simple. In the prisoners' dilemma, for example, player 1 may be committed to playing tit-for-tat rather than either always cooperating or always defecting. When player 2 is short-lived, the reputation bound on player 1's payoff cannot be improved by appealing to commitment types that are not simple. A short-lived player 2 cares only about the action she faces in the period she is active, whether this comes from a simple or more complicated commitment type. Any behavior one could hope to elicit from a short-lived player 2 can then be elicited by having her attach sufficient probability to the appropriate simple commitment type. The result described here is the first of two illustrations of how, when player 2 is long-lived, non-simple commitment types can increase the reputation bound on player 1's payoffs.

The presence of more complicated commitment types allows a stronger reputation result, but also complicates the argument. In particular, we can no longer define a lower bound on player 1's payoff simply in terms of the stage game. Instead, we first consider a finitely repeated game, of length N, in which player 1 does not discount. It will be apparent from the construction that this lack of player 1 discounting simplifies the calculations, but otherwise does not play a role. We fix a pure commitment type σ_1^N for player 1 in this finitely repeated game, and then calculate player 1's average payoff, assuming that player 1 plays the commitment strategy and player 2 plays a best response to this commitment type. Denote this payoff by $v_1^*(\sigma_1^N)$.

Now consider the infinitely repeated game, and suppose that the set of commitment types includes a type who plays σ_1^N in the first N periods, then acts as if the game has started anew and again plays σ_1^N in the next N periods, and then acts as if the game has started anew, and so on. Celentani et al. (1996) establish a lower bound on player 1's payoff in the infinitely repeated game that approaches $v_1^*(\sigma_1^N)$ as player 1 becomes arbitrarily patient. Of course, one can take N to be as large as one would like, and can take the corresponding strategy σ_1^N to be any strategy in the N-length game, as long as one is willing to assume that the corresponding commitment type receives positive probability in the infinitely repeated game. By choosing this commitment type appropriately, one can create a lower bound on player 1's payoff that may well exceed the Stackelberg payoff of the stage game. In this sense, and in contrast to the perfect monitoring results of Sections 4.3.3–4.3.4.1, facing a long-lived opponent can strengthen reputation results.[27]

This argument embodies two innovations. The first is the use of imperfect monitoring to ensure that player 2 is not terrorized by "hidden punishments." The second is the admission of more complicated commitment types, with associated improved bounds on player 1's payoff. It thus becomes all the more imperative to consider the interpretation of the commitment types that appear in the model.

There are two common views of commitment types. One is to work with commitment types that are especially natural in the setting in question. The initial appeal to commitment types, by Kreps and Wilson (1982) and Milgrom and Roberts (1982), in the context of the chain store paradox, was motivated by the presumption that entrants might be especially concerned with the possibility that the incumbent is pathologically committed to fighting. Alternatively, player 2 may have no particular idea as to what commitment types are likely, but may attach positive probability to a wide range of types, sufficiently rich that some are quite close to the Stackelberg type. Both motivations may be less obvious once one moves beyond simple commitment types. It may be more difficult to think of quite complicated commitment types as natural, and the set of

[27] Aoyagi (1996) presents a similar analysis, with trembles instead of imperfect monitoring blurring the distinction between play on and off the equilibrium path, and with player 1 infinitely patient while player 2 discounts.

such types is sufficiently large that it may be less obvious to assume that player 2's prior distribution over this set is sufficiently dispersed as to include types arbitrarily close to any particular commitment type. If so, the results built on the presence of more complicated commitment types must be interpreted with care.

4.3.4.3 Punishing commitment types

This section illustrates more starkly the effects of appropriately chosen commitment types. Fix an action $a_1' \in A_1$, and suppose that player 2's stage-game best response a_2' satisfies $u_2(a_1', a_2') > \underline{v}_2^p := \min_{a_1} \max_{a_2} u_2(a_1, a_2)$, so that 2's best response to a_1' gives her more than her pure action minmax payoff. Let $\hat{a}_1^2$ be the action for player 1 that (pure-action) minmaxes player 2.

Consider a commitment type for player 1 who plays as follows. Play begins in phase 0. In general, phase k consists of k periods of $\hat{a}_1^2$, followed by the play of a_1'. The initial k periods of $\hat{a}_1^2$ are played regardless of any actions that player 2 takes during these periods. The length of time that phase k plays a_1', and hence the length of phase k itself, depends on player 2's actions, and this phase may never end. The rule for terminating a phase is that if player 2 plays anything other than a_2' in periods $k + 1, \ldots$ of phase k, then the strategy switches to phase $k + 1$.

We can interpret phase k of player 1's strategy as beginning with a punishment, in the form of k periods of minmaxing player 2, after which play switches to (a_1', a_2'). The beginning of a new phase, and hence of a new punishment, is prompted by player 2's not playing a_2' when called upon to do so. The commitment type thus punishes player 2, in strings of ever-longer punishments, for not playing a_2'.

Let $\hat{\sigma}_1(a_1')$ denote this strategy, with the a_1' identifying the action played by the commitment type when not punishing player 2. Evans and Thomas (1997) show that for any $\eta > 0$, if player 2 attaches positive probability to the commitment type $\hat{\sigma}_1(a_1')$, for some action profile a' with $u_2(a') > \underline{v}_2^p$, then there exists a $\underline{\delta}_2 \leq 1$ such that for all $\delta_2 \in (\underline{\delta}_2, 1)$, there in turn exists a $\underline{\delta}_1$ such that for all $\delta_1 \in (\underline{\delta}_1, 1)$, player 1's equilibrium payoff is at least $u_1(a') - \eta$. If there exists such a commitment type for player 1's Stackelberg action, then this result gives us player 1's Stackelberg payoff as an approximate lower bound on his equilibrium payoff in the game of incomplete information, as long as this Stackelberg payoff is consistent with player 2 earning more than her pure-strategy minmax.

This technique leads to reputation bounds that can be considerably higher than player 1's Stackelberg payoff. Somewhat more complicated commitment types can be constructed in which the commitment type $\hat{\sigma}_1$ plays a sequence of actions during its nonpunishment periods, rather than simply playing a fixed action a_1, and punishes player 2 for not playing an appropriate sequence in response. Using such constructions, Evans and Thomas (1997, Theorem 1) establish that the limiting lower bound on player 1's payoff, in the limit as the players become arbitrarily patient, with player 1 arbitrarily patient relative to player 2, is given by $\max\{v_1 : v \in \mathcal{F}^{\dagger p}\}$ (where $\mathcal{F}^{\dagger p}$ is that portion of the convex hull of the pure stage-game payoff profiles in which player 2 receives at least her pure minmax payoff). Hence, player 1 can be assured of the largest feasible payoff consistent with player 2's individual rationality.[28]

Like the other reputation results we have presented, this requires the uncertainty about player 1 contain appropriate commitment types. As in Section 4.3.4.2, however, the types in this case are more complicated than the commitment types that appear in many reputation models, particularly the simple types that suffice with short-lived player 2s.[29] In this case, the commitment type not only repeatedly plays the action that brings player 1 the desired payoff, but also consistently punishes player 2 for not fulfilling her role in producing that payoff. The commitment involves behavior both on the path of a proposed outcome and on paths following deviations. Once again, work on reputations would be well served by a better-developed model of which commitment types are likely.

4.4. PERSISTENT REPUTATIONS

In this section, we consider recent work that modifies the basic reputation model to obtain convenient characterizations of nontrivial long run behavior (including cycles of reputation building and destruction) and persistent reputations. We consider three classes of models. To organize our ideas, note that the fundamental force behind the disappearing reputations result of Section 4.2.6 is that a Bayesian with access to an unlimited number of informative signals will eventually learn a fixed parameter. The work examined in this section relaxes each element in this statement (though not all of this work was motivated by the disappearing reputations result). Section 4.4.1 examines models in which the short-lived players have access to only a limited amount of data. Section 4.4.2 considers the case in which the short-lived players are not perfect Bayesians, working instead with a misspecified model of the long-lived player's behavior. Section 4.4.3 considers models in which the short-lived players must learn a moving target.

[28] The idea behind the argument is to construct a commitment type consisting of a phase in which payoffs at least $(\max\{v_1 : v \in \mathcal{F}^{\dagger p}\} - \varepsilon, \bar{v}_2^p + \varepsilon)$ are received, as long as player 2 behaves appropriately, with inappropriate behavior triggering ever-longer punishment phases. Conditional on seeing behavior consistent with the commitment type, a sufficiently patient player 2 must then eventually find it optimal to play the appropriate response to the commitment type. For fixed, sufficiently patient player 2, making player 1 arbitrarily patient then gives the result.

[29] As Evans and Thomas (1997) note, a commitment type with punishments of arbitrary length cannot be implemented by a finite automaton. Hence, one cannot ensure that such commitment types appear in the support of player 2's prior distribution simply by assuming that this support includes the set of all strategies implementable by finite automata. Evans and Thomas (2001), working with two infinitely patient long-lived players, argue that commitment strategies capable of imposing arbitrarily severe punishments are necessary if reputation arguments are to be effective in restricting attention to efficient payoff profiles.

In each case, reputation considerations persist indefinitely. Notice, however, that the focus in this literature, especially the work discussed in Section Section 4.4.3, shifts from identifying bounds on the payoffs of all equilibria to characterizing particular (persistent reputation) equilibria.

4.4.1 Limited observability

We begin with a model in which player 2 cannot observe the entire history of play. Our presentation is based on a particularly simple special case of a model due to Liu (2011). Our point of departure is the product-choice game of Figure 4.1. Player 1 is either a normal type, whose payoffs are those of the product-choice game, or is a commitment type referred to as the "good" type, who invariably plays H. The innovation is that when constructing the associated repeated game, we assume that each successive player 2 can observe only the action taken by player 1 in the previous period. The observation of the previous-period action is perfect.

For $\delta > \frac{1}{2}$, we construct an equilibrium characterized by two states, ω^L and ω^H, determined by the action played by player 1 in the previous period. The play of the normal player 1 and player 2 is described by the automaton in Figure 4.11. Intuitively, we think of player 2 as being exploited in state ω^H, while player 1 builds the reputation that allows such exploitation in state ω^L.

It remains to confirm that these strategies constitute an equilibrium. A player 2 who observes L in the preceding period knows that she faces the normal player 1, who will mix equally between H and L in the next period. Player 2 is then indifferent between c and s, and hence 2's specified mixture is optimal. A player 2 who observes either nothing (in the case of period 0) or H in the previous period attaches a probability to player 1 being good that is at least as high (but no higher in the case of period 0) as the prior probability that player 1 is good. Player 2's actions are thus optimal if and only if the prior probability of a good type is at least 1/2. We will assume this is the case, though note that this takes us away from the common convention in reputation models that

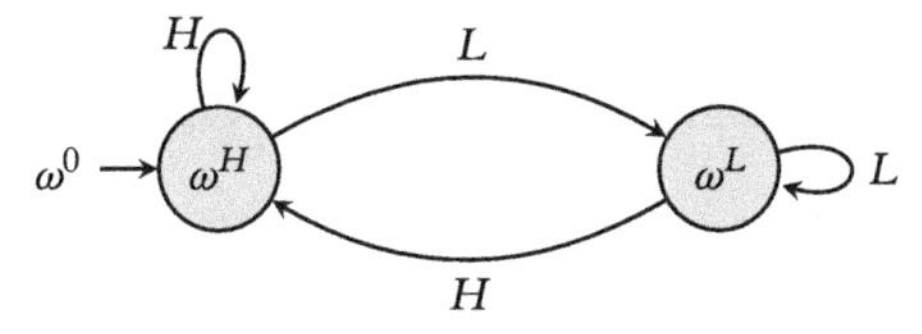

Figure 4.11 The play of the normal player 1 and player 2. In state ω^H, the normal player 1 plays L and player 2 plays c. In state ω^L, players randomize, with the normal player 1 playing $\frac{1}{2} \circ L + \frac{1}{2} \circ H$ and player 2 playing $p \circ c + (1-p) \circ s$, where $p := (2\delta - 1)/(2\delta)$. The transitions between states are determined by the action a_1 of player 1 in the previous period, with a_1 leading to ω^{a_1}. Play begins in ω^H.

the probabilities of the commitment types can be taken to be small. Liu's (2011) more general analysis does not require such a bound on the prior.

We now turn to the normal player 1's incentives. Let $V^L(a_1)$ be player 1's payoff when the current state is ω^L and player 1 plays a_1, with $V^H(a_1)$ being the analogous payoff at the current state ω^H. Since player 1 is supposed to mix between H and L at state ω^L, he must be indifferent between the two actions, and so we must have $V^L(H) = V^L(L)$.

The values satisfy

$$\begin{aligned} V^L(H) &= (1-\delta)2p + \delta V^H(L), \\ V^L(L) &= (1-\delta)(2p+1) + \delta V^L(L), \\ V^H(L) &= (1-\delta)3 + \delta V^L(L), \\ \text{and} \qquad V^H(H) &= (1-\delta)2 + \delta V^H(L). \end{aligned}$$

Solving the second equation yields

$$V^L(L) = 2p + 1,$$

and so

$$\begin{aligned} V^H(L) &= (1-\delta)3 + \delta(2p+1) \\ \text{and} \qquad V^L(H) &= (1-\delta)2p + \delta[(1-\delta)3 + \delta(2p+1)]. \end{aligned}$$

We now choose p so that player 1 is indeed willing to mix at state ω^L, i.e., so that $V^L(H) = V^L(L)$. Solving for p gives

$$p := \frac{2\delta - 1}{2\delta}.$$

Finally, it remains to verify that it is optimal for player 1 to play L at state ω^H. It is a straightforward calculation that the normal player 1 is indifferent between L and H at ω^H, confirming that the proposed strategies are an equilibrium for $\delta \in (\frac{1}{2}, 1)$.

In this equilibrium, player 1 continually builds a reputation, only to spend this reputation once it appears. Player 2 understands that the normal player 1 behaves in this way, with player 2 falling prey to the periodic exploitation because the prior probability of the good type is sufficiently high, and player 2 receives too little information to learn player 1's actual type.

Liu and Skrzypacz (2011) consider the following variation. Once again a long-lived player faces a succession of short-lived players. Each player has a continuum of actions, consisting of the unit interval, but the game gives rise to incentives reminiscent of the product-choice game. In particular, it is a dominant strategy in the stage game for player 1 to choose action 0. Player 2's best response is increasing in player 1's action. If player 2 plays a best response to player 1's action, then 1's payoff is increasing in his action.

To emphasize the similarities, we interpret player 1's action as a level of quality to produce, and player 2's action as a level of customization in the product she purchases. Player 1 may be rational, or may be a (single) commitment type who always chooses some fixed quality $q \in (0, 1]$. The short-lived players can observe only actions taken by the long-lived player, and can only observe such actions in the last K periods, for some finite K.

Liu and Skrzypacz (2011) examine equilibria in which the normal long-lived player invariably chooses either quality q (mimicking the commitment type) or quality 0 (the "most opportunistic" quality level). After any history in which the short lived-player has received K observations of q and no observations of 0 (or has only observed q, for the first $K - 1$ short-lived players), the long-lived player chooses quality 0, effectively burning his reputation. This pushes the players into a reputation-building stage, characterized by the property that the short-lived players have observed at least one quality level 0 in the last K periods. During this phase the long-lived player mixes between 0 and q, until achieving a string of K straight q observations. His reputation has then been restored, only to be promptly burned. Liu and Skrzypacz (2011) establish that as long as the record length K exceeds a finite lower bound, then the limiting payoff as $\delta \to 1$ is given by the Stackelberg payoff. Moreover, they show that this bound holds after every history, giving rise to reputations that fluctuate, but are long lasting.

Ekmekci (2011) establishes a different persistent reputation result in a version of the repeated product-choice game. As usual, player 1 is a long-lived player, facing a succession of short-lived player 2's. The innovation in the model is in the monitoring structure. Ekmekci (2011) begins with the repeated product-choice game with imperfect public monitoring, and then assumes that the public signals are observed only by a mediator, described as a ratings agency. On the basis of these signals, the ratings agency announces one of a finite number of ratings to the players. The short-lived players see only the most recently announced rating, thus barring access to the information they would need to identify the long-lived player's type.

If the game is one of complete information, so that player 1 is known to be normal, then player 1's payoff with the ratings agency can be no higher than the upper bound that applies to the ordinary repeated imperfect monitoring game. However, if player 1 might also be a Stackelberg type, then there is an equilibrium in which player 1's payoff is close (arbitrarily close, for a sufficiently patient player 1) to the Stackelberg payoff after every history. The fact that this payoff bound holds after every history ensures that reputation effects are permanent. If the appropriate (mixed) Stackelberg type is present, then player 1's payoff may exceed the upper bound applicable in the game of complete information. Reputation effects can thus permanently expand the upper bound on 1's payoff.

In equilibrium, high ratings serve as a signal that short-lived players should buy the custom product, low ratings as a signal that the long-lived player is being punished and short-lived players should buy the standard product. The prospect of punishment creates the incentives for the long-lived player to exert high effort, and the long-lived player exerts high effort in any non-punishment period. Punishments occur, but only rarely. Short-lived players observing a rating consistent with purchasing the custom object do not have enough information to determine whether they are facing the Stackelberg type or a normal type who is currently not being punished.

4.4.2 Analogical reasoning

This section considers a model, from Jehiel and Samuelson (2012), in which the inferences drawn by the short-lived players are constrained not by a lack of information, but by the fact that they have a misspecified model of the long-lived players' behavior. As usual, we consider a long-lived player 1 who faces a sequence of short-lived player 2's. It is a standard result in repeated games that the discount factor δ can be equivalently interpreted as reflecting either patience or a continuation probability, but in this case we specifically assume that conditional upon reaching period t, there is a probability $1-\delta$ that the game stops at t and probability δ that it continues.

At the beginning of the game, the long-lived player is chosen to either be normal or to be one of K simple commitment types. Player 2 assumes the normal player 1 chooses in each period according to a mixed action α_0, and that commitment type $k \in \{1, 2, \ldots, K\}$ plays mixed action α_k. Player 2 is correct about the commitment types, but in general is not correct about the normal type, and this is the sense in which player 2's model is misspecified. In each period t, player 2 observes the history of play h^t (though it would suffice for player 2 to observe only the frequencies with which player 1 has played his past actions), and then updates her belief about the type of player 1 she is facing. Player 2 then chooses a best response to the expected mixed action she faces, and so chooses a best response to

$$\sum_{k=0}^{K} \mu_k(h^t)\alpha_k,$$

where $\mu_0(h^t)$ is the posterior probability of the normal type after history h^t and $\mu_k(h^t)$ (for $k \in \{1, 2, \ldots, K\}$) is the posterior probability of the kth commitment type.

The normal long-lived player 1 chooses a best response to player 2's strategy. The resulting strategy profile is an equilibrium if it satisfies a consistency requirement. To formulate the latter, let σ_1 and σ_2 denote the strategies of the normal player 1 and of the short-lived players 2. We denote by $\mathbf{P}^{\sigma}(h)$ the resulting unconditional probability that history h is reached under $\sigma := (\sigma_1, \sigma_2)$ (taking into account the probability of breakdown after each period). We then define

$$A_0 := \frac{\sum_h \mathbf{P}^{\sigma}(h)\sigma_1(h)}{\sum_h \mathbf{P}^{\sigma}(h)}.$$

We interpret A_0 as the empirical frequency of player 1's actions.[30] The consistency requirement is then that

$$A_0 = \alpha_0.$$

The resulting equilibrium notion is a sequential analogy-based expectations equilibrium (Ettinger and Jehiel, 2010; Jehiel, 2005).

We interpret the consistency requirement on player 2's beliefs as the steady-state result of a learning process. We think of the repeated game as itself played repeatedly, though by different players in each case. At the end of each game, a record is made of the frequency with which player 1 (and perhaps player 2 as well, but this is unnecessary) has played his various actions. This in turn is incorporated into a running record listing the frequencies of player 1's actions. A short-lived player forms expectations of equilibrium play by consulting this record. In general, each repetition of the repeated game will leave its mark on the record, leading to somewhat different frequencies for the various player 1 actions in the record. This in turn will induce different behavior in the next game. However, we suppose that the record has converged, giving a steady state in which the expected frequency of player 1 play in each subsequent game matches the recorded frequency. Hence, empirical frequencies recorded in the record will match $\alpha_0, \alpha_1, \ldots, \alpha_K$, leading to the steady state captured by the sequential analogy-based expectations equilibrium.

We illustrate this model in the context of the familiar product-choice game of Figure 4.1. Let p^* $(= 1/2)$ be the probability of H that makes player 2 indifferent between c and s. We assume we have at least one type playing H with probability greater than p^* and one playing H with probability less than p^*. The sequential analogy-based equilibria of the product-choice game share a common structure, which we now describe.

We begin with a characterization of the short-lived players' best responses. For history h, let $n_H(h)$ be the number of times action H has been played and let $n_L(h)$ be the number of times action L has been played. The short-lived player's posterior beliefs after history h depend only on the "state" $(n_L(h), n_H(h))$, and not on the order in which the various actions have appeared.

Whenever player 2 observes H, her beliefs about player 1's type shift (in the sense of first-order stochastic dominance) toward types that are more likely to play H, with the reverse holding for an observation of L. Hence, for any given number n_L of L actions, the probability attached by player 2 to player 1 playing H is larger, the larger the number n_H of H actions. Player 2 will then play c if and only if she has observed enough H

[30] The assumption that δ reflects a termination risk plays a role in this interpretation. The denominator does not equal 1 because $\mathbf{P}^{\sigma}(h)$ is the probability that h appears as either a terminal history, or as the initial segment of a longer history.

actions. More precisely, there exists an increasing function $N_H : \{0, 1, 2, \ldots\} \to \mathbb{R}$, such that for every history h, at the resulting state $(n_L, n_H) = (n_L(h), n_H(h))$,

- player 2 plays c if $n_H > N_H(n_L)$ and
- player 2 plays s if $n_H < N_H(n_L)$.

We now describe the equilibrium behavior of player 1, which also depends only on the state (n_L, n_H). There exists a function $\underline{N}_H(n_L) \leq N_H(n_L)$ such that, for sufficiently large δ and for any history h, at the resulting state $(n_L, n_H) = (n_L(h), n_H(h))$,

- player 1 plays L if $n_H > N_H(n_L + 1)$,
- player 1 plays H if $\underline{N}_H(n_L) < n_H < N_H(n_L + 1)$,
- player 1 plays L if $n_H < \underline{N}_H(n_L)$, and
- $\lim_{\delta \to 1} \underline{N}_H(n_L) < 0$ for all n_L.

Figure 4.12 illustrates these strategies. Note that whenever player 1 chooses H, player 2 places higher posterior probability on types that play H. Doing so will eventually induce player 2 to choose c. Player 1 is building his reputation by playing H, and then enjoying the fruits of that reputation when player 2 plays c. However, it is costly for player 1 to play H when player 2 plays s. If the number of H plays required to build a reputation is too large, player 1 may surrender all thoughts of a reputation and settle for the continual play of Ls. The cost of building a reputation depends on how patient is player 1, and a sufficiently patient player 1 inevitably builds a reputation. This accounts for the last two items in the description of player 1's best response.

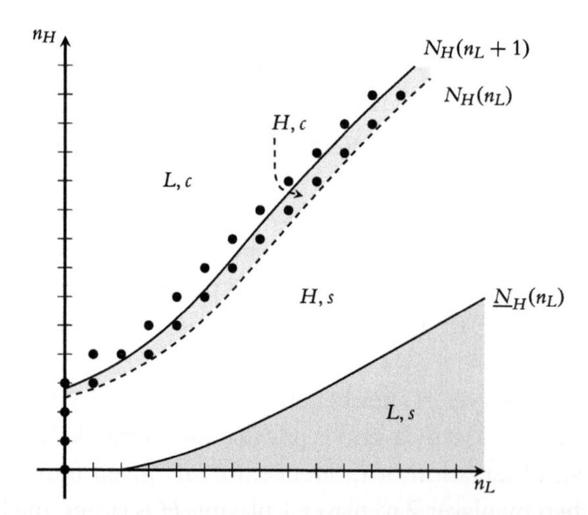

Figure 4.12 Behavior in the product-choice game. The state space (n_L, n_H) is divided into four regions with the players behaving as indicated in each of the four regions. The equilibrium path of play is illustrated by the succession of dots, beginning at the origin and then climbing the vertical axis in response to an initial string of H actions from player 1, with player 1 then choosing H and L so as to induce player 2 to always choose c, but to always be just at the edge of indifference between c and s.

If player 1 encounters a state above $\underline{N}_H$, whether at the beginning of the game or after some nontrivial history, player 1 will choose H often enough to push player 2's beliefs to the point that c is a best response. Once player 1 has induced player 2 to choose c, 1 ensures that 2 thereafter always plays c. The state never subsequently crosses the border $N_H(n_L)$. Instead, whenever the state comes to the brink of this border, 1 drives the state away with a play of H before 2 has a chance to play s.

The functions illustrated in Figure 4.12 depend on the distribution of possible types for player 1 and on the payoffs of the game. Depending on the shapes of these functions, the best responses we have just described combine to create three possibilities for equilibrium behavior in the product-choice game, that correspond to three possibilities for the intercepts of the functions $\underline{N}_H$ and N_H in Figure 4.12. First, it may be that $\underline{N}_H(0) > 0$. In this case, the equilibrium outcome is that the normal player 1 chooses L and player 2 chooses s in every period. Player 1 thus abandons any hope of building a reputation, settling instead for the perpetual play of the stage-game Nash equilibrium Ls. This is potentially optimal because building a reputation is costly. If player 1 is sufficiently impatient, this current cost will outweigh any future benefits of reputation building, and player 1 will indeed forego reputation building. By the same token, this will not be an equilibrium if δ is sufficiently large (but still less than one).

Second, it may be that $\underline{N}_H < 0 < N_H$, as in Figure 4.12. In this case, play begins with a reputation-building stage, in which player 1 chooses H and the outcome is Hs. This continues until player 2 finds c a best response (until the state has climbed above the function N_H). Thereafter, we have a reputation-manipulation stage in which player 1 sometimes chooses L and sometimes H, selecting the latter just often enough to keep player 2 always playing c.

Alternatively, if $N_H(0) < 0$, then play begins with a reputation-spending phase in which player 1 chooses L, with outcome Lc, in the process shifting player 2's beliefs towards types that play L. This continues until player 2 is just on the verge of no longer finding c a best response (intuitively, until the state just threatens to cross the function N_H). Thereafter, we again have a reputation-manipulation stage in which player 1 sometimes chooses L and sometimes H, again selecting the latter just often enough to keep player 2 always playing c.

Which of these will be the case? For sufficiently patient players, whether one starts with a reputation-building or reputation-spending phase depends on the distribution of commitment types. If player 2's best response conditional on player 1 being a commitment type is s, then the rational player 1 must start with a reputation building phase, and we have the first of the preceding possibilities. Alternatively, If player 2's best response conditional on player 1 being a commitment type is c, then the rational player 1 must start with a reputation spending phase, and we have the second of the preceding possibilities.[31] In either case, player 2 remains perpetually uncertain as to which type of agent she faces, with her misspecified model of player 1 obviating the arguments of Cripps et al. (2004, 2007). Player 1's equilibrium payoff approaches (as δ gets large) the mixed Stackelberg payoff, even though there may be no commitment type close to the mixed Stackelberg type.

4.4.3 Changing types

Perhaps the most obvious route to a model in which reputation considerations persist is to assume that player 1's type is not fixed once-and-for-all at the beginning of the game, but is subject to persistent shocks. Intuitively, one's response upon having an uncharacteristically disappointing experience at a restaurant may be not "my previous experiences must have been atypical, and hence I should revise my posterior," but rather "perhaps something has changed." In this case, opponents will never be completely certain of a player's type. At first glance, it would seem that this can only reinforce the temporary-reputation arguments of Cripps et al. (2004, 2007), making it more difficult to find persistent reputation effects. However, this very transience of reputations creates incentives to continually invest in reputation building, in order to assure opponents that one's type has not changed, opening the door for persistent reputations.

This section sketches several recent papers built around the assumption that types are uncertain but not permanently fixed. A critical assumption in all the models described in this section is that the uninformed agents do not observe whether the replacement event has occurred. These papers in turn build on a number of predecessors, such as Holmström (1982), Cole et al. (1995), and Mailath and Samuelson (2001), in which the type of the long-lived player is governed by a stochastic process.

4.4.3.1 Cyclic reputations

We first consider, as a special case adapted from Phelan (2006), the product-choice game under perfect monitoring of actions, assuming that player 1 is constantly vulnerable to having his type drawn anew. The probability of such a replacement in each period is fixed, and we consider the limit in which the discount factor approaches 1.

Player 1 is initially drawn to be either a normal type or the pure Stackelberg type, who always plays H. In each period, player 1 continues to the next period with probability $\lambda > 1/2$, and is replaced by a new player 1 with probability $1 - \lambda$. In the event player 1 is replaced, player 1 is drawn to be the commitment type with probability $\hat{\mu}$, and with

[31] This characterization initially sounds obvious, but it is not immediate that (for example) player 2 will open the game by playing s if her best response to the commitment types is s, since 2's initial best response is an equilibrium phenomenon that also depends on the normal player 1's play. Jehiel and Samuelson (2012) fill in the details of the argument.

complementary probability is the normal type. To simplify the presentation, we assume that player 1 is initially drawn to be the commitment type with probability $(1-\lambda)\hat{\mu}$ (see (4.17)).

Consider first the trigger profile: the normal player 1 always plays H if H has always been observed, plays L if ever L had been played, and player 2 plays c as long as H has been played, and plays s if ever L had been played. This profile is not an equilibrium under replacement, since after the punishment has been triggered by a play of L, the first play of H leads player 2 to believe that the normal type has been replaced by the commitment type and so 2's best response in the next period is to play c. But then the normal player 1 plays H immediately L, destroying the optimality of the punishment phase.

We construct an equilibrium when $2\delta\lambda > 1$. Let μ^t be the period t posterior probability that player 2 attaches to the event that player 1 is the commitment type. In each period t in which μ^t is less than or equal to $\frac{1}{2}$, the normal player 1 plays H with probability

$$\alpha_1(\mu^t) := \frac{1-2\mu^t}{2(1-\mu^t)}.$$

This implies that, in such a period, player 2 faces the mixed action $\frac{1}{2} \circ H + \frac{1}{2} \circ L$, since

$$\mu^t + (1-\mu^t)\frac{1-2\mu^t}{2(1-\mu^t)} = \frac{1}{2}, \qquad [4.15]$$

and hence is indifferent between c and s. When $\mu^t \le \frac{1}{2}$, player 2 mixes, putting probability $\alpha_2(\mu^t)$ (to be calculated) on c, with player 2's indifference ensuring that this behavior is a best response. For values $\mu^t > \frac{1}{2}$, player 2 plays c, again a best response, and the normal player 1 chooses L. Notice that the actions of the customer and the normal player 1 depend only on the posterior probability that player 1 is the Stackelberg type, giving us a profile in Markov strategies.

It remains only to determine the mixtures chosen by player 2, which are designed so that the normal player 1 is behaving optimally. Let $\varphi(\mu \mid H)$ be the posterior probability attached to player 1 being the commitment type, given a prior $\mu \le \frac{1}{2}$ and an observation of H. If the normal type chooses H with probability $\alpha_1(\mu)$, we have

$$\varphi(\mu \mid H) = \lambda\frac{\mu}{\mu + (1-\mu)\alpha_1(\mu)} + (1-\lambda)\hat{\mu} = 2\lambda\mu + (1-\lambda)\hat{\mu}, \qquad [4.16]$$

using [4.15] for the second equality, while the corresponding calculation for L is

$$\varphi(\mu \mid L) = (1-\lambda)\hat{\mu}. \qquad [4.17]$$

Let $\mu(0), \mu(1), \ldots, \mu(N)$ be the sequence of posterior probabilities satisfying $\mu(0) = (1-\lambda)\hat{\mu}$ and $\mu(k) = \varphi(\mu(k-1) \mid H)$ for $(k = 1, \ldots, N)$, with $\mu(N)$ being the first such probability to equal or exceed $\frac{1}{2}$.

We now attach a player 2 action to each posterior in the sequence $(\mu(k))$. Let $V(k)$ be the value to the normal player 1 of continuation play, beginning at posterior $\mu(k)$, and let $\hat{V}$ be player 1's payoff when $\mu = (1-\lambda)\hat{\mu}$. Player 2 must randomize so that the normal type of player 1 is indifferent between L and H, and hence, for $k = 0, \ldots, N-1$,

$$V(k) = (1-\delta\lambda)(2\alpha_2(\mu(k)) + 1) + \delta\lambda\hat{V} \qquad [4.18]$$

$$= (1-\delta\lambda)2\alpha_2(\mu(k)) + \delta\lambda V(k+1), \qquad [4.19]$$

where the right side of the top line is the value if L is played at posterior $\mu(k)$ and the second line is the value if H is played, and where we normalize by $(1-\delta\lambda)$. Finally,

$$V(N) = (1-\delta\lambda)3 + \delta\lambda\hat{V},$$

since the normal player 1 chooses L with certainty for $\mu > \frac{1}{2}$.

Solving [4.18] for $k = 0$ gives

$$\hat{V} = 2\alpha_2(\mu(0)) + 1.$$

We then solve the equality of the right side of [4.18] with [4.19] to obtain, for $k = 1, \ldots, N$,

$$V(k) = \hat{V} + \frac{1-\delta\lambda}{\delta\lambda}.$$

These equations tell us a great deal about the equilibrium. The value $V(0)$ is lower than the remaining values, which are all equal to another, $V(1) = V(2) = \ldots = V(N)$. This in turn implies (from [4.18]) that $\alpha_2(\mu(0))$ is lower than the remaining probabilities $\alpha_2(\mu(1)) = \ldots = \alpha_2(\mu(N-1)) =: \bar{\alpha}_2$. These properties reflect the special structure of the product-choice game, most notably the fact that the stage-game payoff gain to player 1 from playing L rather than H is independent of player 2's action.

It is straightforward to calculate $\alpha_2(\mu(0)) = 1 - (2\delta\lambda)^{-1}$ and $\bar{\alpha}_2 = 1$, and to confirm that a normal player 1 facing posterior $\mu(N)$ prefers to play L. All other aspects of player 1's strategy are optimal because he is indifferent between H and L at every other posterior belief.

Phelan (2006) constructs a similar equilibrium in a game that does not have the equal-gain property of the product-choice game, namely that player 1's payoff increment from playing H is independent of player 2's action. The equilibrium then has a richer dynamic structure.

The breakdown of player 1's reputation in this model is unpredictable, in the sense that the probability of a reputation-ending exertion of low effort is the same regardless of the current posterior that player 1 is good. A normal player 1 who has labored long and hard to build his reputation is just as likely to spend it as is one just starting. Once the reputation has been spent, it can be rebuilt, but only gradually, as the posterior probability of a good type gets pushed upward once more.

4.4.3.2 Permanent reputations

Ekmekci et al. (2012) study a model in which persistent changes in the type of the long-lived player coexist with permanent reputation effects. The model of the preceding section is a special case of that considered here, in which the probability of a replacement is fixed. In this section, we consider the limits as the probability of a replacement approaches zero and the discount factor approaches one. As one might have expected, the order in which these limits is taken is important.

Let us once again consider the product-choice game of Figure 4.1. Player 1 can be one of two possible types, a normal type or a simple commitment type who invariably plays H. At the beginning of the game, player 1's type is drawn from a distribution that puts positive probability on each type. We assume that monitoring is imperfect but public; Ekmekci et al. (2012) allow private monitoring (including perfect and public monitoring as special cases).

So far, this gives us a standard imperfect monitoring reputation game. Now suppose that at the end of every period, with probability $1 - \lambda$ the long-lived player is replaced, and with probability λ the long-lived player continues until the next period. Replacement draws are independent across periods, and the type of each entering player is an independent draw from the prior distribution over types, regardless of the history that preceded that replacement.

Using entropy arguments similar to those described in Sections 4.2.4, 4.2.6, and 4.2.8, Ekmekci et al. (2012) show that the normal type's ex ante equilibrium payoffs in any equilibrium must be at least

$$\underline{w}_H \left(-(1-\delta) \log \mu(H) - \log \lambda \right), \tag{4.20}$$

and the normal type's continuation equilibrium payoffs (after any history) in any equilibrium must be at least

$$\underline{w}_H \left(-(1-\delta) \log(1-\lambda)\mu(H) - \log \lambda \right). \tag{4.21}$$

We consider first [4.20] and the limit as $\lambda \to 1$ and $\delta \to 1$. Player 1 is becoming more patient, and is also becoming increasingly likely to persist from one period to the next. The lower bound on player 1's equilibrium payoff approaches $\underline{w}_H(0)$, the pure Stackelberg payoff of 2.

This payoff bound is unsurprising. One would expect that if replacements are rare, then they will have little effect on ex ante payoffs, and hence that we can replicate the no-replacement Stackelberg payoff bound by making replacements arbitrarily rare. However, we can go beyond this bound to talk about *long-run* beliefs and payoffs. Doing so requires us to consider the lower bound in [4.21], which reveals the necessity of being more precise about the relative rates at which the replacement probability goes to zero and the discount factor gets large.

The bound [4.21] suggests that a particularly interesting limiting case arises if $(1-\delta)\ln(1-\lambda) \to 0$. For example, this is the case when $\lambda = 1-(1-\delta)^{\beta}$ for some $\beta > 0$. Under these circumstances, the replacement probability is not going to zero too fast, compared to the rate at which the discount factor goes to one. The lower bound on player 1's continuation payoff, in any Nash equilibrium and after any history, then approaches (as λ and δ get large) the pure Stackelberg payoff of 2. Notice that this reputation-based lower bound on player 1's payoff is effective after *every* history, giving a long run reputation result. Unlike the case examined by Cripps et al. (2004, 2007), the probability that player 1 is the commitment type can never become so small as to vitiate reputation effects.

Some care is required in establishing this result, but the intuition is straightforward. The probability of a replacement imposes a lower bound on the probability of a commitment type. This lower bound might be quite small, but is large enough to ensure that reputation arguments never lose their force, as long as player 1 is sufficiently patient. Recalling our discussion from Section 4.2.7, the lower-bound arguments from Section 4.2.4 require that limits be taken in the appropriate order. For any given probability of the commitment type, there is a discount factor sufficiently large as to ensure that player 1 can build a reputation and secure a relatively large payoff. However, for a fixed discount factor, the probability of the commitment type will eventually drop so low as to be essentially irrelevant in affecting equilibrium payoffs. In light of this, consider how the limits interact in Ekmekci et al. (2012). As λ goes to one, the lower bound on the probability that player 1 is the commitment type gets very small. But if in the process δ goes to one sufficiently rapidly, then for each value of λ, and hence each (possibly tiny) probability that player 1 is the commitment type, player 1 is nonetheless sufficiently patient that reputation effects can operate. In addition, the constant (if rare) threat of replacement ensues that the lower bound on the probability of a commitment type, and hence the relevance of reputation building, applies to every history of play.

As usual, we have described the argument for the case in which there is a single commitment type, fortuitously chosen to be the Stackelberg type. A similar argument applies to the case in which there are many commitment types. The lower bound on player 1's payoff would then approach the payoff player 1 would receive if 1 were known to be the most advantageous of the possible commitment types.

4.4.3.3 Reputation as separation

This section examines a stylized version of Mailath and Samuelson's (2001) model of "separating" reputations. The prospect of replacements again plays a role. However, the reputation considerations now arise not out of the desire to mimic a good commitment type, but to avoid a bad one.[32]

The underlying game is an imperfect monitoring variant of the product-choice game. In each period, the long-lived player 1 chooses either high effort (H) or low effort (L). Low effort is costless, while high effort entails a cost of c. Player 2 receives either a good outcome $\bar{y}$ or a bad outcome $\underline{y}$. The outcome received by player 2 is drawn from a distribution with probability ρ_H on a good outcome (and probability $1 - \rho_H$ on a bad outcome) when the firm exerts high effort, and probability $\rho_L < \rho_H$ on a good outcome (and probability $1 - \rho_L$ on a bad outcome) when the firm exerts low effort.

At the beginning of the game, player 1 is chosen to be either a normal type or an "inept" type who invariably exerts low effort. The reputation-based incentives for player 1 arise out of player 1's desire to convince player 2 that he is not inept, rather than that he is a good type. We accordingly refer to this as a model of "reputation as separation," in contrast with the more typical reputation-based incentive for a normal player 1 to pool with a commitment type.

As usual, we think of player 1 as a firm and player 2 as a customer, or perhaps as a population of customers. The customer receives utility 1 from outcome $\bar{y}$ and utility 0 from outcome $\underline{y}$. In each period, the customer purchases the product at a price equal to the customer's expected payoff: if the firm is thought to be normal with probability $\tilde{\mu}$, and if the normal firm is thought to choose high effort with probability α, then the price will be

$$p(\tilde{\mu}\alpha) := \tilde{\mu}\alpha\rho_H + (1 - \tilde{\mu}\alpha)\rho_L.$$

It is straightforward to establish conditions under which there exist equilibria in which the normal firm frequently exerts high effort. Suppose the normal firm initially exerts high effort, and continues to do so as long as signal $\bar{y}$ is realized, with the period t price given by $p(\tilde{\mu}_h)$, where $\tilde{\mu}_h$ is the posterior probability of a normal firm after history h. Let signal $\underline{y}$ prompt some number $K \geq 1$ periods of low effort and price ρ_L, after which play resumes with the normal firm exerting high effort. We can choose a punishment length K such that these strategies constitute an equilibrium, as long as the cost c is sufficiently small.

Given this observation, our basic question of whether there exist high-effort equilibria appears to have a straightforward, positive answer. Moreover, uncertainty about player 1's type plays no essential role in this equilibrium. The posterior probability that player 1 is normal remains unchanged in periods in which his equilibrium action is L, and tends to drift upward in periods when the normal player 1 plays H. Player 2's belief

[32] Morris (2001) examines an alternative model of separating reputations, and the bad reputation models discussed in Section 4.5.1 are similarly based on a desire to separate from a bad type.

that player 1 is normal almost certainly converges to 1 (conditional on 1 being normal). Nonetheless, the equilibrium behavior persists, with strings in which player 1 chooses H interspersed with periods of punishment. Indeed, the proposed behavior remains an equilibrium even if player 1 is known to be normal from the start. We have seemingly accomplished nothing more than showing that equilibria can be constructed in repeated games in which players do not simply repeat equilibrium play of the stage game.

However, we prefer to restrict attention to Markov strategies, with the belief about player 1's type as the state variable. This eliminates the equilibrium we have just described, since under these strategies the customer's behavior depends upon the firm's previous actions as well as the customer's beliefs.

Why restrict attention to Markov equilibria? The essence of repeated games is that continuation play can be made to depend on current actions in such a way as to create intertemporal incentives. Why curtail this ability by placing restrictions on the ability to condition actions on behavior? How interesting would the repeated prisoners' dilemma be if attention were restricted to Markov equilibria?

While we describe player 2 as a customer, we have in mind cases in which the player 2 side of the game corresponds to a large (continuum) population of customers. It is most natural to think of these customers as receiving idiosyncratic noisy signals of the effort choice of the firm; by idiosyncratic we mean that the signals are independently drawn and privately observed.[33] A customer who receives a bad outcome from a service provider cannot be sure whether the firm has exerted low effort, or whether the customer has simply been unlucky (and most of the other customers have received a good outcome).

The imperfection in the monitoring per se is not an obstacle to creating intertemporal incentives. There is a folk theorem for games of imperfect public monitoring, even when players only receive signals that do not perfectly reveal actions (Fudenberg et al., 1994). If the idiosyncratic signals in our model were public, we could construct equilibria with a population of customers analogous to the equilibrium described above: customers coordinate their actions in punishing low effort on the part of the firm. Such coordination is possible if the customers observe each other's idiosyncratic signals (in which case the continuum of signals precisely identifies the firm's effort) or if the customers all receive the same signal (which would not identify the firm's effort, but could still be used to engineer coordinated and effective punishments).

However, we believe there are many cases in which such coordinated behavior is not possible, and so think of the idiosyncratic signals as being *private*, precluding

[33] There are well-known technical complications that arise with a continuum of independently distributed random variables (see, for example, Al-Najjar, 1995). Mailath and Samuelson (2006, Remark 18.1.3) describes a construction that, in the current context, avoids these difficulties.

such straightforward coordination. Nonetheless, there is now a significant literature on repeated finite games with private monitoring (beginning with Sekiguchi,1997; Piccione, 2002; and Ely and Välimäki, 2002) suggesting that private signals do not preclude the provision of intertemporal incentives. While the behavioral implications of this literature are still unclear, it is clear that private signals significantly complicate the analysis. Mailath and Samuelson (2006, Section 18.1) studies the case of a continuum of customers with idiosyncratic signals.

In order to focus on the forces we are interested in without being distracted by the technical complications that arise with private signals, we work with a single customer (or, equivalently, a population of customers receiving the same signal) and rule out coordinated punishments by restricting attention to Markov equilibria.

At the end of each period, the firm continues to the next period with probability λ, but with probability $1 - \lambda$, is replaced by a new firm. In the event of a replacement, the replacement's type is drawn from the prior distribution, with probability $\tilde{\mu}^0$ of a normal replacement.

A *Markov strategy* for the normal firm can be written as a mapping $\alpha : [0, 1] \to [0, 1]$, where $\alpha(\tilde{\mu})$ is the probability of choosing action H when the customer's posterior probability of a normal firm is $\tilde{\mu}$. The customers' beliefs after receiving a good signal are updated according to

$$\varphi(\tilde{\mu} \mid \bar{y}) = \lambda \frac{[\rho_H \alpha(\tilde{\mu}) + \rho_L(1 - \alpha(\tilde{\mu}))]\,\tilde{\mu}}{[\rho_H \alpha(\tilde{\mu}) + \rho_L(1 - \alpha(\tilde{\mu}))]\,\tilde{\mu} + \rho_L(1 - \tilde{\mu})} + (1 - \lambda)\tilde{\mu}^0,$$

with a similar expression for the case of a bad signal. The strategy α is a *Markov equilibrium* if it is maximizing for the normal firm.

Proposition 4.4. *Suppose $\lambda \in (0, 1)$ and $\tilde{\mu}^0 > 0$. There exists $\bar{c} > 0$ such that for all $0 \le c < \bar{c}$, there exists a Markov equilibrium in which the normal firm always exerts high effort.*

Proof. Suppose the normal firm always exerts high effort. Then given a posterior probability $\tilde{\mu}$ that the firm is normal, firm revenue is given by $p(\tilde{\mu}) = \tilde{\mu}\rho_H + (1 - \tilde{\mu})\rho_L$. Let $\tilde{\mu}_y := \varphi(\tilde{\mu} \mid y)$ and $\tilde{\mu}_{xy} := \varphi(\varphi(\tilde{\mu} \mid x) \mid y)$ for $x, y \in \{\underline{y}, \bar{y}\}$. Then $\tilde{\mu}_{\bar{y}\bar{y}} > \tilde{\mu}_{\bar{y}} > \tilde{\mu} > \tilde{\mu}_{\underline{y}} > \tilde{\mu}_{\underline{y}\underline{y}}$ and $\tilde{\mu}_{\bar{y}y} > \tilde{\mu}_{\underline{y}y}$ for $y \in \{\underline{y}, \bar{y}\}$. The value function of the normal firm is given by

$$V(\tilde{\mu}) = (1 - \delta\lambda)(p(\tilde{\mu}) - c) + \delta\lambda\left[\rho_H V(\tilde{\mu}_{\bar{y}}) + (1 - \rho_H)V(\tilde{\mu}_{\underline{y}})\right].$$

The payoff from exerting low effort and thereafter adhering to the equilibrium strategy is

$$V(\tilde{\mu}; L) := (1 - \delta\lambda)p(\tilde{\mu}) + \delta\lambda\left[\rho_L V(\tilde{\mu}_{\bar{y}}) + (1 - \rho_L)V(\tilde{\mu}_{\underline{y}})\right].$$

Thus, $V(\tilde{\mu}) - V(\tilde{\mu}; L)$ is given by

$$\begin{aligned} &- c(1 - \delta\lambda) + \delta\lambda(1 - \delta\lambda)(\rho_H - \rho_L)(p(\tilde{\mu}_{\bar{y}}) - p(\tilde{\mu}_{\underline{y}})) \\ &\quad + \delta^2\lambda^2(\rho_H - \rho_L)\{\rho_H[V(\tilde{\mu}_{\bar{y}\bar{y}}) - V(\tilde{\mu}_{\underline{y}\bar{y}})] + (1 - \rho_H)[V(\tilde{\mu}_{\bar{y}\underline{y}}) - V(\tilde{\mu}_{\underline{y}\underline{y}})]\} \\ &\qquad \geq (1 - \delta\lambda)\{-c + \delta\lambda(\rho_H - \rho_L)[p(\tilde{\mu}_{\bar{y}}) - p(\tilde{\mu}_{\underline{y}})]\}, \quad [4.22] \end{aligned}$$

where the inequality is established via a straightforward argument showing that V is increasing in μ.

From an application of the one shot deviation principle, it is an equilibrium for the normal firm to always exert high effort, with the implied customer beliefs, if and only if $V(\tilde{\mu}) - V(\tilde{\mu}; L) \geq 0$ for all feasible $\tilde{\mu}$. From [4.22], a sufficient condition for this inequality is

$$p(\tilde{\mu}_{\bar{y}}) - p(\tilde{\mu}_{\underline{y}}) \geq \frac{c}{\delta\lambda(\rho_H - \rho_L)}.$$

Because there are replacements, the left side of this inequality is bounded away from zero. There thus exists a sufficiently small c such that the inequality holds.

In equilibrium, the difference between the continuation value of choosing high effort and the continuation value of choosing low effort must exceed the cost of high effort. However, the value functions corresponding to high and low effort approach each other as $\tilde{\mu} \to 1$, because the values diverge only through the effect of current outcomes on future posteriors, and current outcomes have very little affect on future posteriors when customers are currently quite sure of the firm's type. The smaller the probability of an inept replacement, the closer the posterior expectation of a normal firm can approach unity. Replacements ensure that $\tilde{\mu}$ can never reach unity, and hence there is always a wedge between the high-effort and low-effort value functions. As long as the cost of the former is sufficiently small, high effort will be an equilibrium.

The possibility of replacements is important in this result. In the absence of replacements, customers eventually become so convinced the firm is normal (i.e., the posterior $\tilde{\mu}$ becomes so high), that subsequent evidence can only shake this belief very slowly. Once this happens, the incentive to choose high effort disappears (Mailath and Samuelson, 2001, Proposition 2). If replacements continually introduce the possibility that the firm has become inept, then the firm cannot be "too successful" at convincing customers it is normal, and so there is an equilibrium in which the normal firm always exerts high effort.

To confirm the importance of replacements, suppose there are no replacements, and suppose further that $\rho_H = 1 - \rho_L$. In this case, there is a unique Markov equilibrium in pure strategies, in which the normal firm exerts low effort after every history.

Under this parameter configuration, an observation of $\bar{y}$ followed by $\underline{y}$ (or $\underline{y}$ followed by $\bar{y}$) leaves the posterior at precisely the level it had attained before these observations. More generally, posterior probabilities depend only on the number of $\bar{y}$ and $\underline{y}$ observations in the history, and not on their order. We can thus think of the set of possible posterior probabilities as forming a ladder, with countably many rungs and extending infinitely in each direction, with a good signal advancing the firm one rung up and a bad signal pushing the firm down one rung. Now consider a pure Markov equilibrium, which consists simply of a prescription for the firm to exert either high effort or low effort at each possible posterior about the firm's type. If there exists a posterior at which the firm exerts low effort, then upon being reached the posterior is never subsequently revised, since normal and inept firms then behave identically, and the firm receives the lowest possible continuation payoff. This in turn implies that if the equilibrium ever calls for the firm to exert high effort, than it must also call for the firm to exert high effort at the next higher posterior. Otherwise, the next higher posterior yields the lowest possible continuation payoff, and the firm will not exert costly effort only to enhance the chances of receiving the lowest possible payoff.

We can repeat this argument to conclude that if the firm ever exerts high effort, it must do so for every higher posterior. But then for any number of periods T and $\varepsilon > 0$, we can find a posterior (close to one) at which the equilibrium calls for high effort, and with the property that no matter what effort levels the firm exerts over the next T periods, the posterior that the firm is normal will remain above $1 - \varepsilon$ over those T periods. By making T large enough that periods after T are insignificant in payoff calculations and making ε small, we can ensure that the effect of effort on the firm's revenue are overwhelmed by the cost of that effort, ensuring that the firm will exert low effort. This gives us a contradiction to the hypothesis that high effort is ever exerted.

Mailath and Samuelson (2001) show that the inability to support high effort without replacements extends beyond this simple case. These results thus combine to provide the seemingly paradoxical result that it can be good news for the firm to have customers constantly fearing that the firm might "go bad." The purpose of a reputation is to convince customers that the firm is normal and will exert high effort. As we have just seen, the problem with maintaining a reputation in the absence of replacements is that the firm essentially succeeds in convincing customers it is normal. If replacements continually introduce the possibility that the firm has turned inept, then there is an upper bound, short of unity, on the posterior $\tilde{\mu}$, and so the difference in posteriors after different signals is bounded away from zero. The incentive to exert high effort in order to convince customers that the firm is still normal then always remains.

A similar role for replacements was described by Holmström (1982) in the context of a signal-jamming model of managerial employment. The wage of the manager in his model is higher if the market posterior over the manager's type is higher, even if the manager chooses no effort. In contrast, the revenue of a firm in the current model is higher for higher posteriors only if customers also believe that the normal firm is choosing high effort. Holmström's manager always has an incentive to increase effort, in an attempt to enhance the market estimation of his talent. In contrast to the model examined here, an equilibrium then exists (without replacements) in which the manager chooses effort levels that are higher than the myopic optimum. In agreement with spirit of our analysis however, this overexertion disappears over time, as the market's posterior concerning the manager's type approaches one.[34]

4.5. DISCUSSION

Even within the context of repeated games, we have focused on a particular model of reputations. This section briefly describes some alternatives.

4.5.1 Outside options and bad reputations

We begin with an interaction that we will interpret as involving a sequence of short-lived customers and a firm, but we allow customers an outside option that induces sufficiently pessimistic customers to abandon the firm, and so not observe any further signals.

The ability to sustain a reputation in this setting hinges crucially on whether the behavior of a firm on the brink of losing its customers makes the firm's product more or less valuable to customers. If these actions make the firm more valuable to customers, it is straightforward to identify conditions under which the firm can maintain a reputation. This section, drawing on Ely and Välimäki (2003), presents a model in which the firms' efforts to avoid a no-trade region destroy the incentives needed for a nontrivial equilibrium. We concentrate on a special case of Ely and Välimäki (2003). Ely et al. (2008) provide a general analysis of bad reputations.

There are two players, referred to as the firm (player 1) and the customer (player 2). We think of the customer as hiring the firm to perform a service, with the appropriate nature of the service depending upon a diagnosis that only the firm can perform. For example, the firm may be a doctor who must determine whether the patient needs to take two aspirin daily or needs a heart transplant.

The interaction is modeled as a repeated game with random states. There are two states of the world, θ_H and θ_L. In the former, the customer requires a high level of service, denoted by H, in the latter a low level denoted by L.

The stage game is an extensive form game. The state is first drawn by Nature and revealed to the firm but not the customer. The customer then decides whether to hire the firm. If the firm is hired, he chooses the level of service to provide.

[34] Neither the market nor the manager knows the talent of the manager in Holmström's (1982) model. The manager's evaluation of the profitability of effort then reflects only market beliefs. In contrast, our normal firms are more optimistic about the evolution of posterior beliefs that are customers. However, the underlying mechanism generating incentives is the same.

	Hire	Not hire
H	u, u	$0, 0$
L	$-w, -w$	$0, 0$

State θ_H

	Hire	Not hire
H	$-w, -w$	$0, 0$
L	u, u	$0, 0$

State θ_L

Figure 4.13 Payoffs for the terminal nodes of the extensive form bad-reputation stage game, as a function of the state (θ_H or θ_L), the customer's decision of whether to hire the firm, and (if hired) the firm's choice of service. We assume $w > u > 0$.

The payoffs attached to each terminal node in the extensive form game are given in Figure 4.13. The firm and the customer thus have identical payoffs. Both prefer that high service be provided when necessary, and that low service be provided when appropriate. If the firm is not hired, then both players receive 0.

Given that interests are aligned, one would think there should be no difficulty in the customer and the firm achieving the obviously efficient outcome, in which the firm is always hired and the action is matched to the state. It is indeed straightforward to verify that the stage game presents no incentive problems. Working backwards from the observation that the only sequentially rational action for the firm is to provide action H in state θ_H and action L in state θ_L, we find a unique sequential equilibrium, supporting the efficient outcome.

Suppose now that the (extensive form) stage game is repeated. The firm is a long-lived player who discounts at rate δ. The customer is a short-lived player. Each period features the arrival of a new customer. Nature then draws the customer's state and reveals its realization to the firm (only). These draws are independent across periods, with the two states equally likely in each case. The customer decides whether to hire the firm, and the firm then chooses a level of service. At the end of the period, a public signal from the set $Y \equiv \{\emptyset, H, L\}$ is observed, indicating either that the firm was not hired ($\emptyset$) or was hired and provided either high (H) or low (L) service. Short-lived players thus learn nothing about the firm's stage-game strategy when the firm is not hired.

For large δ, the repeated game has multiple equilibria. In addition to the obvious one, in which the firm is always hired and always takes the action that is appropriate for the state, there is an equilibrium in which the firm is never hired.[35] Can we restrict attention to a subset of such equilibria by introducing incomplete information? The result of incomplete information is indeed a bound on the firm's payoff, but it is now an upper bound that consigns the firm to a surprisingly low payoff.

[35] The structure of this equilibrium is similar to the zero-firm-payoff equilibrium of the purchase game (Section 4.2.5).

With probability $1 - \hat{\mu} > 0$, the firm is normal. With complementary probability $\hat{\mu} > 0$, the firm is "bad" and follows a strategy of always choosing H. A special case of Ely and Välimäki's (2003) result is then:

Proposition 4.5. *Assume that in any period in which the firm is believed to be normal with probability 1, the firm is hired. Then if the firm is sufficiently patient, there is a unique Nash equilibrium outcome in which the firm is never hired.*

Ely and Välimäki (2003) dispense with the assumption that the firm is hired whenever thought to be normal with probability 1, but this assumption is useful here in making the following intuition transparent. Fix an equilibrium strategy profile and let $\hat{\mu}^{\dagger}$ be the supremum of the set of posterior probabilities of a bad firm for which the firm is (in equilibrium) hired with positive probability. We note that $\hat{\mu}^{\dagger}$ must be less than 1 and argue that $\hat{\mu}^{\dagger} > 0$ is a contradiction. If the firm ever is to be hired, there must be a significant chance that the normal firm chooses L (in state θ_L), since otherwise his value to the customer is negative. Then for any posterior probability $\hat{\mu}'$ sufficiently close to $\hat{\mu}^{\dagger}$ at which the firm is hired, an observation of H must push the posterior of a bad firm above $\hat{\mu}^{\dagger}$, ensuring that the firm is never again hired. But then no sufficiently patient normal firm, facing a posterior probability $\hat{\mu}'$, would ever choose H in state θ_H. Doing so gives a payoff of $(1 - \delta)u$ (a current payoff of u, followed by a posterior above $\hat{\mu}^{\dagger}$ and hence a continuation payoff of 0) while choosing L reveals the firm to be normal and hence gives a higher (for large δ) payoff of $-(1 - \delta)w + \delta u$. The normal firm thus cannot be induced to choose H at posterior $\hat{\mu}'$. But this now ensures that the firm will not be hired for any such posterior, giving us a contradiction to the assumption that $\hat{\mu}^{\dagger}$ is the supremum of the posterior probabilities for which the firm is hired.

The difficulty facing the normal firm is that an unlucky sequence of θ_H states may push the posterior probability that the firm is bad disastrously high. At this point, the normal firm will choose L in both states in a desperate attempt to stave off a career-ending bad reputation. Unfortunately, customers will anticipate this and not hire the firm, ending his career even earlier. The normal firm might attempt to forestall this premature end by playing L (in state θ_H) somewhat earlier, but the same reasoning unravels the firm's incentives back to the initial appearance of state θ_H. We can thus never construct incentives for the firm to choose H in state θ_H, and the firm is never hired.

4.5.2 Investments in reputations

It is common to speak of firms as investing in their reputations. This section presents a discrete-time version of Board and Meyer-ter-Vehn's (2012) continuous-time model, which centers around the idea that investments are needed to build reputations.

In each period, a product produced by a firm can be either high quality or low quality. As in Section 4.4.3.3, the customers on the other side of the market are not strategic, and simply pay a price for the good in each period equal to the probability that it is high quality.

In each period, the firm chooses not the quality of the good, but an amount $\eta \in [0, 1]$ to invest in high quality. The quality of the firm's product is initially drawn to be either low or high. At the end of each period, with probability λ, there is no change in the firm's quality. However, with probability $1 - \lambda$, a new quality draw is taken before the next period. In the event of a new quality draw at the end of period t, the probability of emerging as a high-quality firm is given by η^t, the investment made by the firm in period t.

An investment of level η costs the firm $c\eta$, with $c > 0$. The firm's payoff in each period is then the price paid by the customers in that period minus the cost of his investment. The firm has an incentive to invest because higher-quality goods receive higher prices, and a higher investment enhances the probability of a high-quality draw the next time the firm's quality level is determined.

Customers do not directly observe the quality of the firm. Instead, the customers observe either signal 0 or signal 1 in each period. If the firm is high quality in period t, then with probability ρ_H the customer receives signal 1. If the firm is low quality, then with probability ρ_L the customer receives signal 1. If $\rho_H > \rho_L$, then the arrival of a 1 is good news and pushes upward the posterior of high quality. If $\rho_H < \rho_L$, then the arrival of a 1 is bad news, and pushes downward the posterior that the good is high quality. We say that the signals give perfect good news if $\rho_L = 0$, and perfect bad news if $\rho_H = 0$.

For one extreme case, suppose $\rho_L = 0$, giving the case of perfect good news. Here, the signal 1 indicates that the firm is certainly high quality (at least until the next time its quality is redrawn). We might interpret this as a case in which the product may occasionally allow a "breakthrough" that conveys high utility, but can do so only if it is indeed high quality. For example, the product may be a drug that is always ineffective if low quality and may also often have no effect if it is high quality, but occasionally (when high quality) has dramatic effects.

Conversely, it may be that $\rho_H = 0$, giving the case of perfect bad news. Here, the product may ordinarily function normally, but if it is of low quality it may occasionally break down. The signal 1 then offers assurances that the product is low quality, at least until the next quality draw.

We focus on four types of behavior. In a full-work profile, the firm always chooses $\eta = 1$, no matter the posterior about the product's quality. Analogously, the firm chooses $\eta = 0$ for every posterior in a full-shirk profile. In a work-shirk profile, the firm sets $\eta = 1$ for all posteriors below some cutoff, and chooses $\eta = 0$ for higher posteriors. The firm thus works to increase the chance that its next draw is high quality when its posterior is relatively low, and rides on its reputation when the latter is relatively high.

Turning this around, in a shirk-work profile the firm shirks for all posteriors below a cutoff and works for all higher posteriors. Here, the firm strives to maintain a high reputation but effectively surrenders to a low reputation.

In Board and Meyer-ter-Vehn's (2012) continuous-time model, equilibria for the cases of perfect good news and perfect bad news are completely characterized in terms of these four behavioral profiles. The discrete-time model is easy to describe but cumbersome to work with because it is notoriously difficult to show that the value functions are monotonic in beliefs. For the continuous-time model, Board and Meyer-ter-Vehn (2012) show the following (among other things):

- If signals are perfect good news, an equilibrium exists and this equilibrium is unique if $1 - \lambda \geq \rho_H$. Every equilibrium is either work-shirk or full-shirk. The induced process governing the firm's reputation is ergodic.
- If signals are perfect bad news again equilibrium exists, but $1 - \lambda \geq \rho_L$ does not suffice for uniqueness. Every equilibrium is either shirk-work, full-shirk or full-work. In any nontrivial shirk-work equilibrium, the reputation dynamics are not ergodic.

What lies behind the difference between perfect good news and perfect bad news? Under perfect good news, investment is rewarded by the enhanced probability of a reputation-boosting signal. This signal conveys the best information possible, namely that the firm is certainly high quality, but the benefits of this signal are temporary, as subsequent updating pushes the firm's reputation downward in recognition that its type may have changed. The firm thus goes through cycles in which its reputation is pushed to the top and the then deteriorates, until the next good signal renews the process. The firm continually invests (at all posteriors, in a full-work equilibrium, or at all sufficiently low posteriors, in a work-shirk equilibrium) in order to push its reputation upward.

In the shirk-work equilibrium under perfect bad news, bad signals destroy all hope of rebuilding a reputation, since the equilibrium hypothesis that the firm then makes no investments precludes any upward movement in beliefs. If the initial posterior as to the firm's type exceeds the shirk-work cutoff, the firm has an incentive to invest in order to ward off a devastating collapse in beliefs, but abandons all hope of a reputation once such a collapse occurs. The fact that it is optimal to not invest whenever customers expect no investment gives rise to multiple equilibria.

Some ideas reminiscent of the bad reputation model of Section 4.5.1 reappear here. In the case of a shirk-work equilibrium and perfect bad news, the firm can get trapped in the region of low posteriors. The customers do not literally abandon the firm, as they do in the bad-reputation model, but the absence of belief revision ensures that there is no escape. In this case, however, the firm's desire to avoid this low-posterior trap induces the firm to take high effort, which customers welcome. As a result, the unraveling of the bad reputation model does not appear here.

4.5.3 Continuous time

Most reputation analyses follow the standard practice in repeated games of working with a discrete-time model, while examining the limit as the discount factor approaches one. Since we can write the discount factor as

$$\delta = e^{-r\Delta},$$

where r is the discount rate and Δ the length of a period, there are two interpretations of this limit. On the one hand, this may reflect a change in preferences, with the timing of the game remaining unchanged and the players becoming more patient ($r \to 0$). One sees shadows of this interpretation in the common statements that reputation results hold for "patient players." However, the more common view is that the increasing discount factor reflects a situation in which the players' preferences are fixed, and the game is played more frequently ($\Delta \to 0$).

If one is examining repeated games of perfect monitoring, these interpretations are interchangeable. It is less obvious that they are interchangeable under imperfect monitoring. If periods are shrinking and the game is being played more frequently, one would expect the imperfect signals to become increasingly noisy. A firm may be able to form a reasonably precise idea of how much its rivals have sold if it has a year's worth of data to examine, but may be able to learn much less from a day's worth of data. We should then seek limiting reputations results for the case in which discount factors become large and the monitoring structure is adjusted accordingly.

One convenient way to approach this problem is to work directly in the continuous-time limit. This section describes some recent work on continuous-time reputation games.

It is not obvious that continuous-time games provide a fertile ground for studying reputations. Several examples have appeared in the literature under which intertemporal incentives can lose their force as time periods shrink in complete information games, as is the case in Abreu et al. (1991), Sannikov and Skrzypacz (2007), and Fudenberg and Levine (2007). The difficulty is that as actions become frequent, the information observed in each period provides increasingly noisy indications of actions, causing the statistical tests for cheating to yield too many false positives and trigger too many punishments, destroying the incentives.

The effectiveness of reputation considerations in continuous time hinges on order-of-limits considerations. If we fix the period length and let the discount factor approach one, then we have standard reputation results, with the underlying reasoning reproducing standard reputation arguments. Alternatively, if we fix the discount rate and let the period length go to zero, these reputation effects disappear. Here, we encounter the same informational problems that lie behind the collapse of intertemporal incentives in repeated games of complete information.

This gives us an indication of what happens in the two extreme limit orders. What happens in intermediate cases? Faingold (2005) shows that if we work with the limiting continuous-time game, then there exists a discount factor such that reputation effects persist for all higher discount factors. Hence, we can choose between discrete-time or continuous-time games to study reputations, depending on which leads to the more convenient analysis.

4.5.3.1 Characterizing behavior

Reputation models have typically produced stronger characterizations of equilibrium payoffs than of equilibrium behavior. However, Faingold and Sannikov (2011) have recently exploited the convenience of continuous time to produce a complete characterization of reputation-building behavior. We confine our presentation to the description of an example, taken from their paper.

We consider a variation of the product-choice game. A long-lived firm faces a continuum of customers. At each time t, the firm chooses an effort level $a^t \in [0, 1]$, while each customer chooses a level of service $b^t \in [0, 3]$. The firm observes only the average service level B^t chosen by the continuum of customers, while the customers observe only the current quality level of the firm, X^t. The latter evolves according to a Brownian motion, given by $dX^t = a^t dt + dZ^t$.

The firm maximizes the discounted profits

$$\int_0^{\infty} e^{-rt}(B^t - a^t)dt,$$

so that the firm prefers the customers buy a higher level of service, but finds effort costly. In each period each customer chooses her level of service to maximize her instantaneous payoff, with the maximizing service level increasing in the quality of the firm and decreasing in the average service level chosen by customers. We might interpret the latter as reflecting a setting in which there is congestion in the consumption of the service.

The unique equilibrium outcome in the stage game is that the firm chooses zero effort. In the repeated game, the firm's effort levels are statistically identified—different levels of effort give rise to different processes for the evolution of the firm's quality. This opens the possibility that nontrivial incentives might be constructed in the repeated game. However, the unique equilibrium of the continuous-time repeated game features the relentless play of this stage-game Nash equilibrium.

This result is initially counterintuitive, since it is a familiar result that the long-lived player could be induced to take actions that are not myopic best responses in a discrete-time formulation of this game. One would need only arrange the equilibrium so that "bad" signals about his actions trigger punishments. However, as we have noted, the resulting incentives can lose their force as time periods shrink, as is the case in Abreu et al. (1991), Sannikov and Skrzypacz (2007), andFudenberg and Levine (2007).

Now suppose that with some probability, the firm is a commitment type who always takes the maximal investment, $a = 1$. Faingold and Sannikov (2011) show that there is a unique sequential equilibrium. As the long-lived player becomes arbitrarily patient ($r \to 0$), the long-lived player's payoff converges to the Stackelberg payoff of 2.

Behind these payoffs lie rich equilibrium dynamics. The unique sequential equilibrium is a Markov equilibrium, with the posterior probability μ^t that the firm is a commitment type serving as the state variable. As μ^t approaches unity, the aggregate service level demanded by the short-lived players approaches the best response to the commitment type of 3. Smaller average service levels are demanded for smaller posteriors, though in equilibrium these levels approach 3 as the long-lived player gets more patient. The long-lived player exerts her highest effort levels at intermediate posteriors, while taking lower effort at very low or very high posteriors, with the function specifying the effort level as a function of the posterior converging to unity as the long-lived player gets more patient.

We thus have a reputation version of "number two tries harder." Firms with very high reputations rest on their laurels, finding the cost of high effort not worth the relatively small effect on customer beliefs. Firms with very low reputations abandon all hope of building their reputation. Firms in the middle labor mightily to enhance their reputations. The more patient the firm, the higher the payoff to reputation building, and hence the higher the effort profile as a function of the posterior.

4.5.3.2 Reputations without types

The standard model of reputations is centered around incomplete information concerning the long-lived player's type, with the long-lived player's reputation interpreted in terms of beliefs about his type. In contrast, Bohren (2011) shows that interesting reputation dynamics can appear without any uncertainty as to the type of the long-lived player.

Suppose that at each instant of time, a firm is characterized by a current quality, and chooses an investment. The current quality is a stock variable that is observed, while customers receive only a noisy signal of the investment. There is a continuum of customers, whose individual actions are unobservable. The model builds on Faingold and Sannikov (2011), but without incomplete information and with quality playing the role of posterior beliefs.

In each period, the current quality stock is observed, and then the firm and the customers simultaneously choose an investment level and a quantity to purchase. Higher investment levels tend to give rise to higher signals. The customer's payoff is increasing in the quality stock and the signal, and depends on the customer's quantity. The higher is the stock and the signal, the higher is the customer's payoff maximizing quantity.

The firm's payoff is increasing in the aggregate purchase decision, but is decreasing in the level of investment, which is costly. In a static version of this game, equilibrium would call for an investment level of zero from the firm. The incentive for the firm to undertake investments in the current period then arises out of the prospect of generating larger future quality stocks. In particular, the stock increases in expectation when investment exceeds the current quality level, and decreases in expectation when investment falls short of this level.

Bohren (2011) shows that there is a unique public prefect equilibrium, which is a Markov equilibrium. Once again, this equilibrium gives rise to "number two tries harder" incentives. The firm's investments are highest for an intermediate range of quality stocks, at which the firm works hard to boost its quality. The firm undertakes lower investments at lower quality levels, discouraged by the cost of building its reputation and waiting for a fortuitous quality shock to push it into more favorable territory. Similarly, the firm invests less for high quality levels, content to coast on its reputation. The result is a progression of product quality cycles.

ACKNOWLEDGMENTS

We have benefited from our collaboration and discussions with many coauthors and colleagues over the course of many years, and we are deeply grateful to them. We thank Martin Cripps, Olivier Gossner, Yuichi Yamamoto, and especially Ju Hu and Zehao Hu for their helpful comments. We thank the National Science Foundation (SES-0961540 and SES-1153893) for the financial support. We thank the Stockholm, Toulouse, and UNSW Schools of Economics and the Universities of Cambridge, Chicago, and York for the opportunity to teach some of this material.

REFERENCES

Abreu, D., Milgrom, P., Pearce, D., 1991. Information and timing in repeated partnerships. Econometrica 59(6), 1713–1733.

Al-Najjar, N.I., 1995. Decomposition and characterization of risk with a continuum of random variables. Econometrica 63(5), 1195–1224.

Aoyagi, M., 1996. Evolution of beliefs and the nash equilibrium of normal form games. J. Econ. Theory 70(2), 444–469.

Atakan, A.E., Ekmekci, M., 2012. Reputation in long-run relationships. Rev. Econ. Stud. 79(2), 451–480.

Atakan, A.E., Ekmekci, M., 2013. A two-sided reputation result with long-run players. J. Econ. Theory 148(1), 376–392.

Benabou, R., Laroque, G., 1992. Using privileged information to manipulate markets: insiders, gurus, and credibility. Q. J. Econ. 107(3), 921–958.

Board, S., Meyer-ter-Vehn, M., 2012. Reputation for quality. Econometrica 81(6), 2381–2462.

Bohren, J.A., 2011. Stochastic games in continuous time: persistent actions in long-run relationships. UCSD.

Celentani, M., Fudenberg, D., Levine, D.K., Pesendorfer, W., 1996. Maintaining a reputation against a long-lived opponent. Econometrica 64(3), 691–704.

Cesa-Bianchi, N., Lugosi, G., 2006. Prediction, learning, and games. Cambridge University Press, Cambridge.

Chung, K.L., 1974. A Course in Probability Theory. Academic Press, New York.

Cole, H.L., Dow, J., English, W.B., 1995. Default, settlement, and signalling: lending resumption in a reputational model of sovereign debt. Int. Econ. Rev. 36(2), 365–385.

Conlon, J., 2003. Hope springs eternal: learning and the stability of cooperation in short horizon repeated games. J. Econ. Theory 112(1), 35–65.
Cover, T.M., Thomas, J.A., 2006. Elements of information theory, second ed. John Wiley & Sons, Inc., New York.
Cripps, M.W., Dekel, E., Pesendorfer, W., 2005. Reputation with equal discounting in repeated games with strictly conflicting interests. J. Econ. Theory 121(2), 259–272.
Cripps, M.W., Mailath, G.J., Samuelson, L., 2004. Imperfect monitoring and impermanent reputations. Econometrica 72(2), 407–432.
Cripps, M.W., Mailath, G.J., Samuelson, L., 2007. Disappearing private reputations in long-run relationships. J. Econ. Theory 134(1), 287–316.
Cripps, M.W., Schmidt, K.M., Thomas, J.P., 1996. Reputation in perturbed repeated games. J. Econ. Theory 69(2), 387–410.
Ekmekci, M., 2011. Sustainable reputations with rating systems. J. Econ. Theory 146(2), 479–503.
Ekmekci, M., Gossner, O., Wilson, A., 2012. Impermanent types and permanent reputations. J. Econ. Theory 147(1), 162–178.
Ely, J., Fudenberg, D., Levine, D.K., 2008. When is reputation bad?. Games Econ. Behav. 63(2), 498–526.
Ely, J.C., Välimäki, J., 2002. A robust folk theorem for the prisoner's dilemma. J. Econ. Theory 102(1), 84–105.
Ely, J.C., Välimäki, J., 2003. Bad reputation. Q. J. Econ. 118(3), 785–814.
Ettinger, D., Jehiel, P., 2010. A theory of deception. Am. Econ. J.: Microecon. 2(1), 1–20.
Evans, R., Thomas, J.P., 1997. Reputation and experimentation in repeated games with two long-run players. Econometrica 65(5), 1153–1173.
Evans, R., Thomas, J.P., 2001. Cooperation and punishment. Econometrica 69(4), 1061–1075.
Faingold, E., 2005. Building a reputation under frequent decisions. University of Pennsylvania, Pennsylvania.
Faingold, E., Sannikov, Y., 2011. Reputation in continuous-time games. Econometrica 79(3), 773–876.
Fudenberg, D., Levine, D.K., 1989. Reputation and equilibrium selection in games with a patient player. Econometrica 57(4), 759–778.
Fudenberg, D., Levine, D.K., 1992. Maintaining a reputation when strategies are imperfectly observed. Rev. Econ. Stud. 59(3), 561–579.
Fudenberg, D., Levine, D.K., 2007. Continuous time limits of repeated games with imperfect public monitoring. Rev. Econ. Dyn. 10(2), 173–192.
Fudenberg, D., Levine, D.K., Maskin, E., 1994. The folk theorem with imperfect public information. Econometrica 62(5), 997–1039.
Gossner, O., 2011. Simple bounds on the value of a reputation. Econometrica 79(5), 1627–1641.
Holmström, B., 1982. Managerial incentive problems: a dynamic perspective. In: Essays in Economics and Management in Honour of Lars Wahlbeck. Swedish School of Economics and Business Administration, Helsinki, pp. 209–230. Reprinted from: Rev. Econ. Stud. 66(1), 1999, 169-182.
Hu, J., 2014. Reputation in the presence of noisy exogenous learning. J. Econ. Theory 153(1), 64–73.
Jehiel, P., 2005. Analogy-based expectation equilibrium. J. Econ. Theory 123(2), 81–104.
Jehiel, P., Samuelson, L., 2012. Reputation with analogical reasoning. Q. J. Econ. 127(4), 1927–1969.
Kalai, E., Lehrer, E., 1995. Subjective games and equilibria. Games Econ. Behav. 8(1), 123–163.
Kreps, D., Milgrom, P.R., Roberts, D.J., Wilson, R., 1982. Rational cooperation in the finitely repeated prisoner's dilemma. J. Econ. Theory 27(2), 245–252.
Kreps, D., Wilson, R., 1982. Reputation and imperfect information. J. Econ. Theory 27(2), 253–279.
Liu, Q., 2011. Information acquisition and reputation dynamics. Rev. Econ. Stud. 78(4), 1400–1425.
Liu, Q., Skrzypacz, A., 2011. Limited records and reputation. Columbia University and Graduate School of Business, Stanford.
Mailath, G.J., Samuelson, L., 2001. Who wants a good reputation? Rev. Econ. Stud. 68(2), 415–441.
Mailath, G.J., Samuelson, L., 2006. Repeated Games and Reputations: Long-Run Relationships. Oxford University Press, New York, NY.
Milgrom, P.R., Roberts, D.J., 1982. Predation, reputation and entry deterrence. J. Econ. Theory 27(2), 280–312.
Morris, S., 2001. Political correctness. J. Polit. Econ. 109(2), 231–265.

Phelan, C., 2006. Public trust and government betrayal. J. Econ. Theory, 130(1), 27–43.
Piccione, M., 2002. The repeated prisoner's dilemma with imperfect private monitoring. J. Econ. Theory 102(1), 70–83.
Sannikov, Y., Skrzypacz, A., 2007. Impossibility of collusion under imperfect monitoring with flexible production. Am. Econ. Rev. 97(5), 1794–1823.
Schmidt, K.M., 1993. Reputation and equilibrium characterization in repeated games of conflicting interests. Econometrica 61(2), 325–351.
Sekiguchi, T., 1997. Efficiency in repeated prisoner's dilemma with private monitoring. J. Econ. Theory 76(2), 345–361.
Sorin, S., 1999. Merging, reputation, and repeated games with incomplete information. Games Econ. Behav. 29(1/2), 274–308.
Wiseman, T., 2009. Reputation and exogenous private learning. J. Econ. Theory 144(3), 1352–1357.

CHAPTER 5

Coalition Formation

Debraj Ray*, Rajiv Vohra†

*Department of Economics, New York University, New York, NY, USA
†Department of Economics, Brown University, Providence, RI, USA

Contents

Handbook of game theory
http://dx.doi.org/10.1016/B978-0-12-420056-2.00005-7

Abstract

This chapter surveys a sizable and growing literature on coalition formation. We refer to theories in which one or more groups of agents ("coalitions") deliberately get together to jointly determine within-group actions, while interacting noncooperatively across groups. The chapter describes a variety of solution concepts, using an umbrella model that adopts an explicit real-time approach. Players band together, perhaps disband later and re-form in shifting alliances, all the while receiving payoffs at each date according to the coalition structure prevailing at the time. We use this model to nest two broad approaches to coalition formation, one based on cooperative game theory, the other based on noncooperative bargaining. Three themes that receive explicit emphasis are agent farsightedness, the description of equilibrium coalition structures, and the efficiency implications of the various theories.

Keywords: Coalition formation, Blocking, Bargaining, Farsightedness, Coalition structures, Core, Stable set

JEL Codes: C71, C72, C78, D71

5.1. INTRODUCTION

This chapter surveys the sizable and growing literature on coalition formation. We refer to theories in which one or more groups of agents ("coalitions") deliberately get together to jointly determine their actions. The defining idea of a coalition, in this chapter, is that of a group which can enforce agreements among its members, while it interacts noncooperatively with other nonmember individuals and the outside world in general.

It is hard to overstate the importance of coalition formation in economic, political, and social analysis. Ray (2007) gives several examples in which such a framework comes to life: cartel formation, lobbies, customs unions, conflict, public goods provision, political party formation, and so on. Yet as one surveys the landscape of this area of research, the first feature that attracts attention is the fragmented nature of the literature. The theories that bear on our questions range from collusive behavior in repeated games, to models of bargaining, to cooperative game-theoretic notions of the core, or notions of coalition proofness in noncooperative games. To unravel the many intricacies of this literature would take far more than a survey. To prevent our terms of inquiry from becoming unmanageably large, we impose a basic restriction.

Two fundamental notions are involved (usually separately) in the many theories that coexist. One has to do with the *formation* of groups, the "process" through which a coalition comes together to coordinate its subsequent actions. The other aspect involves the *enforcement* of group actions, say as an equilibrium of an appropriate game. In this survey, we deliberately omit the latter issue. We presume that the problem of enforcement is solved, once coalitions have chosen to form and have settled on the joint actions they intend to take. Of course, that does not necessarily mean that we are in a frictionless world such as the one that Coase (1960) envisaged. After all, the negotiations that lead up to an agreement are fundamentally noncooperative. In addition, it is entirely possible that a negotiation once concluded will be renegotiated. Such considerations generate enough analytical demands that we exclude the additional question of *implementing* an agreement, perhaps via dynamic considerations, from this survey. On the other hand, we are centrally interested in the former issue, which also involve "no-deviation" constraints as the implementation of an agreement would, but a different set of them. Just because a coalition—once formed—is cooperative, does not mean that the creation of that coalition took place in fully cooperative fashion.

An example will make this clear. Suppose that a market is populated by several oligopolists, who contemplate forming a monopolistic cartel. Once formed, the cartel will charge the monopoly price and split the market. A standard question in repeated games (and in this particular case, in the literature on industrial organization) is whether such an outcome can be sustained as an "equilibrium." If a player deviates from the cartel arrangement by taking some other action, there will be punitive responses. One asks that all such play (the initial action path as well as subsequent punishments) be sustainable as an equilibrium of noncooperative play. This is the problem of enforcement. Ignore it by assuming that an agreement, *once made*, can be implemented. In contrast, we emphasize the question of formation. Say that the terms of the cartel are being negotiated. A firm might make proposals to the others about cost and revenue sharing. Other firms might contemplate alternative courses of action, such as the possibility of standing alone. In that case, they would have to predict what their compatriots would do; for instance, whether they would form a smaller cartel made up of all the remaining firms, thus effectively converting the situation into a duopoly. This sort of reasoning in the negotiations process is as noncooperative as the enforcement problem, but it is a different problem. It can exist *even if* we assume that an agreement, once made, can be enforced without cost. And indeed, that is just what we do here.

So questions of enforcement may be out of the way, but a variety of models and theories remain. The literature on coalition formation embodies two classical approaches that essentially form two parts of this chapter:

(i) *The blocking approach*, in which we require the immunity of a coalitional arrangement to "blocking," perhaps subcoalitions or by other groups which intersect the coalition in question. Traditionally, the notion of blocking has been deployed in a negative way, as something that undermines or destroys proposed arrangements. As we shall show, blocking can also be viewed as part of the "negotiation process" that leads up to an agreement.[1]

There is, of course, an entire area of cooperative game theory devoted to such matters, beginning with the monumental work of von Neumann and Morgenstern (1944). This literature includes notions such as the stable set, the core, and the bargaining set (Aumann and Maschler, 1964; Gillies 1953; Shapley 1953; von Neumann and Morgenstern, 1944). Extensions of these ideas to incorporate notions of farsighted behavior were introduced by Harsanyi (1974) and later by Aumann and Myerson (1988). The farsightedness notion—one that is central to this chapter—was further developed by Chwe (1994), Ray and Vohra (1997), Diamantoudi and Xue (2003), and others.

(ii) *Noncooperative bargaining*, in which individuals make *proposals* to form a coalition, which can be accepted or rejected. Rejection leads to fresh proposal-making. The successive rounds of proposal, acceptance, and rejection take the place of blocking. Indeed, on the noncooperative front, theories of bargaining have served as the cornerstone for most (if not all) theories of coalition formation. The literature begins with the celebrated papers of Ståhl (1977), Rubinstein (1982), and Baron and Ferejohn (1989), the subsequent applications to coalition formation coming from Chatterjee et al. (1993), Okada (1996), Seidmann and Winter (1998), and several others. There is also a literature on coalition formation that has primarily concerned itself with situations that involve pervasive externalities across coalitions, such as Bloch (1996), Yi (1996), and Ray and Vohra (1999). In several of these papers bargaining theory explicitly meets coalition formation.

We mean to survey these disparate literatures. This is no easy task. After all, the basic methodologies differ—apparently at an irreconcilable level—over cooperative and noncooperative game approaches, and even with each methodological area there is a variety of contributions. Moreover, this is by no means the first survey of this literature: Bloch (2003), Mariotti and Xue (2003), and Ray (2007) are other attempts and while the survey component varies across these references, all of them provide a perspective on the literature. It is, therefore, important to explain how we approach the current task, and in particular why our survey is so very different from these other contributions.

[1] While we recognize that the term "block" has not been in favor since Shapley (1973), "the blocking approach" seems to us to be preferable to "the improve-upon approach," or to the coining of a new term.

We proceed by suggesting a unifying way of assessing the literature and perhaps taking it further. We propose a framework for coalition formation that has the following properties:

A. It nests the blocking and bargaining approaches under one umbrella model, and in particular, it permits a variety of existing contributions to be viewed from a single perspective.

B. It allows for players to be farsighted or myopic, depending on the particular model at hand.

C. It deals with possible cycles in chains of blocking, a common problem in cooperative game theory.

D. It allows for the expiry or renegotiation of existing agreements or insists that all deals are irreversible, depending on the context at hand.

The chapter is organized as follows. In the next section, we present an abstract, dynamic model of coalition formation, followed by a definition of an equilibrium process of coalition formation (EPCF). This framework will be shown to be general enough to unify the various strands of the literature, and to suggest interesting directions for further research in the area. Section 5.3 concerns the blocking approach to coalition formation. Here we review some of the basic concepts in classical cooperative game theory, which are based on notions of coalitional objections or "blocking." We show how some of the standard notions of coalitional stability, such as the core of characteristic function games, can be subsumed under our general notion of an EPCF, despite the fact that these standard cooperative models are static while our general framework is a dynamic one. We then illustrate some of the limitations of the blocking approach in environments with externalities and argue that our explicitly dynamic model provides a way to resolve some of these difficulties.

Section 5.4 is devoted to a review of coalitional bargaining in noncooperative games. Here, "coalitional moves" are replaced by individual proposals to a group of agents, and acceptance of such a proposal signifies the formation of a coalition. Externalities can be incorporated by considering partition functions rather than characteristic functions, and an equilibrium of the resulting bargaining game now implies the formation of a coalition structure. There are three distinct branches of this literature depending on whether agreements are permanently binding or temporary and whether renegotiation of existing agreements is possible. A suitable specialization of our model seems well suited to encompass all of the bargaining models. We show that in general the process of coalition formation corresponding to an equilibrium of a bargaining game conforms to our notion of an EPCF. We then describe some results on coalition formation from this literature.

The general framework that we use lays bare a large degree of incompleteness in the literature, something that's evident in the asymmetry of exposition between Sections 5.3 and 5.4. In terms of the general framework that we lay out, the existing literature falls short on several counts, and in particular, different aspects receive disparate attention in the bargaining and blocking approaches. For instance, the blocking approach has been concerned with questions of "chains of coalitional objections" so that farsightedness has appeared as a natural component, as in Harsanyi (1974) or Chwe (1994). In contrast, the bargaining approach with its insistence on a well-defined game-theoretic structure has invariably been more explicit about the structure of moves, which then lends itself more naturally to considerations such as the renegotiation of agreements. Unfortunately, these uneven developments are mirrored in the varying emphasis we lay on these matters at different points of the text, and one can only hope that a survey 10 years hence would be far more balanced.

Finally, Section 5.5 concerns one of the most important questions in coalition formation; namely, the possibility of achieving efficiency when there is no impediment to the formation of coalitions. For the clearest understanding of this issue, we assume away the two most commonly recognized sources of inefficiency. First, decentralized or noncooperative equilibria may quite naturally yield inefficient outcomes, as is the case for Nash equilibria in games or competitive equilibria in the context of "market failure" resulting from incomplete markets or externalities. These problems are explicitly assumed away when we allow coalitions to write binding agreements.[2] Second, inefficiency may arise due to incompleteness of information, which we also assume away. Thus, there may be a presumption that in our framework Pareto efficiency will obtain in equilibrium. Indeed, much of cooperative theory is built on this presumption, and the Coaseian idea that efficiency is inevitable in a world of complete information and unrestricted contracting is very much part of the economics folklore. As the recent literature shows, however, efficiency in the presence of externalities, even in a world with complete information and no restrictions on coalition formation, may be more elusive. If coalitional agreements are permanently binding, then the possibility of inefficiency in equilibrium is a robust phenomenon. The ability to renegotiate agreements can restore efficiency under certain conditions, e.g., in the absence of externalities, but not in general. A detailed examination of these issues is provided in Section 5.5.

5.2. THE FRAMEWORK

In this section, we describe a general framework for coalition formation that serves as an umbrella for a variety of different models in the literature. Central to our approach is a description of coalition formation in real time, one which allows for both irreversible and reversible agreements, and yields payoffs as the process unfolds.

[2] For instance, in our framework inefficiency cannot arise in a two-player game even if there are externalities; full cooperation (formation of the grand coalition) must be the equilibrium outcome if it Pareto dominates the "noncooperative" outcome represented by singleton coalitions.

5.2.1 Ingredients

Our framework contains the following components.

[1] A finite set N of *players*, a compact set X of *states*, an infinite set $t = 0, 1, 2 \ldots$ of *dates*, and an initial state x_{-1} at the start of date 0.

[2] For each player i, a continuous one-period payoff function u_i defined on X, and a (common) discount factor $\delta \in (0, 1)$.

[3] For every pair of states x and y, a collection of coalitions $E(x, y)$ that are *effective* in moving the state from x to y, with $E(x, x)$ being the collection of all coalitions.

[4] A *protocol*, ρ, which defines a probability over the choice of an "active coalition" S at each date.

[5] Along with S, a (possibly empty) set of *potential partners* $P \subseteq N \setminus S$, also chosen by the protocol. The interpretation is that S has the exclusive right to propose to a (possibly empty) set of partners $Q \subseteq P$ a move to a new state for which $S \cup Q$ is effective.

[6] An *order of responses*, again given by the protocol, for every set of partners Q included by S in its proposed move. If any individual in Q rejects the proposal, the state remains unchanged. The first such rejector is "responsible" for that move not occurring.

[7] At each date t, possible *histories* h_t that begin at x_{-1} and list all active coalitions, partners, and moves up to period $t - 1$, as well as any individual "responsible" for the refusal of past moves. The protocol ρ is conditioned on this history.

The basic idea is both general and simple. Each date t begins under the shadow of a "going history" h_t, as described in [7]. If a coalition S_t becomes active, as in [4], then it moves to a possibly new state x_t with the help of partners Q_t chosen from the potential set P_t (see [5]); by "move," we refer to $m_t = (x_t, S_t, Q_t)$. Of course, $S_t \cup Q_t$ must be *effective* in implementing the new state x_t.[3] Note that by [3], an active coalition always has the option not to call upon any partners. It can also keep the state unchanged at the status quo (that, too, is a move, by convention). Each player receives a return at each state, with discounted returns added over time to obtain overall payoffs or values, see [2]. The one additional feature we need to embed into histories is a record of the "responsible" individual (if any) who rejected a proposed move in any previous period in which the state did not change, see [6]. There will be no such individual if the active coalition in that period suggested no change, or chose no partners, or if all its partners accepted the proposed move. (The reason we track the rejectors is that in some bargaining models, the choice of future active proposers may depend on the identity of past rejectors.)

It should be noted that implicit in the existence of a "pivotal" or "responsible" individual for each rejected move is that a proposed move must be *unanimously* accepted

[3] The importance of employing a notion of effectiveness in cooperative games was emphasized by Rosenthal (1972).

by partners who have been invited by the active coalition and who respond sequentially given the protocol. As we discuss later in Section 5.4.1.4, there is a little loss of generality in making the unanimity assumption, provided we redefine coalitional worths in an appropriate way.

The process formally continues *ad infinitum*, but all movements in states may or may not cease at a particular date. That will depend on the specification of the model. For instance, it is possible that for some states x, $E(x, y) = \emptyset$ for all $y \neq x$. If such an *end state* is reached, the process ends, though our formalism requires payoffs to be received from that final state "forever after." The notion of end states will be useful in the blocking approach. Another possibility is that the protocol shuts down after certain histories and may cease to choose active coalitions. This is useful in models in which the renegotiation of an existing agreement is not permitted.

The two leading applications of this general framework are, of course, to theories of coalition formation based on cooperative game theory, or what we will refer to here as the *blocking approach*, and a parallel theory based on noncooperative bargaining, what we will call here the *bargaining approach*. In the blocking approach, active coalitions do not make proposals to additional partners; formally, this is captured by having an empty partner set at every history. In the bargaining approach, active coalitions are invariably *individuals*, and proposals made by them to other players (the "partners") will occupy center stage. This allows us to nest a variety of solution concepts under a common descriptive umbrella.

A schematic representation of the model is shown in Figure 5.1. Time goes from left to right. The squares in the upper panel depict implemented states $x_{-1}, \ldots, x_t$, while the ovals in the lower panel denote various individuals. Following every history, the dark ovals denote members of an active coalition and the light ovals are partners. Together, these create histories for later dates. The process continues, possibly indefinitely, and payoffs are received at every date.

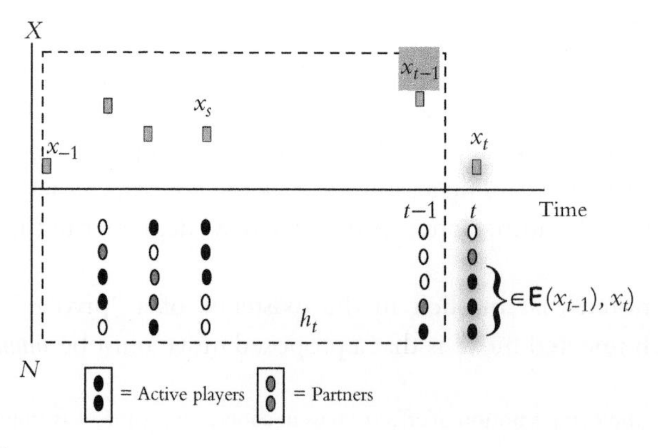

Figure 5.1 Timeline.

5.2.2 Process of coalition formation

A *process of coalition formation* (PCF) is a stochastic process λ that takes existing histories (beginning with $h_0 \equiv \{x_{-1}\}$) to subsequent histories, which obeys the following restrictions:

(i) At every history h_t, λ is consistent with the probability measure induced by the protocol in selecting an active coalition S_t and potential partner set P_t.

(ii) Every move $m_t = (x_t, S_t, Q_t)$ generated by λ has the properties that $Q_t \subseteq P_t$ and $S_t \cup Q_t \in \mathcal{E}(x_{t-1}, x_t)$.

Conditions (i) and (ii) describe obvious feasibility restrictions.

Why might the process be stochastic? One reason is that the protocol that selects an active coalition may be stochastic; for instance, it may choose an active coalition equiprobably from the set of all possible coalitions. In addition, agents and coalitions may randomize their decisions, e.g., the choice of partners or moves.

A PCF defines *values* for every person i at every history h_t:

$$V_i(h_t, \lambda) \equiv \mathbb{E}_\lambda \left(\sum_{s=t}^{\infty} \delta^{s-t} u_i(x_s) | h_t \right),$$

where all states from t onwards are generated by λ conditioned on the history h_t. Note that the expectation is taken prior to the choice of active coalition, which is "about to happen" following that history.

5.2.3 Equilibrium process of coalition formation

We now define an equilibrium process of coalition formation. Suppose that a PCF λ is in place and a particular coalition S is active at some date, along with a set of potential partners. Both S and its chosen set of partners must "willingly participate in the law of motion" of the process. We judge such participation by two criteria: (a) the move to a new state—which includes the possibility of no move—must be "profitable" for both S and any partners that are called upon to implement the move, and (b) of all the profitable moves that can be entertained, there is no other move that S can make that would be better for all the members of S.

Formally, a move $m = (x, S, Q)$ at history h_t and going state x_{t-1} is *profitable* for $S \cup Q$ if $S \cup Q \in \mathcal{E}(x_{t-1}, x)$ and

$$V_i(h_{t+1}, \lambda) \geq V_i(h^i_{t+1}, \lambda) \quad \text{for every } i \in S \cup Q,$$

where for each $i \in S \cup Q$, h^i_{t+1} is a subsequent history consistent with the state remaining unaltered ($x_t = x_{t-1}$). For $i \in Q$, it is the history resulting from i's rejection of the proposed move, under the presumption that all earlier respondents accepted the proposal. For $i \in S$, it is the history resulting from the status quo move. That is, a move is profitable if no member of the active coalition would have been better off with the status quo move and no member of the partner set would have been better off rejecting the move.

Given active coalition S and potential partner set P, a move $m = (x, S, Q)$ at history h_t with state x_{t-1} is *maximal* for S if there is no other move, say $m' = (z, S, Q')$, that is profitable for $S \cup Q'$ for some $Q' \subseteq P$, such that

$$V_i(h'_{t+1}, \lambda) > V_i(h_{t+1}, \lambda) \quad \text{for every } i \in S,$$

where, as before, h_{t+1} is the history created by implementing m and h'_{t+1} is the history created by the alternative profitable move m'. The interpretation is that S can always form the coalition $S \cup Q'$ and make the move m', so no such move must strictly dominate what S actually does under the going PCF.

Note well that profitability and maximality are defined relative to an ongoing process: the value functions will vary with the process in question.

An *equilibrium process of coalition formation* (EPCF) is a PCF with the property that at every history, every active coalition, faced with a given set of potential partners, makes a profitable and maximal move. That is, all movers (weakly) prefer their "prescribed" move to inaction and that move is not dominated for members of S by some other profitable move that S can engineer.

Our formulation is related to the equilibrium definition introduced by Konishi and Ray (2003) and extended by Hyndman and Ray (2007). It is also related to the solution concept used in Gomes and Jehiel (2005) and Gomes (2005). As in these papers, the definition allows for a fully dynamic model of widespread externalities in which coalitions are farsighted in "blocking" a status quo. But there are some important differences. The current approach is designed to allow for a level of generality substantial enough to encompass not only the blocking approach but also the bargaining approach. We include the possibility that an active coalition can enlist the help of others to form an "approval committee" that is capable of carrying the proposed move. As we shall see by example, bargaining is a special case of our formulation in which S is always a singleton asking for acceptance of a proposal from other players (the "partners"). On the other hand, the blocking approach corresponds to the polar case in which the partner set is always empty. This formulation also opens up intermediate and unexplored possibilities, in which a (nonsingleton) coalition can choose partners.

Setting this innovation aside, Konishi and Ray also define profitable and maximal moves. However, there is no protocol that determines which coalition is active at a given date. Konishi and Ray require that if there is *some* coalition for which a *strictly* profitable move exists from the current state, then the state *must* change. That is, some such coalition must perforce become active. Under our definition, it is possible that some coalition has a profitable move, but is not active. This distinction is blurred if different "dates" are very closely bunched together, for then (under a natural full-support assumption) *every* coalition must sooner or later become active. Moreover, as we shall see, having an explicit protocol has the distinct advantage of overcoming coordination problems that can arise with the blocking approach in a dynamic context.[4]

Since we will be showing how various solution concepts in the literature correspond to an EPCF of a suitably specialized version of our general model, existence of an EPCF in each of these cases will follow from corresponding existence results in the literature. While it would be desirable to provide a general result on the existence of an EPCF, we shall not attempt to do so here.

5.2.4 Some specific settings

The framework presented so far is quite general and potentially amenable to a wide range of applications. It will be useful to present some concrete illustrations of how this overall structure can be specialized to cover various models of coalition formation. In particular, we shall discuss state spaces, protocols, and effectivity correspondences in a more explicitly game-theoretic setting.

5.2.4.1 State spaces

So far we have let the state space be an entirely abstract object. Given our specific interest in coalition formation, however, it should be the case that *at a minimum*, a state must describe the coalitions that have formed as well as the (one-period) payoffs to each of the agents.

But states can be suitably "tagged" to encode more information. For instance, in a situation in which all previous actions are irreversible, it is important to keep track of precisely which agents are "free" to engage in further negotiation. In such cases it suffices to tag every agent as "committed" or "uncommitted." When agreements are temporary or renegotiable, every agent is open in principle to further negotiation. However, the concept of effectivity will need to be suitably altered; for instance, with renegotiation, existing signatories to an agreement will need to be part of an effective coalition in any move to a new state even if they are physically not required in the implementation of that state.

In other models, a description of the state in terms of ongoing *action vectors* may be necessary. This will be the case if such actions (rather than the formation of coalitions) are viewed as irreversible, at least for a certain length of time. In yet others, there might be limits on the number of times a particular coalition is permitted to move, so that a state will need to keep track of this information.

[4] Both our definition and that of Konishi and Ray's share the feature that they implicitly rely on dynamic programming; in particular, on the use of the one-shot deviation principle. Because coalitions have vector-valued payoffs, it is possible that the one-shot deviation principle fails when every member of a coalition is required to be better off from a deviation. This issue is discussed in Konishi and Ray (2003) and Ray (2007).

Finally, in other settings that we do not emphasize in this survey, a state might describe a network structure, in which the links of each agent to every other agent are fully described.[5]

5.2.4.2 Characteristic functions and partition functions

A particularly important object that is used to create the state space in a variety of situations is a mapping that assigns each coalition structure (or partition of the player set) to a set of payoffs for each of its member coalitions. The simplest version of such a mapping comes from a cooperative game in characteristic function form as defined by von Neumann and Morgenstern (1944). Such a game is defined as (N, V), where N denotes a finite set of players and for every coalition S (a nonempty subset of N), the set of feasible payoff profiles is denoted $V(S) \subseteq \mathbb{R}^S$.[6] In this setting, a state may consist simply of a coalition structure and a feasible payoff profile. Typically, for a state x, we will denote by $\pi(x)$ the corresponding coalition structure and $u(x)$ the profile of payoffs, and by $u(x)_S$ the restriction of $u(x)$ to coalition S. In some applications, we will find it more convenient to restrict payoffs to be efficient, i.e., $u(x)_S \in \bar{V}(S)$ for each $S \in \pi(x)$, where $\bar{V}(S) = \{w \in V(S) \mid \text{there is no } w' \in V(S) \text{ with } w' \gg w\}$.[7]

The presumption that the feasible payoffs for coalition S can be described independently of the players in $N \setminus S$ is easily justified if there are no external effects across coalitions. In many interesting models of coalition formation, however, the feasible payoffs to a coalition depend on the behavior of outsiders. As we shall see, externalities can be very effectively incorporated into the analysis by generalizing the notion of a characteristic function to a *partition function* (see Thrall and Lucas, 1963), and this is the formulation we adopt for this chapter.[8] In such a game, the feasible payoff profile for a coalition depends on just how the "outsiders" are organized into coalitions of their own. For a partition function game (N, V), the feasible set for S is, therefore, written as $V(S, \pi) \subseteq \mathbb{R}^S$, where π is a coalition structure that contains S. The interpretation is that a coalition S embedded in an ambient structure π can freely choose from the set of payoffs $V(S, \pi)$. In this setting, again, at a minimum, a state x will refer to a pair consisting of a coalition structure $\pi(x)$ and a payoff profile $u(x)$ such that $u(x)_S \in \bar{V}(S, \pi(x))$ for every $S \in \pi(x)$.

[5] There is a literature on networks that we do not address in this survey, but it is worth pointing out that our general framework incorporates the case of network formation as well. See Jackson (2010) for a comprehensive review of the literature on networks. For network formation in particular, see for example, Jackson and Wolinsky (1996), Dutta et al. (2005), Page et al. (2005), and Herings et al. (2009).

[6] $\mathbb{R}^S$ denotes the $|S|$ dimensional Euclidean space with coordinates indexed by members of S.

[7] We use the convention $\gg, >, \geq$ to order vectors in $\mathbb{R}^S$.

[8] For an extended discussion of the historical background on characteristic functions and partition functions see Chapter 11.2.1 of Shubik (1982) and Chapter 2.2 of Ray (2007).

Note, however, that all externalities in a partition function are fed through the existing coalition structure rather than on specific *actions* that noncoalitional members might take. How might a partition function be then compared to the more familiar setting of a game in which players choose actions from their respective strategy sets and the payoff for players in a coalition depends on the actions of all players? The answer is that the partition function, while a primitive object for our purposes, is often derived from just such a normal form setting. The idea is that each coalition in the structure (never mind for the moment how they have formed) can freely coordinate the actions of their members (they can write binding agreements), but they cannot commit to a binding course of play *vis-à-vis* other coalitions and so play noncooperatively "against" them. More precisely, given a coalition structure for a normal form game, equilibrium noncooperative play across coalitions can be defined by first defining coalitional best responses as those which generate vector-valued maximal payoffs for each coalition, and then imposing Nash-like equilibrium conditions, see e.g., Ichiishi (1981), Ray and Vohra (1997), Zhao (1992), and Haeringer (2004). The resulting set of equilibrium payoffs may then be viewed as a partition function.[9] In this way, a partition function can be constructed from a normal form game, with the understanding that an equilibrium concept is already built into the function to begin with.

Here are four examples of partition functions; for others, see Ray (2007).

Example 5.1. (Cournot oligopoly) *A given number, n, of Cournot oligopolists produce output at a fixed unit cost, c. The product market is homogeneous with a linear demand curve: $p = A - bz$, where z is aggregate output. Standard calculations tell us that the payoff to a single firm in an m-firm Cournot oligopoly is*

$$\frac{(A-c)^2}{b(m+1)^2}.$$

If each "firm" is actually a cartel of firms, the formula is no different as long as each cartel attempts to maximize its total profits (and then allocate those profits among its members). Define

$$v(S,\pi) = \frac{(A-c)^2}{b\,[m(\pi)+1]^2},$$

where $m(\pi)$ is the number of cartels in the coalition structure π. Then for every $S \in \pi$, $V(S,\pi)$ is the collection of all payoff vectors for S that add up to no more than $v(S,\pi)$.

[9] In general, when the payoffs to members of a coalition depend on the actions of outsiders, this procedure will yield a partition function rather than a characteristic function. It is possible, though in our view not advisable, to make enough (heroic) assumptions on the behavior of outsiders to go all the way from a normal form to a characteristic function. For example, assuming the worst in terms of outsiders' actions leads to the α-characteristic function. See Section 5.3.5.5 for a critique of this approach.

This partition function is particularly interesting in that it does not depend on S at all, but only on the ambient coalition structure π.[10]

Example 5.2. (Public goods) *Consider the provision of pollution control by n agents (regions, countries). Suppose that r units of control by any agent generates a convex utility cost* $c(r)$ *for that agent. If sidepayments in utils are possible, a coalition of s agents will contribute a per-capita amount* $r(s)$ *to maximize*

$$sr(s) + \bar{R} - c(r(s)),$$

where $\bar{R}$ *is the amount contributed by noncoalition agents.*

It is easy to check that this formulation gives rise to a partition function.[11]

Example 5.3. (Customs unions) *There are n countries, each specialized in the production of a single good. There is a continuum of consumers equally dispersed through these countries. They all have identical preferences. Impose the restriction that no country or coalition can interfere with the workings of the price system except via the use of import tariffs. Then for each coalition structure—a partition of the world into customs unions—there is a coalitional equilibrium, in which each customs union chooses an optimal tariff on goods imported into it.*

In particular, the grand coalition of all countries will stand for the free-trade equilibrium: a tariff of zero will be imposed if lump-sum transfers are permitted within unions.

A specification of trade equilibrium for every coalition structure generates a partition function for the customs union problem.[12]

Example 5.4. (Conflict) *There are several individuals. Each coalition in a coalition structure of individuals expends resources to obtain a reward (perhaps the pleasures of political office). Resources may be spent on lobbies, finance campaigns, or cross-coalitional conflict, depending on the particular application. Suppose that the probability* p_S *that coalition S wins depends on the relative share of resources r expended by it:*

$$p_S = \frac{r_S}{r_S + r_{-S}}. \quad [5.1]$$

The per-capita value of the win will generally depend on the characteristics of the coalition (for instance, coalitional size, $|S| = s$*); write this value as* w_S*. The coalition then chooses resource contributions from its members to maximize*

$$sp_S w_S - \sum_{i \in S} c(r_i),$$

[10] For literature related to this example, see Salant et al. (1983), Bloch (1996), and Ray and Vohra (1997).

[11] See Ray and Vohra (2001).

[12] For literature that relies on a similar formulation, see Krugman (1993), Yi (1996), Krishna (1998), Ray (1998, Chapter 18), Aghion et al. (2007), and Seidmann (2009).

where c is the individual cost function of contributions, $r_S = \sum_{i \in S} r_i$, and r_{-S} is taken as given.

This generates a well-defined transferable-utility partition function.[13]

5.2.4.3 Remarks on protocols and effectivity correspondences

Protocols, along with the specification of coalitions that are effective for each move, can capture various levels of complexity and detail.

When a protocol chooses a nonsingleton coalition to be active, we are typically in the world of cooperative game theory, in which that coalition takes the opportunity to "block" an existing state: here, to be interpreted here in more positive light as moving the current state to a new one. The reinterpretation is important. Under the classical view, the issue of "what happens after" a state is blocked is sidestepped; blocking is viewed more as a negation, as the imposition of a constraint that must be respected (think of the notion of the core, for instance). Here, a "block" is captured by a physical move to a new state made by the active coalition in question. Partner sets, while formally easy enough to incorporate in the definition, generally do not exist in the theory of cooperative games.

In contrast, in the world of noncooperative coalition formation, as captured (for instance) by Rubinstein bargaining, the protocol simply chooses a proposer: a singleton active coalition.

Here are two examples of protocols.

Uniform Protocol: A coalition (or proposer) is randomly selected to be active from the set of all possible coalitions (or proposers) at any date.

Rejector Proposes: The first rejector of the previous proposal is chosen to be the new proposer, while if the previous proposal passed, a new proposer is chosen equiprobably from the set of all "uncommitted" players.

Later, we adopt a generalization of both these well-known protocols.

Protocols that choose active coalitions must also address the question of potential partner selection. As we've discussed, in the blocking approach, the potential partner set is typically empty, or so it is taken to be in existing literature. An active coalition must move on its own. In the bargaining approach, the potential partner set is typically the set of all individuals who are free to receive proposals.

This last item depends intimately on the situation to be analyzed. In models of irreversible agreements, every player who has previously moved (including one who has committed to do so on her own) is no longer free to entertain new offers, or to make them. In models of reversible agreements, *every* player is a potential partner at every date. To be sure, what they can achieve will depend on whether their pre-existing agreements have simply expired or are binding but renegotiable. That is a matter dealt with by the effectivity correspondence, and some remarks on this are in order.

[13] For related literature, see Esteban and Ray (1999), Esteban and Sákovics (2004), and Bloch et al. (2006).

There are two broad sets of considerations that are involved in specifying effectivity. The first has to do with which coalitions can move at a particular state. For instance, as we have already seen, issues of renegotiation (or its absence) can preclude coalitions from moving twice, or perhaps moving only with the blessings of ancillary players who must then be incorporated into the specification of effectivity.

Other conditions that determine which coalitions can move might come from the rules that describe the situation at hand. For instance, in multilateral bargaining with unanimity, as in the Rubinstein model or its multiplayer extensions, only the grand coalition of all players can be effective in making a move. On the other hand, in models in which a majority of players are needed in determining if a new proposal can be implemented, effective coalitions must be a numerical majority.

The second broad consideration that determines effectivity is an understanding of just what a coalition can implement when it does move. After all, the new state specifies payoffs not just for the coalition in question, but for *every* player. For instance, even when the game is described by the relatively innocuous device of a characteristic function, a coalition is presumably "effective" over moves that preserve existing payoffs to other coalitions untouched by its formation, while implementing for itself any payoff vector in its own characteristic function. But there still remains the question of what happens to members of the "coalitional fragments" left behind as the coalition in question forms. Often, the issue is settled by presuming that one can assign any payoff vector in the characteristic function of those fragments, which are after all, coalitions in their own right.

In summary, the effectivity correspondence describes the power that each set of players possesses over the implementation of a new state. This entails the use of several considerations that we discuss in greater detail as we go along. Effectivity specifies both physical feasibility as well as the rules of the game: legalities, constitutions, and voting rules. In addition, and provided the state is described carefully, effectivity correspondences allow easily for situations with irreversible, temporary, or renegotiable agreements. The protocol and the effectivity correspondence play complementary roles.

Various combinations of protocol and effectivity correspondence can be used to serve different descriptive needs. For instance, one might wish to capture the possibility that a coalition which does not move at a particular state remains inactive until the state changes:

A Single Chance at Any State. For any history h_t, let m be the maximal date at which the state last changed, i.e., m is the largest date τ such that $x_\tau \neq x_{\tau-1}$. Then, if $m < t-1$, exclude all active coalitions chosen between $m+1$ and t, and choose a coalition at random from the set of remaining players (no coalition is ever chosen again if the remaining set is empty). This restriction guarantees that at any state, a chosen coalition that does not "actively move" is debarred from being active again, until the state changes.

As another example, the protocol might only choose *doubleton* coalitions to be active, as in Jackson and Wolinsky (1996) on networks:

Link Formation in Networks. The protocol chooses a *pair* of players (perhaps randomly, and perhaps with restrictions depending on what has already transpired); such a pair can form a link in a network. In addition, the protocol can be augmented to choose singleton players to unilaterally sever existing links.

5.2.5 Remarks on the response protocol

We remark on two aspects of the response protocol, which requires individuals to respond in some fixed order, the proposal being rejected as soon as a single individual turns down the offer.

Sequential Responses. When an active individual or coalition has the opportunity to make a proposal to a potential set of partners, as in item [5] of our description, that set of individuals is given the opportunity to respond. This procedure is described in [6]. We assume that such responses are sequential so as to rule out rather uninteresting equilibria with coordination failures ("I say no, because I anticipate that you will say no, and you think likewise."). This is not a major issue, as long as we are prepared to refine the set of equilibria to eliminate such dominated outcomes.

Unanimity. We've presumed that unanimity is needed in order to get a proposal passed. It is not at all difficult to extend the definition to include majority or supermajority rules for obtaining agreement. But some modeling decisions would need to be confronted. Briefly, once unanimity is dropped at the level of the partner set, one might ask whether it can (or should) be dropped at the level of the active coalition: does every member of the active coalition need to agree? Furthermore, does the *renegotiation* of proposals require unanimity, or does it suffice that some given fraction of the original coalition is willing to tear up an existing agreement? To cut down the resulting proliferation of cases, we have decided to avoid these extensions in our exposition.

That said, our framework accommodates some important nonunanimity protocols without any alteration at all. A leading example is the bargaining game of Baron and Ferejohn (1989), in which a majority is enough to force agreement. These cases can be easily covered by defining appropriate minimal winning coalitions, and then regarding the original *majority* game as one in which a proposal requiring *unanimous* support is made to a minimal winning subcoalition. See Section 5.4.1.4 for more details.

An explicit consideration of nonunanimity protocols is warranted in other situations. For example, in coalitional bargaining games with majority protocols, it may be important to record the fact that it is a particular coalition that has actually concluded an agreement via majority, rather than simply track some minimal winning subcoalition of it. An explicit use of the protocol will also be needed in situations in which renegotiation of existing agreements is a possibility.

5.3. THE BLOCKING APPROACH: COOPERATIVE GAMES

The classical concepts in cooperative game theory are based on coalitions as the primary units of decision making. These concepts rely on the notion of coalitional "objections" or "blocking." A proposed allocation is *blocked* by some coalition if there is an alternative allocation, feasible for that coalition, that improves the payoffs of each of its members. Blocking is used as a test to rule *out* certain allocations from membership in the solution. What happens "after" a block is typically not part of the solution, which consists only of those allocations that are *not* blocked. The notion of the core is the leading example of the use of blocking as a negation.[14] Various notions of the bargaining set (for example, Aumann and Maschler, 1964, and Dutta et al. 1989), look beyond the immediate consequence of an objection by considering the possibility of counterobjections. However, these considerations are really meant to refine the logic of negotiations underlying the stability of an allocation in the grand coalition rather than to describe coalition formation. While we will not be reviewing this literature, the interested reader is referred to Maschler (1992) for an extensive survey.

But the concept of blocking can also be used in a more "positive" way, as a generator of actual moves that actively involve the formation of coalitions or coalition structures. In these approaches, it becomes especially necessary to explicitly model what follows a blocking action. Often, there will be repercussions: the blocking of an allocation may be followed by additional moves. In such contexts, the question of farsightedness becomes focal: do players only derive payoffs from the immediate results of their blocking activities, or must they consider the ongoing implications of their initial actions as further blocks are implemented in turn?

When taken to their logical limit, considerations such as these naturally provoke a view of coalition formation as one that occurs in real time, on an ongoing basis, with blocking translated into moves, and with the discount factor as a yardstick for judging just how farsighted the players are.

With these notions in mind, we review the blocking approach to coalition formation. First, we discuss several solution concepts that rely on blocking. We then show how some of these solutions can be usefully subsumed in the general concept of an EPCF introduced in the previous section. Finally, we show how the EPCF serves to both

[14] The formal concept of the core for characteristic function games was introduced by Gillies (1953) and Shapley (1953). See Chapter 6 of Shubik (1982) for additional background.

illustrate and deal with some of the pitfalls of the blocking approach, in which a sequence of moves is viewed more as an abstract shorthand rather than an actual course of actions. The essence of the difficulties stems from the fact that the blocking approach does not incorporate an analog of the *maximality* of profitable moves. We will argue that in many contexts, the simplicity afforded by an abstract notion of blocking can become too restrictive. In summary, in many special cases of interest, the findings from a fully dynamic model parallel those from the classical, "timeless" theory of coalitional blocking. At the same time, in the more general setting, the dynamic model yields a resolution to some of the difficulties that arise with the traditional approach.

5.3.1 The setting

Much of the analysis that follows can be conducted in the setting of our general framework in Section 5.2. However, both from an expositional perspective and in an attempt to link directly to existing literature, it will be useful to work with partition functions. (See Section 5.2.4.2 where we've already introduced them.)

Our state space, then, will explicitly track two objects. A state x contains information about the going coalition structure $\pi(x)$ and a vector of payoffs $u(x)$. It is natural to suppose that these arise from an underlying partition function $V(S, \pi)$ defined for every π and every coalition $S \in \pi$. That is, $u(x)_S \in \bar{V}(S, \pi)$ for each $S \in \pi(x)$.

We will augment this traditional setting with an effectivity correspondence, $E(x, y)$, that specifies the set of coalitions that have the power to change x to y. It is sometimes more convenient to use the notation $x \rightarrow_S y$ to denote $S \in E(x, y)$. Denote by $\Gamma = (N, V, E)$ the (extended) partition function game.

While coalition structures directly generate externalities and affect payoffs in partition function games, they are of interest (as outcomes) even in characteristic function games. For an extensive treatment of various solution concepts in the context of exogenously given coalition structures, see Aumann and Dreze (1974). Our interests are closer to contributions such as Shenoy (1979) and Hart and Kurz (1983) that study the *endogenous* formation of coalition structures. Although this literature is mostly concerned with characteristic function games, thereby assuming away externalities, the payoff to members of a coalition can still depend on the entire coalition structure for strategic reasons.[15] Greenberg (1994) provides an excellent review of this literature.

Note that the traditional cooperative game setting has no explicit notion of time. Yet solution concepts abound that take stock of "farsightedness" and, therefore, implicitly involve time. We will argue below that our abstract dynamic setting allows us to naturally incorporate such farsightedness and thereby integrate different solution concepts in this literature.

[15] See in particular the Owen value; Owen (1977) and Hart and Kurz (1983).

5.3.2 Blocking

We begin with the standard notion of blocking applied to Γ. A pair (T, y), where T is a coalition and y is a state, is an *objection* to state x (or equivalently, T *blocks* x with y) if $T \in E(x, y)$ and $u(y)_T \gg u(x)_T$.

This notion of an objection leads to the fundamental concept of the core: a state is in the *core* of Γ if there does not exist an objection to it:

$$C(N, V, E) = \{x \in X \mid \text{ there does not exist an objection to } x\}.$$

The blocking notion also leads to the equally fundamental (abstract) stable set due to von Neumann and Morgenstern (1944): a set $Z \subseteq X$ of states is *stable* if no state in Z is blocked by any other state in Z (internal stability) and if every state *not* in Z is indeed blocked by some state in Z (external stability).

It is important to emphasize that we have defined blocking in the context of a partition function game, with a state defined in terms of both a coalition structure and a payoff allocation.[16] In the context of a characteristic function game, the standard practice is to take the set of states to be the set of imputations, which are individually rational payoffs in $\bar{V}(N)$, along with the implicit effectivity correspondence $E^*(u, w) = \{T \subseteq N \mid w_T \in V(T)\}$. The core of a characteristic function game is then defined as:

$$C(N, V) = \{u \in \bar{V}(N) \mid \nexists T \subseteq N, w_T \in V(T) \text{ and } w_T \gg u_T\}.$$

As we will see (especially in Section 5.3.6.1), it is sometimes useful, even in characteristic function games, to retain our general definition of a state as a partition as well as a feasible payoff profile, instead of just an imputation. For now, we note that in our model, $C(N, V, E)$ is a reformulation of the standard notion of the core, $C(N, V)$, to account for this difference in the domain: states rather than imputations. To see this, suppose (N, V) is a characteristic function game and, with some abuse of notation, let E^* refer to the corresponding effectivity correspondence over the domain of states, X, i.e., for every $w_T \in V(T)$ there exists a state y where $u(y)_T = w_T$ and $T \in E^*(x, y)$. Then, if $x \in C(N, V, E^*)$ and $\pi(x) = N$, $u(x) \in C(N, V)$. And if $u \in C(N, V)$, then $(u, N) \in C(N, V, E^*)$. For coalition structures other than the grand coalition, $C(N, V, E^*)$ is similarly related to the *coalition structure core of* (N, V), namely $\{x \in X \mid \nexists T \subseteq N$, and $w_T \in V(T)$ with $w_T \gg u(x)_T\}$.

Finally, a similar relationship holds between the stable sets defined here and the *vNM stable sets* for characteristic function games; see Section 5.3.6.1.

[16] It is possible to define blocking in a still more general setting (see, for example, Lucas, 1992). An *abstract game* is defined as $(X, \succ)$, where X is the set of states and $\succ$, referred to as a dominance relation, is a binary irreflexive relation on X. Say that y blocks x if $y \succ x$. In this general formulation, the restrictions imposed by an effectivity correspondence are implicit in the definition of dominance.

5.3.3 Consistency and farsightedness

There are two (related) drawbacks of this notion of blocking, which have both been studied in the recent literature. The first is that of consistency. The traditional notion of the core has been criticized for not being consistent, see Ray (1989) and Greenberg (1990). While allocations for the grand coalition are tested against possible objections from subcoalitions, the objections are not similarly tested for further objections. In some situations, this turns out only to be a conceptual issue that doesn't affect the set of states ultimately judged to be stable (see the discussion following Proposition 5.1 later). However, in general, the issue becomes impossible to ignore. We illustrate this with the following example.

Example 5.5. (Cournot oligopoly and farsightedness) *Consider Example 5.1 with three identical firms, each with a constant average cost of 2. Suppose the inverse demand function is* $p = 14 - z$*, where* z *denotes aggregate output. Suppose that all firms within a coalition are required to share profits equally. We will generally use* π_N *to denote the coalition structure containing the grand coalition alone,* π_i *the coalition structure in which i is a singleton and the other two are together, and* π_0 *the finest partition of three singletons. With minor abuse of notation, we will use* $i, j \ldots$ *to denote singleton coalitions as well as agents, and* ij, $ijk \ldots$ *to denote multi-agent coalitions. Thus,* $\pi_N = \{123\}$, $\pi_i = \{i, jk\}$, *and* $\pi_0 = \{1, 2, 3\}$. *Standard computation yields the following partition function:*

$$
\begin{aligned}
x_N : \pi_N &= \{123\}, & u(x_N) &= (12, 12, 12)\\
x_1 : \pi_1 &= \{1, 23\}, & u(x_1) &= (16, 8, 8)\\
x_2 : \pi_2 &= \{2, 13\}, & u(x_2) &= (8, 16, 8)\\
x_3 : \pi_3 &= \{12, 3\}, & u(x_3) &= (8, 8, 16)\\
x_0 : \pi_0 &= \{1, 2, 3\}, & u(x_0) &= (9, 9, 9)
\end{aligned}
$$

Clearly, any single player i can move from the grand coalition to π_i and, in turn, either j or k can move from π_i to π_0. So every singleton coalition i has a myopic objection to the state corresponding to the grand coalition; it can get 16 rather than 12 by unilaterally moving to π_i. However, this is not a sustainable gain. The intermediate structure π_i is itself unstable: either player in the two-player coalition jk will do better by moving to the finest coalition structure. The myopic blocking notion fails to take such repercussions into account.

The von Neumann-Morgenstern stable set resolves the consistency issue by only taking seriously those objections that are themselves stable. But that brings us to the second drawback of the traditional blocking notion: it is myopic. Such myopia creates problems with the notion of the stable set, as Harsanyi (1974) first pointed out. The following example, due to Xue (1998), provides a simple illustration of this problem.

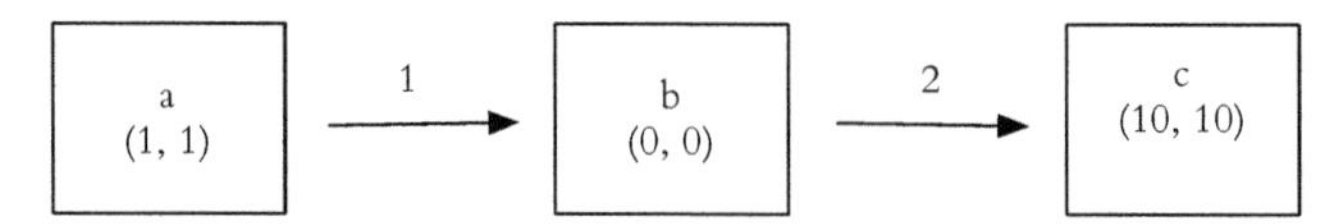

Figure 5.2 Illustration of Example 5.6.

Example 5.6. (Stability and farsightedness) *Consider two players and three states. Suppose that only player 1 is effective in moving from state a to b and only player 2 is effective from b to c. The payoffs to the two players in each of the states are in parentheses in Figure 5.2.*

The core consists of states a and c. These two states also constitute the unique stable set. The stability of a, however, is based on myopic reasoning. Farsightedness on the part of player 1 would surely cause her to move to b, anticipating that player 2's self-interest will then lead to c as the final outcome. Clearly, stability in the sense of von Neumann and Morgenstern does not imply farsightedness.

This discussion motivates the concept of "farsighted blocking." A coalition moves, not necessarily because it has an immediate objection, but because its move can trigger further changes, leading *eventually* to a favorable outcome.

(T, y) is a *farsighted objection* to x if there is a collection of states, $y_0, y_1, \ldots, y_m$ (with $y_0 = x$ and $y_m = y$), and a corresponding collection of coalitions, $T_1, \ldots, T_m$, where $T = T_1$, such that for all $k = 1, \ldots, m$, $y_{k-1} \to_{T_k} y_k$ and $u(y)_{T^k} \gg u(y_{k-1})_{T_k}$. We will often refer to y as farsightedly blocking x, leaving the chain of coalitions implicit.

A farsighted objection pays no attention to what might transpire "immediately" after the objection is made. The first coalition to move may induce an "intermediate" state, in the anticipation that there may well be other states on the way to the "final" state. The definition asks that the objecting coalition be better off at the "end" of this process. Furthermore, it is required that every participant at every intermediate step, namely the coalitions T_k, be better off "pushing" the process a step further at the corresponding states y_{k-1}, once again with the "final state" y in mind. Note that the coalition that initiates the sequence of moves has an optimistic view of the ensuing path. After all, there may be multiple potential continuations from the first step, but it is enough to find *some* sequence of moves that makes all the participating coalitions better off at the "final state."

The notion of farsighted blocking was suggested by Harsanyi (1974) in his critique of the stable set. It was formalized by Chwe (1994) in developing his notion of the largest consistent set and introduced as "sequential blocking" in the context of equilibrium binding agreements by Ray and Vohra (1997).

As we will discuss in Section 5.3.6, this definition of farsighted blocking is not without its own drawbacks. For now, it is imperative to note that the definition cannot make sense unless it is intimately tied to consistency. A farsighted objection y to x has not much meaning unless matters indeed terminate at y. A state that acts as a "credible" farsighted objection must itself have immunity with respect to the same kind of objections that can in turn be leveled at it. The "farsighted stable set" comes close to addressing these issues.

5.3.4 The farsighted stable set

The marriage of farsightedness and consistency leads us to investigate the notion of a farsighted stable set (Harsanyi, 1974).[17] Say that a set Z^* of states is a *farsighted stable set* if no state in Z^* is farsightedly blocked by any other state in Z^* (internal stability), and if every state *not* in Z^* is farsightedly blocked by some state in Z^* (external stability). Put another way, if we attach a description to the states in (or not in) Z^*—call them stable (or unstable)—then no stable state has a farsighted objection that terminates in another stable state, while every unstable state does have such an objection.

Observe that at first sight, this appears to add very little at a conceptual level, simply replacing the blocking relation used for von Neumann-Morgenstern stability by its farsighted analog. But that is not the case. Recall from Example 5.6 that a coalition might "exploit" the von Neumann-Morgenstern stable set by blocking some element in it, while profiting from that block even if the initial objection is "counter-objected" to by some other element in the stable set. In other words, *a coalition could be better off even by moving to an unstable state.*

That exploitation is not possible any more in the farsighted stable set. Suppose that a coalition T replaces a "stable point" $x \in Z^*$ by a new state w. If $w \in Z^*$, then w is "stable" and, by the internal stability property, T cannot be better off, so w cannot serve as a farsighted objection to x. If $w \notin Z^*$, then by external stability, there is a farsighted objection to w that leads to some $y \in Z^*$, which is "stable." But then T cannot be better off under y, for if it were, the entire sequence of states starting with w and terminating in y would act as a farsighted objection to x, which is ruled out by internal stability.

Thus, as we wrote earlier, the farsighted stable set captures the joint imposition of consistency and farsightedness. But there is a certain degree of bootstrapping implicit in the notion of a stable set. In the discussion earlier, a farsighted objection was treated as "credible" if it terminates in a "stable" state, i.e., a state in Z^*. But "stability" does not automatically guarantee that no further farsighted objection exists, only that such an objection must itself terminate in an "unstable" state, defined to be a state *not* in Z^*. Thus, stability and instability need to be simultaneously defined.

It might be thought that a recursive definition of stability can get around this problem. Indeed, a recursive approach might help to some extent when the blocking

[17] It is of interest to note that Harsanyi originally provided a definition that does not conform to the one given here, insisting in addition to farsightedness that each step of the blocking chain results in an instant improvement. However, in the last section of his paper, Harsanyi eliminates—correctly, in our opinion—this extraneous requirement; see also Chwe (1994) which makes this point.

relation is myopic, see, for instance the iterative procedure introduced in von Neumann and Morgenstern (1944, Section 65.7) and extended by Roth (1976). To describe it informally, say that a state is "admissible" if there exists no objection to it. Let A_0 be the set of all such states. (This is just the core). We should certainly label as "inadmissible" all states to which there exists an objection emanating from A_0. Let the set of such states be denoted I_0. Now proceed recursively, labeling additional states as admissible if the only objections to such states stem from I_0. Label all such states A_1. (The set A_0 is included.) Let I_1 be the set of all states to which there is an objection from A_1. Now continue this process until it terminates. The resulting "admissible limit set" will typically extend beyond its starting point A_0. Roth (1976) refers to these limits as "supercores." It is possible, but by no means guaranteed, that supercores will pick out stable sets. Examples when this will *not* happen are abundant, but at least the recursion might offer some traction beyond its initial stage.

Note that the general definition of a farsighted stable set in this section applies to any effectivity correspondence. However, as we consider various applications of farsighted stability, it will become clear that there may be natural restrictions on the effectivity correspondence. And such restrictions may have an important bearing on the properties of the farsighted stable set, see in Section 5.3.6.1. For instance, in the context of a characteristic function game, this recursive procedure is doomed at the outset. Unless restrictions are placed on the coalitions that can participate in such objections, it is often the case that *no* state is immune from a farsighted objection: more on this in Section 5.3.6.1. In short, the analog of A_0 is empty, and the recursion will simply fail to get off the ground. Yet, as we shall see, farsighted stable sets can dramatically narrow the set of outcomes. It is only that such sets are beyond the reach of recursive computability.

One restriction on coalition formation that does permit a recursive description is that only subsets of already formed coalitions are allowed to move. We turn to this in the next section.

5.3.5 Internal blocking

5.3.5.1 A recursive definition

A recursive approach to farsighted stability *does* work well when the effectivity correspondence only permits subsets of existing coalitions to form. We refer to this as *internal blocking*.[18] Such situations may be important when communication becomes impossible across already formed coalitions. In these cases, the potential circularity of farsighted stability is easily avoided.

[18] This is by no means the only situation in which the recursive approach works. More generally, if bounds are placed on the number of times a coalition can move at any node, then one can carry out the same recursive procedure provided that states are appropriately defined to keep track of those bounds. Mariotti (1997) is an instance in which "cycles" of coalitions along a blocking path are ruled out.

For a partition π and a subcoalition T of some $S \in \pi$, denote by $\pi_{|T}$ the partition obtained from π by dividing S into T and $S \setminus T$, leaving all other elements of π unchanged. If $\pi = (S, S_1, \ldots, S_m)$, and $T \subset S$, then $\pi_{|T} = (T, S \setminus T, S_1, \ldots, S_m)$. A partition π' is said to be a *refinement* of π if every $T \in \pi'$ is a subset of some $S \in \pi$, and at least one is a strict subset. It is said to be an *immediate refinement* of π if $\pi' = \pi_{|T}$ for some $T \subset S \in \pi$.

Now we can describe internal blocking by placing a necessary restriction on the effectivity correspondence: if $T \in E(x, y)$ and $x \neq y$, then $T \subset S$ for some $S \in \pi(x)$, $T \in \pi(y)$, and $\pi(y)$ is a refinement of $\pi(x)$. Thus, only a subset of an existing coalition is effective in making a nontrivial change to an existing state.

We will now describe a recursive procedure for constructing the farsighted stable set with internal blocking.[19]

With the restriction to internal blocking (and the use of efficient payoffs), the state corresponding to the finest partition, π_0, is an end state and therefore trivially stable. Consider a state x such that the only refinement of $\pi(x)$ is π_0. Since payoffs in each state are required to be efficient, and objections can only make the coalition structure finer, any objection to x must lead to π_0. Now, say that x is *inadmissible* if there is a (farsighted) objection to it, and it is *admissible* otherwise. Recursively, suppose that all states with associated partitions of cardinality $k + 1$ or greater (where $k < n$) have been labeled either admissible or inadmissible. Consider any state x with associated coalition structure of cardinality k. Since objections can only lead to states that have already been labeled admissible or inadmissible, x can now be assigned a unique label: admissible if no farsighted objection terminates in an admissible state and inadmissible otherwise. Continuing all the way to the coalition structure π_N, every state can therefore be identified as either admissible or inadmissible. It is now easy to see that the admissible states taken together constitute the unique farsighted stable set.

As we shall see, this set corresponds to different solution concepts depending on the context. To explore those connections it will be useful to first draw a general connection between the farsighted stable set and an equilibrium process of coalition formation.

5.3.5.2 EPCF and the farsighted stable set with internal blocking

An important test of the versatility of our dynamic process of coalition formation is its connection to the (static) notion of the farsighted stable set. To study this connection, we consider a process of coalition formation that inherits some of the simplicity of the static model. In particular, a state will refer to a partition and feasible payoffs to coalitions within the partition, and the partner set for any active coalition will be taken to be empty.

[19] The definition will resemble that of a coalition-proof Nash equilibrium of Bernheim et al. (1987). But that concept is purely noncooperative and is not directly related to the theme of this chapter.

In additional to the effectivity correspondence, the only additional ingredient that will be necessary in defining a process of coalition formation is a protocol.

Recall that a protocol chooses an active coalition depending on the history. Here, we will seek to condition the protocol on the history in a minimal way and require that at any given state a coalition has at most one chance to make a move. Suppose $x_{t-1} \neq x_{t-2}$. In other words, the immediate history at date t is not a result of inaction on the part of some coalition. In such a case, the protocol will depend only on the current partition, i.e., on $\pi(x_{t-1})$. Every subcoalition T of a coalition $S \in \pi(x_{t-1})$ will be active with strictly positive probability $\rho(T \mid x_{t-1}) > 0$. Given the restriction to internal blocking $\rho(T' \mid x_{t-1}) = 0$ for any T' that is not a subset of a coalition in $\pi(x_{t-1})$.[20] Suppose, on the other hand, $x_{t-i} = x$ for all $i = 1, \ldots, k$ for some $k \geq 2$ and $x_{t-k} \neq x_{t-k-1}$. The protocol will then assign a zero probability to any coalition that has declined to change the state, i.e, any coalition that was active at dates $t - i$ for $i = 2, \ldots, k$. If all allowable coalitions in $\pi(x_{t-1})$ decline to move, the state remains unchanged and becomes an absorbing state.

We will now explore some conditions under which the absorbing states of an EPCF coincide with the farsighted stable set, and thereby identify situations in which the dynamic process predicts the same set of stable outcomes as those emerging from the static model of internal blocking. To be consistent with the blocking definitions, we will assume throughout the rest of this section that any nonstatus-quo profitable move is *strictly* profitable for all members of the active coalition, i.e., there is a strict inequality in the corresponding definition of Section 5.2.3.

A state x is said to be an *absorbing state* of an EPCF λ if the process does not move from x whatever the history. Formally, x is an absorbing state of λ if $\lambda(h_{t+1} \mid h_t) = 1$ for all histories h_t, h_{t+1} such that $x_{t-1} = x_t = x$.

An EPCF is said to be *absorbing* if from any state it leads to an absorbing state in a finite number of steps.

Given our restrictions on the protocol and because every change in a state only serves to refine the coalition structure, it follows that every process defined on (N, V, E, ρ) is absorbing.

We will now show that if the steps of a farsighted objection and those of a profitable move can be compressed into a single step, then there is a tight connection between the absorbing states of an EPCF and the farsighted stable set. This abstract result will be useful in Sections 5.3.5.4 and 5.3.5.5.

We say that an EPCF is *immediately absorbing* if whenever x is a transient state, there exists an objection (T, y) to x such that y is an absorbing state.

We say that an extended partition function (N, V, E) has the *one-step objection property* if whenever x is not in the farsighted stable set there exists an objection, (T, y) such that

[20] Strictly speaking, this is not necessary. The protocol could assign a positive probability to such a coalition, but the only state this coalition would be effective for is x_{t-1}, and the only change would be that one unit of time would go by.

y belongs to the farsighted stable set. Thus, the initiating coalition achieves a higher payoff in the very first step.[21] These situations are important not only because of the added simplicity of the farsighted stable set but also because there is then no ambiguity stemming from the possibility of multiple continuation paths in a farsighted objection. The latter is crucial in drawing a connection with EPCFs. Otherwise, a farsighted objection, while being profitable, may not be maximal. And then, as we will see in Section 5.3.6, there may be good reason for not expecting the farsighted stable set to be related to an EPCF.

We can now present a result connecting the farsighted stable set to absorbing states of the dynamic model.

Lemma 5.1. *Suppose (N, V, E) has the one-step objection property and λ is an immediately absorbing EPCF of (N, V, E, ρ). Then all absorbing states of λ coincide with the farsighted stable set of (N, V, E).*

Proof. Consider an EPCF λ. The state corresponding to the finest coalition structure is clearly an absorbing state. It is also by definition in the farsighted stable set. Thus, the equivalence between absorbing states of an EPCF and the farsighted stable set holds for the finest coalition structure. We now use an induction argument to prove the result. Accordingly, assume that the result holds for all states with at least $k + 1$ coalitions in the coalition structure.

Suppose $\pi(x)$ consists of k coalitions and x is not an absorbing state. Since λ is immediately absorbing this means that there exists an objection (T, y) to x such that y is an absorbing state for T. Note that T must be a strict subset of some $S \in \pi(x)$ for it to have an objection. Thus, $\pi(y)$ is a refinement of $\pi(x)$, and it follows from the induction hypothesis that y is in the farsighted stable set. This implies that x is not in the farsighted stable set and completes the proof that the farsighted stable set is contained in the set of absorbing states of λ.

Next, we show that if x is an absorbing state, with $\pi(x)$ consisting of k coalitions, it must be in the farsighted stable set. Suppose not. Then, by the one-step objection property, there exists y and a coalition $T \subset S \in \pi(x)$, where $T \in E(x, y)$ and y is in the farsighted stable set. Since $\pi(y)$ is a strict refinement of $\pi(x)$, it follows from the induction hypothesis that y is an absorbing state. By hypothesis, no coalition moves from x regardless of the history. There must be some history for which the protocol chooses T to be active when the current state is x. (Recall that, subject to internal blocking, every subcoalition of an existing coalition has a strictly positive probability of being chosen, as long as it has not already exhausted its single chance to move). Coalition T receives $u(x)_T$ in perpetuity by not moving and $u(y)_T$ in perpetuity by moving to y. Since $u(y)_T \gg u(x)_T$, this is a strictly profitable move, and a contradiction to the hypothesis that x is an absorbing state.

[21] Under this condition farsightedness reduces to consistency. The original definition of the modified vNM set in Harsanyi (1974) satisfies this property.

It is of course important to identify assumptions on the primitive model that will allow us to appeal to Lemma 5.1. That we shall do in Sections 5.3.5.4 and 5.3.5.5.

5.3.5.3 Characteristic functions

Suppose the partition function is actually a characteristic function, so there are no externalities. We shall now impose some restrictions on the effectivity correspondence, which are natural yet implicit in this setting. Throughout this section, in addition to internal blocking, it is assumed that[22]:

(i) [Noninterference] If $T \in E(x, y)$, $S \in \pi(x)$, and $T \cap S = \emptyset$, then $S \in \pi(y)$ and $u(x)_S = u(y)_S$.

(ii) [Full Support] For every state $x \in X$, $T \subseteq N$, and $v \in \bar{V}(T)$, there is $y \in X$ such that $T \in E(x, y)$, $T \in \pi(y)$, and $u(y)_T = v$.

Condition (i) grants coalitional sovereignty to the untouched coalitions: the formation of T cannot "interfere" with the membership of coalitions that are entirely unrelated to T in the original coalition structure, nor can it influence the going payoffs to such coalitions. Condition (ii) grants coalitional sovereignty to the deviating coalition: it can choose not to break up, and it can choose its *own* payoff allocation from the "full support" of its feasible set.

Noninterference acquires its present force because the situation in hand is described by a characteristic function: T truly influences neither the composition of an "untouched" coalition nor the payoffs it can achieve. (In games with externalities, the condition would need to be suitably modified to allow those payoffs to change, but not in a way that is deliberately dictated by the deviating coalition.) These conditions will also be useful when the internal blocking restriction is dropped; see Section 5.3.6.1.

Given internal blocking, it is easy to see that $x \in C(N, V, E)$ if and only if $u(x)_S \in C(S, V)$ for every $S \in \pi(x)$.[23] As our next result shows, under these conditions, the farsighted stable set coincides with the core.

Proposition 5.1. *Suppose (N, V) is a characteristic function game, and E is restricted to internal blocking and satisfies noninterference and full support. Then the farsighted stable set coincides with $C(N, V, E)$, i.e., x belongs to the farsighted stable set if and only if $u(x)_S \in C(S, V)$ for all $S \in \pi(x)$.*

Proof. Suppose $x \in C(N, V, E)$ but there exists a farsighted objection (T, y) to x. Given internal blocking, the coalition structure corresponding to y, $\pi(y)$, must contain a coalition $T' \subseteq T$. Since (T, y) is a farsighted objection to x, $u(y)_{T'} \gg u(x)_{T'}$ and

[22] These conditions appear explicitly in Konishi and Ray (2003), Kóczy and Lauwers (2004), and Ray and Vohra (2013).

[23] For $S \subseteq N$, $C(S, V)$ refers to the core of the characteristic function (N, V) restricted to S.

$u(y)_{T'} \in \bar{V}(T')$. So the full support condition implies that T' is effective in moving from x to y', where $u(y')_{T'} = u(y)_{T'} \gg u(x)_{T'}$. But this contradicts the hypothesis that $x \in C(N, V, E)$.

Suppose x is in the farsighted stable set. We now claim that it must be in $C(N, V, E)$. Suppose not. Then there exists $S \in \pi(x)$ such that $u(x)_S \notin C(S, V)$. Let (S_1, u_1) be an objection to $u(x)_S$ such that $u_1 \in C(S_1, V)$. This can always be assured by taking S_1 to be one of the smallest subcoalitions of S with an objection to $u(x)_S$. By full support, it follows that $S_1 \in E(x, y_1)$, where $u(y_1)_{S_1} = u_1$ and $\pi(y_1) = \pi(x)_{|S_1}$. If $u(y_1)_{S \setminus S_1} \in C(S \setminus S_1, V)$, then (S_1, y_1) is a farsighted objection to which there cannot be any objection from a subset of S_1 or $S \setminus S_1$. If $u(y_1)_{S \setminus S_1} \notin C(S \setminus S_1, V)$ we can find some subcoalition in $S \setminus S_1$, say S_2, with an objection from the core of S_2. Note that by noninterference, a move by S_2 will not affect the payoffs of coalition S_1 (or of any other coalition disjoint from S_2). Continuing in this way it is possible to construct a partition $(S_1, \ldots, S_m)$ of S and $(u_1, \ldots, u_m)$ such that $u_i \in C(S_i, V)$ for every i. Clearly, there is no farsighted objection to $((S_1, \ldots, S_m), (u_1, \ldots, u_m))$ from any allowable coalition in $(S_1, \ldots, S_m)$.

This procedure can be applied to any $S' \in \pi(x)$ for which $u(x)_{S'} \notin C(S', V)$. All such objections can be collected into one farsighted objection that culminates in x' where $\pi(x')$ is a refinement of $\pi(x)$ and $u(x')_T \in C(T, V)$ for all $T \in \pi(x')$. Of course, this must mean that there is no further farsighted objection to x'. But this contradicts the hypothesis that x is in the farsighted stable set.

Note that one of the steps in the above proof relies on the property that if $u \notin C(N, V)$ then there exists an objection (S, u') such that $u' \in C(S, V)$. The only reason this doesn't imply the one-step objection property is because coalitions other than S may also need to "move" in order to arrive at a stable outcome. If we ignore the rest of the coalition structure, then farsighted blocking becomes equivalent to myopic blocking and Proposition 5.1 can we be seen as the coalition structure analog of Ray (1989) and Proposition 6.1.4, Greenberg (1990).

We can now turn to a formal connection between the core and the absorbing states of an EPCF.

Proposition 5.2. *Suppose (N, V) is a superadditive characteristic function game such that $V(S)$ is convex for all $S \subseteq N$. If $x \in C(N, V, E)$, then x is an absorbing state of every EPCF corresponding to (N, V, E, ρ).*

Proof. Suppose x is in the core. Fix a history h at which x has just been arrived at, and suppose that subsequently x is not an absorbing state for EPCF λ. Then, there is a coalition that executes a profitable move when given a chance to move, possibly following other coalitions having given up their opportunity to move from x. Consider the subset of all histories which all begin with h, such that no subsequent coalition called by that history moves at x, and a new active coalition has just been chosen, which is about to announce a move. Because each coalition has a single chance to move, this set is finite. Let h^* be a maximal history of this type. Suppose T is a coalition that moves from x to y following such an history. Clearly, T could have earned $u(x)_T$ forever it chose *not* to move. If y is an absorbing state, the fact that it is a profitable move for T contradicts the hypothesis that x is in the core. Thus, with positive probability, there is a further move from y.

All possible paths leading from y must reach an absorbing state in a finite number of steps. Let m be the maximum number of steps along any such path before an absorbing state is reached. For every step $i = 1, \ldots, m$, following T's move, let μ^i be the probability measure on the states generated by λ. Denote by u^i the corresponding expected utility profile at stage i: $u^i = \int u(x) d\mu^i$. Since T has a strictly profitable move,

$$u_T^1 + \delta u_T^2 + \delta^2 u_T^3 + \cdots + \frac{\delta^{m-1}}{(1-\delta)} u_T^m \gg \frac{1}{(1-\delta)} u(x)_T. \quad [5.2]$$

It follows from superadditivity and the convexity of $V(S)$ that

$$u_T^j \in V(T) \quad \text{for all } j = 1, \ldots, m.$$

Letting

$$\hat{u} = (1-\delta)\left[u^1 + \delta u^2 + \delta^2 u^3 + \cdots + \frac{\delta^{m-1}}{(1-\delta)} u^m\right]$$

(5.2) can be rewritten as:

$$\hat{u}_T \gg u(x)_T. \quad [5.3]$$

Note that $\hat{u}$ is a convex combination of $u^1, \ldots, u^m$. Since $u_T^j \in V(T)$ for all j, it follows that $\hat{u}_T \in V(T)$, but then (5.3) contradicts the hypothesis that x is in the core.

Proposition 5.3. *Suppose x is an absorbing state of an EPCF of (N, V, E, ρ). If E satisfies noninterference and full support, then x belongs to the farsighted stable set of (N, V, E).*

Proof. Suppose x is an absorbing state but does not belong to the farsighted stable set. By Proposition 5.1, there exists $S \in \pi(x)$ such that $u(x) \notin C(S, V)$. Moreover, there exists $T \subset S$ and $u' \in C(T, V)$ such that $u' \gg u(x)_T$. Since x is an absorbing state, T receives $u(x)_T$ in perpetuity by not moving. However, by full support, it could move to a state y in which it receives u'. Since $u' \in C(T, V)$, we know from the previous proposition that no subcoalition of T can move to a higher payoff. The only possible moves from y must come from coalitions in $N \setminus T$. By noninterference, that has no effect on the payoff to T, so we conclude that by moving to y coalition T can receive $u' \gg u(x)_T$. But this contradicts the hypothesis that x is an absorbing state.

Combining Propositions 5.2 and 5.3, we have:

Proposition 5.4. *Suppose* (N, V) *is a superadditive characteristic function game such that* $V(S)$ *is convex for all* $S \subseteq N$ *and* E *satisfies noninterference and full support. Then all absorbing states of every EPCF of* (N, V, E, ρ) *coincide with the core (or farsighted stable set) of* (N, V, E).

There is also an earlier literature that studies processes converging to core allocations. Green (1972), Feldman (1972), and Sengupta and Sengupta (1996) show how in a characteristic function game a process of recontracting can be constructed to lead from any noncore allocation to a core allocation. Recontracting refers to a process in which every active coalition makes a (myopic) improving move, without any guarantee of gaining at the end of the process. In contrast, Proposition 5.4 applies to farsighted behavior.

Konishi and Ray (2003) show how for a farsighted dynamic process can be constructed so as to have any particular core allocation as its absorbing state. They also provide conditions under which a deterministic process converges to a core allocation. Proposition 5.4 provides a stronger connection between the core and absorbing states since it concerns a coincidence of the set of core allocations and absorbing states of *any* EPCF. The key features of our model that make this possible are internal blocking and the specification of a protocol.

5.3.5.4 Effectivity without full support

It is important to stress that the full support property is not always natural. One class of restrictions emanate from the possibility that the *additional* disintegration of a newly formed coalition may be legally or politically impossible, thereby negating full support for the finer subcoalitions. While we will have more to say about such "irreversible agreements" in Sections 5.4 and 5.5, in the current section we discuss a model due to Acemoglu et al. (2008) in which the full support property does not hold for a very different reason. This is a model of a political game of coalition formation in which coalitions are farsighted and their ability to make a nontrivial move depends in an important way on the current state.

The political power of player i is described as $\gamma_i > 0$. A coalition $T \subseteq S$ is said to be *winning within* S if $\gamma_T > \alpha\gamma_S$, where $\gamma_T = \sum_{i \in T} \gamma_i$ and $\alpha \in [0.5, 1)$ denotes the degree of weighted supermajority required to win.

The payoff to players depends on the ruling coalition. If S is the ruling coalition, $w(S)$ denotes the unique profile of utilities for members of S. Players outside the ruling coalition receive 0. A specific functional form, which we assume for convenience, is $w(S) = (\gamma_i/\gamma_S)_{i \in S}$.

In this model, the only coalition of interest at each state is the ruling coalition and it will be useful therefore to define a state as $x = (w(S), S)$ with the interpretation that S is the ruling coalition.[24] We will use $R(x)$ to refer to the ruling coalition at x. Of course, coalition S can enforce such a state only from states in which it is a winning coalition.

$$\text{For } x \neq y, \quad E(x, y) = \begin{cases} R(y) & \text{if } R(y) \text{ is winning in } R(x) \\ \emptyset & \text{otherwise} \end{cases}$$

The effectivity correspondence does not satisfy the full support property because whether or not a coalition can effect a nontrivial move depends on the current state.[25] The model implicitly assumes internal blocking because a winning coalition must necessarily be a subset of the current ruling coalition. Given internal blocking, the recursive procedure described in Section 5.3.5.1 can be applied to determine the farsighted stable set. We illustrate this with the next example.

Example 5.7. *There are three players with $\gamma = (\gamma_1, \gamma_2, \gamma_3) = (4, 5, 6)$. A coalition $T \subseteq S$ is said to be winning within S if $\sum_{i \in T} \gamma_i > 0.5 \sum_{j \in S} \gamma_j$. The payoff profile for a ruling coalition is described as follows:*

$$\begin{aligned} w(123) &= (4/15, 5/15, 6/15), \\ w(12) &= (4/9, 5/9), \\ w(13) &= (0.4, 0.6), \\ w(23) &= (5/11, 6/11), \\ w(i) &= 1, \quad \textit{for all } i. \end{aligned}$$

Each two-player coalition is winning in N and can therefore move to become a ruling coalition and improve upon the status quo. However, within each two-player coalition, the more powerful player can win to become a singleton ruling coalition and earn 1. Thus, although there exist objections to the state corresponding to the grand coalition, none of them is credible because the weaker of the two players in the objecting coalition will be abandoned by the more powerful player at the next stage, ultimately doing worse than at the grand coalition. For example, player 3, who is winning within any two-player

[24] Although we have departed from our earlier formulation in replacing a coalition structure with a ruling coalition, this difference is not substantive. In particular, notions of the farsighted stable set and EPCF, as well as Lemma 5.1 are easily translated into the present model. Alternatively, we could retain the original formalism by associating with each ruling coalition the coalition structure in which all other players are singletons and, in addition, keeping track of the ruling coalition corresponding to every coalition structure of this form. The latter consideration is important for the coalition structure consisting of all singletons because in that case whether a singleton gets 0 or 1 depends on the identity of the ruling coalition.

[25] Pillage games, studied in Jordan (2006), are another interesting model in which the effectivity correspondence depends on the current state. There, the ability of a coalition to move depends on the payoffs, rather than the coalition structure, at the current state.

coalition is effective in becoming a singleton ruling coalition if the current state is (13) or (23), but not if the current state is N. It is easy to see that N belongs to the farsighted stable set even though it is not in the core.

Notice that since players who are not in a ruling coalition earn 0, it is always better to belong to a ruling coalition than not. This implies that every farsighted objection must immediately end in a stable state. Otherwise, there must be at least one member of the initiating coalition who is left out of the final ruling coalition, and any such player would have been better off not participating in the objection. In other words, every farsighted objection that ends in the farsighted stable set must be a one-step objection: the one-step objection property holds.

Now consider an EPCF for this model and assume that the protocol is deterministic: at every state, there is a fixed order in which eligible coalitions are called upon to move. (We will presently specialize the protocol even further.)

Suppose x is a transient state. Thus, there is a coalition T, winning in $R(x)$, with a profitable move. Without loss of generality, let T be the last such coalition given the protocol. If T does not move it is guaranteed a payoff of $u(x)$ forever, which must be less than the discounted payoff from moving. Note that all members of T experience an instantaneous gain by forming a winning coalition. However, if there is a further move (from a subcoalition), those left out of the second move then receive 0. In other words, if T is not an absorbing state, there is some $i \in T$ who receives 0 in all subsequent periods. The discounted payoff for i is therefore $w_i(T)$. For the move to be profitable, it must be the case that

$$w_i(T) > \frac{1}{1-\delta} u_i(x).$$

Since $w_i(T) \leq 1$ and $u_i(x)$ is bounded below by $\min_{i \in N} w_i(N)$, δ can be chosen close enough to 1 so that this is impossible. Thus, for δ high enough, every EPCF in this model must be immediately absorbing. As we've already observed, the one-step objection property holds. We can therefore appeal to Lemma 5.1 to assert:

Proposition 5.5. *In the model of political coalition formation, for δ sufficiently close to 1, all absorbing states of an EPCF coincide with the farsighted stable set.*

With some additional assumptions, it becomes possible to provide a sharper characterization of an EPCF in this model. Observe that for a winning coalition which can induce a state in the farsighted stable set, there is never any advantage in forgoing such an opportunity; a profitable move is maximally profitable. Given the protocol, there is a unique move from any transient state: it is the move by the first coalition according to the protocol which is both winning and moves immediately to a state in the farsighted stable set. A lot depends, therefore, on the protocol.

Let

$$\mathcal{T}(S) = \{T \subset S \mid T \text{ is winning within } S \text{ and } (w(T), T) \text{ is in the farsighted stable set}\}$$

and let the first coalition in $\mathcal{T}(S)$ according to the protocol be denoted $\mathcal{T}^*(S)$. We can now describe the equilibrium process by a mapping ϕ, where

$$\phi(S) = \begin{cases} S \text{ if } (w(S), S) \text{ is in the farsighted stable set} \\ \mathcal{T}^*(S) \text{ otherwise.} \end{cases}$$

Starting from the grand coalition as the initial state, the process moves, in at most one step, to $\phi(N)$ as the absorbing state.

Acemoglu et al. (2008) provide a characterization of the subgame perfect equilibria of their extensive form game of political coalition formation by assuming that the power mapping γ, is generic in the sense that $\gamma_S \neq \gamma_T$ for $S \neq T$. By adopting this assumption and imposing a restriction on the protocol, we can obtain precisely their characterization through our framework. Suppose the protocol is such that among all winning coalitions, relative to the current state, priority is given to those with lower aggregate power. In other words, winning coalitions are arranged in ascending order of aggregate power: if S and S' are both winning, $\gamma_S < \gamma_{S'}$ implies that the protocol chooses S before S'. Given the genericity assumption, this means that $\mathcal{T}^*(S) = \arg\min_{A \in \mathcal{T}(S)} \gamma_A$. Now $\phi(S)$ can be written inductively as follows. Suppose $\phi(.)$ has been defined for all coalitions with fewer than k players. Then, for S with k players let

$$\phi(S) = \arg\min_{A \in \mathcal{T}(S) \cup S} \gamma_A,$$

where

$$\mathcal{T}(S) = \{T \subset S \mid T \text{ is winning within } S \text{ and } T \in \phi(T)\}.$$

This is precisely the mapping ϕ defined by Acemoglu et al. (2008). They prove that every subgame perfect equilibrium of their extensive form game leads to $\phi(N)$ as the ultimate ruling coalition.

5.3.5.5 Internal blocking in the presence of externalities

The characteristic function has proved to be a very useful construction in studying coalitional behavior. It was derived by von Neumann and Morgenstern (1944) from a more general specification of a game by taking the feasible payoffs for a coalition to be those it can achieve by assuming (from its point of view) the worst possible strategy choices of the complementary coalition. While von Neumann and Morgenstern (1944) adopted this conversion to a characteristic function mainly to study zero-sum games, it was subsequently applied to more general (normal form) games by Aumann and Peleg (1960) and Aumann (1961).[26] For example, the α-characteristic function defines $V(S)$

[26] See Chapter 2.2 of Ray (2007) for a historical background.

for a coalition S as the set of payoffs S can achieve through some joint strategy *regardless of the actions of players outside* S.[27]

In some settings, such as exchange economies without externalities, or zero-sum games, this conversion involves no loss of generality. However, in the presence of externalities, the standard construction is *ad hoc*, if not unreasonable. For example, in the Cournot oligopoly, it is hard to see why a cartel should fear that the complementary coalition will flood the market and drive profits to zero, as is implicit in the extreme pessimism embodied in the α-core. It may even be argued that coalition formation should be studied directly through a normal form game. The essence of the problem, however, can usually be captured through a partition function game which makes explicit the manner in which the feasible set of payoffs for a coalition depend on other coalitions. To be sure, this does not completely eliminate the complexities stemming from externalities. A blocking coalition must now predict how players outside the coalition will organize themselves into coalitions. In what follows we will refer to this as the prediction problem. Of course, one could again cut though these complexities by making an assumption about how outsiders will organize themselves in response to a move by a blocking coalition. For example, one could assume that all outsiders will immediately break up into singletons or that all players left behind by a deviating coalition will continue to stay together. The former is related to the notion of γ-stability in Hart and Kurz (1983) and the latter to δ-stability.[28] The question is how to replace such assumptions with predictions about the equilibrium behavior of outsiders. Cooperative game theory has traditionally eschewed such considerations, but it is hard to see how coalitional behavior in the presence of externalities can be studied without making the response of outsiders endogenous to the theory. As we shall discuss, considerable progress has been made in resolving the prediction problem.[29]

Restricting attention to internal blocking, Ray and Vohra (1997) formulate the notion of equilibrium binding agreements (EBA), which can be seen as a variant of the farsighted stable set. In what follows, we shall consider EBA for a special class of *transferable utility* (TU) partition function games. These are partition function games in which the "worth" of coalition S in coalition structure π is denoted $v(S,\pi)$: $V(S,\pi) = \{w \in \mathbb{R}^S \mid \sum_{i\in S} w_i \leq v(S,\pi)\}$. (We will sometimes use (N,v) to denote a

[27] The core of the α-characteristic function is referred to as the α-core. In the β-characteristic function $u \in V(S)$ if for every strategy of players outside S there is some joint strategy in S that yields at least u.

[28] See also Carraro and Siniscalco (1993), Dutta and Suzumura (1993), Chander (2007), and Chander and Tulkens (1997).

[29] It bears mentioning that Aumann and Myerson (1988) tackled the prediction problem head on. We do not discuss this paper here only because its axiomatic emphasis does not fit either the blocking or the bargaining approach which we have confined this chapter to. Maskin (2003) is another important contribution to this issue which doesn't fall within the purview of the present chapter. See also de Clippel and Serrano (2008a,b). For an axiomatic treatment of this issue within the blocking approach, see Bloch and van den Nouweland (2013).

TU partition function.) In a symmetric game, the worth of a coalition depends only on the number of players in the coalition and the numerical coalition structure (number of players in each of the coalitions). Suppose $S \in \pi = \{S_1, \ldots, S_k\}$. Now $v(S, \pi)$ can simply be denoted $v(s, q)$, where s is the cardinality of S and $q = (s_1, \ldots, s_k)$, where s_i is the cardinality of S_i. A game is said to have positive externalities if a coalition's worth is higher when the other coalitions are merged. It is worth noting that the symmetric Cournot oligopoly is one example that satisfies all of these assumptions.

It is reasonable to assume that if $S \in \pi$ and a subcoalition T of S splits from S, in the first step, it induces an immediate refinement, $\pi_{|T}$, i.e., all the remaining players in S stay together and all the coalitions in π remain unchanged. (Of course, there is no presumption that $\pi_{|T}$ will remain unchanged.) When a process of coalition formation refines a partition, there are some coalitions that are active movers, or perpetrators, in splitting from a larger coalition while others can be thought of as residuals, players left behind. If a coalition breaks into k new coalitions, $k - 1$ of them must be perpetrators.

Equilibrium binding agreements (EBA) are defined recursively. The state x_0, corresponding to the finest coalition structure is an EBA. If x is a state with a coalition structure made up of singletons and one coalition, S, consisting of two players, it is said to be blocked by x_0 if one of the players in S prefers x_0 to x. We can now proceed recursively. Suppose EBA and the associated notion of blocking has been defined for all states with coalition structures that are refinements of $\pi(x)$. Then, x is said to be blocked by x' if x' is an EBA and there exists a collection of perpetrators in $\pi'(x)$ such that one of them, a leading perpetrator, prefers x' to x. Moreover, any re-merging of the other perpetrators with their respective residuals is blocked by x', with one of these perpetrators as the leading perpetrator.[30] Note that "blocking" is well defined in the previous sentence because any re-merging of the other perpetrators results in a coalition structure which is a refinement of $\pi(x)$. A state x is an EBA if it is not blocked.

As shown in Ray and Vohra (1997), in a symmetric TU game with positive externalities, a state is an equilibrium binding agreement if no coalition can obtain higher aggregate utility in a binding agreement of *some* refinement of the original coalition structure. In other words, EBA are simply all the states in the farsighted stable set when the effectivity correspondence allows a coalition to move to *any* refinement. Throughout this section, we assume that the effectivity correspondence has this form. Of course, this immediately implies that the one-step objection property holds.

In this setting, additional simplicity comes from the fact that in identifying equilibrium coalition structures, there is no loss of generality in assuming that coalitional worth is divided equally among all members of a coalition, see Proposition 6.3 in Ray and Vohra (1997). In other words, we can assume that states are restricted to satisfy the

[30] The extreme optimism in the notion of a farsighted objection is somewhat tempered by allowing subsequent coalitions to in any arbitrary order or, in fact, to move simultaneously.

property that for any $S \in \pi(x)$, for all $i \in S$, $u_i(x) = a(S, \pi) = \frac{v(S,\pi(x))}{|S|}$. Thus, we can take the number of states to be finite.

Let $a(s, q)$ denote the average worth of a coalition of size s in a numerical structure q, where $q = (q_1, \ldots, q_k)$ describes the sizes of the various coalitions in the partition. We will assume that the distribution of average worths satisfies the genericity assumption in the sense that $a(s, q) \neq a(s', q')$ if $s' \neq s$ or $q' \neq q$.

Proposition 5.6. *Suppose (N, V) is a symmetric, TU, partition function game with positive externalities and the genericity assumption holds. Then, for δ close enough to 1, all absorbing states of every EPCF of (N, V, E, ρ) coincide with EBA.*

Proof. Suppose x is not an absorbing state of an EPCF. This means that for some history at least one coalition has a profitable move. As in the proof of Proposition 5.2, let T be a coalition that could have earned $u(x)_T$ in perpetuity by declining to move, but chooses instead to move from x to y. Either y is an absorbing state or, with positive probability, there is a further move from y. Consider the latter case.

All possible paths leading from y must reach an absorbing state in a finite number of steps. Let m be the maximum number of steps along any such path before an absorbing state is reached. Let u^i denote the expected payoff at each step, $i = 1, \ldots, m$, with $u^1 = u(y)$. Let Z be the set of states in the support of u^m and $p(z)$ be the probability of z being the state at stage m, i.e., $u^m = \sum_{z \in Z} p(z)u(z)$.

The fact that T has a strictly profitable move means that

$$u_T^1 + \delta u_T^2 + \delta^2 u_T^3 + \cdots + \frac{\delta^{m-1}}{(1-\delta)} u_T^m \gg \frac{1}{(1-\delta)} u(x)_T. \quad [5.4]$$

Let $a^* = \max_{z \in Z, S \in \pi(z), S \subset T} a(S, \pi(z))$ be the maximum average worth across all subcoalitions of T in any of the coalition structures corresponding to states in Z. Let z^* be a state in which a^* is attained and S^* the corresponding subcoalition of T. Let $\bar{u}$ be the maximum utility that any player gets in any state. From (5.4) we obtain the following inequality for coalition S^*.

$$(1 + \delta + \cdots + \delta^{m-2})\bar{u} + \frac{\delta^{m-1}}{1-\delta}(a^* - u_i(x)) > 0 \quad \text{for all } i \in S^*. \quad [5.5]$$

Given that $\pi(z^*)$ is a refinement of $\pi(x)$ it follows that $u_i(x) = u_j(x)$ for all $i, j \in S^*$. We now claim that

$$a^* > u_i(x) \quad \text{for all } i \in S^*. \quad [5.6]$$

Suppose not. Then $a^* < u_i(x)$ for all $i \in S^*$ because, by the genericity assumption, $a^* \neq u_i(x)$. Since the number of states is finite, there exists $\epsilon > 0$ which denotes the minimum (absolute) difference, between $a(s, q)$ and $a(s', q')$ for $q \neq q'$. This means that if $a^* < u_i(x)$, then $a^* - u_i(x) \leq -\epsilon$. Substituting this in (5.5), we have

$$(1 + \delta + \cdots + \delta^{m-2})\bar{u} - \left(\frac{\delta^{m-1}}{1-\delta}\right)\epsilon > 0,$$

which is impossible if δ is close enough to 1. This establishes (5.6).

To summarize, we have shown that if x is a transient state, one of the following must be true:

1. [(i)] there is a move by T to y which is an absorbing state and $u(y)_T \gg u(x)_T$,

(ii) there is a coalition $S^* \subset T$ and an absorbing state z^* such that $S^* \in \pi(z^*)$ and $u(z^*)_{S^*} \gg u(x)_{S^*}$.

Since a coalition is effective in moving to any refinement of the current coalition structure, it follows that in case (ii) S^* can move directly from x to z^*. Thus, in any event, there is a profitable move to an absorbing state, and the process must be immediately absorbing. The proof now follows from Lemma 5.1 and the fact that the farsighted stable set is the set of EBA.

Note that in the dynamic model, it is possible for a strictly profitable move to be one in which the absorbing state results in the same utility profile as the current state, with all the gains being reaped in the intervening, transitory periods. Such a move in the dynamic model cannot possibly serve as an objection in the static model. To tie the dynamic model to the static one, therefore, we have to impose additional assumptions. In Proposition 5.6, this is achieved through the genericity assumption.

5.3.6 Beyond internal blocking

To extend the theory beyond internal blocking a recursive definition of farsighted stability will not suffice. Despite the sometimes obscure nature of abstract stable sets, which are typically defined on a nonrecursive basis, they do offer a promising approach at this level of generality. And much of the literature we shall discuss in this section is inspired, directly or indirectly, by farsighted stable sets.

5.3.6.1 *Farsighted stability for characteristic functions*

Some of the issues can be usefully discussed in the simpler setting of characteristic function games. Recall that a set of feasible outcomes Z is a stable set if it is both internally and externally stable (see Section 5.3.2). Consider a characteristic function game (N, V). Let the set of states be the set of all imputations and let the effectivity correspondence be E^* where, for imputations u, w, $E^*(u, w) = \{T \subseteq N \mid w_T \in V(T)\}$. An application of external and internal stability now yields the *vNM stable set.*

A stable set includes the core. But it will generally have other elements: $u \in Z$ may well be dominated by u' as long as $u' \notin Z$. To be sure, external stability guarantees that u' in turn can be "blocked" by some other profile $u'' \in Z$. The presumption, then, is that u should still be considered "stable," because u' does not represent a lasting benefit.

Harsanyi (1974) took issue with this presumption. He observed that this argument is only valid if u'' isn't preferred by the coalition which caused u to be replaced by u'. The vNM stable set is based on a myopic notion of dominance and does not address this concern. Harsanyi replaced the dominance relationship by farsighted dominance.

However, Harsanyi followed von Neumann and Morgenstern in continuing to define effectivity in exactly the same way as they did (with E^* as the effectivity correspondence). That is, S has the power to replace u with w *as long as* w_S *is feasible for* S. In particular, S dictates the complementary allocation w_{N--S}. That allocation need not even be feasible for the complementary set of players. Moreover, even if it is feasible, the presumption is that S has unlimited power in rearranging payoffs for $N - -S$. In effect, the Harsanyi definition violates the noninterference condition of Section 5.3.5.3.

It so happens that with myopic blocking as in vNM, this makes no difference to the stable set. But with farsighted blocking, it is entirely possible that with the vNM (and Harsanyi) notion of effectivity, a coalition may be able to engineer a farsighted block *only* by arranging the payoffs to outsiders in some particular way. To see an example of the violation of coalitional sovereignty implied by this approach, consider a four-player TU game with the following characteristic function. $v(i) = 0$ for all i, $v(12) = v(13) = v(23) = 2$, $v(123) = 6$ and for all S, $v(S \cup 4) = v(S)$. Although Player 4 is a dummy player, there is a singleton Harsanyi stable set in which the dummy receives a positive payoff: $u = (1, 1, 0, 4)$. As an illustration of how external stability is satisfied in this case, consider $w = (2, 2, 2, 0)$, a core allocation. There is a farsighted objection to w culminating in u. It begins with a move by player 4 to $w^1 = (6, 0, 0, 0)$, followed by a move by player 2 to $w^2 = (0, 0, 0, 6)$ and, finally, by coalition 12 to u. The logic depends crucially on player 4 assigning 0's to two of the other three players, and then on player 2 assigning 0's to all players other than player 4.

In fact, the situation illustrated by the previous example is not an accident; the problem runs deeper.

Proposition 5.7. *(Béal et al., 2008) Suppose* (N, v) *is a TU-characteristic function in which* $v(T) > 0$ *for some* $T \subset N$. H *is a Harsanyi stable set if and only if it is a singleton imputation,* u*, such that* $u_S \gg 0$ *and* $\sum_{i \in S} u_i \leq v(S)$ *for some* $S \subset N$.

This proposition implies that *every interior core allocation*[31] *must be excluded from any Harsanyi stable set.*[32] This is an odd implication of stability, to say the least.

[31] u is in the interior of the core if $u_S \notin V(S)$ for any $S \subset N$.

[32] This proposition can also be read as an existence theorem for the Harsanyi stable set. In an earlier contribution, Diamantoudi and Xue (2005) showed that in Lucas's (1968) celebrated example of

To appreciate the extent to which such a result emerges from any lack of restrictions on effectivity, it is important to specify what a coalition can (or cannot) do with some care. The advantage of our overall approach, which explicitly employs the notion of effectivity, is that it permits us to do this. As a first step, move back to the notion of a state: it consists of a coalition structure and an associated feasible payoff profile, rather than simply an imputation. As minimal conditions of coalitional sovereignty, we assume that the effectivity correspondence satisfies the conditions of noninterference and full support of Section 5.3.5.3. That is, when a coalition forms, it leaves unchanged the payoffs to coalitions that are "untouched" by its departure, but it can choose any payoff from its own feasible set.

As Ray and Vohra (2013) show, these restrictions dramatically alter the Harsanyi stable set. The following proposition is a corollary of their Theorem 2:

Proposition 5.8. *(Ray and Vohra, 2013) Consider any characteristic function, and suppose that the effectivity correspondence satisfies noninterference and full support.*

(a) *If u belongs to the interior of the coalition-structure core, then there exists a farsighted stable set with the single-payoff allocation u.*

(b) *If u is not in the coalition-structure core, there is no farsighted stable set containing the single allocation u.*

Contrast Proposition 5.8 with Proposition 5.7. The collection of singleton farsighted sets for the two scenarios could not be more different. See Ray and Vohra (2013) for additional discussion, examples and an application to simple games.

Notice that Propositions 5.7 and 5.8 also imply that no matter whether we consider the Harsanyi stable or the modification discussed here, there is no hope of recursively characterizing these sets along the lines described in Section 5.3.4. Because a multiplicity of single-payoff stable sets exists in both cases, every state has a farsighted objection. Consequently, the recursion cannot get off the ground.

Finally, Proposition 5.8 implies that the core has interesting farsighted stability properties, even without the restriction to internal blocking. This is reminiscent of the Konishi and Ray (2003) results mentioned earlier in Section 5.3.5.3.

A precursor to Proposition 5.8 was obtained by Diamantoudi and Xue (2003) for the class of *hedonic games* with strict preferences. A hedonic game is one in which for every coalition S, $\bar{V}(S)$ is a singleton.[33] This simplifies matters because there is no ambiguity about the payoffs to outsiders, and noninterference and full support are automatically satisfied. For matching games, a special case of hedonic games, Mauleon et al. (2012)

nonexistence of a vNM stable set, the Harsanyi stable set does exist. Bhattacharya and Brosi (2011) extend the Béal et al. (2008) existence theorem to NTU games.

[33] See, for example, Banerjee et al. (2001), Barberà and Gerber (2003, 2007), and Bogomolnaia and Jackson (2002).

prove that if preferences are strict, farsighted stable sets can be characterized as singleton allocations in the core.

5.3.6.2 Farsighted stability for games with externalities

A definitive extension of the results in the previous section is yet to be obtained for games with externalities. However, in the case of partition functions, more can be said provided we restrict attention to hedonic games, and side-step the problem of coalitional sovereignty discussed in the previous section. In this setting, Diamantoudi and Xue (2007) show that the notion of blocking used in defining EBA can be reformulated in terms of farsighted blocking, where the extreme optimism implicit in farsighted blocking is modified to make it robust to the precise manner in which perpetrators move. And this makes it possible, in the context of internal blocking, to interpret EBA as a stable set with an appropriate notion of dominance. They then go on to argue that a suitable extension of EBA to the general case, extended equilibrium binding agreements (EEBA), is the farsighted stable set. Although neither the existence of EEBA nor its efficiency when it does exist can be guaranteed in general, extending the notion of blocking beyond internal blocking can help sustain efficiency. Their positive result on efficiency includes the important example of a Cournot oligopoly; see Section 5.5.

As Greenberg (1990) has shown, it is possible to formulate various stable standards of behavior along the lines of the stable set by assuming conservative rather than optimistic behavior on the part of a blocking coalition. Many of the solution concepts based on farsightedness can be comprehensively viewed through Greenberg's theory of social situations. As a complement to the current section, we refer the reader to Mariotti and Xue (2003) for an excellent review of this approach.

The *largest consistent set* of Chwe (1994) is a prime example of the blocking approach with farsightedness assuming pessimism on the part of a blocking coalition. Chwe considers a more general game than an extended partition function in the sense that the set of states, X, is an abstract set, not necessarily based on a partition function. A set $Z \subset X$ is said to be *consistent* if $x \in Z$ if and only if for all y, S such that $x \rightarrow_S y$, there exists $z \in Z$ such that $z = y$ or z is a farsighted objection to y and there exists some $i \in S$ such that $u(z)_i \leq u_i(x)$. Chwe proves that there is a unique consistent set, the *largest consistent set* (LCS), which contains all consistent sets.[34]

At this stage, one may be tempted to leave well enough alone and accept the idea, as in Knightian uncertainty, that stable outcomes can be modeled either with optimistic beliefs or conservative beliefs or perhaps some combination of the two. However, this is serious drawback of the blocking approach. It is no less ad hoc than making some exogenous assumption about remaining players will organize themselves into a coalition

[34] Chwe also shows how LCS can be related to a conservative stable standard of behavior in Greenberg's theory of social situations.

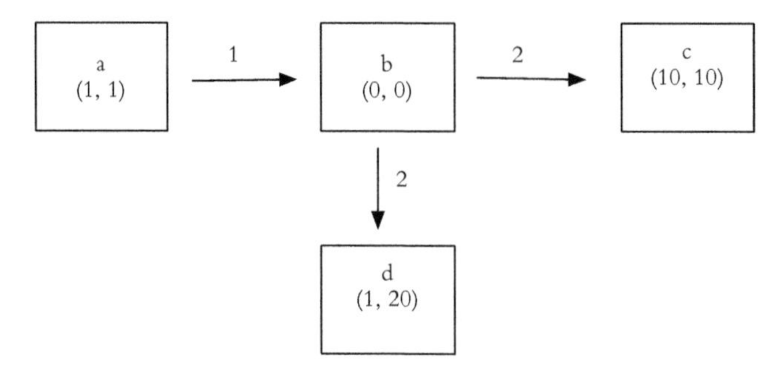

Figure 5.3 Illustration of Example 5.8.

structure in response to a coalitional deviation. The problem is easily shown through the following example.[35]

Example 5.8. *Modify Example 5.6 so that player 2 can move to either state c or d; see Figure 5.3.*

The unique farsighted stable set is $\{c, d\}$. State a is not in the farsighted stable set because player 1 expects player 2 to replace b with c. But now this is too optimistic a prediction. From b player 2 has the choice to move to either c or d, and obliviously the rational decision for player 2 is to move to d. Thus, the (farsighted) reasoning for discarding a now appears to be flawed. In this example, the LCS resolves this problem. For a state to be in the LCS, it is enough that every initial move have *some* continuation that results in a stable outcome which would deter the initial "objection." By this criterion, the LCS is $\{a, c, d\}$; unlike the farsighted stable set, LCS considers a to be stable. In this example, the path from b should lead only to d since that is the only rational move by coalition 2. It so happens that this corresponds to coalition 1 being conservative in its evaluation of the move from a to b. In general, however, a conservative forecast by the initiating coalition need not be "rational," as the next example shows.

Example 5.9. *Modify Example 5.8 by interchanging player 2's payoffs in c and d as shown in Figure 5.4.*

The farsighted stable set and LCS remain unchanged (the former is $\{c, d\}$ and the latter is $\{a, c, d\}$). But now, it should be clear that an "optimistic" view by player 1 is indeed the correct one; player 2 in his own interest will replace b by c, not d. Thus, in this example, it is the LCS which comes to the wrong conclusion.

[35] Many of the examples in this section are based on those in Xue (1998) and Herings et al. (2004). They can all be transformed (with the addition of players) into hedonic partition function games (with appropriately defined effectivity correspondences) and still retain the features discussed in the text.

Mauleon and Vannetelbosch (2004) introduce "cautiousness" in the notion of the LCS by assigning strictly positive probability to all farsighted objections that may follow from a coalitions initial move. In Examples 5.8 and 5.9, the cautious LCS yields the same prediction as the farsighted stable set, which is reasonable in Example 5.9, but not in 5.8. The important point illustrated by these examples is that the prediction of player 2's move at state b should not be based not on optimism or pessimism, or some combination thereof. Instead, it should rely on what is in player 2's best interest. In the language of our dynamic model, player 2 has two profitable moves, only one of which is maximally profitable, and that's the one which ought to be predicted.

It is easy to see that in both these examples, the notion of an EPCF provides a simple resolution. But how far can one go with the traditional blocking approach? Xue (1998) argues, persuasively, that a resolution to this issue requires us to define stability not for allocations, such as a, b, c, and d in the above examples, but for *paths* such as (a, b, c) or (a, b, d). He then considers a stable standard of behavior for a situation with perfect foresight as a collection of paths satisfying internal and external stability.

For a stable standard of behavior σ, let $\sigma(a)$ denote all stable paths originating from a. If $\alpha \in \sigma(a)$, internal stability requires that there not exist a node $b \in \alpha$, a coalition S, which is effective form moving from b to c such that S "prefers" $\sigma(c)$ to α. The same idea is used in defining external stability. The term "prefers" leads to two versions of stability: one based on optimism and other on conservatism.

It is easy to show that this notion of stability yields the correct answer for both Examples 5.8 and 5.9. However, in the next example the problem reappears.

Example 5.10. *Modify Example 5.9 by adding one more player and one more state, as shown in Figure 5.5.*

First consider the stable paths originating from d. Player 2 would like to move to e' while 3 would prefer to move to e. Neither can satisfy internal stability and therefore $\sigma(d) = \emptyset$. Note that while Xue (1998) shows that an acyclicity condition is sufficient for σ to be nonempty valued, this condition is violated at d. (This is similar to Figure 4

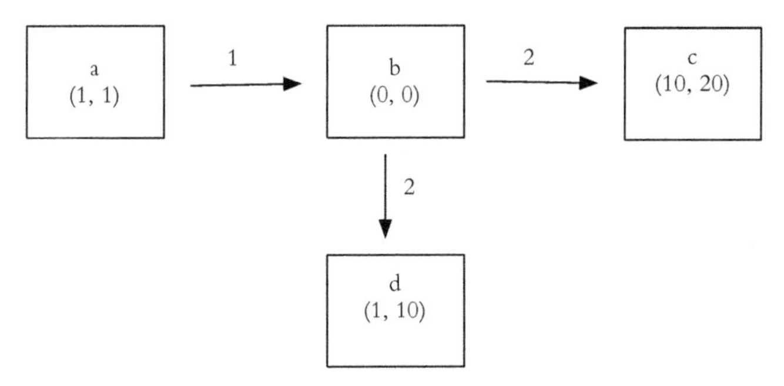

Figure 5.4 Illustration of Example 5.9.

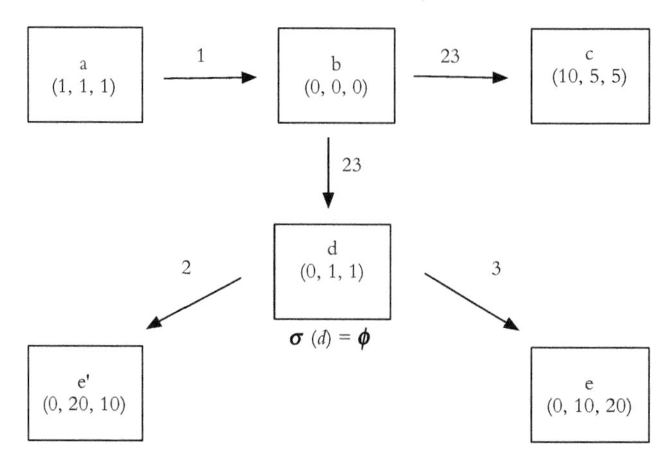

Figure 5.5 Illustration of Example 5.10.

in Xue, 1998). The problem arises because neither of the two objecting coalitions has a priority to make a move. Because $\sigma(d) = \emptyset$, the stability of (a, b, c) is vacuously satisfied; it is not possible to test if it can be deterred by a move from b to d and some $\beta \in \sigma(d)$. Indeed, (a, b, c) is a stable path. But this doesn't seem reasonable because for players 2 and 3 both e and e' dominate c, so coalition 23 should move from b to d, not c.[36]

Unfortunately, this kind of situation is rather common, and it seriously limits the applicability of this solution concept. Note that if there was a protocol, even a probabilistic one, to decide who has the first move, the problem wouldn't arise.[37] For example, if upon arriving at state d, the protocol selects each of two permissible singleton coalitions with equal probability, then the expected payoff to players 2 and 3 (following state d) is 15 each. From state b, therefore, the move by coalition 23 to d is maximally profitable (in the terminology of our dynamic model) but to c is not. And this implies that a is an absorbing state of every EPCF.

Taken together, these examples demonstrate a serious drawback of the blocking approach in dealing with farsightedness, especially in environments with externalities. A key requirement that is missing in the notion of farsighted blocking, is that of constraining objecting coalitions to make maximal moves among their profitable alternatives. As Proposition 5.8 shows, there are conditions under which there exist singleton farsighted stable sets in characteristic function games (without externalities).

[36] For a further discussion of these, and other related issues, see also Herings et al. (2004) and Bhattacharya and Ziad (2012).

[37] Recall that one major difference between our dynamic model and that of Konishi and Ray (2003) is that the latter operates without a protocol. This can result in inefficiencies arising from a "coordination failure." Simply having a protocol of the kind we outlined in Section 5.3.5.2 can help avoid these kinds of inefficiencies.

In such situations, all farsighted blocking paths lead to the same terminal state, and therefore, there is no ambiguity of the kind that plagues the examples discussed above.[38] But, in general, there are limits to how effectively one can capture farsightedness in a static concept of stability. The traditional approach in cooperative game theory has emphasized the virtues of abstracting away from the details of the negotiation process to highlight the essential features of cooperative behavior. In many situations that has indeed been a very fruitful approach, but in the present context, it seems too confining not to introduce some details (as well as explicit dynamics). As the previous examples demonstrate, simply adding the notion of a protocol, and postulating rational behavior on the part of coalitions, can provide a way out of the conundrum that the standard approach produces. The dynamic process of coalition formation described in Section 5.2 is a direct way of studying farsighted coalitional behavior, especially in the presence of externalities. If this framework seems more complex than, say, the characteristic function form, it nevertheless seems to be necessary for the questions at hand. It could even be argued while there is additional structure on the model, the equilibrium concept is much more straightforward that ones we have discussed earlier, for example, EBA, LCS, or farsighted stable sets. The fact that in some of the simpler cases, such as characteristic function games and symmetric TU games with positive externalities, we get the standard conclusions makes it possible to see this approach as a conceptual generalization rather than an alternative. At the same time, as we discuss in Section 5.6, it may be possible, and desirable, to incorporate the principle of maximality into the blocking approach, at least in some classes of games, and thereby resolve this issue without the apparatus of a dynamic model.

5.4. THE BARGAINING APPROACH: NONCOOPERATIVE GAMES

In this section, we study an approach to coalition formation based on noncooperative bargaining. Attention now shifts from active *coalitions* to active *individuals*, and the notion of blocking is replaced by a direct emphasis on proposals and responses. In summary, all negotiations are expressed formally as a *bargaining game*, for which we draw inspiration from Ståhl (1977), Rubinstein (1982), Chatterjee et al. (1993), Okada (1996), and several others.[39]

Throughout, the primitive object on which our analysis is based is the partition function, with the idea that underlying this function is a game in strategic form.

[38] It is plausible to conjecture that, under some reasonable conditions, if a state is a singleton farsighted stable set then it must be the absorbing state of any EPCF. It will be desirable to formally establish such a connection to justify a singleton farsighted stable set as the "right" solution.

[39] For related literature on bargaining, see Selten (1981), Binmore (1985), Baron and Ferejohn (1989), Gul (1989), Winter (1993), Perry and Reny (1994), Krishna and Serrano (1995), Moldovanu and Winter (1995), Hart and Mas-Colell (1996), Bloch and Gomes (2006), and Compte and Jehiel (2010).

As already discussed, a partition function has the virtue of incorporating a number of different situations. Two-person or multiperson bargaining is, of course, included quite trivially. So is coalitional bargaining over a characteristic function, where different coalitions have access to different worths which could each be divided. Over and above these instances, partition functions can also accommodate externalities across coalitions in the determination of coalitional surplus.

We define a noncooperative bargaining game on the partition function. A history-dependent protocol selects an active proposer at each date, perhaps stochastically. In the language of our general framework, active coalitions are now restricted to be singletons. A proposer is free to propose a particular (feasible) payoff allocation to any subset of a partner set specified by the protocol. For each proposal, the protocol also specifies the order in which the responders respond, either to accept or to reject the proposal.[40] Observe that a partner set was of no importance in the blocking approach (at least as it currently appears in the literature), but in the present context, it is crucial: proposers must have others to make proposals *to*. Coalitions form through the course of this bargaining process as proposals are made and either accepted or rejected by responders.

The several ingredients of a bargaining model can be combined in different ways to generate distinct branches of the literature. In the next section, we show how a combination of the effectivity correspondence and protocol can be used to cover a diverse set of models.

5.4.1 Ingredients of a coalitional bargaining model

5.4.1.1 The protocol

When a proposer is chosen by the protocol, she makes a proposal to a subset of eligible partners. The proposal concerns a division of the worth of that subset among its members. If all respondents unanimously accept, the proposed coalition forms, and the process shifts to the set of eligible players remaining in the game. The *rejection* of a proposal creates a bargaining friction: payoffs are delayed by the passage of some time, which is discounted by everybody using the discount factor δ.

The probability with which different individuals are chosen to be proposers will generally depend on the history of events up to that date. Perhaps the simplest protocol is one in which an agent is chosen with uniform probability to be a proposer. This is known as the *uniform protocol*, see, e.g., Baron and Ferejohn (1989) and Okada (1996). Another simple protocol recognizes different individuals in a given order and anoints the first rejector of the previous proposal (if any) as the next proposer. This is the *rejector-proposes protocol*, see, e.g., Rubinstein (1982) and Chatterjee et al. (1993). One way to combine these two protocols while retaining each of them as special cases is to suppose

[40] This avoids the possibility of coordination failure, and is consistent with the rest of the literature.

that an active individual (from the set of free individuals) is chosen randomly at any stage if the previous proposal has been accepted. Otherwise, if the previous proposal has been rejected by some individual, the rejector gets to be the next proposer with probability μ. With probability $1 - \mu$ an exogenous randomization chooses one of the remaining eligible agents to be active. Observe that if $\mu \simeq 1$, the rejector is likely to be the next proposer, while if $\mu \simeq 0$, the rejector is excluded from making a proposal in the next round. This class of protocols accommodates a wide variation in rejector power, including models studied by Eraslan (2002), Okada (2011), and Eraslan and McLennan (2013).

A central feature of our protocol concerns the recognition of those players who are *eligible*, either to make new proposals or to entertain them from others. Which agent is eligible will depend, of course, on the situation to be modeled. In the case of irreversibly binding agreements, a player once included in some previous coalitional agreement is never again recognized by the protocol. When agreements are reversible, so that renegotiation is permitted, the partner set could include a player who belongs to a previously formed coalition, but only if all other members of that coalition are also included as responders. (Their agreement is needed to "free" the player to sign a new deal.) Such restrictions on the protocol can often be substituted by corresponding restrictions on the effectivity correspondence; recall footnote 20.

In short, our description of protocols can be construed as an attempt to model the essential elements of the situation at hand: how easy it is for a rejector to seize the initiative and make a fresh proposal, or how constrained previous signatories are in participating in further bargaining. In this survey, we do not concern ourselves with other approaches to "design" a proposal to deliberately "implement" some known solution concept. For instance, Gul (1989) shows how the Shapley value can be implemented as a stationary perfect equilibrium of a game with pairwise meetings. Hart and Mas-Colell (1996) do so in a more general context, through a bargaining game in which proposers are *required* to make proposals to the complete set of available players. Moreover, the rejection of a proposal leads, with some positive probability, to the proposer being entirely eliminated from future rounds of bargaining. These restrictions are clearly in the spirit of implementing a particular solution concept, as the description makes it clear that there is no particular attempt to identify the protocol with any observed bargaining situation.[41] Indeed, incorporating strategic coalition formation in these models remains an important direction for future work. For a review of this literature we refer the reader to Winter (2002).

[41] It should be added, however, that the above depiction makes more descriptive sense in a two-person bargaining context; see Binmore et al. (1986), and Sutton (1986). There these restrictions reduce, more realistically, to the possibility that the entire bargaining process might break down following any round of negotiation.

5.4.1.2 Possibilities for changing or renegotiating agreements

Whether or not a coalitional agreement once made is subject to future revision is a fundamental issue.

Bargaining with Binding or Irreversible Agreements: In much of the literature, all agreements to form a coalition are "fully" binding, in the sense that they are irreversible. A coalition once formed cannot disintegrate or be subsequently absorbed into a larger group.[42] The protocol responds to histories by choosing only players (as proposers or respondents) who were not part of any previously formed coalition. It follows, therefore, that once all players are included in some formed coalition, the process of coalition formation must come to an end, though payoffs continue to be received as per the various agreements. In particular, the protocol ceases to choose new proposers. For examples of this sort of model, see Chatterjee et al. (1993) and Okada (1996) for characteristic function games, and Bloch (1996) and Ray and Vohra (1999) for partition function games.

Bargaining with Reversible Agreements: Situations in which agreements are only in force for a limited period of time can be modeled by suitably specifying the protocol. For example, it can be supposed that agreements only last for one period, and active proposers are chosen from the *entire* population of players, so that partner sets are never restricted by the history of past coalition formation. For examples, see Stole and Zwiebel (1996), Gomes and Jehiel (2005), and Konishi and Ray (2003).

The possibility of renegotiating existing agreements yields another situation in which agreements may be reversible. In this case, an existing agreement may be changed, but only with the blessings of existing signatories. Such signatories must include all individuals who are party to *any* ongoing agreement that may need to be modified as a consequence of the new proposal. This is captured by an appropriate restriction on the effectivity correspondence or on the protocol. For examples, see Seidmann and Winter (1998), Okada (2000), Gomes and Jehiel (2005), Gomes (2005), and Hyndman and Ray (2007).

5.4.1.3 Payoffs in real time or not

A substantial part of the literature studies models in which payoffs are only experienced after all coalitions have formed. This includes Rubinstein (1982), Bloch (1996), Chatterjee et al. (1993), Ray and Vohra (1999) for binding agreements and Seidmann and Winter (1998) and Okada (2000) for renegotiable agreements. A more recent literature, e.g., Konishi and Ray (2003), Hyndman and Ray (2007), Gomes (2005), Gomes and Jehiel (2005), and Xue and Zhang (2011), considers situations in which payoffs are realized continually, as coalitions can form and continue to renegotiate or discard previous agreements.

It is of course only natural for reversible agreements to be cast in a real-time framework. It should be clear that our general framework is well suited to cover such

[42] In the special case of n-person bargaining, in which only the grand coalition has a surplus to divide, this is hardly an assumption as there is no collective incentive to alter any agreed-upon division of the surplus.

situations and, as we will see below, it can also encompass models in which payoffs are realized at the end of the coalition formation process.

5.4.1.4 *Majority versus unanimity*

There is also a distinction to be drawn using the rules of the game that determine when a proposal is to be passed. Two major candidates are unanimity, as in the Ståhl-Rubinstein model and its descendants, and majority vote, as in the Baron-Ferejohn model (and political economy models more generally). Once again, it is often easy to accommodate a variety of such rules as part of the general framework. When agreements are irreversible, we can often adopt the unanimity approach without any loss of generality, by absorbing other decision rules into the description of the partition function and the form of the effectivity function.

As an example, consider three-person bargaining with majority. The "true" function that describes this example sets the worth of the grand coalition equal to the surplus at stake (say 1 unit), while setting the worth of all subcoalitions to zero. Yet it is possible to use instead the characteristic function

$$v(S) = 1 \quad \text{if and only if } |S| > \frac{n}{2},$$

and use the unanimity protocol. What is altered is essentially a matter of interpretation: a proposal is never actually made to a subcoalition S, but it's *as if* it is: the proposal is in fact made to the grand coalition, with the implicit strategic presumption that the "targeted" majority subgroup S is effective for the change and will approve it.

In short, bargaining models that require majority approval can be easily embedded in coalitional bargaining models in which subcoalitions have power. In this sense, there is little loss of generality in studying unanimity games, *provided* we are general enough to accommodate subcoalitional worths. That said, we hasten to add that the use of majority rules in coalitional settings—especially with reversible agreements—is an interesting and not well-studied research topic.[43]

5.4.2 Bargaining on partition functions

5.4.2.1 *Equilibrium in a real-time bargaining model*

We begin with a baseline model for bargaining in real time. As in the blocking approach, we consider a partition function (N, V) which assigns to each partition π and to each coalition S in that partition, a set of payoff allocations $V(S, \pi)$.

[43] The absorption of the rules into the description of the partition function can sometimes be inadequate when agreements are reversible or can be renegotiated. It may be that coalition {12} is enough to implement an agreement for {123}, but it is not enough to track the situation with coalition {12} alone, when agreements are reversible. Later, when renegotiation might occur, we must recognize that it is coalition {123}, and not {12}, which is bound by an ongoing agreement.

A state is denoted $x = (\pi, u, C)$ where $u_S \in V(S, \pi)$ for every $S \in \pi$ and C is the collection of "committed" players ($N \setminus C$ being the set of "uncommitted" or "free" players). We use the convention that all free players consist of singletons, while committed players belong to formed coalitions (including possibly stand-alone singletons). In other words, $i \notin C$ implies that $\{i\} \in \pi$ and every $j \in S \in \pi$ such that $|S| \geq 2$ implies that $j \in C$. Uncommitted players can be proposers or potential partners in all variations of the model; the others may be included to different degrees if agreements are reversible immediately or later.

At each stage of the bargaining process, we keep track of past proposers, proposals, rejectors (if any), and all formed coalitions as well as free agents.

A *history* at some stage is a list of such objects up to, but not including, the events that will occur at that stage. Such stages may be of various kinds: a proposer is about to be chosen, or a proposal about to be made, or a responder about to respond, or—such matters concluded—a state about to be implemented. Obvious nomenclature may be employed to distinguish between the histories leading up to different stages: "proposer histories," "responder histories," and "implementation histories."

At proposer or responder histories players must either make proposals or react to them. A full listing of a particular player's actions—proposals and responses—for all such histories is a *strategy* for that player. To describe strategies more formally, consider an individual k. For a proposer history h at which k is meant to propose, she must choose a payoff vector and a coalition S that can implement the payoff in question. In standard bargaining theory such a vector would be given by a division of aggregate surplus among the individuals (see, e.g., Rubinstein, 1982, and Baron and Ferejohn, 1989). In coalitional bargaining theory, it would be a division of *coalitional* surplus among the members of that coalition (see, e.g., Chatterjee et al., 1993, Seidmann and Winter, 1998, and Okada, 1996). In bargaining theory with externalities, the payoff vector must come from a "conditional proposal": if coalition structure π forms, we divide in this way, and if π' forms, we divide in that way, and so on (see, e.g., Ray and Vohra, 1999). In our real-time model, the payoffs at each date are feasible given the going coalition structure. For already formed coalitions, they must also reflect the agreed-upon payoff allocation corresponding to this coalition structure. (This can be seen as a restriction on the effectivity correspondence.)

An active agent proposes a new state to one or more of her available partners. She will employ a behavior strategy, which would imply at each history a probability distribution over (y, S), where y represents the new state and S a coalition containing i and a subset of available partners jointly capable of implementing that state. Denote by $\mu_k(h)$ the probability distribution that the active agent employs at proposer history h.

Likewise, at a responder history h at which k is meant to respond, denote by $\lambda_k(h)$ the probability that k will accept the going proposal under that history. The full collection $\sigma = \{\mu_k, \lambda_k\}$ over all players k is a *strategy profile*.

A strategy profile σ induces *value functions* for each player. These are defined at all histories of the game, but the only ones that we will need to track are those just prior to the implementation of a fresh state, or the unaltered continuation of a previous state. Call these *implementation histories*. On the space of such histories, every strategy profile σ (in conjunction with the given protocol) defines a stochastic process P^{σ} as follows. Begin with an implementation history. Then a (possibly new) state is "implemented" at that history. Subsequently, a new proposer is determined. The proposer proposes a state. The state is then accepted or rejected. (The outcome in each of these last three events may be stochastic.) At this point, a new implementation history h' is determined. The entire process is summarized by the transition P^{σ} on implementation histories.

For each person i and given an implementation history h, the *value* for i at that date is given by

$$V_i^{\sigma}(h) = u_i(x) + \delta \int V_i^{\sigma}(h')P^{\sigma}(h, dh'), \quad [5.7]$$

where x is the state implemented at h. Given any transition P^{σ}, a standard contraction mapping argument ensures that V_i^{σ} is uniquely defined for every i.

Say that a strategy profile σ is an *equilibrium* if two conditions are met for each player i:

(a) At every proposer history h for i, $\mu_i(h)$ has support within the set of proposals that maximize the expected value $V_i^{\sigma}(h')$ of i, where h' is the subsequent implementation history induced by i's actions and the given responder strategies.

(b) At every responder history for i, $\lambda_i(h)$ equals 1 if $V_i^{\sigma}(h') > V_i^{\sigma}(h'')$, equals 0 if the opposite inequality holds, and lies in $[0, 1]$ if equality holds, where h' is the implementation history induced by acceptance, and h'' is the implementation history induced by rejection.

5.4.2.2 Two elementary restrictions

To ease the exposition, we impose two simple restrictions on equilibrium. First, equilibria might involve delay: a proposal could be rejected on the equilibrium path. To be sure, it is natural for delays to arise in bargaining with incomplete information.[44] But in complete information models such delays are more artificial and stem from two possible sources. The first is a typical folk theorem-like reason in which history-dependent strategies are bootstrapped to generate inefficient outcomes, including equilibria with delay. More subtly, equilibria may involve delay because an unacceptable proposal is made to deliberately affect the identity of the rejector and subsequently the choice of the next proposer. For examples, see Chatterjee et al. (1993) and Ray and Vohra (1999). This will *only* happen for protocols that are sensitive to the identity of previous rejectors. In several models such as Rubinstein (1982) and random proposer models as in Baron

[44] For a recent example that attempts to get around the Coase conjecture in models of one-sided incomplete information; see Abreu et al. (2012).

and Ferejohn (1989) and Okada (1996, 2006, 2011), this phenomenon is impossible. Moreover, even in situations in which the protocol is sensitive to past rejections, the literature provides reasonable conditions under which all equilibria involve no delay; see in particular Chatterjee et al. (1993) and Ray and Vohra (1999). Accordingly, in what follows, we shall restrict ourselves to *no-delay equilibria*, in which at each proposer history, a proposer makes an acceptable proposal.

We will also restrict ourselves to equilibria which satisfy a minor additional restriction, which we call "compliance." Say that an individual is *compliant* if, whenever she responds to a proposal, she takes an action that makes the proposer better off, provided that this does not harm her in any way. The terms "better off" and "harm" are defined with respect to equilibrium value functions, in just the same way as equilibrium payoffs are. This refinement is of a lexicographic nature: it only applies when there is no danger to the payoff of the individual concerned. Alternatively, one could just as easily think of compliance as an equilibrium refinement rather than as a lexicographic restriction on individual preferences. Thus, an equilibrium strategy profile is *compliant* if for no individual and no history is there a deviation by a responder which increases the payoffs of a proposer, while not decreasing the responder's payoff.

To the extent that we are aware, there is no serious departure from compliance in any part of the literature, so we have no hesitation in imposing this requirement.[45]

5.4.2.3 EPCF and bargaining equilibrium

In this section, we link up bargaining equilibrium (which will nest classical models of two-player noncooperative bargaining but contain much more) with the general solution concept of this chapter, that of an EPCF.

The central feature of such a connection is the link between an acceptable proposal on the one hand, and the notion of a maximally profitable move on the other.

Proposition 5.9. *Consider any bargaining game. Then the equilibrium process P^{σ} corresponding to any no delay, compliant, bargaining equilibrium σ is an EPCF.*

Proof. Pick any bargaining equilibrium σ. It generates a process P^{σ} on all implementation histories. This is a PCF, once we identify every implementation history with the notion of a "history" under the EPCF. Specifically, retain the list of all active coalitions, partners, moves, and previous rejectors up to period $t-1$.

Fix a history, as just identified, and consider the choice of any active (singleton) coalition—the proposer—together with a set of potential partners, as dictated by the

[45] In many cases, compliance is a freely obtained equilibrium property and so imposes no additional restriction. This is true, for instance, of Markovian equilibrium coalitional bargaining with fully transferable utility. Otherwise the proposer could slightly increase his offer to all responders and guarantee acceptance.

protocol of the bargaining game. Consider the equilibrium proposal made. Since, by the no-delay hypothesis each partner accepts the proposal, doing so must yield a value that is at least as high as the value following a rejection. So it is immediate that the proposal is profitable for all the partners concerned. We now establish maximality (and therefore profitability[46]). Suppose, on the contrary, that there is an alternative profitable proposal that makes the proposer strictly better off (in terms of value under P^{σ}). That proposal must involve at least one partner, otherwise the proposer can unilaterally achieve his alternative, which is impossible in equilibrium. Consider the equilibrium responses to this proposal in the order given by the protocol, up to all the respondents or the first equilibrium rejection, if any.[47] By compliance, each respondent (working back recursively from this point) will take the action that benefits the proposer, in case the respondent is indifferent between the equilibrium action and some other. Therefore, the alternative move can be implemented by a proper deviation. This makes him better off relative to the putative equilibrium, a contradiction.

Compliance is needed for this result. As an example, consider a discrete two-player bargaining model in which a unit surplus can be divided in just two ways: $(3/4, 1/4)$ and $(1/4, 3/4)$, where the first entry belongs to player 1 and the second to player 2. Suppose that only player 1 makes offers, and that player 2 has an outside option of $1/4$ which she can enjoy instantly on rejection of an offer. Suppose, moreover, that all agreements are irreversible, and the agreed-on payoffs are enjoyed for all time. There are two equilibria: one in which the division $(3/4, 1/4)$ is proposed and accepted, and the other in which this proposal is rejected, leading to an equilibrium division of $(1/4, 3/4)$. The first can be identified with the unique EPCF of the model, while the second equilibrium is not an EPCF. It is also not compliant.

In addition, there could be more complex history-dependent equilibria of fully transferable-utility bargaining games which are not compliant (and not identifiable with EPCFs). As we do not find noncompliant equilibria particularly interesting, we do not pursue this line any further.

5.4.3 Some existing models of noncooperative coalition formation

In this section, we describe some models of coalition formation and embed these into the real-time setup developed in the previous section. In most cases, these existing models are not real-time theories, and the resulting embedding is perforce somewhat unnatural. That, to us, is a virtue: not only will we be able to describe the positive and useful features of these models, but—to the extent that a real-time description is actually called

[46] With a singleton proposer, verification of maximality suffices to guarantee profitability for the proposer.

[47] Our assumption of no-delay equilibrium does not rule out the possibility that a *deviating* proposal must be accepted even if the responder is indifferent between doing and not doing so. This is where compliance is used.

for in some situations—we will also be able to point out potential inadequacies in the existing literature.

5.4.3.1 The standard bargaining problem

Much of what we do relies on the solution to a standard bargaining problem, due to Ståhl (1977) and Rubinstein (1982). In this section, we briefly recapitulate that problem. A group of n persons divide a cake of size 1; there are no subcoalitions of any value, and there are no externalities. A protocol chooses a proposer in every round, and everyone else responds sequentially to the proposal in some given order. If the proposal is rejected, a new round begins. Future rounds are discounted by a common discount factor.

To cover both the uniform-proposer and rejector-proposes protocols, we consider a general protocol in which the first rejector of a proposal is chosen to be the next proposer with probability $\mu \in [0, 1]$, and with probability $1 - \mu$, *another* uncommitted agent is equiprobably chosen to be the new proposer.

As we will see, in the pure bargaining problem, the (Markov) equilibrium will be one in which the grand coalition forms immediately. Since there are no intervening states before the end of the coalition formation process, there is no distinction between a real-time model and one in which payoffs are received at the end of the process.

If $n = 2$, we have two-person bargaining. A remarkable property of this two-person model is that subgame perfection fully pins down equilibrium payoffs. The proposition that follows extends Rubinstein (1982) to a broader class of protocols:

Proposition 5.10. *There is a unique subgame perfect equilibrium payoff vector in the two-person bargaining model.*

Proof. Existence will be shown below; assume it for now and prove uniqueness. Let M and m be the supremum and infimum equilibrium payoff to either player as a *responder*, conditional on her rejecting the current offer but before the next proposer has been decided.[48] Then, because a proposer can always assure herself an infimum of at least $1 - M$, and because a responder must be given at least m,

$$m \geq \delta[\mu(1 - M) + (1 - \mu)m],$$

where μ is the probability that a current rejector gets to propose next. That implies

$$m \geq \frac{\delta\mu(1 - M)}{1 - \delta(1 - \mu)}. \quad [5.8]$$

[48] When discount factors are not the same, these values vary across the players but the proof follows exactly the same lines.

But no proposer can obtain more than $1 - m$, so it is *also* true that

$$M \leq \delta[\mu(1 - m) + (1 - \mu)M]$$

or

$$M \leq \frac{\delta\mu(1 - m)}{1 - \delta(1 - \mu)}. \qquad [5.9]$$

Combining (5.8) and (5.9), it is easy to see that

$$m \geq \frac{\delta\mu\left(1 - \frac{\delta\mu(1-m)}{1-\delta(1-\mu)}\right)}{1 - \delta(1 - \mu)},$$

and simplifying this yields the inequality

$$m \geq \frac{\delta\mu}{1 - \delta(1 - 2\mu)}. \qquad [5.10]$$

Following an analogous line of reasoning,

$$M \leq \frac{\delta\mu}{1 - \delta(1 - 2\mu)}. \qquad [5.11]$$

and together (5.10) and (5.11) show that

$$M = m = \frac{\delta\mu}{1 - \delta(1 - 2\mu)} \equiv m^*, \qquad [5.12]$$

which establishes uniqueness.

Existence can now be shown by construction. Have each player accept an offer if it yields her at least m^* (defined in (5.12)) and always make the proposal $(1 - m^*, m^*)$ when it is her turn to propose. It is easy to verify that this strategy profile constitutes a perfect equilibrium.

This proposition and its accompanying proof reveal that the equilibrium involves immediate agreement, with the proposer and the responder receiving

$$\frac{1 - \delta(1 - \mu)}{1 - \delta(1 - 2\mu)} \quad \text{and} \quad \frac{\delta\mu}{1 - \delta(1 - 2\mu)},$$

respectively. It is worth noting that no matter how small μ is, as long as it is strictly positive, the division of the cake must converge to an equal split as "bargaining frictions" vanish; i.e., as δ converges to 1. It is true that the first individual to propose may acquire a lot of power, especially if μ is small, but the value of that added power becomes negligible provided both players are extremely patient.

To extend this analysis of the bargaining problem to $n > 2$, we restrict attention to equilibria with stationary strategy profiles. It is easy to see that in equilibrium each player must make an acceptable proposal to the grand coalition. Let m_i be the amount that i

will accept as a responder, provided that all responders *after* her in the response order are planning to accept that proposal.[49]

In equilibrium, (m_i) must be built from an expectation about payoffs conditional on rejection; these would be a probabilistic combination of i's payoff as a proposer $(1-\sum_{j\neq i} m_j)$ and as a responder (m_i). Therefore, m_i must solve the following equation:

$$m_i = \delta\{\mu[1-\sum_{j\neq i} m_j] + (1-\mu)m_i\}$$

or

$$(1-\delta)m_i = \delta\mu\left[1-\sum_{j=i}^{n} m_j\right],$$

which tells us that $m_i = m$ for all i, and

$$m = \frac{\delta\mu}{(1-\delta)+\delta\mu n}. \qquad [5.13]$$

This solution extends the two-person case and once again, convergence occurs to equal division as bargaining frictions disappear, provided that $\mu > 0$.

While our derivation of (5.13) assumed that all players other than the first rejector have an equal chance of becoming the active proposer, the uniqueness of stationary equilibrium payoffs appears to be a more general feature. Eraslan (2002) and Eraslan and McLennan (2013) establish this for a class of bargaining games that includes the Baron and Ferejohn (1989) model and for a class of protocols in which the active proposer is chosen with an exogenously given probability. A characterization of uniqueness for general games and general protocols is yet to be fully explored.[50] However, uniqueness certainly fails to survive with three or more players if we allow for nonstationary equilibrium strategies. The argument, due to Herrero (1985), Shaked (see Sutton, 1986), and Baron and Ferejohn (1989) can be generalized to the full class of protocols we consider here; see Ray (2007) for details. Moreover, when we consider general coalitional games, as we do next, the exact protocol has a bearing on the actual equilibrium payoffs; see Examples 5.11 and 5.13 below.

5.4.3.2 Coalitional bargaining with irreversible agreements

In this section, we study a model of coalitional bargaining in which an agreement once made is irreversible. The protocol is therefore restricted so that no agent who has made

[49] It is unnecessary to describe here what happens if a later responder is planning to reject, as such a proposal will be rejected anyway and that is all that matters for our argument.

[50] For instance, with the pure rejector-proposes protocol, stationary equilibrium payoffs are not unique in Example 5.11. See also Ray (2007, Section 5.3.4.3), which uses a wider class of protocols.

a deliberate decision to join a coalition can participate in future negotiations.[51] This is indeed the framework for most of the bargaining literature that we have already cited. The real-time model developed in Section 5.4.2, which in turn is a special case of an EPCF, can be further simplified to set up a canonical example of irreversible coalitional bargaining with externalities. On still further specialization, the latter yields the bulk of the models used in the literature.

Recall that a proposal refers to a division of the worth of a coalition among its members. In a characteristic function game, for coalition S, this is simply a point in $V(S)$. But in a partition function, the worth of a coalition will vary with the going coalition structure. Therefore, a proposal must consist of a set of *conditional statements* that describe a proposed division of coalitional worth for every contingency, i.e., for every conceivable coalition structure that finally forms. More precisely, a proposal is a pair $(S, \mathbf{v})$, where $\mathbf{v}$ is a collection of allocations $\{v(\pi)\}$, one for each partition π that contains S, feasible in the sense that for every coalition S in π,

$$v_S(\pi) \in V(S, \pi).$$

We continue with the protocol in which the first rejector of a proposal becomes the next proposer with probability μ. Strategies and equilibrium are exactly as defined in the general model.

The literature we will be describing in this section concerns bargaining with payoffs assigned only after all coalitions have formed. However, it is easy enough to describe a general procedure for embedding such models into our real-time framework. The first step is to specify a way of assigning payoffs to the players each time a coalition gets formed (leading to a new state). We do this by presuming that at date 0 all agents are free and the initial coalition structure is one of singletons.[52] From then on, payoffs are received in every period in a perfectly well-defined fashion. For individuals i who have yet to form a coalition or have deliberately decided to stand alone, they are given by $V(i, \pi_t)$ at date t, where π_t is the coalition structure prevailing at that date. For individuals i who are part of some coalition S, they are given by $v_S(\pi_t)$ at date t, which comes from the agreement $\mathbf{v}$ that members of S have entered into at some earlier date.

Next, we need to show that an equilibrium of a standard bargaining game, say σ, is a also an equilibrium in the real-time framework with payoffs defined at and added over all dates. We know that no player has a deviation that can result in higher payoff *after all coalitions have formed*. So the only possibility of a profitable deviation must come from capturing some gains *before* the coalition formation process comes to an end. If the long-

[51] That includes agents who have deliberately made the decision to stand on their own.

[52] In the irreversible agreements model, this assumption is without any serious loss of generality, especially if we assume that already formed coalitions at the start of the game have made their agreements to begin with.

run payoff to the deviating player is *lower*, then for δ close to one, the transitory payoffs cannot make up for the long-run loss. The only possibility that remains for a profitable deviation is that it reaps some transitory gains but eventually results in the same payoff as σ. Moreover, because agreements are irreversible, any such deviation must result in a final coalition structure different from the one under σ. But this can be ruled out by a mild genericity assumption that a different coalition structure cannot yield the deviating player *exactly the same* payoff as the equilibrium payoff. In many models, the equilibrium payoff to a player is closely related to the average worth of her coalition, and this genericity assumption can be then be imposed directly on the partition function. Notice that this genericity argument is very similar to the one we employed in Proposition 5.6 to connect the static notion of an EBA to an EPCF.

It is well known that the perfect equilibria of such a model can generate a huge multiplicity of outcomes. We already know, in fact, that the n-person bargaining game is prey to multiplicity when $n \geq 3$, and that game is a special case of the irreversible agreements model described here. In what follows, then, we retreat to the use of *stationary Markovian strategies.* They depend on a small set of state variables, and do so in a way that's insensitive to the passage of calendar time. The current proposal or response (while permitted to be probabilistic in nature) is not permitted to depend on "past history." Of course, it must be allowed to depend on the current set of free players, on the coalition structure that is currently in place and—in the case of a response—on the going proposal; after all, these are all payoff-relevant objects.[53]

Note that under our initial condition, the coalition structure must steadily coarsen (or remain unchanged) as time wears on. Thus, payoffs are received in real time, and they must finally settle down to a limit value for all concerned. We have therefore successfully, and at little cost, embedded a model of irreversible agreements into the real-time setting of a PCF.

The existence of a stationary equilibrium (possibly with randomization) can be proved along the lines of Ray and Vohra (1999) and Okada (2011).

It is instructive to see how the protocol affects the equilibrium payoffs. We illustrate this with a simple three-player characteristic function.

Example 5.11. (N, v) *is a TU-characteristic function with* $N = \{1, 2, 3\}$, $v(i) = 0$ *for all* i, $v(S) = 1$ *for every nonsingleton* S.

If the protocol chooses the first rejector to become the next proposer ($\mu = 1$), it is easy to see that in any stationary equilibrium, the first proposer offers $\delta/(1 + \delta)$

[53] We will also permit proposers to condition their new proposals on the identity of the last rejector (in the current round of negotiations) and for respondents to condition their responses on the identity of the proposer.

to one of other player and obtains $1/(1+\delta)$. The equilibrium payoffs are therefore exactly as in Rubinstein bargaining with two players; as δ approaches 1, the two players share the aggregate surplus approximately equally. If, however, a proposer is chosen with equal probability ($\mu = 1/3$), regardless of any previous rejection, again a two-player coalition forms in equilibrium. However, the surplus is not shared equally between the two players who form the "winning" coalition. The proposer offers $\delta/3$ to one of the other players and any such offer is accepted. For δ approaching 1, the proposer receives approximately 2/3 and the other player in the winning coalition receives approximately 1/3. This conforms, of course, to the Baron and Ferejohn (1989) characterization of equilibrium with majority voting.

We will have more to say about the efficiency of equilibrium in Section 5.5.

5.4.3.3 Equilibrium coalition structure

The central question of interest is the prediction of equilibrium coalition structure. As the reader might imagine, this is an ambitious and complex undertaking, especially in an ambient environment which allows for a variety of strategic situations and alternative protocols. While we do not have a comprehensive understanding of this problem and highlight it as a fundamental open question, it is possible to make progress in specific cases. We outline some available results.[54]

In what follows, we consider only symmetric TU partition functions though the analysis extends to nonsymmetric games and also to nontransferable payoffs under some restrictions; see the end of this section.

A *symmetric* (TU) partition function has the property that the worth of a coalition depends only on its size and the ambient *numerical* coalition structure it is embedded in; namely, the collection of coalition *sizes* in the coalition structure. Use the notation $\mathbf{n}$ to refer to both numerical coalition structures and substructures, the latter being collections of positive integers (including the "null collection" ϕ) that add up to any number strictly less than the total number of players. In the sequel, a substructure is to be interpreted as a collection of coalitions that has "already formed" in a subgame. Define the *size of a substructure* to be the number of all players in it; that is, the sum of coalition sizes in the substructure.

With some abuse of notation, then, $v(T, \pi)$ may be written as $v(t, \mathbf{n})$, where t is the size of coalition T and $\mathbf{n}$ is the numerical coalition structure corresponding to π. Define the *average worth* of t in $\mathbf{n}$ by

$$a(t, \mathbf{n}) \equiv \frac{v(t, \mathbf{n})}{t}.$$

We may interpret this number as the average worth of any coalition of size T embedded in a coalition structure π with associated numerical structure $\mathbf{n}$. In what follows, we

[54] More detail on the discussion that follows can found in Ray (2007).

impose for expository convenience the genericity condition

$$a(t, \mathbf{n}) \neq a(s, \mathbf{n}') \text{ for all } t \neq s. \quad [5.14]$$

We now present an algorithm that calculates a particular coalition structure. Specifically, for each substructure $\mathbf{n}$ with size less than n, the algorithm assigns a positive integer $t(\mathbf{n})$, to be interpreted in the sequel as the size of the coalition that forms *next*. By applying this rule repeatedly starting from the "null structure" with no coalitions, we will generate a particular numerical coalition structure.

Step 1. For all $\mathbf{n}$ of size $n-1$, define $t(\mathbf{n}) \equiv 1$.

Step 2. Suppose that $t(\mathbf{n})$ is defined for all substructures $\mathbf{n}$ of size greater than m, for some $m \geq 0$. For all such $\mathbf{n}$, define

$$c(\mathbf{n}) \equiv \mathbf{n}.t(\mathbf{n}).t(\mathbf{n}.t(\mathbf{n}))\ldots,$$

where the notation $\mathbf{n}.t_1.\ldots.t_k$ simply refers to the numerical coalition structure obtained by concatenating $\mathbf{n}$ with the integers $t_1, \ldots, t_k$.

Step 3. For any $\mathbf{n}$ of size m, let $t(\mathbf{n})$ be the integer in $\{1, \ldots, n-m\}$ that maximizes the expression $a(t, c(\mathbf{n}.t))$.

Step 4. Once the recursive definition is completed for all structures including the null substructure ϕ, define a numerical coalition structure (by

$$\mathbf{n}^* \equiv c(\phi).$$

This completes the description of the algorithm. Its connection to equilibria is striking and direct under the rejector-proposes protocol, through the links are wider-reaching as we shall see subsequently:

Proposition 5.11. *Assume the rejector-proposes protocol (that is, $\mu = 1$ in our class of protocols). Then under the genericity condition on average worths, there exists $\delta^* \in (0, 1)$ such that if $\delta \in (\delta^*, 1)$, every no-delay equilibrium must yield $\mathbf{n}^*$ as the numerical coalition structure.*

We omit the proof, but it is easy to construct one along the lines of Ray and Vohra (1999). The main argument is based on the following steps. To begin with, when the partition function is symmetric and the rejector gets to propose with probability one, every compatriot to whom an individual makes offers has the same options conditional on refusal as the proposer currently does. It follows that when the discount factor is close to 1, the same argument used to show equal division in Rubinstein bargaining applies, and any formed coalition must exhibit (roughly) equal division of their worth. It follows from an inductive argument that at every stage with a substructure already formed, it pays to form a coalition that maximizes "predicted" average worth, the qualifier "predicted" coming from the fact that "final" average worths are not fully defined until the coalition structure has fully formed. (That is where the induction is used.) That leaves a small item to be verified. Even though coalition formation is occurring in real time and these final payoffs won't be received until every equilibrium coalition has formed, this does not worry the early-forming coalitions provided that they are sufficiently patient, a requirement that will be picked up by the construction of the threshold δ^*. The real-time structure is of little importance in a model with irreversible agreements (our genericity condition (5.14) helps to substantially simplify matters here).

It should be noted that the proposition does not assert that an equilibrium implementing $\mathbf{n}^*$ actually exists. Sometimes it may not (see Ray and Vohra, 1999), but these cases are easily ruled out by a mild restriction on average worths. Define *algorithmic average worth* $\hat{a}(\mathbf{n})$ to be the maximized average worth $a(t, c(\mathbf{n}.t))$ achieved by choosing t at any stage of the algorithm (indexed by the substructure $\mathbf{n}$). Say that algorithmic average worth is *nonincreasing* (or that the partition function satisfies NAW, for "nonincreasing average worth") if $\hat{a}(\mathbf{n}) \geq \hat{a}(\mathbf{n}.t(\mathbf{n}))$ for every substructure $\mathbf{n}$ such that $\mathbf{n}.t(\mathbf{n})$ has size smaller than n.

NAW has bite *only* for partition functions. For all characteristic functions, in which the worth of a coalition depends only on the coalition itself, NAW must be trivially satisfied.[55] Whether or not NAW applies more generally (i.e., when externalities are present) is a less transparent question, and the answer will largely depend on the application at hand. But we haven't come across an interesting economic or political application where NAW isn't satisfied. Both the Cournot oligopoly and the public goods model satisfy NAW. It is also important to note that in both cases the algorithm yields an equilibrium coalition structure which is typically *not* the grand coalition, resulting in inefficient outcomes.

Now we apply NAW by showing that in its presence, $\mathbf{n}^*$ is achieved under a variety of protocols. Recall that in our class of protocols, a rejector counterproposes with probability μ, while another uncommitted player is chosen uniformly otherwise. Say that a protocol from this class is *rejector-friendly* if the rejector gets to counterpropose with better than even probability: $\mu > 1/2$. Of course, the familiar rejector-proposes protocol is a special case.

Proposition 5.12. *Under NAW and genericity, there is a discount factor* $\delta^* \in (0, 1)$ *such that if* $\delta \in (\delta^*, 1)$, *there exists an equilibrium that yields the numerical coalition structure* $\mathbf{n}^*$.

We note once again that $\mathbf{n}^*$ is singled out, because the equilibrium behavior we identify is connected closely to *equal* division of the available worth as bargaining frictions go to zero. For a more nuanced discussion of related issues and some qualifications, see Ray and Vohra (1999) and Ray (2007).

[55] The reason is simple. Our algorithm involves the stepwise maximization of average worth, setting each maximizing coalition aside as the algorithm proceeds. If there are no externalities across coalitions, such a process must result in a sequence of (maximal) average worths that can never increase; for if they did, such coalitions would have been chosen *earlier* in the algorithm, not *later*. This simple observation also assures us that NAW does not demand restrictions such as superadditivity: *all* (symmetric) characteristic functions satisfy it.

An uncomfortably familiar feature of bargaining models is that their predictions are often sensitive to the finer points of procedure—to the *protocol*, in the language of this chapter. This is why we are taking care to present results that cover a broad class of protocols. One might worry, though, that the class isn't broad enough. For instance, the two propositions in this sections have been stated for the class of rejector-friendly protocols, procedures in which the rejector has quite a bit of power. One can show, however (see Ray 2007) that Proposition 5.12 can be extended to *all* the protocols we consider, provided that the rejector always has a strictly positive probability of making the next proposal and can also unilaterally exit with no time delay. The condition that is needed is a strengthening of NAW to one in which average worth *strictly* declines along the path of the algorithm[56]; see Ray (2007, p. 64, Proposition 5.3). It is also possible to provide additional restrictions on algorithmic average with that guarantee that $\mathbf{n}^*$ is the *unique* equilibrium structure under the rejector-proposes protocol (see Ray and Vohra, 1999, Theorem 3.4). As shown in Ray and Vohra (1991, 2001), both the Cournot oligopoly and the public goods model satisfy this additional condition for uniqueness. Thus, $\mathbf{n}^*$ appears to be a focal prediction.

We end our discussion by remarking that it is possible to incorporate nontransferable utility, as well as nonsymmetric cases. First, suppose that we retain all the symmetry assumptions, but replace the TU worth $v(S,\pi)$ by some *symmetric* set of payoffs $V(S,\pi)$. Nothing of substance will change as long as we are willing to assume that each such set is convex. It is easy to obtain an intuition of why the same arguments go through. Average worth will now need to be replaced by the symmetric utility obtained "along the diagonal" for each $V(S,\pi)$, and the same algorithm may be written down with average worth replaced by this symmetric utility. Next, return to the TU case but suppose that the partition function is nonsymmetric. Then the analysis is considerably more complicated, but nevertheless a similar algorithm can be constructed. Ray and Vohra (1999) and Ray (2007) contain the details of this procedure.

5.4.4 Reversibility

So far, we have assumed that a commitment to form a coalition, once made, cannot be undone. In many situations, this isn't a bad assumption. But there are numerous scenarios in which agreements may be reversed freely (or at little cost). A free-trade area or customs union may initially exclude certain countries and later incorporate them. Two groups might form an alliance, or a multiproduct firm may also spin off divisions into sub-firms. Political coalitions may form and reform.

Reversibility comes in two flavors. An agreement may be viewed as indefinitely in place, unless signatories to that agreement voluntarily agree to dissolve it. We will call

[56] That is, $a(\mathbf{n}) > a(\mathbf{n}.t(\mathbf{n}))$ for every substructure $\mathbf{n}$ such that $\mathbf{n}.t(\mathbf{n})$ has size smaller than n.

this the case of *renegotiable agreements*. Or an existing agreement may simply come with an expiry date, after which new options can be freely explored. One might call this the case of *temporary agreements*.[57]

What is the role played by a renegotiable commitment? Why would a commitment to form a group first be made, then reversed? Why not simply eschew the making of that commitment in the first place? As a concrete instance, consider a specific case of the public goods model introduced in Example 5.2:

Example 5.12. (Public goods revisited) *Consider Example 5.2 with three symmetric agents and $c(r) = (1/3)r^3$. As in the Cournot example earlier, let us construct a partition function for this game. Then, it is easy to calculate the partition function. Writing $\mathbf{v}(\pi)$ to be the (ordered) vector of coalitional worths in the coalition structure π, we have, following Ray (2007),*

$$\begin{aligned}
\mathbf{v}(123) &= \{6\sqrt{3}\},\\
\mathbf{v}(\{1,2,3\}) &= \left\{2\frac{2}{3}, 2\frac{2}{3}, 2\frac{2}{3}\right\},\\
\mathbf{v}(\{i,jk\}) &= \left\{2\sqrt{2} + \frac{2}{3}, 2\left[1 + \frac{2}{3}\sqrt{8}\right]\right\}.
\end{aligned}$$

Two features are to be noted. First, the per-capita worth of the grand coalition is smaller than the payoff to i in the coalition structure $\{i,jk\}$. Second, the per-capita payoff to j and k in the coalition structure $\{i,jk\}$ exceeds their corresponding payoff in the coalition structure of singletons. So if one agent commits to (irreversibly) exit the negotiations, it is in the interest of the remaining two players to stay together.

For concreteness, assume that an initial proposer is drawn randomly, that proposals must be universally acceptable to the players involved, and that the first rejector of a going proposal gets to make a new proposal. Then, if group formation is irreversible, there is only one (numerical) equilibrium structure. Player i stands alone, and players j and k band together; the outcome is inefficient.

Now suppose that fresh proposals can always be made. Then, there are two possibilities, both leading to an efficient outcome. First, if a player moves off on her own, the other two players disband as well, incurring a temporary loss of payoff but thereby getting into position to enforce a symmetric, efficient outcome with the all three players coming together. If this path indeed constitutes credible play, then no player will move off in the first place, and the outcome is efficient to begin with.

(This isn't even a possibility if commitments are irreversible. Once player i moves off, there is no bringing her back, so players j and k will never disband.)

[57] To be sure, agreements may be both temporary and renegotiable (within the period for which the agreement is in place), but here we only look at the two features separately.

The second possibility concerns a situation in which once player i moves off, players j and k do not find it worthwhile to disband. For instance, this could happen if player i can make a commitment which is irreversible for some length of time, a situation which can be readily modeled by lowering the discount factor of all players. In this case, the outcome will still be efficient, but the path to efficiency as well as the final outcome will look very different. Some player i *must* initially move off. Thereafter, players j and k must cajole her back to the grand coalition with an offer that gives her more than what she gets in the structure $\{i, jk\}$. So we are ultimately at an efficient outcome, one that is "skewed" in favor of the individual who was lucky enough to be the first to make a commitment. Notice that *the commitment must have been made* for her to take advantage of it, and so the equilibrium path involves a transitory phase of inefficiency, followed by a Pareto-superior outcome.

This example may be easily modified to take account of temporary agreements. For instance, suppose that the signatories to the agreement to bring player i back into the fold cannot commit to honor this agreement in the future. If—in that future—some other player j were to unilaterally desert the agreement and take up the same stance as player i did, then there may be little in the situation to induce player i to take up the conciliatory offer in the first place. That could result in a permanent failure to achieve an efficient outcome, a theme that we return to in Section 5.5.

Both renegotiable and temporary agreements (and several other variants) can be defined using the effectivity correspondence. Suppose that a pair $x = (\pi, u)$ represents a going state, embodying certain agreements, and a move is contemplated to a new state $y = (\pi', u')$, in which one or more of those agreements are disrupted. When agreements are renegotiable, we would like to describe the coalitions that are effective in moving the state from x to y. First, if a player's coalitional *membership* is affected as a consequence of a proposed move, the move *must* be disrupting some previous agreement to which that player was a signatory. That player must be included in any group that is effective for the proposed move. Second, the proposed move might affect the (ongoing) *payoff* to a particular agent, without altering her coalitional membership. Must consent be sought from that agent? The situation here is more subtle. It may be that the payoff is affected because a fellow member of a coalition wishes to reallocate the worth of that coalition. In that case—given that the existing allocation is in force—it is only reasonable that our agent be on the approval committee for the move. On the other hand, our agent's payoff may be affected because of a coalitional change elsewhere in the system, which then affects our agent's coalition via an externality. Our agent is "affected," but need not be on the approval committee because she wasn't part of the agreement "elsewhere" in the first place.[58]

[58] Notice that we wouldn't insist that our player should *not* be on that approval committee; it's just that our definition is silent on the matter.

More formally, for any move from x to y, let $M(x, y)$ denotes the set of individuals whose coalitional membership is altered by the move, and $W(x, y)$ the set of individuals j whose one-period payoffs are altered by the move: $u_j(x) \neq u_j(y)$. Say that agreements are *binding but renegotiable* if the following restrictions are met:

1. [[B.1]] For every state x and proposed move y, $M(x, y) \subseteq S$ whenever $S \in E(x, y)$.
2. [[B.2]] Suppose that $T \cap W(x, y)$ is nonempty for some existing coalition T. Then if the proposed move involves no change in coalition structure, *or* if payoffs are described by a characteristic function, $T \cap W(x, y) \subseteq S$ whenever $S \in E(x, y)$.

[B.1] is obvious. To understand [B.2], note that the payoffs in coalition T have been affected. But if there has been no change in the coalition structure, or if the situation is describable by a characteristic function to begin with, how could that happen? It can only happen if there is a deliberate reallocation within that coalition, and then [B.2] demands that all individuals affected by that reallocation must approve it. It is in this sense that [B.1] and [B.2] together formalize the notion of binding yet renegotiable agreements.

Restrictions such as [B.1] and [B.2] placed on the effectivity correspondence permit us to explore all sorts of other variants. For instance, a theory of "temporary agreements" can characterized as follows: a coalition that's effective for moving x to y must contain all members of at least $m - 1$ of the m new coalitions that form, and the coalition not included must be a subset of an erstwhile coalition.[59] As a second variant, allow a coalition to break up or change if some given fraction (say a majority) of the members in that coalition permit that change. Some political voting games or legislative bargaining would come under this category. Now any effective coalition must consist of at least a majority from *every* coalition affected by the move from one state to another. In the reverse direction, [B.1] and [B.2] could be further strengthened: for instance, one might require that a coalition once formed can never break up again. This would lead us back to the model with irreversible commitments.

In the next section, we turn to the efficiency properties of models with both irreversible and reversible agreements.

5.5. THE WELFARE ECONOMICS OF COALITION FORMATION

A central question in the theory of coalition formation has to do with the attainment of efficiency. The Coaseian idea that efficiency is inevitable in the absence of informational frictions and "unrestricted contracting" is deeply ingrained in the economics literature. The fundamental impediments to efficiency are generally seen as arising from adverse selection or moral hazard, these stemming from deeper asymmetries of information.

[59] It is to be interpreted as a "residual" left by the other "perpetrating coalitions."

That incentive compatibility constraints may rule out first-best efficiency is, of course, well understood. So it is no surprise that notions of second-best efficiency, such as the concept of incentive efficiency of Holmström and Myerson (1984), are well studied.[60] First-best efficiency can sometimes be restored by cleverly designing mechanisms, or rules of the game, that align individual incentives with the social goal of efficiency.[61] In our complete information framework, of course, these complications do not arise; it is trivial to design an efficient mechanism. On the other hand, by granting agents full freedom to form coalitions of their choice we implicitly rule out certain kinds of mechanisms. In effect, every coalition is permitted to adopt an efficient mechanism of its own.[62]

As we shall seek to explain in this section, there are many situations in which the very possibility that groups can form serves as an impediment to efficiency. It isn't a question of incomplete information, though of course there must be some limits to contracting. But what are these limits, and how exactly do these manifest themselves?

5.5.1 Two sources of inefficiency

There are two sources of inefficiency that we seek to make explicit in this section. The first can cause inefficiency even when there are no externalities across coalitions. The second is fundamental to situations with intercoalitional externalities, those typically captured by partition functions. The heart of the first inefficiency is that the "correct" coalition is often not formed, because the active set of players responsible for forming the group seeks to maximize its *own* payoff (or more accurately, to find a maximal payoff vector for itself), but in doing so it will generally need to enlist partners *who have to be suitably compensated.* The nature and amount of that compensation depends crucially on the protocol that governs the process of coalition formation. At the heart of the second inefficiency is a more classical concept: that of externalities. In a typical coalition formation problem, the two effects are often intertwined, but it is instructive to see them separately. We shall begin this discussion by assuming that agreements are irreversible.

Example 5.13. *Consider a three-player TU-characteristic function with* $v(123) = 1 + \epsilon$, *where* $\epsilon \in (0, 0.5)$, $v(ij) = 1$ *for all* i, j, *and* $v(i) = 0$ *for all* i.

[60] For cooperative theory, though, the problem runs even deeper. Restricting all coalitions to incentive compatible contracts does not necessarily yield core stability. Even in otherwise classical environments without externalities, such as exchange economies, the incentive compatible core may be empty; see for example, Vohra (1999) and Forges et al. (2002). For reviews of this literature we refer the reader to Forges et al. (2002) and Forges and Serrano (2013).

[61] See, for example, Palfrey (2002) for a review of the implementation literature.

[62] Recall the discussion in Section 5.4.1.1. See also Ray and Vohra (2001) for further elaboration on this point in the context of the free-riding problem.

While efficiency requires that the grand coalition be formed, the blocking approach does not yield this an equilibrium outcome. This is not a balanced game and its core is empty; for any division of $v(N)$ among the three players, there is a two-player coalition with an objection. If only internal objections are permitted, the coarsest coalition structure that is an EBA consists of a two-player coalition and a singleton. One way to explain why no player will try to form the grand coalition and capture the additional ϵ is that this is too small relative to the dangers posed by the possibility of a two-player secession. Presumably, efficiency could be restored if ϵ is high enough (at least 0.5), or if some player were given more power. To examine the latter possibility, suppose coalitions form as follows: a person is chosen at random to be the "ringleader," and she chooses any coalition she pleases. Once the coalition forms, a fraction k of its worth must be equally divided among the members. The ringleader gets to keep the rest. At one extreme, $k = 1$ and all worths must be equally divided, a case emphasized by Farrell and Scotchmer (1988) and Bloch (1996). At the other extreme, $k = 0$ and the ringleader is a perfect dictator.

It is easy to see in this example that efficiency obtains if and only if k is *below* a certain threshold k^*, given by

$$k^* = \frac{6\epsilon}{1 + 4\epsilon} < 1.$$

The intuition is straightforward. When k is small, the ringleader picks up almost the entire worth and will therefore seek to maximize coalitional worth. The division of payoffs may be distasteful to an inequality-averse social planner, but the outcome is efficient in the Pareto sense. On the other hand, when k exceeds the threshold k^*, the amount that the ringleader has to share with her chosen compatriots becomes an obstacle to efficiency. It is easy to see that in the equal division limit with $k = 1$, the ringleader will seek to maximize average worth, which results in a two-player coalition being formed.

To see how the presence of externalities creates a distinct source of inefficiency, we consider a partition function version of the ringleader example.

Example 5.14. *There are six players in a symmetric TU partition function game. The only (numerical) coalition structures that matter (all others result in 0 to each player) are* $\pi(a) = (3, 3), \pi(b) = (3, 2, 1)$, *and* $\pi(c) = (2, 2, 1, 1)$, *corresponding to three kinds of states, described as follows.*

$$\begin{aligned}
a &: \pi(a) = (3,3), && v(3,\pi(a)) = 4 \\
b &: \pi(b) = (3,2,1), && v(3,\pi(b)) = 2, \quad v(2,\pi(b)) = 5, \quad v(1,\pi(b)) = 0.5 \\
c &: \pi(c) = (2,2,1,1), && v(2,\pi(c)) = 3, \quad v(1,\pi(c)) = 0.5
\end{aligned}$$

Suppose the player with the lowest index in any coalition is the ringleader who can capture the entire worth of the coalition ($k = 0$). Although $(3, 3)$ is the only efficient coalition structure, it is not an "equilibrium." If a three-person coalition were to form, the next ringleader will form a two-player coalition rather than three. But this results in the first ringleader receiving only 2. The equilibrium strategy for the first ringleader will therefore be to form a two-player coalition, which leads to state c, an inefficient outcome.

In summary, we have shown two things so far. First, the ability to internalize the marginal gains from coalition formation is crucial to the formation of the "right" coalitions. Second, when there are externalities, that ability may not be enough to guarantee efficiency. We proceed now to examine a variety of models of coalition formation and their associated implications for efficiency.

5.5.2 Irreversible agreements and efficiency

We begin by studying irreversible agreements, and we draw on both the bargaining and blocking approaches as needed. We address both sources of inefficiency. For the first, it suffices to study characteristic functions.

We have already seen that the stationary equilibria of any reasonable bargaining game will impose some restrictions on how far the payoff to the proposer can diverge from the average worth of the coalition she seeks to form.[63] If this divergence is small enough (effectively giving very little extra power to the ringleader of Example 5.13), the first coalition to form in equilibrium will be the one which maximizes average worth among all coalitions. (In Example 5.13, a two-player coalition will form.) Indeed, as we saw in Section 5.4.3.3, for a wide variety of protocols, the unique equilibrium coalition structure in symmetric games is given by an application of the algorithm that recursively maximizes average worth. In Example 5.13, we can say more. For *any* general protocol in which μ, the probability of the first rejector being the next proposer, is greater than ϵ, there is a unique equilibrium which results in the formation of a two-player coalition[64]; see Ray (2007), p. 143 for details.

As the next proposition shows, inefficiency is even more pervasive than this simple example would suggest. The formation of the grand coalition in equilibrium implies that no other coalition has higher average worth, a condition that may not hold even in a balanced (but non-symmetric) game.

Proposition 5.13. *Suppose (N, V) is a characteristic function game and the first rejector of a proposal is chosen to be the next proposer with probability $\mu \in (0, 1]$. If, for all discount*

[63] Implicitly, the blocking approach also imposes similar restrictions.

[64] The division of the surplus between the proposer and her partner will, however, depend on the precise form of the protocol. In the rejector-proproses protocol, the proposer receives a little more than 1/2, whereas in the random proposer protocol, she receives a little more than 2/3, exactly as in Example 5.11. The difference, of course, is that in the present example the equilibrium outcome is inefficient.

factors sufficiently close to one, there is an equilibrium in which grand coalition forms immediately, regardless of the identity of the first proposer, then $\frac{v(N)}{|N|} \geq \frac{v(S)}{|S|}$ *for all* $S \subseteq N$.

Proof. Suppose in equilibrium each proposer makes an acceptable proposal to the grand coalition. This must mean that the proposer cannot do better by making an acceptable proposal to a subcoalition of N. Letting m_i denote the minimum amount that i will accept from a proposer, provided all remaining responders plan to accept, this implies

$$v(N) - \sum_{j \in N, j \neq i} m_j \geq v(S) - \sum_{j \in S, j \neq i} m_j, \quad \text{for all } S \subseteq N. \qquad [5.15]$$

Since the grand coalition forms immediately in equilibrium, regardless of the proposer's identity, we can apply the same argument used in Section 5.4.3.1 to show that $m_i = m$ for all i, where

$$m = \frac{\mu v(N)}{(1-\delta) + \delta\mu|N|},$$

and rewrite (5.15) as:

$$v(N) - (|N| - 1)m \geq v(S) - (|S| - 1)m$$

or

$$v(N) - (|N| - |S|)m \geq v(S).$$

Substituting for m, this is equivalent to

$$v(N)\frac{(1-\delta) + \delta\mu|S|}{(1-\delta) + \delta\mu|N|} \geq v(S)$$

or

$$\frac{v(N)}{(1-\delta) + \delta\mu|N|} \geq \frac{v(S)}{(1-\delta) + \delta\mu|S|}.$$

As δ converges to 1, this yields $\frac{v(N)}{|N|} \geq \frac{v(S)}{|S|}$.

Thus, if agreements are irreversible, inefficiency cannot be ruled out even in simple characteristic function games for any reasonable bargaining process that doesn't artificially restrict the coalitions that can form.[65,66] A converse of Proposition 5.13 holds for

[65] Bargaining games that implement the Shapley value deliver efficiency *a fortiori*, but they do not allow a proposal to be made to any coalition of the proposer's choosing. For example, Hart and Mas-Colell (1996) require proposals to be made to the full set of available players, and Gul (1989) considers a process of pairwise meetings in which one player buys out the other's resources and continues to bargain with the remaining players. In this respect Gul's model is similar in spirit to models of renegotiation such as Okada (2000) and Seidmann and Winter (1998) in which coalitions form gradually.

[66] Xue and Zhang (2011) suggest another modification of the bargaining model to establish the existence of an efficient equilibrium for a partition function game with irreversible agreements. In their model, the choice of an individual proposer through a protocol is replaced by a bidding mechanism as in Perez-Castrillo and Wettstein (2002). *All players* bid simultaneously on each of the feasible moves from a given

the rejector-proposer protocol (Chatterjee et al. 1993) as well as the random proposer protocol (Okada 1996, 2011). Since we are focusing on inefficiency, we refrain from trying to establish the converse for the more general protocol, but settling this remains an interesting open question.

We now turn to the question of externalities across coalitions. As we observed in Section 5.4.3.3, both the Cournot oligopoly and the public goods model typically yield inefficient equilibria. To explain the nature of the problem, we return to the simple three-player public goods model described in Examples 5.2 and 5.12. The highest social surplus requires full cooperation, and the complete breakdown of cooperation, with each player acting as a singleton, is the worst outcome (in terms of aggregate surplus). However, it is profitable for a single player to break away from the grand coalition *if* the other two stay together. Let us mimic this example with some simple numbers:

Example 5.15. (Another variation on public goods) *Define the per-capita payoffs of a three-player partition function as follows:*

$$\begin{aligned} x_N : \pi_N &= \{123\}, & a(x_N) &= 12 \\ x_1 : \pi_1 &= \{1, 23\}, & a(x_1) &= (16, 7) \\ x_2 : \pi_2 &= \{2, 13\}, & a(x_2) &= (16, 7) \\ x_3 : \pi_3 &= \{12, 3\}, & a(x_3) &= (7, 16) \\ x_0 : \pi_0 &= \{1, 2, 3\}, & a(x_0) &= (6, 6, 6) \end{aligned}$$

The effectivity relations for a three-player hedonic game are as follows. For internal blocking, where $i \neq j \neq k$:

$$\pi_N \rightarrow_i \pi_i, \quad \pi_N \rightarrow_{jk} \pi_i, \quad \pi_i \rightarrow_j \pi_0.$$

When external blocking is permitted we also have:

$$\pi_0 \rightarrow_N \pi_N, \quad \pi_0 \rightarrow_{jk} \pi_i.$$

With internal blocking, it is easy to see that each π_i is an EBA since none of the two players in the larger coalition would gain by precipitating π_0. In fact, these are the coarsest coalition structures corresponding to EBA. It is easiest to see this (though not at all necessary, following Ray and Vohra, 1997) by presuming that there is no transferable utility and that all aggregate payoffs are equally divided. Then, it is clear that the grand coalition is not an EBA because player i gains by moving to π_i and free riding on the other two players. Of course, this conclusion depends critically on the restriction to

state, with the "winning move" being one that attracts the maximum aggregate bid. In particular, player i has an influence on the move, even if the move involves a change in the coalition structure that leaves i's coalition unchanged. As they show, there is at least one stationary equilibrium in which this turns out to be sufficient to internalize the gains from forming the "correct" coalition structure.

internal blocking. Indeed, with internal blocking, the absorbing states of an EPCF yield precisely the same set of outcomes (Proposition 5.6). Which player is able to gain the free-riding advantage starting from the grand coalition depends on the protocol: the first player given an opportunity to move will decide to stand alone. It is natural now to ask how this may change if we depart from the assumption of internal blocking.

Consider the notion of farsighted blocking as embodied in the notion of EEBA introduced by Diamantoudi and Xue (2007) for a hedonic partition function. Recall that this is the set of farsighted stable sets in the present context. Interpret Example 5.15 as a hedonic partition function in which payoffs correspond to the average worth of each coalition. It is not difficult to see that π_N is an EEBA. Since it is a singleton set, it obviously satisfies internal stability. All other coalition structures have farsighted objections culminating in π^N: $\pi_0 \rightarrow_N \pi_N$ and $\pi_i \rightarrow_j \pi_0 \rightarrow_N \pi_N$. Each π_i also constitutes an EEBA. Clearly, $\pi_N \rightarrow_i \pi_i$ and $\pi_0 \rightarrow_N \pi_N \rightarrow_i \pi_i$. Moreover, for $j \neq i$, $\pi_j \rightarrow_i \pi_0 \rightarrow_N \pi_N \rightarrow_i \pi_i$. Note that in the last step, the (optimistic) presumption is that player i will have the opportunity to move from π_N to π_i. In our dynamic model, this presumption will be justified only if the protocol selects player i to be the first potential mover from π_N.

The observation in Example 5.15, that efficiency can be supported through EEBA even when it's not possible to do so through EBA, can be extended to a class of hedonic partition function games in which π_N dominates π_0 and every other coalition structure π contains a player in a non-singleton coalition for whom π_N dominates π.

Proposition 5.14. *(Diamantoudi and Xue, 2007). Consider a hedonic partition function and suppose π_N is a Pareto efficient coalition and Pareto dominates π_0. Then π_N is an EEBA if for all $\pi \neq \pi_N$ and $\pi \neq \pi_0$ there is a coalition $S \in \pi$ such that $|S| \geq 2$ and $u_i(\pi_N) > u_i(\pi)$ for some $i \in S$.*

This is an important positive result because its assumptions are satisfied in symmetric games with positive externalities, e.g., pure public goods economies and the Cournot oligopoly. Thus, efficiency can be restored in such games if we remove the restriction to internal blocking and adopt EEBA as the equilibrium concept. As Diamantoudi and Xue (2007) show, if the assumptions of Proposition 5.14 are not satisfied, *all* EEBAs may be inefficient.

The intuition for Proposition 5.14 should be clear from our discussion of Example 5.15. If π_N Pareto dominates π_0 a coalition seeking to reach π_N only needs to engineer a chain of moves that lead to π_0 as the penultimate step in a far sighted objection terminating in π_N. This argument relies of course on an optimistic view of the world which, as we saw in Example 5.8, can be problematic. Moreover, matters can be more complicated if we properly take account of payoffs on the path to equilibrium. This will be become clear from our continued discussion of Example 5.15 in the next section.

5.5.3 Reversible agreements and efficiency

Agreements may be reversible either because they are temporary or because they can, in principle, be renegotiated. We consider each of these cases in turn.

5.5.3.1 *Temporary agreements*

Suppose that agreements are only valid for one period of time (assuming them to last for some other fixed period of time would make no difference to the analysis). We shall, i.e. allowing for several interim moves, illustrate the efficiency issue by re-examining Example 5.15 through our dynamic model. In doing so, we assume that the protocol has the form described in Section 5.3.5.2. In particular, at each state, a coalition is given at most one chance to make a move. For simplicity, we also assume that when the state is π_N each of the singletons who have not yet been given a chance to move are chosen with equal probability. Also assume that at state π_0 the first coalition chosen to make a potential move is N.

For δ sufficiently high, there does exist an EPCF with π_N as the unique absorbing state. This EPCF has the grand coalition moving from π_0 to π_N and a player $j \neq i$ moving from π_i to π_0. The grand coalition represents an absorbing state because the only coalition that could possibly gain by moving away from this is state is a singleton, say i, hoping to obtain 16 by moving to π_i. However, this is immediately followed by a payoff of 6 and then a return to 12 forever. For $\delta > 2/3$, this is not profitable since $12 + \delta 12 > 16 + \delta 6$. In fact, it can be shown that π_N is the only possible absorbing state for any EPCF in this example. In particular, π_i cannot be an absorbing state. Suppose it is. This means of course that π_0 is not an absorbing state because jk would then have a profitable move to the absorbing state π_i. Nor can π_N be an absorbing state because then jk has an efficient and profitable move to π_N via π_0. (Note that from π_0 player i cannot move to π_i). Now consider a move by player j from π_i to π_0 followed by a move by N to π_N. From here, the process will either get absorbed into π_i or move to π_j or π_k. The worst possible outcome for player j is that it moves immediately to π_i. Thus, the worst that this yields to player j for these steps is $6 + 12\delta + \delta^2 7$ compared to $7 + \delta 7 + \delta^2 7$. For $\delta > 1/5$, this is a profitable move, contradicting the hypothesis that π_i is an absorbing state. Thus, for $\delta > 2/3$, the unique absorbing state is π_N. In particular, unlike EEBA, it is no longer possible to sustain π_i as an absorbing state of an EPCF. This difference results from the explicit accounting of temporary gains which can make it worthwhile for players to move even if there is an eventual return to the status quo. As we shall next show, this reasoning can make it impossible to achieve efficiency through the dynamic process even under the assumptions of Proposition 5.14

Now change Example 5.15 so that in the intermediate coalition structure π, the singleton, i, receives 19 rather than 16 (all other payoffs remain unchanged). Proposition

5.14 continues to apply, and π_N therefore remains an EEBA. However, we will now show that π_N cannot be an absorbing state of an EPCF. In fact, there are no absorbing states, and it is impossible to achieve Pareto efficiency in any EPCF.[67] We claim that π_N cannot be an absorbing state. Consider the case in which player 1 is selected to make a move at π_N and she moves to π_1. The worst that can happen for player 1 from that state is a move to π_0 followed by a move to π_N. This yields player 1 the payoff $19 + \delta 6$ compared to $12 + \delta 12$ in the next two periods (with no change in future periods). Since $\delta < 1$, this is a profitable move, contradicting the hypothesis that π_N is an absorbing state. In fact, the unique EPCF is one in which there is no absorbing state and the transitions between states are the following:

$$\pi_0 \rightarrow_N \pi_N,$$
$$\pi_i \rightarrow_j \pi_0 \text{ whenever } j \text{ is selected to move,}$$
$$\pi_N \rightarrow_i \pi_i \text{ whenever } i \text{ is selected to move.}$$

The equilibrium process immediately moves from π_0 to π_N; from π_N to each of the intermediate coalition structures with probability 1/3 and then immediately to π_0. Thus, the process visits π_0 and π_N one-third of the time, and the remainder is equally divided between the other three intermediate states. The expected payoff to each player, ignore discounting, is therefore $(1/3)12 + (1/3)(11) + (1/3)6 = 9.67$ which is clearly inefficient.

5.5.3.2 Renegotiation

We've seen that proposer incentives are often distorted by the potential loss of control that accompanies a rejected proposal. In short, a proposer must always give some fraction of the surplus away, and a wedge is driven between socially and privately optimal actions. On the other hand, intuition suggests that if outcomes can be renegotiated, then the already agreed-upon arrangements safeguard *existing* payoffs against any loss of control from making a fresh proposal. This suggests two things. First, if there is surplus left on the table, then that surplus should eventually be seized and divided in some way among all parties. Second—and somewhat in contrast to the first point—the seizure of that surplus won't generally happen at the very first round. The safeguards may have to be put in place in earlier rounds, necessitating step-by-step progress toward efficiency (and hence a sacrifice of full dynamic efficiency). These ideas lie at the heart of contributions by Seidmann and Winter (1998), Okada (2000), Gomes and Jehiel (2005), Gomes (2005), and Hyndman and Ray (2007).

To make the point about gradualism completely explicit, recall Example 8. It describes a three-player symmetric characteristic function with $v(123) = 1 + \epsilon$, where

[67] Konishi and Ray (2003) provide other examples of abstract games with similar features in their model.

$\epsilon \in (0, 0.5)$, $v(ij) = 1$ for all i, j, and $v(i) = 0$ for all i. Apply to this the rejector-proposes protocol. Then, if only irreversible arrangements are possible, and the discount factor is close enough to unity, an (inefficient) two-person coalition forms and a valuable third player is omitted, for reasons already discussed. With renegotiation, matters are different. A two-person coalition will still form at first, but the *eventual* outcome is efficient, as the third person can be taken in without any fear of dilution to the already committed players in the two-person coalition. The formation of an "intermediate coalition" essentially protects the parties to that agreement. The agents included in the intermediate coalition can block any attempt by the excluded agent to undercut them, because they are already signatory to a binding agreement that can only be abolished with the consent of both players. That reduces the power of the excluded agent to extract surplus, and the grand coalition can finally form.

One feature of this example is that ultimately all renegotiation ceases and the economy "settles down." More importantly, are those limit payoffs efficient? Consider the following example of a four-person characteristic function, with $v(S) = 3$, if $S = N$, $v(12) = v(34) = 1$, and $v(S) = 0$ otherwise. Suppose that the protocol is "rejector proposes." Provided that the discount factor is close enough to unity, there is an equilibrium in which the coalition structure $\{12, 34\}$ forms but no further progress is made: all proposals to the grand coalition are rebuffed and the rejector demands the entire surplus net of existing payoffs to the other three agents).

Notice, however, that the failure to achieve efficiency is based on rather knife-edge considerations. An efficiency-enhancing proposal may be rejected, true, but events postrejection cannot hurt our existing players by too much, *because ongoing agreements are binding*. If a proposer does not mind being rejected as long as subsequent play benefits others *and* does not hurt her, such history-dependent inefficiencies can be broken provided that the status quo agreements are binding. That motivates the following concept: say that an individual is *benign* if she prefers an outcome in which some other individuals are better off, provided that she (and every individual) is just as well off. The benignness "refinement" is of a lexicographic nature. Our individual first and foremost maximizes her own payoff, and benignness only kicks in when comparisons are made over outcomes in which her payoff is unaffected. There is no danger to the payoff of the individual concerned.

Benignness has found support in a number of different experimental settings (including bargaining); see, e.g., Andreoni and Miller (2002), Charness and Grosskopf (2001), and Charness and Rabin (2002), among others. Indeed, these studies suggest something stronger: people are sometimes willing to *sacrifice* their own payoff in order to achieve a socially efficient outcome. Given its lexicographic insistence on maximizing one's own payoff, benignness certainly doesn't go that far.

Under benignness, asymptotic efficiency must be attained under every possible equilibrium:

Proposition 5.15. *Assume* [B.1] *and* [B.2]. *In characteristic function games all equilibria are absorbing. Moreover, if the set of states is finite, every pure strategy benign equilibrium is asymptotically efficient: every limit payoff is static efficient.*

This proposition, taken from Hyndman and Ray (2007), is to be contrasted with the folk theorem-like results obtained in Herrero (1985) and Chatterjee et al. (1993) for the case of irreversible agreements. With repeated negotiation, no amount of history dependence in strategies can hold players away from an (ultimately) efficient outcome. In this sense, Proposition 5.15 represents a substantial extension of Okada (2000) and Seidmann and Winter (1998), who showed that renegotiation achieves efficiency in superadditive characteristic functions when equilibria are restricted to be Markovian. (Neither superadditivity nor the Markovian assumption is needed here.) The issue of how far these results can be pushed by restricting attention to Markovian equilibria is addressed by Proposition 5.16 later.

While we omit a formal proof, it is easy to see the intuition behind this result. Suppose, contrary to our assertion, that convergence occurs to an inefficient limit. Then a proposer will have the incentive to propose a payoff vector that Pareto dominates this payoff. This follows from two observations. First, because agreements are binding, the proposer cannot be hurt by making such a proposal. She can always continue to enjoy her going payoff.[68] Second, the proposer is benign. She certainly gains from the proposal *if it is accepted*, and there is no reason to invoke benignness. But the point is that she prefers to make the proposal *even if it is rejected*. For rejection must entail that all the rejectors are better off by *not* accepting the proposal, while the assumption that agreements are binding ensures that no one is strictly hurt (see previous paragraph). A benign proposer would therefore prefer the resulting outcome to the presumed equilibrium play, which is continued stagnation at the inefficient payoff vector.

It must be reiterated, though, that the ability to write binding agreements cannot guarantee *full* efficiency in the dynamic sense. As we've seen, absorption will generally require time—i.e., the formation of intermediate coalition structures—before a final outcome is finally settled upon. These intermediate outcomes may well be inefficient. So the path taken as a whole cannot be dynamically efficient.[69]

Most importantly, with externalities across coalitions, matters can be very different. To be sure, renegotiation might restore efficiency, as is easily illustrated by allowing for renegotiation in Example 5.15. Once the two partners j and k in π_i willingly break up to precipitate π_0, and all three join forces to go to π_N, no further changes can occur

[68] To make this argument work, we must "already" be at the limit payoff, otherwise the proposer may do some (small, but positive) damage to her own prospects by the very act of making the proposal. This is why we assume a finite set of states, though the finiteness can be dropped; see Hyndman and Ray (2007).

[69] There is another reason for the failure of dynamic efficiency: such efficiency may necessitate ongoing cycles across different states. See Hyndman and Ray (2007) for an extended discussion.

because player i can no longer trigger π_i without the consent of the other two. Thus, renegotiation yields efficiency in this example. Unfortunately, this is not generally the case. The ubiquitous absorption results reported for characteristic functions break down when externalities are present. Equilibrium payoffs may cycle, and even if they don't, inefficient outcomes may arise and persist. Finally—and in sharp contrast to characteristic functions—such outcomes are not driven by the self-fulfilling contortions of history dependence. They occur even for Markovian equilibria.

It is tempting to think of inefficiencies as entirely "natural" equilibrium outcomes when externalities exist. Such an observation is generally true, of course, for games in which there are no binding agreements. When agreements can be costlessly written, however, no such presumption can and should be entertained. These are models of *binding* agreements, a world in which the so-called "Coase theorem" is relevant. For instance, all two-player games invariably yield efficiency, quite irrespective of whether there are externalities across the two players. This is not to say that the "usual intuition" plays no role here. It must, because the process of negotiation is itself modeled as a noncooperative game. But that is a very different object from the "stage game" over which agreements are sought to be written.

Consider a three-player example.

Example 5.16. (The Failed Partnership) *There are three agents, any two of whom can become "partners." The outsider to the partnership gets a "low" payoff: zero, say. A three-player partnership is assumed not to be feasible (or has very low payoffs)*

$$\begin{aligned}
x_N &: \pi_N = \{123\}, & u(x_N) &= (0,0,0) \\
x_1 &: \pi_1 = \{1,23\}, & u(x_1) &= (0,10,10) \\
x_2 &: \pi_2 = \{2,13\}, & u(x_2) &= (5,0,5) \\
x_3 &: \pi_3 = \{12,3\}, & u(x_3) &= (5,5,0) \\
x_0 &: \pi_0 = \{1,2,3\}, & u(x_0) &= (6,6,6)
\end{aligned}$$

and use any effectivity correspondence that satisfies (B.1) and (B.2), yet allows all free players to get together without any further consultation or clearance.

The crucial feature of this example is that player 1 is a bad partner, or—for the purposes of better interpretation—a *failed partner*. Partnerships between him and any other individual are dominated—both for the partners themselves and for the outsider—by all three standing alone. In contrast, the partnership between agents 2 and 3 is rewarding (for those agents).

In this example, and provided δ lies sufficiently close to 1, the outcomes x_2 and x_3—which are inefficient—must be absorbing states *in every equilibrium.*

A formal proof of this observation isn't needed; the discussion to follow will suffice. Why might x_2 and x_3 be absorbing? The reason is very simple. Despite the fact that x_2

(or x_3) is Pareto dominated by x_0, player 1 won't accept a transition to x_0. If she did, players 2 and 3 would initiate a further transition to x_1. Player 1 *might* accept such a transition if she is very myopic and prefers the short-term payoff offered by x_0, but if she is patient enough she will see ahead to the infinite phase of "outsidership" that will surely follow the short-term gain. That is why it is impossible—in the game as described—to negotiate one's way out of x_2 or x_3. This inefficiency persists in *all* equilibria, history dependent or otherwise.

This example raises four important points:

1. *The Nature of Agreements.* Notice that the players *could* negotiate themselves out of x_2 or x_3 if 2 and 3 could credibly agree never to write an agreement while at x_0. Are such promises reasonable in their credibility? It may be difficult to imagine that from a legal point of view, player 1, who has voluntarily relinquished all other contractual agreements between 2 and 3, could actually hold 2 and 3 to such a meta-agreement. Could one interpret the stand-alone option (x_0) as an *agreement* from which further deviations require universal permission? Or does "stand-alone" mean freedom from all formal agreement, in which case further bilateral deals only need the consent of the two parties involved? It is certainly possible to take the latter view. One might even argue that this is the only compelling view. In that case, efficiency will need to be sacrificed.

But there are situations in which the former view might make sense. No-competition clauses that require an ex-employee not to join a rival firm, at least for some length of time, may be interpreted as an example. Such clauses allow a firm to let a top-level executive or partner go when it is in the interest of both of them (and therefore efficient) to do so, but at the same time prevent—perhaps temporarily—another move in which the executive gets absorbed by a third party. While that latter move is, in itself, also efficient, it might prevent the former move from even being entertained.

2. *The Efficiency Criterion: From Every State, or Some?* Observe that the inefficient states x_2 or x_3 wouldn't be *reached* starting from any other state. This is why the interpretation, the "failed partnership," is useful. The example makes sense in a situation in which players have been locked in with 1 on a past deal, on expectations which have failed since. The game "begins" with the failed partnership, so to speak. Nevertheless, that raises the question of whether there invariably exists *some* initial condition for which efficiency obtains. (That answer is trivially yes in the current example.)

For all three-player games, and provided we are willing to make some minimal assumptions, the answer is in the affirmative. That is, for all δ close enough to 1, there exists an initial state and a stationary Markov equilibrium with efficient absorbing payoff limit from that state (see Proposition 4 in Hyndman and Ray, 2007). But it is also true that a general result in this direction is elusive: there is a *four*-player partition function such that for δ sufficiently close to 1, *every* stationary Markov equilibrium is inefficient starting from *any* initial state (see the four-player example in Hyndman and Ray, 2007).

3. *More on Transfers*. Recall that upfront transfers are not permitted in the failed partnership. Were they allowed in unlimited measure, players 2 and 3 could reimburse player 1 for the present discounted value of his losses in relinquishing his partner. Depending on the discount factor, the amounts involved may be considerable and might strain the presumption of deep pockets or perfect credit markets needed to carry such transfers out. But they would break the deadlock.

How does the ability to make transfers feed into efficiency? It is important to distinguish between two kinds of transfers. Coalitional or partnership worth could be freely transferred between the players within a coalition. Additionally, players might be able to make large upfront payments in order to induce certain coalitions to form. In all cases, of course, the definition of efficiency should match the transfer environment.[70]

Within-coalition transferability often does nothing to remove inefficiency. For instance, nothing changes in the failed partnership of Example 5.16. On the other hand, upfront transfers *across* coalitions have an immediate and salubrious effect in that example. Efficiency is restored from every initial state. The reason is simple. If player 1 is offered any (discount normalized) amount in excess of 5, he will "release" player 2. In view of the large payoffs that players 2 and 3 enjoy at state x_1, they will be only too pleased to make such a payment. The final outcome, then, from any initial condition is the state x_1, and we have asymptotic efficiency.

But transfers can be a double-edged sword. The discussion that follows is based on Gomes and Jehiel (2005).

Example 5.17. (Ubiquitous Bad Partnerships) *Consider the following three-player game:*

$$
\begin{aligned}
x_N &: \pi_N = \{123\}, & u(x_N) &= (0,0,0) \\
x_1 &: \pi_1 = \{1,23\}, & u(x_1) &= (0,a,a) \\
x_2 &: \pi_2 = \{2,13\}, & u(x_1) &= (a,0,a) \\
x_3 &: \pi_3 = \{12,3\}, & u(x_3) &= (a,a,0) \\
x_0 &: \pi_0 = \{1,2,3\}, & u(x_0) &= (b,b,b)
\end{aligned}
$$

Assume that $b > a > 0$.

As in Example 5.16, use any effectivity correspondence that satisfies (B.1) and (B.2), yet allows all free players to get together.

In this symmetric example, there is a unique efficient state by any criterion. It is state x_0. It Pareto dominates every other state. In particular, every two-player partnership is an unambiguous disaster. It is obvious that in any reasonable description of this game that precludes upfront transfers, there is a unique absorbing state, which is the state x_0.

[70] For instance, if transfers are not permitted, it would be inappropriate to demand efficiency in the sense of aggregate surplus maximization.

But the introduction of upfront transfers changes this rather dramatically. Under the uniform proposer protocol, *every stationary Markov equilibrium is inefficient starting from any initial state: the state x_0 can never be absorbing.*

This remarkable observation highlights very cleanly the negative effects of upfront transfers. The usefulness of a transfer is that it frees agents from inefficient outcomes, as in the case of the failed partnership in Example 5.16. But its potential danger lies in the possibility that individuals may *deliberately* generate inefficient outcomes to seek such transfers. This notion of transfers as ransom turns out to be particularly vivid in this example. Whenever it is the turn of an agent to move at state x_0, she creates a bad partnership with another agent and then waits for a transfer to unlock the partnership. That situation recurs again and again. Can we still be sure that transfers will actually be paid, and that all of these actions properly lock together as an equilibrium? The answer is yes; for details, see Gomes and Jehiel (2005) and the detailed exposition in Ray (2007).

In short, the deviating players do suffer a loss in current payoff when they move away from the efficient state. But the prospect of inflicting a *still* greater loss on the outsider raises the possibility that the outsider will pay to have the state moved back—albeit temporarily—to the efficient point. Thus, the presumption that unlimited transfers act to restore or maintain efficiency is wrong.

Notice how the example stands on the presumption (just as in the failed partnership) that two players can always form a partnership starting from the situation in which all three players stand alone. If this contractual right can be eliminated in the act of making an upfront transfer, then efficiency can be restored: once state x_0 is regained, there can be no further deviations from it. This line of discussion is exactly the same as in the failed partnership and there is nothing further to add here.

More generally, the efficient state in this example has the property that a subset of agents can move away from that state, leaving other agents worse off in terms of current payoffs. Whenever this is possible, there is scope for collecting a ransom, and the potential for a breakdown in efficiency. Gomes and Jehiel (2005) develop this idea further.

4. *What About Superadditive Games?* Both the failed partnership, as well as the four-player game mentioned earlier, involve situations that are not superadditive. If, for instance, the grand coalition can realize the Pareto improvement, then player 1 can control any subsequent shenanigans by 2 and 3 (he will need to be part of any coalition that is effective for further change), and he will therefore permit the improvement, thereby restoring efficiency.

There are subtle issues that need to be addressed here. First, in games with externalities, superadditivity is generally not to be expected. For instance, in Example 5.1 (the Cournot oligopoly), it is easy to see that if there are just three firms, firms 1 and 2 do worse together than apart, provided that firm 3 stands separately in both cases. At the same time, this argument does not apply to the grand coalition of all firms. Indeed, it is not hard to show that every partition function derived from a game in strategic form must satisfy *grand coalition superadditivity* (GCS):

[GCS] For every state $x = (u, \pi)$, there is $x' = (u', \{N\})$ such that $u' \geq u$.

Is GCS a reasonable assumption? It may or may not be. One possible interpretation of GCS is that it is a "physical" phenomenon; e.g., larger groups organize transactions more efficiently or share the fixed costs of an enterprise such as a business or public good provision. But such superadditivities are often the exception rather than the rule. After all, the entire doctrine of healthy competition is based on the notion that physical superadditivity, after a point, is not to be had. In general, too many cooks do spoil the broth: competition among groups can lead to efficiency gains not possible when there is a single, and perhaps larger, group attempting to act cooperatively. To be sure, in all of the cases, the argument must be based on some noncontractible factor, such as the creativity or productivity created by the competitive urge, or ideological differences, or the possible presence of stand-alone players who are outside the definition of our set of players but nevertheless have an effect on their payoffs (such as the stand-alone third player in the Cournot example).

But there is a different notion of GCS, summarized in the notion of the *superadditive cover*. After all, the grand coalition can write a contract which exactly replicates the payoffs obtainable in some other coalition structure. For instance, companies do spin off certain divisions, and organizations do set up competing R&D groups. In principle, the grand coalition can agree not to cooperate, if need be, and yet write agreements that bind across all players. For instance, in the failed partnership of Example 5.17, the ability to insert no-compete clauses at will effectively converts the game into its superadditive cover. However, to the extent that such clauses cannot be enforced for an infinite duration, the model without grand-coalition superadditivity can be viewed as a simplification of this and other real-world situations.

In fact, Gomes and Jehiel (2005) demonstrate that in many situations, efficiency can be restored if there exists an efficient state that is negative-externality free (ENF) in the sense that no coalition, or collection of coalitions, can move away from it and hurt a player who is not party to such a change.

[ENF] There exists an efficient state, $x = (u, \pi)$, such that for all $i \in N$ and $x \rightarrow_{S_1} x_1 \rightarrow_{S_2} x_2 \cdots \rightarrow_{S_k} y$, such that $i \notin S_k$ for all k, $u_i(y) \geq u_i(x)$.

Note that ENF is not satisfied in Example 5.15. It is easy to see that ENF is weaker than GCS since a move from the grand coalition requires unanimous consent. And it holds trivially in characteristic function games since there are no externalities.

Proposition 5.16. *Consider a TU partition function. Every Markovian equilibrium is asymptotically efficient under either one of the following set of conditions:*

1. *[(i)] Upfront transfers are permitted, the set of states is finite (modulo transfers) and there exists an ENF state; Gomes and Jehiel (2005).*
2. *[(ii)] There are no upfront transfers and GCS is satisfied; Hyndman and Ray (2007).*

Gomes (2005) establishes a related efficiency result under GCS, without assuming a finite number of states but by assuming that coalitions once formed cannot become smaller. He also provides an example to show that GCS cannot be weakened to ENF without allowing upfront transfers across coalitions.

5.6. COALITION FORMATION: THE ROAD AHEAD

This chapter confronts a plethora of solution concepts in the literature on coalition formation with one umbrella model. That model adopts an explicit real-time approach to the formation of coalitions. That is, a group of players band together, perhaps disband later and re-form in shifting alliances, all the while receiving payoffs at each date (or instant) according to the coalition structure prevailing at the time.

What constrains or disciplines such an approach? Perhaps the most important feature is a specification of the nature of commitment. For instance, one might posit that a coalition, once formed, and an agreement among its members, once reached, is frozen for ever after, in which case the dynamic game in question acquires a new focus, not so much on the *implementation* of an agreement (which, after all, is guaranteed by the rules of such a game), but on the *negotiation* of an agreement. Or, turning to the opposite extreme, one might posit that all agreements have force for just a single period. In that case we are back, not to a standard game (after all, a binding agreement can be written if only for a single period), but to one in which the implementation of existing agreements also acquire focus. Or one might be interested in a situation in which agreements are binding unless renegotiated with the permission of all concerned. The choice of which environment is relevant will be dictated by the particular application that the researcher has in mind, but one thing is common to all of them: that to some degree, *binding* agreements can be written.

When this "equilibrium process of coalition formation" is taken back to particular models, such as characteristic function games, partition function games or models of noncooperative bargaining, it will make predictions. It is of interest to compare such predictions with existing solution concepts designed for such situations, but not explicitly based on dynamic considerations. This is what we have done throughout the chapter. It has been our hope that a common umbrella model serves to illuminate both the strengths and the limitations of existing solution concepts.

Yet our approach is, at present, just that: an approach and not much more. We do not claim to have found a general theory, in the sense of having overarching theorems of existence and behavior that transcend the specific scenarios that have been traditionally dealt with. At the same time, we believe that progress along these lines is possible, perhaps in ways that parallel the modern theory of dynamic noncooperative games. That defines a large and fundamental open area of research, but it is just one half of the program.

The second half of the program would consist in applying those general principles to particular situations, and using those principles either to validate existing solution concepts, or to modify them in useful ways. In a sense this second part of the program is like the reverse of the Nash program. Recall that program asks for noncooperative "implementation" of well-known solution concepts in cooperative game theory, such as the Nash bargaining solution. The Nash program is, we feel, a good statement of implementation theory or mechanism design, but our goal here is different. It is to have a broad *positive* theory of coalition formation in real time, and then let the cards fall where they may in specific situations, without necessarily tailoring the theory to "fit" specific solution concepts. Rather, the theory would be used to sift through the existing concepts, perhaps discarding some and replacing them by the new "projections" of the theory on to the specific situation at hand.

Two aspects of this approach appear to be applicable to a wide range of situations. First, there is the question of intracoalitional division of surplus, once a coalition has formed. Our approach is general enough to encompass several existing theories, but viewed from a broader perspective is quite primitive. We have emphasized two routes to intracoalitional division: the blocking approach, in which any allocation is a potential candidate as long as it is not "objected to" by other coalitions, and the bargaining approach, in which a chosen proposer, cognizant of the "outside options" of her partners, makes a proposal to them that they prefer not to refuse.

But there are other approaches. For instance, a social norm, or a voting process, might dictate how a coalition allocates its surplus. The simplest example of a social norm is unrestricted equal division of the surplus of any coalition that happens to form, as in Farrell and Scotchmer (1988). The simplest example of a voting process is majority rule within a coalition, as in the recent work of Barberà et al. (2013). The impact of such norms or processes on equilibrium coalition structures is nontrivial, interesting, and largely unexplored. Such restrictions placed on division increase the vulnerability of a coalition to threats from other coalitions, but at the same time reduce the flexibility of other coalitions to issue threats. The resulting sets of equilibrium coalition structures are therefore unlikely to be nested.

There are also conceptual issues to be addressed. Perhaps the most fundamental of these is whether a coalitional norm must respect the participation constraints of individuals when it is applied. To return to the example of egalitarian division, it is entirely possible that unrestricted equal division will be blocked by "high-ability" individuals refusing to form coalitions with "low-ability" individuals for fear of expropriation. The resulting outcome will display a high degree of within-coalition equality, but possibly significant amounts of cross-coalitional inequality. If, on the other hand, the egalitarian norm takes into account the participation constraints of individuals, one obtains the *egalitarian solution* of Dutta and Ray (1989), in which the "most egalitarian" division subject to participation constraints is chosen. To be sure, the same considerations apply to all processes and norms. For instance, within-coalition voting in Barberà et al. (2013) is carried out in a manner that parallels unrestricted equal division: the possibility that the coalition could lose its viability is not considered. If it were, that would lead to a new equilibrium set of coalition structures. We should emphasize that we do not take a stand on these fundamental issues, and in particular, we do not claim that one option is inevitably superior. But we ask for a general theory that is cognizant of these alternatives.

The second aspect of the real-time approach is that it naturally contends with the question of farsightedness. The static counterparts of farsightedness (that are dominant in the literature) do a mixed job in capturing its various nuances. They work well in some special situations, e.g., the core in characteristic function games under the assumption of internal blocking, or singleton farsighted stable sets in characteristic function games. But in general, as we discussed in Section 5.3, farsighted stability brings up complications that are not always fully resolved by solution concepts such as the farsighted stable set, the largest consistent set, or equilibrium binding agreements. In contrast, the explicitly dynamic notion of an EPCF has several advantages. It can handle cycles and incorporate explicitly the idea that coalitions make not just profitable moves but *maximally* profitable moves. At the same time, it is an abstract concept and many of the details, such as the protocol and the effectivity correspondence have to be carefully specified in each particular case. These are unwelcome, but they appear to be unavoidable as well.

In going beyond the special cases in which existing static concepts work well, it would be very useful to see if the general framework can provide a "reduced form" solution concept that can be recast in terms of the traditional blocking approach. Indeed, as soon as we take this approach, we see that a key ingredient is missing in the standard models. This is the condition of maximality, which comes so naturally to the real-time formulation: the idea that coalitions will not just make profitable moves but will choose moves that are maximal, or Pareto optimal, for it. In contrast, in the static theory, although a farsighted objection corresponds to a profitable move for the coalitions involved, it does not necessarily ensure that coalitional moves are *maximally* profitable. The reason is that the counterfactuals against which maximality is to be defined are generally ill-defined in the static model. This neglect of counterfactuals has a history. As we noted in Section 5.3, blocking has traditionally been used as a way of ruling *out* a particular state, without comparing the (possibly) multiple objections that a coalition may possess. In short, it is traditionally a negation: a block is enough to rule out an allocation, why worry about the "maximal block"? There are two reasons why maximality may be important. First, when blocking is used in a theory of coalition formation, it is no longer to be used as a negative concept: the block itself determines which coalition forms, and what they do. Second, if blocking is defined in a farsighted sense, imposing maximality becomes important even in testing the stability of a given state. It is possible that a state has a farsighted objection, but every such objection involves some coalition (in the chain of objections) making a profitable move that is *not* maximal. Consequently, neglecting the maximality requirement may lead to farsighted stability concepts that are suspect (see the examples in Section 5.3.6.2). We hope that future work based on the EPCF, or on the principles of profitability and maximality embodied in it, will provide a more satisfactory blocking concept, at least in the absence of cycles.

We end by reminding the reader that there are several aspects of coalition formation that have been beyond the purview of this chapter, e.g., network formation, coalition formation under incomplete information, and axiomatic approaches to coalition formation. For a discussion of future directions for research in some of these areas see Chapter 14 in Ray (2007).

ACKNOWLEDGMENTS

We thank Francis Bloch, Mert Kimya, Hideo Konishi, Licun Xue and two referees for very helpful comments. Ray's research was supported by the National Science Foundation under grant no. SES-0962124.

REFERENCES

Abreu, D., Pearce, D., Stacchetti, E., 2012. One-sided uncertainty and delay in reputational bargaining, mimeo, Princeton University.

Acemoglu, D., Egorov, G., Sonin, K., 2008. Coalition formation in non-democracies, Rev. Econ. Stud., 75, 987–1009.

Aghion, P., Antràs, P., Helpman, E., 2007. Negotiating free trade, J. Inter. Econ., 73, 1–30.

Andreoni, J., Miller, J., 2002. Giving according to GARP: an experimental test of the consistency of preferences for altruism, Econometrica, 40, 737–753.

Aumann, R., 1961. The core of a cooperative game without sidepayments, Trans. Am. Math. Soc. 98, 539–552.

Aumann, R., Dreze, J. H., 1974. Cooperative games with coalition structures, Inter. J. Game Theory, 3, 217–237.

Aumann, R., Maschler, M., 1964. The bargaining set for cooperative games, In: Dresher, M., Shapley, L., Tucker, A., (eds.), Advances in Game Theory, Annals of Mathematical Studies No. 52; Princeton University Press, Princeton, NJ.

Aumann, R., Myerson, R., 1988. Endogenous formation of links between players and of coalitions, an application of the shapley value, In: Roth, A., (ed.), The Shapley Value: Essays in Honor of Lloyd Shapley, pp. 175–191. Cambridge University Press, Cambridge.

Aumann, R., Peleg, B., 1960. von Neumann-Morgenstern solutions to cooperative games without side payments, Bull. Am. Math. Soc., 66, 173–179.

Banerjee, S., Konishi, H., Sönmez, T., 2001. Core in a simple coalition formation game, Soc. Choice Welf. 18, 135–153.

Barberà, S., Gerber, A., 2003. On coalition formation: durable coalition structures, Math. Soc. Sci., 45, 185–203.

Barberà, S., Gerber, A., 2007. A note on the impossibility of a satisfactory concept of stability for coalition formation games, Econ. Le., 95, 85–90.

Barberà, S., Beviá, C., Ponsatí, C., 2013. Meritocracy, egalitarianism and the stability of majoritarian organizations, Working Paper.

Baron, D., Ferejohn, J., 1989. Bargaining in legislatures, Am. Polit. Sci. Rev., 83, 1181–1206.

Béal, S., Durieu, J., Solal, P., 2008. Farsighted coalitional stability in TU-games, Math. Soc. Sci. 56, 303–313.

Bernheim, D., Peleg, B., Whinston, M., 1987. Coalition-proof nash equilibria. I. concepts, J. Econ. Theory, 42, 1–12.
Bhattacharya, A., Brosi, V., 2011. An existence result for farsighted stable sets of games in characteristic function form, Inter. J. Game Theory, 40, 393–401.
Bhattacharya, A., Ziad, A., 2012. On credible coalitional deviations by prudent players, Soc. Choice Welf., 39, 537–552.
Binmore, K., 1985. Bargaining and coalitions, In: Roth, A., (ed.), Game-Theoretic Models of Bargaining, Cambridge University Press, Cambridge.
Binmore, K., Rubinstein, A., Wolinsky, A., 1986. The nash bargaining solution in economic modelling, RAND J. Econ., 17, 176–188.
Bloch, F., 1996. Sequential formation of coalitions in games with externalities and fixed payoff division, Games Econ. Behav., 14, 90–123.
Bloch, F., 2003. Non-cooperative models of coalition formation in games with spillovers, In: Carraro, C., Siniscaclo, D., (eds.), The Endogenous Formation of Economic Coalitions, pp. 311–352. Edward Elger.
Bloch, F., Gomes, A., 2006. Contracting with externalities and outside options, J. Econ. Theory, 127, 172–201.
Bloch, F., Soubeyran, R., Sánchez-Pagés, 2006. When does universal peace prevail? Secession and group formation in conflict, Econ. Govern. 7, 3–29.
Bloch, F., van den Nouweland, A., 2013. Expectation formation rules and the core of partition function games, mimeo.
Bogomolnaia, A., Jackson, M. O., 2002. The stability of hedonic coalition structures, Games Econ. Behav., 38, 201–230.
Carraro, C., Siniscalco, D., 1993. Strategies for the international protection of the environment, J. Public Econ., 52, 309–328.
Chander, P., 2007. The gamma-core and coalition formation, Inter. J. Game Theory, 35, 539–556.
Chander, P., Tulkens, H., 1997. The core of an economy with multilateral environmental externalities, Inter. J. Game Theory, 26, 379–401.
Charness, G., Grosskopf, B., 2001. Relative payoffs and happiness: an experimental study, J. Econ. Behav. Org., 45, 301–328.
Charness, G., Rabin, M., 2002. Understanding social preferences with simple tests, Quar. J. Econ., 117, 817–869.
Chatterjee, K., Dutta, B., Ray, D., Sengupta, K., 1993. A noncooperative theory of coalitional bargaining, Rev. Econ. Stud., 60, 463–477.
Chwe, M., 1994. Farsighted coalitional stability, J. Econ. Theory, 63, 299–325.
Coase, R., 1960. The problem of social cost, J. Law Econ., 3, 1–44.
Compte, O., Jehiel, P., 2010. The coalitional nash bargaining solution, Econometrica, 78, 1593–1623.
de Clippel, G., Serrano, R., 2008a. Bargaining, coalitions and externalities: a comment on maskin, Brown University, Working Paper 2008-16.
de Clippel, G., Serrano, R., 2008b. Marginal contributions and externalities in the value, Econometrica, 76, 1413–1436.
Diamantoudi, E., Xue, L., 2003. Farsighted stability in hedonic games, Soc. Choice Welf., 21, 39–61.
Diamantoudi, E., Xue, L., 2005. Lucas' counter example revisited, Discussion paper, McGill University.
Diamantoudi, E., Xue, L., 2007. Coalitions, agreements and efficiency, J. Econ. Theory, 136, 105–125.
Dutta, B., Ray, D., 1989. A concept of egalitarianism under participation constraints, Econometrica, 57, 615–635.
Dutta, B., Ray, D., Sengupta, K., Vohra, R., 1989. A consistent bargaining set, J. Econ. Theory, 49, 93–112.
Dutta, B., Suzumura, K., 1993. On the sustainability of R&D through private incentives, Indian Statistical Institute, Discussion Paper No. 93-13.
Dutta, B., Ghosal, S., Ray, D., 2005. Farsighted network formation, J. Econ. Theory, 122, 143–164.
Eraslan, H., 2002. Uniqueness of stationary equilibrium payoffs in the Baron-Ferejohn model, J. Econ. Theory, 103, 11–30.
Eraslan, H., McLennan, A., 2013. Uniqueness of stationary equilibrium payoffs in coalitional bargaining, J. Econ. Theory, 148, 2195–2222.
Esteban, J., Ray, D., 1999. Conflict and distribution, J. Econ. Theory, 87, 379–415.

Esteban, J., Sákovics, J., 2004. Olson vs. coase: coalitional worth in conflict, Theory Decis. 55, 339–357.
Farrell, J., Scotchmer, S., 1988. Partnerships, Quart. J. Econ., 103, 279–297.
Feldman, A., 1974. Recontracting stability, Econometrica, 42, 35–44.
Forges, F., Mertens, J.-F., Vohra, R., 2002. The ex ante incentive compatible core in the absence of wealth effects, Econometrica, 70, 1865–1892.
Forges, F., Minelli, E., Vohra, R., 2002. Incentives and the core of an exchange economy: a survey, J. Math. Econ., 38, 1–41.
Forges, F., Serrano, R., 2013. Cooperative games with incomplete information: some open problems, Inter. Game Theory Rev., 15, 1340009-1-17.
Gillies, D., 1953. Locations of solutions In: Kuhn, H.W., (ed.) Report of an Informal Conference on the Theory of N-Person Games, Princeton University, mimeo.
Gomes, A., 2005. Multilateral contracting with externalities, Econometrica, 73, 1329–1350.
Gomes, A., Jehiel, P., 2005. Dynamic processes of social and economic interactions: on the persistence of inefficiencies, J. Polit. Econ., 113, 626–667.
Green, J., 1974. The stability of Edgeworth's recontracting process, Econometrica, 42, 21–34.
Greenberg, J., 1990. The Theory of Social Situations, Cambridge University Press, Cambridge, MA.
Greenberg, J., 1994. Coalition structures, In: *Vol. 2*, Aumann, R. J., Hart, S., (eds.) Handbook of Game Theory, pp. 1305–1337.
Gul, F., 1989. Bargaining foundations of shapley value, Econometrica, 75, 81–95.
Gul, F., 1999. Efficiency and immediate agreement: a reply to Hart and Levy, Econometrica, 67, 913–917.
Haeringer, G., 2004. Equilibrium binding agreements: a comment, J. Econ. Theory, 117, 140–143.
Harsanyi, J., 1974. An equilibrium-point interpretation of stable sets and a proposed alternative definition, Manag. Sci., 20, 1472–1495.
Hart, S., Kurz, M., 1983. Endogenous formation of coalitions, Econometrica, 51, 1047–1064.
Hart, S., Levy, Z., 1999. Efficiency does not imply immediate agreement, Econometric, 67, 909–912.
Hart, S., Mas-Colell, A., 1996. Bargaining and value, Econometrica, 64, 357–380.
Herings, P., Mauleon, A., Vannetelbosch, V., 2004. Rationalizability for social environments, Games Econ. Behav., 49, 135–156.
Herings, P., Mauleon, A., Vannetelbosch, V., 2009. Farsightedly stable networks, Games Econ. Behav., 67(2), 526–541.
Herrero, M., 1985. *n*-player bargaining and involuntary unemployment, Ph.D. Dissertation, London School of Economics.
Holmström, B., Myerson, R., 1983. Efficient and durable decision rules with incomplete information, Econometrica, 51, 1799–1819.
Hyndman, K., Ray, D., 2007. Coalition formation with binding agreements, Rev. Econ. Stud., 74, 1125–1147.
Ichiishi, T., 1981. A social coalitional equilibrium existence lemma, Econometrica, 49, 369–37.
Jackson, M., 2010. Social and Economic Networks, Princeton University Press, Princeton, NJ.
Jackson, M., Wolinsky, A., 1996. A strategic model of social and economic networks, J. Econ. Theory, 71, 44–74.
Jordan, J., 2006. Pillage and property, J. Econ. Theory, 131, 26–44.
Kóczy, L., Lauwers, L., 2004. The coalition structure core is accessible, Games Econ. Behav., 48(1), 86–93.
Konishi, H., Ray, D., 2003. Coalition formation as a dynamic process, J. Econ. Theory, 110, 1–41.
Krishna, P., 1998. Regionalism vs multilateralism: a political economy approach, Quart. J. Econ., 113, 227–250.
Krishna, V., Serrano, R., 1995. Perfect equilibria of a model of n-person noncooperative bargaining, Inter. J. Game Theory, 24, 259–272.
Krugman, P., 1993. Regionalism versus multilateralism: analytical notes, In: de Melo, J., Panagariya, A., (eds.), New Dimensions in Regional Integration, Cambridge University Press, Cambridge, UK.
Lucas, W., 1968. A game with no solution, Bull. Am. Math. Soc., 74, 237–239.
Lucas, W. F., 1992. von Neumann-Morgenstern stable sets, In: Aumann, R. J., Hart, S., (eds.), Handbook of Game Theory, *Vol. 1*, North-Holland, pp. 543–590.
Mariotti, M., 1997. A model of agreements in strategic form games, J. Econ. Theory, 74, 196–217.

Mariotti, M., Xue, L., 2003. Farsightedness in coalition formation, In: Carraro, C., Siniscalco, D., (eds.), *The Endogenous Formation of Economic Coalitions*, Edward Elger, pp. 128–155.
Maskin, E., 2003. Bargaining, coalitions and externalities, Presidential address of the Econometric Society.
Mauleon, A., Vannetelbosch, V., 2004. Farsightedness and cautiousness in coalition formation games with positive spillovers, Theory Decis., 56, 291–324.
Mauleon, A., Vannetelbosch, V., Vergote, W., 2011. von Neumann-Morgenstern farsightedly stable sets in two-sided matching, Theor. Econ., 6, 499–521.
Moldovanu, B., Winter, E., 1995. Order independent equilibria, Games Econ. Behav., 9, 21–34.
Okada, A., 1996. A noncooperative coalitional bargaining game with random proposers, Games Econ. Behav., 16, 97–108.
Okada, A., 2000. The efficiency principle in non-cooperative coalitional bargaining, Jap. Econ. Rev., 51, 34–50.
Okada, A., 2011. Coalitional bargaining games with random proposers: theory and application, Games Econ. Behav., 73, 227–235.
Owen, G., 1977. Values of games with a priori unions, In: Henn, R., Moschlin, O., (eds.), Essays in Mathematical Economics and Game Theory, Springer Verlag, New York, NY.
Page, F., Wooders, M., Kamat, S., 2005. Networks and farsighted stability, J. Econ. Theory, 120, 257–269.
Palfrey, T., 2002. Implementation theory, In: Aumann, R. J., Hart, S., (eds.), Handbook of Game Theory, *Vol. 3*, North-Holland, pp. 2271–2326.
Perez-Castrillo, D., Wettstein, D., 2002. Choosing wisely: a multibidding approach, Am. Econ. Rev., 92, pp. 1577–1587.
Perry, M., Reny, P., 1994. A non-cooperative view of coalition formation and the core, Econometrica, 62, 795–817.
Ray, D., 1989. Credible coalitions and the core, Inter. J. Game Theory, 18, 185–187.
Ray, D., 1998. Development Economics, Princeton University Press, Princeton, NJ.
Ray, D., 2007. A Game-Theoretic Perspective On Coalition Formation, Oxford University Press, New York, NY.
Ray, D., Vohra, R., 1997. Equilibrium binding agreements, J. Econ. Theory, 73, 30–78.
Ray, D., Vohra, R., 1999. A theory of endogenous coalition structures, Games Econ. Behav., 26, 286–336.
Ray, D., Vohra, R., 2001. Coalitional power and public goods, J. Polit. Econ., 109, 1355–1384.
Ray, D., Vohra, R., 2013. The farsighted stable set, Brown University Working Paper, 2013-11.
Rosenthal, R., 1972. Cooperative games in effectiveness form, J. Econ. Theory, 5, 88–101.
Rubinstein, A., 1982. Perfect equilibrium in a bargaining model, Econometrica, 50, 97–109.
Salant, S., Switzer, S., Reynolds, R., 1983. Losses from horizontal merger: the effects of an exogenous change in industry structure on Cournot–Nash equilibrium, Quart. J. Econ., 93, 185–199.
Seidmann, D., 2009. Preferential trading arrangements as strategic positioning, J. Inter. Econ., 79, 143–159.
Seidmann, D., Winter, E., 1998. Gradual coalition formation, Rev. Econ. Stud., 65, 793–815.
Selten, R., 1981. A non-cooperative model of characteristic function bargaining, In: Böhm, V., Nachtkamp, H., (eds.), Essays in Game Theory and Mathematical Economics in Honor of Oscar Morgenstern, Bibliographisches Institut, Mannheim.
Sengupta, A., Sengupta, K., 1996. A property of the core, Games Econ. Behav., 12, 266–73.
Shapley, L., 1953. Open questions In: Kuhn, H.W., (ed.), *Report of an Informal Conference on the Theory of N-Person Games*, Princeton University, mimeo.
Shapley, L., 1973. Let's Block "Block", Econometrica, 41, 1201–1202.
Shenoy, P., 1979. On coalition formation: a game theoretic approach, Inter. J. Game Theory, 8, 133–164.
Shubik, M., 1983. Game Theory in the Social Sciences, MIT Press, Cambridge, MA.
Ståhl, I., 1977. An *n*-person bargaining game in the extensive form, In: Henn, R., Moeschlin, O., (eds.), Mathematical economics and game theory, Springer-Verlag, Berlin.
Stole, L., Zwiebel, J., 1996. Intra-firm bargaining under non-binding contracts, Rev. Econ. Stud., 63, 375–410.
Sutton, J., 1986. Non-cooperative bargaining theory: an introduction, Rev. Econ. Stud., 53, 709–724.
Thrall, R., Lucas, W., 1963. *n*-person games in partition function form, Naval Res. Log. Quart., 10, 281–298.

Vohra, R., 1999. Incomplete information, incentive compatibility, and the core, J. Econ. Theory, 86, 123–147.
von Neumann, J., Morgenstern, O., 1944. *Theory of Games and Economic Behavior*, Princeton University Press, Princeton, NJ.
Winter, E., 1993. Mechanism robustness in multilateral bargaining, Theory Deci., 40, 131–47.
Winter, E., 2002. The shapley value, In: Aumann, R. J., Hart, S., (eds.), Handbook of Game Theory, *Vol. 3*, North-Holland, pp. 2025–2054.
Xue, L., 1998. Coalitional stability under perfect foresight, Econ. Theory, 11, 603–627.
Xue, L., Zhang, L., 2011. Bidding and sequential coalition formation with externalities, Inter. J. Game Theory, 41, 49–73.
Yi, S-S., 1996. Endogenous formation of customs unions under imperfect competition: open regionalism is good, J. Inter. Econ., 41, 153–177.
Zhao, J., 1992. The hybrid solutions of an n-person game, Games Econ. Behav., 4, 145–160.

CHAPTER 6

Stochastic Evolutionary Game Dynamics

Chris Wallace[*], **H. Peyton Young**[†]
[*]Department of Economics, University of Leicester, Leicester, UK
[†]Department of Economics, University of Oxford, Oxford, UK

Contents

Abstract

Traditional game theory studies strategic interactions in which the agents make rational decisions. Evolutionary game theory differs in two key respects: the focus is on large populations of individuals who interact at random rather than on small numbers of players; and individuals are assumed to employ simple adaptive rules rather than to engage in perfectly rational behavior. In such a setting, an equilibrium is a rest point of the population-level dynamical process rather than a form of consistency

Handbook of game theory
http://dx.doi.org/10.1016/B978-0-12-420056-2.00006-9

between beliefs and strategies. This chapter shows how the theory of stochastic dynamical systems can be used to characterize the equilibria that are most likely to be selected when the evolutionary process is subject to small persistent perturbations. Such equilibria are said to be *stochastically stable*. The implications of stochastic stability are discussed in a variety of settings, including 2 × 2 games, bargaining games, public-goods games, potential games, and network games. Stochastic stability often selects equilibria that are familiar from traditional game theory: in 2 × 2 games one obtains the risk-dominant equilibrium, in bargaining games the Nash bargaining solution, and in potential games the potential-maximizing equilibrium. However, the justification for these solution concepts differs between the two approaches. In traditional game theory, equilibria are justified in terms of rationality, common knowledge of the game, and common knowledge of rationality. Evolutionary game theory dispenses with all three of these assumptions; nevertheless, some of the main solution concepts survive in a stochastic evolutionary setting.

Keywords: Stochastic evolutionary dynamics, Stochastically stable equilibrium, Equilibrium selection, Population games, Risk-dominance, Nash bargaining solution, Potential

JEL Codes: C72, C73

6.1. EVOLUTIONARY DYNAMICS AND EQUILIBRIUM SELECTION

Game theory is often described as the study of interactive decision-making by rational agents.[1] However, there are numerous applications of game theory where the agents are not fully rational, yet many of the conclusions remain valid. A case in point is biological competition between species, a topic pioneered by Maynard Smith and Price (1973). In this setting the "agents" are representatives of different species that interact and receive payoffs based on their strategic behavior, whose strategies are hard-wired rather than consciously chosen. The situation is a game because a given strategy's success depends upon the strategies of others. The dynamics are not driven by rational decision-making but by mutation and selection: successful strategies increase in frequency compared to relatively unsuccessful ones. An equilibrium is simply a rest point of the selection dynamics. Under a variety of plausible assumptions about the dynamics, it turns out that these rest points are closely related (though not necessarily identical) to the usual notion of Nash equilibrium in normal form games (Nachbar, 1990; Ritzberger and Weibull, 1995; Sandholm, 2010; Weibull, 1995, Chapter 5).

Indeed, this evolutionary approach to equilibrium was anticipated by Nash himself in a key passage of his doctoral dissertation.

> *"We shall now take up the 'mass-action' interpretation of equilibrium points...*
> *[I]t is unnecessary to assume that the participants have full knowledge of the total structure of the game, or the ability and inclination to go through any complex reasoning processes. But the*

[1] For example, Aumann (1985) puts it thus: "game [...] theory [is] concerned with the interactive behaviour of *Homo rationalis*—rational man."

participants are supposed to accumulate empirical information on the relative advantages of the various pure strategies at their disposal.

To be more detailed, we assume that there is a population (in the sense of statistics) of participants for each position of the game. Let us also assume that the 'average playing' of the game involves n participants selected at random from the n populations, and that there is a stable average frequency with which each pure strategy is employed by the 'average member' of the appropriate population...Thus the assumptions we made in this 'mass-action' interpretation lead to the conclusion that the mixed strategies representing the average behaviour in each of the populations form an equilibrium point.... Actually, of course, we can only expect some sort of approximate equilibrium, since the information, its utilization, and the stability of the average frequencies will be imperfect."

(Nash, 1950b, pp. 21–23)

The key point is that equilibrium does not *require* the assumption of individual rationality; it can arise as the *average behavior* of a population of players who are less than rational and operate with "imperfect" information.

This way of understanding equilibrium is in some respects less problematic than the treatment of equilibrium as the outcome of a purely rational, deductive process. One difficulty with the latter is that it does not provide a satisfactory answer to the question of which equilibrium will be played in games with multiple equilibria. This is true in even the simplest situations, such as 2×2 coordination games. A second difficulty is that pure rationality does not provide a coherent account of what happens when the system is out of equilibrium, that is, when the players' expectations and strategies are not fully consistent. The biological approach avoids this difficulty by first specifying how adjustment occurs at the individual level and then studying the resulting aggregate dynamics. This framework also lends itself to the incorporation of stochastic effects that may arise from a variety of factors, including variability in payoffs, environmental shocks, spontaneous mutations in strategies, and other probabilistic phenomena. The inclusion of persistent stochastic perturbations leads to a dynamical theory that helps resolve the question of which equilibria will be selected, because it turns out that persistent random perturbations can actually make the long-run behavior of the process more predictable.

6.1.1 Evolutionarily stable strategies

In an article in *Nature* in 1973, the biologists John Maynard Smith and George R. Price introduced the notion of an *evolutionarily stable strategy* (or ESS).[2] This concept went on to have a great impact in the field of biology; but the importance of their contribution was also quickly recognized by game theorists working in economics and elsewhere.

Imagine a large population of agents playing a game. Roughly put, an ESS is a strategy σ such that, if most members of the population adopt it, a small number of "mutant"

[2] Maynard Smith and Price (1973). For an excellent exposition of the concept, and details of some of the applications in biology, see the beautiful short book by Maynard Smith (1982).

players choosing another strategy σ' would receive a lower payoff than the vast majority playing σ.

Rather more formally, consider a two-player symmetric strategic-form game $\mathcal{G}$. Let S denote a finite set of pure-strategies for each player (with typical member s), and form the set of mixed strategies over S, written Σ. Let $u(\sigma, \sigma')$ denote the payoff a player receives from playing $\sigma \in \Sigma$ against an opponent playing σ'. Then an ESS is a strategy σ such that

$$u(\sigma, \varepsilon\sigma' + (1-\varepsilon)\sigma) > u(\sigma', \varepsilon\sigma' + (1-\varepsilon)\sigma), \qquad [6.1]$$

for all $\sigma' \neq \sigma$, and for $\varepsilon > 0$ sufficiently small. The idea is this: suppose that there is a continuum population of individuals each playing σ. Now suppose a small proportion ε of these individuals "mutate" and play a different strategy σ'. Evolutionary pressure acts against these mutants if the existing population receives a higher payoff in the post-mutation world than the mutants themselves do, and vice versa. If members of the population are uniformly and randomly matched to play $\mathcal{G}$ then it is as if the opponent's mixed strategy in the post-mutation world is $\varepsilon\sigma' + (1-\varepsilon)\sigma \in \Sigma$. Thus, a strategy might be expected to survive the mutation if [6.1] holds. If it survives all possible such mutations (given a small enough proportion of mutants) it is an ESS.

Definition 6.1a. *$\sigma \in \Sigma$ is an ESS if for all $\sigma' \neq \sigma$ there exists some $\bar{\varepsilon}(\sigma') \in (0,1)$ such that* [6.1] *holds for all $\varepsilon < \bar{\varepsilon}(\sigma')$.*[3]

An alternative definition is available that draws out the connection between an ESS and a Nash equilibrium strategy. Note that an ESS must be optimal against itself. If this were not the case there necessarily would be a better response to σ than σ itself and, by continuity of u, a better response to an ε mix of this strategy with σ than σ itself (for small enough ε). Therefore an ESS must be a Nash equilibrium strategy.

But an ESS requires more than the Nash property. In particular, consider an alternative best reply σ' to a candidate ESS σ. If σ is not also a better reply to σ' than σ' is to itself then σ' must earn at least what σ earns against *any* mixture of the two. But then this is true for an ε mix and hence σ cannot be an ESS. This suggests the following definition.

Definition 6.1b. *σ is an ESS if and only if* (i) *it is a Nash equilibrium strategy, $u(\sigma,\sigma) \geq u(\sigma',\sigma)$ for all σ'; and* (ii) *if $u(\sigma,\sigma) = u(\sigma',\sigma)$ then $u(\sigma,\sigma') > u(\sigma',\sigma')$ for all $\sigma' \neq \sigma$.*

Definitions 6.1a and 6.1b are equivalent. The latter makes it very clear that the set of ESS is a subset of the set of Nash equilibrium strategies. Note moreover that if a Nash equilibrium is strict, then its strategy must be evolutionarily stable.

[3] This definition was first presented by Taylor and Jonker (1978). The original definition (Maynard Smith and Price, 1973; Maynard Smith, 1974) is given below.

One important consequence of the strengthening of the Nash requirement is that there are games for which no ESS exists. Consider, for example, a non-zero sum version of the "Rock–Paper–Scissors" game in which pure strategy 3 beats strategy 2, which in turn beats strategy 1, which in turn beats strategy 3. Suppose payoffs are 4 for a winning strategy, 1 for a losing strategy, and 3 otherwise. The unique (symmetric) Nash equilibrium strategy is $\sigma = \left(\frac{1}{3}, \frac{1}{3}, \frac{1}{3}\right)$, but this is not an ESS. For instance, a mutant playing Rock (strategy 1) will get a payoff of $\frac{8}{3}$ against σ, which is equal to the payoff received by an individual playing σ against σ. As a consequence, the second condition of Definition 6.1b must be checked. However, playing σ against Rock generates a payoff of $\frac{8}{3} < 3$, which is less than what the mutant would receive from playing against itself: there is no ESS.[4]

There is much more that could be said about this and other static evolutionary concepts, but the focus here is on stochastic dynamics. Weibull (1995) and Sandholm (2010) provide excellent textbook treatments of the deterministic dynamics approach to evolutionary games; see also Sandholm's chapter in this volume.

6.1.2 Stochastically stable sets

An ESS suffers from two important limitations. First, it is guaranteed only that such strategies are stable against single-strategy mutations; the possibility that multiple mutations may arise simultaneously is not taken into account (and, indeed, an ESS is not necessarily immune to these kinds of mutations). The second limitation is that ESS treats mutations as if they were isolated events, and the system has time to return to its previous state before the next mutation occurs. In reality however there is no reason to think this is the case: populations are *continually* being subjected to small perturbations that arise from mutation and other chance events. A series of such perturbations in close succession can kick the process out of the immediate locus of an ESS; how soon it returns depends on the global structure of the dynamics, not just on its behavior in the neighborhood of a given ESS. These considerations lead to a selection concept known as *stochastic stability* that was first introduced by Foster and Young (1990). The remainder of this section follows the formulation in that paper. In the next section, we shall discuss the discrete-time variants introduced by Kandori et al. (1993) and Young (1993a).

As a starting point, consider the *replicator dynamics* of Taylor and Jonker (1978). These dynamics are not stochastic—but are meant to capture the underlying stochastic nature of evolution. Consider a continuum of individuals playing the game $\mathcal{G}$ over (continuous)

[4] The argument also works for the standard zero-sum version of the game: here, when playing against itself, the mutant playing Rock receives a payoff exactly equal to that an individual receives when playing σ against Rock—the second condition of Definition 6.1b fails again. The "bad" Rock–Paper–Scissors game analyzed in the above reappears in the example of Figure 6.9, albeit to illustrate a different point.

time. Let $p_s(t)$ be the proportion of the population playing pure strategy s at time t. Let $p(t) = [p_s(t)]_{s \in S}$ be the vector of proportions playing each of the strategies in S: this is the state of the system at t. The simplex $\Sigma = \left\{p(t) : \sum_{s \in S} p_s(t) = 1\right\}$ describes the state space.

The replicator dynamics capture the idea that a particular strategy will grow in popularity (the proportion of the population playing it will increase) whenever it is more successful than average against the current population state. Since $\mathcal{G}$ is symmetric, its payoffs can be collected in a matrix A where $a_{ss'}$ is the payoff a player would receive when playing s against strategy s'. If a proportion of the population $p_{s'}(t)$ is playing s' at time t then, given any individual is equally likely to meet any other, the payoff from playing s at time t is $\sum_{s' \in S} a_{ss'} p_{s'}(t)$, or the sth element of the vector $Ap(t)$, written $[Ap(t)]_s$. The average payoff in the population at t is then given by $p(t)^T Ap(t)$. The replicator dynamics may be written

$$\dot{p}_s(t)/p_s(t) = [Ap(t)]_s - p(t)^T Ap(t), \qquad [6.2]$$

the proportion playing strategy s increases at a rate equal to the difference between its payoff in the current population and the average payoff received in the current population.

Although these dynamics are deterministic they are meant to capture the underlying stochastic nature of evolution. They do so only as an approximation. One key difficulty is that typically there will be many rest-points of these dynamics. They are history dependent: the starting point in the state space will determine the evolution of the system. Moreover, once p_s is zero, it remains zero forever: the boundaries of the simplex state space are absorbing. Many of these difficulties can be overcome with an explicit treatment of stochastic evolutionary pressures. This led Foster and Young (1990) to consider a model directly incorporating a stochastic element into evolution and to introduce the idea of *stochastically stable sets* (SSS).[5]

Suppose there is a stochastic dynamical system governing the evolution of strategy play and indexed by a level of noise ε (e.g., the probability of mutation). Roughly speaking, a state p is stochastically stable if, in the long run, it is nearly certain that the system lies within a small neighborhood of p as $\varepsilon \to 0$. To be more concrete, consider a model of evolution where the noise is well approximated by the following Wiener process

$$\mathrm{d}p_s(t) = p_s(t)\left\{[Ap(t)]_s - p(t)^T Ap(t)\right\}\mathrm{d}t + \varepsilon[\Gamma(p)\mathrm{d}W(t)]_s. \qquad [6.3]$$

We assume that: (i) $\Gamma(p)$ is continuous in p and strictly positive for $p \neq 0$; (ii) $p^T \Gamma(p) = 0^T$; and (iii) $W(t)$ is a continuous white noise process with zero mean and unit variance-covariance matrix. In order to avoid complications arising from boundary

[5] They use dynamical systems methods which build upon those found in Freidlin and Wentzell (1998).

behavior, we shall suppose that each p_i is bounded away from zero (say owing to a steady inflow of migrants).[6] Thus we shall study the behavior of the process in an interior envelope of the form

$$S_\Delta = \{p \in S : p_i \geq \Delta > 0 \text{ for all } i\}. \quad [6.4]$$

We remark that the noise term can capture a wide variety of stochastic perturbations in addition to mutations. For example, the payoffs in the game may vary between encounters, the number of encounters may vary in a given time period. Aggregated over a large population, these variations will be very nearly normally distributed.

The replicator dynamic in [6.3] is constantly affected by noise indexed by ε; the interior of the state space is appropriate since mutation would keep the process away from the boundary so avoiding absorption when a strategy dies out ($p_s(t) = 0$). The idea is to find which state(s) the process spends most time close to when the noise is driven from the system ($\varepsilon \to 0$). For any given ε, calculate the limiting distribution of $p(t)$ as $t \to \infty$. If every neighborhood of p^* has strictly positive probablility in the limit as $\varepsilon \to 0$, then p^* is said to be stochastically stable. The SSS is simply the collection of such states.

Definition 6.2. *The state p^* is* stochastically stable *if, for all $\delta > 0$,*

$$\limsup_{\varepsilon \to 0} \int_{N_\delta(p^*)} f_\varepsilon(p)\,dp > 0,$$

where $N_\delta(p^*) = \{p : |p - p^*| < \delta\}$, and $f_\varepsilon(p)$ is the limiting density of $p(t)$ as $t \to \infty$, which exists because of our assumptions on $\Gamma(p)$. The SSS is the minimal set of p^* for which this is true.

In words, "a *stochastically stable set* (SSS) is the minimal set of states S such that, in the long run, it is nearly certain that the process lies within every open set containing S as the noise tends slowly to zero" (Foster and Young, 1990, p. 221). As it turns out the SSS is often a single state, say p^*. In this case the process is contained within an arbitrarily small neighborhood of p^* with near certainty when the noise becomes arbitrarily small.

Consider the symmetric 2×2 pure coordination game A:

$$A = \begin{bmatrix} 1 & 0 \\ 0 & 2 \end{bmatrix}$$

This game has two ESS, in which everyone is playing the same strategy (either 1 or 2). It also has a mixed Nash equilibrium $(\frac{2}{3}, \frac{1}{3})$ that is not an ESS. Let us examine

[6] The boundary behavior of the process is discussed in detail in Foster and Young (1990, 1997). See also Fudenberg and Harris (1992).

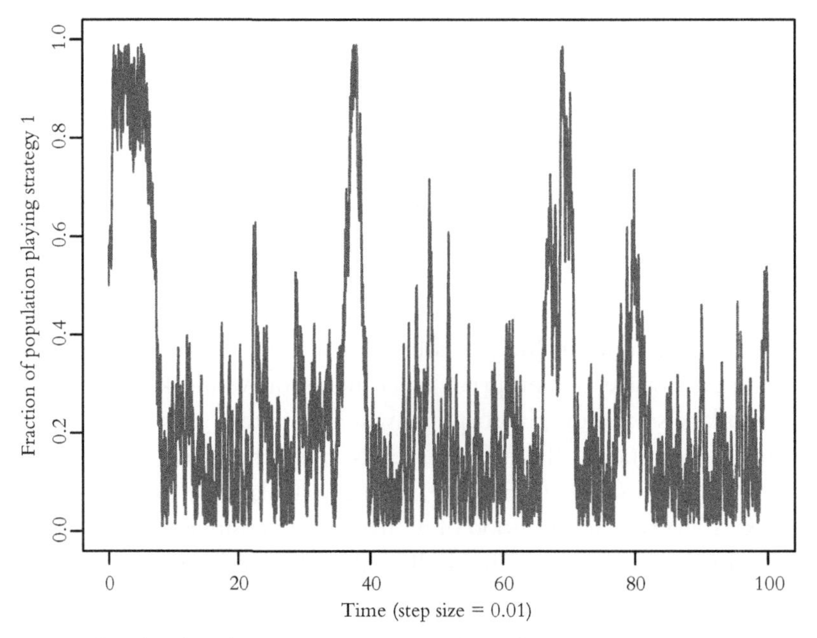

Figure 6.1 A simulated path with $\varepsilon = 0.3$, $\Delta = 0.01$, and initial state $p(0) = 0.5$.

the behavior of the dynamics when a small stochastic term is introduced. Let $p(t)$ be the proportion playing strategy 1 at time t. Assume for simplicity that the stochastic disturbance is uniform in space and time. We then obtain a stochastic differential equation of form

$$\forall p \in S_\Delta, \quad dp(t) = p(t)[p(t) - p^2(t) - 2(1 - p(t))^2]dt + \varepsilon dW(t), \qquad [6.5]$$

where $W(t)$ is $N(0, t)$. Figure 6.1 shows a simulated path with $\varepsilon = 0.3$, the initial condition is $p(0) = 0.5$, and the process is constrained to lie in the interval $[0.01, 0.99]$. Notice that, on average, the process spends more time near the all-2 state than it does near the all-1 states, but it does not converge to the all-2 state. In this simulation the noise level, ε is actually quite large. This illustrates the general point that the noise level does not need to be taken to zero for a significant selection bias toward the stochastically stable state (in this case all-2) to be revealed. When $\varepsilon = 0.05$, for example, the process is very close to all-2 with very high probability in the long run.

In general, there is no guarantee that the stochastically stable equilibrium of a 2×2 coordination game is Pareto dominant. Indeed, under fairly general conditions the dynamics favor the risk-dominant equilibrium, as we shall see in the next section. Furthermore, in larger games there is no guarantee that the dynamics select any equilibrium (ESS or otherwise): even when the game possesses an ESS, the SSS may consist of a cycle. In the next few sections, we illustrate these points using a discrete-time finite version of the selection process. This approach was introduced in two papers that appeared back-to-back in an issue of *Econometrica* in 1993 (Kandori et al., 1993; Young, 1993a). It has the advantage of avoiding the boundary issues that arise in the continuous-time approach and it is also easier to work with analytically. We shall outline the basic framework in the next two sections. Following that we shall show how to apply the analytical machinery to a variety of concrete examples including bargaining, public goods games, and games played on networks. The concluding section addresses the question of how long it takes to converge to the stochastically stable states from arbitrary initial conditions.

6.2. EQUILIBRIUM SELECTION IN 2 × 2 GAMES

At the beginning of this chapter, the problem of multiple Nash equilibria in elementary strategic-form games was used (at least partially) to motivate the study of stochastic evolutionary systems in game theory. When there is more than one strict Nash equilibrium, an equilibrium selection problem arises that cannot be solved with many of the traditional refinement tools of game theory. This section illustrates how the concept of stochastic stability can provide a basis for selection between equilibria in a simple symmetric 2×2 game.

6.2.1 A simple model

The basic idea is to consider a finite population of agents, each of whom must play a game against a randomly chosen opponent drawn from the same population. They do so in (discrete) time. Each period some of the players may revise their strategy choice. Since revision takes place with some noise (it is a stochastic process), after sufficient time any configuration of strategy choices may be reached by the process from any other.

To illustrate these ideas, consider a simple symmetric coordination game in which two players have two strategies each. Suppose the players must choose between **X** and **Y** and that payoffs are as given in the game matrix in Figure 6.2.

Suppose that $a > c$ and $d > b$, so that the game has two pure Nash equilibria, $(\mathbf{X}, \mathbf{X})$ and $(\mathbf{Y}, \mathbf{Y})$. It will also have a (symmetric) mixed equilibrium, and elementary calculations show that this equilibrium requires the players to place probability p on pure action **X**, where

$$p = \frac{(d-b)}{(a-c)+(d-b)} \in (0,1).$$

	X	**Y**
X	a a	c b
Y	b c	d d

Figure 6.2 A 2×2 coordination game.

Now suppose there is a finite population of n agents. At each period $t \in \{0, \ldots, \infty\}$ one of the agents is selected to update their strategy. Suppose that agent i is chosen from the population with probability $\frac{1}{n}$ for all i.[7] In a given period t, an updating agent plays a best reply to the mixed strategy implied by the configuration of other agents' choices in period $t-1$ with high probability. With some low probability, however, the agent plays the strategy that is not a best reply.[8] The state at time t may be characterized by a single number, $x_t \in \{0, \ldots, n\}$: the number of agents at time t who are playing strategy $\mathbf{X}$. The number of agents playing $\mathbf{Y}$ is then simply $y_t = n - x_t$.

Suppose an agent i who in period $t-1$ was playing $\mathbf{Y}$ is chosen to update in period t. Then x_{t-1} other agents were playing $\mathbf{X}$ and $n - x_{t-1} - 1$ other agents were playing $\mathbf{Y}$. The expected payoff for player i from $\mathbf{X}$ is larger only if

$$\frac{x_{t-1}}{n-1}a + \frac{n - x_{t-1} - 1}{n-1}b > \frac{x_{t-1}}{n-1}c + \frac{n - x_{t-1} - 1}{n-1}d \quad \Leftrightarrow \quad \frac{x_{t-1}}{n-1} > p.$$

In this case, player i is then assumed to play the best reply $\mathbf{X}$ with probability $1 - \varepsilon$, and the non-best reply $\mathbf{Y}$ with probability ε. Similarly, a player j who played $\mathbf{X}$ in period $t-1$ plays $\mathbf{X}$ with probability $1 - \varepsilon$ and $\mathbf{Y}$ with probability ε if $(x_{t-1} - 1)/(n-1) > p$. It is then possible to calculate the transition probabilities between the various states for this well-defined finite Markov chain, and examine the properties of its ergodic distribution.

6.2.2 The unperturbed process

Consider first the process when $\varepsilon = 0$ (an "unperturbed" process). Suppose a single agent is selected to revise in each period t. In this case, if selected to revise, the agent will play a best reply to the configuration of opponents' choices in the previous period with probability one. The particularly simple state space in this example can be illustrated on a line ranging from 0 (everyone plays $\mathbf{Y}$) to n (everyone plays $\mathbf{X}$).

Let x^* be the natural number such that $(n-1)p < x^* < (n-1)p + 1$.[9] Then for any state $x_{t-1} > x^*$, no matter the current choice of the revising agent i, the best reply for i is to play $\mathbf{X}$. Hence either the state moves up one: $x_t = x_{t-1} + 1$ (if player i is switching from $\mathbf{Y}$ to $\mathbf{X}$) or remains where it is: $x_t = x_{t-1}$. On the other hand, for any $x_{t-1} < x^*$, the best reply for the revising agent is $\mathbf{Y}$ and the state either moves down or does not change.

So long as any agent might receive the revision opportunity in every period t, it is easy to see that the process will eventually either reach n or 0 depending on which side

[7] The argument would not change at all so long as each agent is chosen with some strictly positive probability. For notational simplicity the uniform case is assumed here.

[8] This process would make sense if, for example, the agent was to play the game $n-1$ times against each other agent in the population at time t, or against just one of the other agents drawn at random. The low probability "mutation" might then be interpreted as a mistake on the part of the revising agent.

[9] We assume for simplicity that n is chosen so that $(n-1)p$ is not an integer.

Figure 6.3 The state space.

of x^* it starts. The process is *history dependent*. If the process starts at exactly x^*, then the probability it moves up or down is simply the probability the agent chosen to update is currently playing **Y** or **X** respectively. The point is that the unperturbed process does not deliver a definitive selection argument: depending on the starting point, the process might eventually reach n or might eventually reach 0. All agents end up playing the same strategy, and thus a pure Nash equilibrium is played if any two agents in the population meet and play the game; but without knowledge of initial conditions the analyst cannot say which equilibrium it will be.

The states to the right of x^* in Figure 6.3 therefore can be interpreted as the "basin of attraction" for the state n. Once in that basin, the process moves inexorably toward n, and once it reaches n, never leaves (it is absorbed). Likewise, the states to the left of x^* are the basin of attraction for the absorbing state 0. The problem that the analyst faces is that the long-run outcome is determined by the initial conditions, which cannot be known a priori. This difficulty disappears once we introduce stochastic perturbations, which are natural in most applications. Once there is (even a very small) probability of moving from the basin of attraction for 0 to that for n and back again, history dependence may be overcome, the Markov process becomes irreducible, and a unique (ergodic) distribution will characterize long-run play.

Recall the objective is to identify the stochastically stable states of such a process. This is equivalent to asking which state(s) are played almost all of the time as the stochastic perturbations introduced to the system are reduced in size. For such vanishingly small perturbations (in this model, $\varepsilon \to 0$) the process will spend almost all time local to *one* of the equilibrium states 0 or n: this state is stochastically stable and the equilibrium it represents (in the sense that at 0 all players are playing **Y**, and at n all the players are playing **X**) is said to have been "selected."

6.2.3 The perturbed process

Consider now the Markov process described above for small but positive $\varepsilon > 0$. The transition probabilities for an updating agent may be calculated directly[10]:

$$\mathbb{P}[x_t = x_{t-1} + 1 \mid x_{t-1} > x^*] = (1 - \varepsilon)\frac{n - x_{t-1}}{n},$$

[10] For the sake of exposition it is assumed that agents are selected to update uniformly, so that the probability that agent i is revising at time t is $\frac{1}{n}$. The precise distribution determining the updating agent is largely irrelevant, so long as it places strictly positive probability on each agent i.

$$\mathbb{P}[x_t = x_{t-1} \,|\, x_{t-1} > x^*] = (1-\varepsilon)\frac{x_{t-1}}{n} + \varepsilon\frac{n - x_{t-1}}{n}, \qquad [6.6]$$
$$\mathbb{P}[x_t = x_{t-1} - 1 \,|\, x_{t-1} > x^*] = \varepsilon\frac{x_{t-1}}{n}.$$

The first probability derives from the fact that the only way to move one step to the right is if first a **Y**-playing agent is selected to revise, and second the agent chooses **X** (a best reply when $x_{t-1} > x^*$) which happens with high probability $(1-\varepsilon)$. The second transition requires either an **X**-playing revising agent to play a best reply, or a **Y**-playing reviser to err. The final transition in [6.6] requires an **X**-player to err. Clearly, conditional on the state being above x^*, all other transition probabilities are zero.

An analogous set of transition probabilities may be written down for $x_{t-1} < x^*$ using exactly the logic presented in the previous paragraph. For $x_{t-1} = x^*$, the process moves down only if an **X**-player is selected to revise and with high probability selects the best reply **Y**. The process moves up only if a **Y**-player is selected to revise and with high probability selects the best reply **X**. The process stays at x^* with low probability, only if whichever agent is selected fails to play a best reply (so this transition probability is simply ε).

As a result, it is easy to see that any state may be reached from any other with positive probability, and every state may transit to itself. These two facts together guarantee that the Markov chain is irreducible and aperiodic, and therefore that there is a unique "ergodic" long-run distribution governing the frequency of play. The $\pi = (\pi_0, \ldots, \pi_n)$ that satisfies $\pi = P\pi$ where $P = [p_{i\to j}]$ is the matrix of transition probabilities is the ergodic distribution.

Note that P takes a particularly simple form: the probability of transiting from state i to state j is $p_{i\to j} = 0$ unless $i = j$ or $i = j \pm 1$ (so P is tridiagonal). It is algebraically straightforward to confirm that for such a process $p_{i\to j}\pi_i = p_{j\to i}\pi_j$ for all i, j.

Combining these equalities for values of i and j such that $p_{i\to j} > 0$, along with the fact that $\sum_i \pi_i = 1$, one obtains a (unique) solution for π. Indeed, consider the expression,

$$\frac{\pi_n}{\pi_0} = \left(\frac{p_{0\to 1}}{p_{1\to 0}}\right) \times \ldots \times \left(\frac{p_{(n-1)\to n}}{p_{n\to(n-1)}}\right), \qquad [6.7]$$

which follows from an immediate algebraic manipulation of $p_{i\to j}\pi_i = p_{j\to i}\pi_j$. The (positive) transition probabilities in [6.7] are given by the expressions in [6.6]. Consider a transition to the left of x^*: the probabilities of moving from state i to state $i+1$ and of moving from state $i+1$ to state i are

$$p_{i\to i+1} = \varepsilon\frac{n-i}{n} \quad \text{and} \quad p_{i+1\to i} = (1-\varepsilon)\frac{i+1}{n}.$$

To the right of x^*, these probabilities are

$$p_{i\to i+1} = (1-\varepsilon)\frac{n-i}{n} \quad \text{and} \quad p_{i+1\to i} = \varepsilon\frac{i+1}{n}.$$

Combining these probabilities and inserting into the expression in [6.7] yields

$$\frac{\pi_n}{\pi_0} = \prod_{i=0}^{x^*-1}\left(\frac{\varepsilon}{1-\varepsilon}\right)\left(\frac{n-i}{i+1}\right) \times \prod_{i=x^*}^{n-1}\left(\frac{1-\varepsilon}{\varepsilon}\right)\left(\frac{n-i}{i+1}\right)$$
$$= \left(\frac{\varepsilon}{1-\varepsilon}\right)^{x^*}\left(\frac{1-\varepsilon}{\varepsilon}\right)^{n-x^*} = \varepsilon^{2x^*-n}(1-\varepsilon)^{n-2x^*}. \quad [6.8]$$

It is possible to write down explicit solutions for π_i for all $i \in \{0, \ldots, n\}$ as a function of ε. However, the main interest lies in the ergodic distribution for $\varepsilon \to 0$. When the perturbations die away, the process becomes stuck for longer and longer close to one of the two equilibrium states. Which one? The relative weight in the ergodic distribution placed on the two equilibrium states is π_n/π_0. Thus, we are interested in the limit:

$$\lim_{\varepsilon\to 0}\frac{\pi_n}{\pi_0} = \lim_{\varepsilon\to 0}\varepsilon^{2x^*-n}(1-\varepsilon)^{n-2x^*} = \lim_{\varepsilon\to 0}\varepsilon^{2x^*-n} = \begin{cases} 0 & \text{if } x^* > \frac{n}{2}, \\ \infty & \text{if } x^* < \frac{n}{2}. \end{cases}$$

It is straightforward to show that the weight in the ergodic distribution placed on all other states tends to zero as $\varepsilon \to 0$. Thus if $x^* > \frac{n}{2}$, $\pi_n \to 0$ and $\pi_0 \to 1$: all weight congregates at 0. Every agent in the population is playing **Y** almost all of the time: the equilibrium $(\mathbf{Y}, \mathbf{Y})$ is selected. Now, recall that $(n-1)p < x^* < (n-1)p + 1$. For n sufficiently large, the inequality $x^* > \frac{n}{2}$ is well approximated by $p > \frac{1}{2}$. $(\mathbf{Y}, \mathbf{Y})$ is selected if $p > \frac{1}{2}$; that is if $c + d > a + b$. If the reverse strict inequality holds then $(\mathbf{X}, \mathbf{X})$ is selected.

Looking back at the game in Figure 6.2, note that this is precisely the condition in a symmetric 2×2 game for the equilibrium to be risk-dominant (Harsanyi and Selten, 1988). A stochastic evolutionary dynamic of the sort introduced here selects the risk-dominant Nash equilibrium in a 2×2 symmetric game. This remarkable selection result appears in both Kandori et al. (1993) and Young (1993a), who nevertheless arrive at the result from quite different adjustment processes.

The reason why the specifics of the dynamic do not matter so much comes from the following intuition: consider Figure 6.4. Suppose the process is currently in state 0. In order to escape the basin of attraction for 0 a selected agent needs to "make a mistake." This happens with low probability ε. Following this, another selected agent (with high probability a **Y**-player) must revise and make a mistake; this also happens with low probability ε. The argument continues for state 2, 3, up to x^*. When ε is

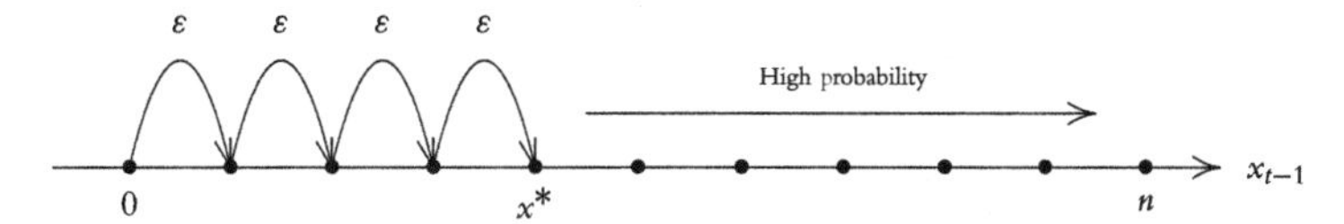

Figure 6.4 Escaping state 0.

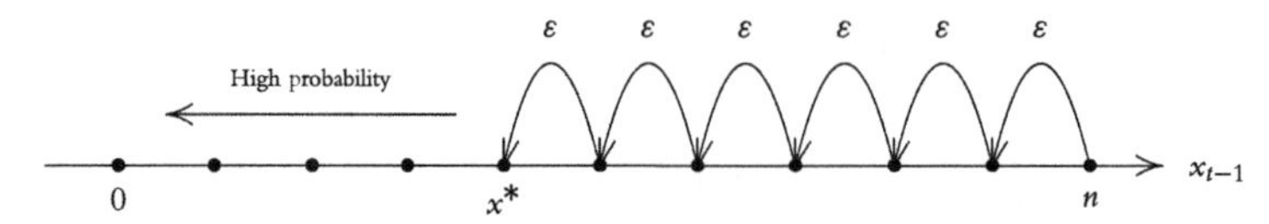

Figure 6.5 Escaping state n.

extremely small, it alone will determine the likelihood of this extremely unlikely event (the non-vanishing probability of a **Y**-player revising is irrelevant for very small ε). Thus to reach the edge of the basin of attraction, x^* errors are required, which will, for small enough ε, have probability well approximated by ε^{x^*}.

Now consider Figure 6.5. The very same argument as in the previous paragraph applies in reverse. Each step toward the edge of the basin of attraction (so that the process can escape into the other basin) takes an extremely unlikely event with probability ε and $n - x^*$ such steps are required. The combined probability of escaping is therefore ε^{n-x^*}.

Thus, whichever equilibrium state has the narrower basin of attraction is easier to escape from. As $\varepsilon \to 0$, the process spends more and more time around the one that is more difficult to escape from (the one with the wider basin of attraction). Equilibrium selection amounts to considering which equilibrium state x^* is further from; and this is determined by p: the one with the wider basin of attraction is risk-dominant.

Notice that the intuition provided above is quite robust to the particularities of the stochastic evolutionary dynamics. For instance, an alternative updating procedure makes little difference: whether the agents update their strategies simultaneously or one-at-a-time will not affect the thrust of the arguments above. Essentially what matters is the rate at which the mutation rates tend to zero: in the simple model described here these are independent of the state (the rate at which ε vanishes does not change depending upon the current state). Bergin and Lipman (1996) made this point and showed that any equilibrium may be selected by *some* model with state-dependent mutations. This is easy to see: if the probability of escaping one or other of the basins remains bounded away from zero (for example) as the probability of escaping the other vanishes, any selection result may be obtained.

Nevertheless, many reasonable specifications of perturbations do return the selection result described above. Blume (2003) characterizes the set of noise processes that work in this way, and shows that a symmetry property of the mutation process is sufficient to preserve the stochastic stability of the risk-dominant equilibrium.

6.3. STOCHASTIC STABILITY IN LARGER GAMES

The previous section focused on the case of 2×2 games. In order to go further and tackle larger games it is convenient to adopt a slightly different approach. The "adaptive learning" model introduced by Young (1993a) has been applied in a variety of contexts, some of which will receive attention later in this chapter: for instance, bargaining games (Section 6.4) and games played on networks (Section 6.6). Therefore, in this section, Young's model is developed, along with the requisite techniques and results from Markov chain theory before applying these methods to larger (coordination) games.[11]

6.3.1 A canonical model of adaptive learning

Consider a general n-player game $\mathcal{G}$ with a typical finite strategy set X_i for each player i. Payoffs are given by $u_i : X \to \mathcal{R}$ where $X = \prod_i X_i$. Suppose there is a population of agents C_i from which player i may be drawn. In each period one player is drawn at random from each population to play the game.[12] At each period t the actions taken by the players selected to play $\mathcal{G}$ may be written $x^t = (x_1^t, \ldots, x_n^t)$, where $x_i^t \in X_i$ is the action taken by the agent occupying player i's position at time t. The state, or history, of play at time t is a sequence of such vectors

$$h^t = \left(x^{t-m+1}, \ldots, x^t\right),$$

where m is the length of the players' *memory*. It represents how far back players are maximally able to recall the actions of other agents. Let the process start at an arbitrary h^0 with m action profiles. The state space is then H^m, the collection of all such feasible h^t.

At each time $t+1$, each agent selected to play $\mathcal{G}$ draws $n-1$ independent random samples of size s from the previous choices in h^t of the other agents. With high probability the agent plays a best reply to the strategy frequencies in these samples, and with low probability he chooses an action uniformly at random. The probability an action is taken at random is written ε. Together, these rules define a Markov process on H^m.

Definition 6.3. *The Markov process $\mathcal{P}^{m,s,\varepsilon}$ on the state space H^m described in the text above is called* adaptive play *with memory m, sample size s, and error rate ε.*

Consider a history of the form $h^* = (x^*, \ldots, x^*)$ where x^* is a strict Nash equilibrium of the game. If the state is currently h^* then a player at $t+1$ will certainly receive s copies of the sample x_{-i}^* of the other players' actions. Since these were strict Nash strategies, player i's best reply is of course x_i^*. Thus, for example, when $\varepsilon = 0$, then $h^{t+1} = h^*$. In other words, once the (deterministic) process $\mathcal{P}^{m,s,0}$ reaches h^* it

[11] The presentation here, including the notation used, for the most part follows that found in Young (1998). Of course, other methods and approaches have been proposed in the literature. A key contribution in this regard is the "radius co-radius" construction of Ellison (2000). Although this approach identifies stochastically stable equilibria in some settings, it provides only a sufficient condition for stochastic stability, unlike the approach described here.

[12] This is an instance of a recurrent game, that is, a repeated game with a changing cast of characters (Jackson and Kalai, 1997).

will never leave. For this reason, h^* is called a "convention." Moreover, it should be clear that all conventions consist of states of the form $(x^*, \ldots, x^*)$ where x^* is a strict Nash equilibrium of $\mathcal{G}$. This fact is summarized in the following proposition.

Proposition 6.1. *The absorbing states of the process $\mathcal{P}^{m,s,0}$ are precisely the conventions $h^* = (x^*, \ldots, x^*) \in H^m$, where x^* is a strict Nash equilibrium of $\mathcal{G}$.*

When $\varepsilon > 0$ the process will move away from a convention with low probability. If there are multiple Nash equilibria, and hence multiple conventions, the process can transit from any convention to any other with positive probability. The challenge is to characterize the stationary distribution (written $\mu^{m,s,\varepsilon}$) for any such error rate. This distribution is unique for all $\varepsilon > 0$ because the Markov process implicit in adaptive play is irreducible, hence ergodic. Thus, in order to find the stochastically stable convention, $\lim_{\varepsilon \to 0} \mu^{m,s,\varepsilon}$ may be found. This is the goal of the next section.

6.3.2 Markov processes and rooted trees

H^m is the state space for a finite Markov chain induced by the adaptive play process $\mathcal{P}^{m,s,\varepsilon}$ described in the previous section. Suppose that P^ε is the matrix of transition probabilities for this Markov chain, where the (h, h')th element of the matrix is the transition probability of moving from state h to state h' in exactly one period: $p_{h \to h'}$. Assume $\varepsilon > 0$.

Note that although many such transition probabilities will be zero, the probability of transiting from any state h to any other $\tilde{h}$ in a finite number of periods is strictly positive. To see this, consider an arbitrary pair $(h, \tilde{h})$ such that $p_{h \to \tilde{h}} = 0$. Starting at $h^t = h$, in period $t + 1$, the first element of h disappears and is replaced with a new element in position m. That is, if $h^t = h = (x^1, x^2, \ldots, x^m)$ then $h^{t+1} = (x^2, \ldots, x^m, y)$ where y is the vector of actions taken at $t + 1$. Any vector of actions may be taken with positive probability at $t + 1$. Therefore, if $\tilde{h} = (\tilde{x}^1, \tilde{x}^2, \ldots, \tilde{x}^m)$, let $y = \tilde{x}^1$. Furthermore, at $t + 2$, any vector of actions may be taken, and in particular $\tilde{x}^2$ can be taken. In this way the elements of h can be replaced in m steps with the elements of $\tilde{h}$. With positive probability h transits to $\tilde{h}$.

This fact means that the Markov chain is irreducible. An irreducible chain with transition matrix P has a unique invariant (or stationary) distribution μ such that $\mu P = \mu$. The vector of stationary probabilities for the process $\mathcal{P}^{m,s,\varepsilon}$, written $\mu^{m,s,\varepsilon}$, can in principle be found by solving this matrix equation. This turns out to be computationally difficult however. A much more convenient approach is the method of rooted trees, which we shall now describe.

Think of each state h as the node of a complete directed graph on H^m, and let $|H^m|$ be the number of states (or nodes) in the set H^m.

Definition 6.4. *A* rooted tree *at $h \in H^m$ is a set T of $|H^m| - 1$ directed edges on the set of nodes H^m such that for every $h' \neq h$ there is a unique directed path in T from h' to h. Let $\mathcal{T}_h$ denote the set of all such rooted trees at state (or node) h.*

For example, with just two states h and h', there is a single rooted tree at h (consisting of the directed edge from h' to h) and a single rooted tree at h' (consisting of the directed edge from h to h'). With three states, h, h', h'', there are three rooted trees at each state. For example, the directed edges from h'' to h' and from h' to h constitute a rooted tree at h, as do the edges from h'' to h and from h' to h, as do the edges from h' to h'' and from h'' to h. Thus $\mathcal{T}_h$ consists of these three elements.

As can be seen from these examples, a directed edge may be written as a pair $(h, h') \in H^m \times H^m$ to be read "the directed edge from h to h'." Consider a subset of such pairs $S \subseteq H^m \times H^m$. Then, for an irreducible process $\mathcal{P}^{m,s,\varepsilon}$ on H^m, write

$$p(S) = \prod_{(h,h') \in S} p_{h \to h'} \quad \text{and} \quad \eta(h) = \sum_{T \in \mathcal{T}_h} p(T) \quad \text{for all } h \in H^m. \qquad [6.9]$$

$p(S)$ is the product of the transition probabilities from h to h' along the edges in S. When S is a rooted tree, these edges correspond to paths along the tree linking every state with the root h. $p(S)$ is called the *likelihood* of such a rooted tree S. $\eta(h)$ is then the sum of all such likelihoods of the rooted trees at h. These likelihoods may be related to the stationary distribution of any irreducible Markov process. The following proposition is an application of a result known as the Markov Chain Tree Theorem.

Proposition 6.2. *Each element $\mu^{m,s,\varepsilon}(h)$ in the stationary distribution $\mu^{m,s,\varepsilon}$ of the Markov process $\mathcal{P}^{m,s,\varepsilon}$ is proportional to the sum of the likelihoods of the rooted trees at h:*

$$\mu^{m,s,\varepsilon}(h) = \eta(h) \Big/ \sum_{h' \in H^m} \eta(h'). \qquad [6.10]$$

Of interest is the ratio of $\mu^{m,s,\varepsilon}(h)$ to $\mu^{m,s,\varepsilon}(h')$ for two different states h and h' as $\varepsilon \to 0$. From the expression in [6.10], this ratio is precisely $\eta(h)/\eta(h')$. Consider the definitions in [6.9]. Note that many of the likelihoods $p(T)$ will be zero: transitions are impossible between many pairs of states h and h'. In the cases where $p(T)$ is positive, what matters is the rate at which the various terms vanish as $\varepsilon \to 0$. As the noise is driven from the system, those $p(T)$ that vanish quickly will play no role in the summation term on the right-hand side of the second expression in [6.9]. Only those that vanish slowly will remain: it is the ratio of these terms that will determine the relative weights of $\mu^*(h)$ and $\mu^*(h')$. This observation is what drives the results later in this section; and indeed those used throughout this chapter.

Of course, calculating these likelihoods may be a lengthy process: the number of rooted trees at each state can be very large (particularly if s is big, or the game itself has many strategies). Fortunately, there is a shortcut that allows the stationary distribution of the limiting process (as $\varepsilon \to 0$) to be characterized using a smaller (related) graph, where each node corresponds to a strict pure Nash equilibrium of $\mathcal{G}$.[13] Again, an inspection of [6.9] and [6.10] provides the intuition behind this step: ratios of non-Nash (non-convention) to Nash (convention) states in the stationary distribution will go to zero very quickly, hence the ratios of Nash-to-Nash states will determine equilibrium selection for ε vanishingly small.

Suppose there are K such Nash equilibria of $\mathcal{G}$ indexed by $k = 1, 2, \ldots K$. Let h_k denote the kth convention: $h_k = (x_k^*, \ldots, x_k^*)$, where x_k^* is the kth Nash equilibrium. With this in place, the Markov processes under consideration can be shown to be *regular perturbed* Markov processes. In particular, the stationary Markov process $\mathcal{P}^{m,s,\varepsilon}$ with transition matrix P^{ε} and noise $\varepsilon \in [0, \bar{\varepsilon}]$ is a regular perturbed process if first, it is irreducible for every $\varepsilon > 0$ (shown earlier); second, $\lim_{\varepsilon \to 0} P^{\varepsilon} = P^0$; and third, if there is positive probability of some transition from h to h' when $\varepsilon > 0$ ($p_{h \to h'} > 0$) then there exists a number $r(h, h') \geq 0$ such that

$$\lim_{\varepsilon \to 0} \frac{p_{h \to h'}}{\varepsilon^{r(h,h')}} = \kappa \quad \text{with } 0 < \kappa < \infty. \qquad [6.11]$$

The number $r(h, h')$ is called the *resistance* (or *cost*) of the transition from h to h'. It measures how difficult such a transition is in the limit as the perturbations vanish. In particular, note that if there is positive probability of a transition from h to h' when $\varepsilon = 0$ then necessarily $r(h_k, h_k) = 0$. On the other hand, if $p_{h \to h'} = 0$ for all $\varepsilon \geq 0$ then this transition cannot be made, and we let $r(h, h') = \infty$.

To measure the difficulty of transiting between any two conventions we begin by constructing a complete graph with K nodes (one for each convention). The directed edge (h_j, h_k) has *weight* equal to the least resistance over all the paths that begin at h_j and end at h_k.

In general the resistance between two states h and h' is computed as follows. Let the process be in state h at time t. In period $t + 1$, the players choose some profile of actions x^{t+1}, which is added to the history. At the same time, the first element of h, x^{t-m+1} will disappear from the history (the agents' memories) because it is more than m periods old. This transition involves some players selecting best replies to their s-length samples (with probability of the order $(1 - \varepsilon)$) and some players failing to play a best reply to

[13] This is appropriate only when dealing with games that possess strict pure Nash equilibria, which will be the focus of this section. For other games the same methods may be employed, but with the graph's nodes representing the recurrence classes of the noiseless process. (A recurrence class of $\mathcal{P}^{m,s,0}$ is a subset of states H^* such that for $h, h' \in H^*$, there is positive probability of moving from h to h' and vice versa and for every $h \in H^*$ and $h' \notin H^*$, $p_{h \to h'} = 0$.)

any possible sample of length s (with probability of the order ε). Therefore, each such transition takes place with probability of the order $\varepsilon^{r(h,h')}(1-\varepsilon)^{n-r(h,h')}$, where $r(h,h')$ is the number of errors (or mutations) required for this transition. It is then easy to see that this "mutation counting" procedure will generate precisely the resistance from state h to h' as defined in [6.11].

Now sum such resistances from h_j to h_k to yield the total (minimum) number of errors required to transit from the jth convention to the kth along this particular path. Across all such paths, the smallest resistance is the *weight* of the transition from h_j to h_k, written r_{jk}. This is the easiest (highest probability) way to get from j to k. When $\varepsilon \to 0$, this is the only way from j to k that will matter for the calculation of the stationary distribution.

Now consider a particular convention, represented as the kth node in the reduced graph with K nodes. A rooted tree at k has resistance $r(T)$ equal to the sum of all the weights of the edges it contains. For every such rooted tree $T \in \mathcal{T}_{h_k}$, a resistance may be calculated. The minimum resistance is then written

$$\gamma_k = \min_{T \in \mathcal{T}_{h_k}} r(T),$$

and is called the *stochastic potential* of convention k. The idea is that for very small but positive ε the most likely paths the Markov process will follow are those with minimum resistance; the most likely traveled of these are the ones that lead into states with low stochastic potential; therefore the process is likely to spend most of its time in states with the lowest values of γ_k: the stochastically stable states are those with the lowest stochastic potential. This is stated formally in the following proposition.

Proposition 6.3. (Young, 1993a) *Suppose $\mathcal{P}^{m,s,\varepsilon}$ is a regular perturbed Markov process. Then there is a unique stationary distribution $\mu^{m,s,\varepsilon}$ such that $\lim_{\varepsilon \to 0} \mu^{m,s,\varepsilon} = \mu^*$ where μ^* is a stationary distribution of $\mathcal{P}^{m,s,0}$. The stochastically stable states (those with $\mu^*(h) > 0$) are the recurrence classes of $\mathcal{P}^{m,s,0}$ that have minimum stochastic potential.*

The next section investigates the consequences of this theorem for the stochastically stable Nash equilibria in larger games by studying a simple example.

6.3.3 Equilibrium selection in larger games

In 2×2 games, the adaptive play process described in this section selects the risk-dominant Nash equilibrium. This fact follows from precisely the same intuition as that offered for the different stochastic process in Section 6.2. The minimum number of errors required to move away from the risk-dominant equilibrium is larger than that required to move away from the other. Because there are only two pure equilibria in 2×2 coordination games, moving away from one equilibrium is the same as moving

	a	b	c
a	6 / 6	3 / 0	0 / 2
b	0 / 3	5 / 5	4 / 1
c	2 / 0	1 / 4	4 / 4

Figure 6.6 A game with no risk-dominant equilibrium.

	a	b	c
a	60 / 5	0 / 0	0 / 0
b	0 / 0	40 / 7	0 / 0
c	0 / 0	0 / 0	1 / 100

Figure 6.7 The three-strategy game $\mathcal{G}_3$.

toward the other. The associated graph for such games has two nodes, associated with the two pure equilibria. At each node there is but one rooted tree. A comparison of the resistances of these edges is sufficient to identify the stochastically stable state in the adaptive play process, and this amounts to counting the number of errors required to travel from one equilibrium to the other and back.

However, in larger games, the tight connection between risk-dominance and stochastic stability no longer applies. First, in larger games there may not exist a risk-dominant equilibrium (whereas there will always exist a stochastically stable set of states); and second, even if there does exist a risk-dominant equilibrium it may not be stochastically stable. To see the first point, consider the two-player three-strategy game represented in Figure 6.6.

In this game, the equilibrium (b, b) risk-dominates the equilibrium (a, a), whilst (c, c) risk-dominates (b, b), but (a, a) risk-dominates (c, c). There is a cycle in the risk-dominance relation. Clearly, since there is no "strictly" risk-dominant equilibrium, the stochastically stable equilibrium cannot be risk-dominant.

Even when there is an equilibrium that risk-dominates all the others it need not be stochastically stable. The following example illustrates this point.[14]

The game $\mathcal{G}_3$ in Figure 6.7 is a pure coordination game (the off-diagonal elements are all zero) with three Nash equilibria. As a result, the risk-dominant equilibrium may be found simply by comparing the products of the payoffs of each of the equilibria. Therefore, (a, a) is strictly risk-dominant (it risk-dominates both (b, b) and (c, c)).

To identify the stochastically stable equilibrium, it is necessary to compute the resistances between the various states in the reduced graph with nodes h_a, h_b, and h_c cor-

[14] This example is taken from Young (1998).

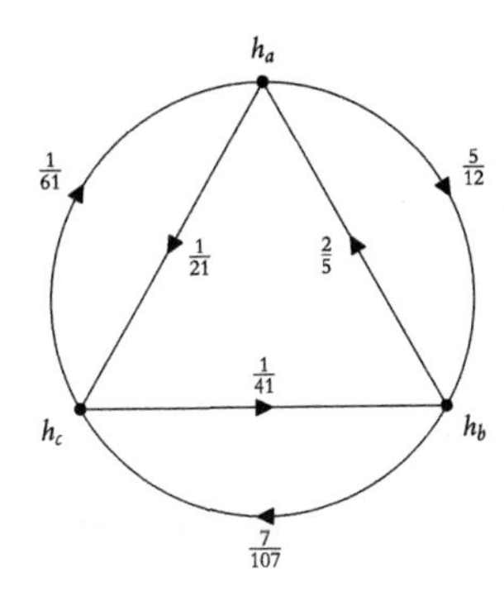

Figure 6.8 Reduced resistances in the game $\mathcal{G}_3$.

responding to the three equilibria. Node h_a represents the state $h_a = ((a, a), \ldots, (a, a))$ and so on. There is a directed edge between each of these nodes. The graph is drawn in Figure 6.8.

The numbers on the directed edges in Figure 6.8 represent the "reduced" resistances of transiting to and from the various conventions. These are calculated by considering the least costly path between the states. Consider for example the resistance of transiting from h_a to h_b. In h_a players 1 and 2 will certainly receive a sample containing s copies of (a, a). Thus, to move away from h_a requires at least one of the players to err (with probability ε). Suppose this is player 1, and that player 1 plays b instead. In the next period player 2 may receive a sample of length s that contains 1 instance of player 1 playing b. For large s this will not be enough for player 2 to find it a best reply to play b. Suppose player 1 again errs and plays a further b. To optimally play b, player 2 requires at least a proportion p^* of the s-length sample to contain b choices, where p^* is found from $60(1 - p^*) = 40p^*$. Thus $p^* = \frac{3}{5}$. Given an s-length sample, there need to be at least $3s/5$ errors by player 1 for player 2 ever to choose b as a best reply to some sample. Of course there are other routes out of convention h_a and into convention h_b. For example, there could be a string of errors by player 2. Player 1 finds it optimal to play b if the sample s contains at least $5s/12$ errors where player 2 has played b. Equally there could be a combination of player 1 and player 2 errors in each period. The key is to find the *least* costly route: clearly these latter paths from h_a to h_b involve at least as many errors as the first two "direct" paths, and so play no role as $\varepsilon \to 0$. Rather (ignoring integer issues), the resistance from h_a to h_b is given by

$$r_{ab} = \min\left\{\frac{5s}{12}, \frac{3s}{5}\right\} = \frac{5s}{12}.$$

Ignoring the sample size s, the "reduced" resistance is $\frac{5}{12}$ as illustrated in Figure 6.8. Similar calculations can be made for each of the other reduced resistances. The next step is to compute the minimum resistance rooted trees at each of the states. Consider $\mathcal{T}_{h_a}$

for example:

$$\mathcal{T}_{h_a} = \{[(h_b, h_a), (h_c, h_b)], [(h_b, h_a), (h_c, h_a)], [(h_b, h_c), (h_c, h_a)]\}.$$

Label these rooted trees T_1, T_2, and T_3 respectively. Then $r(T_1) = \frac{1}{41} + \frac{2}{5}$, $r(T_2) = \frac{1}{61} + \frac{2}{5}$, and $r(T_3) = \frac{7}{107} + \frac{1}{61}$. Hence

$$\gamma_a = \min\{r(T_1), r(T_2), r(T_3)\} = \frac{7}{107} + \frac{1}{61}.$$

In the same way, the stochastic potential for states h_b and h_c may be calculated: $\gamma_b = \frac{1}{21} + \frac{1}{41}$ and $\gamma_c = \frac{1}{21} + \frac{7}{105}$. Now $\gamma_b = \min_{i \in \{a,b,c\}} \gamma_i$, so (b, b) is stochastically stable. It is clear therefore, that the stochastically stable equilibrium need not be risk-dominant, and that the risk-dominant equilibrium need not be stochastically stable.

Nevertheless, a general result does link risk-dominance with stochastic stability in two-player games. Maruta (1997) shows that if there is a *globally* risk-dominant equilibrium, then it is stochastically stable. Global risk-dominance requires more than strict risk-dominance: in particular, a globally risk-dominant equilibrium consists of strategies (a_1, a_2) such that a_i is the unique best reply to any mixture that places at least probability $\frac{1}{2}$ on a_j, for $i, j = 1, 2$.[15] Results can be established for wider classes of games as the next proposition illustrates.

Proposition 6.4. *Suppose $\mathcal{G}$ is an n-player pure coordination game. Let $\mathcal{P}^{m,s,\varepsilon}$ be the adaptive process. Then if s/m is sufficiently small, the process $\mathcal{P}^{m,s,0}$ converges with probability one to a convention from any initial state h^0, and the coordination equilibrium (convention) with minimum stochastic potential is stochastically stable.*

Similar results are available for other classes of game, including potential games and, more generally, weakly acyclic games (Young, 1998).

This framework does not always imply that the dynamics select among the pure Nash equilibria. Indeed there are quite simple games in which they select a cycle instead of a pure Nash equilibrium. We can illustrate this possibility with the following example.

	A	*B*	*C*	*D*
A	3, 3	4, 1	1, 4	−1, −1
B	1, 4	3, 3	4, 1	−1, −1
C	4, 1	1, 4	3, 3	−1, −1
D	−1, −1	−1, −1	−1, −1	0, 0

Figure 6.9 A game with a best-reply cycle.

[15] This is equivalent to the notion of p-dominance described in Morris et al. (1995) with $p = \frac{1}{2}$.

In the game in Figure 6.9, (D, D) is the unique pure Nash equilibrium. It also has the following best reply cycle:

$$(C, A) \to (C, B) \to (A, B) \to (A, C) \to (B, C) \to (B, A) \to (C, A).$$

We claim that the adaptive process $P^{m,s,\varepsilon}$ selects the cycle instead of the equilibrium when the sample size s is sufficiently large and s/m is sufficiently small. The reason is that it takes more errors to move from the cycle to the basin of attraction of the equilibrium than the other way around. Indeed suppose that the process is in the convention where (D, D) is played m times in succession. To move into the basin of the cycle requires that someone choose an action other than D, say C, $\lceil s/6 \rceil$ times in succession. Assuming that s is small enough relative to m, the process will then move into the cycle with positive probability and no further errors. By contrast, to move from the cycle back to the equilibrium (D, D), someone must choose D often enough by mistake so that D becomes a best reply for someone else. It can be verified that it is easiest to escape from the cycle when A, B, and C occur with equal frequency in the row (or column) player's sample, and D occurs 11 times as often as A, B, or C. In this case it takes at least $\lceil 11s/14 \rceil$ mistaken choices of D to transit from the cycle to (D, D). Hence, there is greater resistance to moving from the cycle to the equilibrium than the other way around, from which one can deduce that the cycle is stochastically stable.

More generally this example shows that selection can favor *subsets* of strategies rather than single equilibria; moreover, these subsets take a particular form known as *minimal curb sets* (Basu and Weibull, 1991). For a further discussion of the relationship between stochastic stability and minimal curb sets see Hurkens (1995) and Young (1998, Chapter 7).

6.4. BARGAINING

We now show how the evolutionary framework can be used to derive Nash's bargaining solution. The reader will recall that in his original paper, Nash derived his solution from a set of first principles (Nash, 1950a). Subsequently, Ståhl (1972) and Rubinstein (1982) demonstrated that the Nash solution is the unique subgame perfect equilibrium of a game in which players alternate in making offers to one another. Although many would regard the noncooperative model as more persuasive than Nash's axiomatic approach, it is not entirely satisfactory. A major drawback of the noncooperative model is the assumption that the players' utility functions are common knowledge, and that they fully anticipate the moves of their opponent based on this knowledge. This seems rather far-fetched as an explanation of how people would behave in everyday bargaining situations.

In this section, we present an alternative approach that requires no common knowledge and much less than full rationality. Instead of assuming that two players bargain "face to face" in a repeated series of offers and counteroffers, we shall suppose that bargaining occurs between different pairs of individuals that are drawn from a large population. Thus, even though the protagonists are constantly changing, there is a linkage between periods because the outcomes of earlier bargains act as precedents that shape the expectations of later bargainers. The result is a stochastic dynamical process which (under certain regularity conditions) leads to the Nash bargaining solution, thereby providing an argument for the solution that is quite different from the traditional subgame-perfection-based justification.

6.4.1 An evolutionary model of bargaining

Consider two disjoint populations of agents (men and women, employers and employees, lawyers and clients) who periodically bargain pairwise over their shares of a fixed "pie." One of these populations consists of row players and the other of column players. We shall assume that the players have von Neumann-Morgenstern utility functions that capture their degree of risk aversion. For simplicity let us assume that the row players have the same utility function $u : [0, 1] \mapsto \mathcal{R}$, while the column players have the utility function $v : [0, 1] \mapsto \mathcal{R}$. We shall suppose that u and v are strictly increasing, concave, and that $u(0) = v(0) = 0$. In fact, the analysis generalizes readily to the situation where the agents are fully heterogeneous in their utilities (Young, 1993b).

The basic building block of the evolutionary process is the following one-shot Nash demand game: whenever a row player and column player engage in a bargain, *Row* "demands" a positive share x, *Column* "demands" a positive share y, and they get their demands if $x + y \leq 1$; otherwise they get nothing.

In order to apply the machinery developed in Section 6.3, we shall need to work with a finite state space. To this end we shall assume that the shares are rounded to the nearest d decimal places, that is, the demands are positive integer multiples of $\delta = 10^{-d}$ where $d \geq 1$. Thus the strategy space for both players is $D_\delta = \{\delta, 2\delta, 3\delta, ..., 1\}$, and the payoffs from the one-shot game are as follows:

Nash demand game

$$\begin{array}{ll} u(x), v(y) & \text{if } x + y \leq 1 \\ 0, 0 & \text{if } x + y > 1 \end{array}$$

Assume that at the start of each period a row and column player are drawn at random and they play the Nash demand game. The *state* at the end of time t is the sequence of demands made in the last m periods up to and including t, where m is the memory of the process. We shall denote such a state by $h^t = ((x^{t-m+1}, y^{t-m+1}), \ldots, (x^t, y^t))$.

Fix an integer s such that $0 < s < m$. At the start of period $t + 1$ the following events occur:

(1) A pair is drawn uniformly at random,

(2) *Row* draws a random sample of s demands made by column players in the history h^t,

(3) *Column* draws a random sample of s demands made by row players in the history h^t.

Let $g^t(y)$ denote the relative frequency of demands y made by previous column players in *Row*'s sample, and let $G^t(y) = \int_0^1 g^t(z)dz$ be its cumulative distribution function. Similarly let $f^t(x)$ denote the relative frequency of demands x made by previous row players in *Column's* sample, and let $F^t(x) = \int_0^1 f^t(z)dz$ be its cumulative distribution function.

- With probability $1 - \varepsilon$ *Row* chooses a best reply, $x^{t+1} = \arg\max u(x)G^t(1-x)$. With probability ε he chooses x^{t+1} uniformly at random from D_δ.
- With probability $1 - \varepsilon$ *Column* chooses a best reply, $y^{t+1} = \arg\max v(y)F^t(1-y)$. With probability ε he chooses y^{t+1} uniformly at random from D_δ.

This sequence of events defines a Markov chain $P^{\varepsilon,\delta,s,m}$ on the finite state space $(D_\delta)^m$.

A *bargaining norm* is a state of form $h_x = ((x, 1-x), \ldots, (x, 1-x))$ where $x \in D_\delta$. In such a state, all row players have demanded x, and all column players have demanded $1 - x$, for as long as anyone can remember. The *Nash bargaining norm* is the state h_{x^*} where

$$x^* = \arg\max_{x\in[0,1]} u(x)v(1-x).$$

Proposition 6.5. (Young, 1993b) *When δ is sufficiently small, s and m are sufficiently large and $s \leq m/2$, the stochastically stable bargaining norms are arbitrarily close to the Nash bargaining norm.*

Proof sketch. Fix an error rate ε, precision δ, sample size s, and memory length m satisfying $s \leq m/2$. For expositional simplicity we shall suppose that all errors are *local*, that is, whenever a player makes an error it is always within δ of the true best reply. Proposition 6.5 continues to hold without this simplifying assumption, as shown in Young (1993b).

The first step is to calculate the number of errors it takes to exit from one norm and to enter the basin of attraction of another. We shall illustrate this calculation with a specific example, then pass to the more general case. Suppose that $\delta = 0.1$ and the current norm is .3 for the row players and .7 for the column players. We shall refer to this as the norm (.3, .7). We wish to calculate the minimum number of errors required to transit to the norm (.4, .6). One way for such a transition to occur is that a sequence of row players demands .4 for k periods in succession. If in the next period the current column player happens to sample all k of these deviant demands, and if k is large enough relative to s, then she will choose .6 as a best reply to her sample information. Indeed .6 is a best reply if $v(.6) \geq (1 - k/s)v(.7)$, that is,

$$k \geq [1 - v(.6)/v(.7)]s \tag{6.12}$$

Once .6 is a best reply by the column players for *some* sample, the process can evolve *with no further errors* to the new norm (.4, .6). (This is where the assumption $s \leq m/2$ is used.)

Alternatively, a change of norm could also be induced by a succession of errors on the part of the column players. If a succession of column players demand .6 for k' periods, and if in the next period the current row player samples all of these deviant demands, his best reply is .4 provided that

$$k' \geq [u(.3)/u(.4)]s \qquad [6.13]$$

Once such an event occurs the process can evolve with no further errors to the new norm (.4, .6). Thus, the resistance to transiting from norm (.3, .7) to norm (.4, .6) is the smallest number (k or k') that satisfies one of these inequalities. More generally, the resistance of the transition is $(x, 1-x) \rightarrow (x+\delta, 1-x-\delta)$ is

$$r(x, x+\delta) = \lceil s(1 - v(1-x-\delta)/v(1-x)) \rceil \wedge \lceil s(u(x)/u(x+\delta)) \rceil. \qquad [6.14]$$

(In general, $\lceil z \rceil$ denotes the least integer greater than or equal to z.) Notice that when δ is small, the left-hand term is the smaller of the two; moreover the expression $1 - v(1-x-\delta)/v(1-x)$ is well approximated by $\delta(v'(1-x)/v(1-x))$. Hence, to a good approximation, we have

$$r(x, x+\delta) \approx \lceil \delta s(v'(1-x)/v(1-x)) \rceil. \qquad [6.15]$$

Similiar arguments show that

$$r(x, x-\delta) \approx \lceil \delta s u'(x)/u(x) \rceil. \qquad [6.16]$$

By assumption, the utility functions u and v are concave and strictly increasing. It follows that $r(x, x+\delta)$ is strictly increasing in x and that $r(x, x-\delta)$ is strictly decreasing in x.

We can now construct a least-resistant rooted tree as follows. Let x_δ be a value of $x \in D_\delta$ that maximizes $r(x, x+\delta) \wedge r(x, x-\delta)$ as shown in Figure 6.10.[16] Owing to our assumption that agents only make local errors, we can restrict our attention to transitions between adjacent norms. We shall call such transitions *edges*. The required tree has the

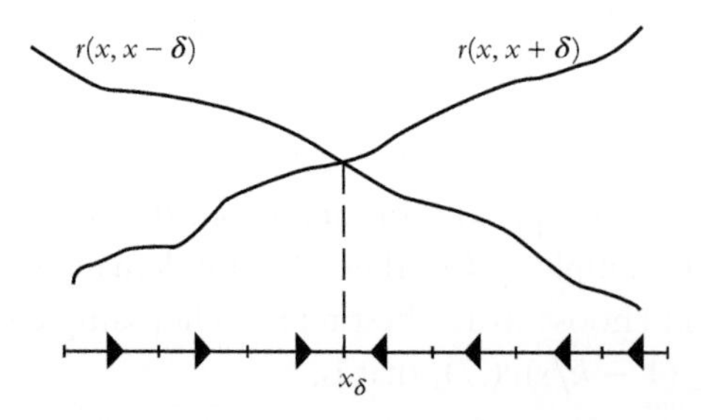

Figure 6.10 A least resistant rooted tree.

[16] Owing to the discrete nature of the state space it can happen that two distinct values $x \in D_\delta$ maximize $f(x) = r(x, x+\delta) \wedge r(x, x-\delta)$. In this case they lie on either side of the unique real-valued maximum of $f(x)$ on the interval [0, 1].

form shown on the x-axis of Figure 6.10: edges to the *left* of x_δ point to the *right*, and edges to the *right* of x_δ point to the *left*. The resistance of each such edge is the *smaller* of the two values $r(x, x+\delta), r(x, x-\delta)$. It is straightforward to check that, among all rooted trees, this one has least total resistance. It follows from Proposition 6.3 that the norm h_{x_δ} is stochastically stable.

When δ is very small and s is very large, the value(s) x_δ are very close to the unique real-valued maximum of $f(x) = u'(x)/u(x) \wedge v'(1-x)/v(1-x)$ on the interval $[0, 1]$. This maximum is precisely the point x^* where the decreasing function $u'(x)/u(x)$ crosses the increasing function $v'(1-x)/v(1-x)$, that is, x^* is the solution to

$$u'(x^*)/u(x^*) = v'(1-x^*)/v(1-x^*). \tag{6.17}$$

Notice that [6.17] is just the first-order condition for maximizing the function $ln\ u(x) + ln\ v(1-x)$ on the interval $x \in [0, 1]$, which is equivalent to maximizing $u(x)v(1-x)$ for $x \in [0, 1]$. It follows that x^* is the Nash bargaining solution. In other words, the stochastically stable norms can be made as close as we like to the Nash bargaining solution by taking δ to be small and s to be large.

6.4.2 The case of heterogeneous agents

The preceding argument can be generalized to the situation where people have different sample sizes. Suppose that the row players all have sample size $s = \alpha m$, whereas the column players all have sample size $s' = \beta m$, where $0 < \alpha, \beta < 1/2$. Then expressions [6.15] and [6.16] have the following analogs:

$$r(x, x+\delta) \approx \lceil \delta\beta m(v'(1-x)/v(1-x)) \rceil, \tag{6.18}$$

$$r(x, x-\delta) \approx \lceil \delta\alpha m u'(x)/u(x) \rceil. \tag{6.19}$$

The stochastically stable norm is determined by the crossing point of these two functions. When m is large and δ is small this crossing point is close to the solution of the equation

$$\alpha u'(x)/u(x) = \beta(v'(1-x)/v(1-x)),$$

which is the solution of

$$\arg\max_{x \in [0,1]} u(x)^\alpha v(1-x)^\beta. \tag{6.20}$$

This is the *asymmetric Nash bargaining solution.* A particular implication is that, for given utility functions, the more information that agents have (i.e., the larger their sample sizes), the better off they are.

6.4.3 Extensions: Sophisticated agents and cooperative games

The preceding framework can be extended in several directions. In the version outlined above, players do not try to anticipate what their opponents are going to do; they implicitly assume that the opponents' behavior is stationary. Suppose, instead, that some positive fraction of each population uses level-2 reasoning (Sáez-Martí and Weibull, 1999): these "clever" players attempt to estimate what their opponent's best reply will be, and choose a best reply to that. (Thus level-2 players act as if their opponents are level-1 players.) To make such an estimate a clever player samples the previous demands of his own population. It turns out that, if the sample size of the clever players is at least as large as the sample size of the opponents, then the clever agents gain nothing in the long run, that is, the stochastically stable solutions are the same as if there were no clever agents. If, however, the clever agents' sample size is smaller than that of the opponents, then they do gain: the smaller sample size of the clever agents has the same effect as reducing the sample size of the opposite population, which reduces the latter's relative share.

Another extension is to learning dynamics in multi-person bargaining games (Agastya, 1997) and in cooperative n-person games (Agastya, 1999). Consider a set $N = \{1, 2, \ldots, n\}$ and a value function v defined on the subsets of N that is convex and contains no dummies, that is, for every two coalitions S and T,

$$v(S) + v(T) \leq v(S \cup T) + v(S \cap T)$$

and for every $i \in N$, there is some coalition S_i containing i such that

$$v(S_i - \{i\}) < v(S_i).$$

There is a population of potential players for each "role" $i \in N$. In each period, one agent is drawn at random from each of the n populations. Simultaneously they make "demands" and a maximal subcoalition forms that is consistent with the demands of its members. Players not in this coalition receive their individual payoffs $v(\{i\})$. As in the bargaining model, players draw independent random samples from the recent history and choose best replies subject to error. The main result is that the stochastically stable demands are in the core of the game, and these demands maximize a real-valued function that is closely related (though not identical) to maximizing the Nash product of the players' utilities subject to their demands being in the core (Agastya, 1999). As in the two-person bargaining set-up, the size of a player's sample determines the exponent to which his utility is raised in the Nash product, thus favoring agents with larger samples.

6.5. PUBLIC GOODS

Public-good games often exhibit multiple Nash equilibria. The simplest example of this can be found in any economics textbook: two agents simultaneously deciding whether to pay for the provision of a public good. Suppose the value of the public good once provided is positive and greater than the costs of its provision by either agent. If it takes only one of them to provide the good, there is a coordination problem: who will provide it? If it takes both agents to successfully generate the value associated with the public good then there may be a different coordination problem: will it be provided or not?

Stochastic evolutionary dynamics may be applied to examine the robustness of these various equilibria, and this section examines several such applications in precisely this context. First, in Section 6.5.1, a very simple worked example is constructed in order to set the scene. This requires little in the way of mathematical complexity and yields insight into the more general results discussed in the remainder of the section. Palfrey and Rosenthal (1984) public-good games, volunteer's dilemmas, and more general public-good games are presented in what follows.

6.5.1 Teamwork

Consider an n-player symmetric game in which each player i can take one of two actions: $z_i = 1$, interpreted as contributing toward the production of a public good, or $z_i = 0$ (not contributing).[17] Suppose that it takes a "team" of $m \leq n$ individual contributions to successfully produce the good, and generate a value of v for each player. Any player contributing pays a cost $c > 0$. Thus payoffs may be written

$$u_i(z) = v \times \mathcal{I}[|z| \geq m] - c \times z_i, \qquad [6.21]$$

where z is the vector of actions $(z_1, \ldots, z_n)$, $|z| = \sum_i z_i$, and $\mathcal{I}$ is the indicator function taking a value 1 when its argument is true, and zero otherwise. When $v > c$ this is a discrete public-good provision game of the kind introduced in Palfrey and Rosenthal (1984).

There are many pure-strategy Nash equilibria. For $m \geq 2$, there is a Nash equilibrium in which $z_i = 0$ for all i: the "no-contribution" equilibrium. No player can unilaterally deviate and do any better: the payoff received in equilibrium is zero, but any unilateral deviation would not result in provision, and thus the deviating player would simply incur the cost c. Any strategy profile where exactly m players contribute (so that $|z| = m$) is also a Nash equilibrium: the contributing players have no incentive to deviate (each contributor is currently receiving $v - c > 0$, the payoff from deviation) and every non-contributor is receiving $v > v - c$, the payoff from contributing.[18]

Focusing on the case where $v > c$ and $m \geq 2$, there are $\binom{n}{m} + 1$ pure equilibria. Which of these are stochastically stable? In the symmetric game considered in this section, there is no difference between any of the m-contributor equilibria, so this amounts to asking the question: is public-good provision stochastically stable?

[17] This section follows the simple example presented in Myatt and Wallace (2005) but here with the rather more convenient logit quantal response employed in the strategy-revision process.

[18] There are also many mixed equilibria, for a full description see Palfrey and Rosenthal (1984).

To answer this question, consider the following "one-step-at-a-time" dynamic strategy-revision process over the state space Z with typical member $z \in Z$. At time t a player is chosen at random (suppose for now with uniform probability $\frac{1}{n}$) to revise their current strategy. Suppose player i receives a revision opportunity at t and the state is currently $z^t = z$. Player i plays a best-reply to the current state with high probability, but plays the other strategy with some non-zero but small probability. Concretely, suppose that conditional on z the log odds of choosing $z_i = 1$ versus $z_i = 0$ is linear in the payoff difference between the two[19,20]:

$$\log \frac{\mathbb{P}[z_i^{t+1} = 1 | z^t = z]}{\mathbb{P}[z_i^{t+1} = 0 | z^t = z]} = \beta \Delta u_i(z), \qquad [6.22]$$

where the payoff difference is simply $\Delta u_i(z) = u_i(z_i = 1; z_{-i}) - u_i(z_i = 0; z_{-i})$. Note that for $\beta = 0$ choices are entirely random, but as $\beta \to \infty$ choices become simple best-replies to the current strategy profile employed by the other players z_{-i}. In the game considered here, $\Delta u_i(z) = v - c$ if $|z_{-i}| = m - 1$, and $\Delta u_i(z) = -c$ otherwise.

It is useful to partition the state space into "layers." Write $Z_k = \{z \in Z : |z| = k\}$, the kth layer where exactly k players are contributing. The pure-strategy Nash equilibria then are the states contained in $Z_0 \cup Z_m$. We are interested in the transition probabilities between the various layers. Write $p_{j \to k} = \mathbb{P}[z^{t+1} \in Z_k | z^t \in Z_j]$. Given that a single player may conduct a strategy revision at any time t, this probability is zero unless k and j differ by at most 1. Suppose the state is currently in layer $m - 1$. The process transits to the state m only if (a) a non-contributing player is selected to revise and (b) this player chooses to contribute. Given that $m - 1$ players currently contribute, $n - m + 1$ players do not. Thus the probability of (a) occurring is $(n - m + 1)/n$. The probability that (b) occurs is then

$$\mathbb{P}[z_i^{t+1} = 1 | z_i = 0 \text{ and } z^t \in Z_{m-1}] = \frac{\exp[\beta(v - c)]}{1 + \exp[\beta(v - c)]}, \qquad [6.23]$$

calculated directly from [6.22].[21] Combining these facts, the probability that the process transits from layer $m - 1$ to m is

$$p_{(m-1) \to m} = \frac{n - m + 1}{n} \times \frac{\exp[\beta(v - c)]}{1 + \exp[\beta(v - c)]}. \qquad [6.24]$$

The other transition probabilities may be calculated in a similar way. The resulting Markov process is ergodic: there is a unique distribution governing the frequency of

[19] This is a logit quantal response (McKelvey and Palfrey, 1995), and admits a random-utility interpretation. In particular, if payoffs were drawn from an i.i.d. extreme-value distribution with scale parameter β, then the payoff differences would be logistically distributed, and [6.22] would follow from a simple best-reply.

[20] This is a common modeling choice. Indeed, later in this chapter the very same log linear construction is used in the context of games played on a network. In particular, Section 6.6 contains further exploration of this model in the discussion leading up to (and including) Proposition 6.7.

[21] Note that this is a special case of the expression given in [6.38].

long-run play when $t \to \infty$. To see this note that first, although many transitions have probability zero, any layer may be reached from any other in finitely many positive-probability transitions, and second, each layer may transit to itself (guaranteeing aperiodicity). Transition probabilities such as [6.24] may be used to calculate the ergodic distribution. Write $\pi_k = \lim_{t\to\infty} \mathbb{P}[z^t \in Z_k]$ for the long-run probability of being in layer k. Balance equations apply to these kinds of finite Markov chain processes (Ross, 1996), that is,

$$\pi_k p_{k\to j} = \pi_j p_{j\to k} \quad \text{for all } j \text{ and } k. \tag{6.25}$$

The ergodic distribution over layers $\pi = (\pi_1, \ldots, \pi_n)$ can then be characterized explicitly (and for any level of β). To understand the properties of the ergodic distribution as $\beta \to \infty$, in order to characterize the stochastically stable state(s) as noise is driven from the process, it is enough to consider the ratio π_m/π_0. The Nash equilibria lie in layers 0 and m, and these are the absorbing states for the noiseless process. This ratio can be calculated from a chain of expressions such as [6.25] as follows

$$\frac{\pi_m}{\pi_0} = \left(\frac{p_{(m-1)\to m}}{p_{m\to(m-1)}}\right) \times \ldots \times \left(\frac{p_{0\to 1}}{p_{1\to 0}}\right). \tag{6.26}$$

Notice the similarity between this exercise and the one conducted in Section 6.2.3. The only difference is with the transition probabilities themselves: here they are formed from the log linear modeling assumption made explicit in [6.22], whereas in the earlier section the probabilities arose from a simple perturbed best-reply process.

Substituting in the transition probabilities found in the step above in [6.24] gives

$$\frac{\pi_m}{\pi_0} = \frac{n-m+1}{n}\frac{\exp[\beta(v-c)]}{(1+\exp[\beta(v-c)])} \times \ldots \times \frac{n}{n}\frac{\exp[-\beta c]}{(1+\exp[-\beta c])} \bigg/ \frac{m}{n}\frac{1}{(1+\exp[\beta(v-c)])} \times \ldots \times \frac{1}{n}\frac{1}{(1+\exp[-\beta c])}. \tag{6.27}$$

Canceling terms and tidying up this expression gives the following result

$$\frac{\pi_m}{\pi_0} = \binom{n}{m} \exp[\beta(v - mc)]. \tag{6.28}$$

The next step is to consider what will happen when noise is driven from the system (that is $\beta \to \infty$ so that players choose best-replies with probability arbitrarily close to 1). The parameter β appears only in the exponent, and therefore the limit of the ratio in [6.28] is zero if $mc > v$ and infinite otherwise. This result is summarized in the following proposition.[22]

[22] To complete the proof, it remains to check that $\pi_k \to 0$ for all $k \neq 0, m$. This follows immediately from similar calculations to those presented above in [6.26]–[6.28]. When $k > m$ find π_k/π_m, and show it converges to zero as $\beta \to \infty$; when $0 < k < m$ find π_k/π_0 and show it again converges to zero.

Proposition 6.6. *Consider a symmetric n-player binary action public-good provision game in which $m \geq 2$ costly contributions are required to successfully provide the public good. Full provision is stochastically stable if and only if the value generated to each player from the good is greater than m times the cost to a contributing individual. When the reverse is true the stochastically stable state involves no contributions.*

All weight in the ergodic distribution is in the layer Z_m whenever $v > mc$. The good is provided only when the value to each individual is sufficiently large (and in particular larger than the social cost being paid for its provision $m \times c$). The *private* value needs to be larger than the *social* cost. This is a point returned to in Section 6.5.4.

To see why this is true, rewrite this condition as $v - c > (m - 1)c$. In order for the process to transit from the "bottom" layer where no-one is contributing, $m - 1$ agents must first choose to contribute when called upon to revise *even though this is not a best reply to the current state.* The final mth contributor is of course playing a best reply to the current state. The cost to each of these players when making their choices is c. Hence the total cost of moving up from Z_0 to Z_m is $(m - 1) \times c$.

The cost of moving down is simpler: it takes but one (contributing) player to choose $z_i = 0$. The payoff lost to this player is $v - c$ (0 is received instead); but once in the $(m - 1)$th layer, it is now a best-reply for revising players to choose not to contribute. Thus there is just one cost involved in moving from Z_m to Z_0 and it is $v - c$. Comparing the cost of reaching the 0th layer from the mth and vice-versa yields the result.

In this case, the ergodic distribution is analytically available. In fact, because of the connection between logit quantal-response dynamics of the kind introduced here and potential (Monderer and Shapley, 1996) the explicit form for the ergodic distribution can be used to say much more about rather more general games (which nest the example of this section). This will be discussed in more detail in Section 6.5.4.

First, though, note that in the example presented so far, symmetry was of great use during much of the analysis; furthermore $m \geq 2$ was maintained throughout. The symmetry means the model automatically remains silent on the very first question raised in this section—who provides the public good? Moreover, when $m = 1$ the game is a volunteer's dilemma, with a different structure (the no-contribution state no longer constitutes a Nash equilibrium) and it is *only* a question of "who volunteers?" The next section relaxes the symmetry assumption in the general $m \geq 2$ case, whilst Section 6.5.3 examines the special case of $m = 1$.

6.5.2 Bad apples

Abandoning the symmetric specification of Section 6.5.1, let each player i receive a potentially different benefit v_i from the provision of the public good, and allow the cost a player faces when choosing $z_i = 1$ to vary with i also (written c_i). Adapting [6.21],

$$u_i(z) = v_i \times \mathcal{I}[|z| \geq m] - c_i \times z_i. \qquad [6.29]$$

This seemingly insignificant change in the model results in a game that does not admit a potential function, the ergodic distribution is no longer explicitly available for non-zero noise, and a rooted-tree analysis (see Section 6.3) is required. However, the intuition behind the resulting characterization for the ergodic distribution as noise is driven from the stochastic process is analogous to that described below Proposition 6.6.

Assuming that $v_i > c_i$ for all i, the Nash equilibria again lie in the mth layer (where exactly m players choose to contribute) and in the 0th layer, where the public good is not provided. These states once again are the absorbing states of a deterministic process where in each period a player is selected at random, and chooses to play a best-reply to the current population state. This process will eventually enter one of these two layers and, once there, never leave. Now suppose that agents only probabilistically play a best-reply: with some (high) probability an updating player selects the strategy corresponding to a best-reply against the current population state; with some (low) probability the other strategy is selected. Once again the question is, as the probability that a non-best-reply is selected tends to zero, in which state does the system spend most of its time?

Myatt and Wallace (2008b) precisely characterize the ergodic distribution for vanishing noise in such a scenario.[23] To convey an idea of the content of this characterization, consider starting in a state where all players are not contributing (the 0th layer). In order to reach a state in which m players contribute (and thus the good is provided), the system needs to transit "upward" to the mth layer. In order to do so, $m-1$ players need to take actions which are non-best-replies in the underlying zero-noise game. Once the $(m-1)$th layer is reached, the final step involves an updating non-contributor playing $z_i = 1$ which, of course, is a best-reply in the zero-noise game, and so has high probability. The first $m-1$ steps each have low probability.

In the symmetric world, each of these high cost/low probability steps entailed a cost of c on the part of a revising player. Here, if player i revises, the associated cost is c_i. Thus, the total cost of moving up from the 0th layer to the mth layer will depend on *which* players bear the cost. In line with the intuition of earlier sections it is the *cheapest* route which matters for the limiting result. What is the cheapest way to reach Z_m from Z_0? Clearly, this will involve the $m-1$ lowest-cost contributors. Ordering the players (without loss of generality) so that $c_1 < c_2 < \ldots < c_n$, the total cost of moving from Z_0 to Z_m is simply $\sum_{j=1}^{m-1} c_j$.[24]

[23] Myatt and Wallace (2008b) also allow for a more general class of revision dynamic, including probit-style noise. In the ensuing discussion here the convenient logistic specification of Section 6.5.1 is maintained.

[24] Note that this leaves the process in a subset of states in Z_m. In fact, it is the set defined by $z \in Z_m$ such that $z_i = 1$ for the $m-1$ lowest cost contributors $(i = 1, \ldots, m-1)$. One other player $j > m-1$ is also contributing—the identity of this player does not affect the cost of transiting out of Z_0.

The cheapest way to transit from Z_m to Z_0 is more complicated. Recall that, in the symmetric model, the cost of such a transition is simply the utility foregone from not contributing for a revising contributor: $v - c$. What is the easiest way to exit from the mth layer when players have different vs and cs? It might be conjectured initially that it is for the player with the lowest $v_i - c_i$ who is currently contributing to stop contributing. Certainly this player finds it easiest to play $z_i = 0$ if currently $z_i = 1$. Two issues arise however: first, the player with the lowest $v_i - c_i$ in the population may or may not be part of the currently functioning team; and second, there is also the possibility of exiting "upward" (a non-contributing player could start contributing). This could be cheaper if the cost of doing so (which is greater or equal to c_m) is smaller than the cost of exiting downward ($\min_{i:z_i=1} v_i - c_i$).

These two problems have a related solution. Whereas before, in a symmetric game, it was clear that it could not possibly be cheaper to exit via Z_{m+1}; here that is no longer the case. Before, such a route would inevitably involve a further transition through Z_m in order to reach Z_0, and therefore involve an additional (but identical) cost $v - c$ of transiting down to Z_{m-1}; here there may be a cheaper route involving a current non-contributor choosing $z_i = 1$ before another player (costlessly) ceases contributing. The point is that, in this new state in Z_m, player i has replaced one of the other contributors: this could make it cheaper to transit to Z_0 if $v_i - c_i$ is particularly low. In particular, it may be possible to (cheaply) "shoehorn" in a player from outside the current set of contributors who has a low $v_i - c_i$ before making the necessary transition down to Z_{m-1}.

To be concrete, suppose the process is in a state $z^\dagger \in Z_m$. The cheapest "direct" route to Z_0 involves a player i with $z_i = 1$ ceasing to contribute (moving the state to Z_{m-1}). $v_i - c_i$ indexes the difficulty of this move. Now consider the "indirect" route via Z_{m+1}. Player j, currently a non-contributor in state $z^\dagger$, chooses to contribute (at a cost of c_j); another player ceases contributing at no cost; from the new state in Z_m, player j stops contributing at a cost of $v_j - c_j$. The total cost of this route out of $z^\dagger$ is therefore $c_j + (v_j - c_j) = v_j$. This is cheaper than the direct route out of $z^\dagger$ if $v_j < v_i - c_i$. It is now clear that if $c_i = c$ and $v_i = v$ for all i this inequality can never hold. But in the absence of such symmetry there can be a cheaper route out of Z_m than the obvious direct route.

Thus, in the asymmetric version of the game, a new and interesting feature arises. The successful provision of a public good will depend not only on the private costs and values of those contributing, but also on the costs and values that exist in the population as a whole. Suppose there is a player with a particularly low valuation for the public good. Even if that player is not directly involved in public-good production (perhaps because their cost of doing so is relatively high) their presence in the population alone may destabilize an otherwise successfully operating team—this player is a "bad apple."

Whilst the stochastic stability of the production equilibria versus the no-production equilibrium will turn on a comparison of the exit costs from Z_0 and those from Z_m, there is also the question of *which* team of contributors is most likely to produce the public good when it is successfully provided. In the symmetric game, this question was moot. Here, it would seem intuitive that the $m - 1$ lowest-cost contributors would be involved (these players pay the costs involved in building the "cheapest team")[25]; but this leaves open the identity of the mth contributor. As might be guessed from the above discussion, the answer to this question depends on the distribution of costs and valuations across the whole set of players. Whilst the full details are described in Myatt and Wallace (2008b, Theorem 1), a similar intuition emerges from the slightly different setting of the volunteer's dilemma, discussed next.

6.5.3 The volunteer's dilemma

So far the public good games discussed have involved a coordination problem of "no provision" versus "provision." In the payoff structure given in [6.29], this requires $m \geq 2$. When $m = 1$ the game is a volunteer's dilemma. It takes one, and only one, player to contribute toward the production of the public good in order for it to be successfully provided. So long as the maintained assumption that $v_i > c_i$ for all i continues to hold, there are n (pure) Nash equilibria, each involving $z_i = 1$ for precisely one player i, and $z_j = 0$ for all $j \neq i$. There is no equilibrium in Z_0 any longer, as (any) single player would wish to deviate and receive $v_i - c_i > 0$. Thus all the (pure) equilibria lie in the first layer, Z_1. The only issue is "who volunteers?" Therefore the volunteer's dilemma provides a simple framework in which to analyze this particular element of the coordination problem (which is nonetheless present in the case of $m \geq 2$ discussed in Section 6.5.2).

Myatt and Wallace (2008a) provide a full analysis: once again, here the intuition behind the results will be discussed without introducing too much formality. Again the stochastically stable state will depend upon the ease with which various equilibrium states are exited. These costs of exit are relatively straightforward to calculate in the $m = 1$ case. In each equilibrium there are two different exit routes available. The single player who is contributing (say i) might choose to cease doing so (at a cost of $v_i - c_i$ in foregone utility). Alternatively another player $j \neq i$ may choose to contribute, bearing an additional cost c_j. The cost of the cheapest exit from the equilibrium in which i contributes is therefore $\min[(v_i - c_i), \min_{j \neq i} c_j]$.[26] Identifying the stochastically stable states boils down to a comparison of these exit costs across the n states in Z_1: whichever of these is the most expensive to exit is played with probability arbitrarily close to 1 as $\beta \to \infty$.

[25] It is not quite this obvious however, as it may be that these $m - 1$ include some very low-valuation (and hence relatively likely to stop contributing) players. Nevertheless, as Myatt and Wallace (2008b) show, these players are in fact always involved in any successful team.

[26] Again, the focus here is on the logit best-reply dynamic introduced in Section 6.5.1. Myatt and Wallace (2008a) allow for a general range of state-dependent (Bergin and Lipman, 1996) updating rules characterized by a pair of probabilities (indexed by some noise parameter ε) for each player.

It is tempting to think that, in a volunteer's dilemma, the player with the lowest cost of contribution ought to be the one providing the good. After all, in every equilibrium the total gross social value generated is the same, and equal to $\sum_j v_j$. The cost paid is simply the cost associated with the lone contributor c_i. Thus net social benefit is maximized when the player with the smallest c_i (the most *enthusiastic* player) contributes. Indeed, if $v_i = v$ for all i this will be the case: recall that (without loss of generality) $c_1 < c_2 < \ldots < c_n$, then

$$\max_i[v_i - c_i] = \max_i[v - c_i] = v - c_1 \quad \text{and} \quad \min_{j \neq i} c_j = c_1 \text{ if } i \neq 1 \quad \text{and} \quad \min_{j \neq 1} c_j = c_2. \tag{6.30}$$

Escaping from the equilibrium in which the most enthusiastic player (i.e., player 1) contributes has a cost of $C_1 = \min[(v - c_1), c_2]$; escaping from any other equilibrium with player $i \neq 1$ contributing has cost $C_i = \min[(v - c_i), c_1]$. The larger of these two numbers is always C_1. Suppose $v - c_1 > c_2$. Now $c_2 > c_1$ by definition, so c_2 is certainly bigger than the minimum of c_1 and any other number. Suppose $c_2 > v - c_1$. Now $v - c_1 > v - c_2$ by [6.30], and hence $v - c_1 > \min[v - c_2, c_1]$. Thus $C_1 > C_i$ for all $i \neq 1$. The stochastically stable equilibrium involves the contribution of player 1, the most enthusiastic player.

This need not be the case however. When $v_i \neq v_j$ for all i and j, the most enthusiastic player (player 1) need not be the most "reliable": the player with the lowest c_i does not also necessarily have the highest $v_i - c_i$. Suppose that player r is the most reliable: $r = \arg\max_i[v_i - c_i]$. Then the stochastically stable state certainly involves $z_r = 1$ if $v_r - c_r < c_1$. To see this, note that in such a circumstance, all the costs of moving up a layer are larger than all the costs (foregone utilities) of moving down a layer: the easiest way to exit any equilibrium involves the contributor ceasing to contribute. It is therefore hardest to leave the state where r contributes.[27]

More enthusiastic players will contribute in the stochastically stable equilibrium the more strongly (negatively) associated are the terms c_i and $v_i - c_i$ across the player set. If it were possible to shift value from high cost players to low cost players, then doing so would reduce the cost paid in the stochastically stable state (by ensuring that a more enthusiastic player is contributing in the selected equilibrium).

A similar feature arises in the case when $m \geq 2$ analyzed in Section 6.5.2. The mth player's identity in a selected provision equilibrium will depend not only on the costs of provision in the pool of players, but also on the "reliability" of the players in the population.

[27] The remaining more complex parameter configurations are dealt with in Myatt and Wallace (2008a), which also contains a formal statement of the result in the following paragraph and other comparative statics.

6.5.4 General public-good games and potential

Further progress can be made for more general public-good games when restricting (a) to the logit dynamics which have been the focus of this section so far and (b) to games that exhibit potential. Blume (1997) provided an early discussion of the relationship between potential games and log-linear choice rules. He observed that balance equations of the kind used in [6.25], for example, are satisfied if and only if the game admits an exact potential function.

To that end consider a general n-player public-good game in which each player i picks an action z_i from a finite set. Suppose that there is a player-specific cost associated with each z_i, written $c_i(z_i)$. Moreover, a payoff $G(z)$ which depends upon the vector z of every player's action accrues to all players. Thus, a player i's payoff may be written

$$u_i(z) = G(z) - c_i(z_i), \qquad [6.31]$$

so that $G(z)$ represents the common payoff obtained by each player and $c_i(z_i)$ represents a private payoff to i alone. The additive separability of these two components is important, as such a game has exact potential (Monderer and Shapley, 1996). To see this, recall that a game has exact potential if and only if there exists a real-valued function ψ such that $u_i(z) - u_i(z') = \psi(z) - \psi(z')$ for all z and z' that differ by only the action of player i. In other words, ψ is a single function which captures all of the essential strategic properties of the game—whenever a single player i deviates from a given strategy profile z, the change in i's payoff is given by the change in potential.

The game specified in [6.31] has potential. Indeed, for this game,

$$\psi(z) = G(z) - \sum_{i=1}^{n} c_i(z_i). \qquad [6.32]$$

As observed earlier, the log-linearity of the choice rule implies that

$$\log\left(\frac{p_{z\to z'}}{p_{z'\to z}}\right) = \beta[u_i(z') - u_i(z)] = \beta[\psi(z') - \psi(z)], \qquad [6.33]$$

where the second equality follows from the definition of the potential function ψ. It follows that the ergodic distribution has a particularly simple representation, known as a Gibbs-Boltzmann distribution (and see Proposition 6.7 in Section 6.6):

$$\pi_z \equiv \lim_{t\to\infty} \mathbb{P}(z^t = z) = \frac{\exp[\beta\psi(z)]}{\sum_{z'} \exp[\beta\psi(z')]}. \qquad [6.34]$$

This is presented in Myatt and Wallace (2009, Lemma 1), who provide a detailed analysis of the game specified by [6.31]. Rather than reproduce the discussion in that paper here, a couple of immediate consequences may be drawn.

First, note that as $\beta \to \infty$ only the states z that maximize potential will have weight in the ergodic distribution. The potential maximizing Nash equilibria are selected as noise vanishes. As a result, in this class of games, the state that maximizes the difference between $G(z)$, the private benefit, and $\sum_i c_i(z_i)$, the social cost, is stochastically stable. In general, this will differ from the social-welfare maximizing state, z^*, where (with additive welfare),

$$z^* = \arg\max_z \left\{ nG(z) - \sum_{i=1}^{n} c_i(z_i) \right\}. \qquad [6.35]$$

Second, note that to make progress with this model there is actually no need to take the limit $\beta \to \infty$. The ergodic distribution is provided in a convenient closed form in [6.34], and analysis can be performed away from the limit. Myatt and Wallace (2009) make full use of this feature to examine the properties of social welfare in [6.35] evaluated at the ergodic limit. Many of the messages discussed in Sections 6.5.1–6.5.3 are echoed in this analysis; particularly the fact that successful public-good provision is dependent upon the relationship between the *private* benefits versus the *social* costs of provision.

6.6. NETWORK GAMES

Up until now we have considered environments where players interact with each other on a purely random basis, that is, the population is uniformly mixed and all pairs of individuals have the same probability of interacting. In practice, it is reasonable to assume that people interact more frequently with others who are "close" in a geographical or social sense. In this section we shall show how to model such situations using the concept of network games. There is an extensive literature on this topic, some of which takes the network as given and the choice of actions as the strategic variable, and some of which takes the choices of links and actions as strategic variables. For comprehensive surveys, see Vega-Redondo (2007), Goyal (2007), Jackson (2008), and Jackson and Zenou (Chapter 3 of this volume).

Here, we shall treat the first situation: the network is fixed, and there is one agent at each node who chooses an action given the actions of his neighbors. We shall assume that the edges of the network are undirected, and each edge has a weight (a nonnegative real number) that measures the importance of that particular interaction. For example, agents who are geographically close might have higher weights than those who are far apart. Let V denote the set of n vertices, and let $i \in V$ denote a particular vertex, which we shall identify with the agent located at i. Let $w_{ij} \geq 0$ be the weight corresponding to the pair $\{i, j\}$. (If $w_{ij} = 0$ there is no edge between i and j, that is, they do not interact.)

We shall assume that the importance of any interaction is weighted equally for the two individuals involved, that is, $w_{ij} = w_{ji}$.[28]

Let X be a finite set of strategies or actions, which we shall suppose is the same for each agent. Let $\mathcal{G}$ be a symmetric two-person game with utility function $u : X \times X \to \mathcal{R}$, that is, $u(x, y)$ is the payoff to an x-player when his opponent plays y. Let Γ be an undirected weighted graph with n vertices. Such a network is fully described by a vector of weights $\vec{w} \in \mathcal{R}_+^{\binom{n}{2}}$. Given $\mathcal{G}$ and $\vec{w}$, the associated *network game* is the n-player game defined as follows. The joint action space is X^n. Given a profile of actions $\vec{x} \in X^n$ the *interactive payoff* to i is defined to be

$$I_i(\vec{x}) = \sum_{j \neq i} w_{ij} u(x_i, x_j). \qquad [6.36]$$

In other words, the payoff to i is the sum of the payoffs when i interacts once with every other player, weighted by the importance of these interactions. (We remark that if the weights at each vertex sum to one, then w_{ij} can be interpreted as the *probability* of the interaction $i \leftrightarrow j$, and $I_i(\vec{x})$ is the expected payoff to i.)

This framework can be extended to include idiosyncratic differences in agents' tastes for particular actions, a generalization that is quite useful in applications. Let $v_i : X \mapsto \mathcal{R}$ denote the *idiosyncratic payoff* that agent i would receive from taking each action in isolation, irrespective of what the others are doing. Agent i's total payoff from playing action x_i is the sum of the idiosyncratic payoff from x_i plus the interactive payoff from playing x_i given the choices of all the other agents:

$$U_i(\vec{x}) = \sum_{j \neq i} w_{ij} u(x_i, x_j) + v_i(x_i). \qquad [6.37]$$

Consider the following example: let X be a set of communication technologies, such as different types of cellphones. The weight w_{ij} is the frequency with which i and j communicate per unit time period. The number $u(x, y)$ is the payoff to someone using technology x when communicating once with someone using technology y. The number $v_i(x)$ is the per-period utility to agent i from owning technology x, which is determined by ease of use, cost, and other factors that may be particular to i. Thus in state $\vec{x}$, the total utility to i per unit time period is given by expression [6.36].

Let us consider how such a system of n interacting agents might evolve over time. Assume that time is discrete: $t = 0, 1, 2, 3, \ldots$ The *state* at time t is denoted by $\vec{x^t}$, where $x_i^t \in X$ represents agent i's choice at that point in time. At the start of each

[28] In practice influence may be asymmetric, that is, agent i may weight j's actions more heavily than j weights i's actions. This situation substantially complicates the analysis and we shall not pursue it here.

period one agent is drawn at random and he reconsiders his choice. This is called a *revision opportunity*.[29] As in Section 6.5 we assume that whenever agent i has a revision opportunity, he chooses an action according to a log linear response function, that is, the probability that i chooses action x in period $t+1$ given that the current state is $\overrightarrow{x^t}$, is given by the expression

$$\mathbb{P}(x_i^{t+1} = x \mid x^t) = \frac{\exp[\beta U_i(x, x_{-i}^{t-1})]}{\sum_{y \in X} \exp[\beta U_i(y, x_{-i}^{t-1})]} \quad \text{for some } \beta \geq 0. \qquad [6.38]$$

The number β is the *response parameter*. The larger β is, the more likely it is that the agent chooses a (myopic) best response given the state $\overrightarrow{x^t}$.

Models of this type were pioneered by Blume (1993, 1995), and have been applied in various settings by Young (1998), Young and Burke (2001), Montanari and Saberi (2010), and Kreindler and Young (2013), amongst others. This type of model is also standard in the discrete choice literature (McFadden, 1976) and can be estimated empirically from a regression model of form[30]

$$\log \mathbb{P}(x_i^t = x \mid x^{t-1}) - \log \mathbb{P}(x_i^t = y \mid x^t) = \beta[U_i(x, x_{-i}^{t-1}) - U_i(y, x_{-i}^{t-1})] + \varepsilon_i^t. \quad [6.39]$$

The stochastic adjustment process described above can be represented as a finite Markov chain on the state space X^n. This process is irreducible, hence it has a unique stationary distribution μ^β. For each $\vec{x} \in X$, $\mu^\beta(x)$ represents the long-run relative frequency with which state x is visited starting from any initial state. In the present case this distribution takes a very simple form. Define the function

$$\rho(\vec{x}) = (1/2) \sum_{1 \leq i,j \leq n} w_{ij} u(x_i, x_j) + \sum_{1 \leq i \leq n} v_i(x_i).$$

We claim that $\rho(\cdot)$ is a potential function for the network game. To establish this it suffices to show that for every agent i, the change in i's utility from a unilateral change in strategy is identical to the induced change in potential. Fix an agent i and a set of choices by the other agents, say $\vec{x}_{-i}$. The difference in i's utility between choosing x and choosing y is

[29] An analogous process can be defined in continuous time as follows. Suppose that each agent i's revision opportunities are governed by a Poisson random variable ω_i with arrival rate $\lambda = 1$, and that the ω_t's are i.i.d. With probability one no two agents revise at exactly the same time. Hence the distinct times at which agents revise define a discrete-time process as assumed in the text.

[30] An alternative model is that each agent chooses a non-best response (given his neighbors' choices) with a uniform error probability $\varepsilon > 0$, as in the model discussed in Sections 6.2 and 6.3. This leads to somewhat different selection results (for more on this topic see Jackson and Zenou, this volume, Chapter 3, Section 3.3).

$$U_i(x, \vec{x}_{-i}) - U_i(y, \vec{x}_{-i}) = \sum_{j \neq i} w_{ij}(u(x, x_j) - u(y, x_j)) + v_i(x) - v_i(y)$$
$$= \rho(x, \vec{x}_{-i}) - \rho(y, \vec{x}_{-i}).$$

It follows that $\rho(\cdot)$ acts as a potential function. Note that the potential of a state is not the same as the total welfare: indeed the latter is

$$W(\vec{x}) = \sum_{1 \leq i,j \leq n} w_{ij} u(x_i, x_j) + \sum_{1 \leq i \leq n} v_i(x_i)$$

In particular, the potential function "discounts" the interactive payoffs by 50% whereas it counts the non-interactive payoffs at full value.

As noted in the previous section, the ergodic distribution takes the following simple form, known as a Gibbs-Boltzmann distribution (Blume, 1993, 1995).

Proposition 6.7. *The ergodic distribution of the log linear updating process is*

$$\mu^\beta(\vec{x}) = \frac{\exp[\beta\rho(\vec{x})]}{\sum_{y \in X^n} \exp[\beta\rho(\vec{y})]}.$$

Corollary 6.1. *The stochastically stable states are the states with maximum potential.*

Some of the implications of this result can be illustrated through the following example. Let A and B be two communication technologies, and suppose that the payoffs from pairwise communications are as shown in the game matrix in Figure 6.11. Suppose further that the population consists of two types of individuals: *hipsters* and *squares*. Let the players' idiosyncratic payoffs be as shown in Figure 6.12.

	A	B
A	c, c	1, 1
B	1, 1	c, c

Figure 6.11 A simple two technology example, with $c > 1$.

Hipsters		Squares	
A	1	A	0
B	0	B	1

Figure 6.12 Idiosyncratic payoffs for the two technology example.

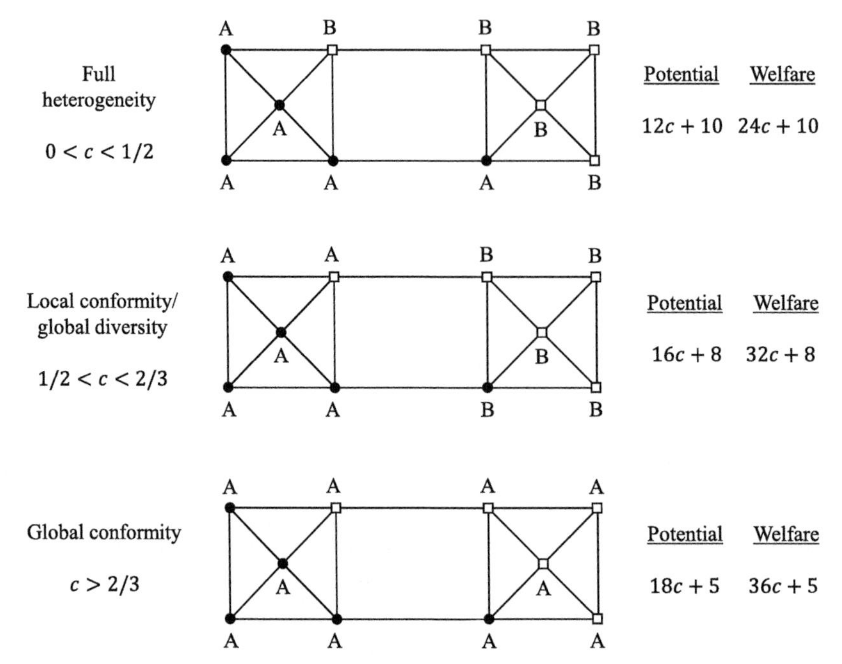

Figure 6.13 A network with two weakly linked clusters. Hipsters are represented by dots, squares by squares. The stochastically stable configurations depend on the coordination payoff c.

Consider the network in Figure 6.13, where the weight on each edge is 1. A consideration of cases shows that the state with maximum potential can take one of three forms depending on the value of c:

(i) *Full heterogeneity*: each person uses his preferred technology.

(ii) *Local conformity/global diversity*: everyone in the left group uses A and everyone in the right group uses B.

(iii) *Global conformity*: Everyone uses the same technology irrespective of his personal tastes.

When the benefits from coordination are low ($c < 1/2$), the most likely state is full heterogeneity. Note that this is not optimal: more conformity would lead to higher total utility. However, when the benefits from coordination are sufficiently large ($c > 2/3$) full coordination is the most likely state and it also maximizes total utility.

Perhaps the most interesting case is the intermediate one where partial coordination results. The logic is that society consists of two groups—left and right—who interact mainly, though not exclusively, with members of their own group. The left group (mostly hipsters) uses technology A even though the minority of its members prefer B, while the opposite holds for the right group. More generally, the model predicts that within-group behavior will be fairly uniform but between-group heterogeneity may exist when

	A	B
A	2, 2	0, 0
B	0, 0	1, 1

Figure 6.14 A simple coordination game.

the groups interact only weakly with each other. In particular, within-group norms may overcome heterogeneity of preferences by members of the group when the conformity motive is sufficiently strong. This is known as the *local conformity/global diversity* effect (Young, 1998, Chapter 10). This effect has been documented empirically in a variety of settings, including contractual norms in agriculture (Young and Burke, 2001), and norms of medical practice (Burke et al., 2010; Wennberg and Gittelsohn, 1973, 1982).

6.7. SPEED OF CONVERGENCE

A common criticism of stochastic models of equilibrium selection is that it can take an extremely long time for a population of agents to reach the stochastically stable state. Indeed, by definition, the stochastically stable states are those whose probability is bounded away from zero when the noise in the adjustment process is vanishingly small. When the noise is extremely small, however, it will take an extremely long time in expectation for enough agents to go against the flow (choose non-best replies) in order to tip the process into the basin of a stochastically stable state.

To take a concrete case, consider a population of 101 agents in which everyone plays everyone else once per period, and the payoffs are given by the game shown in Figure 6.14. Each period one player is selected at random, and he revises his choice using a log linear response rule with parameter β. The stochastically stable state is the one in which everyone chooses A. Now suppose that the process starts in the "bad" equilibrium where everyone chooses B. In this state, a player with a revision opportunity will switch to A with probability $e^0/(e^0 + e^{100\beta}) \approx e^{-100\beta}$. Even for moderate values of β this is an extremely improbable event.

Nevertheless, we cannot conclude that evolutionary selection is impossibly slow in large populations. A special feature of this example is the assumption that everyone plays everyone else in every period. A more realistic assumption would be that, in any given period, an agent interacts only with a few people who are "close" in a geographical or social sense. It turns out that in this situation equilibrium selection can be quite rapid (Young, 1998, 2011).

To be specific, let us assume that agents are located at the nodes of a network Γ, and that in each period every agent plays each of his neighbors exactly once using a 2 x 2 symmetric coordination game. Let us also assume that, whenever they have a revision

	A	B
A	$1+\alpha$, $1+\alpha$	0, 0
B	0, 0	1, 1

Figure 6.15 A 2×2 coordination game.

opportunity, agents use log linear learning with response parameter β. We shall focus on games with the following payoff structure (Figure 6.15):

In this case the risk-dominant and Pareto-dominant equilibria coincide, which simplifies the interpretation of the results.[31] The issue that we wish to examine is how long it takes for the evolutionary process to transit from the all-B state to a neighborhood of the all-A state as a function of: (i) *the size of the advantage* α, (ii) the *degree of rationality* β, and (iii) the *topological properties* of the social network Γ. A key feature of the dynamics is the existence of critical values of α and β such that the waiting times are bounded *independently of the number of agents.*

Before stating this result, however, we need to clarify what we mean by "how long it takes" for the innovation A to spread through a given population. One possibility is to look at the expected time until *all* agents are playing A. Unfortunately this definition is not satisfactory owing to the stochastic nature of the adjustment process. To appreciate the difficulty, consider the case where β is close to zero, and hence the probability of playing A is only slightly larger than the probability of playing B. No matter how long we wait, the probability is high that a sizable proportion of the population will be playing B at any given future time. Thus the expected waiting time until *everyone* plays A is not the relevant concept. (A similar difficulty arises for any stochastic selection process, not just log linear selection.)

We are therefore led to the following definition. For each state x let $a(x)$ denote the *proportion* of agents playing A in state x. Given a target level of *penetration* $0 < p < 1$, define

$$T(\Gamma, \alpha, \beta, p) = E[\min\{t : a(x^t) \geq p \; \& \; \forall t' \geq t, \mathbb{P}(a(x^{t'}) \geq p) \geq p]. \qquad [6.40]$$

In words, $T(\Gamma, \alpha, \beta, p)$ is the expected waiting time until the first time such that: (i) at least p of the agents are playing A, and (ii) the probability is at least p that at least p of the agents are playing A at all subsequent times. Notice that the higher the value of p, the larger β must be for the expected waiting time to be finite.

To distinguish between fast and slow selection we shall consider families of networks of different sizes, where the *size* of a network Γ is the number of its nodes (equivalently, the number of agents).

[31] In general, evolutionary selection favors the risk-dominant equilibrium in symmetric 2×2 games. Analogous results on the speed of convergence hold in this case.

Fast versus slow selection. Given a family of networks $\mathcal{G}$ and $\alpha > 0$, selection is *fast* for $\mathcal{G}$ and α if, for every $p < 1$ there exists $\beta_p > 0$ such that for all $\beta \geq \beta_p$

$$T(\Gamma, \alpha, \beta, p) \text{ is bounded above for all } \Gamma \in \mathcal{G}. \qquad [6.41]$$

Otherwise selection is *slow*, that is, there is an infinite sequence of graphs of increasing size $\Gamma_1, \Gamma_2, \ldots, \Gamma_n, \ldots \in \mathcal{G}$ such that $\lim_{n\to\infty} T(\Gamma_n, \alpha, \beta, p) = \infty$.

6.7.1 Autonomy

We now formulate a general condition on families of networks that guarantees fast selection. Fix a network $\Gamma = (V, W)$, the log linear process P^β, and an advantage $\alpha > 0$. Given a subset of vertices $S \subseteq V$, define the *restricted selection process* $\widetilde{P}^\beta_S$ as follows: all agents $i \notin S$ are *held fixed* at B while the agents in S update according to the process P^β. Let $(\vec{A}_S, \vec{B}_{V-S})$ denote the state in which every member of S plays A and every member of $V - S$ plays B.

Autonomy. Given a network Γ and a value $\alpha > 0$, a subset S of agents is *autonomous* if $(\vec{A}_S, \vec{B}_{V-S})$ is stochastically stable under the restricted process $\widetilde{P}^\beta_S$. (Recall that a state is *stochastically stable* if it has non-vanishing probability in the limit as $\beta \to \infty$.)

Proposition 6.8. (Young, 2011) *Given a family of networks $\mathcal{G}$ and $\alpha > 0$ suppose that there exists a positive integer s such that for every $\Gamma \in \mathcal{G}$, every member of Γ is contained in an autonomous subset of size at most s. Then selection is fast.*

In other words, given any target level of penetration $p < 1$, if the level of rationality β is high enough, the expected waiting time until at least p of the agents are playing A (and continue to do so with probability at least p in each subsequent period) is bounded above independently of the number of the agents in the network.

Proof sketch. For each agent i let S_i be an autonomous set containing i such that $|S_i| \leq s$. By assumption the state $(\vec{A}_{S_i}, \vec{B}_{V-S_i})$ in which all members of S_i play A is stochastically stable. Given a target level of penetration $p < 1$, we can choose β so large that the probability of being in this state after some finite time $t \geq T_{S_i}$ is at least $1 - (1-p)^2$. Since this holds for the restricted process, and the probability that i chooses A does not decrease when someone else switches to A, it follows that in the *unrestricted* process $\mathbb{P}(x_i^t = A) \geq 1 - (1-p)^2$ for all $t \geq T_{S_i}$.[32] Since $|S_i| \leq s$ for all i, we can choose β and T so that $\mathbb{P}(x_i^t = A) \geq (1-p)^2$ for all i and all $t \geq T$. It follows that the *expected proportion* of agents playing A is at least $1 - (1-p)^2$ at all times $t \geq T$.

[32] The last statement follows from a coupling argument, which we shall not give here. See Young (1998, Chapter 6) for details.

Now suppose, by way of contradiction, that the probability is less than p that at least p of the agents are playing A at some time $t \geq T$. Then the probability is greater than $1 - p$ that at least $1 - p$ of the agents are playing B. Hence the expected number playing A is less than $1 - (1 - p)^2$, which is a contradiction.

6.7.2 Close-knittedness

The essential feature of an autonomous set is that its members interact sufficiently closely with each other that they can sustain all-A with high probability even when everyone else plays B. We can recast this as a topological condition as follows (Young, 1998, Chapter 6). Let $\Gamma = (V, W)$ be a graph and $\alpha > 0$ the size of the payoff gap. For every nonempty subset of vertices $S \subseteq V$ let

$$d(S) = \sum_{i \in S, j \in V} w_{ij}. \quad [6.42]$$

Further, for every nonempty subset $S' \subseteq S$ let

$$d(S', S) = \sum_{\{i,j\}: i \in S', j \in S} w_{ij}. \quad [6.43]$$

Thus $d(S)$ is the weighted sum of edges that have at least one end in S, and $d(S', S)$ is the weighted sum of edges that have one end in S' and the other end in S.

Definition 6.5. *Given any real number* $r \in (0, 1/2]$, *the set* S *is* r-close-knit *if*

$$\forall S' \subseteq S, S' \neq \varnothing, \quad d(S', S)/d(S') \geq r. \quad [6.44]$$

Intuitively, S is r-close-knit if no subset has too large a proportion of its interactions with outsiders.

If S is autonomous, then by definition the potential function of the restricted process is maximized when everyone in S chooses A. Straightforward computations show that this implies the following.

Proposition 6.9. (Young, 2011) *Given any network* Γ *and* $\alpha > 0$, *S is autonomous if and only if S is r-close-knit for some* $r > 1/(\alpha + 2)$.

Corollary 6.2. *Given a family of networks* $\mathcal{G}$ *in which all nodes have degree at most* d, *selection is fast whenever* $\alpha > d - 2$.

The latter follows from the observation that when all degrees are bounded by d, then a tree of sufficiently large size is more than $(1/d)$-close-knit, hence more than $1/(\alpha + 2)$-close knit.

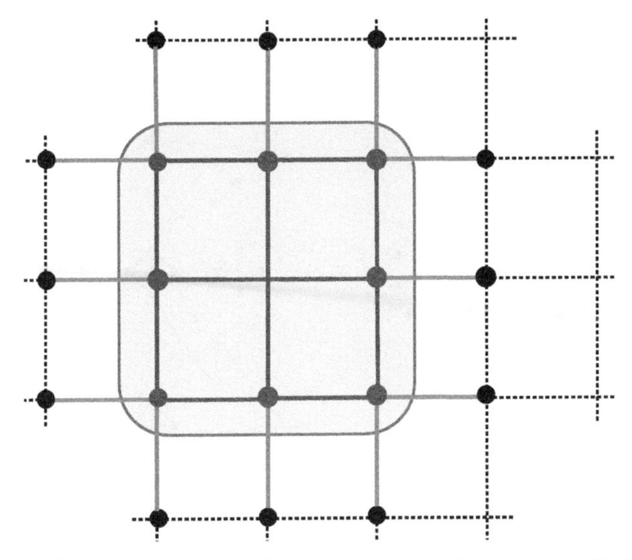

Figure 6.16 A subset of the two-dimensional lattice consisting of nine vertices that is autonomous for any $\alpha > 1$ is highlighted.

Close-knit families. A family of graphs $\mathcal{G}$ is *close-knit* if for *every* $r \in (0, 1/2)$ there exists a positive integer $s(r)$ such that, for every $\Gamma \in \mathcal{G}$, every node of Γ is in an r-close-knit set of cardinality at most $s(r)$.

Corollary 6.3. *Given any close-knit family of graphs $\mathcal{G}$, selection is fast for all $\alpha > 0$.*

To illustrate the latter result, consider a two-dimensional regular lattice (a square grid) in which every vertex has degree 4 (see Figure 6.16). Assume for simplicity that each edge has weight 1. The shaded region in the figure is a square S consisting of nine nodes, which we claim is 1/3-close-knit. Indeed the sum of the degrees is $d(S) = 36$ and the number of internal edges is $d(S, S) = 12$; moreover it can be checked that for every nonempty $S' \subset S$ the ratio $d(S', S)/d(S') \geq 12/36 = 1/3$. It follows from Proposition 6.9 that S is autonomous whenever $\alpha > 1$.

More generally, every square S of side m has $2m(m-1)$ internal edges and m^2 vertices, each of degree 4, hence

$$d(S, S)/d(S) = 2m(m-1)/4m^2 = 1/2 - 1/2m. \qquad [6.45]$$

It is easily checked that for every nonempty subset $S' \subseteq S$,

$$d(S', S)/d(S') \geq 1/2 - 1/2m. \qquad [6.46]$$

Therefore a square of side m is $(1/2 - 1/2m)$-close-knit, so it is autonomous whenever $\alpha > 2/(m-1)$. It follows that, given any $\alpha > 0$, there is an autonomous set of bounded

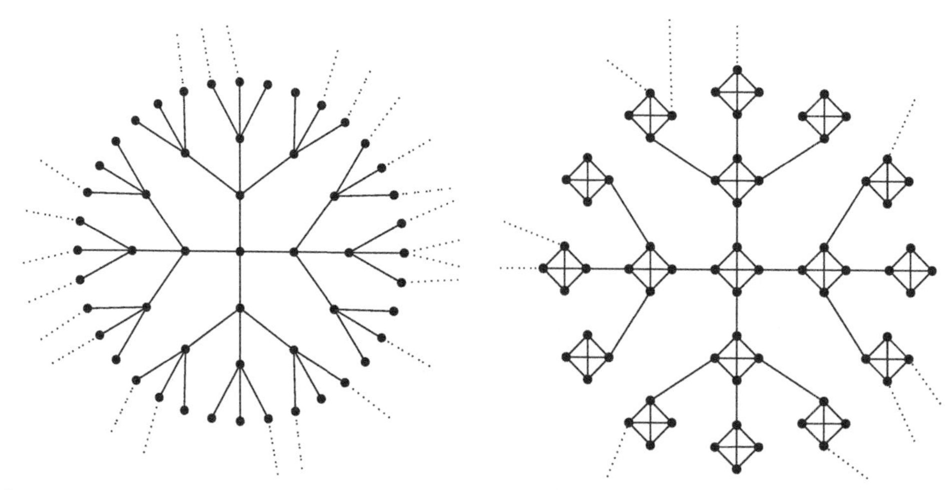

Figure 6.17 Two networks, one with clusters, the other without. Every node other than the end-nodes has degree 4,and there is a hub (not shown) that is connected to every end-node (dashed lines). All edge-weights equal 1.

size (namely a square of side $m > 1 + 2/\alpha$). We have therefore shown that the family of square grids is close-knit, hence selection is fast for any $\alpha > 0$. A similar argument holds for any regular d-dimensional regular lattice: given any $\alpha > 0$, every sufficiently large sublattice is autonomous for α, and this holds independently of the number of vertices in the full lattice.

We remark that in this case fast selection does not arise because neighbors of neighbors tend to be neighbors of one another. In fact, a d-dimensional lattice has the property that no two of the neighbors of a given node are adjacent. Rather, fast selection arises from a basic fact of Euclidean geometry: the ratio of the "surface" to the "volume" of a d-dimensional cube goes to zero as the cube becomes arbitrarily large.

A d-dimensional lattice illustrates the concept of autonomy in a very transparent way, but it applies in many other situations as well. Indeed, one could argue that many real-world networks are composed of relatively small autonomous groups, either because people tend to cluster geographically, or because they tend to interact with people of their own kind (homophily) or for both reasons.

To understand the difference between a network with small autonomous groups and one without, consider the pair of networks in Figure 6.17. The left panel shows a tree in which every node other than the end-nodes has degree 4, and there is a "hub" (not shown) that is connected to all the end-nodes. The right panel shows a graph with a similar overall structure in which every node other than the hub has degree 4; however, in this case everyone (except the hub) is contained in a clique of size 4. In both networks all edges are assumed to have weight 1.

Suppose that we begin in the all-B state in both networks, that agents use log linear selection with $\beta = 1$, and that the size of the advantage is $\alpha > 2/3$. Let each network have n vertices. It can be shown that the expected waiting time to reach a state where at least 99% are playing A is unbounded in n for the network on the left, whereas it is bounded independently of n for the network on the right. In the latter case, simulations show that it takes fewer than 25 periods (on average) for A to penetrate to the 99% level independently of n. The key difference between the two situations is that, in the network with cliques, the innovation can establish a toehold in the cliques relatively quickly, which then causes the hub to switch to the innovation also.

Note, however, that fast selection in the network with cliques does *not* follow from Proposition 6.9, because not every node is contained in a clique. In particular, the hub is connected to all of the leaves, the number of which grows with the size of the tree, so it is not in an r-close-knit set of bounded size for any given $r < 1/2$. Nevertheless selection is fast: any given clique adopts A with high probability in bounded time, hence a sizable proportion of the cliques linked to the hub switch to A in bounded time, and then the hub switches to A also. In other words, fast selection occurs through the combined action of autonomy and contagion (for further details see Young, 2011).

The preceding results show that networks with "natural" topologies frequently exhibit fast selection, but this does not mean that such topologies are *necessary* for fast selection. Indeed, it turns out that fast selection can also occur when everyone is equally likely to interact with everyone else, and they update based on random samples of what others are doing. To be specific, fix a sample size $d \geq 3$ and a game with advantage $\alpha > 0$. Suppose that agents revise their choices according to independent Poisson processes with expectation one. Given a revision opportunity an agent draws a random sample of size d from the rest of the population and chooses a log linear response with parameter β to the distribution of choices in the sample. Let $T(\Gamma, \alpha, \beta, d)$ be the expected waiting time until a majority of agents choose A, starting from the state where everyone chooses B. (Thus we are considering a level of penetration $p = \frac{1}{2}$.) We say that selection is *fast* for (α, β, d) if $T(\Gamma, \alpha, \beta, d)$ is bounded above for all $\Gamma \in \mathcal{G}$. The following result shows that selection is fast whenever α is sufficiently large; in particular, fast selection can occur without the benefit of a particular "topology."

Proposition 6.10. (Kreindler and Young, 2013) *If agents in a large population respond to random samples of size $d \geq 3$ using log linear learning with response parameter β, selection is fast whenever the advantage α satisfies:*

$$\begin{aligned} &\alpha > \min\{(e^{\beta-1} + 4 - e)/\beta, d - 2\} && \text{if } 2 < \beta < \infty, \\ &\alpha > 0 && \text{if } 0 < \beta \leq 2. \end{aligned}$$

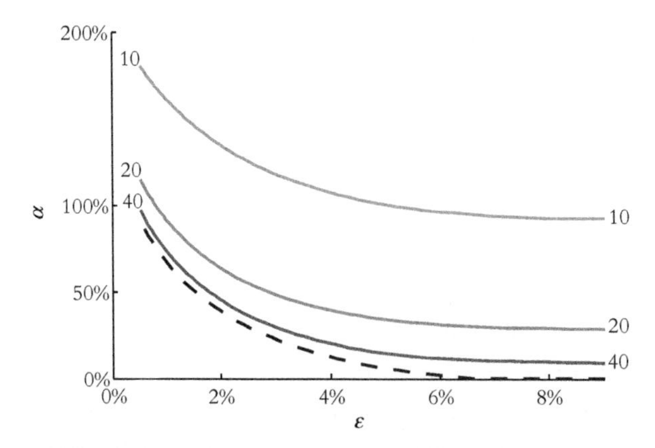

Figure 6.18 Waiting times until 99% of the population plays A. Sample size $d = 4$. The dashed line is the separatrix between bounded and unbounded waiting time.

This result shows that the dynamics exhibit a *phase transition*: for any given $\beta > 2$ there is a critical level of α such that above this level the waiting time is fast, whereas below this level the waiting time grows exponentially with the number of agents. Moreover, if $\beta < 2$ (i.e., the noise is sufficiently large), the waiting time is bounded for *every* positive value of α.

These waiting times are surprisingly short—on the order of 10-40 time periods—for a wide range of parameter values. See Figure 6.18 which shows the waiting times for $d = 4$ and various combinations of α and ε, where $\varepsilon = e^0/(e^0 + e^\beta)$ is the probability that an agent chooses A when everyone in his sample chooses B.

We conclude this section by pointing out some related results in the literature. Morris (2000) studies deterministic best-response dynamics on infinite networks: in each period, each agent myopically best responds to his neighbors' actions. Morris identifies topological conditions such that if the payoff gap between the equilibria is high enough, the high equilibrium spreads by "contagion" from some finite subgroup to the entire population. (This does not address the issue of waiting times as such, but it does identify topological conditions under which the process can escape from the low equilibrium in finite time.) A particularly interesting case arises when the network is connected and all degrees are bounded above by some integer. In this case $\alpha > d - 2$ guarantees that a single adopter is sufficient for the innovation to spread by contagion to the entire population. Note that this same lower bound on α guarantees fast selection in the stochastic models discussed earlier (Proposition 6.9 and Corollary 6.2). For related results on random networks see López-Pintado (2006).

Montanari and Saberi (2010) characterize waiting times in finite networks in terms of certain topological properties drawn from statistical physics. As in the models of Ellison (1993) and Young (1998, 2011) they show that local clustering tends to speed up the selection process. They also show that small-world networks—in which most connections are "local" but there are also some links spanning long distances—tend to have much longer waiting times. It should be noted, however, that these results are concerned with *relative* waiting times for different topologies when the noise level is taken to zero ($\beta \to \infty$), hence the absolute waiting times are unboundedly large. In contrast, Propositions 6.8–6.10 establish conditions under which the absolute values of the waiting times are bounded for *given* values of α and β. Similar results hold for any network irrespective of its topology (Kreindler and Young, 2014).

6.8. CONCLUDING REMARKS

Here we recapitulate some of the main ideas in the preceding sections. The evolutionary framework differs conceptually from that of "classical" game theory in two major respects. First, the focus is on large populations of individuals who interact at random, rather than on a small number of individuals who play a repeated game. Second, individuals are assumed to employ simple adaptive rules rather than to engage in perfectly rational behavior. The argument is that when many players interact randomly over long periods of time, there is little reason to think that any one of them would have enough information to be able to anticipate the system dynamics correctly. Moreover they have little incentive to do so, because no one individual will have a significant influence on the future course of the process. In this setting, an equilibrium is a rest point of the population-level dynamical process rather than a form of consistency between beliefs and strategies, as in traditional game theory.

In spite of this difference in perspective, various solution concepts from traditional theory carry over into the evolutionary setting. A case in point is ESS, which is a refinement of Nash equilibrium. An ESS is a rest point of the dynamics that is robust to small local deviations, that is, when given a small one-time shock, the process returns to the rest point in question. The main contribution of stochastic evolutionary game theory is to show that this theory must be substantially modified when the dynamical system is subjected to *persistent* random perturbations. Such perturbations are quite natural and can arise from a variety of factors, including random utility shocks, heterogeneity in payoffs, mistakes, mutations, and so forth. These persistent perturbations typically lead to an ergodic process whose long-run distribution can be estimated using the theory of large deviations. When these perturbations are "small" in a suitably defined sense, the ergodic distribution places high probability on particular equilibria, which are known as stochastically stable equilibria. Thus stochasticity injects an additional degree of realism into the models, and also leads to a sharp form of equilibrium selection.

In this chapter, we have discussed the consequences of this idea in a variety of settings, including 2×2 games, bargaining games, public goods games, potential games, and network games. Another important application that we did not have the space to treat is stochastic evolution in extensive form games (Nöldeke and Samuelson, 1993;

Hart, 2002). A common theme that emerges from these cases is that stochastic stability often selects equilibria that are well-known in traditional game theory. For example, in 2×2 games one obtains the risk-dominant equilibrium, in bargaining games the Nash bargaining solution, in potential games the potential-maximizing equilibrium, and for some classes of extensive form games the subgame perfect equilibrium. However, the justification for these solution concepts differs between the two approaches. In traditional game theory equilibria are justified in terms of perfect rationality, common knowledge of the game, and common knowledge of rationality, whereas evolutionary game theory dispenses with all three of these assumptions. Nevertheless some of the main solution concepts (or refinements of them) survive in the evolutionary setting.

What then are the major open problems in the subject? Perhaps the single most important challenge is to bring the predictions of theory into closer contact with empirical applications. To date this has been attempted in relatively few cases. One example is the evolution of contract forms in certain types of economic activity, such as agriculture. Stochastic evolutionary models suggest that contract forms will tend to be more homogeneous locally than standard theory would predict owing to positive feedback effects; moreover this prediction is corroborated by data on contract usage in the Midwestern United States (Young and Burke, 2001). Similarly, stochastic evolutionary models can help explain the well-documented tendency of medical treatments to differ substantially by geographical region (Burke et al., 2010; Wennberg and Gittelsohn, 1973, 1982).

More generally, stochastic evolutionary models provide a potentially powerful tool for analyzing the dynamics of social norms. How does the structure of social interactions affect the rate at which norms shift? Do particular interaction topologies tend to encourage local norms or global ones? What intervention strategies are likely to be most effective for instituting a change in norms? These questions can be studied using field data as well as data from controlled experiments. A particularly promising direction is the use of on-line experiments to study the effects of interaction structure, the amount of information available, heterogeneity in payoffs, and a variety of other factors, in a large population setting.

REFERENCES

Agastya, M., 1997. Adaptive play in multiplayer bargaining situations. Rev. Econ. Stud. 64(3), 411–426.

Agastya, M., 1999. Perturbed adaptive dynamics in coalition form games. J. Econ. Theory 89(2), 207–233.

Aumann, R.J., 1985. What is game theory trying to accomplish? In: Arrow, K., Honkapohja, S. (Eds.), *Frontiers of Economics*, Basil Blackwell, Oxford, pp. 28–76.

Basu, K., Weibull, J.W., 1991. Strategy subsets closed under rational behavior. Econ. Lett. 36(2), 141–146.

Bergin, J., Lipman, B., 1996. Evolution with state-dependent mutations. Econometrica 64(4), 943–956.

Blume, L.E., 1993. The statistical mechanics of strategic interaction. Games Econ. Behav. 5(3), 387–424.

Blume, L.E., 1995. The statistical mechanics of best-response strategy revision. Games Econ. Behav. 11(2), 111–145.
Blume, L.E., 1997. Population games. In: Arthur, W.B., Durlauf, S.N., Lane, D.A. (Eds.), *The Economy as an Evolving Complex System II.* Westview Press, Boulder, CO.
Blume, J., 2003. How noise matters. Games Econ. Behav. 44(2), 251–271.
Burke, M.A., Fournier, G.M., Prasad, K., 2010. Geographic variations in a model of physician treatment choice with social interactions. J. Econ. Behav. Organ. 73(3), 418–432.
Ellison, G., 1993. Learning, local interaction and coordination. Econometrica 61(5), 1047–1071.
Ellison, G., 2000. Basins of attraction, long run stochastic stability and the speed of step-by-step evolution. Rev. Econ. Stud. 67(1), 17–45.
Foster, D., Young, H.P., 1990. Stochastic evolutionary game dynamics. Theor. Popul. Biol. 38(2), 219–232.
Foster, D., Young, H.P., 1997. A correction to the paper 'Stochastic Evolutionary Game Dynamics'. Theor. Popul. Biol. 51(1), 77–78.
Freidlin, M.I., Wentzell, A.D., 1998. Random Perturbations of Dynamical Systems, Second ed. Springer-Verlag, Berlin/New York.
Fudenberg, D., Harris, C., 1992. Evolutionary dynamics with aggregate shocks. J. Econ. Theory 57(2), 420–441.
Goyal, S., 2007. Connections: An Introduction to the Economics of Networks. Princeton University Press, Princeton, NJ.
Harsanyi, J.C., Selten, R., 1988. A General Theory of Equilibrium Selection in Games. MIT Press, Cambridge, MA.
Hart, S., 2002. Evolutionary dynamics and backward induction. Games Econ. Behav. 41(2), 227–264.
Hurkens, S., 1995. Learning by forgetful players. Games Econ. Behav. 11(2), 304–329.
Jackson, M.O., 2008. Social and Economic Networks. Princeton University Press, Princeton, NJ.
Jackson, M.O., Kalai, E., 1997. Social learning in recurring games. Games Econ. Behav. 21(1–2), 102–134.
Kandori, M., Mailath, G.J., Rob, R., 1993. Learning, mutation and long-run equilibria in games. Econometrica 61(1), 29–56.
Kreindler, G.E., Young, H.P., 2013. Fast convergence in evolutionary equilibrium selection. Games Econ. Behav. 80, 39–67.
Kreindler, G.E., Young, H.P., 2014. Rapid innovation diffusion in social netwroks. Proc. Nat. Acad. Sci. 111(3), 10881–10888.
López-Pintado, D., 2006. Contagion and coordination in random networks. Int. J. Game Theory 34(3), 371–381.
Maruta, T., 1997. On the relationship between risk-dominance and stochastic stability. Games Econ. Behav. 19(2), 221–234.
Maynard Smith, J., 1974. The theory of games and the evolution of animal conflicts. J. Theor. Biol. 47(1), 209–221.
Maynard Smith, J., 1982. Evolution and the Theory of Games. Cambridge University Press, Cambridge.
Maynard Smith, J., Price, G.R., 1973. The logic of animal conflict. Nature 246(5427), 15–18.
McFadden, D.L., 1976. Quantal choice analysis: a survey. Ann. Econ. Soc. Meas. 5(4), 363–390.
McKelvey, R.D., Palfrey, T.R., 1995. Quantal response equilibria for normal form games. Games Econ. Behav. 10(1), 6–38.
Monderer, D., Shapley, L.S., 1996. Potential games. Games Econ. Behav. 14(1), 124–143.
Montanari, A., Saberi, A., 2010. The spread of innovations in social networks. Proc. Nat. Acad. Sci. 107(47), 20196–20201.
Morris, S., 2000. Contagion. Rev. Econ. Stud. 67(1), 57–78.
Morris, S., Rob, R., Shin, H.S., 1995. *p*-Dominance and belief potential. Econometrica 63(1), 145–157.
Myatt, D.P., Wallace, C., 2005. The evolution of teams. In: Gold, N. (Ed.), *Teamwork: An Interdisciplinary Perspective.* Palgrave MacMillan, New York, pp. 78–101 (Chapter 4).
Myatt, D.P., Wallace, C., 2008a. An evolutionary analysis of the volunteer's dilemma. Games Econ. Behav. 62(1), 67–76.

Myatt, D.P., Wallace, C., 2008b. When does one bad apple spoil the barrel? An evolutionary analysis of collective action. Rev. Econ. Stud. 75(1), 499–527.

Myatt, D.P., Wallace, C., 2009. Evolution, teamwork and collective action: production targets in the private provision of public goods. Econ. J. 119(534), 61–90.

Nachbar, J.H., 1990. 'Evolutionary' selection dynamics in games: convergence and limit properties. Int. J. Game Theory 19(1), 59–89.

Nash, J.F., 1950a. The bargaining problem. Econometrica 18(2), 155–162.

Nash, J.F., 1950b. Non-Cooperative Games. Ph.D. thesis, Princeton University.

Nöldeke, G., Samuelson, L., 1993. An evolutionary analysis of backward and forward induction. Games Econ. Behav. 5(3), 425–454.

Palfrey, T.R., Rosenthal, H., 1984. Participation and the provision of discrete public goods: a strategic analysis. J. Public Econ. 24(2), 171–193.

Ritzberger, K., Weibull, J.W., 1995. Evolutionary selection in normal-form games. Econometrica 63(6), 1371–1399.

Ross, S.M., 1996. Stochastic Processes, Second ed. Wiley, New York.

Rubinstein, A., 1982. Perfect equilibrium in a bargaining model. Econometrica 50(1), 97–109.

Sáez-Martí, M., Weibull, J.W., 1999. Clever agents in young's evolutionary bargaining model. J. Econ. Theory 86(2), 268–279.

Sandholm, W.H., 2010. Population Games and Evolutionary Dynamics. MIT Press, Cambridge, MA.

Ståhl, I., 1972. Bargaining Theory. Stockholm School of Economics, Stockholm.

Taylor, P.D., Jonker, L.B., 1978. Evolutionarily stable strategies and game dynamics. Math. Biosci. 40(1-2), 145–156.

Vega-Redondo, F., 2007. Complex Social Networks. Cambridge University Press, Cambridge.

Weibull, J.W., 1995. Evolutionary Game Theory. MIT Press, Cambridge, MA.

Wennberg, J., Gittelsohn, A., 1973. Small area variations in health care delivery: a population-based health information system can guide planning and regulatory decision-making. Science 182(4117), 1102–1108.

Wennberg, J., Gittelsohn, A., 1982. Variations in medical care among small areas. Sci. Am. 246(4), 120–134.

Young, H.P., 1993a. The evolution of conventions. Econometrica 61(1), 57–84.

Young, H.P., 1993b. An evolutionary model of bargaining. J. Econ. Theory 59(1), 145–168.

Young, H.P., 1998. Individual Strategy and Social Structure: An Evolutionary Theory of Institutions. Princeton University Press, Princeton, NJ.

Young, H.P., 2011. The dynamics of social innovation. Proc. Nat. Acad. Sci. 108(4), 21285–21291.

Young, H.P., Burke, M.A., 2001. Competition and custom in economic contracts: a case study of illinois agriculture. Am. Econ. Rev. 91(3), 559–573.

CHAPTER 7

Advances in Auctions

Todd R. Kaplan[*,†], **Shmuel Zamir**[*,‡]
[*]Department of Economics, University of Exeter Business School, UK
[†]Department of Economics, University of Haifa, Israel
[‡]The Center for the Study of Rationality, The Hebrew University of Jerusalem, Israel

Contents

Handbook of game theory
http://dx.doi.org/10.1016/B978-0-12-420056-2.00007-0

Abstract

As a selling mechanism, auctions have acquired a central position in the free market economy all over the globe. This development has deepened, broadened, and expanded the theory of auctions in new directions. This chapter is intended as a selective update of some of the developments and applications of auction theory in the two decades since Wilson (1992) wrote the previous Handbook chapter on this topic.

Keywords: Auctions, Private-value auctions, Multiunit auctions, All-pay auctions, Resale, Position auctions, Dynamic auctions, Spectrum auctions; Monotone equilibrium

JEL Codes: D44, D82, C72, H57

7.1. INTRODUCTION

Auction theory is a prominent and attractive topic of game theory and economic theory.[1] There are two main reasons for this: the first is that despite its simple rules, it is mathematically challenging and often leads to both surprising and elegant results. The second and probably the more important reason is that as a useful market mechanism, auctions have been widely practiced since ancient history.[2]

[1] As an illustration of the importance of auction theory within game theory, see the recent textbook *Game Theory* (Maschler et al., 2013), where auction theory is presented and developed as one of its chapters.

[2] Herodotus describes Babylonian bridal auctions in 500 B.C. where revenue from the most attractive maidens was used to subsidize the less attractive ones (see Baye et al., 2012b, for an analysis). In 200 B.C., Ptolemy IV of Egypt ran an auction for the tax-farming rights for Palestine and Syria. When the highest bid was at 8000 talents of silver (232 tons), Joseph the Tobiad told the Pharaoh that the bidders were colluding and offered 16,000 talents with the condition that he would be lent soldiers to help in the collection. The profits allowed him to be king of Israel (see Montefiore, 2011, and Adams, 1992).

Both reasons have become even more relevant and convincing since the chapter "Strategic Analysis of Auctions" was written in the first volume of this Handbook (Wilson, 1992). The mathematical developments became more challenging as the theory branched in many directions, such as multiunit auctions, dynamic auctions, combinatorial auctions, auctions with externalities, and more general incomplete information frameworks. The embracing of *free market* economic principles around the world made auctions the main vehicle for executing the huge volume of privatization that took place and is still occurring.

The theory and practice of auctions have stimulated one another. The massive use of spectrum auctions, online auctions and the privatization via auctions of big economic units (such as oil refineries and power plants) called for theoretical investigations, and the availability of new theoretical results and tools encouraged and facilitated the practical use of auctions, as reflected in the title of Milgrom's book, *Putting Auction Theory to Work* (Milgrom, 2004). The importance of mechanism design, of which auction theory is a central part, was recognized not only by politicians and regulators but also by the academic community, as expressed by awarding the 2007 Nobel Prize in Economics to Hurwicz, Maskin, and Myerson. Game theorists and economists became more and more involved in the practice of auctions both in helping regulators to design "the right auction" and in consulting competing companies to choose the "right bidding strategy" (for instance, Binmore, Cramton, Klemperer, McAfee, McMillan, Milgrom, Weber, Wilson, and Wolfstetter, to mention only a few, all acted in at least one of those roles). "Econlit" lists 20 books[3] and 3432 works with the word *Auction* or *Auctions* in the title that have been written since 1992 (according to "Google Scholar" the number is much higher—about 20,000).[4]

Given the volume of work done in the past two decades, we cannot hope to provide a fair and comprehensive presentation of all research and applications on auctions since Wilson's chapter in 1992, within the framework of one chapter. It is even clearer that our bibliography cannot be exhaustive. Instead, we made a selection of the theoretical themes of research and a sample of important applications of auction theory. As with any selection, it is prone to be biased by the taste of the selectors. We hope that, nonetheless, a nonexpert reader will get a reasonably good picture, directly or indirectly through the references, of many of the important developments in auctions in the past few decades.

7.2. FIRST-PRICE AUCTIONS: THEORETICAL ADVANCES

Theoretical and empirical studies of auctions are focused almost exclusively on the prominent solution concept of Nash equilibrium or more precisely Bayes-Nash

[3] Among the relevant books are Cramton et al. (2006), Klemperer (2004), Krishna (2002, 2009), Menezes and Monteiro (2005), Milgrom (2004), Paarsch and Hong (2006), Smith (1989), and Wolfstetter (1999).

[4] Econlit lists 1993 academic journal articles, 977 working papers, 254 papers in collective volumes, 185 dissertations, and 20 books.

equilibrium (BNE), as auctions are games of incomplete information. However, an explicit expression of the BNE is mathematically hard to obtain and so far it is available only for simple models, subject to highly restrictive assumptions. For example, Kaplan and Zamir (2012) provide explicit forms of the equilibrium bid functions in an asymmetric first-price auction with two buyers whose values v_1 and v_2 are uniformly distributed in $[\underline{v}_1, \overline{v}_1]$ and $[\underline{v}_2, \overline{v}_2]$, respectively. Furthermore, even the existence of a BNE is a difficult issue since auctions are games with discontinuous payoff functions and hence many of the standard results on the existence of Nash equilibrium cannot be applied in a straightforward way to auctions. Other tools, specifically tailored for this kind of model, are called for. In this section, we highlight the main results on existence of Nash equilibrium in auctions. We will mainly focus on auctions of a single indivisible object.[5]

Let us begin with the basic first-price auction model. Vickrey (1961) is the first to analyze auctions with independent, private values drawn from a uniform distribution. Riley and Samuelson (1981) extend Vickrey's analysis to n symmetric buyers, with values that are independent and identically distributed according to a general distribution F that is strictly increasing and continuously differentiable. They find that there exists a unique equilibrium that is symmetric with bid function $b(v) = v - \int_{\underline{v}}^{v} F^{n-1}(x)dx/F^{n-1}(v)$ for all bidders; that is, each bidder bids the conditional expectation of the other bidders' highest value given that he wins the auction. Since this contribution, there have been various works extending it to auctions in which one or more of the key assumptions of Riley and Samuelson no longer hold. These assumptions are symmetry and independence of the value distributions and smoothness of the common distribution F.

7.2.1 Mixed-strategy equilibria

When one relaxes one of Riley and Samuelson's (1981) three basic assumptions about the distributions (symmetry, independence and smoothness), for a first-price auction, a pure-strategy Nash equilibrium may not exist. A simple example is the case of two buyers each with a value drawn from the discrete distribution with values 0 and 1 each with probability 1/2. The equilibrium here is unique and in mixed strategies. In a classic example, Vickrey (1961) shows that when one buyer's value is commonly known there is a unique equilibrium where that a buyer uses a mixed strategy.[6]

Jackson and Swinkels (2005) demonstrate that without independence even when both buyers have private values drawn from a continuous distribution, there may not exist a pure-strategy monotonic equilibrium in a first-price auction. As an example consider an auction with two buyers with values v_1 and v_2 uniformly distributed on the triangle $v_1 \geq 0$, $v_2 \geq 0$, and $v_1 + v_2 \leq 1$. The intuition is that a buyer with value

[5] For a recent survey of existence and characterization results, see de Castro and Karney (2012).

[6] See Martínez-Pardina (2006) for a generalization of this situation.

1 knows that the other buyer has value 0 and hence in equilibrium he should not bid above 0. By monotonicity, each buyer should bid 0 for all his values ,which cannot be an equilibrium.

Jackson and Swinkels then prove the existence of equilibrium in mixed strategies in a general model that covers in particular both first-price and all-pay auctions as special cases. More specifically, they consider an environment with multiple identical items of the same indivisible object. Each buyer has an additive, but not necessarily linear, utility for multiple items, he is endowed with a finite (possibly zero) number of items, and he may want to sell or buy some items. Formally, the type $\theta_i = (e_i, v_i)$ of buyer i consists of the number e_i of items that he is endowed with and his values $v_i = (v_{i1}, v_{i2}, \ldots, v_{i\ell})$, where v_{ik} is his marginal utility from the kth item and ℓ is the total number of items in the economy. The key assumptions of the model are that the type space Θ is compact and the distribution F on Θ is continuous and absolutely continuous with respect to the product of its marginal distributions (which they call *imperfect correlation*). Finally, the utility functions over the payoff are assumed to be continuous.

7.2.2 Asymmetric buyers: Existence of mixed and pure-strategy equilibria

The symmetry of buyers was central and crucial for the results of Milgrom and Weber (1982). Existence results for models with asymmetric buyers were obtained by several authors, primarily by Lebrun (1996, 2006) and by Maskin and Riley (2000b).

An early general existence result for equilibria in first-price asymmetric auctions is due to Lebrun (1996) who proved that an *equilibrium in mixed (distributional) strategies* exists in an n-buyer first-price auction for an indivisible single good, under the following assumptions:

- The value distributions $F_1, \ldots, F_n$ are independent, with compact supports contained in $[c, K]$ for some finite c and K $(c < K)$.
- None of the distributions F_i has an atom at c.
- The allowed bidding range is $[c, K]$.

The condition of no atoms at c is needed to avoid payoff discontinuities at c, with the standard tie-breaking rule (all the highest bidders have equal probability of winning). This difficulty can be avoided by "augmenting" the game to include messages to be sent by the bidders and to be used in case of a tie. In this augmented first-price auction, the existence is then proved without the assumptions of no atoms at c. However, this is not very appealing: not only is this "augmentation" rather artificial, but the resulting equilibrium may include buyers using weakly dominated strategies.

This variant of an "augmented" first-price auction, along with a second variant in which in case of a tie, the object is given to the bidder with the highest value, was used in a subsequent paper (Lebrun, 2002) to prove (under the assumptions in Lebrun, 1996) the upper-hemicontinuity of the Nash-equilibrium correspondence with respect to the valuation distributions. This implies the continuity of the Nash-equilibrium under the conditions that guarantee its existence and uniqueness (that is, when the Nash-equilibrium correspondence is single valued). Lebrun views these results as a proof of robustness of the theoretical existence results and the usefulness of the numerical approximations of the Nash equilibrium.

Lebrun (2006) provides results on the existence and uniqueness of the equilibrium in asymmetric first-price auctions with n risk-neutral bidders, reserve price r, and the fair tie-breaking rule. Lebrun makes the following assumptions:

- The value distributions $F_1, \ldots, F_n$ are independent with supports $[c_i, d_i]$; $c_i < d_i$, $i = 1 \ldots n$.
- For each $i = 1, \ldots, n$, the cumulative distribution F_i is differentiable over $(c_i, d_i]$ with derivative (density) f_i locally bounded away from zero over this interval. Assume further that $F_i(c_i) = 0$ for all i.

Assume without loss of generality that $c_1 \geq c_2 \geq c_i$ for all $i \geq 2$; then the main result is:

Theorem 7.1. *If the above assumptions hold and no bidder bids above his value, then under any of the following conditions, the first-price auction has one and only one equilibrium: (i) $r > c_1$; (ii) $c_1 > c_2$; (iii) There exists $\delta > 0$ such that F_i is strictly log-concave over $(c_1, c_1 + \delta) \bigcap (c_i, d_i)$, for all $i \geq 1$, that is, f_i/F_i, the reverse hazard rate, is strictly decreasing over this interval.*[7]

7.2.3 Relaxation of symmetry and independence

Athey (2001) proves existence of a Bayes-Nash equilibrium in nondecreasing pure strategies for a large class of games of incomplete information. Her general result was widely used and in particular it was adapted and used for auctions by Athey herself and other authors (e.g., Reny and Zamir, 2004, to be discussed below). The novelty of Athey's work is that for studying the question of existence of pure-strategy Nash equilibrium (PSNE), which is basically a fixed-point problem, she introduces the *single-crossing condition* (SCC) which plays a central role in the proof of existence. Verifying this condition requires comparative statics analysis for a single-player problem (which is a simpler problem than showing the existence of a fixed point). The SCC of Athey is built on a related notion of *single-crossing property* due to Milgrom and Shannon (1994), who developed comparative statics analysis using only conditions that are ordinal (that is, invariant under order-preserving transformations of the utilities). Let X be a lattice, Θ a partially ordered set, and $h : X \times \Theta \to \mathbb{R}$ a real function on the product set.

Definition 7.1. *The function $h : \mathbb{R}^2 \to \mathbb{R}$ satisfies the (Milgrom-Shannon)* single-crossing property of incremental returns *(SCP-IR) in $(x; \theta)$ if, for all $x_H > x_L$ and all $\theta_H > \theta_L$, $h(x_H, \theta_L) - h(x_L, \theta_L) \geq (>) 0$ implies $h(x_H, \theta_H) - h(x_L, \theta_H) \geq (>) 0$. The function*

[7] For more about the role of log-concave probability in auction theory, see Bergstrom and Bagnoli (2005).

h satisfies weak SCP-IR if for all $x_H > x_L$ *and all* $\theta_H > \theta_L$, $h(x_H, \theta_L) - h(x_L, \theta_L) > 0$ *implies* $h(x_H, \theta_H) - h(x_L, \theta_H) \geq 0$.

This condition requires that the incremental return to x, $h(x_H, \cdot) - h(x_L, \cdot)$, as a function of θ, crosses zero at most once, from below.

Athey considers a game of incomplete information with a finite set of players $I = \{1, \ldots, I\}$. The players' type sets are intervals in the real line, $T_i = [\underline{t}_i, \bar{t}_i]$, and the action sets are compact convex sets in the real line, $A_i \subset \mathbb{R}$. The prior probability on the types is assumed to have density (with respect to the Lebesgue measure). Finally a technical integrability condition is assumed on the utility functions (henceforth Assumption (A1)) to ensure that the objective function $U_i(a_i, t_i; \alpha_{-i})$ of player i of type t_i, when all other players are using nondecreasing strategies α_{-i}, is well defined and finite. For this model, she defines the following single-crossing condition.

Definition 7.2. *The Single-Crossing Condition for games of incomplete information (SCC) is satisfied if for each player* $i \in I$, *whenever every other player* $j \neq i$ *uses a strategy* α_j *that is nondecreasing (in his type), player i's objective function,* $U_i(a_i, t_i; \alpha_{-i})$, *satisfies single crossing of incremental returns (SCP-IR) in* $(a_i; t_i)$.

The method adopted by Athey to prove that the SCC is sufficient for the existence of PSNE is to prove it first for the case of finite action sets and then, for the case of a continuum of actions, to use a sequence of appropriately designed grid approximations to prove that a certain selection of a sequence of monotone PSNE for the finite approximations converges to a monotone PSNE in the original game. The result for finite action sets is:

Theorem 7.2. *If Assumption (A1) is satisfied, SCC holds, and the action set* A_i *is finite for all* $i \in I$, *then the game has a PSNE where each player's equilibrium strategy,* $\beta_i(\cdot)$, *is nondecreasing.*

The results for continuous actions sets are derived from Theorem 7.2 and the following theorem.

Theorem 7.3. *If Assumption (A1) is satisfied and the following hold: (i) for all* $i \in I$, $A_i = [\underline{a}_i, \bar{a}_i]$, *(ii) for all* $i \in I$, $u_i(a, t)$ *is continuous in a, and (iii) for any finite* $A' \subset A = \times_{i \in I} A_i$, *then a PSNE exists in nondecreasing strategies in the game where players choose actions from* A.

Athey applies these results to *supermodular* and *log-supermodular* utility functions and *affiliated types*. Then she shows that the results can be extended with additional assumptions to certain cases of discontinuous utility functions that include some auction games. However, a more general result on the existence of PSNE in asymmetric first-price auctions using Athey's result was given by Reny and Zamir (2004) (henceforth RZ).

RZ notice that the SCC can fail in two possible ways. First, if there are ties at winning bids and, second, if a buyer uses a strategy that yields a negative payoff. They define a single-crossing condition that states that SCC holds whenever we avoid these two possibilities.

Definition 7.3. *The Individually Rational, Tieless, Single-Crossing Condition (IRT-SCC) is satisfied if for each player $i \in I$, actions a_i, a_i', and nondecreasing strategies α_j such that $Pr(\max_{j \neq i} \alpha_j(\cdot) = a_i \text{ or } a_i') = 0$, $U_i(a_i', t_i; \alpha_{-i}) \geq 0$, then $U_i(a_i', t_i; \alpha_{-i}) \geq U_i(a_i, t_i; \alpha_{-i})$ is maintained when t_i rises if $a_i' > a_i$ and when t_i falls if $a_i' < a_i$.*

Using their definition IRT-SCC, RZ are able to prove the existence of PSNE for asymmetric buyers, interdependent values, and affiliated one-dimensional signals. They note, by way of a counterexample, that these results cannot extend to multidimensional signals.

A further significant generalization was obtained by Reny (2011), who proved the existence of monotone pure-strategy equilibria in Bayesian games with locally complete metric semilattices action spaces and type spaces that are partially ordered probability spaces. The novel idea that enabled this generalization was to use the Eilenberg and Montgomery (1946) fixed-point theorem rather than the Kakutani's fixed-point theorem used in Athey (2001). The point is that while Kakutani's theorem requires that the sets of monotone pure-strategy best replies be convex, which may be hard to establish, the Eilenberg and Montgomery fixed-point theorem requires that these sets be contractible. This condition turns out to be rather easy to verify for the general class of Bayesian games studied by Reny. Finally, we note that in a recent unpublished work Zheng (2013) proves the existence of a monotone PSNE in a first-price auction with (endogenous) resale.

7.2.4 Monotonicity and the role of tie-breaking rules

The study of Nash equilibrium in first-price auctions has concentrated primarily on *monotone pure-strategy equilibria* (MPSE). Whenever the existence of such MPSE is established, the natural questions to be asked are: Is there only one MPSE? Are there any *nonmonotone* pure-strategy equilibria? Are there mixed-strategy equilibria (MSE)? These questions, which turned out to be related to each other, were partially or fully answered for various models (under various sets of assumptions), and some are not answered to date. The answers to these questions also often hinge on the *tie-breaking rule* used. For example, in the *general symmetric model* of Milgrom and Weber (1982) where they established a unique symmetric MPSE, it was only 24 years later that McAdams (2006) proved that there are no asymmetric MPSE and in a subsequent paper, McAdams

(2007a), he ruled out nonmonotone equilibria in this model (including mixed-strategy equilibria).[8] In these works, McAdams considered two tie-breaking rules:

- The standard *coin-flip rule* according to which the winner of the object is chosen by a uniform probability distribution over the set of buyers who submitted the highest bid.
- The *priority rule* according to which the winner is the buyer with the highest rank in a prespecified (prior to the bidding) order (permutation) of the buyers.

The results of McAdams are for a general asymmetric model with any (finite) number of buyers with affiliated types and interdependent values. More precisely, he considers the general model of Reny and Zamir (2004) restricted by the additional assumption that the utility functions are strictly decreasing in the bids. In addition he considered two types of bidding sets: the *continuum price grid* where $b_i \in [0, \infty)$ and the *general price grid* where the bids are restricted to an arbitrary subset of $b_i \in [0, \infty)$. For this model, McAdams proves that:

- Given a continuum price grid and the coin-flip rule, every MSE of the first-price auction with no ties is outcome-equivalent to some MPSE.
- Given a general price grid and the priority rule, every MSE of the first-price auction is outcome-equivalent to some MPSE.
- Given the coin-flip rule, every MPSE of the first-price auction has no ties.

Thus, nonmonotone equilibria can exist under the coin-flip rule but they are distinguishable: all nonmonotone equilibria have positive probability of ties whereas all monotone equilibria have zero probability of ties. McAdams provided an example of a nonmonotone pure-strategy equilibrium in a first-price auction (such an example requires three buyers with affiliated types and interdependent values).

In yet another paper, McAdams (2007b) addressed the issue of the uniqueness of the MPSE proved to exist in the RZ model. He proves this uniqueness under considerable restriction of the RZ model. In particular he assumes a symmetric model (roughly: the distribution and the utilities are invariant to any permutation of the buyers). For this he proves:

- There is a unique MPSE in the symmetric first-price auction, up to the bids made by a set of measure zero of types.

7.2.5 Revenue comparisons

An important property of symmetric auctions with independent private values that is lost as soon as we relax the symmetry condition is *revenue equivalence* among a large class of auctions. Although symmetry is not a condition in the Revenue Equivalence

[8] Earlier and weaker results in this direction were obtained in Rodriguez (2000).

Theorem of Myerson (1981), asymmetry implies different equilibrium allocations for different types of auctions (e.g., first-price and second-price auctions), which violates the requirement that equilibrium allocations be the same. Now, the investigation of revenue comparisons between selling mechanisms in general and types of auctions in particular naturally follows. This comparison turns out to be rather difficult, and so far, no strong general results are available. Even the question of which of the two auctions, first price or second price, generates a higher revenue in an asymmetric setting is only very partially answered. In this subsection, we review some of these results.

Maskin and Riley (2000a) study an asymmetric environment with two buyers: a strong one and a weak one. This ranking of the buyers means that the distribution of the strong buyer's value, F_s, conditionally stochastic dominates (CSD) the distribution of the weak buyer's value, F_w. Formally, CSD is defined as follows. For all $x < y$ in the common support of F_s and F_w,

$$Pr\{v_s < x|v_s < y\} = \frac{F_s(x)}{F_s(y)} < \frac{F_w(x)}{F_w(y)} = Pr\{v_w < x|v_w < y\}.$$

Under CSD,[9] in a first-price auction, the weak buyer bids more aggressively than a strong buyer. Furthermore, a strong buyer would prefer a second-price auction while a weak buyer would prefer a first-price auction. Revenue comparisons between the two formats are mixed. When there is a *shift* ($F_s(x+\alpha) = F_w(x)$, $\alpha > 0$) or *stretch* (e.g., for distributions that start at zero, $F_s(x) = \alpha \cdot F_w(x)$, $0 < \alpha < 1$, holds on the common support) a first-price auction generates higher revenue, while when there is a move of probability to the lower end of the support, a second-price auction generates higher revenue ($F_w(x) = \alpha \cdot F_s(x) + \delta$, $0 < \alpha < 1$, $\delta = 1 - \alpha$).[10]

Kirkegaard (2012) expands Maskin and Riley's (2000a) revenue comparison between first-price and second-price auctions. He finds that under CSD, no atoms, and the condition $\int_x^{F_s^{-1}(F_w(x))} (f_w(x) - f_s(y))dy \geq 0$ for all x, a first-price auction generates higher revenue than a second-price auction. This also holds in the presence of a reserve price. Kirkegaard (2014) provides additional sufficient conditions for revenue comparisons of first-price and second-price auctions. Outside the independent private value model, de Castro (2012) finds that with certain cases of dependence, revenue in a first-price auction may be higher than that in a second-price auction even under symmetry, in contrast to Milgrom and Weber (1982), who find the opposite holds in the case of affiliated values.

[9] Maskin and Riley (2000a) use a weaker definition.

[10] For more precise (and general) definitions of shifts, stretches, and probability moves, see Maskin and Riley (2000a).

7.3. MULTIUNIT AUCTIONS

The literature on multiunit auctions for environments in which buyers exhibit demand for multiple units of the same good is of great practical importance. The U.S. government treasury auctions three trillion dollars worth of securities annually (to finance public debts). The underpricing of one cent per $100 would lead to a loss of $300 million!

Treasury auctions are typically sealed-bid auctions in which each buyer submits a demand function (either as the total price $P(x)$ that he is willing to pay for x units, or equivalently his marginal bid $p(x)$ for the x-th unit). The submitted demands determine a clearing price and there are two commonly used pricing methods: uniform pricing, in which all units are sold at the highest price that clears the market, and discriminatory pricing, in which each buyer pays his own bid (for the units that he won). In practice, both pricing methods are used in treasury auctions of various countries and empirically it is not clear which one generates higher revenue (see, for example, Bartolini and Cottarelli, 1997, Binmore and Swierzbinski, 2000, and Brenner et al., 2009). Theoretically, it is known that both types of auctions are inefficient (see, for example, Wilson, 1979, Back and Zender, 1993, and Ausubel et al., 2013) and both are vulnerable to manipulation and cooperative behavior among buyers (see Goswami et al., 1996, Kremer and Nyborg, 2004). In addition to sealed-bid auctions, ascending auctions are commonly used in wine auctions (see Février et al., 2005), timber auctions (see Athey and Levin, 2001, and Athey et al., 2011), spectrum auctions (see Section 7.10), and other instances where multiple units of the same object or similar objects are for sale. These can be run either simultaneously or sequentially.

7.3.1 Efficient ascending-bid auctions

Effective auction design is guided by two principles; both are related to information. The first principle is that the price paid by the winner should be independent of his own bid, inducing the buyer to reveal his information (his true value of the good).[11] This was achieved brilliantly by Vickrey in the second-price auction and was later generalized to the Vickrey-Clarke-Groves (VCG) mechanism. The second principle is to maximize the information of each bidder when placing his bid, as in the English ascending auction. This principle is important if the objective of the design is revenue maximization. When the buyers' values for the object are interdependent and signals are affiliated (roughly speaking, positively correlated), gathering more information about the other bidders'

[11] The independence of the price of the bidder's own bid is not a sufficient condition for bidding the true value in equilibrium. As an example, in the equilibrium a third-price auction (of a single-unit indivisible object and symmetric setting), every bidder bids more than his value (see Kagel and Levin, 1993). It is also not a necessary condition since in an environment having two buyers with values independently drawn from the uniform distribution, a first-price auction where the winning bidder pays half his bid will induce truthful bidding.

signals will typically induce buyers to bid more aggressively and hence increase seller revenue (increasing the expectation of the second-highest bid).[12]

Ausubel (2004) proposes an efficient auction design with these two features for multiunit auctions. He calls it a "dynamic auction" but it is not dynamic in the same sense as in the dynamic auction literature where the dynamic component is either that of the buyers' population or that of the information available to a fixed population of buyers. The dynamic in Ausubel's auction is that of the auction mechanism: information is revealed (implicitly) during the process as the buyers change their demand with the change of prices. It is a dynamic auction in the same sense that the ascending English auction is dynamic. It could therefore be called an *open multiunit auction.*[13]

The starting point of Ausubel is what he calls the "static multiunit Vickrey auction," which is actually the VCG mechanism for selling homogeneous items. This is a sealed-bid auction that generalizes the second-price auction to multiple units: each buyer submits a sealed-bid demand function, which with discrete items is a list of prices that he is willing to pay for the first, second, third item, etc. A market clearing price is determined (where the number of bids above or equal to that price equals the number of units available). Each bid exceeding the clearing price is winning and for each unit won the buyer pays the opportunity cost of assigning this unit to him, that is, the bid of the buyer (other than himself) that was rejected as a result of him receiving the unit. For example, if a buyer receives two items, and the highest rejected prices other than his own were 5 and 3, he pays 5 for his first unit and 3 for his second. Note that these are neither the uniform nor the discriminatory prices and, just as in the second-price auction for a single unit, the prices paid by the buyer are independent of his bids for these two units (which are each of course at least 5).[14] The declared objective of Ausubel in this work was to design an open ascending auction that generalizes the single-unit English clock auction and is analogous to the multiunit Vickrey auction in the same way that the English auction is analogous to the second-price auction for a single unit.

[12] However, it is not always true that the seller's revenue increases monotonically as more information is given to the buyers. A counterexample is provided by Landsberger et al. (2001), who show that when the (two) buyers (in addition to privately knowing their own values) commonly know (only) the ranking of their values, the revenue is higher than the case of complete information when the values are commonly known. This was also shown in Kaplan and Zamir (2000) in the context of the strategic use of seller information in a private-value first-price auction: revealing only part of his information may yield the seller higher revenue than revealing all of it.

[13] In the auction literature, it is also referred to as the Ausubel auction.

[14] For the pure independent private-value setting, the Vickrey multiunit auction achieves an (ex-post) efficient outcome. This need not be the case when the values are interdependent. Perry and Reny (2002) modify the Vickrey auction to obtain efficiency also in the case of interdependent values (while maintaining the assumptions that the objects for sale are homogeneous and that each bidder's demand is downward sloping). Their mechanism consists of a collection of second-price auctions between each pair of bidders conducted over at most two rounds of bidding.

For a homogeneous multiunit auction of M units for sale, the auction proceeds as follows: starting at a low per-unit price, the auctioneer raises the price continuously (or in discrete steps) while each of the buyers posts his demand (i.e., the number of units he is willing to buy) at the current price. The auction ends with the price at which the total demand equals the number of units offered for sale and each buyer receives the number of items that he demanded at this price. The payment made by each buyer is calculated as follows. As the process of increasing the price proceeds, at each price p_t for which the demand $d(p_t)$ at that price drops (by one unit or more), and $d(p_t) \geq M$, it is determined whether for buyer i the demand $d_{-i}(p_t)$ of all other buyers at that price is strictly less than M, that is, whether $d_{-i}(p_t) = M - k$ where $k > 0$. If so, then buyer i has *clinched* winning k units. If his last clinch at a previous stage was at price $p_{t'}$ when the demand of all other buyers was $d_{-i}(p_{t'}) = M - k'$, then buyer i pays a price p_t for each of the last $(k - k')$ units. In other words, the buyer buys "the newly clinched" units for the price p_t at which they were clinched.[15]

Illustrating the process is the following example of Ausubel (2004), loosely patterned after the first U.S. Nationwide Narrowband Auction in which there were five spectrum licenses and five bidders with the limitation that each can buy only up to three licenses. The bidders marginal values for the licenses (their inverse demand functions) are given as columns in Table 7.1 in millions of dollars.

Table 7.1 Marginal values of the buyers in Ausubel's (2004) example.

Unit	A	B	C	D	E
1	123	75	125	85	45
2	113	5	125	65	25
3	103	3	49	7	5

When the auctioneer starts from a low price, say 10, the demands will be $(3, 1, 3, 2, 2)$ for bidders (A,B,C,D,E), respectively. As the price goes up (say continuously), the demands become $(3, 1, 3, 2, 1)$ at price $p = 25$ (a bidder chooses not to buy when indifferent), then $(3, 1, 3, 2, 0)$ at price $p = 45$, then $(3, 1, 2, 2, 0)$ at price $p = 49$, and then $(3, 1, 2, 1, 0)$ at price $p = 65$. At this price, the total demand is 7, which still exceeds the supply of 5. Yet, the total demand of all bidders but A is $4 = 5 - 1$. In the language of Ausubel, "bidder A has *clinched* winning one unit" at price 65 (expressing the fact that

[15] If the demand drops from $d(p_{t-1}) > M$ to $d(p_t) < M$, then the total number of units clinched will exceed the supply M and the process ends at time t. (For example, if the supply is 10, there are two symmetric buyers, there are 0 clinched units at $t - 1$, and the demand drops to $d(p_t) = 8$. Here, each buyer would clinch $10 - (8/2) = 6$ units.) In this case, there may be many possible methods to assign a market clearing allocation of the objects, including random allocation.

at that point it was guaranteed that he would end up with at least one unit). At price $p = 75$, the demand drops to $(3, 0, 2, 1, 0)$ and bidder A clinches winning two items, paying 75 for his second item, and bidder C clinches one item at price 75. The total demand is still higher than the supply and so the price goes on increasing until it reaches $p = 85$, where the demand becomes $(3, 0, 2, 0, 0)$ and hence supply equals demand. At this price 85, bidder A clinches his third unit $(3 = 5 - 2)$ and bidder C clinches his second unit $(2 = 5 - 3)$. The outcome of the auction is therefore: bidder A wins three licenses and pays $65 + 75 + 85 = 225$ million, and bidder C wins two licenses and pays $75 + 85 = 160$ million.

This mechanism yields an efficient allocation: the process allocates the items to the buyers who value them the most, and for this homogeneous discrete multi-item environment with independent values it yields the same allocation as the sealed-bid multiunit Vickrey auction described previously.

For comparison, in a uniform-price ascending-clock auction, the closing price in this example would be 75 (assuming full information), buyer A would reduce his demand and ask for two items at price 75 and end the auction (preferring to get two items at price 75 each rather than three items at price 85 each) with the inefficient allocation $(2, 0, 2, 1, 0)$. Furthermore, the example can be slightly perturbed to make this inefficient outcome the unique outcome of iterated elimination of weakly dominated strategies.

The Ausubel model is for independent values but otherwise it is quite general: it allows for both discrete and divisible objects, both complete information when demand functions are commonly known, and incomplete information when demand functions are only privately known. Ausubel considers two versions of the model, one with a continuous-time clock and one with a discrete-time clock. For private values, the main result about this mechanism which he calls "the alternative ascending-bid auction" is that sincere bidding by all bidders is an ex-post perfect equilibrium yielding the efficient outcome of the multiunit Vickrey auction.

To provide a generalization of the Milgrom-Weber model of interdependent values, Ausubel needs to make some assumptions and limit his general model. He considers a continuous-time game with symmetric interdependent values. A seller offers M discrete and indivisible homogeneous goods. Each buyer i has constant marginal utility v_i for each unit up to a *capacity* of λ_i units and zero marginal utility for any additional unit beyond this capacity. The marginal values v_i are interdependent: they are derived from affiliated private signals. The main result for this model is that if all buyers have the same capacity λ, and M/λ is an integer, then under certain assumptions (analogous to those in the Milgrom-Weber model) both the multiunit Vickrey auction and the alternative ascending-bid auction attain full efficiency. However, the alternative ascending-bid auction yields the same or higher expected revenue than the multiunit Vickrey auction. This is analogous to the Milgrom and Weber result for a single-unit auction: with affiliation, the ascending English auction generates higher revenue than its sealed-bid analog, the second-price auction.

7.3.2 Multiple heterogeneous items

In a subsequent paper, Ausubel (2006) extends his multiunit open ascending auction to K heterogeneous commodities with available supply $S = (S^1, \ldots, S^K)$ among n buyers. The starting point of Ausubel's mechanism is the Walrasian Tâtonnement process (also known as *dynamic clock auction*). One application of such auctions is to sell spectrum rights where many licenses are sold simultaneously and may vary slightly (see the subsequent section of this chapter for further discussion). In this selling mechanism, a "fictitious" auctioneer announces a price vector for the K commodities and the buyers report their demand (vectors), the auctioneer then adjusts the prices by increasing the prices of commodities with positive excess demand and decreasing the prices of commodities with negative excess demand. The process continues until a price is reached in which demand equals supply for all commodities. The trade takes place only at the final (market clearing) price and the buyers' payments are linear in quantities; all buyers pay the same price for all units of the same commodity that was allocated to them.

This process is vulnerable to strategic manipulation as the linear payment method provides an incentive for the bidders to underreport their true demand at the announced price (see Ausubel, 2004, for the ascending auction, and Ausubel and Cramton, 2002, for the sealed-bid auction.) Consequently the process typically does not result in a Walrasian outcome. An extreme empirical demonstration of this is the GSM spectrum auction in Germany (see Grimm et al., 2003, and Riedel and Wolfstetter, 2006). To overcome this drawback, Ausubel replaces the linear pricing by the nonlinear pricing in "clinches" introduced in his above-described one-commodity multiunit ascending auction. Similarly to the process in the homogeneous case, the price vector $P(t)$ is changing (according to an adapted Walrasian Tâtonnement process), and units of each commodity ℓ are allocated (clinched) to buyer i at that price $P_\ell(t)$ when the demand of all other buyers for that commodity is less than the available supply S^ℓ. Thus, in Ausubel's dynamic mechanism, payments are made along the process and they are determined by the prices at various points in time. Hence, buyers pay different prices for units of the same commodity clinched to them at different stages in the process. As a result, there are technical issues to ensure that payments are well defined, i.e., do not depend on the trajectory of the price vector $p(t)$ provided it satisfies certain conditions.

When a bundle of heterogeneous objects is purchased by a bidder in an auction, there is the issue of *complementarity* and *substitutability* of different commodities that affects the efficiency of the auction. For his open ascending auction, Ausubel assumes a *substitutes condition* (or *gross substitutes*) which is needed for the existence of Walrasian equilibrium. This condition requires that if the prices of some commodities are increased while the prices of the remaining commodities are held constant, then a bidder's sincere demand weakly increases for each of the commodities whose prices were held constant.

The issue of complementarity was raised in environments where heterogeneous objects are sold in separate auctions. As it was pointed out in Milgrom (2000), this situation may lead to inefficiency due to what is called *exposure*: a bidder may purchase object A while paying more than his value for A alone or a bidder may purchase bundle AB while paying more than his value for the bundle. This situation may occur, for instance, if, while bidding for A, the price of the complementary object B is not yet determined. The price of B may eventually be too high for it to be worthwhile to purchase B, or the marginal benefit of purchasing B may be worthwhile but the total purchase price of bundle AB is higher than its value to the bidder.[16]

Zheng (2012) considers a sale of two objects A and B with two types of bidders: one type is interested in only one object and another type has an added value for the bundle AB. The author designs a mechanism consisting of two simultaneous ascending English auctions modified so as to allow for *jump bidding* (and other rules accompanying it). The author shows that this modified auction can avoid exposure and restore efficiency since the jump bids can serve as signals.

For divisible goods and strictly concave utility functions and mandatory participation, Ausubel proves that sincere bidding by every buyer is an ex-post perfect equilibrium of the auction game. With sincere bidding, the price vector converges to a Walrasian equilibrium price vector and the outcome is that of a modified VCG mechanism with the same initial price vector $P(0)$.

For the algorithmic aspects of multiunit auctions see the chapter "Algorithmic Mechanism Design" in this Handbook.

7.4. DYNAMIC AUCTIONS

The literature on dynamic auctions provides natural extensions to the well-established static auctions theory to common economic environments: situations where the populations of sellers and buyers, the amount of goods for sale, and the state of information of the various agents are changing dynamically. These changes present the agents and the market designers with dynamic decision problems. Examples are airlines managing their prices when customers enter the market at different times, government surplus auctions where the amount being sold is stochastically changing, owners of internet search engines managing their advertisements when the number of people searching for items varies, and selling a start-up company when the actual value becomes clearer over time.

The framework of modeling and analyzing such economic environments is that of *dynamic mechanism design* within which *dynamic auctions* are a special case of such

[16] See Szentes and Rosenthal (2003a,b) for the exposure problem in the chopsticks auction.

mechanisms. There is no clear line between the two bodies of literature and often a paper with the title "Dynamic Auctions" is actually a paper on "Dynamic Mechanism Design." This is the case with the survey by Bergemann and Said (2011). This paper surveys the literature on dynamic mechanism design by grouping it in a two-dimensional way. In the first dimensions the distinction is between models in which the population of agents change over time while their private information is fixed and models in which the population of agents is fixed but the information of the agents is changing over time. In the second dimension, they group the works according to whether they aim to find mechanisms maximizing social welfare or maximizing revenue.

7.4.1 Dynamic population

Due to the mathematical difficulties involved, each work studies a very special case resulting from restrictions and assumptions of various degrees of plausibility. For this reason, it is hard to compare results obtained by different authors. The common approach to model a changing population of bidders is to have potential bidders enter (or possibly depart from) the market by some exogenous or endogenous process. Crémer et al. (2007) study a mechanism design problem in which a seller wishes to sell a single unit of an indivisible object to one of a finite set I of potential buyers. The problem becomes dynamic since the seller has to contact prospective bidders, and bring the auction to their attention. He incurs a cost of c_i (which they call a search cost) to inform bidder i about the rules of the auction, the identity of the other bidders, and the distribution of their valuations. Bidder i then privately learns his type x_i and decides whether to participate or not. If he agrees to participate, he signs a binding contingency contract. While the types of the bidders are independent, there are interdependent valuations of the object: bidder i's value is $u_i(x_1, \ldots, x_I) = x_i + \sum_{j \neq i} e_{ij}(x_j)$. All bidders have the same discount factor $\delta \in (0, 1]$.

At the start of each period t the seller approaches a set of potential entrants. Those who decide to enter (after learning their type) and sign a contract send a message to the seller who decides whether to stop the mechanism and sell the object to one of the participants (entrants and incumbents) according to the contracts or to approach new bidders. There are variants of information disclosure policies, whether or not the seller reveals to the new entrants the messages received in previous stages. In principle, bidders may send a message at each stage they participate in but there is no loss of generality in limiting the mechanism to a single message at the entering stage. The authors apply a revelation principle argument to observe that any perfect Bayes equilibrium (PBE) outcome can be obtained by an *incentive feasible mechanism* in which the bidders communicate their types truthfully. The main result is that under some technical differentiability conditions, the optimal (seller's profit-maximizing) mechanism is an *optimal search procedure* relative to the *ex-post virtual utility* functions $V_i(x_1, \ldots, x_I) =$

$x_i - \frac{1-F_i(x_i)}{f_i(x_i)} + \sum_{j \neq i} e_{ij}(x_j)$. An interesting part of the result is that the seller's optimal revenue does not depend on the information disclosure policy adopted.

For the special case of private values ($e_{ij}(\cdot) = 0$ for all i, j) and $\delta = 1$, the optimal mechanism can be implemented by a sequence of Myerson's optimal auctions where a new entrant joins at each period as long as the object is not sold. In the symmetric case where the bidders have the same distribution of values, the mechanism consists of a sequence of second-price auctions with a reserve price that declines over time. Crémer et al. show that in this special case the optimal design problem is similar to the Weitzman (1979) pandora problem of searching for the highest reward box when only one box can be opened (at a cost) at each period.

In a subsequent paper, Crémer et al. (2009) consider a variant of the model where the costly recruitment of bidders by the seller is replaced by a cost for a bidder to find out his valuation. The seller can force the bidder to pay an entry fee before finding out his type. They show that the seller can obtain the same profit as if he had full control over the bidders' acquisition of information and could directly observe their valuations once they are informed (intuitively because the bidders are ex-ante identical).

As we move to multiple units for sale, Vulcano et al. (2002) and Pai and Vohra (2013) have similar models in which a seller with K identical items for sale faces a random arrival of buyers with a demand for one unit. The value for the item and the latest time that it has to be purchased by are the private information of the arriving buyer. Both papers address the revenue maximization problem. Board and Skrzypacz (2013) consider a seller with K identical indivisible items for sale in a discrete time span of $\{1, \ldots, T\}$. At time t, a random number N_t of buyers arrive to the market where $\{N_t\}_{t=1}^{\infty}$ are i.i.d. random variables. The number of buyers is observed by the seller but not by the buyers. Each buyer is interested in a single unit for which he has value v_i (his private information). The values are also i.i.d. There is a discount factor δ for the buyers' utilities.[17] Compared to the Crémer et al. model, in the Board and Skrzypacz model the seller has multiple units of the same object rather than one, the set of potential buyers is infinite, and the process of entering the market is random and exogenously given (while it is controlled strategically by the seller in the Crémer et al. model). The values are assumed to be i.i.d. while Crémer et al. allow for interdependent values.

The authors design a selling mechanism in which a buyer, when entering the market and observing his value v_i, declares his value to be $\tilde{v}_i$ and the mechanism determines (probabilistically) the allocation and the transfers. They prove that the optimal (profit-maximizing) allocation is obtained by the following mechanism: at time t with k units left to sell, the seller sells the next unit to the highest-valued agent with a value exceeding a certain *cutoff* x_t^k. Interestingly, the cutoff level x_t^k depends on the time remaining and

[17] For a model of allocating heterogeneous objects to impatient agents arriving sequentially (with privately known characteristics) see Gershkov and Moldovanu (2009).

the number of items for sale but not on the number of agents who have entered in the past and their values. The endogenously determined cutoffs x_t^k are decreasing both in t and k. The cutoff x_t^k is determined by an equation asserting that the seller is indifferent between selling to the agent with the cutoff value today and waiting one more period.

The implementation is achieved by setting, for $t < T$, prices p_t^k such that the "cutoff agent" with value x_t^k will be indifferent between buying and not buying. At the last period $t = T$, the seller allocates the items to the k-highest-value buyers, subject to these values exceeding the static monopoly price. Note that it is sufficient to base these prices only on the items remaining. For example, the price for an item in period 3 would be the same if three items were allocated in period 1 and one item in period 2 or two items were allocated in each period.

For the case of a single item ($K = 1$), this allocation can be implemented by a sequence of second-price auctions with pre-determined reserve prices R_t. For the multiple-item case ($K > 1$), a sequence of second-price auctions cannot implement the optimal allocation since the optimal allocation may have more than one item allocated in the first period. The optimal allocation in period 2 depends upon how many items were allocated in period 1. Hence, if a second-price auctions with reserve prices were to be used to implement the optimal allocation, then the second-period reserve prices would need to be a function of how many items were sold in the first period (or equivalently the number of items remaining to be sold). Thus, as with prices in the optimal allocation, the mechanism needs to have reserve prices that depend upon the period and the number of remaining items, that is, R_t^k.

Board and Skrzypacz also consider the continuous time version of the model in which buyers enter the market according to a Poisson process at arrival rate λ, and instantaneous discount rate r. The optimal mechanism becomes simpler to implement in this case since at any time instant t at most one unit is sold. The cutoff values x_t^k are determined as before but now by the continuous form of the indifference condition for the seller. The implementation is made by setting a take-it-or-leave-it selling price p_t^k that makes the agent with value equal to the cutoff x_t^k indifferent between buying at time t and waiting.

7.4.2 Repeated ascending-price auctions

Said (2008) highlights the difference between static and dynamic auctions by demonstrating that even with a single object for sale and a random arrival of buyers (similar to the Board and Skrzypacz model) repeated sealed-bid second-price auctions may not lead to an efficient allocation.

In his leading example, two buyers are present in the market with values $v_1,\ v_2 \in [0, 1]$, and w.l.o.g. assume $v_1 > v_2$. A third potential buyer with value $v_3 \sim F$, where F is the uniform distribution on $[0, 1]$, may enter the market with probability $q \in [0, 1]$. Assume that v_1 and v_2 are commonly known by the players but the value v_3 of the new entrant is his private information. Each buyer wishes to purchase exactly one unit of the object. There are three units that are sold via a sequence of three second-price auctions in which the buyers' bids are revealed after each round and subsequent rounds are discounted by δ per period. Note that at the first round, buyers 1 and 2 do not know whether buyer 3 has entered or not but, as the bids of the first round are announced, this becomes known to them in the second and third rounds. The efficient allocation should have the highest-valued buyer receiving the unit in the first round, the second-highest-valued buyer receiving the unit in round 2, and the third-highest-valued buyer (if there is one) receiving the last unit.

To see why this situation may lead to inefficiency, let us sketch the equilibrium of the specific case (adapted from Said, 2008) in which $v_1 = \frac{2}{3}$, $v_2 = \frac{1}{3}$, $\delta = \frac{9}{10}$, and $q > 0$. Buyer 1 knows that even if he loses the first round there is a probability of $1 - q$ that he will be alone in the next round, will receive an item at price zero, and obtain a utility of δ times his value. Therefore, his bid in the first round will be at most $(1 - (1 - q)\delta)v_1 = (0.1 + 0.9q)v_1$, which is considerably lower than his value for small q. The same holds for buyer 2.

However, unlike buyers 1 and 2, buyer 3 knows whether or not he has entered the market in round 1. If entering in round 1, buyer 3 knows that he cannot pay zero for the item in round 2 since he would face another buyer. Since in any equilibrium, in round 1, buyer 1 outbids buyer 2, in round 2 buyer 3 faces buyer 2 who bids $(1 - \delta)v_2$ (since the round 3 price is zero). From this, if buyer 3's value is larger than $\frac{1}{3}$, then his potential round 2 profit is $v_3 - \frac{1}{3}(1 - \delta)$. Therefore, in round 1, buyer 3 will bid $v_3 - (v_3 - \frac{1}{3}(1 - \delta))\delta = 0.1v_3 + 0.03$. Consequently, for small enough q, a buyer 3 entering the market with value slightly less than $v_1 = 2/3$ will win the first item, which is inefficient. This happens when $0.1v_3 + 0.03 > (0.1 + 0.9q)v_1$, for example, $v_3 = 0.6$ and $q = 0.03$.

The author suggests that this result is driven first by the fact that the future is discounted and hence the order in which objects are allocated matters. But more importantly, there is a fundamental information asymmetry: while the values of buyers 1 and 2 and their presence are commonly known, the presence of the new entrant and his value are his private information. With this insight, Said shows that efficiency is recovered if the sealed-bid second-price auction is replaced by its open-auction counterpart, namely, the open ascending price auction (while there is still one object per round). The intuition is that with this auction more information is revealed to make up for the initial asymmetry of information.

In a subsequent paper, Said (2012), building on this insight, proves similar results; namely, the repeated sealed-bid second-price auction is inefficient while an appropriate repeated ascending auction is efficient. However, this cannot be viewed as a "more general" result since the model here differs from the one in Said's previous work. In Said's (2012) model, the author considers an infinite-horizon discrete-time process with a single seller. In each period $t \in \mathbb{N}$, the seller has K_t units of a homogeneous and indivisible good available for sale. The amounts K_t are independently distributed according to μ_t. Objects are perishable: at each period, any object that is not allocated or sold perishes at the end of that period. Each period t begins with the arrival of N_t buyers. The variables $\{N_t\}_{t=1}^{\infty}$ are independent with distributions λ_t; $t = 1, 2, \ldots$. Each buyer i present in the market wishes to obtain a single unit and is endowed with a privately known value v_i for that single unit. The values v_i are independent and with distributions F_i. In addition, buyers may exogenously depart from the market (and never return) after each period, where the (common) probability of any buyer i surviving from period t to $t+1$ is $\gamma_t \in [0, 1]$. Otherwise, buyers remain present on the market until they obtain an object. Buyers are risk neutral, with quasilinear and time-separable preferences. All buyers, as well as the seller, discount the future with the common discount factor $\delta \in (0, 1)$. Finally, it is assumed that every buyer is aware of the presence of all other buyers in the market.

In this model, Said first shows that a sequence of second-price sealed-bid auctions with no bid disclosure does not admit an efficient equilibrium. Again, the main reason is that the dynamics creates asymmetry of beliefs between the incumbents and the entrants: the incumbents of a certain round are the losers of the previous round, and each knows his bid and the information disclosed at the end of the previous round (e.g., selling prices). Thus, an incumbent has information on the other incumbents that the new entrant at this period does not have. Also, even when private values are independent, market dynamics and repeated competition generate interdependence: the value of winning an object in a round compared to not winning it in that round depends upon the opportunity cost of not participating in subsequent rounds. This opportunity cost depends upon the competitors' values. Thus, the net values are interdependent.

Here again, efficiency can be restored, by replacing the sealed-bid second-price auction by its open-auction counterpart, namely, the ascending auction. In such an auction, a price clock rises continuously from zero and buyers drop out of the auction at various points. The auction ends as soon as the number of active buyers is K_t or less. Each one will then receive a unit and pay the closing price. If at the beginning of the round there are less than K_t buyers, the auction ends immediately at price zero. Here again, in principle, losers from the previous period seem to have an informational advantage over the new entrants of that period. However, it turns out that the information content of the dropping time in the previous round will be available again in the present period bidding strategy, and this time to all competitors: incumbents and entrants. In other words, in the ascending auction bidders are engaged in a revelation process and at each round a player reveals his information again. Technically, the game has an equilibrium in memory-free strategies. Indeed, the author provides such equilibrium-bidding strategies in the sequential ascending auction and proves that they generate the same allocations and transfers as in the truthful equilibrium of the *dynamic pivot mechanism* developed by Bergemann and Välimäki (2010) as the dynamic version of the VCG mechanism.

7.5. EXTERNALITIES IN SINGLE-OBJECT AUCTIONS

Externalities in auctions are where a player cares not only about winning the object but also about who wins it in case of losing.[18] Two well-suited examples are given by Jehiel et al. (1996). After the breakup of the Soviet Union, Ukraine was left with 176 intercontinental nuclear missiles. Although neither the United States nor Russia had any interest in these old-fashioned missiles, they each paid Ukraine (in various ways) about one billion dollars to dismantle this arsenal. The second example is when China agreed in 1994 not to sell its M-9 and M-11 missiles to Arab countries; as a reward, the United States agreed to lift its one-year-old embargo on satellite exports to China. Although these are not proper auctions, they are cases in which a party is willing to pay, not to have an object, but to prevent a third party from having it. More generally, these are cases in which "whoever wins" may affect the downstream interaction between the players. Typical economic examples are auctioning of a cost-reducing patent to oligopolists, auctioning spectrum licenses to incumbents, and a retailer competing in an auction for a neighboring plot of land against a competitor who, if he wins, intends to build a polluting factory on it.

7.5.1 A general social choice model

Major contributions to this topic are by Jehiel and Moldovanu in numerous papers including a review (Jehiel and Moldovanu, 2006). In this review, they make the distinction between *allocative externality* due to the final allocation of the object and *informational externality* due to information held by the other competitors in the auction. They provide a general social choice mechanism in which multi-object auctions with both types of externalities, as well as combinations of the two and complementarities, are special cases. In this model there are $N+1$ agents, indexed by $i = 0, \ldots, N$ (agent 0 is the seller) and K social alternatives, indexed by $k = 1, \ldots, K$. Each agent gets a private ℓ-dimensional signal about the state of the world $\theta^i \in \Theta^i \subset \mathbb{R}^\ell$. The vector of all private signals is $\theta = (\theta^0, \ldots, \theta^N) \in \Theta = \times_{i=0}^{N} \Theta^i$. Agents have quasi-linear utility functions $u^i(k; \theta; t^i) = v_k^i(\theta) + t^i$ that depends on the chosen alternative k, on the vector of private signals θ, and on monetary payments t^i.

For an auction allocating a set M of (heterogeneous) objects, a partition of M is $P = (P_1, \ldots, P_N)$, where P_i is the bundle of objects allocated to agent i. In this model, the number of alternatives K is equal to the number of possible partitions. Thus, we can

[18] Externalities can extend beyond the identity of the winner such as in the case where there is a positive spillover from the expenditure (see D'Aspremont and Jacquemin, 1988).

write utility as $v^i_P(\theta) + t^i$. In addition, agents receive a signal for each possible partition, and so $\ell = K$. Hence, we can also write, for each partition P, agent i's signal as θ^i_P.

The pure private value is obtained when both the signal θ^i_P and the value $v^i_P(\theta)$ depend only on the bundle P_i allocated to the agent; that is, for any two partitions P and P', $P_i = P'_i$ implies $\theta^i_P = \theta^i_{P'} \equiv \theta^i_{P_i}$ and $v^i_P(\theta) = v^i_{P'}(\theta) \equiv v^i_P(\theta^i_{P_i})$. The pure allocative externalities case is obtained when $v^i_P(\theta_P) \equiv v^i_P(\theta^i_P)$; that is, while the value of an agent depends on the whole distribution of bundles (allocative externalities), it depends only on his own signal and not on the signals of the other agents (no informational externalities). In the case of pure informational externalities and no allocative externalities, the agent cares only about his bundle and about the information of other agents about this bundle; that is, for any two partitions P and P', $P_i = P'_i$ implies $v^i_P(\theta) = v^i_{P'}(\theta) \equiv v^i_{P_i}(\theta)$ for all θ. In addition, since $v^i_{P_i}(\theta)$ depends on the signals of other agents only to the extent that they concern information "about" the bundle P_i, the signals of any agent j can be partitioned into equivalence classes (with respect to their effect on $v^i_{P_i}(\theta)$) indexed by P_i; $\theta^j = (\theta^j_{P_i})_{P_i \subseteq M}$ and we can write $v^i_{P_i}(\theta)$ as $v^i_{P_i}(\theta^0_{P_i}, \ldots, \theta^N_{P_i})$. Finally, the dependence of the value v^i_P on the partition P accommodates complementarities and substitutabilities in the usual way.

7.5.2 Complete information

Considering first the issue of allocative externalities, it turns out that traditional auction formats need not be efficient, and they may give rise to multiple equilibria and strategic nonparticipation. The first observation is that even in the simplest case of a single object and complete information, the very notion of value becomes *endogenous*: if we denote by v^i_i the value of agent i when i wins the object (e.g., a license of a cost reducing patent) and by v^i_j (typically negative) the externality exerted on agent i if agent j wins the object, then the net value for agent i for winning the object compared to losing the auction is $v^i_i - v^i_j$ if he expects agent j to win or $v^i_i - v^i_k$ if he expects agent k to win in the case where he loses the auction. Jehiel and Moldovanu (1996) demonstrate this point and show that, even in a simple second-price auction, not only is there no dominant strategy but there can be multiple equilibria with different allocations of the good. They illustrate this possibility in the following example: $N = 3$ and $v^i_i = v$ for all i, the externality terms are $v^2_1 = v^1_2 = -\alpha$, $v^3_1 = v^3_2 = -\gamma$, and $v^1_3 = v^2_3 = -\beta$, where $\alpha > \gamma > \beta > 0$. In one equilibrium, agents 1 and 2 compete (being "afraid" of each other winning the object) and one of them wins and pays $v + \alpha$. In a second equilibrium, agent 3 competes with agent 1 (or with 2) and wins the object paying $v + \beta$ (since $\gamma > \beta$, agent 3 is more afraid from 1 than 1 is afraid of 3).

Another phenomenon pointed out by Jehiel and Moldovanu (1996) is that with the presence of allocative externalities there is *strategic nonparticipation*. By staying out, an agent may induce an outcome that turns out to be more favorable to him than the outcome that would have arisen if he had participated. An example where firms are the agents (taken from Jehiel and Moldovanu, 2006) has $N = 3$, where firms 1 and 2 are incumbents, while firm 3 is a potential entrant. The incumbents do not value the object per se (such as an innovation that is irrelevant for their production technologies): $v_1^1 = v_2^2 = 0$. Moreover, $v_1^2 = v_2^1 = 0$. The entrant has value $v_3^3 = v$, and it creates a negative externality of α on each of the incumbents, which is greater than his value; $v_3^1 = v_3^2 = -\alpha$ and $\alpha > v$. Consider a second-price auction. Clearly it is the interest of the incumbents to avoid entry by bidding slightly above v. However, there is a free-rider problem between the two incumbents: each one would prefer to let the other deter the entrant. Indeed, in this example, in any equilibrium (there are three) at least one incumbent firm chooses not to participate with positive probability.

7.5.3 Incomplete information

Models with allocative externalities and incomplete information give rise to various situations that depend on the nature of the private information of the agents. When the private information of agent i is the allocative externality v_j^i caused to him by other agents, the resulting situation has no informational externalities. When the private information of agent i is the allocative externality v_i^j that he causes to others, the environment becomes that of private interdependent values. (Of course, having only partial information about these two types of externalities is also conceivable.)

As we just said, models of incomplete information and allocative externalities are closely related to the research on auctions with interdependent values. For example, Jehiel and Moldovanu (1999) consider a two-bidder second-price auction in which each of the private signals is the corresponding private value, that is, $\theta^i = v_i^i$, and the externalities are functions of both signals, that is, $v_j^i = v_j^i(\theta^i, \theta^j)$. With no reserve prices, this is analoguous to the Milgrom and Weber (1982) model with interdependent values $v^i = \theta^i - v_j^i(\theta^i, \theta^j)$, $i = 1, 2$, $j = 3 - i$. However, with the presence of a reserve price r, there is the possibility the seller keeping the object and hence the "net value" of bidder i for winning the object is either θ^i if the alternative is that the seller keeps it, or $\theta^i - v_j^i(\theta^i, \theta^j)$ if the alternative is that bidder j wins it. Jehiel and Moldovanu (1999) find that in this case there is discontinuity in the bidding range in equilibrium. This can happen when the reserve price is binding, that is, when the value including the externalities may be less than the reserve price. For example, take the symmetric case of two buyers with values in $[\underline{v}, \overline{v}]$ with constant (negative) externality $-e$. In a second-price auction with a reserve price r satisfying $\underline{v} + e \leq r \leq \overline{v} + e$, a (basically unique) symmetric equilibrium strategy is bidding $v + e$ when $v \geq r$ and bidding 0 when $v < r$. In other words, there is no relevant bid in the interval $(r, r + e)$. In such a case the optimal reserve price may well be below the seller's value for the object. For positive externalities they show that entry fees and reserve prices need not lead to the same revenue, in contrast to the case with no externalities. Variants of this model were studied by Moldovanu and Sela (2003), Das Varma (2003), and Molnar and Virag (2008).

Fan et al. (2013, 2014) have a model where a patent is auctioned between two firms engaged in a Cournot duopoly where the winner of the auction receives the patent for a fixed fee and then collects royalties from the loser. Since the fixed fee and the royalties depend upon information obtained by bidding in the auction, the bids in this auction have a signaling content that affects the downstream competition. Hence, the externality is bid-dependent.

On the issue of revenue-maximizing auctions, Jehiel et al. (1999) consider a model in which the private information of bidder i is his value v_i^i and all externalities v_j^i caused to him (he receives a multidimensional signal). They find that a second-price auction with an appropriately determined entry fee is the revenue-maximizing mechanism in a class of mechanisms where the object is always sold. Similar models are studied by Das Varma (2002) and Figueroa and Skreta (2004). Caillaud and Jehiel (1998) study the possibility of collusion in the presence of negative externalities.

Informational externalities were studied at early stages by Wilson (1967) and Milgrom and Weber (1982) in single-object auctions with symmetric bidders. The focus was later shifted to multiobject auctions with asymmetric bidders. The main approach of most studies is that of mechanism design, investigating the issues of incentive compatibility, revenue equivalence, and revenue and welfare maximization in *direct-revelation mechanisms*. The breadth, depth, and technical aspects of this literature make a systematic detailed presentation of the main results not feasible within this section. For further reference, we suggest the works of Maskin (1992), Krishna and Maenner (2001), Jehiel and Moldovanu (2001), Hoppe et al. (2006), and the numerous references listed there.

7.6. AUCTIONS WITH RESALE

In the standard private-value auction model, if a buyer wins the auction it is for him worth his private value. This is even the case if another buyer has a higher private value. In this section, we look at what happens when there is the possibility of resale by the winner of the auction to another buyer with a higher value. For this analysis, there must be two stages in the game. In the first stage, the object is sold by the seller via an auction. In the second stage, the winning buyer has the option of selling the object to one of the other buyers. Here, we need to specify both what information is revealed after the first stage and the selling procedure of the second stage. For example, are bids revealed after the first stage and does the winning buyer have full bargaining power in the second stage? The main issue to address is the effect of the introduction of resale on the outcome of the auction, specifically, on the allocation of the object and on the revenue.

7.6.1 First-price and second-price auctions

Under symmetry, adding this possibility of resale does not change the equilibrium in a first-price auction or the symmetric equilibrium (bid your value) in a second-price auction. This is shown when the bids are revealed (Haile, 2003), when the values are revealed (Gupta and Lebrun, 1999), and when the losing bids are not revealed (Hafalir and Krishna, 2008). Under asymmetry, this is no longer the case and revenue equivalence between first-price and second-price auctions breaks down. Hence, it is important to study resale in asymmetric auctions.

We begin with an example adapted from Hafalir and Krishna (2008). A weak buyer has a value drawn from the uniform distribution on $[0, 1]$ and a strong buyer has a value drawn from the uniform distribution on $[0, 4]$. Consider the first-price auction with resale where the winner in the auction in the first stage, without seeing the losing bid, can make a take-it-or-leave-it offer to the loser. The equilibrium bid functions are $\frac{5}{4}v_w$ for the weak buyer and $\frac{5}{16}v_s$ for the strong buyer. If the weak buyer wins in the first stage, he would make a take-it-or-leave-it offer to the strong buyer of double what he paid, $\frac{5}{2}v_w$. Notice that for both buyers the distribution of bids is uniform on $[0, \frac{5}{4}]$. Revenue is thus $\frac{2}{3} \cdot \frac{5}{4} = \frac{5}{6}$. In a second-price auction with resale there is an equilibrium where buyers bid their values (and, hence, there is no resale), revenue is $\frac{3}{4} \cdot \frac{1}{2} + \frac{1}{4} \cdot \frac{1}{3} = \frac{3}{8} + \frac{1}{12} = \frac{11}{24}$, which is less than in the first-price auction with resale. By comparison, in a first-price auction without resale, the equilibrium bid functions are:[19]

$$b_w(v_w) = \frac{16 - 4\sqrt{16 - 15v_w^2}}{15v_w}, \quad b_s(v_s) = \frac{4\sqrt{16 + 15v_s^2} - 16}{15v_s}.$$

Revenue in this equilibrium is approximately 0.59, which is lower than the revenue in the equilibrium of a first-price auction with resale.

Hafalir and Krishna show that these results generalize to any two distributions of values. Namely, in the first-price auction with resale the bid distributions are identical for the two buyers and the revenue of the first-price auction with resale is always higher than that of the second-price auction where the buyers bid their values. Thus, with the introduction of resale, there is a ranking between a first-price auction and a second-price auction that doesn't exist without resale (provided we only consider the bid-one's-value equilibrium in the second-price auction).[20]

The model of Hafalir and Krishna was studied earlier by Garrat and Troger (2006) who consider the special case in which the value distribution of one of the buyers is

[19] These were from Griesmer et al. (1967). Also, see Kaplan and Zamir (2012) for the general solution to the uniform case.

[20] See Section 7.2.5 for examples by Maskin and Riley (2000a) where the revenue between a first-price auction and a second-price auction can be ordered in either way.

degenerate at 0. In other words, one of the buyers is a speculator who has no value for the object per se and this is known to the other buyer. Their main finding is that speculators can affect the equilibrium in auctions with resale. In the first-price auction with resale they find that the speculative bidder uses a mixed-strategy equilibrium and makes zero profit but influences the equilibrium, compared to a first-price auction without resale, by increasing the selling price (and hence the revenue) and changing the allocation to a possibly inefficient one. In a second-price auction there are multiple equilibria where sometimes the speculator makes a profit. Again, the speculator changes the equilibrium by increasing revenue and decreasing efficiency. The focus of Garrat and Troger is on the role of speculation in auctions with resale while the focus of Hafalir and Krishna is on the ranking of the seller's revenue, but as we see in our above example of Hafalir and Krishna, the weak buyer bids $\frac{5}{4}v_w$, which is above his value, and so speculation is present in first-price auctions even in the nondegenerate case.

7.6.2 Seller's optimal mechanism

Zheng (2002) and Lebrun (2012) search for the optimal mechanism when resale is permitted. Myerson (1981) finds the optimal mechanism (i.e., maximizing seller revenue) when resale can be prevented. For example, assume that buyer 1 has a value uniform on $[0, 1]$ and buyer 2 has a value that is drawn uniformly from $[0, 2]$. The optimal mechanism should allocate the object to the buyer with the highest virtual surplus,[21] which means that buyer 1 should get the object if $v_1 \geq v_2 - \frac{1}{2}$. This can be implemented by running a second-price auction with a minimum bid of 1 but where buyer 1 receives a coupon that will reimburse him for his payment up to $\frac{1}{2}$. For instance, if buyer 1 bids $\frac{3}{2}$ and buyer 2 bids $\frac{5}{4}$, then buyer 1 will win the auction at a price of $\frac{5}{4}$, yet pay $\frac{5}{4} - \frac{1}{2} = \frac{3}{4}$. For such a mechanism, buyer 2 will bid his value (or drop out if his value is below 1) and buyer 1 will bid his value plus $\frac{1}{2}$ (or drop out if his value is below $\frac{1}{2}$).

Once resale is introduced, this mechanism will no longer work since if buyer 1 wins, he will then try to sell the object to buyer 2, increasing his surplus. Zheng (2002) proposes a mechanism that will work under resale. The buyers bid for the object and the winner pays some function of his bid (a form of a first-price auction). This function should be set such that buyer 1 bids $v_1 + 1$ and buyer 2 bids v_2. If buyer 1 wins, then he will believe that buyer 2's value is uniform on $[0, v_1 + 1]$ and will make a take-it-or-leave-it offer at $v_1 + \frac{1}{2}$ which would result in the buyer with the highest virtual surplus getting the object as in the optimal mechanism without resale (since the allocation is the same as in the optimal mechanism, the revenue is equivalent). The key insight into this mechanism is that the buyer winning the auction sets a resale auction such that the object is

[21] Virtual surplus for a buyer with value v drawn from the distribution $F(v)$ is $v - \frac{1-F(v)}{f(v)}$.

resold whenever the current owner does not have the highest virtual surplus. The winner of the resale auction will do likewise (as will any subsequent winners). While the necessary conditions for an optimal mechanism in Zheng (2002) are restrictive, Mylovanov and Tröger (2009) are able to relax them such that even for three or more bidders there are several examples of distributions that satisfy them (besides the uniform distribution).

Lebrun (2012) shows that optimal allocation for revenue is an equilibrium of a standard second-price auction with resale (albeit with bidder-specific entry fees). This seemingly contradicts Hafalir and Krishna (2008), who prove that a second-price auction is inferior to a first-price auction with resale; however, rather than focusing on the bid-your-value equilibrium, Lebrun takes one of the inefficient equilibria present in Garratt and Troger (2006). In order for this to be optimal, the beliefs of the winner about the value of the loser must be very specific. In order for them to be so, the buyers must use particular mixed-strategies in the first period.

7.6.3 Further results

Gupta and Lebrun (1999) consider a different model of two-buyer first-price auctions with resale in which they make the assumption that the values are revealed after the first stage.[22] Under this rather strong assumption, the authors provide an equilibrium for any second-stage sale price function $\pi(v_1, v_2)$, for example, $max\{v_1, v_2\}$. If the asymmetric value distributions are F_1 and F_2, then the first-price auction with resale will have the same bid distribution as the symmetric first-price auction (without resale) with value distribution G given by $G^{-1}(q) = \pi(F_1^{-1}(q), F_2^{-1}(q))$, $q \in [0, 1]$. As a consequence, under symmetry the equilibrium doesn't change under resale.

Bikhchandani and Huang's (1989) early contribution to auctions with resale is one of the few that deal with common values. They theoretically investigate treasury bill auctions where the securities can be resold, and find that revenue is higher using a discriminatory auction (similar to first price) rather than a uniform-price auction (similar to second price).

Garratt et al. (2009) show that resale can help reach a collusive equilibria in second-price auctions. With two bidders with values drawn independently from the uniform distribution on $[0, 1]$, bidders can flip a coin (or use a public sunspot) where, depending upon the outcome, one of the bidders bids 1 and the other bids 0. A bidder with a value of 1 receives on average $1/2$. However, with a different distribution on $[0, 1]$, the expectation can be lower than $1/2$, meaning that a bidder with value 1 would prefer the bid-one's-value equilibrium. With resale, the bidder with a value of 1 will receive more than $1/2$ on average since when he bids 0, he will still receive the object by buying it on the resale market. The extra profit is enough to make this collusive equilibrium worthwhile.

[22] Gupta and Lebrun also assume that the support of the value distribution of both buyers is the same.

In this section, we mentioned just some of the recent results on auctions with resale. In Haile (2003), the reason motivating a resale is not the inefficiency in the primary auction but rather information gained afterwards. Additional contributions of note are Bose and Deltas (2007), Pagnozzi (2007, 2009), Harfalir and Krishna (2009), Cheng and Tan (2010), Lebrun (2010a, 2010b), Cheng (2011), Che et al. (2013), Virág (2013), and Xu et al. (2013).

Finally, we note new work on an environment in which the seller cannot commit to not reattempt to sell the object if he fails to do so in the auction (see Vartiainen, 2013, and Skreta, 2013). This topic is somewhat related to auctions with resale since in both cases, the seller cannot prevent post-auction trade.

7.7. ALL-PAY AUCTIONS

A contest is a situation in which players exert effort in an attempt to win a prize. Greater effort increases the chances of winning. All the efforts are sunk while only the winner gets the prize (hence the name *all-pay* as all participants pay a cost). In the literature, contests have been used to describe environments including patent races, sport competitions, court cases, lobbying, political campaigns, promotions, military, and rent-seeking. Explicitly, contests have been used as a tool to spur innovation since the 1700s. The Longitude Prize of £20,000 established by the British Parliament in 1714 induced John Harrison to invent the marine chronometer, for which he won the prize in 1765 (see Sobel, 1996). Motivated by the Orteig Prize, Charles Lindbergh became the first person to fly non-stop solo across the Atlantic in 1927 (Berg, 1998). The self-financed Feymnann prize inspired nanotechnology. Such contests have gained more traction in recent years. For example, the X-Prize Foundation created a number of high-profile prizes (the *Ansari Space Prize* to privately launch a reusable manned spacecraft, the *Lunar Prize* to privately land an unmanned probe on the moon, and the *Tricorder Prize* for a palm-sized instant medical evaluation tool), a number of companies offer (or offered) design prizes (*Topcoder* uses computer programming contests to generate needed code, and *Netflix* held a contest to help design a better movie recommendations system), and the US military holds the DARPA challenges (races) to improve robotic designs. These real-world contests have sparked increased academic interest in the field.

An important ingredient in describing a contest with the set of participants $I = \{1, \ldots, n\}$ is the contest success function, which takes the efforts $\{x_i\}_{i=1}^n$ of the agents and converts them into each agent's probability of winning: $P_i : \mathbb{R}^n \to [0, 1]$. The (expected) utility of a risk-neutral player i with a value of winning the prize v_i and a cost of effort $c_i(x_i)$ is

$$P_i(x_1, \ldots, x_n)v_i - c_i(x_i).$$

There are several popular contest success functions in the literature. The generalized Tullock (1980) success function is

$$P_i(x_1,\ldots,x_n) = \frac{x_i^r}{\sum_j x_j^r} \qquad [7.1]$$

where $r > 0$ is a parameter.

The Lazear-Rosen (1981) success function is

$$P_i(x_1,\ldots,x_n) = P(x_i + \epsilon_i \geq \max_{j\neq i}\{x_j + \epsilon_j\})$$

where ϵ_i are random noise variables with the same distribution.

Perhaps the most "natural" is the auction success function where the prize is awarded (with probability 1) to the competitor that exhibited the greatest effort, and in case of tie, the prize is awarded randomly to one of the maximizers with equal probability to each. A contest with this success function is equivalent to an all-pay (first-price) auction and the auction success function is mathematically the limit of the Generalized Tullock success function [7.1] when $r \to \infty$:

$$P_i(x_1,\ldots,x_n) = \begin{cases} \frac{1}{\#\{j|x_j=\max_j\{x_j\}\}} & \text{if } x_i = \max_j\{x_j\}, \\ 0 & \text{otherwise.} \end{cases}$$

In this section, we will focus on the theoretical literature on all-pay auctions, which is where the contest and auction theory literature overlap. This literature uses two main environments: complete information and incomplete information about the bidders' values of the prize or the bidders' costs. We will begin by describing core developments in these two environments before delving into further advances. Unlike first-price auctions, many research questions have nontrivial results under complete information and hence many topics are covered in the literature using both environments' tools. Intuition gained from the complete-information environment usually carries over to the incomplete-information environment. Therefore, in many cases we cover the complete-information case in more detail. When using the auction terminology we will refer to competitors also as bidders or players and to the effort x_i as the bid of player i.

7.7.1 Complete information

Baye et al. (1996) build upon the work of Hilman and Samet (1987), Hillman (1989), and Hillman and Riley (1989) to characterize the all-pay auction with complete information with two or more players. They allow the values for the prize to be asymmetric (it is possible that $v_i \neq v_j$). In the complete-information models, all values v_i, $i \in I$, are common knowledge among all participants. For each player i, $c_i(x_i) = x_i$.[23]

Before describing the equilibria, it is useful to describe several properties that hold in equilibrium, and to see why. First, the equilibrium involves mixed strategies. In a

[23] For the case of $c_i(x_i) = c_i \cdot x_i$, the same analysis can be applied by redefining values such that $\tilde{v}_i = v_i/c_i$.

pure-strategy equilibrium each player i chooses an effort level x_i with probability 1. If in such an equilibrium there are no ties with the winning bid, then the winning bidder i can profitably deviate to $(x_i + x_j)/2$ where x_j is the second-highest bid. If there is a tie at the winning bid, then a winning bidder i can choose $x_i + \epsilon$ and gain at least $\frac{v_i}{n} - \epsilon$; hence, there is no pure-strategy equilibrium.

Second, in equilibrium any player that chooses zero with positive probability earns zero profits. This is because there must be at least one player who chooses a strictly positive bid with probability one. If not, a player bidding zero would be able to have a discrete jump in profits by increasing his bid to some $\epsilon > 0$.

Third, under symmetry (when $v_i = v$ for all i) all players make zero profits in equilibrium (which is not necessarily symmetric). If one player makes a positive profit, then all players must make a positive profit since any other player can bid at the top of the support of the player making the positive profit (which must be less than v for him since he is making a positive profit). All players cannot be making a positive profit since at least one player bidding at the bottom of the union of supports of all strategies must be making zero profit (because if all players choose this point with positive probability a player will be able to discretely increase profit by increasing his bid by an arbitrarily small amount).

The following describes the equilibria in more detail.

If $v_1 = \ldots = v_m > v_{m+1} \geq \ldots \geq v_n$ with $m \geq 2$, then there is an equilibrium where the first m bidders bid symmetrically according to $F(x) = \left(\frac{x}{v_1}\right)^{\frac{1}{m-1}}$ and the remaining bidders bid zero and all bidders make zero profit. For $m \geq 3$, there is also a continuum of equilibria that are revenue equivalent in which a subset of at least two of the first m bidders randomize continuously on $[0, v_1]$. The remaining of the first m bidders randomize continuously on $[b_i, v_1]$ and bid 0 with positive probability if $b_i > 0$. If two or more players randomize continuously over a common interval, their CDFs are identical over that interval.

If $n = 2$ and $v_1 > v_2$ or if $n > 2$ and $v_1 > v_2 > v_3 \geq \ldots \geq v_n$, then there is a unique equilibrium. In that equilibrium, player i chooses his bid randomly according to the cumulative distribution F_i where

$$F_1(x) = \frac{x}{v_2},\ 0 \leq x \leq v_2, \quad \text{and } F_2(x) = \frac{x + v_1 - v_2}{v_1},\ 0 \leq x \leq v_2, \qquad [7.2]$$

$$F_i(x) = 1 \quad \text{for } x \geq 0,\ i \geq 3.$$

Notice that only the two players with the highest two values are active, and all players except player 1 are making zero profit, since all of them bid zero with positive probability (including player 2, since $F_2(0) > 0$). Since each player must be indifferent among all bids in his support and all players $i > 1$ bid zero with positive probability while player 1 bids strictly above zero with certainty, player i (> 1) bidding zero always loses and thus has no positive profit anywhere in his support.

If $v_1 > v_2 = \ldots = v_m > v_{m+1} \geq \ldots \geq v_n$ with $m \geq 3$, then there exists a continuum of equilibria that are not necessarily revenue equivalent to each other and these are parameterized by $\{b_i\}_{i=2}^{m} \in [0, v_2]^{m-1}$ where $b_i = 0$ for a least one i. Player 1 randomizes continuously on the interval $[0, v_2]$ and players $i \in \{m+1, \ldots, n\}$ bid zero. The remaining players $\{2, \ldots, m\}$ choose 0 with probability α_i and randomize continuously on $[b_i, v_2]$ where $\Pi_{i=2}^{m}\alpha_i = (v_1 - v_2)/v_1$. Each of these equilibria has the property that for any $i, j \in \{2, \ldots, m\}$ and $x \geq \max\{b_i, b_j\}$, we have $F_i(x) = F_j(x)$; that is, the CDFs of i and j are the same in the intersection of their supports.

For an example where revenue is not equivalent between two equilibria, take the case of $m = n = 3$ with $v_1 = 2$ and $v_2 = 1$. One equilibrium (corresponding to $b_2 = 0$ and $b_3 = 1$) has

$$F_1(x) = x,\ F_2(x) = \frac{x+1}{2},\ 0 \leq x \leq 1,\ F_3(0) = 1.$$

While another equilibrium (corresponding to $b_2 = b_3 = 0$) has

$$F_1(x) = x\left(\frac{2}{x+1}\right)^{\frac{1}{2}},\ F_2(x) = F_3(x) = \left(\frac{x+1}{2}\right)^{\frac{1}{2}},\ 0 \leq x \leq 1.$$

The first generates revenue of 0.75 while the second generates revenue of $\frac{5-2\sqrt{2}}{3} \approx$ 0.724.

7.7.2 Incomplete information

In the incomplete information models of all-pay auctions, values are drawn from commonly known distributions, while a player's value is his private information. Thus, this is a game of incomplete information (a Bayesian game) in which the type of a player i is his value v_i.

The symmetric all-pay auction with i.i.d. values drawn from distribution F has a unique equilibrium in which each player bids

$$b(v) = F(v)^{n-1}v - \int_0^v F(\tilde{v})^{n-1}\, d\tilde{v}.$$

The asymmetric two-bidder case with common support was shown by Amann and Leininger (1996) also to have a unique equilibrium. They also showed uniqueness and existence for more general winning payments of the form $(1-\lambda)x_i + \lambda x_{-i}$ for $\lambda \in [0, 1)$ (while the losing payment is still x_i).

Parreiras and Rubinchik (2010) examine the incomplete information case where there are three or more heterogeneous bidders. They find that (unlike the two-bidder case) there can be cases where an individual bidder will either not bid the entire range of equilibrium bids (including having gaps in his support) or even completely drop out of the contest and bid 0.

7.7.3 Multiple prizes

In this section, we consider contests where the prize is a divisible amount of money and address the issue of how to divide the prize among the contestants according to their place. Galton (1902) proposed that one should divide prize money between first and second places in a ratio of 3 to 1. His objective (presumably out of fairness) was to make the ratio of the prizes equal to the ratio of the difference between the first-place and third-place competitors' efforts to the difference between the second-place and third-place competitors' efforts, notably ignoring any incentive effect of the prizes. This is a ratio of the difference between the first-order and the third-order statistic to the difference between the second-order and the third-order statistic, and when efforts are i.i.d. from a normal distribution this ratio is equal to 3. (Lazear and Rosen, 1981, note the loose connection to the concept of marginal product.)

Glazer and Hassin (1988) is the first paper to analyze this problem of the allocation of prize money in all-pay auctions with complete information and find that with symmetry and risk aversion, the expected sum of the efforts is maximized by having homogeneous prizes: the prize money is divided evenly among all but the last-place contestant, who should not get a prize. Under the complete-information environment, Clark and Riis (1998) study homogeneous prizes when the prizes are given either simultaneously or sequentially with the restriction that each participant can win at most one prize. They find that both formats generate the same expected effort. Also under complete information, Barut and Kovenock (1998) characterize the equilibria for heterogeneous prizes and give the expected revenue.[24]

Moldovanu and Sela (2001) readdress Galton's question of how to divide prize money for an all-pay auction with incomplete information about costs when a designer wishes to maximize total effort. More specifically, they use a cost function of the form $c(x) \cdot \theta_i$ where $1/\theta_i$ is the ability of player i and is private information. If $c(x)$ is linear or concave then a designer should use one prize. If $c(x)$ is convex, then using multiple prizes may be optimal. The general intuition is that when $c(x)$ is linear, then the incentive problem is equivalent to one obtained if we divide by θ_i and transform the problem to one with incomplete information about the value of the prize, but with complete information about ability (which is then the same for all players). But revenue equivalence holds in this new formulation. Moreover, we can think of this as auctioning an object (the prize money) where the probability of winning the prize for the n-th highest bid is p_n rather

[24] While Clark and Riis (1998) assume all prizes are the same, they allow the prize value to differ among players. In contrast, Barut and Kovenock (1998) assume that prizes may be different, but each particular prize has the same value for all players.

than giving a fraction f_n of the prize to the n-th place bidder, where $p_n = f_n$. We know from standard optimal auction results that it is revenue-maximizing for the object to be allocated (if it is allocated at all) to the highest bidder (it is optimal to set $p_1 = 1$.) This also holds when the object must be allocated. Hence, we should only have one prize in an all-pay auction (since then $f_1 = p_1 = 1$). Having a concave cost function in an all-pay auction further favors having just one prize. With convex costs, the marginal cost of effort increases in effort. Thus, revenue-maximizing design would provide incentives that increase effort of the lower types at the expense of the effort of higher types. One way to do this is to have multiple prizes such that a lower type has a higher chance of winning at least a part of the prize. This is also the intuition for having bid caps under convex costs of bidding (see Gavious et al., 2002, and Section 7.7.7 of this chapter).

Glazer and Hassin (1988) is also the first paper to analyze the multiple prize problem under incomplete information. As with Moldovanu and Sela (2001), they use the cost function $c(x) \cdot \theta_i$ but they have the additional assumption that $1/\theta_i$ is uniformly distributed on $(0, 1]$. They find that when the players are risk-neutral, then a revenue-maximizing designer would want one prize (which Moldovanu and Sela proved more generally). Under risk-aversion, they find that multiple prizes are indeed optimal (with three or more players). Here, the intuition is that multiple prizes are superior for incentives since they reduce the chance that an agent exerting a high effort will receive any reward. Overall, combining these results one sees that in general with either risk-aversion and/or convex costs it is possible that multiple prizes will enhance revenue. For a longer review of the multiple-prize literature see Sisak (2009).

7.7.4 Bid-dependent rewards

In many of the examples of contests, the reward for winning depends upon the winning bid. This may occur naturally such as in patent races where an earlier patent (obtained from a higher level of effort) is worth more in present-value terms. It may also be set by the designer of the contest. A race director may set a higher prize for breaking a record time. The Longitude prize, described above, had higher prizes set for more accurate methods. Kaplan et al. (2003) analyze a complete-information all-pay auction with a reward for winning of $v(x)$ and with the same cost of effort, $c(x)$, for all firms. We have a symmetric equilibrium where each firm chooses x randomly according to the distribution F given by

$$F(x) = \max_{x' \leq x} \left(\frac{c(x')}{v(x')} \right)^{\frac{1}{n-1}}.$$

When $\frac{c(x)}{v(x)}$ is nondecreasing in x, this simplifies to $F(x) = \left(\frac{c(x)}{v(x)} \right)^{\frac{1}{n-1}}$.

Kaplan et al. (2003) also analyze this environment with two asymmetric contestants. Here, the equilibrium is more complicated. Siegel (2009) defines a player i's reach as the highest bid at which his utility of winning is 0, that is, the largest x such that $v_i(x) = c_i(x)$. Designate player 1 as the player with the largest reach. Denote by π_1 the profit player 1 makes in equilibrium, which is equal to $\max_{x \geq x^*} v_1(x) - c_1(x)$, where x^* is the reach of player 2. We can then define $q_1(x) = \frac{c_1(x)+\pi_1}{v_1(x)}$ and $q_2(x) = \frac{c_2(x)}{v_2(x)}$. Subsequently, we denote $g_1(x) = \min_{x' \geq x} q_1(x')$ and $g_2(x) = \min_{x' \geq x} q_2(x')$ and G_1 and G_2 denote the sets where g_1 and g_2, respectively, are strictly increasing. We then have the equilibrium distribution functions as $F_i(x) = Max\{1 - g_j(\sup_{x' < x} G_i), 0\}$.

This analysis is extended to three or more heterogeneous players in Siegel (2009, 2010, 2014b,c). In addition to the analysis being an important and difficult technical achievement, Siegel finds that the equilibrium with three or more heterogeneous players can have qualitatively different behavior than in the two-heterogeneous-players model.

Kaplan et al. (2002) analyze a variant of this environment with incomplete information about the value of the reward when it is bid-dependent. When types θ_i are i.i.d. from the uniform distribution and the reward is multiplicatively separable in effort and type θ_i, that is, total reward is $\theta_i \cdot R(x)$, where $R(x)$ is a function of effort, the equilibrium effort is given by $b(\theta) = u^{-1}(\theta^n)$ where $u(x) = R(x)^{-\frac{n}{n-1}} \int_0^x \frac{n}{n-1} c'(t) R(t)^{\frac{1}{n-1}} dt$. Interestingly, they find that an increase in the costs or a reduction in rewards may increase the expected sum of the efforts and/or the maximum effort. This is because the information rents (the equilibrium expected profit given one's private information) may be lowered by such an increase.

To see this, by the envelope theorem, the information rent of a type θ is given by

$$\int_0^{\theta} F(\hat{\theta})^{n-1} R(x(\hat{\theta})) d\hat{\theta}.$$

From this equation, we can see why a decrease in the rewards may increase bids. If there is a decrease only for low values of the reward function R, keeping bids constant, this will decrease the rents for all types, not just low types. This requires that the equilibrium profit decrease for all types, even those whose reward function for their equilibrium bid is unaffected by the change. Thus, those types must see an increase in their bid in order for the profits to decrease. Thus, the overall effect can be an increase in bids.

An optimal design approach has been applied to such rewards. Cohen et al. (2008) determine the optimal bid-dependent reward in an all-pay auction under incomplete information when a designer cares about the sum of the efforts or the maximum effort. Kaplan and Wettstein (2013) find the optimal bid-dependent reward under complete information when the designer cares about the maximum effort.

7.7.5 Contests versus lotteries

A contest may also be used as a method to allocate good. For instance, colleges routinely use waiting in line as a means to distribute the right to buy basketball tickets. In other cases, instead of a contest, tickets are distributed by means of a lottery, such as with tickets to Michael Jackson's funeral. In both of these examples, the objective was to maximize welfare rather than revenue. Chakravarty and Kaplan (2013) find the optimal mechanism when an agent of type θ, exerting effort x, has a utility $v(\theta) - c(x)g(\theta)$ where $g(\theta)$ represents the agent's waiting cost and $v(\theta)$ is the agent's value and the type θ drawn i.i.d. from the uniform distribution on $[0, 1]$.[25] The agent's willingness to pay in terms of cost of effort is then $\frac{v(\theta)}{g(\theta)}$, that is, how high an agent would be willing to set cost $c(x)$ in order to obtain the object. They find that if $\left(\frac{v(\theta)}{g(\theta)}\right)'' \leq 0$, then a lottery is optimal. On the other hand, the condition $\left(\frac{v(\theta)}{g(\theta)}\right)'' \geq 0$ is necessary for a contest to be optimal. Still, the optimal mechanism does not depend only upon the willingness to pay. For instance, when $v(\theta) = \theta$ and $g(\theta) = (1 - \theta)^2$, a contest is optimal. When $v(\theta) = \theta(1 - \theta)$ and $g(\theta) = (1 - \theta)^3$, a lottery is optimal while $\frac{v(\theta)}{g(\theta)}$ is the same in both cases.

Hartline and Roughgarden (2009), Condorelli (2012), and Yoon (2011) also made contributions in the analysis of maximizing the bidder's surplus as opposed to the revenue. This objective is closely related to the objective of the work of McAfee and McMillian (1992) in which a bidding ring wants to maximize expected surplus of its members. In earlier literature, Taylor et al. (2003) and Koh et al. (2006) compare lotteries to contests rather than looking at when either of these is an optimal mechanism. Boyce (1994) made an early contribution comparing lotteries to auctions and to queues in both the case where (after the lottery) postallocation selling is permitted and in the case where it is not permitted.

7.7.6 All-pay auctions with spillovers

Baye et al. (2012a) generalize the complete information all-pay auctions allowing for spillovers. If player 1 bids x and player 2 bids y, then player 1's payoff for winning is $v_1 - W(x, y)$ and for losing is $-L(x, y)$ (with ties broken by a coin toss). The degree to which y affects these two payoffs is the spillover. They assume that $W(x, y)$ and $L(x, y)$ are linear functions of x and y. Even with the restriction of linearity, this model covers many economic environments: the all-pay auction that corresponds to $W(x, y) = L(x, y) = x$, the first-price auction that corresponds to $W(x, y) = x$ and $L(x, y) = 0$, and the second-price auction that corresponds to $W(x, y) = y$ and $L(x, y) = 0$. A case with a negative spillover is when $W(x, y) = L(x, y) = x + \frac{y}{2}$. Other cases include: R&D with spillovers (D'Aspremont and Jacquemin, 1988), Varian's model of sales (Varian, 1980), and competition with price-matching policies (Baye and Kovenock, 1994). Baye et al. characterize all the possible equilibria, both pure and mixed.

[25] Following Milgrom (2004, page 111), there is no loss of generality.

Baye et al. (2005) analyze an incomplete information version of this framework in application to litigation systems where the value from winning is private information. Here, they consider the following spillover functions:

$$W(x, y) = \beta x + (1 - \alpha)y,$$
$$L(x, y) = \alpha x + (1 - \beta)y.$$

They classify the legal systems according to the values of α and β.

In the American legal system each side pays his own cost; thus $\beta = \alpha = 1$. In the British system $\alpha = 1$, $\beta = 0$, the loser pays both costs. In the (Dan) Quayle system (a system suggested by the former vice president of the United States, 1989-1993), we have $\alpha = 2$, $\beta = 1$; that is, the loser pays his own costs and reimburses the winner up to his (the loser's) costs. In the Matthew system (from the Gospel of Matthew 5:39-41) the winner should pay not only his own costs but reimburse the loser for a fraction of them (the winner's costs) too: $\alpha = 1$, $\beta > 1$. Of these systems, the British yields the highest legal costs and the lowest litigant utility while the Matthew system yields the lowest legal costs and the highest litigant utility.

7.7.7 Bid caps

In the two-player all-pay auction under complete information, Che and Gale (1998a) analyze the results of the imposition of a bid cap. In many situations that are modeled as all-pay auctions such caps are in place. For instance, in sports many leagues including four major U.S. sports leagues impose a salary cap.[26] In many political contests, there are spending caps in place on campaign spending. The application in their paper was lobbying. In their model, the value of the prizes are v_1, v_2 such that $v_1 > v_2$ and there is a bid cap of m imposed where $m < v_2$. Without a bid cap, the equilibrium is given by [7.2]. The expected revenue in this case is $\frac{v_2(v_1+v_2)}{2v_1}$, which is less than v_2.

When $m \leq \frac{v_2}{2}$, both players bidding at the cap of m is an equilibrium. (When $m = \frac{v_2}{2}$, there are also equilibria where player 1 bids m with probability of $\frac{2m}{v_1}$ or more.) Thus, for m such that $\frac{v_2(v_1+v_2)}{4v_1} < m < \frac{v_2}{2}$, in equilibrium there is higher revenue (equal to $2m$) with a cap than without a cap.

Kaplan and Wettstein (2006) argue that in many contests a designer lacks the ability to impose a hard cap. They show that in the Che and Gale (1998a) model imposing a flexible cap decreases revenues. While Che and Gale (2006) show that adding asymmetry of costs to their model (in addition to asymmetry of values), imposing a flexible cap (resulting in raised costs) may also increase revenue. Gavious et al. (2002) show that with incomplete information about values and a convex bid function, imposing a rigid cap

[26] In professional sports, teams can be thought of as competing with each other in their spending for players. Higher spending increases the team's chances of winning the league championship.

may increase revenues. Pastine and Pastine (2010) study bid caps when the seller prefers to sell to one of the bidders (as can happen in the context of political lobbying). Pastine and Pastine (2013) extend their analysis to include soft (flexible) bid caps. In general, there is also a strong connection between bid caps and financially contained bidders, which have been studied by Che and Gale (1996) and Pai and Vohra (2014) in all-pay auctions.

7.7.8 Research contests

A significant motivation provided by many authors for the study of contests (as we do in this chapter) is the use of contests to spur innovation. There have been a number of papers that have specifically tailored contests to this application. Dasgupta (1986) and Kaplan et al. (2003) specifically use an all-pay auction with complete information, while Pérez-Castrillo and Wettstein (2012) use an all-pay auction with incomplete information. Taylor (1995) considers an innovation contest where a firm attempting an innovation takes a draw from a distribution $F(q)$ for a quality q at cost c. For cost $x \cdot c$, the firm can take x independent draws and use the highest quality. (This is equivalent to one draw from the distribution $F^x(q)$.) The buyer gives a prize to the firm with the highest quality. Interestingly, the success function is mathematically equivalent to a Tullock success function with $r = 1$ (see Equation [7.1] on page 410).[27] Fullerton and McAfee (1999) solve this for asymmetric costs of draws. Fullerton et al. (2002) replace the prize in a Taylor (1995) style tournament with a first-price auction using a "scoring rule" that takes both price and quality into account.[28] Schöttner (2008) uses the Lazear-Rosen success function to model the innovation draw. Ding and Woffstetter (2011) have each firm only take one draw (for a fixed cost of entry) but model the case where firms that have too high a quality will not enter the contest and try to bargain separately.

In an interesting variant of an all-pay auction, Che and Gale (2003) have a scoring rule similar to Fullerton et al. (2002). In their paper, a buyer wishes to purchase an innovation from one of several potential firms. Each firm i expends effort $c(x_i)$ to create innovation with a quality x_i (measured in monetary units). Afterwards, each firm offers its design to the buyer for a price p_i (chosen by the firm). The buyer chooses the firm offering highest surplus s_i which equals $x_i - p_i$ (the quality minus the price). Denote by $G_i(s_i)$ the probability that firm i with surplus s_i will be offering the highest surplus,

[27] To see this for discrete xs: if firm 1 chooses x_1 draws and firm 2 chooses x_2 draws, then there are $x_1 + x_2$ draws in all. Each draw has an equal chance of having the highest quality. Thus, the chance of the highest quality being among the x_1 draws is $x_1/(x_1 + x_2)$.

[28] See Che (1993) and Asker and Cantillon (2008, 2010) for design competitions with a "scoring rule" but without an all-pay component.

that is, $G_i(s_i) = P(s_i > \max\{s_{-i}\})$. Then, firm i chooses effort x_i, surplus s_i, and price p_i to solve

$$\max_{x_i, s_i, p_i} G_i(s_i)p_i - c(x_i) \text{ s.t. } x_i - p_i = s_i.$$

Substituting the constraint into the maximand and the first-order conditions w.r.t. x_i and p_i yields $G_i'(s_i)p_i = c'(x_i)$ and $G_i'(s_i)p_i = G_i(s_i)$, respectively. Combining the two yields: $G_i(s_i) = c'(x_i)$. For arguments similar to the all-pay auction with complete information, firms also make zero profits. Hence $G_i(s_i)p_i = c(x_i)$. Together, we find that firms set prices $p_i = c(x_i)/c'(x_i)$ and choose surpluses by the cumulative distribution set by $G_i(s_i) = c'(x_i)$.

For example, if $c(x) = x^2$ and $n = 2$, then $p_i = \frac{x_i}{2}$. This implies $s_i = \frac{x_i}{2}$ and $G_i(s_i) = F_i(s_i) = 2x_i = 4s_i$. Hence, in the equilibrium each firm chooses a surplus s_i according to the uniform distribution on $[0, 4]$, and sets $x_i = 2s_i$ and $p_i = s_i$. Note that an all-pay auction with a bid-dependent reward of $\frac{x}{2}$ with a cost function of x^2 has a symmetric equilibrium of each firm choosing x according to $F(x) = 2x$. Thus, it has an equivalent equilibrium. Kaplan and Wettstein (2013) prove that the profit from the Che-Gale innovation contest is also the highest expected profit the buyer can earn in a bid-dependent all-pay auction.[29]

7.7.9 Blotto games

Introduced by Borel (1921), a Colonel Blotto game has two players (Colonels) each deciding how to divide their respective army across $n > 2$ battles; that is, they each must choose the size of the force x_i to place in battle i such that the sum of their forces in each battle is less than or equal to the size of their total army. The player with the largest force in a particular battle wins that battle. Winning each battle is worth the same and a tie in a battle is worth half as much as a win. Note that the winner of each battle can also be determined by different success functions (such as Tullock) and overall payoffs can be determined by a winner-take-all for the player who wins most of the battles.

The Blotto game is in essence n all-pay auctions run simultaneously when the cost of using part of one's army in one battle is the opportunity cost of using the troops in another battle. For now assume that both armies are the same size (and equal to one). For similar reasons to those in the all-pay auction with complete information, there is no pure-strategy Nash equilibrium and there cannot be a specific size of force in a battle strictly greater than zero chosen with positive probability (also, at most one player can place an atom at zero for each particular battle). Denote F_i as the distribution played in battle i. Given the other player's choice of F_is, we can now write the Lagrangian of a player's problem where the Lagrange multiplier λ is the shadow price that represents

[29] This is not the case when there is asymmetry among firms.

the opportunity cost of using the troops. Hence, a player will choose his strategy to maximize:

$$\max_{x_1,x_2,\ldots,x_n,\lambda\geq 0} F_1(x_1)+F_2(x_2)+\cdots+F_n(x_n)+\lambda(1-\sum_i x_i). \qquad [7.3]$$

The first-order condition is $F_i'(x_i)=\lambda$ and must hold for all x_i in the support of F_i. Integrating, yields $F_i(x_i)=\lambda x_i+c_i$. However, for the player to be indifferent to shifting forces between two different battles, we must have $c_i=c_j$ for all i,j. Thus, expectation of this uniform distribution is $(1-c)\lambda/2$. Hence for $\sum_i x_i=1$ to hold, we must have $\lambda=\frac{2}{n(1-c)}$. However, since the support of each distribution must be the same for both players and they both cannot have an atom at the bottom of the support, we must have $c=0$ and each player chooses, for each battle, the x_i according to a uniform distribution on $[0,\frac{2}{n}]$.

While each player deploys an army in each battle according to a uniform distribution, there must be a connection between each battle since the total size of the army equals one for each realization and not just in expectation. This joint distribution is what is called a copula in mathematics, but it requires the additional property that the sum of the armies in each battle always be equal. Borel and Ville (1938) provide a solution for $n=3$ by choosing $x_1\sim U\left[0,\frac{2}{3}\right]$ and half the time setting $x_2=\frac{2}{3}-\frac{x_1}{2}$ and $x_3=\frac{1}{3}-\frac{x_1}{2}$, while the other half of the time swapping how one sets x_2 and x_3. Gross and Wagner (1950) extend Borel and Ville's result to n armies and provide an additional method for finding a joint distribution.[30]

Roberson (2006) gives the complete set of equilibrium marginal distributions for general n and asymmetric army sizes. Roberson shows that when the ratio r of army sizes is between $\frac{2}{n}$ and $\frac{n}{2}$, the equilibrium is similar to asymmetric complete-information all-pay auctions given by [7.2].[31] For intermediate asymmetries in army sizes where $\frac{1}{n-1}\leq r<\frac{2}{n}$ or $\frac{n}{2}<r\leq n-1$, the equilibrium is similar to Che and Gale (1998a) but with a bid cap placed only on the weaker firm. He also provides a specific method for finding a joint distribution (making use of n-copulas).

Notice that having battles worth different amounts is just a matter of adding different weights in front of the F_is. This yields a solution where the upper bound of each uniform distribution of each battle is proportional to its weight. This case was studied for symmetric army sizes and general n by Gross (1950), Laslier (2002), and Thomas (2013). Also, Gross and Wagner (1950) solved the case of $n=2$ with heterogeneous values of battles and asymmetric army sizes. The case for $n>2$ with heterogeneous values of battles and asymmetric army sizes has not been generally solved. Another version of the game has

[30] Laslier and Picard (2002) measure the inequality of the equilibria in Gross and Wagner (1950) using the Gini coefficient. Weinstein (2012) provides another method of finding a joint distribution and examines Laslier and Picard's analysis.

[31] The difference is that v_i is replaced by the respective army size times $\frac{n}{2}$.

all the utility going to the player winning the majority of the battles. This version for the $n = 3$ case with symmetric army sizes was solved in Borel and Ville (1938).

Adamo and Matros (2009) solve the Blotto game when the army size is incomplete information. Namely, if the distribution of each army size G is concave, then equally dividing one's army is an equilibrium. Hart (2008) solves the Blotto game when the armies can be divided only discretely. Roberson and Kvasov (2012) extend the analysis of Blotto games to an environment where a player saves an expense proportional to the portion of his army that is not used in a battle.

There is a reason for the strong connection to the all-pay auction with complete information (Roberson, 2006). While the size of the army is fixed and it does not cost more to field the whole army than to field half the army, there is an opportunity cost in that using part of the army in one battle means that it cannot be used in a different battle. This is represented by the shadow price equal to λ in [7.3]. Hence, it is no surprise that both the all-pay auction and the Colonel Blotto game have similar equilibria since they both consist of strategies that are uniform distributions.

7.7.10 Other topics

There are many topics on all-pay auctions not covered in detail in this chapter. For instance, several papers use the basic contest building block in a more complex environment: Moldovanu and Sela (2006) analyze if it is optimal to run one contest or break it down into playoff contests. Konrad and Kovenock (2009) study a sequence of contests where there is a prize for the winner of each individual contest and another for the majority winner of the sequence.[32] There are also studies where the prize is not exactly a monetary prize but rather the utility for winning depends upon the number of prizes given out, as with As in a course (see Modovanu et al., 2007, and Dubey and Geanakoplos, 2010), or losing the contest can come at an additional loss of utility (beyond wasted effort) at a cost to the designer of such a punishment (see Moldovanu et al., 2012, and Thomas and Wang, 2013). Under incomplete information, there is also progress beyond the IPV case (see Lizzeri and Persico, 2000, and Siegel, 2014a). There is also a large body of literature on experiments in all-pay auctions (see Dechenaux et al., 2012, for a review). For a more detailed theoretical review on contests, see Konrad (2009).

7.8. INCORPORATING BEHAVIORAL ECONOMICS

In recent years both the literature and the public have paid increased attention to behavioral economics. Since in many auction experiments, subjects do not bid according to standard theory, auction theory has begun to follow the trend. Probably the most natural behavioral factor to incorporate is a utility for winning the auction. Mathematically this

[32] For other papers on sequential contests where the values of winning each contest are not independent, see Sela (2011, 2012).

is equivalent to shifting the distribution of values (see Cox et al., 1988). Other more complex ways in which behavioral economics has been incorporated into auction theory are explored in this section.

7.8.1 Regret

In an auction, there are two types of regret: winner's regret and loser's regret. With winner's regret, the winning bidder regrets if he overpaid and could have won with a lower bid. With loser's regret, a losing bidder regrets if he could have profitably won the auction by submitting a higher bid.

For a single-object first-price auction, Engelbrecht-Wiggans (1989) (henceforth EW) expresses the utility of the buyer in an auction as the expected profit from winning minus α_1 times the expectation of an increasing function of the overpayment (winner's regret) minus α_2 times the expectation of an increasing function of the missed surplus (loser's regret):

$$P(b > B) \cdot (v - b) - \alpha_1 \cdot E[h(b - B)] - \alpha_2 \cdot E[g(v - B)] \qquad [7.4]$$

where B is the maximum bid of the other buyers and $g(x)$ and $h(x)$ are two functions that are both positive (and typically monotone increasing) for $x > 0$ and 0 for $x \leq 0$. For the special case $g(x) = h(x) = x^+$ and $\alpha_1 = \alpha_2 = \alpha$,[33] EW proved that the equilibrium bid function is the same with and without regret.

To see this, let F be the cumulative distribution of B and let $f = F'$ be its density. Now, in the special case we can rewrite [7.4] as:

$$F(b)(v - b) - \alpha\left(\int_0^b (b - B)f(B)dB + \int_b^v (v - B)f(B)dB\right).$$

The first-order condition for the maximization of this function of b is:

$$f(b)(v - b) - F(b) + \alpha\left(\int_0^b f(B)dB - (v - b)f(b)\right) = 0$$

or

$$f(b)(v - b) - F(b) + \alpha\left(f(b)(v - b) - F(b)\right) = 0. \qquad [7.5]$$

Equation [7.5] is the same whether $\alpha = 0$ (without regret) or $\alpha > 0$ (with regret).

EW also shows that when $\alpha_1 > \alpha_2$ (and $g(x) = h(x) = x^+$), the equilibrium bid function is lower than without regret, and vice versa when $\alpha_1 < \alpha_2$. Logically, when the costs of increasing his bid go up relative to the benefits, a buyer will decrease his bid. Engelbrecht-Wiggans and Katok (2008) (henceforth EWK) show that with n buyers

[33] The notation x^+ denotes x if $x > 0$ and 0 otherwise.

having values drawn from the uniform distribution starting at 0, the equilibrium bid function will be

$$b(v) = \frac{n-1}{n + \frac{\alpha_1 - \alpha_2}{1+\alpha_2}} \cdot v.$$

With the restrictions that $h(x) = g(x) = 0$ for $x \leq 0$ and $h(x)$, $g(x)$, $h'(x)$, and $g'(x)$ positive for $x > 0$, Filiz-Ozbay and Ozbay (2007) (henceforth FOO) prove the rather intuitive result that when there is only loser's regret, $\alpha_1 = 0$ and $\alpha_2 > 0$, the equilibrium bid functions are higher than in the no-regret equilibrium and when there is only winner's regret, $\alpha_1 > 0$ and $\alpha_2 = 0$, the equilibrium bid functions are lower than in the no-regret equilibrium. Consequently, revenue is higher with only loser's regret and lower with only winner's regret than in the no-regret equilibrium.

FOO run a one-shot first-price auction experiment with these three regret conditions as the three treatments by varying the feedback. For the no-regret treatment, there is no feedback except that one has won or lost. For the winner's-regret treatment, in addition the winner sees the second-highest bid. For the loser's-regret treatment, the winning bid is announced. Using groups of four buyers having values drawn from the uniform distribution, they find that indeed the relative ordering of bidding among treatments matches the theoretical predictions.

EWK also run an experiment but with subjects playing repeatedly against computerized strategies. They go further in developing the feedback treatments by telling the winner explicitly the difference between his bid and the second-highest for winner's regret (rather than leaving it up to them to calculate), or telling the losers the exact amount of the money left on the table (the difference between their value and the winning bid for loser's regret). They also add a treatment with both types of regret. As with FOO, they find the bidding among the treatments rank according to theory and that bids with no regret are similar to those with both types of regret, which is consistent with the theory under the assumptions that $\alpha_1 = \alpha_2$ and $g(x) = h(x) = x^+$.

These results suggest that a seller would have higher revenue by announcing the winning bid and keeping the losing bids secret as opposed to only announcing the winner's identity or publicizing all bids. Regret need not be purely psychological. A situation with regret can also exist when an agent is bidding on behalf of a client. The agent wishes to avoid the displeasure of his client. This possibility also holds true for many of the other topics in this section.

7.8.2 Impulse balance

Impulse Balance Theory was introduced by Ockenfels and Selten (2005) (henceforth OS). It is similar to regret except that rather than a utility function being explicitly specified, bidders have an impulse to adjust their bid functions upwards or downwards whenever ex post it is rational to do so. This impulse is proportional to the potential gains (if measurable). For instance, if in a first-price auction one bids 10 and the second-highest bid is 7, then the downwards impulse of the winner is proportional to 3. If instead one loses to a bid of 12 when one's value is 14, then his upwards impulse is proportional to 2. These impulses are in balance if the upward impulse is equal to λ times the downward impulse, where λ is a weight constant that is specific to the agent. If half the time one faces a bid of 7 and half the time one faces a bid of 12 (always with a value of 14), then bidding 10 would be impulse-balanced if $\lambda = 2/3$.

OS analyze impulse balance in an n-bidder first-price auction when values are drawn from the uniform distribution on $[0, 1]$. The approach is not that of best reply and equilibrium. Rather, they define the notion of a bidder's strategy to be impulse-balanced given the strategy of the other bidders. More specifically, they assume that all bidders use a linear bid function $b(v) = av$, and determine for which value of the parameter a each bidder's strategy is impulse-balanced (for a given impulse weight λ).

As a theory attached to an experimental work, OS consider two repeated auction treatments: the full-feedback treatment (denoted F) in which all bids (in particular, the highest and second-highest) are announced at the end of the auction, and the no-feedback treatment (denoted NF) in which only the winning bid is announced. The F treatment is similar to the both-regrets treatments of EWK and the NF treatment is similar to the loser's-regret treatments of OO and EWK. The main differences in design is that OS is repeated with random matching (unlike the one-shot OO) and against human bidders (unlike EWK). (Note also that OS have two bidders rather than four bidders in OO.)

In the F treatment, the downward and upward impulses are defined as the expected forgone profits in case of winning and losing respectively and they are given by:

$$I_{-}^{F} = E[(b - B)^{+}] = \frac{a}{n(n+1)}, \qquad [7.6]$$

and

$$I_{+}^{F} = E[(v - B)^{+} \cdot \mathbf{1}_{b<B}] = \frac{(1-a)^2(n-1)}{2(n+1)}. \qquad [7.7]$$

In the NF treatment, as the foregone profit is not observable for the winner (he does not know how much further he could have lowered his bid and still have won), the impulses are driven just by the fact of winning or losing the auction. Therefore, OS define the impulses to be the probability of these events respectively; that is, the downwards impulse is the probability of winning and the upwards impulse is the probability that a bidder loses when it would be profitable to increase his bid so as to win the auction. For this model, these impulses are given by:

$$I_-^{NF} = P(b > B) = \frac{1}{n}, \quad [7.8]$$

$$I_+^{NF} = P(b < B \,\&\, v > B) = \frac{(1-a)(n-1)}{n}. \quad [7.9]$$

The condition that the strategy $b(v) = av$ be impulse balanced is $I_+^F = \lambda I_-^F$ in the F treatment and $I_+^{NF} = \lambda I_-^{NF}$ in the NF treatment. Solving these equalities shows that if $0 < \lambda < \frac{1}{2}$, then $a^{NF} > a^F > \frac{n-1}{n}$. Recalling that $b(v) = \frac{n-1}{n} v$ is the equilibrium bid of this auction without impulse biases, this means that there is overbidding in both conditions, and there is more overbidding without feedback. This is in agreement with both regret theory and OS's experimental results (which are consistent with the experimental results of OO and EWK).

7.8.3 Reference points

Rosenkranz and Schmitz (2007) (henceforth RS) analyze a model in which the utility of the buyer may depend upon a reference point r; any price paid above this number decreases the buyer's utility and any price below it increases his utility. Formally, a buyer with value v purchasing the object at price p has utility

$$u(v, p, r) = v - p + z(r - p) \quad [7.10]$$

where z is a function satisfying $z' \geq 0$ and $z(0) = 0$.

While RS analyze this model with a linear z, their solution can be generalized, using the envelope theorem, to any nondecreasing function z. In a second-price auction, for the usual reasons, a buyer will set a bid $b_s(v, r)$ such that $u(v, b_s(v, r), r) = 0$. If $z(r - v) > 0$ then $b_s(v, r) > v$. Likewise, if $z(r - v) < 0$, then $b_s(v, r) < v$.

In a first-price auction, if we define the strictly increasing function $w(b, r) := b - z(r - b)$ and its inverse w.r.t. b by $w_r^{-1}(\cdot)$, then the equilibrium bid function is

$$b(v, r) = w_r^{-1}\left(v - \frac{\int_0^v F^{n-1}(\tilde{v})d\tilde{v}}{F^{n-1}(v)}\right). \quad [7.11]$$

RS show that if the reference point r is affected by the minimum bid, then unlike the standard case without reference points, the optimal minimum bid would depend upon the number of bidders. RS also show that in some cases revenue increases if the minimum bid is kept secret and, in which case the buyers have only an exogenous reference point. Finally, RS prove revenue equivalence for the case of linear z (which does not hold for nonlinear z). This equivalence does not hold if the reserve price is kept secret.

7.8.4 Buy-it-now options

A buy-it-now option allows a bidder to buy an object for a specified price before the auction finishes. This option is typically available only for a limited time (for instance, until someone bids on the item or until the reserve price is reached), after which the option vanishes and the auction proceeds normally. The buy-it-now option is available in auctions on eBay (since 1999) and accounts for a large portion of their sales.[34]

Reynolds and Wooders (2009) show that theoretically under risk-neutrality a buy-it-now option cannot improve revenue, but there can be gains if the buyers are risk averse. Shunda (2009) shows that if the buy-it-now option can influence the reference point (in the utility function), then it can improve revenue even under risk-neutrality. The intuition is that a buy-it-now option can push up the reference point and thereby increase revenue.

7.8.5 Level-k bidding

The level-k model was introduced by Stahl and Wilson (1994, 1995) and Nagel (1995). In Nagel's (1995) experimental study of the guessing game, subjects choose a number from 0 to 100 and the number that was closest to 2/3 the average won a prize (ties were broken randomly). Many guesses were bunched around 33 and 22. According to Nagel, a subject who chooses 33 believes that other subjects are choosing a number randomly is classified as L_0 level. Thus, the subjects who choose 33 think one level ahead of L_0 subjects and are classified as level L_1. Those who choose 22 are level L_2 since they think one level ahead of level L_1; that is, if all other subjects are level L_1 players and thus choose 33, it is best to choose 22. The limit of this iterative process goes to the equilibrium of 0.

Crawford and Iriberri (2007) use this type of analysis to derive behavior in auctions. A level L_0 player will bid randomly by choosing a bid uniformly over the whole range of possible bids (from 0 to the highest possible value).[35] Like in the guessing game, a level L_k player ($k > 0$) will best-react under the assumption that all other players are level L_{k-1} players. As an example, in a first-price auction with two buyers and a minimum bid of m, a level L_1 player with a value above m will choose a bid $b = \max\{\arg\max_{\tilde{b}} \tilde{b}(v - \tilde{b}), m\} = \max\{\frac{v}{2}, m\}$. Note that this is independent of the distribution of the players' values since level L_0 is assumed to bid uniformly. In particular, if the values are drawn independently from a distribution on $[0, 1]$ and $m = 0$, level L_1 bidding coincides with equilibrium bidding under the uniform distribution and hence bidding is the same for all levels L_k for $k \geq 1$. On the other hand, if the values of the two players are affiliated, as the level

[34] Before being discontinued, Yahoo! Auctions also had a buy-it-now option that allowed the seller to change the price during the auction as opposed to eBay that only permits the price to be set at the beginning.

[35] Also considered by Crawford and Iriberri (2007) is a "truthful" L_0 that bids his value.

L_1 bidding will still be $\frac{v}{2}$ if both players are level L_1 players, they will not take into account the winner's curse and potentially overbid in the auction (and in expectation lose money).

Crawford et al. (2009) explore the possibility of a different choice of the level L_0 player, letting such a player bid uniformly on the range from the minimum bid m to the highest possible value. Hence, now a level L_1 player with a value above m will choose a bid $b = \arg\max_{\tilde{b}} (\tilde{b} - m) \cdot (v - \tilde{b}) = \frac{v+m}{2}$. (And a level L_2 player with $v > m$ will choose $b = \arg\max_{\tilde{b}} F(2\tilde{b} - m)(v - \tilde{b})$.) Their assumption is somewhat unpalatable in that such an L_0 player will always bid more than m (even if $v < m$). We suggest that perhaps a more reasonable L_0 player for a first-price auction will be one that bids only above the minimum bid if $v > r$ and never bids above his value. In this case for $m = 0$ and uniform values on $[0, 1]$, an L_1 player will choose $b = \arg\max_{\tilde{b}} (\tilde{b} + \int_{\tilde{b}}^{1} \frac{\tilde{b}}{\tilde{v}} d\tilde{v})(v - \tilde{b})$. An L_1 player's inverse bid function would be $v(b) = \frac{-b + 2b\ln(b)}{\ln(b)}$. However, an L_2 player will then in turn bid the highest possible bid from an L_1 player (approximately 0.34) for a range of possible values.

For experimental studies on level-k models see the chapter "Behavior Game Theory Experiments and Modeling" in this Handbook.

7.8.6 Spite

The behavioral considerations thus far have looked only at individual biases, traits, and reasoning. Yet bidders may also have preferences about the surpluses of the other bidders. In particular, they may have spiteful preferences; namely, they are willing to sacrifice some of their own surplus in order to lower the surplus of the other bidders.

Morgan et al. (2003) present a two-bidder model in which a bidder's utility when losing is inversely related to the surplus of the winner. More specifically, for some $\alpha > 0$, in a first-price auction the utility equals

$$P(b > B) \cdot (v_i - b) - \alpha \cdot E[v_{-i} - B | b < B] \qquad [7.12]$$

and in a second-price auction, the utility equals:

$$P(b > B) \cdot (v_i - E[b_{-i} | b > B) - \alpha P(b < B) \cdot E[v_{-i} - b | b < B]. \qquad [7.13]$$

The equilibrium bid function in the first-price auction is:

$$b(v) = v - \frac{\int_0^v F(\tilde{v})^{1+\alpha} d\tilde{v}}{F(v)^{1+\alpha}}.$$

By taking the derivative w.r.t. α, it is straightforward to show that the second term is decreasing in α and thus, for $\alpha > 0$, the bid function is higher than the bid function without spite ($\alpha = 0$) and bidding is increasing in spite. The equilibrium bid function in the second-price auction is

$$b(v) = v + \frac{\int_v^1 (1 - F(\tilde{v}))^{\frac{1+\alpha}{\alpha}}\, d\tilde{v}}{(1 - F(v))^{\frac{1+\alpha}{\alpha}}}.$$

Since this is clearly greater than v and is increasing in α (by straightforward verification), there is overbidding in the second-price auction. Again, the bid function is higher than the bid function without spite and it is increasing in the spite α. It is thus possible that the winner in fact loses in the sense that he has negative utility and would be better off not having won. Morgan et al. (2003) also demonstrates a similar result for English auctions.

Let us now look at the case where there is asymmetry in values. Take the same preferences but where one bidder has a value equal to 9 and the other bidder has a value equal to 0 and there is complete information about values. Assume that $\alpha = 1/2$ and consider a second-price auction in which when both bidders submit the same bid, the high-value bidder wins. In this case, we claim that both bidders submitting the same bid b^*, where $3 < b^* < 6$, is an equilibrium. The high-value buyer will win at price b^* and have utility $9 - b^*$. The low-value bidder will have utility $-(9 - b^*)/2$. To see that this is an equilibrium, the only candidate for a profitable deviation of the high-value bidder is to bid just below b^* and lose the auction in which case his utility will be $b^*/2$. For the low-value bidder, the only candidate for a profitable deviation is to bid just above b^*, win the auction, and have a utility of $-b^*$. For $3 < b^* < 6$ we have $9 - b^* \geq b^*/2$ and $-(9 - b^*)/2 \geq -b^*$; hence, both deviations are not profitable.

In this simple example, we have the low-value bidder overbidding ($b^* > 0$) because of spite and the high-value underbidding ($b^* < 9$) to essentially counterbalance this spite. Nishimura et al. (2011) show that such spite and counter-spite hold more generally in both English and second-price auctions.

7.8.7 Ambiguity

In analyzing auction models, we usually assume that bidders know the probability of winning given a certain bid. In many cases, the bidder may be uncertain about this probability. This can be due to not knowing the distribution of values or strategies of the other bidders. Uncertainty about the relevant probabilities is known as *ambiguity* and its effects on behavior have been demonstrated and studied in a large number of papers starting from the seminal work of Ellsberg (1961). In general, decision makers are, as pointed out by Ellsberg, ambiguity-averse. Theoretical models for studying ambiguity and explaining its effects were laid out by Gilboa (1987), Gilboa and Schmeidler (1989), Schmeidler (1989), Sarin and Wakker (1992), and others.

One main method for incorporating ambiguity into theory is to replace the additive probability of the agent with a non-additive probability measure known as *capacity*. This is a real function $c(\cdot)$ defined on the subsets of the probability space Ω and satisfying

$c(\emptyset) = 0$, $c(\Omega) = 1$, and $A \subseteq B \Rightarrow c(A) \leq c(B)$ for any $A \subseteq \Omega$ and $B \subseteq \Omega$. The additivity condition of a probability measure: $P(A) + P(B) = p(A \cup B) + P(A \cap B)$ is replaced by the *convexity* condition: $c(A) + c(B) \leq c(A \cup B) + c(A \cap B)$ for any $A \subseteq \Omega$ and $B \subseteq \Omega$. The degree of the convexity (that is, the difference between the two sides of the inequality) reflects the degree of ambiguity aversion.

To see this, let A and B be two complementary events, that is, $A \cup B = \Omega$ and $A \cap B = \emptyset$. Now the convexity condition is $c(A) + c(B) \leq 1$. If the decision maker has symmetric information about the occurrence of the two events, then he assigns to them equal probabilities, i.e., $c(A) = c(B)$. But, as Schmeidler says, this equal probability need not be $\frac{1}{2}$ each if the information about the occurrence of these events is meager. This is, for example, the case for the events of drawing a red (R) or a black (B) ball from the Ellsberg urn containing 100 balls with an unknown composition of red and black balls. If $c(R) = c(B) = \frac{3}{7}$ then $c(A) + c(B) < c(A \cup B) + c(A \cap B) = 1$ and the difference between the two sides ($\frac{1}{7}$ in this case) is a measure of the uncertainty, or the *ambiguity*, of the decision maker about the probability of the events.

A convex capacity can be conveniently represented as a composition $c = \phi \circ P$ where P is an additive probability measure and the *probability transformation* ϕ is a function $\phi : [0, 1] \rightarrow [0, 1]$, where the degree of ambiguity aversion is reflected in the convexity of ϕ. In the absence of ambiguity, an agent with utility function $u : \Omega \rightarrow \mathbb{R}$ defined on the state space Ω on which he has subjective (additive) probability P considers the *expected utility*, which can be written as $\int_\tau P(\{\omega \in \Omega | u(\omega) \geq \tau\}) d\tau$. In the presence of ambiguity, the agent considers what is called the Choquet expected utility (CEU), which is the expectation according to the transformed probability $c = \phi \circ P$, which can be written as:

$$\int_\tau \phi(P(\{\omega \in \Omega | u(\omega) \geq \tau\})) d\tau. \qquad [7.14]$$

Instead of maximizing expected utility, an ambiguity-averse agent will maximize this Choquet Expected Utility. Salo and Weber (1995) study the effects of ambiguity in first-price auctions. In their model, an ambiguity-neutral bidder assumes that the other bidders' values are independent and uniformly distributed on $[0, \overline{v}]$. The bidder's CEU profit $\pi_i(v)$ can thus be written as:

$$\pi_i(v) = \phi\left[\left(\frac{B^{-1}(b_i)}{\overline{v}}\right)^{N-1}\right](v_i - b_i) \qquad [7.15]$$

where the random variable B is the highest bid of the other bidders. Using the envelope theorem to find $\pi_i'(v)$ and then integrating and setting $\pi_i(0) = 0$, we find the equilibrium bid function is

$$B(v_i) = v_i - \frac{\int_0^{v_i} \phi([\frac{\tau}{\overline{v}}]^{N-1}) d\tau}{\phi([\frac{v_i}{\overline{v}}]^{N-1})}. \qquad [7.16]$$

If one restricts the ratio $\frac{\phi(\alpha x)}{\phi(x)}$ to only depend upon α, then it would have the form $\phi(x) = x^{\frac{1}{K}}$. Here, $K < 1$ expresses ambiguity aversion ($\phi(x) < x$) and the smaller K implies more ambiguity aversion. $K > 1$ expresses ambiguity-loving ($\phi(x) > x$). The bid function now reduces to:

$$B(v_i) = \frac{N-1}{N-1+K} v_i. \qquad [7.17]$$

We see here the somewhat intuitive result that increasing ambiguity aversion increases the bid function in a first-price auction (and vice versa for ambiguity-loving). Other models of ambiguity have been applied to auctions with similar theoretical results (see Lo, 1998, Ozdenoren, 2002, Chen et al., 2007). However, experimentally, Chen et al. (2007) find that bidding behavior is consistent with ambiguity-loving behavior (lower bids when faced with ambiguity).[36]

Ambiguity could exist about the number of bidders in the auction. Salo and Weber (1995) also examine this ambiguity with CEU and find that the bids are higher (indicating the seller should not reveal the number). Levin and Ozdenoren (2004) revisit this question with maxmin expected utility and also find that bidding should be higher in a first-price auction than in a second-price auction. They also find that if the buyers are more pessimistic than the seller, the seller will want to maintain the ambiguity of the bidders and to refrain from revealing his information about the number of bidders.

When looking at ambiguity aversion of a seller, Turocy (2008) finds that an ambiguity-averse seller will prefer a first-price auction to a second-price auction if bidders make small payoff errors, that is, errors in choosing strategies that result in lowering payoffs by ϵ or smaller.

For detailed discussion of nonexpected utility theory and its implications on auctions when the object itself is a risky prospect (such as a lottery ticket), see the chapter "Ambiguity and Nonexpected Utility" in this Handbook.

7.9. POSITION AUCTIONS IN INTERNET SEARCH

Here we discuss research on a fairly recent auction mechanism that has been the main revenue-driving force behind several large internet companies: the auctioning off of advertisements based upon the keywords used in an internet search. In such a mechanism, several advertisement placements per search are sold. The placements intrinsically vary in quality with the better positions going to the higher bidders.

In 1998, Goto.com introduced the first successful position auction for search results (see Davis et al., 2001, U.S. Patent 6,269,361). When a particular term was searched, Goto.com listed the results in the order of the advertisers' willingness to pay per click

[36] See Dickhaut et al., 2011, for a first-price auction experiment with ambiguity and a common value.

(the advertiser's bid was listed along with the results). This company founded by Bill Gross's Idealab was eventually rebranded as Overture and sold to Yahoo! in 2003 for $1.6 billion. In 2002, Google used the advertisers' quality along with their bids to determine the position of advertisements in their AdWords auction (while listing search results independent of bids). This generated $42 billion in revenues in 2012. In 2004, Google agreed to pay Yahoo! 2.7 million shares (worth $300 million at the time) for patent infringement.

The main driving force behind these auctions is that not only is having an ad posted or link listed valuable, but placement matters too. Being listed first makes it more likely to be chosen. In politics, Orr (2002) argues that in Australia voters were more likely to choose the candidates the higher they were on the ballot.

7.9.1 First-price pay-per-click auctions

In its typical form, a position auction is somewhat of a misnomer: the advertisers do not bid on positions but rather on clicks; that is, the bid represents an amount to be paid when a user is directed to their site.

Assume there are n advertisers. Each advertiser a submits a bid b_a for a click whose value for that advertiser is v_a. There are also a limited number of positions where each position i gets x_i clicks (note that at this stage we assume that the number of clicks per ad depends only upon ad position and not also on ad quality). The lower i indicates the better location, that is, $x_i > x_{i+1}$. Thus, the value for position i for advertiser a is $v_a x_i$. Assume that there is complete information about the bidders' values.

The initial Goto.com position auction was run using the natural extension of first-price auctions. The advertisers are listed in decreasing order of their bids. The advertiser with the highest bid gets the most-preferred position of 1. The advertiser with the second-highest bid gets the second-highest position of 2. This is continued until we run out of advertisers or positions. Ties are broken randomly. The price paid for each click is equal to the bid of the advertiser receiving that click. Advertiser i (that is, the one in position i) will receive $(v_i - b_i)x_i$.

Since many users searched using the same keywords, the same keywords were auctioned off repeatedly. Bids were solicited for a number of searches in advance with an option of changing one's bid at a later time. This created a problem with first-price position auctions. To see this, assume that there are two advertisers and two positions. When first place has an $x > 0$ chance of a click and second place has a $y > 0$ chance of a click (and $x > y$), then under complete information, there is no pure-strategy Nash equilibrium. The advertiser who is listed second will either wish to lower his bid to zero (if it is not already at zero) and stay in second place or raise his bid just enough to come in first. If the second-place advertiser submitted a bid of zero, then the first-place advertiser would wish to lower his bid to $0.01. This cannot be an equilibrium since then the second-place advertiser would like to bid $0.02. Through similar logic, ties can be ruled out. Although this example resembles a first-price auction for the right to be listed first (which would have a pure-strategy Nash equilibrium), it differs in that the loser pays also for the second-place listing.

In position auctions, mixed-strategy equilibria are problematic in that the equilibrium is not efficient (sometimes the low-valued advertiser will be listed higher than an advertiser with a higher value), and effort is then wasted in altering bids and strategizing. Edelman and Ostrovsky (2007) show a sawtooth pattern of cycles in prices in data from June 2002 to June 2003 on Overture.[37] For this reason, Yahoo! switched to using a second-price position auction, which is referred to by Edelman et al. (2007) (henceforth, EOS) as a generalized second-price auction.

7.9.2 Second-price pay-per-click auctions

In second-price position auctions, advertisers are ordered according to their bids and assigned positions in the same manner as in the first-price position auction. For all positions strictly less than n, the price paid per click for the advertiser assigned to position i is now $p_i = b_{i+1}$. The price per click of the advertiser ordered last is zero.

Varian (2007) (henceforth, VAR) shows that the Nash equilibria of the second-price position auctions (under complete information) must satisfy constraints consisting of an advertiser in position i not wanting to move up to a higher position $j, j < i$ (recall that a higher position corresponds to a lower index) and not wanting to move down to a lower position $j, j > i$. To move higher, an advertiser would need to bid slightly above the bid of the advertiser in position j and pay that bid, b_j, which is equal to p_{j-1} (we define $p_0 = b_1$). To move to a lower position, an advertiser needs to pay the price of the advertiser in position j, p_j (by bidding slightly under his bid). Hence, the incentive constraints are:

$$(v_i - p_i)x_i \geq (v_i - p_{j-1})x_j \quad \text{for } j < i, \tag{7.18}$$

$$(v_i - p_i)x_i \geq (v_i - p_j)x_j \quad \text{for } j > i. \tag{7.19}$$

VAR also shows that in the case of complete information (about the values) there are multiple equilibria, as with the standard second-price auction under complete information. However, EOS show that, unlike in the standard second-price auction, truthtelling is no longer a dominant strategy nor even always an equilibrium. To see this let us look at an example where the highest position has a 60% chance of a click, second place has a 50% chance of a click, and third place has a 20% chance of a click. There are three advertisers: one with a value of 3 per click, another with a value of 2, and the last with a value of 1. If everyone bid truthfully, then the advertiser with a value of 3 would

[37] This sawtooth pattern occurs for a similar reason in Edgeworth prices (see Maskin and Tirole, 1988 and Noel, 2008) and empirically is found elsewhere in gasoline markets (see Noel, 2007a,b).

be in first place with utility $(v_1 - p_1)x_1 = (3 - 2)0.6 = 0.6$. However, by bidding 1.9, he could be in second place with utility $(v_1 - p_2)x_2 = (3 - 1)0.5 = 1$. Hence, there is an incentive for this advertiser to deviate.

If we define envy as desiring to swap positions and prices paid with another advertiser, then we can define the envy-free equilibrium as an outcome in which there is no envy between any two advertisers,[38] that is,

$$(v_i - p_i)x_i \geq (v_i - p_j)x_j \quad \text{for all } i, j. \qquad [7.20]$$

Note that, as the name suggests, this concept is indeed a refinement of the notion of equilibrium, since [7.20] is equivalent to [7.19] for $j > i$, and implies [7.18] for $j < i$ since $p_j < p_{j-1}$.

An envy-free equilibrium is in particular locally envy-free; that is, an advertiser does not envy the advertiser located one position above or below his position, that is,

$$(v_i - p_i)x_i \geq (v_i - p_j)x_j \quad \text{for all } i, j = i \pm 1. \qquad [7.21]$$

This concept defined by EOS was proved by VAR to be equivalent to the (global) envy-free equilibrium defined by [7.20]. Indeed, [7.20] implies [7.21], while [7.21] implies [7.20] using the "directional" transitivity of the envy-free conditions (that is, the transitivity of "not envying someone above you" and "not envying someone below you").[39]

VAR shows that the equilibrium with the highest revenue is the highest revenue envy-free equilibrium given by $b_{n+1} = 0$ and $b_{i+1} = (b_{i+2}x_{i+1} + v_i(x_i - x_{i+1}))/x_i$ (in this case, [7.21] holds with equality for $j = i + 1$). In our example, we would thus have $b_3 = 1.2$, $b_2 = 1.5$, and $b_1 > 1.5$ with prices $p_1 = 1.5$, $p_2 = 1.2$, and $p_3 = 0$. Notice that while this is envy-free, it does require advertiser 3 to bid above his value (of 1).

The Vickrey-Groves-Clark (VCG) mechanism with position auctions will induce truthtelling as dominant strategies. With this mechanism, each advertiser i has to pay the externality that he imposes on the other advertisers. Without advertiser i, the advertisers above him would not change their position while each advertiser j positioned below him ($j > i$) would move up one position gaining $v_j(x_{j-1} - x_j)$, which is the advertiser j's value times the

[38] This is called symmetric nash equilibrium by Varian (2007). Note that there are other equilibria. As Borges et al. (2013) mention, if $x_1 = 3, x_2 = 2, x_3 = 1$, and $v_1 = 16, v_2 = 15, v_3 = 14$, then under complete information one advertiser bidding 11, another 9, and another 7 will be a Nash equilibrium. Thus, a nonsymmetric equilibrium could be inefficient since in this example the one valuing a click at 14 could be in the first position.

[39] If i prefers position i to position j (i.e., i does not envy j) and j prefers position j to position k (i.e., j does not envy k) where either $i > j > k$ or $i < j < k$, then $(v_i - p_i)x_i \geq (v_i - p_j)x_j = (v_j - p_j)x_j + (v_i - v_j)x_j \geq (v_j - p_k)x_k + (v_i - v_j)x_j = (v_i - p_k)x_k + (v_i - v_j)(x_j - x_k) \geq (v_i - p_k)x_k$. Thus, i also prefers position i to position k (i.e., i does not envy k).

difference in position clicks. Hence, the entire gain would be $\sum_{j=i+1}^{n} v_j(x_{j-1}-x_j)$. Thus, the VCG mechanism would ask advertisers their true valuations $\tilde{v}$ and sort them into positions according to the reported valuations. It would then require the advertiser in position i to pay a total of $P_i = \sum_{j=i+1}^{n} \tilde{v}_j(x_{j-1} - x_j)$. That is, $P_{i-1} = P_i + v_i(x_{i-1} - x_i)$ where $P_n = 0$. Notice that these payments would cause [7.21] to hold with equality (where $p_i x_i = P_i$ and $j = i - 1$). As shown by both VAR and EOS, this corresponds to the envy-free equilibrium and Nash equilibrium with the lowest revenue to the seller. In our example, this is where $p_1 = 0.833$, $p_2 = 0.6$, and $p_3 = 0$.

7.9.3 Other formats

An English position auction (called a generalized English auction by EOS) is one in which the auctioneer increases the price per click. Advertisers decide when to drop out. The first advertiser to drop out when the remaining advertisers are fewer than the number of positions pays zero and receives the last position. The advertisers dropping out after that pay the price at which the previous advertiser dropped out and get the position one higher than that advertiser. EOS shows that under incomplete information this has a unique perfect Bayesian equilibrium that is equivalent to the VCG mechanism.

Another way to obtain the VCG outcome is for the advertisers to use a mediator. Ashlagi et al. (2009) give conditions under which the VCG outcome will be implemented as an equilibrium of a mediated second-price position auction, that is, a mechanism in which a mediator bids on behalf of the advertisers based upon messages received from them. The advertisers also have the option of ignoring mediation and bidding independently but they have incentives to agree to mediation. Because of this option, the mediator cannot simply utilize the English position auction (which EOS showed leads to the VCG outcome) and use the results to bid in the second-price position auction resulting in the VCG outcome. Reaching the VCG outcome through a mediator is more difficult, since the mediator must be able to induce the advertisers bidding in the auction to use his service.

So far we have discussed the current pay per click. There are two other methods for collecting payments: pay per action, meaning pay based upon the business resulting from the click, and pay for impression (position). For the advertiser, pay per action is ideal as it eliminates several problems that arise in pay per click: the expected value of a click is still uncertain to the advertiser, learning the distribution is costly, and even if the expectation is known, many small advertisers prefer to pass the risk on to the search engine. Moreover, pay per click is subject to click fraud (where owners of websites click fictitiously). However, Agarwal et al. (2009) demonstrate a weakness in pay per action for the seller. Namely, the advertisers know better the potential actions of the searcher and exploit this. For instance, they can use a bait-and-switch technique where the searcher clicks on an ad for a 50-inch TV and winds up buying a 60-inch TV since the other is out of stock. The advertiser could offer less in return for the purchase of the 60-inch TV than the 50-inch TV while the seller would think that the action of purchasing the 50-inch TV is likely.

Some websites, rather than pay per click, use pay for impression. With paying for impression, an advertiser in position i with bid B_i would pay B_{i+1} independently of the number of clicks obtained in that position. Note the effective bid per click would be B_i/x_i and the effective payment per click would be B_{i+1}/x_i, which is different from the effective bid per click of the advertiser in position $i+1$. However, any envy-free equilibrium under pay for click is equivalent to one in pay per impression with $B_i = b_i x_{i-1}$. If the $\{b_i\}$ are an envy-free equilibrium under pay per click, by [7.20] we have:

$$(v_i - b_{i+1})x_i \geq (v_i - b_{j+1})x_j \text{ for all } i, j. \qquad [7.22]$$

By substitution, this implies that

$$v_i x_i - B_{i+1} \geq v_i x_j - B_{j+1} \text{ for all } i, j. \qquad [7.23]$$

These are sufficient conditions for an envy-free equilibrium in the pay-for-impression auction.

The method Google actually uses for position auctions is to take into account the ad quality of the advertiser in addition to his bid per click. Two different advertisers in the same position may receive a different number of clicks. Varian (2007) models this such that advertiser a in position j will have a click-through rate of $x_j e_a$ where e_i is the advertiser's quality. Rather than being ordered by his pay-per-click bid of b_i, an advertiser is ordered by his quality-adjusted bid-per-click of $b_i e_i$. Thus, whereas before an advertiser wanting to move up from position i to position $j < i$ would need to raise his bid to slightly above b_{i-i}, the advertiser now needs to raise his quality-adjusted bid from $b_i e_i$ to slightly above $b_{i-1}e_{i-1}$. This entails raising his bid to slightly above $b_{i-1}e_{i-1}/e_i$ and paying $b_{i-1}e_{i-1}/e_i$ per click as compared to $b_{i+1}e_{i+1}/e_i$ per click in his current position i. Likewise, whereas before an advertiser wanting to move down from position i to position $j > i$ needed to lower his bid from b_i to slightly below b_{i+1}, the advertiser now needs to lower his quality-adjusted bid from $b_i e_i$ to slightly below $b_{i+1}e_{i+1}$. This entails lowering his bid to slightly below $b_{i+1}e_{i+1}/e_i$ and paying $b_{i+2}e_{i+2}/e_i$ per click as compared to $b_{i+1}e_{i+1}/e_i$ per click. The incentive constraints for not wanting to move up or down are then:

$$x_i e_i(v_i - b_{i+1}e_{i+1}/e_i) \geq x_j e_i(v_i - b_j e_j/e_i) \qquad \text{for } j < i, \qquad [7.24]$$

$$x_i e_i(v_i - b_{i+1}e_{i+1}/e_i) \geq x_j e_i(v_i - b_{j+1}e_{j+1}/e_i) \qquad \text{for } j > i. \qquad [7.25]$$

Bringing the e_i inside the parentheses yields:

$$x_i(v_i e_i - b_{i+1}e_{i+1}) \geq x_j(v_i e_i - b_j e_j) \qquad \text{for } j < i, \qquad [7.26]$$

$$x_i(v_i e_i - b_{i+1}e_{i+1}) \geq x_j(v_i e_i - b_{j+1}e_{j+1}) \qquad \text{for } j > i. \qquad [7.27]$$

This is the same structure as before in [7.18] and [7.19] except that, in the pay-per-click auction, the value v_i is replaced by $v_i e_i$ and the bid b_i is replaced by $b_i e_i$. Hence, all the properties stated for the pay per click hold.

Borges et al. (2013) generalize the second-price position auction to include values that are position-dependent and click-through-rates that are advertiser-dependent. Advertiser a in position k will have a value of v_a^k and receive c_a^k clicks. Being in a particular position may also enhance advertiser utility independent of clicks received. This enhancement, called impression value, depends upon advertiser a and position k and is denoted by w_a^k. Hence, the advertiser paying price p would have utility

$$(v_a^k - p)c_a^k + w_a^k. \qquad [7.28]$$

Borges et al. find that as long as $v_a^k c_a^k > v_a^{k+1} c_a^{k+1}$, a symmetric Nash equilibrium exists. They also collected data from 2004 Yahoo! auctions that used this method and found empirical support for the hypothesis that $v_a^1 > v_a^2 > \ldots > v_a^n$.

Unlike auctioning off inanimate objects such as cars, the consumer viewing these advertisements is a player in the game and his surplus and strategy are of concern. An early business model of Goto.com assumed that consumers would prefer to have search results displayed in the order of those willing to pay the most for a click. While Goto.com's business model did not survive (against Google's page rank method), the question of which design benefits the consumer the most is important. Furthermore, once the consumer is considered to be a player, the value of each position and the number of ads that each player clicks should be endogenously determined. Athey and Ellison (2011) do just that (for ads rather than search results) and find that click-weighted auctions are socially optimal as search costs go to zero, whereas this need not be the case with strictly positive search costs. They also find that advertisers hide information about themselves in the click-weighted auction but not in the pay-per-click auction.

Jerath et al. (2011) show that once consumers' actions are taken into account, the position of the advertisers may no longer be sorted by the one with the highest value first. A clear example is two brands of a product: a well-known brand and a lesser-known brand. A well-known brand may get the same number of clicks per position in all positions while the number of clicks of the lesser-known brand may vary significantly more based upon position. In this case, the lesser-known brand may be willing to pay more for the improvement of position and thus be listed in a higher position. For example, when typing "JFK Car rentals" the advertised list on the RHS listed a site called Jfkcarrentals.net first and Avis second. Likewise, when searching for "shoes" in the UK, it listed Schuh first and Clarks second.

7.10. SPECTRUM AUCTIONS

One of the most celebrated achievements of auction theory is the widespread use of spectrum auctions.[40] Many auction theorists are hired either to advise governments on design or help companies on bidding strategies (see McMillan, 1994, for those hired just in the earlier auctions).

In the early spectrum auctions (pre-2000) several issues arose: the exposure problem (analogous to the danger of buying a left shoe without buying a right shoe), strategic demand reduction, and tacit collusion (see Section 7.3 of this chapter for a detailed description of the former two). The exposure problem was particularly problematic in the U.S. where the spectrum was divided up into more than 200 geographical regions and there were complementarities in owning spectrum in neighboring regions. This pushed economists to develop solutions that were eventually implemented in the 4G auctions.

Strategic demand reduction was a major concern in European spectrum auctions, in particular, where no entry was allowed or occurred. For example, in the 1999 German GSM auction, there were four incumbents competing essentially for ten identical blocks of spectrum. The auction was over after just three rounds of bidding with jump-bidding occurring in the first round followed by strategic reduction. While this may be considered a case of collusion, Grimm et al. (2003) (and more generally Riedel and Wolfstetter, 2006) demonstrate that such an auction has a unique subgame-perfect equilibrium where a drastic demand reduction occurs immediately. (Essentially, bidders immediately reduce demand to the number of blocks they would acquire if everyone bid truthfully.) Goeree et al. (2013) confirmed this behavior in an experiment.

7.10.1 3G auctions

In April 2000, the 3G rights were sold off in the UK for 39 billion Euros (£22 billion) (630 Euros per capita). This was coined at the time "the biggest auction ever" (Binmore and Klemperer, 2002). This record lasted all of four months until in Germany the rights were sold off for 51 billion Euros (615 euros per capita).

Besides the magnitude of the revenue, the European 3G auctions are a particularly interesting case study to examine since all the countries had roughly the same spectrum to auction off at roughly the same time. They varied only slightly in the number of incumbents and in GDP per capita.[41] Also, there were only three different designs

[40] The idea of selling spectrum rights by auction dates back to Coase (1959). There is a large literature on the design of spectrum auctions including Banks et al. (2003), Cramton (1997), Klemperer (2004), Plott (1997), Plott and Salmon (2004), Porter and Smith (2006), and Weber (1997). See Hoffman (2011) and Wolf (2012) for a recent summaries.

[41] The incumbents were firms that were already in the telecom industry that owned 2G licenses.

used. Furthermore, there were no combinatoric considerations and networks could successfully operate solely in one country. Each country had roughly 120 MHz of spectrum to auction off. This can be divided into six blocks of 2×10 MHz, four blocks of 2×15 MHz, or three small blocks of 2×10 MHz and two large blocks of 2×15 MHz.

The three basic designs used were: UK-Ascending, Anglo-Dutch, and German-Endogenous. In the UK design, licenses were specified ahead of time and grouped into categories. Initially, the exact spectrum was not specified within a category. An ascending auction was used. Rounds continued as long as there were more active bidders than licenses. Bidders could remain active and continue bidding if they held or beat the previous high bid for a license. Once license ownership was determined for licenses within the same category, there would be an additional sealed-bid auction to determine who gets which specific frequencies.[42]

In the Anglo-Dutch design, the auction begins the same way as the UK design, but switches once the number of active bidders dropped to one plus the number of licenses. At this point, a sealed-bid auction occurs. If K was the number of licenses, then the K top bidders each would get one license for the same price equal to the lowest winning bid. For instance, if there were four licenses, the price would be set to the fourth-highest bid. For determining the specific blocks, the last stage again resembles the UK design.

The German design had an endogenous number of licenses in that the number of owners could range from four to six. The spectrum was divided into 12 lots. Firms were permitted to bid on either two or three lots. If a firm did not remain active on at least two lots, they were considered dropped from the auction. If the auction stopped with a firm owning a single lot, then that firm is dropped and there is an auction for the single lot. Afterwards, a specific spectrum is determined in a way similar to the UK design.

Results from six countries are displayed in Table 7.2.[43] Of these, the two successful auctions were Germany and the UK. Between them, Germany appears more successful since it not only generated higher revenue overall, but did so adjusting for GDP.[44] Furthermore, it granted six licenses rather than five, which means a more competitive market afterwards. While it had the same number of incumbents as the UK, it had fewer entrants, making the outcome more impressive in its competitiveness given that overall there were fewer bidders (although, as Klemperer, 2004, points out, this smaller number of entrants could have been a function of the auction design).

[42] This stage is also useful in ameliorating the exposure problem where there is an additional advantage of holding the same specific frequencies in neighboring countries.

[43] The table is from Wolfstetter (2003) with added data from Klemperer (2004).

[44] In the UK, it was $630/25{,}415 = 0.0248$ Euros per Euro of GDP, while in Germany it was $615/23{,}020 = 0.0267$ Euros per Euro of GDP. See Table 7.2.

Other 3G auctions did not fare so well. Using the UK design, Switzerland with the highest GDP per capita and the lowest number of incumbents was poised for a high per capita sales price (Wolfstetter, 2003). There were also ten firms vying for the four licenses. However, in the end, there were only four entrants competing for the four licenses that went for the reserve price, which was set too low (despite a governmental attempt to change the rules at the last minute). This drop in numbers was in part due to the government allowing joint bidding (Klemperer, 2004).

In Austria, the use of the German design failed to push bids much beyond the reserve price. The six bidders each received a license. Apparently, the stronger bidders preferred to have a smaller license with a market size of six to trying to get a larger license (and with fewer competitors) by bidding on a second smaller license since that would run the risk of increasing all the prices without changing the outcome.

The main weakness of the UK ascending auction is that if there is little private information, even if the number of potential bidders is higher than the number of licenses, then the weaker bidders will realize when they cannot win and will choose not to enter. We then have the same number of bidders as licenses and this leads to the licenses going at the reserve prices. This is particularly problematic if the number of licenses equals the number of incumbents since who is strong is clear. Determining who is strongest among the entrants can take more time. Thus, the fact that the UK ran first may have helped it attract more entrants, while in Switzerland this was not the case.

The Anglo-Dutch design gives a weaker bidder a chance of winning and hence can attract more entrants (at the cost of efficiency). After the failures of the UK ascending design, with four incumbents and four licenses, Denmark used the Anglo-Dutch design (albeit only at the second stage).[45] The auction not only attracted an entrant, but the newcomer replaced one of the incumbents. While the price was 95 Euros per capita, given the drop in the stock market due to the 2000 dot.com crash, this result was successful.

Using an ascending auction for at least part of the auction is problematic in that bidders can communicate indirectly through the auction. In the German auction, bidding at the end of day 11, after 138 rounds, had two top bids by T-Mobile and one by Mannesmann Mobilfunk equal to DM6.666 billion:[46] potentially implying that they should split six ways. Although it was expensive to communicate in such a manner, it is conceivable that it was an attempt to collude. It is also possible to communicate one's strength by entering a jump bid and thereby increasing the bid significantly.

The flexibility of the German design garnered criticism in that it left more combinations for which the bidders could collude. There were also complaints about the complexity of its rules. Also, it was not clear whether the government had the

[45] In the end, it did not matter since there were only five bidders.

[46] While bidding was in Deutsch Marks, which was the currency in circulation at the time, the exchange rate was fixed to the Euro.

Table 7.2 The European 3G auctions in 2000.

Country	Date	Bidders	Licenses	Incum.	Euro/Capita	GDP per Capita
UK	Mar.-Apr. 2000	13	5	4	630	25,415
Netherlands	July 2000	9/6	5	5	170	24,250
The Germany[a]	July-Aug. 2000	12/7	4-6	4	615	23,020
Italy	Oct. 2000	8/6	5	4	210 R	19,451
Austria[a]	Oct. 2000	6	4-6	4	103 R	24,045
Switzerland	Nov.-Dec. 2000	10/4	4	3	19 R	35,739
Denmark[b]	Sept. 2001	-/5	4	4	95	30,034

[a] indicates used the German Design.

[b] indicates using the second phase of the Anglo-Dutch auction.

All others used the UK design (Italy had the additional rule that they would reduce the number of licenses if there was not an excess of serious bidders). R indicates sold at or close to reserve price. For bidders x/y, represents that x bidders were thinking of entering bids but in the end only y bidders entered. Both the German and Austrian auction ended with six licenses awarded. GDP per capita is in 2000 dollars (from the IMF) using then current exchange rates.

knowledge to optimally set the number of different companies owning licenses at the end of the auction. On the other hand, allowing the auction to endogenously determine this number, runs the risk of resulting in too small a number of companies, and hence hurting the competitiveness of the industry. Limiting the number of licenses that one firm can own ensures a minimum level of after-auction firm participation. The advantage of the German design is that when the auction would normally end in a typical ascending auction where each bidder was permitted only one unit, there is potential for the stronger bidders to push the price up further by bidding for an additional unit (and keeping entrants out). As we saw with the German auction, this can end in higher prices without reducing the after-auction firm participation.[47]

See Illing and Klüh (2003) for a collection of papers that interpret and debate the 3G auctions in Europe.

7.10.2 4G auctions

After the 3G auctions, there was the start of a new wave of spectrum auctions, the 4G auctions. These were different from the 3G auctions in many ways. For one, the market had matured. While smart phones were in widespread use, the expectations had been reduced from the dot.com boom. The competition was more Bertrand like resulting in lower profit for the firms. For these reasons, revenue was markedly smaller. Also, the number and type of licenses up for sale was much more varied across countries. In such a mix, there were many complementarities in the blocks. Finally, the number of firms

[47] Ironically, the two entrants returned their licenses within one year (although they could not recover the money spent) in order to avoid the obligation of building infrastructure to provide nationwide coverage. Telefonica was one of those firms and even more ironically took over an incumbent and in 2010 bought back its returned license.

considering entry and incumbents in the market were better known in advance. The need to attract new entrants seemed less of an issue than the need to ensure efficiency in fitting the new licenses to existing ownerships.

The two main designs used are the simultaneous multiround auction (SMRA) and the combinatorial clock auction (CCA).[48] These designs combine elements from both the Anglo-Dutch design and the German design. The CCA is significantly more complex than even the German 3G design[49] and allows for combinatoric bids. Like the German design, the CCA has a built-in flexibility about the amount of spectrum each winning firm can obtain but many implementations have caps imposed on how much spectrum a single bidder can win.

The SMRA is, in essence, an enhanced version of the UK-ascending auction (although a version of it proposed by McAfee, Milgrom, and Wilson has been used by the FCC since 1994). It has been used recently in Germany, Hong Kong, Belgium, Spain, Norway, Sweden, Finland, and other countries. In this format, each specific block of spectrum is bid on separately. In each round, bidders are allowed to bid in each block in which they are active. Bids can be increased in only specified increments (to avoid information transmitted in jump bids). One remains active in round t by either having the leading bid in round $t-2$ or submitting a higher bid than the $t-1$ leading bid in round $t-1$. In some versions, in earlier rounds of the auction one can also remain active for x blocks by satisfying those conditions in only a fraction of the blocks (such as half) for which one was active in round $t-1$. Depending upon the specific rules chosen, either the highest bid or all the bids are shown after each round. The overall auction ends when there is no improvement on the bids in the previous round. The main differences from the UK auction are that the activity rules are block-specific, the bidders are allowed to win more than one block, and the bid increases are restricted. As mentioned, there is a possibility of a cap on the number of blocks one can be active on.

The CCA was introduced by Cramton (2009, 2013), which incorporated many elements of the Proxy-Clock auction proposed by Ausubel et al. (2006). Maldoom (2007) describes an implementation by a private company, dotEcon, that ran many of the European 4G auctions using this format (UK, Ireland, Holland, Denmark, Switzerland).

[48] Among the other designs, Rothkopf et al. (1998) propose using Hierarchical package bidding. This combinatorial auction limits the combinations that can be bid upon to hierarchies of blocks. For instance, one can bid for blocks A, B, C, and D separately, A and B together, C and D together, or A, B, C, and D all together, but one cannot bid on B and C together. This restriction makes computing the maximum revenue significantly easier and allows subjects to better understand the pricing tradeoffs. It was used in practice by the FCC (see Goeree and Holt, 2010). One drawback is that the packages must be predetermined by the auctioneer. The FCC also tried using modified package bidding (also known as resource allocation design) designed by Kwasnica et al. (2005). This design sets shadow prices for the blocks such that the revenue-maximizing bidders are willing to pay the prices (given their bids) and the other bidders are not. This is helpful in price discovery.

[49] Indeed, in the recent auctions in Switzerland and Austria, a bidder could submit more than 1500 bids in the last stage.

Here, bidders are allowed to bid on lots: combinations of spectrum blocks (with a potential cap on the total spectrum size a bidder can own). The CCA has two main stages: the clock stage, during which bidders can bid on just one package in each round, followed by the sealed-bid stage, in which bidders can bid on all possible combinations, restricted by their activity during the clock stage.[50]

Rather than asking the bidders to choose the bids, the clock increases the price and the bidders decide which package of blocks (if any) to bid on. Each block is worth a number of eligibility points and buyers can choose only licenses whose total points are within their eligibility point budget (and spectrum cap). The initial budgets are set by the financial considerations of the bidders and the budgets are reduced to the number of points used in the previous round. The price increase for various lots is determined by the interest in that lot. The clock stops when the demand is less than or equal to the supply. This can lead to overshooting where the demand is strictly less than the supply or it can lead to the highest bid on a single lot below the set threshold. To fix this, there is sealed-bid auction where bids must be consistent with the initial pattern of bids. (The sealed-bid phase also serves to fight strategic demand reduction and collusion.) The auctioneer then chooses the combination of bids (subject to feasibility) that maximizes revenue (which is also efficient if bids are truthful) from both the sealed-bid stage and all rounds of the clock stage. The prices paid are determined by what is called Vickrey-nearest-core pricing. While in Vickrey pricing each individual must pay the externality that they impose on other bidders, in Vickrey-nearest-core pricing each collection of winning bidders must pay the joint externality that is imposed on the other nonwinning bidders.

To demonstrate Vickrey-nearest-core pricing, we use an example from Cramton (2013). There are two lots: A and B. Bidder 1 values A at 28. Bidder 2 values B at 20. Bidder 3 values the package of A and B together at 32. Bidder 4 values A at 14. Bidder 5 values B at 12. The efficient allocation gives A to bidder 1 and B to bidder 2. In Vickrey-Clark-Groves (VCG) pricing, the price of A is 14 and B is 12.[51] However, this is below what the value of A and B together for bidder 3. In order to get a set of prices in the core, the sum of the prices for A and B must be at least 32 (their combined value for bidder 3). This constraint is satisfied by the set of prices that has the minimum revenue

[50] In order to reduce the number of combinations, as with the 3G auctions some blocks are classified as equivalent and bid upon as abstract blocks. After the two main stages, there is a subsequent stage in which (as similar to the UK 3G auction) specific blocks are bid upon among winners of the auction in each equivalent class of blocks.

[51] With VCG, the efficient allocation is chosen and a bidder pays the externality he imposes on others. If bidder 1 does not receive the allocation of A, the next-best allocation is bidder 4 gets A and bidder 2 gets B. This has a total value of 34. This is 14 above the value that the other bidders get when bidder 1 gets A (and bidder 2 gets B). Hence the VCG price for package A is 14.

within the core (MRC). The closest set of prices (geometrically) from the MRC to the Vickrey prices is 17 for bidder 1 and 15 for bidder 2. This also has the advantage of higher revenue than the Vickrey pricing.

Using Vickrey-nearest-core pricing is motivated by the fact that using Vickrey pricing may lead not only lead to outcomes that are not in the core, but to lower revenue, as shown by Day and Milgrom (2008). Day and Milgrom also show that pure VCG pricing may give the bidders an advantage of using a shill bidder (entering as two bidders), and may be nonmonotonic revenue-wise in the number of bidders. They provide an example with two licenses: A and B. Bidder 1 values them together at 10 and each separately at 0. Bidder 2 values the first license at 10 and the second at 0. Bidder 3 values the first license at 0 and the second at 10. With all 3 bidders, the VCG price for either item is 0. With just bidders 1 and 3, the price of item A is 0 but that of item B is 10. Hence in this case, bidder 3 can gain by entering a bidder to bid for item A, thereby, reducing B's price. (Remember that in VCG pricing, a bidder pays the externality he causes by receiving his allocation. With all three bidders, bidders 2 and 3 receive A and B, respectively, each receiving 10 in value. Without bidder 3, either bidder 2 gets A or bidder 1 gets A and B. In either case, the value is 10. Hence, bidder 3 does not impose an externality on the other bidders.)

In support of using Vickrey-nearest pricing, Day and Raghavan (2007) show that the Vickrey-nearest-core prices minimize the sum of the maximum gains of deviating from truthful bidding. Day and Milgrom (2008) suggest that the Vickrey-nearest-core prices maximize the incentive for truth telling among the prices in the core and Day and Cramton (2012) show that the Vickrey-nearest-core prices are unique. However, Goeree and Lien (2012) show that it still distorts incentives for truthful bidding and Erdil and Klemperer (2010) claim that incentives for truthful bidding are superior with a reference point other than VCG, one that does not depend upon the bids of the bidders involved. This would be minimizing incentives locally for small deviations from truthful bidding.

The CCA consistency rules (see Bichler et al., 2013, for a description) are essentially that the final-round bids must be consistent by revealed preferences from the earlier rounds. For instance, if lot A and lot B are the same price in a preliminary round and a bidder chooses to bid on *A* and not on *B*, then in the final, sealed-bid round, that bidder cannot bid more for lot *B* than for lot *A*. (Similar rules apply for quantity reductions.)

While neither design has emerged as dominant in recent auctions, there are some concerns. For the SMRA auction, when there are complementaries, larger bidders may worry about being left exposed by purchasing only part of a complementary lot of blocks. Because of this the SMRA can result in smaller bidders winning at low prices and thus excluding bidders may be beneficial for seller revenue (see Goeree and Lien, 2014).

There are other potential problems as well, as in the case in which a large bidder that has complementarities over a large number of lots is held up by a small bidder. For instance, if bidder A values ten specific licenses together at 100 and each individually only at 1, then a bidder B can speculate by buying one of the ten and then demand a large amount from bidder A. In the advanced wireless services (AWS) auction run by the FCC (see Cramton, 2013), blocks of spectrum were divided up differently geographically. Because of this holdup problem, a block of spectrum sold for 12 times the price of the combined price of a similar block that could be created by buying several geographically divided blocks. Despite these problems, Bichler et al. (2013) experimentally find that the CCA does worse than the SMRA in both efficiency and revenue. Due to its complexity, one may question the feasibility of subjects comprehending the CCA design,[52] but experiments done on teams of bidders recruited from a class on auction theory, who also had two weeks to prepare, yielded similar results. Cramton (2013) claims that part of the problem is that bidders were not provided with the right bidding tools that would have enabled them to easily place bids on the relevant packages.

7.11. CONCLUDING REMARKS

As mentioned in the Introduction, we could not hope to provide a fair representation of all the work going on in auctions in the past two decades. There are many important topics that we did not cover, among them, collusion and corruption, budget constraints, and experiments.[53]

Even in the topics covered in the chapter, we were not exhaustive, but we hope to have given the reader a place to start. Some topics are covered more deeply and formally in other chapters of this handbook: combinatorial auctions and algorithmic mechanism design.

ACKNOWLEDGMENTS

We thank René Kirkegaard, Dan Kovenock, Pricilla Marimo, Brian Roberson, Bradley Ruffle, Aner Sela, David Wettstein, Elmar Wolfstetter, and Charles Zheng for useful comments and Michael Borns for editing the chapter.

[52] See Guala (2001) for a discussion on how experiments are a useful part of spectrum auction design.

[53] We refer the reader to the book of Marshall and Marx (2012) and papers by Skrzypacz and Hopenhayn (2004), Che and Kim (2006, 2009), Pavlov (2008), Garratt et al. (2009), and Rachmilevitch (2013a,b, 2014) for collusion in auctions, Che and Gale (1996, 1998b), Zheng (2001), Pitchik (2009), and Kotowski (2013) for budget constraints, and Kagel (1995) and Kagel and Levin (2014) for surveys of auction experiments.

REFERENCES

Adamo, T., Matros, A. 2009. A Blotto game with incomplete information. Econ. Lett. 105(1), 100–102.

Adams, C. 1993. For Good and Evil: The Impact of Taxes on the Course of Civilization. Rowan & Littlefield, Lanham, Maryland, USA.

Agarwal, N., Athey, S., Yang, D. 2009. Skewed bidding in pay-per-action auctions for online advertising. Am. Econ. Rev. 99(2), 441–447.

Amann, E., Leininger, W. 1996. Asymmetric all-pay auctions with incomplete information: the two-player case. Games Econ. Behav. 14(1), 1–18.

Ashlagi, I., Monderer, D., Tennenholtz, M. 2009. Mediators in position auctions. Games Econ. Behav. 67(1), 2–21.

Asker, J., Cantillon, E. 2008. Properties of scoring auctions. RAND J. Econ. 39(1), 69–85.

Asker, J., Cantillon, E. 2010. Procurement when price and quality matter. RAND J. Econ. 41(1), 1–34.

Athey, S. 2001. Single crossing properties and the existence of pure strategy equilibria in games of incomplete information. Econometrica 69(4), 861–889.

Athey, S., Ellison, G. 2011. Position auctions with consumer search. Q. J. Econ. 126(3), 1213–1270.

Athey, S., Levin, J. 2001. Information and competition in US Forest Service timber auctions. J. Polit. Econ. 109(2), 375–417.

Athey, S., Levin, J., Seira, E., 2011. Comparing open and sealed-bid auctions: evidence from timber auctions. Q. J. Econ. 126(1), 207–257.

Ausubel, L.M. 2004. An efficient ascending-bid auction for multiple objects. Am. Econ. Rev. 94(5), 1452–1475.

Ausubel, L.M. 2006. An efficient dynamic auction for heterogeneous commodities. Am. Econ. Rev. 96(3), 602–629.

Ausubel, L.M., Cramton, P. 2002. Demand reduction and inefficiency in multi unit auctions. University of Maryland Department of Economics, Working Paper.

Ausubel, L.M., Cramton, P., Milgrom, P. 2006. The clock-proxy auction: a practical combinatorial auction design. In: Cramton, R.S.P., Shoham, Y. (Eds.), Combinatorial Auctions. Cambridge: MIT Press, pp. 147–176.

Ausubel, L.M., Cramton, P., Pycia, M., Rostek, M., Weretka, M. 2013. Demand reduction and inefficiency in multi-unit auctions. Working Paper.

Back, K., Zender, J.F. 1993. Auctions of divisible goods: on the rationale for the treasury experiment. Rev. Financial Stud. 6(4), 733–764.

Banks, J., Olson, M., Porter, D., Rassenti, S., Smith, V. 2003. Theory, experiment and the federal communications commission spectrum auctions. J. Econ. Behav. Org. 51(3), 303–350.

Bartolini, L., Cottarelli, C. 1997. "Designing effective auctions for treasury securities." Current Issues in Economics and Finance 3.9.

Barut, Y., Kovenock, D. 1998. The symmetric multiple prize all-pay auction with complete information. Eur. J. Polit. Econ. 14(4), 627–644.

Baye, M.R., Kovenock, D. 1994. How to sell a pickup truck: 'Beat-or-Pay' advertisements as facilitating devices. Int. J. Ind. Org. 12(1), 21–33.

Baye, M.R., Kovenock, D., de Vries, C. 1996. The all-pay auction with complete information. Econ. Theory 8(2), 291–305.

Baye, M.R., Kovenock, D., de Vries, C. 2005. Comparative analysis of litigation systems: an auction-theoretic approach. Econ. J. 115(505), 583–601.

Baye, M.R., Kovenock, D., de Vries, C. 2012a. Contests with rank-order spillovers. Econ. Theory 51(2), 315–350.

Baye, M.R., Kovenock, D., de Vries, C. 2012b. The herodotus paradox. Games Econ. Behav. 74(1), 399–406.

Berg, A.S. 1998. Lindbergh. GP Putnam's Sons. New York.

Bergemann, D., Said, M. 2011. Dynamic Auctions. Wiley Encyclopedia of Operations Research and Management Science. John Wiley & Sons, Hoboken, NJ, USA.

Bergemann, D., Välimäki, J. 2010. The dynamic pivot mechanism. Econometrica 78(2), 771–789.

Bergstrom, T., Bagnoli, M. 2005. Log-concave probability and its applications. Econ. Theory 26(2), 445–469.

Bichler, M., Shabalin, P., Wolf, J. 2013. Do core-selecting combinatorial clock auctions always lead to high efficiency? An experimental analysis of spectrum auction designs. Exp. Econ. 16(4), 511–545.
Bikhchandani, S., Huang, C. 1989. Auctions with resale markets: an exploratory model of treasury bill markets. Rev. Financ. Stud. 2(3), 311–339.
Binmore, K., Klemperer, P. 2002. The biggest auction ever: the sale of the British 3G telecom licences. Econ. J. 112(478), C74–C96.
Binmore, K., Swierzbinski, J. 2000. Treasury auctions: uniform or discriminatory?. Rev. Econ. Des. 5(4), 387–410.
Board, S., Skrzypacz, A. 2013. Revenue management with forward-looking buyers. Working Paper.
Borel, E. 1921. La théorie du jeu et les équations intégrales à noyau symétrique. *Comptes Rendus de l'Académie des Sciences*, 173(1304–1308), 58, English translation by Savage, L. 1953. The theory of play and integral equations with skew symmetric kernels. Econometrica 21(1), 97–100.
Borel, E., Ville, J. 1938. Applications de la théorie des probabilités aux jeux de hasard. Original edition by Gauthier-Villars, Paris; reprinted in Theorie mathematique du bridge a la portee de tous, by E. Borel and A. Cheron, Editions Jacques Gabay, Paris, 1991.
Borges, T., Cox, I., Pesendorfer, M., Petricek, V. 2013. Equilibrium bids in sponsored search auctions. Am. Econ. J. Microecon. 5(4), 163–187.
Bose, S., Deltas, G. 2007. Exclusive versus non-exclusive dealing in auctions with resale. Econ. Theory 31(1), 1–17.
Boyce, J.R. 1994. Allocation of goods by lottery. Econ. Inquiry 32(3), 457–476.
Brenner, M., Galai, D., Sade, O. 2009. Sovereign debt auctions: uniform or discriminatory?. J. Monetary Econ. 56(2), 267–274.
Caillaud, B., Jehiel, P. 1998. Collusion in auctions with externalities. RAND J. Econ. pp. 680–702.
Chakravarty, S., Kaplan, T.R. 2013. Optimal allocation without transfer payments. Games Econ. Behav. 77(1), 1–20.
Che, X., Lee, P., Yang, Y. 2013. The impact of resale on entry in second price auctions. Math. Social Sci. 66(2), 163–168.
Che, Y.-K. 1993. Design competition through multidimensional auctions. RAND J. Econ. 24, 668–680.
Che, Y.-K., Gale, I. 1996. Expected revenue of all-pay auctions and first-price sealed-bid auctions with budget constraints. Econ. Lett. 50(3), 373–379.
Che, Y.-K., Gale, I. 1998a. Caps on political lobbying. Am. Econ. Rev. 88(3), 643–651.
Che, Y.-K., Gale, I. 1998b. Standard auctions with financially constrained bidders. Rev. Econ. Stud. 65(1), 1–21.
Che, Y.-K., Gale, I. 2003. Optimal design of research contests. Am. Econ. Rev. 93(3), 646–671.
Che, Y.-K., Gale, I. 2006. Caps on political lobbying: reply. Am. Econ. Rev. 96(4), 1355–1360.
Che, Y.-K., Kim, J. 2006. Robustly collusion-proof implementation. Econometrica 74(4), 1063–1107.
Che, Y.-K., Kim, J. 2009. Optimal collusion-proof auctions. J. Econ. Theory 144(2), 565–603.
Chen, Y., Katuscak, P., Ozdenoren, E. 2007. Sealed bid auctions with ambiguity: Theory and experiments. J. Econ. Theory 136(1), 513–535.
Cheng, H. 2011. Auctions with resale and bargaining power. J. Math. Econ. 47(3), 300–308.
Cheng, H.H., Tan, G. 2010. Asymmetric common-value auctions with applications to private-value auctions with resale. Econ. Theory 45(1-2), 253–290.
Clark, D.J., Riis, C. 1998. Competition over more than one prize. Am. Econ. Rev. 88(1), 276–289.
Cohen, C., Kaplan, T., Sela, A. 2008. Optimal rewards in contests. RAND J. Econ. 39(2), 434–451.
Condorelli, D. 2012. What money can't buy: efficient mechanism design with costly signals. Games Econ. Behav. 75(2), 613–624.
Cox, J.C., Smith, V.L., Walker, J.M. 1988. Theory and individual behavior of first-price auctions. J. Risk Uncertainty 1(1), 61–99.
Cramton, P. 1997. The FCC spectrum auctions: an early assessment. J. Econ. Manage. Strat. 6(3), 431–495.
Cramton, P. 2009. Spectrum auctions. Discussion paper, University of Maryland, College Park, MD, USA.
Cramton, P. 2013. Spectrum auction design. Rev. Ind. Org. 42(2), 161–190.
Cramton, P., Shoham, Y., Steinberg, R. 2006. Combinatorial Auctions. MIT Press, Cambridge, MA.

Crawford, V.P., Iriberri, N. 2007. Level-k auctions: can a nonequilibrium model of strategic thinking explain the winner's curse and overbidding in private-value auctions? Econometrica 75(6), 1721–1770.
Crawford, V.P., Kugler, T., Neeman, Z., Pauzner, A. 2009. Behaviorally optimal auction design: examples and observations. J. Eur. Econ. Assoc. 7(2-3), 377–387.
Crémer, J., Spiegel, Y., Zheng, C.Z. 2007. Optimal search auctions. J. Econ. Theory 134(1), 226–248.
Crémer, J., Spiegel, Y., Zheng, C.Z. 2009. Auctions with costly information acquisition. Econ. Theory 38(1), 41–72.
Das Varma, G. 2002. Standard auctions with identity-dependent externalities. RAND J. Econ. 689–708.
Das Varma, G. 2003. Bidding for a process innovation under alternative modes of competition. Int. J. Ind. Org. 21(1), 15–37.
Dasgupta, P. 1986. The theory of technological competition. In: Stiglitz, J., Mathewson, G. (Eds.), New Developments in the Analysis of Market Structure (Chapter 17), MIT Press, Cambridge, MA, pp. 519–547.
d'Aspremont, C., Jacquemin, A. 1988. Cooperative and noncooperative R&D in duopoly with spillovers. Am. Econ. Rev. 78(5), 1133–1137.
Davis, D.J., Derer, M., Garcia, J., Greco, L., Kurt, T.E., Kwong, T., Lee, J.C., Lee, K.L., Pfarner, P., Skovran, S., et al. 2001. System and method for influencing a position on a search result list generated by a computer network search engine. US Patent 6,269,361.
Day, R., Cramton, P. 2012. The quadratic core-selecting payment rule for combinatorial auctions. Oper. Res. 60(3), 588–603.
Day, R., Milgrom, P. 2008. Core-selecting package auctions. Int. J. Game Theory 36(3-4), 393–407.
Day, R., Raghavan, S. 2007. Fair payments for efficient allocations in public sector combinatorial auctions. Manage. Sci. 53(9), 1389–1406.
De Castro, L. 2012. A new approach to correlation of types in Bayesian games. Discussion paper.
De Castro, L., Karney, D.H. 2012. Equilibria existence and characterization in auctions: achievements and open questions. J. Econ. Surv. 26(5), 911–932.
Dechenaux, E., Kovenock, D., Sheremeta, R.M. 2012. A survey of experimental research on contests, all-pay auctions and tournaments. Discussion paper, Social Science Research Center Berlin (WZB).
Dickhaut, J., Lunawat, R., Pronin, K., Stecher, J. 2011. Decision making and trade without probabilities. Econ. Theory 48(2-3), 275–288.
Ding, W., Wolfstetter, E.G. 2011. Prizes and lemons: procurement of innovation under imperfect commitment. RAND J. Econ. 42(4), 664–680.
Dubey, P., Geanakoplos, J. 2010. Grading exams: 100, 99, 98,? or A, B, C?. Games Econ. Behav. 69(1), 72–94.
Edelman, B., Ostrovsky, M. 2007. Strategic bidder behavior in sponsored search auctions. Decis. Support Syst. 43(1), 192–198.
Edelman, B., Ostrovsky, M., Schwarz, M. 2007. Internet advertising and the generalized second-price auction: Selling billions of dollars worth of keywords. Am. Econ. Rev. 97(1), 242–259.
Eilenberg, S., Montgomery, D. 1946. Fixed point theorems for multi-valued transformations. Am. J. Math. 68(2), 214–222.
Ellsberg, D. 1961. Risk, ambiguity, and the savage axioms. Q. J. Econ. 75(4), 643–669.
Engelbrecht-Wiggans, R. 1989. The effect of regret on optimal bidding in auctions. Manage. Sci. 35(6), 685–692.
Engelbrecht-Wiggans, R., Katok, E. 2008. Regret and feedback information in first-price sealed-bid auctions. Manage. Sci. 54(4), 808–819.
Erdil, A., Klemperer, P. 2010. A new payment rule for core-selecting package auctions. J. Eur. Econ. Assoc. 8(2–3), 537–547.
Fan, C., Jun, B.H., Wolfstetter, E.G. 2013. Licensing process innovations when losers' messages determine royalty rates. Games Econ. Behav. 82(c), 388–402.
Fan, C., Jun, B.H., Wolfstetter, E.G. 2014. Licensing a common value innovation when signaling strength may backfire. Int. J. Game Theory 43(1), 215–244.
Février, P., Roos, W., Visser, M. 2005. The buyer's option in multi-unit ascending auctions: the case of wine auctions at Drouot. J. Econ. Manage. Strategy 14(4), 813–847.

Figueroa, N., Skreta, V. 2004. Optimal auction design for multiple objects with externalities. Discussion paper, Citeseer.
Filiz-Ozbay, E., Ozbay, E. 2007. Auctions with anticipated regret: theory and experiment. Am. Econ. Rev. 97(4), 1407–1418.
Fullerton, R.L., Linster, B.G., McKee, M., Slate, S. 2002. Using auctions to reward tournament winners: theory and experimental investigations. RAND J. Econ. 33, 62–84.
Fullerton, R.L., McAfee, R.P. 1999. Auctioning entry into tournaments. J. Polit. Econ. 107(3), 573–605.
Galton, F. 1902. The most suitable proportion between the value of first and second prizes. Biometrika 1, No. 4, 385–399.
Garratt, R., Tröger, T. 2006. Speculation in standard auctions with resale. Econometrica 74(3), 753–769.
Garratt, R.J., Tröger, T., Zheng, C.Z. 2009. Collusion via resale. Econometrica 77(4), 1095–1136.
Gavious, A., Moldovanu, B., Sela, A. 2002. Bid costs and endogenous bid caps. RAND J. Econ. 33(4), 709–722.
Gershkov, A., Moldovanu, B. 2009. Dynamic revenue maximization with heterogeneous objects: a mechanism design approach. Am. Econ. J. Microecon. 1(2), 168–198.
Gilboa, I. 1987. Expected utility with purely subjective non-additive probabilities. J. Math. Econ. 16(1), 65–88.
Gilboa, I., Schmeidler, D. 1989. Maxmin expected utility with non-unique prior. J. Math. Econ. 18(2), 141–153.
Glazer, A., Hassin, R. 1988. Optimal contests. Econ. Inquiry 26(1), 133–143.
Goeree, J.K. 2003. Bidding for the future: signaling in auctions with an aftermarket. J. Econ. Theory 108(2), 345–364.
Goeree, J.K., Holt, C.A. 2010. Hierarchical package bidding: a paper & pencil combinatorial auction. Games Econ. Behav. 70(1), 146–169.
Goeree, J.K., Lien, Y. 2012. On the impossibility of core-selecting auctions. Discussion paper, University of Zurich, Switzerland.
Goeree, J.K., Lien, Y. 2014. An equilibrium analysis of the simultaneous ascending auction. J. Econ. Theory, forthcoming.
Goeree, J.K., Offerman, T., Sloof, R. 2013. Demand reduction and preemptive bidding in multi-unit license auctions. Exp. Econ. 16(1), 52–87.
Goswami, G., Noe, T., Rebello, M. 1996. Collusion in uniform-price auction: experimental evidence and implications for treasury auctions. Rev. Fin. Stud. 9(3), 757–785.
Griesmer, J.H., Levitan, R.E., Shubik, M. 1967. Toward a study of bidding processes part IV-games with unknown costs. Naval Res. Logist. Q. 14(4), 415–433.
Grimm, V., Riedel, F., Wolfstetter, E. 2003. Low price equilibrium in multi-unit auctions: the GSM spectrum auction in Germany. Int. J. Ind. Org. 21(10), 1557–1569.
Gross, O.A. 1950. The symmetric Blotto game. RAND Working Paper, RM-424.
Gross, O.A., Wagner, R. 1950. A continuous Colonel Blotto game. Mimeo, Rand Corporation.
Guala, F. 2001. Building economic machines: the FCC auctions. Stud. His. Philos. Sci. Part A 32(3), 453–477.
Gupta, M., Lebrun, B. 1999. First price auctions with resale. Econ. Lett. 64(2), 181–185.
Hafalir, I., Krishna, V. 2008. Asymmetric auctions with resale. Am. Econ. Rev. 98(1), 87–112.
Hafalir, I., Krishna, V. 2009. Revenue and efficiency effects of resale in first-price auctions. J. Math. Econ. 45(9-10), 589–602.
Haile, P.A. 2003. Auctions with private uncertainty and resale opportunities. J. Econ. Theory 108(1), 72–110.
Hart, S. 2008. Discrete Colonel Blotto and general lotto games. Int. J. Game Theory 36(3-4), 441–460.
Hartline, J.D., Roughgarden, T. 2009. Simple versus optimal mechanisms. In: Proceedings of the 10th ACM conference on Electronic commerce. ACM, pp. 225–234.
Hillman, A.L. 1989. The Political Economy of Protection, Harwood Academic Publishers, Chur, London and New York.
Hillman, A.L., Riley, J.G. 1989. Politically contestable rents and transfers. Econ. Polit. 1(1), 17–39.
Hillman, A.L., Samet, D. 1987. Dissipation of contestable rents by small numbers of contenders. Public Choice 54(1), 63–82.

Hoffman, K. 2011. Spectrum auctions. In: Kennington, D.R.J., Olinick, E. (Eds.), Wireless Network Design, International Series in Operations Research and Management Science, vol. 158. New York, Springer, pp. 147–176.
Hoppe, H.C., Jehiel, P., Moldovanu, B. 2006. License auctions and market structure. J. Econ. Manage. Strat. 15(2), 371–396.
Illing, G., Klüh, U. 2003. Spectrum Auctions and Competition in Telecommunications. CESifo Seminar Series, MIT Press, Cambridge, MA, USA.
Jackson, M.O., Swinkels, J.M. 2005. Existence of equilibrium in single and double private-value auctions. Econometrica 73(1), 93–139.
Jehiel, P., Moldovanu, B. 1996. Strategic nonparticipation. RAND J. Econ. 27, 84–98.
Jehiel, P., Moldovanu, B. 1999. Resale markets and the assignment of property rights. Rev. Econ. Stud. 66(4), 971–991.
Jehiel, P., Moldovanu, B. 2000. Auctions with downstream interaction among buyers. RAND J. Econ. 31(4), 768-791.
Jehiel, P., Moldovanu, B. 2001. Efficient design with interdependent valuations. Econometrica 69(5), 1237–1259.
Jehiel, P., Moldovanu, B. 2006. Allocative and informational externalities in auctions and related mechanisms. In: Richard Blundell, W.N., Persson, T. (Eds.), The Proceedings of the 9th World Congress of the Econometric Society, Cambridge University Press.
Jehiel, P., Moldovanu, B., Stacchetti, E. 1996. How (not) to sell nuclear weapons. Am. Econ. Rev. 86(4), 814–829.
Jehiel, P., Moldovanu, B., Stacchetti, E. 1999. Multidimensional mechanism design for auctions with externalities. J. Econ. Theory 85(2), 258–293.
Jerath, K., Ma, L., Park, Y.-H., Srinivasan, K. 2011. A position paradox in sponsored search auctions. Market. Sci. 30(4), 612–627.
Kagel, J. 1995. Auctions: a survey of experimental research. In: Kagel, J., Roth, A. (Eds.), The Handbook of Experimental Economics, Princeton University Press, pp. 501–586.
Kagel, J., Levin, D. 1993. Independent private value auctions: Bidder behaviour in first-, second-and third-price auctions with varying numbers of bidders. Econ. J. 103(419), 868–879.
Kagel, J., Levin, D. 2014. Auctions: a survey of experimental research, 1995–2010. In: Kagel, J., Roth, A. (Eds.), The Handbook of Experimental Economics, *vol. 2.*, forthcoming.
Kaplan, T.R., Luski, I., Sela, A., Wettstein, D. 2002. All-pay auctions with variable rewards. J. Ind. Econ. 50(4), 417–430.
Kaplan, T.R., Luski, I., Wettstein, D. 2003. Innovative activity and sunk cost. Int. J. Ind. Org. 21(8), 1111–1133.
Kaplan, T.R., Wettstein, D. 2006. Caps on political lobbying: comment. Am. Econ. Rev. 96(4), 1351–1354.
Kaplan, T.R., Wettstein, D. 2013. The optimal design of rewards in contests. Working Paper.
Kaplan, T.R., Zamir, S. 2000. The strategic use of seller information in private-value auctions. Center for the Study of Rationality, Hebrew University of Jerusalem, Discussion Paper 221.
Kaplan, T.R., Zamir, S. 2012. Asymmetric first-price auctions with uniform distributions: analytic solutions to the general case. Econ. Theory 50(2), 269–302.
Kirkegaard, R. 2012. A mechanism design approach to ranking asymmetric auctions. Econometrica 80(5), 2349–2364.
Kirkegaard, R. 2014. Ranking asymmetric auctions: filling the gap between a distributional shift and stretch. Games Econ. Behav. 85(1), 60–69.
Klemperer, P. 2004. Auctions: Theory and Practice. Princeton University Press, Princeton, NJ, USA.
Koh, W.T., Yang, Z., Zhu, L. 2006. Lottery rather than waiting-line auction. Soc. Choice Welfare 27(2), 289–310.
Konrad, K.A. 2009. Strategy and Dynamics in Contests. Oxford University Press, Oxford, UK.
Konrad, K.A., Kovenock, D. 2009. Multi-battle contests. Games Econ. Behav. 66(1), 256–274.
Kotowski, M.H. 2013. First-price auctions with budget constraints. Discussion paper.
Kremer, I., Nyborg, K.G. 2004. Underpricing and market power in uniform price auctions. Rev. Financ. Stud. 17(3), 849–877.

Krishna, V. 2002. Auction Theory. Academic Press, San Diego, USA.
Krishna, V. 2009. Auction Theory, second Ed. Academic Press.
Krishna, V., Maenner, E. 2001. Convex potentials with an application to mechanism design. Econometrica 69(4), 1113–1119.
Kwasnica, A.M., Ledyard, J.O., Porter, D., DeMartini, C. 2005. A new and improved design for multi-object iterative auctions. Manage. Sci. 51(3), 419–434.
Landsberger, M., Rubinstein, J., Wolfstetter, E., Zamir, S. 2001. First-price auctions when the ranking of valuations is common knowledge. Rev. Econ. Des. 6(3), 461–480.
Laslier, J.-F. 2002. How two-party competition treats minorities. Rev. Econ. Des. 7(3), 297–307.
Laslier, J.-F., Picard, N. 2002. Distributive politics and electoral competition. J. Econ. Theory 103(1), 106–130.
Lazear, E.P., Rosen, S. 1981. Rank-order tournaments as optimum labor contracts. J. Polit. Econ. 89(5), 841–864.
Lebrun, B. 1996. Existence of an equilibrium in first-price auctions. Econ. Theory 7(3), 421–443.
Lebrun, B. 2002. Continuity of the first price auction Nash equilibrium correspondence. Econ. Theory 20(3), 435–453.
Lebrun, B. 2006. Uniqueness of the equilibrium in first-price auctions. Games Econ. Behav. 55(1), 131–151.
Lebrun, B. 2010a. First-price auctions with resale and with outcomes robust to bid disclosure. RAND J. Econ. 41(1), 165–178.
Lebrun, B. 2010b. Revenue ranking of first-price auctions with resale. J. Econ. Theory 145(5), 2037–2043.
Lebrun, B. 2012. Optimality and the English and second-price auctions with resale. Games Econ. Behav. 75(2), 731–751.
Levin, D., Ozdenoren, E. 2004. Auctions with uncertain numbers of bidders. J. Econ. Theory 118(2), 229–251.
Lizzeri, A., Persico, N. 2000. Uniqueness and existence of equilibrium in auctions with a reserve price. Games Econ. Behav. 30, 83–114.
Lo, K. 1998. Sealed bid auctions with uncertainty averse bidders. Econ. Theory 12(1), 1–20.
Maldoom, D. 2007. Winner determination and second pricing algorithms for combinatorial clock auctions. Discussion paper, 07/01, dotEcon.
Marshall, R.C., Marx, L.M. 2012. The Economics of Collusion. The MIT Press, Cambridge, MA, USA.
Martínez-Pardina, I. 2006. First-price auctions where one of the bidders? valuations is common knowledge. Rev. Econ. Des. 10(1), 31–51.
Maschler, M., Solan, E., Zamir, S. 2013. Game Theory. Cambridge University Press, Cambridge, UK.
Maskin, E. 1992. Auctions and privatization. In: Siebert, H. (Ed.), Privatization: Symposium in Honor of Herbert Giersch. J.C.B. Mohr, Tuebingen, pp. 115–136.
Maskin, E., Riley, J. 2000a. Asymmetric auctions. Rev. Econ. Stud. 67(3), 413–438.
Maskin, E., Riley, J. 2000b. Equilibrium in sealed high bid auctions. Rev. Econ. Stud. 67(3), 439–454.
Maskin, E., Tirole, J. 1988. A theory of dynamic oligopoly, II: price competition, kinked demand curves, and edgeworth cycles. Econometrica 56(3), 571–599.
McAdams, D. 2006. Monotone equilibrium in multi-unit auctions. Rev. Econ. Stud. 73(4), 1039–1056.
McAdams, D. 2007a. Monotonicity in asymmetric first-price auctions with affiliation. Int. J. Game Theory 35(3), 427–453.
McAdams, D. 2007b. Uniqueness in symmetric first-price auctions with affiliation. J. Econ. Theory 136(1), 144–166.
McAfee, R., McMillan, J. 1992. Bidding rings. Am. Econ. Rev. 82(3), 579–599.
McMillan, J. 1994. Selling spectrum rights. J. Econ. Perspectives 8(3), 145–162.
Menezes, F.M., Monteiro, P.K. 2005. An Introduction to Auction Theory. Oxford University Press, Oxford, UK.
Milgrom, P. 2000. Putting auction theory to work: the simultaneous ascending auction. J. Polit. Econ. 108(2), 245–272.
Milgrom, P.R. 2004. Putting Auction Theory to Work. Cambridge University Press, Cambridge, UK.
Milgrom, P.R., Shannon, C. 1994. Monotone comparative statics. Econometrica 62(1), 157–180.

Milgrom, P.R., Weber, R.J. 1982. A theory of auctions and competitive bidding. Econometrica 50(5), 1089–1122.
Moldovanu, B., Sela, A. 2001. The optimal allocation of prizes in contests. Am. Econ. Rev. pp. 542–558.
Moldovanu, B., Sela, A. 2003. Patent licensing to Bertrand competitors. Int. J. Ind. Org. 21(1), 1–13.
Moldovanu, B., Sela, A. 2006. Contest architecture. J. Econ. Theory 126(1), 70–96.
Moldovanu, B., Sela, A., Shi, X. 2007. Contests for status. J. Polit. Econ. 115(2), 338–363.
Moldovanu, B., Sela, A., Shi, X. 2012. Carrots and sticks: prizes and punishments in contests. Econ. Inquiry 50(2), 453–462.
Molnár, J., Virág, G. 2008. Revenue maximizing auctions with market interaction and signaling. Econ. Lett. 99(2), 360–363.
Montefiore, S.S. 2011. Jerusalem: The Biography. Weidenfeld & Nicolson, London, UK.
Morgan, J., Steiglitz, K., Reis, G. 2003. The spite motive and equilibrium behavior in auctions. B.E. J. Econ. Anal. Policy 2(1).
Myerson, R.B. 1981. Optimal auction design. Math. Oper. Res. 6(1), 58–73.
Mylovanov, T., Tröger, T. 2009. Optimal auction with resale—a characterization of the conditions. Econ. Theory 40(3), 509–528.
Nagel, R. 1995. Unraveling in guessing games: an experimental study. Am. Econ. Rev. 85(5), 1313–1326.
Nishimura, N., Cason, T.N., Saijo, T., Ikeda, Y. 2011. Spite and reciprocity in auctions. Games 2(3), 365–411.
Noel, M.D. 2007a. Edgeworth price cycles, cost-based pricing, and sticky pricing in retail gasoline markets. Rev. Econ. Stat. 89(2), 324–334.
Noel, M.D. 2007b. Edgeworth price cycles: evidence from the Toronto retail gasoline market. J. Ind. Econ. 55(1), 69–92.
Noel, M.D. 2008. Edgeworth price cycles and focal prices: computational dynamic Markov equilibria. J. Econ. Manage. Strat. 17(2), 345–377.
Ockenfels, A., Selten, R. 2005. Impulse balance equilibrium and feedback in first price auctions. Games Econ. Behav. 51(1), 155–170.
Orr, G. 2002. Ballot order: donkey voting in Australia. Elect. Law J. 1(4), 573–578.
Ozdenoren, E. 2002. Auctions and bargaining with a set of priors. Unpublished manuscript, University of Michigan.
Paarsch, H.J., Hong, H. 2006. An Introduction to Structural Econometrics of Auctions. MIT Press, Cambridge, MA.
Pagnozzi, M. 2007. Bidding to lose? Auctions with resale. RAND J. Econ. 38(4), 1090–1112.
Pagnozzi, M. 2009. Resale and bundling in auctions. Int. J. Ind. Org. 27(6), 667–678.
Pai, M., Vohra, R. 2013. Optimal dynamic auctions and simple index rules. Math. Oper. Res. 38, 682–697.
Pai, M., Vohra, R. 2014. Optimal auctions with financially constrained buyers. J. Econ. Theory 150, 383–425.
Parreiras, S., Rubinchik, A. 2010. Contests with three or more heterogeneous agents. Games Econ. Behav. 68(2), 703–715.
Pastine, I., Pastine, T. 2010. Politician preferences, law-abiding lobbyists and caps on political contributions. Public Choice 145(1-2), 81–101.
Pastine, I., Pastine, T. 2013. Soft Money and Campaign Finance Reform, Int. Econ. Rev. 54(4) 1117–1131.
Pavlov, G. 2008. Auction design in the presence of collusion. Theor. Econ. 3(3), 383–429.
Pérez-Castrillo, D., Wettstein, D. 2012. Innovation contests. Barcelona GSE WP, 654.
Perry, M., Reny, P.J. 2002. An efficient auction. Econometrica 70(3), 1199–1212.
Pitchik, C. 2009. Budget-constrained sequential auctions with incomplete information. Games Econ. Behav. 66(2), 928–949.
Plott, C.R. 1997. Laboratory experimental testbeds: application to the PCS auction. J. Econ. Manage. Strat. 6(3), 605–638.
Plott, C.R., Salmon, T.C. 2004. The simultaneous, ascending auction: dynamics of price adjustment in experiments and in the UK3G spectrum auction. J. Econ. Behav. Org. 53(3), 353–383.

Porter, D., Smith, V. 2006. FCC license auction design: a 12-year experiment. J. Law Econ. Policy 3(1), 63–80.
Rachmilevitch, S. 2013a. Bribing in first-price auctions. Games Econ. Behav. 77(1), 214–228.
Rachmilevitch, S. 2013b. Endogenous bid rotation in repeated auctions. J. Econ. Theory 148(4), 1714–1725.
Rachmilevitch, S. 2014. First-best collusion without communication. Games Econ. Behav. 83, 224–230.
Reny, P., Zamir, S. 2004. On the existence of pure strategy monotone equilibria in asymmetric first-price auctions. Econometrica 72(4), 1105–1125.
Reny, P.J. 2011. On the existence of monotone pure-strategy equilibria in Bayesian games. Econometrica 79(2), 499–553.
Reynolds, S.S., Wooders, J. 2009. Auctions with a buy price. Econ. Theory 38(1), 9–39.
Riedel, F., Wolfstetter, E. 2006. Immediate demand reduction in simultaneous ascending-bid auctions: a uniqueness result. Econ. Theory 29(3), 721–726.
Riley, J.G., Samuelson, W.F. 1981. Optimal auctions. Am. Econ. Rev. 71(3), 381–392.
Roberson, B. 2006. The Colonel Blotto game. Econ. Theory 29(1), 1–24.
Roberson, B., Kvasov, D. 2012. The non-constant-sum Colonel Blotto game. Econ. Theory 51(2), 397–433.
Rodriguez, G.E. 2000. First price auctions: monotonicity and uniqueness. Int. J. Game Theory 29(3), 413–432.
Rosenkranz, S., Schmitz, P.W. 2007. Reserve prices in auctions as reference points. Econ. J. 117(520), 637–653.
Rothkopf, M.H., Pekeč, A., Harstad, R.M. 1998. Computationally manageable combinational auctions. Manage. Sci. 44(8), 1131–1147.
Said, M. 2008. Information revelation and random entry in sequential ascending auctions. In: Proceedings of the 9th ACM conference on Electronic commerce. ACM, pp. 98–98.
Said, M. 2012. Auctions with dynamic populations: efficiency and revenue maximization. J. Econ. Theory 147(6), 2419–2438.
Salo, A., Weber, M. 1995. Ambiguity aversion in first-price sealed-bid auctions. J. Risk Uncertainty 11(2), 123–137.
Sarin, R., Wakker, P. 1992. A simple axiomatization of nonadditive expected utility. Econometrica 60(6), 1255–1272.
Schmeidler, D. 1989. Subjective probability and expected utility without additivity. Econometrica 57(3), 571–587.
Schöttner, A. 2008. Fixed-prize tournaments versus first-price auctions in innovation contests. Econ. Theory 35(1), 57–71.
Sela, A. 2011. Best-of-three all-pay auctions. Econ. Lett. 112(1), 67–70.
Sela, A. 2012. Sequential two-prize contests. Econ. Theory 51(2), 383–395.
Shunda, N. 2009. Auctions with a buy price: the case of reference-dependent preferences. Games Econ. Behav. 67(2), 645–664.
Siegel, R. 2009. All-pay contests. Econometrica 77(1), 71–92.
Siegel, R. 2010. Asymmetric contests with conditional investments. Am. Econ. Rev. 100(5), 2230–2260.
Siegel, R. 2014a. Asymmetric all-pay auctions with interdependent valuations. J. Econ. Theory, forthcoming.
Siegel, R. 2014b. Asymmetric contests with head starts and non-monotonic costs. Am. Econ. J. Microecon., forthcoming.
Siegel, R. 2014c. Contests with productive effort. Int. J. Game Theory, forthcoming.
Sisak, D. 2009. Multiple-prize contests—the optimal allocation of prizes. J. Econ. Surv. 23(1), 82–114.
Skreta, V. 2013. Optimal auction design under non-commitment. Working Paper.
Skrzypacz, A., Hopenhayn, H. 2004. Tacit collusion in repeated auctions. J. Econ. Theory 114(1), 153–169.
Smith, C.W. 1989. Auctions: The Social Construction of Value. The Free Press, New York, USA.
Sobel, D. 1996. Longitude: The Story of a Lone Genius Who Solved the Greatest Scientific Problem of His Time. HarperCollins UK.

Stahl, D.O., Wilson, P.W. 1994. Experimental evidence on players' models of other players. J. Econ. Behav. Org. 25(3), 309–327.
Stahl, D.O., Wilson, P.W. 1995. On players' models of other players: theory and experimental evidence. Games Econ. Behav. 10(1), 218–254.
Szentes, B., Rosenthal, R.W. 2003a. Beyond chopsticks: symmetric equilibria in majority auction games. Games Econ. Behav. 45(2), 278–295.
Szentes, B., Rosenthal, R.W. 2003b. Three-object two-bidder simultaneous auctions: chopsticks and tetrahedra. Games Econ. Behav. 44(1), 114–133.
Taylor, C.R. 1995. Digging for golden carrots: an analysis of research tournaments. Am. Econ. Rev. pp. 872–890.
Taylor, G.A., Tsui, K.K., Zhu, L. 2003. Lottery or waiting-line auction? J. Public Econ. 87(5), 1313–1334.
Thomas, C.D. 2013. N-dimensional Blotto game with asymmetric battlefield values. Discussion paper.
Thomas, J.P., Wang, Z. 2013. Optimal punishment in contests with endogenous entry. J. Econ. Behav. Org. 91, 34–50.
Tullock, G. 1980. Efficient rent seeking. In: Buchanan, J.M., Tollison, R.D., Tullock, G. (Eds.), Toward a Theory of the Rent-Seeking Society (Chapter 4). Texas A & M Univ Pr, pp. 97–112.
Turocy, T.L. 2008. Auction choice for ambiguity-averse sellers facing strategic uncertainty. Games Econ. Behav. 62(1), 155–179.
Varian, H. 2007. Position auctions. Int. J. Ind. Org. 25(6), 1163–1178.
Varian, H.R. 1980. A model of sales. Am. Econ. Rev. 70(4), 651–659.
Vartiainen, H. 2013. Auction design without commitment. J. Eur. Econ. Assoc. 11(2), 316–342.
Vickrey, W. 1961. Counterspeculation, auctions, and competitive sealed tenders. J. Finance 16(1), 8–37.
Virág, G. 2013. First-price auctions with resale: the case of many bidders. Econ. Theory 52(1), 129–163.
Vulcano, G., Van Ryzin, G., Maglaras, C. 2002. Optimal dynamic auctions for revenue management. Manage. Sci. 48(11), 1388–1407.
Weber, R.J. 1997. Making more from less: strategic demand reduction in the FCC spectrum auctions. J. Econ. Manage. Strat. 6(3), 529–548.
Weinstein, J. 2012. Two notes on the Blotto game. BE J. Theor. Econ. 12(1).
Weitzman, M.L. 1979. Optimal search for the best alternative. Econometrica 47(3), 641–654.
Wilson, R.B. 1967. Competitive bidding with asymmetric information. Manage. Sci. 13(11), 816–820.
Wilson, R.B. 1979. Auctions of shares. Q. J. Econ. pp. 675–689.
Wilson, R.B. 1992. Strategic analysis of auctions. In: Aumann, R.J., Hart, S. (Eds.), Handbook of Game Theory with Economic Applications volume 1, Elsevier Science Publishers (North-Holland).
Wolf, J. 2012. Efficiency, auctioneer revenue, and bidding behavior in the combinatorial clock auction-an analysis in the context of European spectrum auctions. Ph.D. thesis, Technische Universität München, München.
Wolfstetter, E. 1999. Topics in Microeconomics: Industrial Organization, Auctions, and Incentives. Cambridge University Press, Cambridge, UK.
Wolfstetter, E. 2003. The Swiss UMTS spectrum auction flop: bad luck or bad design?. In: Nutzinger, H.G. (Ed.), Regulation, Competition, and the Market Economy. Festschrift for Carl Christian von Weizsäcker. Vandenhoeck & Ruprecht, pp. 281–293.
Xu, X., Levin, D., Ye, L. 2013. Auctions with entry and resale. Games Econ. Behav. 79, 92–105.
Yoon, K. 2011. Optimal mechanism design when both allocative inefficiency and expenditure inefficiency matter. J. Math. Econ. 47(6), 670–676.
Zheng, C.Z. 2001. High bids and broke winners. J. Econ. Theory 100(1), 129–171.
Zheng, C.Z. 2002. Optimal auction with resale. Econometrica 70(6), 2197–2224.
Zheng, C.Z. 2013. Existence of monotone equilibria in first-price auctions with resale. Working Paper.

CHAPTER 8

Combinatorial Auctions

Rakesh V. Vohra[*,†]
[*]Department of Economics, University of Pennsylvania, Philadelphia, PA, USA
[†]Department of Electrical and Systems Engineering, University of Pennsylvania, Philadelphia, PA, USA

Contents

Abstract

Many auctions involve the sale of heterogenous indivisible objects. Examples are wireless spectrum, delivery routes and airport time slots. Because of complementarities or substitution effects between the objects, bidders have preferences not just over individual items but over subsets of them. For this reason, economic efficiency is enhanced if bidders are allowed to bid on bundles or combinations of different assets. This chapter surveys the state of knowledge about combinatorial auctions.

Keywords: Combinatorial auctions, Core, Vickrey auctions, Clock auctions, Package bids

JEL Codes: D44, D47, D82

8.1. INTRODUCTION

The study of combinatorial auctions is concerned with the allocation of heterogenous goods amongst bidders with preferences over subsets of the objects that are quasi-linear. Typically, the goods are indivisible and this is what is assumed here. Because bidders have preferences over subsets of the objects, auctioning each object separately is generally not efficient or revenue maximizing. Thus, bidders should be permitted to bid directly on combinations or bundles of different goods. Auctions where bidders are allowed to submit bids on combinations of items are called combinatorial auctions. Auctions of this type were proposed as early as 1976 for radio spectrum rights (see Jackson, 1976).

Handbook of game theory
http://dx.doi.org/10.1016/B978-0-12-420056-2.00008-2

Rassenti et al. (1982), a little later, proposed such auctions to allocate airport time slots. A number of commercial enterprises now offer software to support the execution of combinatorial auctions. These are used primarily in procurement settings. Examples are CombineNet (US), Tradeslot (Australia) and MinMax Groupe Conseil (Canada).[1] There are a number of others that were subsequently acquired by larger firms (like logistics.com which was acquired by Manhattan Associates).

The complexity of processing and evaluating bids over subsets of objects imposes a substantial computational burden on bidders and auctioneer. Recent computational advances have made such auctions practical. Ledyard et al. (2000), for example, describes the design and use of a combinatorial auction that was employed by Sears in 1993 to select carriers. The objects bid upon were delivery routes (called **lanes**). Since a truck must return to its depot, it was more profitable for bidders to have their trucks full on the return journey. Being allowed to bid on bundles gave bidders the opportunity to construct routes that utilized their trucks as efficiently as possible. Since 1995, London Transport has been auctioning off bus routes using a combinatorial auction. About once a month, existing contracts to service some routes expire and these are offered for auction. Bidders can submit bids on subsets of routes, winning bidders are awarded a five-year contract to service the routes they win (see Cantillon and Pesendorfer, 2006). Epstein et al. (2004) describe a combinatorial auction to assign meal contracts for Chilean schools.

This survey will be organized around three questions. The first is existence of "supporting prices." Auctions determine an allocation via prices. Given a desired allocation, do there exist prices that support (to be made precise later) the allocation and what form do they take? Second, preferences are usually private information and must be elicited. Do the supporting prices identified in the answer to the first question give bidders the incentive to reveal the information necessary to produce the desired allocation? Third, what are the computational constraints that limit the allocations that can be chosen?

Attention is confined to independent private values. This is a huge restriction, but the reader will agree, one hopes, by chapter's end that there is plenty of meat still on the bone. Another important assumption is that the auctions to be examined take place in a vacuum. Thus, we ignore the costly investments that bidders must make before participation, the decision to participate and what happens in the "after market" when the auction ends. The reader left unsated by this survey should consult Cramton et al. (2006).

Assume a finite set of bidders, N, and a finite set of indivisible objects (or goods), G. For each set of objects $S \subseteq G$, each bidder $i \in N$ has a (monetary) non-negative valuation $v_i(S) \geq 0$ (where $v_i(\emptyset) = 0$). To avoid minor technical issues we assume that each $v_i(S)$ is integral. We also assume that $v_i(S)$ is monotone, that is

[1] Note the author was for a period an advisor to CombineNet.

$S \subseteq T \Rightarrow v_i(S) \leq v_i(T)$. A bidder i who consumes $S \subseteq G$ and makes a payment of $p \in \mathbb{R}$ receives a net payoff of $v_i(S) - p$.

8.2. SUPPORTING PRICES

Assume first that preferences, i.e., the v_i's, are common knowledge. In this case, the two objectives an auctioneer can choose from, efficiency and revenue maximization, coincide. Thus we suppose the goal of the auctioneer is to maximize efficiency. In fact for the entire paper we make this assumption. We do so for the usual "drunk in search of his keys" reason.[2]

A straightforward way to find an efficient allocation of the goods is via an integer linear program. This will have the advantage that the dual variables of the underlying linear programing relaxation can be interpreted as (possibly) supporting prices. A set of prices supports an efficient allocation, if at those prices, each bidder is assigned in the efficient allocation a subset of goods that yields both non-negative surplus and maximizes his surplus.[3]

Let $y(S,j) = 1$ mean that the bundle $S \subseteq G$ is allocated to $i \in N$ ...and zero otherwise.[4]

$$V(N) = \max \sum_{j \in N} \sum_{S \subseteq G} v_j(S) y(S,j) \qquad [8.1]$$

$$\text{s.t.} \sum_{S \ni i} \sum_{j \in N} y(S,j) \leq 1 \quad \forall i \in G$$

$$\sum_{S \subseteq G} y(S,j) \leq 1 \quad \forall j \in N$$

$$y(S,j) = 0, 1 \quad \forall S \subseteq G, \forall j \in N$$

The first constraint ensures that each object is assigned at most once. The second ensures that no bidder receives more than one subset. Call this formulation $(CAP1)$. Denote the optimal objective function value of the linear programing relaxation of $(CAP1)$ by $Z_1(N)$.

Even though the linear relaxation of $(CAP1)$ might admit fractional solutions let us write down its linear programing dual. To each constraint

$$\sum_{S \ni i} \sum_{j \in N} y(S,j) \leq 1 \quad \forall i \in G$$

[2] When v_i's are private information, and one takes the incentive problem seriously, one has a problem of mechanism design with multidimensional types. See Vohra (2011).

[3] The exposition here is based on de Vries et al. (2007).

[4] It is important to make clear the notation that, e.g., if bidder i consumes the pair of distinct objects $g, h \in G$, then $y(\{g,h\}, i) = 1$, but $y(\{g\}, i) = 0$. A bidder consumes exactly one set of objects.

associate the variable p_i, which will be interpreted as the price of object i. To each constraint

$$\sum_{S \subseteq G} y(S,j) \leq 1 \quad \forall j \in N$$

associate the variable π_j, which will be interpreted as the surplus of bidder j. The dual is

$$Z_1(N) = \min \sum_{i \in G} p_i + \sum_{j \in N} \pi_j$$

$$\text{s.t. } \pi_j + \sum_{i \in S} p_i \geq v_j(S) \quad \forall S \subseteq G$$

$$\pi_j \geq 0 \quad \forall j \in N$$

$$p_i \geq 0 \quad \forall i \in G$$

Let (π^*, p^*) be an optimal solution to this linear program. It is easy to see that

$$\pi_j^* = \max_{S \subseteq G} \left[v_j(S) - \sum_{i \in S} p_i^* \right]^+$$

which has a natural interpretation. For price vector p^*, π_j^* is the maximum surplus that bidder j will enjoy. If S is a bundle that does not maximize the surplus of bidder j at price vector p^*, then $\pi_j^* > v_j(S) - \sum_{i \in S} p_i^*$. It follows by complementary slackness that $y(S,j) = 0$ in an optimal solution to $(CAP1)$. Equivalently, $y(S,j) > 0$ in an optimal primal solution means that $\pi_j^* = v_j(S) - \sum_{i \in S} p_i^*$. That is, bundle S maximizes bidder j's surplus at prices p^*.

If $V(N) = Z_1(N)$ we can conclude that the efficient allocation is supported by single item prices. Notice, the converse is also true. If the efficient allocation can be supported by single item prices, then $V(N) = Z_1(N)$ (see Bikhchandani and Mamer, 1997). One instance when $V(N) = Z_1(N)$ is when bidders' preferences satisfy the (gross) substitutes property (see Section 8.4 for a definition).

In general $V(N) \neq Z_1(N)$, so we cannot hope to support an efficient allocation using item prices. This can be interpreted as an impossibility result: no auction using only item prices is guaranteed to deliver the efficient allocation. In this case, it is natural to look for a stronger formulation. A standard way to do this is by inserting additional variables. Such formulations are called extended formulations. Bikhchandani and Ostroy (2002) offer one.

To describe this extended formulation, let Π be the set of all possible partitions of the objects in the set G. If σ is an element of Π, we write $S \in \sigma$ to mean that the set $S \subset G$ is a part of the partition σ. Let $z_\sigma = 1$ if the partition σ is selected and zero otherwise. These are the auxiliary variables. Using them we can reformulate $(CAP1)$ as follows:

$$V(N) = \max \sum_{j \in N} \sum_{S \subseteq G} v_j(S) y(S,j)$$

$$\text{s.t.} \sum_{S \subseteq G} y(S,j) \leq 1 \quad \forall j \in N$$

$$\sum_{j \in N} y(S,j) \leq \sum_{\sigma \ni S} z_\sigma \quad \forall S \subseteq G$$

$$\sum_{\sigma \in \Pi} z_\sigma \leq 1$$

$$y(S,j) = 0, 1 \quad \forall S \subseteq G, \ j \in N$$

$$z_\sigma = 0, 1 \quad \forall \sigma \in \Pi$$

Call this formulation ($CAP2$). In words, ($CAP2$) chooses a partition of G and then assigns the sets of the partition to bidders in such a way as to maximize efficiency. Let $Z_2(N)$ denote the optimal objective function value of the linear programming relaxation of ($CAP2$). While ($CAP2$) is stronger than ($CAP1$), it is still not the case that $V(N) = Z_2(N)$ always.

The dual of the linear relaxation of ($CAP2$) involves one variable for every constraint of the form:

$$\sum_{S \subseteq G} y(S,j) \leq 1 \ \forall j \in N,$$

call it π_j, which can be interpreted as the surplus that bidder j obtains. There will be a dual variable π_s associated with the constraint $\sum_{\sigma \in \Pi} z_\sigma \leq 1$ which can be interpreted to be the revenue of the seller. The dual also includes a variable for every constraint of the form:

$$\sum_{j \in N} y(S,j) \leq \sum_{\sigma \ni S} z_\sigma \ \forall S \subseteq G$$

which we denote $p(S)$ and interpret as the price of the bundle S.

The dual will be

$$Z_2(N) = \min \sum_{j \in N} \pi_j + \pi_s$$

$$\text{s.t. } \pi_j \geq v_j(S) - p(S) \quad \forall j \in N, \ S \subseteq G$$

$$\pi_s \geq \sum_{S \in \sigma} p(S) \quad \forall \sigma \in \Pi$$

$$\pi_j, p(S), \pi_s \geq 0 \quad \forall j \in N, \ S \subseteq G$$

and has the obvious interpretation: minimizing the bidders' surplus plus seller's revenue. Notice that prices are now (possibly nonlinear) functions of the bundle consumed.

Such bundle prices support the efficient allocation if and only if $V(N) = Z_2(N)$. Informally, we can interpret this as an impossibility result: in general there can be no auction using only non-linear prices that is guaranteed to deliver the efficient allocation. One instance when $V(N) = Z_2(N)$ is when the v_j's are all subadditive.[5]

To obtain an even stronger formulation (again due to Bikhchandani and Ostroy, 2002), let μ denote both a partition of the set of objects and an assignment of the elements of the partition to bidders. Thus μ and μ' can give rise to the same partition, but to different assignments of the parts to bidders. Let Γ denote the set of all such partition-assignment pairs. We will write μ_j to mean that under μ, bidder j receives the set μ_j. Let $\delta_\mu = 1$ if the partition-assignment pair $\mu \in \Gamma$ is selected, and zero otherwise. Using these new variables the efficient allocation can be found by solving the formulation we will call $(CAP3)$.

$$V(N) = \max \sum_{j \in N} \sum_{S \subseteq G} v_j(S) y(S,j)$$

$$\text{s.t.} \quad y(S,j) \le \sum_{\mu:\mu_j = S} \delta_\mu \quad \forall j \in N, \forall S \subseteq G$$

$$\sum_{S \subseteq G} y(S,j) \le 1 \quad \forall j \in N$$

$$\sum_{\mu \in \Gamma} \delta_\mu \le 1$$

$$y(S,j) = 0,1 \quad \forall S \subseteq G, \forall j \in N$$

Bikhchandani and Ostroy (2002) showed that this formulation's linear programing relaxation has an optimal integer solution.

Theorem 8.1. *Problem (CAP3) has an optimal integer solution.*

Proof. We show that the linear relaxation of $(CAP3)$ has an optimal integer solution. Let (y^*, δ^*) be an optimal fractional solution and $V_{LP}(N)$ its objective function value. To each partition-assignment pair μ there is an associated integer solution: $y^\mu(S,j) = 1$ if S is assigned to bidder j under μ, and zero otherwise.

With probability δ^*_μ select μ as the partition-assignment pair. The expected value of this solution is

$$\sum_{\mu} \left[\sum_{j \in N} \sum_{S \subseteq G} v_j(S) y^\mu(S,j) \right] \delta^*_\mu.$$

[5] See Parkes and Ungar (2000).

But this is equal to

$$\sum_{j\in N}\sum_{S\subseteq G} v_j(S)\left[\sum_{\mu:\mu_j=S}\delta^*_\mu\right] \geq \sum_{j\in N}\sum_{S\subseteq G} v_j(S)y^*(S,j) = V_{LP}(N)$$

The last inequality follows from

$$y^*(S,j) \leq \sum_{\mu:\mu_j=S}\delta^*_\mu \quad \forall j \in N, \forall S \subseteq G$$

Thus, the expected value of the randomly generated integer solution is at least as large as the value of the optimal fractional solution. Therefore, at least one of these integer solutions must have a value at least as large as the optimal fractional solution. It cannot be strictly larger as this would contradict the optimality of (y^*, δ^*). So all y^μ have same objective function value as (y^*, δ^*) or there is no optimal fractional extreme point.

To write down the dual of ($CAP3$), we associate with each constraint $y(S,j) \leq \sum_{\mu:\mu_j=S}\delta_\mu$ a variable $p_j(S) \geq 0$ which can be interpreted as the price that bidder j pays for the set S. This means that different bidders will "see" different prices for the same set of objects. To each constraint $\sum_{S\subseteq G} y(S,j) \leq 1$ we associate a variable $\pi_j \geq 0$ which can be interpreted as bidder j's surplus. To the constraint $\sum_{\mu\in\Gamma}\delta_\mu \leq 1$ we associate the variable π_s which can be interpreted as the seller's revenue. The dual DP3 becomes

$$V(N) = \min \sum_{j\in N}\pi_j + \pi_s$$

$$\text{s.t. } p_j(S) + \pi_j \geq v_j(S) \quad \forall j \in N, \forall S \subseteq G$$

$$-\sum_{\mu:\mu_j=S} p_j(S) + \pi_s \geq 0 \quad \forall \mu \in \Gamma$$

$$p_j(S) \geq 0 \quad \forall j \in N, \forall S \subseteq G$$

$$\pi_i \geq 0 \quad \forall i \in N \cup \{s\}$$

Because ($CAP3$) has the integrality property, the dual above is exact. Hence, an efficient allocation is supported by prices that are nonlinear and nonanonymous, meaning different bidders see different prices for the same bundle.[6]

[6] One could ask whether it is possible to support the efficient allocation by prices that can be decomposed into a nonlinear and nonanonymous part. The answer is yes, see Bikhchandani et al. (2002). There is also a connection between supporting prices and communication complexity, see Nisan and Segal (2006). One can interpret these communication results as saying something about the dimension of the dual polyhedron.

8.2.1 The core

With each instance of a combinatorial auction environment one can associate a co-operative game. Let $N^* = N \cup \{s\}$ where s denotes the seller, be the "players" of the co-operative game. For each $K \subseteq N^*$ set $u(K) = V(K \setminus \{s\})$ if $s \in K$ and zero otherwise. In words, only coalitions containing the seller generate value. The set of vectors $\pi \in \mathbb{R}^{|N^*|}$ that satisfy

$$\sum_{i \in N^*} \pi_i = u(N^*)$$

and

$$\sum_{i \in K} \pi_i \geq u(K) \ \forall K \subset N^*$$

is called the core of the cooperative game (N^*, u). The core of (N^*, u) is nonempty because $\pi_s = u(N^*)$ and $\pi_i = 0$ for all $i \in N$, lies in the core. In fact, the core and the set of optimal solutions to $(DP3)$ are the same.

Theorem 8.2. *Let (π^*, p^*) be an optimal solution to (DP3). Then π^* is an element of the core. If π^* is an element of the core, there exists p^* such that (π^*, p^*) is an optimal solution to (DP3).*

Proof. Suppose first that (π^*, p^*) is an optimal solution to $(DP3)$. We show that π^* lies in the core. The core constraints for sets K disjoint from $\{s\}$ are clearly satisfied. Next, for any $R \subseteq N$, $\{\pi_j^*\}_{j \in N}$, π_s^* and $\{p_j^*(S)\}_{j \in R}$ are a feasible dual solution to the primal program $(CAP3)$ when restricted to the set R. Hence, by weak duality

$$\sum_{j \in R} \pi_j^* + \pi_s^* \geq v(R) = u(R \cup s).$$

Finally, by strong duality

$$\sum_{j \in N} \pi_j^* + \pi_s^* = V(N) = u(N^*).$$

Now suppose π^* is an element of the core. Each element of the core is associated with some allocation of the goods designed to achieve π^*. Let μ^* be the partition assignment associated with π^*. If S is assigned to bidder $j \in N$ under μ^*, set $p_j^*(S) = \max\{v_j(S) - \pi_j^*, 0\}$. It suffices to check that (π^*, p^*) is feasible in $(DP3)$ to verify optimality since, the core condition tells us that the objective function value of the solution is $V(N)$.

An outcome in the core is attractive for the usual reasons. No subset of bidders can approach the seller with an offer that would make each of them better off than what they would get in a core outcome. In this setting, call the marginal product of bidder j, $V(N) - V(N \setminus j)$. This is the maximum added value a bidder contributes to the economy consisting of the one seller and the bidders in N. In a sense this is the largest surplus a bidder is justified in obtaining. There is a relationship between the optimal solutions to ($DP3$) and the marginal product of a bidder that is straightforward to obtain.

Lemma 8.1. *If $\{\pi_i\}_{i \in N \cup s}$ is an optimal solution to (DP3) then $\pi_j \leq V(N) - V(N \setminus j)$ for all $j \in N$.*

Among the set of optimal solutions to ($DP3$) there is one that gives bidder $j \in N$ her marginal product. This is the optimal solution with largest π_j value. Can we expect more? Specifically, is there an optimal solution to ($DP3$) that simultaneously gives to each bidder in N their marginal product? In general, no. But, when the bidders are substitutes condition (BSC) holds, the answer is yes. The definition of the BSC follows: For any subset K of bidders

$$V(N) - V(N \setminus K) \geq \sum_{i \in K} [V(N) - V(N \setminus i)] \qquad [8.2]$$

The definition is due to Shapley and Shubik (1972).[7] In words, the marginal product of the set K is at least as large as the sum of the marginal products of each of its members. In a sense, the whole is greater than the sum of its parts.

Theorem 8.3. *If the BSC holds, the point π^* in the core of (N^*, u) that minimizes π_s satisfies $\pi_j^* = V(N) - V(N \setminus j)$ for all $j \in N$.*

Proof. If $\pi_j^* = V(N) - V(N \setminus j)$ for all $j \in N$ then, to be in the core, we need

$$\pi_s^* = V(N) - \sum_{j \in N} [V(N) - V(N \setminus j)].$$

For any $K \subseteq N$,

$$\sum_{j \in K} \pi_j^* + \pi_s^* = \sum_{j \in K} [V(N) - V(N \setminus j)] + V(N) - \sum_{j \in N} [V(N) - V(N \setminus j)]$$

$$= V(N) - \sum_{j \in N \setminus K} [V(N) - V(N \setminus j)] \geq V(K) = u(K \cup s).$$

The last inequality follows from BSC.

[7] BSC is weaker than submodularity of V.

Let π be any point in the core. Recall that

$$\sum_{j \in N} \pi_j + \pi_s = V(N)$$

and $\sum_{j \in N \setminus i} \pi_j + \pi_s \geq V(N \setminus i)$. Negating the last inequality and adding it to the previous equation, yields $\pi_i \leq V(N) - V(N \setminus i)$. Since π^* achieves this upper bound simultaneously for all $j \in N$, it follows that π^* must have been chosen to be a point in the core that maximizes $\sum_{j \in N} \pi_j$. Equivalently, the point in the core that minimizes π_s.

Thus, when the BSC holds, there exist supporting prices that give to each bidder in N their marginal product. A drawback is that the BSC is a condition not on the primitives of the model but a joint condition on bidder preferences.

8.3. INCENTIVES

As preferences of bidders are private information the price they pay must not only support the desired allocation but tempt them to reveal the information necessary to identify the desired allocation. Hence, when is it the case that supporting prices have this property? An answer to this question can be found in the literature on mechanism design.

In a direct mechanism, the auctioneer announces how the goods will be allocated and the payments set as a function of the private information reported by the bidders. It remains then for the bidders to report (not necessarily truthfully) their private information. Once the reports are made, the auctioneer "applies" the announced function and the resulting allocations made and payments executed. One can think of a direct mechanism as a sealed bid auction, but with the bids being complex enough to communicate the private information of the bidders. If the direct mechanism has the property that it is in the bidders interests, suitably defined, to report their private information truthfully, call it a direct revelation mechanism.

One can imagine a mechanism that is not "direct." Perhaps it involves multiple rounds where participation in one round is conditional on bids or other information revealed in previous rounds. These indirect mechanisms can be quite baroque. It is here that the "revelation principle" comes to our aid. Given any mechanism (and equilibrium outcome of that mechanism), there is a corresponding direct revelation mechanism which will produce the same outcome as the given mechanism. Thus, for the purposes of analysis, one can restrict attention to direct revelation mechanisms. The revelation principle does pose a problem. If a direct mechanism suffices why do we observe indirect mechanisms in practice? We defer a discussion of this to Section 8.3.2.

It is well known that amongst all ex-post individually rational, dominant strategy mechanisms that return the efficient allocation, the one that maximizes the auctioneer's revenue is the Vickrey-Clarke-Groves mechanism or VCG for short.[8] The VCG mechanism achieves these properties by determining the maximum rent bidder j could gain from misreporting without altering the efficient allocation and gives her those rents, so removing the incentive to misreport. The result is that each bidder walks away with a surplus equal to their marginal product. Specifically, if we apply the VCG mechanism, bidder $i \in N$ receives a surplus of $V(N) - V(N \setminus i)$. Given Theorem 8.3 we see that when the BSC holds, there exist supporting prices that coincide with the prices charged by the VCG mechanism. Phrased, differently, when the BSC condition holds, the VCG outcome lies in the core of (N^*, u).

8.3.1 When VCG is not in the core

The VCG auction is criticized for being counter-intuitive (see for example Ausubel and Milgrom, 2006, and Rothkopf, 2007). There are examples where the revenue from the VCG auction can decline as the number of bidders increase. In procurement settings the VCG auction can end up paying many times more for an asset than it is worth. Third, VCG is vulnerable to false name bidding (see Yokoo et al., 2001), that is a single bidder submitting bids under many different identities. We will argue that these instances are neither counter-intuitive or odd. The examples used to make this point will all be set in a procurement context.

Consider two problems defined on graphs. Fix a connected graph G with edge set E and vertex set V. Every edge in the graph is owned by a single bidder and no bidder owns more than one edge.[9] The length/cost of an edge is private information to the bidder who owns the edge. The first problem is the minimum cost spanning tree problem. If A is any subset of bidders, denote by $M(A)$ the length of the minimum spanning tree using edges in A alone.[10]

The second problem is the shortest path problem. Designate a source, s and sink t and let $L(A)$ be the length of the shortest $s - t$ path using edges only in A.[11] Also to make things interesting, assume that the minimum $s - t$ cut is at least two.

In the first case the auctioneer would like to buy edges necessary to build the minimum spanning tree (MST). If the auctioneer uses a VCG auction, what does she end up paying for an edge, e, say, in the MST?

Recall, that in a VCG auction each bidder must obtain their marginal product. To determine the marginal product of bidder e we must ask what would the length of the minimum spanning tree be without edge e. Let T be the MST and f the cheapest edge that forms a cycle with e in $T \cup f$. The increase in cost from excluding bidder e

[8] See Clarke (1971), Groves (1973) and Vickrey (1961).

[9] For this reason we use the words edge and bidder interchangeably.

[10] If A does not span the graph, take $M(A) = \infty$.

[11] If A does not contain a $s - t$ path set $L(A) = \infty$.

would be the difference in cost between f and e. Thus, if the number of edges in G were to increase, the price that the auctioneer would have to pay for an edge can only go down.

Second, if the ratio of the most expensive edge to the cheapest were α, say, each bidder in T would receive at most α times their cost.

Third, could bidder e benefit by introducing a dummy bidder? Only if the dummy bidder serves to increase bidder e's marginal product. If this dummy bidder does not form a cycle through e it cannot make a difference to the outcome. If it does, it must have a length smaller than edge f, so reducing bidder e's marginal product.

In short everything about the VCG auction in this case conforms to our intuition about how auctions should behave.

Now consider the shortest path problem.[12] Consider a graph G with two disjoint $s - t$ paths each involving K edges. The edges in the "top" path have cost 1 each while in the "bottom" path they have cost $1 + r$ each. The VCG auction applied to this case would choose the top path. Each bidder on the top path has a marginal product of rK and so receives a payment of $1 + rK$. The total payment made by the auctioneer is $K + rK^2$, which, for appropriate r and K is many times larger than the length of the shortest path. A clear case of the "generosity" of the VCG auction.

Furthermore, if we increase the number of bidders by subdividing a top edge and assigning each segment a cost of $1/2$, the payments increase without an increase in the length of the path! In fact an individual bidder could increase her return by pretending to be two separate edges rather than one. So, we have two environments, one where the VCG confirms with "intuition" and the other where it does not. What causes these anomalies to occur in the second case? It is that the VCG outcome is not in the core.

In the spanning tree setup BSC holds (see Bikhchandani et al., 2008). Intuitively, a bidder is included in the MST if it has low cost and there is no other bidder with a lower cost that will perform the same "function." That function is to "cover a cycle."[13] Thus, no bidder depends on another to make it into the MST, but another bidder could crowd it out of the MST. This pits the bidders against each other and benefits the auctioneer. Alternatively, the bidders profit by banding together, this in effect is what the BSC means.

In the shortest path case, the BSC is violated. Bidders on the same path complement each other, in that no bidder on the top path can be selected unless all other bidders on the top path are selected. The marginal product of the bidders on the top path is less than the sum of their individual marginal products.

[12] See Bikhchandani et al. (2002), Hershberger and Suri (2001), Elkind et al. (2004).

[13] Recall that one way to phrase to the minimum spanning tree problem is as follows: find the least weight collection of edges that intersect every cycle of the graph at least once.

To summarize, when BSC holds, the VCG auction exhibits the expected properties. When BSC is violated, economic power is in the hands of the bidders, unless, one is prepared to use an auction that does not share some of the properties of the VCG auction.

As an illustration suppose strong complementarities among the bidders and your goal is to reduce the payments made by you to them. Is there another auction that is incentive compatible and individually rational that one might use? If yes, it must give up on efficiency. This is a line taken by Archer and Tardos (2002) who relax the efficiency requirement in the shortest path context. They replace efficiency by a collection of "monotonicity" type conditions and show that the same "odd" behavior ascribed to VCG persists.[14] This is not surprising because it is the complementarity among bidders coupled with incentive compatibility that raises the costs of the VCG auction rather than the desire for efficiency.

To see why, return to the simple graph with two disjoint $s - t$ paths. Now that we understand the complementary nature of the bidders, we recognize that the auctioneer wants $s - t$ paths to compete with each other rather than edges. How can this be arranged? One thing the auctioneer could do is tell the bidders on the top path to band together and submit a single bid and likewise for the ones on the bottom. This forces competition between the two paths and should lower payments. Now think about the winning group of bidders. They receive an amount of money to divide between them. How do they do it? The winning group of bidders face a classical public goods problem. The only way to induce each bidder to truthfully reveal his cost is to throw money at them. If the auctioneer does not provide the money, there is no incentive for the bidders to band together.

8.3.2 Ascending implementations of VCG

Despite the attractive features of the VCG auction, it is argued that an indirect implementation of the VCG mechanism in the form of an ascending auction would be superior. For example, it has long been known that an ascending auction for the sale of a single object to bidders with independent private values exists that duplicates the outcomes of the VCG mechanism. It is the English clock auction, in which the auctioneer continuously raises the price for the object. A bidder is to keep his hand raised until the price exceeds the amount the bidder is willing to pay for the object. The price continues to rise only as long as at least two bidders have raised hands. When the price stops, the remaining bidder wins the object at that price.

In such an auction, no bidder can benefit by lowering his hand prematurely, or by keeping his hand aloft beyond the price he is willing to pay, as long as all other bidders

[14] One can think of these additional conditions as being properties of a path that would be generated by optimizing some "nice" function on paths other than just length.

behave in a way consistent with some valuation for the object. The ascending auction, therefore, shares some of the incentive properties as its direct revelation counterpart.

There are at least three reasons why an ascending version of the VCG auction would be preferable to its direct revelation counterpart. The first is experimental evidence (Kagel and Levine, 2001) that suggests that bidders do not recognize that bidding truthfully is a dominant strategy in the direct version, but do behave according to the equilibrium prescription of the ascending counterpart. The second has to do with costs of identifying and communicating a bid. These costs have bite when what is being sold is a collection of heterogenous objects. The VCG auction requires a bidder to report their type for each possible allocation. This could be a prohibitively large number. An ascending version of the VCG auction potentially avoids this expense in two ways. First, given an appropriate design, a bidder only needs to decide if an allocations value exceeds some threshold at each stage. Second, bidders need focus only on the allocations that "matter" to them.[15] Lastly, an ascending auction allows "winning" bidders to conceal their actual value for some (but not all) combinations of objects. This may be a concern for bidders who suspect that this information might be used against them in some future interaction with the auctioneer.[16]

The English clock auction is an instance of a Walrasian tâtonnement. As imagined by Walras, an auctioneer announces prices for each good and service in the economy and bidders sincerely report their demand correspondences for the announced prices. By comparing supply and reported demand, the auctioneer can adjust prices so as to match the two. In this way, the auctioneer would, in Keynes' words grope, "like a peregrination through the catacombs with a guttering candle," her way to equilibrium prices. Here is what is known about tâtonnement processes.

1. In general no tâtonnement process relying solely on the excess demand at the current price is guaranteed to converge to market clearing prices.
2. Even when such a tâtonnement process exists, the prices can oscillate up and down.
3. It is not necessarily in a bidder's interest to sincerely report her demand correspondence at each price.

The allocation of a single object is remarkable because it is an exception on all three counts. It is natural to ask how far one may go beyond the sale of a single indivisible object in this regard.[17] One way to approach this question is to observe that an ascending auction which returns an efficient allocation must be an algorithm for finding the efficient allocation. Therefore, by studying appropriate algorithms for finding the efficient allocation one will identify ascending auctions. The idea that

[15] Sometimes, in moving from a direct mechanism to an indirect mechanism, the set of equilibrium outcomes is enlarged. Restricting attention to ascending versions serves to limit this.

[16] This last concern we ignore because it can be circumvented through the use of a trusted intermediary or cryptographic protocols.

[17] This is a line of inquiry pioneered by Ausubel (2004).

certain optimization algorithms can be interpreted as auctions goes back at least as far as Dantzig (1963). In particular many of the ascending auctions proposed in the literature are either primal-dual or subgradient algorithms for the underlying optimization problem for determining the efficient allocation. This is formalized in de Vries et al. (2007).

Following Ausubel and Milgrom (2002) (see also Parkes, 1999, and Parkes and Ungar, 2000) we outline a description of an ascending auction that can be interpreted in terms of a procedure for finding an outcome in the core of (N^*, u). At each stage the auctioneer announces non-negative prices $\{p_j(S)\}_{j \in N, S \subseteq G}$. Each bidder then reports their demand correspondence for these prices. The auctioneer selects an allocation of the goods to satisfy two conditions.

1. It maximizes the seller's revenue at the announced prices.
2. Each bidder receives an element in their reported demand correspondence or the empty set.

Conditions 1 and 2 determine an allocation. If it is in the core, the auction terminates. Notice that it terminates in an efficient allocation. If the allocation is outside the core, there must be a blocking coalition. In words, there is a group of bidders, K, each of whom could point to a bundle they would be willing to pay ϵ more. Increase the price of these bundles by ϵ.

1. There is a feasible allocation that assigns these bundles to bidders in K.
2. The revenue of the seller from picking this allocation would be larger than before.

This auction terminates in a point in the core that minimizes the revenue to the seller. In general there can be more than one such point.[18] However, if the BSC condition holds, this point is unique and corresponds to the Vickrey outcome. From this one can argue that it is an ex-post equilibrium for bidders to bid sincerely in the auction.[19] When the BSC is violated, a weaker claim obtains under the assumption that the bidders have common knowledge of each others valuations. In that case, sincere bidding in each round is an ex-post equilibrium (see Day and Milgrom, 2008).

8.3.3 What is an ascending auction?

We introduced the term ascending auction with an example and no definition. A definition is required. Without it, the following could be considered an ascending auction. Bidders step onto an elevator on the ground floor of a tall building. They submit their valuations to the auctioneer. The elevator rises. As it does, the auctioneer

[18] There is some work on which of these points should be selected. See Parkes et al. (2001) and Erdil and Klemperer (2009). For a comparison between these kinds of auctions and the VCG auction see Lamy (2010) and Beck and Ott (2013).

[19] There is a caveat. The only deviations considered are those consistent with BSC. That is, the bids made by a bidder in each round must be consistent with a valuation that would not lead to a violation of the BSC.

computes the Vickrey outcome. Once the elevator reaches the top, the bidders alight with the allocations and payments prescribed by the VCG auction.

Following de Vries et al. (2007) we define an ascending auction that generalizes Gul and Stacchetti's (2000) definition to allow for nonadditive, nonanonymous prices. However, we strengthen the definition by requiring that prices seen during the auction must represent actual payments which could be made by the bidders.

A price path is a function $P\colon [0,1] \to \mathbb{R}^{2^G \times N}$. For each bundle of goods $H \subseteq G$, interpret $P_i^t(H)$ to be the price seen by bidder i for bundle H, at "time" t. A price path is ascending if for all $H \subseteq G$ the function $P_i^t(H)$ is nondecreasing in t.

An ascending auction adjusts prices based only on the reported demands of the bidders. The following definition captures that idea.

An ascending auction assigns to each profile of bidder valuation functions $v \in \mathbb{R}_+^{2^G \times N}$ both an ascending price path P^v and a final assignment y^v satisfying the following two conditions. First, for all valuation profiles $v, v' \in \mathbb{R}_+^{2^G \times N}$,

$$\left[\forall t \in [0,1],\ \forall j \in N,\ D_j(P^{v,t}; v) = D_j(P^{v,t}; v')\right] \Longrightarrow \left[P^{v'} = P^{v}\right] \qquad [8.3]$$

where $D_j(P; v)$ is j's demand correspondence under prices P when his valuations are v_j. That is, if a change to valuation functions v' does not change the reported demands of bidders, then it does not change the resulting price path. Information is revealed only through demand revelation in the auction.[20] Second, the final assignment gives to each bidder a bundle that is in their demand correspondence evaluated at the terminal prices.

An ascending auction assigns goods such that each bidder j receives a bundle H that is in his demand correspondence at prices $P(1)$, and charges that bidder $P_i^1(H)$. In this sense, we require prices in an auction not to be merely artificial constructs.

Unfortunately, when at least one bidder has a valuation function that does not satisfy the gross substitutes condition, an ascending auction cannot always implement VCG outcomes on some class of problems.

Theorem 8.4. *Suppose that there are two objects $G = \{a, b\}$ and that $|N| \geq 3$. Suppose one bidder's valuation function, say v_1, fails the gross substitutes condition. Then there exists a class of gross substitutes valuation functions for the other bidders, $(\mathcal{V}_j)_{j>1}$, such that no ascending auction yields VCG payments for each profile from $\{v_1\} \times \mathcal{V}_2 \times \cdots \times \mathcal{V}_n$.*

The intuition behind Theorem 8.4 can be given with a simple example. For two objects $G = \{a, b\}$, suppose that three bidders have the valuations given in Table 8.1

[20] Naturally, the empty set may be demanded, so information may be revealed when prices get "too high" for a bidder.

Table 8.1 An example violating gross substitutability.

Bidder	Bundle a	b	ab
1	4	6	16
2	2	10	12
3	$8+\alpha$	$2+\alpha$	$8+\alpha$

which violates the gross substitutes assumption. Restricting attention to the cases where $\alpha \in \{-1, 0, 1\}$, it is efficient to give b to Bidder 2, and a to Bidder 3. Their respective VCG payments are $8 - \alpha$ and 6 (while Bidder 1 pays and receives nothing). Observe that the BSC fails in this example (which happens with respect to $M = \{2, 3\}$).

If an ascending auction uses only "real price information" to determine assignments and payments (as we defined above), then it must do two things. First, it must conclude by offering good a to Bidder 3 at a price of 6. Second, it must determine the value of α in order to offer good b to Bidder 2 at the correct price of $8 - \alpha$. However, for the class of Bidder 3's valuation functions obtained by varying $\alpha \in \{-1, 0, 1\}$, the value of α cannot be inferred from his demand correspondence until Bidder 3 "demands" the empty set, which occurs only when his price for a exceeds $8 + \alpha > 6$. This contradicts the fact that his price for that object ascends throughout the auction and ends at 6.

One can circumvent Theorem 8.4 by modifying the definition of an ascending auction. Obviously one can drop the requirement that prices be ascending. See for example Mishra and Parkes (2007).

8.3.4 The clock-proxy auction

Ausubel et al. (2006) have proposed a hybrid of an ascending and sealed bid auction called the clock-proxy auction that attempts to capture the benefits of both. The auction has two phases. The first is a separate English clock auction for each good run simultaneously. In this first phase, called the clock phase, a bidder can participate in multiple single object auctions. As long as the excess demand for any object is positive, the price of that object rises. The clock phase ends when the excess demand for any object is zero. The clock phase is followed by a last-and-final round in which sealed bids on bundles of objects are submitted (the proxy phase). Ausubel et al. suggest the use of a core selecting auction in the proxy phase to determine final allocations and prices. The two phases are linked by the fact that bids in the proxy phase are constrained by terminal bids in the clock phase. For example, a bid on a single object by agent i in the proxy phase must be at least as large as the highest bid agent i submitted in the clock phase. The clock phase is simple, transparent and is hypothesized to encourage "price" discovery. However, because prices in this phase are linear, they cannot support an efficient allocation. Hence, the proxy phrase.

8.4. COMPLEXITY CONSIDERATIONS

A hurdle to implementing the VCG auction is identifying the efficient allocation. This amounts to solving an integer program, called the winner-determination problem. Here, we focus on the formulation ($CAP1$) that is discussed in Rothkopf et al. (1998) and Sandholm (1999). ($CAP1$) is an instance of what is known as the set-packing problem.[21]

It is natural to think that because ($CAP1$) is an instance of the set packing problem it must be NP-hard. One must be careful in drawing such a conclusion, because it depends on how the input to an instance of ($CAP1$) is encoded. If one takes the number of bids as a measure of the size of the input and this number is exponential in $|M|$, any algorithm for ($CAP1$) that is polynomial in the number of bids but exponential in the number of items would, formally, be polynomial in the input size but impractical for large $|M|$.[22] In particular, an instance of ($CAP1$) in which bids were submitted on every possible combination of items would be polynomial in the number of bids. On the other hand, if input size is measured by $|N|$, $|G|$ and the number of bids is polynomial in $|M|$, then ($CAP1$) is indeed NP-hard.

The upshot is that effective solution procedures for the ($CAP1$) must rely on two things. First, the number of distinct bids is not large. Second, the underlying set packing problem can be solved reasonably quickly. For a list of solvable instances of the set packing problem and their relevance to ($CAP1$) see de Vries and Vohra (2003). Solvability of these instances arise from restrictions on the set of combinations that can be bid upon. Indirectly these amount to restrictions on the kinds of bundles bidders can express preferences over. Nisan (2000) discusses how various bid restrictions prevent certain kinds of preferences being expressed. The designer thus faces a tradeoff. Restricting the bids could make the underlying winner determination problem easy. Doing so limits the preferences that can be expressed by bidders. This will produce allocations that are economically inefficient.

The most general class of preferences for which a positive computational result is available is the gross substitutes class (Kelso and Crawford, 1982). To describe it let the value that bidder j assigns to the set $S \subseteq M$ of objects be $v_j(S)$. Given a vector of prices p on objects, let the collection of subsets of objects that maximize bidder j's utility be denoted $D_j(p)$, and defined as

$$D_j(p) = \{S \subseteq M : v_j(S) - \sum_{i \in S} p_i \geq v_j(T) - \sum_{i \in T} p_i \quad \forall T \subseteq M\}.$$

[21] For this reason, the extensive literature on the set packing problem and its relatives is relevant for computational purposes. Had we cast the auction problem in procurement rather than selling terms the underlying integer program would be an instance of the set partitioning problem. The auctions used in the transport industry are of this set-partitioning type.

[22] It is not hard to construct such algorithms.

The gross-substitutes condition requires that for all price vectors p, p' such that $p' \geq p$, and all $A \in D_j(p)$, there exists $B \in D_j(p')$ such that $\{i \in A : p_i = p'_i\} \subseteq B$. A special case of the gross-substitutes condition is when bidders are interested in multiple units of the same item and have diminishing marginal utilities.

In the case when each of the $v^j(\cdot)$ have the gross-substitutes property, the linear-programing relaxation of $(CAP1)$ has an optimal integer solution. This is proved in Kelso and Crawford (1982) as well as Gul and Stacchetti (2000). Murota and Tamura (2000) point out the equivalence between gross substitutes and $M^\natural$-concavity.[23] From this connection it follows from results about $M^\natural$-concave functions that $(CAP1)$ can be solved in time polynomial in the number of objects under the assumption of gross substitutes by using a proper oracle.

8.4.1 Exact methods

Exact methods for solving $(CAP1)$ come in three varieties: branch and bound, cutting planes, and a hybrid called branch and cut. Fast exact approaches to solving the $(CAP1)$ require algorithms that generate both good lower and upper bounds on the maximum objective-function value of the instance. Because even small instances of the $(CAP1)$ may involve a huge number of columns (bids) the techniques described above need to be augmented with column generation. Exact methods for $(CAP1)$ have been proposed by Rothkopf et al. (1998), Fujishima et al. (1999), Sandholm (1999), and Andersson et al. (2000).

8.4.2 Approximation

There is an extensive literature devoted to approximation algorithms for $(CAP1)$. These algorithms have a complexity that is polynomial in $|N|$ and $|G|$ but are not guaranteed to return an optimal solution. The objective of this line of work is to bound in a worst case sense the percentage deviation between the value of the approximate solution and the optimal solution in terms of $|N|$ and $|G|$. For an example, see Dobzinski et al. (2005). Such approximation algorithms are allocation rules and one can ask if they can be implemented in an incentive compatible way. In general no, which reduces their applicability. Thus, a more interesting question is to identify approximation algorithms that are implementable in an incentive compatible way (see Nisan and Ronen, 2000).

Many of the approximation algorithms that are implementable are said to be maximum in range (MIR). The set of feasible allocations is restricted to a subset of all possible feasible allocations. The restriction is selected so that solving $(CAP1)$ over this restricted set is polynomial in $|N|$ and $|G|$. Running the VCG auction restricted to this subset of allocations is now both incentive compatible and computationally efficient. A simple example is the following: allocate the entire set G to the agent whose value for

[23] See also chapter 11 of Murota (2003).

G is largest. In other words, all goods are collapsed into a single bundle and the single bundle is allocated efficiently. The value of this allocation is always at least as large as $|N|^{-1}$ as the value of the efficient allocation. Attempts to identify MIR algorithms with better worst case bounds have not succeeded.

Inspired by ideas in Mossel et al. (2009), Dughmi et al. (2009) and Buchfuhrer and Umans (2011) have shown that no MIR algorithm can achieve a bound better than $|N|^{-1}$ unless $P = NP$. The idea behind these arguments is natural but tricky to execute. Let Γ be the set of all feasible allocations and Γ' be some subset of allocations such that finding an efficient allocation over Γ' can be found in polynomial time. We need the value of the efficient allocation over Γ' to be close to the value of the efficient allocation over Γ. In a sense this forces Γ' to be a sufficiently "rich" subset of Γ. However, if Γ' is sufficiently rich, one can embed within it an instance of a "hard" problem, so that optimizing over Γ' is equivalent to solving that hard problem.

REFERENCES

Andersson, A., Tenhunen, M., Ygge, F. 2000. Integer programming for combinatorial auction winner determination. Proceedings of the Fourth International Conference on Multi-Agent Systems, pp. 39–46.

Archer, A., Tardos, E. 2002. Frugal path mechanisms. Proceedings of the Thirteenth Annual ACM-SIAM Symposium on Discrete Algorithms, pp. 991–999.

Ausubel, L.M. 2004. An efficient ascending-bid auction for multiple objects. Am. Econ. Rev. 94(5), 1452–1475.

Ausubel, L.M., Milgrom, P. 2002. Ascending auctions with package bidding, Front. Theor. Econ. 1, 1–43.

Ausubel, L., Milgrom, P. 2006. The lovely but lonely vickrey auction. In: Cramton, P., Shoham, Y., Steinberg, R. (Eds.), Combinatorial Auctions. MIT Press, Cambridge, MA.

Ausubel, L., Cramton, P., Milgrom, P. 2006. The clock-proxy auction: a practical combinatorial auction design. In: Cramton, P., Shoham, Y., Steinberg, R. (Eds.), Combinatorial Auctions. MIT Press, Cambridge, MA. Available at: http://works.bepress.com/cramton/37.

Beck, M., Ott, M. 2013. Unrelated goods in package auctions-comparing vickrey and core-selecting outcomes, manuscript.

Bikhchandani, S., de Vries, S., Schummer, J., Vohra, R.V. 2002. Linear programming and Vickrey auctions. In: Dietrich, B., Vohra, R.V. (Eds.), Mathematics of the Internet: E-Auction and Markets. The IMA Volumes in Mathematics and its Applications. vol. 127. Springer-Verlag, New York, NY. pp. 75–116.

Bikhchandani, S., Mamer, J.W. 1997. Competitive Equilibrium in an exchange economy with indivisibilities. J. Econ. Theory 74, 385–413.

Bikhchandani, S., Ostroy, J.M. 2002. The package assignment model. J. Econ. Theory 107(2), 377–406.

Bikhchandani, S., de Vries, S., Schummer, J., Vohra, R.V. 2008. Ascending auctions for integral (poly)matroids with concave nondecreasing separable values. Proceedings of the Nineteenth Annual ACM-SIAM Symposium on Discrete Algorithms, 864–873.

Buchfuhrer, D., Umans, C. 2011. Limits on the social welfare of maximal-in-range auction mechanisms, manuscript, 2011.

Cantillon, Estelle, Pesendorfer, M. 2006. Auctioning bus routes: the london experience. In: Cramton, P., Shoham, Y., Steinberg, R. (Eds.), Combinatorial Auctions. MIT Press, Cambridge, MA.

Clarke, E. 1971. Multipart pricing of public goods. Public Choice 8, 19–33.

Cramton, P., Shoham, Y., Steinberg, R. 2006. Combinatorial Auctions. MIT Press. Cambridge, MA.

Dantzig, G.B. 1963. Linear Programming and Extensions. Princeton University Press, Princeton, NJ.

Day, R., Milgrom, P. 2008. Core-selecting package auctions. Int. J. Game Theory 36, 393–407.

de Vries, S., Vohra, R. 2003. Combinatorial auctions: a survey. INFORMS J. Comput. 15, 284–309.
de Vries, S., Schummer, J., Vohra, R. 2007. On ascending vickrey auctions for heterogeneous objects. J. Econ. Theory 132, 95–118.
Dobzinski, S., Nisan, N., Schapira, M. 2005. Approximation algorithms for combinatorial auctions with complement-free bidders. Proceedings of the 37th Symposium on Theory of Computing, pp. 610–618.
Dughmi, S., Fu, H., Kleinberg, R. 2009. Amplified hardness of approximation for vcg based mechanisms, manuscript.
Elkind, E., Sahai, A., Steiglitz, K. 2004. Frugality in path auctions. Proceedings of the 15th annual ACM-SIAM symposium on Discrete algorithms, pp. 701–709.
Epstein, R., Henrquez, L., Cataln, J., Weintraub, G., Martnez, C., Espejo, F. 2004. A combinatorial auction improves school meals in Chile: a case of operations research in developing countries. Int. Trans. Oper. Res. 11, 593–612.
Erdil, A., Klemperer, P. 2009. A new payment rule for core-selecting package auctions, manuscript.
Fujishima, Y., Leyton-Brown, K., Shoham, Y. 1999. Taming the computational complexity of combinatorial auctions: Optimal and approximate approaches. Proceeding of the Sixteenth International Joint Conference on Artificial Intelligence, IJCAI'99, pp. 548–553.
Gul, F., Stacchetti, E. 2000. English and double auctions with differentiated commodities. J. Econ. Theory 92, 66–95.
Groves, T. 1973. Incentives in teams. Econometrica 41(4), 617–631.
Hershberger, J., Suri, S. 2001. Vickrey prices and shortest paths: what is an edge worth? 42nd Annual Symposium on Foundations of Computer Science, pp. 129–140.
Jackson, C. 1976. Technology for spectrum markets. Ph.D. Thesis, Department of Electrical Engineering, Massachusetts Institute of Technology, Cambridge, MA.
Kagel, J.H., Levin, D. 2001. Behavior in multi-unit demand auctions: experiments with uniform price and dynamic Vickrey auctions. Econometrica 69, pp. 413–454.
Kelso, A.S., Crawford, V.P. 1982. Job matching, coalition formation and gross substitutes. Econometrica 50, 1483–1504.
Lamy, L. 2010. Core selecting package auctions: a comment on revenue monotoicity. Int. J. Game Theory 39(3), 503–510.
Ledyard, J.O., Olson, M., Porter, D., Swanson, J.A., Torma, D.P. 2000. The first use of a combined value auction for transportation services. Social Science Working Paper No. 1093, Division of the Humanities and Social Sciences, California Institute of Technology, Pasadena, CA.
Mishra, D., Parkes, D.C. 2007. Ascending price Vickrey auctions for general valuations. J. Econ. Theory 132, 335–366.
Mossel, E., Papadimitriou, C.H., Schapira, M., Singer, Y. 2009. VC vs. VCG: inapproximability of combinatorial auctions via generalizations of the VC dimension, manuscript.
Murota, K. 2003. Discrete Convex Analysis. SIAM Monographs, Philadelphia.
Murota, K., Tamura, A. 2000. New characterizations of M-convex functions and connections to mathematical economics. RIMS Preprint 1307, Research Institute for Mathematical Sciences, Kyoto University, Kyoto, Japan.
Nisan, N. 2000. Bidding and allocation in combinatorial auctions. Proceedings of the 2nd ACM Conference on Electronic Commerce, pp. 1–12.
Nisan, N., Ronen, A. 2000. Computationally feasible VCG mechanisms. Proceedings of the 2nd ACM Conference on Electronic Commerce, pp. 242–252.
Nisan, N., Segal, I. 2006. The communication complexity of efficient allocation problems. J. Econ. Theory 129, 192–224.
Parkes, D.C. 1999. *i*Bundle: an efficient ascending price bundle auction. Proceedings of the 1st ACM Conference on Electronic Commerce, pp. 148–157.
Parkes, D.C., Ungar, L. 2000. Iterative combinatorial auctions: theory and practice. Proceedings 17th National Conference on Artificial Intelligence, pp. 74–81.
Parkes, D.C., Kalagnanam, J., Eso, M. 2001. Achieving budget-balance with Vickrey-based payment schemes in exchanges. Proceedings 17th International Joint Conf. Artificial Intelligence, pp. 1161–168.

Rassenti, S.J., Smith, V.L., Bulfin, R.L. 1982. A combinatorial auction mechanism for airport time slot allocation. Bell J. Econ. 13, 402–417.
Rothkopf, M.H., Pekec, A., Harstad, R.M. 1998. Computationally manageable combinatorial auctions. Manage. Sci. 44, 1131–1147.
Rothkopf, M. 2007. Thirteen reasons why the Vickrey-Clarke-Groves process is not practical. Oper. Res. 55, 191–197.
Shapley, L.S., Shubik, M. 1972. The assignment game I: the core. Int. J. Game Theory 1, 111–130.
Sandholm, T. 1999. An algorithm for optimal winner determination in combinatorial auctions. Proceedings of IJCAI-99, pp. 542–547.
Vickrey, W. 1961. Counterspeculation, auctions, and competitive sealed tenders. J. Finance 16, 8–37.
Vohra, R.V. 2011. Mechanism Design: A Linear Programming Approach (Econometric Society Monographs), Cambridge University Press, New York.
Yokoo, M., Sakurai, Y., Matsubara, S. 2001. Robust combinatorial auction protocol against false-name bids. Arti. Intel. J. 130, 167–181.

CHAPTER 9

Algorithmic Mechanism Design

Through the lens of Multiunit auctions

Noam Nisan[*,†]
[*]Microsoft Research
[†]Hebrew University of Jerusalam, Jerusalam, Israel

Contents

Abstract

Mechanism design is a subfield of game theory that aims to *design games* whose equilibria have desired properties such as achieving high efficiency or high revenue. *Algorithmic mechanism design* is a subfield that lies on the border of mechanism design and computer science and deals with mechanism

Handbook of game theory
http://dx.doi.org/10.1016/B978-0-12-420056-2.00009-4

design in algorithmically complex scenarios that are often found in computational settings such as the Internet.

The central challenge in algorithmic mechanism design is the tension between the computational constraints and the game-theoretic ones. This survey demonstrates both the tension and ways of addressing it by focusing on a single simple problem: multiunit auctions. A variety of issues will be discussed: representation, computational hardness, communication, convexity, approximations, Vickrey-Clarke-Groves (VCG) mechanisms and their generalizations, single-parameter settings vs. multiparameter settings, and the power of randomness.

Keywords: Mechanism design, Algorithms, Bidding languages, Approximation, Truthfulness, Communication complexity, Auctions

JEL Codes: D44, D47, D82, D83, Asymmetric and private information, Mechanism design

9.1. INTRODUCTION

As the Internet was becoming ubiquitous in the 1990s it became increasingly clear that its understanding required combining insights from computer science with those from game theory and economic theory. On one hand, much economic activity was moving to the Internet, and on the other hand many computational systems started involving computers with different owners and goals. Computational implementations of familiar economic institutions such as catalogs, markets, auctions, or stores opened up unprecedented new opportunities—and challenges—in terms of scale, size, speed, and complexity. As familiar computerized protocols were adapted to the Internet environment, the economic aspects of the resulting cooperation and competition needed to be increasingly taken into account. A new field of study for handling such combinations of computerized and economic considerations was born, a field which is often termed *algorithmic game theory*. A textbook for this field is Nisan et al. (2007).

Perhaps the most natural subfield of economics to be combined with computational considerations is that of *mechanism design*. The field of mechanism design aims to *design* economic institutions as to achieve various economic goals, most notably revenue and social welfare. Combining the points of view of social choice and of game theory, it aims to design games whose outcome will be as the game designer desires—whenever rational players engage in them. The field was initiated by the seminal, Nobel prize winning, work of Vickrey (1961), with an extensive amount of further work done over the last half century. As argued in Roth (2003), the field of mechanism design may be viewed as *Engineering*, aiming to create *new* "mechanisms" using its tools of the trade.

In the mid 1990s researchers from several disciplines recognized that mechanism design could be relevant to computation over the Internet. Examples of influential discussions along these lines include "Economic mechanism design for computerized agents" written by an economist (Varian, 1995), "Rules of encounter: designing conventions for automated negotiation among computers" written by Artificial Intelligence researchers (Rosenschein and Zlotkin, 1994), and "Pricing in computer networks: Reshaping the research agenda" written by computer network researchers (Shenkar et al., 1996). By the end of the millennium, formal theoretical models that combine mechanism design with computation were introduced in Nisan and Ronen (2001) where the term *algorithmic mechanism design* was coined.

Before we dive into the special challenges of the *combination* of "algorithmic" with "mechanism design," let us briefly present the basic point of view of each of these two disciplines separately. The field of mechanism design considers a set of *rational players*, where "rational" is defined to be—as usual in economic theory—utility-maximizing. A *social outcome* that affects all these players needs to be chosen by a *planner* who does *not have full information*, information needed for determining the "preferred" social outcome. The planner's goal is to design a mechanism—a protocol—that when played by the rational agents results in an equilibrium that satisfies the planner's goals.

In classical computer science there are no utilities. A problem describes an input-output relationship and the goal is to produce the correct output for every given input. The difficulty is that of computation: we would like to *efficiently* find the correct output, where efficiency is measured in terms of computational resources: mostly computation time, but also memory, communication, etc. Much work in computer science focuses on distributed problems, where the input and output are distributed among multiple computers who must communicate with each other. Such work on distributed systems became very central as the Internet became the default computational platform.

The usual treatment of the distributed processors in computer science was that they were either "obedient"—programmed and controlled by the designer-engineer—or that they were "faulty"—in which case we have no control over their behavior within the bounds of our fault models. However, for many Internet applications, none of these points of view seemed to capture the essence of the situation. What seems to be the essence is that the computers are simply trying to further the goals of their owners. While such behavior is the bread and butter of economics and game-theory, it was certainly new for computer scientists. Hence the *rational* modeling of participants was adopted into computational settings and mechanism design was embraced as a paradigm for the design of distributed computational protocols over the Internet.

9.2. ALGORITHMIC MECHANISM DESIGN AND THIS SURVEY

9.2.1 The field of algorithmic mechanism design

By now, the field called algorithmic mechanism design is quite large and varied. Moreover, the dividing lines between it and "classical" economic fields of mechanism design, market design, and auction design are very blurry. Some dividing line can be put according to whether complexity and algorithms play a crucial role in the problem.

We can say that a problem is part of algorithmic mechanism design whenever we have an economic design challenge where a central essence of the problem is the multitude of "input pieces" or "outcome possibilities." In other words, whenever some combinatorial complexity is a key issue interacting with incentive issues. Thus, for example, while auctioning a single item is the paradigmatic problem of mechanism design, auctioning *multiple* (related) items would be the paradigmatic problem of algorithmic mechanism design. Market or auction design problems whose domain of application is computerized (e.g., ad auctions) most usually fall into this category and considered to be part of algorithmic mechanism design. Finally, as the theory related to such scenarios has developed, some "approaches" to mechanism design that are characteristic to computer scientists have emerged, in particular a preference to "worst-case" analysis and a willingness to accept approximate solutions.

This survey will not attempt to give a general overview of the field of mechanism design, for which surveys can be found in textbooks such as Mas-Collel et al. (1995) or Osborne and Rubistein (1994) or in the more comprehensive books Krishna (2002) and Milgrom (2004). This survey will not even attempt covering the whole field of algorithmic mechanism design, a nearly impossible task due to its size, ill-defined borders and rapid pace of change. Instead this survey will focus on the single key issue driving the field: *the clash between algorithmic considerations and incentive constraints*. To this effect, we will be limiting ourselves to the following choices:

1. **Private values:** We will only be talking about scenarios where each participant has full information about his own value for each possible outcome. In contrast, a significant amount of classical mechanism design (see Krishna, 2002; Milgrom, 2004) as well as some recent algorithmic mechanism design (e.g., Roughgarden and Talgam-Cohen, 2013) considers situations where bidders have only partial knowledge about their values.
2. **Quasi-linear utilities:** We will assume that bidders' values can be measured in monetary units (money) that can be transferred to and from the auctioneer. This assumption of quasi-linearity is quite common in general mechanism design, although a very large body of work on matching markets does not require this assumption, as do some other works in "algorithmic mechanism design without money" (see, e.g., Procaccia and Tennenholtz, 2009; Schummer and Vohra, 2007).
3. **Risk neutrality:** We will assume that all participants are risk neutral. There is of course much work in mechanism design that concerns effects of risk aversion.
4. **Focus on efficiency:** The two central goals of both mechanism design and algorithmic mechanism design are revenue and efficiency (social welfare). In this survey, we will completely ignore all revenue issues and exclusively consider the goal of efficiency. See Hartline and Karlin (2007) and Hartline (2012) for surveys of work in algorithmic mechanism design that is focused on revenue.

5. **Dominant strategy analysis:** The standard way of modeling lack of information in economics (and most scientific fields) is using a Bayesian prior, and looking at probabilities and expected values over the possible "states of the world." In computer science, however, there is a general mistrust in our ability to correctly capture real world distributions, and the standard approach is to use worst-case analysis. While being more limited, worst case analysis is more robust as well as usually technically simpler. In our context, the worst-case point of view favored by computer scientists translates to a focus on dominant strategy equilibria that makes no distributional assumptions.[1] While much of the classical work in mechanism design is in the Bayesian setting, this survey will stick to the simpler dominant strategy analysis. For a survey of the growing body of recent work in algorithmic mechanism design that uses Bayesian notions or intermediate ones, see Hartline (2012).

All these limiting choices are made as to allow us to focus on the central issue in the field of algorithmic mechanism design, the very basic clash between incentives and computation. We should perhaps say explicitly that we consider the algorithmic issues to include many facets of the complexity of implementation like communication or description costs.

9.2.2 Our example: multiunit auctions

Even after we make all the restrictions specified above, and after deciding to focus on the clash between incentives and computation, we are still left with quite a large body of work which is hard to survey in an organized manner. The choice we have made is to take a single representative problem and demonstrate all the basic issues just on it. The problem we have chosen for this is "multiunit auctions"—an auction of a multiple homogenous goods. Beyond its importance as of itself, it allows us to go over many of the common themes in algorithmic mechanism design, covering issues that arise when addressing many other problems such as the paradigmatic *combinatorial auctions* covered from various aspects in Cramton et al. (2006), Milgrom (2004), Blumrosen and Nisan (2007), or de Vries and Vohra (2003). There are two main reasons for the choice of this specific problem. The first is its relative combinatorial simplicity: almost all the issues that we deal with are technically simpler in the case of multiunit auctions than they are in most other problems, and thus allow us to focus in the cleanest way on the basic concepts. The second reason is personal: the author has simply worked much on this problem and thus finds it easier to demonstrate general concepts with it. The reader should be warned however that this problem is not the most striking one in practical terms: the computational issues become a real world bottleneck only for very large numbers of items; this is in contrast to other applications such as combinatorial auctions for which

[1] Or, more generally, ex post Nash equilibria, which in independent private-value settings is essentially equivalent to dominant strategy equilibria (see Nisan, 2007).

the computational bottlenecks are biting starting with the modest numbers of items that occur in most practical settings.

So here is the multiunit auction problem:

- We have m identical indivisible units of a single good to be allocated among n strategic players also called *bidders*.
- Each bidder i has a privately known *valuation function* $v_i : \{0, \ldots, m\} \rightarrow \Re^+$, where $v_i(k)$ is the value that this bidder has for receiving k units of the good. We assume that $v_i(0) = 0$ and free disposal: $v_i(k) \leq v_i(k+1)$ (for all $0 \leq k < m$).
- Our goal (as the auctioneer) is to allocate the units among the bidders in a way that optimizes *social welfare*: each bidder i gets m_i units and we aim to maximize $\sum_i v_i(m_i)$, where the feasibility constraint is that $\sum_i m_i \leq m$.

Our goal is to develop a *computationally efficient mechanism* for this problem. *Efficiency* in the computational sense here should imply a protocol that can be realistically implemented even for large numbers of bidders and goods. There may be different quantitative notions of computational-efficiency, but we will see that (for our problem as well as many others) the answer will be quite robust to the exact choice. A *mechanism* here means a protocol that ensures that the required outcome is achieved *when the bidders act rationally*. Again, as mentioned in the problem description, the required outcome is the socially efficient allocation of resources.

There are basically three types of difficulties we need to overcome here:

1. **Representation:** The representation of the "input" to this problem, i.e., the vector of bidders' valuations, contains m numbers. We are interested in large values of m, where communicating or handling such a large amount of data is infeasible. We will thus need to cleverly find more succinct representations.
2. **Algorithmic:** Even for the simplest types of valuations, it turns outs that just computing the optimal allocation is computationally intractable. This will lead us to considering also relaxations of our economic efficiency goals, settling for *approximate* optimization.
3. **Strategic:** In order to achieve our social efficiency goal, the auctioneer would need to know the valuations of the bidders. However, the bidders' valuations are *private information*. So how can the auctioneer access them?

Combinatorial Auctions and Beyond

As we present the issues for our problem of multiunit auctions, we will also shortly point out the corresponding issues for combinatorial auctions. In a *combinatorial auction* we are selling m *heterogenous* items to n bidders. The valuation function of each bidder thus needs to assign a value for each *subset* of the items and so is captured by $2^m - 1$ numbers $v_i(S)$ for every non empty set S of the items (we assume $v(\emptyset) = 0$). Free disposal is usually assumed, i.e., $S \subseteq T$ implies $v_i(S) \leq v_i(T)$. The first goal considered is, as in the multiunit auction case, the socially efficient allocation of resources. These same three issues are the ones that arise in the treatment of combinatorial auctions. As we go along we will point out some early or central references, but for more information on these issues in the context of combinatorial auctions, the reader is referred to the surveys Cramton et al. (2006) or Blumrosen and Nisan (2007). We will also very shortly mention the situation for other algorithmic mechanism design problems.

9.2.3 Where are we going?

Before taking the journey into the analysis of our problem, let us first take a high-level view of the map of this coming journey.

In the first part of the journey—Sections 9.3 and 9.4—we will just be laying the foundations for our representational and algorithmic treatment. These are far from trivial due to the large "input" size. We start this part of the journey by describing the two basic approaches to handling such "large inputs": via *bidding languages* and via *queries*, the most general of which are usefully viewed in terms of *communication*. We then continue by looking at algorithms for the problem, starting with one that it is not computationally efficient. While for *convex settings* a computationally efficient algorithm can be designed, we demonstrate that this is not possible in general. We present two different *intractability proofs* of this last fact, that correspond to the choices made in input representation. At this point we reach the culmination of the algorithmic part of the trip, a computationally efficient algorithm for *approximating* the optimal allocation arbitrarily well.

Up to this point we have not really touched any strategic or incentive issues, and this is what we do in the second part of the journey—Section 9.5—where we study the introduction of *incentives* that will motivate rational bidders to interact with the algorithms of the first part in a way that will lead to the desired results. We start by presenting the classic *VCG mechanisms* which would solve the problem, had there not been any computational concerns. We then show how this solution clashes with the computational constraints, and study in detail the subclass of *maximum-in-range mechanisms* where no such clash exists. For the special *single parameter* case we present a good non-VCG mechanism, and then discuss the difficulties in obtaining such for general *multi parameter* settings. We conclude by demonstrating that *randomization* in mechanisms can help.

9.3. REPRESENTATION

Any process that solves a problem, must certainly get access to the data specifying the problem requirements. In our case this is to the valuation functions of the n players. Now, each valuation function v is specified by m numbers, $v(1), v(2), \ldots, v(m)$. In some cases, where m is of moderate size, these numbers can just be listed, and we have a full representation of the problem (consisting of mn real numbers), but in other cases we may want to consider huge values of m, in which case a full listing of m

numbers is infeasible. Economic examples of this flavor may include auctions of bonds or other financial instruments, where often billions of identical units are sold. Examples in computerized systems may include auctions for communication bandwidth in a computer network where we may imagine selling the billions of communication packets over a communication link in a single second or selling millions of online ad slots of a certain type. This case, of a large value of m, is the one that we deal with here. Alternatively, and essentially equivalently, one may wish to model a huge number of identical items as a single infinitely divisible good, in which case m would just correspond to the precision that we can handle of specifying the fractions of the good.

Before we continue, let us get one hurdle out of the way: the representation of real numbers. In principle we would need infinite precision to represent even a single real number. From such a point of view we can never implement anything dealing with real numbers over any computer system. In practice this is never an issue: the standard 64-bit precision in a standard personal computer is more than enough for representing the world Gross domestic product (GDP) in micro-penny precision. The usual formal treatment in computer science is to allow finite representations in the form of rational numbers and count the number of bits in the representation as part of the input size. Almost equivalently, and as we do here, is to ignore the issue and assume that real numbers are represented directly at the required precision. Usually, and as is the case here, this merely modifies all complexity calculations by a factor that is the number of bits of precision that we require, a modification that does not matter significantly.

So how can we succinctly represent a valuation when m is too large? The simple answer is that we can't: in an information-theoretic sense the valuation cannot be compressed to less than m numbers.[2] However, there are two useful approaches for such a representation: the first one is by means of fixing a "bidding language"—a formalism for succinctly representing "useful" types of valuations, and the second is by means of "queries" that specify the type of interactions possible with the valuation. We will describe these two approaches below.

9.3.1 Bidding languages

The first approach for specifying valuations is to fix some syntactic formalism for doing so. Such a formalism is a "language" where we need to specify the syntax as well as the semantics, the latter being a mapping from strings of characters to valuations. Specifying languages is a standard concept in computer science, and the usual tradeoff is between the dual goals of expressiveness and simplicity. On one hand, we would like to be able to *succinctly* express as many as possible "interesting" types of valuations, and on the other

[2] One may always encode an arbitrary finite number of real numbers by a single real number since $\Re^m$ and $\Re$ have the same cardinality. This however completely breaks down in any finite representation. When working formally in a model of direct representation of real numbers, some technical dimension-preserving condition is taken to rule out such unrealistic encodings. See, e.g., Nisan and Segal (2006).

hand we want to make sure that it is "easy" to handle valuations expressed in such a language. Here "ease" may have multiple meanings, both regarding conceptual ease for humans and regarding technical ease for computers. Here are three common bidding languages, in increasing order of expressiveness:

1. **Single-minded bids:** This language allows only representing "step functions," valuations of the form $v(k) = 0$ for $k < k^*$ and $v(k) = w^*$ for $k \geq k^*$. Clearly to represent such a valuation we only need to specify two numbers: k^* and w^*. Clearly, also, this is a very limited class of valuations.
2. **Step functions:** This language allows specifying an arbitrary sequence of pairs $(k_1, w_1), (k_2, w_2), \ldots, (k_t, w_t)$ with $0 < k_1 < k_2 \cdots < k_t$ and $0 < w_1 < w_2 < \cdots < w_t$. In this formalism $v(k) = w_j$ for the maximum j such that $k \geq k_j$. Thus, for example, the bid $((2, 7), (5, 23))$ would give a value of \$0 to 0 or 1 items, a value of \$7 to 2, 3, or 4 items, and a value of \$23 to 5 or more items. Every valuation can be represented this way, but most valuations will take length m. Simple ones—ones that have only a few "steps"—will be succinctly represented.
3. **Piece-wise linear:** This language allows specifying a sequence of marginal values rather than of values. Specifically, a sequence $(k_1, p_1), (k_2, p_2), \ldots, (k_t, p_t)$ with $0 < k_1 < k_2 \cdots < k_t$ and $p_j \geq 0$ for all j. In this formalism p_j is the marginal value of item k for $k_j \leq k < k_{j+1}$. Thus $v(k) = \sum_{l=1}^{k} u_l$ with $u_l = p_j$ for the largest j such that $l \geq k_j$. In this representation, the bid $((2, 7), (5, 23))$ would give a value of \$7 to 1 item, \$14 to 2 items, \$37 to 3 items, \$60 to 4 items, and \$83 to 5 or more items.

Languages can often be compared to each other in terms of their expressiveness. It is clear that the step function language is more expressive than the single-minded bid language since it contains the latter as a special case but may represent a larger class of valuations. However, we may also say that the Piece-wise linear language is more expressive than the step function language: every valuation expressed as a step function $((k_1, w_1) \ldots (k_t, w_t))$ can be expressed also in the Piece-wise linear form by converting each step (k_j, w_j) into two marginal values: $(k_j, w_j - w_{j-1})$ and $(k_j + 1, 0)$. This only mildly increases the size of expressing the valuation (by a factor of 2), but a similar mild increase would not be possible had we wanted the opposite emulation. The Piece-wise linear bid $(m, 1)$ which represents the valuation $v(k) = k$ may be represented as a step function but only by listing all possible m steps (k, k), blowing up the size of the representation by an unacceptable factor of m.

The choice of a language is somewhat arbitrary and there are various reasonable ways to strike the expressiveness vs. simplicity tradeoff, depending on the application. For maximum bite, we would like to prove positive results for the strongest language that we can handle, while proving negative results for the weakest language for which they still hold. This will be the case for the results we show in this chapter.

Of course, one may easily think of even more expressive (and less simple) languages such as allowing Piece-wise quadratic functions, or arbitrary expressions, etc. The extreme expressive language in this sense is to allow the expression of an arbitrary computation. To be concrete, one may take a general purpose programming language and allow expressing a valuation v by a program in this programming language. While it would seem that it would not be easy to manipulate valuations presented in such general form, we will see below that even this is not hopeless, and in fact there is much we can do even with such "black box" access that provides the value $v(k)$ for any given "query" k.

Combinatorial Auctions

Taking the wider perspective of combinatorial auctions, there is very large space of possible bidding languages, striking various balances between simplicity and power. The simplest "Single-minded bid" language, allows the specification of a value for a single target subset of the items. More complicated variants allow various ways of combining such simple bids. A survey of bidding languages for combinatorial auctions appears in Cramton et al. (2006, Chapter 9).

9.3.2 Query access to the valuations

A more abstract way of representing the input valuations is by treating them as opaque "black boxes." In this approach, we do not concern ourselves with how the valuations are represented but rather only with the interface they provide to the algorithm or mechanism. This has the advantage of leaving significant freedom in the actual representation of the valuation, decoupling the algorithm design from the representation issue. Of course, the choice of interface—the set of "queries" that the black box needs to answer—is important here. If this interface is too weak then we will have a hard time constructing computationally efficient algorithms using this weak set of queries while if the set of allowed queries is too strong then we will have difficulties in implementing them with realistic representations. nevertheless, even unrealistically strong sets of queries are interesting for proving "lower bounds"—limits on what computationally efficient algorithms or mechanisms may achieve: any limitation of mechanisms or algorithms that are allowed to use even strong unrealistic queries will certainly apply also to mechanisms that only use more realistic ones. In general, one usually attempts using the strongest possible set of queries when proving "lower bounds" and attempts using the weakest possible set of queries when proving "upper bounds"—actual algorithms or mechanisms. This is what we will also do here, use the realistic and weak "value queries" for upper bounds and use the unrealistically strong and general "communication queries" for lower bounds.

9.3.2.1 Value queries

A *value query* to a valuation $v : \{0, \ldots, m\} \rightarrow \Re_+$ gets a number of items k and simply returns the value $v(k)$.

The requirement that a certain bidding language be able to answer value queries essentially means that it has an effective implementation. This is, that there is a computationally efficient algorithm that takes as input is a valuation represented in the language as well as a query k and produces as output the value $v(k)$. That the algorithm is computationally efficient means that it runs in time that is polynomial in its input size, i.e., in the *representation length*. Note that all the bidding languages suggested in the previous section can indeed answer such queries easily.

Take for instance the strongest one suggested above, the Piece-wise linear bidding language. Given a valuation v represented in this language as $((k_1, p_1), (k_2, p_2), \ldots, (k_t, p_t))$ and a query k, how can we efficiently compute the value of $v(k)$? Directly applying the definition of Piece-wise linear valuations requires summing the k marginal values, an operation whose running time may be significantly larger than the input size that is proportional to t (rather than k which may be as big as the number of items m). However, a quick inspection will reveal that one can replace the repeated addition of equal marginal values by multiplication, calculating $v(k) = \sum_{j=1}^{l-1} k_j p_j + (k - \sum_{j=1}^{l-1} k_j) \cdot p_l$, where l is the largest integer such that $\sum_{j=1}^{l-1} k_j \leq k$. This now takes time that is polynomial in the input size, as required.

9.3.2.2 General communication queries

The strongest imaginable set of queries to a valuation is to simply allow all possible queries. This means that we could allow all functions on the set of valuations.

Even in this general setting two key restrictions remain: the first is that a query may only address a single valuation function and may not combine several. Thus for example a query "find the odd k that maximized $v_i(k) - k^2$" would be allowed, but "find the k that maximizes $v_1(k) + v_2(m - k)$" can not be a single query since it combines two separate valuations. The second is that the reply from a query needs to be small, we shouldn't for example be able to ask for an extremely long string that encodes the whole valuation. To be completely precise we should require that the reply to a query is always a single bit (i.e., a yes/no answer), but equivalently we can allow arbitrary answers, counting the number of answer bits instead of the number of queries.

When thinking about an algorithm or mechanism that makes such general queries to the valuations, it is best to think about it in the completely equivalent model of *communication*. In such a model the algorithm is really a protocol for communication between the valuation black boxes which we now think of as the bidders themselves. An algorithmic query to a valuation is interpreted as the protocol specifying that the appropriate bidder communicate (to the algorithm) the answer to the query—an answer that obviously depends on the valuation that the bidder has, but not on any other information (beyond the state of the protocol). Under this interpretation, the total number of queries that an algorithm makes is exactly the total number of communicated "numbers" made by the protocol.

Combinatorial Auctions

For combinatorial auctions, there are a variety of natural specific query models, the simplest of which is that of value queries, and the strongest of which is the communication model. Another very natural query model is that of "demand queries," where a query specifies a price vector $(p_1, \ldots, p_m)$ and the response is the demanded set of items under these prices, i.e., a set S that maximizes the utility $v(S) - \sum_{i \in S} p_i$. A systematic discussion of query models for combinatorial auctions appears in Blumrosen and Nisan (2010), while a discussion of communication appears in Nisan and Segal (2006).

9.4. ALGORITHMS

By now we have only discussed how our "input" may be represented or accessed by the algorithm or mechanism. At this point we can start discussing the algorithmic challenge of optimally allocating the items:

The Allocation Problem

Input: The sequence of valuations of the players: $v_1, \ldots, v_n$

Output: An allocation $m_1, \ldots, m_n$ with $\sum_{i=1}^{n} m_i \leq m$ that maximizes $\sum_{i=1}^{n} v_i(m_i)$

Combinatorial Auctions

For the case of combinatorial auctions, the valuations v_i provide a value for each subset of the items, and the required output is an allocation $(S_1 \ldots S_n)$ that maximizes $\sum_i v_i(S_i)$, where the constraint is that $S_i \cap S_j = \emptyset$ for $i \neq j$.

9.4.1 Algorithmic efficiency

Our basic requirement from the algorithm is that it be "computationally efficient." Specifically, for every input we would want the algorithm to provide an output "quickly." As usual, the running time is a function of the input size parameters n and m. Also as usual, the running time may depend on the precise definition of the model of computation, but this dependence is very minor, and the running time will be almost the same under all reasonable definitions. Thus, as usual, we will just formally require that the running time is polynomial in the *input size*. In all cases in this survey a polynomial running time will actually have reasonably small degrees and small hidden constants while a non-polynomial running time will actually be exponential.

In the case where we are dealing with the input presented in a certain bidding language the interpretation of this is the completely standard one: we want an algorithm whose running time is polynomial in the sum of lengths of the representations of the input valuations. In the case where the input valuations are given by query access to "black boxes" there is no explicit representation of the input, so what we will require is that the running time is polynomial in the numbers of bidders, n, and the number of bits required to represent a basic quantity in the system: $\log m$ bits to represent a number of items, and the maximum number of bits of precision of a value, denoted by t.

Let us reflect further on why we do not view a running time that is polynomial in m as computationally efficient. First, consider that the black box query model is intended to give an abstract way of considering a wide class of representations. For none of these representations do we envision its size to be "trivial"—of an order of magnitude similar to m—but rather we expect it to be significantly smaller, and thus a run time that is polynomial in m will not be polynomial in the input size under this representation. Second, consider the query model as an abstraction of some protocol between the players. A run time that is polynomial in m allows sending complete valuations, completely losing the point of the abstraction. Third, it is not uncommon to have multiunit auctions of millions or billions of items (e.g., treasury bonds or web ad impressions). While all of the algorithms and mechanisms presented below that have a run time that is polynomial in $\log m$ and n can be practically applied in these cases, algorithms whose run time is polynomial in m will probably be too slow.

Let us start with a *non-computationally efficient* but still not-trivial dynamic programming[3] algorithm for solving this problem. We will present the algorithm in the value-queries model. Presenting it in this weak model ensures that it applies also to all the other models presented above. The basic idea is to use dynamic programming to fill up a table that for any $0 \leq i \leq n$ and $0 \leq k \leq m$ holds the optimal total value that can be achieved by allocating k items among the first i bidders. An entry in this table can be filled using the values of previously filled entries, and this information suffices for actually reconstructing the optimal allocation itself.

General Allocation Algorithm

1. Fill the $(n+1) * (m+1)$ table s, where $s(i,k)$ is the maximum value achievable by allocating k items among the first i bidders:
   (a) For all $0 \leq k \leq m$: $s(0,k) = 0$
   (b) For all $0 \leq i \leq n$: $s(i,0) = 0$
   (c) For $0 < i \leq n$ and $0 < k \leq m$: $s(i,k) = max_{0 \leq j \leq k}[v_i(j) + s(i-1, k-j)]$
2. The total value of the optimal allocation is now stored in $s(n,m)$
3. To calculate the m_i's themselves, start with $k = m$ and for $i = n$ down to 1 do:
   (a) Let m_i be the value of j that achieved $s(i,k) = v_i(j) + s(i-1, k-j)$
   (b) $k = k - m_i$

---

[3] Dynamic programming is a common algorithmic paradigm that builds up a table of partial results, each of which is computed using previous ones.

The running time of this algorithm is $O(nm^2)$: the calculation of each of the mn cells in the table s requires going over at most m different possibilities. Note that while the dependence of the running time on n is acceptable, the dependence on m is not so since we required a polynomial dependence on $\log m$. Thus, we do not consider this algorithm to be computationally-efficient and indeed it would likely not be practical when handling even as few as a million items. We have thus proved:

Theorem 9.1. *There exists an algorithm for the multiunit allocation problem (in any of the models we considered) whose running time is polynomial in n and m (rather than in $\log m$).*

Combinatorial Auctions

In the case of combinatorial auctions, the valuations hold information that is exponential in m. In this case computationally efficient algorithms are allowed a running time that is polynomial in both n and m, rather than grow exponentially in m. Mirroring the case for multiunit auctions, it is not difficult to come up with a dynamic programming algorithm for combinatorial auctions whose running time is polynomial in n and 2^m.

9.4.2 Downward sloping valuations

At this point, it would perhaps be a good idea to see an example of a computationally efficient algorithm. We can provide such an algorithm for the special case of "downward sloping valuations," those that exhibit diminishing marginal values, i.e., that satisfy $v_i(k+1) - v_i(k) \leq v_i(k) - v_i(k-1)$ for all bidders i and number of items $1 \leq k \leq m-1$. This is the case of a "convex economy," a case which turns out to be algorithmically easy. A simple intuitive algorithm will allocate the m items one after another in a greedy fashion, at each point giving it to the player that has highest marginal value for it (given what it was already allocated). As stated, this algorithm takes time that increases linearly in m rather than in $\log m$ as desired, but its simple greedy nature allows us to use binary search[4] to convert it to a computationally efficient one.

The idea here is that with downward sloping valuations we have a market equilibrium: a clearing price p, and an allocation $(m_1, \ldots, m_n)$ with $\sum m_i = m$ that has the property that for each i, m_i is i's demand under price p, i.e., that $v_i(m_i) - v_i(m_{i-1}) \geq p > v_i(m_{i+1}) - v_i(m_i)$ which implies that $v_i(m_i) - m_i p \geq v_i(k) - kp$ for all k. *The First Welfare Theorem* in this context states that such a market equilibrium must maximize $\sum_i v_i(m_i)$, since for any other allocation $(k_1, \ldots, k_n)$ with $\sum k_i \leq m$ we have that $\sum_i v_i(m_i) - mp = \sum_i (v_i(m_i) - m_i p) \geq \sum_i (v_i(k_i) - k_i p) \geq \sum_i v_i(k_i) - mp$.

[4] Binary search is a common algorithmic paradigm used to find a target value within a given range by repeatedly splitting into two subranges and determining which one contains the desired target. The key point is that this process requires only time that is only logarithmic in the range size.

Algorithmically, given a potential price p, we can calculate players' demands using binary search to find the point where the marginal value decreases below p. This allows us to calculate the total demand for a price p, determining whether it is too low or too high, and thus search for the right price p using binary search. For clarity of exposition we will assume below that all values of items are distinct, $v_i(k) \neq v_{i'}(k')$ whenever $(i, k) \neq (i', k')$. (This is without loss of generality since we can always add values $\epsilon_{i,k}$ to $v_i(k)$ to achieve this.) This assumption also makes it evident that a clearing price p exists: for the proof imagine starting with price $p = 0$ and then increasing p gradually. Since the total demand starts high and decreases by a single unit at each price-point $v_i(k)$, at some point it will reach exactly m units and the market will clear.

Allocation Algorithm for Downward Sloping Valuations

1. Using binary search, find a clearing price p in the range $[0, V]$, where $V = max_i[v_i(1)]$:
   a. For each $1 \leq i \leq n$, use binary search over the range $\{0, 1, \ldots, m\}$, to find m_i such that $v_i(m_i) - v_i(m_{i-1}) \geq p > v_i(m_{i+1}) - v_i(m_i)$
   b. If $\sum_i m_i > m$ then p is too low; if $\sum_i m_i < m$ then p is too high, otherwise we have found the right p
2. The optimal allocation is given by $(m_1, \ldots, m_n)$

The important point is that this algorithm is computationally efficient: the binary search for p requires only t rounds, where t is the number of bits of precision used to represent a value. Similarly the binary search for m_i requires only $\log m$ rounds. We have thus shown:

Theorem 9.2. *There exists an algorithm for the multiunit allocation problem among downward sloping valuations (in any of the models we considered) whose running time is polynomial in n, $\log m$, and the number of bits of precision.*

Combinatorial Auctions

In the case of combinatorial auctions, there is a corresponding class of "convex" valuations, called "substitutes" valuations. For this class, equilibrium prices are known to exist (Gul and Stacchetti, 1999) and the optimal allocation among such valuations can be found computationally efficiently (Blumrosen and Nisan, 2010; Gul and Stacchetti, 2000).

9.4.3 Intractability

Now the natural question is whether one can get a computationally efficient algorithm even for the general case, not just the downward sloping one. It turns out that it is quite easy to show that this is impossible in the value queries model.

Consider such a hypothetical algorithm even just for the special case of two players. For any two valuations v_1 and v_2 it outputs the optimal allocation (m_1, m_2). Now let us follow this algorithm when it accepts as input the two linear valuations $v_1(k) = k$ and $v_2(k) = k$ for all k. The algorithm keeps querying the two valuations at various values and for each query k gets as a response the value k. After some limited amount of queries it must produce an answer (m_1, m_2). Now the critical thing is that if the algorithm was computationally efficient and thus in particular made less than $2m - 2$ queries then it had to produce the answer without querying some value (even excluding m_1, m_2 which it may or may not have queried). Without loss of generality let us say that it never queried $v_1(z)$ (with $z \notin \{m_1, m_2\}$). Now we may consider a different input where we increase the value of $v_1(z)$ by one, i.e., $v'_1(k) = k$ except for $v'_1(z) = z + 1$. Note that this v'_1 is still a proper valuation, as monotonicity was maintained. Note also that now the unique optimal allocation is $(z, m - z)$ which gives value $v'_1(z) + v_2(m - z) = m + 1$, while any other allocation only gives a total value of m. Our poor algorithm, however, has no way to know that we are now in the case of (v'_1, v_2) rather than (v_1, v_2) since the only difference between the two cases is for the query z, a query which it did not make, so it will provide the same answer (m_1, m_2) as before, an answer which is not optimal. We have thus shown:

Theorem 9.3. *Every algorithm for the multiunit allocation problem, in the value-queries model, needs to make at least* $2m - 2$ *queries for some input.*

This argument concludes the impossibility result for the weakest black-box model, the one that only allows value queries. It is already quite interesting, exhibiting a significant computational gap between the "convex economy" case where a computationally efficient algorithm exists and the general case for which it does not. It turns out that the same impossibility result is true even in stronger (and more interesting) models. We now bring two different intractability results, one for the case where the valuations are given in a bidding language—even the weakest possible such language, and the other one for query models—even for the strongest one that allows communication queries.

9.4.3.1 NP-Completeness

Let us consider the simplest possible bidding language, that of single-minded bids. In this language each valuation v_i is represented as (k_i, p_i) with the meaning that for $k < k^*$ we have $v_i(k) = 0$ and for $k \geq k_i$ we have $v_i(k) = p_i$. The optimal allocation will thus need to choose some subset of the players $S \subseteq \{1 \ldots n\}$ which it would satisfy. This would give total value $\sum_{i \in S} p_i$ with the constraint that $\sum_{i \in S} k_i \leq m$. This optimization problem is a very well known problem called the knapsack problem, and it is one of a family of problems known to be "NP-complete." In particular, no computationally efficient algorithm exists for solving the knapsack problem unless "P=NP," which is generally conjectured not to be the case (the standard reference is still Garey and Johnson, 1979). It follows that no computationally efficient algorithm exists for our problem when the input valuations are presented in any of the bidding languages mentioned above all of which contain the single-minded one as special cases. We thus have:

Theorem 9.4. *Assuming $P \neq NP$, there is no polynomial-time algorithm for the multi-unit allocation problem, in any of the bidding language models.*

We should mention that this intractability result does rely on the unproven assumption of $P \neq NP$. On the other hand the allocation problem—for any effective bidding language—is clearly an optimization variant of an NP problem. Thus, if it turns out that $P = NP$ then a computationally efficient allocation algorithm does exists.

9.4.3.2 Communication complexity

When the valuations are given in any effective bidding language, our allocation problem is a "normal" computational problem and we cannot hope for unconditional answers as our problem is equivalent to the $P = NP$ problem. However, when the input to our problem is given as black boxes, we are really looking for algorithms whose running time should be sublinear in "true" input size, and for this we have techniques for proving unconditional impossibility results. In particular, we have already seen that there is no sublinear algorithm that uses only value queries to access the input valuations. Here we extend this result and show that it holds even when allowing arbitrary queries to each of the input valuations, i.e., in the communication model. The "proper way" of proving this is by using the well known results from the theory of communication complexity (surveyed in Kushilevitz and Nisan, 1997) and deducing from them impossibilities in this situation (as was done in Nisan and Segal, 2006). Here we give, however, a self-contained treatment that uses the simplest technique from the field of communication complexity ("fooling sets") directly on our problem.

Let us consider the following set of valuations: for every set $S \subseteq \{1, \ldots m-1,\}$ let $v_S(k) = k + 1_{k \in S}$, i.e., for $k \notin S$ we have $v_S(k) = k$ while for $k \in S$ we get one more, $v_S(k) = k + 1$. Now let us consider a two-player situation when player 1 gets v_S and player 2 gets the "dual" valuation v_{S^*}, where $S^* = \{0 < k < m | m - k \notin S\}$. The dual valuation was defined so that any allocation between v_S and v_{S^*} would have only value exactly $v_S(k) + v_{S^*}(m-k) = m$. Now there are 2^{m-1} possible pairs of input valuations (v_S, v_{S^*}) that we are considering, and the main point of the proof will show that for no two of these pairs can the algorithm see exactly the same sequence of answers from the two black-boxes (i.e., in terms of communication, exactly the same sequence of bits communicated between the two parties). It follows that the total number of possible sequences of answers is at least 2^{m-1} and thus at least one of them requires at least $m-1$ bits to write down. As each answer to a query is "short"—say at most t bits, for some small t—the total number of queries made by the algorithm must be at least $(m-1)/t$ for at least one of these input pairs (in fact, for most of them), and this is a lower bound on the number of queries made by the algorithm (communication protocol) and thus on its running time.

We are left with showing that indeed for no two pairs of valuations (v_S, v_{S^*}) and (v_T, v_{T^*}) with $S \neq T$ the algorithm cannot receive the same sequence of answers. The main observation is that, had this been the case, then the algorithm would also see the same sequence of answers on the "mixed" pairs (v_S, v_{T^*}) and (v_T, v_{S^*}). Consider the algorithm when running on the input pair (v_S, v_{T^*}). It starts by making some query, a query which is also the first query it made on both (v_S, v_{S^*}) and (v_T, v_{T^*}). If it queries the first player then it gets the same answer as it did on input (v_S, v_{S^*}) (since it has the same valuation for the first player) while if it queries the second player then it gets the same answer as in (v_T, v_{T^*}). In both cases it gets the same information as in the original cases (v_S, v_{S^*}) and (v_T, v_{T^*}) and thus continues to make the next query exactly as it has made the next query in those cases. Again, it must get the same answer for this new query, and as it continues getting the same answers and thus making the same queries as in the original cases, it never observes a deviation from these two cases and thus ends up with the same final outcome as it did in the original cases. The same is true for the other mixed pair (v_T, v_{S^*}) for which it gives again the same final outcome. This however contradicts the optimality of the algorithm since an outcome allocation that is optimal for (v_S, v_{T^*}) is never optimal for (v_T, v_{S^*}). To see this fact just note that since $S \neq T$ there exists some k in the symmetric difference, say, wlog, $k \in S - T$. In this case notice that $v_S(k) + v_{T^*}(m-k) = m+1$ and thus the optimal allocation would give at least (actually, exactly) this total value. However, any k which gives an optimal allocation for (v_S, v_{T^*}), i.e., $v_S(k) + v_{T^*}(m-k) = m+1$ would give a suboptimal allocation for (v_T, v_{S^*}), since $v_S(k) + v_{S^*}(m-k) + v_T(k) + v_{T^*}(m-k) = 2m$, and thus we would have $v_T(k) + v_{S^*}(m-k) = m-1$. We have thus shown:

Theorem 9.5. *(Nisan and Segal, 2006)Every algorithm for the multiunit allocation problem, in any query model, needs to ask queries whose total answer length is at least $m-1$ in the worst case.*

The argument presented above also holds in the stronger model where the algorithm is given "for free" the optimal allocation and only needs to *verify* it, a model known as "non-deterministic protocols". This can be viewed as a lower bound on the "number of prices" needed to support an equilibrium. Informally, and taking a very broad generalization, a generalized equilibrium specifies some "generalized pricing information" for a market situation, information that allows each player to *independently of the other players* decide on the set of goods that is allocated to him. The above argument then actually provides a lower bound on the number "generalized prices" required in any general notion of equilibrium. See Nisan and Segal (2006) for more details.

Combinatorial Auctions

Going back again to combinatorial auctions, similar NP-hardness results are known for bidding language models (Sandholm, 1999) as well as query and communication models

(Nisan and Segal, 2006), showing that a running time that is polynomial in m is not possible.

9.4.4 Approximation

At this point we seem to have very strong and general impossibility results for algorithms attempting to find the optimal allocation. However, it turns out that if we just slightly relax the optimality requirement and settle for "approximate optimality" then we are in much better shape. Specifically, an *approximation scheme* for an optimization problem is an algorithm that receives, in addition to the input of the optimization problem, a parameter ϵ which specifies how close to optimal do we want to solution to be. For a maximization optimization problem (like our allocation problem) this would mean a solution whose value is at least $1 - \epsilon$ times the value of the optimal solution. It turns out that this is possible to do, computationally efficiently, for our problem using dynamic programming.

The basic idea is that there exists an optimal algorithm whose running time is polynomial in the *range of values*. (This is a different algorithm than the one presented in Section 9.4.1 whose running time was polynomial in the number of items m.) Our approximation algorithm will thus truncate all values $v_i(k)$ so that they become small integer multiples of some δ and run this optimal algorithm on the truncated values. Choosing $\delta = \epsilon V/n$ where $V = \max_i[v_i(m)]$ strikes the following balance: on one hand, all values are truncated to an integer multiple $w\delta$ for an integer $0 \leq w \leq n/\epsilon$, and so the range of the total value is at most the sum of n such terms, i.e., at most n^2/ϵ. On the other hand, the total additive error that we make is at most $n\delta \leq \epsilon V$. Since V is certainly a lower bound on the total value of the optimal allocation this implies that the fraction of total value that we lose is at most $(\epsilon V)/V = \epsilon$ as required.

The allocation algorithm among the truncated valuations uses dynamic programming, and fills a $(n+1) * (W+1)$-size table K, where $W = n^2/\epsilon$. The entry $K(i, w)$ holds the minimum number of items that, when allocated optimally between the first i players, yields total value of at least $w\delta$. Each entry in the table can be computed efficiently from the previous ones, and once the table is filled we can recover the actual allocation.

Approximation Scheme

1. Fix $\delta = \epsilon V/n$ where $V = max_i[v_i(m)]$, and fix $W = n^2/\epsilon$
2. Fill an $(n+1) * (W+1)$ table K, where $K(i, w)$ holds the minimum number of items that, when allocated optimally between the first i players, yields total value of at least $w\delta$:
   (a) For all $0 \leq i \leq n$: $K(i, 0) = 0$
   (b) For all $w = 1, \ldots, W$: $K(0, w) = \infty$

(c) For all $i = 1, \ldots, n$ and $w = 1, \ldots, W$ compute $K(i, w) = min_{0 \leq j \leq w} val(j)$, for $val(j)$ defined by:
   i. Use binary search to find minimum number of items k such that $v_i(k) \geq j\delta$
   ii. $val(j) = k + K(i-1, w-j)$
3. Let w be the maximum index such that $K(n, w) \leq m$
4. For i going from n down to 1:
   (a) Let j be so that $K(i, w) = val(j)$ (as computed above)
   (b) Let m_i be the minimum value of k such that $v_i(k) \geq j\delta$
   (c) Let $w = w - j$

Our algorithm's running time is dominated by the time required to fill the table K which is of size $n \times n^2/\epsilon$ where filling each entry in the table requires going over all possible n/ϵ values j and for each performing a binary search that takes $O(\log m)$ queries. Thus the total running time is polynomial in the input size as well as in the precision parameter ϵ which is often called a fully polynomial time approximation scheme. For most realistic optimization purposes such an arbitrarily good approximation is essentially as good as an optimal solution. We have thus obtained:

Theorem 9.6. *There exists an algorithm for the approximate multiunit allocation problem (in any of the models we considered) whose running time is polynomial in n, $\log m$, and ϵ^{-1}. It produces an allocation $(m_1, \ldots, m_n)$ that satisfies $\sum_i v_i(m_i) \geq (1 - \epsilon) \sum_i v_i(opt_i)$, where $(opt_1, \ldots, opt_n)$ is the optimal allocation.*

Combinatorial Auctions

When going to the wider class of combinatorial auctions, such a good approximation scheme is not possible. The best possible polynomial time approximations lose a factor of $O(\sqrt{m})$ and no better is possible. Significant work has been done on obtaining improved approximation factors for various subclasses of valuations. See Blumrosen and Nisan (2007) for survey and references.

9.5. PAYMENTS, INCENTIVES, AND MECHANISMS

Up to this point we have treated the bidders in the auction as doing no more than providing their inputs to the allocation algorithm. A realistic point of view would consider them as economic agents that act rationally to maximize their own profits. To formalize such a setting we would need to make assumption about what exactly is it that the players are aiming to maximize. The simplest, most natural, such model will having bidder i rationally maximize his net utility: $u_i = v_i(m_i) - p_i$ where m_i is the number of units he is allocated and p_i is the price that he is asked to pay for these items. Notice how we already went beyond the pure algorithmic question that only involved the allocation $(m_1 \ldots m_n)$ and went into an economic setting which explicitly deals with the payments $(p_1 \ldots p_n)$ for these items. The assumption that bidders make payments p_i and that bidders' utilities (which each aims to maximize) are the simple differences $u_i = v_i(m_i) - p_i$ (rather than some more complicated function of m_i and p_i) is often called "quasi-linear utilities" and is essentially equivalent to the existence of a common currency that allows comparison and transfer of utilities between players.

At this point we will apply the basic notions of mechanism design and define a *(direct revelation) mechanism* for our setting: this is just an algorithm that is adorned with payment information. In our case the input to a mechanism will be the n-tuple of valuations $(v_1, \ldots, v_n)$ and the output will consist of both the allocation $(m_1, \ldots, m_n)$ and the vector of payments $(p_1, \ldots, p_m)$. So here is the strategic setup that we are considering: the auctioneer gets to specify a mechanism, which is public knowledge ("power of commitment"). Each bidder i *privately* knows his valuation functions v_i. The bidders then interact with the mechanism transmitting to it their "bids" and are hence put in a game whose outcome is the resulting allocation and payment. In this game the players are acting strategically as to optimize their resulting utility. In particular the "bids" that they will report to the mechanism need not necessarily be their true valuations but rather whatever optimizes their utility. The *equilibrium* outcome of a given mechanism for the input valuations $(v_1, \ldots, v_n)$ will be its allocation $(m_1, \ldots, m_n)$ and payments $(p_1, \ldots, p_n)$ as calculated for the equilibrium bids (rather than as calculated for the true valuations).

At this point we need to further specify our equilibrium notion which will of course depend on our informational assumptions: what do players know about each other's valuation and behavior. There are two main notions in the literature here. The Bayesian setup assumes that bidder's valuations v_i come from some known distribution F on valuation tuples. The players (and mechanism designer) have common knowledge of F and of each others' rationality, but each player i only knows the realization of his own v_i and not the realization of the other v_j's. The second notion is more robust as well as technically simpler, and makes a very strong requirement: that the equilibrium be robust to any information that players may have. This is the definition that we will use here.

Notations: A mechanism $M = (m(\cdot), p(\cdot))$ accepts as input an n-tuple of valuations $(v_1, \ldots, v_n)$ and outputs an allocation $(m_1, \ldots, m_n)$ and payments $(p_1, \ldots, p_n)$. Both the allocation and payments are functions of the n-tuple of v_i's and are denoted using $m_i(v_1, \ldots, v_n)$ and $p_i(v_1, \ldots, v_i)$. We will also use the abbreviation (v_{-i}, w_i) to denote the n-tuple of valuations where i's valuation is w_i and j's valuation for $j \neq i$ is v_j.

Definition: We say that a mechanism $M = (m(\cdot), p(\cdot))$ is *truthful* (in dominant strategies) if for every $(v_1, \ldots, v_n)$, every i, and every $\tilde{v}_i$ we have that $u_i \geq u_i'$ where $u_i = v_i(m_i(v_i, v_{-i})) - p_i(v_i, v_{-i})$ and $u_i' = v_i(m_i(\tilde{v}_i, v_{-i})) - p_i(\tilde{v}_i, v_{-i})$. Truthful mechanisms are equivalently called *strategy-proof* or *incentive compatible*.

What this means is that in no case will a player have any motivation to report to the mechanism anything other than his true valuation v_i; any other bid $\tilde{v}_i$ that he may attempt to report instead can never give him higher utility. For such mechanisms we may realistically assume that all bidders will indeed report their private valuations truthfully to the mechanism who may then come up with a good outcome.

This requirement may seem quite excessive and even unrealistic. First, the informational assumptions are quite strict: players strategic behavior does not depend on any information about the others, i.e., strategically all players have dominant strategies. Second, this dominant strategy is not arbitrary but is to report their true valuations. While the first assumption has significant bite and is what gives the high robustness to this notion relative to Bayesian notions mentioned above, the second point turns out to be without loss of generality. One can convert any mechanism with dominant strategies to an equivalent incentive compatible one by internalizing the dominant strategies into the mechanism. This is also true for general mechanisms with arbitrary protocols (not just getting the v_i's as input): as long as dominant strategies exist they can be incorporated into the mechanism, with the result being a truthful mechanism; this is known as the "revelation principle."

9.5.1 Vickrey-Clarke-Groves mechanisms

The basic general positive result of mechanism design applies to our situation. This result, originally suggested by Vickrey (1961) and then generalized further by Clarke (1971) and Groves (1973), elucidates the following simple idea: if we charge each participant an amount that is equal to his *externality* on the others then we effectively align all users' interests with the total social welfare, and thus ensure that they behave truthfully as long as our mechanism maximizes social welfare. Specifically, here is this mechanism applied to multiunit auctions:

VCG Mechanism

1. Each player reports a valuation $\tilde{v}_i$
2. The mechanism chooses the allocation $(m_1, \ldots, m_n)$ that maximizes $\sum_j \tilde{v}_j(m_j)$ and outputs it
3. For each player i:
   (a) The mechanism finds the allocation $(m'_1, \ldots, m'_n)$ that maximizes $\sum_{j \neq i} \tilde{v}_j(m'_j)$
   (b) Player i pays $p_i = \sum_{j \neq i} \tilde{v}_j(m'_j) - \sum_{j \neq i} \tilde{v}_j(m_j)$

The main property of this mechanism is that it is incentive compatible. To see this, first notice that the term $\sum_{j \neq i} \tilde{v}_j(m'_j)$ in the calculation of p_i has no strategic implications since its value does not depend in any way on the valuation $\tilde{v}_i$ reported by i and thus it does not effect in any way i's incentives (for any fixed value of v_{-i}). So to analyze i's strategic behavior we can simply assume that the first term is 0, i.e., that he is getting *paid* the amount $\sum_{j \neq i} \tilde{v}_j(m_j)$ and thus trying to maximize $v_i(m_i) + \sum_{j \neq i} \tilde{v}_j(m_j)$. However, notice that as long as i indeed reports $\tilde{v}_i = v_i$ the mechanism is performing exactly this optimization! Thus i can not do better than by being truthful. The reader may wonder why is the term $\sum_{j \neq i} \tilde{v}_j(m'_j)$ used at all in the calculation of payments since, if we only care about truthfulness, it can be eliminated or replaced by an arbitrary expression that does not depend on v_i. However, this term is usually added as it ensures that the payments are always "in the right ballpark," satisfying "no positive transfers," $p_i \geq 0$, as well as "individual rationality," $v_i - p_i \geq 0$. We have thus proved:

Theorem 9.7. *(Groves, 1973; Vickrey, 1961) There exists a truthful mechanism for the multiunit allocation problem.*

The central problem with this mechanism is its computational efficiency. The allocation algorithm it uses is supposed to find the optimal allocation, and this we have already seen is impossible to do computationally in an efficient way. Now, for the special case of downward sloping valuations we did see, in Section 9.4.2, a computationally efficient optimal allocation algorithm, and thus for this special case we get a truthful computationally efficient mechanism. This is the case that was dealt with in the seminal paper Vickrey (1961). This however does not hold for the general problem. The basic difficulty in the field of algorithmic mechanism design is to *simultaneously* achieve both incentive compatibility and computational efficiency. We will now continue trying to do so for multiunit auctions.

9.5.2 The clash between approximation and incentives

As we have already seen in Section 9.4.3 that no computationally efficient algorithm can always find the optimal allocation, adding yet another requirement—that of truthfulness—would seem totally hopeless. However, we have also seen in Section 9.4.4 that one may computationally efficiently obtain approximately optimal allocations, and we could similarly hope for arbitrarily good *truthful approximation mechanisms*. Specifically, we would desire a truthful computationally efficient mechanism that for every given $\epsilon > 0$ produces an allocation whose total value is at least $1 - \epsilon$ fraction of the optimal one. We may even settle for weaker approximation guarantees, settling for an allocation whose total value is fraction α, for as large as possible fixed value of α.

The natural approach to this goal would be to simply replace in the previous algorithm the step of "find an allocation that maximizes $\sum_j \tilde{v}_j(m_j)$" by "find an allocation that *approximately* maximizes $\sum_j \tilde{v}_j(m_j)$." This we would do for all $n + 1$ optimizations problems encountered by the mechanism: the global optimization to determine the allocation, as well as the n personalized ones, needed to compute the payments of the

n bidders. We can call such a mechanism *VCG-based*. It would seem natural to assume that this would maintain truthfulness, but this is not the case.

Let us present a simple example of an approximation algorithm where if we use its results instead of the optimal ones we would lose truthfulness. Consider the allocation of two items between two players, both of whose valuations are: $v(1) = 1.9$ and $v(2) = 3.1$. The optimal allocation would allocate one item to each bidder, giving each a value of 1.9, and requesting from each a payment of $1.2 = 3.1 - 1.9$ which is the difference for the other between getting a single item or both. Now consider the case where we use an approximation algorithm that first truncates the values down to integer values and then finds the optimal allocation. Let us first consider the case where both players bid truthfully, $(1.9, 3.1)$. The approximation algorithm would truncate these values to $v(1) = 1$ and $v(2) = 3$, for which the optimal allocation is to give both items to one of the players. His payment will be the loss of the other player from not getting both items which is 3.1, giving our player zero (real) utility. Had our player been non-truthful bidding $\tilde{v}(1) = \tilde{v}(2) = 3$ then the approximation algorithm would choose the allocation that assigns a single item to each player, charging our non-truthful one $1.2 = 3.1 - 1.9$ giving him a net positive utility. (This calculation used the reported (unrounded) values for calculations of payments; using the rounded values will not restore truthfulness either.)

This example is in no way unusual. Not only is such a truncation algorithm essentially the approximation algorithm of Section 9.4.4, but also it turns out that essentially *every* approximation algorithm will exhibit this non-truthfulness if we just plug it into the VCG payment calculation (Dobzinski and Nisan, 2007; Nisan and Ronen, 2007). There is a small, almost trivial, set of approximation algorithms for which this is not the case and we study them in the next section.

Combinatorial Auctions and Beyond

Taking a wider look at the whole field of algorithmic game theory, this clash between incentives and computation, is a main occurring theme. For many other algorithmic mechanism design problems we have reasonable approximation algorithms even though finding the optimal solution is not computationally feasible. In all of these cases, we are not able to apply the VCG technique as to obtain a computationally efficient truthful mechanism. The central problem of this form is combinatorial auctions that has been the focal point such investigations since the early Lehmann et al. (2002). There is yet another well-studied type of problems in algorithmic mechanism design where the main challenge is the lack of applicability of VCG mechanisms. These are problems where the optimization goal itself is not the maximization of social welfare, but rather some other computational goal. A central problem of this form is that of machine scheduling where the bidders correspond to machines that can execute jobs for a fee and a standard goal is the minimization of the the "makespan"—time that the last job is finished (Nisan and Ronen, 2001).

At this point one may naturally attempt to deal with the non-truthfulness that approximation algorithms cause by slightly relaxing the notion of truthfulness. One approach would be to require only "approximate truthfulness" in some sense. A major problem with this approach is that players' utilities and payments are *differences* between quantities. A good approximation of each quantity does not imply a good approximation of the difference.[5] Another relaxation approach would be to notice that lying is worthwhile only if it improves the overall allocation, something which is supposedly computationally hard. However, it is not completely clear how to define such "bounded-computational"-truthfulness, sensibly quantifying over different running times of different algorithms on different inputs. None of these approaches seems to be "clean" enough in general, so we will continue with the basic definition of exact truthfulness and continue studying whether it may be achieved.

9.5.3 Maximum-in-range mechanisms

The optimal VCG mechanism requires maximization of $\sum_i v_i(m_i)$ over all possible allocations $(m_1, \ldots, m_n)$. One can think about a "restricted" version of VCG which does not aim to maximize over all possible allocations but rather only over some pre-specified range of possible allocations. The smaller the range, the easier, potentially, it will be to optimize over it, but the less good will the best allocation in this limited range be. To take an extreme case, we could limit ourself in advance to bundle all the m items. This means that we will only maximize over the range of n possible allocations that have $m_i = m$ for a single agent i and $m_j = 0$ for all other j. This optimization is easy to do computationally (just check which i has the largest $v_i(m)$), but clearly it may only recover a $1/n$ fraction of possible welfare (e.g., when each bidder is willing to pay 1 for a single item, but no more than 1 even if he gets all items).

Mechanisms that take this approach are called maximum-in-range: such a mechanism fixes in advance a range of possible allocations; completely optimizes over this range; and then uses a VCG-like payments where everything is computed just over this range. The important point is that such mechanisms will always be truthful: the proof of truthfulness of the usual VCG mechanisms as given above directly applies also to every VCG-based mechanism. To see this, note the truthfulness of VCG mechanism was due to the fact that it aligned the utility of each player with that of the social welfare. This remains, and nothing that a bidder can do will take the result outside the range of allocations

[5] An example of this phenomena exists in the recent FCC "incentive auctions" for buying back spectrum from TV broadcasters. This auction requires sophisticated optimization and while the FCC's optimization programs were able to provably reach 97% efficiency, when these were plugged into the VCG formula, nonsensical results were obtained, including, e.g., negative payments.

over which the mechanism already optimized. As mentioned, these turn out to be that the *only* approximation mechanisms that become truthful when simply using the VCG payment rule.

While the family of VCG-based mechanisms seems to be a trivial extension of VCG mechanisms we can still get non-trivial mileage from it. The trick would be to carefully choose the range so that on one hand it is simple enough so we can effectively optimize over it, and on the other hand is rich enough so that optimizing over it guarantees some non-trivial level of approximation. For multiunit auctions such a range will be obtained by pre-bundling the items into a small number of "bundles." For simplicity of presentation we will assume without loss of generality that m is an integer multiple of n^2:

Approximation Mechanism

1. Split the m items into n^2 equi-sized bundles of size m/n^2 each
2. Run the optimal VCG mechanism with the items being the n^2 bundles, using the general allocation algorithm from Section 9.4.1

Note that this mechanism is maximum-in-range and hence truthful since be are completely optimizing over allocations that correspond to the pre-bundling. The point is that this mechanism is also computationally efficient since it is just an auction with n^2 items (the bundles), which we have already seen in Section 9.4.1 can be done in time that is polynomial in the number of items, which is now just polynomial in n.

The quality of allocations produced by this mechanism is that they always have at least $1/2$ the total value of the optimal one: $\sum_i v_i(m_i) \geq \frac{1}{2}\sum_i v_i(m'_i)$, where $(m_1, \ldots, m_n)$ is the optimal allocation that respects this bundling (which is the one the mechanism produces) and $(m'_1, \ldots, m'_n)$ is any other allocation. To see this, start with the optimal allocation $(m'_1, \ldots, m'_n)$ and consider the bidder i that got the largest number of items, $m'_i \geq m/n$. Now there are two possibilities. If at least half the total value comes just from this i, $v_i(m'_i) \geq \sum_{j \neq i} v_j(m'_j)$, then clearly the allocation that gives everything to him (which is part of our restricted optimization range) gets at least $\frac{1}{2}\sum_i v_i(m'_i)$ so our allocation is only better. If on the other hand $v_i(m'_i) < \sum_{j \neq i} v_j(m'_j)$ then we can take away i's allocation, and use these $m'_i \geq m/n$ items to complete the allocations of all other bidders to exact multiples of m/n^2 (which may require at most m/n^2 for each bidder), now resulting in a bundling allocation that is in our restricted optimization range but with value of at least $\sum_{j \neq i} v_j(m'_j) \geq \frac{1}{2}\sum_i v_i(m'_i)$ which is thus a lower bound on $\sum_i v_i(m_i)$ as needed. We thus have:

Theorem 9.8. *(Dobzinski and Nisan, 2007) There exists a truthful mechanism for the 2-approximate multiunit allocation problem (in any of the models we considered) whose running* *time is polynomial in n and $\log m$. It produces an allocation $(m_1, \ldots, m_n)$ that satisfies $\sum_i v_i(m_i) \geq (\sum_i v_i(opt_i))/2$, where $(opt_1, \ldots, opt_n)$ is the optimal allocation.*

So we have obtained a truthful computationally efficient 2-approximation mechanism. This is still far from the arbitrarily good approximation that is possible when we just take into account one of the two constraints of efficient computation or of truthfulness. Can we do better? We will now see that no maximum-in-range mechanism (with any range) can get a better approximation ratio. We will prove the impossibility of a better approximation even for the special case of two players for which we have already seen in Section 9.4.3 that finding the exactly optimal allocation is intractable even in the general queries (communication) model.

Consider an algorithm that guarantees strictly better than $1/2$ of the maximum possible value. If the range is complete, i.e., contains all possible allocations $(m_1, m - m_1)$ then we are doing perfect optimization which as we have seen is hard, and thus the algorithm is not computationally efficient. Otherwise, our range is missing at least one possible outcome $(m_1, m - m_1)$. Consider the input valuations given by the single-minded bids where the first player is willing to pay 1 for at least m_1 items and the second is willing to pay 1 for at least $m - m_1$ items. Clearly the optimal allocation for this case would be $(m_1, m - m_1)$ with $v_1(m_1) + v_2(m - m_1) = 2$. However, this allocation is not in our range, so our maximum-in-range mechanism will output a different allocation, and all other allocations get value of at most 1, which is exactly half of optimal. This argument assumes a black-box query model and, in fact, Dobzinski and Nisan (2007) also show that better approximations are possible in the bidding language models discussed above. We have thus shown:

Theorem 9.9. *(Dobzinski and Nisan, 2007) There is no maximum-in-range mechanism for approximate multiunit allocation (in any query model) whose running time is polynomial in n and $\log m$ and always produces an allocation $(m_1, \ldots, m_n)$ that satisfies $\sum_i v_i(m_i) > (\sum_i v_i(opt_i))/2$, where $(opt_1, \ldots, opt_n)$ is the optimal allocation.*

So, as we have reached the limits of the VCG technique, are there other new techniques for the construction of truthful mechanisms? Can any of these techniques give an efficiently computable mechanism that gets a better approximation ratio?

Combinatorial Auctions

When looking at the wider class of combinatorial auctions, even the algorithmic problem can not be approximated arbitrarily well. As significant algorithmic work has been done on obtaining various approximation factors for various subclasses of valuations, for each of these subclasses, the corresponding challenge is to find a truthful mechanism that achieves similar approximation ratios. Maximum-in-range mechanisms achieve various partial results in this respect—see Blumrosen and Nisan (2007) for survey and references.

9.5.4 Single parameter mechanisms

There is one class of situations where classical mechanism design provides another, non-VCG, technique for constructing truthful mechanisms. This is the case where buyers' private valuations are *single dimensional*. For the single dimensional case, Myerson (1981) essentially gives a complete characterization of incentive compatible mechanisms. Our presentation here essentially utilizes these results, but slightly differs from the classic treatment in two respects. First, we will limit our discussion to deterministic mechanisms, as we have done so far. Second, our scenario considers *single-minded bids*, which are really slightly more than single dimensional as they contain an additional discrete parameter (the desired number of items).

So, in this section we restrict ourselves to bidders whose valuations are single minded: each bidder i desires a specific number of items k_i and is willing to pay v_i if he gets at least k_i items (and gets value zero otherwise). This is a very limited class of valuations for which we have no representation problems. However, in this section, we will rely on their *strategic simplicity* which is beyond the representational one. In particular, the results in this section do not apply when we allow even double-minded bidders that are still simple in the representational sense.

A bidder that is single minded may either lose or win the auction—winning gives value v_i and losing gives value 0. We can always assume without loss of generality that a mechanism never allocate to any winner i more than exactly the $\tilde{k}_i$ items it requested and never allocates anything to a losing bidder. As a normalization, we may also assume without loss of generality that losing bidders always pay 0. Under these conventions it turns out that, when bidders are limited to being single minded, truthful mechanisms are exactly the ones that satisfy the following two properties:

1. **Monotonicity:** If a player wins with bid (k_i, v_i) then it will also win with any bid which offers at least as much money for at most as many items. That is, i will still win if the other bidders do not change their bids and i changes his bid to some (k_i', v_i') with $k_i' \leq k_i$ and $v_i' \geq v_i$.
2. **Critical payment:** The payment of a winning bid (k_i, v_i) is the smallest value needed in order to win k_i items, i.e., the infimum of v_i' such that (k_i, v_i') is still a winning bid, when the other bidders do not change their bids.

Let us see why these are sufficient conditions: First let us see why for a bidder whose real valuation is (k_i, v_i) bidding a quantity $k_i' \neq k_i$ is always dominated by k_i. If $k_i' < k_i$ then even if i wins he gets only $k_i' < k_i$ items and thus his value from winning is 0 and thus his utility is not positive. On the other hand, a bid $k_i' > k_i$ can not help him since monotonicity implies that anything that he wins with bid k_i' he could also win with k_i and furthermore more, given the the critical payment condition, the payment in the latter case is never higher than in the former case. We now show that misreporting v_i is not beneficial either. If v_i is winning and paying the critical value $p_i \leq v_i$ then every possible bid $v_i' \geq p_i$ is still winning and still paying the same amount p_i leading to the same utility as when bidding v_i. A bid $v_i' < p_i$ in this case will lose, getting utility 0, which is certainly not better than for bidding v_i. If, on the other hand, v_i is a losing bid, and the smallest winning bid is $p_i \geq v_i$ then bidding less than p_i will still lose, and bidding at least p_i will win but will yield negative utility.

This characterization is quite fruitful: first, it decouples the algorithmic issue from the strategic one: algorithmically we only need to devise a good *monotone* approximation algorithm. Incentive compatibility will follow once we add the critical payment rule. Second, it opens up a wide design space of monotone algorithms. At first, the limitation of being monotone does not seem to be too challenging as it actually is hard to imagine any useful algorithmic idea that is not monotone. Then, at second thought, one finds that any interesting algorithmic idea attempted breaks monotonicity. Finally, one learns how to be quite careful in the design of algorithms and ensure monotonicity. Let us look at these three stages as they relate to our problem.

Let us start with the approximation algorithm presented in Section 9.4.4 and consider it in our single-minded setting of inputs (k_i, v_i). This gives the following mechanism:

Single-Minded Additive Approximation Mechanism

1. Let $\epsilon > 0$ be the required approximation level, and let δ be a truncation precision parameter
2. Truncate each v_i to $j\delta$ where $0 \leq j \leq n/\epsilon$ is an integer
3. Find the optimal allocation among the truncated v_i's, using the algorithm from Section 9.4.4
4. Charge the critical payments

It is not difficult to see that for each fixed ϵ and δ, the δ-scale approximation algorithm is indeed monotone: the truncation is clearly a monotone operation, and the optimal solution is clearly monotone. However, what should the choice of δ be? If we just take a fixed choice of δ then we do get an *additive* approximation error of $n\delta$, however this will not translate to any approximation *ratio* since for very small valuations we may lose everything. In our original description of the algorithm in Section 9.4.4 our choice was $\delta = \epsilon V/n$ where $V = \max_i v_i$. This choice strikes a balance ensuring on one hand that an additive error of $n\delta$ which the truncation causes translates to an approximation factor of $1 - \epsilon$ and on the other hand keeps the range of rounded values small (i.e., n/ϵ) enough for computational efficiency. However, this choice of δ is not monotone: it is possible that a bidder who has the largest bid $v_i = V$ is a winner, but when he slightly increases his bid, then δ will also increase, leading to a different rounding in which he loses. Consider selling $m = 12$ items to $n = 7$ bidders: each of three bidders are willing to each pay \$7 for four items, while each of four bidders are willing to pay \$5.9 for three items. Now for $\epsilon = 1$ we get $\delta = \$1$ and the first group of three bidders all win. Had one of the first group of bidders slightly increased his bid to \$7.1, then δ would increase slightly above 1, and the new rounding would favor the second group of bidders over the first group, breaking monotonicity hence implying that the mechanism is not truthful.

It is not at all clear whether there is any monotone way of choosing δ in a way that will maintain both the approximation ratio and the computational efficiency. The solution proposed in Briest et al. (2011) is to simultaneously try multiple values of δ as all different orders of scale:

Single-Minded Approximation Mechanism:

1. Let $\epsilon > 0$ be the required approximation level
2. For all $\delta = 2^t$ where t is an integer (possibly negative) with $\epsilon V/(2n^2) \leq \delta \leq V$, where $V = \max_i v_i$ do:
   - **(a)** Run the additive approximation algorithm with precision parameter δ obtaining a "δ-scale allocation"
3. Choose the δ-scale allocation that achieved the highest value (as counted with the truncated values) to determine the final allocation
4. For each winner i in the final allocation: Find, using binary search, the lowest value that would still have bidder i win when the others' values are unchanged, and set this critical value as i's payment

---

It would seem that we have only made things worse: previously we had a single δ that depended on V, now we have a range of such δ's. However, the main observation is that in fact the range of δ's does not depend on V: for $\delta > V$ the truncation would anyway result in rounding everything to 0's, while for $\delta < \epsilon V/(2n^2)$, the δ-scale approximation will truncate each v_i to at most $V/(2n)$, yielding a total value of at at most $V/2$, which is less than that obtained for the largest δ in the range checked by allocating to the highest bidder. Thus to strategically analyze this mechanism we can imagine that we really tried *all* (infinitely many) possible values of $\delta = 2^t$ for integer t and chose the best allocation encountered. This imaginary infinite search would yield exactly the same outcome as our finite one does. However, it is clear that in this imaginary mechanism the choice of each δ does not depend in any way on the bids so we maintain monotonicity.

Or do we? While it is true that for every single δ we have a monotone algorithm, it does not follow that an algorithm that runs several monotone algorithms and then chooses the best among them is monotone itself. It could be that a winner, that wins when the optimum was achieved via a certain value of δ, increases his bid, causing a different value of δ to be the winning choice, a value of δ in which our original bidder loses. However, in our case it is possible to show that this does not happen: a nice property of the additive approximation is that the only way that the value of a player influences the solution quality is by winning or not. Thus if an increase in v_i causes a new value of δ to achieve the highest value then v_i must be winning for the new value of δ and thus global monotonicity is maintained. The same is true for manipulations in the requested number of items k_i. We have thus obtained:

Theorem 9.10. *(Briest et al., 2011) There exists a truthful mechanism for the approximate multiunit allocation problem among single-minded bidders whose running time is polynomial in n, $\log m$, and ϵ^{-1}. It produces an allocation $(m_1, \ldots, m_n)$ that satisfies $\sum_i v_i(m_i) \geq (1 - \epsilon) \sum_i v_i(opt_i)$, where $(opt_1, \ldots, opt_n)$ is the optimal allocation.*

So we have shown a truthful polynomial time approximation mechanism for the single-minded case. What about general valuations?

Combinatorial Auctions and Beyond

A similar phenomena appears in many others algorithmic mechanism design problems, where good approximation mechanisms are known for single-parameter settings. This was demonstrated effectively in the context of combinatorial auctions in Lehmann et al. (2002) where a truthful $O(\sqrt{m})$-approximation mechanism was designed, matching the algorithmic possibilities. Such mechanisms can be applied also to problems that do not maximize social welfare, as has been done, for example in the context of scheduling in early work of Archer and Tardos (2001).

9.5.5 Multiparameter mechanisms beyond VCG?

Once we have seen that for the strategically simple case of single-minded bidders we can design arbitrarily good computationally efficient truthful approximation mechanisms it is natural to try to get the same type of result for general bidders. Is that possible? While this is still open, it is known that truthfulness implies rather severe constraints in multiparameter settings. This section will demonstrate these type of constraints by proving a partial impossibility result. We will show that a truthful mechanism for $n = 2$ players that is required to *always allocate all items* which obtains an approximation factor that is strictly better than 2 requires exponentially many queries, in the value queries model. So let us fix such a mechanism $M = (m_1(\cdot), m_2(\cdot), p_1(\cdot), p_2(\cdot))$, where m_i denotes the number of items allocated to player i, where we always have $m_1 + m_2 = m$.

We start by deducing a necessary condition on the allocation function itself $(m_1(\cdot), m_2(\cdot))$ (independently of the payment functions). This condition will then allow us to analyze the behavior of truthful mechanisms and bound their power. Our analysis starts with stating the basic structure of payments induced by fixing one of the two input valuations and varying only the other.

The taxation principle: For every fixed value of v_2, there exists prices $\{p_a\}$ for $0 \le a \le m$ such that $p_1(v_1, v_2) = p_a$ whenever $a = m_1(v_1, v_2)$ (we will denote $p_a = \infty$ if there is no v_1 with $a = m_1(v_1, v_2)$). Moreover, $m_1(v_1, v_2) = argmax_a\{v_1(a) - p_a\}$.

The taxation principle states that, once v_2 is fixed, the prices that the first bidder face may only depend on the allocation it gets rather than arbitrarily on his full valuation. Moreover, under this pricing rule, his allocation maximizes his utility. Otherwise, if $m_1(v_1, v_2) = m_1(v_1', v_2)$ but $p_1(v_1, v_2) > p_1(v_1', v_2)$ then clearly a bidder with valuation v_1 will gain by misreporting v_1'. Similarly if $m_1(v_1, v_2) \neq argmax_a\{v_1(a) - p_a\} = m_1(v_1', m_2)$ then a bidder with valuation v_1 will gain by misreporting v_1'.

Let us now consider a situation where, for some fixed value of v_2, when the first bidder changes his valuation from v_1 to v_1' then his allocation changes from a to a'. By the taxation principle this means that $v_1(a) - p_a \ge v_1(a') - p_{a'}$ while $v_1'(a') - p_{a'} \ge v_1'(a) - p_a$. Combining these two inequalities allows us to cancel p_a and $p_{a'}$, showing that the allocation rule of a truthful mechanism satisfies the following property, a property that also turns out to be sufficient (Lavi et al., 2003):

Weak monotonicity: The allocation function $m_1(\cdot)$ is called *weakly monotone* if whenever $m_1(v_1, v_2) = a \neq a' = m_1(v_1', v_2)$ then we have that $v_1(a) - v_1'(a) \ge v_1(a') - v_1'(a')$.

The heart of our proof will use this property to show that for some fixed value of v_2, the range of $m_1(\cdot, v_2)$ is full, i.e., for every $0 \le a \le m$ there exists v_1 such that $m_1(v_1, v_2) = a$ and, moreover, each of these v_1's is strictly monotone, $0 < v_1(1) < \cdots < v_1(m)$. Using the taxation principle, this in particular implies $p_0 < p_1 < \cdots < p_m < \infty$. This allows to use the following adversary argument. Fix v_2 and our adversary will choose v_1 adaptively, as it answers value queries, as follows: for each query a it will answer $v_1(a) = p_a - p_0$. Note that as long as the algorithm is missing even a single query $0 < a \le m$, it is still possible that for the missing a, we have that $v_1(a) > p_a - p_0$ which would give strictly higher utility than all previous queries, and so the truthful mechanism may need to output it. Thus in order for the algorithm to be truthful it must query $v_1(a)$ for all m possible values of a.

We will choose v_2 to be the linear valuation $v_2(k) = 2k$ (there is nothing special about this valuation and any other strictly monotone one will do as well). For every $0 \le a \le m$ we will exhibit a valuation v_1 such that for the input (v_1, v_2), the output must be $m_1 = a$ and $m_2 = m - a$. We will argue this by starting with some other valuations (v_1', v_2') for which we can know the output and "modifying" them to become the desired (v_1, v_2) using weak monotonicity to control how the results may change.

Let us denote by $1_{\ge a}$ the valuation that gives value ϵ for each item with a large bonus of 1 for at least a items (i.e., $1_{\ge a}(k) = k\epsilon$ for $k < a$ and $1_{\ge a}(k) = 1 + k\epsilon$ for $k \ge a$) and consider the allocation on the input $v_1' = 1_{\ge a}$ and $v_2' = 1_{\ge m-a}$. The fact that the mechanism is a strictly better than 2 approximation implies that for this input $m_1 = a$ and $m_2 = m - a$.

We now consider a "stretched" version v_1 of v'_1 where for $k < a$ we define $v_1(k) = v'_1(k)$, for $k > a$ we define $v_1(k) = c + v'_1(k)$ and $v_1(a) = c + v'_1(a) + \epsilon/2$, where c is a large constant. We claim that for input (v_1, v'_2) our mechanism must still produce the allocation $m_1 = a$ and $m_2 = m - a$. This is true due to weak monotonicity as in our construction $v_1(a) - v'_1(a) \leq v_1(a') - v'_1(a')$ only holds for $a' = a$ (notice that the roles of a and a' are switched from their roles in the definition).

Let us now consider the output of our mechanism on the input (v_1, v_2). Since c is large, to maintain an approximation ratio the first player must be allocated at least a items and thus the second must be allocated at most $m - a$ items. Can the second player be allocated strictly less items, $m - a'$ for $a' > a$? If so, then using the weak monotonicity of m_2 for the fixed choice of v_1, we have that $v_2(m - a) - v'_2(m - a) \leq v_2(m - a') - v'_2(m - a')$ which in our construction only holds for $a' \leq a$. Thus $m_2 = m - a$ and $m_1 = a$ as required, completing the proof of:

Theorem 9.11. *(Lavi et al., 2003) Every truthful mechanism for approximate multiunit allocation with two players that always allocates all items, $m_1 + m_2 = m$, and for some fixed $\alpha > 1/2$ always satisfies $v_1(m_1) + v_2(m_2) \geq \alpha(v_1(opt_1) + v_2(opt_2))$, where (opt_1, opt_2) is the optimal allocation, requires at least m queries in the value queries model.*

So how "biting" are the constraints induced by truthfulness? Can we extend this theorem to hold without the requirement of always allocating all items? This is still not known. To gain a perspective about what we may be able to achieve, it will be useful to consider a strong characterization result due to Roberts (1979) that holds in a more general mechanism design setting. In this abstract non-restricted setting there are n agents that have valuations that assign an arbitrary values to possible abstract outcomes in the set A, i.e., $v_i : A \rightarrow \Re$. The mechanism in such a case has to choose a single outcome $a \in A$ (as well as payments) in a truthful way. It is not difficult to verify that the VCG mechanism that we have seen in Section 9.5.1 directly generalizes to this setting yielding a truthful mechanism whose outcome maximizes $\sum_i v_i(a)$ over all $a \in A$. It is also not difficult to generalize the VCG mechanism by assigning weights $\alpha_i \geq 0$ to the players as well as weights β_a to the outcomes, obtaining a truthful mechanism that maximizes $\beta_a + \sum_i \alpha_i v_i(a)$ over all $a \in A$. A mechanism whose output always maximizes this expression (for some choice of $\{\alpha_i\}, \{\beta_a\}$) is called an *affine maximizer.* The payment rule to make an affine maximizer truthful is given by a natural weighted variant of the VCG payments. Roberts' theorem states that these are the *only* truthful mechanisms for the general unrestricted mechanism design problem when there are at least three outcomes. This is in stark contrast to the single-parameter case where Myerson's technique provides many truthful non-affine-maximizing (non-VCG) mechanisms.

Now back to our setting of multiunit auctions. The case of single-minded bidders is close to being single parameter and as we saw in Section 9.5.4 there are good, non-VCG, mechanisms for this case. The more general case, where bidders have more general valuations is far from being single parameter and we basically lack any techniques that go beyond (weighted versions of) VCG, which are very limited in their power. For example, as shown in Section 9.5.3, they cannot guarantee, in polynomial time, more than half of the social welfare. (It is not difficult to generalize the proof there to weighted versions of VCG, i.e., to any affine maximizer.) Had Roberts' theorem applied to multiunit auctions and not just to the unrestricted case then we would have a general impossibility result of going beyond (weighted) VCG. However, the setting of multiunit auctions is more restricted in two respects. First, we are assuming free disposal (v_i is monotone) and, second, we are implicitly assuming no externalities. (v_i is only a function of m_i rather than of the full allocation). The case analyzed here of two players where all items must always be allocated, voids the "no externalities" restriction and becomes sufficiently close to the unrestricted case as to allow Lavi et al. (2003) to extend (a version of) the Roberts characterization to it. On the other hand, without this restriction, several non-VCG truthful mechanisms are known for multiunit auctions, including some that guarantee more than half of the social welfare (Dobzinski and Nisan, 2011), although not in polynomial time. An impossibility result is given in Dobzinski and Nisan (2011) for the subclass of "scalable mechanisms"—those where multiplying all values by the same constant factor does not change the allocation.

The general question remains open: are there any *useful* truthful non-VCG mechanisms for multiunit auctions? This is open for most interpretations of "useful," including the one we embrace in this survey: computationally efficient and guaranteeing strictly more than half the optimal social welfare.[6]

Combinatorial Auctions and Beyond

The basic question of existence of "useful" non-VCG computationally efficient truthful combinatorial auctions is open for general valuations as well as for most interesting subclasses. There is one case where a truthful non-VCG mechanism is known (Bartal et al., 2003), as well one case where impossibility results are known (Dobzinski, 2011; Dobzinski and Vondrák, 2012). A similar situation holds for other mechanism design problems where again essentially no non-VCG mechanisms are known for multiparameter settings. For a problem that is "close" to being unrestricted, there is a strong characterization extending Roberts (Papadimitriou et al., 2008). There is also a restricted impossibility results for scheduling due to Ashlagi et al. (2009).

[6] It is probably good to remark that if we agree to relax our notion of equilibria to Bayesian, then a multitude of computationally efficient Bayesian-truthful mechanisms exist (e.g., Hartline et al., 2011).

9.5.6 Randomization

We now slightly widen our scope of mechanisms and allow also randomized ones. We will see that these indeed do provide us with increased capabilities. A randomized mechanism will still accept as input the valuations of the players but will now output a *distribution over allocations and payments*. A randomized mechanism will be called truthful (in expectation) if for all v_i, v_{-i}, v_i' we have that $E[v_i(M_i) - P_i] \geq E[v_i(M_i') - P_i']$ where M_i and P_i are the random variables indicating the number of items allocated to i and the payment of i when he bids v_i, and M_i' and P_i' are these random variables when he bids v_i'. We will concentrate on a specific class of randomized mechanisms that first *deterministically* choose a *distribution* on allocations, and a payment and then just output a random realization of an allocation from the chosen distribution.

We can now generalize the idea of maximum-in-range mechanisms. We can think of a mechanism that pre-decides on a range of *distributions on allocations*. For any such possible distribution D on allocations $(m_1 \ldots m_n)$, we can look at the expected value of each v_i on this distribution: $v_i(D) = E_{(m_1,\ldots,m_n)\sim D}[v_i(m_i)]$. From this point of view we can look at any such distribution as a new possible allocation. We will call a randomized mechanism *maximum in distributional range* if it always maximizes the expected social value over such a pre-chosen range of distributions, i.e., always chooses the distribution that maximizes $\sum_i v_i(D) = E_{(m_1\ldots m_n)\sim D}[\sum_i v_i(m_i)]$ over all D in the range. Clearly, we can apply the VCG payment rule to such a mechanism by simply viewing the range of distributions as a new range of "deterministic" outcomes, this way obtaining a randomized truthful (in expectation) mechanism.

Can such mechanisms be helpful? In particular can we build a maximum-in-distributional-range mechanism that provides a good approximation while being computationally efficient? The approximation ratio of a randomized mechanism is defined to be the worst case ratio obtained over all possible inputs, where for each input we measure ratio between the expected social welfare produced and the optimal social welfare for this input. As shown by Dobzinski and Dughmi (2009), it turns out that the answer is yes. In fact, a very basic randomized range already suffices for this. The distributions we will use in our range will all have support size of 2: the empty allocation ($m_i = 0$ for all i) and a single other allocation $(m_1, \ldots, m_n)$. We will have such a distribution for every possible allocation $(m_1, \ldots, m_n)$; we will define a "penalty probability" $0 \leq z(m_1, \ldots, m_n) < \epsilon$ for each $(m_1, \ldots, m_n)$, and will choose the empty allocation with probability $z(m_1, \ldots, m_n)$ and allocate m_i to each i with probability $1 - z(m_1, \ldots, m_n)$. The expected social welfare of such a distribution is simply $(1 - z(m_1, \ldots, m_n)) \sum_i v_i(m_i)$, and it is this expression that our algorithm maximizes (over all $(m_1, \ldots, m_n)$.) It should immediately be clear that this allocation will also approximately maximize the non-penalized social welfare, i.e., $\sum_i v_i(m_i) \geq$

$(1-\epsilon)\sum_i v_i(m_i')$ for every other allocation $(m_1' \ldots m_n')$. The wonder is that we will be able to define these penalties in a way that will allow the computationally efficient maximization of $(1-z(m_1,\ldots,m_n))\sum_i v_i(m_i)$ over all allocations.

The idea is to penalize "non-roundness." Let us define the weight of an integer m_i, $w(m_i)$, to be the largest power of two that divides m_i (thus the weight of 8 is 3, the weight of 9 is 0, and the weight of 10 is 1). Let us also define the weight of an allocation to be the sum of the weights of the m_i's, $w(m_1,\ldots,m_n)=\sum_i w(m_i)$. Our penalty function will be defined as $z(m_1,\ldots,m_n)=(C-w(m_1,\ldots,m_n))\delta$, where C is a fixed upper bound on the weight of any allocation ($C=n\log m$) and δ chosen to be small enough so that z is in the required range, ($\delta=\epsilon/C$). The point is that these penalties for "non-round" allocations will suffice to significantly trim down the search space.

Let us look at a single player: do we need to consider all m possible values of m_i to allocate to him or can we significantly trim down the search space and consider only a smaller number of possible m_is? Consider two possible values $m_i < m_i'$ with $w(m_i) > w(m_i')$. The point is that if their values are close, $v_i(m_i)/v_i(m_i') > (1-\delta)$, then m_i' can clearly be trimmed from the search space since the tiny addition to $v()$ when increasing m_i to m_i' can not offset the loss of welfare due to the decrease in $w_i()$ (which increases z). It turns out that this suffices for the construction of an computationally efficient algorithm.

Maximum-in-Distributed-Range Mechanism

1. For each player i, compute a list $List_i$ of possible allocations m_i to player i by calling $List_i = List_i(0,\log m)$:

   For an integer $0\le t\le \log m$, and $0\le L\le m$ with $w(L)\le t$, $LIST_i(L,t)$, will output the list of all non-trimmable values of m_i in the range $L < m_i \le L+2^t$:
   (a) If $t=0$ then return the singleton $\{L+1\}$
   (b) If $v_i(L)/v_i(L+2^{t-1}) < 1-\delta$ then return $LIST_i(L,t-1)$ concatenated with $LIST_i(L+2^{t-1},t-1)$
   (c) Else return $LIST_i(L+2^{t-1},t-1)$
2. Find the allocation $(m_1,\ldots,m_n)$ that maximizes $(1-z(m_1\ldots m_n))\sum_i v_i(m_i)$, over all $m_i\in List_i$
3. With probability $z(m_1,\ldots,m_n)$ output the empty allocation, otherwise output $(m_1,\ldots,m_n)$
4. Compute VCG payments for each bidder i by re-running the optimization without i

The point is that the subroutine LIST branches to two recursive calls only if $v(L)/v(L+2^{t-1}) < 1-\delta$, otherwise it only makes a single recursive call. In terms of correctness this is OK since if the gap between $v(L)$ and $v(L+2^{t-1})$ is too small then we have seen that the whole range $L < m_i \le L+2^{t-1}$ can be trimmed away. To bound the running time, we will bound the number of invocations of LIST where a branch can be made, and will do that for each fixed value of the $1 + \log m$ possible levels t's. Since the subranges on which *LIST* is called at level t are a partition, and since a branch is undertaken only if there is a $(1 - \delta)$-gap between the value of v_i at the two end points of the subrange, then there can be at most $O(T/\delta)$ invocations of LIST that branch at any given level t of the recursion, where T is the number of bits of precision in the representation of v_i. Thus the total number of recursive invocations of $LIST_i$ is bounded by $O(T \log m/\delta)$ giving a total polynomial bound on the running time as well as on the length of $LIST_i$.

If we just worry about information considerations, i.e., the number of queries made, then we are already done: Finding the optimal allocation only requires querying the values of $v_i(m_i)$ only for these polynomially many $m_i \in LIST_i$. To completely specify a polynomial time algorithm we need to explicitly describe how to use this polynomial-length information to find the optimal allocation. For the case of a constant number of players, this is immediate by exhaustive search over all possible combinations. The general case is more complex and appears in Dobzinski and Dughmi (2009). We have thus obtained:

Theorem 9.12. *(Dobzinski and Dughmi, 2009) There exists a randomized truthful (in expectation) mechanism for the approximate multiunit allocation problem (in any of the models we considered) whose running time is polynomial in n, $\log m$, and ϵ^{-1}. It produces an allocation $(m_1, \ldots, m_n)$ that satisfies $E[\sum_i v_i(m_i)] \geq (1 - \epsilon) \sum_i v_i(opt_i)$, where $E[]$ denotes expectation over the random choices of the mechanism and $(opt_1, \ldots, opt_n)$ is the optimal allocation.*

A slightly stronger result in terms of truthfulness is also known (Vöcking, 2012). This mechanism randomly chooses among deterministic truthful mechanism for the problem. The guarantee on the quality of the output is still probabilistic, but the truthfulness holds universally for each random choice rather than only in expectation.

Combinatorial Auctions and Beyond

Looking at the wider picture of algorithmic mechanism design, the power of randomized mechanisms is a general phenomena and there are many cases where we have randomized mechanisms that are better than the deterministic ones. This, in particular, is so for combinatorial auctions where a randomized mechanism is known that achieves an $O(\sqrt{m})$-approximation ratio which matches the ratio possible due to algorithmic constraints alone (Dobzinski et al., 2006). Similarly, for many subclasses of valuations we have randomized mechanisms whose approximation ratios are much better than those of the known truthful deterministic ones. See Blumrosen and Nisan (2007) for references.

9.6. CONCLUSION

The challenge of designing good truthful efficiently computable (deterministic) mechanisms for welfare maximization in multiunit auctions remains open.

The same is true for most other interesting algorithmic mechanism design problems such as combinatorial auctions.

ACKNOWLEDGMENTS

I would like to thank Moshe Babaioff, Shahar Dobzinski, Michal Feldman, and Renato Paes Leme for extensive comments on a previous draft. I would like to thank Shahar for the proof of Theorem 9.11. Research partially supported by a grant from the Israeli Science Foundation.

REFERENCES

Archer, A., Tardos, E., 2001. Truthful mechanisms for one-parameter agents. In: FOCS'01.
Ashlagi, I., Dobzinski, S., Lavi, R., 2009. An optimal lower bound for anonymous scheduling mechanisms. In: ACM Conference on Electronic Commerce, pp. 169–176.
Bartal, Y., Gonen, R., Nisan, N., 2003. Incentive compatible multi unit combinatorial auctions. In: TARK.
Blumrosen, L., Nisan, N., 2007. Combinatorial auctions (a survey). In: Nisan, N., Roughgarden, T., Tardos, E., Vazirani, V. (Eds.), Algorithmic Game Theory. Cambridge University Press.
Blumrosen, L., Nisan, N., 2010. On the computational power of demand queries. SIAM J. Comput. 39(4), 1372–1391.
Briest, P., Krysta, P., Vcking, B., (2011). Approximation techniques for utilitarian mechanism design. SIAM J. Comput. 40(6), 1587–1622.
Clarke, E.H., 1971. Multipart pricing of public goods. Public Choice, 17–33.
Cramton, P., Shoham, Y., Steinberg, R. (Eds.), 2006. Combinatorial Auctions. MIT Press.
de Vries, S., Vohra, R., 2003. Combinatorial auctions: a survey. INFORMS J. Comput., 15(3), 284–309.
Dobzinski, S., 2011. An impossibility result for truthful combinatorial auctions with submodular valuations. In: STOC, pp. 139–148.
Dobzinski, S., Dughmi, S., 2009. On the power of randomization in algorithmic mechanism design. In: FOCS, pp. 505–514.
Dobzinski, S., Nisan, N., 2007. Mechanisms for multi-unit auctions. In: Proceedings of the 8th ACM Conference on Electronic Commerce, EC '07, pp. 346–351.
Dobzinski, S., Nisan, N., 2011. Multi-unit auctions: beyond roberts. In: Proceedings of the 12th ACM Conference on Electronic Commerce, EC '11, pp. 233–242.
Dobzinski, S., Vondrák, J., 2012. The computational complexity of truthfulness in combinatorial auctions. In: ACM Conference on Electronic Commerce, pp. 405–422.
Dobzinski, S., Nisan, N., Schapira, M., 2006. Truthful randomized mechanisms for combinatorial auctions. In: STOC.
Garey, M.R., Johnson, D.S., 1979. Computers and Intractability: A Guide to the Theory of NP-Completeness. W. H. Freeman & Co., New York, NY, USA. ISBN 0716710447.
Groves, T., 1973. Incentives in teams. Econometrica, pp. 617–631.
Gul, F., Stacchetti, E., 1999. Walrasian equilibrium with gross substitutes. J. Econ. Theory 87, 95–124.
Gul, F., Stacchetti, E., 2000. The english auction with differentiated commodities. J. Econ. Theory 92(3), 66–95.
Hartline, J., 2012. Approximation in Economic Design. Lecture Notes.
Hartline, J.D., Karlin, A.R., 2007. Profit maximization in mechanism design. In: Nisan, N., Roughgarden, T., Tardos, E., Vazirani, V. (Eds.), Algorithmic Game Theory. Cambridge University Press.

Hartline, J.D., Kleinberg, R., Malekian, A., 2011. Bayesian incentive compatibility via matchings. In: Proceedings of the Twenty-Second Annual ACM-SIAM Symposium on Discrete Algorithms, pp. 734–747.

Krishna, V., 2002. Auction Theory. Academic Press.

Kushilevitz, E., Nisan, N., 1997. Communication Complexity. Cambridge University Press.

Lavi, R., Mu'alem, A., Nisan, N., 2003. Towards a characterization of truthful combinatorial auctions. In: FOCS.

Lehmann, D., O'Callaghan, L.I., Shoham, Y., 2002. Truth revelation in approximately efficient combinatorial auctions. JACM 49(5), 577–602.

Mas-Collel, A., Whinston, W., Green, J., 1995. Microeconomic Theory. Oxford University Press.

Milgrom, P., 2004. Putting Auction Theory to Work. Cambridge University Press.

Myerson, R.B., 1981. Optimal auction design. Math. Oper. Res., 6(1), 58–73.

Nisan, N., 2007. Introduction to mechanism design (for computer scientists). In: Nisan, N., Roughgarden, T., Tardos, E., Vazirani, V. (Eds.), Algorithmic Game Theory. Cambridge University Press.

Nisan, N., Ronen, A., 2001. Algorithmic mechanism design. Games Econ. Behav. 35, 166–196 (A preliminary version appeared in STOC 1999).

Nisan, N., Ronen, A., 2007. Computationally feasible vcg mechanisms. J. Artif. Int. Res. 29(1), 19–47. ISSN 1076-9757.

Nisan, N., Segal, I., 2006. The communication requirements of efficient allocations and supporting prices. J. Econ. Theory 129(1), 192–224.

Nisan, N., Roughgarden, T., Tardos, E., Vazirani, V.V., 2007. Algorithmic Game Theory. Cambridge University Press, New York, NY, USA. ISBN 0521872820.

Osborne, M.J., Rubistein, A., 1994. A Course in Game Theory. MIT press.

Papadimitriou, C.H., Schapira, M., Singer, Y., 2008. On the hardness of being truthful. In: FOCS, pp. 250–259.

Procaccia, A.D., Tennenholtz, M., 2009. Approximate mechanism design without money. In: Proceedings of 10th EC, pp. 177–186.

Roberts, K., 1979. The characterization of implementable choise rules. In: Laffont, J.-J. (Ed.), Aggregation and Revelation of Preferences. Papers presented at the first European Summer Workshop of the Economic Society. North-Holland, pp. 321–349.

Rosenschein, J.S., Zlotkin, G., 1994. Rules of Encounter: Designing Conventions for Automated Negotiation Among Computers. MIT Press.

Roth, A.E., 2003. The economist as engineer: game theory, experimentation, and computation as tools for design economics. Econometrica 70(4), 1341–1378.

Roughgarden, T., Talgam-Cohen, I., 2013. Optimal and near-optimal mechanism design with interdependent values. In: Proceedings of the Fourteenth ACM Conference on Electronic Commerce, pp. 767–784.

Sandholm, T., 1999. An algorithm for optimal winner determination in combinatorial auctions. In: IJCAI-99.

Schummer, J., Vohra, R.V., 2007. Mechanism design without money. In: Nisan, N., Roughgarden, T., Tardos, E., Vazirani, V. (Eds.), Algorithmic Game Theory. Cambridge University Press.

Shenkar, S., Clark, D.E., Hertzog, S., 1996. Pricing in computer networks: reshaping the research agenda. ACM Comput. Comm. Rev., 19–43.

Varian, H.R., 1995. Economic mechanism design for computerized agents. In: Proceedings of the First Usenix Conference on Electronic Commerce, New York.

Vickrey, W., 1961. Counterspeculation, auctions and competitive sealed tenders. J. Finance 8–37.

Vöcking, B., 2012. A universally-truthful approximation scheme for multi-unit auctions. In: Proceedings of the Twenty-Third Annual ACM-SIAM Symposium on Discrete Algorithms, SODA '12, pp. 846–855.

CHAPTER 10

Behavioral Game Theory Experiments and Modeling

Colin F. Camerer*, **Teck-Hua Ho**†
*California Institute of Technology, Pasadena, CA, USA
†University of California, Berkeley, Berkeley, CA, USA

Contents

Handbook of game theory
http://dx.doi.org/10.1016/B978-0-12-420056-2.00010-0

Abstract

This chapter reviews recent experimental data testing game theory and behavioral models that have been inspired to explain those data. The models fall into four groups: in cognitive hierarchy or level-*k* models, the assumption of equilibrium is relaxed by assuming agents have beliefs about other agents who do less reasoning (i.e., some are non-strategic, and others are more strategic and understand they are playing non-strategic players). A different approach, quantal response equilibrium, retains equilibrium expectations but adds stochastic response (of which players are aware). Learning theories explain choices in games played repeatedly as a result of past actions and payoffs that were received (as in classical reinforcement) or foregone payoffs (model-directed learning). Finally, many studies reject the joint hypothesis of equilibrium expectations and optimization, along with self-interest in valuing outcomes. Social preference models have emerged to explain these data, capturing concepts like inequity-aversion, reciprocity, and social image.

Keywords: Behavioral game theory, Cognitive hierarchy, Level-k, Learning, Nonequilibrium, Bounded rationality

JEL Codes: C7, C9, D7, D8

10.1. INTRODUCTION

For explaining individual decisions, rationality—in the sense of accurate belief and optimization—may still be an adequate approximation even if a modest percentage of players violate the theory. But game theory is different. Players' fates are intertwined. The presence of players who do not have accurate belief or optimize can dramatically change what rational players should do. As a result, what a population of players is likely to do when some do not have accurate belief or optimize can only be predicted by an analysis which explicitly accounts for bounded rationality as well, preferably in a precise way. This chapter is about what has been learned about boundedly rational strategic behavior from hundreds of experimental studies (and some field data).

In the experiments, equilibrium game theory is almost always the benchmark model being tested. However, the frontier has moved well beyond simply comparing actual behaviors and equilibrium predictions, because that comparison has inspired several types of behavioral models. Therefore, the chapter is organized around precise behavioral models of boundedly rational choice, learning and strategic teaching, and social preferences. We focus selectively on several example games which are of economic interest and explain how these models work.

We focus on five types of models

1. Cognitive hierarchy (CH) model which captures players' beliefs about steps of thinking, using one parameter to describe the average step of strategic thinking (level-*k* (LK) models are closely related). These models are designed to predict one-shot games or initial conditions in a repeated game.

2. A noisy optimization model called quantal response equilibrium (QRE). Under QRE, players are allowed to make small mistakes but they always have accurate beliefs about what other players will do.
3. A learning model (called experience-weighted attraction learning (EWA)) to compute the path of equilibration. The EWA learning algorithm generalizes both fictitious play and reinforcement models. EWA can predict the time path of play as a function of the initial conditions (perhaps supplied by CH model).
4. An extension of the learning models to include sophistication (understanding how others learn) as well as strategic teaching, and nonequilibrium reputation-building. This class of models allows the population to have sophisticated players who actively influence adaptive players' learning paths to benefit themselves.
5. Models of how social preferences map monetary payoffs (controlled in an experiment) into utilities and behavior.

Our approach is guided by three stylistic principles: Precision; generality; and empirical accuracy. The first two are standard desiderata in equilibrium game theory; the third is a cornerstone in empirical economics.

Precision: Because game theory predictions are sharp, it is not hard to spot likely deviations and counterexamples. Until recently, most of the experimental literature consisted of documenting deviations (or successes) and presenting a simple model, usually specialized to the game at hand. The hard part is to distill the deviations into an alternative theory that is similarly precise as standard theory and can be widely applied. We favor specifications that use one or two free parameters to express crucial elements of behavioral flexibility. This approach often embeds standard equilibrium as a parametric special case of general theory. It also allows sensible measures of individual and group differences, through parameter variation. We also prefer to let data, rather than intuition, specify parameter values.

Generality: Much of the power of equilibrium analyses, and their widespread use, comes from the fact that the same principles can be applied to many different games. Using the universal language of mathematics enables knowledge to cumulate worldwide. Behavioral models of games are also meant to be general, too, in the sense that the same models can be applied to many games with minimal customization. Keep in mind that this desire for universal application is not held in all social sciences. For example, many researchers in psychology believe that behavior is so context-specific that it is might be too challenging to create a common theory that applies to all contexts. Our view is that we can't know whether general theories are hopeless until we try to apply them broadly.

Empirical discipline: Our approach is heavily disciplined by data. Because game theory is about people (and groups of people) thinking about what other people and groups will do, it is unlikely that pure logic alone will tell us what will happen. As the Nobel Prize-winning physicist Murray Gell-Mann said, "Imagine how difficult physics would be if electrons could think." It is even harder if we don't watch what 'electrons' do when interacting.

Our insistence on empirical discipline is shared by others. Van Damme (1999) thought the same:

> Without having a broad set of facts on which to theorize, there is a certain danger of spending too much time on models that are mathematically elegant, yet have little connection to actual behavior. At present our empirical knowledge is inadequate and it is an interesting question why game theorists have not turned more frequently to psychologists for information about the learning and information processes used by humans.

Experiments play a privileged role in testing game theory. Game-theoretic predictions are notoriously sensitive to what players know, when they move, and how they value outcomes. Laboratory environments provide crucial control of all these variables (see Camerer, 2003; Crawford, 1997). As in other lab sciences, the idea is to use lab control to sort out which theories work well and which don't, then later use them to help understand patterns in naturally occurring data. In this respect, behavioral game theory resembles data-driven fields like labor economics or finance more than equilibrium game theory.

Many behavioral game theory models also circumvent two long-standing problems in equilibrium game theory: refinement and selection. These models "automatically" refine sets of Bayesian-Nash equilibria (BNE) because they allow all events to occur with positive probability, and hence Bayes' rule can be used in all information sets. Some models (e.g., CH and LK) also automatically avoid multiplicity of equilibria by making a single statistical prediction. Surprisingly, assuming *less* rationality on players therefore can lead to *more* precision (as noted previously by Lucas, 1986).

10.2. COGNITIVE HIERARCHY AND LEVEL-K MODELS

CH and LK models are designed to predict behavior in one-shot games and to provide initial conditions for models of learning.

We begin with notation. Strategies have numerical attractions that determine the probabilities of choosing different strategies through a logistic response function. For player i, there are m_i strategies (indexed by j) which have initial attractions denoted $A_i^j(0)$. Denote i's jth strategy by s_i^j, chosen strategies by i and other players (denoted $-i$) in period t as $s_i(t)$ and $s_{-i}(t)$, and player i's payoffs of choosing s_i^j by $\pi_i(s_i^j, s_{-i}(t))$.

A logit response rule is used to map attractions into probabilities:

$$P_i^j(t+1) = \frac{e^{\lambda \cdot A_i^j(t)}}{\sum_{j'=1}^{m_i} e^{\lambda \cdot A_i^{j'}(t)}} \qquad [10.1]$$

where λ is the response sensitivity.[1] Since CH and LK models are designed to predict strategic behaviors in only one-shot games, we focus mainly on $P_i^j(1)$ (i.e., no learning). We shall use the same framework for learning models too and the model predictions there will go beyond one period.

We model thinking by characterizing the number of steps of iterated thinking that subjects do, their beliefs, and their decision rules.[2] Players using zero steps of thinking, do not reason strategically at all. We think these players are using simple low-effort heuristics, such as choosing salient strategies (cf. Shah and Oppenheimer, 2008). In games which are physically displayed (e.g., battlefields) salience might be based on visual perception (e.g., Itti and Koch, 2000). In games with private information, a strategy choice that matches an information state might be salient (e.g., bidding one's value or signal in an auction). Randomizing among all strategies is also a reasonable step-0 heuristic when no strategy is particularly salient.

Players who do one step of thinking believe they are playing against step 0 types. Proceeding inductively, players who use k steps think all others use from zero to $k-1$ steps. Since they do not imagine same- and higher-step types there is missing probability; we assume beliefs are normalized to adjust for this missing probability. Denote beliefs of level-k players about the proportion of step h players by $g_k(h)$. In CH, $g_k(h) = f(h)/\sum_{k'=0}^{k-1} f(k')$ for $h \leq k-1$ and $g_k(h) = 0$ for $h \geq k$. In LK, $g_k(k-1) = 1$. The LK model is easier to work with analytically for complex games.

It is useful to ask why the number of steps of thinking might be limited. One possible answer comes from psychology. Steps of thinking strain "working memory," because higher step players need to keep in mind more and more about what they believe lower steps will do as they form beliefs. Evidence that cognitive ability is correlated with more thinking steps is consistent with this explanation (Camerer et al., 2004; Devetag and Warglien, 2003; Gill and Prowse, 2012).

However, it is important to note that making a step k choice does not logically imply a limit on level of thinking. In CH, players are essentially defined by their beliefs about others, not by their own cognitive capabilities. For example, a step 2 player believes others use 0 or 1 steps. It is possible that such players are capable of even higher-step reasoning (e.g., choosing Nash equilibria) if they thought their opponents were of higher steps. CH and level-k models are not meant to permit separation of beliefs and "reasoning skill," per se (though see Kneeland, 2013 for evidence).

An appealing intuition for why strategic thinking is limited is that players endogeneously choose whether to think harder. The logical challenge in such an approach is this: A player who fully contemplated the benefit from doing one more thinking

[1] Note the timing convention—attractions are defined *before* a period of play; so the initial attractions $A_i^j(0)$ determine choices in period 1, and so forth.

[2] This concept was first studied by Stahl and Wilson (1995) and Nagel (1995), and later by Ho et al. (1998).

step would have to derive the optimal strategy after the additional thinking. Alaoui and Penta (2013) derive an elegant axiomatization of such a process by assuming that players compute the benefit from an extra step of thinking based on its maximum payoff. They also calibrate their approach to data of Goeree and Holt (2001) with some success.

Of course, it is also possible to allow players to change their step k choice. Indeed, Ho and Su (2013) and Ho et al. (2013) make CH and level-k models dynamic by allowing players to update their beliefs of other players' steps of thinking in sequential games. In their dynamic level-k model, players not only choose rules based on their initial guesses of others' steps (like CH and level-k) but also use historical plays to improve their guesses. Their dynamic level-k model captures two systematic violations of backward induction in centipede games, limited induction (i.e., people violate backward induction more in sequential games with longer decision trees) and repetition unraveling (i.e., choices unravel towards backward induction outcomes over time) and explains learning in p-beauty contests and sequential bargaining games.

The key challenge in thinking steps models is pinning down the frequencies of players using different numbers of thinking steps. A popular nonparametric approach is to let $f(k)$ be free parameters for a limited number of steps k; often the three steps 0, 1, and 2 will fit adequately. Instead, we assume those frequencies have a Poisson distribution with mean and standard deviation τ $\left(\text{the frequency of step } k \text{ types is } f(k) = \frac{e^{-\tau}\tau^k}{k!}\right)$. Then τ is an index of aggregate bounded rationality.

The Poisson distribution has only one free parameter (τ); it generates "spikes" in predicted distributions reflecting individual heterogeneity (some other approaches do not reflect heterogeneity[3]); and for typical τ values the Poisson frequency of step types is roughly similar to estimates of nonparametric frequencies (see Ho et al., 1998; Nagel et al., 1999; Stahl and Wilson, 1995).

Given this assumption, players using $k > 0$ steps are assumed to compute expected payoffs given their adjusted beliefs, and use those attractions to determine choice probabilities according to

$$A_i^j(0|k) = \sum_{h=1}^{m_{-i}} \pi_i(s_i^j, s_{-i}^h) \cdot \left\{ \sum_{c=0}^{k-1} [\frac{f(c)}{\sum_{k'=0}^{k-1} f(c)} \cdot P_{-i}^h(1|k')] \right\} \qquad [10.2]$$

where $A_i^j(0|k)$ and $P_i^j(1|k')$) are the attraction of step k in period 0 and the predicted choice probability of lower step k' in period 1. Hence, the predicted probability of level K in period 1 is given by:

$$P_i^j(1|K) = \frac{e^{\lambda \cdot A_i^j(0|K)}}{\sum_{j'=1}^{m_i} e^{\lambda \cdot A_i^{j'}(0|K)}}. \qquad [10.3]$$

[3] Alternatives include the theory of noisy expectation by Capra (1999) and the closely related theory of "noisy introspection" by Goeree and Holt (1999). Both models do not accommodate heterogeneity.

Next we apply CH to three types of games: dominance-solvable p-beauty contests, market entry, and LUPI lottery choices. Note that CH and LK have also been applied to literally hundreds of other games of many structures (including private information) (see Crawford et al., 2013).

10.2.1 *P*-beauty contest

In a famous passage, Keynes (1936) likens the stock market to a newspaper contest in which people guess which faces others will judge to be the most beautiful. He writes:

> It is not the case of choosing those which, to the best of one's judgment, are really the prettiest, nor even those which average opinion genuinely thinks the prettiest. We have reached the third degree, where we devote our intelligences to anticipating what average opinion expects the average opinion to be. And there are some, I believe, who practice the fourth, fifth, and higher degrees *[p. 156]*.

The essence of Keynes's observation is captured in a game in which players are asked to pick numbers from 0 to 100, and the player whose number is closest to p times the average number wins a prize. Suppose $p = 2/3$ (a value used in many experiments). Equilibrium theory predicts each contestant will reason as follows: "Even if all the other players guess 100, I should guess no more than 2/3 times 100, or 67. Assuming that the other contestants reason similarly, however, I should guess no more than 45..." and so on, finally concluding that the only rational and consistent choice for all the players is zero. When the beauty contest game is played in experimental settings, the group average is typically between 20 and 35. Apparently, some players are not able to reason their way to the equilibrium value, or they assume that others are unlikely to do so. If the game is played multiple times with the same group, the average moves close to 0.

Table 10.1 shows data from 24 p-beauty contest games, with various p and conditions. Estimates of τ for each game were chosen to minimize the (absolute) difference between the predicted and actual mean of chosen numbers. The table is ordered from top to bottom by the mean number chosen. The first seven lines show games in which the equilibrium is not zero; in all the others the equilibrium is zero.[4]

The first four columns describe the game or subject pool, the source, group size, and total sample size. The fifth and sixth columns show the Nash equilibrium and the difference between the equilibrium and the average choice. The middle three columns show the mean, standard deviation, and the mode in the data. The mean choices are generally far off from the equibrium; they choose numbers that are too low when the equilibrium is high (first six rows) and the numbers that are too high when the equibrium is low (lower rows). The rightmost six columns show the estimate of τ from the Poisson-CH model, and the mean, prediction error, standard deviation, and mode predicted by

[4] In games in which the equilibrium is not zero, players are asked to choose a number that is closest to p times the average number + a nonzero constant.

Table 10.1 Data and ch estimates of τ in various p-beauty contest games.

Subject pool		Group	Sample	Nash	Pred'n	Data				Fit of CH model			Bootstrapped	
or game	Source[1]	size	size	equil'm	error	Mean	Std Dev	Mode	τ	Mean	Error	Std Dev	Mode	90% C.I.
$p=1.1$	HCW (98)	7	69	200	47.9	152.1	23.7	150	**0.10**	151.6	−0.5	28.0	165	(0.0,0.5)
$p=1.3$	HCW (98)	7	71	200	50.0	150.0	25.9	150	**0.00**	150.4	0.5	29.4	195	(0.0,0.1)
High $	CHW	7	14	72	11.0	61.0	8.4	55	**4.90**	59.4	−1.6	3.8	61	(3.4,4.9)
Male	CHW	7	17	72	14.4	57.6	9.7	54	**3.70**	57.6	0.1	5.5	58	(1.0,4.3)
Female	CHW	7	46	72	16.3	55.7	12.1	56	**2.40**	55.7	0.0	9.3	58	(1.6,4.9)
Low $	CHW	7	49	72	17.2	54.8	11.9	54	**2.00**	54.7	−0.1	11.1	56	(0.7,3.8)
$0.7(M+18)$	Nagel (98)	7	34	42	−7.5	49.5	7.7	48	**0.20**	49.4	−0.1	26.4	48	(0.0,1.0)
PCC	CHC (new)	2	24	0	−54.2	54.2	29.2	50	**0.00**	49.5	−4.7	29.5	0	(0.0,0.1)
$p=0.9$	HCW (98)	7	67	0	−49.4	49.4	24.3	50	**0.10**	49.5	0.0	27.7	45	(0.1,1.5)
PCC	CHC (new)	3	24	0	−47.5	47.5	29.0	50	**0.10**	47.5	0.0	28.6	26	(0.1,0.8)
Caltech board	Camerer	73	73	0	−42.6	42.6	23.4	33	**0.50**	43.1	0.4	24.3	34	(0.1,0.9)
$p=0.7$	HCW (98)	7	69	0	−38.9	38.9	24.7	35	**1.00**	38.8	−0.2	19.8	35	(0.5,1.6)
CEOs	Camerer	20	20	0	−37.9	37.9	18.8	33	**1.00**	37.7	−0.1	20.2	34	(0.3,1.8)
German students	Nagel (95)	14–16	66	0	−37.2	37.2	20.0	25	**1.10**	36.9	−0.2	19.4	34	(0.7,1.5)
70 yr olds	Kovalchik	33	33	0	−37.0	37.0	17.5	27	**1.10**	36.9	−0.1	19.4	34	(0.6,1.7)
US high school	Camerer	20–32	52	0	−32.5	32.5	18.6	33	**1.60**	32.7	0.2	16.3	34	(1.1,2.2)
Econ PhDs	Camerer	16	16	0	−27.4	27.4	18.7	N/A	**2.30**	27.5	0.0	13.1	21	(1.4,3.5)
1/2 mean	Nagel (98)	15–17	48	0	−26.7	26.7	19.9	25	**1.50**	26.5	−0.2	19.1	25	(1.1,1.9)
Portfolio mgrs	Camerer	26	26	0	−24.3	24.3	16.1	22	**2.80**	24.4	0.1	11.4	26	(2.0,3.7)
Caltech students	Camerer	17–25	42	0	−23.0	23.0	11.1	35	**3.00**	23.0	0.1	10.6	24	(2.7,3.8)
Newspaper	Nagel (98)	3696, 1460, 2728	7884	0	−23.0	23.0	20.2	1	**3.00**	23.0	0.0	10.6	24	(3.0,3.1)
Caltech	CHC (new)	2	24	0	−21.7	21.7	29.9	0	**0.80**	22.2	0.6	31.6	0	(4.0,1.4)
Caltech	CHC (new)	3	24	0	−21.5	21.5	25.7	0	**1.80**	21.5	0.1	18.6	26	(1.1,3.1)
Game theorists	Nagel (98)	27–54	136	0	−19.1	19.1	21.8	0	**3.70**	19.1	0.0	9.2	16	(2.8,4.7)
								mean	**1.30**					
								median	**1.61**					

Note 1: HCW (98) is Ho et al. (1998); CHC are new data from Camerer, Ho, and Chong; CHW is Camerer et al. (unpublished); Kovalchik is unpublished data collected by Stephanie Kovalchik

the best-fitting estimate of τ, and the 90% confidence interval of τ estimated from a randomized resampling (bootstap) procedure.

There are several interesting patterns in Table 10.1. The prediction errors of the mean (column 13, "error") are extremely small, less than .6 in all but two cases. This is no surprise since τ is estimated (separately in each row) to minimize this prediction error. The pleasant surprise is that the predicted standard deviations and the modal number choices which result from the error-minimizing estimate of τ are also fairly close (across rows, the correlation of the predicted and actual standard deviation is 0.72) even though τ's were not chosen to match these moments. The values of τ have a median and mean across rows of 1.30 and 1.61. The confidence intervals have a range of about one in samples of reasonable size (50 subjects or more).

Note that nothing in the Poisson-CH model, per se, requires τ to be fixed across games or subject pools, or across details of how games are presented or choices are elicited. Outlying low and high values of τ are instructive about how widely τ might vary, and why. Estimates of τ are quite low (0-0.1) for the p-beauty contest game when $p > 1$ and, consequently, the equilibrium is at the upper end of the range of possible choices (rows 1-2). In these games, subjects seem to have trouble realizing that they should choose very large numbers when $p > 1$ (though they equilibriate rapidly by learning; see Ho et al., 1998). Low τ's are also estimated among the PCC (Pasadena City College) subjects playing two- and three-player games (row 8 and 10). High values of τ ($\approx$ 3-5) appear in games where the equilibrium in the interior, 72 (row 7-10)—small incremental steps toward the equilibrium in these games produce high values of τ. High τ values are also estimated in games with an equilibrium of zero when subjects are professional stock market portfolio managers (row 19), Caltech students (row 20), game theorists (row 24), and subjects self-selecting to enter newspaper contests (row 25). The latter subject pools show that in high analytical and educated subject pools (especially with self-selection) τ can be much higher than in other subject pools.

A sensible intuition is that when stakes are higher, subjects will use more steps of reasoning (and may think others will think harder too). Rows 3 and 6 compare low stakes ($1 per person per period) and high stakes ($4) in games with an interior equilibrium of 72. When stakes are higher, τ is estimated to be twice as large (5.01 versus 2.51), which is a clue that some sort of cost-benefit analysis may underlie steps of reasoning (cf. Alaoui and Penta, 2013).

Notwithstanding these interesting outliers, there is also substantial regularity across very diverse subject pools and payoff steps. About half the samples have confidence intervals that include $\tau = 1.5$. Subsamples of corporate CEOs, high-functioning 80-year-old spouses of memory-impaired patients, and high school students (rows 13, 15-16) all have τ values from 1.1-1.7.

An interesting question about CH modeling is how persistent a player's thinking step is across games. The few studies that have looked carefully found fairly stable estimated steps within a subject across games of similar structure (Costa-Gomes et al., 2001; Stahl and Wilson, 1995). For example, Chong et al. (2004) estimated separate τ values for each of 22 one-shot games with mixed-equilibria. An assignment procedure then chose a most-likely step for each subject and each game (given the 22 τ estimates). For each subject, an average of their first 11 estimated steps and last 11 estimated steps were computed. The correlation coefficient of these two average steps was 0.61. This kind of within-subject stability is a little lower than many psychometric traits (e.g., intelligence, extraversion) and is comparable to econographic traits such as risk-aversion.

Two studies looked most carefully at within-subject step stability. Georgiadis et al. (2013) had subjects play variants of two structurally different games. They found modest stability of step choices within each game type, and low stability across the two game types. Hyndman et al. (2013) did the most thorough study; they measured both choices and beliefs across several baseline payoff-isomorphic games. They found that 30% of subjects appear to maintain a stable type belief (mostly step 1) across games, and others fluctuate between lower and higher belief between games. These studies bracket what we should expect to see about stability of step types (i.e., a mixture of stability). Since nothing in these theories commits to how stable "types" are, this empirical result is not surprising at all.

Finally, whether subjects maintain a single step type across games is neither predicted by the theory nor important for most applications. The most basic applications involve aggregate prediction, and sensitivity of predicted results to comparative static changes in game structure. It is very rare that the scientific goal is to predict what a specific subject will do in one game type, based on their behavior in a different game type.

10.2.2 Market entry games

Consider binary entry games in which there is a market with demand c (expressed as a percentage of the total number of potential entrants). Each of the many potential entrants decides simultaneously whether or not to enter the market. If a potential entrant thinks that fewer than c% will enter she will enter; if she thinks more than c% will enter she stays out.

There are three regularities in many experiments based on entry games like this one (see Camerer, 2003; Ochs, 1999; Seale and Rapoport, 2000, chapter 7): (1) Entry rates across different levels of demand c are closely correlated with entry rates predicted by (asymmetric) pure equilibria or symmetric mixed equilibria; (2) players slightly over-enter at low levels of demand and under-enter at high levels of demand; and (3) many players use noisy cutoff rules in which they stay out for levels of demand below some cutoff c^* and enter for higher levels of demand.

In Camerer et al. (2004), we apply the thinking model with best response (i.e., $\lambda = \infty$) to explain subject behaviors in this game. Step zero players randomize and enter half the time. This means that when $c < 0.5$ one step thinkers stay out and when $c > 0.5$ they enter. Players doing two steps of thinking believe the fraction of zero steppers is $f(0)/(f(0) + f(1)) = 1/(1 + \tau)$. Therefore, they enter only if $c > 0.5$ and $c > \frac{0.5+\tau}{1+\tau}$, or when $c < 0.5$ and $c > \frac{0.5}{1+\tau}$. To make this more concrete, suppose $\tau = 2$. Then two-step thinkers enter when $c > 5/6$ and $1/6 < c < 0.5$. What happens is that more steps of thinking "iron out" steps in the function relating c to overall entry. In the example, one-step players are afraid to enter when $c < 1/2$. But when c is not too low (between 1/6 and 0.5) the two-step thinkers perceive room for entry because they believe the relative proportion of zero-steppers is 1/3 and those players enter half the time. Two-step thinkers stay out for capacities between 0.5 and 5/6, but they enter for $c > 5/6$ because they know half of the (1/3) zero-step types will randomly stay out, leaving room even though one-step thinkers always enter. Higher steps of thinking smooth out steps in the entry function even further.

The surprising experimental fact is that players can coordinate entry reasonably well, even in the first period. ("To a psychologist," Kahneman (1988) wrote, "this looks like magic.") The thinking steps model provides a possible explanation for this magic and can account for the second and third regularities discussed above for reasonable τ values. Figure 10.1 plots entry rates from the first block of two studies for a game similar to the one above (Seale and Rapoport, 2000; Sundali et al., 1995). Note that the number of actual entries rises almost monotonically with c, and entry is above capacity at low c and below capacity at high c.

Figure 10.1 also shows the thinking steps entry function $N(all|\tau)(c)$ for $\tau = 1.25$. The function reproduces monotonicity and the over- and under-capacity effects. The thinking-steps models also produce approximate cutoff rule behavior for all higher thinking steps except two. When $\tau = 1.25$, step 0 types randomize, step 1 types enter for all c above 0.5, step 2-4 types use cutoff rules with one "exception," and steps 5-above use strict cutoff rules. This mixture of random, cutoff and near-cutoff rules is roughly what is observed in the data when individual patterns of entry across c are measured (e.g., Seale and Rapoport, 2000).

This example makes a crucial point about the goal and results from CH modeling. The goal is *not* (just) to explain nonequilibrium behavior. Another goal is to explain a *lack* of nonequilibrium behavior—that is, when is equilibration remarkably good, even with no special training or experience, and no opportunities for learning or communication? Note that in p-beauty contest games, if some players are out of equilibrium then even sophisticated players will prefer to be out of equilibrium. However, in entry games if some players over- or under-react to the capacity c then the sophisticated players will behave oppositely, leading to aggregate near-equilibrium. More generally, in games with strategic complementarity a little irrationality (e.g., step 0) will be multiplied; in games with strategic substitutes a little irrationality will be mitigated (Camerer and Fehr, 2006).

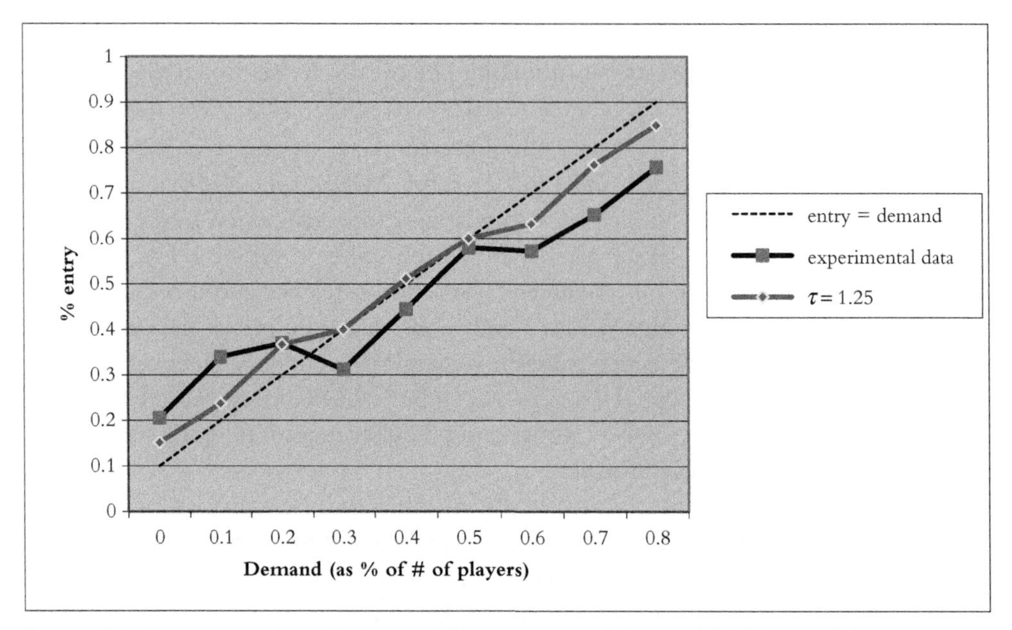

Figure 10.1 How entry varies with demand (D), experimental data and thinking model.

10.2.3 LUPI lottery

A unique field study used a simple lottery played in Sweden by an average of 53,783 players per day, over 7 weeks (Ostling et al., 2011). Participants in this lottery paid 1 euro to pick an integer from 1 to 99,999. The participant who chose the lowest unique positive integer won a prize of 10,000 euros (hence, the lottery is called LUPI). Interestingly, solving for the Nash equilibrium for a fixed number of n players is computationally intractable for large n. However, if the number of players is random and Poisson distributed across days, then the methods of Poisson games (Myerson, 1998) can be applied. The symmetric Poisson-Nash equilibrium (PNE) has an elegant simple structure where n is the mean number of players and p_k is the probability of playing integer k. The symmetric PNE has a striking nonlinear shape: numbers from 1 to around 5500 are chosen almost equally often, but with slightly declining probability (i.e., 1 is most likely, 2 is slightly less likely, etc.) (see Figure 10.2). A bold property of the PNE is that numbers above 5513—a range that includes 95% of all available numbers—should rarely be chosen.

Figure 10.3 shows the full distribution of number choices in the first week, along with the PNE and a best-fitting quantal CH model (i.e., λ is finite). Compared to the PNE, players chose too many low numbers, below 2000, and too many high numbers. As a result, not enough numbers between 2000 and 5000 are chosen. CH can fit these deviations from PNE with a value of $\tau = 1.80$, which is close to estimates from many experimental games. Intuitively, step 1 players choose low numbers because they think

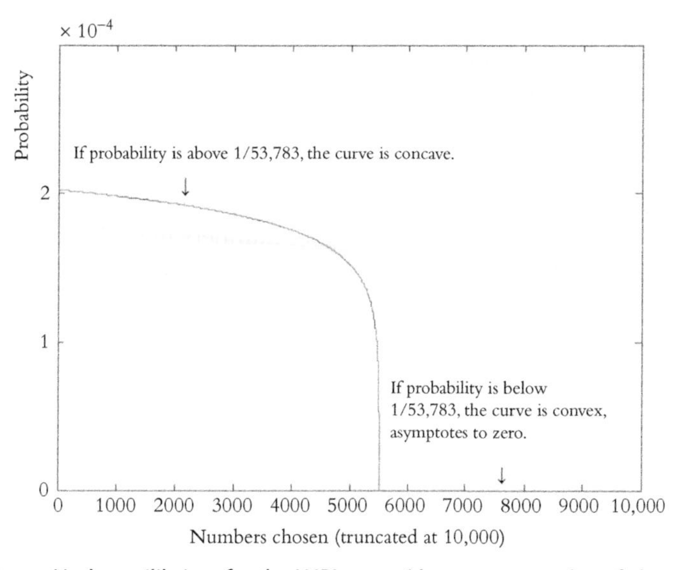

Figure 10.2 Poisson-Nash equilibrium for the LUPI game *(the average number of players is 53,783).*

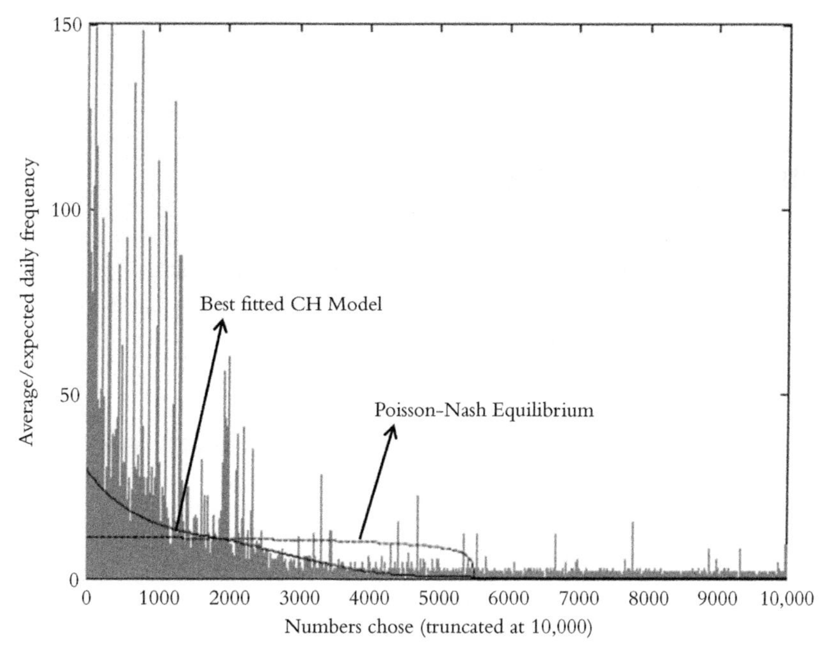

Figure 10.3 The full distribution of actual choices.

step 0 randomizers will pick too many high numbers, step 2 numbers choose above the step 1 choices, and so on.

Despite that CH can explain the deviations of the number choices in LUPI lottery, the actual behavior is not far from the PNE prediction in general, given how precisely bold the predicted distribution is. Recall that PNE predicts only numbers below 5513 will be played—excluding about 95% of the strategies—and overall, 93.3% of the numbers are indeed below that threshold. Furthermore, over seven weeks almost every statistical feature of the empirical distribution of numbers moved toward the PNE. For example, PNE predicts the average number will be 2595. The actual averages were 4512 in the first week and 2484 in the last week (within 4% of the prediction). A scaled-down laboratory version of LUPI also replicated these basic patterns.

10.2.4 Summary

Simple models of thinking steps attempt to predict choices in one-shot games and provide initial conditions for learning models. CH and level-k approaches incorporate discrete steps of iterated thinking. In Poisson-CH, the frequencies of players using different numbers of steps is Poisson-distributed with mean τ. While these models have been applied to hundreds of experimental data sets, and several field settings, in this chapter we focussed on data from dominance-solvable p-beauty contests, market entry, a LUPI field lottery, and a private-information auction (see Section 10.3). For Poisson-CH, estimates of τ are typically around 1-2. Importantly, CH can typically explain large deviations from equilibrium *and* surprising cases in which equilibration occurs without communication or experience (in entry games).

10.3. QUANTAL RESPONSE EQUILIBRIUM

The CH and LK models explain nonequilibrium behavior by mistakes in beliefs about what other subjects will do, due to bounded rationality in strategic thinking. A different approach is to assume that beliefs are accurate, but responses are noisy. This approach is called QRE (McKelvey and Palfrey, 1995). QRE relaxes the assumption that players always choose the best actions given their beliefs by incorporating "noisy" or "stochastic" response. The theory builds in a sensible principle that actions with higher expected payoffs are chosen more often; that is, players "better-respond" rather than "best-respond." If the response function is logit, QRE is defined by

$$A_i^j(0|K) = \sum_{h=1}^{m_{-i}} \pi_i(s_i^j, s_{-i}^h) \cdot P_{-i}^h(1)$$

$$P_i^j(1) = \frac{e^{\lambda \cdot A_i^j(0)}}{\sum_{h=1}^{m_i} e^{\lambda \cdot A_i^h(0)}} \qquad [10.4]$$

Mathematically, the QRE nests Nash equilibrium as a special case. Specifically, when the noise parameter λ goes to infinity, QRE converges to Nash equilibrium.

The errors in the players' QRE best-response functions are usually interpreted as decision errors in the face of complex situations or as unobserved latent disturbances to the players' payoffs (i.e., the players are optimizing given their payoffs, but there is a component of their payoffs that only they understand). In other words, the relationship between QRE and Nash equilibrium is analogous to the relationship between stochastic choice and deterministic choice models. Note that QRE is different from trembling hand perfect equilibrium in that the noise parameter λ in QRE is part of the equilibrating process and players use it to determine an action's expected payoff as well as its choice probability.

QRE has been used successfully in many applications to explain deviations from Nash equilibrium in games (Goeree and Holt, 2001; McKelvey and Palfrey, 1995; Rosenthal, 1981).[5] The key feature of the approach is that small mistakes can occur, but if a small mistake by player i has a large impact on player j, QRE prediction can be far from Nash equilibrium. QRE can also be used as a tool to check robustness in designing institutions or predicting how structural changes will change behavior.

10.3.1 Asymmetric hide-and-seek game

Let's see how QRE can improve equilibrium prediction by considering a game with an unique mixed-strategy equilibrium (see Table 10.2 for its payoff matrix). The row player's strategy space consists of actions A1 and A2, while the column player's chooses between B1 and B2. The game is a model of "hide-and-seek" in which one player wants to match another player's numerical choice (e.g., A1 responding to B1), and another player wants to mismatch (e.g., B1 responding to A2). The row player earns either 9 or 1 from matching on (A1, B1) or (A2, B2) respectively. The column player earns 1 from mismatching on (A1, B2) or (A2, B1).

The empirical frequencies of each of the possible actions, averaged across many periods of an experiment conducted on this game, are also shown in Table 10.2 (McKelvey and Palfrey, 1995). What is the Nash equilibrium prediction for this game? We start by observing that there is no pure-strategy Nash equilibrium for this game so we look for a mixed-strategy Nash equilibrium. Let us suppose that the Row player chooses A1 with probability p and A2 with probability $1 - p$, and the Column player chooses B1 with probability q and B2 with probability $1 - q$. In a mixed-strategy equilibrium, the players actually play a probabilistic mixture of the two strategies. If their valuation of

[5] To use QRE to predict behaviors ex ante, one must know the noise parameter (λ). This can be accomplished by first estimating λ in a similar game. Previous studies appear to show that λ does vary from game to game. Hence, an open research question is to develop a theory that maps features of a game to the value of λ so that one can then use QRE to predict behavior in any new game.

Table 10.2 Asymmetric hide-and-seek game

	B1 q	B2 1-q	Empirical frequency ($N = 128$)	Nash equilibrium	QRE
A1 p	9, 0	0, 1	0.54	0.50	0.65
A2 1-p	0, 1	1, 0	0.46	0.50	0.35
Empirical Frequency	0.33	0.67			
Nash Equilibrium	0.10	0.90			
QRE	0.35	0.65			

outcomes is consistent with expected utility theory, they only prefer playing a mixture if they are indifferent between each of their pure strategies. This property gives a way to compute the equilibrium mixture probabilities p and q. The mixed-strategy Nash equilibrium for this game turns out to be $[(0.5A1, 0.5A2), (0.1B1, 0.9B2)]$. Comparing this with the empirical frequencies, we find that Nash prediction is close to actual behavior by the Row players, whereas it under-predicts the choice of B1 for the Column players.

If one player plays a strategy that deviates from the prescribed equilibrium strategy, then according to the optimization assumption in Nash equilibrium, the other player must best-respond and deviate from Nash equilibrium as well. In this case, even though the predicted Nash equilibrium and actual empirical frequencies almost coincide for the Row player, the players are not playing a Nash equilibrium jointly, because the Row player should have played differently given that the Column player has deviated quite far from the mixed-strategy Nash equilibrium (playing B1 33% of the time rather than 10%).

We now illustrate how to derive the QRE for a given value of λ in this game. Again, suppose the Row player chooses A1 with probability p and A2 with probability $1 - p$, while the Column player chooses B1 with probability q and B2 with probability $1 - q$. Then the expected payoffs from playing A1 and A2 are $q * 9 + (1 - q) * 0 = 9q$ and $q * 0 + (1 - q) * 1 = 1 - q$ respectively. Therefore, we have $p = \frac{e^{\lambda \cdot 9q}}{e^{\lambda \cdot 9q} + e^{\lambda \cdot (1-q)}}$. Similarly, the expected payoffs to B1 and B2 for the Column player are $1 - p$ and p respectively, so we have $q = \frac{e^{\lambda \cdot (1-p)}}{e^{\lambda \cdot (1-p)} + e^{\lambda \cdot p}}$. Notice that q is on the right hand side of the first equation, which determines p, and p is on the right hand side of the second equation, which determines q. For any value of λ, there is only one pair of (p, q) values that solves the simultaneous equations and yields a QRE. If $\lambda = 2$, for example, the QRE predictions are $p* = 0.646$ and $q* = 0.343$ which are closer to the empirical frequencies than the Nash equilibrium predictions are. If $\lambda = \infty$, the QRE predictions are $p* = 0.5$ and $q* = 0.1$, which are identical to Nash equilibrium predictions.

Using the actual data, a precise value of λ can be estimated using maximum-likelihood methods. The estimated λ for the QRE model for the asymmetric "hide-and-seek" game is 1.95. The negative of the log likelihood of QRE (an overall measure of goodness of fit) is 1721, a substantial improvement over a random model benchmark ($p = q = 0.5$) which has a fit of 1774. The Nash equilibrium prediction has a fit of 1938, which is worse than random (because of the extreme prediction of $q = 0.1$).

10.3.2 Maximum value auction

We describe a private-information auction which illustrates another application of QRE. In the maximum-value second price auction, two players observe private integer signals x_1 and x_2 drawn independently from a commonly known distribution in the interval [0,10] (Bulow and Klemperer, 2002; Carrillo and Palfrey, 2011). They bid for an object which has a common value equal to the maximum of those two signals, $\max(x_1, x_2)$. The highest bidder wins the auction and pays a price equal to the second-highest bid.

How should you bid? Bidding less than your signal is a mistake because your bid just determines whether you win, not what you pay; so bidding less never saves money. In fact, you might miss a chance to win at a price below the object's value if you bid too low and get outbid, so underbidding is weakly dominated. Second, bidding above your signal could be a mistake: If the other bidder is also overbidding, either of you may get stuck overpaying for an item with a low maximum-value. In the unique symmetric (weak) BNE, therefore, players simply bid their values. In fact, the symmetric equilibrium where both players bid their signal can be solved for with two rounds of elimination of weakly dominated strategies.

Note that the unique symmetric BNE in which bidders bid their signals, but the equilibrium is weak; indeed it is just about as weak as an equilibrium can possibly be. If the other bidder is bidding according to the Nash equilibrium, i.e., bidding their signals, then *every bid greater than or equal to your signal* is a (weak) best response to Nash equilibrium.

QRE impose the assumption that beliefs about choices of other players are accurate, but allow imperfect (noisy) response. A natural conjecture is that (symmetric) regular QRE (Goeree et al., 2005) in these auctions will typically entail frequent overbidding, because bidding a little too high is a very small mistake.

In these experiments (Ivanov et al., 2010) subjects first participated in 11 independent auctions, once for each private signal 0-10, *with no feedback*. For each possible value, the rectangles in Figures 10.4ab show the median bid (thick horizontal line), and the vertical boundaries of the rectangle are the first and third quartiles. The main result is that there is substantial overbidding which has a "hockey stick" shape: Bids when signals are below \$5 are around \$4-\$5 and do not change much as the signal increases. Bids when signals are above \$5 are a little above the signals, and increase as signals increase. The BNE is just to bid the signal value (shown by a dotted 45-degree line). To fit the data, hierarchical QRE (HQRE) and CH-QR (with quantal response and only levels 0-2) were first fit to

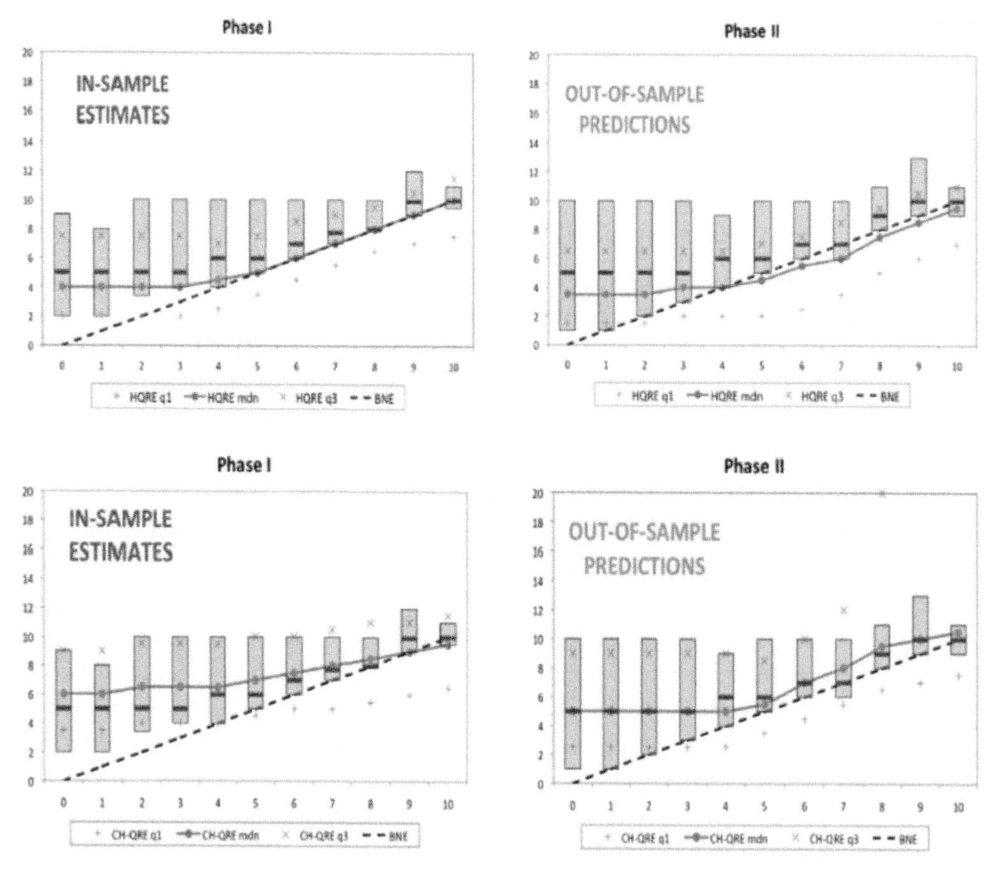

Figure 10.4 Bids by Signal, MinBid Treatment. *Data from Nunnari et al. (2012).*

the bidding data from phase I. Those best-fitting parameters were then used to predict bids in phase II. Note that HQRE and CH-QR make probabilistic predictions—that is, they predict the distribution of likely bids given each signal. The predictions are therefore represented by a dot for the predicted median and small x's and +s for the first and third quartiles. Visually, a good fit is indicated if, for each signal, the dot is near the horizontal line and the x's and +s are near the rectangle boundaries.

Both models do a reasonable job of picking up the major regularity in how bids deviation from BNE, which is the hockey stick pattern. However, HQRE predicts bids which are a little too low (including a substantial rate of *under*bidding). CH-QR fits very nicely, except at the highest signals \$8-10, where it predicts too many high bids.

In this case, the essential feature of QRE approaches is that small mistakes are expected to occur, but can predict bids far from BNE. Since low signal values are unlikely to determine the object's value (the other signal is likely to be the maximum), there is only a small loss from bidding too high on low signals. However, with high signals the bidder's own signal is likely to be the maximum, so overbidding is a bigger mistake and should be less common. These two intuitions generate the hockey stick shape that is evident in the data.

This example illustrates how different best-response equilibrium analysis can be from either CH or QRE. Curiously, Ivanov et al. (2010) conclude from their data that "Overall, our study casts a serious doubt on theories that posit the WC [winner's curse] is driven by beliefs [i.e., by CH and cursed equilibrium]."(p. 1435) However, as Figure 10.4b shows, the nonequilibrium belief theory CH fits the data quite well.

Finally, recall the LUPI lottery game described previously, in which the player with a lowest unique positive integer wins a fixed prize. It happens that LUPI is well suited for distinguishing CH and QRE approaches. For LUPI games with a small number of players, for which both QRE and CH can be solved numerically, the QRE distribution approaches NE from below, while the CH distribution predicts more low number choices than NE. Thus, QRE cannot explain the large number of low number choices occuring in the data, as CH can.

However, in a comparison of several other games, we have typically found that QRE and CH make about equally accurate predictions (e.g., Rogers et al., 2009). Moinas and Pouget (2013) use a remarkable bubble investment game to carefully compare models and find good fit from a heterogeneous form of QRE. A careful meta-analysis by Wright and Leyton-Brown (2013) shows that Poisson-CH generally fits a little better than other versions (given its parsimony) and all limited-thinking approaches do better than equilibrium in predicting behavior.

10.4. LEARNING

By the mid-1990s, it was well-established that simple models of learning could explain some movements in choice over time in specific game and choice contexts.[6] The challenge taken up since then is to see how well a specific parametric model can account for finer details of the equilibration process in a wide range of classes of games.

This section describes a one-parameter theory of learning in decisions and games called functional EWA (or fEWA for short; also called "EWA Lite" to emphasize its 'low-calorie' parsimony) (Ho et al., 2007). fEWA predicts the time path of individual behavior in any normal-form game given the initial conditions. Initial conditions can be imposed or estimated in various ways. We use the CH model described in the previous

[6] To name only a few examples, see Camerer (1987) (partial adjustment models); Smith et al. (1988) and Camerer and Weigelt (1993) (Walrasian excess demand); McAllister (1991) (reinforcement); Roth and Erev (1995) (reinforcement); Ho and Weigelt (1996) (reinforcement and belief learning); Cachon and Camerer (1996) (Cournot dynamics).

section to specify the initial conditions. The goal is to predict both initial conditions and equilibration in new games in which behavior has never been observed, with minimal free parameters (the model uses two, τ and λ).

10.4.1 Parametric EWA learning: interpretation, uses, and limits

fEWA is a relative of a parametric model of learning called experience-weighted attraction (EWA) (Camerer and Ho, 1998, 1999). As in most theories, learning in EWA is characterized by changes in (unobserved) attractions based on experience. Attractions determine the probabilities of choosing different strategies through a logistic response function. For player i, there are m_i strategies (indexed by j) which have initial attractions denoted $A_i^j(0)$. The best-response CH model is used to generate initial attractions given parameter value τ.

Denote i's j'th strategy by s_i^j, chosen strategies by i and other players (denoted $-i$) by $s_i(t)$ and $s_{-i}(t)$, and player i's payoffs by $\pi_i(s_i^j, s_{-i}(t))$.[7] Define an indicator function $I(x, y)$ to be zero if $x \neq y$ and one if $x = y$. The EWA attraction updating equation is

$$A_i^j(t) = \frac{\phi N(t-1)A_i^j(t-1) + [\delta + (1-\delta)I(s_i^j, s_i(t))]\pi_i(s_i^j, s_{-i}(t))}{N(t-1)\phi(1-\kappa)+1} \qquad [10.5]$$

and the experience weight (the "EW" part) is updated according to $N(t) = N(t-1)\phi(1-\kappa)+1$.

Notice that the term $[\delta + (1-\delta)I(s_i^j, s_i(t))]$ implies that a weight of one is put on the payoff term when the strategy being reinforced is the one the player chose ($s_i^j = s_i(t)$), but the weight on foregone payoffs from unchosen strategies ($s_i^j \neq s_i(t)$) is δ.

Attractions are mapped into choice probabilities using a logit response function given by:

$$P_i^j(t+1) = \frac{e^{\lambda \cdot A_i^j(t)}}{\sum_{k=1}^{m_i} e^{\lambda \cdot A_i^k(t)}} \qquad [10.6]$$

where λ is the response sensitivity.[8] The subscript i, superscript j, and argument $t+1$ in $P_i^j(t+1)$ are reminders that the model aims to explain every choice by every subject in every period.[9]

[7] To avoid complications with negative payoffs, we rescale payoffs by subtracting them by the minimum payoff so that rescale payoffs are always weakly positive.

[8] Note that we can use the same parameter λ in both [10.3] and [10.6]. The parameter maps attractions into probabilities.

[9] Other models aim to explain aggregated choices at the population level. Of course, models of this sort can sometimes be useful. But our view is that a parsimonious model which can explain very fine-grained data can probably explain aggregated data well too.

Each EWA parameter has a natural interpretation:

- The parameter δ is the weight placed on foregone payoffs. It presumably is affected by imagination (in psychological terms, the strength of counterfactual reasoning or regret, or in economic terms, the weight placed on opportunity costs and benefits) or reliability of information about foregone payoffs.
- The parameter ϕ decays previous attractions due to forgetting or, more interestingly, because agents are aware that the learning environment is changing and deliberately "retire" old information (much as firms junk old equipment more quickly when technology changes rapidly).
- The parameter κ controls the rate at which attractions grow. When $\kappa = 0$ attractions are weighted averages and grow slowly; when $\kappa = 1$ attractions cumulate. We originally included this variable because some learning rules used cumulation and others used averaging. It is also a rough way to capture the distinction in machine learning between "exploring" an environment (low κ), and "exploiting" what is known by locking in to a good strategy (high κ) (e.g., Sutton and Barto, 1998).
- The initial experience weight $N(0)$ is like a strength of prior beliefs in models of Bayesian belief learning. It plays a minimal empirical role so it is set to one in our current work.

EWA is a hybrid of two widely studied models, reinforcement and belief learning. In reinforcement learning, only payoffs from chosen strategies are used to update attractions and guide learning. In belief learning, players do not learn about which strategies work best; they learn about what others are likely to do, then use those updated beliefs to change their attractions and hence what strategies they choose (see Brown, 1951; Fudenberg and Levine, 1998). EWA shows that reinforcement and belief learning, which were often treated as fundamentally different, are actually related in a non-obvious way, because both are special kinds of a general reinforcement rule[10]. When $\delta = 0$ the EWA rule is a simple reinforcement rule.[11] When $\delta = 1$ and $\kappa = 0$ the EWA rule is equivalent to belief learning using weighted fictitious play.[12]

Foregone payoffs are the fuel that runs EWA learning. They also provide an indirect link to "direction learning" and imitation. In direction learning players move in the direction of observed best response (Selten and Stoecker, 1986). Suppose players follow EWA but don't know foregone payoffs, and believe those payoffs are monotonically

[10] See also Cheung and Friedman (1997, pp. 54–55), Fudenberg and Levine (1998, pp. 184–185); and Hopkins (2002).

[11] See Bush and Mosteller (1955), Harley (1981), Cross (1983), Arthur (1991), McAllister (1991), Roth and Erev (1995), and Erev and Roth (1998).

[12] When updated fictitious play beliefs are used to update the expected payoffs of strategies, precisely the same updating is achieved by reinforcing all strategies by their payoffs (whether received or foregone). The beliefs themselves are an epiphenomenon that disappear when the updating equation is written expected payoffs rather than beliefs.

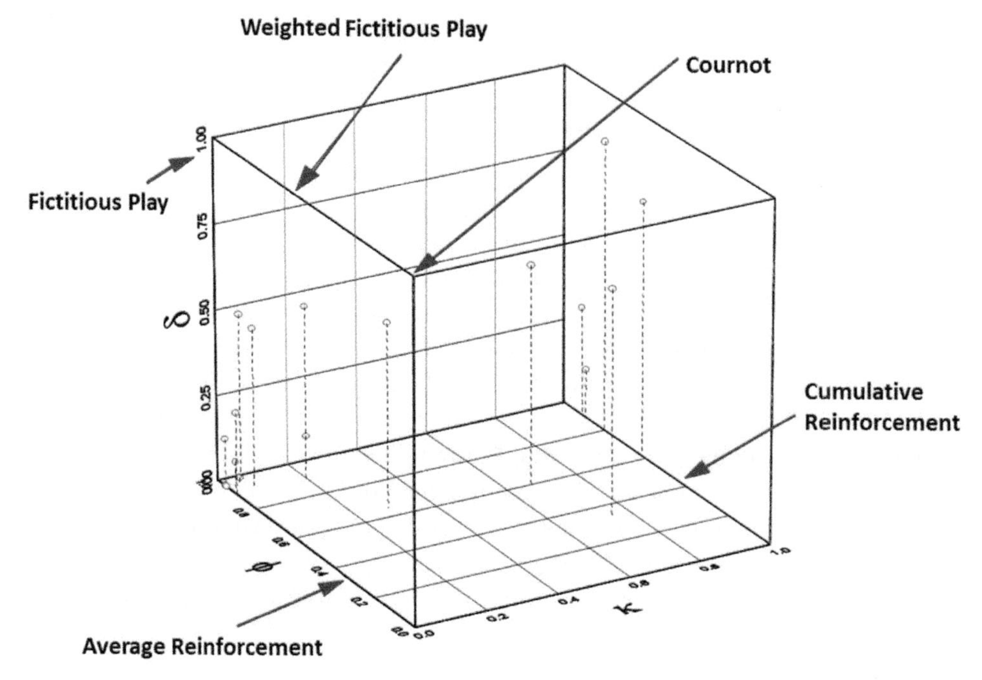

Figure 10.5 EWA cube.

increasing between their choice $s_i(t)$ and the best response. If they also reinforce strategies near their choice $s_i(t)$ more strongly than strategies that are further away, their behavior will look like direction learning. Imitating a player who is similar and successful can also be seen as a way of heuristically inferring high foregone payoffs from an observed choice and moving in the direction of those higher payoffs.

The relation of various learning rules can be shown visually in a cube showing configurations of parameter values (see Figure 10.5). Each point in the cube is a triple of EWA parameter values which specifies a precise updating equation. The corner of the cube with $\phi = \kappa = 0, \delta = 1$, is Cournot best-response dynamics. The corner $\kappa = 0, \phi = \delta = 1$, is standard fictitious play. The vertex connecting these corners, $\delta = 1, \kappa = 0$, is the class of weighted fictitious play rules (e.g., Fudenberg and Levine, 1998). The vertices with $\delta = 0$ and $\kappa = 0$ or 1 are averaging and cumulative choice reinforcement rules (Erev and Roth, 1998; Roth and Erev, 1995).

The EWA model has been estimated by ourselves and many others on about 40 data sets (see Camerer et al., 2002). The hybrid EWA model predicts more accurately than the special cases of reinforcement and weighted fictitious in most cases, except in games with mixed-strategy equilibrium where reinforcement does equally well.[13] In our

[13] In mixed games no model improves much on Nash equilibrium (and often don't improve on QRE at all), and parameter identification is poor; see Salmon, 2001.

model estimation and validation, we always penalize the EWA model in ways that are known to make the *adjusted* fit *worse* if a model is too complex (i.e., if the data are actually generated by a simpler model).[14] Furthermore, econometric studies show that if the data were generated by simpler belief or reinforcement models, then EWA estimates would correctly identify that fact for most games and reasonable sample sizes (see Cabrales and Garcia-Fontes, 2000; Salmon, 2001). Since EWA is capable of identifying behavior consistent with special cases, when it does not then the hybrid parameter values are improving fit.

Figure 10.5 also shows estimated parameter triples from 20 data sets. Each point is an estimate from a different game. If one of the special case theories is a good approximation to how people generally behave across games, estimated parameters should cluster in the corner or vertex corresponding to that theory. In fact, parameters tend to be sprinkled around the cube, although many (typically mixed-equilibrium games) cluster in the averaged reinforcement corner with low δ and κ. The dispersion of estimates in the cube raises an important question: Is there regularity in which games generate which parameter estimates? A positive answer to this question is crucial for predicting behavior in brand new games.

10.4.2 fEWA functions

This concern is addressed by a version of EWA, fEWA, which replaces free parameters with specific values and functions that are then used in the EWA updating equation to determine attractions, which then determine choices probabilistically. Since the functions also vary across subjects and over time, they have the potential to inject heterogeneity and time-varying "rule learning," and to explain learning better than models with fixed parameter values across people and time. And since fEWA has only one parameter which must be estimated (λ),[15] it is especially helpful when learning models are used as building blocks for more complex models that incorporate sophistication (some players think others learn) and teaching, as we discuss in the section below.

As shown in Figure 10.5, the front face of the cube ($\kappa = 0$) captures almost all familiar special cases except for the cumulative reinforcement model. The cumulative model has been supplanted by the averaging model (with $\kappa = 0$) because the latter seems to be more robust in predicting behavior in some games (see Roth and Erev, 1995). This sub-class of EWA learning models is the simplest model that nests averaging reinforcement and weighted ficititous play (and hence Cournot and simple fictitious play) as special cases.

[14] We typically penalize in-sample likelihood functions using the Akaike and Bayesian information criteria, which subtract a penalty of one, or $\log(n)$, times the number of degrees of freedom from the maximized likelihood. More persuasively, we rely mostly on out-of-sample forecasts which will be *less* accurate if a more complex model simply appears to fit better because it overfits in-sample.

[15] Note that if your statistical objective is to maximize hit rate, λ does not matter and so fEWA is a zero-parameter theory given initial conditions.

It can also capture a weighted fictitious play model using time-varying belief weights (such as the stated-beliefs model explored by Nyarko and Schotter, 2002), as long as subjects are allowed to use a different weight to decay lagged attractions over time (i.e., move along the top edge of the side in Figure 10.5). There is also an empirical reason to set κ to a particular value. Our prior work suggests that κ does not seem to affect fit much (e.g., Ho et al., 2008). The initial experience $N(0)$ was included in the original EWA model so that Bayesian learning models are nested as a special case—$N(0)$ represents the strength of prior beliefs. We restrict $N(0) = 1$ here because its influence fades rapidly as an experiment progresses and most subjects come to experiments with weak priors anyway.

Consequently, we are left with three free parameters– ϕ, δ, and λ. To make the model simple to estimate statistically, and self-tuning, the parameters ϕ and δ are replaced by deterministic functions $\phi_i(t)$ and $\delta_{ij}(t)$ of player i's experience with strategy j, up to period t. These functions determine numerical parameter values for each player, strategy, and period, which are then plugged into the EWA updating equation above to determine attractions in each period. Updated attractions determine choice probabilities according to the logit rule, given a value of λ. Standard maximum-likelihood methods for optimizing fit can then be used to find which λ fits best.[16]

10.4.2.1 The change-detector function $\phi_i(t)$

The decay rate ϕ which weights lagged attractions is sometimes called "forgetting" (an interpretation, which is carried over from reinforcement models of animal learning). While forgetting obviously does occur, the more interesting variation in $\phi_i(t)$ across games, and across time within a game, is a player's perception of how quickly the learning environment is changing. The function $\phi_i(t)$ should therefore "detect change." When a player senses that other players are changing, a self-tuning $\phi_i(t)$ should dip down, putting less weight on distant experience. As in physical change detectors (e.g., security systems or smoke alarms), the challenge is to detect change when it is really occurring, but not falsely mistake small fluctuations for real changes too often.

The core of the $\phi_i(t)$ change-detector function is a "surprise index," which is the difference between other players' most recently chosen strategies and their chosen strategies in all previous periods. First define a cumulative history vector, across the other players' strategies k, which records the historical frequencies (including the last period t) of the choices by other players. The vector element $h_i^k(t)$ is $\frac{\sum_{\tau=1}^{t} I(s_{-i}^k, s_{-i}(\tau))}{t}$.[17]

[16] If one is interested only in the hit rate—the frequency with which the predicted choice is the same as what a player actually picked—then it is not necessary to estimate λ. The strategy that has the highest attraction will be the predicted choice. The response sensitivity λ only dictates how *frequently* the highest-attraction choice is actually picked, which is irrelevant if the statistical criterion is hit rate.

[17] Note that if there is more than one other player, and the distinct choices by different other player's matter to player i, then the vector is an $n-1$-dimensional matrix if there are n players.

The immediate 'history' $r_i^k(t)$ is a vector of 0's and 1's which has a one for strategy $s_{-i}^k = s_{-i}(t)$ and 0's for all other strategies s_{-i}^k (i.e., $r_i^k(t) = I(s_{-i}^k, s_{-i}(t))$). The surprise index $S_i(t)$ simply sums up the squared deviations between the cumulative history vector $h_i^k(t)$ and the immediate history vector $r_i^k(t)$; that is,

$$S_i(t) = \sum_{k=1}^{m_{-i}} (h_i^k(t) - r_i^k(t))^2. \qquad [10.7]$$

In other words, the surprise index captures the degree of change of the most recent observation[18] from the historical average. Note that it varies from zero (when the last strategy the other player chose is the one they have always chosen before) to two (when the other player chose a particular strategy 'forever' and suddenly switches to something brand new). When surprise index is zero, we have a stationary environment; when it is two, we have a turbulent environment. The change-detecting decay rate is:

$$\phi_i(t) = 1 - \frac{1}{2} \cdot S_i(t). \qquad [10.8]$$

Because $S_i(t)$ is between zero and two, ϕ is always (weakly) between one and zero.

Some numerical boundary cases help illuminate how the change-detection works. If the other player chooses the strategy she has always chosen before, then $S_i(t) = 0$ (player i is not surprised) and $\phi_i(t) = 1$ (player i does not decay the lagged attraction at all, since what other players did throughout is informative). The opposite case is when an opponent has previously chosen a single strategy in every period, and suddenly switches to a new strategy. In that case, $\phi_i(t)$ is $\frac{2t-1}{t^2}$. This expression declines gracefully toward zero as the string of identical choices up to period t grows longer. (For $t = 2$, 3, 5 and 10 the $\phi_i(t)$ values are 0.75, 0.56, 0.36, and 0.19.) The fact that the ϕ values decline with t expresses the principle that a new choice is bigger surprise (and should have an associated lower ϕ) if it follows a *longer* string of *identical* choices which are different from the surprising new choice. Note that since the observed behavior in period t is included in the cumulative history $h_i^k(t)$, $\phi_i(t)$ will never dip completely to zero. (which could be a mistake because it erases all the history embodied in the lagged attraction). For example, if a player chose the same strategy for each of 19 periods and a new strategy in period 20, then $\phi_i(t) = 39/400 = 0.0975$.

Another interesting special case is when unique strategies have been played in every period up to $t - 1$, and another unique strategy is played in period t. (This is often true in games with large strategy spaces.) Then $\phi_i(t) = 0.5 + \frac{1}{2t}$, which starts at 0.75

[18] In games with mixed equilibria (and no pure equilibria), a player should expect other players' strategies to vary. Therefore, if the game has a mixed equilibrium with W strategies which are played with positive probability, the surprise index defines recent history over a window of the last W periods (e.g., in a game with four strategies that are played in equilibrium, $W = 4$). Then $r_i^k(t) = \sum_{k=1}^{m_{-i}} \left[\frac{\sum_{\tau=t-W+1}^{t} I(s_{-i}^k, s_{-i}(\tau))}{W} \right]$.

and asymptotes at 0.5 as t increases. Comparing the case where the previous strategy was the same, and the previous strategies were all different, it is evident that if the choice in period t is new, the value of $\phi_i(t)$ is higher if there were more variation in previous choices, and lower if there were less variation. This mimics a hypothesis-testing approach in which more variation in previous strategies implies that players are less likely to conclude there has been a regime shift, and therefore do not lower the value of $\phi_i(t)$ too much.

Note that the change-detector function $\phi_i(t)$ is individual and time specific and it depends on information feedback. Nyarko and Schotter (2002) show that a weighted fictitious play model that uses stated beliefs (instead of empirical beliefs posited by the fictitious play rule) can predict behavior better than the original EWA model in games with unique mixed-strategy equilibrium. One way to intrepret their result is that their model allows each subject to attach a different weight to previous experiences over time. In the same vein, the proposed change-detector function allows for individual and time heterogeneity by positing them theoretically.

10.4.2.2 The attention function, $\delta_{ij}(t)$

The parameter δ is the weight on foregone payoffs. Presumably this is tied to the attention subjects pay to alternative payoffs, ex-post. Subjects who have limited attention are likely to focus on strategies that would have given higher payoffs than what was actually received, because these strategies present missed opportunities, which show that such a regret-driven rule converges to correlated equilibrium (Hart and Mas-Colell, 2001). To capture this property, define[19]

$$\delta_{ij}(t) = \begin{cases} 1 & \text{if } \pi_i(s_i^j, s_{-i}(t)) \geq \pi_i(t), \\ 0 & \text{otherwise.} \end{cases} \qquad [10.9]$$

That is, subjects reinforce chosen strategies and all unchosen strategies with (weakly) better payoffs by a weight of one. They reinforce unchosen strategies with strictly worse payoffs by zero.

Note that this $\delta_{ij}(t)$ can transform the self-tuning rule into special cases over time. If subjects are strictly best-responding (ex post), then no other strategies have a higher ex-post payoff. Hence $\delta_{ij}(t) = 0$ for all strategies j which were not chosen, reducing the model to choice reinforcement. However if they always choose the worst strategy, then $\delta_{ij}(t) = 1$, which corresponds to weighted fictitious play. If subjects neither choose

[19] In games with unique mixed-strategy equilibrium, we use $\delta_{ij}(t) = \frac{1}{W}$ if $\pi_i(s_i^j, s_{-i}(t)) \geq \pi_i(t)$ and 0 otherwise. This modification is driven by the empirical observation that estimated δ's are often close to zero in mixed games (which might also be due to misspecified heterogeneity). Using only $\delta_{ij}(t)$ without this adjustment produces slightly worse fits in the two mixed-equilibrium games examined below where the adjustment matters (patent-rate games and the Mookerjhee-Sopher games).

the best nor the worst strategy, the updating scheme will push them (probabilistically) towards those strategies that yield better payoffs, as is both characteristic of human learning and normatively sensible.

The updating rule is a natural way to formalize and extend the "learning direction" theory of Selten and Stoecker (1986). Their theory consists of an appealing property of learning: Subject move in the direction of ex-post best-response. Broad applicability of the theory has been hindered by defining "direction" only in terms of numerical properties of ordered strategies (e.g., choosing 'higher prices' if the ex-post best response is a higher price than the chosen price). The self-tuning $\delta_{ij}(t)$ defines the "direction" of learning set-theoretically, as shifting probability toward the set of strategies with higher payoffs than the chosen ones.

The self-tuning $\delta_{ij}(t)$ also creates the "exploration-exploitation" shift in machine learning (familiar to economists from multi-armed bandit problems). In low-information environments, it makes sense to explore a wide range of strategies, then gradually lock in to a choice with a good historical relative payoffs. In self-tuning EWA, if subjects start out with a poor choice, many unchosen strategies will be reinforced by their (higher) foregone payoffs, which shifts choice probability to those choices and captures why subjects "explore." As equilibration occurs, only the chosen strategy will be reinforced, thereby producing an "exploitation" or "lock-in." This is behaviorally very plausible. The updating scheme also helps to detect any change in environment. If a previously optimal response becomes inferior because of an exogenous change, other strategies will have higher ex-post payoffs, triggering higher $\delta_{ij}(t)$ values (and reinforcement of superior payoffs) and guiding players to re-explore better strategies.

The self-tuning $\delta_{ij}(t)$ function can also be seen as a reasonable all-purpose rule which conserves a scarce cognitive resource—attention. The parametric EWA model shows that weighted fictitious play is equivalent to generalized reinforcement in which all strategies are reinforced. But reinforcing many strategies takes attention. As equilibration occurs, the set of strategies which receive positive $\delta_{ij}(t)$ weights shrinks so attention is conserved when spreading attention widely is no longer useful. When an opponent's play changes suddenly, the self-tuning $\phi_i(t)$ value drops. This change reduces attractions (since lagged attractions are strongly decayed) and spreads choice probability over a wider range of strategies due to the logit response rule. This implies that the strategy chosen may no longer be optimal, leading $\delta_{ij}(t)$ to allocate attention over a wider range of better-responses. Thus, the self-tuning system can be seen as procedurally rational (in Herbert Simon's language) because it follows a precise algorithm and is designed to express the basic features of how people learn—by exploring a wide range of options, locking in when a good strategy is found, but re-allocating attention when environmental change demands such action.

A theorist's instinct is to derive conditions when flexible learning rules choose parameters optimally, which is certainly a direction to explore in future research.

However, broadly optimal rules will likely depend on the set of games an all-purpose learning agent encounters, and also may depend sensitively on how cognitive costs are specified (and should also jibe with data on the details of neural mechanisms, which are not yet well-understood). So it is unlikely to find a universally optimal rule that can always beat rules which adapt locally.

Our approach is more like the exploratory work in machine learning. Machine learning theorists try to develop robust heuristic algorithms which learn effectively in a wide variety of low-information environments. Good machine learning rules are not provably optimal but perform well on tricky test cases and natural problems like those which good computerized robots need to perform (navigating around obstacles, hill-climbing on rugged landscapes, difficult pattern recognition, and so forth).

Before proceeding to estimation, it is useful to summarize the properties of the self-tuning model. First, the use of simple fictitious play and reinforcement theories in empirical analysis is often justified by the fact that they have only a few free parameters. The self-tuning EWA is useful by this criterion because it requires estimating only one parameter, λ (which is difficult to do without in empirical work). Second, the functions in self-tuning EWA naturally vary across time, people, games, and strategies. The potential advantage of this flexibility is that the model can predict across new games better than parametric methods. Whether this advantage is realized will be examined below. Third, the self-tuning parameters can endogenously shift across rules. Early in a game, when opponent choices vary a lot and players are likely to make ex-post mistakes, the model automatically generates low values of $\phi_i(t)$ and high $\delta_{ij}(t)$ weights—it resembles Cournot belief learning. As equilibration occurs and behavior of other players stabilizes, $\phi_i(t)$ rises and $\delta_{ij}(t)$ falls—it resembles reinforcement learning. The model therefore keeps a short window of history (low ϕ) and pays a lot of attention (high δ) when it should, early in a game, and conserves those cognitive resources by remembering more (high ϕ) and attending to fewer foregone strategies (low δ) when it can afford to, as equilibration occurs.

fEWA has three advantages. First, it is easy to use because it has only one free parameter (λ). Second, parameters in fEWA naturally vary across time and people (as well as across games), which can capture heterogeneity and mimic "rule learning" in which parameters vary over time (e.g., Salmon, 1999, and Stahl, 1996, 2000). For example, if ϕ rises across periods from 0 to 1 as other players stabilize, players are effectively switching from Cournot-type dynamics to fictitious play. If δ rises from 0 to 1, players are effectively switching from reinforcement to belief learning. Third, it should be easier to theorize about the limiting behavior of fEWA than about some parametric models. A key feature of fEWA is that as a player's opponents' behavior stabilizes, $\phi_i(t)$ goes toward one and (in games with pure equilibria) $\delta_i(t)$ does too. Since $\kappa = 0$, fEWA automatically turns into fictitious play; and a lot is known about theoretical properties of fictitious play.

10.4.3 fEWA predictions

In this section we compare in-sample fit and out-of-sample predictive accuracy of different learning models when parameters are freely estimated, and check whether fEWA functions can produce game-specific parameters similar to estimated values. We use seven games: Games with unique mixed strategy equilibrium (Mookerjhee and Sopher, 1997); R&D patent race games (Rapoport and Amaldoss, 2000); a median-action order statistic coordination game with several players (Van Huyck et al., 1990); a continental-divide coordination game, in which convergence behavior is extremely sensitive to initial conditions (Van Huyck et al., 1997); dominance-solvable p-beauty contests (Ho et al., 1998); and a price-matching game (called "travelers' dilemma" by Capra et al., 1999).

The estimation procedure for fEWA is sketched briefly here (see Ho et al., 2007 for details). Consider a game where N subjects play T rounds. For a given player i, the likelihood function of observing a choice history of $\{s_i(1), s_i(2), \ldots, s_i(T-1), s_i(T)\}$ is given by:

$$\Pi_{t=1}^{T} P_i^{s_i(t)}(t) \qquad [10.10]$$

The joint likelihood function L of observing all players' choice is given by

$$L(\lambda) = \Pi_i^N \left[\Pi_{t=1}^{T} P_i^{s_i(t)}(t) \right] \qquad [10.11]$$

Most models use first-period data to determine initial attractions. We also compare all models with burned-in attractions with a model in which the thinking steps model from the previous section is used to create initial conditions and combined with fEWA. Note that the latter hybrid uses only two parameters (τ and λ) and does not use first-period data at all.

Given the initial attractions and initial parameter values,[20] attractions are updated using the EWA formula. fEWA parameters are then updated according to the functions above and used in the EWA updating equation. Maximum likelihood estimation is used to find the best-fitting value of λ (and other parameters, for the other models) using data from the first 70% of the subjects. Then the value of λ is frozen and used to forecast behavior of the entire path of the remaining 30% of the subjects. Payoffs were all converted to dollars (which is important for cross-game forecasting).

In addition to fEWA (one parameter), we estimated the parametric EWA model (five parameters), a belief-based model (weighted fictitious play, two parameters) and the two-parameter reinforcement models with payoff variability (Erev et al., 1999; Erev et al., 2002), and QRE.

[20] The initial parameter values are $\phi_i(0) = 0.5$ and $\delta_i(0) = \phi_i(0)/W$. These initial values are averaged with period-specific values determined by the functions, weighting the initial value by $\frac{1}{t}$ and the functional value by $\frac{t-1}{t}$.

The first question we ask is how well models fit and predict on a game-by-game basis (i.e., parameters are estimated separately for each game). For out-of-sample validation we report both hit rates (the fraction of most-likely choices which are picked) and log likelihood (LL). Recall that these results forecast a holdout sample of subjects *after* model parameters have been estimated on an earlier sample and then "frozen" for holdout. If a complex model is fitting better within a sample purely because of spurious overfitting, it will predict more poorly out of sample. Results are summarized in Table 10.3.

The best fits for each game and criterion are printed in bold; hit rates that are statistically indistinguishable from the best (by the McNemar test) are also in bold. Across games, parametric EWA is as good as all other theories or better, judged by hit rate, and has the best LL in four games. fEWA also does well on hit rate in six of seven games. Reinforcement is competitive on hit rate in five games and best in LL in two. Belief models are often inferior on hit rate and never best in LL. QRE clearly fits worst.

Combining fEWA with CH model to predict initial conditions (rather than using the first-period data) is only a little worse in hit rate than EWA and slightly worse in LL.

The bottom line of Table 10.3, "pooled," shows results when a single set of common parameters is estimated for all games (except for game-specific λ). If fEWA is capturing parameter differences across games effectively, it should predict especially accurately, compared to other models, when games are pooled. It does: when all games are pooled, fEWA predicts out-of-sample better than other theories, by both statistical criteria.

Some readers of our functional EWA paper were concerned that by searching across different specifications, we may have overfitted the sample of seven games we reported. To check whether we did, we announced at conferences in 2001 that we would analyze all the data people sent us by the end of the year and report the results in a revised paper. Three samples were sent and we have analyzed one so far—experiments by Kocher and Sutter (2005) on *p*-beauty contest games played by individuals and groups. The KS results are reported in the bottom row of Table 10.3. The game is the same as the beauty contests we studied (except for the interesting complication of group decision making, which speeds equilibration), so it is not surprising that the results replicate the earlier findings: Belief and parametric EWA fit best by LL, followed by fEWA, and reinforcement and QRE models fit worst. This is a small piece of evidence that the solid performance of fEWA (while worse than belief learning on these games) is not entirely due to overfitting on our original seven-game sample.

The table also shows results (in the column headed "Thinking+fEWA") when the initial conditions are created by the CH model rather than from first-period data and combined with the fEWA learning model. Thinking plus fEWA are also a little more accurate than the belief and reinforcement models in five of seven games. The hit rate and LL suffer only a little compared to the fEWA with estimated parameters. When common parameters are estimated across games (the row labelled "pooled"), fixing initial conditions with the thinking steps model only lowers fit slightly.

Table 10.3 Out of sample accuracy of learning models (Ho et al. 2007)

Game	Thinking +fEWA		fEWA		EWA		Weighted Fict. Play		Reinf. with PV		QRE	
	%Hit	LL	%Hit	LL	%Hit	LL	%Hit	LL	%Hit	LL	%Hit	LL
Cont'l divide	**45**	−483	**47**	−470	**47**	**−460**	25	−565	45	−557	5	−806
Median action	71	−112	74	−104	**79**	**−83**	**82**	−95	74	−105	49	−285
p-BC	**8**	−2119	**8**	−2119	**6**	**−2042**	**7**	−2051	**6**	−2504	4	−2497
Price matching	**43**	−507	**46**	−445	**43**	**−443**	36	−465	**41**	−561	27	−720
Mixed games	**36**	−1391	**36**	**−1382**	**36**	−1387	**34**	−1405	33	−1392	**35**	−1400
Patent Race	**64**	−1936	**65**	−1897	**65**	−1878	53	−2279	**65**	**−1864**	40	−2914
Pot Games	**70**	−438	**70**	−436	**70**	−437	66	−471	**70**	**−429**	51	−509
Pooled	**50**	−6986	**51**	**−6852**	49	−7100	40	−7935	46	−9128	36	−9037
KS p−BC			6	−309	3	−279	3	−279	4	−344	1	−346

Note: Sample sizes are 315, 160, 580, 160, 960, 1760, 739, 4674 (pooled), 80.

Note that all learning models have a low hit-rate in p-beauty contests (p-BC). This is so for three reasons. First, the p-BC has many strategies (101 altogether) and as a consequence it is much harder to predict behavior correctly in this game. Second, some subjects may be sophisticated and their behavior may not be captured by learning models (see Section 10.5 for details). Finally, subjects may exhibit increasing sophistication over time (i.e., subjects increase their depth of reasoning) and adaptive learning models do not allow for this dynamic rule shift (see Ho and Su, 2013a,b for a dynamic model that explicitly accounts for such behavior).

10.4.4 Example: mixed strategy games

Some of the first games studied experimentally featured only mixed-strategy equilibria.[21] These games usually have a purely competitive structure in which one person wins and another loses for each strategy combination (although they are not always zero-sum). This section summarizes the discussion in Camerer (2003, Chapter 3) and updates it.

The earliest results were considered to be discouraging rejections of mixed equilibrium. Most lab experiments do show non-independent over-alternation of strategies (i.e., subjects use fewer long runs than independence predicts). However, the conclusion that actual mixture frequencies were far from equilibrum proved to be too pessimistic. A survey of many experimental studies in Camerer (2003, Chapter 3) comparing actual *aggregate* choice frequencies with predicted frequencies shows they are highly correlated across all games ($r = 0.84$); the mean absolute deviation between predictions and data is 0.067. Fitting QRE instead of Nash mixtures improves the actual-predicted correlation a bit ($r = 0.92$) . . . and there is little more room for improvement because the correlation is so high!

Careful experiments by Binmore et al. (2001) and some field data on tennis and soccer (e.g., Palacio-Huerta and Volij, 2008; Walker and Wooders, 2001) draw a similarly positive conclusion about overall accuracy of mixed equilibrium.[22] So does an unusual example we describe next.

[21] The earliest studies were conducted in the 1950s, shortly after many important ideas were consolidated and extended in Von Neumann and Morgenstern's (1944) landmark book. John Nash himself conducted some informal experiments during a famous summer at the RAND Corporation in Santa Monica, California. Nash was reportedly discouraged that subjects playing games repeatedly did not show behavior predicted by theory: "The experiment, which was conducted over a two-day period, was designed to test how well different theories of coalitions and bargaining held up when real people were making the decisions . . . For the designers of the experiment . . . the results merely cast doubt on the predictive power of game theory and undermined whatever confidence they still had in the subject" (Nasar, 1998).

[22] The Palacios-Huerta and Volij study is unique because they are able to match data from actual play on the field from one group of players (in European teams) with laboratory behavior of some players from that group (although not matching the same players' field and lab data). Importantly, PHV also found that high school students as a group behaved less game theoretically than the soccer pros, except that high school students with substantial experience playing soccer were much closer to game-theoretic. However, a reanalysis by Wooders (2010) later showed a higher degree of statistical dependence than shown by PHV.

Table 10.4 Asymmetric mixed-strategy hide-seek games in Martin et al. (2013)

	L	R	Nash mse
L	X,0	0,2	$\frac{1}{2}$
R	0,2	Y,0	$\frac{1}{2}$
Nash	$\frac{Y}{X+Y}$	$\frac{X}{X+Y}$	

In the example study Martin et al. (2013) used three variants of asymmetric matching pennies (Table 10.4).

Subjects simultaneously chose L or R by pressing a touch screen display of a left or right box (like an iPad). The "Matcher" player wins either X or Y if both players match their choices on L or R, respectively. The "Mismatcher" player wins if they mismatch. Subjects participated in multiple sessions with 200 trials in each, switching matcher/mismatcher roles after each session. The games used (X,Y) matcher payoff pairs of (2,2), (3,1) and (4,1). NE makes the highly counterinitutive prediction that while the matcher payoffs change across the three games, the predicted $P_{\text{match}}(R) = 0.50$ does *not* change, but predicted $P_{\text{mismatch}}(R)$ does change, from 0.50 to 0.75 to 0.80. Data averaged over trials are shown in Figure 10.6 shows the averaged results from each of the three games, choosing over several hundred trials. The actual frequencies of $P_{\text{mismatch}}(R)$ The frequencies are remarkably close to the NE predictions. In five of the six comparisons, the absolute deviations are 0-0.03.

The experiments used six chimpanzees housed in the Kyoto PRI facility, playing in three mother-child pairs for apple cube payoffs delivered immediately. Also, we used a no-information protocol in which the chimpanzees were not told what the payoffs were; instead, they learned them from trial and error.

We can compare the chimpanzee behavior with two human groups, which separately played one of the three games, with $(X, Y) = (4, 1)$ (called "inspection game"), also using the low-information protocol. The chimpanzees are clearly closer to NE than the humans. The two human groups' data are close together, despite large differences in the two groups (Japanese vs. African workers at a reserve in Boussou earning the equivalent of US $134). The human subjects' deviations are also similar in magnitude to many earlier experiments.

Why are the chimpanzees so close, and closer than humans? The apparent reason is that they learn better. Martin et al. estimated a weighted fictitious play model in which beliefs update according to $p(t+1)^* = p_t^* + \eta\delta_t^p$ where $\delta_t^p = P_t - p_t^*$ is the prediction error (the difference between observed play, 0 or 1, and belief). The mismatcher choice probability in each trial is $f(V^L - V^R)^{\text{Mismatcher}} = f([\pi(L,L)p^* + \pi(R,L)(1 - p^*)] - [\pi(L,R)p^* + \pi(R,R)(1 - p^*)])$ where $f()$ is the logit function $f(V^a - V^b) = 1/(1 + e^{-\rho(V^a - V^b + \alpha)})$ (where α allows for bias toward L or R). When

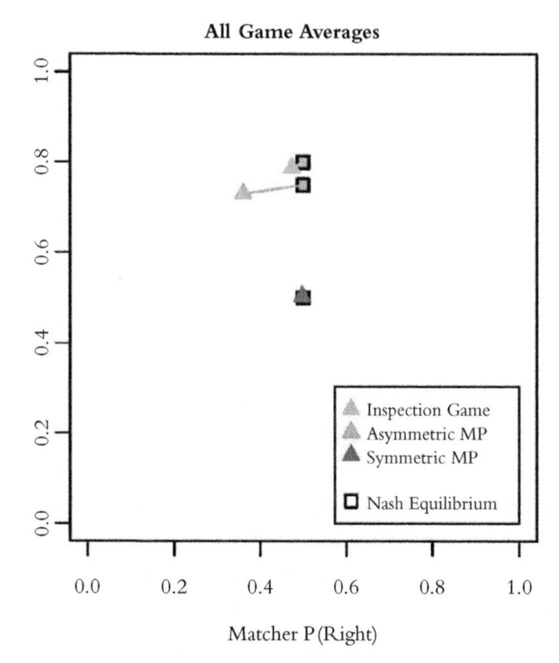

Figure 10.6 Actual relative frequencies of chimpanzee choices and predictions of Nash equilibrium, in three games.

fictitious play (or reinforcement learning[23]) models are fit to the chimpanzee and human data, the chimpanzees show an enhanced sensitivity to past reward and opponent play (higher learning weight η, and to payoff structure (changing behavior between the matcher and mismatcher roles), compared to humans.

Martin et al. (2013) speculate that the chimpanzees are as good as the human subjects in mixing strategically (or perhaps even better in general), because chimpanzees may have retained the cognitive capacity to detect patterns in competitive games, and practice it more often in childhood, than humans do (Matsuzawa, 2007). Winning repeated competitions is important in the chimpanzees' natural ecology, in the form of hide-and-seek (predator-prey) games and winning physical competitions to establish status dominance. The learning model results suggest the chimpanzees learn better than the humans do. In general, the results strikingly supports the interpretation of (Binmore, 1999) of game theory as often "evolutive"—that is, describing the result of an adaptive, perhaps biological process, rather than introspection and analysis.[24]

[23] Note that EWA was not fit because in these types of games reinforcement, fictitious play and EWA are difficult to separate statistically

[24] This is a plausible explanation. However, there are other differences between the chimpanzee and human experiments. The most obvious difference is that chimpanzees receive instant gratification (e.g., they

10.4.5 Summary

Learning is clearly important for economics. Equilibrium theories are useful because they suggest a possible limit point of a learning process and permit comparative static analysis. But if learning is slow, or the time path of behavior selects one equilibrium out of many, a precise theory of equilibration is crucial for knowing which equilibrium will result, and how quickly.

The theory described in this section, fEWA, replaces the key parameters in the parametric EWA learning models with functions that change over time in response to experience. One function is a "change detector" ϕ which goes up (limited by one) when behavior by other players is stable, and dips down (limited by zero) when there is surprising new behavior by others. When ϕ dips down, the effects of old experience (summarized in attractions which average previous payoffs) is diminished by decaying the old attraction by a lot. The second "attention" function δ is one for strategies that yield equal or better than actual payoff and zero otherwise. This function ties sensitivity to foregone payoffs to attention, which is likely to be on strategies that give equal or better than actual payoff ex post. Self-tuning EWA is more parsimonious than most learning theories because it has only one free parameter– the response sensitivity λ.

We report out-of-sample prediction of data from several experimental games using fEWA, the parameterized EWA model, and QRE. Both QRE and self-tuning EWA have one free parameter, and EWA has five. We show that fEWA predicts slightly worse than parametric EWA in these games. Since fEWA generates functional parameter values for ϕ and δ, which vary sensibly across games, it predicts better than other QRE and parametric EWA when games are pooled and common parameters are estimated. fEWA therefore represents one solution to the problem of flexibly generating EWA-like parameters across games.

10.5. SOPHISTICATION AND TEACHING

The learning models discussed in the last section are adaptive and backward-looking: Players only respond to their own previous payoffs and knowledge about what others did. While a reasonable approximation, these models leave out two key features: Adaptive players do not explicitly use information about *other* players' payoffs (though subjects actually do[25]); and adaptive models ignore the fact that when the same players are matched together repeatedly, their behavior is often different than then they are not rematched together, generally in the direction of greater efficiency (e.g., Andreoni and Miller, 1993; Clark and Sefton, 1999; Van Huyck, et al., 1990).

eat the apple cubes right away) and human subjects earn money that can only be traded for a delayed gratification. Such differences between the two experiments could also explain the observed differences in behaviors.

[25] Partow and Schotter (1993), Mookerjee and Sopher (1994), and Cachon and Camerer (1996).

In this section adaptive models are extended to include sophistication and strategic teaching in repeated games (see Camerer et al., 2002; and Stahl, 1999, for details). Sophisticated players believe that others are learning and anticipate how others will change in deciding what to do. In learning to shoot a moving target, for example, soldiers and fighter pilots learn to shoot *ahead*, toward where the target *will be*, rather than shoot at the current target. They become sophisticated.

Sophisticated players who also have strategic foresight will "teach"—that is, they choose current actions which teach the learning players what to do, in a way that benefits the teacher in the long-run. Teaching can be either mutually beneficial (trust-building in repeated games) or privately beneficial but socially costly (entry-deterrence in chain-store games). Note that sophisticated players will use information about payoffs of others (to forecast what others will do) and will behave differently depending on how players are matched, so adding sophistication can conceivably account for effects of information and matching that adaptive models miss.[26]

10.5.1 Sophistication

Let's begin with myopic sophistication (no teaching). The model assumes a population mixture in which a fraction α of players are sophisticated. To allow for possible overconfidence, sophisticated players think that a fraction $(1-\alpha')$ of players are adaptive and the remaining fraction α' of players are sophisticated like themselves.[27] Sophisticated players use the fEWA model to forecast what adaptive players will do, and choose strategies with high expected payoffs given their forecast and their guess about what sophisticated players will do. Denoting choice probabilities by adaptive and sophisticated players by $P_i^j(a,t)$ and $P_i^j(s,t)$, attractions for sophisticates are

$$A_i^j(s,t) = \sum_{k=1}^{m_{-i}} [\alpha' P_{-i}^k(s,t+1) + (1-\alpha') \cdot P_{-i}^k(a,t+1)] \cdot \pi_i(s_i^j, s_{-i}^k) \quad [10.12]$$

Note that since the probability $P_{-i}^k(s,t+1)$ is derived from an analogous condition for $A_i^j(s,t)$, the system of equations is recursive. Self-awareness creates a whirlpool of recursive thinking which means QRE (and Nash equilibrium) are special cases in which all players are sophisticated and believe others are too ($\alpha = \alpha' = 1$).

An alternative model links steps of sophistication to the steps of thinking used in the first period. For example, define zero learning steps as using fEWA; one step is

[26] Sophistication may also potentially explain why players sometimes move in the *opposite* direction predicted by adaptive models (Rapoport et al., 1999), and why measured beliefs do not match up well with those predicted by adaptive belief learning models (Nyarko and Schotter, 2002).

[27] To truncate the belief hierarchy, the sophisticated players believe that the other sophisticated players, like themselves, believe there are α' sophisticates.

Table 10.5 Sophisticated and adaptive learning model estimates for the *P*-beauty contest game (Camerer et al., 2002) (standard errors are in parentheses)

	Inexperienced subjects		Experienced subjects	
	Sophisticated EWA	**Adaptive EWA**	**Sophisticated EWA**	**Adaptive EWA**
ϕ	**0.44**	0.00	**0.29**	0.22
	$(0.05)^2$	(0.00)	(0.03)	(0.02)
δ	**0.78**	0.90	**0.67**	0.99
	(0.08)	(0.05)	(0.05)	(0.02)
α	**0.24**	0.00	**0.77**	0.00
	(0.04)	(0.00)	(0.02)	(0.00)
α'	**0.00**	0.00	**0.41**	0.00
	(0.00)	(0.00)	(0.03)	(0.00)
LL				
(In sample)	**−2095.32**	−2155.09	**−1908.48**	−2128.88
(Out of sample)	**−968.24**	−992.47	**−710.28**	−925.09

best-responding to zero-step learners; two steps is best-responding to choices of one-step sophisticates, and so forth (see Ho et al., 2013). This model can produce results similar to the recursive one we report below, and it replaces α and α' with τ from the theory of initial conditions so it reduces the entire thinking-learning-teaching model to only two parameters. In addition, this model allows players to become more or less sophisticated over time as they learn about others' thinking steps.

We estimate the sophisticated EWA model using data from p-beauty contests introduced above. Table 10.5 reports results and estimates of important parameters (with bootstrapped standard errors). For inexperienced subjects, adaptive EWA generates Cournot-like estimates ($\hat{\phi} = 0$ and $\hat{\delta} = 0.90$). Adding sophistication increases $\hat{\phi}$ and improves LL substantially both in- and out-of-sample. The estimated fraction of sophisticated players is 24% and their estimated perception $\hat{\alpha}'$ is zero, showing overconfidence (as in the thinking-steps estimates from the last section).[28]

Experienced subjects are those who play a second 10-period game with a different p parameter (the multiple of the average which creates the target number). Among experienced subjects, the estimated proportion of sophisticates increases to $\hat{\alpha} = 77\%$. Their estimated perceptions increase too but are still overconfident ($\hat{\alpha}' = 41\%$). The estimates reflect "learning about learning": Subjects who played one 10-period game come to realize an adaptive process is occurring; and most of them anticipate that others are learning when they play again.

[28] The gap between apparent sophistication and perceived sophistication shows the empirical advantage of separating the two. Using likelihood ratio tests, we can clearly reject both the rational expectations restriction $\alpha = \alpha'$ and the pure overconfidence restriction $\alpha' = 0$ although the differences in log-likelihood are not large.

10.5.2 Strategic teaching

Sophisticated players matched with the same players repeatedly often have an incentive to "teach" adaptive players, by choosing strategies with poor short-run payoffs which will change what adaptive players do, in a way that benefits the sophisticated player in the long-run. Game theorists have showed that strategic teaching could select one of many repeated-game equilibria (teachers will teach the pattern that benefits them) and could give rise to reputation formation without the complicated apparatus of Bayesian updating of Harsanyi-style payoff types (see Fudenberg and Levine, 1989; Watson, 1993; Watson and Battigali, 1997). This section of the paper describes a parametric model which embodies these intuitions, and tests it with experimental data. The goal is to show how the kinds of learning models described in the previous section can be parsimoniously extended to explain behavior in more complex games which are, perhaps, of even greater economic interest than games with random matching.

Consider a finitely repeated trust game. A borrower B wants to borrow money from each of a series of lenders denoted L_i ($i = 1, \ldots, N$). In each period a lender makes a single lending decision (*Loan* or *No Loan*). If the lender makes a loan, the borrower either (*repays* or *defaults*). The next lender in the sequence, who observed all the previous history, then makes a lending decision. The payoffs used in experiments are shown in Table 10.6.

There are actually two types of borrowers. As in post-Harsanyi game theory with incomplete information, types are expressed as differences in borrower payoffs which the borrowers know but the lenders do not (though the probability that a given borrower is each type is commonly known). The honest (Y) types actually receive *more* money from repaying the loan, an experimenter's way of inducing preferences like those of a person who has a social utility for being trustworthy (see Camerer, 2003, Chapter 3 and references therein). The normal (X) types, however, earn 150 from defaulting and only 60 from repaying. If they were playing just once and wanted to earn the most money, they would default.[29]

Table 10.6 Payoffs in the Borrower-Lender trust game, Camerer and Weigelt (1988)

Lender strategy	Borrower strategy	Payoffs to lender	Payoffs to borrower normal (X)	Payoffs to borrower honest (Y)
Loan	Default	−100	150	0
	Repay	40	60	60
No loan	(No choice)	10	10	10

[29] Note that player types need not be stage game types. They can also be repeated game types where players are preprogrammed to play specific repeated game strategies. We do not allow repeated game types in this Section.

In the standard game-theoretic account, paying back loans in finite games arises because there is a small percentage of honest types who *always* repay. This gives normal-type borrowers an incentive to repay until close to the end, when they begin to use mixed strategies and default with increasing probability.

Whether people actually play these sequential equilibria is important to investigate for two reasons. First, the equilibria impose consistency between optimal behavior by borrowers and lenders and Bayesian updating of types by lenders (based on their knowledge and anticipation of the borrowers' strategy mixtures); whether reasoning or learning can generate this consistency is an open behavioral question (cf. Selten, 1978). Second, the equilibria are very sensitive to the probability of honesty (if it is too low the reputational equilibria disappear and borrowers should always default), and also make counterintuitive comparative statics predictions which are not confirmed in experiments (e.g., Jung et al., 1994; Neral and Ochs, 1992).

In the experiments subjects play many sequences of eight periods. The eight-period game is repeated to see whether equilibration occurs across many sequences of the entire game.[30] Surprisingly, the earliest experiments showed that the pattern of lending, default, and reactions to default across experimental periods within a sequence is roughly in line with the equilibrium predictions. Typical patterns in the data are shown in Figure 10.7a and b. Sequences are combined into ten-sequence blocks (denoted "sequence") and average frequencies are reported from those blocks. Periods $1, \ldots, 8$ denote periods in each sequence. The figures show relative frequencies of no loan and default (conditional on a loan). Figure 10.7a shows that in early sequences lenders start by making loans in early periods (i.e., there is a low frequency of no-loan), but they rarely lend in periods 7-8. In later sequences they have learned to always lend in early periods and rarely lend in later periods. Figure 10.7b shows that borrowers rarely default in early periods, but usually default (conditional on getting a loan) in periods 7-8. The within-sequence pattern becomes sharper in later sequences.

The teaching model is a boundedly rational model of reputation formation in which the lenders learn whether to lend or not. They do not update borrowers' types and do not anticipate borrowers' future behavior (as in equilibrium models); they just learn. In the teaching model, some proportion of borrowers are sophisticated and teach; the rest are adaptive and learn from experience but have no strategic foresight.

A sophisticated teaching borrower's attractions for sequence k after period t are specified as follows ($j \in \{repay, default\}$ is the borrower's set of strategies):

$$A_B^j(s,k,t) = \sum_{j'=\text{Loan}}^{\text{NoLoan}} P_L^{j'}(a,k,t+1) \cdot \pi_B(j,j') +$$

[30] Borrower subjects do not play consecutive sequences, which removes their incentive to repay in the eighth period of one sequence so they can get more loans in the first period of the next sequence.

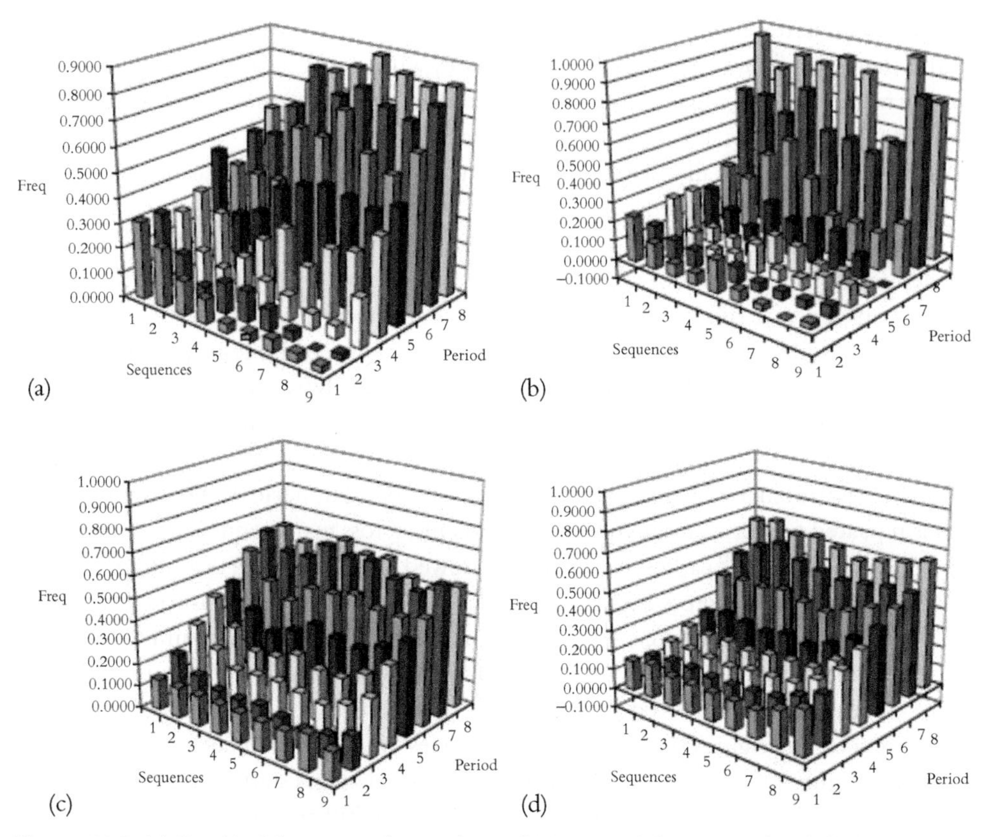

Figure 10.7 (a) Empirical frequency for no loan. (b) Empirical frequency for default conditional on loan. (c) Predicted frequency for no loan. (d) Predicted frequency for default conditional on loan.

$$\max_{J_{t+1}} \left\{ \sum_{v=t+2}^{T} \sum_{j'=\text{Loan}}^{\text{NoLoan}} \hat{P}_L^{j'}(a,k,v|j_{v-1} \in J_{t+1}) \cdot \pi_B(j_v \in J_{t+1}, j') \right\}$$

The set J_{t+1} specifies a possible path of future actions by the sophisticated borrower from round $t+1$ until end of the game sequence. That is $J_{t+1} = \{j_{t+1}, j_{t+2}, \ldots, j_{T-1}, j_T\}$ and $j_{t+1} = j$.[31] The expressions $\hat{P}_L^{j'}(a,k,v|j_{v-1})$ are the overall probabilities of either getting a loan or not in the future periods v, which depends on what happened in

[31] To economize in computing, we search only paths of future actions that always have default following repay because the reverse behavior (repay following default) generates a lower return.

the past (which the teacher anticipates).[32] $P_B^j(s, k, t+1)$ is derived from $A_B^j(s, k, t)$ using a logit rule.

The updating equations for adaptive players are the same as those used in fEWA with two twists. First, since lenders who play in later periods know what has happened earlier in a sequence, we assume that they learned from the experience they saw as if it had happened to them.[33] Second, a lender who is about to make a decision in period 5 of sequence 17, for example, has *two* relevant sources of experience to draw on—the behavior she saw in periods 1-4 in sequence 17, *and* the behavior she has seen in the period 5's of the previous sequences (1-16). Since *both* kinds of experience could influence her current decision, we include both using a two-step procedure. After period 4 of sequence 17, for example, attractions for lending and not lending are first updated based on the period 4 experience. Then attractions are partially updated (using a degree of updating parameter σ) based on the experience in period 5 of the previous sequences.[34] The parameter σ is a measure of the strength of "peripheral vision"—glancing back at the "future" period 5's from previous sequences to help guess what lies ahead.

Of course, it is well-known that repeated-game behavior can arise in finite-horizon games when there are a small number of "unusual" types (who act like the horizon is unlimited), which creates an incentive for rational players to behave as if the horizon is unlimited until near the end (e.g., Kreps and Wilson, 1982). But specifying why some types are irrational, and how many they are, makes this interpretation difficult to test. In the teaching approach, which "unusual" type the teacher pretends to be arises endogenously from the payoff structure: They are Stackelberg types, who play the strategy they would choose if they could commit to it. For example, in trust games, they would like to commit to repaying; in entry-deterrence, they would like to commit to fighting entry.

The model is estimated using repeated game trust data from Camerer and Weigelt (1988). In Ho et al. (2007), we used parametric EWA to model behavior in trust games. That model allows two different sets of EWA parameters for lenders and borrowers. In this paper we use fEWA to model lenders and adaptive borrowers so the model has fewer

[32] Formally, $\hat{P}_L^{j'}(a, k, v|j_{v-1}) = \hat{P}_L^{\text{Loan}}(a, k, v-1|j_{v-1}) \cdot P_L^{j'}(a, k, v|(\text{Loan}, j_{v-1})) + \hat{P}_L^{\text{NoLoan}}(a, k, v-1|j_{v-1}) \cdot P_L^{j'}(a, k, v|(\text{NoLoan}, j_{v-1}))$.

[33] This is called "observational learning" (see Duffy and Feltovich, 1999). Without this assumption the model learns far slower than the lenders do so it is clear that they are learning from observing others.

[34] The idea is to create an "interim" attraction for round t, $B_L^j(a, k, t)$, based on the attraction $A_L^j(a, k, t-1)$ and payoff from the round t, then incorporate experience in round $t+1$ from previous sequences, transforming $B_L^j(a, k, t)$ into a final attraction $A_L^j(a, k, t)$. See Ho et al. (2007) for details.

Table 10.7 Model parameters and fit in repeated trust games

	Statistic	Model: fEWA+ Teaching	Model: Agent QRE
In-sample	Hit rate (%)	76.5%	73.9%
Calibration ($n = 5757$)	Log-likelihood	−2975	−3131
Out-of-sample	Hit rate (%)	75.8%	72.3%
Validation ($n = 2894$)	Log-likelihood	−1468	−1544
parameters		**estimates**	
Cross-sequence learning	σ	0.93	-
% of teachers	α	0.89	-
Homemade prior p(honest)	θ	-	0.91

parameters.[35] Maximum likelihood estimation is used to estimate parameters on 70% of the sequences in each experimental session, then behavior in the holdout sample of 30% of the sequences is forecasted using the estimated parameters.

As a benchmark alternative to the teaching model, we estimated an agent-based version of QRE suitable for extensive-form games (see McKelvey and Palfrey, 1998). Agent-QRE is a good benchmark because it incorporates the key features of repeated-game equilibrium—strategic foresight, accurate expectations about actions of other players, and Bayesian updating—but assumes stochastic best-response. We use an agent-based form in which players choose a distribution of strategies at each node, rather than using a distribution over all history-dependent strategies. We implement agent QRE with four parameters—different λ's for lenders, honest borrowers, and normal borrowers, and a fraction θ, the percentage of players with normal-type payoffs who are thought to act as if they are honest (reflecting a "homemade prior" which can differ from the prior induced by the experimental design[36]). (Standard equilibrium concepts are a special case of this model when λ's are large and $\theta = 0$, and fit much worse than AQRE does.)

The models are estimated separately on each of the eight sessions to gauge cross-session stability. Since pooling sessions yields similar fits and parameter values, we report only those pooled results in Table 10.7 (excluding the λ values). The interesting parameters for sophisticated borrowers are estimated to be $\hat{\alpha} = 0.89$ and $\hat{\sigma} = 0.93$, which means most subjects are classified as teachers and they put a lot of weight on previous sequences. The teaching model fits in-sample and predicts better out-of-sample than AQRE by a modest margin (and does better in six of eight individual experimental

[35] We use four separate λ's, for honest borrowers, lenders, normal adaptive borrowers, and teaching borrowers, an initial attraction for lending $A(0)$, and the spillover parameter σ and teaching proportion α.

[36] See Camerer and Weigelt (1988), Palfrey and Rosenthal (1988), and McKelvey and Palfrey (1992).

sessions), predicting about 75% of the choices correctly. The AQRE fits reasonably well too (72% correct) but the estimated "homemade prior" θ is 0.91, which is absurdly high. (Earlier studies estimated numbers around 0.1-0.2.) The model basically fits best by assuming that *all* borrowers simply prefer to repay loans. This assumption fits most of the data but it mistakes teaching for a pure repayment preference. As a result, it does not predict the sharp upturn in defaults in periods 7-8, which the teaching model does.

Figures 10.7c and d show average predicted probabilities from the teaching model for the no-loan and conditional default rates. No-loan frequencies are predicted to start low and rise across periods, as they actually do, though the model underpredicts the no-loan rate in general. The model predicts the increase in default rate across periods reasonably well, except for underpredicting default in the last period.

The teaching approach as a boundedly rational alternative to type-based equilibrium models of reputation-formation.[37] It has always seemed improbable that players are capable of the delicate balance of reasoning required to implement the type-based models, unless they learn the equilibrium through some adaptive process. The teaching model is one parametric model of that adaptive process. It retains the core idea in the theory of repeated games—namely, strategic foresight—and consequently, respects the fact that matching protocols matter. And since the key behavioral parameters (α and σ) appear to be near one, restricting attention to these values should make the model workable for doing theory.

10.5.3 Summary

In this section we introduced the possibility that players can be sophisticated (i.e., they believe others are learning). Sophistication links learning theories to equilibrium ones if sophisticated players are self-aware. Adding sophistication also improves the fit of data from repeated beauty-contest games. Interestingly, the proportion of estimated sophisticates is around a quarter when subjects are inexperienced, but rises to around three-quarters when they play an entire 10-period game a second time, as if subjects learn about learning. Sophisticated players who know they will be rematched repeatedly may have an incentive to "teach," which provides a boundedly rational theory of reputation formation. We apply this model to data on repeated trust games. The model adds only two behavioral parameters, representing the fraction of teachers and how much "peripheral vision" learners have (and some nuisance λ parameters), and predicts substantially better than a quantal response version of equilibrium.

[37] One direction we are pursuing is to find designs or tests which distinguish the teaching and equilibrium updating approaches. The sharpest test is to compare behavior in games with types that are fixed across sequences with types that are independently "refreshed" in each period within a sequence. The teaching approach predicts similar behavior in these two designs but type-updating approaches predict that reputation formation dissolves when types are refreshed.

10.6. SOCIALITY

Some of the simplest economic games have proved to be useful as ways to measure sociality—departures from pure self-interest—and test specific models of social preferences. By far the most popular examples are prisoner's dilemma (or multiplayer "commons" dilemmas) and public goods contribution games (PGG). While these games are trivial from a strategic point of view, at first blush, explaining behavioral regularities has required fundamental developments and applications in game theory. Since the literature is large, we will discuss selective results from PGG, along with ultimatum and dictator games.

Before proceeding, however, it is crucial to clarify what sociality means in game theory. As every reader should know, game theory makes predictions about behavior conditional on a proper specification of utilities for outcomes. Rejecting money in an ultimatum game, for example, is definitely *not* a rejection of game theoretic rationality if the utility of having nothing is greater than the utility of having a small amount when a proposer has a much larger amount (due to fairness preferences, for example). However, since we typically do not have independent measurement of utilities for outcomes, in testing game theory predictions we are almost always testing the joint hypothesis of a particular utility specification—usually, selfishness—along with a theory of behavior given those specified utilities. The evidence discussed in this section shows systematic ways in which people do not appear to be selfish, and new ideas about what utility specifications can explain the data. In a sense, the goal of this approach is very "conservative"—the hope is that a reasonable utility specification will emerge so that data can be reasonably explained by standard analysis given that new type of utility.

It is also notable that the tendency of economists to be skeptical about human prosociality (e.g., Stigler, 1981; List quoted in Stanley, 2011) conflicts with the prevailing view of exceptional human prosociality, compared to other species, in biology and anthropology. Scientists in those fields have amassed much evidence that humans are more prosocial toward non-kin than all other species. For example, Boyd and Richerson wrote: "Humans cooperate on a larger scale than most other mammals.....The scale of human cooperation is an evolutionary puzzle" (Boyd and Richerson, 2009, p. 3281).

10.6.1 Public goods

In a typical linear public goods game experiment, four players are endowed with 20 tokens. A number c_i of tokens can be contributed to a public pool where they earn M, which is distributed equally. The rest of the tokens, $20 - c_i$, earn a private return of 1. The fraction $\frac{M}{4}$ is called the marginal per capita return (MCPR). Assumptions $1 < M < 4$ are imposed so that M is above the private return but the MCPR is below the private return. Player i's payoff is $u_i(c_i, c_k) = 20 - c_i + \frac{M}{4}\sum_{j=1}^{4} c_j$. The collective payoff of $80 - (M - 1)\sum_{j=1}^{4} c_j$ is maximized if $c_i = 20\forall i$. Then each person earns $(20M * 4)/4 = 20M$. If everyone is selfish they contribute $c_i = 0$ and everyone earns 20.

Notice that under selfish preferences, keeping all tokens is a dominant strategy, so the predictions do not depend on the depth of strategic thinking, Therefore, the PGG is mostly interesting as a way to study social (nonselfish) preferences, and the interaction of such preferences with structural variables.

In most experiments, contributions begin with a bimodal distribution of many 100% and 0% contributions, averaging around 50% of the tokens. Given feedback about overall contributions, within 10 periods or so the average contribution erodes to about 10%. Contribution is substantially higher in a "partner" protocol in which the same people are matched each period, compared to a random-matching "stranger" protocol. If subjects are surprised and the entire experimental sequence is repeated, initial contributions jump back up near 50%. This "restart effect" (Andreoni, 1988) is clear evidence that subjects who contribute are not just confused but need experience to learn to give nothing, since they would give more after a restart.

What seems to be going on in the PGG is that some people are selfish, and give nothing no matter what. Another large fraction of people are "conditionally cooperative," and give in anticipation that others will give. If they are too optimistic, or choose to punish the selfish free-riders, these conditional cooperators only recourse is to give less. These combined forces lead to a steady decline in contribution. The best evidence for conditional cooperation is an elegant experiment that asked subjects how much they would contribute, depending on how much other subjects did (and enforced actual contribution in this way; Fischbacher and Gachter, 2010). In this protocol, a large number of subjects choose to give only when others give.

10.6.2 Public goods with punishment

An interesting variant of PGG includes costly punishment after players contribute. The idea is that in natural settings players can often find out how much others contributed to a public good, and can then punish or reward other people based on their observed contributions, typically at a cost to themselves.

These punishments range from gossip, scolding, and public shaming (Guala, 2010)—now magnified by the internet—to paralegal enforcement (vigilantism, family or neighborhood members sticking up for each other), or organized legal enforcement through policing and taxation.

Yamagishi (1986) was the first to study punishment in the form of sanction. Since then, a paradigm introduced by Fehr and Gachter (2000) has become standard. In their method, after contributions are made a list of the contribution amounts is made public (without identifying identity numbers or other information about contributors, to avoid repeated-game spillovers). Each player can then spend 1 token to subtract 3 tokens from one other participant, based only on the punished participants contribution.

The first wave of studies showed that punishment was effective in raising contributions close to the socially optimal step (although total efficiency suffers a bit since punishments reduce total payoff). When punishment is available, early punishment of free riders causes them to increase their contributions, and punishment is more rare in later trials as a result. Punishment works when some of those who contribute greatly to the public good also act as "social police," using personal resources to goad low contributors to give more.

However later studies complicated this simple finding in several ways:

1. Some of the effect of punishment can be created by symbolic shaming, without any financial cost or penalty (Masclet et al., 2013).
2. Players have a choice with whom they wish to play the game (Page et al., 2005), what type of sanctions are allowed (Markussen et al., 2014), or whether or not to allow punishment in a community (Gurerk et al., 2006).
3. What happens if those who are punished can then punish the people who punish them? (Note that in natural settings, there is likely to be no way to allow informal punishment without retaliation against punishers.) One study showed that in this scenario, it is easy for efficiency-reducing punishment "feuds" to arise (Nikoforakis and Engelmann, 2011).
4. Herrmann et al. (2008) and Gachter et al. (2010) found an empirical continuum of punishment in an amazing study across 16 societies, sampling a wider variety of cultures than previously studied. In countries with English-speaking, Confucian, or Protestant histories, those who punish a lot are typically high contributors punishing low contributors (punishment is "prosocial" in intent). However, in several Arabic, Mediterranean, and former Communist block countries they sampled, there were also high rates of "antisocial" punishment (i.e., low contributors often punish high contributors)! When antisocial punishment is common, giving rates do not increase across trials, and efficiency is reduced because punishments are costly for both parties involved.

10.6.3 Negative reciprocity: ultimatums

In a typical ultimatum game, a proposer offers a specific division of a known amount of money to a responder. The responder can accept it, and the money amounts are paid to both sides, or reject it, and both sides get nothing.

This simple game was first studied experimentally by Guth et al. (1982) and has been replicated hundreds of times since then (including offers of water to very thirsty people; Wright et al., 2012). If players are self-interested, then the subgame perfect equilibrium is to offer zero (or the smallest divisible amount) and for responders to accept. This joint hypothesis about strategic thinking is typically rejected because average offers are around 37%, and offers below 20% or so are often rejected. (The overall rejection rate was 16%, in a survey of 75 studies by Ooesterbeek et al., 2004.) Since the responders rejection choice does not involve any strategic thinking at all, these stylized facts are clear evidence that many players deciding whether to reject or accept offers are not purely self-interested.

One explanation that emerged prominently early on is that rejections are simple monetary mistakes which selfish players learn to avoid over time (e.g., Erev and Roth, 1998). This learning explanation is clearly incomplete: It does not specify the time scale over which learning takes place, and cannot explain why fewer rejection "mistakes" are made when unequal offers are generated exogenously by a computer algorithm or when rejections do not hurt the proposer. Evidence for learning across repeated experimental trials is also not strong (e.g., Cooper and Stockman, 2002).

The leading current explanation for ultimatum rejections is "negative reciprocity": Some players trade off money with a desire to deprive a player who has treated them unfairly from earning money. Summarized from strongest to weakest, evidence for a negative reciprocity explanation comes from the facts that: players reject offers less frequently if an unequal offer is created by a computerized agent (Blount, 1995; Sanfey et al., 2003); rejections are rare if rejecting the offer does not deprive the proposer of her share (Bolton and Zwick, 1995); low offers and rejections are associated with brain activity in insula, a bodily discomfort region (Sanfey et al., 2003), and with wrinkling facial muscles associated with felt anger and disgust (Chapman et al., 2009) or exogeneously induced disgust (Moretti and di Pellegrino, 2010); and rejections are associated with self-reported anger (Pillutla and Murnighan, 1996).

Many studies have changed the amount of money being divided, over substantial ranges including days or weeks of wage equivalents (beginning with Cameron, 1995). Offers generally go down as the stakes go up (Oosterbeek et al., 2004). As stakes go up, the absolute amounts players will reject increase, but the percentages they reject decrease. Note that this pattern does *not* imply responders are becoming "more rational" as stakes increase, because the money amounts they will reject increase in magnitude. In fact, such a pattern is predicted to arise if responders care both about their own earnings and about their relative share (or if equality is an inferior good).

Several formalizations of social preference have been developed to explain rejected offers. The simplest of these propose utility functions that include aversion to unequal outcome differences (Fehr and Schmidt, 1999) or unequal shares (Bolton and Ockenfels, 2000). However, these theories cannot explain acceptance of low offers generated exogeneously (which create equal inequity). Therefore, other theories model ultimatums as psychological games in which beliefs enter the utility function (Rabin, 1993).

An interesting empirical strategic question is whether proposers have accurate beliefs about how much responders are willing to accept, and make payoff-maximizing offers given those beliefs. Early findings in four countries from Roth et al. (1991) support this "rational proposer" hypothesis, but without formal statistical testing or adjustment for risk-aversion. This hypothesis is an attractive one for game theory because it suggests players have equilibrium beliefs and choose best responses (or noisy quantal responses) given their beliefs. If so, after adjusting for social preferences or negative reciprocity, equilibrium analysis is an accurate description of offers and rejection rates.

However, several later studies found domains in which proposers appear to offer too much or too little, as evidenced by apparent gaps between actual offers and profit-maximizing offers given accurate beliefs.

For example, many small-scale societies show generous near-equal offers along with extremely low rejection rates at all offer steps, suggesting that proposer beliefs are pessimistic (Henrich et al., 2001). Offers in a study in the Khasi-speaking part of India (Andersen et al., 2011) are half as large in most samples (12-23%) but rejection rates are higher than in any country ever studied, around (34%). Either the Khasi proposers' beliefs about acceptance are too optimistic, or they are extremely risk-seeking.

10.6.4 Impure altruism and social image: dictator games

A dictator game is a simple decision about how to allocate unearned money between the dictator and some recipient(s). It is an ultimatum game with no responder rejection. Dictator games were first used to see whether ultimatum offers are altruistic (if so, dictator and ultimatum offers will be the same) or are selfishly chosen to avoid rejection (if so, case dictator offers will be lower). Dictator offers are clearly lower.

However, dictator game allocations are sensitive to many details of procedure and (likely) personality. In psychological terms, the dictator game is a "weak situation" because norms of reasonable sharing can vary with context, may not be widely shared, and compete with robust self-interest.

In initial dictator game experiments some subjects give nothing but a small majority do allocate some money to recipients; the average allocation is typically 10-25%. Subjects clearly give much less when the money to be allocated was earned by the dictator rather than unearned and randomly allocated by the experimenter (Camerer and Thaler, 1995, p. 216; cf. Hoffman and Spitzer, 1982). Therefore, it should be clear that experimental dictator giving in lab settings is likely to be much more generous than natural charitable giving out of wealth. Allocations are also sensitive to the range of what can be given or taken (Bardsley, 2008; List, 2007), and to endogenous formation of norms about what others give (Krupka and Weber, 2009).

In the canonical dictator game, the total allocation amount is fixed, so there is no tradeoff between giving more and giving more equally. However, if the possible total allocations vary, then it is useful to know how much people will sacrifice efficiency (choosing a higher total allocation) for equity (choosing a more equal allocation). Several experiments explored this tradeoff (Charness and Rabin, 2002; Engelmann and Strobel, 2006). There is substantial evidence for efficiency-enhancing motivations, which appears in some studies to depend on the subjects' academic background and training (Fisman et al., 2007).

The last several years have seen a correction to the language used to describe dictator game allocations imply. Some early discussions of these results called dictator giving "altruistic," because of the contrast to strategic offers in ultimatum games (which are not generous, per se, but instead meant to avoid rejections). Of course, the term "altruism" means something specific in public finance: Pure altruism is a desire to make another person better-off on that person's own terms. Impure altruism means having a private motive to help another person, such as a "warm glow" from giving.) In addition, Camerer and Thaler (1995) described dictator allocations as reflecting "manners," meaning adherence to an accepted social norm of giving.

The idea of "manners" was made precise in formal theories of "social image." In these theories, people give in dictator games because they have a utility for believing that others think they are a generous type (i.e., have good manners) or to avoid disappointing the expectations that they think other people have (guilt-aversion; Hoffman et al., 1996). Andreoni and Bernheim adapted a strategic theory of rational conformity to this setting. It predicts that if there is a P chance a dictator allocation will be stopped, so the recipient gets zero, then dictators will be more inclined to give 0 if P is higher. (The intuition is that if recipients are likely to get nothing anyway, the dictators won't feel as unfairly judged if they give zero also.) Indeed, they find such an effect in their experiment. Berman and Small (2012) also find that when the selfish choice is imposed exogeneously, subjects are happier. Tonin and Vlassopoulos (2013) found that a third of subjects who initially gave a positive amount change their minds when asked if they want to keep all the money. Malmendier et al. (2012) found that many subjects would rather opt out of a $10 ultimatum game, to get $9, provided the potential recipient subject would not know about their choice.[38]

These studies indicate that in simple dictator allocations, a major motivation for dictator giving is wanting to *appear* generous, in order to believe that a recipient subject think the dictator is being generous. (In formal terms, people have a preference for an iterated belief that another person believes their type is generous.) Disrupting any element of this complicated chain of perception and cognition can easily push giving toward zero. Interesting biological evidence also supports the social image view: Contributing when others are watching activates reward areas (Izuma et al. 2008), and autistic adults do not adjust giving when being watched (which control subjects do).

It is always useful to ask how well findings from lab experiments correlate with field data when relevant features such as incentives, subject pools, choice sets etc. are closely matched (Camerer, in press). For prosociality, lab-field correlation is over-

[38] There are also studies showing that dictators may take away other player's money in order to equalize payoffs between them and others (Zizzo, 2003; List, 2007).

whelmingly positive and often very impressive in magnitude (comparable to other kinds of psychometric reliability). For example, Brazilians who catch fish and shrimp exhibit lab prosociality and patience which is associated with more cooperative behavior in their fishing (Fehr and Liebbrandt, 2011). Boats of Japanese fishermen who disapprove of how much others keep selfishly tend to catch more collectively (Carpenter and Seki, 2011). Students who are generous in the (lab) dictator game are also more inclined to return a (field) misdirected letter containing money (Franzen and Pointner, 2013).

10.6.5 Summary

There is now massive evidence of various kinds of prosocial behavior. The challenge is to distill this evidence into a minimal set of workable, psychologically plausible models that can be used in public finance, game theory, political science and other applications. Leading theories assume simple kinds of adjustment to outcome utility based on inequity of payoff differences or shares, reciprocity, or social image. All theories are backed by various kinds of data. Notably, the data comparing lab measures of prosociality and field measures (holding as many lab-field factors constant as possible) shows a substantial correspondence of behavior in the lab and field.

10.7. CONCLUSION

This chapter presented experimental evidence motivating behavioral models based on limited rationality, stochastic choice, learning and teaching, and prosocial preferences. These models are intended to explain behavior in many different games with a single general specification (sometimes up to one or more free parameters).

Please note that not all examples illustrate how deviations from conventional equilibrium are explained by behavioral game theory. In some cases, the fact that equilibrium concepts predict surprisingly well can be explained by behavioral models. One example is entry games, which approximately equilibrate with no learning at all because, we suggest, higher-step thinks fill in deviations from equilibrium that result from lower-step thinkers. Another example is how responsive learning by highly trained chimpanzees in hide-and-seek games (perhaps a conserved cognitive capacity due to its important for fitness in their competitive societies) leads to near-perfect frequencies of Nash play.

While these behavioral models are actively researched and applied, we are struck by how little attention is currently paid to them in typical game theory courses and textbooks. If a student asks, "What happens when people actually play these games?" equilibrium analysis should be part of any explanation, but it will typically not provide the best currently available answer to the student's question. Many of the behavioral ideas are also easy to teach and easily grasped by students—partly because the intuitions underlying the models correspond to students' intuitions about they would do or expect. And there are plenty of experimental data to contrast equilibrium predictions with intuitions (or to show they may match surprisingly well).

Many interesting behavioral and experimental topics were excluded in this chapter to save space. One set of models are boundedly rational solutions concepts which predict stationary strategy distributions (typically using one free parameter.) These include action sampling, payoff sampling, and impulse-balance theory (Chmura et al., 2008) Note that those solution concepts were not derived a priori from any specific evidence about cognitive limits (and have not been tested by eyetracking or fMRI data). There is also no robust evidence that these concepts improve on QRE when errors in Chmura et al. (2008) were corrected (see Brunner et al., 2011).

We also acknowledge cursed equilibrum (CE, Eyster and Rabin, 2008) and analogy-based expectational equilibrum (ABEE, Jehiel, 2005). There are very few new data testing these theories and, despite their potential, their current scope and applicability are limited. CE is specifically designed only to explain deviations in BNE with private information, so it coincides with Nash equilibrium in games with complete information. As a result, it is not a serious candidate for a general theory of deviation from Nash equilibrium across all types of games. ABEE predictions depend on how analogies are created, which requires auxiliary assumptions.

Frontiers of behavioral and experimental game theory are pushing far out from conventional lab choices, to tests with field data. They are also pushing far in and down, to show cognitive and neural mechanisms. We believe it is possible to have theories, such as CH, which makes specific predictions about mental processes *and* can be abstracted upward to make predictions about price, quantity and quality in markets.

REFERENCES

Alaoui, L., Penta, A., 2013. Depth of reasoning and incentives. Econ. Wisc working paper.

Andersen, S., Erta, S., Gneezy, U., Hoffman, M., List, J.A., 2011. Stakes matter in ultimatum games. Am. Econ. Rev. 101, 3427–3439.

Andreoni, J., 1988. Why free ride? Strategies and learning in public goods experiments. J. Public Econ. 37, 291–304.

Andreoni, J., Miller, J., 1993. Rational cooperation in the finitely repeated prisoner's dilemma: experimental evidence. Econ. J. 103, 570–585.

Arthur, B., 1991. Designing economic agents that act like human agents: a behavioral approach to bounded rationality. Am. Econ. Rev. 81 (2), 353–359.

Bardsley, N., 2008. Dictator game giving: altruism or artifact. Exp. Econ. 11, 122–133.

Berman, J.Z., Small, D.A., 2012. Self-interest without selfishness: the hedonic benefit of imposed self-interest. Psychol. Sci. 23 (10), 1193–1199.

Binmore, K., 1999. Why experiment in economics? The Econ. J. 109 (453), 16–24.

Binmore, K., Swierzbinski, J., Proulx, C., 2001. Does maximin work? An experimental study. Econ. J. 111, 445–464.

Blount, S., 1995. When social outcomes aren't fair: the effect of causal attributions on preferences. Org. Behav. Human Decis. Processes 63, 131–144.
Bolton, G.E., Ockenfels, A., 2000. ERC: a theory of equity, reciprocity, and competition. Am. Econ. Rev. 90 (1), 166–193.
Bolton, G.E., Zwick, R., 1995. Anonymity versus punishment in ultimatum bargaining. Games Econ. Behav. 10 (1), 95–121.
Boyd, R., Richerson, P.J., 2009. Culture and the evolution of human cooperation. Philos. Trans. R. Soc. B 364, 3281–3288.
Brown, G., 1951. Iterative Solution of Games by Fictitious Play. In Activity Analysis of Production and Allocation. John Wiley & Sons, New York.
Brunner, C., Camerer, C., Goeree, J., 2011. Stationary concepts for experimental 2 × 2 games: comment. Am. Econ. Rev. 101, 1029–1040.
Bulow, J., Klemperer, P., 2002. Prices and the winner's curse. RAND J. Econ. 33 (1), 1–21.
Bush, R., Mosteller, F., 1955. Stochastic Models for Learning. John Wiley & Sons, New York.
Cabrales, A., Garcia-Fontes, W., 2000. Estimating learning models with experimental data. University of Pompeu Febra Working Paper.
Cachon, G.P., Camerer, C.F., 1996. Loss-avoidance and forward induction in experimental coordination games. Q. J. Econ. 111 (1), 165–194.
Camerer, C.F., 1987. Do biases in probability judgment matter in markets? Experimental Evidence. Am. Econ. Rev. 77, 981–997.
Camerer, C.F., 1995. Individual decision making. In: Handbook of Experimental Economics. Princeton University Press, Princeton, pp. 587–703.
Camerer, C.F., 2003. Behavioral Game Theory: Experiments on Strategic Interaction. Princeton University Press, Princeton.
Camerer, C.F., The promise and success of lab-field generalizability in experimental economics: a critical reply to Levitt and List. In: Frechette, G., Schotter, A. (Eds.), The Methods of Modern Experimental Economics. Oxford University Press, in press.
Camerer, C.F., Fehr, E., 2006. When does "economic Man" Dominate Social Behavior?. Science 311 (5757), 47–52.
Camerer, C.F., Weigelt, K., 1988. Experimental tests of a sequential equilibrium reputation model. Econometrica 56, 1–36.
Camerer, C.F., Weigelt, K., 1993. Convergence in experimental double auctions for stochastically lived assets. In: Friedman, D., Rust, J. (Eds.), The Double Auction Market: Theories, Institutions and Experimental Evaluations. Addison-Wesley, Redwood City, CA, pp. 355–396.
Camerer, C.F., Thaler, R.H., 1995. Anomalies: ultimatums, dictators and manners. J. Econ. Perspect. 9 (2), 209–219.
Camerer, C.F., Ho, T.-H., 1998. EWA learning in normal-form games: probability rules, heterogeneity and time variation. J. Math. Psychol. 42, 305–326.
Camerer, C.F., Ho, T.-H., 1999. Experience-weighted attraction learning in normal-form Games. Econometrica 67, 827–874.
Camerer, C.F., Ho, T.-H., Chong, J.K., 2002. Sophisticated EWA learning and strategic teaching in repeated games. J. Econ. Theory 104 (1), 137–188.
Camerer, C.F., Hsia, D., Ho, T.-H., 2002. EWA learning in bilateral call markets. In: Rapoport, A., Zwick, R. (Eds.), Experimental Business Research, pp. 255–284.
Camerer, C., Ho, T.-H., Chong, J.K., 2004. Behavioural game theory: thinking, learning and teaching. In: Advances in Understanding Strategic Behaviour: Game Theory, Experiments, and Bounded Rationality: Essays in Honor of Werner Güth. Palgrave Macmillan, New York, pp. 120–180.
Capra, M., 1999. Noisy expectation formation in one-shot games. Unpublished dissertation, University of Virginia.
Capra, M., Goeree, J., Gomez, R., Holt, C., 1999. Anomalous behavior in a traveler's dilemma. Am. Econ. Rev. 89, 678–690.
Carpenter, J., Seki, E., 2011. Do social preferences increase productivity? Field experimental evidence from fishermen in Toyama bay. Econ. Inquiry 49 (2), 612–630.

Carrillo, J.D., Palfrey, T.R., 2011. No trade. Games Econ. Behav. 71 (1), 66–87.
Chapman, H.A., Kim, D.A., Susskind, J.M., Anderson, A.K., 2009. In bad taste: evidence for the oral origins for moral disgust. Science, 27, 1222–1226.
Charness, G., Rabin, M., 2002. Understanding social preferences with simple tests. Q. J. Econ. 117 (3), 817–869.
Cheung, Y.-W., Friedman, D., 1997. Individual learning in normal form games: some laboratory results. Games Econ. Behav. 19, 46–76.
Chmura, T., Sebastian, G., Selten, R., 2007. Stationary concepts for experimental 2x2-games. Am. Econ. Rev. 98(3), 938–66.
Chong, J.K., Camerer, C., Ho, T.H., 2004. Cognitive Hierarchy: A Limited Thinking Theory of Games, Experimental Business Research, Vol. II. In: Zwick, R., Rapoport, A. (Eds.), Kluwer Academic Publishers.
Clark, K., Sefton, M., 1999. Matching protocols in experimental games. University of Manchester working paper.
Cooper, D.J., Stockman, C.K., 2002. Learning to punish: experimental evidence from a sequential step-level public goods game. Exp. Econ. 5 (1), 39–51.
Costa-Gomes, M., Crawford, V., Broseta, B., 2001. Cognition and behavior in normal-form games: an experimental study. Econometrica 69, 1193–1235.
Crawford, V., 1997. Theory and experiment in the analysis of strategic interactions. In: Kreps, D., Wallis, K. (Eds.), Advances in Economics and Econometrics: Theory and Applications. Seventh World Congress, Vol. I. Cambridge University Press, Cambridge.
Crawford, V., Costa-Gomes, M.A., Iriberri, N., 2013. Structural models of nonequilibrium strategic thinking: theory, evidence, and applications. J. Econ. Lit. 51, 5–62.
Cross, J., 1983. A theory of adaptive learning economic behavior. Cambridge University Press, New York.
Devetag, G, Warglien, M., 2003. Games and phone numbers: Do short-term memory bounds affect strategic behavior? J. Econ. Psych. 24, 189–202.
Duffy, J., Feltovich, N., 1999. Does observation of others affect learning in strategic environments? An experimental study. Int. J. Game Theory 28, 131–152.
Engelmann, D., Strobel, M., 2006. Inequality aversion, efficiency, and maximin preferences in simple distribution experiments: reply. Am. Econ. Rev. 96 (5), 1918–1923.
Erev, I., Roth, A., 1998. Predicting how people play games: reinforcement learning in experimental games with unique, mixed-strategy equilibria. Am. Econ. Rev. 88, 848–881.
Erev, I., Bereby-Meyer, Y., Roth, A., 1999. The effect of adding a constant to all payoffs: experimental investigation, and a reinforcement learning model with self-adjusting speed of learning. J. Econ. Behav. Org. 39, 111–128.
Erev, I., Roth, A.E., Slonim, R.L., Barron, G., 2002. Predictive value and the usefulness of game theoretic models. Int. J. Forecast. 18 (3), 359–368.
Eyster, E., Rabin, M., 2005. Cursed equilibrium. Econometrica 73, 1623–1672.
Fehr, E., Schmidt, K.M., 1999. A theory of fairness, competition, and cooperation. Q. J. Econ. 114 (3), 817–868.
Fehr, E., Gachter, S., 2000. Cooperation and punishment in public goods experiments. Am. Econ. Rev. 90 (4), 980–994.
Fehr, E., Leibbrandt, A., 2011. A field study on cooperativeness and impatience in the tragedy of the commons. J. Public Econ. 95 (9–10), 1144–1155.
Fischbacher, U., Gachter, S., 2010. Social preferences, beliefs, and the dynamics of free riding in public goods experiments. Am. Econ. Rev. 100 (1), 541–556.
Fishman, R., Kariv, S., Markovits, D., 2007. Individual preferences for giving. Am. Econ. Rev. 97, 1858–1876.
Franzen, A., Pointner, S., 2013. The external validity of giving in the dictator game: a field experiment using the misdirected letter technique. Exp. Econ. 16, 155–169.
Fudenberg, D., Levine, D., 1989. Reputation and equilibrium selection in games with a patient player. Econometrica 57, 759–778.
Fudenberg, D., Levine, D., 1998. The Theory of Learning in Games. MIT Press, Boston.

Gachter, S., Herrmann, B., Thoni, C., 2010. Culture and cooperation. Phil. Trans. R. Soc. B 365 (1553), 2651–2661.

Georgiadis, S., Healy, P.J., Weber, R., 2013. On the persistence of strategic sophistication. University of London Working Paper.

Gill, D., Prowse, V., 2012. Cognitive ability and learning to play equilibrium: a level-k analysis. Munich Personal RePEc Archive, Paper No. 38317, posted 23 April, 2012. http://mpra.ub.uni-muenchen.de/38317/

Goeree, J.K., Holt, C.A., 1999. A theory of noisy introspection. University of Virginia Department of Economics.

Goeree, J.K., Holt, C.A., 2001. Ten little treasures of game theory, and ten contradictions. Am. Econ. Rev. 91 (5), 1402–1422.

Goeree, J.K., Holt, C.A., Palfrey, T.R., 2005. Regular quantal response equilibrium. Exp. Econ. 8, 347–367.

Guala, F., 2010. Reciprocity: weak or strong? What punishment experiments do (and do not) demonstrate. University of Milan Department of Economics, Business and Statistics Working Paper No. 2010-23.

Gurerk, O., Irlenbusch, B., Rockenbach, B., 2006. The competitive advantage of sanctioning institutions. Science 312 (5770), 108–111.

Guth, W., Schmittberger, R., Schwarze, B., 1982. An experimental analysis of ultimatum bargaining. J. Econ. Behav. Org. 3 (4), 367–388.

Harley, C., 1981. Learning the evolutionary stable strategies. J. Theor. Biol. 89, 611–633.

Hart, S., Mas-Colell, A., 2001. A reinforcement procedure leading to correlated equilibrium. In: Debreu G., Neuefeind W., Trockel (Eds.), Economics Essay, Springer Berlin Heidelberg, 181–200.

Henrich, J., Boyd, R., Bowles, S., Camerer, C.F., Fehr, E., 2001. In search of homo economicus: behavioral experiments in 15 small-scale societies. Am. Econ. Rev. 91 (2), 73–78.

Herrmann, B., Thoni, C., Gachter, S., 2008. Antisocial punishment across societies. Science 319 (5868), 1362–1367.

Ho, T.-H., Weigelt, K., 1996. Task complexity, equilibrium selection, and learning: an experimental study. Manage. Sci. 42, 659–679.

Ho, T.-H., Su, X., 2013. A dynamic level-k model in sequential games. Manage. Sci. 59, 452–469.

Ho, T.-H., Camerer, C.F., Weigelt, K., 1998. Iterated dominance and iterated best response in experimental 'p-beauty contests'. Am. Econ. Rev. 88, 947–969.

Ho, T.-H., Camerer, C.F., Chong, J.K., 2007. Self-tuning experience-weighted attraction learning in games. J. Econ. Theory 133, 177–198.

Ho, T.-H., Wang, X., Camerer, C.F., 2008. Individual differences in the EWA learning with partial payoff information. Econ. J. 118, 37–59.

Ho, T.-H., Park, S.-E., Su, X., 2013. Level-r model with adaptive and sophisticated learning. University of California, Berkeley Working Paper.

Hoffman, E., Spitzer, M.L., 1982. The coase theorem: some experimental tests. J. Law Econ. 25 (1), 73–98.

Hoffman, E., McCabe, K., Smith, V.L., 1996. Social distance and other regarding behavior in dictator games. Am. Econ. Rev. 86, 653–660.

Hopkins, E., 2002. Two competing models of how people learn in games. Econometrica 70 (6), 2141–2166.

Hyndman, R.J., Booth, H., Yasmeen, F., 2013. Coherent mortality forecasting: the product-ratio method with functional time series models. Demography, 50 (1), 261–283.

Itti, L., Koch, C., 2000. A saliency-based search mechanism for overt and covert shifts of visual attention. Vision Res. 40 (10–12), 1489–1506.

Ivanov, A., Levin, D., Niederle, M., 2010. Can relaxation of beliefs rationalize the winner's curse? An experimental study. Econometrica 78 (4), 1435–1452.

Izuma, K., Saito, D.N., Sadato, N., 2008. Processing of social and monetary rewards in the human striatum. Neuron 58 (2), 284–294.

Jehiel, P., 2005. Analogy-based expectation equilibrium. J. Econ. Theory 123, 81–104.

Jung, Y.J., Kagel, J.H., Levin, D., 1994. On the existence of predatory pricing: an experimental study of reputation and entry deterrence in the chain-store game. RAND J. Econ. 25, 72–93.

Kahneman, D., 1988. Experimental economics: a psychological perspective. In: Tietz, R., Albers, W., Selten, R., (Eds.), Bounded Rational Behavior in Experimental Games and Markets. Springer-Verlag, New York. pp. 11–18.

Keynes, J.M., 1936. The General Theory Of Employment, Interest And Money. Macmillan Cambridge University Press, for Royal Economic Society.
Kneeland, T., 2013. Rationality and consistent beliefs: theory and experimental evidence. University of British Columbia Working Paper.
Kocher, M.G., Sutter, M., 2005. When the 'decision maker' matters: individual versus team behavior in experimental 'beauty-contest' games. Econ. J. 115, 200–223.
Kreps, D., Wilson, R., 1982. Reputation and imperfect information. J. Econ. Theory 27, 253–279.
Krupka, E., Weber, R.A., 2009. The focusing and informational effects of norms on pro-social behavior. J. Econ. Psychol. 30 (3), 307–320.
List, J.A., 2007. On the interpretation of giving in dictator games. J. Polit. Econ. 115 (3), 482–493.
Lucas, R.G., 1986. Adaptive behavior and economic theory. J. Bus. 59, S401–S426.
Malmendier, U., Lazear, E., Weber, R., 2012. Sorting in experiments with application to social preferences. Am. Econ. J. Appl. Econ. 4 (1), 136–163.
Markussen, T., Putterman, L., Tyran, J.-R., 2014. Self-organization for collective action: an experimental study of voting on sanction regimes. Rev. Econ. Stud. 81(1), 301-324.
Martin, C., Bossaerts, P., Matsuzawa, T., Camerer, C., 2014. Experienced chimpanzees play according to game theory. Nature Scientific Reports, 4(5182).
Masclet, D., Noussair, C.N., Villeval, M., 2013. Threat and punishment in public goods experiments. Econ. Inquiry 51 (2), 1421–1441.
Matsuzawa, T., 2007. Comparative cognitive development. Dev. Sci. 10, 97–103.
McAllister, P.H., 1991. Adaptive approaches to stochastic programming. Ann. Operat. Res. 30, 45–62.
McKelvey, R.D., Palfrey, T.R., 1992. An experimental study of the centipede game. Econometrica 60, 803–836.
McKelvey, R.D., Palfrey, T.R., 1995. Quantal response equilibria for normal-form games. Games Econ. Behav. 7, 6–38.
McKelvey, R.D., Palfrey, T.R., 1998. Quantal response equilibria for extensive-form games. Exp. Econ. 1, 9–41.
Moinas, S., Pouget, S., 2013. The bubble game: an experimental study. Econometrica 81 (4), 1507–1539.
Mookerjee, D., Sopher, B., 1994. Learning behavior in an experimental matching pennies game. Games Econ. Behav. 7, 62–91.
Mookerjee, D., Sopher, B., 1997. Learning and decision costs in experimental constant-sum games. Games Econ. Behav. 19, 97–132.
Moretti, L., Di Pellegrino, G., 2010. Disgust selectively modulates reciprocal fairness in economic interactions. Emotion 10 (2), 169–180.
Myerson, R.B., 1998. Population uncertainty and poisson games. Int. J. Game Theory 27, 375–392.
Nagel, R., 1995. Experimental results on interactive competitive guessing. Am. Econ. Rev. 85, 1313–1326.
Nagel, R., Bosch-Domenech, A., Satorra, A., Garcia-Montalvo, J., 1999. One, two, (three), infinity: newspaper and lab beauty-contest experiments. Universitat Pompeu Fabra.
Nasar, S., 1998. A beautiful mind: a biography of John Forbes Nash Jr. Simon and Schuster, New York.
Neral, J., Ochs, J., 1992. The sequential equilibrium theory of reputation building: a further test. Econometrica 60, 1151–1169.
Nikiforakis, N., Engelmann, D., 2011. Altruistic punishment and the threat of feuds. J. Econ. Behav. Org. 78 (3), 319–332.
Nunnari, S., Camerer, C.F., Palfrey, T., 2012. Quantal response and nonequilibrium beliefs explain overbidding in maximum-value auctions. Caltech Working Paper.
Nyarko, Y., Schotter, A., 2002. An experimental study of belief learning using elicited beliefs. Econometrica 70 (3), 971–1005.
Ochs, J., 1999. Entry in experimental market games. In: Budescu, D., Erev, I., Zwick, R. (Eds.), Games and Human Behavior: Essays in Honor of Amnon Rapoport. Lawrence Erlbaum Assoc. Inc., New Jersey.
Oosterbeek, H., Sloof, R., van de Kuilen, G., 2004. Differences in ultimatum game experiments: evidence from a meta-analysis. Exp. Econ. 7 (2), 171–188.
Ostling, R., Wang, J.T.-y., Chou, E.Y., Camerer, C.F., 2011. Testing game theory in the field: Swedish LUPI lottery games. Am. Econ. J. Microecon. 3 (3), 1–33.

Palacios-Huerta, I., Volij, O., 2008. Experientia docet: professionals play minimax in laboratory experiments. Econometrica 76 (1), 71–115.
Palfrey, T.R., Rosenthal, H., 1988. Private incentives in social dilemmas: the effects of incomplete information and altruism. J. Public Econ. 35, 309–332.
Page, T., Putterman, L., Unel, B., 2005. Voluntary association in public goods experiments: reciprocity, mimicry, and efficiency. Econ. J. 115, 1032–1053.
Partow, J., Schotter, A., 1993. Does game theory predict well for the wrong reasons? An experimental investigation. C.V. Starr Center for Applied Economics Working Paper, New York University, pp. 93–46.
Pillutla, M.M., Murnighan, J.K., 1996. Unfairness, anger, and spite: emotional rejections of ultimatum offers. Org. Behav. Human Decis. Processes 68 (3), 208–224.
Rabin, M., 1993. Incorporating fairness into game theory and economics. Am. Econ. Rev. 83 (5), 1281–1302.
Rapoport, A., Amaldoss, W., 2000. Mixed strategies and iterative elimination of strongly dominated strategies: an experimental investigation of states of knowledge. J. Econ. Behav. Org. 42, 483–521.
Rapoport, A., Lo, A., K.-C., Zwick, R., 1999. Choice of prizes allocated by multiple lotteries with endogenously determined probabilities. University of Arizona, Department of Management and Policy Working Paper.
Rogers, B.W., Palfrey, T.R., Camerer, C.F., 2009. Heterogeneous quantal response equilibrium and cognitive hierarchies. J. Econ. Theory 144 (4), 1440–1467.
Rosenthal, R.W., 1981. Games of perfect information, predatory pricing and the chain-store paradox. J. Econ. Theory 25, 92–100.
Roth, A., Erev, I., 1995. Learning in extensive-form games: experimental data and simple dynamic models in the intermediate term. Games Econ. Behav. 8, 164–212.
Roth, A., Prasnikar, V., Okuno-Fujiwara, M., Zamir, S., 1991. Bargaining and market behavior in Jerusalem, Ljubljana, Pittsburgh, and Tokyo: an experimental study. Am. Econ. Rev. 81 (5), 1068–1095.
Salmon, T., 1999. Evidence for 'learning to learn' behavior in normal-form games. Caltech Working Paper.
Salmon, T., 2001. An evaluation of econometric models of adaptive learning. Econometrica 69, 1597–1628.
Sanfey, A.G., Rilling, J.K., Aronson, J.A., Nystrom, L.E., Cohen, J.D., 2003. The neural basis of economic decision-making in the ultimatum game. Science 300, 1755–1758.
Seale, D.A., Rapoport, A., 2000. Elicitation of strategy profiles in large group coordination games. Exp. Econ. 3, 153–179.
Selten, R., 1978. The chain store paradox. Theory Decis. 9, 127–159.
Selten, R., Stoecker, R., 1986. End behavior in sequences of finite prisoner's dilemma supergames: a learning theory approach. J. Econ. Behav. Org. 7, 47–70.
Shah, A.K., Oppenheimer, D.M., 2008. Heuristics made easy: an effort-reduction framework. Psychol. Bull. 134 (2), 207–222.
Smith, V.L., Suchanek, G., Williams, A., 1988. Bubbles, crashes and endogeneous expectations in experimental spot asset markets. Econometrica 56, 1119–1151.
Stahl, D.O., 1996. Boundedly rational rule learning in a guessing game. Games Econ. Behav. 16, 303–330.
Stahl, D.O., 1999. Sophisticated learning and learning sophistication. University of Texas at Austin Working Paper.
Stahl, D.O., 2000. Local rule learning in symmetric normal-form games: theory and evidence. Games Econ. Behav. 32, 105–138.
Stahl, D.O., Wilson, P., 1995. On players models of other players: theory and experimental evidence. Games Econ. Behav. 10, 213–254.
Sundali, J.A., Rapoport, A., Seale, D.A., 1995. Coordination in market entry games with symmetric players. Org. Behav. Human Decis. Processes 64, 203–218.
Sutton, R.S., Barto, A.G., 1998. Introduction to Reinforcement Learning. MIT Press, Cambridge.
Tonin, M., Vlassopoulos, M., 2013. Experimental evidence of self-image concerns as motivation for giving. J. Econ. Behav. Org. 90 (C), 19–27.
Van Damme, E., 1999. Game theory: the next stage. In: Gerard-Varet, L.A., Kirman, A.P., Ruggiero, M. (Eds.), Economics Beyond the Millennium, Oxford University Press, 184–214.

Van Huyck, J., Battalio, R., Beil, R., 1990. Tacit cooperation games, strategic uncertainty, and coordination failure. Am. Econ. Rev. 80, 234–248.
Van Huyck, J., Cook, J., Battalio, R., 1997. Adaptive behavior and coordination failure. J. Econ. Behav. Org. 32, 483–503.
Von Neumann, J., Morgenstern, O., 1944. Theory of games and economic behaviour. Bull. Amer. Math. Soc. Princeton University Press, 18–625.
Walker, M., Wooders, J., 2001. Minimax play at Wimbledon. Am. Econ. Rev. 91 (5), 1521–1538.
Watson, J., 1993. A 'reputation' refinement without equilibrium. Econometrica 61, 199–205.
Watson, J., Battigali, P., 1997. On 'reputation' refinements with heterogeneous beliefs. Econometrica 65, 363–374.
Wooders, J., 2010. Does experience teach? Professionals and minimax play in the lab. Econometrica 78 (3), 1143–1154.
Wright, J., Leyton-Brown, K., 2013. Evaluating, understanding, and improving behavioral game theory models for predicting human behavior in unrepeated normal-form games. Working Paper.
Wright, N.D., Hodgson, K., Fleming, S.M., Symmonds, M., Guitart-Masip, M., Dolan, R.J., 2012. Human responses to unfairness with primary rewards and their biological limits. Scientific Reports, Wellcome Trust.
Yamagishi, T., 1986. The provision of a sanctioning system as a public good. J. Pers. Social Psychol. 51 (1), 110–116.
Zizzo, D.J., 2003. Empirical evidence on interdependent preferences: nature or nurture?, Cambridge J. Econ., 27 (6), 867–880.

CHAPTER 11

Evolutionary Game Theory in Biology

Peter Hammerstein[*], Olof Leimar[†]
[*]Institute for Theoretical Biology, Humboldt University Berlin, 10115 Berlin, Germany
[†]Department of Zoology, Stockholm University, 10691 Stockholm, Sweden

Contents

Handbook of game theory
http://dx.doi.org/10.1016/B978-0-12-420056-2.00011-2

Abstract

This chapter reviews the origin and development of game-theoretic ideas in biology. It covers more than half a century of research and focuses on those models and conceptual advancements that are rooted in fundamental biological theory and have been exposed to substantial empirical scrutiny. The different areas of research—ranging from molecules and microbes to animals and plants—are described using informative examples rather than attempting an all-encompassing survey.

Keywords: Plant defenses, Sex ratios, Evolutionary dynamics, Animal fighting, Genetic conflict, Symbiosis, Cooperation, Biological markets, Animal signaling

JEL Codes: C73

11.1. STRATEGIC ANALYSIS—WHAT MATTERS TO BIOLOGISTS?

Biology is a natural science and as such it aims to understand the properties of real life. Its philosophy, therefore, differs fundamentally from that of conventional normative game theory (e.g., Hammerstein and Boyd, 2012), where central concepts rely on axioms of rationality and the consistency with these axioms matters more than empirical support. Neither rationality nor the consistency with any axiomatic system is a fundamental issue in biology. It is fundamental though that evolution acting on genes has shaped the biological properties of organisms on Earth.

Many of these properties can be conceived as strategies by which organisms interact adaptively with their living environment. To illustrate this, consider the morphological development of water fleas. In principle, these small crustaceans are capable of building a helmet-like structure that protects them from predators. Helmets are very costly to produce, however, and predation pressure can be large or small. When growing up, water fleas screen their aquatic environment for chemical traces left by predators. Based on the presence or absence of these chemical cues they grow large helmets only when predation is likely (Agrawal et al., 1999). What evolution has invented for these small crustaceans is more than just a protective device; it is the *strategy* to grow a helmet if there are predators around or else save this growth effort.

As demonstrated by the water flea example, the notion of a strategy can help biologists to simultaneously conceptualize traits and their contingencies. This creates an important interdisciplinary link with game theory, which can be quite fruitful for the broader understanding of phenomena involving conflict, cooperation, or signaling. It must be emphasized, however, that strategic analysis in biology often focuses on those details of interactions that do not fall within the traditional scope of game theory. The key to understanding strategic interaction in animals, plants, and microorganisms lies frequently in the sophisticated mechanisms and machineries involved. Where this is the case, the wisdom of biochemistry, molecular biology, or cognitive science can be of greater importance than that of game theory.

All identification of evolved strategies requires a thorough understanding of the biological context in which they occur. To give a simple example, falling coconuts sometimes kill humans that are relaxing under a palm tree. The dropping of coconuts, however, has not evolved as the tree's strategy to kill large mammals by throwing a solid object on their neck. It occurs instead as part of a seed dispersal strategy. Coconuts serve the seeds as vehicles and enable them to travel long distances in the sea. These vehicles need a sturdy design to act as a boat and let their passengers survive in saltwater. Hamilton and May (1977) were inspired by game theory when they argued that plants are usually under selection to send some of their propagules on a journey even if the "travel fee" is high and if habitats are stable and saturated. Launching sturdy vehicles is a response to this selection pressure and produces human casualties as a side effect.

In contrast, when nicotine kills insect larvae feeding on a tobacco plant, the larvae fall victim to a highly sophisticated strategy of plant defense against herbivores. Nicotine is one of the most powerful insecticides known today and it is also very toxic to mammals. Three tiny droplets of this substance would suffice to kill an adult human. It is an ingenious strategy of the tobacco plant to use nicotine for defenses because this toxin interferes with neuronal signaling systems that are widespread in the animal world. By partially mimicking the neurotransmitter acetylcholine and binding to receptors for this chemical signal, nicotine can very effectively disrupt vital regulatory processes in a multitude of animal species.

Plants in general have evolved a great variety of chemical defenses that manipulate animal nervous systems. Different plant compounds interfere with nearly every step in neuronal signaling (Wink, 2000), including neurotransmitter synthesis, storage, release, binding, and re-uptake. They also interfere with signal receptor activation, and with key enzymes involved in signal transduction. In many cases, plant compounds achieve these effects because they have evolved to resemble endogenous neurotransmitters of the targeted set of animal species.

Obviously, it would have been difficult for herbivores to stop such antagonistic interference through evolutionary modification of their own complex communication systems. They found other countermeasures instead (Karban and Agrawal, 2002;

Petzinger and Geyer, 2006). In turning the table, herbivores have evolved a number of compounds that interfere antagonistically with endogenous plant signaling. These compounds prevent or attenuate the induction of plant chemical defenses. In addition, herbivores evolved detoxification mechanisms—even through symbiotic relationships with microorganisms—and possess aversive learning mechanisms that permit feeding on less toxic tissues.

Has game theory played a significant role in revealing these strategic aspects of plant-herbivore interactions where conflict of interest is almost universal? The answer is no, and it is easy to explain why. In a figurative sense, plants act like programmers who "hack" the computers (neural systems) of target individuals (see Hagen et al., 2009, for a review). Conversely, the targets try to construct a firewall against these hacking attempts and may even "hack back." Needless to say that successful computer manipulation requires mainly the understanding of operating systems. Neither operating systems nor neural systems fall within the subject domain of traditional game theory but the latter have been studied extensively within the realm of biology.

There are many research areas, however, where game theory matters to biology. These include the study of sex ratios and sex allocation, animal fighting, intragenomic conflict, cooperation—or the frequent lack thereof—in microbes and higher organisms, market-like biological interactions, and honest versus deceptive signaling. In the following, we discuss these areas in turn, describe their affinities with game theory and highlight the lessons about nature biologists have learnt under the influence of game theory. Interestingly, the spirit of game theory appeared in biology already in the early work by Düsing (1883, 1884a,b) and Fisher (1930) on sex ratios. They foreshadowed evolutionary game theory in a way that is comparable to Cournot's well-known anticipation of classical game theory. We will address sex-ratio theory in the following to introduce some fundamentals of evolutionary game theory and describe the unexpected impact this theory had on empirical research and groundbreaking ideas in biology.

11.2. SEX RATIOS—HOW THE SPIRIT OF GAME THEORY EMERGED IN BIOLOGY

Many animal species produce male and female offspring in almost equal proportions. Intriguingly, this often happens even if males have lower chances than females to reach the age of sexual maturity, if they do not care for their young, and if competition among males is so strong that many of them will never reproduce. Shouldn't these aspects matter to sex-ratio evolution? No less than Charles Darwin (1874) felt that the frequent occurrence of approximate 1:1 sex ratios posed a serious challenge to his theory of adaptation through natural selection. He had to admit, however, that he himself could not master this challenge. What was so difficult about it?

11.2.1 There is a hitch with fitness

To understand the deeper nature of Darwin's difficulty, let us look at the relationship between a mother's offspring sex distribution $s = (s_f,\ s_m)$ and her fitness defined as the number u of her surviving adult offspring:

$$u = (s_f a_f + s_m a_m)n. \qquad [11.1]$$

Here, a_f and a_m are the sex-specific viabilities for females, males, respectively; n is the total number of the mother's newborn offspring, s_f is the relative frequency of her daughters, and $s_m = 1 - s_f$.

In order to demonstrate the hitch with [11.1], suppose that, as it is often the case, females are more likely than males to reach the age of sexual maturity ($a_f > a_m$). Mothers who produce only females would then have the largest number of surviving offspring—i.e., the highest fitness.

Here, we see a conceptual problem that relates to the question of how to define a biological notion of utility. Simple views on Darwinian evolution would suggest that in an evolving population where there are two competing traits, the one with the higher fitness u would be selected. From this perspective and with the assumptions made above, natural selection should be expected to create highly female-biased sex ratios and perhaps eradicate males from many populations on earth. Furthermore, in populations where there are no viability differences between males and females, the population sex ratio would be subject to random drift and could never stabilize. So, obviously there is something problematic about using the number of surviving offspring as a measure of success (utility, payoff) relevant to natural selection.

11.2.2 Düsing's solution—the first biological game

Only a few years after Darwin had given up his attempt to explain even sex ratios, the demographer Düsing (1883, 1884a,b) resolved Darwin's puzzle. Instead of looking at the number of surviving children, Düsing calculated a mother's reproductive success as the number of her grandchildren. This step toward a more far-sighted notion of reproductive success (utility) made it possible to identify the selective forces at work. Düsing realized that the expected number of grandchildren obtained through children of a given sex would depend on the overall sex ratio produced in a population. If the population on the whole produces one sex excessively, children of this sex will—on average—reward their mothers with fewer grandchildren than those of the opposite sex.

Düsing's insight follows from a *balance equation* which states that in a population with discrete nonoverlapping generations, the class of all males produces exactly the same number of offspring as the class of all females. This balance equation makes sense for genetically diploid organisms like mammals and birds because in this case individuals have exactly one genetic father and one genetic mother. It does not hold, however, for haplodiploid organisms like ants, bees, and wasps, where males have no genetic father.

Using the balance equation, it is easy to calculate the number of grandchildren a mother would obtain through sons and daughters. This number depends on the population sex ratio and thus on the aggregate sex ratio "decisions" made by other mothers in the population. Obviously, the "flavor" of a game then appears if one considers the individual mother as a player in a population of players.

To be more concrete, let F_0, F_1, and F_2 denote the generations of a mother, her offspring, and her grandchildren. Suppose that for the entire population, the F_1-generation consists of N individuals, counted at the newborn stage, with fractions s_f, s_m of females, males, respectively, and that the F_2 population consists of M newborn individuals. On average, mothers would then obtain grandchildren as follows:

Expected number of a mother's grandchildren

$$(i)\ \textit{per daughter} = \frac{M}{s_f N}, \qquad [11.2]$$

$$(i)\ \textit{per son} = \frac{M}{s_m N}. \qquad [11.3]$$

From [11.2] and [11.3], it follows readily that a mother who wishes to maximize the number of her grandchildren would play a best response to the population by producing only sons if $s_f > s_m$ and only daughters if the inverse strict inequality holds. This game-theoretic reasoning serves biologists today as an important heuristic step toward a full-fledged dynamic theory of sex-ratio evolution (presented further later). It gives us the crucial hint about how natural selection pushes the offspring sex distribution in the direction of the equilibrium state s^* with $s_m^* = s_f^* = 0.5$, where according to [11.2] and [11.3] sons and daughters would on average be equally successful in producing grandchildren.

Düsing (1883, 1884a,b) phrased his sex-ratio arguments in a way very similar to the way we present them here, although he could of course not use the language of game theory. This language still had to be born but the spirit of game theory was definitely visible in his work. Sex-ratio theory can be regarded as both a precursor of evolutionary game theory and its current flagship if one considers the strong influence it has had for many decades on biological thought, experimentation, and field studies. We will describe this impact in Section 11.3.

11.2.3 Fisher's treatment of sex-ratio theory

The pioneering work by Düsing in the 19th century was long forgotten, partly because Fisher (1930), one of the most influential theoreticians of evolutionary biology, did not mention Düsing when he gave his own view on sex ratios in a classic monograph. Fisher draws primary attention to the "parental expenditure" made by parents when producing their young. To sketch his arguments informally, Fisher considers a "resource cake" limiting this expenditure and asks how parents would divide the cake among sons and daughters if their resource allocation took place under the influence of natural selection. His answer is surprisingly simple. Evolved parents would split the cake in two equal halves, one for all the sons, and the other for all the daughters. Obviously, if the sexes differ in how big a slice of cake they need for development, this unbiased division of the cake requires a sex ratio biased toward the "cheaper" sex.

The derivation of Fisher's result requires no more than a tiny modification of [11.2] and [11.3]. Suppose that on average daughters and sons will obtain slices of cake with sizes c_f, c_m, respectively. The expected number of grandchildren per resource unit invested in a daughter is then expression (*i*) divided by c_f. The analogous calculation can be made for investment in sons, using (*ii*). At equilibrium, the return per invested resource unit must be equal for male and female offspring:

$$\frac{s_f}{s_m} = \frac{c_m}{c_f} \quad \textit{The Fisherian sex ratio.} \qquad [11.4]$$

This equation captures one of the earliest quantitative predictions made in theoretical biology. It implies the equal split of the resource cake and is testable wherever one can measure the slices of the cake required by the sexes.

11.2.4 Does it suffice to count grandchildren—what is the utility?

In his reasoning about sex ratios Fisher (1930) did not explicitly talk of grandchildren as the measure of a parent's success. He suggested instead counting the children weighted by their reproductive value. In principle, this way of defining utility has a recursive flavor because in order to define the reproductive value of an offspring we would have to know the value of each of its offspring, etc. Taylor (1990), Charlesworth (1994a), McNamara and Houston (1996), and Grafen (2006) discuss the mathematical and conceptual intricacies of defining reproductive value. In Grafen's view, Fisher had the intuition that using reproductive value appropriately over one generation is the equivalent of looking an infinite number of generations ahead in order to obtain an asymptotic measure of fitness. This intuition—though often correct—needs careful examination (see Section 11.2.6).

If one does not fully trust the game-theoretic treatment of sex ratios or other biological phenomena, there are of course ways to invoke more mathematical rigor. This rigor is usually achieved at the expense of intuitive appeal. It seems to be the fate of game theory in biology that while it helps us tremendously in shaping our intuition and in generating insights, there is often the need to check these insights in a mathematical framework that captures more explicitly the dynamics of natural selection. In the following, we will give an example how this can be done for sex-ratio theory.

11.2.5 Evolutionary dynamics

In sex-ratio theory, populations are class structured because there are males and females. In principle, there could be more classes. Let $x_i(t)$ denote the number of individuals in class i at generation t. In a deterministic model with nonoverlapping generations, the evolutionary dynamics can be written as $x_i(t+1) = \sum_j a_{ij}x_j(t)$. The matrix A with elements a_{ij} is the so-called population projection matrix (Caswell, 2001), and if x is the vector of the x_i, we have $x(t+1) = Ax(t)$. Since we are dealing here with game-theoretic problems, the a_{ij} will depend on x and on the strategies present in the population. The elements a_{ij} are the per capita contributions of individuals in class j to class i in terms of offspring.

A common practice in evolutionary biology is to look for a stationary x for the case where all individuals use a strategy s. For such a stationary population one then examines whether initially rare mutant strategies s' would increase in number. Following this analytical program, we can now re-analyze the sex-ratio problem. Suppose that a mother has control over the sex ratio and produces a son with probability s_m and a daughter with probability $s_f = 1 - s_m$. Looking as before at nonoverlapping generations, and counting individuals at the newborn stage, the dynamics of x can be written as

$$\begin{pmatrix} x_f(t+1) \\ x_m(t+1) \end{pmatrix} = \begin{pmatrix} s_f b/2 & s_f bq/2 \\ s_m b/2 & s_m bq/2 \end{pmatrix} \begin{pmatrix} x_f(t) \\ x_m(t) \end{pmatrix}, \qquad [11.5]$$

where b is the expected number of offspring of a newborn female, bq is the corresponding reproductive output of a newborn male, and the factor $1/2$ accounts for the genetic shares of the parents in their offspring for diploid inheritance

Using the above mentioned balance equation again, the reproductive output of all males must equal that of all females. This implies $q = x_f(t)/x_m(t)$ and thus that $q = s_f/s_m$. In a stationary population $b = 1/s_f$ must hold, which could come about through a dependence of b on the total population size. Introducing now a mutant strategy s' and the matrix

$$B(s', s) = \frac{1}{2} \begin{pmatrix} s'_f/s_f & s_f/s_m \\ s'_m/s_f & 1 \end{pmatrix}, \qquad [11.6]$$

the population projection matrix for a stationary population is $A = B(s, s)$. The number of females and males, $(x_f,\ x_m)$, in the stationary population is proportional to the leading eigenvector, $w = (s_f, s_m)$ of $B(s, s)$.

We now consider a mutant gene that causes a female to adopt the sex-ratio strategy s', but has no effect in a male. If the mutant strategy is rare, only the strategy of heterozygous mutant females needs to be taken into account, and the dynamics of the mutant subpopulation can be written as $x'(t+1) = A'x'(t)$ with $A' = B(s', s)$. Computation of the leading eigenvalue of A' shows that a mutant with $s'_m > s_m$ can invade if $s_m < 0.5$ and one with $s'_m < s_m$ can invade if $s_m > 0.5$, so there is an evolutionary equilibrium at $s^* = (0.5,\ 0.5)$.

11.2.6 Reproductive value

Intuitively, the reproductive value of an individual of class i measures the individual's expected genetic contribution to future generations. For the case of nonoverlapping generations, the value is proportional to the ith component of the leading "left eigenvector" v of the population projection matrix $A = B(s, s)$, i.e., the leading eigenvector of the transpose of A. This follows from the Perron-Frobenius theorem (we assume A is a primitive matrix with leading eigenvalue $\lambda = 1$ and leading left and right eigenvectors v and w, normalized such that $\sum_i v_i w_i = 1$). The theorem entails that for a dynamics $x(t+1) = Ax(t)$ for $t = 0, 1, \ldots, x(t)$ will asymptotically approach $V_0 w$, where $V_0 = \sum_i v_i x_i(0)$ is the starting total reproductive value of the population.

For the sex-ratio problem, with $A = B(s, s)$ and B from [11.6], the leading left and right eigenvectors are $v = \frac{1}{2}(1/s_f, 1/s_m)$ and $w = (s_f, s_m)$. In this case, the reproductive value of a female is proportional to her expected number of offspring, $1/s_f$, and the same is true for a male. We see here the reason why Düsing's counting of grandchildren led to the correct conclusion, but his approach is not valid in general. For a mutant strategy s', the expected reproductive value per offspring is

$$V(s', s) = \frac{1}{2}\left(\frac{s'_f}{s_f} + \frac{s'_m}{s_m}\right), \qquad [11.7]$$

and this is sometimes used as a fitness or utility function, first introduced by Shaw and Mohler (1953). The fitness function gives the same conclusion, of an equilibrium for $s^* = (0.5, 0.5)$, as an examination of the leading eigenvalue. Although $V(s', s)$ differs from the leading eigenvalue $\lambda(s', s)$ of the mutant population projection matrix $A' = B(s', s)$, it is easy to see that V and λ have the same first order dependence on deviations of the mutant from the resident strategy, so V can be used to search for evolutionary equilibria. While reproductive value has the advantage of being a guide to intuition, it cannot fully replace a direct analysis of the evolutionary dynamics.

11.2.7 Haplodiploid sex-ratio theory

Let us repeat the analysis for haplodiploid insects, where haploid males develop from unfertilized eggs laid by their mother and diploid females develop from fertilized eggs. Given this mechanism of sex determination, a mother can readily influence the sex ratio among her offspring, by a choice of whether or not to fertilize an egg. In insects, eggs are fertilized one by one, right before they are laid. Corresponding to the dynamics [11.5], we then have

$$\begin{pmatrix} x_f(t+1) \\ x_m(t+1) \end{pmatrix} = \begin{pmatrix} s_f b/2 & s_f bq/2 \\ s_m b & 0 \end{pmatrix} \begin{pmatrix} x_f(t) \\ x_m(t) \end{pmatrix} \qquad [11.8]$$

for haplodiploids. Because each female has one mother and one father, $q = s_f/s_m$ must hold, and for a stationary population, we have $b = 1/s_f$, just as for the diploid case. For a rare mutant strategy s' and the matrix

$$B(s', s) = \begin{pmatrix} \frac{1}{2}s'_f/s_f & \frac{1}{2}s_f/s_m \\ s'_m/s_f & 0 \end{pmatrix}, \qquad [11.9]$$

the population projection matrix for the mutant subpopulation, consisting of female mutant heterozygotes and male haploid mutants, is $A' = B(s', s)$. The leading eigenvalue of $A = B(s, s)$ is $\lambda = 1$, with leading left and right eigenvectors $v = \frac{1}{2}(2/s_f, 1/s_m)$ and $w = (s_f, s_m)$. The expected number of offspring is $1/s_f$ for a female and s_f/s_m for a male, so counting grandchildren is not an appropriate measure of utility here. Instead, one needs to extend the accounting into the future, keeping track of genetic contributions at each step. Even so, the expected reproductive value per offspring, weighted by the genetic share, is proportional to $V(s', s)$ in [11.7], so the equilibrium sex-ratio strategy for the haplodiploid case is $s^* = (0.5, 0.5)$, just as for the diploid case. It is in a sense fortuitous that diploid and haplodiploid patterns of inheritance lead to the same sex-ratio equilibrium. In general, we should expect that evolutionary equilibria depend on the details of how genes are passed on to future generations.

11.3. THE EMPIRICAL SUCCESS OF SEX-RATIO THEORY

As persuasive as the Düsing-Fisher theory of sex ratios and sex allocation may be, some caution always needs to be exercised when evaluating functional explanations. In their famous "pamphlet" criticizing the "adaptationist program," Gould and Lewontin (1979) argued that many apparent adaptations may not have evolved for their seeming benefits, but may instead exist as part or by-product of another trait that has its own evolutionary logic. For example, in species with sex chromosomes like our human X and Y, where males are typically XY and females XX, the even sex ratio might simply occur as an unavoidable consequence of the sex determination system. Mendelian segregation of chromosomes implies an equal production of X- and Y-bearing sperm, so sex-ratio evolution could be constrained by an XY sex-determination mechanism if there are no genes that can modify it. Toro and Charlesworth (1982) showed indeed for an outbred population of the fruit fly *Drosophila melanogaster* that genetic variation in the sex ratio was effectively absent in that population.

11.3.1 Experimental evolution

Such observations do not decide the issue, however. The absence of genetic variation may actually result from natural selection for an even sex ratio (Bull and Charnov, 1988). By starting from a situation where there is genetic variation in the sex-ratio trait and observing evolution in the laboratory, it is possible to investigate empirically whether frequency-dependent selection stabilizes the 1:1 ratio as soon as it has genetic material to work on. Carvalho et al. (1998) and Blows et al. (1999) created lab populations with genetic variation in the sex ratio and an initial mean deviation from an even sex ratio, using species of *Drosophila*. By tracing the evolution of their lab populations, both were able to demonstrate an approach towards an even sex ratio. In addition, the specific design of their experiments made it possible to examine the selective forces hypothesized in the Düsing-Fisher model.

Conover and Van Voorhees (1990) conducted a similar experiment with fish. They studied the Atlantic silverside, which has temperature-dependent sex determination, and kept several populations for half a decade in artificial constant-temperature environments, either constant low or constant high temperature. Initially this caused the sex ratio to be strongly biased—toward females in low and males in high temperatures—but the proportion of the minority sex increased over the years until a balanced sex ratio was established.

11.3.2 Measuring the slices of the cake

In order to test Fisher's cake model for sex allocation [11.4], one needs to measure the different amounts of resources invested in sons and daughters. For many solitary wasps, it is possible to literally see these slices of Fisher's cake. These wasps produce an individual brood cell for each of their offspring, fill the cell almost completely with captured insect larvae, and lay one egg in it. The captured larvae serve as food provisioning for the wasp larva emerging from the egg. Cell size can thus be used to quantify the slice of cake. Krombein (1967) investigated this size and its relation to offspring sex for a substantial number of wasp species. His data enabled Trivers and Hare (1976) to provide empirical support for Fisher's result [11.4], namely that the sex ratio is the inverse of the investment ratio. Furthermore, Trivers and Hare analyzed sex-allocation data for a variety of ant species where, in principle, there is an evolutionary conflict between the queen and her workers over the investment ratio. Their study shows that workers often provide a larger portion of the cake to the queen's daughters than what would have been optimal from the queen's point of view. This is what haplodiploid sex-ratio theory (introduced in 11.2.7) predicts under appropriate conditions (Bulmer and Taylor, 1981)

Torchio and Tepedino (1980) also present support for the Fisherian sex ratio, but in the solitary bee they studied the investment pattern varies through the season and requires, in principle, a more detailed life-history model. More generally, it has to be said that the sex ratio can be biased for various reasons other than those captured by [11.4], some of these will be given in the following. More information can, for example, be found in Charnov's (1982) classic monograph and in West's (2009) detailed evaluation of various sex-ratio results and observations.

11.3.3 Local mate competition

The theory outlined so far tacitly assumes random mating and an unstructured population. In a publication now considered as one of the founding works of evolutionary game theory, Hamilton (1967) highlighted the fact that mating often occurs locally in more or less isolated groups. He suggested that local mate competition (LMC) would favor the evolution of female-biased sex ratios and substantiated his theoretical argument with a long list of insect examples where sib mating is frequent and the typical batch of offspring strongly female biased. The basic idea is seen most readily in the extreme case where mating only occurs between sibs. The mother's genetic contribution to future generations is then directly proportional to the number of her daughters that go on to produce new batches of offspring, so the mother ought to produce the minimum number of sons needed to fertilize her daughters. This is true for Hamilton's perhaps most exotic case, which is not an insect but the live-bearing mite *Acarophenax tribolii*. As Hamilton points out, males of this species usually die before they are born, and they mate with their sisters in the mother's body. Hamilton lists a mother's typical batch of offspring as 1 son and 14 daughters, which is perfectly in line with the extreme form of LMC.

Many empirical studies have supported Hamilton's theory of LMC. Herre (1985), for example, studied varying degrees of LMC in fig-pollinating wasps. Here, a few females—referred to as foundresses—enter a fig nearly simultaneously, pollinate the flowers, lay eggs, and die. As the fruit ripens, the wasps' offspring emerge and mate inside the fruit. The number of foundresses can be used to estimate the intensity of LMC. As this number decreases, LMC increases, and one should expect the female bias to increase as well. Herre's empirical findings are qualitatively and quantitatively in good agreement with a sex-ratio model designed for this kind of biological system.

11.3.4 Environmental sex determination and the logic of randomization

The performance of males and females is sometimes differentially influenced by environmental conditions, such as local temperatures. If so, environmental sex determination has a chance to evolve (Charnov and Bull, 1977). This is nicely illustrated by Pen et al. (2010), who analyzed data from snow skink lizard populations from low- and highland regions of Tasmania. In the warmer lowlands, there is temperature-dependent sex determination (TSD), and females with good local basking opportunities give birth to mostly female offspring, usually born earlier in the season. This could be adaptive, because snow skink females benefit more than males from being born early and having a longer period to grow. A possible drawback of TSD, on the other hand, is that large-scale temperature fluctuations, for instance between years, might result in population-level fluctuations in the sex ratio, which can be detrimental to members of the majority sex.

If the fluctuations are large enough, it would be better for an individual, or its mother, to ignore the environment and instead randomize the sex. Mendelian segregation is a kind of random number generator, so genetic sex determination (GSD) might then evolve if temperature fluctuations are large (Bulmer and Bull, 1982; Van Dooren and Leimar, 2003). In agreement which this idea, Pen et al. (2010) found GSD in highland snow skink populations, for which temperatures are lower and much more variable from year to year. Using a model that included details of snow skink life history and plausible sex-determination mechanisms, they found good agreement between field data on sex ratios and the model output.

We can compare such evolved strategies of environmental sex determination with Harsanyi's (1973) famous conception of the "purification of mixed strategies." If each decision maker, be it the mother or the developing individual itself, experiences a random and independent deviation in temperature from an overall mean, and this deviation influences the relative advantages of becoming a male or a female in a monotone manner, the equilibrium strategy is evidently to develop as male on one side of a threshold and as a female on the other. Harsanyi showed that, in the limit of the payoff perturbations conditional on the experience or observation of a decision maker being small, these pure threshold strategies approach a mixed equilibrium of the unperturbed game. For the sex-ratio case, the limit would be a 50-50 male-female randomization. The snow skinks, however, seem not to approximate the mixed equilibrium through such pure strategies with sharp thresholds. In the lowlands, there is instead a gradual change of the chance of becoming female with temperature (Pen et al., 2010), and in the highlands Mendelian segregation imposes randomization even though the relative payoff effects most likely still are present. So what can one say about these other ways of approximating a mixed strategy equilibrium of an "unperturbed game"?

Leimar et al. (2004) suggest a reason for the frequent lack of bang-bang like solutions in nature. In the "purification of mixed strategies" paradigm, the players of a game each observe an independent, private variable that has some influence on their payoff. Instead of, for example, tossing a coin to play a random mixture of two pure strategies, the players make their actions dependent on the realization of the private variable and play one option for low values, the other for high values—bang-bang. As long as there is statistical independence of private information, this could possibly work in nature. Returning to the snow skinks, however, we see a case of statistical dependence among the private variables. Why? The snow skink's private variable is its estimate of the temperature during a sensitive period. Obviously, this information lacks statistical independence because of large-scale temperature fluctuations. Leimar et al. (2004) point out that the applicability of Harsanyi's purification idea to biology is limited to the case of statistically independent private variables. When these variables of interacting individuals are correlated, randomized, or even complex, oscillating strategies might instead be the outcome.

11.4. ANIMAL FIGHTING AND THE OFFICIAL BIRTH OF EVOLUTIONARY GAME THEORY

For more than half the 20th century, a number of empirical biologists were guided by the intuition that natural selection favors essentially those animal traits that increase the "well-being" of the species. As a child of his time, Lorenz (1963, 1966) aimed to explain the biological patterns of aggression with this idea in mind. He emphasized three ways in which aggression would benefit the species: it causes its members to be widely distributed in space, it ensures that the strongest individuals propagate the species, and it serves to protect the young. He even went as far as to postulate for animals an innate inhibition to kill members of their own species. Impressive fights such as occur in stags, appeared to him like "ritualized contests" where the opponents respect hidden rules to avoid serious injury, while still selecting the better fighter. It could be said that Lorenz looked at fighting animals as if they were humans in a fencing tournament.

There are two major problems with this view of aggression. First, stags are not as well behaved as Lorenz described them. When given a chance, they can target the vulnerable parts of their opponents and even kill them. Second, natural selection tends to act far more efficiently at the individual level than at the level of the species. With some important exceptions, individual rather than species benefits dominate the selection pressures on phenotypic traits. Methodological individualism is indeed of great importance to biology and here lies the potential for exchange of ideas with game theory.

Maynard Smith and Price (1973) realized this potential and explored how animal conflict can be studied in a formal game-theoretic model. Their aim was to explain the frequent observation of "limited war" in animal contests, without appealing to group or species benefits. The work triggered an avalanche of research activities in biology and founded evolutionary game theory as a field. Their paper centers on a model that depicts a conflict over a resource between two animals. The simplest, unburdened version of this model is now well known as the "Hawk-Dove" game, in which either player has two extreme behavioral options, namely to "be ready for escalation, use weapons, and fight it out" or "avoid escalation, display your interest and go for a peaceful conflict resolution." These strategies were later called "escalate" and "display" by Maynard Smith (1982). The model looks as follows:

	Escalate	Display
Escalate	$(V-C)/2,\ (V-C)/2$	$V,\ 0$
Display	$0,\ V$	$V/2,\ V/2$

[11.10]

Parameter V is the benefit of winning and C the cost of being wounded.

11.4.1 The basic idea of an evolutionary game

How does a game like Hawk-Dove relate to a dynamic model of evolution? The idea is that strategies are heritable traits that evolve in a given population and are subject to natural selection. For a matrix game, for instance, the trait would be a mixed strategy. The game is played every generation between animals who find themselves in a conflict depicted by the game. The animals play inherited strategies and their reproductive success depends at least partly on the game payoff. In a sense, natural selection makes the choice of strategy, not the animal that we can observe as an actor in the game.

An important question is, who meets whom in the population. Maynard Smith and Price (1973) and more explicitly Maynard Smith (1982) make a 'mass action assumption' and assume that strategies are randomly paired so that the probability of encountering a strategy q is its frequency x_q. This assumption would, of course, be violated in games between close genetic relatives.

Suppose now that two strategies p and q occur at population frequencies x_p and x_q so that the population distribution of strategies is $x = (x_p, x_q)$. Let $E(p, q)$ denote the expected game payoff for playing a strategy p against strategy q. In an otherwise simple biological world, we are justified to define the fitness of strategy p as follows:

$$w_p(x) = w_0 + x_p E(p,p) + x_q E(p,q) \quad \text{Frequency-dependent fitness} \qquad [11.11]$$

Here, w_0 is the animal's basic fitness expectation that is altered by the game payoff. Maynard Smith and Price were not explicit about how this payoff shows up in a selection equation but the simplest such equation fitting their approach is the following, which describes exact inheritance in a population with discrete, nonoverlapping generations:

$$x'_p = x_p w_p(x)/\bar{w}(x) \quad \text{Discrete replicator equation} \qquad [11.12]$$

Here, $\bar{w}(x) = x_p w_p(x) + x_q w_q(x)$ is the mean population fitness and x'_p is the frequency of p in the next generation.

11.4.2 The concept of an evolutionarily stable strategy

As already discussed in Section 11.2, a common practice in evolutionary biology is to look at a population state where all individuals play the same strategy and to examine whether initially rare mutant strategies would be able to invade. Maynard Smith and Price called a strategy p evolutionarily stable if according to [11.12] no initially rare mutant strategy q can invade. It is easy to show that—for the given model of frequency-dependent selection—this corresponds to the following characterization of an Evolutionarily Stable Strategy (ESS).

A strategy p is evolutionarily stable if and only if it fulfills the following two criteria: [11.13]

(i) $E(p,p) \geq E(q,p)$ for all strategies q.
(ii) All strategies q with $q \neq p$ and $E(q,p) \geqq E(p,p)$ satisfy $E(p,q) > E(q,q)$.

To gain some perspective on this characterization, let us consider a finite, symmetric matrix game with n pure strategies, defined by the matrix A and its transpose, and define frequency-dependent fitness by [11.11] with $E(p,q) = \sum_{ij} p_i a_{ij} q_i$. If the population frequency of p is $1-\varepsilon$ and that of q is ε, the difference ΔW in fitness between strategies p and q can be written as follows:

$$\begin{aligned} \Delta w &= w_p(x) - w_q(x) \\ &= (1-\varepsilon)[E(p,p) - E(q,p)] + \varepsilon[E(p,q) - E(q,q)]. \end{aligned} \qquad [11.14]$$

Looking at the two square brackets and their weights we see conditions (i) and (ii) emerge. On the one hand, (i) means that the pair of strategies (p,p) is a symmetric Nash equilibrium, and condition (ii) can be seen as a refinement of that equilibrium. This creates a well-founded link between biology and classical game theory. On the other hand, the fitness [11.11] and dynamics [11.12] relate to a rather simple world of models, which could limit the scope of the link.

The ESS concept addresses primarily the uninvadability of a population that is monomorphic for the strategy p, and the versatility of this general idea, linking game theory to evolutionary dynamics, might be the most important reason for its success. In biology, polymorphic populations are also of great interest, and the monomorphism/polymorphism distinction is important. Seen in this light, the combination of a mass action assumption and bilinear payoff is a special, degenerate case, where the ESS condition characterizes population states that are averages of distributions of strategies. Maynard Smith and Price (1973) were aware of this degeneracy.

What Maynard Smith and practically everyone else did not know in the 1970s is that Nash himself had offered a mass action interpretation of game-theoretic equilibrium in his PhD thesis. Unfortunately, this did not appear in the publication (Nash, 1951) of his doctoral work. As a result, it was buried for more than 40 years in the archives of Princeton University. The scientific community noticed Nash's mass action argument only in the 1990s when he was considered for a prestigious award. Once a Nobel laureate, Nash (1996) finally published the missing bit.

11.4.3 What does the Hawk-Dove game tell us about animal fighting?

If escalated fights are costly in relation to resource value (i.e., if $C > V$), the Hawk-Dove game has a mixed evolutionarily stable strategy, which is to "escalate with probability V/C and display otherwise." The main wisdom gained from this toy model is, therefore, that individual selection will limit the amount of aggression when only low fitness gains are at stake. So far, the model goes some way toward what Lorenz (1963) wanted to explain but without resorting to a group selection or species benefit argument. For high fitness gains (i.e., $V > C$), however, the model makes nature look "red in tooth and claw," in sharp contrast with the ideas expressed by Lorenz.

Opponents may differ in size, age, sex, or status, as well as in many other aspects that possibly influence the costs and benefits of a biological game. Some asymmetries can be payoff irrelevant, such as the random order of arrival at a site of conflict. Maynard Smith and Parker (1976) studied an asymmetric version of the Hawk-Dove game in which one player is the first to be present at this site, the other second. They showed that if $V < C$, the so-called Bourgeois strategy would be an ESS, which is to "escalate when first" and "display when second." Hammerstein (1981) extended their model and investigated contests that are asymmetric in more than one aspect. He decomposed the genuinely symmetric evolutionary game into asymmetric role games (subgames) where, for example, the role "weaker owner" is confronted with the opposite role "stronger intruder." With this decomposition, and using a result by Selten (1980), the ESS can be characterized as a combination of strict Nash equilibria for the role games. Hammerstein's analysis of the Hawk-Dove game with two asymmetric aspects shows that—within limits—the Bourgeois strategy remains evolutionarily stable even if this means that a weaker owner escalates and a stronger intruder avoids escalation (see Hammerstein and Riechert, 1988, for a possible example in the world of spiders). So here could be a second explanation for limited aggression: asymmetries might lead to conflict settlement without dangerous fighting but based on the logic of deterrence (see Hammerstein and Boyd, 2012, for the role of emotions in this context and Kokko et al. 2006 for the role of feedback between individual level behavior and population dynamics).

A third explanation (Parker, 1974), not well represented by the Hawk-Dove game, is that opponents gain information about their fighting abilities (size, strength, weapons, etc.) during a contest and use this information in deciding whether to withdraw or persist (Enquist and Leimar, 1983).

11.4.4 The current view on limited aggression

The qualitative conclusion by Maynard Smith and Price (1973) that a low resource value relative to the cost of escalating fighting limits aggression, still stands and has strong empirical support. There is also now an understanding that the value of a contested resource should be put into a life-history perspective. If a large part of the remaining lifetime opportunities for reproduction are at stake in a single contest, dangerous and even fatal fights are the expected consequence (Grafen, 1987, Enquist and Leimar, 1990), whereas limited aggression is expected when there are many reproductive opportunities beyond the current contest. As an example of the former, males of certain species of fig wasps have evolved into extreme fighters, with armored bodies and huge mandibles, and they often fight to the death inside the figs where they emerge. A comparison of a range of species of fig wasps showed that limited future mating opportunities for the males was the most important factor explaining this extreme form of fighting (West et al., 2001).

Over the years, a great number of studies of animal contests have accumulated, and overall it is the case that winners differ from losers in morphological and physiological characteristics that might indicate fighting ability, with higher values for winners (Vieira and Peixoto, 2013). It is also established that contestants generally acquire information, both in the current contest and from previous contests, and use this in deciding whether to withdraw or persist (Hsu et al., 2006), so some assessment of fighting abilities is widespread. This acts to limit the seriousness of aggression, because information is effectively used to predict the outcome of all-out fights, without paying the cost of entering those serious fights.

The importance of clear-cut asymmetries in contest settlement, as suggested by the Bourgeois strategy for the Hawk-Dove game, is a more controversial topic in the study of animal contests. There is no doubt that "owners" of territories or other resources often win contests, but the reason could at least partly be that these individuals have higher fighting ability or place a higher value on the resource (Leimar and Enquist, 1984). So, for instance, the conclusion from a classical study on contests in male Speckled wood butterflies over sunspot territories (Davies, 1978), supporting the Bourgeois strategy, is now in doubt. Effects of the motivation of a male to persist in the contest, for instance because he has private information about the chances of encountering females in the territory, could instead be the main factor explaining contest outcomes (Bergman et al., 2010). Even so, the possibility that owners and intruders could use different strategies also when they have identical information about fighting abilities and resource values cannot be ignored, but it is not yet settled how important this effect might be.

11.5. EVOLUTIONARY DYNAMICS

The strongest and most productive link between game theory and evolutionary dynamics lies in the study of the invasion of a rare mutant into a resident population. In many cases, the dynamics of the mutant subpopulation is approximately linear in the mutant frequencies, as long as the mutant is rare. The asymptotic rate of growth of the mutant subpopulation can then be employed as a measure of the fitness of the mutant strategy in the environment given by the resident strategies. If this linear problem allows a description of the rate of growth in terms of reproductive value, which we saw in Section 11.2 is connected to the Perron-Frobenius theorem, there is an even closer link involving the concept of utility or payoff in game theory. This approach is a general recipe that can be applied in many situations. For instance, the mutant could be an allele at some particular genetic locus, with the analytical machinery of population genetics, involving Mendelian segregation and recombination with genes at other loci, put to use to study the change in mutant frequency. Alternatively, one might make the simplifying assumption of exact, asexual inheritance of strategies. As soon as the situation of the mutant and the resident population is specified, one needs essentially to bring to bear the relevant mathematical tools to the problem. A main goal of the analysis is to find resident strategies such that no mutant can invade. An equally important question is in which directions and towards which strategies such evolution through mutant invasion leads.

11.5.1 The replicator equation

Instead of the discrete-generation replicator equation [11.12], a continuous-time version is sometimes used to establish a link between game theory and evolutionary dynamics that goes beyond the question of the invasion of a rare mutant into a resident population. Assuming the population has strategies p_k with frequencies x_k, $k = 1, \ldots, N$, we define the fitness of p_k from the game payoffs as

$$f_k(x) = \sum_l E(p_k, p_l)x_l. \qquad [11.15]$$

The continuous-time replicator equation is then

$$dx_k/dt = x_k[f_k(x) - \bar{f}(x)], \qquad [11.16]$$

with $\bar{f}(x) = \sum_k x_k f_k(x)$. This equation is extensively discussed in an impressive book by Hofbauer and Sigmund (1998). For matrix games, the dynamics [11.16] specifies the change in the mean strategy, $\bar{p} = \sum_k x_k p_k$ and the payoff $E(p, q)$ can be expressed in terms of a payoff matrix. Hofbauer and Sigmund (1998) show that a strategy $\hat{p}$ is an ESS of the matrix game if and only if it is strongly stable, in the sense that under the dynamics [11.16], the mean strategy converges to $\hat{p}$ provided that it starts out at a point sufficiently close to $\hat{p}$. So for the continuous-time replicator equation, an ESS corresponds to a dynamically stable population strategy. As we shall see, however, for other kinds of evolutionary dynamics this form of stability is not a general property of uninvadable strategies.

11.5.2 Adaptive dynamics and invasion fitness

Any evolutionary dynamics where natural selection plays a role could be described as an adaptive dynamics, but the term has come to be used in particular for evolutionary sequences of steps where in each step a mutant strategy has a chance to invade and replace a resident strategy, or possibly, invade without ousting an alternative, thus creating polymorphism (Geritz et al., 1998; Hofbauer and Sigmund, 1998; Metz et al., 1996). A basic tool for the study of adaptive dynamics is the concept of invasion fitness, which plays the role of determining the probability with which a mutant can invade. Let s denote a strategy or trait, which we take to be an n-dimensional real vector. For a population monomorphic for s, the invasion fitness $F(s', s)$ of a mutant trait s' has the interpretation of a dominant Lyapunov exponent (Metz et al., 1992), corresponding to the mean rate of change of the logarithm of the size of the (small) mutant gene subpopulation. So, when the mutant is the same as the resident, we have $F(s, s) = 0$. In the examples of sex-ratio evolution discussed in Section 11.2, invasion fitness would be the logarithm of the leading eigenvalue of the mutant population projection matrix. A mutant with $F < 0$ has no chance of invading (in a large population), whereas one with $F > 0$ might invade. If one takes into account that this mutant could go extinct during an initial phase of low mutant copy number, one finds that the probability of invasion is proportional to $F(s', s)$ (Dieckmann and Law, 1996).

If a population is at an ESS, the adaptive dynamics cannot lead away from it, but it is also of interest to ask if trait-substitution sequences will lead toward the ESS. For one-dimensional trait spaces, this question of so-called convergence stability has been much studied (e.g., Eshel, 1983; Taylor, 1989). A basic idea is that the selection gradient

$$\partial F(s', s)/\partial s'|_{s'=s} \qquad [11.17]$$

should point toward a convergence stable point, resulting in the criterion that

$$\frac{\partial^2 F(s', s)}{\partial^2 s'^2} + \frac{\partial^2 F(s', s)}{\partial s' \partial s} < 0 \quad \text{at } s' = s \qquad [11.18]$$

should hold for a convergence stable s. This criterion is not equivalent to uninvadability, so there can be an uninvadable s that evolutionary change does not approach but rather moves away from. Furthermore, there can be local fitness minima that are convergence stable. These are called branching points (Metz et al., 1996; Geritz et al., 1998) and are associated with the evolutionary emergence of polymorphism.

In multidimensional strategy spaces, the distinction between uninvadability and evolutionary convergence is even greater (Leimar, 2009), in that genetic correlations between strategy components can influence evolutionary trajectories. It is a general phenomenon that evolutionary change is guided by, but not fully determined by fitness. The possible influence of the mutation process in directing evolutionary change will be greater in higher-dimensional phenotype spaces, simply because with genetic correlations there are more possibilities for bigger gains in some trait component to trade off against smaller gains or even losses in other components.

11.5.3 Gene-frequency dynamics and the multilocus mess

Short-term evolutionary change is often not the result of natural selection operating on new mutations, but rather the consequence of selection on already existing genetic variation, sometimes called standing genetic variation. This variation is also important in artificial selection, so the matter of responses to selection that only involve gene- and genotype-frequency dynamics, without new mutations entering a population, is of great practical importance. In practical applications, quantitative genetics (Falconer and Mackay, 1996) is often used to investigate and predict the response to selection, and the approach has also been extended to study long-term evolution (e.g., Lande, 1981). The basic idea is that the response to selection depends on the presence of additive genetic variation in the traits in question. Over the longer term, new mutations can replenish the supply of additive variation. Evolutionary quantitative genetics is an often used modeling alternative to adaptive dynamics.

There is, however, a difficulty that appears if one wants to base a theory of adaptive change solely on gene- and genotype-frequency dynamics, without considering new mutations. Quite some time ago, Moran (1964) and Karlin (1975) showed that one can readily find two-locus systems where the mean fitness of a population decreases over time. As a sort of resolution of this seeming dilemma, Eshel (1996), Hammerstein (1996), and Weissing (1996) pointed out that if one allows for a sufficiently wide range of mutations, long-term evolution is not constrained by the idiosyncrasies of multilocus genotype-frequency dynamics.

11.5.4 Finite populations and stochasticity

In finite populations, the randomness of reproduction and survival can have a noticeable influence on gene-frequency dynamics. The study of genetic drift at selectively neutral loci is a major endeavor in population genetics. Random changes in gene frequencies can also modify and even act in opposition to natural selection. This could be particularly important if, for instance, genetic drift allows a population to move from a local fitness peak to another higher peak, with chance events driving it across the fitness valley between peaks. There are many variants on this theme in the history of evolutionary thinking, with Wright's (1932, 1977) shifting balance theory being the most prominent. There is no consensus in biology about the importance for long-term evolution of random movements between selective equilibria (e.g., Coyne et al., 1997, question the importance of the shifting balance theory), but even so one should keep in mind what these processes might achieve.

In game theory, the idea that external shocks or an intrinsically noisy learning process can correspond to equilibrium selection (Young, 1993) has been put forward, and more recently this kind of thinking gained support in biologically oriented game theory (e.g., Nowak et al., 2004). Lacking decisive empirical studies, the contribution of random shifts to long-term evolution remains an open question. One should however keep in mind that the probability of a random shift from one equilibrium to another will depend very strongly on population sizes, so for these processes to play a major role, important evolutionary changes should sometimes happen in very small populations, and this is a significant constraint.

11.6. INTRAGENOMIC CONFLICT AND WILLFUL PASSENGERS

Although most genes probably have a function that serves the organism as a whole, some genetic elements (genes or larger parts of DNA) have properties that enhance their own transmission relative to the rest of an individual's genome. These are called selfish genetic elements and are excellently reviewed by Werren et al. (1988) and Burt and Trivers (2006). Selfish elements are in conflict with the rest of the genome and thus fall within the domain of evolutionary game theory. Conceptually speaking now, what is a genetic conflict? Hurst et al. (1996) proposed the following definition: "Genes are in conflict if the spread of one gene creates the context for the spread of another gene, expressed in the same individual, and having the opposite effect." Evolutionary game theory can be helpful in identifying genetic conflict as such. For instance, applying the concept of reproductive value as a measure of Darwinian utility (as was done in Section 11.2) to a genetic element that is transmitted to a mother's daughters but not to her sons, it is evident that sons have zero utility, seen from the strategic perspective of the element. Even so, it usually makes little sense to offer anything like a "solution concept" for genetic conflict. The power of the players in genetic games depends too much on mechanistic detail (*cf.* Section 11.1), and there is often a never-ending arms race.

11.6.1 Meiotic drive

In organisms that carry pairs of chromosomes, each of these chromosomes usually has a 50% chance of transmission to offspring. Some chromosomes contain "selfish regions" that enhance this percentage at the expense of the paired chromosome's transmission, a phenomenon called meiotic drive. These regions may contain "killer genes" and "target genes," which protect the selfish element from killing itself. An example is the so-called t-haplotype found in in natural populations of the house mouse. Males carrying a chromosome with this selfish genetic element transmit the chromosome to approximately 90% of their offspring. Males carrying two copies of the t-haplotype are sterile. Furthermore, both males and females have reduced viability if they carry two copies of the t-haplotype As a consequence, this particular selfish element cannot spread to fixation in a population (e.g., Charlesworth, 1994b)

Meiotic drive genes are known from many animal species, including fruit flies and mosquitos. The vast majority of driving genetic elements are located on sex chromosomes, in which case they bias the sex ratio (see Jaenike, 2001, for a review). Many theory-guided sex-ratio studies have therefore led to the discovery of previously unknown selfish genetic elements—an unforeseen triumph for Düsing, Fisher, Charnov, and other developers of sex-ratio theory.

11.6.2 Example of an ultraselfish chromosome

Werren (1991) made an exciting discovery that even tops the story of meiotic drive. In the parasitoid wasp *Nasonia vitripennis*, he found small parasitic chromosome fragments—referred to as *PSR*, which stands for "paternal sex ratio"—and investigated their basic properties and population dynamics. These fragments are so-called B-chromosomes and exhibit the maximal degree of genetic selfishness found in nature. Werren and

Stouthamer (2003), therefore, called them "the ultimate selfish genetic elements." In a male, *PSR* first coexists peacefully with the wasp's regular chromosomes. But the peace is hollow. As soon as these regular chromosomes are passed on to a female's egg, a "magic hand" destroys them all with great efficacy. The presence of *PSR* is thus detrimental to the male's entire genome.

As a game theorist one has to wonder about the strategic advantage of this massive destruction. There is a hidden benefit. *PSR*—not being a regular chromosome—has problems with transmission through the female's germ line. The obstacle lies in the detailed mechanics of how ova are generated, and *PSR* would often get stranded during meiosis. By destroying the paternal chromosome after fertilization it acts as a "gender bender" and transforms the diploid female organism into a haploid organism that—according to the sex-determination system of wasps—develops into a male. This individual's sperm will carry *PSR* into the next generation and *PSR* will again transform daughters into sons, starting a new reproductive cycle of this selfish "player."

11.6.3 Endosymbionts in conflict with their hosts

Many organisms have microbial passengers. Humans, for example, have more bacterial symbionts than their bodies have cells. Our individual survival depends strongly on microbial cooperation. Even mitochondria, now known as "organelles" and as the "power plants" of animal and plant cells, used to be microbial passengers. About two billion years ago, they still were "autonomous" bacterial organisms. At some point, they became endosymbiotic partners of other kinds of bacteria that evolved into nucleated cells. The endosymbionts subsequently lost much of their genetic autonomy and are now largely controlled by genes located in the nucleus of the eukaryotic cell.

Mitochondria and some intracellular bacteria are transmitted via eggs from mother to offspring but are not transmitted via sperm. Such uniparental transmission has a remarkable theoretical consequence, highlighted by Cosmides and Tooby (1981). Whenever mitochondria are passed on to a male organism, this results in a dead end for their own reproduction. They should thus be under strong selection to manipulate the reproductive system of the host and increase as much as possible the fraction of females in host offspring. Taking into account, however, that mitochondria have very few genes left in their own genomes, it seems hardly possible for them to respond evolutionarily to the selection pressure just described. Yet, in plants, such as corn and rye, mitochondrial variants do exist that target the pollen-producing anthers and disrupt their development (Burt and Trivers, 2006). The otherwise highly cooperative mitochondria act here as enemies of male tissue—a striking example of how the sex ratio conflict described by Cosmides and Tooby can manifest itself in nature.

For us humans, the conflict between mitochondria and their hosts has an irritating aspect. Since men do not transmit mitochondria, the latter would not be punished by natural selection if they harmed the former. Frank and Hurst (1996) suspect that this could play an important role in the understanding of human male diseases, such as the widely observed aberrations in sperm motility and male fertility. They opened a still underexplored research area where evolutionary game theory meets medicine and thus takes part in a broader scientific endeavor initiated by Nesse and Williams (1994) as "evolutionary medicine."

11.6.4 *Wolbachia*—master manipulators of reproduction

As manipulators of their hosts, mitochondria look like beginners compared to *Wolbachia*, their closest living relatives in the bacterial world. *Wolbachia* are intracellular passengers of insects, mites and other arthropods. They share their mode of inheritance with that of mitochondria and have found several impressive ways to alter the sex ratio, which puts them in conflict with the nuclear genome of their hosts. Their repertoire of manipulative skills (Werren et al., 2008) includes *feminization,* the transformation of genetic males into egg producers, *induction of parthenogenesis,* a mode of reproduction that leads to the exclusive production of daughters, and *male killing*, a destructive interference with male embryonic development. With respect to evolutionary medicine, it is of interest that the most typical manipulation exerted by *Wolbachia* is a modification of sperm causing male infertility in matings with uninfected females.

Koehncke et al. (2009) explore the strategic options for hosts to counter the adverse action of their bacterial passengers. It turns out from their analysis that - in an evolutionary arms race - host populations would have a fair chance of driving *Wolbachia* to extinction. This seems inconsistent with the empirical estimate that 40 percent (Zug and Hammerstein, 2012) or even more of all arthropod species are infected. Zug et al. (2012) resolve this puzzle by considering an epidemiological model in evolutionary time, where *Wolbachia* can infect new species before becoming extinct in a given one, and where resistance against *Wolbachia* attenuates over time in an uninfected species. This way the master manipulator can save its neck, and the host-parasite game can finally be understood on a long time scale in a species-overarching context. The games *Wolbachia* play happen at different time scales.

11.7. COOPERATION IN MICROBES AND HIGHER ORGANISMS

Game theorists have used models like the Prisoner's dilemma and related empirical paradigms to demonstrate the obstacles to cooperation in a world where players act entirely according to their self-interest. For animals, plants, and microbes, however, these obstacles are often removed through the effect of genetic relatedness. When participants in a cooperative endeavor are genetic relatives, they may individually obtain a substantial share of their partners' success through indirect transmission of genes. This is the core argument of classical kin selection theory (Hamilton, 1964a,b) and enables us, for example, to understand the high degrees of sociality found in ants, wasps, and bees. In fact, kin selection often "steals the show" from game theory when cooperative phenomena are to be explained. With this caveat in mind, let us now focus on the wisdom of game theory.

11.7.1 Reciprocal altruism

Evolutionary biologists use the term "altruism" for acts that are costly to the actor and beneficial to others. Their notion of altruism is solely based on the fitness consequences of behavior and leaves open the psychological question of whether there is a true concern for others. Consider now the case of an altruistic "donor" who provides support to a "receiver" in need of help. When there is some degree of genetic relatedness between donor and recipient, kin selection might elegantly explain this altruism by showing that the indirect fitness benefits compensate the direct fitness losses of the donor. In contrast, when there is no genetic relatedness, altruism remains a puzzle.

Can game theory help us resolve this puzzle? Trivers (1971) created in biology the paradigm of "reciprocal altruism," which closely relates to the theory of repeated games also addressed by Axelrod and Hamilton (1981), and more critically by Selten and Hammerstein (1984) as well as Boyd and Lorberbaum (1987). Trivers recognized that altruism has a chance to evolve in the context of repeated interactions if the roles of donor and recipient alternate between interacting, unrelated organisms. He considered the strategy of "reciprocal altruism," which is to help "by default" the partner in a repeated game but mirror refusal of help with a refusal to help. Assuming that the cost of a donor's help is smaller than the benefit to a receiver, that there are many repeated interactions, and that a sufficient share of a population plays reciprocal altruism, this very strategy will not be punished by negative selection. It simply pays donors to help because help begets help—this is how Trivers made helping lose its altruistic flavor.

Wilkinson (1984) presented a vampire bat example of reciprocal altruism that received much attention in biology and the social sciences. As the name indicates, vampire bats feed on blood, mostly from mammals. They live in colonies and hunt with varying success in the dark. After about three days of fasting a bat would die from starvation but hungry bats often receive help through a "blood donation," in which a donor regurgitates food to the hungry receiver. In his experiment, Wilkinson used bats from two colonies and controlled experimentally the access to food (blood). He showed that donations were more likely to come from those individuals that had already received a donation from the receiver than from those that had not.

The vampire bat example may look like reciprocal altruism. There are major problems, however, with Wilkinson's otherwise fascinating study. He did not provide any evidence for internal "bookkeeping" of help refusals, and we do not know from his experiment to what extent a refusal to help would be mirrored at all. Furthermore, the bats live in kin groups and help occurred mainly within the kin groups used in the experiment. Perhaps, this study provided evidence for kin selection rather than reciprocal altruism. Based on a recent experiment, Carter and Wilkinson (2012) argue that kin selection may not be as important as it looked in the original study but they do not tackle the crucial bookkeeping issue. Hammerstein (2003) and Clutton-Brock (2009) criticize Wilkinson's example and come to the more general conclusion that the evidence for Trivers's reciprocal altruism is extremely scarce in nature. We think that the theory of repeated games has largely failed in biology. There are possible exceptions in the world of primates were the evidence for reciprocal altruism is so mixed and imprecise that it would be too cumbersome to present them here.

11.7.2 Indirect reciprocity

Nowak and Sigmund (1998) studied a model in which the entire population consists of only 100 individuals. Each generation 500 interactions take place where a randomly chosen "recipient" asks a randomly chosen "donor" for help. The donor can refuse to help but the entire community observes everyone's behavior. Individuals each have a reputation, which increases whenever the individual helps and decreases whenever help is refused. Given the high degree of stochasticity in their model, Nowak and Sigmund asked how much helping would on average be seen if one lets this population evolve on a long time scale. They found the average helping rate to be high and this was due to the frequent occurrence of a so-called "image-scoring strategy," which is to help those with sufficiently high reputation and not to help otherwise.

Leimar and Hammerstein (2001) pointed out that in the given model the image-scoring strategy contains an unexplained altruistic element because every time an individual refuses help it loses some of its reputation. They investigated a similar model with a more realistic, sizeable population that consists of 100 groups of 100 individuals, and where migration between groups occurs. As expected, the image-scoring strategy does not evolve in this model. The study by Nowak and Sigmund demonstrates a basic insight from population genetics, namely that selection works differently in populations with a small compared to a large "effective population size."

Sugden (1986) made the earliest suggestion of how indirect reciprocity might actually work. His approach differs from Nowak and Sigmund (1998) mainly in how reputation is updated. Sugden's players do not lose reputation when they fail to help those with a bad reputation. Engelmann and Fischbacher (2003) conducted a human experiment where Sugden's reputational logic seemed to work. This logic might be cognitively too demanding, however, to be realistic for animals. Wedekind and Milinski (2000) and Milinski et al. (2001) showed in experiments with humans that image scoring can occur but it remains an open question why this is so. In a theoretical study, Panchanathan and Boyd (2004) explored cost-free ways of punishing that would allow indirect reciprocity to work. Yet, outside the human world and perhaps with some primate exceptions, indirect reciprocity has not been shown to be broadly significant in biology.

11.7.3 Tragedy of the commons

In nature, iron is essential for elementary processes such as DNA synthesis. Bacteria have found an efficient method by which they can supply themselves with this metal. They secrete compounds that have a chemical affinity to iron and "grab" it. These so-called siderophores enable the bacteria to "fish" for the resource under discussion. A catch is then consumed via reuptake of the iron-catching molecules. If anything in biology looks like a "tragedy of the commons" in the sense of Hardin (1968), this is it. The production of siderophores is costly and serves producers as well as scroungers as a "meal."

How can this iron fishery work if it offers a strong incentive for free riding? This is an interesting conceptual question. As Ostrom (1990) explained so beautifully in her work on the tragedy of the commons, human societies have found many ways to counter free riding through institutions that change the structure of the game. Needless to say, for example, that an efficient police controlling resource use will make free riding unprofitable. But bacteria are different from human societies. They cannot easily design institutions. Kin selection, however, is a candidate process that might restrict the amount of free riding.

Griffin et al. (2004) studied a soil bacterium, *Pseudomonas aeruginosa*, to examine the role of kin selection in limiting the amount of free riding. They used an experimental evolution approach and used lab populations with different compositions of clones, so that the degree of relatedness varied between evolving populations. Griffin et al. showed that higher levels of siderophore production evolved in the higher relatedness treatments.

As in many other bacterial systems, kin selection plays here the dominant role in explaining cooperation. Without the influence of game theory, however, biologists would have been much less inclined to engage in experimental activities like the one presented here.

11.7.4 Common interest

Kin selection is very important but the biological theory of cooperation has more "tricks up its sleeve," and one is the effect of common interest. Following now Leimar and Hammerstein (2010), the basic idea of common interest is that individual organisms have a stake in the success of others or in the success of a joint project. This idea is translatable into the world of genes or larger genetic elements. Joint projects can be as simple as cooperative hunting in carnivores or as elaborate as the contribution of different genes and regulatory elements to the development of a multicellular organism. The common interest perspective plays an important role in biological attempts to understand the evolution of organismality (Queller and Strassmann, 2009; Strassmann and Queller, 2010).

How can common interest as such evolve? First, if partners are likely to stay together in the future, for instance because of costs associated with partner change, common interest would typically become greater. The coupling of reproduction is another general reason for greater common interest. The extreme form of cooperation found in the genes operating in an organism provides an example. A comprehensive survey on selfish genetic elements (Burt and Trivers, 2006) confirms the view that most genes act most of the time to the benefit of the organism. This high degree of intragenomic cooperation is based on the genes' shared interest to "keep the organism running" that carries them into the next generation and "delivers" its passengers in a rather fair way. Fair transmission of genes in animals and plants occurs partly because gene movements are strictly regulated during mitotic and meiotic cell divisions. Molecular devices, such as the "spindle check point," supervise these "passenger movements." The alignment of gene interests through molecular control mechanisms represents one of the major transitions in evolution (Maynard Smith and Szathmáry, 1995).

It is important to understand how evolution can increase the dependence of partners on each other. The following plant-animal symbiosis documents this impressively. *Acacia* plants house mutualistic ants as their "body guards" and possess specific glands to provide them with extrafloral nectar, a sucrose-free nutritive product (Heil et al., 2005; Kautz et al., 2009). Due to the absence of sucrose the *Acacia's* nectar is unattractive to ants in general. The mutualistic *Pseudomyrmex* ants, however, are specialized to live on *Acacia* and depend strongly on this plant's specific nectar. They have even lost the capability to digest sucrose (Kautz et al., 2009). The dependence most likely evolved because the plant was under selection to become less attractive to nonmutualistic ants, and the mutualistic ants were forced to adapt to the evolutionarily changing *Acacia* to be efficient in their main habitat.

11.7.5 Common interest through lifetime monogamy

In the evolutionary view of the animal world, a female and her mate share an interest in their joint offspring but interests typically diverge when it comes to the question of how to split the work involved in raising that offspring (see Houston et al., 2005, for a review of modeling approaches to conflict between parents over care). Males may even benefit from actively harming female partners. The fruit fly *Drosophila melanogaster* is a naturally promiscuous species where males add toxic components to their seminal fluid. When they mate, these toxic components increase the fertilization chances of their sperm at the expense of the female's long-term reproduction and survival (Chapman et al., 1995). In the absence of mate fidelity, it is indeed possible for such harmful behavior to evolve because the male's reproductive success does not depend on the female's long-term survival. This makes *D. melanogaster* an interesting study object for experimental evolution of cooperation.

What if the fruit flies showed more partner fidelity? Holland and Rice (1999) conducted an evolutionary experiment to explore how monogamy can alter the harm males inflict on their female partners. In the evolving lab population, they imposed monogamy by letting each female house individually with one randomly assigned male. After 47 generations, males had evolved to be less harmful to their mates and females where less resistant to male-induced harm. This demonstrates impressively the role monogamy can play in the evolution of cooperation.

Lifetime monogamy is likely to have played a key role in the evolution of social insects. As Boomsma (2009) reports, "all evidence currently available indicates that obligatory sterile eusocial castes only arose via the association of lifetime monogamous parents and offspring." This is extremely interesting because lifetime parental monogamy induces a reliable common interest between parents and their offspring. This is so because the average genetic relatedness between siblings is then equal to the relatedness between parent and offspring. Children may then stay with their parents as helpers at the nest if their positive effect on parental reproductive success is a little higher than the gains possible through own reproductive attempts (a standard kin selection argument). After presenting this simple but highly consequential argument, Boomsma discusses the possible counterargument that his line of reasoning makes eusociality "too easy to evolve." The rare occurrence of lifetime monogamy in nature and various other facts weaken this counterargument.

11.8. BIOLOGICAL TRADE AND MARKETS

As described by Leimar and Hammerstein (2010), social partner choice (Bull and Rice, 1991) and the concept of a biological market (Noë et al., 1991; Noë and Hammerstein, 1994, 1995) have gained increasing attention of researchers who explore the evolutionary stability of cooperation among microbes, animals, and plants. This may seem surprising because conventional market models in economics rely on idealizations that biologists cannot refer to (Bowles and Hammerstein, 2003). For example, many market models describe interactions in which the goods and services traded are subject to complete contracts that are enforceable at no cost. These models do not apply to biology since nonhuman organisms do not sign contracts and even if they did, these contracts could not be enforced. Under such circumstances, one cannot expect biological markets to clear in the sense that supply matches demand at evolutionary equilibrium.

The most basic aspect of a biological market is the possibility to choose between offers in trade-like interactions where some exchange of "commodities" takes place. There is ample evidence for such choices, and this is important for the understanding of cooperation. Trades in nature often take place in mutually beneficial interactions between members of different species (interspecific mutualism). These interactions tend to follow the pattern of "hosts" offering food or shelter to "visitors" while gaining benefits from the latter (Cushman and Beattie, 1991). If visitors are sufficiently mobile, this enables them to exert choice between potential hosts.

11.8.1 Pollination markets

Pollination biology has a long tradition of thinking in terms of market analogies. Perhaps the first use of this analogy was made by Von Frisch (1967), who wrote about "regulation of supply and demand on the flower market." A large body of work now supports the view that, on a rather short time scale, insect-flower systems may approach an equilibrium with approximately equal profitability for visitors to different hosts (Schaffer et al., 1983). We know in addition that the choosiness of pollinators can strongly influence the composition of host communities. Dramatic changes in community structure occur, for example, when an invading plant outcompetes the residents, offering a particularly rich nectar to lure pollinators away from the native plants (Chittka and Schürkens, 2001). There is also evidence from experiments with artificial flowers showing the dramatic impact rewards can have on pollinator visitation rates (Internicola et al., 2007). Many other systems exist where visitors choose among hosts. These include ants visiting aphids (Fischer et al., 2001; Völkl et al., 1999) and client reef fish visiting cleaner wrasse stations (Bshary and Noë, 2003; Bshary and Schäffer, 2002).

11.8.2 Principal-agent models, sanctioning and partner choice

The reverse choice occurs when a stationary individual can accept or reject incoming "applicants" for a "position." For example, the power of strong males to keep weak males off their territory enables them to act as if they were "principals" in trades with the less strong "agents" applying for entry to their territory. Bowles and Hammerstein (2003) have therefore likened certain biological scenarios to principal-agent problems in economics. Among possible examples are the males of a songbird called lazuli bunting (the name alludes to a well-known gemstone). Bright-plumaged males, the principals, allow less competitive dull-plumaged males, the agents, to settle at the periphery of their high-quality territories (Greene et al., 2000). This deal is beneficial to both principal and agent because the latter will attract a new female to the territory and both will mate with her. Note that the principal can enforce his share of the mating activities but if he goes too far and monopolizes access to this female, the agent will have no reason to care for her offspring. In this way, the trade works without an enforceable contract.

Other possible examples are large male fiddler crabs accepting smaller territorial neighbors for help with territorial defense (Backwell and Dennions, 2004; Detto et al., 2010). There are also indications of choosiness by hosts in legume-rhizobium mutualism (Heath and Tiffin, 2009). *Rhizobium* is a soil bacterium that forms an association with the roots of legumes and fixes nitrogen. Admittedly, the mechanisms by which a plant could recognize and choose a beneficial *Rhizobium* are currently unknown. Termination of interactions with unprofitable visitors is frequently thought of as a form of sanction resulting in partner choice (Bull and Rice, 1991; Kiers et al., 2003; Kiers and Denison, 2008; Simms et al., 2006).

From the range of empirical data, it seems that choice in biological markets is a widespread and basic mechanism that stabilizes cooperation in nature. This mechanism can operate in situations where there is little or no common interest and where partners only meet once. Note, however, that choosiness will often be costly and there must be sufficient variability to choose from in order to offset such costs (McNamara and Leimar, 2010; McNamara, 2013).

11.8.3 Supply and demand

Even if biological markets cannot be generally expected to clear, there are a number of examples that show some similarity to the influence of variation in supply and demand in idealized markets (Noë and Hammerstein, 1994, 1995). This is nicely demonstrated by mutualisms where ants protect the larvae of lycaenid butterflies from enemies and receive food in return. It has been found that a larva will sharply increase its delivery of food when it perceives itself to be under attack from enemies or when ants return to it after an interruption in attendance, i.e., when the larva's need for ant protection is particularly high (Agrawal and Fordyce, 2000; Axén et al., 1996; Leimar and Axén, 1993).

This kind of influence of the availability and value of partners may also be present in cleaning mutualisms, where members of the cleaner fish species "run cleaning stations" and a number of client fish species visit these stations to be purged from skin parasites. As reviewed by Bshary and Noë (2003), local clients for which long-distance moves are very costly receive less thorough cleaning than visitors from further away, who might more readily switch between cleaning stations and thereby exert partner choice. This is what the economic theory of monopolistic competition would predict: buyers with few alternative sources of supply will have less advantageous transactions than those who can shop around (Bowles and Hammerstein, 2003). There is also data showing that cleaner service quality improves when clients become scarce (Soares et al., 2008). A similar phenomenon has also been found in the social behavior of vervet monkeys, where the providers of food items were found to receive more grooming when the food was scarce (Fruteau et al., 2009). Such adjustments are potentially general properties of biological markets and are therefore of broad interest.

11.9. ANIMAL SIGNALING—HONESTY OR DECEPTION?

Honest communication sometimes looks like a good in limited supply. Generations of school children, for example, were tricked by their teachers who, perhaps unwittingly, used a fake citation from the classical roman literature to underscore the importance of their educational efforts. In the fake, it is said that education serves to prepare us for our future lives and is not intended to just get us through school. What Seneca really said is the exact opposite: *Non vitae, sed scholae discimus.*

11.9.1 Education and the Peacock's tail

Spence (1973) elaborated on Seneca's caricature of education. In an article that laid the foundation of signaling games in economics, he asked the following question about job markets: Can the acquisition of higher levels of education lead to higher wages even if education fails to improve a person's productivity? Spence's model hinges on the idea that those who apply for a job know more about themselves than the hiring company can easily observe. In real life, this private information can be on anything like health, learning ability, talent, or the efficiency in getting long-term projects done. In the model, there are simply more or less talented "types of persons." A random move assigns the talent to a person. The person can then choose a level of education conditional upon its talent, it being less costly for the more talented to continue in school. Following completion of schooling, firms observe the person's education and make wage offers.

As is well-known today, Spence found that a game-theoretic equilibrium can exist in which education signals talent and higher education implies higher wage. The reason is that only the talented persist in long years of schooling, so employers use years of schooling as a signal of the unobservable trait, talent. This result is remarkable because education is costly in the model and does not increase a person's productivity.

There is a striking parallel between Spence's original signaling game in economics and the so-called "handicap principle" in biology. Without knowing signaling games, the behavioral biologist Zahavi (1975) claimed that animals acquire costly handicaps, such as the peacock's tail, just to impress others. He emphatically maintained that the credibility of signals hinges on their costliness. Zahavi first failed to convince the community of theoretical biologists (e.g., Maynard Smith, 1976). Subsequently, however, Pomiankowski (1987), Grafen (1990a,b), and others showed that Zahavi's verbal idea can be expressed in coherent mathematical models.

In these models—like in Spence's job market—the animal (e.g., a peacock) chooses the size of its handicap (e.g., a giant tail) depending on private information (e.g., body condition). The size of the handicap is the signal perceived by a receiver who aims to gain information about hidden qualities of the sender (e.g, a female checking a potential mate). Bowles and Hammerstein (2003) demonstrated how a few lines of "heuristic algebra" lead to the core of the biological handicap principle—circumventing a number of important technical issues. Their shortcut of Grafen's (1990a,b) approach makes the link with Spence's theory particularly transparent. Grafen himself confronted certain biological aspects of signaling theory that economists are usually unfamiliar with. For example, it is known from Fisher's (1930) model of "runaway selection" that in the context of sexual selection a genetic covariance can build up between (a) genes relevant to mate choice and (b) genes relevant to the traits that are favored through mate choice. Such covariance is of importance because if males carry a trait that is preferentially chosen, their sons will propagate the genes for the attractive trait as well as the genes for the preference for this trait. This alone can drive the evolution of exaggerated male characteristics (Lande, 1981; Kirkpatrick, 1982).

11.9.2 Does the handicap principle work in practice?

To confront the handicap principle with reality, let us now have a closer look at the signaling games played between male and female widowbirds. Male long-tailed widowbirds are rather small but have an extreme sexual ornament; their tails are about half a meter long. In a famous field experiment, Andersson (1982) truncated the tails of some males and glued the obtained snippets to the tails of others. He showed that males with more than normal-sized tails attract more females. More recently, Pryke and Andersson (2005) did a similar experiment and manipulated the tails of red-collared widowbirds. They demonstrated in particular that (a) females use tail length as the primary mate choice cue, (b) tail length is an indicator of body condition, and (c) tails are costly. Their results confirm the role of mate choice and quality advertising as the main selection pressures behind elongated tails in widowbirds. So here, the handicap principle might indeed be at work and the widowbirds' tails lend empirical support to the basic theory of signaling games.

The analysis of widowbird signaling does not end here though. Do we, for example, understand the evolutionary origin of this system? In their open grassland and savanna habitats, male widowbirds make themselves conspicuous by displaying their impressive tails. Pryke and Andersson (2005) argue that the evolution of elongated tails probably started because it increased conspicuousness (visibility of the male) and that simultaneously or subsequently the exaggerated tail might have become an indicator of male condition and quality. This is an important argument because only those handicaps can evolve that are not a handicap to begin with. For example, female birds would never prefer males that chop off one of their legs just to please Spence, Grafen, and Zahavi. The set of handicaps that can evolve is limited but signaling games are blind to this issue.

Female mate choice in stalk-eyed flies is another possible application of Zahavi's handicap principle (David et al., 2000). Males of these flies have long eyestalks, so their eyes are far apart. Because flies in general do not have this trait, the eyestalks most likely come at some cost. Nevertheless, high phenotypic quality, either from high nutrition or from genes, cause stalk-eyed fly males to develop longer eye stalks. Female stalk-eyed flies prefer to mate with these males thus providing a selective force that might maintain the long eyestalks.

11.9.3 The rush for ever more handicaps, has it come to an end?

Powerful ideas are seductive. Many researchers were mining for examples of Zahavian handicaps like diggers in search for a vein of gold. During this rush, efforts were also made to scrutinize the applicability of the handicap principle. From these efforts, we know that apparently costly signals can be surprisingly cheap and far from a serious handicap. To illustrate this, consider the so-called parent-offspring conflict (Trivers, 1974). This is a conflict about the allocation of resources from parent to offspring. From a mother's genetic point of view, she is equally related to all her children. The children, however, are more related to themselves than to their brothers and sisters. As a consequence, one would expect individual offspring to always want a little more provisioning from a mother than the mother would want to give—an evolutionary conflict within the family that, in principle, always exists. Signaling plays a role in this conflict because the need for food varies over time, and this need is to some extent private information. It is tempting to interpret the begging behavior of nestling birds as a costly signal because it may, for example, attract predators.

Haskell (1994) tested this idea by playing back begging calls from radios placed in artificial nests. This way he could measure changes in predation rates with changes in begging. Haskell found no effect of begging on predation for nests positioned at sites where the recorded calls might have naturally occurred. McCarty (1996) measured energetic expenditure associated with begging in starlings and tree swallows and found this expenditure surprisingly low. These and many other behavioral studies led various biologists to emphasize the limited scope of the handicap principle (Zollman, 2013). Recent advances in experimental methodology enabled researchers, however, to demonstrate a cost of begging for a few bird species. Haff and Magrath (2011) and similarly Ibáñez-Álamo et al. (2012) showed that elevated nestling calling attracts predators to active, natural nests. Moreno-Rueda and Redondo (2011) found that intensive begging decreases simultaneously the nestlings' growth rate and intensity of immune reactions.

11.9.4 Warning signals and mimicry

Certain animals who, based on their size and general life style, would be regarded as prey by some predators are actually unpalatable to these predators. Larvae of the monarch butterfly (*Danaus plexippus*), for instance, feed on milkweed plants that contain cardiac glycosides, which are toxic to many animals. The monarch larvae, however, do not suffer from ingesting these substances, but sequester them in their bodies. The defense substances are passed on to the adult butterflies, who otherwise would be suitable prey for insectivorous birds. The toxin in the butterflies instead turns them into a powerful emetic, producing extreme nausea when ingested (Brower, 1969). Both the larvae and the adults of the monarch have a distinctive and conspicuous coloration, serving as warning to predators. This type of warning, or aposematic coloration is widespread in nature (Poulton, 1890; Ruxton et al., 2004). An interesting question is then if, and for what reason, the coloration is an honest signal of unpalatability.

The traditional view on warning coloration dates back to discussions between Charles Darwin and Alfred Russel Wallace in the 1860s and proposes that aposematism works because predators learn through experience to associate the coloration with unpalatability. Naturally, such learning could only happen if, on the whole, there is a statistical correlation between the appearance and the palatability of prey. The correlation need not be perfect, as illustrated by the phenomenon of Batesian mimicry, in which a palatable species is similar in appearance to an aposematic model species, and gains protection through identification mistakes by predators. For the monarch butterfly, the viceroy (*Limenitis archippus*) is a well-studied example of a Batesian mimic (Van Zandt Brower, 1958).

Both Zahavi (1991) and Grafen (1990b) interpret warning coloration as handicap signaling. Grafen develops the idea that palatable prey would often suffer costs if they were conspicuous, because predators find them more readily and learn that they are good to eat, so the honesty of the signal is tested. While this might well be true, the idea that aposematism is best understood as handicap signaling is not widely accepted (e.g., Guilford and Dawkins, 1993), basically for two reasons. First, the strongest and most direct application of the handicap principle would state that only potentially very unpalatable animals have the capacity to produce strong warning signals. To a large extent, the empirical evidence rejects this possibility. For monarch butterflies, Brower (1969) showed that they can be reared on plants that do not contain cardiac glycosides, which makes them palatable to birds but does not change their appearance. Also, studying wild-caught monarchs, Brower (1972) found that there is much variation in the concentration of cardiac glycosides in adult butterflies, both between different areas in North America and between individuals sampled from the same area, such that some individuals would be suitable as prey. There is thus no direct link between unpalatability and appearance, so the signal is not honest in this sense. Second, there is the question of how natural selection in fact has molded warning signals. If this process embodied the handicap principle and conspicuousness is the cost of signaling, there ought to be selection for conspicuousness per se when the signal evolves. The general view is instead that warning signals have been shaped by their role in the learning process, for instance to promote rapid learning and high memorability, and conspicuousness appears mainly as a side effect (Ruxton et al., 2004).

Even so, constraints on the success of Batesian mimics, from being sampled and found to be tasty by predators, throw light on why warning signals are not everywhere corrupted by mimicry, and can maintain an overall correlation between unpalatability and appearance. Mimics are often less numerous than their aposematic models, or show mimetic polymorphism, such that only some individuals are mimetic or that different individuals mimic different models (Mallet and Joron, 1999). In this way, the mimics avoid being too common. There is a frequency dependence in Batesian mimicry, with an advantage of being rare, because predators then have less opportunity, and reason, to learn about the palatable mimics.

As an aside, polymorphic Batesian mimicry appears to be the first biological example to which the term "theory of games" was applied. In his discussion of the nature of genetic polymorphism, Fisher (1958) elaborated on his early work (Fisher, 1934) on randomized strategies in card play and suggested that such strategies could evolve in prey, to improve their chances of escaping predation. It is perhaps curious that Fisher, who evidently had a strong interest in game theory (he saw himself as one of the originators of this theory), did not propose a link to sex-ratio theory, which seems so natural today, but instead compared polymorphic Batesian mimicry with a mixed equilibrium of a game.

The study of warning coloration in evolutionary biology shows why one may need to understand particular biological mechanisms and how they shape the evolutionary process (cf. section 11.1), and that the usefulness of game theory is primarily as a tool to sharpen the analysis and guide the intuition. So, knowing about signaling games affords us some understanding of the evolution of aposematism – that some factor must limit the proliferation of palatable mimics – but the understanding remains incomplete without deeper insights into the interaction between predators and prey.

Warning coloration is the result of the evolutionary interaction of two complex systems, the developmental machinery producing prey coloration and the neural, cognitive machinery responsible for the learning and generalization of prey appearances by predators. When predators learn to discriminate between unpalatable and palatable prey, the most salient (i.e., striking) aspects of the differences in appearance will act as a focal point of learning; the importance of the salience of stimuli for learning was investigated already by Pavlov (1927). Having learned to discriminate, predators will tend to generalize to new prey appearances in a biased manner, for instance acting as if prey whose appearance further accentuates the learned characteristics of unpalatable prey are even more unpalatable.

This biased response from learned discrimination was first studied by animal psychologists (Spence, 1937; Hanson, 1959), and is a quite general and carefully investigated phenomenon. Leimar et al. (1986) proposed that it could drive the evolution of warning coloration. Together with similar cognitive properties such as memorability, there is thus a plausible account of how warning colors evolve (Ruxton et al., 2004). From this perspective, we can ascribe an evolutionary function to the appearance of the monarch butterfly. The striking and even beautiful coloration has evolved so that—after some learning by a predator—it is effective in evoking in this predator a feeling of revulsion. This insight derives, at least in terms of the ideas it builds on, from outside of game theory.

More generally, there is a trend in contemporary biology to integrate function and mechanisms (Hammerstein and Stevens 2012; McNamara, 2013; McNamara and Houston, 2009). It is still too early to tell what this trend means for the future of game theory in biology. Perhaps the role of game theory will be diminished, but there is also an alternative, that game theory enriches its framework by incorporating mechanisms. This way it could enter into a more direct confrontation with empirical observation—and with biology.

REFERENCES

Agrawal, A.A., Fordyce, J.A., 2000. Induced indirect defence in a lycaenid-ant association: The regulation of a resource in a mutualism. Proc. R. Soc. B 267, 1857–1861.

Agrawal, A. A., Laforsch, C., Tollrian, R., 1999. Transgenerational induction of defences in animals and plants. Nature 401, 60–63.

Andersson, M., 1982. Female choice selects for extreme tail length in a widowbird. Nature 299, 818–820.

Axelrod, R., Hamilton, W.D., 1981. The evolution of cooperation. Science 211, 1390–1396.

Axén, A.H., Leimar, O., Hoffman, V., 1996. Signalling in a mutualistic interaction. Anim. Behav. 52, 321–333.

Backwell, P.R.Y., Dennions, M.D., 2004. Animal behaviour: Coalition among male fiddler crabs. Nature 430, 417.

Bergman, M., Olofsson, M., Wiklund, C., 2010. Contest outcome in a territorial butterfly: The role of motivation. Proc. R. Soc. B 277, 3027–3033.

Blows, M.W., Berrigan, D., Gilchrist, G.W., 1999. Rapid evolution towards equal sex ratios in a system with heterogamety. Evol. Ecol. Res. 1, 277–283.

Boomsma, J.J., 2009. Lifetime monogamy and the evolution of eusociality. Philos. Trans. R. Soc. B 364, 3191–3207.

Bowles, S., Hammerstein, P., 2003. Does market theory apply to biology? In: Hammerstein, P., (Ed.), *Genetic and Cultural Evolution of Cooperation*, MIT Press, Cambridge, MA, pp. 153–165.

Boyd, R., Lorberbaum, J.P., 1987. No pure strategy is evolutionarily stable in the repeated Prisoner's Dilemma game. Nature 327, 58–59.

Brower, L.P., 1969. Ecological chemistry. Sci. Am. 220, 22–29.

Brower, L.P., 1972. Variation in cardiac glycoside content of monarch butterflies from natural populations in Eastern North America. Science 177, 426–429.

Bshary, R., Noë, R., 2003. Biological markets: The ubiquitous influence of partner choice on the dynamics of cleaner fish–client reef fish interactions. In: Hammerstein, P., (Ed.), Genetic and Cultural Evolution of Cooperation, MIT Press, Cambridge, MA, pp. 167–184.

Bshary, R., Schäffer, D., 2002. Choosy reef fish select cleaner fish that provide high-quality service. Anim. Behav. 63, 557–564.

Bull, J.J., Charnov, E., 1988. How fundamental are Fisherian sex ratios? Oxford Surv. Evol. Biol. 5, 96–135.

Bull, J.J., Rice, W.R., 1991. Distinguishing the mechanisms for the evolution of co-operation. J. Theor. Biol. 149, 63–74.

Bulmer, M.G., Bull, J.J., 1982. Models of polygenic sex determination and sex ratio control. Evolution 326, 13–26.

Bulmer, M.G., Taylor, P.D., 1981. Worker-queen conflict and sex ratio theory in social hymenoptera. Heredity 47, 197–207.

Burt, A., Trivers, R., 2006. Genes in Conflict: The Biology of Selfish Genetic Elements. Harvard University Press, Cambridge, MA.

Carter, G.G., Wilkinson, G.S., 2012. Food sharing in vampire bats: reciprocal help predicts donations more than relatedness or harassment. Proc. R. Soc. B 280, 2012–2573.

Carvalho, A.B., Sampajo, M.C., Varandas, F.R., Klaczko, L.B., 1998. An experimental demonstration of Fisher's principle: Evolution of sexual proportion by natural selection. Genetics 148, 719–731.

Caswell, H., 2001. Matrix Population Models: Construction, Analysis, and Interpretation. Second Ed. Sinauer, Sunderland, MA.

Chapman, T., Liddle, L.F., Kalb, J.M., Wolfner, M.F., Partridge, L., 1995. Cost of mating in *Drosophila melanogaster* females is mediated by male accessory gland products. Nature 373, 241–244.

Charlesworth, B., 1994a. Evolution in Age-Structured Populations. Cambridge University Press, Cambridge, UK.

Charlesworth, B., 1994b. The evolution of lethals in the t-haplotype system of the mouse. Proc. Biol. Sci. 22, 101–107.

Charnov, E.L., 1982. The Theory of Sex Allocation. Princeton University Press, Princeton, NJ.

Charnov, E.L., Bull, J., 1977. When is sex environmentally determined? Nature 266, 828–830.

Chittka, L., Schürkens, S., 2001. Successful invasion of a floral market. Nature 411, 653.

Clutton-Brock, T., 2009. Cooperation between non-kin in animal societies. Nature 462, 51–57.
Conover, D.O., Van Voorhees, D.A., 1990. Evolution of a balanced sex ratio by frequency-dependent selection in a fish. Science 250, 1556–1558.
Cosmides, L.M., Tooby, J., 1981. Cytoplasmic inheritance and intragenomic conflict. J. Theor. Biol. 89, 83–129.
Coyne, J.A., Barton, N.H., Turelli, M., 1997. Perspective: A critique of Sewall Wright's shifting balance theory of evolution. Evolution 51, 643–671.
Cushman, J.H., Beattie, A.J., 1991. Mutualisms: Assessing the benefits to hosts and visitors. Trends Ecol. Evol. 6, 193–195.
Darwin, C., 1871. The Descent of Man and Selection in Relation to Sex, 2nd Ed., John Murray, London; 1874.
David, P., Bjorksten, T., Fowler, K., Pomiankowski, A., 2000. Condition-dependent signalling of genetic variation in stalk-eyed flies. Nature 406, 186–188.
Davies, N.B., 1978. Territorial defence in the speckled wood butterfly (*Pararge aegeria*): The resident always wins. Anim. Behav. 26, 138–147.
Detto, T., Jennions, M.D., Backwell, P.R.Y., 2010. When and why do territorial coalitions occur? Experimental evidence from a fiddler crab. Am. Nat. 175, E119–E125.
Dieckmann, U. Law, R., 1996. The dynamical theory of coevolution: A derivation from stochastic ecological processes. J. Math. Biol. 34, 579–612.
Düsing, C., 1883. Die Factoren welche die Sexualität entscheiden. Jenaische Zeitschrift für Naturwissenschaft 16, 428–464.
Düsing, C., 1884a. Die Regulierung des Geschlechtsverhätlnisses bei der Vermehrung der Menschen, Tiere und Pflanzen. Jenaische Zeitschrift für Naturwissenschaft 17, 593–940.
Düsing, C., 1884b. Die Regulierung des Geschlechtsverhältnisses bei der Vermehrung der Menschen, Tiere und Pflanzen. Gustav Fischer: Jena.
Engelmann, D., Fischbacher, U., 2003. Indirect reciprocity and strategic reputation building in an experimental helping game. Games Econ. Behav. 67, 399–407.
Enquist, M., Leimar, O., 1983. Evolution of fighting behaviour: Decision rules and assessment of relative strength. J. Theor. Biol. 102, 387–410.
Enquist, M., Leimar, O., 1990. The evolution of fatal fighting. Anim. Behav. 39, 1–9.
Eshel, I., 1983. Evolutionary and continuous stability. J. Theor. Biol. 103, 99–111.
Eshel, I., 1996. On the changing concept of evolutionary population stability as a reflection of a changing point of view in the quantitative theory of evolution. J. Math. Biol. 34, 485–510.
Falconer, G.S., Mackay, T.F.C., 1996. Introduction to Quantitative Genetics, Fourth ed. Addison Wesley Longman, Harlow, Essex, UK.
Fischer, M.K., Hoffmann, K.H., Völkl, W., 2001. Competition for mutualists in an ant-homopteran interaction mediated by hierarchies of ant attendance. Oikos 92, 531–541.
Fisher, R.A., 1930. The Genetical Theory of Natural Selection. Clarendon Press, Oxford.
Fisher, R.A., 1934. Randomization, an old enigma of card play. Math. Gaz. 18, 294–297.
Fisher, R.A., 1958. Polymorphism and natural selection. J. Ecol. 46, 289–293.
Frank, S.A., Hurst, L.D., 1996. Mitochondria and male disease. Nature 383, 224.
Fruteau, C., Voelkl, B., van Damme, E., Noë, R., 2009. Supply and demand determine the market value of food providers in wild vervet monkeys. Proc. Natl. Acad. Sci. USA 106, 12007–12012.
Geritz, S.A.H., Kisdi, É., Meszena, G., Metz, J.A.J., 1998. Evolutionarily singular strategies and the adaptive growth and branching of the evolutionary tree. Evol. Ecol. 12, 35–57.
Gould, S.J., Lewontin, R.C., 1979. The spandrels of San Marco and the Panglossian paradigm: A critique of the adaptationist programme. Proc. R Soc. B 205, 581–598.
Grafen, A., 1987. The logic of divisively asymmetric contests – respect for ownership and the desperado effect. Anim. Behav. 35, 462–467.
Grafen, A., 2006. A theory of Fisher's reproductive value. J. Math. Biol. 53, 15–60.
Grafen, A., 1990a. Biological signals as handicaps. J. Theor. Biol. 144, 517–546.
Grafen, A., 1990b. Sexual selection unhandicapped by the Fisher process. J. Theor. Biol. 144, 473–516.

Greene, E., Lyon, B.E., Muehter, V.R., Ratcliffe, L., Oliver, S.J., Boag, P.T., 2000. Disruptive sexual selection for plumage coloration in a passerine bird. Nature 407, 1000–1003.
Griffin, A.S., West, S.A., Buckling, A., 2004. Cooperation and competition in pathogenic bacteria. Nature 430, 1024–1027.
Guilford, T., Dawkins, M.S., 1993. Receiver psychology and the design of animal signals. Trends Neurosci. 16, 430–436.
Haff, T.M., Magrath, R.D., 2011. Calling at a cost: Elevated nestling calling attracts predators to active nests. Biol. Lett. 7, 493–495.
Hagen, E.H., Sullivan, R.J., Schmidt, R., Morris, G., Kempter, R., Hammerstein, P., 2009. Ecology and neurobiology of toxin avoidance and the paradox of drug reward. Neuroscience 160, 69–84.
Hamilton, W.D., 1964a. The genetical evolution of social behaviour. I. J. Theor. Biol. 7, 1–16.
Hamilton, W.D., 1964b. The genetical evolution of social behaviour. II. J. Theor. Biol. 7, 17–52.
Hamilton, W.D., 1967. Extraordinary sex ratios. Science 156, 477–488.
Hamilton, W.D., May, R.M., 1977. Dispersal in stable habitats. Nature 269, 578–581.
Hammerstein, P., 1981. The role of asymmetries in animal contests. Anim. Behav. 29, 193–205.
Hammerstein, P., 1996. Darwinian adaptation, population genetics and the streetcar theory of evolution. J. Math. Biol. 34, 511–532.
Hammerstein, P., 2003. Why is reciprocity so rare in social animals? A protestant appeal. In: Hammerstein, P., (Ed.), Genetic and Cultural Evolution of Cooperation, MIT Press, Cambridge, MA, pp. 84–93.
Hammerstein, P., 2012. Towards a Darwinian theory of decision making: Games and the biological roots of behavior. In: Binmore, K., Okasha, S., (Eds.), Evolution and Rationality, Cambridge University Press, Cambridge, NY, pp. 7–22.
Hammerstein, P., Boyd, R., 2012. Learning, cognitive limitations, and the modeling of social behavior. In: Hammerstein, P., Stevens, J.R., (Eds.), Evolution and the Mechanisms of Decision Making, MIT Press, Cambridge, MA, pp. 319–343.
Hammerstein, P., Riechert, S.E., 1988. Payoffs and strategies in territorial contests: ESS analyses of two ecotypes of the spider *Agelenopsis aperta*. Evol. Ecol. 2, 115–138.
Hammerstein, P., Stevens, J., (Eds.)., 2012. Evolution and the Mechanisms of Decision Making. MIT Press, Cambridge, MA.
Hanson, H.M., 1959. Effects of discrimination training on stimulus generalization. J. Exp. Psychol. 58, 321–333.
Haskell, D., 1994. Experimental evidence that nestling begging behaviour incurs a cost due to nest predation. Proc. R. Soc. B 257, 161–164.
Hardin, G., 1968. The tragedy of the commons: The population problem has no technical solution; it requires a fundamental extension in morality. Science 162, 1243–1248.
Harsanyi, J., 1973. Games with randomly disturbed payoffs: A new rationale for mixed-strategy equilibrium points. Int. J. Game Theory 2, 1–23.
Heath, K.D., Tiffin, P., 2009. Stabilizing mechanisms in a legume-rhizobium mutualism. Evolution 63, 652–662.
Heil, M., Rattke, J., Boland, W., 2005. Postsecretory hydrolysis of nectar sucrose and specialization in ant/plant mutualism. Science 308, 560–563.
Herre, E.A., 1985. Sex ratio adjustment in fig wasps. Science 228, 896–898.
Hofbauer J., Sigmund, K., 1998. Evolutionary Games and Population Dynamics. Cambridge University Press, Cambridge.
Holland, B., Rice, W.R., 1999. Experimental removal of sexual selection reverses intersexual antagonistic coevolution and removes a reproductive load. Proc. Natl. Acad. Sci. USA 96, 5083–5088.
Houston, A.I., Székely, T., McNamara, J.M., 2005. Conflict between parents over care. Trends Ecol. Evol. 20, 33–38.
Hsu, Y., Earley, R.L., Wolf, L.L., 2006. Modulation of aggressive behavior by fighting experience: mechanisms and contest outcomes. Biol. Rev. 81, 33–74.
Hurst, L.D., Atlan, A., Bengtson, B.O., 1996. Genetic conflict. Q. Rev. Biol. 71, 317–364.
Ibáñez-Álamo, J.D., Arco, L., Soler, M., 2012. Experimental evidence for a predation cost of begging using active nests and real chicks. J. Ornithol. 153, 801–807.

Internicola, A.I., Page, P.A., Bernasconi, G., Gigord, L.D.B., 2007. Competition for pollinator visitation between deceptive and rewarding artificial inflorescences: An experimental test of the effects of floral colour similarity and spatial mingling. Funct. Ecol. 21, 864–872.
Jaenike, J., 2001. Sex chromosome meiotic drive. Annu. Rev. Ecol. Syst. 32, 25–49.
Karban, R., Agrawal, A.A., 2002. Herbivore offense. Annu. Rev. Ecol. Syst. 33, 641–644.
Karlin, S., 1975. General two-locus selection models: Some objectives, results and interpretations. Theor. Pop. Biol. 7, 364–398.
Kautz, S., Lumbsch, H.T., Ward, P.S., Heil, M., 2009. How to prevent cheating: A digestive specialization ties mutualistic plant-ants to their ant-plant partners. Evolution 63, 839–853.
Kiers, E.T., Denison, R.F., 2008. Sanctions, cooperation, and the stability of plant-rhizosphere mutualisms. Annu. Rev. Ecol. Syst. 39, 215–236.
Kiers, E.T., Rousseau, R.A., West, S.A., Denison, R.F., 2003. Host sanctions and the legume-rhizobium mutualism. Nature 425, 78–81.
Kirkpatrick, M., 1982. Sexual selection and the evolution of female choice. Evolution 36, 1–12.
Koehncke, A., Telschow, A., Werren, J.H., Hammerstein, P., 2009. Life and death of an influential passenger: *Wolbachia* and the evolution of CI-modifiers by their hosts. PLoS One 4:e4425.
Kokko, H., López-Sepulcre, A., Morell, L.J., 2006. From hawks and doves to self-consistent games of territorial behavior. Am. Nat. 167, 901–912.
Krombein, K., 1967. Trap-Nesting Wasps and Bees: Life Histories, Nest and Associates. Smithonian Press, Washington.
Lande, R., 1981. Models of speciation by sexual selection on polygenic traits. Proc. Natl. Acad. Sci. USA 78, 3721–3725.
Leimar, O., 2009. Multidimensional convergence stability. Evol. Ecol. Res. 11, 191–208.
Leimar, O., Axén, A.H., 1993. Strategic behaviour in an interspecific mutualism: Interactions between lycaenid larvae and ants. Anim. Behav. 46, 1177–1182.
Leimar, O., Enquist, M., 1984. Effects of asymmetries in owner-intruder conflicts. J. Theor. Biol. 111, 475–491.
Leimar, O., Enquist, M., Sillén Tullberg, B., 1986. Evolutionary stability of aposematic coloration and prey unprofitability: A theoretical Analysis. Am. Nat. 128, 469–490.
Leimar, O., Hammerstein, P., 2001. Evolution of cooperation through indirect reciprocity. Proc. R. Soc. B 268, 745–753.
Leimar, O., Hammerstein, P., 2010. Cooperation for direct fitness benefits. Philos. Trans. R. Soc. B 365, 2619–2626.
Leimar, O., Van Dooren, T.J.M., Hammerstein, P., 2004. Adaptation and constraint in the evolution of environmental sex determination. J. Theor. Biol. 227, 561–570.
Lorenz, K., 1963. Das sogenannte Böse. Zur Naturgeschichte der Aggression. Borotha-Schoeler, Wien.
Lorenz, K., 1966. On Aggression. Methuen, London.
Mallet, J., Joron, M., 1999. Evolution of diversity in warning color and mimicry: Polymorphisms, shifting balance, and speciation. Annu. Rev. Ecol. Syst. 30, 201–233.
Maynard Smith, J., 1976. Sexual selection and the handicap principle. J. theor. Biol. 57, 239–242.
Maynard Smith, J., 1982. Evolution and the Theory of Games. Cambridge University Press, Cambridge.
Maynard Smith, J., Parker, G.A., 1976. The logic of asymmetric contests. Anim. Behav. 24, 159–175.
Maynard Smith, J., Price, G., 1973. The logic of animal conflict. Nature 246, 15–18.
Maynard Smith, J., Szathmáry, E., 1995. The Major Transitions in Evolution. Oxford University Press, Oxford.
McCarty, J.P., 1996. The energetic cost of begging in nestling passerines. The Auk 113, 178–188.
McNamara, J.M., 2013. Towards a richer evolutionary game theory. J. R. Soc. Interf. 10, 20130544.
McNamara, J.M., Houston, A.I., 1996. State-dependent life histories. Nature 380, 215–221.
McNamara, J.M., Houston, A.I., 2009. Integrating function and mechanism. Trends Ecol. Evol. 24, 670–675.
McNamara, J.M., Leimar, O., 2010. Variation and the response to variation as a basis for successful cooperation. Philos. Trans. R. Soc. B 365, 2627–2633.

Metz, J.A.J., Geritz, S.A.H., Meszéna, G., Jacobs, F.J.A., Van Heerwaarden, J.S., 1996. Adaptive dynamics: A geometrical study of the consequences of nearly faithful reproduction. In: van Strien, S.J., Verduyn Lunel, S.M., (Eds.), Stochastic and Spatial Structures of Dynamical Systems, Elsevier, North-Holland, pp. 183–231.
Metz, J.A.J., Nisbet, R., Geritz, S.A.H., 1992. How should we define 'fitness' for general ecological scenarios? Trends Ecol. Evol. 7, 198–202.
Milinski, M., Semmann, D., Bakker, T.C.M., Krambeck, H.J., 2001. Cooperation through indirect reciprocity: Image scoring or standing strategy? Proc. R. Soc. B 268, 2495–2501.
Moran, P.A.P., 1964. On the nonexistence of adaptive topographies. Ann. Hum. Genet 27, 383–393.
Moreno-Rueda, G., Redondo, T., 2011. Begging at high level simultaneously impairs growth and immune response in southern shrike (*Lanius meridionalis*) nestlings. J. Evol. Biol. 24, 1091–1098.
Nash, J.F., 1951. Non-cooperative games. Ann. Math. 54, 286–295.
Nash, J.F., 1996. Essays on Game Theory. Edward Elgar, Cheltenham, UK.
Nesse, R.M., Williams, G.C., 1994. Why We Get Sick: The New Science of Darwinian Medicine. Vintage, New York.
Noë, R., Hammerstein, P., 1994. Biological markets: Supply and demand determine the effect of partner choice in cooperation, mutualism and mating. Behav. Ecol. Sociobiol. 35, 1–11.
Noë, R., Hammerstein, P., 1995. Biological markets. Trends Ecol. Evol. 10, 336–339.
Noë, R., van Schaik, C.P., van Hooff, J.A.R.A.M., 1991. The market effect: An explanation for pay-off asymmetries among collaborating animals. Ethology 87, 97–118.
Nowak, M.A., Sasaki, S., Taylor, C., Fudenberg, D., 2004. Emergence of cooperation and evolutionary stability in finite populations. Nature 428, 646–650.
Nowak, M.A., Sigmund, K., 1998. Evolution of indirect reciprocity by image scoring. Nature 393, 573–577.
Ostrom, E., 1990. The Evolution of Institutions for Collective Action. Cambridge University Press, Cambridge, UK.
Panchanathan, K., Boyd, R., 2004. Indirect reciprocity can stabilize cooperation without the second-order free rider problem. Nature 432, 499–502.
Parker, G.A., 1974. Assessment strategy and the evolution of animal conflicts. J. Theor. Biol. 47, 223–243.
Pavlov, I.P., 1927. Conditioned Reflexes: An investigation of the physiological activity of the cerebral cortex. Translated and edited by G. V. Anrep. Oxford University Press, London.
Pen, I., Uller, T., Feldmeyer, B., Harts, A., While, G.M., Wapstra, E., 2010. Climate-driven population divergence in sex-determining systems. Nature 468, 436–438.
Petzinger, E., Geyer, J., 2006. Drug transporters in pharmacokinetics. Naunyn. Schmied. Arch. Pharmacol. 372, 465–475.
Pomiankowski, A., 1987. Sexual selection: The handicap principle does work - sometimes. Proc. R. Soc. B 231, 123–145.
Poulton, E.B., 1890. The Colours of Animals: Their Meaning and Use, Especially Considered in the Case of Insects. Kegan Paul, London.
Pryke, S.R., Andersson, S., 2005. Experimental evidence for female choice and energetic costs of male tail elongation in red-collared widowbirds. Biol. J. Linnean Soc. 86, 35–43.
Queller, D.C., Strassmann, J.E., 2009. Beyond society: The evolution of organismality. Philos. Trans. R. Soc. B 364, 3143–3155.
Ruxton, G., Sherratt, T., Speed, M., 2004. Avoiding Attack: The Evolutionary Ecology of Crypsis, Warning Signals and Mimicry. Oxford University Press, Oxford.
Schaffer, W.M., Zeh, D.W., Buchmann, S.L., Kleinhans, S., Schaffer, M.W., Antrim, J., 1983. Competition for nectar between introduced honey bees and native North American bees and ants. Ecology 64, 564–577.
Selten, R., 1980. A note on evolutionarily stable strategies in asymmetric animal conflicts. J. Theor. Biol. 83, 93–101.
Selten, R., Hammerstein, P., 1984. Gaps in Harley's argument on evolutionarily stable learning rules and in the logic of 'tit for tat'. Behav. Brain Sci. 7, 115–116.
Shaw, R.F., Mohler, J.D., 1953. The selective significance of the sex ratio. Am. Nat. 87, 337–34.

Simms, E.L., Taylor, D.L., Povich, J., Shefferson, R.P., Sachs, J.L., Urbina, M., Tausczik, Y., 2006. An empirical test of partner choice mechanisms in a wild legume-rhizobium interaction. Proc. Biol. Sci. 273, 77–81.
Soares, M.C., Bshary, R., Cardoso, S.C., Côté, I.M., 2008. Does competition for clients increase service quality in cleaning gobies? Ethology 114, 625–632.
Spence, K.W., 1937. The differential response in animals to stimuli varying in a single dimension. Psychol. Rev. 44, 430–444.
Spence, M., 1973. Job market signaling. Q. J. Econ. 87, 355–374.
Strassmann, J.E., Queller, D.C., 2010. The social organism: congresses, parties, and committees. Evolution 64, 605–616.
Sugden, R., 1986. The Economics of Rights, Co-operation and Welfare. Basil Blackwell, Oxford, UK.
Taylor, P.D., 1989. Evolutionary stability in one-parameter models under weak selection. Theor. Pop. Biol. 36, 125–143.
Taylor, P.D., 1990. Allele-frequency change in a class-structured population. Am. Nat. 135, 95–106.
Trivers, R.L., 1971. The evolution of reciprocal altruism. Q. Rev. Biol. 46, 35–57.
Trivers, R.L., 1974. Parent-offspring conflict. Am. Zool. 14, 249–264.
Trivers, R.L., Hare, H., 1976. Haplodiploidy and the evolution of the social insects. Science 191, 249–263.
Torchio, P.F., Tepedino, V.J., 1980. Sex ratio, body size and seasonality in a solitary bee, *Osmia lignaria propinqua* Cresson (Hymenoptera: Megachilidae). Evolution 34, 993–1003.
Toro, M.A., Charlesworth, B., 1982. An attempt to detect genetic variation in sex ratio in *Drosophila melanogaster*. Heredity 49, 199–209.
Van Dooren, T.J. Leimar, O., 2003. The evolution of environmental and genetic sex determination in fluctuating environments. Evolution 57, 2667–2677.
Van Zandt Brower, J., 1958. Experimental studies of mimicry in some North American butterflies: Part I. The monarch, *Danaus plexippus*, and viceroy, *Limenitis archippus archippus*. Evolution 12, 32–47.
Vieira, M.C. Peixoto, P.E.C., 2013. Winners and losers: A meta-analysis of functional determinants of fighting ability in arthropod contests. Funct. Ecol. 27, 305–313.
Völkl, W., Woodring, J., Fischer, M., Lorenz, M.W., Hoffmann, K.H., 1999. Ant-aphid mutualisms: The impact of honeydew production and honeydew sugar composition on ant preferences. Oecologia 118, 483–491.
Von Frisch, K., 1967. The Dance Language and Orientation of Bees. Harvard University Press, Cambridge, MA.
Wedekind, C., Milinski, M., 2000. Cooperation through image scoring in humans. Science 288, 850–852.
Weissing, F.J., 1996. Genetic versus phenotypic models of selection: Can genetics be neglected in a long-term perspective? J. Math. Biol. 34, 533–555.
Werren, J.H., 1991. The paternal-sex-ratio chromosome of *Nasonia*. Am. Nat. 137, 392–402.
Werren, J.H., Baldo, L., Clark, M.E., 2008. Wolbachia: master manipulators of invertebrate biology. Nat. Rev. Microbiol. 6, 741–751.
Werren, J.H., Nur, U., Wu, C., 1988. Selfish genetic elements. Trends Evol. Ecol. 3, 297–302.
Werren, J.H., Stouthamer, R., 2003. PSR (paternal sex ratio) chromosomes: the ultimate selfish genetic elements. Genetica 117, 85–101.
West, S., 2009. Sex Allocation. Princeton University Press, Princeton, NJ.
West, S., Murray, M.G., Machado, C.A., Griffin, A.S., Herre, E.A., 2001. Testing Hamilton's rule with competition between relatives. Nature 409, 510–513.
Wilkinson, G.S., 1984. Reciprocal food sharing in the vampire bat. Nature 308, 181–184.
Wink, M., 2000. Interference of alkaloids with neuroreceptors and ion channels. Stud. Natl. Prod. Chem. 21, 3–122.
Wright, S., 1932. The roles of mutation, inbreeding, crossbreeding and selection in evolution. Proceedings of the 6th International Congress on Genetics 1, 356–366.
Wright, S., 1977. Evolution and the Genetics of Populations. Vol. 3: Experimental Results and Evolutionary Deductions. University of Chicago Press, Chicago.
Young, H.P., 1993. The evolution of conventions. Econometrica 61, 57–84.
Zahavi, A., 1975. Mate selection - a selection for a handicap. J. Theor. Biol. 53, 205–214.

Zahavi, A., 1991. On the definition of sexual selection, Fisher's model, and the evolution of waste and of signals in general. Anim. Behav. 42, 501–503.
Zollman, K.J.S., 2013. Finding alternatives to handicap theory. Biol. Theory 8, 127–132.
Zug, R., Koehncke, A., Hammerstein, P., 2012. Epidemiology in evolutionary time: The case of *Wolbachia* horizontal transmission between arthropod host species. J. Evol. Biol. 25, 2149–2160.
Zug, R., Hammerstein, P., 2012. Still a host of hosts for *Wolbachia*: Analysis of recent data suggests that 40% of terrestrial arthropod species are infected. PLoS ONE 7(6), e38544.

CHAPTER 12

Epistemic Game Theory

Eddie Dekel[*,†], **Marciano Siniscalchi**[*,†]
[*]Tel Aviv University, Tel Aviv, Israel and Northwestern University, Evanston, IL, USA
[†]Northwestern University, Evanston, IL, USA

Contents

Handbook of game theory
http://dx.doi.org/10.1016/B978-0-12-420056-2.00012-4

Abstract

Epistemic game theory formalizes assumptions about rationality and mutual beliefs in a formal language, then studies their behavioral implications in games. Specifically, it asks: what do different notions of rationality and different assumptions about what players believe about. . .what others believe about the rationality of players imply regarding play in a game? Being explicit about these assumptions can be important, because solution concepts are often motivated intuitively in terms of players' beliefs and their rationality; however, the epistemic analysis may show limitations in these intuitions, reveal what additional assumptions are hidden in the informal arguments, clarify the concepts or show how the intuitions can be generalized. A further premise of this chapter is that the primitives of the model—namely, the hierarchies of beliefs—should be elicitable, at least in principle. Building upon explicit assumptions about elicitable primitives, we present classical and recent developments in epistemic game theory and provide characterizations of a nonexhaustive, but wide, range of solution concepts.

Keywords: Epistemic game theory, Interactive epistemology, Solution concepts, Backward induction, Forward induction, Rationalizability, Common-prior assumption, Hierarchies of beliefs, Conditional probability systems, Lexicographic probability systems

JEL Codes: C72, D81

12.1. INTRODUCTION AND MOTIVATION

Epistemic game theory formalizes assumptions about rationality and mutual beliefs in a formal language, then studies their behavioral implications in games. Specifically, it asks: what do different notions of rationality and different assumptions about what players believe about...what others believe about the rationality of players imply regarding play in a game? A well-known example is the equivalence between common belief in rationality and iterated deletion of dominated strategies.

The reason why it is important to be formal and explicit is the standard one in economics. Solution concepts are often motivated intuitively in terms of players' beliefs and their rationality. However, the epistemic analysis may show limitations in these intuitions, reveal what additional assumptions are hidden in the informal arguments, clarify the concepts, or show how the intuitions can be generalized. We now consider a number of examples.

Backward induction was long thought to be obviously implied by "common knowledge of rationality." The epistemic analysis showed flaws in this intuition, and it is now understood that the characterization is much more subtle (Sections 12.7.4.3 and 12.7.5).

Next, consider the solution concept that deletes one round of weakly dominated strategies and then iteratively deletes strictly dominated strategies. This concept was first proposed because it is robust to payoff perturbations, which were interpreted as a way to perturb players' rationality. Subsequent epistemic analysis showed this concept is exactly equivalent to "almost common belief" of rationality and of full-support conjectures—an explicit robustness check of common belief in rationality (see Section 12.5). Thus, the epistemic analysis generalizes and formalizes the connection of this concept to robustness.

The common-prior assumption (Section 12.4.3) is used to characterize Nash equilibrium with $n > 2$ players, but is not needed for two-player games (compare Theorems 12.5 and 12.7). This result highlights the difference between the assumptions implicit in this solution concept across these environments. Furthermore, the common prior is known to be equivalent to no betting when uncertainty is exogenous. We argue that the interpretation of the common-prior assumption and its connection to no-betting results must be modified when uncertainty is endogenous, e.g, about players' strategies (see Example 12.4).

Finally, recent work has shown how forward induction and iterated deletion of weakly dominated strategies can be characterized. These results turn out to identify important, nonobvious, assumptions and require new notions of "belief." Moreover, they clarify the connection between these concepts (see Section 12.7.4.4).

Epistemic game theory may also help provide a rationale, or "justification," for or against specific solution concepts. For instance, in Section 12.6, we identify those cases where interim independent rationalizability is and is not a "suitable" solution concept for games of incomplete information.

We view nonepistemic justifications for solution concepts as complementary to the epistemic approach. For some solution concepts, such as forward induction, we think the epistemic analysis is more insightful. For others, such as Nash equilibrium, learning theory may provide the more compelling justification. Indeed, we do not find the epistemic analysis of objective equilibrium notions (Section 12.4) entirely satisfactory.

This is because the epistemic assumptions needed are often very strong and hard to view as a justification of a solution concept. Moreover, except for special cases (e.g., pure-strategy Nash equilibrium), it is not really possible to provide necessary and sufficient epistemic conditions for equilibrium *behavior* (unless we take the view that mixed strategies are actually available to the players). Rather, the analysis constitutes a fleshing-out of the textbook interpretation of equilibrium as "rationality plus correct beliefs." To us this suggests that equilibrium behavior cannot arise out of strategic reasoning alone. Thus, as discussed earlier, this epistemic analysis serves the role of identifying where alternative approaches are required to justify standard concepts.

While most of the results we present are known from the literature, we sometimes present them differently, to emphasize how they fit within our particular view. We have tried to present a wide swath of the epistemic literature, analyzing simultaneous-move games as well as dynamic games, considering complete and incomplete-information games, and exploring both equilibrium and non-equilibrium approaches. That said, our choice of specific topics and results is still quite selective and we admit that our selection is driven by the desire to demonstrate our approach (discussed next), as well as our interests and tastes. Several insightful and important papers could not be included because they did not fit within our narrative. More generally, we have ignored several literatures. The connection with the robustness literature mentioned earlier (see Kajii and Morris, 1997b, for a survey) is not developed. Nor do we study self-confirming based solution concepts (Battigalli, 1987; Fudenberg and Levine, 1993; Rubinstein and Wolinsky, 1994).[1] Moreover, we do not discuss epistemics and k-level thinking (Crawford et al., 2012; Kets, 2012) or unawareness (see Schipper, 2013, for a comprehensive bibliography). We find all this work interesting, but needed to narrow the scope of this paper.

12.1.1 Philosophy/Methodology

The basic premise of this chapter is that the primitives of the model should be observable, at least in principle. The primitives of epistemic game theory are players' beliefs about the play of the game, their beliefs about players' beliefs about play, and so on; these are called *hierarchies of beliefs*. Obviously, these cannot be observed directly, but we can ask that they be *elicitable* from observable choices, e.g., their betting behavior, as is standard in decision theory (De Finetti, 1992; Savage, 1972).

However, there are obvious difficulties with eliciting a player's beliefs about his own behavior and beliefs. Our basic premise then requires that we consider hierarchies of

[1] The concept of RPCE (Fudenberg and Kamada, 2011) is a recent example where epistemics seem to us useful. Its definition is quite involved, and, while examples illustrate the role of various assumptions, the epistemic analysis confirms the equivalence of the solution concept to the assumptions used in its intuitive description.

beliefs over *other* players' beliefs and rationality, thereby ruling out "introspective" beliefs (see also Section 12.2.6.3). With this stipulation, it *is* possible to elicit such hierarchies of belief; see Section 12.2.6.2.

By contrast much of the literature, following Aumann's seminal developments, allows for introspective beliefs (Aumann, 1987). This modeling difference does have implications, in particular in characterization results that involve the common-prior assumption (Theorems 12.4 and 12.8).

Rather than working with belief hierarchies directly, we use a convenient modeling device due to Harsanyi (1967), namely *type structures*. In the simple case of strategic-form games, these specify a set of "types" for each player, and for each type, a belief over the opponents' strategies and types. Every type generates a hierarchy of beliefs over strategies, and conversely, every hierarchy can be generated in some type structure; details are provided in Sections 12.2.3 and 12.2.4.

We emphasize that we use type structures solely as a modeling device. Types are not real-world objects; they simply represent hierarchies, which are. Therefore, although we will formally state epistemic assumptions on types, we will consider only those assumptions that can also be stated as restrictions on belief hierarchies, and we will interpret them as such. In particular, our assumptions cannot differentiate between two types that generate the same belief hierarchy. One concrete implication of this can be seen in the analysis of solution concepts for incomplete-information games (Section 12.6.1).

To clarify this point further, note that type structures can be used in a different way. In particular, they can be used to represent an information structure: in this case, a type represents the hard information a player can receive—for example, a possible outcome of some study indicating the value of an object being auctioned. Here, it makes perfect sense to distinguish between two types with different hard information, even if the two pieces of information lead to the same value for the object, and indeed the same belief hierarchy over the value of the object. However, in this chapter, types will *only* be used to represent hierarchies of beliefs, without any hard information.[2]

Finally, it is important to understand how to interpret epistemic results. One interpretation would go as follows. Assume we have elicited a player's hierarchy of beliefs. The theorems identify testable assumptions that determine whether that player's behavior is consistent with a particular solution concept. We do not find this interpretation very interesting: once we have elicited a player's hierarchy, we know her best replies, so it is pointless to invest effort to identify what assumptions are satisfied. Instead our preferred interpretations of the results are as statements about play that follow without knowing the exact hierarchy. That is, the theorems we present answer the following question: if all we knew about the hierarchies of beliefs was that they satisfied certain assumptions, what would we be able to say about play? Naturally, we cannot identify necessary conditions:

[2] We can add hard information to our framework, at the cost of notational complexity: see Footnote 49.

a player might play a Nash-equilibrium strategy "just because" he wanted to. (There is, however, a sense in which the results we present provide necessary conditions as well: see the discussion in Section 12.3.2.)

12.2. MAIN INGREDIENTS

In this section, we introduce the basic elements of our analysis. We begin with notation and a formal definition of strategic-form games, continue with hierarchies of beliefs and type structures, and conclude with rationality and beliefs.

12.2.1 Notation

For any finite set Y, let $\Delta(Y)$ denotes the set of probability distributions over Y and any subset E of Y is an event. For $Y' \subset Y$, $\Delta(Y')$ denotes the set of probabilities on Y that assign probability 1 to Y'. The support of a probability distribution $p \in \Delta(Y)$ is denoted by supp p. Finally, we adopt the usual conventions for product sets: given sets X_i, with $i \in I$, we let $X_{-i} = \prod_{j \neq i} X_j$ and $X = \prod_{i \in I} X_i$.

All our characterization theorems include results for which infinite sets are not required. However, infinite sets are needed to formally present hierarchies of beliefs, their relationship to type structures and for part of the characterization results. To minimize technical complications, infinite sets are assumed to be compact metric spaces endowed with the Borel sigma algebra. We denote by $\Delta(Y)$ the set of Borel probability measures on Y and endow $\Delta(Y)$ with the weak convergence topology.[3] Cartesian product sets are endowed with the product topology and the product sigma algebra. Events are a measurable subsets of Y.

12.2.2 Strategic-form games

We define finite strategic-form games and best replies.

Definition 12.1. *A (finite)* ***strategic-form game*** *is a tuple* $G = (I, (S_i, u_i)_{i \in I})$*, where* I *is finite and, for every* $i \in I$*,* S_i *is finite and* $u_i : S_i \times S_{-i} \to \mathbb{R}$*.*

As is customary, we denote expected utility from a mixed strategy of i, $\sigma_i \in \Delta(S_i)$, and a belief over strategies of opponents, $\sigma_{-i} \in \Delta(S_{-i})$, by $u_i(\sigma_i, \sigma_{-i})$. We take the view that players always choose *pure* strategies. On the other hand, certain standard solution concepts are defined in terms of mixed strategies. In the epistemic analysis, mixed strategies of i are replaced by strategic uncertainty of i's opponents, that is, their beliefs about i's choice of a pure strategy. We allow for mixed strategies as actual choices only when there is an explicit mixing device appended to the game.

[3] For detailed definitions see, e.g., Billingsley (2008).

Definition 12.2. *Fix a game $(I, (S_i, u_i)_{i \in I})$. A strategy $s_i \in S_i$ is a* ***best reply*** *to a belief $\sigma_{-i} \in \Delta(S_{-i})$ if, for all $s_i' \in S_i$, $u_i(s_i, \sigma_{-i}) \geq u_i(s_i', \sigma_{-i})$; the belief σ_{-i} is said to* ***justify*** *strategy s_i.*

12.2.3 Belief hierarchies

The essential element of epistemic analysis is the notion of hierarchies of belief. These are used to define rationality and common belief in rationality, which are then used to characterize solution concepts. A belief hierarchy specifies a player's belief over the basic space of uncertainty (e.g., opponents' strategies), her beliefs over opponents' beliefs, and so on.

To formally describe belief hierarchies, we first specify the basic space of uncertainty X_{-i} for each player i. In the epistemic analysis of a strategic-form game, the basic uncertainty is over the opponents' strategies, so $X_{-i} = S_{-i}$. More generally, we will allow for *exogenous* uncertainty as well, which is familiar from the textbook analysis of incomplete-information games. For instance, in a common-value auction, each player i is uncertain about the value of the object, so X_{-i} includes the set of possible values.

Once the sets X_{-i} have been fixed, each player i's hierarchy of beliefs is a sequence of probability measures $(p_i^1, p_i^2, \ldots)$. It is simpler to discuss these beliefs in the case of two players. Player i's *first-order belief* p_i^1 is a measure over the basic domain X_{-i}: $p_i^1 \in \Delta(X_{-i})$. Player i's *second-order belief* p_i^2 is a measure over the Cartesian product of X_{-i} and the set of all possible first-order beliefs for player $-i$: that is, $p_i^2 \in \Delta(X_{-i} \times \Delta(X_i))$, where X_i is the domain of $-i$'s first-order beliefs. The general form of this construction is as follows. First, let $X_{-i}^0 = X_{-i}$ for each player $i = 1, 2$; then, inductively, for each $k = 1, 2, \ldots$, let

$$X_{-i}^k = X_{-i}^{k-1} \times \Delta(X_i^{k-1}). \qquad [12.1]$$

Then, for each $k = 1, 2, \ldots$, the domain of player i's kth order beliefs is X_{-i}^{k-1}. Consequently, the set of all belief hierarchies for player i is $H_i^0 = \prod_{k \geq 0} \Delta(X_{-i}^k)$; the reason for the superscript "0" will be clear momentarily.

Note that, for $k \geq 2$, the domain of i's kth order beliefs includes the domain of her $(k-1)$th order beliefs. For instance, $p_i^2 \in \Delta(X_{-i} \times \Delta(X_i))$, so the marginal of p_i^2 also specifies a belief for i over X_{-i}, just like p_i^1. For an arbitrary hierarchy $(p_i^1, p_i^2, \ldots)$, these beliefs may differ. The reader may then wonder why we did not define i's second-order beliefs just over her opponent's first-order beliefs, i.e., as measures over $\Delta(X_i)$ rather than $X_{-i} \times \Delta(X_i)$.

Intuitively, the reason is that we need to allow for correlation in i's beliefs over X_{-i} and $-i$'s beliefs over X_i. Specifically, consider a simple 2×2 coordination game, with strategy sets $S_i = \{H, T\}$ for $i = 1, 2$. Suppose that the analyst is told that player 1 (i) assigns equal probability to player 2 choosing H or T, and (ii) also assigns equal probability to the events "player 2 believes that 1 chooses H" and "player 2 believes that 1 chooses T"

(where by "believes that" we mean "assigns probability one to the event that"). Can the analyst decide whether or not player 1 believes that player 2 is rational? The answer is negative. Given the information provided, it may be the case that player 1 assigns equal probability to the events "player 2 plays H and believes that 1 plays T" and "player 2 plays T and believes that 1 plays H."

To sum up, i's second-order belief p_i^2 must be an element of $\Delta(X_{-i} \times \Delta(X_i))$. Hence, we need to make sure that its marginal on X_{-i} coincides with i's first-order belief p_i^1. More generally, we restrict attention to *coherent* belief hierarchies, i.e., sequences $(p_i^1, p_i^2, \ldots) \in H_i^0$ such that, for all $k \geq 2$,

$$\text{marg}_{X_{-i}^{k-2}} p_i^k = p_i^{k-1}. \qquad [12.2]$$

Let H_i^1 denote the subset of H_i^0 consisting of coherent belief hierarchies.

Brandenburger and Dekel (1993) use Kolmogorov's theorem (see Aliprantis and Border, 2007, Section 15.6, or Dellacherie and Meyer, 1978, p. 68) to show that there exists a homeomorphism

$$\eta_i : H_i^1 \to \Delta\left(X_{-i} \times H_{-i}^0\right) \qquad [12.3]$$

that "preserves beliefs" in the sense that for $h_i = \left(p_i^k\right)_{k=1}^{\infty}$, $\text{marg}_{X_{-i}^k} \eta_i(h_i) = p_i^{k+1}$. To understand this, first note that η_i maps a coherent hierarchy h_i into a belief over i's basic space of uncertainty, X_{-i}, and $-i$'s hierarchies, H_{-i}^0. Therefore, we want this mapping to preserve i's first-order beliefs. In particular, h_i's first-order beliefs should equal the marginal of $\eta_i(h_i)$ on X_{-i}. Now consider second-order beliefs. Recall that $H_{-i}^0 = \prod_{\ell \geq 0} \Delta(X_i^\ell) = \Delta(X_i^0) \times \prod_{\ell \geq 1} \Delta(X_i^\ell)$. Therefore, $X_{-i} \times H_{-i}^0 = X_{-i} \times \Delta(X_i) \times \prod_{\ell \geq 1} \Delta(X_i^\ell) = X_{-i}^1 \times \prod_{\ell \geq 1} \Delta(X_i^\ell)$. Hence, we can view $\eta_i(h_i)$ as a measure on $X_{-i}^1 \times \prod_{\ell \geq 1} \Delta(X_i^\ell)$, so we can consider its marginal on X_{-i}^1. Preserving beliefs means that this marginal is the same as i's second-order belief p_i^2 in the hierarchy h_i. Higher-order beliefs are similarly preserved.

The function η_i in [12.3] maps coherent hierarchies of player i to beliefs about the basic uncertainty X_{-i} and the hierarchies of the other player, H_{-i}^0. Thus, in a sense, η_i determines a "first-order belief" over the expanded space of uncertainty $X_{-i} \times H_{-i}^0$. However, since η_i is onto, some coherent hierarchies of i assign positive probability to *in*coherent hierarchies of $-i$. These hierarchies of $-i$ do *not* correspond to beliefs over $X_i \times H_i^0$. Therefore, there are coherent hierarchies of i for which "second-order beliefs" over the expanded space $X_{-i} \times H_{-i}^0$ are not defined. To address this, we impose the restriction that coherency is "common belief"; that is, we restrict attention to

$$H_i = \cap_{k=0}^{\infty} H_i^k, \qquad [12.4]$$

where for $k > 0$ $H_i^k = \left\{h_i \in H_i^{k-1} : \eta_i(h_i)\left(X_{-i} \times H_{-i}^{k-1}\right) = 1\right\}$. It can then be shown that the function η_i in [12.3] restricted to H_i is one-to-one and onto $\Delta(X_{-i} \times H_{-i})$.[4] In the next subsection, we will interpret the elements of H_i as "types." With this interpretation, that η_i is one-to-one means that distinct types have distinct beliefs over X_{-i} and the opponent's types. That η_i is onto means that any belief about X_{-i} and the opponent's types is held by some type of i.

It is important to note that belief hierarchies are elicitable via bets. We elaborate on this point in Section 12.2.6.

12.2.4 Type structures

As Harsanyi noted, type structures provide an alternative way to model interactive beliefs. A type structure specifies for each player i the space X_{-i} over which i has uncertainty, the set T_i of types of i, and each type t_i's hierarchy of beliefs, $\beta_i(t_i)$.[5]

Definition 12.3. *For every player $i \in I$, fix a compact metric space X_{-i}. An $(X_{-i})_{i\in I}$-based **type structure** is a tuple $\mathcal{T} = (I, (X_{-i}, T_i, \beta_i)_{i\in I})$ such that each T_i is a compact metric space and each $\beta_i : T_i \to \Delta(X_{-i} \times T_{-i})$ is continuous.[6] A type structure is **complete** if the maps β_i are onto.*

We discuss the notion of completeness immediately before Definition 12.7.

An *epistemic type structure* for a strategic-form game of complete information models players' strategic uncertainty: hierarchies are defined over opponents' strategies. This is just a special case of Definition 12.3. However, since epistemic type structures play a central role in this chapter, we provide an explicit definition for future reference. Also, when it is clear from the context, we will omit the qualifier "epistemic."

Definition 12.4. *An **epistemic type structure** for the complete-information game $G = (I, (S_i, u_i)_{i\in I})$ is a type structure $\mathcal{T} = (I, (X_{-i}, T_i, \beta_i)_{i\in I})$ such that $X_{-i} = S_{-i}$ for all $i \in I$.*

Given an (epistemic) type structure $\mathcal{T}$, we can assess the belief hierarchy of each type t_i. As discussed earlier, type t_i's first-order belief is what she believes about S_{-i}; her

[4] For further details on the construction of belief hierarchies, see Armbruster and Böge (1979), Böge and Eisele (1979), Mertens and Zamir (1985), Brandenburger and Dekel (1993), Heifetz (1993), and Heifetz and Samet (1998), among others.

[5] As we discussed in the Introduction, in this definition, players do not have *introspective* beliefs—that is, beliefs about their own strategies and beliefs: see Section 12.2.6 for a discussion of this modeling choice.

[6] The topological assumptions we adopt are for convenience; we do not seek generality. For instance, compactness of the type spaces and continuity of the belief maps β_i provides an easy way to show that sets corresponding to assumptions such as, "Player i is rational," or "Player i believes that Player j is rational," are closed, and hence measurable.

second-order belief is what she believes about S_{-i} and about other player j's beliefs about S_{-j}, and so on. Also recall [12.4] that the set of all hierarchies of beliefs over strategies for a player i is denoted by H_i.

Definition 12.5. *Given a type structure $\mathcal{T}$, the function mapping types into hierarchies is denoted by $\varphi_i(\mathcal{T}) : T_i \to H$. The type structure $\mathcal{T}$ is redundant if there are two types $t_i, t_i' \in T_i$ with the same hierarchy, i.e., such that $\varphi_i(\mathcal{T})(t_i) = \varphi_i(\mathcal{T})(t_i')$; such types are also called redundant.*

When the type structure $\mathcal{T}$ is clear from the context, we will write $\varphi_i(\cdot)$ instead of $\varphi_i(\mathcal{T})(\cdot)$.

Because a type's first-order beliefs—those over S_{-i}—play a particularly important role, it is convenient to introduce specific notation for them:

Definition 12.6. *The **first-order beliefs map** $f_i : T_i \to \Delta(S_{-i})$ is defined by $f_i(t_i) = \text{marg}_{S_{-i}}\beta_i(t_i)$ for all $t_i \in T_i$.*

Example 12.1. *We illustrate these notions using a finite type structure.*

In the type structure on the right-hand side of Figure 12.1, type t_1^1 of player 1 (the row player) assigns equal probability to player 2 choosing L and C: these are type t_1^1's first-order beliefs. Similarly, the first-order beliefs of type t_1^2 of player 1 assign probability one to player 2 choosing L. The second-order beliefs of player 1's types are straightforward, because both t_1^1 and t_1^2 assign probability one to t_2^1, and hence to the event that player 2 is certain that (i.e., assigns probability one to the event that) 1 chooses T. Thus, for example, the second-order beliefs of type t_1^1 are that, with probability $\frac{1}{2}$, player 2 chooses L and believes that 1 chooses T, and with probability $\frac{1}{2}$, player 2 chooses C and believes that 1 chooses T.

Now consider type t_2^2 of player 2, who assigns equal probability to the pairs (M, t_1^1) and (B, t_1^2). This type's first-order beliefs are thus that player 1 is equally likely to play M or B; his second-order beliefs are that, with equal probability, either (i) player 1 plays M and expects player 2 to choose L and C with equal probability or (ii) player 1 plays B and is certain that 2 plays L. We can easily describe type t_2^2's third-order beliefs as well: this type believes that, with equal probability, either (i) player 1 plays M, expects 2 to choose L and C with equal probability, and

	L	C	R
T	2,1	3,1	0,0
M	4,3	0,2	4,0
B	3,0	1,2	2,5

	(L, t_2^1)	(L, t_2^2)	(C, t_2^1)	(C, t_2^2)	(R, t_2^1)	(R, t_2^2)
$\beta_1(t_1^1)$	$\frac{1}{2}$	0	$\frac{1}{2}$	0	0	0
$\beta_1(t_1^2)$	1	0	0	0	0	0

	(T, t_1^1)	(T, t_1^2)	(M, t_1^1)	(M, t_1^2)	(B, t_1^1)	(B, t_1^2)
$\beta_2(t_2^1)$	1	0	0	0	0	0
$\beta_2(t_2^2)$	0	0	$\frac{1}{2}$	0	0	$\frac{1}{2}$

Figure 12.1 A strategic-form game and an epistemic type structure.

is certain that 2 is certain that 1 plays T or (ii) player 1 plays B, is certain that 2 chooses L, and is certain that 2 is certain that 1 plays T.

A number of questions arise in connection with type structures. Is there a type structure that generates *all* hierarchies of beliefs? Is there a type structure into which any other type structure can be embedded?[7] Is a given type structure complete, as in Definition 12.3, i.e., such that every belief over Player *i*'s opponents' strategies and types is generated by some type of Player *i*? These are all versions of the same basic question: is there a rich enough type structure that allows for "all possible beliefs?" We ask this question because we take beliefs as primitive objects; hence, we want to make sure that using type structures as a modeling device does not rule out any beliefs.

Under our assumptions on the sets X_{-i}, the answer to these questions is affirmative. Indeed, we can consider H_i (defined in [12.4]) as a set of type profiles and define $\mathcal{T} = (I, (X_{-i}, T_i, \beta_i)_{i \in I})$, where $T_i = H_i$ and $\beta_i = \eta_i$ (where η_i was defined in [12.3]). This is the "largest" nonredundant type structure, that generates all hierarchies, embeds all other type structures, and is complete.[8]

Once again, type structures are devices and belief hierarchies are the primitive objects of interest. Therefore, asking whether a player's hierarchy of beliefs "resides" in one type structure or another is meaningless. In particular, we cannot ask whether it "resides" in a rich type structure. We state results regarding the implications of epistemic assumptions in both rich and arbitrary type structures. The interest in rich type structures is twofold. First, one of convenience: while rich type structures are uncountable and complex mathematical objects, the fact that they are complete simplifies the statements of our characterization results. The second appeal of rich type structure is methodological: because they generate *all* hierarchies, they impose no implicit assumption on beliefs. Any smaller type structure does implicitly restrict beliefs; we explain this point in Section 12.7.4.4, because it is particularly relevant there. On the other hand, small (in particular, finite) type structures are convenient to discuss examples of epistemic conditions and characterization results.

12.2.5 Rationality and belief

We can now define rationality (expected payoff maximization) and belief, by which we mean "belief with probability one."

[7] We are not going to formally define the relevant notion of embedding. Roughly speaking, it is that each type in one type space can be mapped to a type in the other in such a way as to preserve hierarchies of beliefs.

[8] Because we restrict attention to compact type spaces and continuous belief maps, these notions of "richness" are all equivalent. See the references in Footnote 4 for details, as well as Friedenberg (2010). In particular, it is sufficient that X_{-i} and T_i are compact metrizable and that the sets T_i are nonredundant.

Definition 12.7. *Fix a type structure* $(I, (S_{-i}, T_i, \beta_i)_{i\in I})$ *for a strategic game* $(I, (S_i, u_i)_{i\in I})$. *For every player* $i \in I$:

1. *Strategy* $s_i \in S_i$ *is* ***rational*** *for type* $t_i \in T_i$ *if it is a best reply to* $f_i(t_i)$; *let*
$$R_i = \{(s_i, t_i) \in S_i \times T_i \ : \ s_i \text{ is rational for } t_i\}.$$
2. *Type* $t_i \in T_i$ ***believes*** *event* $E_{-i} \subset S_{-i} \times T_{-i}$ *if* $\beta_i(t_i)(E_{-i}) = 1$; *let*
$$B_i(E_{-i}) = \{(s_i, t_i) \in S_i \times T_i \ : \ t_i \text{ believes } E_{-i}\}.$$

Note that R_i, the set of strategy-type pairs of i that are rational for i, is defined as a subset of $S_i \times T_i$, rather than a subset of $S \times T$. This is notationally convenient and also emphasizes that R_i is an assumption about Player i alone.

For any event $E_{-i} \subset S_{-i} \times T_{-i}$, $B_i(E_{-i})$ represents the types of i that believe E_{-i} obtains. It is convenient to define it as an event in $S_i \times T_i$, but it is clear from the definition that no restriction is imposed on i's strategies.[9] We abuse terminology and for sets $E_i \subset S_i \times T_i$ write "t_i in E_i" if there exists s_i such that $(s_i, t_i) \in E_i$.

The map associating with each event $E_{-i} \subset S_{-i} \times T_{-i}$ the subset $B_i(E_{-i})$ is sometimes called Player i's *belief operator*. While we do not develop a formal syntactic analysis, we do emphasize two important related properties satisfied by probability-one belief, *Monotonicity* and *Conjunction*:[10] for all events $E_{-i}, F_{-i} \subset S_{-i} \times T_{-i}$,

$$E_{-i} \subset F_{-i} \quad \Rightarrow \quad B_i(E_{-i}) \subset B_i(F_{-i}) \quad \text{and} \quad B_i(E_{-i} \cap F_{-i}) = B_i(E_{-i}) \cap B_i(F_{-i}). \qquad [12.5]$$

Finally, we define mutual and common belief. Consider events $E_i \subset S_i \times T_i$ (with $E = \prod_i E_i$ and $E_{-i} = \prod_{j\neq i} E_i$ as usual). Then the events "E is *mutually believed*," "*kth order believed*," and "*commonly believed*" are

$$B^1(E) = B(E) = \prod_{i\in I} B_i(E_{-i}), \quad B^k(E) = B\left(B^{k-1}(E)\right) \text{ for } k > 1,$$
$$CB(E) = \bigcap_{k\geq 1} B^k(E). \qquad [12.6]$$

Note that each of these events is a Cartesian product of subsets of $S_i \times T_i$ for $i \in I$, so we can write $B_i^k(E)$ and $CB_i(E)$ for the ith projection of these events.[11]

[9] That is, if $(s_i, t_i) \in B_i(E_{-i})$ for some $s_i \in S_i$, then $(s_i', t_i) \in B_i(E_{-i})$ for all $s_i' \in S_i$.

[10] We can split the "=" in the Conjunction property into two parts, "$\subset$" and "$\supset$." It is easy to see that the "$\subset$" part is equivalent to Monotonicity for any operator, no matter how it is defined.
The "p-belief," "strong belief," and "assumption" operators we consider in Sections 12.5, 12.7, and 12.8, respectively, do not satisfy Monotonicity, and hence the "$\subset$" part of Conjunction—a fact that has consequences for the epistemic analysis conducted therein.

[11] Thus, $CB(E) = \bigcap_{k\geq 1} B^k(E) = \prod_i B_i(E_{-i}) \cap \bigcap_{k\geq 2} \prod_i B_i(B_{-i}^{k-1}(E)) = \prod_i (B_i(E_{-i}) \cap \bigcap_{k\geq 2} B_i(B_{-i}^{k-1}(E)))$ $\equiv \prod_i CB_i(E)$. Hence, $CB_i(E) = B_i^1(E) \cap \bigcap_{k\geq 2} B_i(B_{-i}^{k-1}(E)) = \bigcap_{k\geq 1} B_i^k(E)$ and also $CB_i(E) = B_i(E_{-i} \cap \bigcap_{k\geq 1} B_{-i}^k(E)) = B_i(E_{-i} \cap CB_{-i}(E))$.

12.2.6 Discussion

The above definitions of a game, type structure, rationality, and belief all incorporate the assumption that players have state-independent expected-utility preferences. This modeling assumption raises three issues, discussed next: relaxing state independence, relaxing expected utility, and eliciting beliefs. Another modeling assumption discussed subsequently is that the type structure in principle allows any strategy to be played by any type. We conclude this discussion section by commenting on our use of semantic models rather than the alternative syntactic approach.

12.2.6.1 State dependence and nonexpected utility

A more general definition of a game would specify a consequence for each strategy profile, and a preference relation over acts that map opponents' strategies into consequences.[12] Maintaining the expected-utility assumption one could allow for state dependence: the ranking of consequences may depend on the opponents' strategies (as in Morris and Takahashi, 2011). One could also allow for a richer model where preferences may be defined over opponents' beliefs (as in Geanakoplos et al., 1989) or preferences (as in Gul and Pesendorfer, 2010), as well as material consequences. All these interesting directions lie beyond the scope of this chapter.

Moreover, type structures can also be defined without making the expected-utility assumption. Some generalizations of expected utility are motivated by refinements: in particular, lexicographic beliefs (Blume et al., 1991) and conditional probability systems (Myerson, 1997; Siniscalchi, 2014).[13] We discuss these, and the type structures they induce, in Sections 12.7 and 12.8. Other generalizations of expected utility are motivated by the Allais and Ellsberg paradoxes; Epstein and Wang (1996) show how type spaces can be constructed for a wide class of non-expected utility preferences.[14]

12.2.6.2 Elicitation

Our analysis puts great emphasis on players' beliefs; thus, as discussed in the Introduction, it is crucial that such beliefs can in fact be elicited from preferences. Indeed one would expect that Player 1's beliefs about 2's strategies can be elicited by asking 1 to bet on which strategy 2 will in fact play, as in Savage (1972) and Anscombe and Aumann (1963). Given this, one can then elicit 2's beliefs about 1's strategies and beliefs by having 2 bet on 1's strategies and bets, and so on.[15] (Similarly, we could elicit utilities over consequences.)

[12] As in Anscombe and Aumann (1963), consequences could be lotteries over prizes.

[13] See also Asheim and Perea (2005). Morris (1997) considers alternative, preference-based definitions of belief.

[14] See also Ahn (2007), Di Tillio (2008), and Chen (2010).

[15] See Morris (2002) and Dekel et al. (2006). More generally, one can in principle elicit Player 1's preferences over acts mapping 2's strategies to consequences, then elicit 2's preferences over acts mapping 1's (strategies and) preferences to consequences, and so on; this underlies the aforementioned construction of Epstein and Wang (1996).

However, adding these bets changes the game, because the strategy space and payoffs now must include these bets. Potentially, this may change the players' beliefs about the opponents' behavior in the original game. Hence, some delicacy is required in adding such bets to elicit beliefs.[16]

12.2.6.3 Introspective beliefs and restrictions on strategies

Our definition of a type structure assumes that each type t_i has beliefs over $S_{-i} \times T_{-i}$. This has two implications. First, players do not hold introspective beliefs, as we noted in the Introduction. Second, by specifying a type space, the analyst restricts the hierarchies the player may hold, but does not restrict play. An alternative popular model (Aumann, 1999a,b) associates with each type t_i a belief on opponents' types *and* a strategy, $\sigma_i(t_i)$. Such a model restricts the strategies that a player with a given hierarchy may choose; moreover, such restrictions are common belief among the players. We can incorporate such assumptions as well, but, in keeping with our view of the goals of the epistemic literature, we make them explicit: see, for example, Section 12.4.5.

12.2.6.4 Semantic/syntactic models

Finally, we note that our modeling approach is what is called *semantic*: it starts from a type structure, and defines the belief operator, B_i, using the elements of the type structure; its properties, such as conjunction and monotonicity, follow from the way it is defined. An alternative approach, called *syntactic*, is to start with a formal language in which a belief operator is taken as a primitive; properties such as the analogs of conjunction and monotonicity are then explicitly imposed as axioms. There is a rich literature on the relation between the semantic and syntactic approaches; see for example, Fagin et al. (1995), Aumann (1999a,b), Heifetz and Mongin (2001), and Meier (2012). Due to its familiarity to economists, we adopt the semantic approach here.

12.3. STRATEGIC GAMES OF COMPLETE INFORMATION

In this section, we study common belief in rationality, as this is a natural starting point. Like the assumptions of perfect competition or rational expectations, common belief in rationality is not meant to be descriptively accurate. However, like those notions, it is a useful benchmark. We present the equivalence of the joint assumptions of rationality and common belief in rationality with iterated deletion of dominated strategies,

[16] Aumann and Dreze (2009) raise this concern and propose a partial resolution, although they do not elicit unique beliefs and only study first-order beliefs. (A related concern was raised by Mariotti, 1995, and addressed by Battigalli, 1996b.) Aumann and Dreze (2009) also note that, by assuming common belief in rationality—as we will through most of this paper—beliefs can also be elicited by adding to the game bets with payoffs that are suitably small. Siniscalchi (2014) adds bets differently, avoiding all these concerns.

i.e., (correlated) rationalizability, and best-reply sets. We also discuss a refinement of rationalizability that allows for additional restrictions on beliefs.

12.3.1 Rationality and common belief in rationality

As noted, we focus on the joint assumptions of rationality and common belief in rationality. There is more than one way of stating this assumption. The immediate definition is

$$RCBR = R \cap B(R) \cap B^2(R) \cap \cdots \cap B^m(R) \cap \cdots = R \cap CB(R).^{17} \qquad [12.7]$$

In words, RCBR is the event that everybody is rational, everybody believes that everyone else is rational, everybody believes that everyone else believes that others are rational, and so on. However, there is an alternative definition. For all $i \in I$, let:

$$R_i^1 = R_i; \qquad [12.8]$$

and for any $m \geq 1$,

$$R_i^{m+1} = R_i^m \cap B_i(R_{-i}^m). \qquad [12.9]$$

Finally, we let

$$RCBR_i = \bigcap_{m \geq 1} R_i^m \quad \text{and} \quad RCBR = \prod_{i \in I} RCBR_i. \qquad [12.10]$$

To see how [12.7] and [12.10] relate consider the case $m = 3$. We have

$$R_1^3 = R_1 \cap B_1(R_2) \cap B_1(R_2 \cap B_2(R_1))$$

whereas

$$R_1 \cap B_1(R_2) \cap B_1^2(R_1) = R_1 \cap B_1(R_2) \cap B_1(B_2(R_1)).$$

Inspecting the last term, the definition of R_1^3 is seemingly more demanding. However, thanks to monotonicity and conjunction (see [12.5]), the two are equivalent. Inductively, it is easy to see that the two definitions of $RCBR_i$ in [12.7] and [12.10] are also equivalent. However, when we consider nonmonotonic belief operators—as we will have to for studying refinements—this equivalence will fail.

Having defined the epistemic assumptions of interest in this section, we now turn to the relevant solution concepts. In general, different (but obviously related) solution concepts characterize the behavioral implications of epistemic assumptions such as RCBR in complete type structures, where nothing is assumed beyond RCBR, and

[17] A typographical note: we write "RCBR" in the text as the acronym for "rationality and common belief in rationality," and "*RCBR*" in equations to denote the event that corresponds to it.

smaller type structures, in which players' beliefs satisfy additional (commonly believed) assumptions. Here, the relevant concepts are rationalizability and best-reply sets.

Definition 12.8. (Rationalizability) *Fix a game* $(I, (S_i, u_i)_{i \in I})$. *Let* $S_i^0 = S_i$ *for all* $i \in I$. *Inductively, for* $m \geq 0$, *let* S_i^{m+1} *be the set of strategies that are best replies to conjectures* $\sigma_{-i} \in \Delta(S_{-i}^m)$. *The set* $S_i^\infty = \bigcap_{m \geq 0} S_i^m$ *is the set of* ***(correlated) rationalizable*** *strategies of Player* i.

Bernheim (1984) and Pearce (1984) propose the solution concept of rationalizability, which selects strategies that are best replies to beliefs over strategies that are themselves best replies, and so on. Intuitively, one expects this to coincide with the iterative deletion procedure in Definition 12.8. Indeed, these authors prove this, except that they focus on beliefs that are product measures, i.e., stochastically independent across different opponents' strategies.

A strategy $s_i \in S_i$ is **(strictly) dominated** if there exists a distribution $\sigma_i \in \Delta(S_i)$ such that, for all $s_{-i} \in S_{-i}$, $u_i(\sigma_i, s_{-i}) > u_i(s_i, s_{-i})$. It is well known (Gale and Sherman, 1950; Pearce, 1984; Van Damme, 1983) that a strategy is strictly dominated if and only if it is not a best reply to any belief about the opponents' play.[18] Therefore, S_i^∞ is also the set of strategies of i that survive **iterated strict dominance**, i.e., the solution concept that selects the iteratively undominated strategies for each player. In the game of Figure 12.1, it is easy to verify that $S^1 = \{T, M\} \times \{L, C, R\}$ and $S^2 = S^\infty = \{T, M\} \times \{L, C\}$.

A best-reply set is a collection of strategy profiles with the property that every strategy of every player is justified by (i.e., is a best response to) a belief restricted to opponents' strategy profiles in the set. A best-reply set is full if, in addition, *all* best replies to each such justifying belief also belong to the set.[19]

Definition 12.9. *Fix a game* $(I, (S_i, u_i)_{i \in I})$. *A set* $B = \prod_{i \in I} B_i \subset S$ *is a* ***best-reply set*** *(or BRS) if, for every player* $i \in I$, *every* $s_i \in B_i$ *is a best reply to a belief* $\sigma_{-i} \in \Delta(B_{-i})$.

B is a ***full BRS*** *if, for every* $s_i \in B_i$, *there is a belief* $\sigma_{-i} \in \Delta(B_{-i})$ *that justifies* s_i *and such that all best replies to* σ_{-i} *are also in* B_i.

Notice that the player-by-player union of (full) BRSs is again a (full) BRS.[20] Thus, there exists a unique, maximal BRS, which is itself a full BRS; it can be shown that it is equal to S^∞.

To clarify the notion of full BRS, refer to the game in Figure 12.1. The profile (T, C) is a BRS, but not a full BRS, because, if player 1 plays T, then L yields the same payoff

[18] This equivalence holds for games with compact strategy sets and continuous payoff functions (in particular, the finite games we consider here). See also Dufwenberg and Stegeman (2002) and Chen et al. (2007).

[19] For related notions, see Basu and Weibull (1991).

[20] That is: if $B = \prod_i B_i$ and $C = \prod_i C_i$ are (full) BRSs, then so is $\prod_i (B_i \cup C_i)$.

to player 2 as C. On the other hand, $\{T\} \times \{L, C\}$ is a full BRS, because T is the unique best reply for player 1 to a belief that assigns equal probability to L and C, and L and C are the only best replies to a belief concentrated on T.

We can now state the epistemic characterization result.

Theorem 12.1. (Brandenburger and Dekel, 1987; Tan and da Costa Werlang, 1988) [21] *Fix a game* $G = (I, (S_i, u_i)_{i \in I})$.

1. *In any type structure* $(I, (S_{-i}, T_i, \beta_i)_{i \in I})$ *for* G, $\text{proj}_S RCBR$ *is a full BRS.*
2. *In any complete type structure* $(I, (S_{-i}, T_i, \beta_i)_{i \in I})$ *for* G, $\text{proj}_S RCBR = S^\infty$.
3. *For every full BRS* B, *there exists a finite type structure* $(I, (S_{-i}, T_i, \beta_i)_{i \in I})$ *for* G *such that* $\text{proj}_S RCBR = B$.

We do not provide proofs in this chapter; they can be found in the cited papers or can be adapted from arguments therein. For some results, we provide the details in an online appendix, Dekel et al. (2014). We discuss this result in Section 12.3.2. For an example of Theorem 12.1, consider the type structure of Figure 12.1. Then, $\text{proj}_S RCBR = \{T, M\} \times \{L, C\}$, which is a full BRS and indeed equals S^∞. Next, consider the smaller type structure $\mathcal{T}'$ containing only type t_1^1 for player 1 and type t_2^1 for player 2. Now $\text{proj}_S RCBR = \{T\} \times \{L, C\}$, which, as noted earlier, is indeed a full BRS.

12.3.2 Discussion

Theorem 12.1 characterizes the implications of RCBR. One could also study the weaker assumption of common belief in rationality (CBR). The latter is strictly weaker because belief in an event does not imply it is true. Hence, CBR only has implications for players' beliefs; we focus on RCBR because it also restricts behavior. Epistemic models that allow for introspective beliefs have the feature that, if a player has correct beliefs about her own strategy and beliefs, then, if she believes that she is rational she is indeed rational. Hence, in such models, CBR is equivalent to RCBR.

The interpretation of part (1) in Theorem 12.1 is that, if the analyst assumes that RCBR holds, but allows for the possibility that the players' beliefs may be further restricted (i.e., something in addition to rationality is commonly believed), then the analyst can only predict that play will be consistent with *some* full BRS.[22] This implies that, unless the analyst knows what further restrictions on players' beliefs hold, he must allow for the player-by-player union of all full BRSs. As we noted, this is equal to S^∞.

Part (2) in Theorem 12.1 is an epistemic counterpart to this. A complete type structure embeds all other type structures; it is therefore "natural" to expect that the predictions of RCBR in a complete structure should also be S^∞. Theorem 12.1 shows that this is the case. This convenient equivalence fails when we consider refinements.

[21] See also Armbruster and Böge (1979) and Böge and Eisele (1979).

[22] It must be a full BRS because we do not restrict play: see the discussion at the end of Section 12.2.

Part (3) confirms that the result in part (1) is tight: *every* full BRS represents the behavioral implications of RCBR in some type structure. If this was not the case, then RCBR would have more restrictive behavioral implications than are captured by the notion of full BRS. Furthermore, the result in part (3) indicates a sense in which RCBR is "necessary" for behavior to be consistent with a full BRS. While, as noted in the Introduction, players may choose strategies in a given full BRS B by accident, or following thought processes altogether different from the logic of RCBR, the latter is always a *possible* reason why individuals may play strategy profiles in B.

12.3.3 Δ-Rationalizability

As we discussed, a type structure encodes assumptions about players' hierarchies of beliefs. This may provide a convenient way to incorporate specific assumptions of interest. For example, one may wish to study the assumption that players' beliefs over opponents' play are independent or that players believe that, for some reason, a particular strategy—even if it is rationalizable—will not be played, and so on. An alternative approach (Battigalli and Siniscalchi, 2003) is to make them explicit. In this subsection, we outline one way to do so.

For every player $i \in I$, fix a subset $\Delta_i \subset \Delta(S_{-i})$. Given a type structure, the event that Player i's beliefs lie in the set Δ_i is

$$[\Delta_i] = \left\{(s_i, t_i) : f_i(t_i) \in \Delta_i\right\}.$$

We wish to characterize RCBR combined with common belief in the restrictions Δ_i.[23]

Definition 12.10. *Fix a game $(I, (S_i, u_i)_{i\in I})$ and a collection of restrictions $\Delta = (\Delta_i)_{i\in I}$. A set $B = \prod_{i\in I} B_i \subset S$ is a* **Δ-best-reply set** *(or Δ-BRS) if, for every player $i \in I$, every $s_i \in B_i$ is a best reply to a belief $\sigma_{-i} \in \Delta(B_{-i}) \cap \Delta_i$; it is a* **full Δ-BRS** *if, for every $s_i \in B_i$, there is a belief $\sigma_{-i} \in \Delta(B_{-i}) \cap \Delta_i$ that justifies s_i and such that all best replies to σ_{-i} are also in B_i.*

Definition 12.11. *Fix a game $(I, (S_i, u_i)_{i\in I})$. For each $i \in I$, let $S_i^{\Delta,0} = S_i$. Inductively, for $m \geq 0$, let $S_i^{\Delta,m+1}$ be the set of strategies that are best replies to conjectures $\sigma_{-i} \in \Delta_i$ such that $\sigma_{-i}(S_{-i}^{\Delta,m}) = 1$. The set $S_i^{\Delta,\infty} = \bigcap_{m\geq 0} S_i^{\Delta,m}$ is the set of* **Δ-rationalizable** *strategies of Player i.*

Obviously, the set of Δ-rationalizable strategies may be empty for certain restrictions Δ. Also, note that $S_i^{\Delta,\infty}$ is a full Δ-BRS.

[23] For related solution concepts (albeit without a full epistemic characterization) see Rabin (1994) and Gul (1996).

Theorem 12.2. *Fix a game* $G = (I, (S_i, \beta_i)_{i \in I}$ *and a collection of restrictions* $\Delta = (\Delta_i)_{i \in I}$.

1. *In any type structure* $(I, (S_{-i}, T_i, \beta_i)_{i \in I})$ *for* G, $\text{proj}_S(RCBR \cap CB([\Delta]))$ *is a full* Δ-*BRS*.
2. *In any complete type structure* $(I, (S_{-i}, T_i, \beta_i)_{i \in I})$ *for* G, $\text{proj}_S(RCBR \cap CB([\Delta])) = S^{\Delta,\infty}$.
3. *For every full* Δ-*BRS* B, *there exists a finite type structure* $(I, (S_{-i}, T_i, u_i)_{i \in I})$ *for* G *such that* $\text{proj}_S(RCBR \cap CB([\Delta])) = B$.[24]

Results (1)-(3) in Theorem 12.2 correspond to results (1)-(3) in Theorem 12.1.

The notion of Δ-rationalizability extends easily to games with incomplete information and is especially useful in that context. We provide an example in Section 12.6.5; more applied examples can be found, e.g., in Battigalli and Siniscalchi (2003). In the context of complete-information games, Bernheim's and Pearce's original definition of rationalizability required that beliefs over opponents' strategies be independent. This can also be formulated using Δ-rationalizability: for every player i, let Δ_i be the set of product measures over S_{-i}. Restrictions on first-order beliefs may also arise in a learning setting, where players observe only certain aspects of play in each stage. We return to this point in Section 12.7.6, where we discuss self-confirming equilibrium in extensive games.

12.4. EQUILIBRIUM CONCEPTS

12.4.1 Introduction

A natural question is what epistemic characterizations can be provided for equilibrium concepts. By this we mean solution concepts where players best reply to opponents' *actual* strategies. This is in contrast with solution concepts like rationalizability, where players best-reply to conjectures about opponents' strategies that may be incorrect.

Before turning to Nash equilibrium, we consider two weaker solution concepts, objective and subjective correlated equilibrium. Somewhat surprisingly, it turns out that the latter equilibrium concept is equivalent to correlated rationalizability. Hence, RCBR does provide a characterization of an equilibrium concept as well. Subsection 12.4.2 develops this point. The main idea is that any incorrect beliefs about opponents' strategies can be "shifted" to incorrect beliefs about a correlating device, thereby maintaining the assumption that players have correct beliefs about the mapping from correlating signals to strategies.

[24] In fact, one can define a type structure in which the restrictions Δ hold for *every* type. In such a structure, $\text{proj}_S(RCBR \cap CB([\Delta])) = \text{proj}_S RCBR$. It is also possible to construct a type structure that is not complete, but is infinite and contains all belief hierarchies that are consistent with it being common belief that players' first-order beliefs satisfy the restrictions Δ. Unlike the type structures constructed to obtain part (3), this contains no other assumptions on beliefs beyond common belief in Δ. Consequently, RCBR characterizes $S^{\Delta,\infty}$ in this structure.

Subsections 12.4.3 and 12.4.4 characterize objective correlated and Nash equilibrium. In contrast to other results in this paper, in which epistemic conditions fully characterize *play*, in Subsections 12.4.3 and 12.4.4, epistemic conditions only imply that *beliefs* correspond to equilibrium. For example, we do not show that under certain conditions players play Nash-equilibrium strategies, only that the profile of their first-order beliefs is an equilibrium profile. Indeed, this is one of the insights that emerges from the epistemic analysis: Nash equilibrium is usefully interpreted as a property of (first-order) beliefs, not play. This point was made by Harsanyi (1973) and Aumann (1987), among others.

A critical assumption in Subsections 12.4.3 and 12.4.4 is the existence of a "common prior" that generates beliefs. While we mostly follow Aumann (1987) and Aumann and Brandenburger (1995), in contrast to their approach we do not allow players to have beliefs over their own strategies.[25] Due to this difference, a direct adaptation of the common-prior assumption to our setting turns out to be weaker than in those papers (indeed betting becomes possible). Hence, we formulate an additional assumption that is needed for a full characterization of these equilibrium concepts. The final subsection presents an alternative sufficient—but not necessary—assumption to obtain these concepts.

12.4.2 Subjective correlated equilibrium

Aumann (1987) defined (subjective) correlated equilibria; these are equivalent to Nash equilibria of a game in which players observe signals from a correlating device prior to choosing their actions. A correlating device consists of a finite set Ω of realizations, and, for each player, a partition Π_i of this finite set, and a conditional probability distribution $\mu_i(\cdot|\pi_i)$ for each cell π_i in the partition.[26] In a correlated equilibrium of a strategic-form game $G = (I, (S_i, u_i)_{i\in I})$, players choose strategies in S_i as a function of their signal $\pi_i \in \Pi_i$, so as to maximize their conditional expected payoff, taking as given the equilibrium behavior of their opponents.

Definition 12.12. *Fix a game* $G = (I, (S_i, u_i)_{i\in I})$.

*A **correlating device** for the game G is a tuple* $\mathcal{C} = (\Omega, (\Pi_i, \mu_i)_{i\in I})$, *where* Ω *is a finite set, for every* $i \in I$, Π_i *is a partition of* Ω *with typical element* π_i, *and* μ_i *is a conditional belief map, i.e.,* $\mu_i : 2^\Omega \times \Pi_i \to [0, 1]$ *satisfies* $\mu_i(\cdot|\pi_i) \in \Delta(\Omega)$ *and* $\mu_i(\pi_i|\pi_i) = 1$ *for all* $\pi_i \in \Pi_i$. *If there exists* $\mu \in \Delta(\Omega)$ *such that, for every* $i \in I$, *and* $\pi_i \in \Pi_i$, $\mu_i(\cdot|\pi_i) = \mu(\cdot|\pi_i)$, *then it is an **objective correlating device**.*

[25] For other approaches, see Tan and da Costa Werlang (1988), Brandenburger and Dekel (1987), and Perea (2007).

[26] Aumann (1987) defines correlating devices slightly differently; for details and to see how this affects the results herein, see his paper (see also Brandenburger and Dekel, 1987).

*A **subjective correlated equilibrium** is a correlating device and a tuple $(\mathbf{s}_i)_{i\in I}$ where, for each $i \in I$, $\mathbf{s}_i : \Omega \to S_i$ is measurable with respect to Π_i and, for every $\pi_i \in \Pi_i$,*

$$\sum_{\omega\in\pi_i} \mu_i(\{\omega\}|\pi_i)u_i(\mathbf{s}_i(\omega), \mathbf{s}_{-i}(\omega)) \geq \sum_{\omega\in\pi_i} \mu_i(\{\omega\}|\pi_i)u_i(s_i, \mathbf{s}_{-i}(\omega)) \quad \forall s_i \in S_i. \quad [12.11]$$

*An **objective correlated equilibrium** is a subjective correlated equilibrium where the correlating device is objective.*

*Given an objective correlated equilibrium $C = (\Omega, \Pi_i, \mu)$, the **objective correlated equilibrium distribution** induced by C is the probability distribution $\sigma \in \Delta(S)$ defined by $\sigma(s) = \mu(\{\omega : \mathbf{s}(\omega) = s\})$ for all $s \in S$.*

There are obvious formal similarities between type structures and correlating devices—although their interpretation is very different. As noted repeatedly, a type structure is merely a mathematical construct used to represent belief hierarchies. On the other hand, a correlating device is meant to represent a real signal structure—something the players and (potentially) the analyst observe. The formal similarities yield the following result.

Theorem 12.3. *[Brandenburger and Dekel, 1987] Fix a game $G = (I, (S_i, u_i)_{i\in I})$.*

1. *For any type structure $(I, (S_{-i}, T_i, \beta_i)_{i\in I})$ for G, there exists a subjective correlated equilibrium $\big((\Omega, \Pi_i), \mathbf{s}\big)$ of G such that, for all $i \in I$, $\mathrm{proj}_{S_i} RCBR_i = \mathbf{s}_i(\Omega)$.*
2. *Given any subjective correlated equilibrium $\big((\Omega, \Pi_i), \mathbf{s}\big)$ of G there exists a type structure $(I, (S_{-i}, T_i, \beta_i)_{i\in I})$ for G such that $\mathrm{proj}_{S_i} RCBR_i \supseteq \mathbf{s}_i(\Omega)$ for all $i \in I$.*

The essence of this theorem is that, under RCBR, the strategic uncertainty in an epistemic type structure (or, equivalently, in rationalizability) is interchangeable with the exogenous uncertainty of a (subjective) correlating device in a correlated equilibrium. The proof of part 1 sets $\Omega = RCBR \subseteq S \times T$, defines the cells in each player's partition Π_i to be of the form $\{(s_i, t_i)\} \times RCBR_{-i}$, and chooses the belief μ_i given the cell $\{(s_i, t_i)\} \times RCBR_{-i}$, so that its marginal on $RCBR_{-i}$ equals $\beta_i(t_i)$. For part 2, let $T_i = \Pi_i$ and, for any $\pi_i \in \Pi_i$, let $\beta_i(\pi_i)(s_{-i}, \pi_{-i}) = \mu_i(\{\omega : \forall j, \mathbf{s}_j(\omega) = s_j \text{ and } \Pi_j(\omega) = \pi_j\}|\pi_i)$, where $\Pi_j(\omega)$ denotes the element of Π_j that contains ω. Note that, in part 2, there may be strategies in $RCBR_i$ that are not played in the correlated equilibrium, but the set of (interim) payoffs under RCBR and in the equilibrium are the same.

12.4.3 Objective correlated equilibrium

To characterize objective correlated equilibrium and Nash equilibrium, we want to define the event that Player i's beliefs are "consistent with a common prior." By this we mean that her belief hierarchy can be generated in some type structure where the beliefs held by each type t_i can be obtained from some probability measure μ (the common prior) over the profiles of strategies and types $S \times T$, by conditioning on the event that i's type is indeed t_i. Note that we state the common-prior assumption as a property of belief hierarchies, rather than type structures; in this, we deviate from the received literature, but are consistent with our premise that the primitives of our analysis should be elicitable.

To make this formal, we proceed in two steps. First, a type structure $\mathcal{T} = (I, (S_{-i}, T_i, \beta_i)_{i \in I})$ admits a common prior μ on $S \times T$ if the belief maps β_i are obtained from μ by conditioning on types. This differs from the standard definition (e.g., Aumann, 1987) because it conditions only on types, not on strategies; we discuss this important point after Example 12.3. However, it is as "close" as possible to the standard definition, given our premise that players do not hold beliefs about their own strategies.

Definition 12.13. *A finite type structure $\mathcal{T} = (I, (S_{-i}, T_i, \beta_i)_{i \in I})$* ***admits*** *(or* ***is generated by****)* ***a common prior*** *$\mu \in \Delta(S \times T)$ if, for all $t_i \in T_i$, $\mu(S_i \times \{t_i\} \times S_{-i} \times T_{-i}) > 0$ and $\beta_i(t_i) = \mathrm{marg}_{S_{-i} \times T_{-i}} \mu(\cdot | S_i \times \{t_i\} \times S_{-i} \times T_{-i})$.*

We now translate Definition 12.13 into one that is stated in terms of hierarchies, rather than types. Given any type structure $\mathcal{T}$, we deem type $t_i \in T_i$ consistent with a common prior μ if its induced hierarchy is the same as the one which would arise in the ancillary type structure $\mathcal{T}^\mu$ which admits μ as a common prior, where the type spaces of $\mathcal{T}^\mu$ are subsets of those in $\mathcal{T}$.[27] The following definition makes this precise.

Definition 12.14. *Fix a finite type structure $\mathcal{T} = (I, (S_{-i}, T_i, \beta_i)_{i \in I})$, a player $i \in I$, and a probability $\mu \in \Delta(S \times T)$. Consider the type structure $\mathcal{T}^\mu = (I, (S_{-i}, T_i^\mu, \beta_i^\mu)_{i \in I})$ that admits μ as a common prior and such that, for every $i \in I$, $T_i^\mu \subseteq T_i$. The event "Player i's beliefs are consistent with a* ***common prior*** *μ" is*

$$CP_i(\mu) = \Big\{ (s_i, t_i) \ : \ \mu(S \times \{t_i\} \times T_{-i}) > 0 \text{ and } \varphi_i(\mathcal{T})(t_i) = \varphi_i(\mathcal{T}^\mu)(t_i) \Big\}. \quad [12.12]$$

The prior μ is ***minimal*** *for t_i if t_i is in $CP_i(\mu)$ and, for all $\nu \in \Delta(S \times T)$ with t_i in $CP_i(\nu)$, $\mathrm{supp}\,\nu \not\subset \mathrm{supp}\,\mu$.*

The following examples illustrate Definition 12.14.

Example 12.2. *Definition 12.14 is stated in terms of hierarchies, which are elicitable. This example shows a further benefit of this formulation. Let $T_1 = \{t_1\}$, $T_2 = \{t_2\}$, $T_3 = \{t_3', t_3''\}$, $S_i = \{s_i\}$ for all i, $\beta_1(t_1)(t_2, t_3', s_2, s_3) = 1$, $\beta_2(t_2)(t_1, t_3'', s_1, s_3) = 1$,*

[27] To clarify, to obtain $\mathcal{T}^\mu$ from $\mathcal{T}$ we may eliminate some types and replace the belief maps β_i with maps β_i^μ derived from μ by conditioning.

$\beta_3(t_3)(t_1, t_2, s_1, s_2) = 1$ *for* $t_3 \in T_3$. *Since all players have a single strategy, all types commonly believe the profile* (s_1, s_2, s_3), *so the hierarchies of beliefs over strategies* should *be deemed consistent with a common prior; indeed, this is the case according to our Definition 12.14. Yet this type space does not have a common prior in the standard sense of Definition 12.13 because of the redundancy. Specifically,* t'_3 *and* t''_3 *induce the same hierarchies, even though they are distinct types: Player 1 is sure 3's type is* t'_3 *while 2 is sure 3's type is* t''_3. *Note that, as an alternative to the definition above, another way around this difficulty is to rule out redundant type spaces.*

Example 12.3. *We illustrate two aspects of Definition 12.14: first, the role of minimality, and second, why we allow the type spaces* T_i^{μ} *in the ancillary structure* $\mathcal{T}^{\mu}$ *to be a strict subset of the type spaces* T_i *in the original structure.*

Let $\mathcal{T}$ *be the type structure with* $T_i = \{t_i^a, t_i^b\}$, $S_i = \{s_i^a, s_i^b\}$, $\beta_i(t_i^k)(s_{-i}^k, t_{-i}^k) = 1$. *This type structure is really the combination of two separate structures,* $\mathcal{T}^a$ *and* $\mathcal{T}^b$: *in each structure* $\mathcal{T}^k$, *for* $k = a, b$, *the type spaces are* $T_i^k = \{t_i^k\}$ *for* $i = 1, 2$ *and the profile* s^k *is commonly believed. The structure* $\mathcal{T}$ *is consistent with any common prior* μ *that assigns probability* $\mu_k > 0$ *to* $(s_i^k, t_i^k)_{i \in I}$ *with* $\mu_a + \mu_b = 1$. *However, focusing on minimal common priors treats the two components* $\mathcal{T}^a$ *and* $\mathcal{T}^b$ *distinctly. In particular, the minimal common prior for both types* t_i^a *assigns probability one to* $(s_i^a, t_i^a)_{i \in I}$; *it generates the beliefs in the ancillary type structure* $\mathcal{T}^a$.

Treating $\mathcal{T}^a$ *as distinct from* $\mathcal{T}^b$ *is important to characterize correlated equilibrium. Assume that* s_i^a *is strictly dominant for* $i = 1, 2$. *Consider type structure* $\mathcal{T}$. *Then at* $(s_i^a, t_i^a)_{i=1,2}$ *RCBR holds and by construction beliefs are consistent with the common prior* μ. *However,* $\text{marg}_S \mu$ *assigns positive probability to the strictly dominated strategies* s_i^b, $i = 1, 2$, *and hence, it is not a correlated equilibrium distribution. On the other hand,* $\text{marg}_S \mu^a$ *is a correlated equilibrium distribution.*

Aumann proved the important result that a common prior together with common belief in rationality implies that the distribution over actions is an objective correlated equilibrium distribution. Aumann's framework is different from ours: in his model, player i's "type" incorporates i's strategy, and hence corresponds to a pair (s_i, t_i) in our framework. As a result, the existence of a common prior in Aumann's framework requires that t_i's beliefs be obtained by conditioning on (s_i, t_i), rather than just t_i as in Definition 12.13. This implies that a common prior in the sense of Definition 12.13 need not be a common prior in Aumann's framework. The implications of this distinction are apparent in the following example (due to Brandenburger and Dekel, 1986).

Table 12.1 A two-player game

	L	C	R
T	$7, 0$	$0, 5$	$0, 7$
M	$5, 0$	$2, 2$	$5, 0$
B	$0, 7$	$0, 5$	$7, 0$

Table 12.2 The common prior

	(L, t_2)	(C, t_2)	(C, t_2')	(R, t_2')
(T, t_1)	0	0	$\frac{1}{4}$	0
(M, t_1)	$\frac{1}{4}$	0	0	0
(M, t_1')	0	0	0	$\frac{1}{4}$
(B, t_1')	0	$\frac{1}{4}$	0	0

Table 12.3 Player 1's beliefs over $S_2 \times T_2$

	(L, t_2)	(C, t_2)	(C, t_2')	(R, t_2')
$\beta_1(t_1)$	$\frac{1}{2}$	0	$\frac{1}{2}$	0
$\beta_1(t_1')$	0	$\frac{1}{2}$	0	$\frac{1}{2}$

Table 12.4 Player 2's beliefs over $S_1 \times T_1$

	(T, t_1)	(M, t_1)	(M, t_1')	(B, t_1')
$\beta_2(t_2)$	0	$\frac{1}{2}$	0	$\frac{1}{2}$
$\beta_2(t_2')$	$\frac{1}{2}$	0	$\frac{1}{2}$	0

Example 12.4. *Consider the game in Table 12.1 (due to Bernheim, 1984) and the type structure $\mathcal{T}$ generated by the common prior μ in Table 12.2 as per Definition 12.13 (adapted from Brandenburger and Dekel, 1986).*[28] *A fortiori, each type's belief hierarchy is consistent with the common prior μ in the sense of Definition 12.14; furthermore, RCBR holds for every strategy-type profile, i.e., RCBR $= S \times T$. For instance, (T, t_1) and (M, t_1) are both rational because T and M are best replies to type t_1's first-order belief that 2 plays L and C with equal probability. Yet, the distribution over strategies induced by μ is* not *an objective correlated equilibrium distribution. (The only objective correlated equilibrium places probability one on the profile (M, C).) Thus, a common prior in the sense of Definition 12.13 and RCBR do not characterize objective correlated equilibrium.*

Moreover, the beliefs in the example permit a form of betting between individual players and an outside observer (or dummy player) whose beliefs are given by the prior μ. Consider the bet described in Table 12.5, where the numbers specify the payments from the outside observer to Player 1.

The outside observer computes the value of this bet using the prior μ from Table 12.2; the marginal on S is given in Table 12.6. The observer expects to receive $\frac{1}{2}$ with probability one. The unusual feature of this example (relative to the literature) is that Player 1 is betting on his own actions as well as those of Player 2. Suppose his type is t_1; in this case he is indifferent between

[28] For convenience, Tables 12.3 and 12.4 indicate the beliefs associated with the types of players 1 and 2, respectively, induced by the common prior μ of Table 12.2.

Table 12.5 A bet between player 1 and an outside observer

	L	C	R
T	1	$-\frac{1}{2}$	0
M	$-\frac{1}{2}$	1	$-\frac{1}{2}$
B	0	$-\frac{1}{2}$	1

Table 12.6 The outside observer's beliefs over S

	L	C	R
T	0	$\frac{1}{4}$	0
M	$\frac{1}{4}$	0	$\frac{1}{4}$
B	0	$\frac{1}{4}$	0

T and M in terms of his payoffs in the game. Moreover, by playing either T or M, he expects to get 1 or $-\frac{1}{2}$ with equal probability from the bet. As type t_1 is indifferent between T and M, and strictly prefers both to B in the game and in the bet, the outside observer has no concern that the bet will affect type t_1's incentives. The same analysis applies to type t_1'. Therefore, the observer and both types of Player 1 expect strictly positive payoffs from the bet.

The preceding example may seem puzzling in light of the so-called "no-trade theorems." These results state that, in a setting in which type structures are used to model hierarchical beliefs about exogenous events, rather than strategic uncertainty, the existence of a common prior is equivalent to the absence of mutually agreeable bets.[29] Example 12.4 instead shows that, in an environment in which the events of interest are endogenous—each player chooses his strategy—certain bets are not ruled out by the existence of a common prior in the sense of Definition 12.14.[30] These bets can be ruled out if we impose a further assumption on the common prior—one that is automatically satisfied in Aumann's model, due to the way "types" are defined. This assumption, condition AI of Definition 12.15, states that conditioning the prior on a player's strategy does not imply more information than conditioning only on his type. Clearly, the prior in Table 12.2 violates this: $\mu(\cdot|(t_1, T)) \neq \mu(\cdot|(t_1, M))$. We conjecture that, in the present setting where uncertainty is strategic, a suitable definition of "no betting" that

[29] That a common prior implies no betting follows from Aumann (1976) and the subsequent no-trade literature (Milgrom and Stokey, 1982; Rubinstein and Wolinsky, 1990). The opposite direction requires a more involved statement; see Morris (1994), Bonanno and Nehring (1999), Feinberg (2000), and Samet (1998a). For different characterizations (not in terms of betting), see Samet (1998b) and Heifetz (2006).

[30] As noted previously, another difference with the received literature is that here players do not have beliefs about their own strategies, whereas no-trade theorems consider environments in which every agent's beliefs are defined over the entire state space.

takes into account the fact that players can choose their own strategies, do not have beliefs about them, but can bet on them as well as on opponents' play, can characterize this additional assumption on the common prior.[31]

Definition 12.15. *A prior $\mu \in \Delta(S \times T)$ satisfies* ***Condition AI*** *if, for every $i \in I$, event $E_{-i} \subset S_{-i} \times T_{-i}$, strategies $s_i, s_i' \in S_i$ and type $t_i \in T_i$ with $\mu(\{(s_i, t_i)\} \times S_{-i} \times T_{-i}) > 0$ and $\mu(\{(s_i', t_i)\} \times S_{-i} \times T_{-i}) > 0$,*

$$\mu(E_{-i} \times S_i \times T_i | \{(s_i, t_i)\} \times S_{-i} \times T_{-i}) = \mu(E_{-i} \times S_i \times T_i | \{(s_i', t_i)\} \times S_{-i} \times T_{-i}). \quad [12.13]$$

Roughly speaking, Condition AI requires that the conditional probability $\mu(E_{-i} \times S_i \times T_i | \{(s_i, t_i)\} \times S_{-i} \times T_{-i})$ be independent of s_i. We discuss this condition further in Subsection 12.4.6.1.

We then obtain a version of Aumann's celebrated result: correlated equilibrium is equivalent to RCBR and hierarchies consistent with a common prior that is minimal and satisfies Condition AI.[32]

Theorem 12.4. *Fix a game $G = (I, (S_i, u_i)_{i \in I})$.*

1. *For every type structure $(I, (S_{-i}, T_i, \beta_i)_{i \in I})$, if $(s_i, t_i) \in CP_i(\mu) \cap RCBR_i$ for some $\mu \in \Delta(S \times T)$, and μ is minimal for t_i and satisfies Condition AI, then $\text{marg}_S \mu$ is an objective correlated equilibrium distribution.*
2. *Conversely, for every objective correlated equilibrium distribution ν of G, there are a type structure $(I, (S_{-i}, T_i, \beta_i)_{i \in I})$ and a prior $\mu \in \Delta(S \times T)$ satisfying Condition AI, such that $\text{marg}_S \mu = \nu$ and, for all states $(s, t) \in \text{supp } \mu$, $(s, t) \in CP(\mu) \cap RCBR$.*

12.4.4 Nash equilibrium

We turn now to Nash equilibrium. We start with two players as the epistemic assumptions required are much weaker. In particular, this is the only objective equilibrium assumption for which no cross-player consistency assumptions (such as a common prior or agreement) must be imposed.

For every $i \in I$, let $\phi_i \in \Delta(S_{-i})$ be a conjecture of Player i about her opponents' play. In any type structure $(I, (S_{-i}, T_i, \beta_i)_{i \in I})$, let $[\phi_i] = \{(s_i, t_i) : f_i(t_i) = \phi_i\}$ be the event that Player i's first-order beliefs are given by ϕ_i. Then, the following theorem says that if 1's and 2's first-order beliefs are (ϕ_2, ϕ_1), their first-order beliefs are mutually believed to be (ϕ_2, ϕ_1) and rationality is also mutually believed, then (ϕ_1, ϕ_2) is a Nash

[31] For a different perspective on obtaining objective correlated equilibrium from no-betting conditions, see Nau and McCardle (1990).

[32] Barelli (2009) shows that a characterization of correlated equilibrium in the Aumann (1987) setting can be obtained using a weaker common-prior assumption that only restricts first-order beliefs.

equilibrium. As discussed in the introduction to this section, this result, as well as the subsequent generalizations, provides conditions under which beliefs—not play—form a Nash equilibrium.

Theorem 12.5. *Assume that $I = 2$. If $[\phi] \cap B(R \cap [\phi]) \neq \emptyset$, then (ϕ_2, ϕ_1) is a Nash equilibrium.*[33]

We can obtain a straightforward extension of Theorem 12.5 to n-player games by explicitly adding the assumptions that players' beliefs over opponents' strategies are independent and that any two players have the same beliefs over any common opponent's strategies. In particular, define the events "i has independent first-order beliefs" and "i's opponents agree":

$$Ind_i = \left\{(s_i, t_i) : f_i(t_i) = \Pi_{j \neq i} \text{marg}_{S_j} f_i(t_i)\right\},$$

$$Agree_{-i} = \left\{(s_{-i}, t_{-i}) : \forall j, k, \ell \in I \text{ s.t. } j \neq i, k \neq i, j \neq \ell, k \neq \ell, \text{marg}_{S_\ell} f_j(t_j) = \text{marg}_{S_\ell} f_k(t_k)\right\}.$$

It is worth emphasizing that $Agree_{-i}$, like the common prior, is a restriction that relates different players' beliefs, in contrast to all the other assumptions throughout this paper.

Theorem 12.6. *If $[\phi] \cap \bigcap_{i \in I} B_i(R_{-i} \cap [\phi]_{-i} \cap Ind_{-i} \cap Agree_{-i}) \neq \emptyset$, then there exist $\sigma_j \in \Delta(S_j)$ for all j such that σ is a Nash equilibrium and $\phi_j = \prod_{k \neq j} \sigma_j$ for all j.*

Aumann and Brandenburger (1995) show that these additional conditions can be derived from arguably more primitive assumptions: the common prior and common belief in the conjectures. For the reasons discussed in the preceding section, we need to add Condition AI.[34]

[33] Aumann and Brandenburger (1995) require only mutual belief in rationality and in the conjectures. This is because, in their framework, players have beliefs about their own strategies and hierarchies, and furthermore, these beliefs are correct. Thus, mutual belief in the conjectures ϕ implies that i's conjecture is ϕ_i. As we do not model a player's introspective beliefs (here beliefs about her own beliefs), we need to explicitly assume that the conjectures are indeed ϕ.

Alternatively, in Theorems 12.5, 12.6, 12.7, and 12.9, we could drop the event $[\phi]$ and replace "B" with "B^2." We could also state these results using only assumptions on one player's beliefs, as in Theorem 12.4. For example, in Theorem 12.5 and for $i = 1$, the assumptions would be $B_1([\phi_2]) \cap B_1(B_2(R_1)) \cap B_1(B_2([\phi_1]))$ and $B_1(B_2([\phi_1])) \cap B_1(B_2(B_1(R_2))) \cap B_1(B_2(B_1([\phi_2])))$.

Finally, Aumann and Brandenburger allow for incomplete information in the sense of Section 12.6, but assume that there is common belief in the game being played. Liu (2010) shows that this assumption can be weakened to second-order mutual belief in the game, but not to mutual belief.

[34] To see why assumption AI is necessary, consider the three-player game in Figure 5 of Aumann and Brandenburger (1995). Let $T_i = \{t_i\}$ for $i = 1, 2, 3$, and define $\mu \in \Delta(S \times T)$ by $\mu(H, t_1, h, t_2, W, t_3) = \mu(T, t_1, t, t_2, W, t_3) = 0.4$, $\mu(H, t_1, t, t_2, W, t_3) = \mu(T, t_1, h, t_2, W, t_3) = 0.1$. Define β_1 and β_2 via μ, as in Example 12.4. The first-order beliefs of types t_1 and t_2 place equal probability on H, T and h, t, respectively; therefore, $R_i = S_i \times T_i = S_i \times \{t_i\}$ for $i = 1, 2$. Furthermore, player 3 assigns a high

Theorem 12.7. *If there is a probability $\mu \in \Delta(S \times T)$ that satisfies Condition AI, and a tuple $(t_1, \ldots, t_I)$ in $CP(\mu) \cap [\phi] \cap CB([\phi]) \cap B(R)$ for which μ is minimal, then there exist $\sigma_j \in \Delta(S_j)$ for all j such that σ is a Nash equilibrium and $\phi_j = \prod_{k \neq j} \sigma_j$ for all j.*

12.4.5 The book-of-play assumption

We now consider a related approach to dealing with the issues pointed out in Example 12.4 and footnote 34. We introduce a "book of play": a commonly believed function from hierarchies into strategies (Brandenburger and Dekel, 1986). The interpretation is that, once we condition on a player's hierarchical beliefs, there is no residual uncertainty about her play. The existence of such a function reflects a (perhaps naive) determinism perspective—a player's hierarchical beliefs uniquely determine his strategy—and hence may be of interest in its own right.

It turns out that common belief in a "book of play" implies that Condition AI in Definition 12.15 holds. Therefore, we can obtain sufficient epistemic conditions for objective correlated and Nash equilibrium by replacing Condition AI with common belief in a "book of play." We do so in Theorems 12.8 and 12.9. The advantage relative to Theorems 12.4 and 12.7 is that common belief in a "book of play" is a more easily interpretable assumption. However, we will see in Example 12.5 later that, in the absence of any exogenous uncertainty, this assumption is restrictive: essentially, it rules out certain forms of randomization.[35]

Consider a game $G = (I, (S_i, u_i)_{i \in I})$, a type structure $\mathcal{T} = (I, (S_{-i}, T_i, \beta_i)_{i \in I})$, and a function

$$n_i : \varphi_i(\mathcal{T})(T_i) \to S_i; \qquad [12.14]$$

this specifies, for each type of Player i, the strategy she is "expected" to play, where any given hierarchy is associated with a unique (pure) strategy.[36] We then define the event that "i's play adheres to the book n_i":

$$[n_i] = \left\{(s_i, t_i) \ : \ s_i = n_i(\varphi_i(\mathcal{T})(t_i))\right\}. \qquad [12.15]$$

probability to players 1 and 2 playing either (H, h) or (T, t), so that $R_3 = \{(W, t_3)\}$. Thus, there is common belief in rationality and the first-order beliefs, as well as a common prior in the sense of Definition 12.14. However, player 3 has a correlated first-order belief, so we do not get a Nash equilibrium. Furthermore, players 1 and 3 could bet on the correlation between 1's and 2's strategies, so once again the common prior does not preclude bets.

[35] One can also explore the implications of this assumption in nonequilibrium contexts. Under RCBR, Brandenburger and Friedenberg (2008) consider weaker conditions that enable them to study the notion of "intrinsic" correlation in games with more than 2 players, which corresponds to it being common belief that there are no *exogenous* unmodeled correlating devices. Peysakhovich (2011) shows that objective correlated equilibrium outcomes are also consistent with RCBR and intrinsic correlation. The converse is false, as Example 12.4 shows.

[36] Here, as is the case throughout this chapter with the exception of Theorem 12.10, players do not have access to randomizing devices. Rather, randomizations reflect opponents' beliefs, as discussed in Section 12.2.2.

Theorem 12.8. *Fix a game* $G = (I, (S_i, u_i)_{i\in I})$. *For every type structure* $(I, (S_{-i}, T_i, \beta_i)_{i\in I})$ *and book of play* n, *if* $(s_i, t_i) \in CP_i(\mu) \cap RCBR_i \cap CB([n])_i$ *for some* $\mu \in \Delta(S \times T)$, *and* μ *is minimal for* t_i, *then* $\text{marg}_S\mu$ *is an objective correlated equilibrium distribution.*

Theorem 12.9. *Fix a type structure* $(I, (S_{-i}, T_i, \beta_i)_{i\in I})$. *If there are a book of play* n, *a profile of conjectures* ϕ, *and a probability* $\mu \in \Delta(S \times T)$ *such that* $[\phi] \cap B(R) \cap CP(\mu) \cap CB([\phi]) \cap CB([n]) \neq \emptyset$, *then there exist* $\sigma_j \in \Delta(S_j)$ *for all* j *such that the profile* σ *is a Nash equilibrium and* $\phi_j = \prod_{k\neq j} \sigma_j$ *for all* j.

Notice that we did not state a converse to the preceding theorems. In fact, as the following example due to Du (2011) shows, the converses are false.

Example 12.5. *Consider Matching Pennies. The unique correlated and Nash equilibrium is of course* $\sigma_1 = \sigma_2 =$ *"*$\frac{1}{2}$ *Heads,* $\frac{1}{2}$ *Tails."*

First, consider the converse to Theorem 12.9. Fix a type t_i. *If there is common belief that the conjectures are* (σ_2, σ_1), *then every type* t_j *to which* t_i *assigns positive probability must have the same belief hierarchy, and hence, by the book-of-play assumption, must play the same strategy, either Heads* or *Tails. But then type* t_i*'s first-order belief cannot be* σ_{-i}.

Next, consider the converse to Theorem 12.8. Fix a type structure, a type t_1, *and a common prior* μ *minimal for* t_1 *and such that its marginal on S is the equilibrium distribution. Because common belief in the book of play holds (in the eyes of* t_1*), we can partition the types of each player i in the support of* μ *into those that play Heads and those that play Tails, say* T_i^H *and* T_i^T. *Consider a type* $t_1' \in T_1^H$; *assuming wlog that he wants to match, (common belief in) rationality requires that* t_1' *assigns probability at least* $\frac{1}{2}$ *to Heads. Repeating the argument for all types in* T_1^H *implies that the common prior must assign probability at least* $\frac{1}{2}$ *to Heads conditional on 1's type being in* T_1^H. *This is equivalent to saying that, conditional on 1 playing Heads, 2 must play Heads with probability at least* $\frac{1}{2}$. *But by assumption, conditional on 1 playing Heads, 2 plays heads with probability exactly* $\frac{1}{2}$. *This implies that all types in* T_1^H *have the same first-order beliefs, i.e.,* σ_2. *Repeating the argument shows that all types of i in the support of* μ *have the same first-order beliefs, for* $i = 1, 2$. *Hence, all types of each player i have the same hierarchy of beliefs, and by common belief in the book of play, they must actually be playing the same strategy.*

The essence of the example is that the book-of-play assumption makes it impossible, in certain games, to attribute mixed strategies to beliefs and not to actual mixing. Thus, this important perspective, highlighted by Harsanyi and Aumann—indeed one of the benefits of the epistemic perspective—is not possible under the book-of-play assumption. Indeed, one way to obtain a converse to Theorems 12.8 and 12.9 is to either explicitly allow for mixing or add extrinsic uncertainty (so as to "purify" the mixing).

Once we allow for mixed strategies to be played, first-order beliefs are over mixed strategies; that is, for player j, they are measures $\phi_j \in \Delta(\prod_{k\neq j}\Delta(S_k))$. It is then useful to have a notation for the "expected belief" over pure strategies; given a first-order belief ϕ_j, let $\mathbb{E}\phi_j \in \Delta(S_{-j})$ be defined by $\mathbb{E}\phi_j(s_{-j}) = \int_{\prod_{k\neq j}\Delta(S_k)} \bar{\sigma}_{-j}(s_{-j})\,\phi_j(d\bar{\sigma}_{-j})$ for all s_{-j}. We can then state the following converse to Theorem 12.9: given a Nash equilibrium, there is a type structure where hierarchies of beliefs are consistent with a common prior, and there is common belief in rationality, the book of play, and first-order beliefs whose expectations are the equilibrium strategy profile.

Theorem 12.10. *For any Nash equilibrium* $(\sigma_i)_{i\in I}$, $\sigma_i \in \Delta(S_i)$, *there is a profile* $\phi = (\phi_i)_{i\in I}$ *of first-order beliefs, with* $\phi_i \in \Delta\left(\prod_{j\neq i}\Delta(S_j)\right)$ *and* $\mathbb{E}\phi_i = \prod_{j\neq i}\sigma_j$ *for every i, and a type structure* $(I, (\prod_{j\neq i}\Delta(S_j), T_i, \beta_i)_{i\in I})$, *such that, for all i,* $T_i = CP_i(\mu) \cap CB(R)_i \cap CB([n]) \cap CB([\phi])$.

12.4.6 Discussion

12.4.6.1 Condition AI

In Theorem 12.4, we assume that a player's hierarchy is consistent with a common prior μ which satisfies Condition AI. This is an elicitable assumption because it is about beliefs, but it is arguably somewhat opaque. As noted earlier, we conjecture that common priors satisfying Condition AI may be characterized via a suitable no-betting condition. This would provide a more transparent behavioral characterization.

We emphasize that we cannot interpret Condition AI, i.e., [12.13], directly as a restriction on Player *i*'s beliefs: in our environment, players do not have beliefs about their own strategies. Instead, it is a restriction on the beliefs of the other players, and perhaps those of an outside observer whose beliefs are given by μ. In Example 12.4, it implies in particular that, conditional on t_1, an outside observer must believe that Player 1's and 2's strategies are stochastically independent. Hence, the name of the condition: *A*umann *I*ndependence.[37]

We observed that Condition AI is implied by the common-prior assumption in Aumann's model, due to the fact that a "type" therein comprises both a belief about the other players and a strategy.[38] Our framework instead requires that we make

[37] Aumann deserves no blame for this definition; the label only indicates that it is inspired by his analysis.

[38] By the common-prior assumption, the beliefs of an Aumann type for player i about the other players' Aumann types is derived from a common prior μ by conditioning on *i*'s Aumann type, and hence by definition on both that type's beliefs *and* strategy. Therefore, if we condition μ on two Aumann types that feature the same beliefs about the other players, but different strategies, the two resulting measures must obviously have the same marginal on the set of other players' Aumann types. This corresponds to [12.13].

this independence assumption explicit and, hence, helps us highlight a key epistemic condition required to characterize objective correlated and Nash equilibrium.

Finally, at the risk of delving too far into philosophy, one can also relate Condition AI to the notion of free will. If there is "free will" then players cannot learn anything from their own choice of strategy. If instead players' choices are predetermined, then it is possible they would learn something from their own choices, unless one explicitly assumes that there is enough independence to rule out any such learning. Condition AI requires that there be no such learning, either because of "free will" or because there is sufficient independence so that there is nothing to learn.

12.4.6.2 Comparison with Aumann (1987)

Our characterization of objective correlated equilibrium in Theorem 12.4 seems more complex than Aumann's original result. Translated to our setting, his result is as follows:

> *Let $\mathcal{T}$ be a type structure that admits a common prior μ (in the sense of Definition 12.13) which satisfies Condition AI. If* supp $\mu \subset R$, *then* $\text{marg}_S \mu$ *is an objective correlated equilibrium distribution.*

While elegant, this formulation involves hypotheses that are not directly verifiable. To begin with, we cannot verify whether "the type structure admits a common prior" because, as we have noted several times, we cannot elicit the type structure that generated the players' actual belief hierarchies. This is why we were led to Definition 12.14 rather than 12.13. Once we have elicited a player's hierarchy, we *can* verify whether that hierarchy is consistent with a common prior. Our Theorem 12.4 translates Aumann's result above into the language of hierarchies. Moreover, Aumann's hypothesis of rationality for every strategy-type pair in the support of the prior implies, but does not explicitly state, that RCBR will also hold. Our focus on hierarchies forces us to make this latter assumption explicit.

12.4.6.3 Nash equilibrium

As discussed, the epistemic analysis leads to the interpretation of Nash equilibria in mixed strategies as descriptions of players' conjectures, rather than their actual behavior. The definition of Nash equilibrium then becomes a mutual consistency requirement: conjectures must be mutually believed and consistent with rationality. In games with more than two players, they must also be independent and suitably consistent across players. Theorems 12.5 and 12.6 formalize the Nash consistency requirement in the language of belief hierarchies.

A separate question is whether these results provide a "justification" for equilibrium concepts. In games with more than two players, the assumptions of independence and agreement (corresponding to the events Ind_i and $Agree_{-i}$ used in Theorem 12.6) appear strong; understandably, the literature has sought more basic conditions. Theorems 12.7-12.10 clarify the need for Condition AI or common belief in the "book of play."

Ultimately, we interpret Theorems 12.6-12.10 as negative: they highlight the demanding epistemic assumptions needed to obtain equilibrium concepts.[39] Naturally, there is room for other types of justification for equilibrium analysis, such as learning and evolution.

12.5. STRATEGIC-FORM REFINEMENTS

In this section, we provide a first introduction to the epistemic analysis of refinements of rationalizability. These are important for two reasons. First, refinements yield tighter predictions, and hence can be useful in applications. Second, the epistemic conditions that yield them turn out to be of interest. In particular, this section introduces the notions of admissibility/weak dominance and common p-belief.

Weak dominance is a strengthening of Bayesian rationality (expected-utility maximization). A strategy $s_i \in S_i$ is **weakly dominated** if there exists a distribution $\sigma_i \in \Delta(S_i)$ such that, for all $s_{-i} \in S_{-i}$, $u_i(\sigma_i, s_{-i}) \geq u_i(s_i, s_{-i})$, and the inequality is strict for at least one $s^*_{-i} \in S_{-i}$. A strategy is **admissible** if it is not weakly dominated. Analogously to strict dominance, a strategy is weakly dominated if and only if it is not a best reply to any full-support belief about the opponents' play (that is, a belief that assigns strictly positive probability to every opponents' strategy profile).

A natural first step to refine the assumption of RCBR might be to consider "admissibility and common belief in admissibility" and try to obtain an analog of Theorem 12.1 in the preceding text. However, there is a tension between the logic of admissibility and that of common belief. Loosely speaking, the former is motivated from the perspective that anything is possible, whereas the latter does restrict what is possible.[40] To address this tension, one can relax the definition of belief.

Monderer and Samet (1989) introduce the notion of p-belief, i.e., belief with probability at least p, to game theory.[41] For $p = 1$, this is the notion of belief we have considered so far. For p close to 1, p-belief has similar behavioral implications, but it enables us to resolve the tension just discussed. It makes it possible to formulate "almost" common belief in admissibility, which *is* consistent with full-support beliefs (anything is possible).

As we just noted, one motivation for our discussion of p-belief is the observation that, while admissibility is an interesting and common strengthening of rationality (expected-payoff maximization), common belief in admissibility leads to difficulties. Common p-belief in admissibility may be viewed as one way to approximate these

[39] The examples here and in Aumann and Brandenburger (1995) demonstrate the extent to which the theorems are tight and these demanding assumptions are needed. Barelli (2009) indicates one way in which the common-prior assumption can be weakened.

[40] See, e.g., Samuelson (1992).

[41] This notion originates in modal logic: see, e.g., Fagin and Halpern (1994), Fagin et al. (1990), and the references therein.

epistemic conditions of interest. However, there are two additional reasons. First, as we shall see momentarily, for p sufficiently high, common p-belief in admissibility characterizes a solution concept that was originally motivated by different considerations. Thus, Theorem 12.11 below provides a different perspective on the derived solution concept. Second, we can employ the notion of p-belief to carry out a robustness check for our analysis of RCBR. It can be shown that, for p sufficiently close to 1, rationality and common p-belief of rationality have the same behavioral implications as RCBR (in stark contrast with admissibility and common p-belief thereof): see Hu (2007).

Definition 12.16. *Fix a game* $G = (I, (S_i, u_i)_{i \in I})$ *and a type structure* $\mathcal{T} = (I, (S_{-i}, T_i, \beta_i)_{i \in I})$ *for* G. *The event that Player* i assigns probability at least $p \in [0, 1]$ to $E_{-i} \subset S_{-i} \times T_{-i}$ *is*

$$B_i^p(E_{-i}) = \left\{(s_i, t_i) \; : \; \beta_i(t_i)(E_{-i}) \geq p\right\}. \quad [12.16]$$

The event that Player i has full-support beliefs *is*

$$FS_i = \left\{(s_i, t_i) \; : \; \operatorname{supp} f_i(t_i) = S_{-i}\right\}. \quad [12.17]$$

We can now define "admissibility" and "mutual and common p-belief of admissibility" as follows. We proceed analogously to the definition of the event $RCBR_i$ in [12.8] and [12.9]. We assume full-support belief in addition to rationality, and weaken belief to p-belief. (As noted, rationality and full support are equivalent to admissibility). For every $i \in I$, let

$$\begin{aligned} ACBA_i^{p,0} &= R_i \cap FS_i, \\ ACBA_i^{p,k} &= ACBA_i^{p,k-1} \cap B_i^p(ACBA_{-i}^{p,k-1}) \text{ and} \\ ACBA_i^p &= \bigcap_{k \geq 0} ACBA_i^{p,k}. \end{aligned} \quad [12.18]$$

Here it *does* matter whether we define mutual and common p-belief as above, or by iterating the p-belief operator as in [12.7].[42] The reason is that p-belief does not satisfy the Conjunction property in [12.5]. On the other hand, given any *finite* type structure, there is a $\pi \in (0, 1)$ such that, for all $p \geq \pi$, Conjunction holds, i.e., player i p-believes events A_{-i} and B_{-i} if and only if she p-believes $A_{-i} \cap B_{-i}$. A similar statement holds for Monotonicity.

[42] Indeed, $B_1^p(R_2 \times S_3 \times T_3) \cap B_1^p(S_2 \times T_2 \times R_3) \neq B_1^p(R_2 \times R_3)$ in general, whereas equality does hold for $p = 1$.

To capture the behavioral implications of ACBA, consider the following adaptation of the notion of best-reply sets (Definition 12.9).[43]

Definition 12.17. *Fix $p \in [0,1]$. A set $B = \prod_{i \in I} B_i \subset S$ is a **p-best-reply set** (or p-BRS) if, for every player $i \in I$, every $s_i \in B_i$ is a best reply to a full-support belief $\sigma_{-i} \in \Delta(S_{-i})$ with $\sigma_{-i}(B_{-i}) \geq p$; it is a **full p-BRS** if, for every $s_i \in B_i$, there is a justifying full-support belief $\sigma_{-i} \in \Delta(B_{-i})$ with $\sigma_{-i}(B_{-i}) \geq p$ such that all best replies to σ_{-i} are also in B_i.*

A basic refinement of rationalizability is to carry out one round of elimination of weakly dominated strategies, followed by the iterated deletion of strictly dominated strategies. This procedure was introduced in Dekel and Fudenberg (1990), who—following Fudenberg et al. (1988)—were motivated by robustness considerations.[44] Let $S^\infty W$ denotes the set of strategy profiles that survive this procedure.

For every game, there exists $\pi \in (0,1)$ such that every p-BRS with $p \geq \pi$ is contained in $S^\infty W$. Furthermore, $S^\infty W$ is itself a p-BRS, for $p \geq \pi$. This inclusion is a consequence of the fact that, as discussed earlier, p-belief satisfies Conjunction and Monotonicity for p large enough.

Theorem 12.11. *Fix a game $G = (I, (S_i, u_i)_{i \in I})$. Then, there is $\pi \in (0,1)$ such that, for $p \geq \pi$:*[45]

1. *In any type structure $(I, (S_{-i}, T_i, \beta_i)_{i \in I})$, $\text{proj}_S ACBA^p$ is a full p-BRS contained in $S^\infty W$;*
2. *In any complete type structure, $(I, (S_{-i}, T_i, \beta_i)_{i \in I})$, $\text{proj}_S ACBA^p = S^\infty W$;*
3. *For every full p-BRS B, there exists a finite type structure $(I, (S_{-i}, T_i, \beta_i)_{i \in I})$ such that $\text{proj}_S ACBA^p = B$.*

Note that π depends upon the game G. Consequently, the epistemic conditions that deliver $S^\infty W$ depend upon the game. We view this as unappealing, because to some extent the assumptions are tailored to the game. We will shortly mention an alternative approach that avoids this issue. However, this approach requires a different notion of type structure. The advantage of Theorem 12.11 is that it can be stated using the machinery developed so far.

[43] For related notions, see Tercieux (2006) and Asheim et al. (2009).

[44] Like Bernheim's perfect rationalizability (Bernheim, 1984), this procedure is a nonequilibrium analog to trembling-hand perfection (Selten, 1975). Borgers (1994) (see also Hu, 2007) provided a characterization using common p-belief. The main difference with perfect rationalizability is that in $S^\infty W$ it is not assumed that players agree about the trembles of other players, and trembles are not required to be independent. For refinements of rationalizability motivated by proper equilibrium (Myerson, 1978), see Pearce (1984) Section 3, Schuhmacher (1999), and Asheim (2002).

[45] If $p < \pi$, then these results are modified as follows. First, (1) holds except for the claim that the p-BRS is contained in $S^\infty W$. Regarding (2), ACBA characterizes the largest p-BRS, which can be computed using the procedure in Borgers (1994). Finally, (3) holds for all p.

An alternative way to characterize $S^\infty W$ builds on Schuhmacher (1999). Instead of weakening belief to p-belief, we weaken rationality to "ϵ-rationality." This is easiest to implement in a model where players can explicitly randomize, as in Theorem 12.10.[46] Consider the mixed extension $(I, (\Delta(S_i), u_i)_{i\in I})$ of the original game $(I, (S_i, u_i)_{i\in I})$, and a type structure $\left(I, \left(\prod_{j\neq i} \Delta(S_j), T_i, \beta_i\right)_{i\in I}\right)$. For given $\epsilon > 0$, a (mixed) strategy-type pair (σ_i, t_i) is **ϵ-rational** if σ_i assigns probability at most ϵ to any pure strategy that is *not* a best reply to type t_i's first-order beliefs;[47] we thus define the event

$$R_i^\epsilon = \left\{(\sigma_i, t_i) \ : \ u_i(s_i, \mathbb{E}f_i(t_i)) < \max_{s_i'\in S_i} u_i(s_i', \mathbb{E}f_i(t_i)) \Rightarrow \sigma_i(s_i) \leq \epsilon. \right\},$$

where, as in Theorem 12.10, $\mathbb{E}f_i(t_i)$ is the reduction of the measure $f_i(t_i)$.

As in Theorem 12.11, we need a full-support assumption in order to obtain admissibility. Given that we consider the mixed extension of the game, as in the discussion preceding Theorem 12.10, player i's first-order beliefs are now a probability measure over profiles of *mixed* strategies of the opponents. The appropriate full-support assumption remains over *pure* strategy profiles. We formalize this using "expected" first-order beliefs: e.g., for type t_i, $\mathbb{E}f_i(t_i) \in \Delta(S_{-i})$ has full support. The event where this is the case is

$$\widehat{FS}_i = \left\{(\sigma_i, t_i) \ : \ [\mathbb{E}f_i(t_i)](s_{-i}) > 0 \ \forall s_{-i} \in S_{-i}\right\}.$$

Theorem 12.12. *Fix a game $G = (I, (S_i, u_i)_{i\in I})$. Then, there is $\bar{\epsilon} \in (0, 1)$ such that, for $\epsilon \leq \bar{\epsilon}$,*

1. *In any complete type structure $(I, (\prod_{j\neq i} \Delta(S_j), T_i, \beta_i)_{i\in I})$, $s \in S^\infty W$ if and only if there is $(\sigma_i, t_i)_{i\in I} \in R^\epsilon \cap \widehat{FS} \cap CB(R^\epsilon \cap \widehat{FS})$ such that $\sigma_i(s_i) > \epsilon$ for each i.*
2. *In any type structure $(I, (\prod_{j\neq i} \Delta(S_j), T_i, \beta_i)_{i\in I})$, if $(\sigma_i, t_i)_{i\in I} \in R^\epsilon \cap \widehat{FS} \cap CB(R^\epsilon \cap \widehat{FS})$ and $\sigma_i(s_i) > \epsilon$ for each i then $s \in S^\infty W$.*

Schuhmacher (1999) uses this approach to define a counterpart to Myerson (1978)'s notion of proper equilibrium. He strengthens ϵ-rationality to "ϵ-properness": σ_i must be completely mixed and, if a strategy s_i is worse than another strategy s_i' given player i's first-order beliefs, then $\sigma_i(s_i) \leq \epsilon\sigma_i(s_i')$, where σ_i is i's mixed strategy.[48]

[46] This characterization (Theorem 12.12) could also be stated without the mixed-strategy extension, but as a result concerning beliefs and not play (see our discussion in the third paragraph of Section 12.4.1).

[47] Note that this is not the same as saying that the strategy obtains within ϵ of the maximal payoff; this is also often called "ϵ-rationality." The definition in the text is in the spirit of Selten (1975) and Myerson (1978), as well as Schuhmacher (1999).

[48] Without the requirement that σ_i be completely mixed, ϵ-properness may lose its bite: if, given i's beliefs, strategy s_i is strictly better than s_i', which in turn is strictly better than s_i'', then a mixed strategy that assigns

A strategy σ_i is then deemed "ϵ-properly rationalizable" if there is a type structure and a type t_i such that the pair (s_i, t_i) is consistent with ϵ-properness and common belief thereof. Notice that this definition is epistemic; Schuhmacher (1999) provides an algorithmic procedure that yields some, but not all properly rationalizable strategies; Perea (2011a) provides a full algorithmic characterization. Finally, Asheim (2002) provides an epistemic definition of proper rationalizability using lexicographic probability systems.

At this point, it would be natural to investigate epistemic conditions leading to iterated admissibility. These would require strengthening common p-belief. It turns out that it is possible to do so by replacing probabilistic beliefs with lexicographic probability systems. This in turn necessitates modifying the notion of a type structure. (This is the different type structure alluded to above in which an alternative version of Theorem 12.11 can be given: see Brandenburger, 1992.) It is convenient to present this material after studying extensive-form refinements: see Section 12.8.

12.6. INCOMPLETE INFORMATION

12.6.1 Introduction

A game has *incomplete information* if the payoff to one or more players is not fully determined by the strategy profile; we, therefore, allow for a parameter $\theta \in \Theta$ that enters players' payoff functions.[49] In this section, we provide an epistemic analysis of such games, focusing mainly on RCBR.

The key issue that arises is to what extent the model is meant to be "complete," that is, to describe all possible aspects of the world that might be relevant to the agents. The alternative (to a complete model) is to adopt a "small worlds" perspective, where we understand that many aspects are not included in our specification. This is a general issue in modeling, that is particularly relevant in this chapter, and that is especially critical in this section. The particular aspect of concern is whether there might be additional uncertainty and information beyond what the model describes. Such information could enable correlations that we might otherwise exclude.

One way to deal with this is to adopt the small-worlds approach and study solution concepts that are "robust" to adding such unmodeled uncertainty explicitly. The other is to insist on the model being complete. We consider both in this section.

probability one to s_i would formally be ϵ-proper, but one could not say that s_i' is "much more likely" than s_i''.

[49] To economize on notation, this formulation does not allow for hard private information (signals) the players may receive. To accommodate private information, one can let $\Theta = \Theta_0 \times \prod_{i \in I} \Theta_i$, where Θ_i is the set of signals that i may receive and Θ_0 represents residual uncertainty. For example, see Battigalli et al. (2011a).

To clarify this issue we consider two distinct solution concepts that embody different degrees of correlation: interim independent and correlated rationalizability (denoted IIR and ICR, respectively). Consider the game of incomplete information in Table 12.7.

The players' hierarchy of beliefs over $\Theta = \{\theta_1, \theta_2\}$ corresponds to it being common belief that the two parameters are equally likely. (Neither player receives any hard information.) These hierarchies can be modeled using two distinct type structures based on Θ (i.e., $X_{-i} = \Theta$), denoted $\mathcal{T}^N = \left\{I, (\Theta, T_i^N, \beta_i^N)_{i\in I}\right\}$ and $\mathcal{T}^R = \left\{I, (\Theta, T_i^R, \beta_i^R)_{i\in I}\right\}$. For both structures, player 1 has a single type: $T_1^N = \{t_1^N\}$ and $T_1^R = \{t_1^R\}$. However, $T_2^N = \left\{t_2^N\right\}$, whereas $T_2^R = \left\{t_2^R, \bar{t}_2^R\right\}$. The belief maps β_i^N and β_i^R are described in Table 12.8. Notice that these type structures describe beliefs about Θ alone, and not also about players' strategies (that is, $X_{-i} = \Theta$). Solution concepts for incomplete-information games typically use such Θ-based type structures. When we turn to the epistemic analysis, we will need to consider type structures that model beliefs about both Θ and players' strategies (i.e., $X_{-i} = \Theta \times S_{-i}$).

Structure $\mathcal{T}^R$ has redundant types (cf. Definition 12.5), since the hierarchies of beliefs of types t_2^R and $\bar{t}_2^R$ are the same. Furthermore, there is no hard private information in this example: types are used solely to model hierarchies of beliefs.[50] In this sense, t_2^R and $\bar{t}_2^R$ are indistinguishable.

We can (iteratively) delete strategies that are dominated (i.e., non-best replies) for a given type. We now present some intuitive arguments about two different deletion procedures; these are formally defined in Sections 12.6.2 and 12.6.3. First consider structure $\mathcal{T}^R$. One might argue that D is not dominated for t_1^R: it is a best reply to the belief that t_2^R plays R and $\bar{t}_2^R$ plays L. This conclusion crucially depends on the fact that type t_1^R believes that 2's type is t_2^R when $\theta = \theta_1$ and $\bar{t}_2^R$ when $\theta = \theta_2$. This induces a correlation between the strategy that t_1^R expects 2 to play and the payoff parameter θ; this correlation is essential for D to be a best reply.

For structure $\mathcal{T}^N$ the analysis is more subtle. One could argue that D is dominated for player 1's sole type t_1^N since, for any belief over S_2 *independently* combined with the belief that θ_1 and θ_2 are equally likely, D is not a best reply. This is the perspective underlying the solution concept of IIR, which we analyze in Subsection 12.6.3. Alternatively, one could argue that D is not dominated: it is a best reply to the belief that, with

Table 12.7 An incomplete-information game

θ_1	L	R
U	1,1	0,0
D	$\frac{1}{4}$,0	$\frac{1}{4}$,0

θ_2	L	R
U	0,0	1,1
D	$\frac{1}{4}$,0	$\frac{1}{4}$,0

[50] If there was private information, we would model it explicitly: see footnote 49.

Table 12.8 A nonredundant (Top) and a redundant (Bottom) type structure for the game in Table 12.7

	(θ_1, t_2^N)	(θ_2, t_2^N)
$\beta_1^N(t_1^N)$	$\frac{1}{2}$	$\frac{1}{2}$

	(θ_1, t_1^N)	(θ_2, t_1^N)
$\beta_2^N(t_2^N)$	$\frac{1}{2}$	$\frac{1}{2}$

	(θ_1, t_2^R)	(θ_1, t_2^R)	$(\theta_2, \bar{t}_2^R)$	$(\theta_2, \bar{t}_2^R)$
$\beta_1^R(t_1^R)$	$\frac{1}{2}$	0	0	$\frac{1}{2}$

	(θ_1, t_1^R)	(θ_2, t_1^R)
$\beta_2^R(t_2^R)$	$\frac{1}{2}$	$\frac{1}{2}$
$\beta_2^R(\bar{t}_2^R)$	$\frac{1}{2}$	$\frac{1}{2}$

probability one-half, player 2's sole type t_2^N plays R and the state is θ_1, and otherwise, t_2^N plays L and the state is θ_2. This corresponds to ICR (Subsection 12.6.2). Is this latter belief "reasonable"? Certainly yes if there is unmodeled uncertainty: player 1 can believe that player 2's actions are correlated with θ through some unmodeled payoff-irrelevant signal.[51] If there is no unmodeled uncertainty however, one might want to exclude such beliefs.

To do so, we introduce an explicit independence assumption into the epistemic model. Intuitively, in the absence of unmodeled hard information, we want to rule out the possibility of "excessive" correlation between the payoff-relevant parameter θ and 2's strategy. However, what is "excessive" needs to be defined with care. It certainly seems reasonable to allow player 1 to believe that 2 plays differently depending on 2's hierarchical beliefs about Θ. However, *conditional on 2's hierarchy of beliefs over* Θ, 1's beliefs about θ and 2's strategies should be independent. Thus, by definition—since types that have the same hierarchy must be treated the same—an epistemic analysis that adopts this independence assumption will not result in different solutions for the two type structures.[52]

In the next two subsections, we develop this formally. First, we show ICR characterizes RCBR. Then we show that IIR corresponds to RCBR plus common belief of a suitable independence assumption *if the type space is non-redundant* (and IIR is a coarser solution concept in general). We then briefly discuss Δ-rationalizability (Section 12.3.3) for incomplete-information games. In the last subsection, we briefly discuss equilibrium concepts. Little has been done here in terms of using the ideas of Section 12.4 above under incomplete information to characterize standard equilibrium concepts,

[51] Indeed, one view of ICR is that it is the same as IIR when certain types of unmodeled correlation are explicitly added. For related ideas see Liu (2009), who discusses this idea in the context of Bayesian Nash equilibrium. See also Sadzik (2011) and Bergemann and Morris (2011).

[52] This formulation of independence is due to Battigalli et al. (2011a). These authors emphasize that epistemic analysis should be carried out solely in terms of *expressible* assumptions about the primitives of the model. For incomplete-information games, the primitives are the payoff states θ and the strategy sets; in our probabilistic setting, the only expressible assumptions are those about each player i's hierarchies of beliefs on $\Theta \times S_{-i}$. We agree with these authors' emphasis on expressibility. In fact, as argued in the Introduction, we take the stronger stand that assumptions should be elicitable.

and in particular the many different versions of correlated equilibrium concepts in the literature.

12.6.2 Interim correlated rationalizability

We begin by formally defining incomplete-information games. We then define ICR and conclude this subsection by relating this solution concept to RCBR.

Definition 12.18. *A (finite)* ***(strategic form) incomplete-information game*** *is a tuple* $G = (I, \Theta, (S_i, u_i)_{i\in I})$, *where* I *and* Θ *are finite and, for every* $i \in I$, S_i *is finite and* $u_i : S_i \times S_{-i} \times \Theta \to \mathbb{R}$.

This description is partial because it does not specify the players' (hierarchies of) beliefs about Θ. One way to address this is to append to the game the players' hierarchies of beliefs over Θ. We model these hierarchies using a Θ-based type structure, i.e., a type structure as in Definition 12.3 where we set $X_{-i} = \Theta$.

As discussed, ICR is the solution concept that iteratively eliminates strategies that are not best replies to beliefs over $\Theta \times S_{-i}$, where beliefs allow for correlation.[53]

Definition 12.19. [54] *Consider a game* $G = (I, \Theta, (S_i, u_i)_{i\in I})$ *and a* Θ*-based type structure* $\mathcal{T}^\Theta = (\Theta, (T_i^\Theta, \beta_i^\Theta)_{i\in I})$. *For every* $t_i^\Theta \in T_i^\Theta$, *let*

- $ICR_i^0(t_i^\Theta) = S_i$;
- *for* $k > 0$, $s_i \in ICR_i^k(t_i^\Theta)$ *if there exists a map* $\sigma_{-i} : \Theta \times T_{-i}^\Theta \to \Delta(S_{-i})$ *such that, for all* $\theta \in \Theta$ *and* $t_{-i}^\Theta \in T_{-i}^\Theta$, $\sigma_{-i}(\theta, t_{-i}^\Theta)(ICR_{-i}^{k-1}(t_{-i}^\Theta)) = 1$ *and*

$$\forall s_i' \in S_i, \sum_{\theta, t_{-i}^\Theta} \beta_i^\Theta(t_i^\Theta)(\theta, t_{-i}^\Theta) \sum_{s_{-i}} \sigma_{-i}(\theta, t_{-i}^\Theta)(s_{-i}) u_i(s_i, s_{-i}, \theta)$$
$$\geq \sum_{\theta, t_{-i}^\Theta} \beta_i^\Theta(t_i^\Theta)(\theta, t_{-i}^\Theta) \sum_{s_{-i}} \sigma_{-i}(\theta, t_{-i}^\Theta)(s_{-i}) u_i(s_i', s_{-i}, \theta).$$

The set $ICR_i^\infty(t_i^\Theta) = \bigcap_{k\geq 0} ICR_i^k(t_i^\Theta)$ *is the set of* ***interim correlated rationalizable*** *strategies for type* t_i^Θ.

[53] With $I > 2$ players, there are two forms of correlation: that between the underlying uncertainty and opponents' strategies, and (as in correlated rationalizability—Definition 12.8—and correlated equilibrium—Definition 12.12) that among opponents' strategies, even conditioning on the underlying uncertainty. One could allow one but not the other, in principle. For simplicity we allow for both.

[54] The definition of ICR we adopt differs in inessential ways from the one originally proposed by Dekel et al. (2007). See also Liu (2014) and Tang (2011).

To understand this definition, recall that, in a Θ-based structure $(\Theta, (T_i^\Theta, \beta_i^\Theta)_{i \in I})$, player i's type t_i^Θ represents her beliefs about $\Theta \times T_{-i}^\Theta$, but not S_{-i}. ICR then assumes that beliefs over S_{-i} are determined by a function $\sigma_{-i} : \Theta \times T_{-i}^\Theta \to \Delta(S_{-i})$. Specifically, the probability that type t_i^Θ assigns to opponents playing a given profile s_{-i} equals

$$\sum_{(\theta, t_{-i}^\Theta)} \beta_i^\Theta(t_i^\Theta)(\theta, t_{-i}^\Theta)) \cdot \sigma_{-i}(\theta, t_{-i}^\Theta)(s_{-i}). \qquad [12.19]$$

The fact that $\sigma_{-i}(\cdot) \in \Delta(S_{-i})$ depends upon both θ and t_{-i}^Θ allows for the possibility of unmodeled correlating information received by i's opponents, as discussed. For example, in the game of Table 12.7 augmented with Θ-based type structure $\mathcal{T}^1$, the strategy D of player 1 is a best reply for type t_1 given the belief on $\Theta \times S_{-i}$ constructed by defining $\sigma_2(\theta_1, t_2')(R) = 1$ and $\sigma_2(\theta_2, t_2'')(L) = 1$. The strategy D is also a best reply for type t_1 of player 1 in the structure $\mathcal{T}^2$ when we define $\sigma_2(\theta_1, t_2)(R) = 1$ and $\sigma_2(\theta_2, t_2)(L) = 1$.

To study the relationship between RCBR and ICR, we start with an epistemic type structure where $X_{-i} = \Theta \times S_{-i}$. It is important to keep track of the difference between this and the Θ-based type structure appended to the game of incomplete information and used in defining ICR (Definition 12.19). Of course, in our epistemic analysis, we will need to relate the belief hierarchies generated by the Θ-based type structure to the $\Theta \times S_{-i}$-based hierarchies in the epistemic type structure.

Thus, consider an epistemic type structure $\mathcal{T} = (I, (\Theta \times S_{-i}, T_i, \beta_i)_{i \in I})$. As in Definition 12.6, continue to denote the first-order belief map for player i by $f_i : T_i \to \Delta(\Theta \times S_{-i})$, defined by $f_i(t_i) = \text{marg}_{\Theta \times S_{-i}} \beta_i(t_i)$. Naturally, first-order beliefs are now over $\Theta \times S_{-i}$. Analogously to Definition 12.7, a strategy s_i is *rational* for type $t_i \in T_i$, written $(s_i, t_i) \in R_i$, iff

$$\forall s_i' \in S_i, \qquad \sum_{\theta \in \Theta, s_{-i} \in S_{-i}} f_i(t_i)(\theta, s_{-i}) u_i(s_i, s_{-i}, \theta) \geq \sum_{\theta \in \Theta, s_{-i} \in S_{-i}} f_i(t_i)(\theta, s_{-i}) u_i(s_i', s_{-i}, \theta).$$

The event, "Player i believes event $E_{-i} \subset \Theta \times S_{-i} \times T_{-i}$" is defined as in Definition 12.7 part (2): $(s_i, t_i) \in B_i(E_{-i})$ if $\beta_i(t_i)(E_{-i}) = 1$. As before, both R_i and $B_i(E_{-i})$ are subsets of $S_i \times T_i$. Recall that we defined mutual belief in a product event $E = \prod_i E_i$, where $E_i \subseteq S_i \times T_i$, as $B(E) = \prod_i B_i(E_{-i})$. Since now beliefs are also about Θ, in order to simplify the definition of $B(B(E))$, and so on, it is convenient to instead have $B(E)$ be a subset of $\Theta \times S \times T$, as follows. For $F = Q \times \prod_i E_i$, where $Q \subset \Theta$ and $E_i \subseteq S_i \times T_i$, let

$$B(F) = \Theta \times \prod_{i \in I} B_i(Q \times E_{-i}).$$

This way, $B(F)$ has the same product structure as F, so common belief can be defined as before by iterating $B(\cdot)$. Correspondingly, we adapt the definition of RCBR:

$$RCBR = (\Theta \times R) \cap CB(\Theta \times R).$$

As noted, in order to provide an epistemic analysis of ICR, we must discuss the relationship between the epistemic type structure and the Θ-based type structure that we append to the game of incomplete information and use to define solution concepts. We start with a type t_i^Θ in a Θ-based type structure—such as t_i^N in $\mathcal{T}^N$, or t_i^R in $\mathcal{T}^R$—which induces a belief hierarchy over Θ. Denote this hierarchy by $\varphi_i^\Theta(t_i^\Theta)$. We then ask whether a type t_i in the epistemic type structure, which induces belief hierarchies over $\Theta \times S_{-i}$, has the same "marginal" hierarchy on Θ, denoted $\varphi_{i,\Theta}(t_i)$. To illustrate this, we relate the type structure $\mathcal{T}^N$ of Table 12.8 to the epistemic type structure defined in Table 12.9 below.

In this epistemic type structure, types t_1 and t_2 both believe that θ_1 and θ_2 are equally likely; furthermore, t_1 assigns probability one to 2's type being t_2, and conversely. So, if players' belief hierarchies over $\Theta \times S_{-i} \times T_{-i}$ are described by t_1 and t_2, respectively, there is common belief that each payoff parameter θ_i is equally likely. This is the same belief hierarchy over Θ held by types t_1^N and t_2^N in the Θ-based type structure $\mathcal{T}^N$. Formally, we have $\varphi_i^\Theta(t_i^N) = \varphi_{i,\Theta}(t_i)$.

To further illustrate how to construct $\varphi_{i,\Theta}(\cdot)$, consider type t_1'. This type also believes that θ_1 and θ_2 are equally likely; however, t_1' has more complex second-order beliefs. Specifically, t_1' assigns probability $\frac{1}{2}$ to the event that the payoff parameter is θ_1 and that 2 thinks θ_1 and θ_2 are equally likely, and probability $\frac{1}{2}$ to the event that the payoff state is θ_2 and 2 thinks that the probability of θ_1 is $\frac{1}{3}$. Iterating this procedure yields the hierarchical beliefs on Θ held by t_1', i.e., $\varphi_{i,\Theta}(t_1')$. Notice that no type in the Θ-based structure $\mathcal{T}^N$ (or $\mathcal{T}^R$) generates this hierarchy over Θ.

Thus, for an epistemic type structure $\mathcal{T}$ and Θ-based type structure $\mathcal{T}^\Theta$, we can define the event that each player i's hierarchy over Θ is the one generated by some type t_i^Θ in $\mathcal{T}^\Theta$:

$$[\varphi^\Theta(t^\Theta)] = \left\{(\theta, s, t) \;:\; \forall i,\; \varphi_{i,\Theta}(t_i) = \varphi_i^\Theta(t_i^\Theta)\right\}.$$

Then, we can ask what strategies are consistent with RCBR and the assumption that hierarchical beliefs on Θ are generated by a given type t_i^Θ in the Θ-based structure $\mathcal{T}^\Theta$.

Table 12.9 An epistemic type structure for the game in Table 12.7.

	(θ_1, L, t_2)	(θ_1, L, t_2')	(θ_1, R, t_2)	(θ_1, R, t_2')	(θ_2, L, t_2)	(θ_2, L, t_2')	(θ_2, R, t_2)	(θ_2, R, t_2')
$\beta_1(t_1)$	0	0	$\frac{1}{2}$	0	$\frac{1}{2}$	0	0	0
$\beta_1(t_1')$	$\frac{1}{2}$	0	0	0	0	$\frac{1}{2}$	0	0

	(θ_1, U, t_1)	(θ_1, U, t_1')	(θ_1, D, t_1)	(θ_1, D, t_1')	(θ_2, U, t_1)	(θ_2, U, t_1')	(θ_2, D, t_1)	(θ_2, D, t_2')
$\beta_2(t_2)$	$\frac{1}{2}$	0	0	0	$\frac{1}{2}$	0	0	0
$\beta_2(t_2')$	0	$\frac{1}{3}$	0	0	0	$\frac{2}{3}$	0	0

The following theorem (due to Battigalli et al., 2011; for a related result, see Dekel et al., 2007) states that these are precisely the strategies in $ICR_i^\infty(t_i^\Theta)$.[55]

Theorem 12.13. *Fix a Θ-based type structure and a complete epistemic type structure for a game $\left(I, (S_i)_{i\in I}, \Theta, (u_i)_{i\in I}\right)$. Then, for any Θ-type profile $t^\Theta \in T^\Theta$,*

$$ICR^\infty(t^\Theta) = \mathrm{proj}_S\left(RCBR \cap [\varphi^\Theta(t^\Theta)]\right).$$

One important implication of this characterization is that the set of ICR strategies for a Θ-based type t_i^Θ depends solely upon the hierarchical beliefs on Θ that it generates. In particular, if two such types $t_i^\Theta, \hat{t}_i^\Theta$ induce the same hierarchies, i.e., if they are redundant, they share the same set of ICR strategies (see Dekel et al., 2007).

12.6.3 Interim independent rationalizability

As noted earlier, a key feature of ICR is the fact that it allows a player to believe that her opponents' strategic choices are correlated with the uncertainty Θ. The definition of ICR does so by introducing maps $\sigma_{-i} : \Theta \times T_{-i}^\Theta \to \Delta(S_{-i})$, which explicitly allow player i's conjecture about her opponents' play to depend upon the realization of θ, in addition to their Θ-based types t_{-i}. Correspondingly, in an epistemic type structure, any correlation between the Θ and S_{-i} components of the first-order beliefs is allowed.

As discussed, if there is no unmodeled uncertainty one might want to rule out such correlations and assume that, conditioning on opponents' hierarchies, opponents' strategies should be uncorrelated with Θ. *Interim Independent Rationalizability*, or IIR, reflects such considerations. Like ICR, this procedure applies to an incomplete-information game augmented with a Θ-based type structure $(I, (\Theta, T_i^\Theta, \beta_i^\Theta)_{i\in I})$, and iteratively eliminates strategies that are not best replies for each player type. The difference is in the way beliefs about S_{-i} are constructed: IIR employs maps $\sigma_{-i} : T_{-i}^\Theta \to \Delta(S_{-i})$ that associate with each profile $t_{-i} \in T_{-i}^\Theta$ a distribution over strategy profiles. This explicitly rules out the possibility that opponents' strategies may be directly correlated with the payoff parameter θ. The probability that a type t_i^Θ attaches to strategy profile s_{-i}, given the function σ_{-i}, is then

$$\sum_{t_{-i}^\Theta} \beta_i^\Theta(t_i^\Theta)(\Theta \times \{t_{-i}^\Theta\}) \cdot \sigma_{-i}(t_{-i}^\Theta)(s_{-i}), \qquad [12.20]$$

in contrast with [12.19]. Any correlation between the payoff parameter θ and opponents' play must thus come from correlation between θ and opponents' types, because direct

[55] For brevity, we only state the analog to part (2) in Theorems 12.1, 12.2, and 12.11. We could also define a notion of best response set and provide analogs to parts (1) and (3) as well. In particular, in any epistemic type structure, $ICR^\infty(t^\Theta) \supseteq \mathrm{proj}_S\left(RCBR \cap [\varphi^\Theta(t^\Theta)]\right)$.

correlation between θ and s_{-i} is ruled out by the definition of the maps $\sigma_{-i}(\cdot)$. This implies that redundant types can matter. Consider the game in Table 12.7 and the type structure $\mathcal{T}^R$ defined in Table 12.8. Strategy D is a best reply for type t_1^R, given the function $\sigma_2 : T_2^R \to \Delta(S_2)$ such that $\sigma_2(t_2^R)(R) = 1 = \sigma_2(\bar{t}_2^R)(L)$; notice how the belief over $\Theta \times S_2$ derived from $\beta_1^R(t_1^R) \in \Delta(\Theta \times T_2^R)$ and σ_2 induces correlation between θ and 2's strategy via correlation between θ and 2's type. If we instead consider the type structure $\mathcal{T}^N$, it is impossible to induce correlation between θ and 2's strategy in this indirect way, because 2 has only one type. As a result, D is not a best reply given this type structure. This indicates that IIR can deliver different predictions in type structures that generate the same hierarchical beliefs about Θ.

The IIR procedure (Ely and Peski, 2006) is formally defined as follows.

Definition 12.20. [56] *Consider a game $G = (I, \Theta, (S_i, u_i)_{i\in I})$ and a Θ-based type structure $\mathcal{T}^\Theta = (\Theta, (T_i^\Theta, \beta_i^\Theta)_{i\in I})$. For every $t_i^\Theta \in T_i^\Theta$, let*

- $IIR_i^0(t_i^\Theta) = S_i$;
- *for $k > 0$, $s_i \in IIR_i^k(t_i^\Theta)$ if there exists a map $\sigma_{-i} : T_{-i}^\Theta \to \Delta(S_{-i})$ such that, for all $t_{-i}^\Theta \in T_{-i}^\Theta$, $\sigma_{-i}(t_{-i}^\Theta)(IIR_{-i}^{k-1}(t_{-i}^\Theta)) = 1$ and*

$$\forall s_i' \in S_i, \sum_{\theta, t_{-i}^\Theta} \beta_i^\Theta(t_i^\Theta)(\theta, t_{-i}^\Theta) \sum_{s_{-i}} \sigma_{-i}(t_{-i}^\Theta)(s_{-i}) u_i(s_i, s_{-i}, \theta)$$
$$\geq \sum_{\theta, t_{-i}^\Theta} \beta_i^\Theta(t_i^\Theta)(\theta, t_{-i}^\Theta) \sum_{s_{-i}} \sigma_{-i}(t_{-i}^\Theta)(s_{-i}) u_i(s_i', s_{-i}, \theta).$$

*The set $IIR_i^\infty(t_i^\Theta) = \bigcap_{k\geq 0} IIR_i^k(t_i^\Theta)$ is the set of **interim independent rationalizable** strategies for type t_i^Θ.*

We now turn to the epistemic characterization of IIR. The key is to formalize the assumption that player i's beliefs about $\Theta \times S_{-i}$ are independent, conditional upon any hierarchical beliefs about Θ that i thinks may be held by her opponents. Thus, fix a finite Θ-based type structure $\mathcal{T}^\Theta$ and an epistemic type structure $\mathcal{T}$. We denote by $T_{i,CI}$ the set of i's types t_i whose beliefs satisfy this assumption: formally, $t_i \in T_{i,CI}$ if

$$\mathrm{marg}_{\Theta\times S_{-i}} \beta_i(t_i) \left(\cdot \middle| [\varphi_{-i}^\Theta(t_{-i}^\Theta)]\right)$$

is the product of its marginals on Θ and S_{-i}, whenever the above conditional probability is well-defined, i.e., for every type t_{-i}^Θ in the Θ-based structure $\mathcal{T}^\Theta$ such that $\beta_i(t_i)([\varphi_{-i}^\Theta(t_{-i}^\Theta)]) > 0$.[57] Finally, let

$$CI = \{(\theta, s, t) \ : \ \forall i, \ t_i \in T_{i,CI}\}.$$

[56] IIR can also be described as "rationalizability in the interim strategic form": see Battigalli et al. (2011a).

[57] For completeness, $[\varphi_{-i}^\Theta(t_{-i}^\Theta)] = \left\{(\theta, s_{-i}, t_{-i}) : \forall j \neq i, \ \varphi_{j,\Theta}(t_j) = \varphi_j^\Theta(t_j^\Theta)\right\}$.

For example, consider the epistemic type structure in Table 12.9 and the Θ-based type structure $\mathcal{T}^R$. Recall that both types t_2^R and $\bar{t}_2^R$ in $\mathcal{T}^R$ generate the same hierarchy of beliefs about Θ, namely that it is common belief that θ_1 and θ_2 are equally likely. Observe that epistemic type t_2 generates precisely this hierarchical belief about Θ, whereas epistemic type t_2' generates a different hierarchy. Therefore, the events $[\varphi_2^\Theta(t_2^R)]$ and $[\varphi_2^\Theta(\bar{t}_2^R)]$ coincide and are equal to $\Theta \times S_2 \times \{t_2\}$ in the epistemic structure of Table 12.9. Now consider player 1's type t_1. This type assigns probability one to 2's type t_2, and hence to the event $[\varphi_2^\Theta(t_2^R)]$. Conditional on this event, t_1's beliefs over $\Theta \times S_2$ assign equal probability to (θ_2, L) and (θ_1, R), which is not an independent product. Therefore, $t_1 \notin T_{1,CI}$. On the other hand, type t_1' assigns positive probability to both types t_2 and t_2' of player 2. Conditional on t_2, i.e., conditional on $[\varphi_2^\Theta(t_2^R)]$, t_2' assigns probability one to (θ_1, L), which is trivially an independent product. Conditional independence does not impose any further restriction on t_1', because the Θ-hierarchy generated by t_2' differs from the hierarchy of any type in the Θ-based structure $\mathcal{T}^R$. Therefore, $t_1' \in T_{1,CI}$.

We then have the following characterization (Battigalli et al., 2011a).

Theorem 12.14. *Fix a finite Θ-based type structure $\mathcal{T}^\Theta$ and a* complete *epistemic type structure $\mathcal{T}$ for the game $\left(I, (S_i)_{i\in I}, \Theta, (u_i)_{i\in I}\right)$. Then, for any profile of Θ-types $t^\Theta \in T^\Theta$,*

$$IIR^\infty(t^\Theta) \supset \text{proj}_S \left(RCBR \cap CI \cap CB(CI) \cap [\varphi^\Theta(t^\Theta)]\right);$$

if, furthermore, the Θ-based type structure is not redundant (see Definition 12.5), then the above inclusion is an equality.

Note that the inclusion in Theorem 12.14 may be strict when the Θ-based structure is redundant, even if the epistemic type structure is complete. This contrasts with the results in Theorems 12.1, 12.2, 12.11, and 12.13, where completeness implies equality. In those four theorems, inclusion may be strict only if some justifying beliefs are simply not present in a given incomplete epistemic type structure.

To see that the inclusion in Theorem 12.14 may be strict in a redundant Θ-based structure, consider the game in Table 12.7 augmented with the type structure $\mathcal{T}^R$. As was argued before Definition 12.20, $IIR^\infty(t_1^R) = \{U, D\}$. Now consider a strategy-type pair $(s_1, t_1) \in RCBR_1 \cap CI_1 \cap CB_1(CI_2) \cap [\varphi_1^\Theta(t_1^R)]$ in a complete epistemic type structure for this game. Since $(s_1, t_1) \in [\varphi_1^\Theta(t_1^R)$, epistemic type t_1 must satisfy common belief that θ_1, θ_2 are equally likely. Hence, t_1 must believe that 2 also commonly believes this. Therefore, $(s_1, t_1) \in CI_1$ implies that t_1's beliefs about Θ and S_2 must be independent conditional on 2 commonly believing this. But then, D cannot be a best reply: that is, $(s_1, t_1) \in RCBR_1 \cap CI_1 \cap CB_1(CI_2 \cap [\varphi_1^\Theta(t_1^R)]$ implies $s_1 = U$. Thus, the epistemic assumptions result in $\{U\}$, a strict subset of the IIR prediction of $\{U, D\}$. On the other hand, repeating the analysis in the structure $\mathcal{T}^N$ leads to an equality: the only IIR strategy for type t_1^N is U.

To summarize, the main point is quite simple: if the type space has redundant types, then the solution concept should treat them symmetrically, since they are decision-theoretically indistinguishable.[58]

12.6.4 Equilibrium concepts

The characterization of Nash equilibrium in games of incomplete information requires significantly stronger assumptions relative to the complete-information case (cf. Theorem 12.5). There are two differences. One is that the first-order belief of a type in an epistemic type structure is a belief about opponents' strategies, $s_{-i} \in S_{-i}$, whereas an equilibrium of an incomplete-information game specifies maps from Θ-types into strategies. Therefore, while Theorem 12.5 obtained Nash equilibrium by assuming that first-order beliefs are mutually believed, now this needs to be modified so that the maps from Θ-hierarchies into strategies are (at least) mutually believed. The other, and more interesting, difference is that the assumption of mutual belief of these maps and of rationality needs to be strengthened to common belief. The following example illustrates this.

Example 12.6. *Consider a 2-person game with payoff irrelevant uncertainty $\Theta = \{\theta, \theta'\}$; player 2 has only one action, L, player 1 has two actions, U, D where U is strictly dominant, and the Θ-based type structure $\mathcal{T}^{\Theta}$ is generated by the common prior in Table 12.10:*

Consider the following maps from types in $\mathcal{T}^{\Theta}$ to strategies: $\psi_1(t_1^{\Theta}) = U$, $\psi_1(\hat{t}_1^{\Theta}) = D$, $\psi_2(t_2^{\Theta}) = \psi_2(\hat{t}_2^{\Theta}) = L$. The pair $\psi = (\psi_1, \psi_2)$ is obviously not a Bayesian Nash equilibrium. However, consider the common-prior epistemic type structure $\mathcal{T}$ obtained from $\mathcal{T}^{\Theta}$ and ψ being common belief: see Table 12.11.

The Θ-based hierarchies generated by the epistemic type structure $\mathcal{T}$ coincide with those generated by $\mathcal{T}^{\Theta}$: for example, $\varphi_{1,\Theta}(t_1) = \varphi_1^{\Theta}(t_1^{\Theta})$. The type profile (t_1, t_2) satisfies rationality

Table 12.10 A Θ-based type structure

θ	t_2^{Θ}	$\hat{t}_2^{\Theta}$
t_1^{Θ}	$\frac{1}{2}$	$\frac{1}{4}$
$\hat{t}_1^{\Theta}$	0	0

θ'	t_2^{Θ}	$\hat{t}_2^{\Theta}$
t_1^{Θ}	0	0
$\hat{t}_1^{\Theta}$	0	$\frac{1}{4}$

[58] In an intriguing and surprising result, Ely and Peski (2006) show that one can associate with each type a hierarchy of beliefs about $\Delta(\Theta)$, rather than Θ, in such a way as to distinguish between types that are redundant in the usual sense. However, we do not understand how their notion of hierarchy can be elicited. For example, the first-order belief of player i in an Ely-Pesky hierarchy is an element of $\Delta(\Delta(\Theta))$, representing i's beliefs about her own beliefs about Θ (for details, see p. 28 in their paper). As we argued in Section 12.2.6, introspective beliefs cannot easily be interpreted behaviorally. Moreover, a probability measure in $\Delta(\Delta(\Theta))$ is meant to represent how i's beliefs about Θ would change *if i were informed of j's type*. This obviously depends on the type space chosen to represent beliefs; we do not see how one could elicit these beliefs.

and mutual, but not common, belief in rationality. Finally, by construction, Player 2's beliefs about 1's strategies, conditional on 1's Θ-hierarchy, are as specified by ψ_1, and this is common belief (the same is trivially true for Player 1's beliefs about 2's sole strategy). This shows that mutual belief in rationality, even with common belief in the maps ψ_i, is not enough to obtain Bayesian Nash equilibrium.[59]

We now show that strengthening the assumptions of Theorem 12.5 as indicated above yields a characterization of Bayesian Nash equilibrium in two-player games (see Pomatto, 2011, and Sadzik, 2011). We believe (but have not verified) that a similar analog to Theorem 12.7 holds for games with more than two players. There exist several incomplete-information versions of correlated equilibrium (Bergemann and Morris, 2011; Forges, 1993, 2006; Liu, 2014). The epistemic characterizations for these concepts may be insightful and have not yet been developed.

Fix a nonredundant Θ-based type structure T^{Θ} on Θ, and maps $\psi_i : T^{\Theta}_{-i} \to \Delta(S_{-i})$. For every i, we interpret this map as Player i's conjecture about the behavioral strategy of her opponents.[60]

Given an epistemic type structure $\mathcal{T}$, let $[s_{-i}] = \Theta \times \{s_{-i}\} \times T_{-i}$ and $[t^{\Theta}_{-i}] = \Theta \times S_{-i} \times \{t_{-i} : \varphi_{-i,\Theta}(t_{-i}) = \varphi^{\Theta}_{-i}(t^{\Theta}_{-i})\}$. These are the events that "the opponents play s_{-i}" and "the opponents' Θ-hierarchies are as specified by t^{Θ}_{-i}." Then, the event that "each player's first-order beliefs are consistent with ψ" is

$$[\psi] = \Big\{(\theta, s, t)\ :\ \forall i,\ \forall t^{\Theta}_{-i} \in T^{\Theta}_{-i} \text{ s.t. } \beta_i(t_i)([t^{\Theta}_{-i}]) > 0, \\ \beta_i(t_i)([s_{-i}] | [t^{\Theta}_{-i}]) = \psi_i(t^{\Theta}_{-i})(s_{-i})\Big\}. \qquad [12.21]$$

Define "Θ-hierarchies are consistent with a given Harsanyi type structure $\mathcal{T}^{\Theta}$":

$$[\mathcal{T}^{\Theta}] = \left\{(\theta, s, t)\ :\ \forall i,\ \varphi_{i,\Theta}(t_i) \in \varphi^{\Theta}_i(T^{\Theta}_i).\right\}.$$

We need one more definition. Θ-based structure $\mathcal{T}^{\Theta}$ is *minimal* if, for every pair of players and types $i, j \in I$, $t^{\Theta}_i \in T^{\Theta}_i$, $t^{\Theta}_j \in T^{\Theta}_j$, there is a finite sequence $t^1_{i(1)}, \ldots, t^N_{i(N)}$

Table 12.11 An epistemic type structure

θ	L, t_2	$L, \hat{t}_2$
U, t_1	$\frac{1}{2}$	$\frac{1}{4}$
$D, \hat{t}_1$	0	0

θ'	L, t_2	$L, \hat{t}_2$
U, t_1	0	0
$D, \hat{t}_1$	0	$\frac{1}{4}$

[59] Furthermore, for any finite $k \geq 2$, we can modify the above example so that there is kth order mutual belief in rationality, and still the conjectures do not form a Bayesian Nash equilibrium. Similarly, the necessity of common belief in ψ can be demonstrated. See Pomatto (2011).

[60] The maps ψ_i resemble, but are distinct from, the "books of play" n_i (see [12.14]) since the former map from Θ-hierarchies, whereas the latter map from S_{-i} hierarchies.

such that $i(1) = 1$, $t^1_{i(1)} = t^\Theta_i$, $i(N) = j$, $t^N_{i(N)} = t^\Theta_j$, and for all $n = 2, \ldots, N$, $\beta^\Theta_{i(n-1)}(t^{n-1})([t^n_{i(n)}]) > 0$. That is, loosely speaking, it is not possible to partition $\mathcal{T}^\Theta$ into two components such that each component is a type structure in and of itself.

Theorem 12.15. *Assume that there are two players. Fix an incomplete-information game G and a nonredundant, minimal Θ-based type structure $\mathcal{T}^\Theta$ and maps ψ_1, ψ_2 as above. If there is an epistemic type structure $\mathcal{T}$ in which $CB([\mathcal{T}^\Theta]) \cap CB([\psi]) \cap CB(\Theta \times R) \neq \emptyset$, then (ψ_2, ψ_1) is a Bayesian Nash equilibrium of the Bayesian game $(G, \mathcal{T}^\Theta)$.*

12.6.5 Δ-Rationalizability

The role of Θ-based type structures in the definition of ICR and IIR is to represent assumptions about players' interactive beliefs concerning exogenous payoff uncertainty, Θ. An alternative approach is to adapt the notion of Δ-rationalizability discussed in Section 12.3.3. Doing so is straightforward: for every player i, let $\Delta_i \subset \Delta(\Theta \times S_{-i})$ represent the restrictions on i's first-order beliefs that we would like to maintain. Notice that these restrictions can also be about i's opponents' strategies, not just the exogenous uncertainty. The set of Δ-rationalizable profiles, which we continue to denote by $S^{\Delta,\infty}$, can then be defined exactly as in Definition 12.11, with the understanding that players best respond to conjectures $\sigma_{-i} \in \Delta(\Theta \times S_{-i})$, rather than in $\Delta(S_{-i})$. The epistemic characterization of $S^{\Delta,\infty}$ via RCBR and common beliefs in the restrictions Δ_i provided in Theorem 12.2 also extends, provided RCBR is defined as in Section 12.6.2.

This framework allows us to model, for example, a situation in which players' ordinal preferences over strategy profiles are fixed (and commonly believed), but their cardinal preferences (i.e., their risk attitudes) are unknown. To study this situation, Börgers (1993) proposes the following notion of rationality: given a complete-information game $(I, (S_i, u_i)_{i \in I})$, s_i is rational if and only if there is a belief $\sigma_{-i} \in \Delta(S_{-i})$ and a function $v_i : S \to \mathbb{R}$ that is a strictly increasing transformation of u_i such that s_i is a best reply to σ_{-i} with utility function v_i. He characterizes this notion of rationality in terms of a novel pure-strategy dominance property and argues that common belief in his notion of rationality corresponds to the iterated deletion of strategies that are pure-strategy dominated in his sense. As Borgers notes, it is straightforward to formalize this by suitably modifying the definition of the events R_i, and hence RCBR.

Instead of modifying the notion of rationality for complete-information games, we can obtain an alternative epistemic characterization of iterated pure-strategy dominance in Borgers' sense by considering a related game in which players have incomplete information about the risk preferences of their opponents, but know their ordinal rankings. To do so, we retain the usual notion of rationality for incomplete-information games and consider the implications of RCBR and common belief of the ordinal rankings. We model common belief of the ordinal rankings by a suitable choice of payoff parameter space Θ and commonly believed restrictions on first-order beliefs Δ. By the incomplete-information analog of Theorem 12.2, RCBR and common belief of the ordinal rankings characterize the set $S^{\Delta,\infty}$ of Δ-rationalizable profiles. Given the choice of Θ and Δ, $S^{\Delta,\infty}$ is precisely the set of iteratively pure-strategy undominated profiles. To sum up, Borgers relates iterated pure-strategy dominance to RCBR in the original complete-information game, but redefines what it means for a player to be rational. The argument described here relates iterated pure-strategy dominance to RCBR and common belief in the ordinal rankings in an associated incomplete-information game, where rationality has the usual meaning.

To make this precise, we specify the appropriate Θ and Δ. Given the complete-information game $G = (I, (S_i, u_i)_{i\in I})$, let Θ_i be the set of all payoff functions $\theta_i : S \to [0, 1]$, and Θ_i^u be the set of utilities $\theta_i^u \in \Theta_i$ that are ordinally equivalent to u_i. Furthermore, let Δ_i be the set of all finite-support probability measures $\sigma_{-i} \in \Delta(\Theta \times S_{-i})$ such that player i (i) is certain of her own *cardinal* utility, and (ii) is certain of her opponents' *ordinal* preferences: i.e., $\sigma_{-i}(\{\theta_i^u\} \times \Theta_{-i}^u \times S_{-i}) = 1$ for some $\theta_i^u \in \Theta_i^u$. Note that (i) says that Player i is certain of her own risk preferences, but these need not coincide with u_i. Finally, we define the incomplete-information game in the obvious way, that is, $(I, \Theta, (S_i, \upsilon_i)_{i\in I})$, where $\upsilon_i(\theta, s) = \theta_i(s)$ for every $\theta \in \Theta$ and $s \in S$. With these definitions, a strategy $s_i \in S_i$ is a best reply for i in the incomplete-information game $(I, \Theta, (S_i, \upsilon_i)_{i\in I})$ if and only if it is a best reply for i given *some* (complete-information) payoff function $\theta_i : S \to \mathbb{R}$ that is ordinally equivalent to u_i, i.e., if it is not pure-strategy dominated in the sense of Borgers. It follows that $S^{\Delta,\infty}$ coincides with iterated pure-strategy dominance; thus, by the incomplete-information analog of Theorem 12.2, iterated pure-strategy dominance in the complete-information game $(I, (S_i, u_i)_{i\in I})$ is characterized by RCBR with the restrictions Δ specified here for the incomplete-information game $(I, \Theta, (S_i, \upsilon_i)_{i\in I})$.

12.6.6 Discussion

The epistemic analysis of games of incomplete information is recent and indeed incomplete. The results above make three points. First, IIR is suitable only when there are no redundant types and no unmodeled correlation. ICR on the other hand is a robust solution concept that corresponds to RCBR. Finally, we saw that equilibrium concepts are difficult to characterize with clean and insightful epistemic conditions. The assumptions required are more demanding than the analogous concepts in complete-information games.[61]

[61] This point, from a different perspective, is consistent with the work on learning in games, which argues for weaker concepts than Nash equilibrium, such as self-confirming equilibrium (see, e.g., Fudenberg and Levine, 1993; Dekel et al., 1999, 2004, and Fudenberg and Kamada, 2012).

As we do throughout this chapter, in this section, we continue to interpret type structures merely as representations of hierarchies of beliefs. As we noted in Section 12.1.1 type structures are also used to model hard information. If one takes this view, then one would want to replace the independence assumption *CI* with the statement that, conditional on *types*, beliefs are independent. Together with the other epistemic assumptions in Theorem 12.14, this would fully characterize IIR. However, we prefer to model such hard information explicitly and distinctly from types. The basic space of uncertainty should include the hard information, and each player's hierarchy of beliefs should be consistent with the hard information received. In such a structure, types associated with different hard information *are* distinguishable—they have different beliefs about the information. In particular, the suitable analogs to type structures $\mathcal{T}^N$ and $\mathcal{T}^R$ are not equivalent in terms of hierarchies, and the types in the latter are not redundant. Consequently, the epistemic assumptions in Theorem 12.14 would yield different answers for the two structures, consistently with IIR.

We do not discuss almost common belief in the payoff structure within the context of games with incomplete information. This is an issue which, starting from Rubinstein's (1989) e-mail game, has led to many interesting developments. Among others, these include Monderer and Samet (1989)'s introduction of p-belief in game theory, and the literature on global games and robustness (see, e.g., Carlsson and Van Damme, 1993; Kajii and Morris, 1997a; Morris and Shin, 2003; Weinstein and Yildiz, 2007). Some of these issues (in particular, almost common belief in rationality for games with incomplete information) may benefit from epistemic analysis.

12.7. EXTENSIVE-FORM GAMES

12.7.1 Introduction

The results in the previous sections apply *verbatim* to the strategic form of a multistage game. However, merely invoking those results disregards an essential implication of the dynamic nature of the interaction: players may be *surprised* in the course of game play. As is familiar from the textbook presentation of extensive-form solution concepts such as sequential equilibrium (Kreps and Wilson, 1982), and refinements such as the Intuitive Criterion of Cho and Kreps (1987), different assumptions about beliefs at unexpected histories can lead to very different predictions. Epistemic game theory provides a rich framework to analyze such assumptions. We now illustrate this using as an example the debate on the relationship between backward induction and "common belief in rationality."

Consider the three-legged centipede game in Figure 12.2. A common, informal—and, it turns out, controversial—argument suggests that "common belief in rationality" implies that the backward-induction (BI) outcome should be played: Player 1, if rational, will choose D at the third node; but then, if Player 2 is rational and believes that 1 is rational, he should choose *d* at the second node; finally, if Player 1 is rational and anticipates all this, she will choose *D* at the initial node.

The problem with this informal argument is that it is not immediately obvious what is meant by "belief" and "rationality" in an extensive game such as the Centipede. Trivially, if one assumes that players commit to their strategies (perhaps by delegating actual play to agents, or machines) at the beginning of the game, and are rational in the sense of Section 12.3, then one reduces the situation to a game with simultaneous moves. It is easy to verify that, in that strategic-form game, RCBR only eliminates strategy *AA* for Player 1. Thus, choosing *D* at the initial node is consistent with RCBR, but so is choosing *A* at the first node and then *D* at the third node. In particular, player 1 will rationally commit to play strategy *AD* if she believes that 2 will play *a* with sufficiently high probability. In turn, player 2 will commit to play *a* if he expects 1 to actually choose *D* at the first node, because in this case, his choice will not matter. So, the profile (*AD*, *a*) is consistent with RCBR.

A possible objection to this argument begins by noting that, if the game actually reaches the second node, player 2 may regret his commitment to *a*. At the second node, he knows that player 1 did not choose *D*. Thus, if he continues to believe that 1 has committed to a strategy consistent with RCBR, he concludes that 1 will play *D* next. In this case, player 2's commitment to *a* results in a net loss of 1 "util." Then, since 2's choice of *a* vs. *d* only matters if play reaches the second node, shouldn't player 2 anticipate all this and commit to *d* instead?

Despite their intuitive appeal, these considerations pertain to player 2's knowledge, beliefs, and expected payoffs at the second node, and as such are irrelevant from the ex-ante point of view. What then if one abandons this ex-ante perspective? Suppose one takes into account the beliefs that players *would actually hold at different points in the game*, and correspondingly adopts "sequential rationality" as the appropriate behavioral principle (Kreps and Wilson, 1982). For player 2, this requires that his choice of *a* or *d* be optimal given the beliefs she would hold at the second node. The preceding argument now seems to apply: at that node, player 2 knows that 1 did not play *D*, and it appears that player 2 should conclude that 1 is rationally playing *AD*, whatever 2's initial beliefs might have been. Thus, it appears that the informal argument supporting the BI solution does in fact apply if one takes the dynamic nature of the game into consideration.

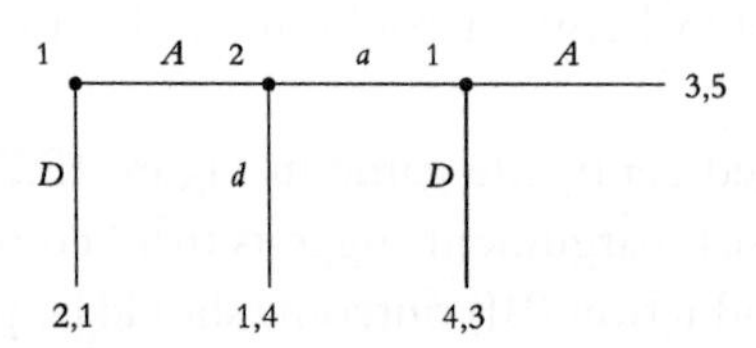

Figure 12.2 A three-legged centipede game.

However, in an important contribution, Reny (1992) questions the argument just given. Reny points out that, while this argument allows for the possibility that player 2's beliefs *about 1's play* may change as the game progresses, it implicitly assumes that 2's beliefs *about 1's rationality* will not. If one instead allows 2's beliefs about 1's rationality to change, the BI prediction need not obtain, even though *initially* there is common belief in (sequential) rationality. The intuition is as follows. Suppose that, *before the game is played*, player 2 expects the BI outcome, with the usual rationalization: that is, 2 expects 1 to choose D at the initial node because 1 expects d, which is in turn justified by 2's belief that 1 will choose D at the third node. Suppose further that, if the second node is reached—an event that 2 does not expect to occur—then player 2 changes his mind about 1's rationality. This is not unreasonable: after all, if 1 expects 2 to choose d, 1 should not play A at the first node, so observing A does provide circumstantial evidence to the effect that 1 is *not* rational. In particular, upon reaching the second node, player 2 revises his beliefs about 1's play and now expects 1 to choose A at the third node as well; this makes his own choice of a sequentially rational. To sum up, player 2's beliefs may *initially* be consistent with BI, and—informally—with RCBR, and yet he may (plan to) play a. Furthermore, suppose player 1 is rational and has correct beliefs about 2's strategy and beliefs. Then, 1 will rationally choose A at the first node, and then D at the third. The point is that, while 2's beliefs about 1's behavior and beliefs are incorrect, both players' *initial* beliefs are consistent with RCBR. Yet, BI does not obtain.[62]

BI can be obtained if we translate into an epistemic model the assumptions that players make conditionally optimal choices at all information sets and that they believe the same is true of opponents at all subsequent decision nodes; see Subsection 12.7.5. We find it more surprising that it also follows from a strengthening of the notion of belief that is motivated by forward-induction ideas, i.e., from requiring that players "believe in rationality as much as possible": see Corollary 12.1 in Subsection 12.7.4.

This example illustrates that a careful, explicit modeling of interactive beliefs is extremely important in extensive games. In particular, Reny's argument points out a hidden assumption in the informal argument relating "common belief in rationality" and BI. These subtle issues suggest that a formal epistemic analysis may be insightful. Indeed, as with simultaneous-move games, epistemic models provide the language and tools to study the implications of common belief in rationality in dynamic games and to study how modifications of the notion of rationality and belief characterize different solution concepts. Moreover, as before, dynamic epistemic models can and should force the theorist to make all assumptions explicit (or, at least, they can make it easier to spot hidden assumptions).

[62] One might further conjecture that assuming RCBR at *every* history may deliver the BI solution. However, in many games, RCBR *cannot* possibly hold at every history: this is the case for centipede games, for instance. See Reny (1993) and Battigalli and Siniscalchi (1999) for details and additional results.

This chapter emphasizes this role of epistemic models. We begin by characterizing initial RCBR in multistage games with observed actions, based on the work of Ben-Porath (1997) and Battigalli and Siniscalchi (1999). We then introduce the notion of *strong belief*, and show that, in complete type structures, rationality and common strong belief thereof (RCSBR) is characterized by *extensive-form rationalizability* (Battigalli, 1997; Pearce, 1984). RCSBR captures a (strong) principle of *forward induction* (Kohlberg and Mertens, 1986), as we indicate by means of examples. At the same time, RCSBR in complete type structures implies that the backward-induction outcome will obtain in generic perfect-information games; thus, it provides sufficient conditions for BI (see Battigalli and Siniscalchi, 2002).

12.7.2 Basic ingredients

For notational convenience, we focus on multistage games with observed actions (Fudenberg and Tirole, 1991a; Osborne and Rubinstein, 1994). These are extensive games in which play proceeds in stages. Perfect-information games are an example, but more generally, two or more players may be active in a given stage, in which case they choose simultaneously. The crucial assumption is that players observe all past choices at the beginning of each stage. Although we shall not do so, it is trivial to add incomplete information to such games (see Subsection 12.7.7). Most extensive games of interest in applications are in fact multistage games with observed actions (henceforth, "multistage game") and possibly incomplete information.

We follow the definition of multistage games in Osborne and Rubinstein (1994, Section 6.3.2), and introduce here only the notation that we need subsequently. We identify a multistage game Γ with the tuple $(I, \mathcal{H}, (\mathcal{H}_i, S_i(\cdot), u_i)_{i \in I})$, where:

- $\mathcal{H}$ is the set of (terminal and non-terminal) histories in Γ. In a perfect-information game, these are (possibly partial) sequences of actions. In the Centipede game of Figure 12.2, $\langle A, a, D \rangle$ is a (terminal) history. In a general multistage game, histories are sequences of action profiles. For simplicity, the profiles only indicate actions of players who have nontrivial choices. In the game of Figure 12.3 in Example 12.8 below, $\langle In \rangle$ and $\langle In, (T, L) \rangle$ are both histories.
- $\mathcal{H}_i$ is the subset of $\mathcal{H}$ where Player i has nontrivial choices, and $\emptyset$ denotes the initial history;
- S_i is the set of strategic-form strategies of Player i: these are maps from histories to actions;
- $\mathcal{H}_i(s_i)$ is the set of histories in $\mathcal{H}_i$ that are *not precluded* if i plays strategy s_i (whether or not any $h \in \mathcal{H}_i(s_i)$ is actually reached thus only depends upon the play of i's opponents);

- $S_i(h)$ is the set of strategies of player i that allow history h to be reached. The Cartesian product $S(h) = \prod_i S_i(h) = S_i(h) \times S_{-i}(h)$ is the set of strategy profiles that reach h.[63] Note also that $S_i(\emptyset) = S_i$;
- $u_i : S \to \mathbb{R}$ is i's strategic-form payoff function.

A history represents a (possibly partial) path of play. In a perfect-information game, a history is an ordered list of actions

In order to analyze players' reasoning at each point in the game, it is necessary to adopt an expanded notion of probabilistic beliefs and correspondingly redefine type structures. Specifically, we need a model of *conditional beliefs*. Following Ben-Porath (1997) (see also Battigalli and Siniscalchi, 1999), we adopt the following notion, originally proposed by Rényi (1955).

Definition 12.21. *Fix a measurable space* (Ω, Σ) *and a countable collection* $\mathcal{B} \subset \Sigma$. *A **conditional probability system**, or CPS, is a map* $\mu : \Sigma \times \mathcal{B} \to [0, 1]$ *such that:*
1. *For each* $B \in \mathcal{B}$, $\mu(\cdot|B) \in \Delta(\Omega)$ *and* $\mu(B|B) = 1$.
2. *If* $A \in \Sigma$ *and* $B, C \in \mathcal{B}$ *with* $B \subset C$, *then* $\mu(A|C) = \mu(A|B) \cdot \mu(B|C)$.

The set of CPSs on (Ω, Σ) *with conditioning events* $\mathcal{B}$ *is denoted* $\Delta^{\mathcal{B}}(\Omega)$.

A conditional probability system is a belief over a space of uncertainty Ω, together with conditional beliefs over a collection $\mathcal{B}$ of conditioning events $B \subset \Omega$.[64] In the simplest application of this definition we shall consider, we take the point of view of Player i: the domain Ω of her uncertainty is the set S_{-i} of strategy profiles that may be played by her opponents, and, roughly speaking, $\mathcal{B}$ is the set of i's information sets. Formally, if play reaches a (nonterminal) history $h \in \mathcal{H}$, Player i can infer that her opponents are playing a strategy profile in $S_{-i}(h)$. Thus, the relevant set of conditioning events is $\mathcal{B} \equiv \{S_{-i}(h) : h \in \mathcal{H}\}$, and $\mu(\cdot|S_{-i}(h))$ denotes the conditional belief held by i at history h.

Condition 2 is the essential property of CPSs: it requires that the standard rule of conditioning, or updating, be applied "whenever possible." Note that player i's initial beliefs are given by $\mu(\cdot|S_{-i}(\emptyset)) = \mu(\cdot|S_{-i})$. Now suppose that, in a given play of the game, the history reached next is h. If Player i initially considered h possible—that is, if $\mu(S_{-i}(h)|S_{-i}) > 0$—then her beliefs $\mu(\cdot|S_{-i}(h))$ must be obtained from $\mu(\cdot|S_{-i})$ via the usual updating formula. If, on the other hand, Player i initially assigned zero probability to the event that h was reached (more precisely, to opponents' strategies that allow h to be reached), then the standard updating rule clearly cannot apply, and $\mu(\cdot|S_{-i}(h))$

[63] For general extensive games, one can define the sets $S(h)$, $S_i(h)$ and $S_{-i}(h)$; perfect recall implies the equality $S(h) = S_i(h) \times S_{-i}(h)$. The class of games we consider do satisfy perfect recall, so this equality holds.

[64] For a decision-theoretic analysis of conditional probability systems, see Myerson (1997), Blume et al. (1991), and Siniscalchi (2014).

is unconstrained (except for the natural requirement that $\mu(S_{-i}(h)|S_{-i}(h)) = 1$). However, suppose that the history reached after h is h', and $\mu(S_{-i}(h')|S_{-i}(h)) > 0$: in this case, Condition 2 requires that Player i derive $\mu(\cdot|S_{-i}(h'))$ from $\mu(\cdot|S_{-i}(h))$ via standard updating. That is, following a surprise event, a player is allowed to revise her beliefs in an essentially unconstrained way; however, once she has done so, she has to conform to standard updating until a new surprise event is observed. Analogous assumptions underlie solution concepts such as sequential equilibrium (Kreps and Wilson, 1982) or perfect Bayesian equilibrium (Fudenberg and Tirole, 1991b); however, these equilibrium concepts add further restrictions on beliefs following surprise events.

For example, consider the centipede game of Fig. 12.2, and suppose that 2's initial beliefs μ are $\mu(\{AD\}|S_1) = \mu(\{D\}|S_1) = 0.5$. Then, conditional upon reaching the second node, player 2 must assign probability one to AD. If instead $\mu(\{D\}|S_1) = 1$, then 2's conditional beliefs at the second node must assign probability zero to strategies that choose D at the initial node, but are otherwise unconstrained.

We now define sequential rationality with respect to a CPS over opponents' strategies. Unlike, e.g., Kreps and Wilson (1982), but as in, e.g., Rubinstein (1991), Reny (1992), and Dekel et al. (1999), we do not require that a strategy s_i of Player i be optimal at *all* histories, but only at those that are not ruled out by s_i itself.[65]

Here and subsequently, we denote by $\mathcal{B}_{-i}$ the conditioning events $S_{-i}(h)$, $h \in \mathcal{H}$.

Definition 12.22. *[(Weak) Sequential Rationality] Fix a player $i \in I$, a CPS $\mu \in \Delta^{\mathcal{B}_{-i}}(S_{-i})$ and a strategy $s_i \in S_i$. Say that s_i is a* ***sequential best response*** *to μ iff, for all $h \in \mathcal{H}_i(s_i)$ and all $s_i' \in S_i(h)$,*

$$\mathbb{E}_{\mu(\cdot|S_{-i}(h))}[u_i(s_i, \cdot)] \geq \mathbb{E}_{\mu(\cdot|S_{-i}(h))}[u_i(s_i', \cdot)].$$

In this case, we say that the CPS μ ***justifies*** *the strategy s_i.*

That is, the strategy specified by s_i at every information set it reaches is optimal given the conditional beliefs at that information set.

Finally, we define type structures for multistage games. The definition is analogous to the one for strategic-form games (Definition 12.3); the key difference is that types are mapped to CPSs (rather than probabilities) over opponents' strategies and types. An essential element of the following definition is the assumption that each type holds beliefs conditional upon reaching every history; thus, the conditioning events are of the form $S_{-i}(h) \times T_{-i}$.

[65] Choices specified by s_i at histories precluded by s_i itself are payoff-irrelevant. They are important in equilibrium notions such as Sequential Equilibrium because they represent other players' beliefs about i's play at counterfactual histories; in particular, the requirement that choices at such histories be optimal reflect the assumption that opponents believe i to be rational. For example, in the game of Figure 12.2, 1's BI strategy DD encodes 2's belief that 1 will choose D at the last node. But, in an epistemic approach such assumptions can and should be modeled explicitly using opponents' CPSs.

Definition 12.23. *A type structure for the multistage game* $\Gamma = (I, \mathcal{H}, (\mathcal{H}_i, S_i(\cdot), u_i)_{i\in I})$ *is a tuple* $\mathcal{T} = (I, (\mathcal{C}_{-i}, T_i, \beta_i)_{i\in I})$, *where each* T_i *is a compact metric space,*
1. $\mathcal{C}_{-i} = \{S_{-i}(h) \times T_{-i} : h \in \mathcal{H}\}$,
2. $\beta_i : T_i \to \Delta^{\mathcal{C}_{-i}}(S_{-i} \times T_{-i})$,
and each β_i *is continuous.*[66] *We also write* $\beta_{i,h}(t_i) = \beta_i(t_i)(\cdot|S_{-i}(h) \times T_{-i})$.

Note that a type t_i for player i specifies conditional beliefs at histories h where i has nontrivial choices to make, and also histories at which i is essentially not active. In particular, this is true for $h = \emptyset$, the initial history. This simplifies the discussion of assumptions such as "common belief in rationality at a history" (e.g., initial CBR). It is sometimes convenient to refer to a tuple $(s, t) = (s_i, t_i)_{i\in I}$ as a **state**.

Battigalli and Siniscalchi (1999) construct a type structure for extensive games that is canonical (types are collectively coherent hierarchies of conditional beliefs), embeds all other structures as a belief-closed subset, and is complete. Their construction extends the one we provided in Section 12.2 for strategic-form games. As in that section, we denote by H_i the set of X_{-i}-based hierarchies of conditional beliefs for player i, and by $\varphi_i : T_i \to H_i$ the **belief hierarchy map** that associates with each type in a type structure $\mathcal{T}$ the hierarchy of conditional beliefs that it generates (cf. Definition 12.5).

As in the previous sections, it is convenient to introduce explicit notation for first-order beliefs. The first-order beliefs of a type t_i in an epistemic type structure for an extensive game is a CPS on S_{-i}. Thus, given a type structure $(I, (\mathcal{C}_{-i}, T_i, \beta_i)_{i\in I})$ for the extensive game $(I, \mathcal{H}, (\mathcal{H}_i, S_i(\cdot), u_i)_{i\in I})$, the **first-order belief map** $f_i : T_i \to \Delta^{\mathcal{B}_{-i}}(S_{-i})$ for Player i is defined by letting $f_i(t_i)(\cdot|S_{-i}(h)) = \text{marg}_{S_{-i}}\beta_{i,h}(t_i)$ for all $h \in \mathcal{H}$. It can be shown that $f_i(t_i)$ is indeed a CPS on S_{-i} with conditioning events $S_{-i}(h)$, $h \in \mathcal{H}$.

We now define the key ingredients of our epistemic analysis. The following is analogous to Definition 12.7 in Section 12.2.

Definition 12.24. *[Rationality and Conditional Belief] The event "Player i is sequentially rational" is*

$$R_i = \{(s_i, t_i) \in S_i \times T_i : s_i \text{ is a sequential best reply to } f_i(t_i)\}.^{67}$$

For every measurable subset $E_{-i} \subset S_{-i} \times T_{-i}$ *and history* $h \in \mathcal{H}$, *the event "Player i would believe that* E_{-i} *if h was reached" is*

$$B_{i,h}(E_{-i}) = \{(s_i, t_i) \in S_i \times T_i : \beta_{i,h}(t_i)(E_{-i}) = 1\}.$$

[66] The set $\Delta^{\mathcal{C}_{-i}}(S_{-i} \times T_{-i})$ is endowed with the relative product topology.

[67] This is a slight abuse of notation, because we have used R_i to denote strategic-form rationality in Definition 12.7.

12.7.3 Initial CBR

We begin with the simplest set of epistemic assumptions that take into account the extensive-form nature of the game, but are still close to strategic-form analysis in spirit.

Following Ben-Porath (1997), we consider the assumption that players are (sequentially) rational and *initially* commonly believe in (sequential) rationality:

$$RICBR_i^0 = R_i,$$
$$RICBR_i^k = RICBR_i^{k-1} \cap B_{i,\phi}(RICBR_{-i}^{k-1}) \quad \text{for} \quad k > 0, \qquad [12.22]$$
$$RICBR_i = \bigcap_{k \geq 0} RICBR_i^k.$$

Except for the fact that rationality is interpreted in the sense of Definition 12.24, these epistemic assumptions are analogous to RCBR in simultaneous-moves games, as defined in Eqs. 12.9 and 12.10. Direct restrictions on beliefs are imposed only at the beginning of the game–the "I" in $RICBR_i^k$ refers to this feature. In particular, following a surprise move by an opponent, Player i's beliefs are not constrained.

We now illustrate the above definitions. Table 12.12 represents a type structure for the Centipede game in Figure 12.2. Because we need to represent beliefs at different points in the game, we adopt a more compact notation than in the preceding sections. For each player, we indicate a numbered (non-exhaustive) list of strategy-type pairs; for each such pair (s_i, t_i), we describe the (conditional) beliefs associated with type t_i as a probability vector over the strategy-type pairs of the opponent. By convention, all strategy-type pairs that are not explicitly listed are assigned zero probability at all histories. Furthermore, we omit beliefs at histories that are not relevant for the analysis.

To interpret, each row in the two tables corresponds to a pair consisting of a strategy and a type for a player. Each vector in such a row is that type's probability distribution over the rows of the other player's table, conditional on reaching the history described by the column label. For example, consider the row numbered "1" in the table on the left, which corresponds to strategy-type pair (D, t_1^1) of player 1. The vector $(1, 0)$ indicates that, at the initial history ϕ, type t_1^1 believes that player 2 would choose d at the second node, and that 2's type is t_2^1; the vector $(0, 1)$ indicates that at the third node, i.e., after history $\langle A, a\rangle$, type t_1^1 believes that player 2 actually chose a at the second node. The interpretation of the other types is similar. Since we do not list strategy-type pair (D, t_1^2) in the table on the left, player 2 assigns probability 0 to it at every history.

Table 12.12 A type structure for the centipede game of Figure 12.2.

	(s_1, t_1)	$\beta_{1,\phi}(t_1)$	$\beta_{1,\langle A,a\rangle}(t_1)$
1	(D, t_1^1)	(1,0)	(0,1)
2	(AD, t_1^2)	(0,1)	(0,1)
3	(AA, t_1^3)	$(\frac{1}{2}, \frac{1}{2})$	(0,1)

	(s_2, t_2)	$\beta_{2,\phi}(t_2)$	$\beta_{2,\langle A\rangle}(t_2)$
1	(d, t_2^1)	(1,0,0)	(0,1,0)
2	(a, t_2^2)	(1,0,0)	(0,0,1)

State (D, d, t_1^1, t_2^1) supports the BI prediction. Player 1 chooses D at the initial node ϕ because she expects 2 to play d at the second node $\langle A \rangle$; Player 2 initially expects 1 to play D at ϕ, but indeed plans to choose d at $\langle A \rangle$ because, should he observe A, he would revise his beliefs and conclude that 1 is actually (rationally) playing AD. Instead, state (AD, a, t_1^2, t_2^2) corresponds to Reny's story, as formalized by Ben-Porath: player 1 initially expects 2 to choose a, and thus best responds with AD; player 2 initially expects 1 to play D and to hold beliefs consistent with backward induction, but upon observing A he revises his beliefs and concludes that 1 is actually irrational and will continue with A at the third node. Both states are consistent with RICBR. We obtain $RICBR_1^1 = \{(D, t_1^1), (AD, t_1^2)\}$ and $RICBR_2^1 = \{(d, t_2^1), (a, t_2^2)\}$. (Recall that subscripts refer to players and superscripts to iterations.) Note that all strategy-type pairs for 2 are in $RICBR_2^1$; hence, every type for 1 in $RICBR_1^1$ trivially assigns probability one to $RICBR_2^1$ at the initial history, so $RICBR_1^2 = RICBR_1^1$. Moreover, every type of 2 *initially* assigns probability one to (D, t_1^1), which is in $RICBR_1^1$; hence, $RICBR_2^2 = RICBR_2^1$. Repeating the argument shows that $RICBR_i^k = RICBR_i^1$ for all $k \geq 1$. Thus, as claimed in the introduction to this section, the BI prediction is consistent with RICBR, but so is the profile (AD, a).

As is the case for RCBR in simultaneous-move games, RICBR can be characterized via an iterative deletion algorithm (Battigalli and Siniscalchi, 1999) as well as a suitable notion of best reply set. Initial rationalizability (Definition 12.25) is like rationalizability, in that it iteratively deletes strategies that are not best replies. In each iteration, players' beliefs are restricted to assign positive probability only to strategies that survived the previous rounds. The differences are that here "best reply" means "sequential best reply" to a CPS, and only beliefs at the beginning of the game are restricted. Definition 12.26 is similarly related to best-reply sets, as in Definition 12.9.

Definition 12.25. (Initial Rationalizability) *Fix a multistage game $(I, \mathcal{H}, (\mathcal{H}_i, S_i(\cdot), u_i)_{i \in I})$. For every player $i \in I$, let $S_{i,\phi}^0 = S_i$. Inductively, for every $k > 0$, let $S_{i,\phi}^k$ be the set of strategies $s_i \in S_i$ that are sequential best replies to a CPS $\mu \in \Delta^{\mathcal{B}_{-i}}(S_{-i})$ such that $\mu(S_{-i,\phi}^{k-1} | S_{-i}) = 1$. Finally, the set of **initially rationalizable** strategies for i is $S_{i,\phi}^\infty = \bigcap_{k \geq 0} S_{i,\phi}^k$.*

Definition 12.26. *Fix a multistage game $(I, \mathcal{H}, (\mathcal{H}_i, S_i(\cdot), u_i)_{i \in I})$. A set $B = \prod_{i \in I} B_i \subset S$ is a **sequential best-reply set** (or SBRS) if, for every player $i \in I$, every $s_i \in B_i$ is a sequential best reply to a CPS $\mu_{-i} \in \Delta^{\mathcal{B}_{-i}}(S_{-i})$ such that $\mu(B_{-i} | S_{-i}) = 1$.*

*B is a **full SBRS** if, for every $s_i \in B_i$, there is a CPS $\mu_{-i} \in \Delta^{\mathcal{B}_{-i}}(S_{-i})$ that justifies it and such that (i) $\mu(B_{-i} | S_{-i}) = 1$, and (ii) all sequential best replies to μ_{-i} are also in B_i.*

One can easily see that, in any extensive game, S_ϕ^∞ is the largest SBRS. We have:

Theorem 12.16.
1. *In any type structure $\mathcal{T}$,* $\text{proj}_S RICBR$ *is a full SBRS;*
2. *in any complete type structure $\mathcal{T}$,* $\text{proj}_S RICBR = S^\infty_\phi$*;*
3. *for every full SBRS Q there exists a type structure $\mathcal{T}$ such that* $Q = \text{proj}_S RICBR$.

Ben-Porath (1997) shows that, in generic perfect-information games, RICBR characterizes the $S^\infty W$ procedure discussed in Section 12.5. Since S^∞_ϕ coincides with $S^\infty W$ in such games, Theorem 12.16 generalizes Ben-Porath's. Thus, for generic perfect-information games, Theorem 12.16 or, equivalently, Ben-Porath's result, provide an alternative, but related, interpretation of the $S^\infty W$ procedure: instead of relying on common p-belief in strategic-form rationality (Definition 12.7), RICBR imposes common 1-belief in sequential rationality (Definition 12.22) at the beginning of the game, but no restrictions on beliefs at other points in the game.

12.7.4 Forward induction

While the literature has considered a wide variety of "forward-induction" notions,[68] a common thread emerges: surprise events are regarded as arising out of purposeful choices of the opponents, rather than mistakes or "trembles." In turn, this implies that a player may try to draw inferences about future play from a past surprising choice made by an opponent. This leads to restrictions on beliefs conditional upon unexpected histories—precisely the beliefs that RICBR does not constrain.

In this section, we consider a particular way to constrain beliefs at unexpected histories, namely iterated strong belief in rationality. We first define strong belief (Section 12.7.4.1) and provide examples showing how strong belief in rationality yields forward and backward induction (Section 12.7.4.2). After providing the characterization results in Section 12.7.4.3, we discuss important properties of the notion of strong belief in Section 12.7.4.4.

12.7.4.1 Strong belief

Stalnaker (1998) and, independently, Battigalli and Siniscalchi (2002) introduce the notion of "strong belief" and argue that it is a key ingredient of forward-induction reasoning:

[68] The expression "forward induction" was coined by Kohlberg and Mertens (1986). The Battle of the Sexes with an outside option is an early example, which Kreps and Wilson (1982) attribute to Kohlberg. See also Cho and Kreps (1987). A recent axiomatic approach is proposed by Govindan and Wilson (2009) (though axioms are imposed on solution concepts, not behavior or beliefs).

Definition 12.27. (Strong Belief) *Fix a type structure $(I, (\mathcal{C}_{-i}, T_i, \beta_i)_{i\in I})$ for an extensive game $(I, \mathcal{H}, (\mathcal{H}_i, S_i(\cdot), u_i)_{i\in I})$. For any player $i \in I$ and measurable subset $E_{-i} \subset S_{-i} \times T_{-i}$, the event "Player i* ***strongly believes*** *that E_{-i}" is*

$$SB_i(E_{-i}) = \bigcap_{h\in\mathcal{H}:[S_{-i}(h)\times T_{-i}]\cap E_{-i}\neq\emptyset} B_{i,h}(E_{-i}).$$

In words: if E_{-i} *could* be true in a state of the world where h can be reached,[69] then, upon reaching h, player i must believe that E_{-i} is in fact true. More concisely: *player i believes that E_{-i} is true whenever possible.*

12.7.4.2 Examples

In this subsection, we discuss the implications of strong belief in rationality for the Centipede game (Figure 12.2) and the Battle of the Sexes with an outside option (Figure 12.3; see Kohlberg and Mertens, 1986).

Example 12.7. (The Centipede Game) *Consider again the type structure in Table 12.12. As noted above, type t_2^2 of player 2 initially believes that 1 is rational, but becomes convinced that Player 1 is irrational in case 1 chooses A at ϕ. However, note that there is a rational strategy-type pair for player 1 that chooses A at ϕ, namely (AD, t_1^2). Strong belief in 1's rationality, $SB_2(R_1)$, then requires player 2 to believe at the second node, i.e., at history $\langle A\rangle$, that 1 is rational. Therefore, type t_2^2 of player 2 is* not *consistent with strong belief in 1's rationality, because, conditional on $\langle A\rangle$, he assigns probability one to 1 playing the irrational strategy AA. On the other hand, consider now type t_2^1: upon seeing $\langle A\rangle$ type t_2^1 assigns probability one precisely to 1's strategy-type pair (AD, t_1^2); therefore, t_2^1 is the only type of 2 that is consistent with strong belief in 1's rationality.*

Since $SB_2(R_1) = \{t_2^1\}$ and t_2^1 expects player 1 to play D at the third node, $R_2 \cap SB_2(R_1) = \{(d, t_2^1)\}$. That is, the joint assumptions that 2 is rational and that he strongly believes in 1's rationality yield the conclusion that 2 should plan to play d. Thus, in the type structure of Table 12.12, rationality and strong belief in rationality eliminate the non-BI outcome (AD, a). Observe that this is achieved not by arguing that, at the second node, player 2 believes that 1's initial choice of A was a mistake—an unintended deviation from her planned strategy; rather, player 2 interprets prior actions as purposeful, insofar as this is possible. If in addition player 1 is rational and strongly believes $R_2 \cap SB_2(R_1)$, one obtains the backward-*induction outcome via* forward-*induction reasoning. We will return to this point in Subsection 12.7.4.3.*

Example 12.8. (Battle of the Sexes with an outside option) *Consider the game in Figure 12.3.*

[69] That is, if there is a profile $(s_{-i}, t_{-i}) \in E_{-i}$ such that $s_{-i} \in S_{-i}(h)$.

An informal forward-induction argument runs as follows: InB is a strictly dominated strategy for Player 1, because Out yields a strictly higher payoff regardless of 2's choice. "Therefore," if the simultaneous-moves subgame is reached, Player 2 should expect Player 1 to play T, and best-respond with L. But then, if Player 1 anticipates this, she will best respond with In, followed by T (i.e., she will choose strategy InT).

The right-hand side of Figure 12.3 displays a type structure, denoted $\mathcal{T}^{FI}$, *where strong belief in rationality reflects this reasoning process. Note that InB is irrational regardless of 1's beliefs, and furthermore InT is irrational for 1's type* t_1^1, *because this type expects 2 to play R. Thus,* $R_1 = \{(Out, t_1^1), (InT, t_1^2)\}$; *moreover, all strategy-type pairs are rational for Player 2. Neither types* t_2^1 *nor type* t_2^2 *are in* $SB_2(R_1)$. *To see this, first note that, conditional on every history, type* t_2^1 *assigns probability one to the irrational strategy-type pair* (InT, t_1^1). *Second, type* t_2^2 *initially believes that Player 1 rationally chooses Out, but upon observing* $\langle In \rangle$, *he switches to the belief that 1 plays the irrational strategy InB. On the other hand,* t_2^3 *is consistent with* $SB_2(R_1)$. *Since furthermore L is rational for type* t_2^3, *we have* $R_2 \cap SB_2(R_1) = \{(L, t_2^3)\}$. *Consequently, if one further assumes that 1 strongly believes that* $R_2 \cap SB_2(R_1)$, *type* t_1^1 *of player 1 must be eliminated (because it assigns probability one to* $(R, t_2^2) \notin R_2 \cap SB_2(R_1)$ *at every history). Thus,* $R_1 \cap B_{1,\phi}(R_2 \cap SB_2(R_1)) = \{(InT, t_1^2)\}$. *We have obtained the forward-induction outcome of this game, as claimed. Notice that the assumption that 1 initially believes that* $R_2 \cap SB_2(R_1)$ *is an assumption on how 1 expects 2 to revise his beliefs in case 2 is surprised: specifically, 1 expects 2 to maintain the belief that 1 is rational as long as possible.*

In the preceding examples, iterated strong belief in rationality selects backward- and forward-induction outcomes. Theorem 12.17 and Corollary 12.1 in Section 12.7.4.3 show that in complete type structures this always holds.[70] The following example shows that, in arbitrary, small type structures, these results need not hold. Section 12.7.4.4 discusses the reasons for the different conclusions reached in Examples 12.8 and 12.9. Theorem 12.17 also provides a characterization of iterated strong belief in rationality for arbitrary type structures.

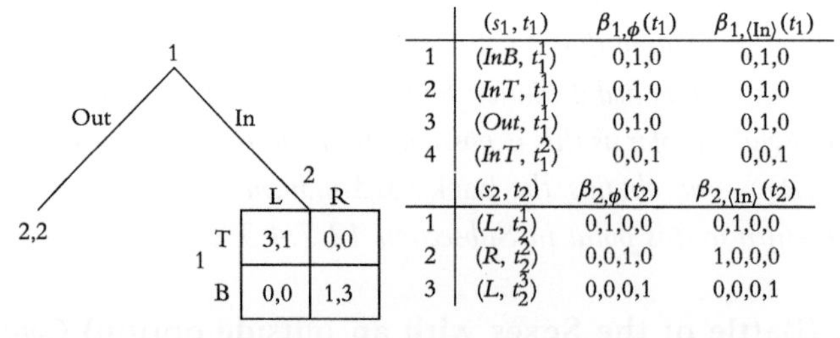

	(s_1, t_1)	$\beta_{1,\phi}(t_1)$	$\beta_{1,\langle In \rangle}(t_1)$
1	(InB, t_1^1)	0,1,0	0,1,0
2	(InT, t_1^1)	0,1,0	0,1,0
3	(Out, t_1^1)	0,1,0	0,1,0
4	(InT, t_1^2)	0,0,1	0,0,1

	(s_2, t_2)	$\beta_{2,\phi}(t_2)$	$\beta_{2,\langle In \rangle}(t_2)$
1	(L, t_2^1)	0,1,0,0	0,1,0,0
2	(R, t_2^2)	0,0,1,0	1,0,0,0
3	(L, t_2^3)	0,0,0,1	0,0,0,1

Figure 12.3 Battle of the Sexes with an outside option and the type structure $\mathcal{T}^{FI}$.

[70] Recall that in a complete type structure the belief maps are onto.

Example 12.9. *Consider the type structure in Table 12.13, denoted $\mathcal{T}^{NFI}$, for the game in Figure 12.3.*

Note that, relative to the type structure $\mathcal{T}^{FI}$ given in the table in Figure 12.3, we have removed type t_1^2 for Player 1. As a consequence, now $R_1 = \{(Out, t_1^1)\}$. This implies that there is no rational strategy-type pair of Player 1 who plays In in the type structure $\mathcal{T}^{NFI}$. Therefore, upon observing In, *Player 2's beliefs are unconstrained by strong belief; thus, type t_2^2* is *consistent with* $SB_2(R_1)$ *in $\mathcal{T}^{NFI}$. Therefore, repeating the analysis in Subsection 12.7.4.2 now leads to a different conclusion: if Player 1 initially (or strongly) believes $R_2 \cap SB_2(R_1)$, and if she is rational, she will choose the action Out at the initial node, contrary to what the standard FI argument for this game predicts.*■

12.7.4.3 RCSBR and extensive-form rationalizability

Battigalli and Siniscalchi (2002) consider the implications of rationality and common strong belief in rationality (RCSBR), which is the strong-belief counterpart of [12.8] and [12.9] for belief and, respectively, [12.18] for p-belief. For every player $i \in I$, let

$$\begin{aligned} RCSBR_i^0 &= R_i, \\ RCSBR_i^k &= RCSBR_i^{k-1} \cap SB_i(RCSBR_{-i}^{k-1}) \quad \text{for} \quad k > 0, \\ RCSBR_i &= \bigcap_{k \geq 0} RCSBR_i^k. \end{aligned} \qquad [12.23]$$

As was the case for iterated p-belief, it *does* matter whether we define mutual and common strong belief as above, or by iterating the mutual strong belief operator as in [12.7]. Once again, the reason is that strong belief does not satisfy the Conjunction property in [12.5]. We discuss this further in Section 12.7.4.4.

To see why RCSBR selects the forward-induction outcome, as illustrated by Example 12.8, we study [12.23] in more detail. Consider the two-player case for simplicity, take the point of view of player 1, and focus on $k = 2$ to illustrate:

$$RCSBR_1^2 = R_1 \cap SB_1(R_2) \cap SB_1\big(R_2 \cap SB_2(R_1)\big). \qquad [12.24]$$

Note that, since $R_2 \cap SB_2(R_1) \subset R_2$, every history h in an arbitrary game can fall into one of three categories: (0) histories inconsistent with R_2 and hence a fortiori with $R_2 \cap SB_2(R_1)$; (1) histories consistent with R_2 but not with $R_2 \cap SB_2(R_1)$; and

Table 12.13 Type structure $\mathcal{T}^{NFI}$ for the battle of the sexes.

	(s_1, t_1)	$\beta_{1,\phi}(t_1)$	$\beta_{1,(\text{In})}(t_1)$
1	(InB, t_1^1)	0,1	0,1
2	(InT, t_1^1)	0,1	0,1
3	(Out, t_1^1)	0,1	0,1

	(s_2, t_2)	$\beta_{2,\phi}(t_2)$	$\beta_{2,(\text{In})}(t_2)$
1	(L, t_2^1)	0,1,0	0,1,0
2	(R, t_2^2)	0,0,1	1,0,0

(2) histories consistent with $R_2 \cap SB_2(R_1)$ and hence a fortiori R_2. Equation [12.24] requires the following: at histories of type 1, player 1 should assign probability one to R_2; at histories of type 2, she should assign probability one to $R_2 \cap SB_2(R_1)$.[71] Interpreting $R_2 \cap SB_2(R_1)$ as a "strategically more sophisticated" assumption about 2's behavior and beliefs than R_2, [12.24] requires that, at any point in the game, players draw inferences from observed play by attributing the highest possible degree of strategic sophistication to their opponents.[72]

Theorem 12.17 below shows that RCSBR is characterized by extensive-form rationalizability (Pearce, 1984) in any complete type structure, and by the notion of "extensive-form best reply set" (Battigalli and Friedenberg, 2012) in arbitrary type structures. Extensive-form rationalizability is similar to initial rationalizability (Definition 12.25), except that beliefs are restricted to assign positive probability to strategies that survive the previous rounds at *all* histories where it is possible to do so: see [12.25]. Extensive-form best-reply sets bear the same relationship to SBRSs (Definition 12.26).

Definition 12.28. *[Extensive-Form Rationalizability] Fix a multistage game* $(I, \mathcal{H}, (\mathcal{H}_i, S_i(\cdot), u_i)_{i \in I})$. *For every player* $i \in I$, *let* $\hat{S}_i^0 = S_i$. *Inductively, for every* $k > 0$, *let* $\hat{S}_i^k$ *be the set of strategies* $s_i \in \hat{S}_i^{k-1}$ *that are sequential best replies to a CPS* $\mu \in \Delta^{\mathcal{B}_{-i}}(S_{-i})$ *such that*

$$\text{for every } h \in \mathcal{H}_i, \qquad S_{-i}(h) \cap \hat{S}_{-i}^{k-1} \neq \emptyset \quad \text{implies} \quad \mu(\hat{S}_{-i}^{k-1} | S_{-i}(h)) = 1. \qquad [12.25]$$

$\hat{S}^\infty = \bigcap_{k \geq 0} \hat{S}^k$ *is the set of* ***extensive-form rationalizable*** *strategy profiles.*

Definition 12.29. *Fix a two-player multistage game* $(\{1,2\}, \mathcal{H}, (\mathcal{H}_i, S_i(\cdot), u_i)_{i=1,2})$.[73] *A set* $B = B_1 \times B_2 \subset S$ *is an* ***extensive-form best-reply set*** *(or EFBRS) if, for every player* $i = 1, 2$, *every* $s_i \in B_i$ *is a sequential best reply to a CPS* $\mu_{-i} \in \Delta^{\mathcal{B}_{-i}}(S_{-i})$ *such that, for every* $h \in \mathcal{H}_i$ *with* $S_{-i}(h) \cap B_{-i} \neq \emptyset$, $\mu(B_{-i}|S_{-i}(h)) = 1$.

B is a ***full EFBRS*** *if, for every* $i = 1, 2$ *and* $s_i \in B_i$, *there is a CPS* $\mu_{-i} \in \Delta^{\mathcal{B}_{-i}}(S_{-i})$ *that justifies* s_i, *and such that (i)* $\mu(B_{-i}|S_{-i}) = 1$ *for every* $h \in \mathcal{H}_i$ *that satisfies* $S_{-i}(h) \cap B_{-i} \neq \emptyset$, *and (ii) all sequential best replies to* μ_{-i} *are also in* B_i.

Theorem 12.17.

1. *In any type structure* $\mathcal{T}$ *for a two-player multistage game,* $\text{proj}_S RCSBR$ *is a full EFBRS;*
2. *in any complete type structure* $\mathcal{T}$ *for an arbitrary multistage,* $\text{proj}_S RCSBR = \hat{S}^\infty$;
3. *for every full EFBRS Q of a two-player multistage game, there exists a type structure* $\mathcal{T}$ *such that* $Q = \text{proj}_S RCSBR$.

[71] Because strong belief does not satisfy conjunction, the right-hand side of [12.24] is *not* equivalent to $R_1 \cap SB_1(R_2 \cap SB_2(R_1))$.

[72] Battigalli (1996a) calls this the *Best Rationalization Principle.*

[73] We restrict attention to the two-player case to avoid issues of correlation (see Battigalli and Friedenberg, 2012, Section 9c for additional discussion).

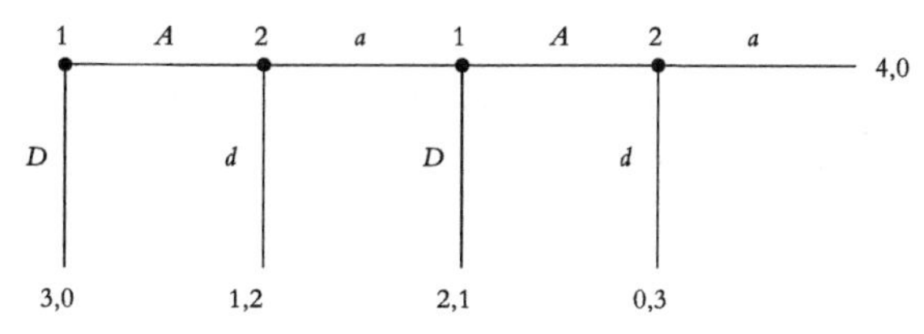

Figure 12.4 Backward and forward induction.

One consequence of Theorem 12.17 and the notion of strong belief is that, if a history h is reached under a strategy profile $s \in \hat{S}^{\infty}$, then there is common belief in rationality at h. Thus, while there may be histories where common belief in rationality may fail to hold, it does hold on the path(s) of play predicted by RCSBR. Though he does not use type structures, Reny (1993) defines iterative procedures motivated by the assumption that rationality is common belief at a given history.

Extensive-form rationalizability yields the BI outcome in generic perfect-information games (Battigalli, 1997; Heifetz and Perea, 2013). Combining this with Theorem 12.17, we obtain the following

Corollary 12.1. *In any complete type structure $\mathcal{T}$ for a generic perfect-information game Γ, any strategy profile $s \in \mathrm{proj}_S RCSBR$ induces the backward-induction outcome.*

Corollary 12.1 thus states that RCSBR in a complete type structure provides a *sufficient* epistemic condition for the BI outcome. Note that Corollary 12.1 does *not* state that $s \in \mathrm{proj}_S RCSBR$ is the (necessarily unique) BI *profile*, but only that such an s induces the BI *outcome*. Indeed, the BI profile may even be inconsistent with RCSBR. Both points are illustrated by the game in Figure 12.4, due to Reny (1992).

The unique BI profile in this game is (DD, dd). However, the extensive-form rationalizable profiles—hence, the profiles supported by RCSBR in complete type structures—are $(\{DD, DA\}, ad)$. To see this, note that strategy AD is strictly dominated by choosing D at the initial node. Hence, if play reaches the second node, player 2 must conclude that player 1 is playing AA, which makes ad strictly better than choosing d at the second node (note that strategy aa is not sequentially rational). In turn, this leads player 1 to choose D at the first node; hence, the backward-induction outcome obtains. However, RCSBR implies that, conditional upon observing A at the first node, and hence reaching the second node, 2 would expect that 1 will continue with A at the third node. Therefore, RCSBR implies that 2 would play ad, which is not his backward-induction strategy.

12.7.4.4 Discussion

The different predictions in Examples 12.8 and 12.9 raise several questions. First, in one case, the event *RCSBR* yields the forward-induction outcome, and in the other case it does not. While we noted that the forward-induction conclusion relies on the type structure being sufficiently rich, this merits further discussion. Relatedly, in Example 12.8 type t_2^2 in $\mathcal{T}^{FI}$ is in $RCSBR$, while in Example 12.9 type t_2^2 in $\mathcal{T}^{NFI}$, which has exactly the same hierarchy of beliefs, is not in $RCSBR$. This raises doubts about whether RCSBR depends only on belief hierarchies, or on the type structure as well, i.e., whether it is an elicitable assumption or not. In this section we address these concerns. It is useful to begin with a discussion of properties of strong belief.

Strong belief is not monotonic, and violates conjunction (cf. [12.5]).[74] To see this, consider again the type structure in Table 12.12 for the Centipede game, and focus on the events

$$SB_1(R_2 \cap SB_2(R_1)) \quad \text{and} \quad SB_1(R_2) \cap SB_1(SB_2(R_1)).$$

As shown in Example 12.7, $R_2 \cap SB_2(R_1) = \{(d, t_2^1)\}$. Now observe that type t_1^1 initially assigns probability one to $\{d, t_2^1)\}$. Furthermore, if player 2 plays d, history $\langle A, a\rangle$ is not reached; hence, strong belief in $R_2 \cap SB_2(R_1) = \{(d, t_2^1)\}$ imposes no restriction on beliefs at $\langle A, a\rangle$. Therefore, type t_1^1 strongly believes $R_2 \cap SB_2(R_1)$, so $SB_1(R_2 \cap SB_2(R_1)) \neq \emptyset$. On the other hand, $SB_1(SB_2(R_1)) = \emptyset$. To see this, note first that $SB_2(R_1) = S_2 \times \{t_2^1\}$. This event is consistent with the third node, i.e. history $\langle A, a\rangle$, being reached; therefore, strong belief in $SB_2(R_1)$ requires that player 1 assign probability one to this event conditional on $\langle A, a\rangle$. However, no type of player 1 in Table 12.12 assigns positive probability to 2's type t_2^1 at that history. Hence, $SB_1(R_2) \cap SB_1(SB_2(R_1)) = \emptyset \neq SB_1(R_2 \cap SB_2(R_1))$.[75] This failure of monotonicity and conjunction plays an important role in the subsequent discussion.

Throughout this chapter, we interpret events such as R_i, $B_i(E_{-i})$ or $SB_i(E_{-i})$ defined in a given type structure $\mathcal{T}$ as "player i is rational," "player i believes E_{-i}," or "player i strongly believes E_{-i}." While convenient, this is not quite accurate. Every type structure defines the set of belief hierarchies that are allowed for each player. For instance, consider type structure $\mathcal{T}^{NFI}$ in Example 12.9 and denote by φ_i^{NFI} its belief hierarchy maps (Definition 12.5). The event R_1 in $\mathcal{T}^{NFI}$ should be interpreted as "player 1 is rational and her belief hierarchy is $\varphi_1^{NFI}(t_1^1)$." Similarly, the event $B_{2,\phi}(R_1)$ should be interpreted as "player 2 initially believes that '1 is rational and her belief hierarchy is $\varphi_1^{NFI}(t_1^1)$' and 2's belief hierarchy is either $\varphi_2^{NFI}(t_2^1)$ or $\varphi_2^{NFI}(t_2^2)$." We typically avoid such convoluted statements, but must recognize that the simpler statements "1 is rational" and "2 initially believes that 1 is rational" are precise interpretations of R_1 and $B_{2,\phi}(R_1)$ only if we define these events in a rich type structure—one that generates all hierarchies of (conditional) beliefs.

[74] We noted in Footnote 10 that monotonicity is equivalent to the "$\subset$" part of conjunction. Indeed, that is the only part of conjunction that strong belief does not satisfy.

[75] It can also be shown that $SB_1(R_2) \cap SB_1(SB_2(R_1))$ is empty in any complete type structure for the centipede game in Figure 12.2.

Fortunately, this is not an issue when interpreting results that use monotonic belief operators. For example, consider Theorem 12.1. On the one hand, the event *RCBR* is accurately interpreted as "rationality and common belief in rationality," RCBR, only in a complete type structure. Indeed, in smaller type structures, as explained earlier, the event *RCBR* should be interpreted as RCBR jointly with additional assumptions about hierarchical beliefs. On the other hand, because of Monotonicity,[76] the hierarchies consistent with the event *RCBR* in a small type structure are also consistent with *RCBR* in a complete type structure. Consequently, every full BRS is a subset of the rationalizable set S^{∞}. Since the latter strategies are unambiguously interpreted as consistent with RCBR (because $S^{\infty} = \text{proj}_S RCBR$ in a complete type structure), so are the former.

Now consider strong belief. A cumbersome but precise interpretation of the event $SB_2(R_1)$ in the type structure $\mathcal{T}^{NFI}$ is as follows: "player 2 believes that 'player 1 is rational and her belief hierarchy is $\varphi_1^{NFI}(t_1^1)$' at every history that can be reached if this assertion is true, and 2's belief hierarchy is either $\varphi_2^{NFI}(t_2^1)$ or $\varphi_2^{NFI}(t_2^2)$." If instead we define event $SB_2(R_1)$ in the type structure $\mathcal{T}^{FI}$ of Figure 12.3, its precise interpretation is, "player 2 believes that 'player 1 is rational and her belief hierarchy is either $\varphi_1^{FI}(t_1^1)$ or $\varphi_1^{FI}(t_1^2)$' at every history that can be reached if this assertion is true, and 2's belief hierarchy is $\varphi_2^{FI}(t_2^1)$,$\varphi_2^{FI}(t_2^2)$ or $\varphi_2^{FI}(t_2^3)$."

Observe that these statements are expressed in terms of strategies and hierarchies of conditional beliefs, and hence, they may be elicited in principle. Thus, there is no conflict between our goal of elicitation and the notion of strong belief. The apparent conflict arises from an imprecise interpretation of strong belief in small type structures.

That said, the interpretation of the event *RCSBR* in small type structures is subtle because strong belief does not satisfy Monotonicity. As above, the event *RCSBR* is accurately interpreted as "rationality and common strong belief in rationality," RCSBR, only in a complete type structure. However, in contrast to the case of RCBR, due to the failure of Monotonicity, the hierarchies consistent with the event *RCSBR* in a small type structure need *not* be consistent with *RCSBR* in a larger (a fortiori, in a complete) type structure.[77] Hence, a full EFBRS need not be a subset of the extensive-form rationalizable set $\hat{S}^{\infty}$.[78] Thus, while the latter strategies can accurately be interpreted

[76] If a hierarchy of player 2 is consistent with belief in 1's rationality under the assumption that 1's beliefs are constrained in some specific way, then by monotonicity it is also consistent with belief in 1's rationality without additional restrictions on 1's beliefs (and obviously this argument can be iterated).

[77] Recall that, in Examples 12.8 and 12.9, the hierarchy $\varphi_2^{NFI}(t_2^2) = \varphi_2^{FI}(t_2^2)$ of player 2 is consistent with strong belief in the statement "1 is rational and 1's hierarchy is $\varphi_1^{NFI}(t_1^1)$," but not with strong belief in the statement "1 is rational and 1's hierarchy is either $\varphi_1^{FI}(t_1^1) = \varphi_1^{NFI}(t_1^1)$ or $\varphi_1^{FI}(t_1^2)$."

[78] Consider for instance the game in Figure 12.3. As we noted above, $\hat{S}^{\infty} = \{InT\} \times \{L\}$. However, consider the set $Q = \{Out\} \times \{R\}$. The strategy *Out* for player 1 is the unique (sequential) best reply to any CPS that initially assigns probability 1 to R. Furthermore, R is the unique sequential best reply to the CPS that assigns probability one to *Out* at the beginning of the game, and to *InB* conditional on reaching the

as consistent with RCSBR, this is not the case for the former. We can say, however, that strategies in an EFBRS are consistent with "rationality plus additional assumptions on hierarchies, and common strong belief thereof."

So, what is the "right" solution concept? If the analyst is interested in the implications of RCSBR, without any additional assumptions, then the answer is $\hat{S}^{\infty}$. (Analogously, for simultaneous-move games, the answer is S^{∞}). If the analyst wants to impose some particular additional assumption about beliefs, then the answer is a particular EFBRS; which EFBRS depends on the assumption. (For simultaneous-move games, the answer is a particular BRS). Finally, if the analyst wants to be "cautious" and consider the predictions that would arise were she to adopt any possible assumption, then the answer is the (player-by-player) union of all EFRBS's. (For simultaneous-move games, the answer is the union of all BRS's, which in this case is again S^{∞}.) The bottom line is that, when interpreting assumptions involving strong belief, one should be careful to specify whether or not additional assumptions are imposed on players' beliefs.

These considerations apply in particular to the relationship between strong belief and forward-induction reasoning. As we noted, the basic intuition underlying forward induction is that players attempt to maintain the assumption that their opponents are rational as long as possible, in the face of unexpected behavior. This suggests that no a priori constraint is placed on players' attempt to rationalize deviations. In other words, a connection can be established between forward induction and (iterated) strong belief in *complete* type structures. Theorem 12.17 confirms this. When strong belief is applied in small type structures, there is an interaction between the rationalization logic of forward induction and whatever assumptions are exogenously imposed on beliefs; this yields EFBRS's, as stated in Theorem 12.17.[79]

12.7.5 Backward induction

As we noted above, RCSBR in a complete type structure yields the backward-induction outcome, but *not* the backward-induction profile. The game of Figure 12.4 provides an example. An alternative way to get backward induction is to simply make explicit the assumption inherent in BI: at all nodes where he is active, i makes conditionally optimal

simultaneous-move subgame. Therefore, Q is an EFBRS, because Q_1 does not reach the subgame and so there are no further restrictions on 2's beliefs. Yet, the sets Q and $\hat{S}^{\infty}$ are disjoint, and induce distinct outcomes.

[79] An equivalent way (in our setting) to incorporate belief restrictions in the analysis is to work in the canonical type structure and explicitly define events C_i that formalize the desired additional assumptions. Then, studying the behavioral implications of events such as $R_i \cap C_i \cap SB_i(R_{-i} \cap C_{-i})$ is the same as studying the implications of the event $R_i \cap SB_i(R_{-i})$ in a type structure that incorporates the desired restrictions on beliefs. A detailed discussion of these issues can be found in Battigalli and Friedenberg (2012) and Battigalli and Prestipino (2012).

choices, i believes that his opponent j does so at subsequent histories where she is active, i believes that j believes that k will also choose optimally at subsequent nodes where k is active, etc.; see Perea (2011b).[80]

Aumann (1995) (see also Balkenborg and Winter, 1997) derives the backward induction profile from "common knowledge of rationality" in a very different epistemic model.[81] As it does not explicitly incorporate belief revision, Aumann's model lies outside our framework.

12.7.6 Equilibrium

The epistemic analysis of equilibrium concepts for extensive games is largely yet to be developed. In this subsection, we briefly describe results on subgame-perfect equilibrium, self-confirming equilibrium in signaling games, and the relationship between EFBRS's and (subgame-perfect) equilibrium.

A basic question is whether sufficient conditions for subgame-perfect equilibrium (Selten, 1975) can be provided by adapting the results of Aumann and Brandenburger (1995). Theorem 12.5, on two-player games, can be easily adapted. As this is not in the literature, we sketch the steps here.

To avoid introducing new notation, we describe i's play in an extensive-form game using a CPS σ_i on S_i, instead of using behavioral strategies.[82] Thus, a *subgame-perfect equilibrium* (SPE) of a multistage game $(I, \mathcal{H}, (\mathcal{H}_i, S_i(\cdot), u_i)_{i \in I})$ is a profile $(\sigma_i)_{i \in I}$, where $\sigma_i \in \Delta^{\mathcal{B}_i}(S_i)$ is a CPS on S_i with conditioning events $\mathcal{B}_i = \{S_i(h) : h \in \mathcal{H}\}$ for each player i, such that, at every history $h \in \mathcal{H}$, $\left(\sigma_i(\cdot|S_i(h))\right)_{i \in I}$ is a Nash equilibrium of the strategic-form game $\left(I, \left(S_i(h), u_i\right)_{i \in I}\right)$. To clarify how this definition is related to the usual ones, consider the profile $(OutB, R)$ in the game of Figure 12.3. Player 1's strategy is represented by any CPS σ_1 such that $\sigma_1(\{OutT, OutB\}|S_1) = 1$ and $\sigma_1(\{InB\}|S_1(h)) = 1$, where h denotes the simultaneous-move subgame. Player 2's strategy is represented by the CPS σ_2 defined by $\sigma_2(\{R\}|S_2) = \sigma_2(\{R\}|S_2(h)) = 1$. It is easy to verify that the profile of CPSs (σ_1, σ_2) satisfies the definition of SPE we have just given.

Turning to the epistemic analysis, we adapt the notation from Section 12.4: given a CPS $\mu_i \in \Delta^{\mathcal{B}_{-i}}(S_{-i})$ and a type structure $(I, (\mathcal{C}_{-i}, T_i, \beta_i)_{i \in I})$, let $[\mu_i]$ be the event that i's first-order belief is μ_i; given a profile $(\mu_i)_{i \in I}$, the event $[\mu]$ is defined as the intersection

[80] Stalnaker (1998) discusses belief revision in dynamic games. In particular, he characterizes backward induction in a similar way to that discussed earlier, but interprets belief in subsequent optimality as a consequence of a suitable independence assumption on beliefs about future and past play. See also Perea (2008).

[81] See also Aumann (1998) and Samet (1996). Samet (2013) extends Aumann (1995)'s analysis from knowledge to belief.

[82] We use σ_i to denote the CPS on S_i that describes i's *play*, as opposed to i's *beliefs*, denoted μ_i below, which are defined on S_{-i}.

of all $[\mu_i]$ for $i \in I$. Finally, define the event "Player i makes a rational choice at history $h \in \mathcal{H}$" as

$$R_{i,h} = \left\{ (s_i, t_i) : s_i \in \arg\max_{s'_i \in S_i(h)} u_i(s'_i, f_{i,h}(t_i)) \right\}.^{83}$$

To see how this is different from event R_i (Definition 12.24), consider strategy *InB* in the game of Figure 12.3. This is strictly dominated, hence (sequentially) irrational in the entire game; however, it does specify a choice in the simultaneous-move subgame that is a best reply to, e.g., the belief that assigns probability one to 2 playing R.

We can now state the counterpart to Theorem 12.5. Fix a CPS $\mu_i \in \Delta^{\mathcal{B}_{-i}}(S_{-i})$ for $i = 1, 2$. If $[\mu] \cap \bigcap_{h \in \mathcal{H}} B_h(R_h \cap [\mu]) \neq \emptyset$, then $(\sigma_1, \sigma_2) = (\mu_2, \mu_1)$ is a SPE.[84]

As is the case for simultaneous-move games, the situation is more delicate if there are more than two players. One approach is to adapt the definitions of agreement and independence of first-order beliefs in Section 12.4.4, thereby obtaining a counterpart to Theorem 12.6. Alternatively (cf. Barelli, 2010), one can adapt the notion of common prior (Definition 12.14) and obtain a counterpart to Theorem 12.7.[85] In order to adapt the arguments used to prove Theorem 12.7, the common "prior" for a type structure $(I, (\mathcal{C}_{-i}, T_i, \beta_i)_{i \in I})$ must be defined as a CPS $\mu \in \Delta^{\mathcal{B}}(S \times T)$, where $\mathcal{B} = \{S(h) : h \in \mathcal{H}\}$ such that $\beta_{i,h}(t_i) = \text{marg}_{S_{-i} \times T_{-i}} \mu(\cdot | S(h) \times \{t_i\} \times T_{-i})$ for *all* histories $h \in \mathcal{H}$.[86]

Asheim and Perea (2005) provide an epistemic characterization of sequential equilibrium (Kreps and Wilson, 1982) in two-player games. In their analysis, beliefs are represented using a generalization of lexicographic probability systems.

A different approach is explored by Battigalli and Siniscalchi (2002) in the context of signaling games. We do not formalize it, because doing so would require introducing notation that we do not use anywhere else in this chapter. Roughly speaking, they show that, in any epistemic model, if there is a state in which players' first-order beliefs are consistent with an outcome of the game (that is, a probability

[83] Recall that $f_i(t_i)$ is t_i's first-order belief, and $f_{i,h}(t_i)$ is the conditional of the CPS $f_i(t_i)$ given history h.

[84] Alternatively, one could get sufficient epistemic conditions for SPE by assuming that conjectures are given by μ and, at every history, there is mutual belief in $R \cap [\mu]$, where R is the event that all players are sequentially rational in the sense of Definition 12.24. This is the approach taken by Barelli (2009). However, these conditions rule out the SPE $(OutB, R)$ in the game of Figure 12.3. More generally, they may preclude certain SPE in games in which some histories can only be reached if a player plays a strictly dominated strategy.

[85] In this approach, since we continue to assume that players only hold beliefs about their opponents, the independence condition in Definition 12.15 would also have to be adapted. Note that Barelli (2009) allows players to hold beliefs about their own strategies, and hence does not require any additional condition.

[86] Barelli (2009) notes that this notion is very demanding, because it requires no betting even conditional upon histories that players do not expect to be reached.

distribution over terminal histories), and there is initial mutual belief in rationality and in the event that first-order beliefs are consistent with the outcome, then there exists a self-confirming equilibrium that induces that outcome. They also provide necessary and sufficient epistemic conditions for the outcome to be supported in a self-confirming equilibrium that satisfies the *Intuitive Criterion* of Cho and Kreps (1987).

Finally, Battigalli and Friedenberg (2012) relate EFBRSs, and hence iterated strong belief in rationality, with Nash and subgame-perfect equilibrium in two-person multistage games with observable actions. Every pure-strategy SPE is contained in some EFBRS. Moreover, under a no-relevant-ties condition (Battigalli, 1997), a pure-strategy SPE profile *is* an EFBRS. In perfect-information games that satisfy the "transfer of decision-maker indifference" of Marx and Swinkels (1997), if a state (s, t) in a type structure is consistent with the event $RCSBR$, then s is outcome equivalent to a Nash equilibrium. Conversely, in games with no relevant ties, for any Nash equilibrium profile (s_1, s_2) such that each strategy s_i is also a sequential best reply to some CPS on S_{-i}, there is a type structure and a profile t of types such that (s, t) is consistent with the event $RCSBR$.[87]

As should be clear from the above, this area is fertile ground for research. For instance, it would be interesting to investigate the implications of strong belief in rationality, and of the best-rationalization principle, in an equilibrium setting. Care is needed; for instance, one cannot assume that—as may appear natural—there is mutual or common belief in the conjectures at *every* information set, because that may be inconsistent with the logic of best rationalization.[88,89]

12.7.7 Discussion

Strategies: The choice-theoretic interpretation of strategies deserves some comment. In a simultaneous-move game, whether or not a player plays a given strategy is easily observed ex-post. In an extensive game, however, a strategy specifies choices at several histories as play unfolds. Some histories may be mutually exclusive, so that it is simply

[87] Proposition 7.1 in Brandenburger and Friedenberg (2010) implies that a similar relationship exists between Nash equilibria of perfect-information games and the epistemic conditions of lexicographic rationality and common assumption thereof, which we analyze in Section 12.8.

[88] For example, consider the game in Figure 12.2. RCSBR suggests that, on the one hand, player 1 believes that player 2 will choose d at his final decision node. On the other hand, if player 2 observes that player 1 has unexpectedly chosen A at her first move, then RCSBR within a complete type structure requires that player 2 explain this event by believing that player 1 believes that player 2 will choose a at his final decision node. In other words, RCSBR in a rich type structure implies that the equilibrium continuation strategies are *not* commonly believed at the third node.

[89] Reny (1992) introduces a notion of *explicable equilibrium* that is similarly motivated, though his analysis does not employ type structures and is thus closer to Pearce (1984)'s definition of extensive-form rationalizability.

not possible to observe whether a player actually follows a given strategy. Furthermore, it may be the case that epistemic assumptions of interest imply that a given history h will not be reached, and at the same time have predictions about what the player on the move at h would do if, counterfactually, h was reached. Consider for instance the Centipede game of Figure 12.2: as we noted above, RCSBR (in the type structure of Table 12.12) implies that player 1 will choose D at the initial node, and that player 2 would choose d if the second node was reached. Verifying predictions about such unobserved objects seems problematic. This is troublesome both in terms of testing the theory, and because it is not obvious how to elicit players' beliefs about such objects.

One obvious way to avoid this difficulty is to assume that players commit to observable contingent plans at the beginning of the game. While this immediately addresses the issue of verifiability, it seems to do so at the cost of turning the extensive game into a strategic-form game. However, one can impose the requirement that players prefer their plans to be conditionally, not just ex-ante optimal, even at histories they do not expect to be reached.[90] In this case, while players commit to specific plans, the extensive-form structure retains its role. Siniscalchi (2014) develops this approach.

An alternative approach, explored in Battigalli et al. (2011b), is to take as primitives the paths of play, rather than strategy profiles. In this case, at any history, player i chooses an action, given her beliefs about possible continuation paths. Notice that these paths include actions by i's opponents as well as actions that i herself takes. In this respect, such a model requires introspective beliefs about one's future play, in conflict with one of our key desiderata (Section 12.2.6.3). However, this approach does resolve the issue of verifiability of predictions, because these are now observable paths of play and not strategy profiles.[91]

It also enables decomposing the assumption of sequential rationality into the assumptions that (i) the player *expects* her (future) actions to be optimal given her (future) beliefs, and (ii) her *actual* choices at a state coincide with her *planned* actions.[92] This more expressive language can be used to elegantly characterize backwards induction and should also be useful to study environments where players do not correctly forecast their own play (including cases where utility depends on beliefs and are hence not necessarily dynamically consistent).[93]

[90] This is related to, but weaker than lexicographic expected-utility maximization (Definition 12.32); for details, see Siniscalchi (2014).

[91] A related model of conditional beliefs in dynamic games is considered by Di Tillio et al. (2012).

[92] See also Bach and Heilmann (2011).

[93] The characterization is elegant in that it obtains backward induction by weakening the assumption of correct forecasting (which is a way to model "trembles").

Incomplete information: The definition of multistage games can be easily extended to incorporate payoff uncertainty. As for simultaneous-move games, we specify a set Θ of payoff states or parameters, and stipulate that each player's payoff function takes the form $u_i : \Theta \times S \to \mathbb{R}$. If one assumes that payoff states are not observed, the analysis in the preceding subsections requires only minimal changes.[94] First-order beliefs are modeled as CPSs on $\Theta \times S$, with i's conditioning events being $\mathcal{B}_{-i} = \{\Theta \times S_{-i}(h) : h \in \mathcal{H}\}$. In an epistemic type structure, the conditioning events are $\mathcal{C}_{-i} = \{\Theta \times S_{-i}(h) \times T_{-i} : h \in \mathcal{H}\}$, and the belief maps are defined as functions $\beta_i : T_i \to \Delta^{\mathcal{C}_{-i}}(\Theta \times S_{-i} \times T_{-i})$. Chen (2011) and Penta (2012) extend the notion of ICR to dynamic games. It is also straightforward to adapt the notion of Δ-rationalizability to allow for incomplete information; one obtains versions of initial or strong rationalizability that incorporate commonly-believed restrictions on first-order beliefs. Epistemic characterizations adapting Theorems 12.16 and 12.17 may be found in Battigalli and Siniscalchi (2007) and Battigalli and Prestipino (2012).

12.8. ADMISSIBILITY

We now return to strategic-form analysis to analyze epistemic conditions and solution concepts related to *admissibility*, i.e., ruling out weakly dominated strategies. In particular, we will discuss epistemic conditions for *iterated admissibility*. This continues the analysis in Section 12.5: as noted therein, there is a conceptual inconsistency between the "everything is possible" logic behind admissibility and common-belief conditions. In Section 12.5, we introduced the notion of p-belief to resolve this inconsistency and weakened the notion of common belief accordingly. This section explores an alternative approach: we replace probabilistic beliefs with the richer concept of a *lexicographic probability system* (LPS). These are related to the CPSs introduced in Section 12.7 to study extensive-form solution concepts; we elaborate on the connection in Section 12.8.4. We saw that common p-belief in rationality yields $S^\infty W$. We shall now see that suitable epistemic conditions characterize iterated admissibility (and its best-reply set analog, "self-admissible sets"). The main idea (Brandenburger et al., 2008) is to introduce an analog to the notion of strong belief (Definition 12.27) for LPSs, called *assumption*.

A lexicographic probability system is a finite array $\mu_0, \ldots, \mu_K$ of probabilistic beliefs over, say, opponents' strategy profiles; μ_k is the *kth level* of the LPS (distinct from a kth *order* belief in a belief hierarchy). The lowest-level beliefs are the most salient, in the sense that, if a strategy s_i yields a strictly higher expected utility than another strategy s_i' with respect to μ_0, then s_i is preferred to s_i'. If, however, s_i and s_i' have the same μ_0-expected utility, then Player i computes μ_1-expected utilities, and so on. Thus, higher-level

[94] Even if there is private information, the changes required to adapt the analysis are only notational.

(less salient) probabilities are used to break ties.[95] In order to formalize the notion of "common assumption in lexicographic rationality," we need to modify our notion of type structure: types will now be mapped to LPSs over opponents' strategies and types.

12.8.1 Basics

We begin by defining LPSs and lexicographic type spaces.

Definition 12.30. (Blume et al., 1991; Brandenburger et al., 2008) *A **lexicographic probability system** (or LPS) $\sigma = (\mu_0, \ldots, \mu_{n-1})$ on a compact metric space Ω is a sequence of length $n < \infty$ in $\Delta(\Omega)$.*

*An LPS $\sigma = (\mu_0, \ldots, \mu_{n-1})$ has **full support** if $\bigcup_\ell \operatorname{supp} \mu_\ell = \Omega$. The set of LPSs on Ω is denoted $\mathcal{L}(\Omega)$. The set of full-support LPSs on Ω is denoted $\mathcal{L}^+(\Omega)$.*

Definition 12.31. *A **lexicographic type structure** for the strategic-form game $G = (I, (S_i, u_i)_{i \in I})$ is $\mathcal{T} = (N, (T_i, \beta_i)_{i \in I})$ where each T_i is a compact metric space and each $\beta_i : T_i \to \mathcal{L}(S_{-i} \times T_{-i})$ is continuous.*

In order to define best replies, we first recall the lexicographic (i.e., "dictionary") order on vectors.

Definition 12.32. *Fix vectors $x = (x_\ell)_{\ell=0}^{n-1}, y = (y_\ell)_{\ell=0}^{n-1} \in \mathbb{R}^n$, write $x \geq^L y$ iff*

$$y_j > x_j \text{ implies } x_k > y_k \text{ for some } k < j.^{96} \qquad [12.26]$$

*Given a strategic-form game $G = (I, (S_i, u_i)_{i \in I})$, a strategy s_i of player i is a **(lexicographic) best reply** to an LPS $\sigma_{-i} = (\mu_0, \ldots, \mu_{n-1})$ on S_{-i} if $\left(\pi_i(s_i, \mu_\ell)\right)_{\ell=0}^{n-1} \geq_L \left(\pi_i(s_i', \mu_\ell)\right)_{\ell=0}^{n-1}$ for all $s_i' \in S_i$.*

It is easy to see that a strategy is admissible if and only if it is a lexicographic best reply to a *full-support* LPS.

Given a type structure, we define "rationality" as usual; we also define "full-support beliefs" analogously to Definition 12.16.

Definition 12.33. *Fix a lexicographic type structure $\mathcal{T} = (I, (T_i, \beta_i)_{i \in I})$ for the game G. The event that player i is* rational *is*

$$R_i = \left\{ (s_i, t_i) \ : \ s_i \text{ is a lexicographic best reply to } \left(\operatorname{marg}_{S_{-i}} \mu_\ell\right)_{\ell=0}^{n-1}, \text{ where } \beta_i(t_i) = (\mu_0, \ldots, \mu_{n-1}) \right\}.^{97} \qquad [12.27]$$

[95] A behavioral characterization of lexicographic expected-utility maximization is provided by Blume et al. (1991).

[96] That is: either $x = y$ or there exists $k \in \{0, \ldots, n-1\}$ such that $x_j = y_j$ for $j = 0, \ldots, k-1$, and $x_k > y_k$.

[97] As in Definition 12.24, the repeated use of R_i is a slight abuse of notation.

The event that player i has full-support beliefs *is*

$$FS_i = \left\{(s_i, t_i) \ : \ \beta_i(t_i) \in \mathcal{L}^+(S_{-i} \times T_{-i})\right\}.^{98} \qquad [12.28]$$

12.8.2 Assumption and mutual assumption of rationality

We can now introduce the notion of assumption.

Definition 12.34. *Fix a lexicographic type structure $\mathcal{T} = (I, (T_i, \beta_i)_{i \in I})$ and an event $E_{-i} \subset S_{-i} \times T_{-i}$. Then (s_i, t_i)* ***assumes*** *E_{-i}, written $(s_i, t_i) \in A_i(E_{-i})$, iff $\beta_i(t_i) = (\mu_0, \ldots, \mu_{n-1})$ has full support and there is $\ell^* \in \{0, \ldots, n-1\}$ such that:*

(i) $\mu_\ell(E_{-i}) = 1$ for $\ell \leq \ell^*$;

(ii) $E_{-i} \subseteq \bigcup_{\ell \leq \ell^*} \operatorname{supp} \mu_\ell$;[99]

(iii) for every $\ell > \ell^*$ there exist numbers $\alpha_1, \ldots, \alpha_{\ell^*} \in \mathbb{R}$ such that, for every event $F_{-i} \subseteq E_{-i}$ such that $\mu_\ell(F_{-i}) > 0$, $\mu_\ell(F_{-i}) = \sum_{k \leq \ell^*} \alpha_k \mu_k(F_{-i})$.

Assumption captures the notion that E_{-i} and all its subsets are infinitely more likely than the complement of E_{-i}. The level-zero measure must assign probability one to E_{-i}, although its support may be a strict subset of E_{-i}. If it is a strict subset, then the remainder of E_{-i} must receive probability one in the next level, and so on, until all of E_{-i} has been "covered." For those measures that assign positive probability outside E_{-i}, i.e., those after level ℓ^*, their restriction to E_{-i} is behaviorally irrelevant. To elaborate, in any LPS on a set Ω, a measure that is a linear combination of lower-level measures can be removed without changing lexicographic expected-utility rankings. Therefore, part (iii) of Definition 12.34 states that, at levels $\ell > \ell^*$, either $\mu_\ell(E_{-i}) = 0$ or $\mu_\ell(\cdot|E_{-i})$ is a linear combination of lower-level conditionals, and hence is irrelevant on E_{-i}. For example, if Ω consists of three points, the LPS given by $(\frac{1}{2}, \frac{1}{2}, 0)$, $(\frac{1}{3}, \frac{1}{3}, \frac{1}{3})$ will assume the event consisting of the first two points; the LPS given by $(\frac{1}{2}, \frac{1}{2}, 0)$, $(\frac{1}{2}, \frac{1}{4}, \frac{1}{4})$ will not.

Strong belief in an event also captures the notion that it is infinitely more likely than its complement; we discuss the connection between assumption and strong belief in Section 12.8.4. In view of this connection, it should come as no surprise that assumption also violates both Monotonicity and Conjunction (cf. [12.5]). As for strong belief, this implies that care must be taken when iterating the assumption operator. Furthermore,

[98] Note that, in Definition 12.16, the event FS_i required full support of the beliefs over opponents' strategies only; here we follow Brandenburger et al. (2008) and require that the beliefs on strategies *and* types have full support. Catonini and De Vito (2014) show that full support of first-order beliefs is enough to obtain the results in this section.

[99] Since the support of a measure is the smallest closed set with measure 1, this condition implies that the notion of "assumption" depends upon the topology; see also Section 12.7.7.

as for RCSBR, the behavioral implications of rationality and common assumption of rationality will not be monotonic with respect to the type structure. The discussion of these and related issues in Section 12.7.4.4 apply here *verbatim*.

We can now define the events "admissibility and mutual or common assumption thereof."

$$ACAA_i^0 = R_i \cap FS_i; \quad [12.29]$$
$$ACAA_i^k = ACAA_i^{k-1} \cap A_i(ACAA_{-i}^{k-1}) \quad \text{for } k > 0.$$

The event that admissibility and common assumption of admissibility hold is $ACAA = \bigcap_{k \geq 0} ACAA^k$.

12.8.3 Characterization

Just like we need sufficiently rich type structures for RCSBR to yield forward induction (more precisely, extensive-form rationalizability: see Section 12.7.4.4), now we need sufficiently rich structures to obtain iterated admissibility from mutual or common assumption of admissibility. Adapting arguments from Brandenburger et al. (2008), one can readily show that there exists a complete lexicographic type structure.[100]

We recall the definitions of admissibility with respect to a Cartesian product of strategy sets and iterated admissibility. We then introduce a suitable analog of best-reply sets. As in Brandenburger et al. (2008), we restrict attention to two-player games.

Definition 12.35. *Fix $B_1 \times B_2 \subset S_1 \times S_2$. An action $s_i \in B_i$ is* ***weakly dominated with respect to*** *$B_1 \times B_2$ if there is $\mu_i \in \Delta(B_i)$ such that $u_i(\mu_i, s_{-i}) \geq u_i(s_i, s_{-i})$ for all $s_{-i} \in B_{-i}$, and $u_i(\mu_i, s_{-i}^*) > u_i(s_i, s_{-i}^*)$ for some $s_{-i}^* \in B_{-i}$. The action $s_i \in B_i$ is* ***admissible with respect to*** *$B_1 \times B_2$ if it is not weakly dominated with respect to $B_1 \times B_2$.*

Definition 12.36. (Iterated Admissibility) *Fix a two-player strategic-form game $(I, (S_i, u_i)_{i \in I})$. For every player $i \in I$, let $W_i^0 = S_i$. For $k > 0$, let $s_i \in W_i^k$ iff $s_i \in W_i^{k-1}$ and s_i is admissible w.r.to $W_1^{k-1} \times W_2^{k-1}$. The set of* ***iteratively admissible*** *strategies is W^∞.*

We need an additional definition. Say that a strategy $s_i' \in S_i$ of player i *supports* $s_i \in S_i$ if there exists a mixed strategy $\sigma_i \in S_i$ for i that duplicates s_i and has s_i' in its support: that is, $u_i(\sigma_i, s_{-i}) = u_i(s_i, s_{-i})$ for all $s_{-i} \in S_{-i}$, and $\sigma_i(s_i') > 0$.

[100] Specifically, one can adapt the proof of Proposition 7.2 in Brandenburger et al. (2008) (p. 341). Their argument goes further because they restrict attention to a subset of LPSs (see Section 12.8.4). We suspect, but have not proved, that a *canonical* construction à la Mertens and Zamir (1985) or Brandenburger and Dekel (1993) is also possible for LPSs. Ganguli and Heifetz (2012) show how to construct a non-topological "universal" type structure for LPSs, such that every other such LPS-based type structure can be uniquely embedded in it.

Definition 12.37. *Fix a two-player strategic-form game* $(I, (S_i, u_i)_{i \in I})$. *A set* $B = \prod_{i \in I} B_i \subset S$ *is a* ***self-admissible set*** *(or SAS) if, for every player* $i \in I$, *every* $s_i \in B_i$ *is admissible with respect to both* $S_i \times S_{-i}$ *and* $S_i \times B_{-i}$;[101] *it is a* ***full SAS*** *if, in addition, for every player* $i \in I$ *and strategy* $s_i \in B_i$, *if* s_i' *supports* s_i *then* $s_i' \in B_i$.

In the definition of full SAS, including in the set B_i a strategy s_i' that supports some other strategy $s_i \in B_i$ plays the same role as including all best replies to a belief that justifies some element of a full BRS. For additional discussion, see Brandenburger et al. (2008).

As is the case for extensive-form rationalizability and EFBRSs, the set W^∞ is a full SAS; however, it is not the largest (full) SAS, and indeed there may be games in which a full SAS is disjoint from the IA set. For example, in the strategic form of the game in Figure 12.3, the unique IA profile is (InT, L); however, $B = \{OutT, OutB\} \times \{R\}$ is also a full SAS.

The characterization result is as follows.

Theorem 12.18. *Fix a two-person game* $G = (I, (S_i, u_i)_{i \in I})$.

1. *In any lexicographic type structure* $(I, (S_i, T_i, \beta_i)_{i \in I})$ *for* G, $\text{proj}_S ACAA$ *is a full SAS.*
2. *In any complete lexicographic type structure* $(I, (S_i, T_i, \beta_i)_{i \in I})$ *for* G, *and for every* $k \geq 0$, $\text{proj}_S ACAA^k = W^{k+1}$.
3. *For every full SAS* B, *there exists a finite lexicographic type structure* $(I, (S_i, T_i, \beta_i)_{i \in I})$ *for* G *such that* $\text{proj}_S ACAA = B$.

12.8.4 Discussion

We start by discussing three issues in the characterization of IA. These are the relationship to the characterization in Brandenburger et al. (2008), the full-support assumption, and common vs. mutual assumption of admissibility. We then discuss the relationship between the current section and the extensive-form analysis of Section 12.7. In particular, we relate LPSs to CPSs, assumption to strong belief, and admissibility to sequential rationality.

[101] To see why we need admissibility with respect to both $S_i \times B_{-i}$ and $S_i \times S_{-i}$, consider the following two-person games (only Player 1's payoffs are indicated).

	L	R
T	0	0
B	0	1

	L	R
T	0	1
B	1	0

In the game on the left, T is admissible with respect to $S_1 \times \{L\}$, but not with respect to $S_1 \times S_2$. On the other hand, in the game on the right, T is admissible with respect to $S_1 \times S_2$, but not with respect to $S_1 \times \{L\}$.

Before turning to these issues, we note that, as is the case for strong belief, since assumption violates Monotonicity and Conjunction, its interpretation in small type structures is somewhat delicate. We do not repeat the discussion of these issues here, as the treatment in Section 12.7.4.4 regarding RCSBR applies verbatim here.

12.8.4.1 Issues in the characterization of IA

Relationship with Brandenburger et al. (2008): Our presentation differs from Brandenburger et al. (2008) in that their main results are stated for LPSs with disjoint supports; following Blume et al. (1991), we call these "lexicographic *conditional* probability systems," or LCPSs. We choose to work with LPSs to avoid certain technical complications that arise with LCPSs (for example, the definition and construction of a complete type structure). The proof of Theorem 12.18 can be found in Dekel et al. (2014).

Full-support beliefs: The characterization of IA focuses on types that commonly assume rationality and full-support beliefs. This raises the question whether one could incorporate the full-support assumption in the definition of lexicographic type structures. That is, could we assume that *all* types have full-support beliefs, or at least full-support first-order beliefs? Recall that, in the characterization of $S^{\infty}W$ in Section 12.5, we also focus on types that commonly p-believe in both rationality and full-support beliefs. There, we could restrict attention to type structures where each type's belief over the opponents' strategies have full support. The following example demonstrates that we cannot do this in the current environment.

Example 12.10. (Figure 2.11 in Battigalli, 1993; see also Figure 2.6 in Brandenburger et al., 2008)

Consider the strategic-form game in Table 12.14. The IA set is $\{U, M, D\} \times \{C, R\}$.

Fix an arbitrary lexicographic type structure. Note first that, since L is strictly dominated for player 2, $(L, t_2) \notin R_2$ *for any type* t_2 *of 2; a fortiori,* $(L, t_2) \notin R_2 \cap FS_2$. *Moreover, C and R always yield a payoff of 1, and hence both* $(C, t_2) \in R_2 \cap FS_2$ *and* $(R, t_2) \in R_2 \cap FS_2$ *hold if and only if type* t_2 *has full-support beliefs.*

Now consider a type t_1 *of player 1 such that* $(D, t_1) \in R_1 \cap FS_1 \cap A_1(R_2 \cap FS_2)$, *and let* $\beta_1(t_1) = (\mu_0, \ldots, \mu_{n-1})$. *Since the definition of assumption (Definition 12.34) requires full-support beliefs, as* t_1 *assumes* $R_2 \cap FS_2$, *this type must have full-support beliefs; in particular, there must be an order k with* $\mu_k(\{L\} \times T_2) > 0$. *Furthermore, since* t_1 *assumes* $R_2 \cap FS_2$, *and L is irrational for 2, it must be the case that* $k > 0$.

Table 12.14 Iterated admissibility and ACAA

	L	C	R
U	4,0	4,1	0,1
M	0,0	0,1	4,1
D	3,0	2,1	2,1

Next, by lexicographic utility maximization, for all $\ell = 0, \ldots, k-1$, *we must have* $\mu_\ell(\{C\} \times T_2) = \mu_\ell(\{R\} \times T_2) = \frac{1}{2}$ *for* $0 \leq \ell \leq k-1$: *otherwise, D could not be a best reply. But then, U and M are also best replies to* $\text{marg}_{S_2}\mu_\ell$, $\ell = 0, \ldots, k-1$. *In other words, D ties with U and M against the beliefs* $\text{marg}_{S_2}\mu_0, \ldots, \text{marg}_{S_2}\mu_{k-1}$. *Then, the optimality of D requires that D also be a best response to the kth level belief* $\text{marg}_{S_2}\mu_k$.

Finally, for this to be the case, we must have $\mu_k(\{R\} \times T_2) > 0$. *Moreover, as* $\mu_k(\{L\} \times T_2) > 0$ *and L is not rational for 2,* $\mu_k(R_2) < 1$, *hence* $\mu_k(R_2 \cap FS_2) < 1$. *However,* t_1 *assumes* $R_2 \cap FS_2$. *Therefore, by the definition of assumption,* $\mu_k(R_2 \cap FS_2)$ *must equal either 1 or 0. Hence, it must be the case that* $\mu_k(R_2 \cap FS_2) = 0$. *On the other hand,* $\mu_k(\{R\} \times T_2) > 0$, *so there must be types* t_2 *of 2 for whom* $(R, t_2) \notin R_2 \cap FS_2$. *That is, because 1's kth-level belief assigns zero probability to 2 being rational and having full-support beliefs, and positive probability to 2 playing R, it must be that 1 expects those types of 2 who are playing R to hold beliefs that either do not justify R, or do not have full support. But, since R is a best reply against any beliefs, the only way this can hold is if 1 expects 2's type to not have full-support beliefs. This means that the type structure under consideration* must *contain types for player 2 that do not have full-support beliefs.*

Common vs. mutual assumption of admissibility: Finally, there is an additional subtlety. Note that Theorem 12.18 does not characterize *common* assumption of admissibility for complete type structures—merely finite-order assumption of admissibility. Indeed, Brandenburger et al. (2008) show that, under completeness and restricting attention to LCPSs, $\bigcap_{k \geq 0} ACAA_i^k$ is empty. Admissibility and common assumption of admissibility thus cannot hold in a complete, LCPS-based type structure. We believe (but have not proved) that the same is true when beliefs are represented by LPSs.

This is a puzzling result. In a recent paper, Lee and Keisler (2011) demonstrate that the problem arises out of the requirement in Definition 12.31 that the belief maps β_i be *continuous*. If one drops this requirement, and merely asks that they be measurable, it is possible to construct a complete, LCPS-based type structure in which $\text{proj}_S ACAA$ equals IA, so that ACAA *is* possible (and characterizes iterated admissibility).[102]

12.8.4.2 Extensive-form analysis and strategic-form refinements

LPSs and CPSs: LPSs and CPSs are clearly similar. CPSs are also collections of probabilities, that also may differ in terms of saliency (lower-saliency beliefs come into play as unexpected events are encountered). However, there are also differences, due to the fact that the former are strategic-form objects, whereas the latter are defined for

[102] Other papers that provide epistemic conditions related to IA include Asheim and Dufwenberg (2003), Barelli and Galanis (2011), Yang (2011), Perea (2012), Lee (2013), and Catonini and De Vito (2014).

extensive-form games.[103] Probabilities in an LPS are completely ordered, whereas in a CPS the order is partial. For example, consider a game in which Player 1 can choose T, M or B, and Player 2 (who moves immediately after 1) is initially certain of T. Then, Player 2's conditional beliefs following M and B are not ranked in terms of their salience, although they are less salient than Player 2's initial belief. Second, the supports of any two probabilities in a CPS are either disjoint, or one is included in the other; in an LPS, the supports can overlap arbitrarily. In addition, a technical distinction in the context of type structures is that, for a finite extensive game, the number of probabilities in a CPS is fixed and equal to the number of nonterminal histories in the game; on the other hand, in general there is no upper bound on the number of levels in an LPS.

Strong belief and assumption: As we noted above, both strong belief and assumption capture the notion that an event and its subsets are infinitely more likely than its complement. Recall that player i assumes E_{-i} if she assigns probability one to it or some subset of it in each of the first ℓ^* levels of her LPS, until all of E_{-i} has been given probability 1 at some level; furthermore, higher-level measures either assign probability zero to E_{-i}, or are behaviorally irrelevant conditional on E_{-i}. Analogously, if player i strongly believes E_{-i} in an extensive game, then her initial beliefs assign probability one to E_{-i} or some subset thereof. Moreover, so long as E_{-i} has not been contradicted by observed play, when player i revises her beliefs,[104] she continues to assign probability one to some subset of E_{-i}. Once E_{-i} has been contradicted, it must receive probability zero. Thus, with strong belief, the "level" at which E_{-i} is no longer believed is objective, while in the case of assumption, the level at which i no longer believes E_{-i} is subjective. Nevertheless, assumption and strong belief are quite similar. Specifically, for finite spaces Ω, there is a one-to-one mapping between LCPSs (but not arbitrary LPSs) and CPSs in which the set of conditioning events consists of all nonempty subsets of Ω. Furthermore, an LCPS λ "assumes" an event E_{-i} (analogously to Definition 12.34) if and only if the corresponding CPS μ "strongly believes" E_{-i}.[105]

Admissibility and sequential rationality: Brandenburger (2007) shows that, in single-person, dynamic choice problems, admissibility is equivalent to sequential rationality in all decision trees that have the same strategic form, up to the addition or deletion of strategies that are convex combinations of other strategies (i.e., trees that have the same *fully reduced* normal form in the sense of Kohlberg and Mertens, 1986). Nevertheless, Brandeburger's result is about single-person problems; adding or deleting convex combinations of existing strategies in an extensive *game* may affect the players' strategic reasoning (see e.g., Hillas, 1994, and Govindan and Wilson, 2009).

[103] In fact, CPSs are no different from regular probabilities for extensive forms of simultaneous-move games.

[104] When we say that she "revises" her beliefs, we allow for both standard belief updating, following positive-probability observations, as well as formulating entirely new beliefs, following zero-probability observations.

[105] For the case of infinite sets Ω, see Brandenburger et al. (2007).

ACKNOWLEDGEMENT

We thank Drew Fudenberg for his detailed comments and Robert Molony and Luciano Pomatto for excellent research assistantship. We also thank Pierpaolo Battigalli, Adam Brandenburger, Yi-Chun Chen, Amanda Friedenberg, Joe Halpern, Qingmin Liu, Andres Perea, two anonymous referees, and the editors, Peyton Young and Shmuel Zamir, for helpful feedback. Eddie Dekel gratefully acknowledges financial support from NSF grant SES-1227434.

REFERENCES

Ahn, D.S., 2007. Hierarchies of ambiguous beliefs. J. Econ. Theory, 136 (1), 286–301.
Aliprantis, C.D., Border, K.C. 2007. Infinite Dimensional Analysis: A Hitchhiker's Guide. Springer Verlag.
Anscombe, F.J., Aumann, R.J., 1963. A definition of subjective probability. Ann. Math. Stat. 34 (1), 199–205.
Armbruster, W., Böge, W., 1979. Bayesian game theory. Game Theory and Related Topics. North-Holland, Amsterdam.
Asheim, G.B., Dufwenberg, M., 2003. Admissibility and common belief. Games Econ. Behav. 42 (2), 208–234.
Asheim, G.B., Perea, A., 2005. Sequential and quasi-perfect rationalizability in extensive games. Games Econ. Behav. 53 (1), 15–42.
Asheim, G.B., 2002. Proper rationalizability in lexicographic beliefs. Inter. J. Game Theory 30, 453–478. ISSN 0020-7276. URL http://dx.doi.org/10.1007/s001820200090. 10.1007/s001820200090.
Asheim, G.B., M. Voorneveld, Weibull, J.W. et al., 2009. Epistemically stable strategy sets. Memorandum.
Aumann, R., Brandenburger, A., 1995. Epistemic conditions for Nash equilibrium. Econometrica: J. Econ. Soc. 63 (5), 1161–1180.
Aumann, R.J., 1976. Agreeing to disagree. Ann. Stat. 4 (6), 1236–1239.
Aumann, R.J., 1987. Correlated equilibrium as an expression of Bayesian rationality. Econometrica 55 (1), 1–18. ISSN 0012-9682.
Aumann, R.J., 1995. Backward induction and common knowledge of rationality. Games Econ. Behav. 8 (1), 6–19. ISSN 0899-8256.
Aumann, R.J., 1998. On the centipede game. Games Econ. Behav. 23, 97–105.
Aumann, R.J., 1999a. Interactive epistemology i: knowledge. Inter. J. Game Theory 28 (3), 263–300.
Aumann, R.J., 1999b. Interactive epistemology ii: Probability. Inter. J. Game Theory 28 (3), 301–314.
Aumann, R.J., Dreze, J.H., 2004. Assessing strategic risk. Mimeo.
Aumann, R.J., Dreze, J.H., 2009. Assessing strategic risk. Am. Econ. J.: Microecon. 1 (1), 1–16.
Bach, C.W., Heilmann, C., 2011. Agent connectedness and backward induction. Int. Game Theory Rev. 13 (02), 195–208.
Balkenborg, D., Winter, E., 1997. A necessary and sufficient epistemic condition for playing backward induction. J. Math. Econ. 27 (3), 325–345. ISSN 0304-4068.
Barelli, P., 2009. Consistency of beliefs and epistemic conditions for nash and correlated equilibria. Games Econ. Behav. 67 (2), 363–375.
Barelli, P., 2010. Consistent beliefs in extensive form games. Games 1 (4), 415–421.
Barelli, P., Galanis, S., 2011. Admissibility and Event-Rationality.
Basu, K., Weibull, J.W., 1991. Strategy subsets closed under rational behavior. Econ. Lett. 36 (2), 141–146.
Battigalli, P., 1993. Restrizioni Razionali su Sistemi di Probabilita' Soggettive e Soluzioni di Giochi ad Informazione Incompleta. E.G.E.A., Milan.
Battigalli, P., 1996a. Strategic rationality orderings and the best rationalization principle. Games Econ. Behav. 13 (2), 178–200.
Battigalli, P., 1996b. Comment to on the decision theoretic foundation of game theory by M. Mariotti. In: Arrow, K.J., Colombatto, E., Perlman, M., Schmidt, C., (Eds.), The Foundations of Rational Economic Behavior. London: McMillan.
Battigalli, P., 1997. On rationalizability in extensive games. J. Econ. Theory 74 (1), 40–61.

Battigalli, P., Friedenberg, A., 2012. Forward induction reasoning revisited. Theoretical Economics, 7(1), 57–98.

Battigalli, P., Prestipino, A., 2012. Transparent restrictions on beliefs and forward induction reasoning in games with asymmetric information. The BE Journal of Theoretical Economics, 13(1), 79–130.

Battigalli, P., Siniscalchi, M., 1999. Hierarchies of Conditional Beliefs and Interactive Epistemology in Dynamic Games. J. Econ. Theory 88 (1), 188–230.

Battigalli, P., Siniscalchi, M., 2002. Strong belief and forward induction reasoning. J. Econ. Theory 106 (2), 356–391.

Battigalli, P., Siniscalchi, M., 2003. Rationalization and incomplete information. Adv. Theor. Econ. 3 (1), 1073–1073.

Battigalli, P., Siniscalchi, M., 2007. Interactive epistemology in games with payoff uncertainty. Res. Econ. 61 (4), 165–184.

Battigalli, P., Di, A. Tillio, Grillo, E., and Penta, A., 2011a. Interactive epistemology and solution concepts for games with asymmetric information. BE J. Theor. Econ. 11 (1).

Battigalli, P., Di Tillio, A., Samet, D., 2011b. Strategies and Interactive Beliefs in Dynamic Games.

Battigalli, P., 1987. Comportamento razionale ed equilibrio nei giochi e nelle situazioni sociali. unpublished undergraduate dissertation, Bocconi University, Milano.

Ben-Porath, E., 1997. Rationality, Nash equilibrium and backwards induction in perfect-information games. Rev. Econ. Stud. 64, pp. 23–46.

Bergemann, D., Morris, S., 2011. Correlated equilibrium in games with incomplete information. Cowles Foundation Discussion Papers, (1822).

Bernheim, B.D., 1984. Rationalizable strategic behavior. Econometrica: J. Econ. Soc. 52 (4), 1007–1028. ISSN 0012-9682.

Billingsley, P., 2008. Probability and Measure. Wiley India.

Blume, L., Brandenburger, A., Dekel, E., 1991. Lexicographic probabilities and choice under uncertainty. Econometrica: J. Econ. Soc. 59 (1), 61–79.

Böge, W., Eisele, T., 1979. On solutions of bayesian games. Inter. J. Game Theory 8 (4), 193–215.

Bonanno, G., Nehring, K., 1999. How to make sense of the common prior assumption under incomplete information. Inter. J. Game Theory 28 (3), 409–434.

Börgers, T., 1993. Pure strategy dominance. Econometrica: J. Econ. Soc. 61 (2): 423–430. ISSN 0012-9682.

Borgers, T., 1994. Weak dominance and approximate common knowledge. J. Econ. Theory 64 (1), 265–276. URL http://econpapers.repec.org/RePEc:eee:jetheo:v:64:y:1994:i:1:p:265-276.

Brandenburger, A., Dekel, E., 1986. "Bayesian Rationality in Games," in Brandenburger, A., "Hierarchies of Beliefs in Decision Theory," PhD dissertation, Cambridge University.

Brandenburger, A., 1992. Lexicographic probabilities and iterated admissibility. In: Dasgupta, P., Gale, D., Hart, O., Maskin, E., (Eds), Econ. Anal. of Markets Games 282–290. MIT Press, Cambridge, Massachusetts, USA

Brandenburger, A., 2007. The power of paradox: some recent developments in interactive epistemology. Inter. J. Game Theory 35 (4): 465–492.

Brandenburger, A., Dekel, E., 1987. Rationalizability and correlated equilibria. Econometrica 55, 1391–1402.

Brandenburger, A., Dekel, E., 1993. Hierarchies of beliefs and common knowledge. J. Econ. Theory 59, 189–189.

Brandenburger, A., Friedenberg, A., 2008. Intrinsic correlation in games. J. Econ. Theory 141 (1): 28–67.

Brandenburger, A., Friedenberg, A., 2010. Self-admissible sets. J. Econ. Theory 145 (2): 785–811. ISSN 0022-0531.

Brandenburger, A., Friedenberg, A., Keisler, J., 2007. Notes on the relationship between strong belief and assumption. Mimeo.

Brandenburger, A., Friedenberg, A., Keisler, H.J., 2008. Admissibility in games. Econometrica 76 (2), 307.

Carlsson, H., Van Damme, E., 1993. Global games and equilibrium selection. Econometrica: J. Econ. Soc. 61 (5): 989–1018.

Catonini, E., De Vito, N., 2014. Common assumption of cautious rationality and iterated admissibility. Mimeo, http://www.hse.ru/data/2014/06/11/1324209270/Common%20assumption%20of%20cautious%20rationality%20and%20iterated%20admissibility.pdf

Chen, Y.-C., 2011. A Structure Theorem for Rationalizability in Dynamic Games. Working paper, National University of Singapore.

Chen, Y.C., 2010. Universality of the epstein-wang type structure. Games Econ. Behav. 68 (1), 389–402.

Chen, Y.C., Long, N.V., Luo, X., 2007. Iterated strict dominance in general games. Games Econ. Behav. 61 (2), 299–315.

Cho, I.K., Kreps, D.M., 1987. Signaling games and stable equilibria. Quart. J. Econ. 102 (2), 179.

Crawford, V.P., Costa-Gomes, M.A., Iriberri, N., 2012. Structural models of nonequilibrium strategic thinking: Theory, evidence, and applications. J. Econ. Lit.

De Finetti, B., 1992. Foresight: Its Logical Laws, Its Subjective Sources, volume Breakthroughs in Statistics: Foundations and Basic Theory, pp. 134–174. Springer-Verlag, New York.

Dekel, E., Fudenberg, D., 1990. Rational behavior with payoff uncertainty. J. Econ. Theory 52 (2), 243–267. ISSN 0022-0531.

Dekel, E., Fudenberg, D., Levine, D.K., 1999. Payoff information and self-confirming equilibrium. J. Econ. Theory 89 (2), 165–185.

Dekel, E., Fudenberg, D., Levine, D.K., 2004. Learning to play bayesian games. Games Econ. Behav. 46 (2), 282–303.

Dekel, E., Fudenberg, D., Morris, S., 2006. Topologies on types. Theor. Econ. 1, 275–309.

Dekel, E., Fudenberg, D., Morris, S., 2007. Interim correlated rationalizability. Theor. Econ. 2 (1), 15–40.

Dekel, E., Friedenberg, A., Siniscalchi, M., 2014, Lexicographic Belief and Assumption, working paper, http://www.public.asu.edu/~afrieden/admissibility.pdf

Dekel, E., Pomatto, L., Siniscalchi, M., 2014. Epistemic Game Theory: Online Appendix. http://faculty.wcas.northwestern.edu/~msi661/EGT-Proofs.pdf

Dellacherie, C., Meyer, P.A., 1978. Probability and Potential. Hermann, Paris.

Di Tillio, A., 2008. Subjective expected utility in games. Theor. Econ. 3 (3): 287–323.

Di Tillio, A., Halpern, J., Samet, D., Conditional Belief Types. January 2012.

Du, S., 2011. Correlated equilibrium and higher order beliefs about play.

Dufwenberg, M., Stegeman, M., 2002. Existence and uniqueness of maximal reductions under iterated strict dominance. Econometrica 70 (5), 2007–2023.

Ely, J.C., Peski, M., 2006. Hierarchies of belief and interim rationalizability. Theor. Econ. 1 (1), 19–65.

Epstein, L.G., Wang, T., 1996. "Beliefs about beliefs" without probabilities. Econometrica 64 (6), 1343–1373. ISSN 0012-9682.

Fagin, R., Halpern, J.Y., 1994. Reasoning about knowledge and probability. J. ACM (JACM) 41 (2), 340–367.

Fagin, R., Halpern, J.Y., Megiddo, N., 1990. A logic for reasoning about probabilities. Information and computation 87 (1), 78–128.

Fagin, R., Halpern, J., Moses, Y., Vardi, M., 1995. Reasoning about knowledge. The MIT Press, Cambridge, Massachusetts.

Feinberg, Y., 2000. Characterizing common priors in the form of posteriors⋆ 1. J. Econ. Theory 91 (2), 127–179.

Forges, F., 1993. Five legitimate definitions of correlated equilibrium in games with incomplete information. Theory Decis. 35 (3), 277–310.

Forges, F., 2006. Correlated equilibrium in games with incomplete information revisited. Theory Deci. 61 (4): 329–344.

Friedenberg, A., 2010. When do type structures contain all hierarchies of beliefs? Games Econ. Behav. 68 (1), 108–129.

Fudenberg, D., Kamada, Y., 2011. Rationalizable partition-confirmed equilibrium. Technical report, mimeo.

Fudenberg, D., Levine, D.K., 1993. Self-confirming equilibrium. Econometrica 61 (3), 523–545.

Fudenberg, D., Tirole, J., 1991a. Game Theory. MIT Press, Cambridge, Massachusetts, USA.

Fudenberg, D., Tirole, J., 1991b. Perfect bayesian equilibrium and sequential equilibrium. J. Econ. Theory 53 (2), 236–260.

Fudenberg, D., Kreps, D.M., Levine, D.K., 1988. On the robustness of equilibrium refinements. J. Econ. Theory 44 (2), 354–380.

Fudenberg, D., Kamada, Y., 2012. Rationalizable partition-confirmed equilibrium. Working paper.

Gale, D., Sherman, S., 1950. Solutions of finite two-person games. In: Kuhn, H., Tucker, A., (Eds.), Contributions to the Theory of Games. Princeton University Press.
Ganguli, J., Heifetz, A., 2012. Universal Interactive Preferences. Available at SSRN 2174371.
Geanakoplos, J., Pearce, D., Stacchetti, E., 1989. Psychological games and sequential rationality. Games Econ. Behav. 1 (1), 60–79.
Govindan, S., Wilson, R., 2009. On forward induction. Econometrica 77 (1), 1–28. ISSN 1468-0262.
Gul, F., 1996. Rationality and coherent theories of strategic behavior. J. Econ. Theory 70 (1), 1–31.
Gul, F., Pesendorfer, W., 2010. Interdependent preference models as a theory of intentions. Technical report, Princeton University Working Paper.
Harsanyi, J.C., 1967. Games with incomplete information played by "Bayesian" players, I-III. Part I. The basic model. Manag. Sci., 14, 159–182.
Harsanyi, J.C., 1973. Games with randomly disturbed payoffs: A new rationale for mixed-strategy equilibrium points. Inter. J. Game Theory 2 (1), 1–23.
Heifetz, A., 1993. The bayesian formulation of incomplete information the non-compact case. Inter. J. Game Theory 21, 329–338. ISSN 0020-7276. URL http://dx.doi.org/10.1007/BF01240148. 10.1007/BF01240148.
Heifetz, A., 2006. The positive foundation of the common prior assumption. Games Econ. Behav. 56 (1), 105–120.
Heifetz, A., Mongin, P., 2001. Probability logic for type spaces. Games Econ. Behav. 35 (1-2), 31–53.
Heifetz, A., Samet, D., 1998. Topology-free typology of beliefs. J. Econ. Theory 82 (2), 324–341.
Heifetz, A., Perea, A., 2013. On the outcome equivalence of backward induction and extensive form rationalizability. Mimeo, Maastricht University.
Hillas, J., 1994. How much of forward induction is implied by backward induction and ordinality. Mimeo, University of Auckland.
Hu, T.W., 2007. On p-rationalizability and approximate common certainty of rationality. J. Econ. Theory 136 (1), 379–391.
Kajii, A., Morris, S., 1997a. The robustness of equilibria to incomplete information. Econometrica: J. Econ. Soc., 1283–1309.
Kajii, A., Morris, S., 1997b. Refinements and higher order beliefs: A unified survey. Northwestern University CMS-EMS discussion paper n. 1197. URL http://www.kellogg.northwestern.edu/research/math/papers/1197.pdf.
Kets, W., 2012. Bounded reasoning and higher-order uncertainty. Available at SSRN 2116626.
Kohlberg, E., Mertens, J.F., 1986. On the strategic stability of equilibria. Econometrica: J. Econ. Soc. 54 (5), 1003–1037.
Kreps, D.M., Wilson, R., 1982. Sequential equilibria. Econometrica: J. Econ. Soc. 50 (4), 863–894.
Lee, B.-S., Keisler, H.J., 2011. Common Assumption of Rationality . Working paper, University of Toronto, Rotman School of Management.
Lee, B.S, 2013. Conditional beliefs and higher-order preferences. Mimeo, http://individual.utoronto.ca/byungsoolee/papers/cbhop.pdf
Liu, Q., 2009. On redundant types and Bayesian formulation of incomplete information. J. Econ. Theory. 144, 2115–2145.
Liu, Q., 2010. Higher-Order Beliefs and Epistemic Conditions for Nash Equilibrium, working paper, Columbia University, https://sites.google.com/site/qingmin/home/epiNash.pdf?attredirects=0
Liu, Q., August 2014. Correlation and Common Prior in Games with Incomplete Information, working paper, Columbia University, https://sites.google.com/site/qingmin/home/MS%202011775%20Revision.pdf?attredirects=0
Mariotti, M., 1995. Is bayesian rationality compatible with strategic rationality? Econ. J. 105, 1099–1109.
Marx, L.M., Swinkels, J.M., 1997. Order independence for iterated weak dominance. Games Econ. Behav. 18 (2), 219–245.
Meier, M. (2012). An infinitary probability logic for type spaces. Israel Journal of Mathematics, 192(1), 1–58.
Mertens, J.F., Zamir, S., 1985. Formulation of Bayesian analysis for games with incomplete information. Inter. J. Game Theory 14 (1), 1–29.

Milgrom, P., Stokey, N., 1982. Information, trade and common knowledge. J. Econ. Theory 26 (1), 17–27.
Monderer, D., Samet, D., 1989. Approximating common knowledge with common beliefs. Games Econ. Behav. 1 (2), 170–190.
Morris, S., 1994. Trade with heterogeneous prior beliefs and asymmetric information. Econometrica: J. Econ. Soc. 62, 1327–1347.
Morris, S., 1997. Alternative definitions of knowledge, volume 20, Epistemic Logic and the Theory of Games and Decisions, Kluwer, Boston, Dordrecht, London 217–233.
Morris, S., 2002. Typical types. Mimeo, Department of Economics. Princeton University.[448, 449, 469].
Morris, S., Shin, H.S., 2003. Global games: theory and applications. In: M. Dewatripont, L.P. Hansen, and S.J. Turnovsky (eds.) Advances in Economics and Econometrics: Theory and applications, Eighth world Congress, vol. 1, pp. 56–114.
Morris, S.E., Takahashi, S., 2011. Common Certainty of Rationality Revisited. SSRN eLibrary. doi: 10.2139/ssrn.1789773.
Myerson, R.B., 1997. Game Theory: Analysis of Conflict. Harvard Univ Pr, Cambridge, Massachusetts, USA.
Myerson, R.B., 1978. Refinements of the nash equilibrium concept. Inter. J. Game Theory 7 (2), 73–80.
Nau, R.F., McCardle, K.F., 1990. Coherent behavior in noncooperative games. J. Econ. Theory 50 (2), 424–444. ISSN 0022-0531.
Osborne, M.J., Rubinstein, A., 1994. A Course in Game Theory. The MIT Press, Cambridge, Massachusetts, USA.
Pearce, D.G., 1984. Rationalizable strategic behavior and the problem of perfection. Econometrica 52 (4), 1029–1050. ISSN 0012-9682.
Penta, A., 2012. Higher order uncertainty and information: Static and dynamic games. Econometrica 80 (2), 631–660.
Perea, A., 2007. A one-person doxastic characterization of nash strategies. Synthese 158 (2), 251–271.
Perea, A., 2008. Minimal belief revision leads to backward induction. Mathematical Social Sciences 56 (1), 1–26.
Perea, A., 2011a. An algorithm for proper rationalizability. Games Econ. Behav. 72 (2), 510–525.
Perea, A., 2011b. Belief in the Opponents' Future Rationality. Mimeo.
Perea, A., 2012. Epistemic Game Theory: Reasoning and Choice. Cambridge University Press.
Peysakhovich, A., 2011. Correlation Without Signals. Technical report.
Pomatto, L., 2011. Epistemic Conditions for Bayesian Nash Equilibrium. Mimeo, Northwestern University.
Rabin, M., 1994. Incorporating behavioral assumptions into game theory. In: Friedman, J.W., Samuels, W.J., Darity, W., (Eds.), Problems of Coordination in Economic Activity, volume 35 of Recent Economic Thought Series, Springer Netherlands, pp. 69–87. ISBN 978-94-011-1398-4.
Reny, P.J., 1992. Backward induction, normal form perfection and explicable equilibria. Econometrica: J. Econ. Soc., 60(3), pp. 627–649.
Reny, P.J., 1993. Common belief and the theory of games with perfect information. J. Econ. Theory 59, 257–257.
Rényi, A., 1955. On a new axiomatic theory of probability. Acta Math. Hung. 6 (3), 285–335. ISSN 0236-5294.
Rubinstein, A., 1989. The Electronic Mail Game: Strategic Behavior Under "Almost Common Knowledge." The American Economic Review 79 (3), 385–391.
Rubinstein, A., 1991. Comments on the interpretation of game theory. Econometrica: J. Econ. Soc. 909–924.
Rubinstein, A., Wolinsky, A., 1990. On the logic of 'agreeing to disagree' type results. J. Econ. Theory 51 (1), 184–193.
Rubinstein, A., Wolinsky, A., 1994. Rationalizable conjectural equilibrium: between nash and rationalizability. Games Econ. Behav. 6 (2), 299–311.
Sadzik, T., 2011. Beliefs Revealed in Bayesian-Nash Equilibrium. New York University.
Samet, D., 1996. Hypothetical knowledge and games with perfect of of information. Games Econ. Behav. 17, 230–251.
Samet, D., 1998a. Common priors and separation of convex sets. Games Econ. Behav. 24, 172–174.

Samet, D., 1998b. Iterated expectations and common priors. Games Econ. Behav. 24, 131–141.
Samet, D., 2013. Common belief of rationality in games of perfect information. Games Econ. Behav., ISSN 0899-8256. doi: 10.1016/j.geb.2013.01.008. URL http://www.sciencedirect.com/science/article/pii/S0899825613000195.
Samuelson, L., 1992. Dominated strategies and common knowledge. Games Econ. Behav. 4 (2), 284–313.
Savage, L.J., 1972. The Foundations of Statistics. Dover Publications.
Schipper, B., 2013. The unawareness bibliography. URL http://www.econ.ucdavis.edu/faculty/schipper/unaw.htm.
Schuhmacher, F., 1999. Proper rationalizability and backward induction. Inter. J. Game Theory 28 (4), 599–615. ISSN 0020-7276.
Selten, R., 1975. Reexamination of the perfectness concept for equilibrium points in extensive games. Inter. J. Game Theory 4 (1), 25–55.
Siniscalchi, M., 2014. Sequential Preferences and Sequential Rationality. Mimeo, Northwestern University.
Stalnaker, R., 1998. Belief revision in games: forward and backward induction. Mathematical Social Sciences 36 (1), 31–56.
Tan, T.C.C., da Costa Werlang, S.R., 1988. The Bayesian foundations of solution concepts of games. J. Econ. Theory 45 (2), 370–391. ISSN 0022-0531.
Tang, Q., 2011. Interim partially correlated rationalizability.
Tercieux, O., 2006. p-best response set. J. Econ. Theory 131 (1), 45–70.
Van Damme, E., 1983. Refinements of the Nash Equilibrium Concept. Springer-Verlag, Berlin, New York.
Weinstein, J., Yildiz, M., 2007. A structure theorem for rationalizability with application to robust predictions of refinements. Econometrica 75 (2), 365–400. ISSN 1468-0262.
Yang, C.-C., 2011. Weak Assumption and Iterative Admissibility.

CHAPTER 13

Population Games and Deterministic Evolutionary Dynamics

William H. Sandholm*
*Department of Economics, University of Wisconsin, Madison, WI, USA

Contents

Handbook of game theory
http://dx.doi.org/10.1016/B978-0-12-420056-2.00013-6

Abstract

Population games describe strategic interactions among large numbers of small, anonymous agents. Behavior in these games is typically modeled dynamically, with agents occasionally receiving opportunities to switch strategies, basing their choices on simple myopic rules called *revision protocols*. Over finite time spans the evolution of aggregate behavior is well approximated by the solution of a differential equation. From a different point of view, every revision protocol defines a map—a *deterministic evolutionary dynamic*—that assigns each population game a differential equation describing the evolution of aggregate behavior in that game.

In this chapter, we provide an overview of the theory of population games and deterministic evolutionary dynamics. We introduce population games through a series of examples and illustrate their basic geometric properties. We formally derive deterministic evolutionary dynamics from revision protocols, introduce the main families of dynamics—imitative/biological, best response, comparison to average payoffs, and pairwise comparison—and discuss their basic properties. Combining these streams, we consider classes of population games in which members of these families of dynamics converge to equilibrium; these classes include potential games, contractive games, games solvable by iterative solution concepts, and supermodular games. We relate these classes to the classical notion of an evolutionarily stable state and to recent work on deterministic equilibrium selection. We present a variety of examples of cycling and chaos under evolutionary dynamics, as well as a general result on survival of strictly dominated strategies. Finally, we provide connections to other approaches to game dynamics, and indicate applications of evolutionary game dynamics to economics and social science.

Keywords: Evolutionary game theory, Learning in games, Population games, Revision protocols, Deterministic evolutionary dynamics, Global convergence, Local stability, Nonconvergence, Stochastic processes, Dynamical systems

JEL Codes: C72, C73

13.1. INTRODUCTION

Consider a population of commuters, each of whom must select a path through a highway network from his home to his workplace. Each commuter's payoff depends not only on the route he chooses, but also on the distribution of the route choices of other drivers, as these will determine the delays the commuter will face.

Population games are a powerful tool for modeling strategic environments like traffic networks, in which the number of agents is large, each agent is small, and agents are anonymous, with each agent's payoffs depending on his own strategy and the distribution of others' strategies. One typically imposes further restrictions on the agents' diversity: there are a finite number of populations, and agents in each population are identical, in that they choose from the same set of strategies and have identical payoff functions. Despite their simplicity, population games offer a powerful tool for applications in economics, computer science, biology, sociology, and other fields that study interactions among large numbers of participants.[1]

The traditional approach to prediction in noncooperative games, equilibrium analysis, is based on strong assumptions about what players know. Such assumptions—that players fully understand the game they are playing, and that they are able to correctly anticipate how others will act—are overly demanding in many applications, particularly those in which the number of participants is large.

An alternative approach, one especially appropriate for recurring interactions with many agents, proceeds through a dynamic, disequilibrium analysis. One assumes that agents occasionally receive opportunities to switch strategies. A modeling device called a *revision protocol* specifies when and how they do so. The definition of the protocol reflects what information is available to agents when they make decisions, and how this information is used. Protocols can capture imitation, optimization, or any other criterion that agents can employ to respond to current strategic conditions.

Together, a population game, a population size, and a revision protocol generate a Markov process on the set of population states—that is, of distributions over pure strategies. While this Markov process can be studied directly, particularly powerful conclusions can be reached by evaluating this process over a fixed time horizon in the

[1] Research on population games is distinguished from the literature on large noncooperative games (Schmeidler, 1973; Khan and Sun, 2002; Balder, 2002; Carmona and Podczeck, 2009) by both the limited diversity assumption and the central role played by disequilibrium dynamics.

large population limit.[2] A suitable law of large numbers implies that as the population size grows large, the sample paths of the Markov process can be approximated arbitrarily well by a deterministic trajectory. This trajectory is obtained as a solution of an ordinary differential equation: the *mean dynamic* induced by the game and the protocol.

We can view this development in a different light by fixing the revision protocol, thereby obtaining a map from population games to differential equations. This map—and by the usual synecdoche, the output of this map—is known as a *deterministic evolutionary dynamic.* Deterministic evolutionary dynamics reflect the character of the protocols that generate them; for example, dynamics based on imitation are readily distinguished from those based on optimization. Nevertheless, in some classes of games, dynamics derived from a variety of choice principles exhibit qualitatively similar behavior.

Evolutionary dynamics are nonlinear differential equations, so explicit formulas for solution trajectories are almost never available. Instead, methods from the theory of dynamical systems are used to establish local stability, global convergence, and nonconvergence results. Software for computing and visualizing numerical solutions offers a powerful tool for understanding the behavior of evolutionary dynamics, particularly in cases of cyclical and chaotic behavior.[3]

Section 13.2 introduces population games, provides some basic examples, and offers geometric intuitions. Section 13.3 introduces revision protocols and mean dynamics, and describes the deterministic approximation theorem that allows one to move from one to the other. Section 13.4 formalizes the notion of deterministic evolutionary dynamics, and introduces criteria that relate their behavior to incentives in the underlying games. Section 13.5 introduces the main families of evolutionary dynamics that have been studied in the literature: imitative dynamics (and biological dynamics), the best response dynamic and its variants, and dynamics based on comparisons to average payoffs and pairwise comparisons.

With these basic ingredients in place, the next four sections present convergence and nonconvergence results for various combinations of games and dynamics. Section 13.6 considers the class of potential games, for which the most general convergence results are available. Section 13.7 introduces the notion of an evolutionarily stable state (ESS), which provides a general sufficient condition for local stability. It also studies the class of contractive games, which are motivated by a version of the ESS condition, and which admit global convergence results. Section 13.8 addresses iterative solution concepts, supermodular games, and deterministic equilibrium selection. Section 13.9

[2] The other main approach to studying this process focuses on its infinite horizon behavior, as either a noise parameter or the population size approaches its limit. This approach, known as *stochastic stability analysis*, is the subject of the chapter by Wallace and Young in this volume, and is discussed in Section 13.10.1 below. For complete treatments, see Young (1998) and Sandholm (2010c, Chapters 11 and 12).

[3] The phase diagrams in this chapter were created using the *Dynamo* software suite (Sandholm et al., 2012). See Franchetti and Sandholm (2013) for an introduction.

provides examples of cycling and chaos under evolutionary dynamics, and presents a general result on survival of strictly dominated strategies.

While the material above is the core of the theory of population games and deterministic evolutionary dynamics, many other topics in the theory have been explored to various degrees. Section 13.10 provides snapshots of work on these topics, connections to other approaches to game dynamics, and a brief summary of applications in economics and social science.

Other treatments of evolutionary game theory with a focus on deterministic dynamics include survey papers by Hofbauer and Sigmund (2003), Sandholm (2009a), Hofbauer (2011) and Cressman (2011), and books by Hofbauer and Sigmund (1988, 1998), Weibull (1995), Cressman (2003), and Sandholm (2010c). The last reference includes detailed treatments of the mathematics used in the analysis of evolutionary game dynamics, and provides full accounts of many of the topics introduced below.

13.2. POPULATION GAMES

This section introduces games played by a single population, with all agents sharing the same strategy set and payoff function. Extending the model to allow multiple populations is simple, but complicates the notation; see Sandholm (2010c) for details.

13.2.1 Definitions

We consider a unit mass of agents, each of whom chooses a pure strategy from the set $S = \{1, \ldots, n\}$. The aggregate behavior of these agents is described by a *population state*. This is an element of the simplex $X = \{x \in \mathbb{R}^n_+ : \sum_{j \in S} x_j = 1\}$, with x_j representing the proportion of agents choosing pure strategy j. The standard basis vector $e_i \in \mathbb{R}^n$ represents the *pure population state* at which all agents choose strategy i.

We identify a *population game* with a continuous vector-valued payoff function $F: X \to \mathbb{R}^n$. The scalar $F_i(x)$ represents the payoff to strategy i when the population state is x.

A population state is a *Nash equilibrium* of F, denoted $x \in NE(F)$, if no agent can improve his payoff by unilaterally switching strategies. More explicitly, x^* is a Nash equilibrium if

$$x_i^* > 0 \text{ implies that } F_i(x^*) \geq F_j(x^*) \text{ for all } j \in S. \qquad [13.1]$$

By representing Nash equilibria as fixed points of the game's best response correspondence (see Section 13.5.2), one can use the Kakutani fixed point theorem to prove that equilibrium exists.

Theorem 13.1. *Every population game admits at least one Nash equilibrium.*

As we argued in the introduction, the direct assumption of equilibrium play rests on strong knowledge assumptions, assumptions that are particularly suspect in games with large numbers of participants. But in large population settings, equilibrium predictions can sometimes be justified by consideration of disequilibrium dynamics. Sections 13.6, 13.7, and 13.8 will show that for certain interesting classes of population games, various dynamic adjustment processes lead to equilibrium play. It is not coincidental that in these classes of games, existence of equilibrium can be established by elementary methods—in particular, without recourse to fixed point theorems. In games for which the full power of fixed point theory is needed to establish existence of equilibrium, the direct assumption of equilibrium play is least convincing.

13.2.2 Examples

We now introduce some basic examples of population games that we revisit throughout the chapter.

Example 13.1. (Matching in two-player symmetric normal form games) *In a* symmetric two-player normal form game, *each of the two players chooses a (pure) strategy from the finite set $S = \{1, \ldots, n\}$. The game's* payoffs *are described by the matrix $A \in \mathbb{R}^{n \times n}$. Entry A_{ij} is the payoff a player obtains when he chooses strategy i and his opponent chooses strategy j; this payoff does not depend on whether the player in question is called player 1 or player 2.*

Suppose that the unit mass of agents are randomly matched to play the symmetric normal form game A (or, alternatively, that each agent is matched once with each possible opponent). The expected payoff to strategy i at population state x is described by the linear function $F_i(x) = \sum_{j \in S} A_{ij} x_j$. The payoffs to all strategies can be expressed concisely as $F(x) = Ax$.

For many reasons, population games based on matching in normal form games are by far the most commonly studied in the literature. Linearity of payoffs in the population state makes these games mathematically simple. Furthermore, some research on population games and evolutionary dynamics is motivated not by a direct interest in large games per se, but rather by the possibility of obtaining equilibrium selection results for (reduced) normal form games by embedding them in a larger context. But if one is genuinely interested in modeling behavior in large populations of strategically interacting agents, focusing on the linear case is unnecessarily restrictive. Indeed, there are some applications in which nonlinearity is essential.

Example 13.2. (Congestion games) *Consider the following model of highway congestion, due to Beckmann et al. (1956). A pair of towns, Home and Work, are connected by a network of* links. *To commute from Home to Work, an agent must choose a* path *$i \in S$ connecting the two towns. The payoff the agent obtains is the negation of the delay on the path he takes. The delay on* *the path is the sum of the delays on its links, and the delay on a link is a function of the number of agents who use that link.*

To formalize this environment as a congestion game, *let $\mathscr{L}$ be the collection of links in the highway network. Each strategy $i \in S$ is a path from Home to Work, and so is identified with a set of links $\mathscr{L}_i \subseteq \mathscr{L}$. Each link ℓ is assigned a* cost function $c_\ell \colon \mathbb{R}_+ \to \mathbb{R}$, *whose argument is link ℓ's* utilization level u_ℓ:

$$u_\ell(x) = \sum_{i \in S(\ell)} x_i, \quad \textit{where } S(\ell) = \{i \in S \colon \ell \in \mathscr{L}_i\}.$$

The payoff of choosing path i is the negation of the total delays on its links:

$$F_i(x) = -\sum_{\ell \in \mathscr{L}_i} c_\ell(u_\ell(x)).$$

Since highway congestion involves negative externalities, cost functions in models of highway congestion are increasing. They are typically convex as well: delays are essentially fixed until a link becomes congested, at which point they increase quickly. Congestion games can also be used to model positive externalities, like the choice between different technological standards; in this case, cost functions are decreasing in the utilization levels.

Population games also provide a useful framework for macroeconomic applications.

Example 13.3. (Search with positive externalities) *Agents in a unit mass population choose levels of search effort from the set $S = \{1, \ldots, n\}$. Stronger efforts increase the likelihood of finding trading partners, so that payoffs are increasing both in own search effort and in aggregate search effort. This search model is represented by a population game with payoffs $F_i(x) = m_i\, b(a(x)) - c_i$, where $a(x) = \sum_{k=1}^{n} k x_k$ represents aggregate search effort, the increasing function $b \colon \mathbb{R}_+ \to \mathbb{R}$ represents the benefits of search as a function of aggregate effort, the increasing function $m \colon S \to \mathbb{R}$ is the benefit multiplier, and the arbitrary function $c \colon S \to \mathbb{R}$ captures search costs.*

13.2.3 The geometry of population games

An important advantage of the population game framework is the possibility of representing games' incentive structure geometrically, at least in low-dimensional cases. Intuition obtained in these cases often carries over to higher dimensions.

To present population games in pictures, we introduce the matrix $\Phi = I - \frac{1}{n}\mathbf{1}\mathbf{1}' \in \mathbb{R}^{n \times n}$, where $\mathbf{1} \in \mathbb{R}^n$ is the vector of ones. This matrix represents the orthogonal projection of $\mathbb{R}^n$ onto the subspace $TX = \{z \in \mathbb{R}^n \colon \sum_{i \in S} z_i = 0\}$, which is the tangent space of the simplex X. For a payoff vector $\pi \in \mathbb{R}^n$, the projected payoff vector

$$\Phi\pi = \pi - \mathbf{1}\left(\frac{1}{n}\sum_{i\in S}\pi_i\right) \equiv \pi - \mathbf{1}\bar{\pi}$$

is obtained by subtracting the population average payoff $\bar{\pi}$ from each component. Thus applying the projection Φ to a payoff vector eliminates information about average payoffs, while preserving information about payoff differences, and hence about the incentives that revising agents face.

Example 13.4. (Drawing two-strategy games) *Figure 13.1 presents the payoff vectors and projected payoff vectors for the two-strategy coordination game F^{C2} and the Hawk-Dove game F^{HD}:*

$$F^{C2}(x) = \begin{pmatrix} 1 & 0 \\ 0 & 2 \end{pmatrix}\begin{pmatrix} x_1 \\ x_2 \end{pmatrix} = \begin{pmatrix} x_1 \\ 2x_2 \end{pmatrix}; \quad F^{HD}(x) = \begin{pmatrix} -1 & 2 \\ 0 & 1 \end{pmatrix}\begin{pmatrix} x_H \\ x_D \end{pmatrix} = \begin{pmatrix} 2x_D - x_H \\ x_D \end{pmatrix}.$$

In each case, the first strategy is represented on the vertical axis, so as to agree with the payoff matrix, and the projected payoff vectors are those running parallel to the simplex. In the coordination game, the payoff vectors push outward, away from the mixed equilibrium $x^* = (\frac{2}{3}, \frac{1}{3})$ and toward the pure equilibria e_1 and e_2. In Hawk-Dove, the payoff vectors push inward, toward the unique Nash equilibrium $x^* = (\frac{1}{2}, \frac{1}{2})$.

We will see that the basic payoff monotonicity condition for disequilibrium dynamics (see Section 13.4.2) requires the vector describing the motion of the population state to agree with the payoff vector, in the weak sense that the angle between the two is acute. In a two-strategy game, this condition and feasibility completely determine the direction in which evolution should proceed: the state should move in the direction indicated by the projected payoff vector. Thus, Figure 13.1 shows that evolution sends the population toward a pure equilibrium in the coordination game, and toward the mixed equilibrium in the Hawk-Dove game.

In games with three strategies, it is no longer possible to draw the payoff vectors directly; only the projected payoff vectors may be drawn. Moreover, while these vectors

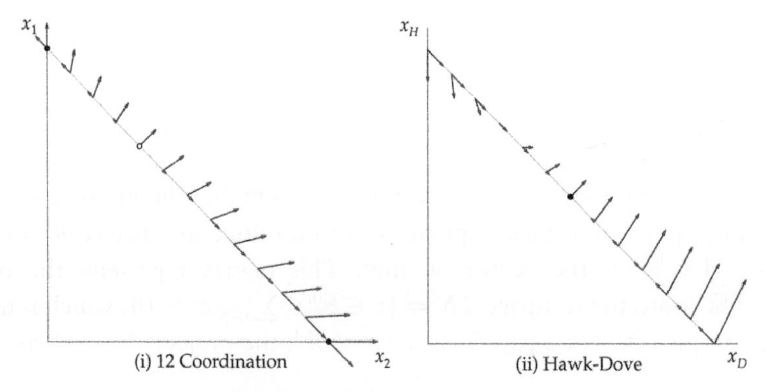

Figure 13.1 Payoffs and projected payoffs in two-strategy games.

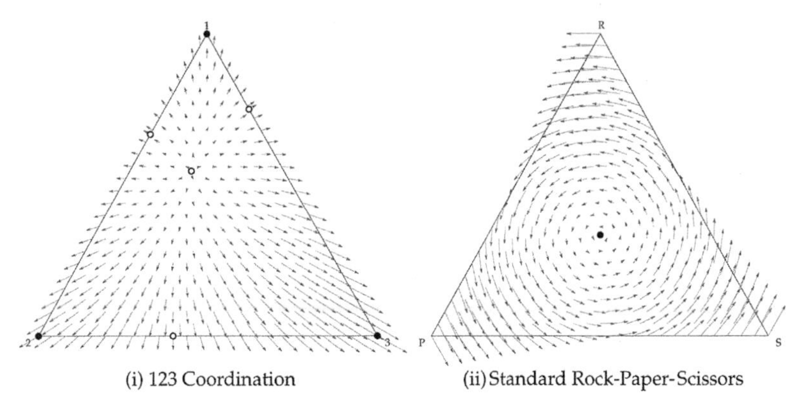

(i) 123 Coordination (ii) Standard Rock-Paper-Scissors

Figure 13.2 Projected payoffs in three-strategy games.

continue to describe incentives, they are no longer enough to determine the direction in which the population state will move, which will depend on the details of the revision procedure.

Example 13.5. (Drawing three-strategy games) *Figure 13.2 illustrates projected payoffs in the three-strategy coordination game F^{C3} and the standard Rock-Paper-Scissors game F^{RPS}.*

$$F^{C3}(x) = \begin{pmatrix} 1 & 0 & 0 \\ 0 & 2 & 0 \\ 0 & 0 & 3 \end{pmatrix} \begin{pmatrix} x_1 \\ x_2 \\ x_3 \end{pmatrix} = \begin{pmatrix} x_1 \\ 2x_2 \\ 3x_3 \end{pmatrix};$$

$$F^{RPS}(x) = \begin{pmatrix} 0 & -1 & 1 \\ 1 & 0 & -1 \\ -1 & 1 & 0 \end{pmatrix} \begin{pmatrix} x_R \\ x_P \\ x_S \end{pmatrix} = \begin{pmatrix} x_S - x_P \\ x_R - x_S \\ x_P - x_R \end{pmatrix}.$$

In the coordination game, the payoff vectors again push outward toward the pure equilibria. We will see in Section 13.6.5 that for a large class of evolutionary dynamics, play will converge to one of these equilibria from most initial states.

In the standard Rock-Paper-Scissors game, the projected payoffs cycle around the simplex. This suggests that evolutionary dynamics in this game need not converge to Nash equilibrium. As we will see below, whether or not convergence occurs depends on the revision protocol agents employ.

In Figures 13.1 and 13.2, the Nash equilibria of the games in question are drawn as dots. While these Nash equilibria can be found by checking definition [13.1], they can also be discovered geometrically.

For each population state x, we define the *tangent cone* of X at x to be the set of directions of motion from x that do not cause the state to leave the simplex X:

$$TX(x) = \left\{ z \in \mathbb{R}^n \colon z = \alpha\,(y - x) \text{ for some } y \in X \text{ and some } \alpha \geq 0 \right\}.$$

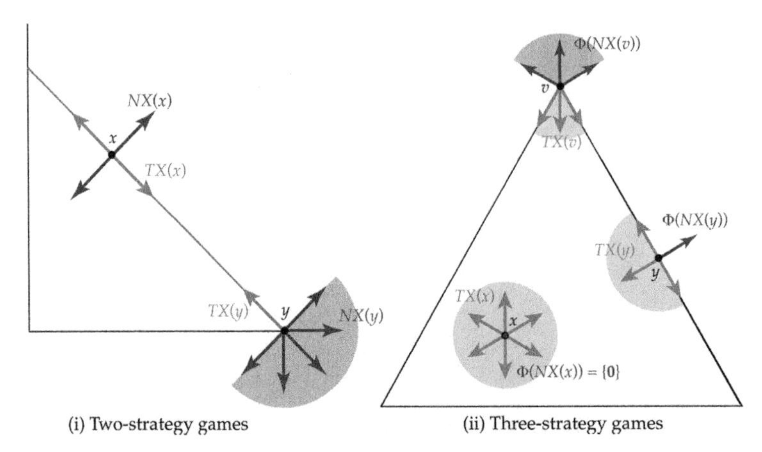

(i) Two-strategy games (ii) Three-strategy games

Figure 13.3 Tangent cones and normal cones.

The *normal cone* of X at x is the polar of the tangent cone of X at x. In other words, it is the set of directions that form an obtuse or right angle with every vector in $TX(x)$:

$$NX(x) = (TX(x))^{\circ} = \left\{ y \in \mathbb{R}^n \colon y'z \leq 0 \text{ for all } z \in TX(x) \right\}.$$

The possible forms of the tangent and normal cones in two- and three-strategy games are presented in Figure 13.3. In the latter case this is done after applying the projection Φ.

Observe that

$$\begin{aligned} x \in NE(F) &\Leftrightarrow [x_i > 0 \Rightarrow F_i(x) \geq F_j(x)] \text{ for all } i, j \in S \\ &\Leftrightarrow x'F(x) \geq y'F(x) \text{ for all } y \in X \\ &\Leftrightarrow (y - x)'F(x) \leq 0 \text{ for all } y \in X. \end{aligned} \qquad [13.2]$$

We thus have the following geometric characterization of Nash equilibrium.

Theorem 13.2. *Let F be a population game. Then $x \in NE(F)$ if and only if $F(x) \in NX(x)$.*

According to the theorem, the Nash equilibria drawn in Figures 13.1 and 13.2 are precisely those states whose payoff vectors lie in the relevant normal cones, as drawn in Figure 13.3.

13.3. REVISION PROTOCOLS AND MEAN DYNAMICS

Evolutionary dynamics for population games are designed to capture two basic assumptions. The first, *inertia*, means that agents only occasionally consider switching strategies. This assumption is natural when the environment being modeled is one of many in which the agents participate, so that the agents only pay limited attention to each. The second assumption, *myopia*, says that agents do not attempt to forecast future behavior, but instead base their decisions on the information they have about the current strategic environment. These two assumptions are mutually reinforcing: myopic behavior is most sensible when opponents' behavior adjusts slowly, so that strategies that perform well now are likely to continue to do so.[4]

To specify the adjustment process most transparently, we consider how individual agents make decisions. Formally, this is accomplished by means of objects called revision protocols. Revision protocols describe when and how agents decide to switch strategies, and they implicitly specify what information agents use to make these decisions.

Together, a population game and a revision protocol generate a differential equation called the mean dynamic. This dynamic describes the evolution of aggregate behavior when the revision protocol is employed during recurrent play of the game.

This section introduces revision protocols and mean dynamics, and justifies the use of the latter to describe the evolution of the population state. This approach to defining deterministic evolutionary dynamics via microfoundations was first developed for imitative dynamics by Björnerstedt and Weibull (1996) and Weibull (1995), and later expanded by Benaïm and Weibull (2003, 2009) and Sandholm (2003, 2010b).

13.3.1 Revision protocols

In the most general case, a *revision protocol* ρ is a map that assigns each population game F a function $\rho^F\colon X \to \mathbb{R}^{n\times n}_{+}$, which maps population states $x \in X$ to collections of *conditional switch rates* $\rho^F_{ij}(x)$. A population game F, a revision protocol ρ, and a finite population size N together define a stochastic evolutionary process—a Markov process—that runs on the discrete grid $\mathscr{X}^N = X \cap \frac{1}{N}\mathbb{Z}^n = \{x \in X : Nx \in \mathbb{Z}^n\}$.[5]

This process, which we define formally below, can be described as follows. Each agent in the society is equipped with a "stochastic alarm clock." The times between rings of an agent's clock are independent, each with a rate R exponential distribution, where $R = R(\rho^F) \geq \max_{x,i} \sum_{j\neq i} \rho^F_{ij}(x)$,[6] and different agents' clocks are independent of one another. The ringing of a clock signals the arrival of a revision opportunity for the clock's owner. If an agent playing strategy $i \in S$ receives a revision opportunity, he

[4] Other approaches to dynamics for population games build on different assumptions. The most notable example, the *perfect foresight dynamics* of Matsui and Matsuyama (1995), are obtained by retaining the assumption of inertia, but assuming that agents are forward looking. While evolutionary dynamics describe disequilibrium adjustment, perfect foresight dynamics represent adjustments that occur within a dynamic equilibrium.

[5] This process can also be formulated in discrete time, and can be adjusted to account for finite-population effects; see Benaïm and Weibull (2003, 2009) and Sandholm (2010c, Section 10.3). One can also incorporate finite-population effects into the definition of population games; see Sandholm (2010c, Section 11.4).

[6] As we will see, the mean dynamic is not affected by the value of the clock rate R.

switches to strategy $j \neq i$ with probability $\frac{1}{R}\rho_{ij}^{F}(x)$, and he continues to play strategy i with probability $1 - \frac{1}{R}\sum_{j\neq i}\rho_{ij}^{F}(x)$. This decision is made independently of the timing of the clocks' rings. If a switch occurs, the population state changes accordingly, from the old state x to a new state y that accounts for the agent's choice.

This interpretation of the stochastic evolutionary process has the advantage of working for any revision protocol. But once a protocol is fixed, one can use the structure provided by that protocol to provide a simpler interpretation of the process—see Example 13.6 below.

According to the description above, the diagonal elements $\rho_{ii}^{F}(x)$ of a revision protocol are irrelevant. An important exception occurs when the protocol has unit row sums, $\sum_{j\in S}\rho_{ij}^{F}(x) = 1$ for all $x \in X$ and $i \in S$, so that $\rho_{i\cdot}^{F}(x) = (\rho_{i1}^{F}(x), \ldots, \rho_{in}^{F}(x))$ is a probability vector. In this case, we set the clock rate R at 1, and refer to $\rho_{ij}^{F}(x)$ as a *conditional switch probability*.

13.3.2 Information requirements for revision protocols

The model above describes the behavior of agents who only occasionally consider switching strategies. To assess how well a particular revision protocol agrees with this general approach, we introduce some restrictions concerning the information that revision protocols may require.

Most of the protocols studied in the literature are of the form $\rho^{F}(x) = \rho(F(x), x)$, so that the current conditional switch rates only depend on the game by way of the current payoff. We call such protocols *reactive*. The remaining protocols, which we call *prospective*, are more demanding, in that they require agents to know enough about the payoff functions to engage in counterfactual reasoning. We will only consider prospective protocols in Section 13.8.3, where we will see that dynamics based on them can differ markedly from those based on reactive protocols.

Protocols can also be distinguished by the *amount and types of data* about the current strategic environment that they require. The least demanding protocols require knowledge of only one payoff, that of the current or the candidate strategy; slightly more demanding ones require both. Still more demanding ones require knowledge of the current payoffs to all strategies. Protocols that require further information—say, information about the average payoffs in the population—may be regarded as too demanding for typical applications.

A third distinction separates *continuous* and *discontinuous* protocols. Under continuous protocols, agents' choices do not change abruptly after small changes in the strategic environment. This property accords well with the evolutionary paradigm. Discontinuous protocols, which require exact information about the current strategic environment, are perhaps less natural under this paradigm, but are used to define important ideal cases.

13.3.3 The stochastic evolutionary process and mean dynamics

Formally, the game F, the protocol ρ, and a finite population size N define a Markov process $\{X_t^N\}_{t\geq 0}$ on the finite state space $\mathscr{X}^N$. Since each of the N agents receives revision opportunities at rate R, revision opportunities arrive in the population as a whole at rate NR, which is thus the expected number of revision opportunities arriving during each unit of clock time. Letting τ_k denote the arrival time of the kth revision opportunity, we can describe the transition law of the process $\{X_t^N\}$ by

$$\mathbb{P}\left(X_{\tau_{k+1}}^N = y \,|\, X_{\tau_k}^N = x\right) = \begin{cases} \dfrac{x_i \rho_{ij}^F(x)}{R} & \text{if } y = x + \dfrac{1}{N}(e_j - e_i), j \neq i, \\ 1 - \displaystyle\sum_{i \in S} \sum_{j \neq i} \dfrac{x_i \rho_{ij}^F(x)}{R} & \text{if } y = x, \\ 0 & \text{otherwise.} \end{cases}$$

While one can analyze this Markov process directly, we instead consider its limiting behavior as the population size N becomes large. Notice that all of the randomness in the process $\{X_t^N\}$ is idiosyncratic: both the assignments of revision opportunities and the randomizations performed by the agents are independent of past events conditional on the current state. Therefore, taking the limit as N grows large enables us to approximate $\{X_t^N\}$ by a deterministic trajectory—namely, a solution of the so-called *mean dynamic* generated by ρ and F. We introduce the mean dynamic next, saving the formal approximation result for the next section.

To derive the mean dynamic, we consider the behavior of the process $\{X_t^N\}$ over the next dt time units, starting from state x. Since each of the N agents receives revision opportunities at rate R, the expected number of opportunities arriving during these dt time units is $NR\,dt$. Each opportunity is equally likely to go to each agent, so the expected number of these opportunities that are received by current strategy i players is $Nx_i R\,dt$; the expected number of these that lead to switches to strategy j is $Nx_i \rho_{ij}^F(x)\,dt$. Hence, the expected change in the proportion of agents using strategy i is

$$\left(\sum_{j \in S} x_j \rho_{ji}^F(x) - x_i \sum_{j \in S} \rho_{ij}^F(x)\right) dt.$$

The *mean dynamic* induced by population game F and revision protocol ρ is thus

$$\dot{x} = V^F(x), \quad \text{where } V_i^F(x) = \sum_{j \in S} x_j \rho_{ji}^F(x) - x_i \sum_{j \in S} \rho_{ij}^F(x). \qquad [13.3]$$

Here $\dot{x} = \dot{x}_t = \frac{\mathrm{d}}{\mathrm{d}t} x_t$ denotes the time derivative of the solution trajectory $\{x_t\}_{t\geq 0}$.

The mean dynamic [13.3] is an ordinary differential equation defined on the simplex X. Evidently, $V_i^F(x) \geq 0$ whenever $x_i = 0$, so $V(x) \in TX(x)$ for all $x \in X$; that is, V^F never points outward from the boundary of X. So long as V^F is Lipschitz continuous,[7] existence and uniqueness of solutions to [13.3] follow from standard results.

Theorem 13.3. *If V^F is Lipschitz continuous, then the mean dynamic* [13.3] *admits a unique forward solution from each $\xi \in X$: that is, a trajectory $\{x_t\}_{t\geq 0}$ satisfying $x_0 = \xi$ and $\frac{d}{dt}x_t = V^F(x_t)$ for all $t \geq 0$.*

Typically, Lipschitz continuity of V^F is ensured by assuming that the game F is Lipschitz continuous and that the protocol $\rho^F(x) = \rho(F(x), x)$ is reactive and Lipschitz continuous.

Example 13.6. (Pairwise proportional imitation and the replicator dynamic) *Helbing (1992) and Schlag (1998) introduce the following reactive protocol, called* pairwise proportional imitation*:*

$$\rho_{ij}^F(x) = \rho_{ij}(F(x), x) = x_j[F_j(x) - F_i(x)]_+. \qquad [13.4]$$

Under protocol [13.4], *an agent who receives a revision opportunity chooses an opponent at random. This opponent is a strategy j player with probability x_j. The agent imitates the opponent only if the opponent's payoff is higher than his own, doing so with probability proportional to the payoff difference.*

Substituting protocol [13.4] *into formula* [13.3], *we find that the mean dynamic associated with this protocol is*

$$\begin{aligned}\dot{x}_i &= \sum_{j\in S} x_j x_i [F_i(x) - F_j(x)]_+ - x_i \sum_{j\in S} x_j [F_j(x) - F_i(x)]_+ \\ &= x_i \sum_{j\in S} x_j (F_i(x) - F_j(x)) \\ &= x_i \left(F_i(x) - \sum_{j\in S} x_j F_j(x) \right).\end{aligned}$$

This is the replicator dynamic *of Taylor and Jonker (1978), the best known dynamic in evolutionary game theory. Under this dynamic, the percentage growth rate $\dot{x}_i/x_i$ of each strategy currently in use is equal to the difference between that strategy's payoff and the average payoff obtained in the population; unused strategies always remain so. We discuss this dynamic further in Example 13.8 below.*

[7] The function $V^F \colon X \to \mathbb{R}^n$ is *Lipschitz continuous* if there exists a constant $K > 0$ such that $|V^F(y) - V^F(x)| \leq K|y - x|$ for all $x, y \in X$.

13.3.4 Finite horizon deterministic approximation

The basic link between the Markov processes $\{X_t^N\}$ and the mean dynamic [13.3] is provided by the following theorem.

Theorem 13.4. *Suppose that the mean dynamic V^F is Lipschitz continuous. Let the initial conditions $X_0^N = x_0^N$ converge to state $x_0 \in X$, and let $\{x_t\}_{t \geq 0}$ be the solution to the mean dynamic* [13.3] *starting from x_0. Then for all $T < \infty$ and $\varepsilon > 0$,*

$$\lim_{N \to \infty} \mathbb{P}\left(\sup_{t \in [0,T]} \left| X_t^N - x_t \right| < \varepsilon \right) = 1. \qquad [13.5]$$

Theorem 13.4, due to Kurtz (1970), says that when the population size N is large, nearly all sample paths of the Markov process $\{X_t^N\}$ stay within ε of a solution of the mean dynamic [13.3] through time T. In particular, by choosing N large enough, we can ensure that with probability close to one, X_t^N and x_t differ by no more than ε at all times t between 0 and T (Figure 13.4).[8]

The intuition for this result comes from the law of large numbers. At each revision opportunity, the increment in the process $\{X_t^N\}$ is stochastic. But if we fix dt, then

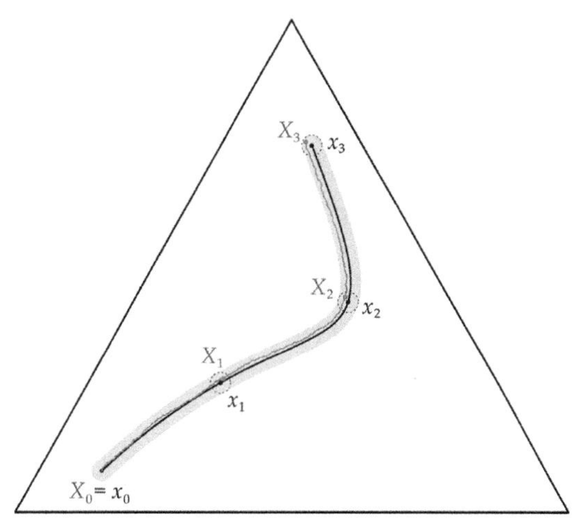

Figure 13.4 Deterministic approximation of the Markov process $\{X_t^N\}$.

[8] Benaïm and Weibull (2003) prove that the rate of convergence in [13.5] is exponential in N. Roth and Sandholm (2013) extend the theorem to allow the mean dynamic to be an upper hemicontinuous differential inclusion, as in the case of the best response dynamic (see Section 13.5.2). Hwang et al. (2013) consider stochastic evolutionary processes for large games on spatial domains with long-range interactions, and establish convergence of these processes to solutions of deterministic integro-differential equations.

when N is large enough, the expected number of revision opportunities arriving during time interval $I = [t, t + dt]$, namely $NR\,dt$, is large as well. Since each opportunity leads to an increment of the state of size $\frac{1}{N}$, the overall change in the state during interval I is only of order $R\,dt$. Thus, during this interval, there are many of revision opportunities, each involving nearly the same transition probabilities. The law of large numbers therefore suggests that the change in $\{X_t^N\}$ during this interval should be almost completely determined by the expected motion of $\{X_t^N\}$, as described by the mean dynamic [13.3].

It is important to note that Theorem 13.4 cannot be extended to an infinite horizon result. If, for instance, the conditional switch rates $\rho_{ij}^F(x)$ are always positive, then the process $\{X_t^N\}$ is irreducible, and thus must visit every state in $\{X_t^N\}$ infinitely often with probability one. Even so, one can use the mean dynamic [13.3] to obtain restrictions on the infinite horizon behavior of the process $\{X_t^N\}$, a point we explain in Section 13.10.1.

13.4. DETERMINISTIC EVOLUTIONARY DYNAMICS

13.4.1 Definition

We are now prepared to state a formal definition of deterministic evolutionary dynamics. Let $\mathscr{F}$ be a set of population games $F\colon X \to \mathbb{R}^n$ (with some fixed number of strategies n). Let $\mathscr{D}$ be the class of Lipschitz continuous ordinary differential equations $\dot{x} = V(x)$ on the simplex X, where the vector field $V\colon X \to \mathbb{R}^n$ satisfies $V(x) \in TX(x)$ for all $x \in X$.[9] A map that assigns each game $F \in \mathscr{F}$ a differential equation in $\mathscr{D}$ is called a *deterministic evolutionary dynamic.*

Every well-behaved revision protocol implicitly defines a deterministic evolutionary dynamic. Specifically, suppose that the revision protocol ρ is such that for each $F \in \mathscr{F}$, the function $\rho^F\colon X \to \mathbb{R}_+^{n\times n}$ is Lipschitz continuous. Then ρ defines an evolutionary dynamic $\dot{x} = V^F(x)$ by way of equation [13.3].

13.4.2 Incentives and aggregate behavior

In order to draw links between deterministic evolutionary dynamics and traditional game-theoretic analyses, we must introduce conditions that relate the evolution of aggregate behavior under the dynamics to the incentives in the underlying game. The two most important conditions are these:

$$\textit{Positive correlation} \qquad V^F(x) \neq \mathbf{0} \;\Rightarrow\; V^F(x)'F(x) > 0. \qquad \text{[PC]}$$

$$\textit{Nash stationarity} \qquad V^F(x) = \mathbf{0} \;\Leftrightarrow\; x \in NE(F). \qquad \text{[NS]}$$

[9] In what follows, we often identify the differential equation $\dot{x} = V(x)$ with the vector field V.

Positive correlation (PC) is the basic restriction on disequilibrium dynamics. In game-theoretic terms, it requires that there be a positive correlation between growth rates and payoffs under the uniform probability distribution on strategies: under this distribution,

$$\mathbb{E}(V^F(x)) = \sum_{k \in S} \tfrac{1}{n} V_k^F(x) = 0, \text{ and so}$$

$$\operatorname{Cov}(V^F(x), F(x)) = \mathbb{E}(V^F(x)\, F(x)) - \mathbb{E}(V^F(x))\, \mathbb{E}(F(x)) = \tfrac{1}{n} V^F(x)' F(x).$$

Geometrically, (PC) requires that whenever the growth rate vector $V^F(x)$ is nonzero, the angle it forms with the payoff vector $F(x)$ is acute. Thus, away from Nash equilibrium (see Proposition 13.1 below), (PC) restricts the direction of motion to a half space. In this sense, it is as weak a condition as one could hope to be useful. We will see in Sections 13.6 and 13.7 that it is useful indeed.[10]

Example 13.7. (Drawing positive correlation) *Figure 13.5 presents projected payoff vectors and vectors of motion under the replicator dynamic in 123 Coordination (Example 13.5). Evidently, the direction of motion makes an acute angle with the payoff vector whenever the*

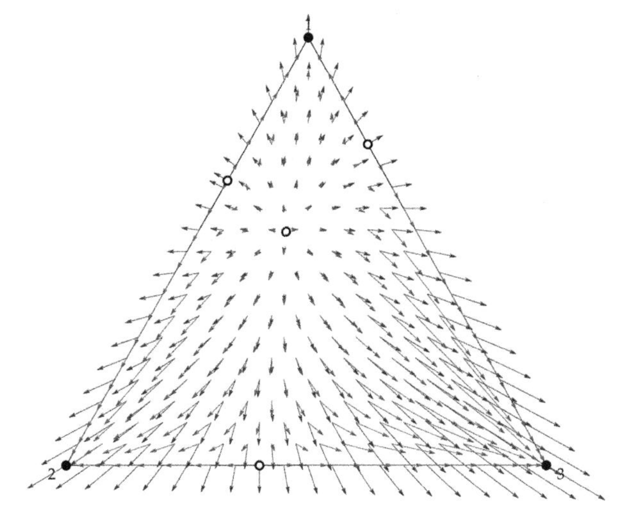

Figure 13.5 Condition (PC): Projected payoff vectors and vectors of motion under the replicator dynamic in 123 Coordination. The latter point within the simplex.

[10] Versions of this condition are considered by Friedman (1991), Swinkels (1993), Sandholm (2001b), and Demichelis and Ritzberger (2003). For other monotonicity conditions, some of which are particular to imitative dynamics, see Nachbar (1990), Friedman (1991), Samuelson and Zhang (1992), Swinkels (1993), Ritzberger and Weibull (1995), Hofbauer and Weibull (1996), and Viossat (2011).

population is not at rest. This illustrates the general fact that the replicator dynamic satisfies positive correlation (PC) (see Section 13.5.1.3).

Nash stationarity (NS) requires that the Nash equilibria of the game F and the rest points of the dynamic V^F coincide. It can be split into two distinct restrictions. First, (NS) asks that every Nash equilibrium of F be a rest point of V^F. If state x is a Nash equilibrium, then no agent benefits from switching strategies; (NS) demands that in this situation, aggregate behavior is at rest under V^F. This does not imply that individual agents' behavior is also at rest—remember that V^F only describes the *expected* motion of the underlying stochastic process.[11]

In any case, this direction of Nash stationarity is implied by positive correlation:

Proposition 13.1. *If V^F satisfies* (PC)*, then $x \in NE(F)$ implies that $V^F(x) = \mathbf{0}$.*

The proof of this result is simple if we use the geometric ideas from Section 13.2.3. If $x \in NE(F)$, then $F(x) \in NX(x)$ (by Theorem 13.2). But $V^F(x) \in TX(x)$ (since it is a feasible direction of motion from x). Thus $V^F(x)'F(x) \leq 0$ (by the definition of normal cones), so (PC) implies that $V^F(x) = \mathbf{0}$.

Second, Nash stationarity asks that every rest point of V^F be a Nash equilibrium of F. If the current population state is not a Nash equilibrium, then there are agents who would benefit from switching strategies. (NS) requires that enough agents avail themselves of this opportunity that aggregate behavior is not at rest.

13.5. FAMILIES OF EVOLUTIONARY DYNAMICS

Evolutionary dynamics are defined in families, with members of a family being derived from qualitatively similar revision protocols. This approach addresses the fact that in practice, one does not expect to know the protocols agents employ with much precision. If we can show that qualitatively similar protocols lead to qualitatively similar aggregate dynamics, then knowing which family a protocol comes from may be enough to draw conclusions about aggregate play.

We present five basic examples of revision protocols and their mean dynamics in Table 13.1. As the protocols in this section are all reactive, we write them as $\rho_{ij}(\pi, x)$, so that $\rho_{ij}^F(x) = \rho_{ij}(F(x), x)$.

We also introduce some additional notation. We define $\hat{F}_i(x) = F_i(x) - \bar{F}(x)$ to be the *excess payoff* to strategy i; this is the difference between strategy i's payoff and the population average payoff $\bar{F}(x) = \sum_{i \in S} x_i F_i(x)$. In addition, we let $M\colon \mathbb{R}^n \Rightarrow X$ denote the *(mixed) maximizer correspondence*, $M(\pi) = \arg\max_{y \in X} y'\pi$.

[11] This distinction is important in local stability analyses of the underlying stochastic process—see Sandholm (2003).

Table 13.1 Five basic deterministic dynamics.

Revision protocol	Mean dynamic	Name
$\rho_{ij} = x_j[\pi_j - \pi_i]_+$	$\dot{x}_i = x_i\hat{F}_i(x)$	Replicator
$\rho_{i\cdot} = M(\pi)$	$\dot{x} \in M(F(x)) - x$	Best response
$\rho_{ij} = \dfrac{\exp(\eta^{-1}\pi_j)}{\sum_{k\in S}\exp(\eta^{-1}\pi_k)}$	$\dot{x}_i = \dfrac{\exp(\eta^{-1}F_i(x))}{\sum_{k\in S}\exp(\eta^{-1}F_k(x))} - x_i$	Logit
$\rho_{ij} = [\pi_j - \sum_{k\in S} x_k\pi_k]_+$	$\dot{x}_i = [\hat{F}_i(x)]_+ - x_i\sum_{j\in S}[\hat{F}_j(x)]_+$	BNN
$\rho_{ij} = [\pi_j - \pi_i]_+$	$\dot{x}_i = \sum_{j\in S} x_j[F_i(x) - F_j(x)]_+ - x_i\sum_{j\in S}[F_j(x) - F_i(x)]_+$	Smith

Figure 13.6 presents phase diagrams for the five basic dynamics when agents are matched to play standard Rock-Paper-Scissors (Example 13.5). The phase diagram of the replicator dynamic displays closed orbits around the unique Nash equilibrium $x^* = \left(\frac{1}{3}, \frac{1}{3}, \frac{1}{3}\right)$. Since this dynamic is based on imitation (or on reproduction), each face and each vertex of the simplex X is an invariant set: a strategy initially absent from the population will never subsequently appear.

The other four dynamics presented in the figure are based on protocols that allow agents to select unused strategies. Under these dynamics, the Nash equilibrium is the sole rest point, and attracts solutions from all initial conditions.[12] Under the best response dynamic, solution trajectories quickly change direction and then accelerate when the best response changes; under the Smith dynamic, solutions approach the Nash equilibrium in a less angular, more gradual fashion.

13.5.1 Imitative dynamics

Imitative dynamics are the most thoroughly studied dynamics in evolutionary game theory. They are the descendants, or more accurately, a reinterpretation of the game dynamics studied in biology,[13] and they predominated in the early economic literature on evolutionary game dynamics.[14]

[12] In the case of the logit dynamic, the rest point happens to coincide with the Nash equilibrium only because of the symmetry of the game—see Section 13.5.2.3.

[13] See, for instance, Taylor and Jonker (1978), Maynard Smith (1982), and Hofbauer and Sigmund (1988, 1998).

[14] See, for example, Nachbar (1990), Samuelson and Zhang (1992), Björnerstedt and Weibull (1996), Weibull (1995), and Hofbauer (1995a).

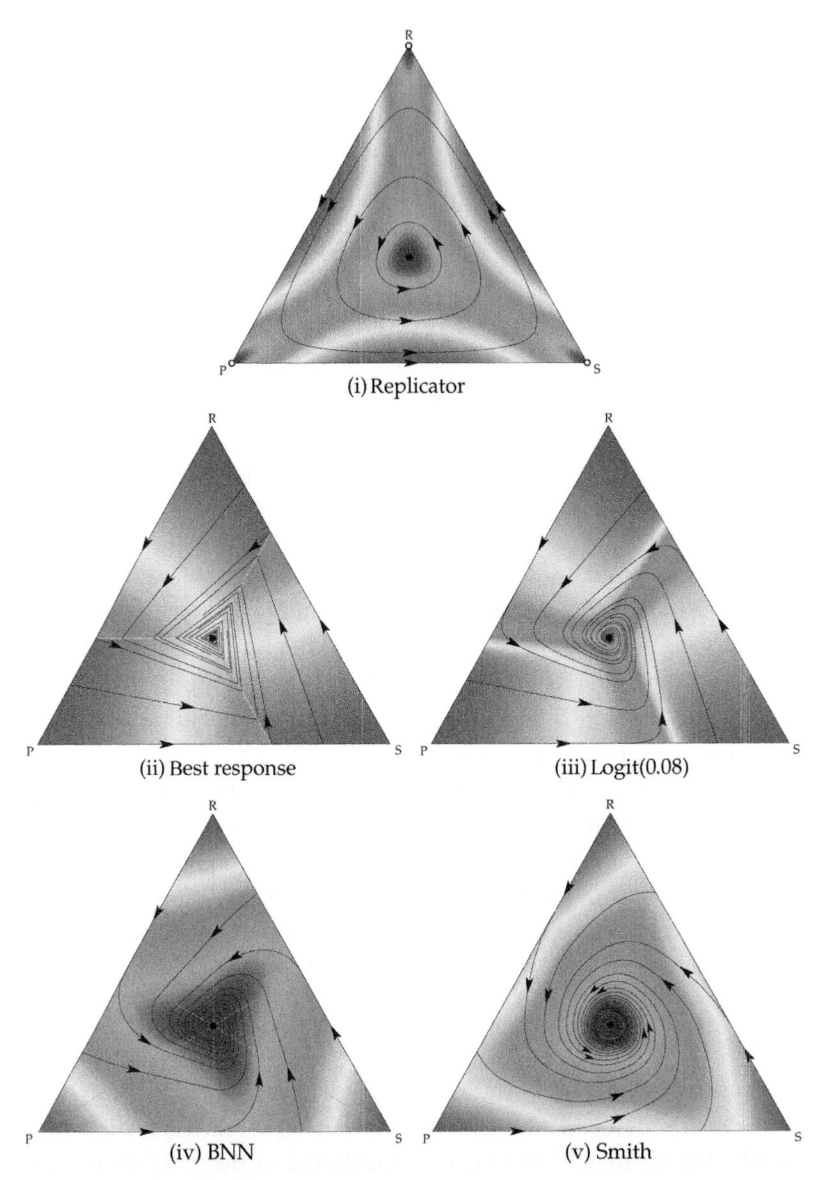

Figure 13.6 Five basic deterministic dynamics in standard Rock-Paper-Scissors. Shading represents speeds.

Imitative dynamics are derived from protocols under which agents consider switching to the strategy of a randomly sampled opponent, with the ultimate decision depending on payoff considerations. This specification leads imitative dynamics to have a particularly convenient functional form, in which strategies' absolute growth rates are proportional to their current levels of use. This functional form is just as apposite from the biological point of view: there the dynamics reflect births and deaths of different genetic types, and so are naturally expressed relative to current population shares.

13.5.1.1 Definition

The Lipschitz continuous revision protocol ρ is an *imitative protocol* if

$$\rho_{ij}(\pi, x) = x_j r_{ij}(\pi, x), \text{ where} \quad [13.6a]$$

$$\pi_j \geq \pi_i \iff [r_{kj}(\pi, x) - r_{jk}(\pi, x) \geq r_{ki}(\pi, x) - r_{ik}(\pi, x) \text{ for all } i, j, k \in S]. \quad [13.6b]$$

The values of r_{ij} are called *conditional imitation rates.* Condition [13.6b], called *net monotonicity of conditional imitation rates,* says that when strategy j has a higher payoff than strategy i, the net rate of imitation from any strategy k to j exceeds the net rate of imitation from k to i. This condition is enough to ensure that the resulting mean dynamic has the monotonicity properties we desire.

Substituting the functional form [13.6a] into the general equation [13.3] for mean dynamics, we obtain

$$\dot{x}_i = x_i \sum_{j \in S} x_j \left(r_{ji}(F(x), x) - r_{ij}(F(x), x) \right). \quad [13.7]$$

When the conditional imitation rates satisfy net monotonicity [13.6b], we call [13.7] an *imitative dynamic.* Under these dynamics, strategies' (absolute) growth rates are proportional to their current levels of use, while unused strategies remain so.

Table 13.2 lists a variety of specifications of imitative protocols consistent with condition [13.6b]. Under the first three, the agent observes the strategy of a single opponent and decides whether to imitate him; under the last, the agent repeatedly draws opponents until he deems one worthy of imitating. The protocols also require different pieces of data. Under imitation via pairwise comparisons, the agent compares his own payoff to that of the opponent he observes; under imitation driven by dissatisfaction, he only observes his own payoff, and under both forms of imitation of success, he only considers the payoffs of those he observes.

Table 13.2 Some specifications of imitative revision protocols.

Formula	Restriction	Interpretation
$\rho_{ij}(\pi, x) = x_j \phi(\pi_j - \pi_i)$	$\text{sgn}(\phi(d)) = \text{sgn}([d]_+)$	Imitation via pairwise comparisons
$\rho_{ij}(\pi, x) = a(\pi_i)\, x_j$	a decreasing	Imitation driven by dissatisfaction
$\rho_{ij}(\pi, x) = x_j\, c(\pi_j)$	c increasing	Imitation of success
$\rho_{ij}(\pi, x) = \dfrac{x_j\, w(\pi_j)}{\sum_{k \in S} x_k\, w(\pi_k)}$	w increasing	Imitation of success with repeated sampling

13.5.1.2 Examples

We now consider some important instances of these protocols and the dynamics they induce.

Example 13.8. (The replicator dynamic) *Example 13.6 introduced the replicator dynamic, which we can express concisely using the notation introduced above as*

$$\dot{x}_i = x_i \hat{F}_i(x). \qquad [13.8]$$

Example 13.6 derived this dynamic from imitation via pairwise comparisons with the semilinear functional form $\rho_{ij}(\pi, x) = x_j[\pi_j - \pi_i]_+$. *It can also be derived from the linear versions of imitation driven by dissatisfaction* ($\rho_{ij}(\pi, x) = (K - \pi_i)x_j$ *with* K *sufficiently large) and of imitation of success* ($\rho_{ij}(\pi, x) = x_j(\pi_j - K)$ *with* K *sufficiently small).*

While in social science contexts the replicator dynamic is best understood as describing the aggregate consequences of imitation, the dynamic first appeared as a biological model: it was introduced by Taylor and Jonker (1978) to provide a dynamic foundation for Maynard Smith and Price's (1973) notion of an evolutionarily stable strategy.[15]

The replicator dynamic has deep connections with other dynamic models from biology. Hofbauer (1981) shows that the replicator dynamic is equivalent after a nonlinear change of variable to the Lotka-Volterra equation, a fundamental model of the dynamics of ecological systems. Schuster and Sigmund (1983) observe that basic models of population genetics (Crow and Kimura (1970)) and of biochemical evolution (Eigen and Schuster (1979)) can be viewed as special cases of the replicator dynamic; they are also the first to refer to the dynamic by this name.[16]

Example 13.9. (The Maynard Smith replicator dynamic) *Suppose that agents employ imitation of success with repeated sampling, that payoffs are positive, and that* $w(\pi_j) = \pi_j$.[17] *In this case, the mean dynamic* [13.7] *becomes*

$$\dot{x}_i = \frac{x_i \hat{F}_i(x)}{\bar{F}(x)}. \qquad [13.9]$$

This is known as the Maynard Smith replicator dynamic, *after Maynard Smith (1982). In the current single-population context, dynamics* [13.8] *and* [13.9] *differ only by a change of speed. With multiple populations this is no longer true, since the changes of speed differ across populations; consequently, the two dynamics have different stability properties in some multipopulation games.*

[15] However, the connection between the two constructions is looser than one might expect: see Sections 13.7.1 and 13.10.3.

[16] For more on the links among these models, see Hofbauer and Sigmund (1988).

[17] In the biology literature, the stochastic evolutionary process generated by this protocol is called the *frequency-dependent Moran process*, after Moran (1962). See Nowak (2006) for references and discussion.

Example 13.10. (The imitative logit dynamic) *If agents again employ imitation of success with repeated sampling, using the exponential transformation* $w(\pi_j) = \exp(\eta^{-1}\pi_j)$ *with* noise level $\eta > 0$, *then* [13.7] *becomes*

$$\dot{x}_i = \frac{x_i \exp(\eta^{-1} F_i(x))}{\sum_{k \in S} x_k \exp(\eta^{-1} F_k(x))} - x_i.$$

This is the imitative logit dynamic *of Björnerstedt and Weibull (1996) and Weibull (1995). When the noise level* η *is small, behavior under this dynamic resembles that under the best response dynamic (Section 13.5.2), at least away from the boundary of the simplex.*

13.5.1.3 Basic properties

We now consider some general properties of imitative dynamics. Later sections will describe their behavior in various classes of games and interesting examples.

Theorem 13.3 showed that in general, Lipschitz continuous mean dynamics admit unique forward solutions $\{x_t\}_{t \geq 0}$ from every initial condition in the simplex. For imitative dynamics [13.7], under which $\dot{x}_i$ is proportional to x_i, more is true: solutions $\{x_t\}_{t \in (-\infty, \infty)}$ exist in both forward and backward time, and along any solution the support of x_t does not change.

Turning to payoff monotonicity properties, let us rewrite the dynamic [13.7] as

$$\dot{x}_i = V_i(x) = x_i G_i(x), \text{ where } G_i(x) = \sum_{k \in S} x_k \left(r_{ki}(F(x), x) - r_{ik}(F(x), x)\right). \quad [13.10]$$

If strategy $i \in S$ is in use, then $G_i(x) = V_i(x)/x_i$ represents the *percentage growth rate* of the number of agents using this strategy. Condition [13.6b] implies that strategies' percentage growth rates are ordered by their payoffs,

$$G_i(x) \geq G_j(x) \text{ if and only if } F_i(x) \geq F_j(x), \quad [13.11]$$

a property called *monotonicity of percentage growth rates*. This property, a strong restriction on strategies' percentage growth rates, can be shown to imply positive correlation (PC), a weak restriction on strategies' absolute growth rates.[18]

It is easy to show that the rest points of any imitative dynamic V_F include the Nash equilibria of F; in fact, this follows from the previous claim and Proposition 13.1. But imitative dynamics may also have non-Nash rest points. For instance, any pure state e_i is a rest point under [13.7]: since everyone is playing the same strategy, imitation leads to stasis. In fact, x is a rest point of [13.7] if and only if it is a *restricted equilibrium* of F, meaning that it is a Nash equilibrium of a restricted version of F in which only strategies in the support of x can be played. Non-Nash rest points of [13.7] are not natural predictions of play: they cannot be locally stable, nor can they be approached

[18] See Sandholm (2010c, Theorem 5.4.9).

by any interior solution trajectory (Bomze, 1986; Nachbar, 1990). Even so, continuous dynamics move slowly near rest points, so escape from the vicinity of non-Nash rest points is necessarily slow.

13.5.1.4 Inflow-outflow symmetry

To introduce a final property, let us compare the general equation for mean dynamics with that for imitative dynamics:

$$\dot{x}_i = \sum_{j\in S} x_j \rho_{ji}^F(x) - x_i \sum_{j\in S} \rho_{ij}^F(x), \qquad [13.3]$$

$$\dot{x}_i = x_i \sum_{j\in S} x_j \left(r_{ji}(F(x), x) - r_{ij}(F(x), x) \right). \qquad [13.7]$$

In general, mean dynamics exhibit an asymmetry between their inflow and outflow terms: the rate of switches from strategy j to strategy i is proportional to x_j, while the rate of switches from strategy i to strategy j is proportional to x_i. But under imitative dynamics, both of these rates are proportional to both x_i and x_j. This property, called *inflow-outflow symmetry*, underlies a number of properties of imitative dynamics that fail for other continuous dynamics. Figure 13.6 illustrates this point in the standard Rock-Paper-Scissors game: while the other dynamics converge, the replicator dynamic exhibits a continuum of closed orbits. A more surprising distinction concerns the behaviors of the dynamics in games with dominated strategies: see Sections 13.8.1 and 13.9.2.

Another dynamic that satisfies inflow-outflow symmetry is the *projection dynamic* (Nagurney and Zhang, 1997; Lahkar and Sandholm, 2008). On the interior of the simplex, this dynamic is defined by

$$\dot{x} = \Phi F(x), \qquad [13.12]$$

so that the direction of motion is given by the projected payoff vector.[19] It can be derived from revision protocols reflecting "revision driven by insecurity," under which the conditional switch rate ρ_{ij} is inversely proportional to x_i. The rate of switches from strategy i to strategy j under [13.12] is proportional to neither x_i nor x_j. By virtue of this inflow-outflow symmetry, the projection dynamic exhibits close connections to the replicator dynamic, at least in the interior of the simplex (Sandholm et al., 2008). While the projection dynamic is mathematically appealing, the discontinuities in the dynamic and its protocols at the boundary of the simplex raise doubts about its appropriateness for applications.

[19] The dynamic is defined globally by $\dot{x} = \mathrm{Proj}_{TX(x)}(F(x))$, where the right hand side represents the closest point projection of $F(x)$ onto $TX(x)$. This definition ensures that unique forward solution trajectories exist from every initial condition in X.

13.5.2 The best response dynamic and related dynamics

Traditionally, choice in game theory is based on optimization. The original evolutionary dynamic embodying this paradigm is the best response dynamic of Gilboa and Matsui (1991) (see also Matsui, 1992, and Hofbauer, 1995b). This dynamic can also be derived as a continuous-time version of the well-known *fictitious play* process of Brown (1949, 1951) (see Section 13.10.2). We now introduce the best response dynamic and some interesting variations, most notably those defined by means of payoff perturbations.

13.5.2.1 Target protocols and target dynamics

Under the revision protocol for the best response dynamic, an agent's conditional switch rates do not depend on his current strategy. Protocols with this feature have identical rows: $\rho_{ij}^F(x) = \rho_{\hat{\imath}j}^F(x)$ for all $x \in X$ and $i, \hat{\imath}, j \in S$. In this case, we use $\tau^F \equiv \rho_{i\cdot}^F$ to refer to the common row of ρ^F, and call τ^F a *target protocol.*

Substitution into equation [13.3] shows that target protocols generate mean dynamics of the form

$$\dot{x}_i = \tau_i^F(x) - x_i \sum_{j \in S} \tau_j^F(x), \qquad [13.13]$$

which we call *target dynamics*. When $\lambda^F(x) = \sum_{j \in S} \tau_j^F(x)$ is not zero, we can define $\sigma^F(x) \in X$ by $\sigma_j^F(x) = \tau_j^F(x)/\lambda^F(x)$ and express the target dynamic [13.13] as

$$\dot{x} = \lambda^F(x)(\sigma^F(x) - x). \qquad [13.14]$$

Geometrically, equation [13.14] says that the population state moves from its current position x toward the target state $\sigma^F(x)$ at rate $\lambda^F(x)$.

A further simple property of the best response dynamic's protocol is that its entries sum to one, $\sum_{j \in S} \tau_j^F(x) = 1$ for all $x \in X$, so that the conditional switch rates τ_j^F are actually conditional switch probabilities (see Section 13.3.1). To highlight this property, we denote the protocol by $\sigma^F(x)$, and express the target dynamic as

$$\dot{x} = \sigma^F(x) - x. \qquad [13.15]$$

13.5.2.2 The best response dynamic

Under the best response protocol, agents receive revision opportunities at a unit rate, and use these opportunities to switch to a current best response. This protocol is reactive ($\rho^F(x) = \rho(F(x), x)$), and it is a target protocol with unit row sum ($\rho_{i\cdot}(F(x), x) = \sigma(F(x), x) \in X$); moreover, it does not condition directly on the population state ($\sigma(F(x), x) = \sigma(F(x))$). It is defined formally by

$$\sigma(\pi) = M(\pi), \tag{13.16a}$$

where the map $M\colon \mathbb{R}^n \Rightarrow X$, defined by

$$M(\pi) = \underset{y \in X}{\arg\max}\, y'\pi, \tag{13.16b}$$

is the (*mixed*) *maximizer correspondence*. Substituting [13.16a] and [13.16b] into the mean dynamic equation [13.15], we obtain the *best response dynamic*,

$$\dot{x} \in M(F(x)) - x. \tag{13.17}$$

This dynamic can also be expressed as

$$\dot{x} \in B^F(x) - x,$$

where $B^F = M \circ F$ is the (*mixed*) *best response correspondence* for F.

Since the maximizer correspondence M is set-valued and discontinuous, the best response dynamic is a *differential inclusion*, and lies outside the framework developed in Section 13.3. Thus, the basic results on existence and uniqueness of solutions (Theorem 13.3) and on deterministic approximation (Theorem 13.4) do not apply here. Fortunately, versions of both of these results are available for the current setting. Since M is a convex-valued and upper hemicontinuous correspondence, results from the theory of differential inclusions imply the existence of a *Carathéodory solution* from every initial condition: a Lipschitz continuous trajectory $\{x_t\}_{t \geq 0}$ that satisfies $\dot{x}_t \in V(x_t)$ at all but a measure zero set of times.[20] But solutions are generally not unique: for instance, starting from a mixed equilibrium of a coordination game, there is not only a stationary solution, but also solutions that head immediately toward a pure equilibrium, as well as solutions that do so after an initial delay. Regarding deterministic approximation, one can define versions of the stochastic evolutionary process $\{X_t^N\}$ that account for the multivaluedness of M, and whose sample paths are approximated by solutions of [13.17].[21]

Despite these technical complications, most solutions of the best response dynamic take a very simple form. To draw a phase diagram of [13.17] for a given game F, one divides the state space X into the best response regions for each strategy. When the state is in the best response region for strategy i, solutions of [13.17] move directly toward pure state e_i. At states admitting multiple best responses, more than one direction of motion is possible in principle, though not always in practice.

Example 13.11. (The best response dynamic for standard RPS) *The best response dynamic for standard Rock-Paper-Scissors (Example 13.5) is illustrated in Figure 13.6ii. Within*

[20] The formulation of the best response dynamic as a differential inclusion is due to Hofbauer (1995b). For an introduction to differential inclusions, see Smirnov (2002).

[21] See Roth and Sandholm (2013).

each best response region, solutions proceed toward the relevant vertex. At the boundaries of these regions, the only continuation path consistent with [13.17] turns instantly toward the next vertex in the best response cycle. In the limit, the process converges to the mixed equilibrium $x^* = \left(\frac{1}{3}, \frac{1}{3}, \frac{1}{3}\right)$ *after an infinite number of turns.*[22]

Despite its absence in this example, multiplicity of solutions under the best response dynamic is quite common. We noted earlier that there are always multiple solutions emanating from mixed equilibria of coordination games. In other cases, multiplicity of solutions can lead to complicated dynamics, including solutions that cycle in and out of equilibrium in perpetuity (Hofbauer, 1995b).

It is easy to verify that the best response dynamic satisfies versions of positive correlation (PC) and Nash stationarity (NS) suitable for differential inclusions. For the former, note that if $y \in M(F(x)) = B^F(x)$ is any best response to x, we have

$$(y - x)'F(x) = \max_{j \in S} F_j(x) - \bar{F}(x) = \max_{j \in S} \hat{F}_j(x) \geq 0,$$

with equality only when x is a Nash equilibrium (see Proposition 13.2 below). And it is clear from [13.17] that stationary solutions coincide with Nash equilibria.

While the best response dynamic is the simplest dynamic based on the idea of exact optimization, it is not the only such dynamic. Balkenborg et al. (2013) consider the *refined best response dynamic*, under which agents only switch to best responses that are robust, in the sense that they are unique best responses at some nearby state. Under the *tempered best response dynamics* of Zusai (2011), all revising agents switch to best responses, but the rate at which an agent revises declines with the payoff of his current strategy. A third alternative, the *sampling best response dynamics* of Oyama et al. (2012), is introduced in Section 13.8.3.

13.5.2.3 Perturbed best response dynamics

For both technical convenience and realism, it is natural to consider variants of the best response dynamic that define smooth dynamical systems. The most important alternative, introduced by Fudenberg and Levine (1998) (see also Fudenberg and Kreps, 1993), supposes that agents optimize after their payoffs have been subject to perturbations. The perturbations ensure that agents' choice probabilities vary smoothly with payoffs, and so lead to differentiable dynamics. This allows us to avail ourselves of the full toolkit for analyzing smooth dynamics, including both linearization around rest points and the theory of cooperative differential equations (Section 13.8.2). The resulting perturbed best response dynamics have important connections with work in stochastic stability theory (Section 13.10.1) and models of heuristic learning in

[22] For a detailed analysis, see Gaunersdorfer and Hofbauer (1995).

games (Section 13.10.2); they also describe the dynamics of aggregate behavior in heterogeneous populations of exact optimizers (Ellison and Fudenberg, 2000; Ely and Sandholm, 2005).

Perturbed best response protocols are target protocols of the form

$$\sigma(\pi) = \tilde{M}(\pi), \tag{13.18}$$

where the *perturbed best response function* $\tilde{M}: \mathbb{R}^n \to \text{int}(X)$ is a smooth approximation of the maximizer correspondence M. This function is defined most conveniently in terms of a smooth *deterministic perturbation* v: $\text{int}(X) \to \mathbb{R}$ of the payoff to each *mixed strategy*:

$$\tilde{M}^v(\pi) = \underset{y \in \text{int}(X)}{\arg\max} \, (y'\pi - v(y)), \quad \text{where} \tag{13.19a}$$

$$z'\nabla^2 v(y) z > 0 \text{ for all } z \in TX \text{ and } y \in \text{int}(X), \text{ and} \tag{13.19b}$$

$$\lim_{k \to \infty} y_k \in \text{bd}(X) \Rightarrow \lim_{k \to \infty} \left| \nabla v(y_k) \right| = \infty. \tag{13.19c}$$

The convexity condition [13.19b] and the steepness condition [13.19c] ensure that the optimal solution to [13.19a] is unique and lies in the interior of X. Indeed, taking the first order condition for [13.19a] shows that

$$\tilde{M}^v(\pi) = (\nabla v)^{-1}(\Phi\pi). \tag{13.20}$$

From a game-theoretic point of view, it is more natural to define perturbed maximization using *stochastic perturbations* of the payoff of each *pure strategy*:

$$\tilde{M}_i^\varepsilon(\pi) = \mathbb{P}\left(i = \underset{j \in S}{\arg\max} \, \pi_j + \varepsilon_j\right), \tag{13.21}$$

where ε is a random vector that admits a density function that is positive throughout $\mathbb{R}^n$. While the explicit expression for $\tilde{M}^\varepsilon$ is quite cumbersome, one can avoid working with it directly: Hofbauer and Sandholm (2002) show that under appropriate smoothness conditions, any maximizer function of form [13.21] can be represented using a deterministic perturbation as in [13,19].

Example 13.12. (Logit choice) *The best known perturbed maximizer function is the* logit choice function *with* noise level $\eta > 0$:

$$\tilde{M}_i(\pi) = \frac{\exp(\eta^{-1}\pi_i)}{\sum_{j \in S} \exp(\eta^{-1}\pi_j)}. \tag{13.22}$$

When η is large, $\tilde{M}(\pi)$ is close to a uniform probability vector. When η is close to 0, $\tilde{M}(\pi)$ is a close approximation to the maximizer $M(\pi)$, but places positive probability on every strategy. The logit choice function can be expressed in form [13.21] using stochastic perturbations that are *i.i.d. with a double exponential distribution. It can also be expressed in form [13,19] by letting the deterministic perturbation v be the negated entropy function $v(y) = \eta \sum_{j \in S} y_j \log y_j$.*

If agents revise using a perturbed best response protocol [13,18], then aggregate behavior evolves according to the *perturbed best response dynamic*

$$\dot{x} = \tilde{M}^v(F(x)) - x. \qquad [13.23]$$

When the perturbations that generate $\tilde{M}^v$ are small in a suitable sense, the dynamic [13.23] is a smooth approximation of the best response dynamic [13.17].

Because its definition uses payoff perturbations, perturbed best response dynamics cannot satisfy positive correlation (PC) and Nash stationarity (NS) exactly. They do, however, satisfy perturbed versions of these conditions. Considering the latter first, observe that the rest points of [13.23] are the fixed points of the *perturbed best response function* $\tilde{B}^v = \tilde{M}^v \circ F$. These *perturbed equilibria* approximate Nash equilibria when the perturbations are small.[23]

To obtain a useful alternative characterization, define the *virtual payoffs* $\tilde{F}^v \colon X \to \mathbb{R}^n$ associated with the pair (F, v) by $\tilde{F}^v(x) = F(x) - \nabla v(x)$. By way of interpretation, note that the convexity and steepness of v ensure that strategies played by few agents have high virtual payoffs. Equation [13.20] implies that the perturbed equilibria for the pair (F, v) are precisely those states for which $\tilde{F}^v(x)$ is a constant vector.

Virtual payoffs are also used to define the appropriate analog of positive correlation. Hofbauer and Sandholm (2002, 2007) show that the perturbed best response dynamics [13.23] satisfy *virtual positive correlation*

$$V^{F,v}(x) \neq 0 \text{ implies that } V^{F,v}(x)'\tilde{F}^v(x) > 0. \qquad [13.24]$$

This condition is just what is needed to extend stability and convergence results for the best response dynamic to perturbed best response dynamics—see Sections 13.6.5 and 13.7.5.

13.5.3 Excess payoff and pairwise comparison dynamics

A basic question addressed by evolutionary game dynamics is whether Nash equilibrium can be interpreted as stationary behavior among agents who employ simple myopic rules. Indeed, a version of this interpretation was offered by Nash himself in his doctoral dissertation.[24] Imitative dynamics fail to satisfy Nash stationarity, and so do not provide an ideal basis for this interpretation. Best response dynamics are also not ideal for this purpose: they are based on discontinuous revision protocols, which require more precise information than simple agents should be expected to possess.

[23] In the experimental literature, perturbed equilibria are known as *quantal response equilibria*; see McKelvey and Palfrey (1995) and Goeree et al. (2008).

[24] See Nash (1950) and Weibull (1996).

In this section, we introduce two classes of continuous dynamics that satisfy Nash stationarity, and so provide the interpretation of Nash equilibrium we seek.

13.5.3.1 Excess payoff dynamics

The excess payoff function $\hat{F}\colon X \to \mathbb{R}^n$ for game F, defined by

$$\hat{F}_i(x) = F_i(x) - \bar{F}(x) = F_i(x) - \sum_{j\in S} x_j F_j(x)$$

describes the performance of each strategy relative to the population average. Clearly, the excess payoff vector $\hat{F}(x)$ cannot lie in the interior of the negative orthant $\mathbb{R}^n_-$, as this would mean that all strategies receive a worse-than-average payoff. In fact, it is not difficult to establish that in this context, the boundary of the negative orthant plays a special role:

Proposition 13.2. *$x \in NE(F)$ if and only if $\hat{F}(x) \in \mathrm{bd}(\mathbb{R}^n_-)$.*

To interpret this proposition, we let $\mathbb{R}^n_* = \mathbb{R}^n \setminus \mathrm{int}(\mathbb{R}^n_-)$ denote the set of vectors in $\mathbb{R}^n$ with at least one non-negative component. If the excess payoff vector lies on the boundary of this set, $\mathrm{bd}(\mathbb{R}^n_*) = \mathrm{bd}(\mathbb{R}^n_-)$, then the maximal component of the excess payoff vector is zero. Viewed in this light, Proposition 13.2 says that the Nash equilibria are just those states at which no strategy receives an above-average payoff.

Excess payoff protocols are target protocols (Section 13.5.2.1) under which conditional switch rates are expressed as functions of the excess payoff vector $\hat{F}(x)$.[25] Specifically, we call a target protocol τ an *excess payoff protocol* if it is Lipschitz continuous and satisfies

$$\tau_j(\pi, x) = \tau_j(\hat{\pi}), \text{ where } \hat{\pi}_i = \pi_i - x'\pi, \text{ and} \qquad [13.25a]$$

$$\hat{\pi} \in \mathrm{int}(\mathbb{R}^n_*) \Rightarrow \tau(\hat{\pi})'\hat{\pi} > 0. \qquad [13.25b]$$

Condition [13.25b], called *acuteness*, requires that away from Nash equilibrium, strategies with higher growth rates tend to be those with higher excess payoffs.

Substituting equation [13.25a] into the mean dynamic [13.13] yields the corresponding class of evolutionary dynamics, the *excess payoff dynamics*:

$$\dot{x}_i = \tau_i(\hat{F}(x)) - x_i \sum_{j\in S} \tau_j(\hat{F}(x)). \qquad [13.26]$$

Example 13.13. (The BNN dynamic) *If the protocol τ takes the semilinear form*

$$\tau_i(\hat{\pi}) = [\hat{\pi}_i]_+, \qquad [13.27]$$

[25] Best response protocols and perturbed best response protocols can also be expressed in this way—a point we return to in Section 13.7.5.

we obtain the Brown-von Neumann-Nash *(*BNN*)* dynamic*:*

$$\dot{x}_i = [\hat{F}_i(x)]_+ - x_i \sum_{j \in S} [\hat{F}_j(x)]_+. \tag{13.28}$$

This dynamic was introduced in the early days of game theory by Brown and von Neumann (1950) in the context of symmetric zero-sum games. Nash (1951) used a discrete-time analog of this dynamic as the basis for a simple proof of existence of equilibrium via Brouwer's theorem. The dynamic was then forgotten for 40 years before being reintroduced by Skyrms (1990), Swinkels (1993), Weibull (1996), and Hofbauer (2000).

It is not difficult to verify that the BNN dynamic satisfies both positive correlation (PC) and Nash stationarity (NS). Sandholm (2005a) shows that these properties are satisfied by all excess payoff dynamics.

13.5.3.2 Pairwise comparison dynamics

Excess payoffs dynamics are not completely satisfactory as a model of behavior in population games. To use an excess payoff protocol, an agent needs to know the vector of excess payoffs, and so, implicitly, the average payoff obtained in the population. Unless this information were provided by a planner, it is not information that agents could easily obtain.

As a more credible alternative, we consider dynamics based on pairwise comparisons. When an agent receives a revision opportunity, he selects an alternative strategy at random. He compares its payoff to that of his current strategy, and considers switching only if the former exceeds the latter.

Formally, *pairwise comparison protocols* are Lipschitz continuous protocols $\rho \colon \mathbb{R}^n \times X \to \mathbb{R}^{n \times n}_+$ that satisfy *sign preservation*:

$$\operatorname{sgn}(\rho_{ij}(\pi, x)) = \operatorname{sgn}([\pi_j - \pi_i]_+) \text{ for all } i, j \in S. \tag{13.29}$$

The resulting evolutionary dynamics, described by equation [13.3], are called *pairwise comparison dynamics*.

Example 13.14. (The Smith dynamic) *Suppose the revision protocol ρ takes the semilinear form*

$$\rho_{ij}(\pi, x) = [\pi_j - \pi_i]_+, \tag{13.30}$$

Inserting this formula into the mean dynamic [13.3], *we obtain the* Smith dynamic*:*

$$\dot{x}_i = \sum_{j \in S} x_j [F_i(x) - F_j(x)]_+ - x_i \sum_{j \in S} [F_j(x) - F_i(x)]_+. \tag{13.31}$$

Table 13.3 Families of revision protocols and evolutionary dynamics, and their properties.

Family	Example	Continuity	Data Req.	(PC)	(NS)
Imitation	Replicator	Yes	Weak	Yes	No
Optimization	Best response	No	Moderate	Yes	Yes
Perturbed optimization	Logit	Yes	Moderate	Approx.	Approx.
Excess payoff	BNN	Yes	Strong	Yes	Yes
Pairwise comparison	Smith	Yes	Weak	Yes	Yes

This dynamic was introduced by Smith (1984), who used it to model disequilibrium adjustment by drivers in highway networks.

Protocol [13.30] is closely related to the pairwise proportional imitation protocol from Example 13.6, $\rho_{ij}(\pi, x) = x_j[\pi_j - \pi_i]_+$. Indeed, the protocols only differ in how candidate strategies are chosen. Under the imitative protocol, a revising agent obtains a candidate strategy by observing the choice of an opponent. Under protocol [13.30], he obtains a candidate strategy by choosing uniformly from a list of all strategies.

This difference between imitative and pairwise comparison protocols has clear consequences for the dynamics of aggregate behavior. Under imitative dynamics, rare strategies are unlikely to be chosen, and unused strategies, even optimal ones, are never chosen. Pairwise comparison dynamics have neither of these properties, and as a consequence satisfy not only positive correlation (PC), but also Nash stationarity (NS): see Smith (1984) and Sandholm (2010b).

In fact, pairwise comparison protocols can be used in combination with imitative protocols to improve the performance of the latter. Sandholm (2005a, 2010b) considers *hybrid protocols* that combine imitation with consideration of unused strategies via protocols of form [13.25] or [13.29]. The resulting *hybrid dynamics* satisfy both positive correlation (PC) and Nash stationarity (NS). It is thus not imitation per se, but rather the exclusive use of imitation, that allows non-Nash rest points to exist.

Our main conclusions about families of revision protocols and evolutionary dynamics are summarized in Table 13.3. For each family, the table describes the continuity and data requirements of the protocols, and the incentive properties of the corresponding dynamics.

13.6. POTENTIAL GAMES

In this section and the following two, we consider classes of population games that have attractive theoretical properties and are useful in applications. Games in these classes admit simple characterizations of Nash equilibrium, and ensure global convergence under various evolutionary dynamics.

The most general convergence results are available for potential games, in which all information about incentives can be captured by a scalar-valued function defined on the set of population states. Dynamics satisfying positive correlation (PC) and Nash stationarity (NS) ascend this function and converge to Nash equilibrium.

The first appearance of potential functions in game theory is in the work of Beckmann et al. (1956), who use potential function arguments to analyze congestion games. Rosenthal (1973) introduced congestion games with finite numbers of players, motivating Monderer and Shapley's (1996) definition of finite-player potential games. Potential function arguments have long been used in models from population genetics that are equivalent to the replicator dynamic in normal form potential games; see Kimura (1958), Shahshahani (1979), Akin (1979), and Hofbauer and Sigmund (1988, 1998). Our presentation of potential games played by continuous populations of agents follows Sandholm (2001b, 2009b).

13.6.1 Population games and full population games

Some of the classes of games introduced in the coming sections can be characterized in terms of the externalities that players of different strategies impose on one another. In discussing these externalities, it is natural to consider the effect of adding new agents playing strategy j on the payoffs of agents currently choosing strategy i. In principle, this effect should be captured by the partial derivative $\frac{\partial F_i}{\partial x_j}$. However, since payoffs in population games are only defined on the simplex, this partial derivative does not exist.

This difficulty can circumvented by considering *full population games*, in which payoffs are defined on the positive orthant $\mathbb{R}^n_+$. We can interpret the extended payoff functions as describing the payoffs that would arise were the population size to change. Although there are some subtleties involved in the use of these extensions, in the end they are harmless, and we employ them here with little further comment.

13.6.2 Definition, characterization, and interpretation

Let $F : \mathbb{R}^n_+ \to \mathbb{R}^n$ be a (full) population game. We call F a *potential game* if there exists a continuously differentiable function $f: \mathbb{R}^n_+ \to \mathbb{R}$, called a *potential function*, satisfying

$$\nabla f(x) = F(x) \text{ for all } x \in \mathbb{R}^n_+, \text{ or equivalently}$$
$$\frac{\partial f}{\partial x_i}(x) = F_i(x) \text{ for all } i \in S \text{ and } x \in \mathbb{R}^n_+.$$

Thus the partial derivatives of the potential function are the payoff functions of the game.

If the payoff function F is continuously differentiable, then well-known results from calculus tell us that F is a potential game if and only if

$$DF(x) \text{ is symmetric for all } x \in \mathbb{R}^n_+, \text{ or equivalently}$$

$$\frac{\partial F_i}{\partial x_j}(x) = \frac{\partial F_j}{\partial x_i}(x) \text{ for all } i, j \in S \text{ and } x \in \mathbb{R}^n_+.$$

This condition, which we call *full externality symmetry*, has a simple game-theoretic interpretation: the effect on the payoffs to strategy i of introducing new agents choosing strategy j always equals the effect on the payoffs to strategy j of introducing new agents choosing strategy i.[26]

For intuition concerning the potential function itself, suppose that some members of the population switch from strategy i to strategy j, so that the population state moves in direction $z = e_j - e_i$. If these switches improve the payoffs of those who switch, then

$$\frac{\partial f}{\partial z}(x) = \nabla f(x)'z = F(x)'z = F_j(x) - F_i(x) > 0.$$

Thus profitable strategy revisions increase potential.

For more general sorts of adjustment, we have the following simple lemma:

Lemma 13.1. *Let F be a potential game with potential function f, and suppose the dynamic V^F satisfies positive correlation* (PC). *Then along any solution trajectory $\{x_t\}$, we have $\frac{\mathrm{d}}{\mathrm{d}t} f(x_t) > 0$ whenever $\dot{x}_t \neq \mathbf{0}$.*

The proof follows directly from the chain rule:

$$\frac{\mathrm{d}}{\mathrm{d}t} f(x_t) = \nabla f(x_t)'\dot{x}_t = F(x_t)' V^F(x_t) \geq 0,$$

where equality holds only if $V^F(x_t) = \mathbf{0}$.

13.6.3 Examples

Potential games admit a number of important applications, three of which are described next.

Example 13.15. (Matching in games with common interests) *Example 13.1 considered matching in the two-player symmetric normal form game $A \in \mathbb{R}^{n\times n}$, which generates the population game $F(x) = Ax$. The game A is said to exhibit* common interests *if A is a symmetric matrix ($A_{ij} = A_{ji}$), so that two matched players always receive the same payoffs. Matching in two-player games with common interests defines a fundamental model from population genetics; the common interest assumption reflects the shared fate of two genes that inhabit the same organism.*[27]

[26] If we only defined payoffs on the simplex, the corresponding condition would be *externality symmetry*, which requires that $DF(x)$ be symmetric with respect to $TX \times TX$ (in other words, that $z'DF(x)\hat{z} = \hat{z}'DF(x)z$ for all $z, \hat{z} \in TX$ and $x \in X$.)

[27] See Hofbauer and Sigmund (1988, 1998).

Since $DF(x) = A$, the population game derived from a game with common interests is a potential game. Its potential function, $f(x) = \frac{1}{2}x'Ax = \frac{1}{2}\bar{F}(x)$, is one-half of the average payoff function. Thus Lemma 13.1 implies that in common interest games, evolutionary dynamics satisfying positive correlation (PC) improve social outcomes.

Example 13.16. (Congestion games) *Example 13.2 introduced congestion games, whose payoff functions are of the form*

$$F_i(x) = -\sum_{\ell \in \mathscr{L}_i} c_\ell\left(u_\ell(x)\right).$$

In the context of highway congestion, each ℓ represents a link in the highway network, and $\mathscr{L}_i$ is the set of links that make up path i.

In a congestion game, the marginal effect of adding an agent to path i on the payoffs to drivers on path j is due to the marginal increases in congestion on the links the two paths have in common. The marginal effect of adding an agent to path j on the payoffs to drivers on path i is the same:

$$\frac{\partial F_i}{\partial x_j}(x) = -\sum_{\ell \in \mathscr{L}_i \cap \mathscr{L}_j} c'_\ell(u_\ell(x)) = \frac{\partial F_j}{\partial x_i}(x).$$

Thus congestion games satisfy externality symmetry, and so are potential games. Their potential functions are of the form

$$f(x) = -\sum_{\ell \in \mathscr{L}} \int_0^{u_\ell(x)} c_\ell(z)\, \mathrm{d}z. \qquad [13.32]$$

The potential function [13.32] of a congestion game is generally unrelated to its average payoff function,

$$\bar{F}(x) = -\sum_{\ell \in \mathscr{L}} u_\ell(x) c_\ell(u_\ell(x)).$$

Dafermos and Sparrow (1969) observed that if each cost function is a monomial of the same degree $\eta \geq 0$, so that $c_\ell(u) = a_\ell u^\eta$, then the potential function is proportional to average payoffs: $f(x) = \frac{1}{\eta+1}\bar{F}(x)$. For general full potential games, it follows from Euler's theorem that the potential function f is proportional to aggregate payoffs if and only if F is a homogeneous function of degree $k > -1$ (Hofbauer and Sigmund, 1988; Sandholm, 2001b).

Congestion games have received considerable attention in the computer science literature. Much of this work focuses on so-called "price of anarchy" results, which bound the ratio of the total delay in the network in Nash equilibrium to the minimal feasible delay. The most basic and best-known result in this literature, due to Roughgarden and

Tardos (2002, 2004) (see also Correa et al. 2004, 2008), shows that in congestion games with non-negative, increasing, affine cost functions, total delays in Nash equilibrium must be within a factor of $\frac{4}{3}$ of the minimal feasible delay. In fact, tight bounds can be obtained for quite general classes of cost functions.

Example 13.17. (Games generated by variable pricing schemes) *Population games can be viewed as models of externalities for environments with many agents. One way to force agents to internalize the externalities they impose upon others is to introduce pricing schemes. Given an arbitrary population game F with average payoff function $\bar{F}$, we define a new game $\tilde{F}$ by*

$$\tilde{F}_i(x) = F_i(x) + \sum_{j \in S} x_j \frac{\partial F_j}{\partial x_i}(x).$$

We interpret the second term as a price (either subsidy or tax) imposed by a planner. It represents the marginal effect that a strategy i player has on the payoffs of his opponents.

Observe that

$$\frac{\partial \bar{F}}{\partial x_i}(x) = \frac{\partial}{\partial x_i} \sum_{j \in S} x_j F_j(x) = F_i(x) + \sum_{j \in S} x_j \frac{\partial F_j}{\partial x_i}(x) = \tilde{F}_i(x).$$

In words, the augmented game $\tilde{F}$ is a full potential game, and its full potential function is the average payoff function of the original game F. Thus when individual agents switch strategies in response to the combination of original payoffs and prices, average payoffs in the population increase. Sandholm (2002, 2005b, 2007b) uses this construction as the basis for an evolutionary approach to implementation theory.

13.6.4 Characterization of equilibrium

The fact that evolutionary processes increase potential suggests a connection between local maximization of potential and Nash equilibrium. We therefore consider the problem of maximizing potential over the set of population states:

$$\max f(x) \quad \text{subject to} \sum_{j \in S} x_j = 1 \text{ and } x_i \geq 0 \text{ for all } i \in S.$$

The Lagrangian for this maximization problem is

$$L(x, \mu, \lambda) = f(x) + \mu \left(1 - \sum_{i \in S} x_i \right) + \sum_{i \in S} \lambda_i x_i \,,$$

so the Kuhn-Tucker first-order necessary conditions for maximization are

$$\frac{\partial f}{\partial x_i}(x) = \mu - \lambda_i \qquad \text{for all } i \in S, \qquad [13.33a]$$

$$\lambda_i x_i = 0, \qquad \text{for all } i \in S, \text{ and} \qquad [13.33b]$$

$$\lambda_i \geq 0 \qquad \text{for all } i \in S. \qquad [13.33c]$$

These conditions characterize the Nash equilibria of a potential game.

Theorem 13.5. *Let F be a potential game with potential function f. Then x is a Nash equilibrium of F if and only if (x, μ, λ) satisfies [13.33a]–[13.33c] for some $\lambda \in \mathbb{R}^n$ and $\mu \in \mathbb{R}$.*

The proof is simple: the multiplier μ is the maximal payoff, and the multiplier λ_i is the payoff deficit of strategy i relative to the optimal strategy.

Theorem 13.5 provides a simple proof of existence of equilibrium in potential games. Since f is continuous and X is compact, the former attains its maximum on the latter, and the theorem implies that any maximizer is a Nash equilibrium.

Example 13.18. (Nash equilibria in a potential game) *The coordination game F^{C3} from Example 13.5 is generated by a common interest game, and so is a potential game. Its potential function is the convex function $f^{C3}(x) = \frac{1}{2}((x_1)^2 + 2(x_2)^2 + 3(x_3)^2)$. Figure 13.7 presents a graph and a contour plot of this function. The three pure states locally maximize potential, and so are Nash equilibria. There are four additional states that do not maximize potential, but that satisfy the Kuhn-Tucker first-order conditions. Geometrically, this means that if we linearize the potential function at these states, the linearized functions admit no feasible direction of increase from these states. By Theorem 13.5, these points too are Nash equilibria, although Theorem 13.7 will show they are not locally stable under typical evolutionary dynamics.*

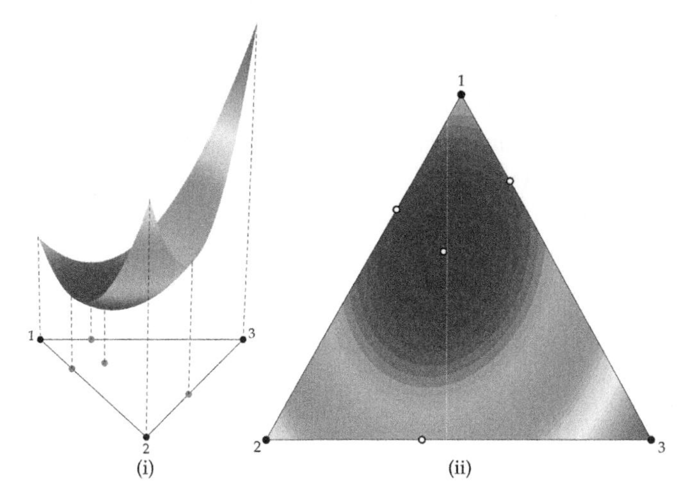

Figure 13.7 Graph and contour plot of the potential function $f^{C3}(x) = \frac{1}{2}((x_1)^2 + 2(x_2)^2 + 3(x_3)^2)$, with Nash equilibria of game F^{C3} marked.

The potential function in the previous example was convex. If instead a game's potential function is concave—as is true, for instance, in congestion games with increasing cost functions—then only the global maximizers of potential are Nash equilibria.

13.6.5 Global convergence and local stability

Lemma 13.1 tells us that evolutionary dynamics satisfying positive correlation (PC) will ascend the potential function whenever they are not at rest. In dynamical systems terminology, the potential function is a *strict (global) Lyapunov function* for all such dynamics.

Stating the consequences of this fact for the global behavior of the dynamics requires a formal definition. Let $\dot{x} = V(x)$ be a Lipschitz continuous differential equation that is forward invariant on X. For each initial condition $\xi \in X$, we define the *ω-limit set* $\omega(\xi)$ to be the set of all points that the solution trajectory $\{x_t\}_{t\geq 0}$ starting from $x_0 = \xi$ approaches arbitrarily closely infinitely often:

$$\omega(\xi) = \left\{ y \in X\colon \text{ there exists } \{t_k\}_{k=1}^{\infty} \text{ with } \lim_{k\to\infty} t_k = \infty \text{ such that } \lim_{k\to\infty} x_{t_k} = y \right\}.$$

In general, ω-limit points include rest points, periodic orbits, and chaotic attractors. However, standard results from dynamical systems theory show that the existence of a strict global Lyapunov function rules out the latter possibilities. We therefore have:

Theorem 13.6. *Let F be a potential game, and let V^F be an Lipschitz continuous evolutionary dynamic that satisfies positive correlation (PC). Then all ω-limit points of V^F are rest points. If in addition V^F satisfies Nash stationarity (NS), then all ω-limit points of V^F are Nash equilibria.*

Lemma 13.1 also suggests that only local maximizers of f should be locally stable. To present such a result, we introduce some formal definitions of local stability for the differential equation $\dot{x} = V(x)$ above. These definitions are provided for a single state $y \in X$; replacing y with a closed set $Y \subset X$ yields the definitions for closed sets of states.

State y is *Lyapunov stable* if for every neighborhood O of y, there exists a neighborhood O' of y such that every solution $\{x_t\}_{t\geq 0}$ that starts in O' is contained in O: that is, $x_0 \in O'$ implies that $x_t \in O$ for all $t \geq 0$. State y is *attracting* if there is a neighborhood O of y such that every solution that starts in O converges to y; if we can choose $O = X$, y is *globally attracting*. Finally, state y is *asymptotically stable* if it is Lyapunov stable and attracting, and it is *globally asymptotically stable* if it is Lyapunov stable and globally attracting.

With these definitions, we have:

Theorem 13.7. *Let F be a potential game with potential function f, and let V^F be an evolutionary dynamic that satisfies positive correlation (PC) and Nash stationarity (NS). Then* *state $x \in X$ is asymptotically stable under V^F if and only if it is a local maximizer of f and is isolated in $NE(F)$.*

Versions of Theorem 13.7 remain true if Nash stationarity is not assumed, or if we consider local maximizer sets instead of points.[28]

The results above are stated for Lipschitz continuous dynamics that satisfy positive correlation (PC). Analogous results hold for the best response dynamic, and also for perturbed best response dynamics. For the analysis of the latter case, one introduces the perturbed potential function $\tilde{f}(x) = f(x) - v(x)$. Since the gradient of this function is the virtual payoff function $\tilde{F}(x) = F(x) - \nabla v(x)$, condition [13.24] implies that perturbed best response dynamics ascend this function and converge to perturbed equilibria.

13.6.6 Local stability of strict equilibria

The existence of a potential function allows one to establish convergence and stability results for all dynamics satisfying positive correlation (PC). Similar reasoning can be used to establish local stability of any *strict equilibrium*: that is, any pure state e_k such that $F_k(e_k) > F_j(e_k)$ for all $j \neq k$.

This result requires a slightly stronger restriction on the dynamics. We say that a dynamic V^F for game F satisfies *strong positive correlation* in $Y \subseteq X$ if

$$\text{There exists a } c > 0 \text{ such that for all } x \in Y,$$
$$V^F(x) \neq 0 \text{ implies that } \operatorname{Corr}(V^F(x), F(x)) = \frac{V^F(x)'\Phi F(x)}{\left|V^F(x)\right| \left|\Phi F(x)\right|} \geq c.$$

That is, the correlation between strategies' growth rates and payoffs, or equivalently, the cosine of the angle between the growth rate and projected payoff vectors, must be bounded away from zero on Y. This condition is satisfied by imitative dynamics, excess payoff dynamics, and pairwise comparison dynamics in a neighborhood of any strict equilibrium.

Sandholm (2014) shows that the function

$$L(x) = (e_k - x)'F(e_k),$$

representing the payoff deficit of mixed strategy x at the strict equilibrium e_k, is a *strict local Lyapunov function* for state e_k: its value decreases along solutions of V^F in a neighborhood of e_k, and it is non-negative in this neighborhood, equaling zero only when $x = e_k$. It thus follows from standard results that if e_k is an isolated rest point of V^F, then it is asymptotically stable under V^F.

[28] See Sandholm (2001b).

The next section introduces a more general criterion for local stability. This criterion is defined for states throughout the simplex, but it requires more structure on the dynamics for stability to be assured.

13.7. ESS AND CONTRACTIVE GAMES

13.7.1 Evolutionarily stable states

The birth of evolutionary game theory can be dated to the definition of an *evolutionarily stable strategy (ESS)* by Maynard Smith and Price (1973). Their model is one of monomorphic populations, whose members all choose the same mixed strategy in a symmetric normal form game. The notion of an evolutionarily stable strategy is meant to capture the capacity of a monomorphic population to resist invasion by a monomorphic mutant group whose members play some alternative mixed strategy.

This framework is quite different from the polymorphic, pure-strategist model we consider here. Nevertheless, if we reinterpret Maynard Smith and Price's (1973) conditions as constraints on population states—and call a point that satisfies these conditions an *evolutionarily stable state*—we obtain a sufficient condition for local stability under a variety of evolutionary dynamics.[29,30]

There are three equivalent ways of defining ESS. The simplest one, and the most useful for our purposes, was introduced by Hofbauer et al. (1979).[31] It defines an ESS to be an *(infinitesimal) local invader*:

$$\text{There is a neighborhood } O \text{ of } x \text{ such that } (y-x)'F(y) < 0 \text{ for all } y \in O \setminus \{x\}. \qquad [13.34]$$

To interpret this condition, fix the candidate state x, and consider some nearby state y. Condition [13.34] says that if the current population state is y, and an infinitesimal group of agents with strategy distribution x joins the population, then the average payoff of the agents in this group, $x'F(y) = \sum_{j\in S} x_j F_j(y)$, exceeds the average payoff in the population as a whole, $y'F(y)$. Of course, forming predictions based on group average payoffs runs counter to the individualistic approach that defines noncooperative game theory. Thus in the present context, the definition of ESS is not of interest directly, but only instrumentally, as a sufficient condition for stability under evolutionary dynamics.

[29] The distinction between evolutionarily stable strategies and ESS is emphasized by Thomas (1984). General references on ESS theory include the survey of Hines (1987) and the monographs of Bomze and Pötscher (1989) and Cressman (1992).

[30] How one ought to extend the ESS definition to multipopulation settings depends on which of the interpretations above one has in mind. For the mixed-strategist environment, the appropriate extension is *Cressman ESS* (Cressman, 1992, 2006; Cressman et al., 2001), while for population games, it is *Taylor ESS* (Taylor, 1979). See Sandholm (2010c, Chapter 8) for further discussion.

[31] See also Pohley and Thomas (1983) and Thomas (1985).

The second definition, introduced by Taylor and Jonker (1978) and Bomze (1991), defines an ESS as a state that possesses *uniform invasion barrier*:

$$\begin{aligned}&\text{There is an } \bar{\varepsilon} > 0 \text{ such that } (y - x)'F(\varepsilon y + (1 - \varepsilon)x) < 0\\ &\text{for all } y \in X \setminus \{x\} \text{ and } \varepsilon \in (0, \bar{\varepsilon}).\end{aligned} \quad [13.35]$$

In contrast to condition [13.34], definition [13.35] looks at invasions by groups with positive mass, and compares the average payoffs of the incumbent and invading groups in the postentry population.

The third definition, the original one of Maynard Smith and Price (1973), shows what restrictions ESS adds to Nash equilibrium.

$$x \text{ is a Nash equilibrium: } (y - x)'F(x) \leq 0 \text{ for all } y \in X. \quad [13.36a]$$

$$\begin{aligned}&\text{There is a neighborhood } O \text{ of } x \text{ such that for all } y \in O \setminus \{x\},\\ &(y - x)'F(x) = 0 \text{ implies that } (y - x)'F(y) < 0.\end{aligned} \quad [13.36b]$$

The stability condition [13.36b] says that if a state y near x is an alternative best response to x, then an incumbent population with strategy distribution y obtains a lower average payoff against itself than an infinitesimal group of invaders with strategy distribution x obtains against the incumbents.

The following theorem confirms the equivalence of these definitions.

Theorem 13.8. *The following are equivalent:*
(i) *x satisfies condition [13.34].*
(ii) *x satisfies condition [13.35].*
(iii) *x satisfies conditions [13.36a] and [13.36b].*
A state that satisfies these conditions is called an evolutionary stable state (ESS).

For certain local stability results, we need a slightly stronger condition than ESS. We call state x a *regular ESS* (Taylor and Jonker, 1978) if

$$x \text{ is a \textit{quasistrict equilibrium}: } F_i(x) = \bar{F}(x) > F_j(x) \text{ whenever } x_i > 0 \text{ and } x_j = 0. \quad [13.37a]$$

$$z'DF(x)z < 0 \text{ for all } z \in TX \setminus \{\mathbf{0}\} \text{ such that } z_i = 0 \text{ whenever } x_i = 0. \quad [13.37b]$$

The quasistrictness condition [13.37a] strengthens the Nash equilibrium condition [13.36a] by requiring unused strategies to be suboptimal. The first-order stability condition [13.37b] strengthens the stability condition [13.36b] by requiring the strict inequality to hold even at the level of a linear approximation. If $x = e_k$ is a pure state, then condition [13.37a] requires x to be a strict equilibrium, and condition [13.37b] is vacuous. At the other extreme, if x is in the interior of the simplex, then condition [13.37a] is equivalent to Nash equilibrium [13.36a], and condition [13.37b] requires that the derivative matrix $DF(x)$ be negative definite with respect to the tangent space TX.

This last requirement is the motivation for our next class of games.

13.7.2 Contractive games

The population game $F\colon X \to \mathbb{R}^n$ is a *(weakly) contractive game*[32] if

$$(y - x)'(F(y) - F(x)) \leq 0 \text{ for all } x, y \in X. \qquad [13.38]$$

If the inequality in condition [13.38] holds strictly whenever $x \neq y$, F is *strictly contractive*, while if this inequality always binds, F is *conservative* (or *null contractive*).

For a first intuition, notice that if F is a potential game, so that $F \equiv \nabla f$, then [13.38] says that the potential function f is concave, ensuring the convexity and global stability of the set of Nash equilibria.

For a more general intuition, consider again the projection dynamic [13.12], defined on int(X) by $\dot{x} = \Phi F(x)$. Solutions to this dynamic "follow the payoff vectors" to the greatest extent possible. If we run this dynamic from two initial states x_0 and y_0, the squared distance between the states changes according to

$$\frac{\mathrm{d}}{\mathrm{d}t}\left|y_t - x_t\right|^2 = 2(y_t - x_t)'(\dot{y}_t - \dot{x}_t) = 2(y_t - x_t)'(F(y_t) - F(x_t)). \qquad [13.39]$$

Thus, in a contractive game, following the payoff vectors brings states (weakly) closer together.[33]

The connection between these games and the notion of ESS follows from the characterization for the differentiable case. By the fundamental theorem of calculus, if F is continuously differentiable, it is contractive if and only if it satisfies *self-defeating externalities:*

$$DF(x) \text{ is negative semidefinite with respect to } TX \text{ for all } x \in X. \qquad [13.40]$$

This condition, which can be rewritten as

$$\sum_{i \in S} z_i \frac{\partial F_i}{\partial z}(x) \leq 0 \;\text{ for all } z \in TX \text{ and } x \in X,$$

provides an economic interpretation of contractive games. The vector $z \in TX$ represents the aggregate effect of revisions by a group of agents on the population state. Condition [13.40] requires that improvements in the payoffs of strategies to which revising agents are switching are always exceeded by the improvements in the payoffs of strategies which revising agents are abandoning. For instance, when $z = e_j - e_i$, this condition becomes

[32] Also known as *stable games* (Hofbauer and Sandholm, 2009) or *negative semidefinite games* (Hopkins, 1999b). In the convex analysis literature, an F satisfying condition [13.38] (sometimes with the inequality reversed) is called a *monotone operator*—see Hiriart-Urruty and Lemaréchal (2001).

[33] This remains true on the boundary of the simplex if the dynamic is defined there via closest-point projection—see Lahkar and Sandholm (2008).

$$\frac{\partial F_j}{\partial(e_j - e_i)}(x) \leq \frac{\partial F_i}{\partial(e_j - e_i)}(x).$$

According to this inequality, any gains that the switches create for the newly chosen strategy j are dominated by gains for the abandoned strategy i.

13.7.3 Examples

Example 13.19. (Matching in symmetric zero-sum games) *A symmetric two-player normal form game A is* symmetric zero-sum *if A is skew-symmetric: $A_{ji} = -A_{ij}$ for all $i, j \in S$. If agents are matched to play this game, then the resulting population game $F(x) = Ax$ satisfies $z'DF(x)z = z'Az = 0$ for all vectors $z \in \mathbb{R}^n$, and so is conservative.*

Example 13.20. (Matching in Rock-Paper-Scissors) *Let $F(x) = Ax$ with*

$$A = \begin{pmatrix} 0 & -l & w \\ w & 0 & -l \\ -l & w & 0 \end{pmatrix}.$$

*Here, $w > 0$ and $l > 0$ represent the benefit from a win and the cost of a loss, respectively. When $w = l$, we refer to A as (*standard*)* RPS*; when $w > l$, we refer to A as* good RPS, *and when $w < l$, we refer to A as* bad RPS. *In all cases, the unique Nash equilibrium of $F(x) = Ax$ is $x^* = (\frac{1}{3}, \frac{1}{3}, \frac{1}{3})$.*

To evaluate condition [13.40] on the derivative matrix $DF(x) = A$, notice that the symmetric matrix $A + A'$ has eigenvalue $l - w$ with geometric multiplicity 2, corresponding to the eigenspace TX. It follows that F is conservative in standard RPS, and strictly contractive in good RPS. In the latter case, Nash equilibrium x^* is a regular ESS.

Example 13.21. ((Perturbed) concave potential games) *A potential game F with a strictly concave potential function satisfies condition [13.38] with a strict inequality, and so is a strictly contractive game. If we slightly perturb F, then the new game is quite unlikely to be a potential game. But [13.38] will continue to hold, so the new game is still a strictly contractive game.*

13.7.4 Equilibrium in contractive games

The contraction condition [13.39] suggests that obedience of incentives will push the population toward some "central" equilibrium state. To work toward a confirmation of this idea, we say that x is a *globally neutrally stable state* of F, denoted $x \in GNSS(F)$, if

$$(y - x)'F(y) \leq 0 \quad \text{for all } y \in X. \tag{13.41}$$

GNSS is the global analog of Maynard Smith's (1982) notion of a *neutrally stable strategy*, which is obtained by replacing the strict inequality in definition [13.34] of ESS with a weak one. Rewriting inequality [13.41] as $(x-y)'F(y) \geq 0$, we obtain a simple geometric interpretation: proceeding from any state y an infinitesimal distance in the direction specified by payoff vector $F(y)$ moves the state (weakly) closer to the GNSS x.

Example 13.22. (GNSS in standard RPS) *Starting from a selection of states $y \in X$, Figure 13.8 presents projected payoff vectors $\Phi F(y)$ from standard RPS, along with vectors $x^* - y$ leading to the Nash equilibrium $x^* = (\frac{1}{3}, \frac{1}{3}, \frac{1}{3})$. All such pairs of vectors are orthogonal, so x^* is a GNSS.*

The set *GNSS*(F) is an intersection of half spaces, and so is convex. A simple geometric argument shows that it only contains Nash equilibria.

Proposition 13.3. *GNSS*(F) $\subseteq$ *NE*(F).

To prove this inclusion, let $x \in GNSS(F)$ and let $y \neq x$. Define $x_\varepsilon = \varepsilon y + (1-\varepsilon)x$. Since x is a GNSS, $(x - x_\varepsilon)'F(x_\varepsilon) \geq 0$ for all $\varepsilon \in (0, 1]$: that is, motion from x_ε in direction $F(x_\varepsilon)$ is weakly toward x (see Figure 13.9). Taking ε to zero yields $(y-x)'F(x) \leq 0$ for all $y \in X$, which is definition [13.2] of Nash equilibrium.

While every GNSS is a Nash equilibrium, not every game has a GNSS. But the next result shows that in contractive games, a GNSS must exist, because every Nash equilibrium is a GNSS.

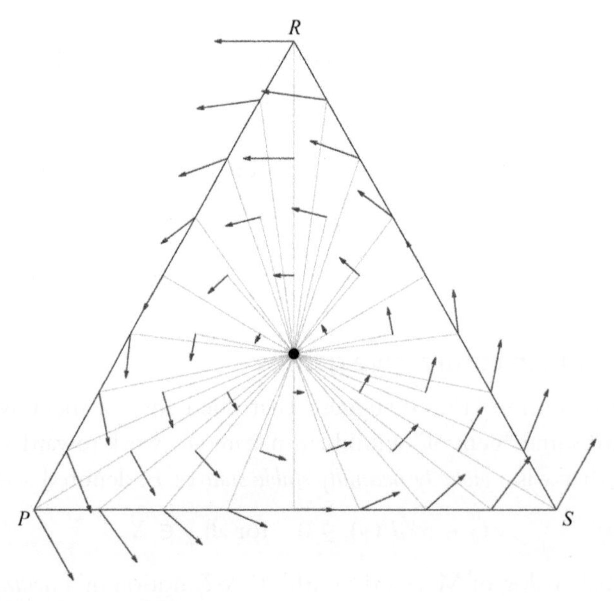

Figure 13.8 The GNSS of standard RPS.

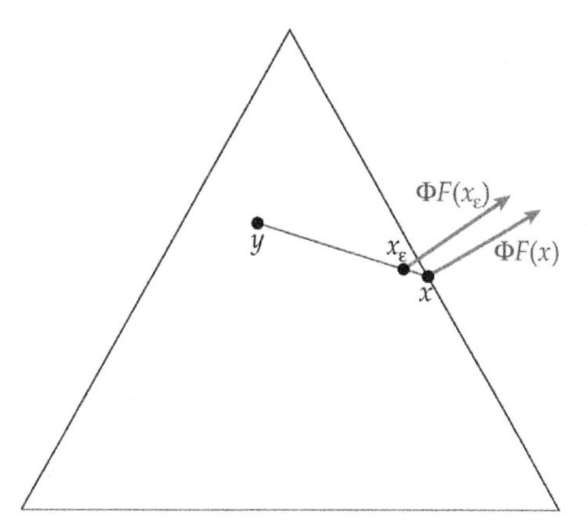

Figure 13.9 Why every GNSS is a Nash equilibrium.

Theorem 13.9. *If F is a contractive game, then $NE(F) \subseteq GNSS(F)$. Thus $NE(F) = GNSS(F)$, and so is a convex set.*

To prove the theorem, add the definition [13.38] of contractive games to definition [13.2] of Nash equilibrium; the sum is the definition [13.41] of GNSS.

If inequality [13.41] is strict whenever $y \neq x$, we call x a *globally evolutionarily stable state*. This condition is the global analog of ESS condition [13.34]. A variation on the arguments above shows that in strictly contractive games, a Nash equilibrium is a GESS, and that the Nash equilibrium is unique.

In light of Theorem 13.9 and the foregoing analysis, it is natural to ask whether one can prove existence of Nash equilibrium in contractive games using tools from convex analysis, avoiding an appeal to fixed point theory. Indeed, adapting an argument of Minty (1967), one can provide a direct proof of existence of equilibrium using the minmax theorem and a compactness argument.[34]

13.7.5 Global convergence and local stability

The discussions so far suggest that contractive games should admit global convergence results under evolutionary dynamics, and that ESS should serve as a sufficient condition for local stability. This section verifies these intuitions.

13.7.5.1 Imitative dynamics

The original global convergence result for strictly contractive games, due to Hofbauer et al. (1979) and Zeeman (1980), is for the replicator dynamic [13.8]. The analysis proceeds as follows: If F is a strictly contractive game, it admits a GESS x^*. We can therefore consider the function

[34] See Sandholm (2010c, Sec. 3.3.5).

$$H_{x^*}(x) = \sum_{i \in \text{supp}(x^*)} x_i^* \log \frac{x_i^*}{x_i},$$

which is defined on the set of states $X_{x^*} = \{x \in X\colon \text{supp}(x^*) \subseteq \text{supp}(x)\}$ whose supports contain the support of x^*. Jensen's inequality implies that H_{x^*} is non-negative, and equals 0 precisely when $x = x^*$. Moreover, the time derivative of this function under the replicator dynamic [13.8] is

$$\dot{H}_{x^*}(x) = \nabla H_{x^*}(x)'\dot{x} = -\sum_{i \in \text{supp}(x^*)} \frac{x_i^*}{x_i} \cdot x_i \hat{F}_i(x) = -\sum_{i \in S} x_i^* \hat{F}_i(x) = -(x^* - x)'F(x).$$

Thus, since x^* is a GESS, the value of H_{x^*} falls over time. Standard results on Lyapunov functions then imply that x^* is asymptotically stable, attracting solutions from every initial condition in X_{x^*}. Whether it is possible to prove global convergence results for contractive games under other imitative dynamics is an open question.

The only property of the game F used in this calculation was the fact that x^* is a GESS. Therefore, virtually the same argument implies that any ESS is locally asymptotically stable under the replicator dynamic.

One can obtain more general conclusions about local stability using linearization. Taylor and Jonker (1978) show that when the derivative matrix $DV(x^*)$ of the replicator dynamic is evaluated at a regular ESS x^*, its eigenvalues corresponding to directions in TX have negative real part, implying that solutions starting near the ESS converge to it at an exponential rate. Remarkably, Cressman (1997b) shows that in hyperbolic cases,[35] the linearizations of all imitative dynamics at a given restricted equilibrium are positive multiples of one another, and consequently have the same stability properties. Among other things, this implies the local stability of any regular ESS under typical imitative dynamics.

13.7.5.2 Target and pairwise comparison dynamics: global convergence in contractive games

We saw earlier that in potential games, the potential function serves as a Lyapunov function for any evolutionary dynamic satisfying positive correlation (PC). Contractive games, by contrast, do not come equipped with candidate Lyapunov functions. In order to construct Lyapunov functions for dynamics in these games, we must look to the dynamics themselves to obtain the necessary structure, following Smith (1984), Hofbauer (1995b, 2000), and Hofbauer and Sandholm (2007, 2009).

[35] A rest point x^* is *hyperbolic* if the eigenvalues of the derivative matrix $DV(x^*)$ have nonzero real part. The Hartman-Grobman theorem (see Robinson, 1995) shows that near a hyperbolic rest point, the flow of a nonlinear dynamic is topologically conjugate to the flow of its linearization.

The best response dynamic, perturbed best response dynamics, and excess payoff dynamics all can be derived from target protocols $\tau : \mathbb{R}^n \to \mathbb{R}^n_+$ that condition on the excess payoff vector $\hat{F}(x)$. Since target protocols are maps from $\mathbb{R}^n$ into itself, it makes sense to ask whether a target protocol is integrable: that is, whether there exists a *revision potential* $\gamma : \mathbb{R}^n \to \mathbb{R}$ satisfying $\tau(\hat{\pi}) = \nabla\gamma(\hat{\pi})$ for all $\hat{\pi} \in \mathbb{R}^n$. In this case, the revision potential provides a building block for constructing a Lyapunov function to analyze the dynamic at hand.[36]

For example, it is easy to verify that the revision protocol [13.27] for the BNN dynamic is integrable:

$$\tau_i(\hat{\pi}) = [\hat{\pi}_i]_+ \implies \gamma(\hat{\pi}) = \frac{1}{2}\sum_{i\in S}[\hat{\pi}_i]_+^2.$$

To make use of the revision potential γ, we introduce the Lyapunov function $\Gamma : X \to \mathbb{R}$, defined by

$$\Gamma(x) = \gamma(\hat{F}(x)). \tag{13.42}$$

This function is non-negative, and by Proposition 13.2, it equals zero precisely at the Nash equilibria of F. A computation shows that along solutions to the BNN dynamic, the value of Γ obeys

$$\dot{\Gamma}(x) = \dot{x}' DF(x)\dot{x} - (\tau(\hat{F}(x))'\mathbf{1})(F(x)'\dot{x}).$$

If x is not a rest point, then the contribution of the second term here is negative by positive correlation (PC), and if F is contractive, the contribution of the first term is nonpositive by condition [13.40]. Thus Γ is a (decreasing) strict global Lyapunov function, and so standard results imply that the set $NE(F)$ is globally asymptotically stable under the BNN dynamic. The specific functional form of τ is not too important here: as long as the protocol is acute [13.25b] and integrable, all solutions to the corresponding excess payoff dynamic converge to the set of Nash equilibria.

Similar results hold for the best response dynamic and all perturbed best response dynamics, as they too can be derived from integrable target protocols:[37]

$$M(\hat{\pi}) = \operatorname*{arg\,max}_{y\in X} y'\hat{\pi} \implies \mu(\hat{\pi}) = \max_{y\in X} y'\hat{\pi},$$

$$\tilde{M}(\hat{\pi}) = \operatorname*{arg\,max}_{y\in \text{int}(X)} y'\hat{\pi} - v(y) \implies \tilde{\mu}(\hat{\pi}) = \max_{y\in \text{int}(X)} y'\hat{\pi} - v(y).$$

[36] For a game-theoretic interpretation of this integrability condition, see Sandholm (2014b).

[37] We can replace actual payoffs with excess payoffs as the arguments of M and $\tilde{M}$, since maximizers of actual payoffs are also maximizers of excess payoffs. Also, since M is a correspondence, the function μ is not a potential function in the usual calculus sense, but in the sense of convex analysis: we have $\partial\mu \equiv M$, where the correspondence $\partial\mu : \mathbb{R}^n \rightrightarrows \mathbb{R}^n$ is the subdifferential of μ.

Table 13.4 Lyapunov functions for five basic deterministic dynamics in contractive games.

Dynamic	Lyapunov function for contractive games
Replicator	$H_{x^*}(x) = \sum_{i \in \mathrm{supp}(x^*)} x_i^* \log \frac{x_i^*}{x_i}$
Best response	$G(x) = \max_{i \in S} \hat{F}_i(x)$
Logit	$\tilde{G}(x) = \max_{y \in \mathrm{int}(X)} \left(y' \hat{F}(x) - \eta \sum_{i \in S} y_i \log y_i \right) + \eta \sum_{i \in S} x_i \log x_i$
BNN	$\Gamma(x) = \frac{1}{2} \sum_{i \in S} [\hat{F}_i(x)]_+^2$
Smith	$\Psi(x) = \frac{1}{2} \sum_{i \in S} \sum_{j \in S} x_i [F_j(x) - F_i(x)]_+^2$

Therefore, Lyapunov functions analogous to [13.42] can be constructed in these cases as well.

Pairwise comparison dynamics cannot be derived from target protocols. Even so, Smith (1984) constructs a more complicated Lyapunov function for his dynamics for contractive games, and Hofbauer and Sandholm (2009) show that a similar construction can be used for a broader class of pairwise comparison dynamics.[38]

The Lyapunov functions for the five basic dynamics in contractive games are summarized in Table 13.4.

13.7.5.3 Target and pairwise comparison dynamics: local stability of regular ESS

These Lyapunov functions for target and pairwise comparison dynamics can also be used to establish local stability of regular ESS. In the case of an interior ESS, the contractive game condition [13.40] holds in a neighborhood of the ESS, so local stability follows immediately from the previous analysis.

The analysis is trickier for a regular ESS on the boundary of the simplex. The relevant contraction condition, [13.37b], only applies in directions tangent to the face on which the ESS lies, reflecting switches among strategies currently in use. For its part, quasistrictness [13.37a] requires that the strategies in use are strictly better than those that are not. Intuitively, one expects that quasistrictness should drive the state toward the face of the simplex, and that once the state is close enough to the face, the contraction condition should be enough to ensure stability. To formalize this argument, Sandholm (2010a) augments the Lyapunov functions introduced above by a term that penalizes mass on strategies outside the support of the ESS. The values of these augmented functions decrease over time, establishing local stability of regular ESS in each case.

[38] The condition that is needed, *impartiality*, requires that the function of the payoff difference $\pi_j - \pi_i$ that describes the conditional switch rate from i to j does not depend on an agent's current strategy i.

13.8. ITERATIVE SOLUTION CONCEPTS, SUPERMODULAR GAMES, AND EQUILIBRIUM SELECTION

This section considers connections between evolutionary game dynamics and traditional iterative solution concepts. In some cases iteration is used directly to study the dynamics, but in the case of supermodular games a rather different tool, the theory of cooperative differential equations, is employed. This section also introduces a new class of dynamics, sampling best response dynamics, which select unique equilibria in certain games with multiple strict equilibria.

13.8.1 Iterated strict dominance and never-a-best-response

To begin, we show that traditional iterative solution concepts are respected by imitative dynamics and the best response dynamic. Although these dynamics are the most commonly studied, the properties described here do not extend to other dynamics, as Section 13.9.2 will show.

For imitative dynamics, we consider strict dominance, following Akin (1980) and Nachbar (1990). Strategy i is *strictly dominated* by strategy j if $F_j(x) > F_i(x)$ for all $x \in X$.[39]

Theorem 13.10. *Let $\{x_t\}$ be an interior solution trajectory of an imitative dynamic in game F. If strategy $i \in S$ is strictly dominated in F, then $\lim_{t\to\infty} (x_t)_i = 0$.*

To prove this result, suppose that strategy i is strictly dominated by strategy j, and write $r = x_i/x_j$. If we express the imitative dynamic in the percentage growth rate form $\dot{x}_i = x_i G_i(x)$ as in [13.10], then monotonicity condition [13.11] and a compactness argument imply that $G_j(x) - G_i(x) \geq c > 0$ for all $x \in X$. Applying the quotient rule yields

$$\frac{\mathrm{d}}{\mathrm{d}t}r = \frac{\mathrm{d}}{\mathrm{d}t}\frac{x_i}{x_j} = \frac{\dot{x}_i x_j - \dot{x}_j x_i}{(x_j)^2} = \frac{x_i G_i(x) x_j - x_j G_j(x) x_i}{(x_j)^2} = r(G_i(x) - G_j(x)) \leq -cr,$$

and so x_i vanishes.

Of course, once we remove a dominated strategy from a game, other strategies may become dominated in turn. A simple continuity argument shows that the conclusion of Theorem 13.10 extends to strategies eliminated via iterated dominance.

It is easy to see that under the best response dynamic, not only does any strictly dominated strategy vanish; so too does any strategy i that is never a best response, in the sense that for every $x \in X$ there is a $j \in S$ such that $F_j(x) > F_i(x)$.[40] In fact, such

[39] For results on strict dominance by mixed strategies, see Samuelson and Zhang (1992), Hofbauer and Weibull (1996), and Viossat (2011).

[40] Any strategy that is strictly dominated by a pure strategy is never a best response, but because we only consider pure strategies here, the converse statement is false, even when $F(x) = Ax$ is linear.

a strategy must vanish at an exponential rate: $(x_t)_i = (x_0)_i e^{-t}$ for any $x_0 \in X$, so that elimination occurs not just from interior initial conditions, but from all initial conditions. And by an argument based on continuity of payoffs, any strategy eliminated by iterative removal of never-a-best-response strategies eventually vanishes at an exponential rate.

It follows that at each stage of this elimination, the face of the simplex corresponding to the surviving strategies is asymptotically stable under the best response dynamic. The next example shows that the corresponding claim for imitative dynamics is false.

Example 13.23. (Dominated strategies and local instability under imitative dynamics) *Consider the game $F(x) = Ax$ with*

$$A = \begin{pmatrix} 1 & 1 & 1 \\ 2 & 2 & 2 \\ 0 & 4 & 0 \end{pmatrix}.$$

Since strategy 1 is strictly dominated in this game, Theorem 13.10 implies that under any imitative dynamic, x_1 vanishes along any interior solution trajectory. However, if strategy 2 is omitted, strategy 1 strictly dominates strategy 3, so on the interior of the e_1e_3 face, solutions of imitative dynamics converge to e_1. We therefore conclude that the e_2e_3 face is neither Lyapunov stable nor attracting.

However, if a game is dominance solvable, with only strategy k surviving, then state e_k is a strict equilibrium, and hence asymptotically stable—see Section 13.6.6.

13.8.2 Supermodular games and perturbed best response dynamics

Supermodular games, introduced in finite player contexts by Topkis (1979), Vives (1990), and Milgrom and Roberts (1990), are defined by the property that higher choices by one's opponents make one's own higher strategies look relatively more desirable. This implies the monotonicity of the best response correspondence, which in turn implies the existence of minimal and maximal Nash equilibria. We consider the large population version of these games. Following Hofbauer and Sandholm (2002, 2007), we use techniques from the theory of cooperative dynamical systems to establish almost global convergence results for perturbed best response dynamics.

To define supermodular games, we introduce the stochastic dominance matrix $\Sigma \in \mathbb{R}^{(n-1)\times n}$, defined by $\Sigma_{ij} = 1_{j>i}$. Then $(\Sigma x)_i = \sum_{j=i+1}^{n} x_j$ equals the total mass on actions greater than i at population state x, and $\Sigma y \geq \Sigma x$ if and only if y stochastically dominates x.

We call F a *supermodular game* if

$$\Sigma y \geq \Sigma x \text{ implies that } F_{i+1}(y) - F_i(y) \geq F_{i+1}(x) - F_i(x) \text{ for all } i < n. \qquad [13.43]$$

That is, if y stochastically dominates x, then for any action $i < n$, the payoff advantage of $i+1$ over i is greater at y than at x. If F is continuously differentiable, then it is supermodular if and only if

$$\frac{\partial(F_{i+1} - F_i)}{\partial(e_{j+1} - e_j)}(x) \geq 0 \text{ for all } i < n, j < n, \text{ and } x \in X. \qquad [13.44]$$

In words, if some agents switch from strategy j to strategy $j+1$, the performance of strategy $i+1$ improves relative to that of strategy i. Conditions [13.43] and [13.44] are both called *strategic complementarity*.

Example 13.24. (Search with positive externalities) *Example 13.3 introduced a population game model of search: $F_i(x) = m_i\, b(a(x)) - c_i$, where $a(x) = \sum_{k=1}^{n} kx_k$ represents aggregate search effort, the increasing function b represents the benefits of search as a function of aggregate effort, the increasing function m is the benefit multiplier, and the arbitrary function c captures search costs. Since*

$$\frac{\partial(F_{i+1} - F_i)}{\partial(e_{j+1} - e_j)}(x) = (m_{i+1} - m_i)\, b'(a(x)) \geq 0,$$

F is a supermodular game.

It is intuitively clear that supermodular games must have increasing best response correspondences: when opponents choose higher strategies, an agent's own higher strategies look relatively better, so his best strategies must be higher as well. The next result makes this observation precise, and presents its implications for the structure of the Nash equilibrium set.

Let $\underline{B}\colon X \to X$ and $\bar{B}\colon X \to X$, defined by $\underline{B}(x) = \min B(x)$ and $\bar{B}(x) = \max B(x)$, be the minimal and maximal best response functions, where the minimum and maximum are defined with respect to the stochastic dominance order. Evidently, these functions always evaluate to pure states. For states $\underline{x}, \bar{x} \in X$ satisfying $\Sigma\underline{x} \leq \Sigma\bar{x}$, we define the interval $[\underline{x}, \bar{x}]$ as the set of states lying between $\underline{x}$ and $\bar{x}$ in the stochastic dominance order: $[\underline{x}, \bar{x}] = \{x \in X : \Sigma\underline{x} \leq \Sigma x \leq \Sigma\bar{x}\}$.

Theorem 13.11. *Suppose F is a supermodular game. Then*

(i) *$\underline{B}$ and $\bar{B}$ are increasing in the stochastic dominance order: if $\Sigma x \leq \Sigma y$, then $\Sigma\underline{B}(x) \leq \Sigma\underline{B}(y)$ and $\Sigma\bar{B}(x) \leq \Sigma\bar{B}(y)$.*

(ii) *The sequences of iterates $\{\underline{B}^k(e_1)\}_{k\geq 0}$ and $\{\bar{B}^k(e_n)\}_{k\geq 0}$ are monotone sequences of pure states, and so converge within n steps to their limits, $\underline{x}^*$ and $\bar{x}^*$.*

(iii) *$\underline{x}^* = \underline{B}(\underline{x}^*)$ and $\bar{x}^* = \bar{B}(\bar{x}^*)$, so $\underline{x}^*$ and $\bar{x}^*$ are pure Nash equilibria of F.*

(iv) *$NE(F) \subseteq [\underline{x}^*, \bar{x}^*]$. Thus if $\underline{x}^* = \bar{x}^*$, then this state is the unique Nash equilibrium of F.*

Given the monotonicity of the best response correspondence in supermodular games, it is natural to look for convergence results under the best response dynamic. It follows from the results in the previous section that this dynamic must converge to the interval $[\underline{x}^*, \bar{x}^*]$ whose endpoints are the minimal and maximal Nash equilibria. But to prove convergence to Nash equilibrium, one must impose additional structure on the game.[41]

General convergence results can be proved using the theory of *(strongly) cooperative differential equations*. These are continuously differentiable differential equations $\dot{\mathscr{X}} = \mathscr{V}(\mathscr{X})$ whose derivative matrices have positive off-diagonal elements, so that increases in any component of increase the growth rates of other components.

Since cooperative differential equations must be smooth, this theory cannot be applied to the best response dynamic. However, Hofbauer and Sandholm (2002, 2007) show that the needed monotonicity is preserved by perturbed best response dynamics, although only those derived from stochastic perturbations.

Specifically, they consider the perturbed best response dynamic

$$\dot{x} = \tilde{M}^{\varepsilon}(F(x)) - x, \qquad [13.45]$$

where the perturbed best response function $\tilde{M}^{\varepsilon}$ is defined in terms of the stochastic perturbation ε, as in equation [13.21]. Equation [13.45] is transformed into a cooperative differential equation using the change of variable defined by the stochastic dominance operator Σ. Let $\mathscr{X} = \Sigma X \subset \mathbb{R}^{n-1}$ denote the image of X under Σ, and let $\bar{\Sigma}: \mathscr{X} \to X$ denote the (affine) inverse of the map Σ. Then the change of variable Σ converts [13.45] into the dynamic

$$\dot{\mathscr{X}} = \Sigma\tilde{M}^{\varepsilon}(F(\bar{\Sigma}\mathscr{X})) - \mathscr{X}, \qquad [13.46]$$

on $\mathscr{X}$.

Combining strategic complementarity [13.44] with properties of the derivative matrix $D\tilde{M}^{\varepsilon}(\pi)$, Hofbauer and Sandholm (2002, 2007) show that equation [13.46] is strongly cooperative. Results from the theory of cooperative differential equations (Hirsch, 1988; Smith, 1995) then yield the following result.

Theorem 13.12. *Let F be a C^1 strictly supermodular game, and let $\dot{x} = V^{F,\varepsilon}(x)$ be a stochastically perturbed best response dynamic for F. Then*

(i) *States $\underline{x}^* \equiv \omega(\underline{x})$ and $\bar{x}^* \equiv \omega(\bar{x})$ exist and are the minimal and maximal elements of the set of perturbed equilibria. Moreover, $[\underline{x}^*, \bar{x}^*]$ contains all ω-limit points of $V^{F,\varepsilon}$ and is globally asymptotically stable.*

[41] For instance, Berger (2007) proves that in two-population games generated by two-player normal form games that are supermodular and satisfy a diminishing returns condition, most solution trajectories of the best response dynamic converge to pure Nash equilibria.

(ii) *Solutions to $\dot{x} = V^{F,\varepsilon}(x)$ from an open, dense, full measure set of initial conditions in X converge to perturbed equilibria.*

13.8.3 Iterated *p*-dominance and equilibrium selection

The evolutionary dynamics we have studied so far have been based on reactive protocols $\rho^F(x) = \rho(F(x), x)$, with switch rates depending on the game only through current payoffs. This formulation leads naturally to dynamics satisfying positive correlation (PC) and Nash stationarity (NS). While this approach has many advantages, it also imposes restrictions on what the dynamics can achieve. For instance, in coordination games, all pure equilibria are locally stable, implying that predictions of play must depend on initial conditions.

Following Oyama et al. (2012) (see also Sandholm, 2001a), we now argue that dynamics based on prospective revision protocols may have quite different properties, and in particular may lead to unique predictions in certain games with multiple strict equilibria.[42]

We consider an analog of the best response dynamic in which agents do not know the population state, and so base their decisions on information from samples. A revising agent obtains information by drawing a sample of size k from the population. The agent then plays a best response to the empirical distribution of strategies in his sample. This is a prospective revision protocol: agents are not reacting to current payoffs, but to the payoffs that would obtain if their sample were representative of behavior in the population at large.

Let $\mathbb{Z}_+^{n,k} = \{z \in \mathbb{Z}_+^n : \sum_{i \in S} z_i = k\}$ be the set of possible outcomes of samples of size k. The *k-sampling best response protocol* is the target protocol

$$\sigma^F(x) = B^{F,k}(x), \text{ where} \tag{13.47a}$$

$$B^{F,k}(x) = \sum_{z \in \mathbb{Z}_+^{n,k}} \binom{k}{z_1 \; \ldots \; z_n} \left(x_1^{z_1} \cdots x_n^{z_n}\right) B^F\left(\frac{1}{k}z\right). \tag{13.47b}$$

is the *k-sampling best response correspondence*.[43] The mean dynamic of this protocol is the *k-sampling best response dynamic*

$$\dot{x} \in B^{F,k}(x) - x. \tag{13.48}$$

Sampling best response dynamics agree with standard dynamics in one basic respect: strict equilibria of F are rest points of [13.48]. However, while strict equilibria are locally

[42] Dynamics based on reactive protocols with a moderate amount of noise can have similar properties—see Oyama et al. (2012) and Kreindler and Young (2013).

[43] While in general $B^{F,k}$ is multivalued, it is single valued when every possible sample of size k generates a unique best response in F. For linear games $F(x) = Ax$, this is true for generic choices of the matrix A.

stable under standard dynamics (Section 13.6.6), this need not be true under sampling best response dynamics.

For $p \in [0, 1]$, we call strategy $i \in S$ a *p-dominant equilibrium* of F (Morris et al., 1995) if it is the unique optimal strategy at any state x satisfying $x_i \geq p$. Thus 1-dominant equilibria are strict equilibria, while 0-dominant equilibria correspond to strictly dominant strategies. The basic selection result for sampling best response dynamics is:

Theorem 13.13. *Suppose that strategy i is $\frac{1}{k}$-dominant in game F, and let V^F be the k-sampling best response dynamic. Then state e_i is asymptotically stable, and attracts solutions from all initial conditions with $x_i > 0$.*

To explain this result, it is sufficient to consider a coordination game with strategy set $S = \{0, 1\}$ and in which strategy 1 is $\frac{1}{k}$-dominant but not $\frac{1}{k+1}$-dominant. An example is the game $F(x) = Ax$ with

$$A = \begin{pmatrix} 1 & 0 \\ 0 & k-1+\varepsilon \end{pmatrix} \text{ and } \varepsilon \in (0, 1).$$

Since strategy 1 is $\frac{1}{k}$-dominant and the sample size is k, an agent will only choose strategy 0 if all of the agents he samples choose strategy 0. Thus if $x_1 \in (0, 1)$,

$$\dot{x}_1 = B_1^{F,k}(x) - x_1 = \left(1 - (1 - x_1)^k\right) - x_1 = (1 - x_1) - (1 - x_1)^k.$$

Since this expression is positive whenever $x_1 \in (0, 1)$, state $x_1 = 1$ attracts all interior initial conditions.

The use of a fixed sample size is not essential: the same result holds so long as enough probability is placed on sample sizes no greater than k. Moreover, the selection result extends to iterated versions of p-dominance (Tercieux, 2006; Oyama and Tercieux, 2009), as we illustrate through an example.

Example 13.25. (Selection of iterated $\frac{1}{2}$-dominant equilibrium) *Consider the following 3×3 coordination game of Young (1993):*

$$F(x) = Ax, \text{ where } A = \begin{pmatrix} 6 & 0 & 0 \\ 5 & 7 & 5 \\ 0 & 5 & 8 \end{pmatrix}.$$

The phase diagram of the best response dynamic for this game, presented in Figure 13.10i, shows that each of the three strict equilibria of F has a non-negligible basin of attraction.

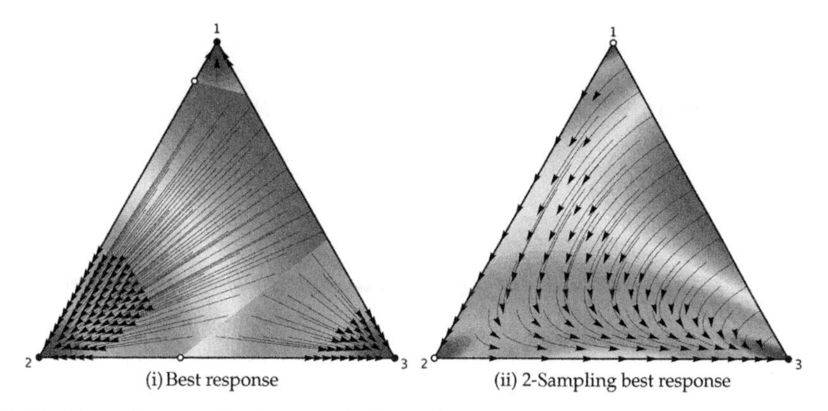

Figure 13.10 Phase diagrams for dynamics in Young's game.

*Examining the best response regions of F as depicted in Figure 13.10i, it is clear that no strategy in this game is $\frac{1}{2}$-dominant. However, the set $\{2, 3\}$ is a $\frac{1}{2}$-*best response set*: if at least half the population plays strategies from this set, then all pure best responses are elements of the set. Moreover, once strategy 1 is removed, strategy 3 is $\frac{1}{2}$-dominant in the game that remains. As Figure 13.10ii shows, state e_3 is asymptotically stable under the 2-sampling best response dynamic in the original game, and attracts solutions from all states x with $x_3 > 0$.*

To establish asymptotic stability in Example 13.25, one uses a transitivity theorem for asymptotic stability due to Conley (1978). In the present context, this theorem says that if edge e_2e_3 is an asymptotically stable invariant set in X, and state e_3 is asymptotically stable with respect to the dynamic restricted to edge e_2e_3, then state e_3 is asymptotically stable in X. This theorem may prove useful elsewhere for transferring the logic of iteration from the analysis of a game to the analysis of a corresponding evolutionary dynamic.

13.9. NONCONVERGENCE OF EVOLUTIONARY DYNAMICS

The analyses in the previous sections have highlighted combinations of classes of games and dynamics for which convergence to equilibrium can be assured. Beyond these combinations, there are few general guarantees of convergence. Indeed, a key reason for studying evolutionary game dynamics is to understand when equilibrium analysis should be augmented or supplanted by an analysis of persistent disequilibrium behavior. In this section, we offer a selection of examples in which convergence to equilibrium fails, and present an implication of this possibility for the survival of strictly dominated strategies.

13.9.1 Examples

Example 13.26. (Bad RPS) *Example 13.20 introduced Rock-Paper-Scissors:*

$$F(x) = Ax, \text{ where } A = \begin{pmatrix} 0 & -l & w \\ w & 0 & -l \\ -l & w & 0 \end{pmatrix}.$$

In bad Rock-Paper-Scissors, *$l > w > 0$, so that the cost of a loss is higher than the benefit of a win. Figure 13.11 presents phase diagrams of the five basic evolutionary dynamics for bad RPS with $l = 2$ and $w = 1$. Under the replicator dynamic, interior solutions approach a heteroclinic cycle along the three edges of the simplex. Under the other four dynamics, solutions approach interior limit cycles, with the position of the cycle depending on the dynamic in question.*

Gaunersdorfer and Hofbauer (1995) provide a full analysis of the replicator and best response dynamics in Rock-Paper-Scissors games. They dub the triangular closed orbit of the best response dynamic in bad RPS a Shapley polygon, *after Shapley (1964), who constructed the first example of cycling under the fictitious play process (Section 13.10.2). Benaïm et al. (2009) establish the existence of attracting Shapley polygons in higher dimensional examples with a similar cyclic structure.*

Since the state space for bad RPS is two-dimensional, the existence of limit cycles for the remaining dynamics can be proved using the Poincaré-Bendixson theorem, *which states that in planar systems, forward invariant regions containing no rest points must contain closed orbits; see Berger and Hofbauer (2006), Sandholm (2007a), and Hofbauer and Sandholm (2011).*

Example 13.27. (Mismatching Pennies) Mismatching Pennies *(Jordan, 1993) is a three-player normal form game in which each player has two strategies, Heads and Tails. Player p receives a payoff of 1 for choosing a different strategy than player $p + 1$ and a payoff of 0 otherwise, where players are indexed modulo 3. The unique Nash equilibrium of this game has each player play each of his strategies with equal probability.*

Figure 13.12 presents phase diagrams of the replicator dynamic and the best response dynamic for the three-population game obtained by matching triples of agents to play Mismatching Pennies. Almost all solutions of the replicator dynamic (Figure 13.12i) converge to a six-sided heteroclinic cycle on the boundary of the state space; this cycle follows the best response cycle of the normal form game. Almost all solutions of the best response dynamic (Figure 13.12ii) converge to a Shapley polygon, here a six-sided closed orbit in the interior of the state space. The behaviors of the two dynamics are intimately related: Gaunersdorfer and Hofbauer (1995) show that the time averages of interior solutions of the replicator dynamic approach the Shapley polygon of the best response dynamic. This link is not accidental: quite general connections between the time-averaged replicator dynamic and the best response dynamic are established by Hofbauer et al. (2009b).

Hart and Mas-Colell (2003) show that the failures of convergence exhibited in Figure 13.12 are not exceptional. They consider three-population games based on normal form games whose

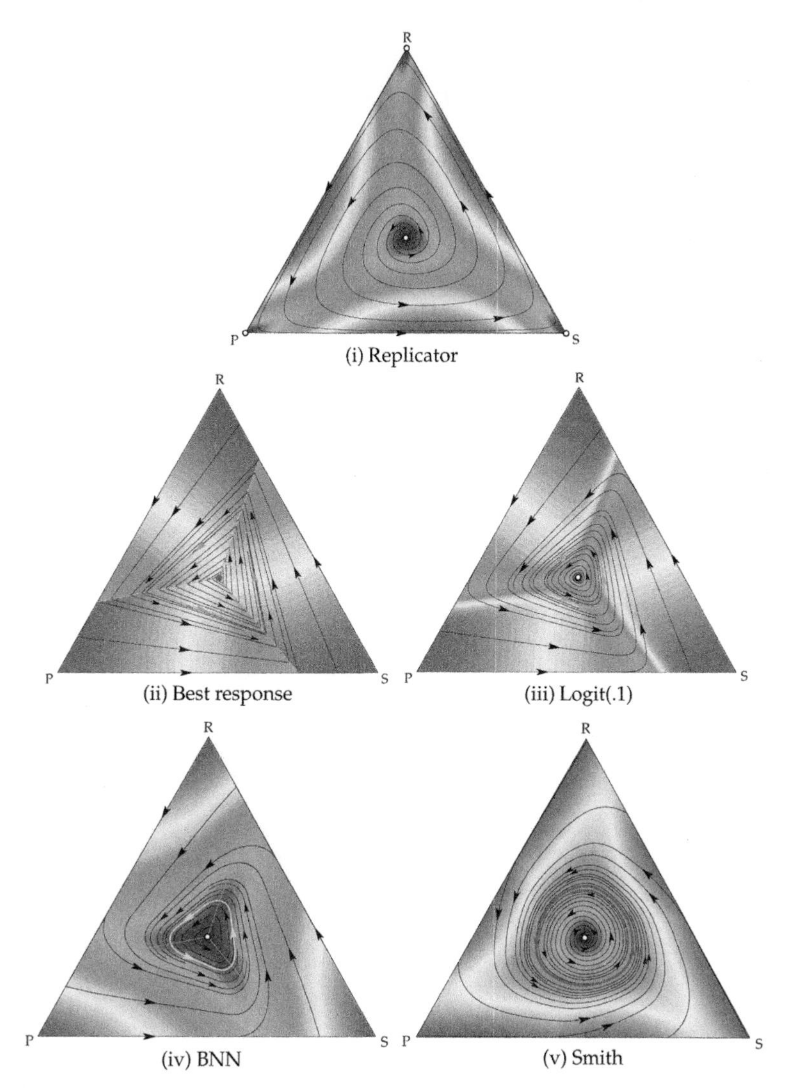

Figure 13.11 Five basic deterministic dynamics in bad Rock-Paper-Scissors.

unique Nash equilibrium is completely mixed. Under the weak assumption that each population's revision protocol does not condition on other populations' payoffs, they prove that any hyperbolic rest point of the resulting evolutionary dynamic must be unstable. It follows that if the corresponding dynamics satisfy Nash stationarity (NS), solutions from almost all initial conditions do not converge.

Example 13.28. (The hypnodisk game) Hypnodisk *(Hofbauer and Sandholm (2011)) is a three-strategy population game with nonlinear payoffs. In a small circle centered at* $x^* = (\frac{1}{3}, \frac{1}{3}, \frac{1}{3})$, *the payoffs are those of the coordination game*

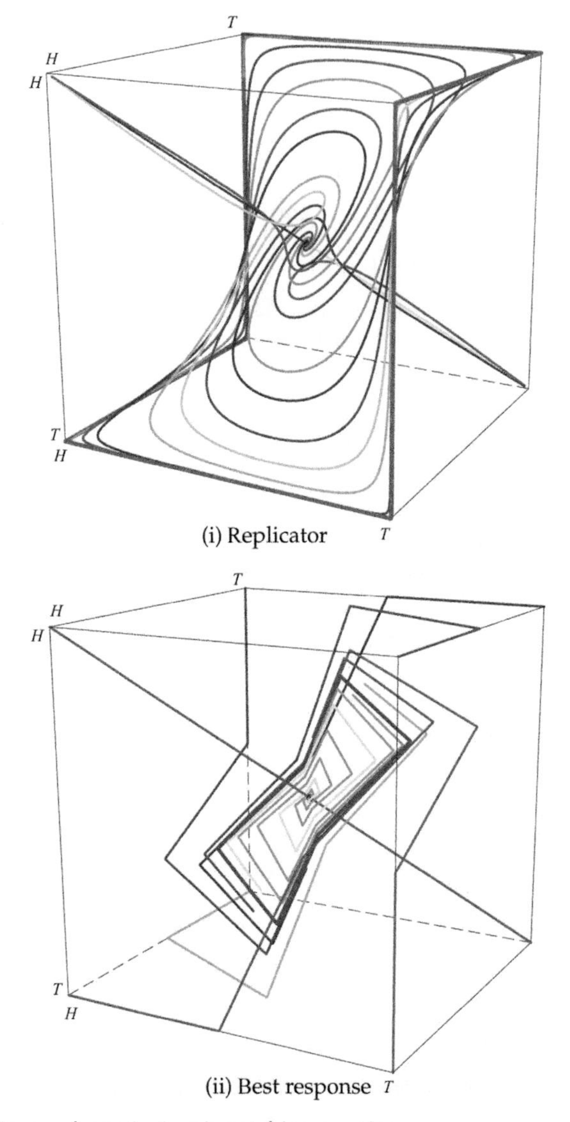

Figure 13.12 Evolutionary dynamics in mismatching pennies.

$$F^C(x) = Cx = \begin{pmatrix} 1 & 0 & 0 \\ 0 & 1 & 0 \\ 0 & 0 & 1 \end{pmatrix} \begin{pmatrix} x_1 \\ x_2 \\ x_3 \end{pmatrix} = \begin{pmatrix} x_1 \\ x_2 \\ x_3 \end{pmatrix},$$

a potential game with convex potential function $f^C(x) = \frac{1}{2}((x_1)^2 + (x_2)^2 + (x_3)^2)$. *Outside a larger circle centered at* x^*, *the payoffs are those of the anticoordination game* $F^{-C}(x) = -Cx$, *a potential game with concave potential function* $f^{-C}(x) = -\frac{1}{2}((x_1)^2 + (x_2)^2 + (x_3)^2)$. *In between, payoffs are defined so that* F *is continuous, and so that* x^* *is the game's unique Nash equilibrium. Geometrically, this is accomplished by starting with the vector field for* F^C, *and then*

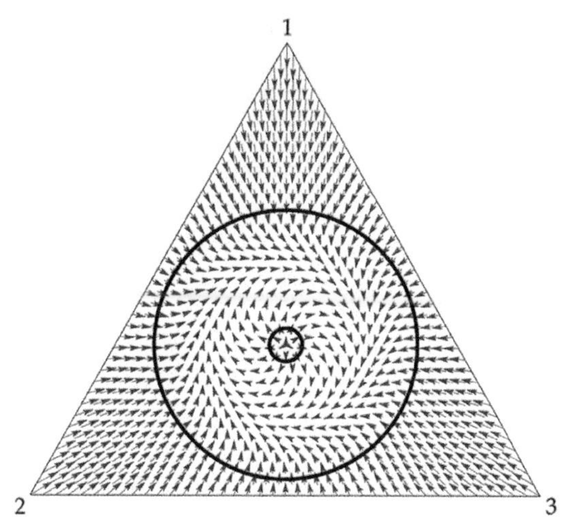

Figure 13.13 Projected payoffs in the hypnodisk game.

twisting the portion of the vector field outside the inner circle in a clockwise direction, excluding larger and larger circles as the twisting proceeds, so that the outer circle is reached when the total twist is 180°. *The projected payoffs of the resulting game are illustrated in Figure 13.13.*

Now consider evolution under an evolutionary dynamic satisfying positive correlation (PC) and Nash stationarity (NS). By the analysis in Section 13.6, solutions starting at states besides x^* *in the inner disk must leave the inner disk. Similarly, solutions from states outside the outer disk must enter the outer disk. Since there are no Nash equilibria, and hence no rest points, in the annulus bounded by the circles, the Poincaré-Bendixson theorem implies that every solution other than the one at* x^* *must converge to a limit cycle in the annulus.*

Example 13.29. (Chaotic dynamics) *In population games with four or more strategies, and hence state spaces with three or more dimensions, solution trajectories of game dynamics can converge to complicated sets called* chaotic attractors. *Central to most definitions of chaos is* sensitive dependence on initial conditions*: solution trajectories starting from close together points on the attractor move apart at an exponential rate. Chaotic attractors can also be recognized in phase diagrams by their intricate appearance.*

Following Arneodo et al. (1980) and Skyrms (1992), we consider evolution in the game

$$F(x) = Ax = \begin{pmatrix} 0 & -12 & 0 & 22 \\ 20 & 0 & 0 & -10 \\ -21 & -4 & 0 & 35 \\ 10 & -2 & 2 & 0 \end{pmatrix} \begin{pmatrix} x_1 \\ x_2 \\ x_3 \\ x_4 \end{pmatrix},$$

whose lone interior Nash equilibrium is $x^* = \left(\frac{1}{4}, \frac{1}{4}, \frac{1}{4}, \frac{1}{4}\right)$. *Figure 13.14 presents a solution to the replicator dynamic for this game from initial condition* $x_0 = (0.24, 0.26, 0.25, 0.25)$. *This solution spirals clockwise about* x^*. *Near the rightmost point of each circuit, where the value of* x_3 *gets close to zero, solutions sometimes proceed along an "outside" path on which the value of*

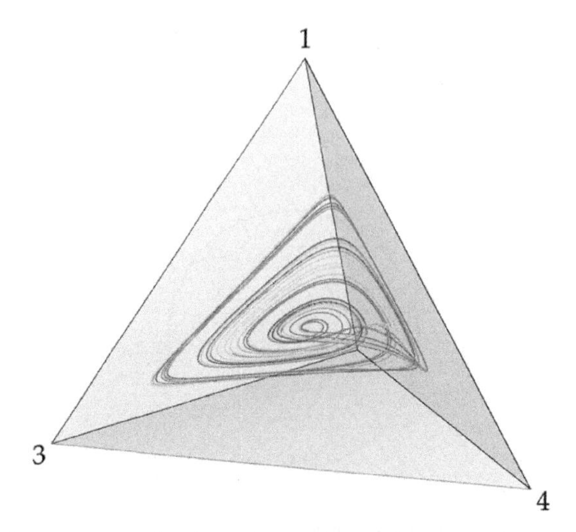

Figure 13.14 A chaotic attractor under the replicator dynamic.

x_3 surpasses 0.6. But they sometimes follow an "inside" path on which x_3 remains below 0.4, and at other times do something in between. Which of these alternatives occurs is difficult to predict from approximate information about the previous behavior of the system. While this game has a complicated payoff structure, in multipopulation contexts one can find chaotic evolutionary dynamics in simple games with three strategies per player.[44]

13.9.2 Survival of strictly dominated strategies

In Section 13.8.1, we argued that that the best response dynamic and all imitative dynamics eliminate dominated strategies, at least along solutions starting from most initial conditions. These conclusions may seem unsurprising given the fundamental role played by dominance arguments in traditional game-theoretic analyses. In fact, these conclusions are quite special, as the following example illustrates.

Example 13.30. (Survival of strictly dominated strategies) *Following Berger and Hofbauer (2006) and Hofbauer and Sandholm (2011), we first consider the Smith dynamic for "bad RPS with a twin" (Figure 13.15i):*

$$F(x) = Ax = \begin{pmatrix} 0 & -2 & 1 & 1 \\ 1 & 0 & -2 & -2 \\ -2 & 1 & 0 & 0 \\ -2 & 1 & 0 & 0 \end{pmatrix} \begin{pmatrix} x_R \\ x_P \\ x_S \\ x_T \end{pmatrix}.$$

The Nash equilibria of F are the states on line segment $\{x^* \in X\colon x^* = (\frac{1}{3}, \frac{1}{3}, c, \frac{1}{3} - c)\}$, which is a repellor under the Smith dynamic. Away from Nash equilibrium, strategies gain players at

[44] See Sato et al. (2002) for examples of chaos under the replicator dynamic, and Cowan (1992), Sparrow et al. (2008), and van Strien and Sparrow (2011) for analyses of chaos under the best response dynamic.

rates that depend on their payoffs, but lose players at rates proportional to their current usage levels. The proportions of players choosing the twin strategies are therefore equalized, with the state approaching the plane $\mathscr{P} = \{x \in X : x_S = x_T\}$. Since F is based on bad RPS, solutions on plane $\mathscr{P}$ approach a closed orbit away from any Nash equilibrium.[45]

Figure 13.15ii presents the Smith dynamic in "bad RPS with a feeble twin,"

$$F^d(x) = A^d x = \begin{pmatrix} 0 & -2 & 1 & 1 \\ 1 & 0 & -2 & -2 \\ -2 & 1 & 0 & 0 \\ -2-d & 1-d & -d & -d \end{pmatrix} \begin{pmatrix} x_R \\ x_P \\ x_S \\ x_T \end{pmatrix}$$

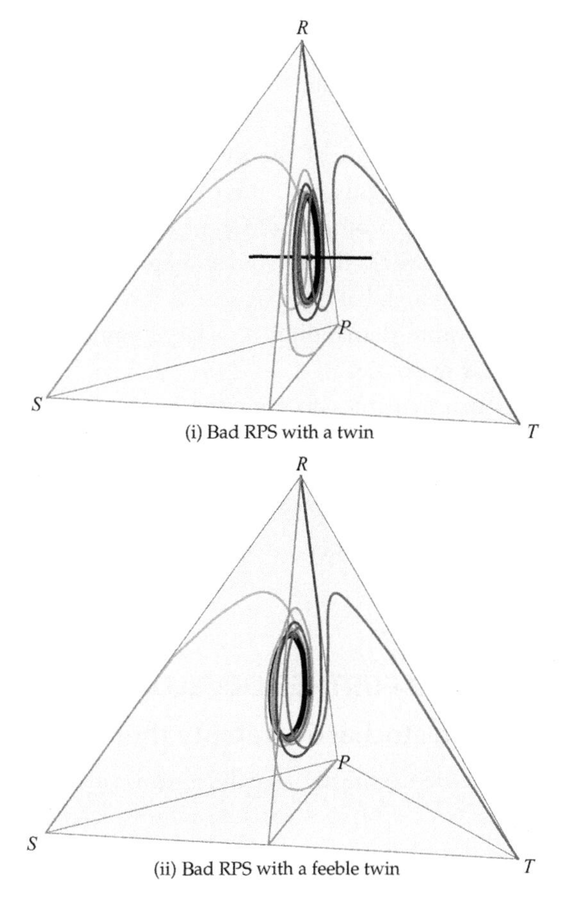

Figure 13.15 Survival of strictly dominated strategies: the Smith dynamic in two games.

[45] Under the replicator dynamic in this game, the state space X foliates into a continuum of invariant planes on which the ratio x_S/x_T is constant. This foliation, which is in part a consequence of inflow-outflow symmetry (Section 13.5.1.4), is structurally unstable, helping to explain why elimination results for imitative dynamics (Theorem 13.10) should be viewed as knife-edge cases.

with $d = \frac{1}{10}$. Evidently, the attractor from Figure 13.15i moves slightly to the left, reflecting the fact that the payoff to Twin has gone down. But since the new attractor is in the interior of X, the strictly dominated strategy Twin is always played by fractions of players bounded away from zero. These fractions need not be small: Twin is recurrently played by at least 10% of the population when $d \leq 0.31$, by at least 5% of the population when $d \leq 0.47$, and by at least 1% of the population when $d \leq 0.66$.

Using the Poincaré-Bendixson theorem and results on continuation of attractors, Hofbauer and Sandholm (2011) show that the phenomenon illustrated in Example 13.30 is quite general. They consider continuous evolutionary game dynamics satisfying positive correlation (PC), Nash stationarity (NS), and a third condition, *innovation*, which requires that away from Nash equilibrium, unused optimal strategies have positive growth rates. They show that for any such dynamic, there are games in which strictly dominated strategies survive in perpetuity.

We should emphasize two features of the evolutionary process that underlie this result. First, it is important that agents base their decisions on the strategies' present performances. If agents were able to compare different strategies' payoff functions in their entirety, they could recognize strictly dominated strategies and avoid them. While the knowledge needed to make such global comparisons is assumed implicitly in traditional analyses, this assumption is quite demanding in a large game. Without it, dominated strategies need not disappear from use.

Second, the possibility that dynamics do not converge is crucial to the survival result. Under dynamics satisfying Nash stationarity (NS), any solution trajectory that converges must converge to a Nash equilibrium, implying that dominated strategies eventually go unplayed. Example 13.30 illustrates that when solutions do not converge, so that payoffs remain in flux, evolutionary dynamics need not eliminate strategies that often perform well, but are never optimal.

13.10. CONNECTIONS AND FURTHER DEVELOPMENTS

13.10.1 Connections with stochastic stability theory

Theorem 13.4 showed that deterministic dynamics can be used to approximate the stochastic evolutionary process $\{X_t^N\}$ in a particular limit, one that fixes a finite time horizon T and takes the population size N to infinity. To focus on very long run behavior, one can consider the opposite order of limits, fixing the population size N and taking the time horizon T to infinity. If the revision protocol is such that all strategies are always chosen with positive probability, then over an infinite time horizon, the process $\{X_t^N\}$ will visit each state in the finite grid $\mathscr{X}^N$ infinitely often. The very long run behavior of the process is summarized by its unique stationary distribution, μ^N, which describes the limiting distribution of the process, as well as its limiting empirical distribution on almost every sample path.

Pioneering work by Foster and Young (1990), Kandori et al. (1993), and Young (1993) showed that over very long time spans, stochastic evolutionary processes often admit unique predictions of play, even in games with multiple strict equilibria. While the stationary distribution μ^N places positive mass on every state in $\mathscr{X}^N$, it typically concentrates its mass on a particular subset of states, often just a single state. To obtain clean results, certain limits are taken to make this concentration as stark as possible. The population states selected in this fashion are said to be *stochastically stable*.

The most common approach to stochastic stability considers revision protocols parameterized by a noise level, whether a fixed mistake probability, or a parameter reflecting the size of payoff perturbations, as in the logit choice rule (Example 13.12). States that retain mass in the stationary distribution as the noise level approaches zero are said to be *stochastically stable in the small noise limit*. This approach to stochastic stability theory is described in the chapter by Wallace and Young in this volume.

Another approach to stochastic stability, introduced by Binmore et al. (1995) and Binmore and Samuelson (1997), considers the limiting behavior of the stationary distributions μ^N as the population size N approaches infinity. This analysis of *stochastic stability in the large population limit* can be viewed as the infinite horizon analog of the finite horizon analysis of $\{X_t^N\}$ provided by deterministic dynamics.

In view of this connection, it is natural to expect the behavior of deterministic evolutionary dynamics to have consequences for stochastic stability in the large population limit. Connections of this sort can be established using tools from *(constant step size) stochastic approximation theory*. Benaïm and Weibull (2003) show that states retaining mass in the stationary distributions μ^N in the large population limit must satisfy a basic notion of recurrence for the mean dynamic [13.3]: they must be elements of its *Birkhoff center*, which contains rest points, closed orbits, and more complicated limit sets.[46] Some states in the Birkhoff center—for instance, unstable rest points—seem unlikely candidates for stochastic stability. Benaïm (1998) provides conditions of a global nature ensuring that mass in μ^N does not accumulate near such states, implying that these states cannot be stochastically stable. To go further than this—to actually identify the component of the Birkhoff center containing the stochastically stable states—requires a detailed analysis of large deviations properties of the stochastic process $\{X_t^N\}$; for work in this direction, see Benaïm et al. (2014).

13.10.2 Connections with models of heuristic learning

This chapter has considered disequilibrium dynamics in games played by large populations of agents. There are close connections between these evolutionary game dynamics

[46] See also Benaïm (1998), Sandholm (2007c), and Roth and Sandholm (2013).

and models of *heuristic learning in games*, which consider disequilibrium dynamics during repeated play of games with small numbers of players—for instance, when there is one individual in each player role of a recurring normal form game.[47]

The original model of disequilibrium dynamics in games is *fictitious play*, introduced by Brown (1949, 1951) as a method of computing equilibria. During each period of the fictitious play process, each agent plays a best response to the time average of his opponents' past choices. The state variable of this process, the time average of play, has increments that become smaller over time, since choices late in the process have small effects on the overall time average. After accounting for this reparameterization of time, the fictitious play process is essentially a discrete-time version of the best response dynamic [13.17]. Indeed, convergence results developed in one context often have analogs in the other.[48]

Since best responses are generically pure, convergence of time averages in the fictitious play process does not imply convergence of players' period-by-period intended play. To generate such a link, Fudenberg and Kreps (1993) introduced *stochastic fictitious play*, in which players play best responses only after their payoffs have been subjected to perturbations. The result is a stochastic process with decreasing step size whose expected motion is described by perturbed best response dynamics [13.23]. By combining analyses of these dynamics with tools from (*decreasing step size*) *stochastic approximation theory* (Benaïm, 1999), one can establish global convergence results for stochastic fictitious play.[49]

The decision rule used in stochastic fictitious play generates an *ε-consistent* strategy for repeated play of a normal form game: a player using this rule could not have improved his average payoff by more than ε by having played a single pure strategy in all previous periods. There are other subtle links between consistency in repeated games and evolutionary game dynamics. For instance, the convergence results for contractive games in Section 13.7.5.2 are directly inspired by a class of consistent repeated game strategies introduced by Hart and Mas-Colell (2001).

There are also connections between evolutionary game dynamics and models of *reinforcement learning*, in which each player's mixed strategy weights are determined by his strategies' aggregated past performances. Börgers and Sarin (1997) show that when players' choices are based on the reinforcement learning rule of Cross (1973), then in the continuous-time limit, the players' mixed strategies evolve according to the replicator dynamic [13.8].[50] Still other connections exist with hybrid models, in which large

[47] For surveys of this literature, see Young (2004) and Hart (2005).

[48] See Hofbauer (1995b), Harris (1998), Berger (2005, 2007, 2008, 2012), and Benaïm et al. (2005).

[49] See Fudenberg and Kreps (1993), Kaniovski and Young (1995), Benaïm and Hirsch (1999), Hofbauer and Sandholm (2002), Benaïm and Faure (2012), and Perkins and Leslie (2013).

[50] Hopkins (2002) (see also Hofbauer et al., 2009b) uses surprising relations between the replicator and logit dynamics to establish strong links between behavior under reinforcement learning and period-by-period

populations of agents employ heuristic learning rules that condition on statistics of the history of play.[51]

13.10.3 Games with continuous strategy sets

Many applications of game theory are naturally modeled using a continuous strategy space. Population dynamics for this setting were first used by Hines (1980) and Zeeman (1981) to study evolution in populations of agents playing mixed strategies, a generalization of the environment in which Maynard Smith and Price (1973) introduced the ESS concept (see Section 13.7.1). Bomze (1990, 1991) continued this line of research by defining the replicator dynamic for general games with continuous strategy sets.

Analyzing population dynamics in this setting introduces novel technical and conceptual issues. If the strategy set is a convex subset of $\mathbb{R}^n$, then a population state is a probability measure on this space. Since the space of these probability measures is a function space, defining evolutionary dynamics requires one not only to write down an ordinary differential equation, but also to specify a norm with respect which to this equation's time derivative is defined.

Moreover, while the meaning of local stability is unambiguous when strategy sets are finite, this is no longer the case with a continuous strategy sets, a point emphasized by Oechssler and Riedel (2002). With finite strategy sets, population states are close to one another if they assign similar probabilities to each pure strategy. With continuous strategy sets, this notion of closeness is captured by using a variational norm on probability measures. But in the latter context, there is a second sense in which population states might be close. For instance, one might want to regard two pure population states as close together if the strategies played in these states are close together in the Euclidean metric on $\mathbb{R}^n$.

Therefore, before studying local stability under evolutionary dynamics, one must first choose the definition of a neighborhood in the space of population states. If one hews closely to the finite case by defining neighborhoods in terms of the variational norm, one can obtain sufficient conditions for local stability under the replicator dynamic using suitable generalizations of the ESS concept.[52] Alternatively, if the definition of a neighborhood captures only the second sort of closeness, allowing large numbers of agents to make small changes in strategy, then quite distinct conditions for local stability

choices under stochastic fictitious play. Kosfeld et al. (2002), Tsakas and Voorneveld (2009), and Laraki and Mertikopoulos (2013) study mixed strategy dynamics for normal form games derived from other heuristic principles.

[51] See Hopkins (1999a), Ellison and Fudenberg (2000), Ramsza and Seymour (2010), Fudenberg and Takahashi (2011), and Lahkar and Seymour (2013).

[52] See Bomze (1990), Oechssler and Riedel (2001), and Norman (2008).

based on comparisons to the performances of nearby strategies become relevant.[53] Finally, one might want to define neighborhoods in a way that captures both sorts of closeness. To accomplish this, Oechssler and Riedel (2002) introduce a condition called *evolutionary robustness*, which uses neighborhoods derived from the weak topology on probability measures, and show that it is sufficient for local stability under a class of imitative dynamics.[54]

Once one allows that certain pairs of strategies are closer together than others, it becomes natural to ask whether not only notions of stability, but also the very definition of the dynamic, should take distances between strategies into account. Friedman and Ostrov (2008, 2010, 2013) consider a framework in which each agent adjusts his strategy continuously in response to the incentives he faces; the use of continuous adjustment is justified as an optimal response to adjustment costs. Rather than obeying an ordinary differential equation, the evolution of aggregate behavior in this framework follows a partial differential equation. This formulation introduces a wide range of new dynamic phenomena, and requires analytical techniques different from those considered in this chapter, leaving many open questions for future research.

13.10.4 Extensive form games and set-valued solution concepts

To describe behavior in populations whose members are engaged in sequential interactions, one can define population games via matching in extensive form games, and evaluate the properties of evolutionary dynamics in these population games.

Even in the traditional theory, both the notion of a strategy for an extensive form game and the modeling of beliefs about events off of the path of play raise conceptual difficulties. Variations on these difficulties persist in large population contexts. But new difficulties arise as well. For instance, it is well known that in extensive form games with perfect recall, there is a many-to-one map between mixed strategies and outcome equivalent behavior strategies, with equivalent mixed strategies exhibiting different correlations among choices at temporally unordered information sets. Chamberland and Cressman (2000) show that under evolutionary dynamics, statistical correlation among choices at different information sets can interfere with the forces of selection, leading to unexpected forms of dynamic instability.

Traditionally, a fundamental idea in the analysis of extensive form games is *sequential rationality*: the requirement that players' equilibrium strategies specify actions that are

[53] Eshel and Motro (1981), Eshel (1983), and Apaloo (1997) introduce such conditions, and Eshel and Sansone (2003), Cressman (2005), Cressman et al. (2006), and van Veelen and Spreij (2009) show them to be sufficient for local stability under the replicator dynamic.

[54] See Cressman and Hofbauer (2005) for further stability results for the replicator dynamic. For definitions and stability analyses of the BNN, logit, and Smith dynamics with continuous strategy sets, see Hofbauer et al. (2009a), Lahkar and Riedel (2013), and Cheung (2013), respectively.

rational from the perspective of the moments the actions would be taken, even if these moments should not occur during equilibrium play. It is a basic question whether the predictions of evolutionary dynamics agree with solution concepts embodying sequential rationality. The strongest agreements occur in games of (stagewise) perfect information, for equilibria in which all information sets are reached: Cressman (1997a) and Cressman and Schlag (1998) show that in such cases, asymptotic stability under the replicator and best response dynamics agree with subgame perfection in the extensive form. Although positive results are available for certain further instances,[55] connections between sequential rationality and asymptotic stability have not been established in great generality.[56]

In extensive form games, the possibility of unreached information sets gives rise to connected sets of outcome-equivalent equilibria, which differ only in the choices they specify off the path of play. For this reason, it is natural to consider the stability of sets of equilibria—for instance, by defining set-valued extensions of the notion of ESS. In single-population contexts, Thomas's (1985) notion of an *evolutionarily stable set* (or *ES set*) provides a sufficient condition for asymptotic stability under the replicator dynamic.[57] In multipopulation contexts, Balkenborg and Schlag (2007) show that Balkenborg's (1994) notion of a *strict equilibrium set* (or *SE set*) plays an analogous role.

Using ideas from differential topology, Demichelis and Ritzberger (2003) develop a powerful approach to evaluating local stability of sets of Nash equilibria.[58] They assign each component of Nash equilibria an integer, the *index* of the component, that is determined by the behavior of the payoff vector field in a neighborhood of the component. For any evolutionary dynamic satisfying positive correlation (PC) and Nash stationarity (NS), a necessary condition for the component to be asymptotically stable is that its index agree with the component's Euler characteristic. In typical (specifically, contractible) cases, this means that a component can only be asymptotically stable if its index is 1. Through this result, one can establish the instability of a component of equilibria under a large class of evolutionary dynamics—the same class in which convergence to Nash equilibrium is ensured in potential games—using only information about the payoffs of the game.

13.10.5 Applications

Population games provide a formal framework for applications in which large numbers of participants make interdependent decisions. As we argued at the start, such applications

[55] See, e.g., Cressman (1996).

[56] See Cressman (2003) for a comprehensive treatment of this literature.

[57] See Balkenborg and Schlag (2001) and van Veelen (2012) for alternative characterizations and related notions.

[58] See also Ritzberger (1994) and Demichelis and Germano (2000, 2002).

are precisely the ones in which traditional equilibrium assumptions seem questionable. By taking evolutionary dynamics to applications, one can both justify equilibrium predictions, and, more interestingly, call attention to settings in which disequilibrium predictions are warranted.

This latter possibility is well illustrated by analyses of price dispersion. While in idealized markets, homogeneous goods sell at a single price, in actual markets one often observes a variety of prices for the same good. Research in information economics has shown that with heterogeneously informed consumers, price dispersion can occur as an equilibrium phenomenon: different sellers choose different prices, and each seller, knowing the choices of the others, is content with his own choice.[59]

Using tools from evolutionary game theory, Hopkins and Seymour (2002), Lahkar (2011), Hahn (2012), and Lahkar and Riedel (2013) argue that equilibrium is not the best explanation for the price dispersion we see. They show that equilibria with price dispersion are unstable, but that price dispersion can persist through disequilibrium cycles. Indeed, empirical and experimental evidence supports the conclusion that price dispersion may be a cyclical phenomenon rather than an equilibrium phenomenon.[60]

Deterministic evolutionary dynamics have been used to study a variety of other topics in economics and social science. A partial list includes auctions (Louge and Riedel, 2012); fiat money (Sethi, 1999); conspicuous consumption (Friedman and Ostrov, 2008); common resource use (Sethi and Somanathan, 1996); cultural evolution (Bisin and Verdier, 2001; Sandholm, 2001c; Kuran and Sandholm, 2008; Montgomery, 2010); the evolution of language (Pawlowitsch, 2008); implementation problems (Cabrales and Ponti, 2000; Sandholm, 2002, 2005b; Fujishima, 2012); international trade (Friedman and Fung, 1996); residential segregation (Dokumacı and Sandholm, 2007); preference evolution (Sandholm, 2001c; Heifetz et al., 2007; Norman, 2012); and theories of mind (Mohlin, 2012). Further applications of deterministic evolutionary game dynamics to these topics and others offer fascinating directions for future research.

ACKNOWLEDGEMENTS

I thank Man Wah Cheung for meticulous proofreading and an anonymous reviewer for helpful comments. Financial support from NSF Grant SES-1155135 is gratefully acknowledged.

REFERENCES

Akin, E., 1979. The Geometry of Population Genetics. Springer, Berlin.
Akin, E., 1980. Domination or equilibrium. Math. Biosci. 50, 239–250.

[59] See Varian (1980) and Burdett and Judd (1983).
[60] See Lach (2002), Eckert (2003), Noel (2007), and Cason et al. (2005).

Apaloo, J., 1997. Revisiting strategic models of evolution: the concept of neighborhood invader strategies. Theor. Popul. Biol. 52, 71–77.
Arneodo, A., Coullet, P., Tresser, C., 1980. Occurrence of strange attractors in three-dimensional Volterra equations. Phys. Lett. 79A, 259–263.
Balder, E.J., 2002. A unifying pair of Cournot–Nash equilibrium existence results. J. Econ. Theory 102, 437–470.
Balkenborg, D., 1994. Strictness and evolutionary stability. Discussion paper 52, Center for the Study of Rationality, Hebrew University.
Balkenborg, D., Hofbauer, J., Kuzmics, C., 2013. Refined best-response correspondence and dynamics. Theor. Econ. 8, 165–192.
Balkenborg, D., Schlag, K.H., 2001. Evolutionarily stable sets. Int. J. Game Theory 29, 571–595.
Balkenborg, D., Schlag, K.H., 2007. On the evolutionary selection of sets of Nash equilibria. J. Econ. Theory 133, 295–315.
Beckmann, M., McGuire, C.B., Winsten, C.B., 1956. Studies in the Economics of Transportation. Yale University Press, New Haven.
Benaïm, M., 1998. Recursive algorithms, urn processes, and the chaining number of chain recurrent sets. Ergodic Theory Dyn. Syst. 18, 53–87.
Benaïm, M., 1999. Dynamics of stochastic approximation algorithms. In: Azéma, J., Émery, M., Ledoux, M., Yor, M., (ed.), Séminaire de Probabilités XXXIII, volume 1709 of Lecture Notes in Mathematics Springer, Berlin, pp. 1–68.
Benaïm, M., Faure, M., 2012. Stochastic approximations, cooperative dynamics, and supermodular games. Ann. Appl. Probab. 22, 2133–2164.
Benaïm, M., Hirsch, M.W., 1999. Mixed equilibria and dynamical systems arising from fictitious play in perturbed games. Games Econ. Behav. 29, 36–72.
Benaïm, M., Hofbauer, J., Hopkins, E., 2009. Learning in games with unstable equilibria. J. Econ. Theory 144, 1694–1709.
Benaïm, M., Hofbauer, J., Sorin, S., 2005. Stochastic approximations and differential inclusions. SIAM J. Control Optim. 44, 328–348.
Benaïm, M., Sandholm, W.H., Staudigl, M., 2014. Large deviations and stochastic stability in the large population limit. Unpublished manuscript, Université de Neuchâtel, University of Wisconsin, and Bielefeld University.
Benaïm, M., Weibull, J.W., 2003. Deterministic approximation of stochastic evolution in games. Econometrica 71, 873–903.
Benaïm, M., Weibull, J.W., 2009. Mean-field approximation of stochastic population processes in games. Unpublished manuscript, Université de Neuchâtel and Stockholm School of Economics.
Berger, U., 2005. Fictitious play in 2 × n games. J. Econ. Theory 120, 139–154.
Berger, U., 2007. Two more classes of games with the continuous-time fictitious play property. Games Econ. Behav. 60, 247–261.
Berger, U., 2008. Learning in games with strategic complementarities revisited. J. Econ. Theory 143, 292–301.
Berger, U., 2012. Non-algebraic convergence proofs for continuous-time fictitious play. Dyn. Games Appl. 2, 4–17.
Berger, U., Hofbauer, J., 2006. Irrational behavior in the Brown-von Neumann-Nash dynamics. Games Econ. Behav. 56, 1–6.
Binmore, K., Samuelson, L., 1997. Muddling through: noisy equilibrium selection. J. Econ. Theory 74, 235–265.
Binmore, K., Samuelson, L., Vaughan, R., 1995. Musical chairs: modeling noisy evolution. Games Econ. Behav. 11, 1–35. Erratum, 21 (1997), 325.
Bisin, A., Verdier, T., 2001. The economics of cultural transmission and the dynamics of preferences. J. Econ. Theory 97, 298–319.
Björnerstedt, J., Weibull, J.W., 1996. Nash equilibrium and evolution by imitation. In: Arrow, K.J. et al., (eds.), The Rational Foundations of Economic Behavior, St. Martin's Press, New York, pp. 155–181.

Bomze, I.M., 1986. Non-cooperative two-person games in biology: a classification. Int. J. Game Theory 15, 31–57.

Bomze, I.M., 1990. Dynamical aspects of evolutionary stability. Monatsh. Math. 110, 189–206.

Bomze, I.M., 1991. Cross entropy minimization in uninvadable states of complex populations. J. Math. Biol. 30, 73–87.

Bomze, I.M., Pötscher, B.M., 1989. Game Theoretical Foundations of Evolutionary Stability. Springer, Berlin.

Börgers, T., Sarin, R., 1997. Learning through reinforcement and the replicator dynamics. J. Econ. Theory 77, 1–14.

Brown, G.W., 1949. Some notes on computation of games solutions. Report P-78, The Rand Corporation.

Brown, G.W., 1951. Iterative solutions of games by fictitious play. In: Koopmans, T.C. et al., (eds), Activity Analysis of Production and Allocation, Wiley, New York, pp. 374–376.

Brown, G.W., von Neumann, J., 1950. Solutions of games by differential equations. In: Kuhn, H.W. and Tucker, A.W., (eds.), Contributions to the Theory of Games I, volume 24 of Annals of Mathematics Studies, Princeton University Press, Princeton, pp. 73–79.

Burdett, K., Judd, K.L., 1983. Equilibrium price dispersion. Econometrica 51, 955–969.

Cabrales, A., Ponti, G., 2000. Implementation, elimination of weakly dominated strategies, and evolutionary dynamics. Rev. Econ. Dyn. 3, 247–282.

Carmona, G., Podczeck, K., 2009. On the existence of pure-strategy equilibria in large games. J. Econ. Theory 144, 1300–1319.

Cason, T.N., Friedman, D., Wagener, F., 2005. The dynamics of price dispersion, or Edgeworth variations. J. Econ. Dyn. Control 29, 801–822.

Chamberland, M., Cressman, R., 2000. An example of dynamic (in)consistency in symmetric extensive form evolutionary games. Games Econ. Behav. 30, 319–326.

Cheung, M.W., 2013. Pairwise comparison dynamics for games with continuous strategy space. J. Econ. Theory, forthcoming.

Conley, C., 1978. Isolated Invariant Sets and the Morse Index. American Mathematical Society, Providence, RI.

Correa, J.R., Schulz, A.S., Stier-Moses, N.E., 2004. Selfish routing in capacitated networks. Math. Oper. Res. 29, 961–976.

Correa, J.R., Schulz, A.S., Stier-Moses, N.E., 2008. A geometric approach to the price of anarchy in nonatomic congestion games. Games Econ. Behav. 64, 457–469.

Cowan, S.G., 1992. Dynamical Systems Arising from Game Theory. PhD thesis, University of California, Berkley.

Cressman, R., 1992. The Stability Concept of Evolutionary Game Theory: A Dynamic Approach. Springer, Berlin.

Cressman, R., 1996. Evolutionary stability in the finitely repeated prisoner's dilemma game. J. Econ. Theory 68, 234–248.

Cressman, R. 1997a. Dynamic stability in symmetric extensive form games. Int. J. Game Theory 26, 525–547.

Cressman, R. 1997b. Local stability of smooth selection dynamics for normal form games. Math. Soc. Sci. 34, 1–19.

Cressman, R., 2003. Evolutionary Dynamics and Extensive Form Games. MIT Press, Cambridge.

Cressman, R., 2005. Stability of the replicator equation with continuous strategy space. Math. Soc. Sci. 50, 127–147.

Cressman, R., 2006. Uninvadability in N-species frequency models for resident-mutant systems with discrete or continuous time. Theor. Popul. Biol. 69, 253–262.

Cressman, R., 2011. Beyond the symmetric normal form: extensive form games, asymmetric games, and games with continuous strategy spaces. In: Sigmund, K., (ed.), Evolutionary Game Dynamics, American Mathematical Society, Providence, RI, pp. 27–59.

Cressman, R., Garay, J., Hofbauer, J., 2001. Evolutionary stability concepts for N-species frequency-dependent interactions. J. Theor. Biol. 211, 1–10.

Cressman, R., Hofbauer, J., 2005. Measure dynamics on a one-dimensional continuous trait space: theoretical foundations for adaptive dynamics. Theor. Popul. Biol. 67, 47–59.

Cressman, R., Hofbauer, J., Riedel, F., 2006. Stability of the replicator equation for a single-species with a multi-dimensional continuous trait space. J. Theor. Biol. 239, 273–288.
Cressman, R., Schlag, K.H., 1998. On the dynamic (in)stability of backwards induction. J. Econ. Theory 83, 260–285.
Cross, J.G., 1973. A stochastic learning model of economic behavior. Q. J. Econ. 87, 239–266.
Crow, J.F., Kimura, M., 1970. An Introduction to Population Genetics Theory. Harper and Row, New York.
Dafermos, S., Sparrow, F.T., 1969. The traffic assignment problem for a general network. J. Res. Nat. Bur. Stand. B 73, 91–118.
Demichelis, S., Germano, F., 2000. On the indices of zeros of Nash fields. J. Econ. Theory 94, 192–217.
Demichelis, S., Germano, F., 2002. On (un)knots and dynamics in games. Games Econ. Behav. 41, 46–60.
Demichelis, S., Ritzberger, K., 2003. From evolutionary to strategic stability. J. Econ. Theory 113, 51–75.
Dokumacı E., Sandholm, W.H., 2007. Schelling redux: an evolutionary model of residential segregation. Unpublished manuscript, University of Wisconsin.
Eckert, A., 2003. Retail price cycles and presence of small firms. Int. J. Ind. Org. 21, 151–170.
Eigen, M., Schuster, P., 1979. The Hypercycle: A Principle of Natural Self-Organization. Springer, Berlin.
Ellison, G., Fudenberg, D., 2000. Learning purified mixed equilibria. J. Econ. Theory 90, 84–115.
Ely, J.C., Sandholm, W.H., 2005. Evolution in Bayesian games I: theory. Games Econ. Behav. 53, 83–109.
Eshel, I., 1983. Evolutionary and continuous stability. J. Theor. Biol. 103, 99–111.
Eshel, I., Motro, U., 1981. Kin selection and strong evolutionary stability of mutual help. Theor. Popul. Biol. 19, 420–433.
Eshel, I., Sansone, E., 2003. Evolutionary and dynamic stability in continuous population games. J. Math. Biol. 46, 445–459.
Foster, D.P., Young, H.P., 1990. Stochastic evolutionary game dynamics. Theor. Popul. Biol. 38, 219–232. Corrigendum 51 (1997), 77–78.
Franchetti, F., Sandholm, W.H., 2013. An introduction to *Dynamo*: diagrams for evolutionary game dynamics. Biol. Theory 8, 167–178.
Friedman, D., 1991. Evolutionary games in economics. Econometrica 59, 637–666.
Friedman, D., Fung, K.C., 1996. International trade and the internal organization of firms: an evolutionary approach. J. Int. Econ. 41, 113–136.
Friedman, D., Ostrov, D.N., 2008. Conspicuous consumption dynamics. Games Econ. Behav. 64, 121–145.
Friedman, D., Ostrov, D.N., 2010. Gradient dynamics in population games: some basic results. J. Math. Econ. 46, 691–707.
Friedman, D., Ostrov, D.N., 2013. Evolutionary dynamics over continuous action spaces for population games that arise from symmetric two-player games. J. Econ. Theory 148, 743–777.
Fudenberg, D., Kreps, D.M., 1993. Learning mixed equilibria. Games Econ. Behav. 5, 320–367.
Fudenberg, D., Levine, D.K., 1998. The Theory of Learning in Games. MIT Press, Cambridge.
Fudenberg, D., Takahashi, S., 2011. Heterogeneous beliefs and local information in stochastic fictitious play. Games Econ. Behav. 71, 100–120.
Fujishima, S., 2012. Evolutionary implementation with estimation of unknown externalities. Unpublished manuscript, Washington University.
Gaunersdorfer, A., Hofbauer, J., 1995. Fictitious play, Shapley polygons, and the replicator equation. Games Econ. Behav. 11, 279–303.
Gilboa, I., Matsui, A., 1991. Social stability and equilibrium. Econometrica 59, 859–867.
Goeree, J.K., Holt, C.A., Palfrey, T.R., 2008. Quantal response equilibrium. In: Blume, L.E., Durlauf, S.N., (eds.), The New Palgrave Dictionary of Economics, Palgrave Macmillan, Basingstoke, U.K.
Hahn, M., 2012. An evolutionary analysis of Varian's model of sales. Dyn. Games Appl. 2, 71–96.
Harris, C., 1998. On the rate of convergence of continuous-time fictitious play. Games Econ. Behav. 22, 238–259.
Hart, S., 2005. Adaptive heuristics. Econometrica 73, 1401–1430.
Hart, S., Mas-Colell, A., 2001. A general class of adaptive strategies. J. Econ. Theory 98, 26–54.
Hart, S., Mas-Colell, A., 2003. Uncoupled dynamics do not lead to Nash equilibrium. Am. Econ. Rev. 93, 1830–1836.

Heifetz, A., Shannon, C., Spiegel, Y., 2007. What to maximize if you must. J. Econ. Theory 133, 31–57.
Helbing, D., 1992. A mathematical model for behavioral changes by pair interactions. In: Haag, G., Mueller, U., and Troitzsch, K.G., (eds.), Economic Evolution and Demographic Change: Formal Models in Social Sciences, Springer, Berlin, pp. 330–348.
Hines, W. G.S., 1980. Strategic stability in complex populations. J. Appl. Probab. 17, 600–610.
Hines, W. G.S., 1987. Evolutionary stable strategies: a review of basic theory. Theor. Popul. Biol. 31, 195–272.
Hiriart-Urruty, J.-B., Lemaréchal, C., 2001. Fundamentals of Convex Analysis. Springer, Berlin.
Hirsch, M.W., 1988. Systems of differential equations that are competitive or cooperative III: competing species. Nonlinearity 1, 51–71.
Hofbauer, J., 1981. On the occurrence of limit cycles in the Volterra-Lotka equation. Nonlinear Anal. 5, 1003–1007.
Hofbauer, J. 1995a. Imitation dynamics for games. Unpublished manuscript, University of Vienna.
Hofbauer, J. 1995b. Stability for the best response dynamics. Unpublished manuscript, University of Vienna.
Hofbauer, J., 2000. From Nash and Brown to Maynard Smith: equilibria, dynamics, and ESS. Selection 1, 81–88.
Hofbauer, J., 2011. Deterministic evolutionary game dynamics. In: Sigmund, K., (ed.), Evolutionary Game Dynamics, American Mathematical Society, Providence, RI, pp. 61–79.
Hofbauer, J., Oechssler, J., Riedel, F. 2009a. Brown-von Neumann-Nash dynamics: the continuous strategy case. Games Econ. Behav. 65, 406–429.
Hofbauer, J., Sandholm, W.H., 2002. On the global convergence of stochastic fictitious play. Econometrica 70, 2265–2294.
Hofbauer, J., Sandholm, W.H., 2007. Evolution in games with randomly disturbed payoffs. J. Econ. Theory 132, 47–69.
Hofbauer, J., Sandholm, W.H., 2009. Stable games and their dynamics. J. Econ. Theory 144, 1665–1693.
Hofbauer, J., Sandholm, W.H., 2011. Survival of dominated strategies under evolutionary dynamics. Theor. Econ. 6, 341–377.
Hofbauer, J., Schuster, P., Sigmund, K., 1979. A note on evolutionarily stable strategies and game dynamics. J. Theor. Biol. 81, 609–612.
Hofbauer, J., Sigmund, K., 1988. Theory of Evolution and Dynamical Systems. Cambridge University Press, Cambridge.
Hofbauer, J., Sigmund, K., 1998. Evolutionary Games and Population Dynamics. Cambridge University Press, Cambridge.
Hofbauer, J., Sigmund, K., 2003. Evolutionary game dynamics. Bull. Am. Math. Soc. (New Series), 40, 479–519.
Hofbauer, J., Sorin, S., Viossat, Y. 2009b. Time average replicator and best reply dynamics. Math. Oper. Res. 34, 263–269.
Hofbauer, J., Weibull, J.W., 1996. Evolutionary selection against dominated strategies. J. Econ. Theory 71, 558–573.
Hopkins, E. 1999a. Learning, matching, and aggregation. Games Econ. Behav. 26, 79–110.
Hopkins, E. 1999b. A note on best response dynamics. Games Econ. Behav. 29, 138–150.
Hopkins, E., 2002. Two competing models of how people learn in games. Econometrica 70, 2141–2166.
Hopkins, E., Seymour, R.M., 2002. The stability of price dispersion under seller and consumer learning. Int. Econ. Rev. 43, 1157–1190.
Hwang, S.-H., Katsoulakis, M., Rey-Bellet, L., 2013. Deterministic equations for stochastic spatial evolutionary games. Theor. Econ. 8, 829–874.
Jordan, J.S., 1993. Three problems in learning mixed-strategy Nash equilibria. Games Econ. Behav. 5, 368–386.
Kandori, M., Mailath, G.J., Rob, R., 1993. Learning, mutation, and long run equilibria in games. Econometrica 61, 29–56.
Kaniovski, Y.M., Young, H.P., 1995. Learning dynamics in games with stochastic perturbations. Games Econ. Behav. 11, 330–363.

Khan, M.A., Sun, Y., 2002. Non-cooperative games with many players. In: Aumann, R.J., Hart, S., (eds.), Handbook of Game Theory with Economic Applications, volume 3, chapter 46, North Holland, Amsterdam, pp. 1761–1808.
Kimura, M., 1958. On the change of population fitness by natural selection. Heredity 12, 145–167.
Kosfeld, M., Droste, E., Voorneveld, M., 2002. A myopic adjustment process leading to best reply matching. J. Econ. Theory 40, 270–298.
Kreindler, G.E., Young, H.P., 2013. Fast convergence in evolutionary equilibrium selection. Games Econ. Behav. 80, 39–67.
Kuran, T., Sandholm, W.H., 2008. Cultural integration and its discontents. Rev. Econ. Stud. 75, 201–228.
Kurtz, T.G., 1970. Solutions of ordinary differential equations as limits of pure jump Markov processes. J. Appl. Probab. 7, 49–58.
Lach, S., 2002. Existence and persistence of price dispersion: an empirical analysis. Rev. Econ. Stat. 84, 433–444.
Lahkar, R., 2011. The dynamic instability of dispersed price equilibria. J. Econ. Theory 146, 1796–1827.
Lahkar, R., Riedel, F., 2013. The continuous logit dynamic and price dispersion. Unpublished manuscript, Institute for Financial Management and Research and Bielefeld University.
Lahkar, R., Sandholm, W.H., 2008. The projection dynamic and the geometry of population games. Games Econ. Behav. 64, 565–590.
Lahkar, R., Seymour, R.M., 2013. Reinforcement learning in population games. Games Econ. Behav. 8, 10–38.
Laraki, R., Mertikopoulos, P., 2013. Higher order game dynamics. J. Econ. Theory 148, 2666–2695.
Louge, F., Riedel, F., 2012. Evolutionary stability in first price auctions. Dyn. Games Appl. 2, 110–128.
Matsui, A., 1992. Best response dynamics and socially stable strategies. J. Econ. Theory 57, 343–362.
Matsui, A., Matsuyama, K., 1995. An approach to equilibrium selection. J. Econ. Theory 65, 415–434.
Maynard Smith, J., 1982. Evolution and the Theory of Games. Cambridge University Press, Cambridge.
Maynard Smith, J., Price, G.R., 1973. The logic of animal conflict. Nature 246, 15–18.
McKelvey, R.D., Palfrey, T.R., 1995. Quantal response equilibria for normal form games. Games Econ. Behav. 10, 6–38.
Milgrom, P., Roberts, J., 1990. Rationalizability, learning, and equilibrium in games with strategic complementarities. Econometrica 58, 1255–1278.
Minty, G.J., 1967. On the generalization of a direct method of the calculus of variations. Bull. Am. Math. Soc. 73, 315–321.
Mohlin, E., 2012. Evolution of theories of mind. Games Econ. Behav. 75, 299–318.
Monderer, D., Shapley, L.S., 1996. Potential games. Games Econ. Behav. 14, 124–143.
Montgomery, J., 2010. Intergenerational cultural transmission as an evolutionary game. Am. Econ. J.: Microecon. 2, 115–136.
Moran, P. A.P., 1962. The Statistical Processes of Evolutionary Theory. Clarendon Press, Oxford.
Morris, S., Rob, R., Shin, H.S., 1995. p-Dominance and belief potential. Econometrica 63, 145–157.
Nachbar, J.H., 1990. Evolutionary selection dynamics in games: convergence and limit properties. Int. J. Game Theory 19, 59–89.
Nagurney, A., Zhang, D., 1997. Projected dynamical systems in the formulation, stability analysis, and computation of fixed demand traffic network equilibria. Transport. Sci. 31, 147–158.
Nash, J.F., 1950. Non-Cooperative Games. PhD thesis, Princeton. Reprinted In: The Essential John Nash, Kuhn, H.W., Nasar, S. (Eds.), Princeton University Press, Princeton (2002), pp. 51–83.
Nash, J.F., 1951. Non-cooperative games. Ann. Math. 54, 287–295.
Noel, M.D., 2007. Edgeworth price cycles: evidence from the Toronto retail gasoline market. J. Ind. Econ. 55, 69–92.
Norman, T.W.L., 2008. Dynamically stable sets in infinite strategy spaces. Games Econ. Behav. 62, 610–627.
Norman, T.W.L., 2012. Equilibrium selection and the dynamic evolution of preferences. Games Econ. Behav. 74, 311–320.
Nowak, M.A., 2006. Evolutionary Dynamics: Exploring the Equations of Life. Belknap/Harvard, Cambridge.

Oechssler, J., Riedel, F., 2001. Evolutionary dynamics on infinite strategy spaces. Econ. Theory 17, 141–162.
Oechssler, J., Riedel, F., 2002. On the dynamic foundation of evolutionary stability in continuous models. J. Econ. Theory 107, 223–252.
Oyama, D., Sandholm, W.H., Tercieux, O., 2012. Sampling best response dynamics and deterministic equilibrium selection. Theor. Econ., forthcoming.
Oyama, D., Tercieux, O., 2009. Iterated potential and robustness of equilibria. J. Econ. Theory 144, 1726–1769.
Pawlowitsch, C., 2008. Why evolution does not always lead to an optimal signaling system. Games Econ. Behav. 63, 203–226.
Perkins, S., Leslie, D.S., 2013. Stochastic fictitious play with continuous action sets. J. Econ. Theory, forthcoming.
Pohley, H.-J., Thomas, B., 1983. Non-linear ESS models and frequency dependent selection. BioSystems 16, 87–100.
Ramsza, M., Seymour, R.M., 2010. Fictitious play in an evolutionary environment. Games Econ. Behav. 68, 303–324.
Ritzberger, K., 1994. The theory of normal form games from the differentiable viewpoint. Int. J. Game Theory 23, 207–236.
Ritzberger, K., Weibull, J.W., 1995. Evolutionary selection in normal form games. Econometrica 63, 1371–1399.
Robinson, C., 1995. Dynamical Systems: Stability, Symbolic Dynamics, and Chaos. CRC Press, Boca Raton, FL.
Rosenthal, R.W., 1973. A class of games possessing pure strategy Nash equilibria. Int. J. Game Theory 2, 65–67.
Roth, G., Sandholm, W.H., 2013. Stochastic approximations with constant step size and differential inclusions. SIAM J. Control Optim. 51, 525–555.
Roughgarden, T., Tardos, É., 2002. How bad is selfish routing? J. ACM 49, 236–259.
Roughgarden, T., Tardos, É., 2004. Bounding the inefficiency of equilibria in nonatomic congestion games. Games Econ. Behav. 49, 389–403.
Samuelson, L., Zhang, J., 1992. Evolutionary stability in asymmetric games. J. Econ. Theory 57, 363–391.
Sandholm, W.H. 2001a. Almost global convergence to p-dominant equilibrium. Int. J. Game Theory 30, 107–116.
Sandholm, W.H. 2001b. Potential games with continuous player sets. J. Econ. Theory 97, 81–108.
Sandholm, W.H. 2001c. Preference evolution, two-speed dynamics, and rapid social change. Rev. Econ. Dyn. 4, 637–639.
Sandholm, W.H., 2002. Evolutionary implementation and congestion pricing. Rev. Econ. Stud. 69, 667–689.
Sandholm, W.H., 2003. Evolution and equilibrium under inexact information. Games Econ. Behav. 44, 343–378.
Sandholm, W.H. 2005a. Excess payoff dynamics and other well-behaved evolutionary dynamics. J. Econ. Theory 124, 149–170.
Sandholm, W.H. 2005b. Negative externalities and evolutionary implementation. Rev. Econ. Stud. 72, 885–915.
Sandholm, W.H. 2007a. Evolution in Bayesian games II: stability of purified equilibria. J. Econ. Theory 136, 641–667.
Sandholm, W.H. 2007b. Pigouvian pricing and stochastic evolutionary implementation. J. Econ. Theory 132, 367–382.
Sandholm, W.H. 2007c. Simple formulas for stationary distributions and stochastically stable states. Games Econ. Behav. 59, 154–162.
Sandholm, W.H. 2009a. Evolutionary game theory. In: Meyers, R.A., (ed.), Encyclopedia of Complexity and Systems Science, Springer, Heidelberg, pp. 3176–3205.
Sandholm, W.H. 2009b. Large population potential games. J. Econ. Theory 144, 1710–1725.
Sandholm, W.H. 2010a. Local stability under evolutionary game dynamics. Theor. Econ. 5, 27–50.

Sandholm, W.H. 2010b. Pairwise comparison dynamics and evolutionary foundations for Nash equilibrium. Games 1, 3–17.

Sandholm, W.H. 2010c. Population Games and Evolutionary Dynamics. MIT Press, Cambridge.

Sandholm, W.H., 2014a. Local stability of strict equilibrium under evolutionary game dynamics. J. Dyn. Games 1, 485–495

Sandholm, W.H., 2014b. Probabilistic interpretations of integrability for game dynamics. Dyn. Games Appl. 4, 95–106.

Sandholm, W.H., Dokumacı, E., Franchetti, F., 2012. Dynamo: diagrams for evolutionary game dynamics. Software. `http://www.ssc.wisc.edu/~whs/dynamo`.

Sandholm, W.H., Dokumacı, E., Lahkar, R., 2008. The projection dynamic and the replicator dynamic. Games Econ. Behav. 64, 666–683.

Sato, Y., Akiyama, E., Farmer, J.D., 2002. Chaos in learning a simple two-person game. Proc. Nat. Acad. Sci. 99, 4748–4751.

Schlag, K.H., 1998. Why imitate, and if so, how? A boundedly rational approach to multi-armed bandits. J. Econ. Theory 78, 130–156.

Schmeidler, D., 1973. Equilibrium points of non-atomic games. J. Sta. Phys. 7, 295–300.

Schuster, P., Sigmund, K., 1983. Replicator dynamics. J. Theor. Biol. 100, 533–538.

Sethi, R., 1999. Evolutionary stability and media of exchange. J. Econ. Behav. Org. 40, 233–254.

Sethi, R., Somanathan, E., 1996. The evolution of social norms in common property resource use. Am. Econ. Rev. 86, 766–788.

Shahshahani, S., 1979. A new mathematical framework for the study of linkage and selection volume 211 of Memoirs of the American Mathematical Society.

Shapley, L.S., 1964. Some topics in two person games. In: Dresher, M., Shapley, L.S., Tucker, A.W., editors, Advances in Game Theory, volume 52 of Annals of Mathematics Studies, Princeton University Press, Princeton, pp. 1–28.

Skyrms, B., 1990. The Dynamics of Rational Deliberation. Harvard University Press, Cambridge.

Skyrms, B., 1992. Chaos in game dynamics. J. Logic Lang. Inform. 1, 111–130.

Smirnov, G.V., 2002. Introduction to the Theory of Differential Inclusions. American Mathematical Society, Providence, RI.

Smith, H.L., 1995. Monotone Dynamical Systems: An Introduction to the Theory of Competitive and Cooperative Systems. American Mathematical Society, Providence, RI.

Smith, M.J., 1984. The stability of a dynamic model of traffic assignment—an application of a method of Lyapunov. Trans. Sci. 18, 245–252.

Sparrow, C., van Strien, S., Harris, C., 2008. Fictitious play in 3×3 games: the transition between periodic and chaotic behaviour. Games Econ. Behav. 63, 259–291.

Swinkels, J.M., 1993. Adjustment dynamics and rational play in games. Games Econ. Behav. 5, 455–484.

Taylor, P.D., 1979. Evolutionarily stable strategies with two types of players. J. Appl. Probab. 16, 76–83.

Taylor, P.D., Jonker, L., 1978. Evolutionarily stable strategies and game dynamics. Math. Biosci. 40, 145–156.

Tercieux, O., 2006. p-best response set. J. Econ. Theory 131, 45–70.

Thomas, B., 1984. Evolutionary stability: states and strategies. Theor. Popul. Biol. 26, 49–67.

Thomas, B., 1985. On evolutionarily stable sets. J. Math. Biol. 22, 105–115.

Topkis, D., 1979. Equilibrium points in nonzero-sum n-person submodular games. SIAM J. Control Optim. 17, 773–787.

Tsakas, E., Voorneveld, M., 2009. The target projection dynamic. Games Econ. Behav. 67, 708–719.

van Strien, S., Sparrow, C., 2011. Fictitious play in 3×3 games: chaos and dithering behavior. Games Econ. Behav. 73, 262–286.

van Veelen, M., 2012. Robustness against indirect invasions. Games Econ. Behav. 74, 382–393.

van Veelen, M., Spreij, P., 2009. Evolution in games with a continuous action space. Econ. Theory 39, 355–376.

Varian, H., 1980. A model of sales. Am. Econ. Rev. 70, 651–659.

Viossat, Y., 2011. Monotonic dynamics and dominated strategies. Unpublished manuscript, Université Paris-Dauphine.

Vives, X., 1990. Nash equilibrium with strategic complementarities. J. Math. Econ. 19, 305–321.
Weibull, J.W., 1995. Evolutionary Game Theory. MIT Press, Cambridge.
Weibull, J.W., 1996. The mass action interpretation. Excerpt from The work of John Nash in game theory: Nobel Seminar, December 8, 1994. J. Econ. Theory 69, 165–171.
Young, H.P., 1993. The evolution of conventions. Econometrica 61, 57–84.
Young, H.P., 1998. Individual Strategy and Social Structure. Princeton University Press, Princeton.
Young, H.P., 2004. Strategic Learning and Its Limits. Oxford University Press, Oxford.
Zeeman, E.C., 1980. Population dynamics from game theory. In: Nitecki, Z., Robinson, C., (eds.), Global Theory of Dynamical Systems (Evanston, 1979), number 819 in Lecture Notes in Mathematics, Berlin. Springer, pp. 472–497.
Zeeman, E.C., 1981. Dynamics of the evolution of animal conflicts. J. Theor. Biol. 89, 249–270.
Zusai, D., 2011. The tempered best response dynamic. Unpublished manuscript, Temple University.

CHAPTER 14

The Complexity of Computing Equilibria

Christos Papadimitriou*
*University of California, Berkeley, USA

Contents

Abstract

In one of the most influential existence theorems in mathematics, John F. Nash proved in 1950 that any normal form game has an equilibrium. More than five decades later, it was shown that the computational task of finding such an equilibrium is intractable, that is, unlikely to be carried out within any feasible time limits for large enough games. This chapter develops the necessary background and formalism from the theory of algorithms and complexity developed in computer science, in order to understand this result, its context, its proof, and its implications.

Keywords: Normal form games, Nash equilibrium, Algorithms, Computational complexity, Polynomial-time algorithms, NP-complete problems, PPAD-complete problems

JEL Codes: C72

14.1. THE TASK

It takes a little calculation to find a mixed Nash equilibrium even in a small two-person game. Take for example the 3×3 game shown below.

	a	b	c
1	7, 3	6, 4	9, 2
2	5, 2	7, 3	7, 2
3	6, 2	12, 2	6, 4

The only Nash equilibrium has supports $\{1, 3\}$ and $\{b, c\}$. Once the right supports are in hand, all one has to do is solve a linear system and check a few inequalities. But

Handbook of game theory
http://dx.doi.org/10.1016/B978-0-12-420056-2.00014-8

how does one guess the right supports? In this example, of course, it is feasible to try all supports, but what if the game was 10×10? 100×100?

The thought has occurred to almost all of us: *Is there a technique*[1] *that scales well with the number of strategies?* This question is our subject.

14.2. PROBLEMS AND ALGORITHMS

If the game is zero-sum, the question has a straightforward answer: it is well known that the min-max strategies (which coincide with the Nash equilibrium in this case) can be found by linear programming. And linear programming is one of the happy computational problems for which we know "a good algorithm."

Ever since we learned long division in elementary school, we have known the concept of an algorithm:

> *An algorithm for a problem is a clear and unambiguous sequence of elementary steps, expressed in an appropriately rigorous language, which, when applied on any input of the problem, it always stops eventually with the correct solution.*[2]

But what is a "good" algorithm? In other words, when should we consider a computational problem solved satisfactorily? Over the past half century, computer scientists have come up with a comprehensive theory on the matter (see Garey and Johnson, 1989; Papadimitriou, 1994a; Kleinberg and Tardos, 2005; Dasgupta et al., 2008, for standard textbooks and expositions, and Roughgarden, 2010, for one addressed to economists). To understand this theory, let us look at a few examples of problems:

Shortest path

Given an undirected graph[3] G with n nodes, and an integer $k < n$, find a path from node A to node B that has at most k edges—or report that no such path exists.

[1] There, of course, the Lemke-Howson algorithm, see Lemke and Howson (1964) and Chapter 45 of this Handbook) for two-player games, a clever method for jumping from support to support until a Nash equilibrium is found. We'll come to that—see Savani and von Stengel (2004) and the chapter's last theorem.

[2] The intuitive concept of algorithm is ancient, known to Euclid and Archimedes in the 3rd century BC, and popularized by the Arab scholar Al Khwarizmi in the 9th century AD. In the mid 1930s, several mathematical formalisms of this concept were developed independently. The best known is the *Turing machine*, but there are others, Church's λ calculus, Kleene's recursive functions, Post's correspondence systems, Markov's algorithms, etc. All these formalisms were eventually shown to be equivalent in power, and therefore they all express the same real-world concept. Thirty years later, these concepts were refined to capture what is a "good algorithm," the best known refinement being the polynomial-time Turing machine. Again, many different mathematical formalisms of the concept of a "good algorithm" end up being mathematically equivalent, showcasing the robustness of the notion and of the class **P** introduced in Section 4.

[3] An undirected graph is just a finite set of points called *nodes* and a set of lines each joining two nodes called *edges*.

Linear programing

Given an $m \times n$ integer matrix A and a integer m-vector b, find a rational n-vector x such that $Ax \geq b$—or report that no such vector exists.

Gale-Shapley matching

Given a set of n hospitals and n residents, and for each hospital its preference ordering of the residents, and similarly for every resident his/her preference ordering of the hospitals, find a *stable matching*, that is, a pairing of each resident with one hospital such that there is no resident-hospital pair that is *unstable*, in that the resident and the hospital are not currently matched together, but they both prefer each other to their current match.

Independent set of edges

Given a graph with n nodes and an integer $k \leq \frac{n}{2}$, find k edges of the graph that are *independent*, in that no two of them share a node—or report that there is no such set of edges.

Two-coloring

Given a graph with n nodes, color each node white or black so that no edge has endpoints with the same color—or report that no such coloring is possible.

These are all *computational problems:* In each, we are given some input data (a matrix, a graph, etc.), and are asked to return a solution meeting certain specification *vis à vis* the input (or report that no solution meeting the specification exists). They are precisely the kind of problems one would hope to solve on a computer.

14.3. GOOD ALGORITHMS

All these problems have *good algorithms.* What does this mean, exactly?

Let us consider the GALE-SHAPLEY MATCHING[4] problem. It requires us that we find, among all possible assignments, one that satisfies a certain criterion. The total number of possible assignments of n residents to n hospitals is of course an astronomical $n!$. For $n = 30$, an extremely modest number in comparison to what is done in the world, this is 2.65×10^{32}. It is impossible to inspect them exhaustively: there are not enough computers in the world and not enough picoseconds in our lives for this. But the Gale-Shapley algorithm finds the required stable matching by a clever iterative construction, instead of by rote enumeration. All told, the total number of steps the algorithm takes is $O(n^2)$—by which we mean at most some constant times n^2.

The SHORTEST PATH problem can be solved in time $O(m)$, where m is the number of edges of the graph, by a simple algorithm called *breadth-first search:* Mark A as "0"

[4] Since in this chapter computational problems are mathematical objects, we shall treat them with a measure of formality; for example, we shall always use SMALL CAPS when mentioning a computational problem.

(meaning that its distance from A is zero), then mark all of its neighbors "1," then mark all of their neighbors "2" (unless a neighbor is already marked, in which case we do nothing), and so on. Then see if B has been marked k or less; if it is, tracing back the labels down to 0 will identify the path.

The TWO-COLORING problem is also easy to solve, and essentially by the same algorithm: Start with a node A and color it white. Color all its neighbors black. Color all *their* neighbors white. And so on. If at some point you must change the color of an already colored node, then you can be sure that there is no way to two-color this graph.

The story with LINEAR PROGRAMING is a little more complicated, but the gist of it is that we have algorithms that solve it in time L^3, where L is the length of the representation of the matrix A.

The INDEPENDENT SET OF EDGES problem (better known as NONBIPARTITE MATCHING) can be solved in $O(n^3)$ steps by a very elegant and highly nontrivial algorithm discovered by Jack Edmonds in 1965. In fact, it was on the occasion of this discovery that Edmonds used for the first time the term "good algorithm," in the same sense we use it today:

> *An algorithm is good if, on an input of size* n*, it always halts with the correct solution after a number of elementary steps bounded by some polynomial in* n.

To arrive at a formal definition, let us start with the concept of a computational problem (of which we have seen several examples already): in a computational problem one is presented with an *input*, a mathematical object such as a graph or a matrix of integers encoded in some finite alphabet, and we are asked to find a *solution* to this input: another mathematical object likewise encoded. We are interested in *algorithms* for solving computational problems. There are many formal models of algorithms, such as the Turing machine (see for example Chapter 2 of Papadimitriou, 1994a, for a treatment); but introducing such a model now would be a needless digression. Instead, we shall evoke the intuitive understanding of the concept shared by all. So, for each algorithm and each input, one can speak of the "time" it takes for the algorithm to produce the solution, measured in terms of elementary steps. What is an elementary step of an algorithm? This is a fair question, and it opens many possibilities for an interesting discussion. But suffice it to say that, at the level of coarseness of interest here—when we are oblivious of multiplicative constants, soon even of distinctions between polynomials—all reasonable definitions are equivalent. Thus, the complexity of an algorithm is denoted as a function $f(n)$ which upper bounds the *maximum* number of steps the algorithm takes in order to solve an input of size n.

But what is the "size" of an input? The input to a problem is presented by its encoding as a string over a finite alphabet, and the size of the input is precisely the length of this encoding. But it is often much more convenient to use instead surrogate quantities, such as the number of nodes in the graph or the dimensions of the matrix, which are closely related to that length.

The four algorithms we have presented are deemed satisfactory because their time requirements are bounded by some polynomial function in the size of the input: n^2, n^3, perhaps n^5. Polynomial functions do grow, of course, but their growth is relatively benign. In contract, exponential functions, such as 2^n and $n!$, quickly overtake any polynomial function and soon enough reach absurd magnitudes.

Importantly, all the problems above seem to call for a search among a population of solutions that is exponential in size. Thus, a naïve enumeration algorithm would be infeasible. Luckily, for each of these problems we have a clever algorithm that avoids such exhaustive search, and quickly zeroes in the desired solution. "Quickly," again, means polynomially fast. This is the concept of "good algorithm" universally accepted among computer scientists.

Definition 14.1. *We say that an algorithm is* polynomial-time *if for any input of size n it produces a solution within time bounded from above by $c \cdot n^d$ for some positive integer constants c and d.*

There are legitimate objections to identifying this notion with that of a "good algorithm." Here are two:

- Why require that an algorithm run in polynomial time on all inputs? Wouldn't a method that works fast for typical inputs, or most inputs, or in expectation, be valuable too?
- Wouldn't an n^{80} algorithm be unrealistic, even though it is polynomial? And wouldn't a $2^{n/1000}$ algorithm, or an $n^{\log n}$ algorithm, be superior for all practical purposes, even though neither of them is, strictly speaking, polynomial?

The first objection marks a well-known ideological divide between economists and computer scientists. Computer scientists are obsessed with worst-case scenaria because, unlike economists, they don't believe in priors—and there is a reason for this, besides research culture. For example, which prior distribution of the inputs of the SHORTEST PATH problem is it reasonable to assume? Random graphs, in which each edge is present with probability $\frac{1}{2}$, independently? Not a good prior, since such graphs are unrealistically dense and tend to have paths of length one or two between any two nodes. Besides, an algorithm for SHORTEST PATH will be typically used as a subroutine in a larger algorithm, and thus its inputs will be the *output* of another part of that algorithm—and it is unnatural to assume that the outputs of a complex process are distributed according to an organic prior. Of course, if for some problem we have a reasonable and realistic prior distribution of the inputs, then it is legitimate and valuable to analyze its average case performance—and in this field we often do.

The second objection is the kind one can raise to all mathematical definitions attempting to capture a real-world concept. Take *smooth functions,* for example. Most mathematicians identify this concept with C_∞ (the class of functions that can be differentiated again and again). But there are functions outside C_∞ that would seem extremely smooth to a casual observer, and there are functions in C_∞ that feel very bumpy.

The value of a mathematical definition lies in the extent to which it enables the development of an elegant and comprehensive theory that is faithful to the definition's worldly motivation. In the 1960s, when this definition of a good algorithm was being tried out, nobody could predict the elegant and unreasonably comprehensive theory which would result.

14.4. P AND NP

The problems we have seen so far represent some of the most glorious moments in the quest for good algorithms. Meet now a few of the greatest embarrassments of the quest:

Integer linear programing

Given an $n \times m$ integer matrix A and an integer n-vector b, find an integer n-vector x such that $Ax \geq b$—or report that no such vector exists.

Longest path

Given an undirected graph G with n nodes, and an integer $k < n$, find a path from node A to node B, which repeats no node more than once, and has at least k edges—or report that no such path exists.

Gale-Shapley matching with couples

We are given a set of n hospitals and n residents, and for each hospital its preference ordering of the residents, and similarly for every resident his/her preference ordering of the hospitals. The hospitals are partitioned into sets called *cities*, and certain residents are known to be married to each other—that is, we are also given a set of disjoint sets of residents of cardinality two. We are asked to find a *stable matching*, that is, a pairing of each resident with one hospital such that there is no resident-hospital pair that is *unstable*, in that the resident and the hospital are not currently matched together, but they both prefer each other to their current match; and furthermore, all married couples of residents are assigned to hospitals in the same city.

Independent set of nodes

Given a graph with n nodes and an integer $k \leq n$, find k nodes of the graph that are *independent*, in that no two of them are adjacent—or report that there is no such set of nodes.

Three-coloring

Given a graph with n nodes, color each node white, black, and red, so that no edge has endpoints with the same color—or report that no such coloring is possible.

Each of these problems is a "twist" on a problem that we saw in the previous section. The twist may seem innocuous to the untrained eye. But this similarity is deceiving. Despite decades of herculean efforts by computer scientists and mathematicians, there is no known good algorithm for any of these problems. For example, LONGEST PATH turns out to be closely related the notorious TRAVELING SALESMAN PROBLEM, which has evaded good algorithms for a century. These repeated failures have convinced most researchers in the area that none of these problems has a good algorithm. This mathematical conjecture is known as

$$\text{“}\mathbf{P} \neq \mathbf{NP}\text{.”}$$

It is one of the deepest and most important open problems in Mathematics today.

What is **P** and **NP**? To answer, let us go back to the concept of a computational problem: A computational problem consists of a set of *possible inputs*—descriptions of a mathematical object such as a matrix or a graph on which the algorithm works. For each input x, we are asked to find a solution y—another mathematical object, such as a set of edges or a vector of integer—or rational—numbers. The point is that *the solution should stand in a particular mathematical relationship to the input:* The vector should satisfy all inequalities, the set of edges should be a path of length at most k, the assignment of residents to hospitals should be stable, etc. If no such solution exists, we are asked to report so.

In other words, a computational problem is essentially a relationship between inputs and solutions—a specification based on x that restricts the desired solution y. Here is a crucial observation: In all examples in this and the previous section, *whether alleged solution y satisfies the specification* viz. *input x can be checked easily.* There is a good—polynomial-time—algorithm for telling whether a sequence of nodes is a path of length at most k on graph G, where G and k are given. Similarly for "at least k." There is a good algorithm for testing whether an assignment of residents to hospitals is stable—just try all n^2 hospital-resident pairs. Ditto for checking in addition whether an assignment accommodates all couples. One can check easily whether a three-coloring (or two-coloring) of a given graph is legitimate. And so on. In addition to this, the solution y in all these problems is small compared to the input. This is the essence of the class **NP**:

Definition 14.2. *We say that a computational problem is in* **NP** *if for each input either there is no solution for this input, or each solution is of size bounded from above by a polynomial in the size of the input; and furthermore there is a polynomial-time algorithm which, given an input and an alleged solution for this input, determines if, indeed, it is one.*

P *on the other hand is the class of all computational problems in* **NP** *which can be solved by a polynomial-time algorithm.*

Thus, GALE-SHAPLEY MATCHING, SHORTEST PATH, INDEPENDENT SET OF EDGES, and TWO-COLORING are all in **P**—and therefore also in **NP**—while GALE-SHAPLEY MATCHING WITH COUPLES, LONGEST PATH, INDEPENDENT SET OF NODES, and THREE-COLORING are in **NP** (and not believed to be in **P**).

P is our goal, our Shangri-La, the state of Nature with which we are happy. We hope and pray that the problems life places on our path are in there. In contrast, **NP** is a seemingly much broader class of problems. In fact, **NP** is arguably the natural limit of our ambition: If you have no way of telling if an alleged solution satisfies the specifications, how can you hope to find quickly a solution that does?[5]

Is **P** = **NP**? If this were the case, it would be a remarkable fact, both of mathematics and of life. It would mean that exponential exhaustive search is never necessary. To solve any computational problem, to discover a solution satisfying any given set of specs, one would only need to be able to verify the specs. For example, it would be possible to discover any mathematical proof with effort that scales well with the proof's length!

Most computer scientists and mathematicians working in the field are convinced that this is not the case. That **P** ≠ **NP**. But, more than half a century after the articulation of this conjecture by Jack Edmonds, there is little progress towards a proof, either way.

But there has been progress on a different front: We have identified many problems which have no good algorithms *under the assumption that* **P** ≠ **NP**.

Optimization

Notice by the way that in our examples we have steered clear of optimization problems. For example, we have not stated Integer programing in its familiar form: "minimize cx subject to $Ax \geq b$," but in the form "find a solution to $Ax \geq b$," in which the objective cx has been absorbed in the inequalities. The reason is that the optimization version does not qualify as a member of **NP**: We know of no good algorithm for verifying whether a solution to an integer linear program is optimum. In complexity theory, optimization problems are treated through their surrogate "search problems," with the added specification that the objective be better than some bound (as in our examples above). If there is a good algorithm for the surrogate, the optimization problem can be solved in polynomial time too.[6] And if the surrogate is difficult, what hope is there for the optimization problem?

14.5. REDUCTIONS AND NP-COMPLETE PROBLEMS

We have seen two groups of problems.

1. Problems in P

These are Linear programing, Shortest path, Gale-Shapley matching, Two-coloring, Independent set of edges.

[5] The problem of determining if a position in a board game, such as the $n \times n$ Go, is winning for White is an example of a problem that is very natural and of interest in Game Theory, but seems to lie outside **NP**. We come back to such problems later.

[6] This entails the process of binary search—guessing the optimum objective value bit by bit—to eventually zero in the optimum solution.

2. Problems in NP believed *not* to be in P

INTEGER LINEAR PROGRAMING, LONGEST PATH, GALE-SHAPLEY MATCHING WITH COUPLES, THREE-COLORING, INDEPENDENT SET OF NODES.

We would love to know for sure that the problems in the second group are not in **P**—so we can stop wasting our energies trying to develop good algorithms for them. But we already know that establishing this would answer a famous question in Math, which has been open for fifty years. What more can there be to say?

During the past forty years, computer scientists have managed to say quite a bit more. They have established that, among all problems in **NP**—which in a sense means "among all problems"—the problems in Group 2 above are *the least likely to be in* **P**. They are **NP**-*complete*. To understand what it means, we need to look at *reductions*.

Reductions are ubiquitous in math, science and life. Reducing problem A to problem B means to establish that, once problem B is solved, then problem A is solved as well—and, hopefully, problem B has already been solved. Game theory is replete with reductions: The proof of Nash's theorem is essentially a reduction from the problem of establishing the existence of mixed Nash equilibria to establishing the existence of Brouwer fixpoints (which, thankfully, had been solved decades before Nash was born). Finding the minmax of a zero-sum two-person games can be reduced to solving a linear program.

So, reduction from A to B is useful for showing that A can be done. There is a more subtle (some say perverse) use of reduction: To establish that *problem B cannot be solved*—this is the case when we already know that problem A cannot. The proof of the Gibbard-Satterthwaite theorem is an example: it establishes that finding a choice rule that selects one order among a given list of preference orders (problem A) can be reduced to the problem of finding a rule for selecting a single winner from a list of preference orders (problem B). Since Arrow's impossibility theorem states that no rule for choosing an order can have good properties (problem A is impossible), it follows that there is no good rule for choosing a winner (problem B is impossible). If problem A cannot be solved, then neither can problem B. This is the type of reduction most useful for establishing the difficulty of problems.

Definition 14.3. *A* reduction *from a computational problem A to a computational problem B consists of two functions f and g, both computable by a polynomial-time algorithm. Function f transforms any input x of problem A to an input $f(x)$ of problem B. Function g takes an solution z of problem B and transforms it into an solution of problem B. Functions f and g must satisfy this: Input x has a solution in problem A if and only if input $f(x)$ has a solution in problem B. And if z is a legitimate solution for input $f(x)$ in problem B, then $g(z)$ must be a legitimate solution for input x in problem A.*

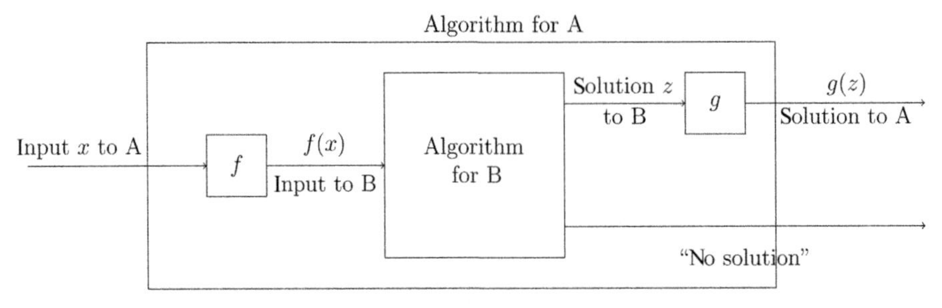

Figure 14.1

Pictorially, we have the situation shown in Figure 14.1. Notice how the algorithm for A together with the algorithms for computing f and g constitute a perfectly legitimate algorithm for problem A.

If there is a reduction from A to B we say that A *reduces* (or *can be reduced*) to B.

We are now ready to define what **NP**-complete means:

Definition 14.4. *A problem is* **NP**-*complete if (a) it is in* **NP** *and (b) all problems in* **NP** *can be reduced to it.*

Some remarks are in order.

1. Reductions *compose*, in that if problem A reduces to problem B, and B reduces to C, then problem A reduces directly to C. This follows from the well-known fact that a polynomial of a polynomial is a polynomial; for example, if $f(x) = x^2$ and $g(x) = x^3 + x$ then $g(f(x)) = x^6 + x^2$.
2. Note that the same fact implies that, if A reduces to B, and B is in **P**, then A must be in **P** as well.
3. Ergo, if an **NP**-complete problem A is in **P**, then all of the problems in **NP** are in **P** (because they all reduce to A, by definition), and **P** = **NP**. Therefore, **NP**-complete problems are indeed the hardest problems of **NP**, its mysterious aristocracy, the ones that are least likely to be in **P**. If one of them is, then all of them are, and **P** = **NP**.
4. At first sight, showing that a problem A is **NP**-complete seems a daunting task: we must somehow establish that, of the infinitely many problems in **NP**, every single one of them reduces to A. In fact, this difficulty has to be confronted only once (Stephen Cook did it for us in 1971, see Cook, 1971). Once one or more problems have been proved **NP**-complete, we can take advantage of the transitivity property of reductions explained in (1) above, and prove that A is **NP**-complete by reducing another problem B, already known to be **NP**-complete, to A.
5. The situation can be described pictorially, as shown below (computational complexity increases going upwards) (Figure 14.2).

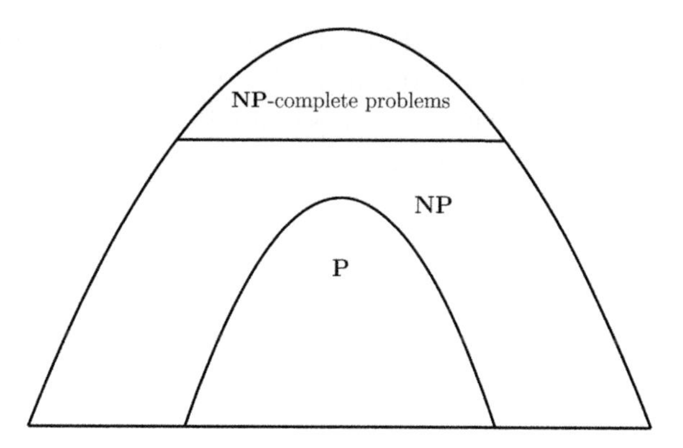

Figure 14.2

One has to bear in mind, however, that this picture is a conjecture: For all we know, all this structure could collapse down to just one bubble: **P**. The third possibility, the one in which **P** $\neq$ **NP**, and all of **NP** is neatly partitioned into problems in **P** or **NP**-complete problems is known to be impossible (see Ladner, 1975). That is, if **P** $\neq$ **NP**, there will be several other problems lying in between. This is worth remembering when, in the next section, we confront the complexity of the Nash equilibrium problem.

6. If a problem A reduces to B, the intuitive understanding is that B is "at least as hard as" A (or that A "is no harder than" B). If in addition B reduces back to A, then we say that the two problems are *polynomially equivalent*. It is easy to see that all **NP**-complete problems are polynomially equivalent, and so are all problems in **P**. That is, the class **P** and the **NP**-complete problems in the figure above are the lowest and highest equivalence classes of problems in **NP**. If **P** $=$ **NP**, of course, then these two classes coincide. But we know that, if **P** $\neq$ **NP**, then there are infinitely many other equivalence classes in between Ladner (1975). Again, this is worth remembering in connection to the Nash equilibrium problem.
7. **NP**-completeness elevates a problem into the status of paradigm. For example, we can now define the important class **NP** as "the set of all problems that reduce to INDEPENDENT SET OF NODES." This again presages our analysis of the complexity of NASH, which comes next.

We shall see examples of reductions soon enough.

14.6. THE COMPLEXITY OF NASH EQUILIBRIUM

Finally, we come back to where we started: *Is there a good algorithm for finding the Nash equilibrium of a given game?* In this section we shall show that the answer is "no" (appropriately qualified, of course): computing a mixed Nash equilibrium is hard in the worst case, even for bimatrix games. "Hard" here means something analogous to "**NP**-complete" but, for technical reasons explained below, the precise notion is a bit different. The importance of this complexity result goes beyond computation of Nash equilibria; it has fundamental implications for using and interpreting the Nash equilibrium concept.

We outline below the background and proof of this theorem. The more technical parts of the argument, which could be skipped at a first reading, are in smaller font.

The Problem Nash

To start, we must first formalize the Nash equilibrium question as a computational problem. It turns out that the best way to do this is in terms of a notion of approximation[7]: recall that a mixed strategy profile is a Nash equilibrium if all strategies in its support are best responses. Fix a small number $\epsilon > 0$, and call a mixed strategy profile with rational probabilities an ϵ-Nash equilibrium if all strategies in its support are best responses *give or take* ϵ. That is, there may be another strategy of the same player that has higher expected payoff, *but not by more than* ϵ.[8]

Nash

Given a game in tabular form and a positive rational number ϵ, find an ϵ-Nash equilibrium.

So, this is the problem whose complexity we shall study. The question now in front of us is this: How does this problem stand in relation to **P**, **NP**, and the **NP**-complete problems?

It is clearly in the class **NP** (given an alleged Nash equilibrium, we just check that each strategy in each support is within ϵ of the best response). But, within **NP**, which of the two is the case: Is NASH in **P** (like TWO-COLORING and LINEAR PROGRAMING) or is it **NP**-complete (like THREE-COLORING and INTEGER LINEAR PROGRAMING)?

[7] In games with more than two players, Nash equilibria can contain irrational numbers, roots of polynomials of very high degree, and describing such numbers exactly is problematic—in fact, there are reasons to believe that computing exactly a Nash equilibrium may not be in **NP** at all (see Etessami and Yannakakis, 2007, and the discussion in the last section)! For all these reasons—and also because in real life there are rarely grounds for objecting to a solution that solves our problem approximately, simply because the problem we are solving by computer is almost always an approximation of the true problem—we deal with an approximate notion of Nash equilibrium

[8] This is a useful definition, but we should note that an ϵ-Nash equilibrium need not be in any way close to a true Nash equilibrium—in the same way that a number x that makes a polynomial $p(x)$ very close to zero may not be near a root of the polynomial.

Unlike the 10 other computational problems that we have seen so far, which could be readily classified as either in **P** or **NP**-complete, problem NASH seems to be in neither class. It shares with the **NP**-complete problems the characteristic that it is a problem seeking one among an exponential number of solutions, for which, despite efforts over decades, no polynomial-time algorithm has been devised. However, it has another property which sets it apart from all these problems: *Every input has a solution.* Whereas it is perfectly possible to be given a graph that is not three-colorable (consider four nodes connected to one another in all six possible ways), or a system of equations that has no integer solution (say, $x + y = 3, x - y = 2$), Nash's theorem tells us that *every finite game has a Nash equilibrium*, and therefore every input of NASH has a solution (since a Nash equilibrium exists, it is easy to see that, by choosing a rational point close enough to it, a solution to NASH also exists. This puts NASH in a very special position: It is very unlikely that it is **NP**-complete. To see why, imagine a reduction from a known **NP**-complete problem A, say THREE-COLORING to NASH. What would this reduction be like? If the input x in this reduction were a graph which is not three-colorable, what would $f(x)$ be? It cannot be a game that has no Nash equilibrium, since there are no such games. And how about $g(f(x))$?

Total problems and Best Nash

Call a computational problem *total* if every input has a solution. NASH is therefore total. There is an agreement among students of complexity theory that total problems are almost as unlikely to be **NP**-complete as problems in **P** are. Totality is strong evidence against **NP**-completeness. And conversely, the only thing that separates NASH from **NP-completeness** is totality. We show next that, if this totality is removed somehow by force, then **NP**-completeness reigns immediately.

The easiest way to do this is define the following problem:

Best Nash

Given a two-person game in tabular form and a rational number P, find a Nash equilibrium in which both the payoffs of players are at least P or determine that no such equilibrium exists.

Theorem 14.1. BEST NASH *is* **NP**-*complete.*

The problem is certainly in **NP** (if we are given a pair of mixed strategies, it is easy to verify whether they are a Nash equilibrium, and, if so, whether their expected payoffs add up to P or more.

We shall next show a reduction from INDEPENDENT SET OF NODES to BEST NASH. Given a graph $G = (V, E)$ and an integer k, we must show how to construct (this is function f in the definition of a reduction) in polynomial time a two-player game and a number P such that G has a set of k independent nodes iff the game has a Nash equilibrium with su of payoffs at least P, and in fact from any such Nash equilibrium we can recover in polynomial time (this is function g) k independent nodes in G.

The game we construct is symmetric. Each player has $2|V|$ strategies, two for each node of G. If v is a node of G, the corresponding strategies are denoted v_C and v_D (C and D for "collaborate" and "defect").

The payoffs are as follows:

- Suppose a player plays v_D (he "defects on v"). Then, if the other player collaborates on v, the defecting player gets a huge payoff $2k^2$ and the collaborating player gets $-2k^2$. If the other player responds to v_D in any other way (also defects, or collaborates on another node), both payoffs are zero.
- So, suppose that both players collaborate: They play u_C and v_C, respectively, for $u, v \in V$. Then they both get:
  - $2k + 1$ if $u = v$;
  - $2k$ if u and v are different vertices not connected by an edge;
  - 0 if u and v are connected by an edge,

  The target sum of payoffs is $P = 2k + \frac{1}{k}$.

We must show that the target is achievable by some Nash equilibrium if G gas k independent nodes. The "if" direction is easy: If G has k independent nodes, then it is a Nash equilibrium for the two players to randomize uniformly between collaborating on these k nodes. The expected payoff for both is exactly P (the average between $k - 1$ $k's$ and one $2k + 1$).

The converse is established by a sequence of observations, each of which is not too difficult to check:

- At equilibrium, no player can collaborate on any node with probability that is too high (higher than $\frac{1}{k}(1 + \frac{1}{2k})$, as it turns out). This is because such exposure would incentivize the other player to defect on the same node.
- On any good Nash equilibrium (by this we mean, one with payoff at least P for both players) no player defects more than $\frac{1}{2k}$ of the time. This is because defections are costly, given the above point.
- Therefore, on a good equilibrium each player collaborates with nonzero probability over at least k nodes; and for each of these nodes, the other player collaborates with probability at least $\frac{1}{k}$.
- It follows that the only good Nash equilibria are of the form we already saw: both players choose a set of k independent nodes, and collaborate over them, uniformly.

To finish the **NP**-completeness proof, notice that, by the same reasoning, it is very easy to transform any good Nash equilibrium into an independent set of k nodes in G, by reading off the strategies of the two players. Thus we obtain the function g, the reduction's last component.

The Class PPAD

Coming back to NASH, the consensus is that, since it is a total problem in **NP**, it cannot be **NP**-complete. On the other hand, it would come as a big surprise to many seasoned researchers if there were a polynomial-time algorithm for NASH. This makes NASH a perfect candidate for belonging to one of the infinitely many classes of polynomially equivalent problems which we know must lie between **P** and the **NP**-complete problems (recall the last figure, assuming of course that $\mathbf{P} \neq \mathbf{NP}$). The question is, which one?

To answer, we must look at the origin of our troubles: the proof of Nash's theorem. By following step by step the argument establishing the existence of a Nash equilibrium, starting from any game, and coming up with good algorithms for every step of the proof, we might even be able to come up with a good algorithm for finding this Nash equilibrium. Alas, as the reader may suspect, it turns out that Nash's proof has a crucial step that is not easy to make constructive: the place where it evokes Brouwer's theorem:

Brouwer's Theorem

Any continuous function from a convex compact set to itself has a fixpoint.

One can associate with this fundamental existence theorem a natural total computational problem:

BROUWER. Given a function $\phi : [0,1]^d \mapsto [0,1]^d$, and two positive rational numbers ϵ and λ, either find a rational point x in the region such that $|\phi(x) - x| \leq \epsilon$, or two rational points x, x' such that $|\phi(x) - \phi(x')| > \lambda \cdot |x - x'|$.

Notice that in this definition, instead of postulating that the function ϕ is continuous (which is necessary for applying Brouwer's theorem but we would not know how to check) we accept as an answer a violation of continuity (if the function is discontinuous, such a violation must exist no matter how large λ is). If there is no discontinuity, an approximate fixpoint must exist.

There is a key technical issue we must address now regarding the problem BROUWER. In all computational problems we had seen so far, the input is a simple combinatorial object (a graph, a list of permutations, a matrix of integers, etc.). But in this one, the input is a real-valued function ϕ. How is this object presented to us? The answer is, *through a program that computes it.* The program has a very simple form: It consists of a finite sequence of statements of the form

$$x = y \circ z,$$

where:

- x is a real variable in $[0, 1]$ which is associated with, and defined by, this statement of the program;
- each of y and z is either a rational constant, or a variable defined in a previous statement, or else it is one of the inputs of the function f. In all cases, all variables and constants take values in $[0, 1]$;
- the operation $\circ$ is one of $+, -, *$, and $>$. The first three are the familiar arithmetic operations, with the twist that the result is guaranteed to be in $[0, 1]$. If the result of the operation is greater than 1, the final result is 1, if it is negative the final result is 0. Thus, for us $\frac{1}{3} + \frac{3}{4} = 1$ and $\frac{1}{3} - \frac{3}{4} = 0$;
- important: if $\circ = *$, then one of y, z must be a constant (this ensures the function ϕ is piecewise linear);
- the operation $x < y$ returns the number 1 if $x < y$, and the number 0 if $x \geq y$. It is the operation which introduces the possibility that f is discontinuous;
- finally, d of the real-valued variables defined in the program are designated as the output, the d components of $\phi(x)$.

For a simple example, take the function ϕ mapping any point (x, y) of the unit square to $(\frac{x+y}{2}, \max\{x - \frac{1}{2}, 0\})$. The corresponding program would be:

$$\begin{aligned} a:&\quad u = x + y \\ b:&\quad v = u * \tfrac{1}{2} \\ c:&\quad z = x - \tfrac{1}{2} \end{aligned} \qquad [14.1]$$

Variables v and z are designated as the output of ϕ.

It may be surprising that this restricted class of functions (basically, piecewise linear) is enough to capture the full generality of Brouwer's theorem. But it does; the reason is, roughly, that such functions can simulate Boolean functions, which in turn can approximate any real-valued function.

The problem BROUWER is a total computational problem that is fundamental and paradigmatic. It is the basis for defining the class **PPAD**,[9] a broad class of total computational problems:

Definition 14.5. PPAD *is the class of all total problems in* **NP** *which reduce to* BROUWER. *A problem is* **PPAD**-*complete if it belongs in* **PPAD** *and all other problems in* **PPAD** *reduce to it.*

[9] The class **PPAD** was defined in Papadimitriou (1994b) in a very different and perhaps more principled way, which however would be awkward to present here: **PPAD** is the class of all total computational problems whose proof of totality is based on the following simple yet powerful principle: *"If a finite directed graph has an unbalanced node (one in which the number of incoming edges is different from the number of outgoing edges), then it must have another unbalanced node."* This is also the origin of the name of the class, an acronym for "Polynomial Parity Argument for Directed graphs."

Many total problems reduce to BROUWER and are therefore in **PPAD**, including the computational version of Sperner's lemma, Tucker's lemma, the Borsuk-Ulam theorem, computing price equilibria in Arrow-Debreu markets, along with, as we shall soon see, NASH Papadimitriou (1994b) Some of these problems have, in fact, been shown **PPAD**-complete.[10] But the problem NASH, arguably the most important problem in the class, and the main motivation for defining it in the first place, had resisted all attempts at proving completeness until rather recently. The remaining part of this section is devoted to the proof of the chapter's main theorem:

Theorem 14.2. NASH *is* **PPAD**-*complete even for games with two players.*

But first, why is the problem in **PPAD**? It turns out that the proof of Nash's Theorem (for example, as in Geanakoplos, 2003) is essentially a reduction from NASH to BROUWER.

From Brouwer to Network Games

The reduction from BROUWER to NASH, and its proof of correctness, are quite technical; here we shall give a rather high-level description illustrating the main ideas involved. For the complete proof, see Daskalakis et al. (2009a).

The reduction employs an important kind of game called *network game*. Let $G = (V, E)$ be a directed graph. A *network game* defined on G is a game in which the players are the vertices in V. Each vertex $v_i \in V$ has its own set of strategies S_i, and a payoff function u_i. The peculiarity is that u_i is not a function mapping $\prod_{j:v_j \in V} S_j$ to the reals, as it would be in a generic game among $|V|$ players, but a function mapping only $S_i \times \prod_{j:(v_j,v_i)\in E} S_j$ to the reals. In other words, the payoff of player v_i depends not on the actions of all players, but only on the actions of itself and all nodes from which there is an edge to v_j.

What makes network games an important model is *succinctness of representation:* While an n-person game with just two strategies per player takes an astronomical total of $n2^n$ numbers to describe, a network game with n players takes at most $n2^{d+1}$ numbers, where d is the maximum in-degree of any node in the network and can be quite small. We shall discuss the important issue of succinctness in more detail in the next section.

Network games are interesting for another reason: Finding a Nash equilibrium in such games is **PPAD**-complete, and this is a key intermediate step in the proof of the Theorem.

Consider for example the network game shown in Figure 14.3, with four players named x, y, z, and a. (The utilities of players x and y are not specified, because they do not depend on the other players.) All players have two strategies which we shall call 0 and 1. For this particular game, let us call x, y, z the probabilities with which the homonymous

[10] Notice that, by the definition of **PPAD**, BROUWER itself is trivially **PPAD**-complete; it turns out that it is so even in its three-dimensional version Papadimitriou (1994b).

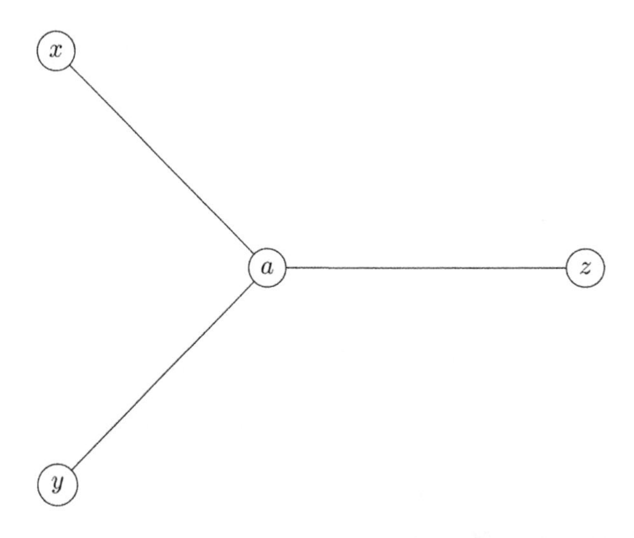

Figure 14.3

if a plays 0	then a's utility is 1 if both x and y play 1, and 0 otherwise
if a plays 1	then a's utility is 1 if z plays 1, and 0 otherwise

if z plays 0	then z's utility is 0
if z plays 1	then z's utility is 1 if a plays 0 and it is 0 otherwise

players play strategy 1—in other words, we deliberately blur the distinction between the name of a player and the player's mixed strategy. Then it is not very hard to check the following remarkable property of this game:

Lemma 14.1. *For every possible combination of values $(x, y) \in [0, 1]^2$ there is a Nash equilibrium, and in all such Nash equilibria $z = x \cdot y$.*

In other words, this network game "behaves" like a statement

$$a : z = x * y$$

in our programing language! It is a "multiplication game," with "inputs" x and y and "output" z. Furthermore, the values of the "inputs" are not affected by the choices of the other players in the game (and so each can legitimately be an output of another such network game). It turns out that there are similar network games which "implement" the $+, -$, and $>$ operations in our programs.[11]

[11] There is a difficulty with the implementation of $>$ in the proof of this theorem in Daskalakis et al. (2009a): If the inputs of this particular network game are equal, then the output is not 1, as it should be, but it can take arbitrary values. In fact, in some sense this difficulty is inherent, in that it can be shown that, if we had a way to faithfully implement $>$ by a network game, then we would be able to construct a network game which has no Nash equilibria, contradicting Nash's theorem! Taking care of this complication is a large part of the proof in Daskalakis et al. (2009a), and one of the many aspects of that proof we omit here.

We can now proceed with the reduction from BROUWER to NASH. Given a program computing some function ϕ from the d-dimensional hypercube to itself, by going through the sequence of statements in the given program implementing the function, and identifying the inputs of each with the outputs of some previous statement (or with one of the inputs of the program), we can create a large network game which implements ϕ. The vertices of this network game are the variables of the program, plus one node per statement labeled with the operation of that statement. The edges of the network are the ones connecting the inputs and outputs of a statement with its operation node to implement the operation ($+$, $-$, $*$ or $>$) of the statement.

We have thus created a network game which computes the function ϕ! d of the players of this game, namely players $x_1, \ldots, x_d$, are the inputs of ϕ—the coordinates of the hypercube—while d other nodes are its output—the results of the computation, call them $y_1, \ldots, y_d$. In the crucial last step of the construction, *we identify each input node x_i with the corresponding output node y_i*. That is, x_i and y_i are one and the same node.

For example, if ϕ is the function computed by (1), the corresponding network game is shown in Figure 14.4. Player a is the middle player of an addition network, b of a multiplication network, and c of a subtraction network. Player d forces player $\frac{1}{2}$'s mixed strategy to be $\left(\frac{1}{2}, \frac{1}{2}\right)$ by playing, say, matching pennies with him; any constant can be implemented this way.

We have thus created a huge game, which at equilibrium captures the computation of ϕ and furthermore—because of the last step of identifying each input with the

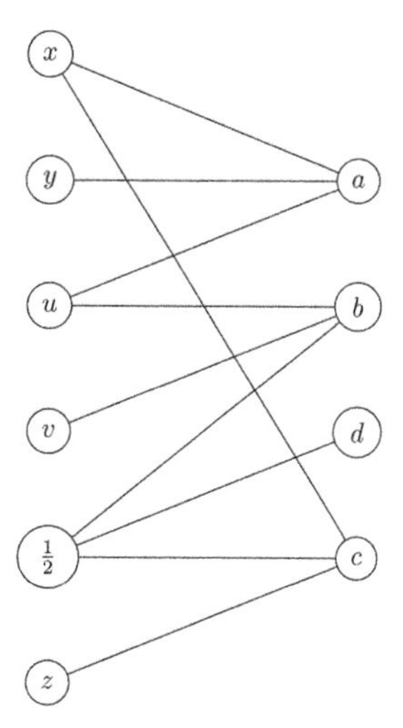

Figure 14.4

corresponding output—we are in effect requiring that, again at equilibrium, ϕ is computed at a fixpoint. That is, we seem to be ready to prove that, from any approximate Nash equilibrium of this game, we can reconstruct either an approximate fixpoint of ϕ, or a break in ϕ's continuity. But this is easier said than done; the actual proof of this statement in Daskalakis et al. (2009a) is quite indirect and technical in ways that would be a long digression here.

From Network Games to Bimatrix Games

We already have an interesting result, namely that finding a Nash equilibrium of a network game is **PPAD**-complete. But this is a far cry from our final goal, proving the same result for bimatrix games. How does one simulate a network game by a bimatrix game?

It turns out that this is possible for *additive bipartite network games*, which means network games with these two properties:

- The underlying network is *bipartite*, that is, its nodes are partitioned into two classes, and no directed edge connects two nodes in the same class. The network game produced in our reduction so far clearly fall into this category, with its nodes divided into *variables* and *operations* (recall the figure);
- the utility of any node while playing one of its strategies must be a *linear* function of the probabilities with which other nodes play theirs; the network game we constructed has this property, since the only multiplication of variables allowed is by a constant.[12]

We shall next sketch informally how to simulate any additive bipartite network game by a bimatrix game. We know that the players of the network game are subdivided into two sets, call them *men* and *women*, and the utility of each man and woman is affected additively by the mixed strategies of several players on the other side. Without loss of generality, assume that there are n men and n women. The two players or the bimatrix game can be thought as two *lawyers*, one who represents the n men, and one who represents the n women. The strategy set of the men's lawyer is the disjoint union of the strategy sets of all men, and similarly for the women's side. The intention is that

[12] Some history: Pinpointing the complexity of NASH was the main motivation behind the definition of the class **PPAD** Papadimitriou (1994b), see also Megiddo and Papadimitriou (1991). The result for four players was proved in a sequence of two papers Goldberg and Papadimitriou (2006), Daskalakis et al. (2006b) some of the techniques in these papers had been anticipated in Bubelis (1979). The improvement to three players came a few months later in Chen and Deng (2005) and independently in Daskalakis and Papadimitriou (2005). Soon after that, the brilliant observation was made in Chen and Deng (2006) that the proof of Goldberg and Papadimitriou (2006), Daskalakis et al. (2006b) also works for two players, thanks to the additivity property of the constructed network game, thus completing the picture. For a detailed proof of the theorem see Daskalakis et al. (2009a), and see Daskalakis et al. (2009b) for another informal exposition.

the mixed strategy profiles of the two lawyers will be in one-to-one correspondence with the mixed strategy profiles of the original game as follows: if the $x_1, \ldots, x_n$ denotes the mixed strategies played by the n men, then the mixed strategy of their lawyer is $\frac{1}{n}(x_1, x_2, \ldots, x_n)$, and similarly for the women. That is, each lawyer subdivides his or her capacity into n equal parts, and then plays in each part exactly what the corresponding client would play. Another way to describe it: each lawyer chooses first a client—uniformly at random—and then a strategy of this client—following in this second choice the same mixed strategy as the client. Because of additivity, the utility of each lawyer is the average of the utilities if his or her clients. From this it would follow that Nash equilibria are preserved.

But there is a problem. Some clients of a lawyer may be more lucrative than others, so to speak, in that they represent a possibility of a larger payoff. The lawyer would then focus on these clients and neglect the others, instead of assigning to all equal probability masses, thus subverting the simulation.

The solution is this: we have the two lawyers play "on the side" a high-stakes game of *generalized rock-paper-scissors.* In generalized rock-paper-scissors the strategies of both players are $1, 2, \ldots, n$ and the payoff of a player playing i when the opponent plays j is (a) 1 if $i = j + 1 \bmod n$; (b) -1 if $i = j + 1 \bmod n$, and (c) zero otherwise. Obviously, the players *must* randomize uniformly.

More formally: we modify the bimatrix game constructed so far by adding to the payoffs of the lawyers the numbers M, $-M$, or zero, for some very large number M, according to the rules of the generalized rock-paper-scissors game, and *depending on their choices of clients.* As a result, each of the two lawyers must now randomize uniformly enough among her clients, and therefore the final payoffs approximate well the payoffs of the original network game.

This concludes the reduction and the proof of the theorem: by setting the approximation bound low enough, and M high enough, we can guarantee that an approximate Nash equilibrium of the final bimatrix game is close enough to a Nash equilibrium of the network game, and therefore to a fixpoint of ϕ.

But what Does it Mean?

We have established something interesting: Finding an (approximate) Nash equilibrium of a game is **PPAD**-complete—which means, as hard as finding a Brouwer fixpoint. But why, exactly, is this bad news?

How do we know that Brouwer is hard?

We do not, but we strongly suspect that it is, simply because nobody has found a good algorithm for it, despite the fact that the challenge has been around for a century. Besides, it has been shown Hirsch et al. (1989) that finding an approximate Brouwer fixpoint of a function ϕ given to us as a "black box" (that is, we can access ϕ only by computing $\phi(x)$ for any x we want) does require exponentially many steps. Thus, any good algorithm for BROUWER must look into the innards of ϕ—the code that computes it—and do something clever with it; nobody has an idea how to accomplish something like that.

But what if P = NP?

In mathematics, we must hold as possible any statement whose negation has not been proved; and nobody has proved that $\mathbf{P} \neq \mathbf{NP}$—or seems to be about to do so. Still, most researchers in the area would be astonished if it turned out that $\mathbf{P} = \mathbf{NP}$. Nash equilibria would be easy to compute—but this will not be the most surprising consequence. A myriad (literally) of important computational problems from all realms of science and mathematics—problems which in many cases have resisted decades of sustained efforts at a good algorithm—would all be susceptible to good algorithms. Cryptography would be impossible, and all design would be essentially trivial. We would be able to discover any mathematical proof with an effort commensurate to its length.[13] All this sounds just too amazing to be true.

So, finding a Nash equilibrium is intractable, in this worst-case sense. So what?

Game theorists should care about the computational complexity of a solution concept. There is something wrong when we cannot calculate an outcome of our prediction. Perhaps such inability is evidence that this may not be a good prediction of what rational agents would do if faced with the situation in hand.[14]

Among solution concepts, the Nash equilibrium is paramount, and therefore its intractability is a serious matter. Roger Myerson has argued convincingly (Myerson, 1999) that the Nash equilibrium concept, and the conception of rationality that it embodies, lie at the foundation of modern economic thought. The universality of the Nash equilibrium (the fact that all finite games have one by virtue of Nash's theorem; Nash, 1951) is a large component of the concept's importance, a sine qua non.[15] And the intractability of the Nash equilibrium (even in this conjectured and worst-case sense) makes this universality suspect: It suggests that there are games—however far-fetched and rare—whose equilibria, even though they exist by Nash's theorem, are not accessible in any realistic way.

[13] Thus, in some far-fetched sense, the continuing absence of a proof that $\mathbf{P} = \mathbf{NP}$ is evidence against its existence...

[14] In the words of Kamal Jain: "If your laptop can't find it, then neither can the market."

[15] Myerson states in Myerson (1991), p. 215, that "it is hard to take a solution concept seriously if it is sometimes empty."

14.7. APPROXIMATION, SUCCINCTNESS, AND OTHER TOPICS

Approximate Nash equilibria

Once a problem has been shown intractable, algorithmic research turns towards developing fast algorithms for finding an approximate solution. This road has been taken for Nash equilibria.

- Algorithms have been developed for finding mixed strategy profiles in bimatrix games which are not quite best responses to each other, but instead have the property that no other strategy is better by more than an additive $\delta > 0$. All payoffs are assumed to be normalized in $[0, 1]$. It has been surprisingly hard to come up with polynomial algorithms which bring δ down to something really small; after several iterations, the state of the art seems to be $\delta \approx 0.3$ (Tsaknakis and Spirakis, 2007).
- If we want to approximate the Nash equilibrium within a *multiplicative* factor $(1 + \delta)$ (this is what is pursued for other intractable problems, see Vazirani, 2003), then the problem is as hard as finding the Nash equilibrium (Daskalakis, 2011).
- Approximation algorithms make sense in the case of a hard optimization problem like the traveling salesman problem: Somebody faced with solving such a problem has little other choice than to settle for the best guarantee available (in the case of the TSP, this is currently 50% above the optimum), because we do not know how to find a better solution. But games are very different. Why would a player settle for playing a strategy that is not a best response, when the best response is easy to find?
- Ergo, approximation in games can only make sense when δ is *very* small. If this is the case, then it is not totally unreasonable to argue that players will not bother to deviate from a suggested play for the sake such a tiny advantage. One very important problem regarding the complexity of Nash equilibria is this: *Is there a polynomial-time algorithm which, given a game and a desired accuracy $\delta > 0$, returns a δ-approximate Nash equilibrium, no matter how small $\delta > 0$ is?* Such family of algorithms, one for each δ, would be called a *polynomial-time approximation scheme*, or PTAS, for NASH. It is known Chen and Deng (2006) that, if such a PTAS exists, it cannot be polynomial also in $\frac{1}{\delta}$; that is to say, $\frac{1}{\delta}$ must somehow appear in the exponent of the running time—unless of course **PPAD** = **P**.
- So, we do not know whether NASH has a PTAS. However, there is a clever algorithm Lipton et al. (2003) which, given a game with an arbitrary but fixed number of players with n strategies each, computes a δ-approximate Nash equilibrium in time growing roughly as $n^{\frac{\log n}{\delta^2}}$; such function is not polynomial, of course, but it is far better than the usual exponential growth one encounters in problems not in **P**. The idea of the algorithm is simple and is based on the following surprising fact: There is always an approximate mixed Nash equilibrium with probabilities which are all multiples of $\frac{1}{k}$, where $k = \frac{\log n}{\delta^2}$, and so the algorithm has to search exhaustively among all these.

> Notice that this fact says something interesting about large games: if you play a game with a huge number N of strategies, you can always find a mixed play reasonably close to equilibrium that employs only $O(\log N)$ strategies...

Exact Nash equilibrium

On the subject of approximation, recall that in the definition of problem NASH we are asked to find a Nash equilibrium within some desired level of accuracy. What if we insisted on finding the precise Nash equilibrium? This is a legitimate question, and in fact there are bona fide applications of Game Theory in Computer Science where exactness is *sine qua non.* See Etessami and Yannakakis (2007) and Yannakakis (2009) for the interesting story of the exact version of NASH. For starters, the exact version is quite a bit harder than NASH—for example, it is not believed to lie within the class **PPAD**, not even within **NP**. For two players, of course, there is no difference between the two versions (the Nash equilibrium is always rational, and therefore precision is not an issue).

Special Classes of Games

Another road traveled frequently after intractability is to explore *easy special cases* of the problem—restrictions on the input that render the problem in **P**. This has also been done in the case of NASH. For example, we have already discussed that anonymous games with few strategies admit PTAS Daskalakis and Papadimitriou (2008b).

The *rank* of a bimatrix game (A, B) is the rank of the matrix $A + B$. If this rank is zero—that is, if the game is zero sum—then we know that finding a Nash equilibrium is in **P**. But what if the rank is, not quite zero, but small? It turns out that there is a nontrivial and interesting polynomial algorithm for finding Nash equilibria in games of rank one Adsul et al. (2011), as well as a PTAS for games of any finite rank Kannan and Theobald (2007).

A second positive result for a special class of games (another generalization of von Neumann's Maxmin Theorem, in a different direction) involves network games. Consider a network in which each edge is a bimatrix game played between the two nodes. In all games corresponding to the edges leaving a node v, player v has the same strategy set, and v's choice of strategy must be consistent across all these games. Player v's payoff is the sum of the payoffs from these games. Such games are called *separable network games.* Suppose now that the game is zero sum. That is, it so happens that, for all pure strategy profiles, all payoffs of all players add up to zero (the constituent bimatrix games may not be zero-sum). For three or more players the zero-sum property does not make the problem easy (just imagine an extra player who absorbs excess payoff). But here it does: A Nash equilibrium for zero-sum separable network games can be found by linear programming Cai and Daskalakis (2011).

Restricting games in other ways does not help: Symmetric two-player games Gale et al. (1950), and two-player games with payoffs in $\{0, 1\}$ are as hard as general games Abbott et al. (2005).

Succinct Representations of Games

The tabular form of a game with a large number of players is a huge affair. To fully represent all payoffs for all possible strategy profiles in a game with n players requires at least $n \cdot 2^n$ numbers (since every player has at least two strategies). This is an exponential amount of information, rendering the problem meaningless in practice when n is large.[16]

Fortunately, there are certain important families of games which *can be represented succinctly*. A good example is the class of *network games* (sometimes called *graphical*, see Kearns et al., 2001), already introduced and used in the **PPAD**-completeness proof: Suppose that players are the nodes of a network, and each is affected by the choices of the neighboring nodes. Furthermore, no node has more than, say, d neighbors, where d is a small number such as three. The number of payoffs one needs to describe this game is no longer astronomical: just ns^{d+1}—a polynomial in n. There are many other important and meaningful classes of succinctly representable games:

- *Multimatrix games*, in which each player plays simultaneously many bimatrix games with other players, on the same strategy set, and the payoffs add up.
- *Congestion games* Rosenthal (1973), in which the strategies of each player are paths (sets of edges, not necessarily realized on a graph), and the (negative) payoff of each player is the sum of the resulting delays over the edges of the path—but these delays depend on how many other players have chosen paths through the same edges through function, one for each edge, mapping number of players to the reals.
- *Symmetric games* over large numbers of players can also be represented in a succinct way, by exploiting symmetry to avoid repeating the same payoffs again and again. A related but much broader class is that of *anonymous* games Blonski (1999), in which each players have a different payoff function, but these payoffs depend on *how many* other players choose each strategy, not *which* other players do. There is a PTAS for finding Nash equilibria of anonymous games (Daskalakis and Papadimitriou, 2007, 2008a,b).
- Well known game formalisms such as *Bayesian games* or games in extensive form (see Chapter 2 of this Handbook), as well as *repeated games* (in which players play many rounds of the same game, called in this context the "stage game," with the payoffs, perhaps discounted over, added up; see Chapter 4 of this Handbook) can all be rendered in tabular form, albeit with huge overhead; therefore, they can all be considered as succinctly represented games.

[16] With astronomically large inputs, even exhaustive algorithms such as enumeration of all supports are suddenly polynomial in the size of the input, just because the size of the input has become so huge that it has rendered the running time less than absurd—in comparison...

Computing Equilibria in Succinct Games

What is the complexity of finding a Nash equilibrium in the above categories on succinctly representable games? There is a sweeping, and rather surprising, result on this subject. It states, roughly, that for almost all of the above categories of succinct games finding a Nash equilibrium is exactly as hard as it is for games in tabular form: **PPAD**-complete (Daskalakis et al., 2006a).

Congestion games are special: finding any Nash equilibrium of a congestion game seems to be quite a bit easier than **PPAD**. The reason is that there are *two independent proofs* of Nash's Theorem for congestion games: One is inherited from finite games, and the other is a potential argument proof due to Rosenthal establishing the existence of *pure* Nash equilibrium (Rosenthal, 1973). Neither proof is constructive, but they are non-constructive in "different ways," and this makes the problem easier. In fact, the problem of defining a Nash equilibrium in a congestion belongs in another complexity class, called **CLS** (Daskalakis and Papadimitriou, 2011), also containing a few other game-theoretic specimens not known to be in **P**.

The *correlated equilibrium* is a well-known and important relaxation of the Nash equilibrium concept (Aumann, 1974): a distribution on strategy profiles such that, if a strategy profile is drawn according to this distribution and each player is given her own strategy in the profile, then the players would have no incentive to deviate from this suggestion. The set of all correlated equilibria is described by a set of linear inequalities, and therefore it is straightforward, given a game, to find a correlated equilibrium in polynomial time through linear programming. But what if the game is succinctly represented as in one of the categories seen above? Again, there is a theorem (Papadimitriou and Roughgarden, 2008) implying that a correlated equilibrium of a succinctly represented game can be computed in polynomial time, provided the succinct representation has the following property: If one is given a mixed strategy profile in this game, there is a polynomial-time algorithm which computes the expected payoff of each player. It turns out that essentially all genres of succinctly representable games have this property.

Incidentally, it is very surprising to a complexity theorist that so many and varied succinct representations of input retain the computational complexity of the original problem. In complexity theory, succinct representations typically usher in an exponential increase in complexity. This contrast is a compliment to game theory: it suggests that the classes of succinctly represented games chosen over the decades by researchers in the field are all well chosen and meaningful.

Extensive Form and Refinements

An extensive form game is given in terms of a game tree. One can find a Nash equilibrium of a two-player zero-sum game in extensive form by formulating the game in tabular form and then using linear programming; unfortunately, the first step is exponential. In games of perfect recall, there is a clever way to do this in polynomial time by a simple alternative representation of the mixed strategies (Koller et al., 1992). This results in a fast algorithm for solving zero-sum extensive form games. Solving games of imperfect recall is **NP**-complete (Koller and Megiddo, 1996).

Extensive form games are also the arena for several important refinements of the Nash equilibrium concept. The concept of subgame perfect equilibrium does not differ much computationally from that of the Nash equilibrium, because it entails finding a Nash equilibrium for all subtrees of the tree (and there are at most n subtrees in a tree with n nodes). But there are many other refinements: sequential equilibrium, trembling hand equilibrium, quasi-perfect equilibrium, proper equilibrium, closed under rational behavior, and many more. These, in general, tend to be considerably harder: see Hansen et al. (2010) for some recent results on this front.

Repeated Games and the Folk Theorem

How hard is it to find a Nash equilibrium in a repeated game—for example, the infinitely repeated prisoners' dilemma with discounted payoffs? There is a classical result known as the *Folk Theorem* (see Chapter 4 of this Handbook), which seems to suggest that the Nash equilibrium problem becomes very easy in the case of repeated games, in that in any repeated game there is a huge set of Nash equilibria with payoffs approximating any point in the interior of the so-called *individually rational region* of the game, consisting of all profiles that dominate the minmax profile (also called the *threat point* for reasons that will soon become clear). For example, in the repeated prisoner's dilemma, the threat point is the "defect-defect" outcome, and thus any payoff combination that dominates it—for example, one arbitrarily close to the "collaborate-collaborate" outcome—can be realized through the Folk Theorem. The Nash equilibria guaranteed by the Folk Theorem have the following structure: The players agree to a periodic sequence of play which approximates in payoff the target point. Any player who deviates from this plan is punished by the others playing the minmax against the defector. Because of this threat, the plan is a Nash equilibrium achieving the target payoffs.

However, if computational complexity is taken into account, there are serious problems with the Folk Theorem (Borgs et al., 2008). First, for three or more players the threat point is **NP**-complete to compute. Furthermore, it can be shown that finding an "algorithmic equilibrium" in a repeated game (a polynomial algorithm for each player that chooses an action at each repetition) is **PPAD**-complete—that is to say exactly as hard as finding a Nash equilibrium in the stage game. (Notice that a trivial Nash equilibrium in the repeated game is to keep playing a Nash equilibrium of the stage game.)

Intriguingly, it has been recently pointed out by Halpern and Pass (2013) that this last negative implication (intractability of finding an algorithmic equilibrium in repeated games) can be remedied by *cryptography:* If the players are allowed to randomize, and communicate via encrypted messages, then they can agree on a favorable *correlated* equilibrium and stick to it under mutual threat.

If there are only two players, then the threat point is essentially computing the minmax in a zero-sum game and can therefore be computed in polynomial time, and the Folk Theorem is restored (Littman and Stone, 2005).

Price Equilibria

The proof of Nash's theorem in 1950 provided inspiration and impetus for Arrow and Debreu to prove their own famed existence theorem for price equilibria in economies. Six decades later the pattern was repeated: inspired by the proof that NASH is intractable (Daskalakis et al., 2009a), several researchers showed that finding price equilibria in several contexts is a **PPAD**-complete problem—it was already known in the 1990s that finding price equilibria for (appropriately restricted) excess demand functions is **PPAD**-complete (Papadimitriou, 1994b). There is a simple reduction from two-player NASH to the problem for finding certain kind of price equilibrium in a market with Leontieff consumer utilities (Codenotti et al., 2006). More elaborate reductions are needed for more realistic consumer utilities (Chen et al., 2009, 2013). For the more general setting of an economy with production, see Papadimitriou and Wilkens (2011) for an account of the complexity phenomena that take hold if the assumption of convex production is violated.

And it gets worse: PSPACE-completeness

We have treated the complexity class **NP** as the limit of our ambition in computation. Are there meaningful computational problems beyond that?[17]

The complexity class **PSPACE** consists of all problems which can be solved with an amount of *memory* that is polynomial in the size of the input. This class contains the class **NP**, because exhaustively trying all possible solutions of a search problem can be done in polynomial memory—because memory can be *reused*... **PSPACE** is broadly believed to be strictly more powerful than **NP**, but here again we have no proof.[18]

There are several connections between Game Theory and this complexity class. Perhaps the most fundamental one concerns *board games*. Define a game of $n \times n$ Go by simply specifying an initial position (a configuration of black and white stones on

[17] The reader is probably aware that there are mathematical problems that are completely unsolvable by computation, such as the *halting problem*, or the problem of finding integer solutions of polynomial equations, see for example Chapter 3 of Papadimitriou (1994a). In fact, there has been a strand of work considering the implications of such impossibility results in economics (Velupillai, 2000).

[18] In fact, there is no proof in sight even that **PSPACE** $\neq$ **P**...

the $n \times n$ grid) and asking whether it is a win for White. Similarly for Checkers.[19] The tabular form of this game would have an astronomical number of strategies (each strategy contains a full list of the player's response to every possible position reachable from the initial position). Notice that such board games can be considered succinct representations of some exponentially large extensive form games. There is a long list of parametrized board games that have been shown **PSPACE**-complete (see Demaine, 2001, for an overview of the broader area).

PSPACE-completeness is considered an even graver grade of intractability than **NP**-completeness—which in turn is more serious than the **PPAD**-completeness plaguing NASH. However, certain variants of NASH turn out, rather surprisingly, to be **PSPACE**-complete. Ironically, they all originate from efforts to combat the problem's intractability:

- The *homotopy method* (Eaves, 1972) is a well known algorithm for finding Nash equilibria in games (and solving other fixpoint problems). It starts with a game whose equilibrium is easy to find, and slowly, continuously transforms it to the game whose equilibrium we wish to find, pivoting from solution to solution along the way. The algorithm is guaranteed to end up with the desired Nash equilibrium.
- The Lemke-Howson algorithm (see Lemke and Howson, 1964, and Chapter 45 of this Handbook) is a very elegant algorithm for finding a Nash equilibrium of a two-player game. The algorithm works on pairs of supports (one support for each player—recall that the Nash equilibrium is trivial to find once we have the right support) and pivots from one support pair to the next very much the same way that simplex pivots from one basis to the other (one important difference is that in the case of Lemke-Howson there is never a choice of next basis). The algorithm is guaranteed to end up at a Nash equilibrium.
- Besides computational intractability, one potentially undesirable aspect of Nash equilibria is their *multiplicity*, the fact that a game can have many. Harsanyi and Selten (1988) proposed a remedy for multiplicity, an algorithm for selecting one of the possibly many Nash equilibria of a game in a principled way, which they called the *linear tracing procedure*.

Theorem 14.3. *(Goldberg et al., 2013) It is a* **PSPACE**-*complete problem, given a two-person game, to find the Nash equilibrium that would be produced by (a) the homotopy method; (b) the Lemke-Howson algorithm; or (c) the Harsanyi-Selten linear tracing procedure.*

[19] Chess (discussed in the first chapter of this Handbook), however, is hard to deal with in this context, because Chess is a finite object—8×8 board, at most 10^{50} positions—and therefore impossible to deal within a theory which proudly sheds constants... If we bend Chess enough to make it an unbounded game similar to $n \times n$ Go and Checkers, then **PSPACE**-completeness kicks in immediately.

ACKNOWLEDGMENTS

Many thanks to Tim Roughgarden for his extensive comments, which resulted in substantial improvements, to Costis Daskalakis and an anonymous reviewer for valuable feedback on an earlier version, and the editors for their support and patience.

REFERENCES

Adsul, B., Garg, J., Mehta, R., Sohoni, M., 2011. Rank-1 bimatrix games: a homeomorphism and a polynomial time algorithm. Proceedings of the 43rd annual ACM symposium on Theory of computing (STOC), pp. 195–204.

Aumann, R.J., 1974. Subjectivity and correlation in randomized strategies. J. Math. Econ. 1, 67–96.

Abbott, T.G., Kane, D., Valiant, P., 2005. On the complexity of two-player win-lose games. In the 46th Annual IEEE Symposium on Foundations of Computer Science (FOCS).

Borgs, C., Chayes, J.T., Immorlica, N., Kalai, A.T., Mirrokni, V.S., Papadimitriou, C.H., 2008. The Myth of the Folk Theorem. In the 40th Annual ACM Symposium on Theory of Computing (STOC).

Blonski, M., 1999. Anonymous games with binary actions. Games Econ. Behav. 28(2), 171–180.

Bubelis, V., 1979. On equilibria in finite games. Int. J. Game Theory 8(2), 65–79.

Cai Y., Daskalakis, C., 2011. On minmax theorems for multiplayer games. Symposium on Discrete Algorithms (SODA), pp. 217–234.

Chen, X., Dai, D., Du, Y., Teng, S.-H., 2009. Settling the complexity of arrow-debreu equilibria in markets with additively separable utilities. In: Proceedings of the 50th Annual IEEE Symposium on Foundations of Computer Science (FOCS).

Chen, X., Paparas, D., Yannakakis, M., 2013. The complexity of non-monotone markets. In: Proceedings of the 45th annual ACM symposium on Symposium on theory of computing (STOC), pp. 181–190.

Chen, X., Deng, X., 2005 3-Nash is PPAD-complete. Electronic colloquium on computational complexity (ECCC) 134.

Chen, X., Deng, X., 2006. Settling the complexity of two-player nash equilibrium. 47th Annual IEEE Symposium on Foundations of Computer Science (FOCS).

Codenotti, B., Saberi, A., Varadarajan, K.R., Ye, Y., 2006. Leontief economies encode nonzero sum two-player games. In the 17th Annual ACM-SIAM Symposium on Discrete Algorithms (SODA).

Cook, S.A., 1971. The complexity of theorem-proving procedures. In: Proceedings of the third annual ACM symposium on Theory of computing, pp. 151–158.

Dantzig, G.B., 1963. Linear Programming and Extensions. Princeton University Press: Princeton, NJ.

Daskalakis, C., 2011. On the complexity of approximating a nash equilibrium. In: Proc. 2011 Symp. on Discrete Algorithms (SODA).

Daskalakis, C., Fabrikant, A., Papadimitriou, C.H., 2006a. The game world is flat: the complexity of nash equilibria in succinct games. In: the 33rd International Colloquium on Automata, Languages and Programming (ICALP).

Daskalakis, C., Papadimitriou, C.H., 2005. Three-player games are hard. ECCC 139.

Daskalakis, C., Papadimitriou, C.H., 2007. Computing equilibria in anonymous games. In: the 48th Annual IEEE Symposium on Foundations of Computer Science (FOCS).

Daskalakis, C., Papadimitriou, C.H., 2008a. Discretized multinomial distributions and nash equilibria in anonymous games. In: the 49th Annual IEEE Symposium on Foundations of Computer Science (FOCS).

Daskalakis, C., Papadimitriou, C.H., 2008b. On oblivious PTAS for nash equilibrium. Manuscript.

Daskalakis, C., Papadimitriou, C., 2011. Continuous local search. In: Proceedings of the Twenty-Second Annual ACM-SIAM Symposium on Discrete Algorithms, pp. 790–804.

Daskalakis, C., Goldberg, P.W., Papadimitriou, C.H., 2006b. The complexity of computing a nash equilibrium. In: the 38th Annual ACM Symposium on Theory of Computing (STOC).

Daskalakis, C., Goldberg, P.W., Papadimitriou, C.H., 2009a. The complexity of computing a nash equilibrium. SIAM J. Comput. 39, 1, 195–259.

Daskalakis, C., Goldberg, P.W., Papadimitriou, C.H., 2009b. The complexity of computing a nash equilibrium. Communi. ACM 52, 2, 89–97.
Dasgupta, S., Papadimitriou, C.H., Vazirani, U., 2008. Algorithms, McGraw-Hill. New York, NY.
Demaine, E.D., 2001. Playing games with algorithms: algorithmic combinatorial game theory. Mathematical Foundations of Computer Science, Lecture Notes in Computer Science Volume 2136, pp. 18–33.
Eaves, B.C., 1972. Homotopies for computation of fixed points. Math. Program. 3, 1–22.
Etessami, K., Yannakakis, M., 2007. On the complexity of nash equilibria and other fixed points (extended abstract). In: the 48th Annual IEEE Symposium on Foundations of Computer Science (FOCS).
Fabrikant, A., Papadimitriou, C.H., Talwar, K., 2004. The complexity of pure nash equilibria. In: the 36th Annual ACM Symposium on Theory of Computing (STOC).
Gale, D., Kuhn, H.W., Tucker, A.W., 1950. On symmetric games. In: Kuhn, H.W., Tucker, A.W. (Eds.), Contributions to the Theory of Games, volume 1. Princeton University Press: Princeton, NJ, pp. 81–87.
Garey, M.R., Johnson, D.S., 1989. Computers and Intractability: A Guide to the Theory of NP-completeness, Freeman: San Francisco, CA.
Geanakoplos, J., 2003. Nash and Walras equilibrium via Brouwer. Econ. Theory 21 2-3, 585–603.
Goldberg, P.W., Papadimitriou, C.H., 2006. Reducibility among equilibrium problems. In: the 38th Annual ACM Symposium on Theory of Computing (STOC).
Goldberg, P.W., Papadimitriou, C.H., Savani, R., 2013. The complexity of the homotopy method, equilibrium selection, and Lemke-Howson solutions. ACM Trans. Econ. Comput. 1, 2.
Halpern, J., Pass, R., 2013. The Truth Behind the Myth of the Folk Theorem. Working paper, Cornell University.
Hansen, K.A., Miltersen, P.B., Srensen, T.B., 2010. The computational complexity of trembling hand perfection and other equilibrium refinements. Algorithmic Game Theory, Lecture Notes in Computer Science, Volume 6386, pp. 198–209.
Harsanyi, J.C., Selten, R., 1988. A General Theory of Equilibrium Selection in Games, MIT Press Books.
Hirsch, M.D., Papadimitriou, C.H., Vavasis, S.A., 1989. Exponential lower bounds for finding brouwer fixed points. J. Complex. 5(4), 379–416.
Kannan, R., Theobald, T., 2007. Games of fixed rank: a hierarchy of bimatrix games. In: the 18th Annual ACM-SIAM Symposium on Discrete Algorithms (SODA).
Kearns, M., Littman, M.L., Singh, S., 2001. Graphical models for game theory. In: Proceedings of the Seventeenth conference on Uncertainty in artificial intelligence, pp. 253–260.
Kleinberg, J., Tardos, É., 2005. Algorithm design, Addison-Wesley, Reading, MA.
Koller, D., Megiddo, N., 1996. The complexity of two-person zero-sum games in extensive form. Games Econ. Behav. 14, 247–259.
Koller, D., Megiddo, N., von Stengel, B., 1992. Efficient computation of equilibria for extensive two-person games. Games Econ. Behav. 4, 528–552.
Ladner, R.E., 1975. On the structure of polynomial time reducibility. J. ACM 22, 155–171.
Lemke, C.E., Howson Jr., J.T., 1964. Equilibrium points of bimatrix games. J. Soc. Ind. Appl. Math. 12(2) 413–423.
Lipton, R.J., Markakis, E., Mehta, A., 2003. Playing large games using simple strategies. In: the 4th ACM Conference on Electronic Commerce (EC).
Littman, M.L., Stone, P., 2005. A polynomial-time Nash equilibrium algorithm for repeated games. Decision Sup. Syst. 39, 1, 55–66.
Megiddo, N., Papadimitriou, C.H., 1991. On total functions, existence theorems and computational complexity. Theor. Comput. Sci. 81(2), 317–324.
Myerson, R.B., 1991. Game Theory, Harvard University Press, Cambridge, MA.
Myerson, R.B., 1999. Nash equilibrium and the history of economic theory. J. Econ. Lit. 37, 3.
Nash, J., 1951. Non-cooperative games. Ann. Math. 54(2), 286–295.
Papadimitriou, C.H., 1994a. Computational Complexity. Addison-Wesley, Reading, MA.
Papadimitriou, C.H., 1994b. On the complexity of the parity argument and other inefficient proofs of existence. J. Comput. Syst. Sci. 48(3), 498–532.
Papadimitriou, C.H., Roughgarden, T., 2008. Computing correlated equilibria in multi-player games. J. ACM (JACM) 55, 3.

Papadimitriou, C.H., Wilkens, C.A., 2011. Economies with non-convex production and complexity equilibria. Proc. 12th ACM conference on Electronic commerce (EC11), pp. 137–146.
Rosenthal, R.W., 1973. A class of games possessing pure-strategy nash equilibria. Int. J. Game Theory 2(1), 65–67.
Roughgarden, T., 2010. Computing equilibria: a computational complexity perspective. Econ. Theory 43, 193–236.
Savani, R., von Stengel, B., 2004. Exponentially many steps for finding a nash equilibrium in a bimatrix game. In: the 45th Symposium on Foundations of Computer Science (FOCS).
Tsaknakis, H., Spirakis, P.G., 2007. An optimization approach for approximate nash equilibria. In: the 3rd International Workshop on Internet and Network Economics (WINE).
Vazirani, V.V., 2003. Approximation Algorithms, Springer, Berlin.
Velupillai, K.V., 2000. Computable Economics, Oxford University Press, Oxford.
Yannakakis, M., 2009. Computational aspects of equilibria. In: Algorithmic Game Theory, Lecture Notes in Computer Science, Volume 5814, pp. 2–13.

CHAPTER 15

Theory of Combinatorial Games

Aviezri S. Fraenkel[*], Robert A. Hearn[†], Aaron N. Siegel[‡]

[*]Department of Computer Science and Applied Mathematics, Weizmann Institute of Science, Rehovot, Israel
[†]H3 Labs LLC, Palo Alto, CA, USA
[‡]Twitter, San Francisco, CA, USA

Contents

Abstract

Aim: To present a systematic development of the theory of combinatorial games from the ground up. **Approach**: Computational complexity. Combinatorial games are completely determined; the questions of interest are efficiencies of strategies. **Methodology**: Divide and conquer. Ascend from Nim to Chess and Go in small strides at a gradient that is not too steep. **Presentation**: Mostly informal; examples

Handbook of game theory
http://dx.doi.org/10.1016/B978-0-12-420056-2.00015-X

of combinatorial games sampled from various strategic viewing points along scenic mountain trails illustrate the theory. **Add-on**: A taste of *constraint logic*, a new tool to prove intractabilities of games.

Keywords: Combinatorial game theory, Partizan games, Misère play, Nim, Chess, Go, Impartial games, Sprague-grundy theory, Computational complexity, Constraint logic

JEL Codes: C72, C79

15.1. MOTIVATION AND AN ANCIENT ROMAN WAR-GAME STRATEGY

The current mainstream of the family of combinatorial games consists of two-person games with perfect information (unlike some card games where information is hidden), without chance moves (no dice), and **outcome** restricted to (lose, win), (tie, tie), and (draw, draw) for the two players who move alternately (no passing).

Instead of the long terminology "combinatorial game(s)," we shall usually simply write "game(s)." In *normal* play, to *win a game* means to make the last move in it. This is the main concern of game theory, covered in Sections 15.2-15.5. But in Section 15.6, we expose the modern theory of *misère* play, where the player making the last move loses. A tie is an end position with no winner and no loser, as may occur in tic-tac-toe, for example. A draw is a "dynamic tie," i.e., a nonend position such that neither player can force a win, but each can find a next nonlosing move. (In "noncombinatorial" game theory, each player receives a **payoff** at the end of the game. For combinatorial games it is natural to assign a payoff of $+1$ to the winner, -1 to the loser and 0 for tying or drawing: once play is in a draw cycle it is abrogated. Our games are **zero-sum** games in this sense.)

The modern theory of combinatorial games is portrayed in the groundbreaking work of Conway (2001), the encyclopedic compilation of Berlekamp et al. (2001–2004), the attractive textbook by Albert et al. (2007), and the authoritative graduate-level book of Siegel (2013) that studies the modern theory of partizan games and misère play.

The primeval and simplest combinatorial game is NIM: Given m piles of finitely many tokens, a move consists of selecting a single nonempty pile and removing from it a positive number of tokens, that is, at least one, and up to and including the entire pile. The player first unable to move loses, the opponent wins (**normal** play). For $m = 1$, player I can win if the pile is nonempty, simply by removing it entirely. For $m = 2$, player I can win if the piles are of unequal size, by a move that equalizes their size, followed by imitating on one pile what player II does on the other. For $m > 2$, the winning strategy, first given in Bouton (1901–1902), is quite surprising, yet simple: compute the eXclusive OR (XOR) of the binary representation of the pile sizes. If the resulting binary **nim-sum** is nonzero, the *next* player (player I) has a move making it zero (a winning move). If it is zero, *every* move will make it nonzero (a losing move). This is shown in Section 15.2 in the more general setting of "NIM-type" games. Thus for $m = 3$ and pile sizes $1, 2, 3$, a simple case analysis shows that the *previous* player (player II) can win. Indeed, the nim-sum $1 \oplus 2 \oplus 3$ is 0.

As an exercise, can you win by beginning to play in a game of NIM with four piles of sizes $2, 3, 5, 7$? If so, do you have a unique winning strategy?

The family of combinatorial games contains simple games such as NIM, as well as seemingly complex games such as CHECKERS, CHESS, and GO. The fundamental question that arises naturally is why some games, such as NIM, are easy to solve, whereas others in the family, such as GO, seem so complex? The quest for answers to this problem motivates this survey.

For throwing some light on the question, a Roman Caesars' motto is adopted:

DIVIDE AND CONQUER .

There are several mathematical differences between NIM-type and CHESS-type games. After identifying them, a concentrated attack is launched on each of them separately, which seems to have a better chance of success than trying in vain to scale the sheer cliff separating NIM from CHESS. Thus, we ascend from NIM towards CHESS and GO at a moderate gradient, by gradually introducing into NIM more and more complications in a natural order of increasing complexity. The adventures occurring on the way comprise the story of this chapter.

In Section 15.2, we review the classical theory of acyclic games, **sum** of games and the Sprague-Grundy function, which is the main tool for solving acyclic games. We also show that complexities of games are normally much higher than those encountered in optimization problems such as the Traveling Salesperson Problem.

An "apparent" difference between NIM and CHESS is the **board** which exists for the latter but not for the former. However, Figure 15.1 shows that also NIM can be considered as a board game: a_i indicates a nim-heap of size i, and the directed edges indicate the permissible moves. Thus placing a token on each of the vertices a_1, a_2, and a_3 and moving them along directed edges, where any number of tokens may reside on any vertex, is isomorphic to NIM with pile sizes 1, 2, 3. Conclusion: this "apparent" difference is not really a mathematical difference.

Here are some more substantive differences:

- **Cycles.** NIM-type games are finite and "acyclic," i.e., there is an underlying "well-ordering principle" which guarantees that no position is assumed twice. This is not the case for CHESS-type games. Applying the Divide And Conquer Principle, we deal with such "cyclic" games separately in Section15.3, where it is shown that cycles indeed destroy the classical theory. A generalized theory is developed there which recovers a polynomial strategy for cyclic games.
- **Token interactions.** Another difference is that in NIM-type games, considered as board games, tokens coexist peacefully on the same vertex (board square), whereas they interact in various ways such as jumping, deflecting, capturing, etc., in

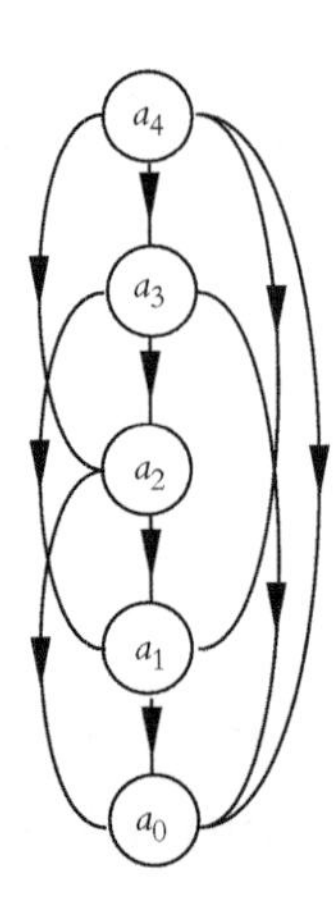

Figure 15.1 NIM as a board game.

CHESS-type games. Many of these interactions cause the games to become PSPACE-hard (notion explained near the end of Section 15.2) even in simplified form, e.g., when played on planar or acyclic or bipartite graphs. However, if both tokens disappear on impact, a "just barely polynomial" strategy can be given for general cyclic digraphs (directed graphs). This topic is studied in Section 15.4.

- **Partizanship.** A game is **impartial** if the set of **options** (positions reached in a single move) of every position is the same for the two players. If this does not necessarily hold, the game is **partizan**. NIM-type games are impartial, whereas CHESS-type games are partizan (the "black" player cannot move a white piece and vice versa). Note that the set of impartial games is a subset of the set of partizan games. It turns out that partizan games, taken up in Section 15.5, are in general PSPACE-hard even on acyclic digraphs; see Yedwab (1985), and Morris (1981). See also Pultr and Morris (1984).
- **Termination set.** Another difference concerns the conventions for ending the play of the game, i.e., the termination set τ. Roughly, the complexity of the strategy seems to increase with the size $|\tau|$ of τ. The simplest games are those played on a digraph G, where τ is the set of *leaves* of G (vertices of outdegree 0), followed by those in which τ consists of all positions whose only options are leaves – such as in **misère** play: the player making the last move loses – to cases where τ is even larger, such as in CHESS and GO. A theory for general τ has yet to be developed, but we treat misère play in Section 15.6.

As we progress from the easy games to the more complex ones, we will develop some understanding for the *poset* of tractabilities and efficiencies of game strategies: in the realm of existential questions, tractabilities, and efficiencies are, by and large, linearly ordered, from polynomial to exponential. However, as explained near the end of Section 15.2, game problems are formulated by an—often unbounded—number of alternating quantifiers. For such problems the notion of a "tractable," "polynomial," or "efficient" computation—defined formally in Definition 15.1, Section 15.2—is much more complex. (Which is more tractable: a game that ends after four moves, but it is undecidable who wins (Rabin, 1957), or a game requiring an Ackermann function of moves to finish but the winner can play randomly, having to pay attention only near the end (Fraenkel et al., 1988; Fraenkel and Nešetřil, 1985) ?) Since we are concerned with game complexities, we present, in Section 15.7, a modern tool for proving game intractabilities conveniently, and efficiently. In Section 15.8, the Conclusion, we briefly illuminate our ascent from NIM to CHESS and GO, and indicate possible further directions of combinatorial game theory.

15.2. THE CLASSICAL THEORY, SUM OF GAMES, COMPLEXITY

In this section, we will see how to play "arbitrary" finite acyclic games such as BEAT DOUG (Figure 15.2). (DOUG—"DAG," Directed Acyclic Graph.)

Place one token on each of the four starred vertices. A move consists of selecting a token and moving it, along a directed edge, to a neighboring vertex on this acyclic digraph. As usual we consider normal play, so the player making the last move wins. Tokens can coexist peacefully on the same vertex. For the given position, how much time does it take to:

(a) compute who can win;
(b) compute an optimal next move; and
(c) consummate the win, that is, actually make the last move?

Denote by $\mathbb{N}$ and $\mathbb{N}^+$ the set of all nonnegative integers and the set of all positive integers, respectively. Following the divide and conquer methodology, let us begin with a more structured digraph, rather than solving immediately the "arbitrary" BEAT DOUG. Given $n \in \mathbb{N}^+$ (the initial score) and $t \in \mathbb{N}^+$ (the maximal step size), a move in the game SCORING consists of selecting $i \in \{1, \ldots, t\}$ and subtracting i from the current score,

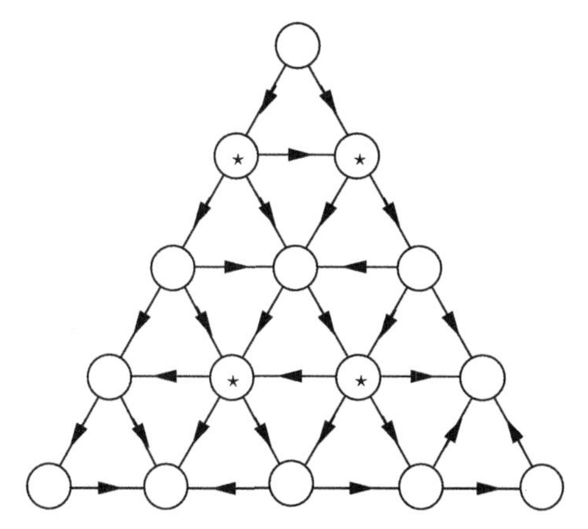

Figure 15.2 Beat DOUG on this directed acyclic graph (DAG).

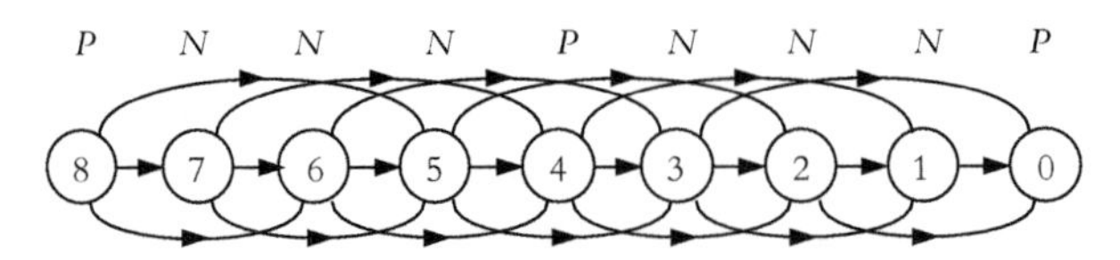

Figure 15.3 The digraph for Scoring, with initial score $n = 8$ and maximal step $t = 3$. Positions marked N are wins and P are losses for the player moving from those positions.

initially n, to generate the new score. Play ends when the score 0 is reached. The player reaching 0 wins (normal play). Notice that Nim is the special case $t = \infty$ of Scoring.

The digraph $G = (V, E)$ for Scoring is shown in Figure 15.3 for $n = 8$ and $t = 3$: it is an acyclic digraph, where V is the set of game positions, and $(u, v) \in E$ if and only if there is a move from u to v (then v is an **option** of u). A position (vertex) $u \in V$ is labeled N (for Next player win) if the player moving from u can win; otherwise it is a P-position (Previous player win). Denote by $\mathcal{P}$ the set of all P-positions, by $\mathcal{N}$ the set of all N-positions, and by $F(u)$ the set of all options of any vertex u. For any acyclic game, the partition of the vertex-set into $\mathcal{P}$, $\mathcal{N}$ exists uniquely and satisfies,

$$u \in \mathcal{P} \quad \text{if and only if} \quad F(u) \subseteq \mathcal{N}\,, \qquad [15.1]$$

$$u \in \mathcal{N} \quad \text{if and only if} \quad F(u) \cap \mathcal{P} \neq \emptyset\,. \qquad [15.2]$$

In words: u is a P-position if and only if all its options (direct followers) are N-positions; and u is an N-position if and only if it has an option in $\mathcal{P}$.

As suggested by Figure 15.3, we have $\mathcal{P} = \{k(t+1) : k \in \mathbb{N}\}$, so $\mathcal{N} = \{\{0, \ldots, n\} \setminus \mathcal{P}\}$. The winning strategy consists of dividing n by $t + 1$. Then $n \in \mathcal{P}$ if and only if the remainder r is zero. If $r > 0$, the unique winning move is from n to $n - r$.

Is this a tractable strategy? ("Tractable"—see Definition 15.1.)

Input size: $\Theta(\log n)$ (**succinct** input).

Strategy computation: $O(\log n)$ (division of n by t).

Length of play: $\lceil n/(t+1) \rceil$.

Thus the computation time is linear in the input size, but the length of play is exponential!

To the "run-of-the-mill-algorithmicians" the latter fact dooms the game as intractable. It may be quite a surprise to them that it does not prevent the strategy from being tractable: whereas we dislike computing in more than polynomial time, we observe that at least some members of the human race relish to see some of its members being tormented for an exponential length of time, from before the era of the Spanish matadors and inquisition, through soccer and tennis, to Chess and Go! But there are other requirements for making a strategy polynomial as we will see presently, so at present let us say that the strategy is *tractable*.

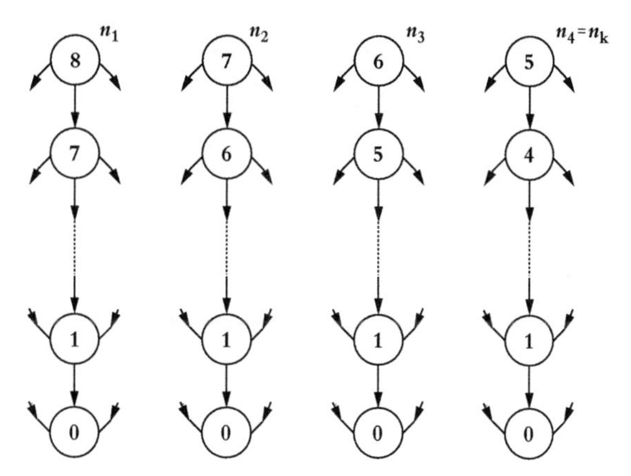

Figure 15.4 A SCORING game consisting of a sum of four SCORING games. Here $k = 4$, $n_1 = 8$, $n_2 = 7$, $n_3 = 6$, $n_4 = 5$, and $t = 3$.

Recapping our story up to now, we have made some progress: we got a tractable strategy for winning in SCORING. But what about the case when we have k scores $n_1, \ldots, n_k \in \mathbb{N}^+$ and $t \in \mathbb{N}^+$? A move consists of selecting one of the current scores and subtracting from it some $i \in \{1, \ldots, t\}$. Play ends when all the scores are zero. Figure 15.4 shows an example ($k = 4$). This is a **sum** of SCORING games, itself also a SCORING game. The notion of sum often permits us to simplify the strategy analysis, if the components of the game are disjoint. For example, NIM is the sum of its piles. It is easy to see that the game of Figure 15.4 is equivalent to the game played on the digraph of Figure 15.5, with tokens on vertices 5, 6, 7, and 8. A move consists of selecting a token and moving it right by not more than $t = 3$ places. Tokens can coexist on the same vertex. Play ends when all tokens reside on 0. What is a winning strategy?

We hit two snags when trying to answer this question:

(i) Though the sum of P-positions is in $\mathcal{P}$, the sum of N-positions is in $\mathcal{P} \cup \mathcal{N}$. Thus a game of two tokens, one on each of 5 and 7, is seen to be an N-position (the move $7 \to 5$ clearly results in a P-position), whereas the sum of a token on 3 and 7 is seen, by inspection, to be a P-position. So the simple P-, N-strategy breaks down for sums, which arise frequently in combinatorial game theory.

(ii) The **game-graph** has exponential size in the input size $\Omega(\Sigma_{i=1}^{k} \log n_i)$ of the "regular" digraph $G = (V, E)$ (with $|V| = n + 1$, where $n = \max_i n_i$) on which the game is played with k tokens (Figure 15.5 in our case). However, G is not the game-graph of the game: each tuple of k tokens on G corresponds to a single vertex of the game-graph, whose vertex-set thus has size $\binom{k+n}{n}$—the number of k-combinations of $n + 1$ distinct objects with at least k repetitions. For $k = n$ this gives $\binom{2n}{n} = \Theta(4^n/\sqrt{n})$, which is doubly exponential in the input size!

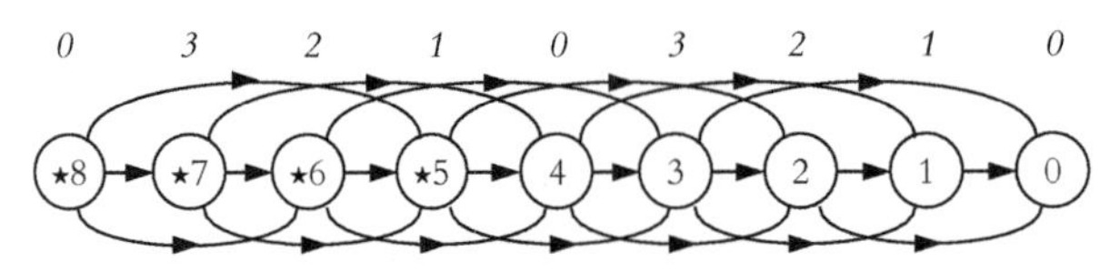

Figure 15.5 A game on a graph, but not a game-graph.

The main contribution of the classical theory is to provide a polynomial strategy for sums despite the exponential size of the game-graph. On G, label each vertex u with the least nonnegative integer not among the labels of the options of u (see top of Figure 15.5). These labels are called the **Sprague–Grundy** function values of the game on G, or the g-function for short (Sprague, 1935–1936; Grundy, 1939). It is a function from the vertices of a digraph into the nonnegative integers, defined recursively by

$$g(u) = \operatorname{mex} g(F(u)),$$

where for any subset $S \subsetneq \mathbb{N}$,

$$\operatorname{mex} S = \min \mathbb{N} \backslash S$$

is the least nonnegative integer not in S. Notice that g of the empty set is 0. The function g exists uniquely on every finite acyclic digraph.

For $u = (u_1, \ldots, u_k)$, a vertex of the game-graph (whose very construction entails exponential effort), we have

$$g(u) = g(u_1) \oplus \cdots \oplus g(u_k)\ , \quad \mathcal{P} = \{u \colon g(u) = 0\}\ , \quad \mathcal{N} = \{u \colon g(u) > 0\}\ ,$$

where $\oplus$ denotes **nim-sum** (summation over GF(2), also known as XOR, which we already met in Section 15.1). To compute a winning move from an N-position, note that there is some i for which $g(u_i)$ has a 1-bit at the binary position where $g(u)$ has its leftmost 1-bit. Reducing $g(u_i)$ appropriately makes the Nim-sum 0, and there is a corresponding move with the ith token. For the example of Figure 15.5 we have

$$g(5) \oplus g(6) \oplus g(7) \oplus g(8) = 1 \oplus 2 \oplus 3 \oplus 0 = 0\ ,$$

a P-position, so every move from this position is losing.

Is this Sprague-Grundy strategy polynomial? For Scoring, the remainders $r_1, \ldots, r_k$ of dividing $n_1, \ldots, n_k$ by $t+1$ are the g-values, as suggested by Figure 15.5. The computation of each r_j has size $O(\log n)$, where $n = \max n_i$. Since $k \log n < (k + \log n)^2$, the strategy computation (items **(a)** and **(b)** at the beginning of this section) is polynomial in the input size (k is a constant). The length of play remains exponential.

Since the strategy for Scoring is tractable for a single game as well as for a sum, we may say that Scoring has a **polynomial** strategy (see Definition 15.1).

Now consider a general nonsuccinct acyclic digraph $G = (V, E)$, that is, the input size is not logarithmic: If the graph has $|V| = n$ vertices and $|E| = m$ edges, the input

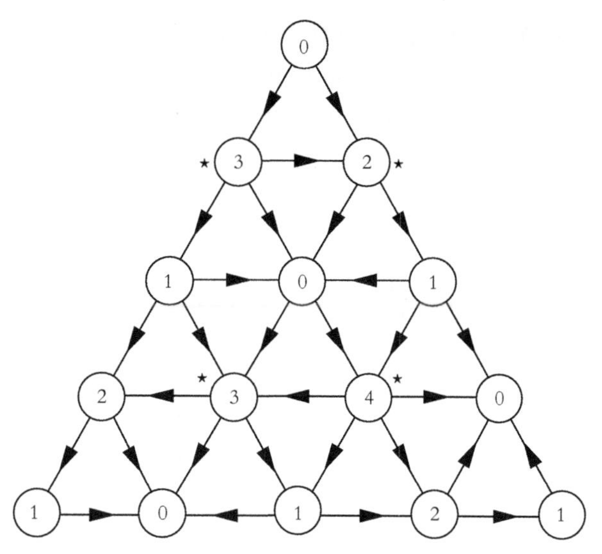

Figure 15.6 The beaten Doug.

size is $\Theta((m+n)\log n)$ (each vertex is represented by its index of size $\log n$, and each edge by a pair of indices), and g can be computed in $O((m+n)\log n)$ steps (by a "depth-first" search; each g-value is at most n, of size at most $\log n$). For a sum of k tokens on the input digraph, the input size is $\Theta((k+m+n)\log n)$, and the strategy computation for the sum can be carried out in $O((k+m+n)\log n)$ steps (nim-adding k summands of g-values). Note also that for a nonsuccinct digraph the length of play is only linear rather than exponential, in contrast to a succinct (logarithmic input size) digraph.

Our original Beat Doug problem is now also solved with a polynomial strategy. Figure 15.6 depicts the original digraph of Figure 15.2 with the g-values added in (we will see later how to compute g). Since $2 \oplus 3 \oplus 3 \oplus 4 = 6$, the given position is in $\mathcal{N}$. Moving $4 \to 2$ is a unique winning move. The winner can consummate a win in polynomial time. Also notice that the strategy for Nim is a special case of the Sprague-Grundy strategy.

However, the strategy of classical games is not very robust: slight perturbations in various directions can make the analysis considerably more difficult. Thus, the theory for Welter, which amounts to Nim with all piles of *distinct* size, is rather complicated (Conway, 2001; Chapter 13).

We point out that there is an important difference between the strategies of Beat Doug and Scoring. In both, the g-function plays a key role. But for the latter, some further property is needed to yield a strategy that is polynomial, since the input graph is (logarithmically) succinct. In this case the extra ingredient is the periodicity modulo $(t+1)$ of g, which was easy to establish. For other succinct games, it may be harder to prove polynomiality, such as for general *octal games* (Berlekamp et al., 2001–2004, vol. 1).

15.2.1 Complexity, hardness, and completeness

What, then, are tractable, polynomial, and efficient games? We abstract some of the properties of NIM, since it has a simple strategy, and it is the sum of its piles.

Definition 15.1. *Let $c > 1$ denote arbitrary constants and denote by n the size of a sufficiently succinct encoding of a digraph $G = (V, E)$. A subset T of combinatorial games with a* ***polynomial*** *strategy has the following properties. For normal play of every $G = (V, E) \in T$, and every position u of G:*

(a) *The P-, N-, D-, and tie-label of u can be computed in time $O(n^c)$ (polynomial time; D denotes draw—see next section).*

(b) *An optimal next move from any N- to a P-position and from any D- to a D-position and from any nonend tie- to a tie-position can be computed in time $O(n^c)$ (polynomial time).*

(c) *The winner can consummate a win in at most $O(c^n)$ moves (exponential time).*

(d) *The subset T is closed under summation, i.e., $G_1, G_2 \in T$ implies $G_1 + G_2 \in T$. (Thus* **(a)-(c)** *hold for $G_1 + G_2$ for every independently chosen position of G_1 and for every independently chosen position of G_2.)*

A subset $T_1 \subseteq T$ for which **(a)-(d)** *hold also for misère play—the player making the last move loses—is a subset of games with an* ***efficient*** *strategy.*

A superset $T^1 \supseteq T$ for which **(a)-(c)** *hold is a superset of games with a* ***tractable*** *strategy.*

A game in some such T or T_1 or T^1 is called ***polynomial*** *or* ***efficient*** *or* ***tractable****, respectively.*

A decidable game[1] *which has no polynomial (tractable) strategy is called* ***nonpolynomial*** *(****intractable****).*

Strictly speaking, in view of **(c)**, the terminology "polynomial" ought to be replaced by something else, such as "adequate." But "polynomial" is so universally used for problems that are computationally reasonable, that "polynomial" is preferred. Ramifications in several directions of Definition 15.1 are considered in Fraenkel (2004).

To prove that a problem is tractable, polynomial, or efficient, the normal procedure is to construct an algorithm that has those properties. But how do we show that, no matter how hard we try, a problem does not have a good solution? We explain briefly a next best way to do something in this direction.

Roughly, NP consists of all problems whose solution can be *verified*—not necessarily *found*, only verified—using an amount of time that is polynomial in a succinct input size of the problem. It is NP-complete if it is among the hardest problems in NP. It is NP-hard if it is NP-complete, except that it needs *at least* a polynomial amount of time. PSPACE consists of all problems that can be solved using a polynomial amount of space (but may need more than polynomial time), and EXPTIME—all problems that can be solved in an

[1] A problem is decidable if there exists an algorithm to solve all its instances. Otherwise it is undecidable.

exponential amount of time. Hardness and completeness are defined analogously to the respective definitions of NP. NP-complete problems share the following idiosyncrasies:

- If any NP-complete problem will be shown to have a polynomial-time algorithm, then all of them are polynomial, and if any is shown to have a lower nonpolynomial bound, then all of them are nonpolynomial.
- It is widely believed that NP-complete problems are nonpolynomial.
- Completeness results are asymptotic. With any NP-complete problem there is associated some parameter n, and the result holds for large n. For games, n is typically the size of a side of the board.

Analogous results hold *a-fortiori* for PSPACE-complete problems. But EXPTIME-completeness is an unconditional *provable* intractability: any EXPTIME-complete problem has a lower exponential time bound for its solution, asymptotically.

Optimization problems, such as Traveling Salesperson Problem (TSP) are typically NP-complete, since there is a single existential quantifier (does there exist a tour of cost $< C$?). In a two-person game, the question whether player I can win involves an alternating number of existential and universal quantifiers: does player I have a move such that for every move of player II there exists a move of player I $\cdots$ such that player I wins? If the number of alternating quantifiers is bounded, the game tends to be PSPACE-complete, such as HEX (Reisch, 1981); if their number is unbounded, it is typically EXPTIME-complete, such as CHESS (Fraenkel and Lichtenstein, 1981).

We do not know of any PSPACE-complete or EXPTIME-complete game problem that has a known polynomial solution for finite boards as encountered in practice, such as 8×8 or 19×19. Thus, though completeness and hardness are asymptotic properties, in practice they seem to say something also about actual games.

15.3. INTRODUCING DRAWS

In this section, we learn how to beat Craig (**C**yclic d**IGRA**ph) efficiently. The four starred vertices in Figure 15.7 contain one token each. The moves are identical to those of BEAT DOUG; tokens can coexist peacefully on any vertex. The only difference is that now the digraph $G = (V, E)$ may have cycles and loops (the latter correspond to passing a move), or may be infinite. In addition to the P- and N-positions, which satisfy [15.1] and [15.2] we now may have also Draw-positions, D.

Definition 15.2. *Given a game Γ, with game-graph $G = (V, E)$, where G may be finite or infinite, acyclic or cyclic. Denote by $\mathcal{O}$ the set of all nonnegative ordinals not exceeding $|V|$. By recursion on $n \in \mathcal{O}$ define the multisets,*

$$P_n = \{u \in V,\ n = \min\ m : F(u) \subseteq \bigcup_{i<m} N_i\},$$

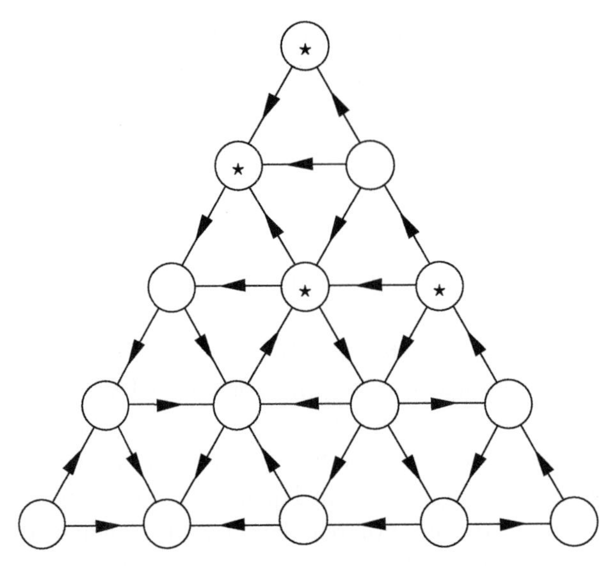

Figure 15.7 Beat CRAIG in this cyclic dIGRAph.

$$N_n = \{u \in V,\ n = \min\ m : F(u) \cap \left(\bigcup_{i<m} P_i\right) \neq \emptyset\}.$$

Finally, let

$$\mathcal{P} = \bigcup_{n \in \mathcal{O}} P_n, \quad \mathcal{N} = \bigcup_{n \in \mathcal{O}} N_n, \quad \mathcal{D} = V \backslash (\mathcal{P} \cup \mathcal{N}),$$

where $\mathcal{D}$ is the set of all D-positions.

The definition implies

$$u \in \mathcal{D} \quad \textit{if and only if} \quad F(u) \cap \mathcal{P} = \emptyset \textit{ and } F(u) \cap \mathcal{D} \neq \emptyset\ .$$

Introducing cycles causes several problems:

- Moving a token from an N-position such as vertex 4 in Figure 15.8 to a P-position such as vertex 5 is a nonlosing move, but does not necessarily lead to a win. A win is achieved only if the token is moved to the *leaf* 3. The digraph might be embedded inside a large digraph, and it may not be clear to which P-option to move in order to realize a win.
- The partition of V into $\mathcal{P}, \mathcal{N}$, and $\mathcal{D}$ is not unique, as it is for $\mathcal{P}$ and $\mathcal{N}$ in the classical case. For example, vertices 1 and 2 in Figure 15.8, if labeled P and N, would still satisfy [15.1] and [15.2], and likewise for vertices 8 and 9 (either can be labeled P and the other N).

Both of these shortcomings can be remedied by introducing a suitable counter function J—see Fraenkel and Yesha (1986).

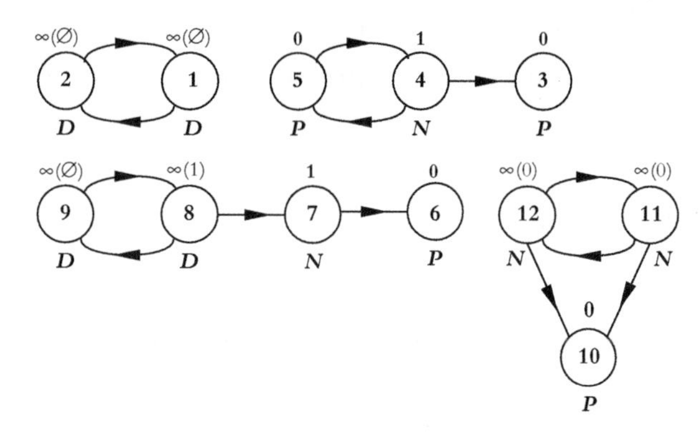

Figure 15.8 P-, N-, D-, and γ-values for simple digraphs.

For handling sums, we would like to use the g-function (Sprague-Grundy function), but there are two problems:

- The question of the existence of g on a digraph G with cycles or loops is NP-complete, even if G is planar and its degrees are ≤ 3, with each indegree ≤ 2 and each outdegree ≤ 2 (Fraenkel, 1981). (NP-completeness without these restrictions, or less restrictions, has been proved in Chvátal, 1973; Fraenkel and Yesha, 1979; van Leeuwen, 1976).
- The strategy of a cyclic game is not always determined by the g-function, even if it exists.

This is one of those rare cases where two failures are better than one! The second failure opens up the possibility that perhaps there is another tool that always works, and if we are optimistic, we might even hope that it is also polynomial. There is indeed such a generalized g-function γ. It was introduced by Smith (1966); an improved version was (re)discovered in Fraenkel and Perl (1975); see also Conway (2001, Chapter 11) and Fraenkel and Yesha (1986).

The γ-function is defined the same way as the g-function, except that it can assume not only values in $\mathbb{N}$, but in $\mathbb{N} \cup \{\infty\}$, where the symbol ∞ denotes a value bigger than every natural number. We also use the notation $\gamma(u) = \infty(K)$, where K is the set of finite γ-values of the options of u. We have $\gamma(u) = \infty(K)$, if there is $v \in F(u)$ with $\gamma(v) = \infty$, and v has no option w with $\gamma(w)$ equal to the least nonnegative integer not in K. The formal definition is given in Fraenkel and Yesha (1986). Figure 15.8 depicts γ-values for some simple digraphs. Every finite digraph with n vertices and m edges has a unique γ-function that can be computed in $O(mn \log n)$ steps. This is a polynomial-time computation, though bigger than the g-values computation.

To get a strategy for sums, define the **generalized nim-sum** as the ordinary nim-sum augmented by:

$$a \oplus \infty(L) = \infty(L) \oplus a = \infty(L \oplus a) = \infty(a \oplus L), \qquad \infty(K) \oplus \infty(L) = \infty(\emptyset),$$

where $a \in \mathbb{N}$ and $L \oplus a = \{l \oplus a\colon l \in L\}$. For a sum of k tokens on a digraph $G = (V, E)$, let $u = (u_1, \ldots, u_k)$. We then have $\gamma(u) = \gamma(u_1) \oplus \cdots \oplus \gamma(u_k)$, and

$$\begin{aligned} \mathcal{P} &= \{u\colon \gamma(u) = 0\}, \\ \mathcal{N} &= \{u\colon 0 < \gamma(u) < \infty\} \cup \{u\colon \gamma(u) = \infty(K) \text{ and } 0 \in K\}, \\ \mathcal{D} &= \{u\colon \gamma(u) = \infty(K) \text{ and } 0 \notin K\}. \end{aligned} \qquad [15.3]$$

Thus a sum consisting of a token on vertex 4 and one on 8 in Figure 15.8 has γ-value $1 \oplus \infty(1) = \infty(1 \oplus 1) = \infty(0)$, which is an N-position (the move $8 \to 7$ results in a P-position). Also one token on 11 or else on 12 is an N-position. But a token on both 11 and 12; or on 8 and 12 is a D-position of their sum, with γ-value $\infty(\emptyset)$. Also a token on 7 and 12 is a D-position, since $\infty(0) \oplus 1 = \infty(0 \oplus 1) = \infty(1)$. A token on 4 and 7 is a P-position of the sum.

With k tokens on a digraph, the strategy for the sum can be computed in $O((k + mn) \log n)$ steps. It is polynomial in the input size $\Theta((k + m + n) \log n)$, since $k + mn \leq (k + m + n)^2$. Also, for certain succinct "linear" graphs, γ provides a polynomial strategy. See Fraenkel and Tassa (1975).

Beat Craig is now also solved with a polynomial strategy. From the γ-values of Figure 15.9 we see that the position given in Figure 15.7 has γ-value $0 \oplus 1 \oplus 2 \oplus$

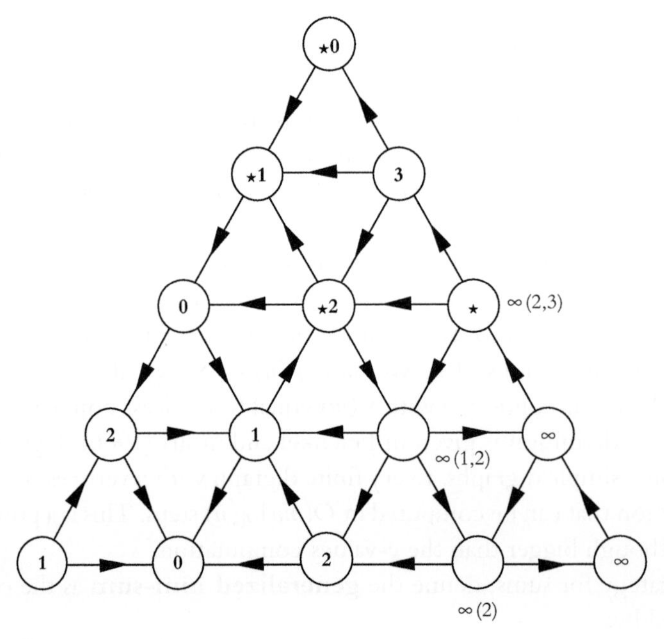

Figure 15.9 Craig has also been beaten.

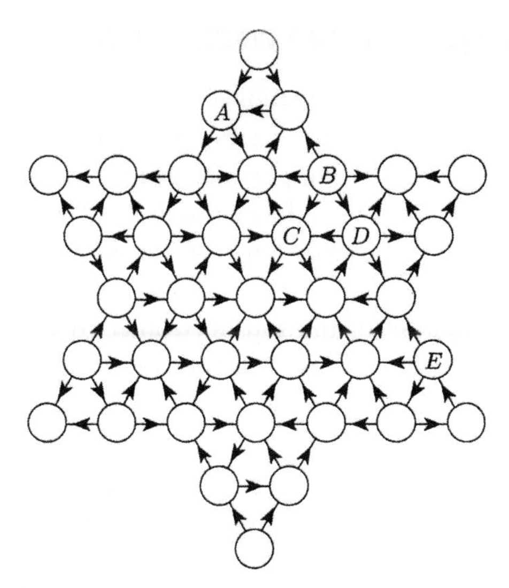

Figure 15.10 Beat an even bigger Craig.

$\infty(2,3) = 3 \oplus \infty(2,3) = \infty(1,0)$, so by [15.3] it is an N-position, and the unique winning move is $\infty(2,3) \to 3$. Again the winner can force a win in polynomial time.

As an exercise, beat an even bigger Craig: compute the labels P, N, D for the digraph of Figure 15.10 with tokens placed on vertices A–E, or for various other initial token placements.

We end this section with the *Fundamental Theorem of Combinatorial Game Theory* for impartial games which may be cyclic.

Theorem 15.1. *Let Γ be a two-person cyclic game with perfect information whose game-graph may be infinite, without chance moves and without ties. Then for every position of Γ there either exists a winning move for precisely one of the two players, or else, both players can maintain a draw.*

Proof. Every position has at least one label from among $\{\mathcal{P}, \mathcal{N}, \mathcal{D}\}$. Indeed, for any position u which is neither in $\mathcal{P}$ nor in $\mathcal{N}$, Definition 15.2 implies $u \in \mathcal{D}$. So suppose that there exists $u_0 \in (\mathcal{P} \cap \mathcal{N})$. Then $u_0 \in (P_{m_0} \cap N_{k_0})$ for some ordinals $k_0, m_0 \in \mathcal{O}$. It then follows that $F(u_0) \subseteq \mathcal{N}$, $F(u_0) \cap \mathcal{P} \neq \emptyset$. By Definition 15.2, there thus is $u_1 \in F(u_0)$ with $u_1 \in (P_{m_1} \cap N_{k_1})$, where $k_1 < m_0$, $m_1 < k_0$. Hence $F(u_1) \subseteq \mathcal{N}$, $F(u_1) \cap \mathcal{P} \neq \emptyset$. Thus there is $u_2 \in F(u_1)$ with $u_2 \in (P_{m_2} \cap N_{k_2})$, where $k_2 < m_1$, $m_2 < k_1$. This leads to two infinite sequences $k_0 > m_1 > k_2 > m_3 > \cdots$ and $m_0 > k_1 > m_2 > k_3 > \cdots$, such that $u_i \in (P_{m_i} \cap N_{k_i})$ for all $i \in \mathbb{N}$. This contradicts the well ordering of the ordinals. Hence $(\mathcal{P} \cap \mathcal{N}) = \emptyset$.

By the definition of $\mathcal{D}$ in Definition 15.2, $\mathcal{N} \cap \mathcal{D} = \mathcal{P} \cap \mathcal{D} = \emptyset$. We have shown that every position of Γ gets a unique label from among $\{P, N, D\}$.

15.4. ADDING INTERACTIONS BETWEEN TOKENS

Here, we learn how to beat Anne (**Ann**ihilation). On the five-component digraph depicted in Figure 15.11, place tokens at arbitrary locations, but at most one token per vertex. A move is defined as in the previous games, but if a token is moved onto an occupied vertex, both tokens are annihilated (removed). The digraph has cycles, and could also have loops (passing positions). Note that the three components with z-vertices are identical, as are the two y-components. The only difference between a z- and a y-component is in the orientation of the top horizontal edge. With tokens on the 12 starred vertices, can the first player win or at least draw, and if so, what is an optimal move? How "good" is the strategy?

The indicated position may be a bit complicated as a starter. So consider first a position consisting of four tokens only: one on z_0 and the other on z_2 in two of the z-components. Secondly, consider the position also consisting of four tokens: a single token on each of y_0 and y_2 in each y-component. It is clear that in both of these games player II can at least draw, simply by imitating on one component what player I does on the other. Can player II actually win in one or both of these games?

Annihilation games were proposed by John Conway. It is easy to see that on a finite *acyclic* digraph, annihilation can affect the length of play, but the strategy is the same as for the classical games: Since $g(u) \oplus g(u) = 0$, the winner does not need to use annihilation, and the loser cannot be helped by it. But the situation is quite different in the presence of cycles. In Figure 15.12a, a token on each of the vertices z_1 and z_3 is clearly a D-position for the nonannihilation case, but it is a P-position when played with annihilation (the second move is a winning annihilation move). In Figure 15.12b, with annihilation, a

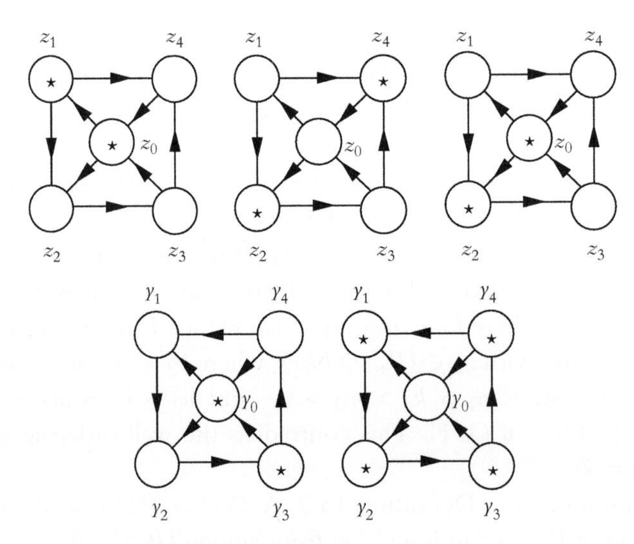

Figure 15.11 Beat Anne in this ANNihilation game.

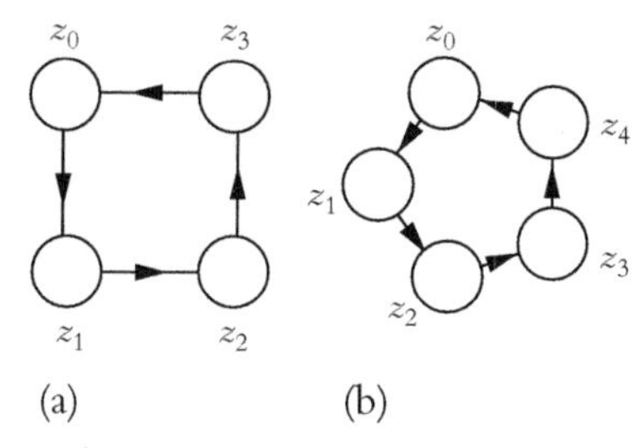

Figure 15.12 Annihilation on simple cyclic digraphs.

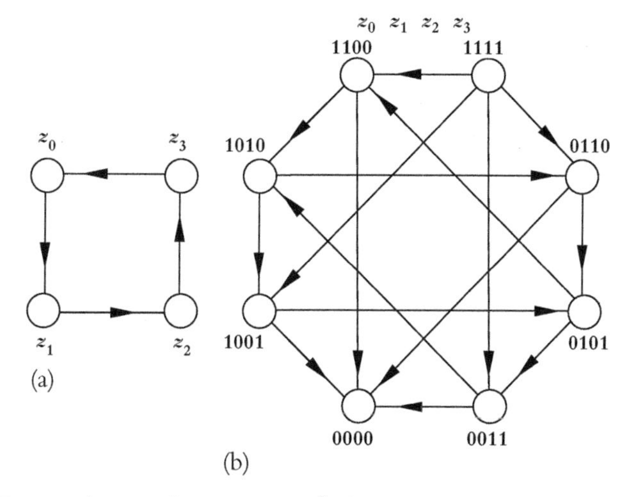

Figure 15.13 (b) Depicts the "even" component G^0 of the annihilation graph G of the digraph (a).

token on each of z_1 and z_2 is an N-position, whereas a token on each of z_1 and z_3 is a D-position. The theory of annihilation games is discussed in depth in Fraenkel and Yesha (1982); see also Fraenkel (1974), Fraenkel and Yesha (1976, 1979), and Fraenkel et al. (1978). Misère annihilation play was analyzed by Ferguson (1984).

The **annihilation graph** is a certain game-graph of an annihilation game. The annihilation graph of the annihilation game played on the digraph of Figure 15.12a consists of two components. One is depicted in Figure 15.13b, namely, the component $G^0 = (V^0, E^0)$ with 8 vertices and an even number of tokens. The "odd" component G^1 also has 8 vertices. In general, a digraph $G = (V, E)$ with $|V| = n$ vertices has an annihilation graph $G = (V, E)$ with $|V| = 2^n$ vertices, namely all n-dimensional binary vectors. The γ-function on G determines whether any given position is in $\mathcal{P}$, N, or $\mathcal{D}$, according to [15.3]; and γ, together with its associated counter function, determines an optimal next move from an N- or D-position.

The only problem is the exponential size of G. We can recover an $O(n^6)$ strategy by computing an **extended γ-function** σ on an induced subgraph of G of size $O(n^4)$,

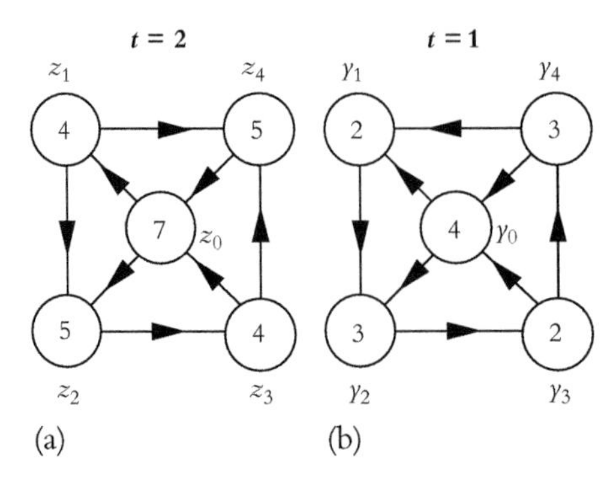

Figure 15.14 The σ-function.

namely, on all vectors of weight ≤ 4 (at most four 1-bits). In Figure 15.14, the numbers inside the vertices are the σ-values, computed by Gaussian elimination over GF(2) of an $n \times O(n^4)$ matrix. This computation also yields the values $t = 2$ for Figure 15.14a and $t = 1$ for Figure 15.14b: If $\sigma(u) \geq 2^t$, then $\gamma(u) = \infty$, whereas $\sigma(u) < 2^t$ implies $\gamma(u) = \sigma(u)$.

Thus for Figure 15.14a we have $\sigma(z_0, z_2) = 5 \oplus 7 = 2 < 4$, so $\gamma(z_0, z_2) = 2$. Hence, two such copies constitute a P-position ($2 \oplus 2 = 0$). (How can player II consummate a win?) In Figure 15.14b we have $\sigma(y_0, y_2) = 3 \oplus 4 = 7 > 2$, so $\gamma(y_0, y_2) = \infty$, in fact, $\infty(0, 1)$, so two such copies constitute a D-position. (How can the two players maintain the draw?) We have thus answered the two questions posed in the second paragraph of the present section.

The position given in Figure 15.11 is repeated in Figure 15.15, together with the σ-values. From left to right we have: for the z-components, $\gamma = 3 \oplus 0 \oplus 2 = 1$; and for the y-components, $\infty(0, 1) \oplus 0 = \infty(0, 1)$, so the γ-value is $\infty(0, 1) \oplus 1 = \infty(0, 1)$. Hence the position is an N-position by [15.3]. There is, in fact, a unique winning move, namely $y_0 \to y_2$ in the first component from the left. Any other move leads to drawing or losing. We have learned how to beat Anne.

For small digraphs, a counter function c is not necessary, but for larger ones it is needed for consummating a win. There is a problem in computing c: our polynomial algorithm produces γ and c only for an $O(n^4)$ portion of G. Whereas γ can then be extended easily to all of G, this does not seem to be the case for c. There is a way out involving a **broad** strategy.

A strategy is **narrow** if it uses only the present position u for deciding whether u is a P-, N-, or D-position, and for computing a next optimal move. It is **broad** (Fraenkel, 1991) if the computation involves any of the possible predecessors of u, whether actually encountered or not. It is **wide** if it uses any ancestor that was actually encountered in the play of the game. Wide strategies were defined by Kalmár (1928) and Smith (1966), but then both authors immediately reverted back to narrow strategies, since both authors

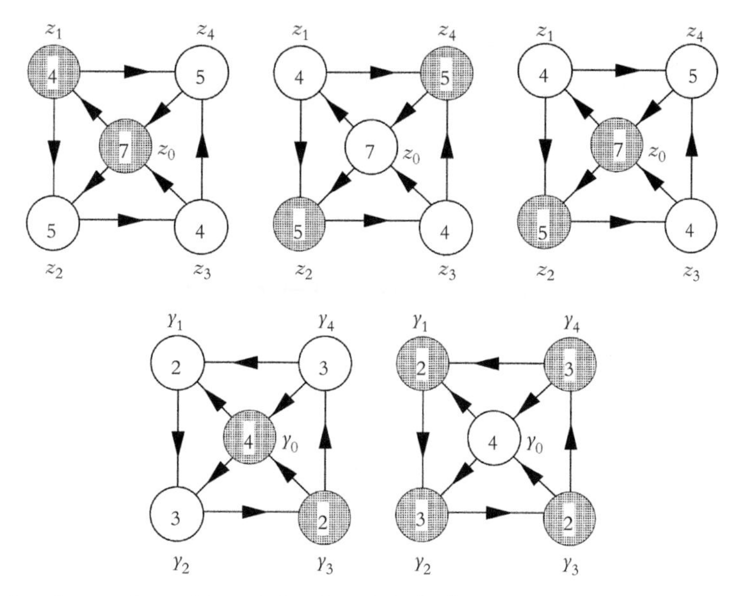

Figure 15.15 Poor beaten Anne. (Gray circles show initial token positions.)

remarked that the former do not seem to have any advantage over the latter. Yet for annihilation games, only a broad strategy was found that is polynomial. For details, see Fraenkel and Yesha (1982).

For certain (Chinese) variations of Go, for CHESS and some other games, there are rules that forbid certain repetitions of positions, or modify the outcome in the presence of such repetitions. Now if all the history is included in the definition of a move, then every strategy is narrow. But the way, Kalmár (1928) and Smith (1966) defined a move—much the same as the intuitive meaning—there is a difference between a narrow and wide strategy for these games.

As an exercise, compute the label $\in \{P, N, D\}$ of the stellar configuration marked by letters in "Interstellar encounter with Jupiter" (Figure 15.16), where J is Jupiter, the other letters are various fragments of the Shoemaker-Levy comet, and all the vertices are "space-stations." A move consists of selecting Jupiter or a fragment, and moving it to a neighboring space-station along a directed trajectory. Any two bodies colliding on a space-station explode and vanish in a cloud of interstellar dust. Whereas in "Beat Anne" there is no leaf, here there are six "black holes," where a body is absorbed and cannot escape. Both players are viciously bent on making the final move to destroy this solar subsystem. Is the given position a win for player I or for player II? Or is it a draw, so that a part of this subsystem will exist forever? And if so, can it be arranged for Jupiter to survive as well? (An encounter of the Shoemaker-Levy comet with Jupiter took place in mid-July, 1994.)

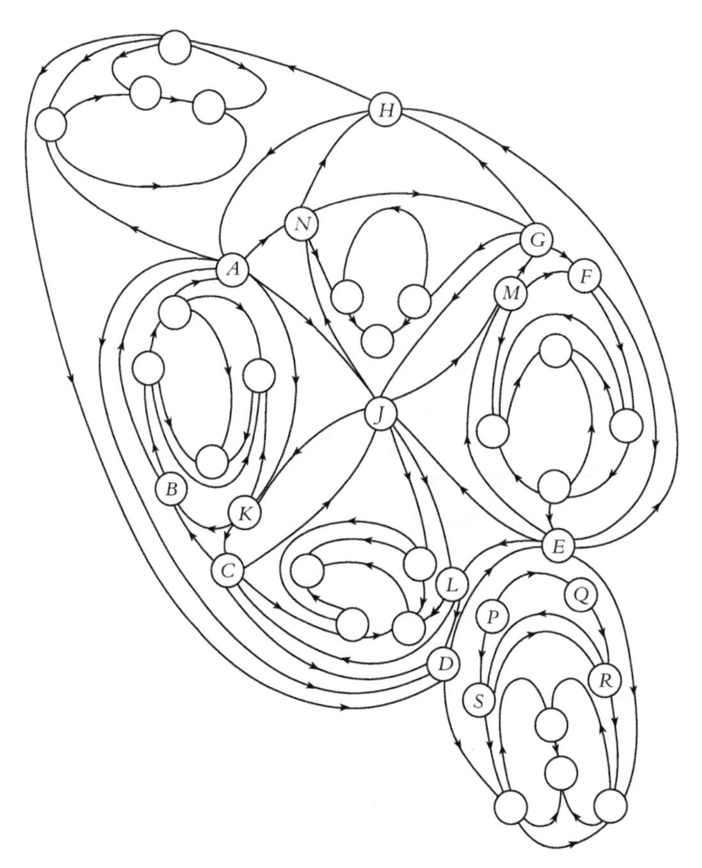

Figure 15.16 Interstellar encounter with Jupiter.

Various impartial and partizan variations of annihilation games were shown to be NP-hard, PSPACE-complete, or EXPTIME-complete (Fraenkel and Goldschmidt, 1987; Goldstein and Reingold, 1995). We mention here only briefly an interaction related to annihilation. Electrons and positrons are positioned on vertices of the game MATTER AND ANTIMATTER (Figure 15.17). A move consists of moving a particle along a directed trajectory to an adjacent station—if not occupied by a particle of the same kind, since two electrons (and two positrons) repel each other. If there is a resident particle, and the incoming particle is of the opposite type, they annihilate each other, and both disappear from the play. It is not very hard to determine the label of any position on the given digraph. But what can be said about a general digraph? About succinct digraphs? Note that the special case where all the particles are of the same type, is the generalization of WELTER played on the given digraph. WELTER, mentioned in the penultimate paragraph of section 15.2, is NIM with the restriction that no two piles have the same size. It has a polynomial strategy, but its validity proof is rather intricate (Conway, 2001, Chapter 13).

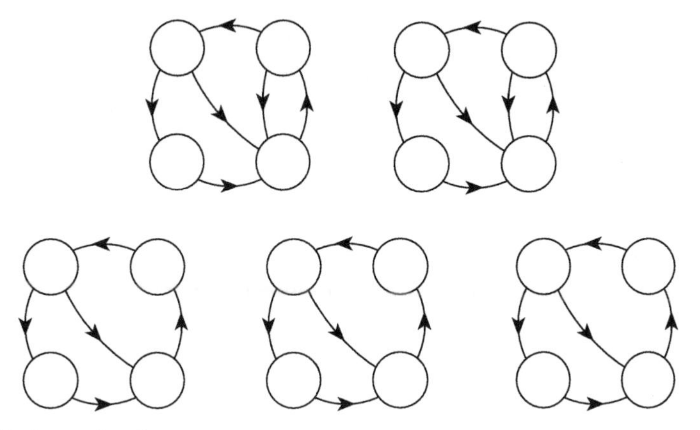

Figure 15.17 Matter and antimatter.

15.5. PARTIZAN GAMES

In a **partizan combinatorial game** there are two players, Left and Right, who have distinct sets of moves available from each position. A game G is **short** if it meets both of the following conditions:

- G is **finite**: it has just finitely many distinct subpositions, and
- G is **acyclic**: there is no infinite sequence of moves proceeding from G.

Formally, a short partizan game G can be represented as an ordered pair $(\mathscr{G}^L, \mathscr{G}^R)$, where $\mathscr{G}^L$ and $\mathscr{G}^R$ are sets of "simpler" games (that is, games with strictly fewer subpositions). Elements of $\mathscr{G}^L$ (respectively, $\mathscr{G}^R$) are called **Left** (respectively, **Right**) **options** of G. We will sometimes write

$$G = \left\{\mathscr{G}^L \mid \mathscr{G}^R\right\},$$

though we will usually list the options of G explicitly:

$$G = \left\{G_1^L, G_2^L, \ldots, G_m^L \mid G_1^R, G_2^R, \ldots, G_n^R\right\}$$

or abuse notation and write simply

$$G = \left\{G^L \mid G^R\right\}$$

to indicate that G^L and G^R range over all the Left and Right options of G. The simplest game is the **empty game** 0, from which there are no options for either player:

$$0 = \{\ \mid\ \}.$$

Then we define the set of short games $\tilde{\mathbb{G}}$ by

$$\tilde{\mathbb{G}}_0 = \{0\}; \quad \tilde{\mathbb{G}}_{n+1} = \left\{ \left\{\mathscr{G}^L \mid \mathscr{G}^R\right\} : \mathscr{G}^L, \mathscr{G}^R \subset \tilde{\mathbb{G}}_n \right\}; \quad \tilde{\mathbb{G}} = \bigcup_{n \geq 0} \tilde{\mathbb{G}}_n.$$

The theory of partizan games was introduced by Berlekamp, Conway and Guy in the 1970s and early 1980s. The classical texts *Winning Ways for Your Mathematical Plays* (Berlekamp et al., 2001–2004) and *On Numbers and Games* (Conway, 2001) remain excellent introductions.

15.5.1 Two examples: HACKENBUSH and DOMINEERING

HACKENBUSH is played on a finite undirected graph with colored edges, such as the one in Figure 15.18a. The solid horizontal line in Figure 15.18a represents a single vertex of the graph, the **ground**. On her turn, Left may remove any bLue (soLid; dark gray in print version) edge; Right may remove any Red (paRallel; parallel line in print version) one. GrEen (dottEd; dotted in print version) edges may be removed by either player. After each move, any edges no longer connected to the ground are also removed from play. HACKENBUSH follows the same normal-play convention as NIM: whoever makes the last move wins.

DOMINEERING is played on an $m \times n$ checkboard, typically 8×8. Left and Right alternately place dominoes on the board. Each domino must cover exactly two adjacent squares, and dominoes may never overlap. Moreover, Left must place verticaLly-oriented dominoes, and Right must place hoRizontally-oriented ones. Eventually, the players will run out of moves (since the board will fill up with dominoes), and whoever makes the last move wins. (Notice that making the last move coincides with placing the most dominoes, with ties broken in favor of the second player.)

Figure 15.19a shows a typical position after each player has made one move: Left made an opening move in the northeast corner of the board, and Right responded in the southeast. Figure 15.19b shows the first 14 moves of a game played between David Wolfe and Dan Calistrate, in the finals of the first (and last) World Domineering Championship. Left 13 was a fatal mistake, and after Right 14 Calistrate went on to win the match and the tournament. A description of the Wolfe-Calistrate DOMINEERING match can be found in West (1996).

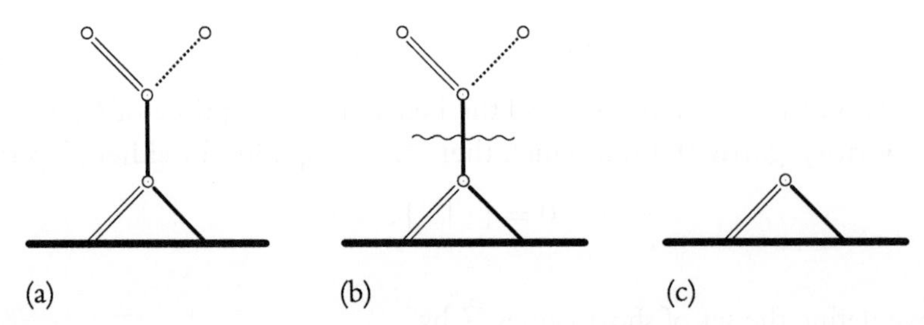

Figure 15.18 (a) A HACKENBUSH position; (b) a typical opening move for Left; and (c) the resulting position after Left's move.

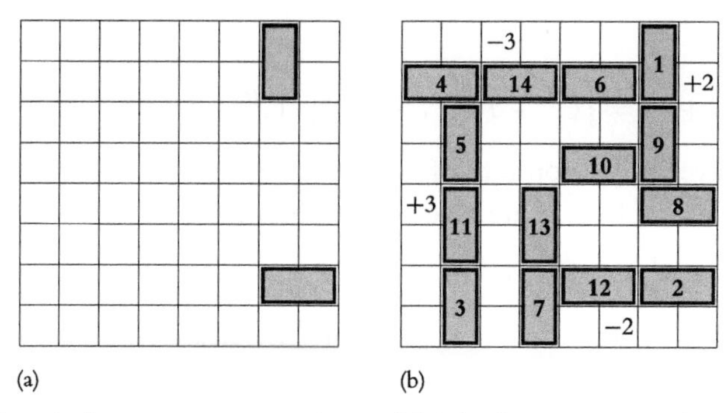

Figure 15.19 (a) A typical DOMINEERING opening and (b) the first 14 moves of Wolfe-Calistrate 1994, Round 3.

Notice that the position in Figure 15.19b can be subdivided into six separate territories, and no single move can affect more than one such component. Subsequent play on the four components labelled +3, +2, −2 and −3 is entirely predictable: Left will place exactly n dominoes on each $+n$ component, and Right will place n dominoes on each $-n$ component. The remaining two regions are more exciting; their resolutions depend on who plays first on which territory. Assigning meaningful mathematical values to such components, and describing their combinatorial interactions, is a central goal of the partizan theory.

15.5.2 Outcomes and sums

If G is a short partizan game, then G belongs to one of four **outcome classes**:

- $\mathscr{N}$ first player (the *N*ext player) can force a win;
- $\mathscr{P}$ second player (the *P*revious player) can force a win;
- $\mathscr{L}$ Left can force a win, no matter who moves first; and
- $\mathscr{R}$ Right can force a win, no matter who moves first.

The proof that every game belongs to one of these four classes is a trivial generalization of Theorem 15.1, the Fundamental Theorem for impartial games. We denote by $o(G)$ the outcome class of G. Figure 15.20 gives examples of HACKENBUSH positions representing all four classes.

The **disjunctive sum** $G + H$ is formed as follows: Place copies of G and H side-by-side; on her turn, a player must move in *exactly one* of the two components. Formally,

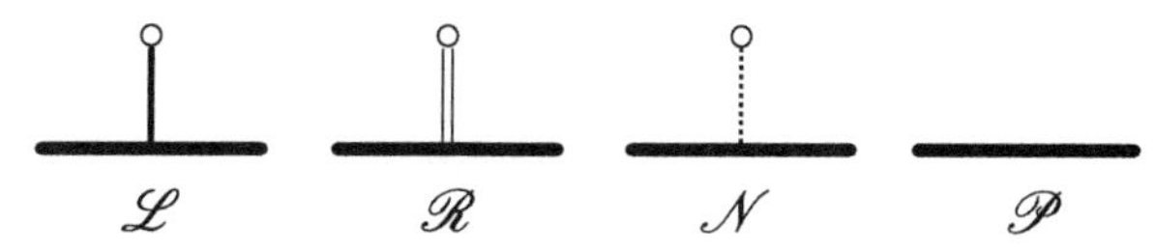

Figure 15.20 Four HACKENBUSH positions with distinct outcome classes.

we may write

$$G+H=\{G^L+H,\ G+H^L \mid G^R+H,\ G+H^R\}. \qquad [\dagger]$$

Here G^L ranges over all Left options of G, and H^L ranges over all Left options of H, so that the Left options of $G+H$ are given by the union

$$\{X+H : X \in \mathcal{G}^L\} \cup \{G+Y : Y \in \mathcal{H}^L\}. \qquad [\ddagger]$$

The notation in the equation marked (†) is generally clearer and more succinct than set notation (‡), and we will use it throughout this article without further comment.

Each game G also has a **negative** $-G$, obtained by interchanging the roles of Left and Right:

$$-G=\{-G^R \mid -G^L\}$$

We write $G-H$ as shorthand for $G+(-H)$.

The definition of disjunctive sum is motivated by examples such as DOMINEERING, in which endgame positions decompose naturally into sums. The position in Figure 15.19b, for example, can be written as the sum of six independent territories. Likewise, positions in NIM and KAYLES can be written as the disjunctive sum of single piles.

This modularity is central to combinatorial game theory. Given a sum of games

$$G=G_1+G_2+\cdots+G_k,$$

it is often impractical to undertake a brute-force analysis of G itself. Instead, we study the components G_i individually, and attempt to extract information that can be pieced back together to determine $o(G)$. In Section 15.2, this "information" took the form of **nim values**; in the context of partizan games, a more general notion of **game value** is needed.

Observe that it's not always sufficient to know the outcomes of each component. For example, let G and H be the following simple HACKENBUSH positions:

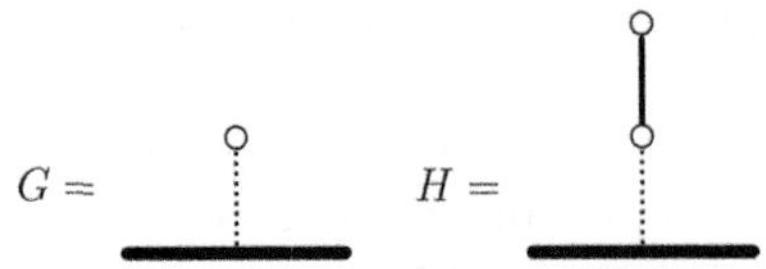

$G =$ $H =$

Then $o(G)=o(H)=\mathcal{N}$: either player can win immediately (on either game, played in isolation) by chopping the unique green (dotted line in print version) edge, moving to 0. Also $o(G+G)=\mathcal{P}$, by the obvious symmetry argument. However, on the sum

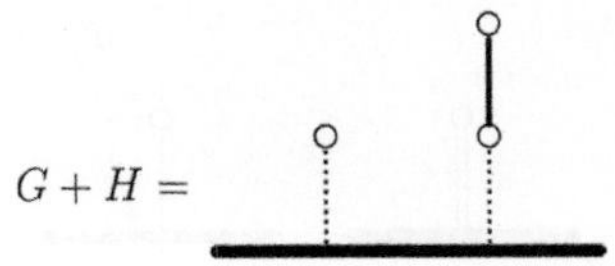

$G+H =$

Left can win no matter who moves first, since she can arrange that *Right* is always first to chop a green (dotted line in print version) edge. So $o(G+H) = \mathscr{L}$, and this shows that G and H have unequal values.

15.5.3 Values

If G and H are partizan games, then we write

$$G = H \quad \text{if} \quad o(G+X) = o(H+X) \text{ for all } X.$$

Here X ranges over *all* short partizan games (that is, all elements of $\tilde{\mathbb{G}}$). In particular, suppose G and H are HACKENBUSH positions. Then X ranges over all HACKENBUSH positions, but *also* over games that are not necessarily representable in HACKENBUSH. This is deliberate: the universal quantifier is essential in order to get a good theory, and as we'll see in a moment it provides a common language for identifying shared structure in combinatorial games.

The **game value** of G is its equivalence class modulo equality. The idea is that given an arbitrary sum

$$G = G_1 + G_2 + \cdots + G_k,$$

the value, and hence the outcome, of G can be computed from the values of each G_i. The set of game values is denoted by $\mathbb{G}$.

Figure 15.21 gives a nontrivial example of two games with the same value.

The outcome classes are naturally partially ordered by *favorability to Left*:

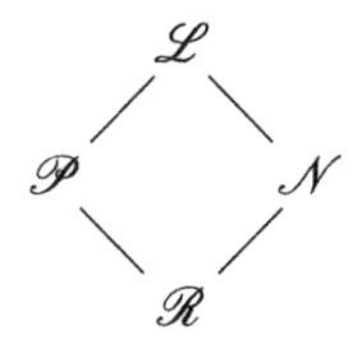

This induces a partial order of $\mathbb{G}$:

$$G \geq H \quad \text{if} \quad o(G+X) \geq o(H+X) \text{ for every all } X.$$

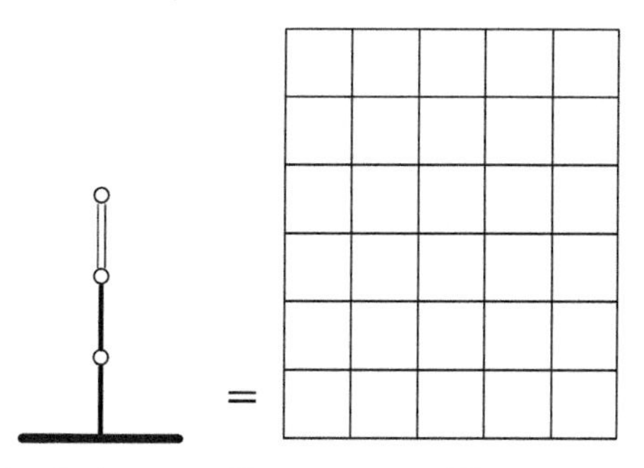

Figure 15.21 A nontrivial identity between HACKENBUSH and DOMINEERING.

If $G \geq H$, then Left will be satisfied to replace the component H with G, in any conceivable sum of games. The basic theorems are as follows:

Theorem 15.2. *$o(G) \geq \mathscr{P}$ if and only if $G \geq 0$, for all short games G.*

Theorem 15.3. *$\mathbb{G}$ is a partially ordered Abelian group under disjunctive sum, with identity 0.*

Note that $o(G) \geq \mathscr{P}$ if and only if Left can force a win on G as second player. So Theorem 15.2 implies that *every second-player win is equal to* 0. This directly generalizes the impartial theory, in which every second-player win has nim value 0.

We will also write $G \cong H$ to mean that G and H are identical (**isomorphic**) games. Certainly $G \cong H$ implies $G = H$, but $G = H$ does *not* imply $G \cong H$ (since in particular, if G is *any* second-player win, then $G = 0$).

15.5.4 Simplest forms

The central result of the partizan theory is the *Simplest Form Theorem*: every game value has a unique simplest representative. The Simplest Form Theorem is obtained through the following explicit construction.

For a given G, we identify several types of "extraneous" options:

- A Left option G^{L_1} is **dominated** (by G^{L_2}) if $G^{L_2} \geq G^{L_1}$ for some other Left option G^{L_2}.
- A Right option G^{R_1} is **dominated** (by G^{R_2}) if $G^{R_2} \leq G^{R_1}$ for some other Right option G^{R_2}.
- A Left option G^{L_1} is **reversible** (through $G^{L_1R_1}$) if $G^{L_1R_1} \leq G$ for some Right option $G^{L_1R_1}$.
- A Right option G^{R_1} is **reversible** (through $G^{R_1L_1}$) if $G^{R_1L_1} \geq G$ for some Left option $G^{R_1L_1}$.

Dominated options can be removed from G without affecting its value: in any sum $G + X$ from which Left would like to move to $G^{L_1} + X$ (with G^{L_1} dominated by G^{L_2}), she is equally satisfied to play $G^{L_2} + X$ instead.

Reversible options are a bit more subtle. If G^{L_1} is reversible through $G^{L_1R_1}$, then G^{L_1} can be replaced with the set of all $G^{L_1R_1L}$, without affecting the value of G. Symbolically:

$$G = \left\{ G^{L_1R_1L}, G^{L'} \mid G^R \right\},$$

with $G^{L'}$ ranging over all Left options of G *except* G^{L_1}. This operation is known as **bypassing** the reversible move G^{L_1} (through $G^{L_1R_1}$).

Any game G can be simplified by repeatedly eliminating dominated options and bypassing reversible ones. Each such operation strictly reduces the number of edges in the game tree of G, so this process necessarily produces a game K with no dominated or reversible options, and such that $K = G$. Such K is called the **canonical form** or **simplest form** of G, and the following theorem shows that it is unique.

Theorem 15.4. (Simplest Form Theorem) *Suppose that $G = H$, and neither G nor H has any dominated or reversible options. Then $G \cong H$.*

The Simplest Form Theorem follows immediately by inductive application of the following lemma:

Lemma 15.1. *Suppose that $G = H$, and neither G nor H has any dominated or reversible options. Then for every H^L, there is a G^L such that $G^L = H^L$, and vice versa; and likewise for Right options.*

Proof. Consider a Left option H^L. Since $G - H \geq 0$, Left must have a winning response to Right's opening move $G - H^L$. In particular, either $G^L - H^L \geq 0$ for some G^L, or else $G - H^{LR} \geq 0$ for some H^{LR}. But the latter would imply

$$H = G \geq H^{LR},$$

contradicting the assumption that H has no reversible options. So necessarily $G^L \geq H^L$ for some G^L. An identical argument now shows that $H^{L'} \geq G^L$ for some $H^{L'}$, so that

$$H^{L'} \geq G^L \geq H^L.$$

But H has no dominated options, so none of the inequalities can be strict, and in particular $G^L = H^L$. Proofs of the other cases are the same.

15.5.5 Numbers

Consider a single blue (dark gray in print version) HACKENBUSH stalk, from which Left can move to 0, and Right has no move at all:

$$= \{ \quad | \ \} = \{0 \mid \ \}$$

This game is denoted by 1, since it behaves like one spare move for Left. Since $1 > 0$, it generates a subgroup of $\mathbb{G}$ isomorphic to $\mathbb{Z}$, and it is customary to identify this subgroup with $\mathbb{Z}$. In particular we have

$$2 = 1 + 1 = \{1 \mid \ \}, \quad 3 = 2 + 1 = \{2 \mid \ \}, \quad \ldots$$

and in general $n + 1 = \{n \mid \ \}$, and $-(n + 1) = \{ \ \mid -n\}$.

In Figure 15.22, we see various other **numbers**, for example

$$= \{ \quad | \ \} = \{0 \mid \ \}$$

The identity $\frac{1}{2} + \frac{1}{2} = 1$ is easily verified by showing that the difference game

$$\tfrac{1}{2} + \tfrac{1}{2} - 1$$

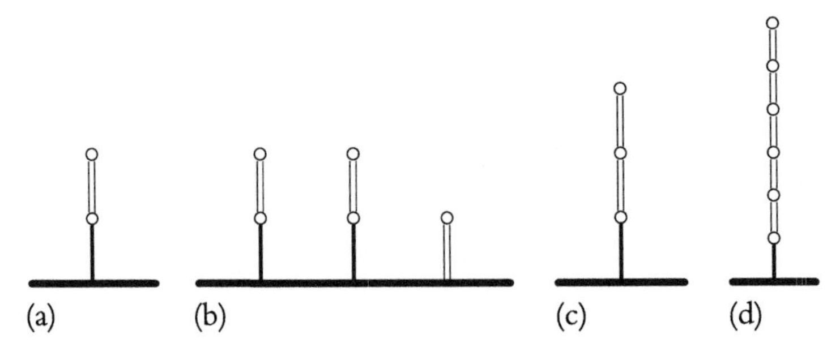

Figure 15.22 HACKENBUSH positions: (a) $\frac{1}{2}$; (b) $\frac{1}{2}+\frac{1}{2}-1$; (c) $\frac{1}{4}$; and (d) $\frac{1}{32}$.

is a second-player win. Larger denominators can be similarly constructed:

$$\frac{1}{2^{n+1}} = \left\{0 \,\middle|\, \frac{1}{2^n}\right\}$$

and such numbers generate a subgroup of $\mathbb{G}$ isomorphic to $\mathbb{D}$, the group of **dyadic rationals**:

$$\mathbb{D} = \left\{q \in \mathbb{Q} : 2^n q \in \mathbb{Z} \text{ for some } n \geq 0\right\}.$$

The canonical form of $m/2^n$ (in lowest terms) is given by

$$\frac{m}{2^n} = \left\{\frac{m-1}{2^n} \,\middle|\, \frac{m+1}{2^n}\right\}.$$

The inductive structure of numbers is neatly visualized in Figure 15.23. For each $n \geq 0$, there are 2^n numbers with **birthday** exactly n.

Now if x is a number, then it is a disadvantage to move on x, in the sense that

$$x^L < x < x^R$$

for every x^L and x^R. Remarkably, this criterion characterizes the dyadic rationals.

Theorem 15.5. *Let x be a short game, and suppose that $y^L < y < y^R$ for every subposition y of x and every y^L and y^R. Then $x \in \mathbb{D}$.*

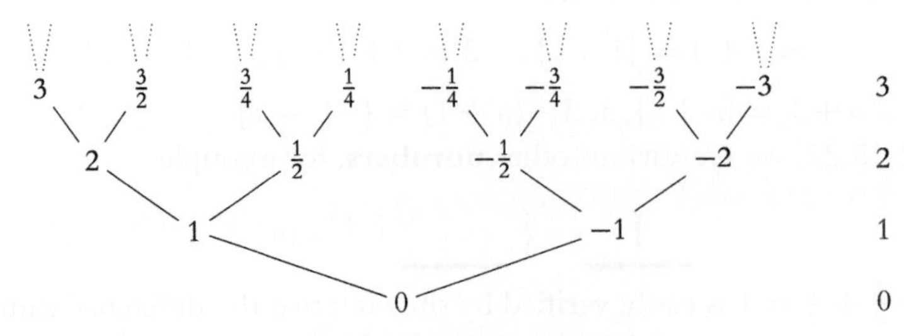

Figure 15.23 The number tree (with birthdays labeled on the right).

This observation has several fundamental consequences.

Theorem 15.6. (Number Avoidance Theorem) *Suppose that x is equal to a number but G is not. If Left (resp. Right) has a winning move on $G + x$, then she can win by playing on G.*

Theorem 15.7. (Number Translation Theorem) *Suppose that x is equal to a number but G is not. Then*

$$G + x = \left\{ G^L + x \mid G^R + x \right\}.$$

15.5.6 Infinitesimals

Numbers provide a natural metric against which other games can be calibrated. In particular, there is a vast hierarchy of games that are **infinitesimal** in the sense that

$$x > G > -x$$

for all positive numbers x.

The simplest nonzero infinitesimal is the game $*$ (pronounced "star"), from which either player can move to 0:

$$* = \{0 \mid 0\} =$$

It's easily checked that $*$ is an infinitesimal, since on the sum

$$* + \frac{1}{2^n} =$$

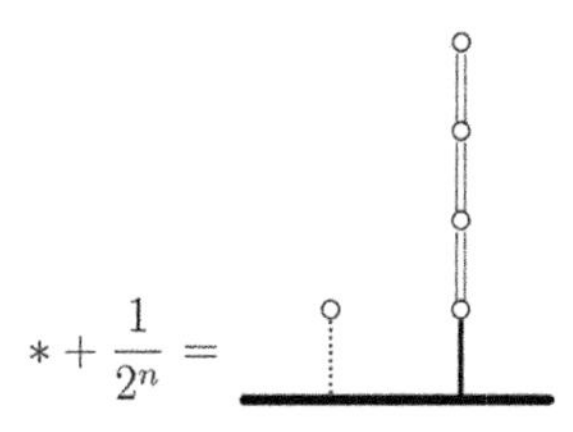

Left can win easily by playing preferentially on $*$, independent of the value of n.

Note that $*$ is isomorphic to a nim-heap of size 1. In the partizan context, a nim-heap of size m is denoted by $*m$ (pronounced "star m"). Symbolically:

$$*m = \{0, *, *2, \ldots, *(m-1) \mid 0, *, *2, \ldots, *(m-1)\}.$$

Each $*m$ (for $m \geq 1$) is a first-player win, and so is confused with 0. The simplest *signed* infinitesimals are

$$\uparrow = \{0 \mid *\} \quad (\text{"up"}) \quad \text{and} \quad \downarrow = -\uparrow = \{* \mid 0\} \quad (\text{"down"})$$

Certainly $\uparrow > 0$, since Left can win no matter who moves first. But $\uparrow$ is infinitesimal, by the same argument used for $*$: on $\downarrow +2^{-n}$ (say), Left can win by playing preferentially on $\downarrow$.

15.5.7 Stops and the mean value

If G is not a number, then its **confusion interval** is given by

$$\mathcal{C}(G) = \{x \in \mathbb{D} : G \ngtrless x\}.$$

The reader is invited to check the following examples:

- $\mathcal{C}(*) = \{0\}$, a singleton.
- $\mathcal{C}(\uparrow) = \emptyset$.
- $\mathcal{C}(\{3 \mid -3\})$ is the closed interval $[-3, 3]$.
- $\mathcal{C}(\{3 + * \mid -3\})$ is the half-open interval $[-3, 3)$.

The endpoints of $\mathcal{C}(G)$ are fundamental invariants of G, known as the **Left stop** $L(G)$ and **Right stop** $R(G)$ of G. Between them lies a third invariant, the **mean value** $m(G)$, which has the following remarkable properties:

$$m(G+H) = m(G) + m(H) \quad \text{for all } G \text{ and } H;$$

and for all G, the difference

$$\big(n \cdot G\big) - \big(n \cdot m(G)\big)$$

is bounded by a constant independent of n. Therefore, $m(G)$ is a number that closely approximates the limiting behavior of many copies of G.

One can think of G as vibrating between its Left and Right stops in such a way that its "center of gravity" lies at $m(G)$.

15.6. MISÈRE PLAY

We now return to the subject of impartial games, but considered under the **misère play convention**, in which the player who makes the last move *loses*. The misère theory was introduced by Plambeck (2005) and Plambeck and Siegel (2008); see Siegel (2013) for a concise overview.

The Fundamental Theorem works in misère play too, with the same proof, so that every impartial game G has a **misère outcome** ($\mathscr{N}$ or $\mathscr{P}$) in addition to its normal outcome. The misère outcome of G is denoted by $o^-(G)$.

The motivating question in misère impartial games is this: What is the misère analog of the Sprague-Grundy Theory? There are several reasonable answers to this question, each relevant in a different set of circumstances.

15.6.1 Misère Nim value

Let G be a NIM position, with heaps of sizes $a_1, \ldots, a_k$. Recall that $o(G) = \mathscr{P}$ if and only if $a_1 \oplus \cdots \oplus a_k = 0$. A similar rule works in misère play, but it is slightly more complicated.

Theorem 15.8. (Bouton) *The* NIM *position G with heaps $a_1, \ldots, a_k$ is a misère $\mathscr{P}$-position if and only if*

$$a_1 \oplus \cdots \oplus a_k = 0,$$

unless every $a_i = 0$ or 1. In that case, G is a $\mathscr{P}$-position if and only if

$$a_1 \oplus \cdots \oplus a_k = 1.$$

In particular, note that $*$ is a misère $\mathscr{P}$-position, but $*m$ is an $\mathscr{N}$-position for all $m \neq 1$. This motivates the following misère analog of nim values. Recall that the *(normal) nim value* of G is given recursively by

$$\mathscr{G}(G) = \begin{cases} 0 & \text{if } G \cong 0; \\ \operatorname*{mex}_{G' \in G} \mathscr{G}(G') & \text{otherwise.} \end{cases}$$

The **misère nim value** is similarly defined, but with a different base case:

$$\mathscr{G}^-(G) = \begin{cases} 1 & \text{if } G \cong 0; \\ \operatorname*{mex}_{G' \in G} \mathscr{G}(G') & \text{otherwise.} \end{cases}$$

The misère nim value of G determines its outcome. In fact we can say something slightly stronger:

Theorem 15.9. *$\mathscr{G}^-(G)$ is the unique value of m such that $o^-(G + *m) = \mathscr{P}$. In particular, G is a misère $\mathscr{P}$-position if and only if $\mathscr{G}^-(G) = 0$.*

The problem with misère nim values is that they are not well-behaved in sums. For example, let $G = *$ and $H = *2 + *2$. Then G and H are both $\mathscr{P}$-positions (by Theorem 15.8), so

$$\mathscr{G}^-(G) = \mathscr{G}^-(H) = 0.$$

However, it is not hard to show (using Theorem 15.9, say) that

$$\mathscr{G}^-(G + *2) = 3, \quad \text{but} \quad \mathscr{G}^-(H + *2) = 2.$$

So the misère nim value of a sum of games cannot be determined from the nim values of its components.

15.6.2 Genus theory

The **genus** of G (plural **genera**), denoted by $\mathscr{G}^{\pm}(G)$, is obtained by conjoining its normal and misère nim values:

$$\mathscr{G}^{\pm}(G) = \big(\mathscr{G}(G), \mathscr{G}^{-}(G)\big).$$

For brevity it is customary to write $\mathscr{G}^{\pm}(G) = a^b$ in place of $\mathscr{G}^{\pm}(G) = (a, b)$.

Remarkably, genus values *are* well behaved in sums, but *only* for a particular class of games known as **tame games**. First, note that one can easily classify all the genera that arise in misère NIM:

- If G has no heaps of size ≥ 2, then $\mathscr{G}^{\pm}(G) = 0^1$ or 1^0, depending on the parity of the number of heaps of size 1.
- Otherwise, $\mathscr{G}^{\pm}(G) = a^a$, where $a = \mathscr{G}(G)$. (This follows from Theorems 15.8 and 15.9.)

So the only genera in misère NIM are 0^1, 1^0, and those of the form a^a for some $a \geq 0$. An arbitrary game G is **tame** if all its subpositions have genus values drawn from this ensemble.

The nim-addition operator $\oplus$ extends to tame genera according to the following addition table:

$$\begin{aligned} 0^1 \oplus 0^1 &= 0^1 & a^a \oplus 0^1 &= a^a \\ 0^1 \oplus 1^0 &= 1^0 & a^a \oplus 1^0 &= (a \oplus 1)^{a\oplus 1} \\ 1^0 \oplus 1^0 &= 0^1 & a^a \oplus b^b &= (a \oplus b)^{a\oplus b} \end{aligned}$$

The main theorem is the following:

Theorem 15.10. (Conway) *If G and H are tame, then so is $G + H$, and moreover*

$$\mathscr{G}^{\pm}(G + H) = \mathscr{G}^{\pm}(G) \oplus \mathscr{G}^{\pm}(H).$$

This provides a reasonably straightforward extension of the theory of misère NIM to arbitrary tame games. In particular, any tame game can be treated as a NIM position *in sums involving other tame games*.

For example, let $G = *$ and $H = *2 + *2$. We noted above that $\mathscr{G}^{-}(G) = \mathscr{G}^{-}(H) = 0$, but $G + *2$ and $H + *2$ have distinct misère nim values. This is explained by the fact that $\mathscr{G}^{\pm}(G) = 1^0$, but $\mathscr{G}^{\pm}(H) = 0^0$. There are two fundamentally different types of tame games with $\mathscr{G}^{-}$-value 0, corresponding to the two cases in the statement of Theorem 15.8.

Likewise, consider $J = *2$ and $K = *2 + *2 + *2$. Here we have $\mathscr{G}^{\pm}(J) = \mathscr{G}^{\pm}(K) = 2^2$. Since J and K have the same genus, Theorem 15.10 implies that $o^{-}(J + X) = o^{-}(K + X)$ for any *tame* X. However, consider the game

$$X = \{0,\ *2 + *3\}$$

whose options are 0 and $*2 + *3$. X is not tame (since its genus is 2^0), and indeed it is not hard to check that

$$o^-(J + X) = \mathscr{N}, \quad \text{whereas} \quad o^-(K + X) = \mathscr{P}.$$

So even though J and K are both tame and have the same genus, they nonetheless behave differently in sums with a suitable *wild* game. The question of how best to extend the genus theory to wild games is an ongoing research problem; the rest of this section will describe the (considerable) advances that have been made in this direction.

15.6.3 Misère canonical form

The most straightforward idea is simply to define *misère equality* for impartial games, the same way we defined equality for partizan games in Section 15.5:

$$G = H \quad \text{if} \quad o^-(G + X) = o^-(H + X) \text{ for all } X,$$

with X ranging over all impartial games. Then the **misère game value** of G is its equivalence class modulo misère equality. This obviously works, in the sense that misère game value is automatically well behaved in sums. The central problem with misère nim values (and genus values for wild games) is therefore definitionally circumvented.

But misère game values suffer from a different problem, which is that there are rather a lot of them. If G is an impartial game, then an option $G' \in G$ is said to be **(misère) reversible** if there is some $G'' \in G'$ such that $G'' = G$. Obviously if G is misère reversible, then it is equal to a simpler game, namely G'', so this is a sort of analog of partizan reversible moves from Section 15.5. The following theorem of Conway is one of the crowning results of the misère theory.

Theorem 15.11. (Conway) *Suppose that $G = H$, and neither G nor H has any reversible moves. Then $G \cong H$.*

Theorem 15.11 says that reversible moves are *only* type of reduction available for impartial games. This is true in both normal and misère play: the "=" sign in Theorem 15.11 can be interpreted to mean *either* normal *or* misère equality (provided the corresponding notion of "reversible" is also used). In normal play, it is essentially a restatement of the Sprague-Grundy Theorem, so here we have a quite clear analog of the normal-play theory.

Sadly, reversible moves in misère play are exceedingly rare. Consider the set of game values with birthday ≤ 6. In normal play, there are just seven of them:

$$0, *, *2, \ldots, *6.$$

Conversely, in misère play Conway has shown that there are more than $2^{4171779}$. In this sense misère game values spectacularly fail to yield a coherent theory.

15.6.4 Misère quotients

The above results suggest that genus values preserve too little information, whereas misère game values preserve too much. The theory of *misère quotients* offers a third approach: rather than aim for a single, fully general extension of the Sprague-Grundy theory, we instead accept a multiplicity of *local* analogs.

Recall the definition of misère equality:

$$G = H \quad \text{if} \quad o^-(G + X) = o^-(H + X) \text{ for all } X.$$

In defining misère game values, we allowed X to range over all impartial games. If instead G, H, and X are restricted to range over *tame* games, then the resulting equivalence classes correspond one-to-one with genus values (and in fact this is just a restatement of the genus theory). So genus values can be viewed as the structure obtained when misère equivalence is localized to the set of tame games.

Along these lines, let $\mathscr{A}$ be any nonempty set of impartial games that is **closed** in the following sense:

- If $G, H \in \mathscr{A}$, then $G + H \in \mathscr{A}$ (additive closure) and
- If $G \in \mathscr{A}$ and $G' \in G$, then $G' \in \mathscr{A}$ (hereditary closure).

Then define

$$G \equiv H \pmod{\mathscr{A}} \quad \text{if} \quad o^-(G + X) = o^-(H + X) \text{ for all } X \in \mathscr{A}.$$

Let $\mathcal{Q}$ be the corresponding set of equivalence classes. The closure assumptions on $\mathscr{A}$ imply that $\mathcal{Q}$ is a commutative monoid, and there is a surjective homomorphism

$$\Phi : \mathscr{A} \to \mathcal{Q}.$$

Denote by $\mathcal{P} \subset \mathcal{Q}$ the subset corresponding to $\mathscr{P}$-positions from $\mathscr{A}$:

$$\mathcal{P} = \{\Phi(G) : G \in \mathscr{A},\ o^-(G) = \mathscr{P}\}.$$

The structure $(\mathcal{Q}, \mathcal{P})$ is the **misère quotient** of $\mathscr{A}$, and is denoted by $\mathcal{Q}(\mathscr{A})$. It serves as a localized analog of the Sprague-Grundy theory, in the following sense. Suppose that we wish to study a game $G \in \mathscr{A}$ that decomposes in $\mathscr{A}$:

$$G = G_1 + G_2 + \cdots + G_k, \quad \text{each } G_i \in \mathscr{A}.$$

Given the Φ-values of each G_i, say $x_i = \Phi(G_i)$, then we can multiply them out in the arithmetic of $\mathcal{Q}$ to determine $\Phi(G)$:

$$\Phi(G) = x = x_1 x_2 \cdots x_k$$

and then check whether $x \in \mathcal{P}$. So far we have not said anything terribly profound. What is surprising (and what makes misère quotients so powerful) is that the monoid $\mathcal{Q}$ often turns out to be finite, even when $\mathscr{A}$ is infinite, and even when $\mathscr{A}$ contains some wild

$$\mathcal{Q} \cong \langle a, b, c, d, e, f, g \mid a^2 = 1,\ b^3 = b,\ bc^2 = b,\ c^3 = c,\ bd = bc,\ cd = b^2,\ d^3 = d,\ be = bc,\ ce = b^2,\ e^2 = de,\ bf = ab,\ cf = ab^2c,\ d^2f = f,\ f^2 = b^2,\ b^2g = g,\ c^2g = g,\ dg = cg,\ eg = cg,\ fg = ag,\ g^2 = b^2 \rangle$$

$$\mathcal{P} = \{a, b^2, ac, ac^2, d, ad^2, e, ade, adf\}$$

Figure 15.24 The misère quotient of KAYLES.

games. In such cases, the problem of determining the outcome of the sum G reduces to a small number of operations on the finite multiplication table $\mathcal{Q}$.

For a simple example, let $\mathscr{A}$ consist of all sums involving $*$ and $*2$. Then every element of $\mathscr{A}$ is tame, so the elements of $\mathcal{Q}$ correspond to genera of games in $\mathscr{A}$, which are restricted to the six possibilities

$$0^1,\ 1^0,\ 0^0,\ 1^1,\ 2^2,\ 3^3.$$

The structure of the corresponding monoid follows directly from the addition table for genus values:

$$\mathcal{Q} \cong \langle a, b : a^2 = 1,\ b^3 = b \rangle,$$

with $\mathcal{P} = \{a, b^2\}$, corresponding to genera 1^0 and 0^0.

A fairly typical misère quotient is shown in Figure 15.24. It is the quotient of the set of positions in the game KAYLES, and therefore succinctly describes the winning strategy for misère KAYLES. It is worth noting that the original solution to misère KAYLES ran 43 pages long. A streamlined proof in *Winning Ways* reduced this to "just" five pages. That the entire proof can be encoded by the succinct monoid presentation in Figure 15.24 nicely illustrates the power of the quotient theory.

15.7. CONSTRAINT LOGIC

While combinatorial game theory seeks efficient algorithms for games, often no efficient algorithm exists. Then, we seek instead to show hardness. In recent years a new tool has emerged for proving hardness of games: **constraint logic** (Demaine and Hearn, 2008; Hearn and Demaine, 2009). With constraint logic, the games we consider are both more specialized and more general than what is traditionally addressed by classical game theory. More specialized, because we are concerned only with determining the winner of a game, and not with other issues such as maximizing payoff, cooperative strategies, etc. More general, because classical game theory is concerned only with the interactions of two or more players, whereas constraint logic addresses, in addition, games with only one player (puzzles) and even with no players at all (simulations). Constraint logic offers, for a variety of types of game, a simple path to hardness reductions; generally a small

Table 15.1 Game categories and their natural complexities. Constraint Logic is complete in each class.

	Zero player (simulation)	One player (puzzle)	Two player	Team, imperfect information
Unbounded length	PSPACE	PSPACE	EXPTIME	Undecidable
Bounded length	P	NP	PSPACE	NEXPTIME

number of constraint logic "gadgets" must be built out of components of the target game.

The starting point of constraint logic is the perspective that *games model computation.* Different types of game model different types of computation. For example, the idea of nondeterministic computation nicely matches the feature of puzzles that a player must choose a sequence of moves or piece placements to satisfy some global property. Thus, puzzles are often NP-complete (see Section 15.2.1). Even more striking is the correspondence between alternation, the natural extension to nondeterminism, and two-player games. Constraint logic is a family of games (played on directed graphs) which model computation ranging from that of monotone Boolean circuits (P-complete) all the way to unrestricted Turing machines (undecidable). For any game to be analyzed, the category of game will suggest a potential complexity, which may be proved by a reduction from the corresponding type of constraint logic. The entire range of constraint-logic games and complexities is shown in Table 15.1.

The chief advantage in showing a game hard by a reduction from constraint logic, rather than from a standard problem such as SAT or QBF, is that constraint logic is very similar in nature to many actual games, often making reductions extremely simple. For example, essentially the entire proof that sliding-block puzzles are PSPACE-complete is contained in Figure 15.30 (Hearn and Demaine, 2005). This problem, originally posed by Gardner (1964), had been open for nearly 40 years. Other games and puzzles shown hard via constraint logic include TipOver (Hearn, 2006a), sliding tokens (a dynamic version of Independent Set) (Hearn and Demaine, 2005), River Crossing (Hearn, 2004), Triangular Rush Hour (Hearn and Demaine, 2009), Push-2-F (Demaine et al., 2002), Amazons (Hearn, 2009), Konane (Hearn, 2009), Cross Purposes (Hearn, 2009), Hitori (Hearn and Demaine, 2009), and Wriggle puzzles (Maxime, 2007). Some games and puzzles with existing hardness proofs have also been shown hard via constraint logic, with simpler constructions (in some cases, also strengthening the existing results), including Sokoban (Hearn and Demaine, 2005), Rush Hour (Hearn and Demaine, 2005), and the Warehouseman's Problem (Hearn and Demaine, 2005). Finally, constraint logic has also been applied to several problems outside the domain of games proper, including showing undecidability of some decision problems for multiport finite-state machines (Hierons, 2010).

15.7.1 The constraint-logic framework

The general model of games we develop is based on the idea of a **constraint graph**; the rules defining legal moves on such graphs are called **constraint logic**. In later sections, the graphs and the rules will be specialized to produce one-player, two-player, etc., games.[2] A game played on a constraint graph is a computation of a sort, and simultaneously serves as a useful problem to reduce to other games to show their hardness.

A **constraint graph** is a directed graph with edge weights among $\{1, 2\}$. An edge is then called **red** (light gray in print version) or **blue** (dark gray in print version), respectively. The **inflow** at each vertex is the sum of the weights on inward-directed edges. Each vertex has a nonnegative **minimum inflow**. A **legal configuration** of a constraint graph has an inflow of at least the minimum inflow at each vertex; these are the **constraints**. A **legal move** on a constraint graph is the reversal of the direction of a single edge that results in a legal configuration. Generally, in any game, the goal will be to reverse a given edge by executing a sequence of (legal) moves. In multiplayer games, each edge is controlled by an individual player, and each player has his own goal edge. In deterministic games, a unique sequence of moves is forced. For the bounded games, each edge may only reverse once.

It is natural to view a game played on a constraint graph as a computation. Depending on the nature of the game, it can be a deterministic computation, or a nondeterministic computation, or an alternating computation, etc. The constraint graph then **accepts** the computation just when the game can be won.

AND/OR constraint graphs; planarity

Certain vertex configurations in constraint graphs are of particular interest. An **AND vertex** (Figure 15.25a) has minimum inflow constraint 2 and incident edge weights of

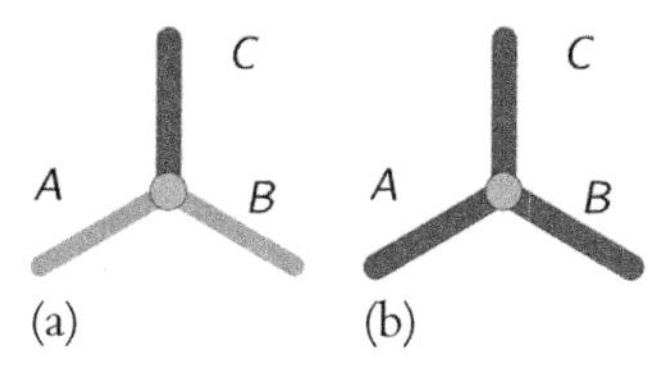

Figure 15.25 AND and OR vertices. Red (light gray, thinner) edges have weight 1, blue (dark gray, thicker) edges have weight 2, and vertices have a minimum in-flow constraint of 2. (a) AND vertex. Edge *C* may be directed outward if and only if edges *A* and *B* are both directed inward. (b) OR vertex. Edge *C* may be directed outward if and only if either edge *A* or edge *B* is directed inward.

[2] In the interest of space, we omit some of the definitions—and all discussion of zero-player games (DETERMINISTIC CONSTRAINT LOGIC)—and refer the reader to Hearn (2006b), Demaine and Hearn (2008), or Hearn and Demaine (2009).

1, 1, and 2. It behaves as a logical AND in the following sense: the weight-2 (blue; dark gray in print version) edge may be directed outward if and only if both weight-1 (red; light gray in print version) edges are directed inward. Otherwise, the minimum inflow constraint of 2 would not be met. An **OR vertex** (Figure 15.25b) has minimum inflow constraint 2 and incident edge weights of 2, 2, and 2. It behaves as a logical OR: a given edge may be directed outward if and only if at least one of the other two edges is directed inward.

It turns out that for all the game categories, it will suffice to consider constraint graphs containing only AND and OR vertices. For some of the game categories, there can be many subtypes of AND and OR vertex, because each edge may have a distinguishing initial orientation (in the case of bounded games), and a distinct controlling player (when there is more than one player). In some cases there are alternate vertex "basis sets" that enable simpler reductions to other problems than do the complete set of ANDs and ORs.

For all but the bounded zero-player case, it also suffices to only consider planar constraint graphs. In practice this makes for much easier hardness reductions; often, crossover gadgets are the most difficult pieces of a reduction to construct. With constraint logic, we get them for free. The most common problem used to show NP-hardness is 3SAT, but in many instances this planarity property makes constraint logic reductions simpler.

Directionality; Fanout

As implied above, although it is natural to think of AND and OR vertices as having inputs and outputs, there is nothing enforcing this interpretation. A sequence of edge reversals could first direct both red (light gray in print version) edges into an AND vertex, and then direct its blue (dark gray in print version) edge outward; in this case, we could say that its "inputs" have "activated," enabling its "output" to "activate." But the reverse sequence could equally well occur. In this case we could view the AND vertex as a splitter, or **FANOUT** gate: directing the blue (dark gray in print version) edge inward allows both red (light gray in print version) edges to be directed outward, effectively splitting a signal.

In the case of OR vertices, again, we can speak of an active input enabling an output to activate. However, here the choice of input and output is entirely arbitrary, because OR vertices are symmetric.

15.7.2 One-player games

The one-player version of constraint logic is called NONDETERMINISTIC CONSTRAINT LOGIC (NCL). The rules are simply that on a turn the player reverses a single edge that results in a legal configuration. The goal is to reverse a particular edge.

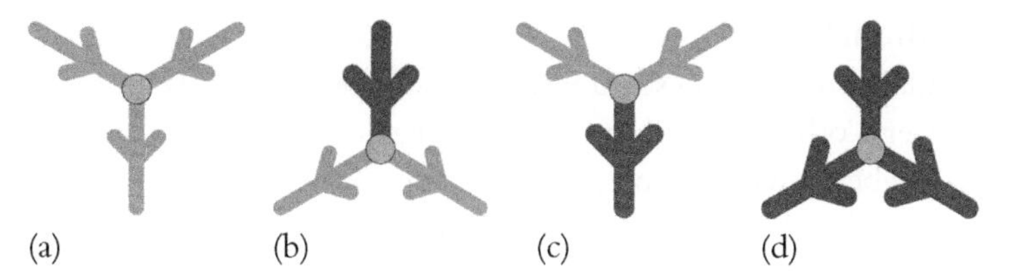

Figure 15.26 Basis vertices for bounded NCL. (a) CHOICE; (b) AND; (c) FANOUT; and (d) OR.

15.7.2.1 Bounded games

BOUNDED NONDETERMINISTIC CONSTRAINT LOGIC (BOUNDED NCL) is formally defined as follows:

BOUNDED NONDETERMINISTIC CONSTRAINT LOGIC (BOUNDED NCL)

1. []
   INSTANCE: Constraint graph G, edge e in G.
2. []
   QUESTION: Is there a sequence of moves on G that eventually reverses e, such that each edge is reversed at most once?

BOUNDED NCL is NP-complete (reduction from 3SAT). It remains NP-complete when the graph G is required to be a planar graph which uses only the vertex types shown in Figure 15.26.[3] It also turns out to be useful to reduce from graphs that have the property that only a single edge can initially reverse; this problem is also NP-complete.

A related problem is CONSTRAINT GRAPH SATISFIABILITY:

CONSTRAINT GRAPH SATISFIABILITY

1. []
   INSTANCE: Unoriented planar constraint graph G using only AND and OR vertices.
2. []
   QUESTION: Does G have a configuration that satisfies all the constraints?

Properly, this problem is not a constraint-logic game, because the moves (assignments of edge orientations) are not reversals from one legal configuration to another. But it is similar in spirit, and can prove useful for reductions. CONSTRAINT GRAPH SATISFIABILITY is NP-complete (reduction from 3SAT). Note that for CONSTRAINT GRAPH SATISFIABILITY, unlike proper BOUNDED NCL, only two types of vertex are needed.

[3] Here we show the initial, "inactivated" orientation of the edges. In an AND, the blue (dark gray in print version) edge may reverse if the red (light gray in print version) edges first reverse; in a FANOUT, the red (light gray in print version) edges may reverse if the blue (dark gray in print version) edge first reverses.

Sample application: Hitori

Hitori was popularized by Japanese publisher Nikoli, along with its more-famous sibling Sudoku, and several other "pencil-and-paper" puzzles. In Hitori, we are given a grid with each square labeled with an integer, and the goal is to paint a subset of the squares so that (1) no row or column has a repeated unpainted label (similar to Sudoku), (2) painted squares are never adjacent, and (3) the unpainted squares are all connected. A simple Hitori puzzle and its solution are shown in Figure 15.27. We give a reduction from Constraint Graph Satisfiability (Section 15.7.2) showing that it is NP-complete to determine whether a given $n \times n$ Hitori puzzle has a solution (Hearn and Demaine, 2009).

Wiring

We represent graph edge orientation with wires, or strings of adjacent squares, consisting of integers $x_1, x_1, x_2, x_2, \ldots, x_{n-1}, x_{n-1}, x_n, x_n$, where the x_i are distinct. If the first x_1 is unpainted, then the next must be painted (by rule 1 above), forcing the first x_2 to be unpainted (by rule 2), etc.; thus the last x_n must be painted. If the first x_1 is painted, the last x_n may be painted or unpainted: we could (for example) have the second x_1 and the first x_2 both unpainted without violating the rules.

Wires may be turned, as in Figure 15.28(a): if the bottom a is unpainted, then the right d must be painted. (We assume that the unlabeled squares all contain distinct integers not otherwise used in the gadgets.)

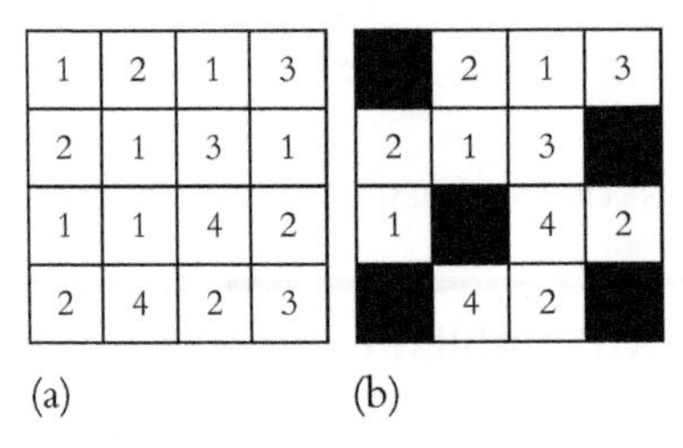

Figure 15.27 A simple Hitori puzzle and its solution. (a) Puzzle and (b) Solution.

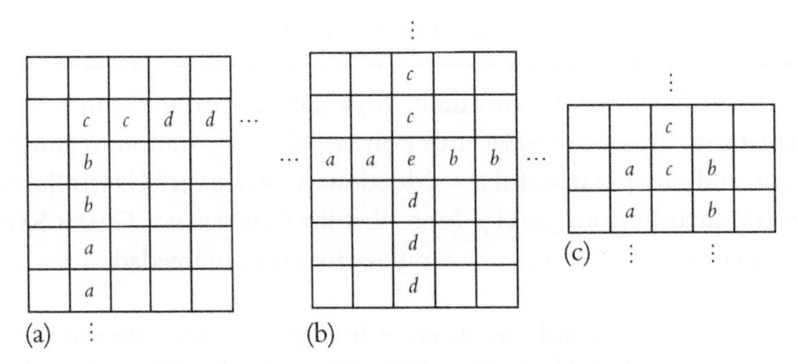

Figure 15.28 Hitori gadgets. (a) Wire, turn; (b) OR, parity; and (c) AND.

OR vertex/parity gadget

In Figure 15.28(b), first consider the *d*s. At most one can be unpainted, but no two adjacent may be painted. Therefore, both the lower and the upper one must be painted, and *e* must be unpainted.

If both the left *a* and the right *b* are unpainted, then the right *a* and the left *b* must be painted. As an unpainted square, *e* must be connected to the other unpainted squares (rule 3); the lower *c* is the only way out. Therefore, the lower *c* is unpainted, and the upper one painted. But if either the left *a* or the right *b* is painted, then the other *a* or *b* will be unpainted, allowing another way out for *e*. Then the lower *c* may be painted, and the upper *c* unpainted. These are the same constraints an OR vertex has, again with an unpainted "port" square (left *a*, right *b*, and top *c*) corresponding to an outward-directed edge, and a painted port square corresponding to an inward-directed edge.

This gadget can also serve to alter the positional parity in wiring, so that the various gadgets can be connected arbitrarily, by using only one input, and blocking the other one (for example, by adding another *b* to the right of the right one).

AND vertex

Similar but simpler reasoning as above shows that the gadget in Figure 15.28(c) satisfies the same constraints as an AND vertex, with the lower *a* and *b* (inputs) corresponding to the red (light gray in print version) edges, and the upper *c* (output) to the blue (dark gray in print version) edge: the output square may be unpainted if and only if both input squares are painted.

Assembly

Given a planar AND/OR constraint graph, we construct a HITORI puzzle by connecting together AND and OR vertex gadgets with wires, adjusting positional parity as needed. If the graph has a legal configuration, then every wire can be painted so as to satisfy all the HITORI constraints, as described. Similarly, if the HITORI puzzle can be solved, then a legal graph configuration can be read off the wires.

15.7.2.2 Unbounded games

NONDETERMINISTIC CONSTRAINT LOGIC (NCL) is the canonical form of constraint logic:

NONDETERMINISTIC CONSTRAINT LOGIC (NCL)
INSTANCE: Constraint graph *G*, edge *e* in *G*.
QUESTION: Is there a sequence of moves on *G* that eventually reverses *e*?

NCL is PSPACE-complete (reduction from QBF), and remains PSPACE-complete when the graph G is required to be a planar graph which uses only AND and OR vertices (Figure 15.25). NCL reductions are often very straightforward, for two reasons. First, only two gadgets must be constructed. Second, one-player games (puzzles) are generally easier to reduce to than multiplayer games. For these reasons, and because there is a large supply of candidate puzzles to analyze, NCL reductions form the largest set of existing constraint-logic reductions.

Sample application: Sliding Blocks

In the usual kind of sliding-block puzzle, one is given a box containing a set of rectangular pieces, and the goal is to slide the blocks around so that a particular piece winds up in a particular place. A popular example is Dad's Puzzle, shown in Figure 15.29; it takes 59 moves to slide the large square to the bottom left. We outline a reduction from Nondeterministic Constraint Logic (Section 15.7.2) showing that it is PSPACE-complete to determine whether a given sliding-block puzzle in an $n \times n$ box has a solution. For a formal proof (which is also stronger, using only 1×2 blocks), see Hearn and Demaine (2005) or Hearn and Demaine (2009).

AND vertex

The gadget shown in Figure 15.30(a) satisfies the same constraints as an AND vertex. Assume that the outer, dark-colored "wall" blocks are fixed. Then, the only way the top "signal" (light-colored) block may slide down is if the left signal block first slides left, and bottom signal block slides down. This allows the other signal blocks to move out of the way.

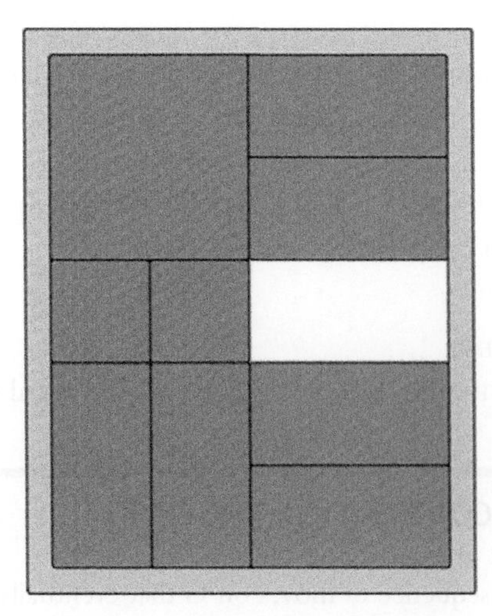

Figure 15.29 Dad's puzzle.

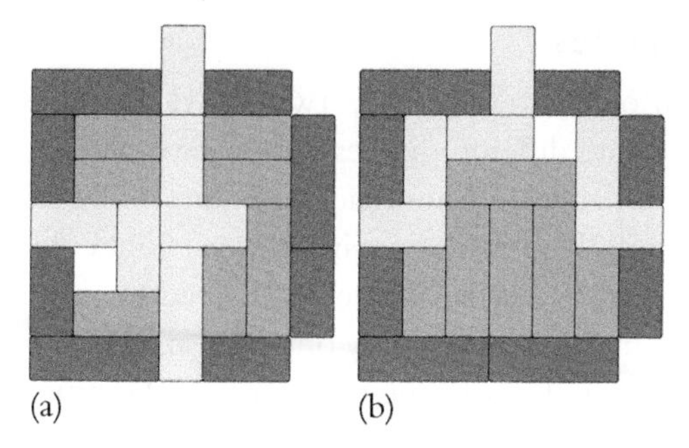

Figure 15.30 Constraint-logic gadgets showing PSPACE-completeness of Sliding Blocks. (a) AND and (b) OR.

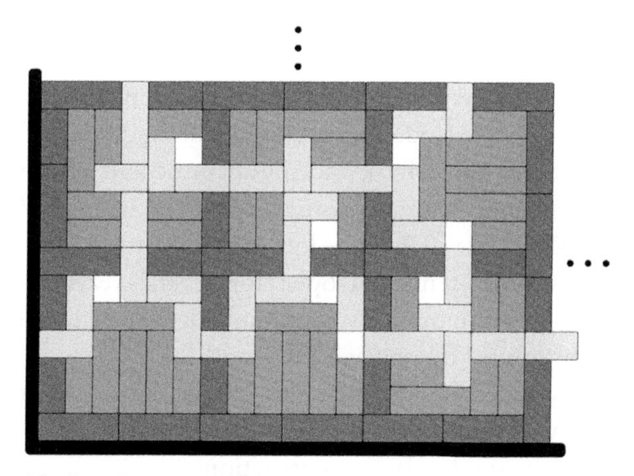

Figure 15.31 Sliding-block gadget assembly.

OR vertex

Similarly, the gadget shown in Figure 15.30(b) satisfies the same constraints as an OR vertex—the top signal block may slide down if and only if either the left or the right signal block first slides out.

Assembly

To use these gadgets to represent arbitrary planar AND/OR constraint graphs, we assemble them as shown in Figure 15.31. The wall blocks are shared between adjacent vertices, as are the signal blocks that act as graph edges. We put a grid of the gadgets inside a box. This keeps the wall blocks from moving, as required. The goal is to slide the particular signal block that corresponds to the target edge in the input constraint graph. Then, the puzzle can be solved just when the constraint-logic problem is solvable.

15.7.3 Two-player games

The two-player version of constraint logic, Two-Player Constraint Logic (2CL), is defined as follows. To create different moves for the two players, Black and White, we label each constraint graph edge as either Black or White. (This is independent of the red (light gray in print version)/blue (dark gray in print version) coloration, which is simply a shorthand for edge weight.) Black (White) is allowed to reverse only Black (White) edges. As before, a move must reverse exactly one edge and result in a valid configuration. Each player has a target edge he is trying to reverse.[4] (We omit the formal definitions here.)

Bounded games

The bounded case permits each edge to reverse at most once. Bounded 2CL is PSPACE-complete (reduction from G_{pos}(POS CNF), a variant of QBF; Schaefer, 1978). It remains PSPACE-complete when the constraint graph is a planar graph using only the vertex types shown in Figure 15.32. Indeed, the actual reduction showing Bounded 2CL PSPACE-complete is almost trivial, and the main benefit of using Bounded 2CL for game reductions, rather than simply using one of the many QBF variants, is that when reducing from Bounded 2CL one does not have to build a crossover gadget. The complexity of Amazons remained open for several years, despite some effort by the game-complexity community; its constraint-logic reduction showing PSPACE-completeness is straightforward (Hearn and Demaine, 2009).

The vertex set in Figure 15.32 is actually almost the same as that for Bounded NCL (Figure 15.26); the only addition is a single vertex type allowing for player interaction. Most of the gadgets can be single-player constructions.

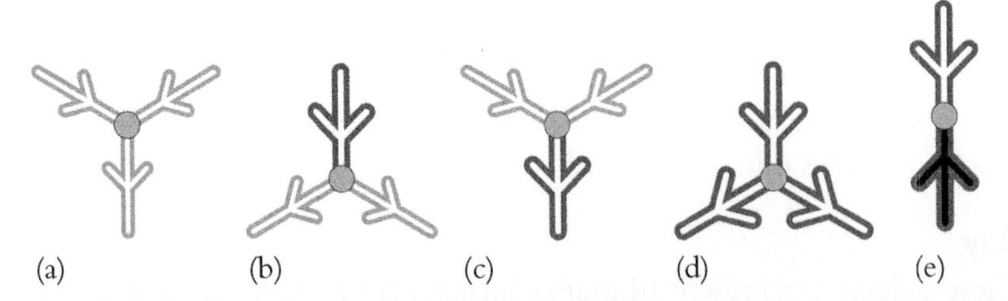

Figure 15.32 Basis vertices for Bounded 2CL. (a) CHOICE; (b) AND; (c) FANOUT; (d) OR; and VARIABLE.

[4] In combinatorial game theory, it is normal to define the loser as the first player unable to move. This definition would work perfectly well for 2CL, rather than using target edges to determine the winner; the hardness reduction would not be substantially altered. However, the given definition is more consistent with the other varieties of constraint logic: always, the goal is to reverse a given edge.

Unbounded games

The unbounded case simply removes the restriction of edges reversing at most once. 2CL is EXPTIME-complete (reduction from G_6, one of the several Boolean formula games shown EXPTIME-complete by Stockmeyer and Chandra (1979)). 2CL remains EXPTIME-complete when the graph is a planar graph using only the vertices shown in Figure 15.33. In principle, this should enable much simpler reductions to actual games than the standard reductions from Boolean formula games. The existing CHESS (Fraenkel and Lichtenstein, 1981), CHECKERS (Robson, 1984b), and GO (Robson, 1983) hardness results are all quite complicated; there could be simpler reductions from 2CL. However, such reductions have not yet been found. Enforcing the necessary constraints in a two-player game gadget is much more difficult than in a one-player game.

15.7.4 Team games

The natural team private-information constraint logic (TPCL) assigns to each player a set of edges he can reverse, and a set of edges whose orientation he can see, in addition to the target edge he aims to reverse. There are two teams, Black and White; a team wins when a player on that team reverses his target edge. (We omit the formal definitions here.)

Bounded games

As usual for bounded games, with BOUNDED TEAM PRIVATE CONSTRAINT LOGIC we allow each edge to reverse at most once. BOUNDED TPCL is NEXPTIME-complete (reduction from the DEPENDENCY QBF problem introduced in Peterson and Reif, 1979). It remains NEXPTIME-complete, even for planar graphs which use only AND and OR vertices, and

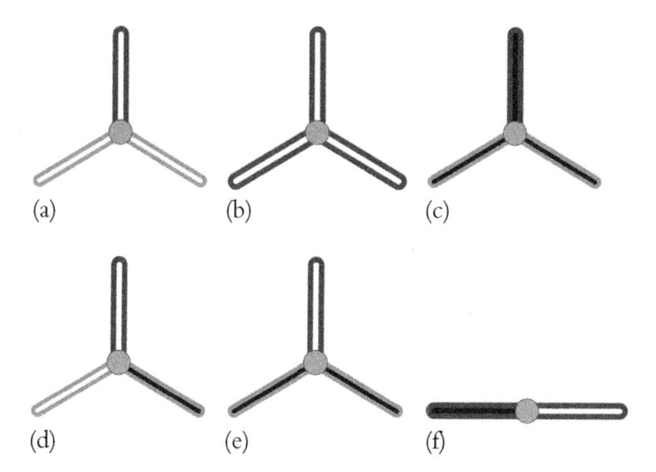

Figure 15.33 Basis vertices for 2CL. (a) White AND; (b) White OR; (c) Black AND; (d) Multiplayer AND 1 ; (e) Multiplayer AND 2 ; and (f) Black-White.

when there is only one Black player and two White players. (Unlike other forms of constraint logic, here we do not enumerate a specific smaller set of basis vertices; note that there are several different types of AND and OR vertices, depending on controlling player, initial edge orientation, and edge visibility.)

Unbounded games

To enable a simpler reduction to an unbounded form of team constraint logic, we allow each player to reverse up to some given constant k edges on his turn, rather than just one, and leave the case of $k = 1$ as an open problem. TPCL is undecidable (shown by a series of reductions beginning with acceptance of a Turing machine on an empty input). It remains undecidable even for planar graphs which use only AND and OR vertices. As with BOUNDED TPCL, several different AND- and OR-subtypes are used in the reduction, which we do not enumerate.

The undecidability here is rather striking, given that this is a game played with a finite number of positions! Essentially, this means that the games with a bounded amount of state can simulate any unbounded Turing computation.

The ability for a player to reverse multiple edges on a turn, and the lack of a small set of basis vertices, would seem to make TPCL a challenging problem to reduce from to show other problems undecidable. However, TPCL has already been applied to show some decision problems for multiport finite-state machines undecidable (Hierons, 2010).

15.8. CONCLUSION

We have now forged a trail from NIM to CHESS and GO. In Section 15.2, we dealt with classical acyclic impartial games. We then presented a polynomial theory of cyclic games in Section 15.3. In Section 15.4, we added interactions between tokens. Next we tackled partizan games in Section 15.5, and misère play in Section 15.6. All of these—cycles, token interactions, misère play—are absent from NIM but very much present in CHESS and GO. This trail is still rather thin: token interaction was restricted to annihilation. Most of the other interactions lead to intractable games. Similarly for partizan games. Misère play was portrayed in Section 15.1 as a special case of "Termination Set," general cases of which occur in CHESS and GO. In this sense, we listed misère play as a road step towards CHESS and GO.

Game intractability results, besides their intrinsic value, serve as trail guides: They indicate the boundary beyond which polynomial strategies are unlikely to exist, where we have to resort to analysis of restricted or special cases. Section 15.7 provides a modern convenient tool for proving intractability, though it does not seem to work for all cases. Intractable games, in the technical meaning of intractability, though it's only an asymptotic result—for $n \times n$ boards as n goes to infinity—are rather unlikely to have a tractable strategy for a finite actual board. In fact, we do not know of any such case.

Therefore, only special cases are likely to be analysable. For misère play, this was done by restricting the universe of the games. For CHESS and GO it was done by treating endgames.

Thus, we have arrived at CHESS and GO from two directions: The former has been showed to be EXPTIME-complete (Fraenkel and Lichtenstein, 1981), which is a provable intractability, and the latter even EXPSPACE-complete, under certain game-rules of GO (Robson, 1984a); and there are some constructive results for their endgames. Elkies (1996), Elkies (2004/2006) has some results about CHESS endgames, and Berlekamp and his students have some spectacular results about GO endgames (Berlekamp, 1991; Berlekamp and Wolfe, 1994). GO play tends to break up into almost independent subgames at the end, so the strong tool of game-sums can be unleashed to attack them. This is not quite the case for CHESS, partly because some of the pieces are so strong that they dominate much of the entire board, rather than only locally.

How can we broaden this still rather thin trail? One direction could be to extend the misère play theory to more general termination sets, as exist for CHESS and GO. Another is to broaden the fledgling theory of scoring games, where scores are accumulated during play. These were and are independently studied by John Milnor, Mark Ettinger, Fraser Stewart, Will Johnson, and Carlos Santos. Related avenues include competitive auctions, incentives (Elwyn Berlekamp), bidding games (Sam Payne) and, more generally, connections between combinatorial games and classical games with applications to economics. In quite a different direction, the yearned for emergence of quantum and biological computing are potential brute force tools to bridge the complexity gap between polynomial and nonpolynomial games.

ACKNOWLEDGMENT

We thank an anonymous referee whose remarks made us explain better concepts from Combinatorial Game Theory to Classical Game Theorists, and improved the exposition in various ways.

REFERENCES

Albert, M., Nowakowski, R.J., Wolfe, D., 2007. Lessons in Play: An Introduction to Combinatorial Game Theory. A K Peters, Wellesley, MA.

Berlekamp, E., 1991. Introductory overview of mathematical Go endgames. In: Combinatorial Games (Columbus, OH, 1990), Proceedings of the Symposia in Applied Mathematics. American Mathematical Society, vol. 43. Providence, RI, pp. 73–100.

Berlekamp, E.R., Conway, J.H., Guy, R.K., 2001–2004. Winning Ways for Your Mathematical Plays, second ed. A K Peters Ltd., Wellesley, MA. Vol. 1: 2001; Vols. 2 and 3: 2003; Vol. 4: 2004.

Berlekamp, E., Wolfe, D., 1994. Mathematical Go. A K Peters Ltd., Wellesley, MA. (Chilling gets the last point, With a foreword by James Davies).

Bouton, C.L., 1901/02. Nim, a game with a complete mathematical theory. Ann. Math. (2) 3(1-4), 35–39.

Chvátal, V., 1973. On the computational complexity of finding a kernel. Report No. crm-300, Centre de Recherches Mathématiques, Université de Montréal.

Conway, J.H., 2001. On Numbers and Games, second ed. A K Peters Ltd., Natick, MA.

Demaine, E.D., Hearn, R.A., 2008. Constraint logic: a uniform framework for modeling computation as games. Computational Complexity, Annual IEEE Conference on, pp. 149–162.
Demaine, E.D., Hearn, R.A., Hoffmann, M., 2002. Push-2-F is PSPACE-complete. In: Proceedings of the 14th Canadian Conference on Computational Geometry (CCCG 2002). Lethbridge, Alberta, Canada, August 12–14, pp. 31–35.
Elkies, N.D., 1996. On numbers and endgames: combinatorial game theory in chess endgames. In: Games of No Chance (Berkeley, CA, 1994), Mathematics Sciences Research Institute Publication, vol. 29. Cambridge University Press, Cambridge, pp. 135–150.
Elkies, N.D., 2004/06. New directions in enumerative chess problems. Electron. J. Combin., 11(2), Article 4, 14 pp. (electronic).
Ferguson, T.S. 1984. Misère annihilation games. J. Combin. Theory Ser. A, 37(3), 205–230.
Fraenkel, A.S., 1974. Combinatorial games with an annihilation rule. In: The Influence of Computing on Mathematical Research and Education (Proc. Sympos. Appl. Math., Vol. 20, Univ. Montana, Missoula, Mont., 1973). American Mathematical Society, Providence, R.I., pp. 87–91.
Fraenkel, A.S., 1981. Planar kernel and Grundy with $d \leq 3$, $d_{\text{out}} \leq 2$, $d_{\text{in}} \leq 2$ are NP-complete. Discrete Appl. Math., 3(4), 257–262.
Fraenkel, A.S., 1991. Complexity of games. In Combinatorial games (Columbus, OH, 1990), volume 43 of Proc. Sympos. Appl. Math., pp. 111–153. American Mathematical Society, Providence, RI.
Fraenkel, A.S., 2004. Complexity, appeal and challenges of combinatorial games. Theoret. Comput. Sci. 313(3), 393–415. (Algorithmic combinatorial game theory).
Fraenkel, A.S., Goldschmidt, E., 1987. PSPACE-hardness of some combinatorial games. J. Combin. Theory Ser. A 46(1), 21–38.
Fraenkel, A.S., Lichtenstein, D., 1981. Computing a perfect strategy for $n \times n$ chess requires time exponential in n. J. Combin. Theory Ser. A 31(2), 199–214.
Fraenkel, A.S., Loebl, M., Nešetřil, J., 1988. Epidemiography. II. Games with a dozing yet winning player. J. Combin. Theory Ser. A 49(1), 129–144.
Fraenkel, A.S., Nešetřil, J., 1985. Epidemiography. Pacific J. Math. 118(2), 369–381.
Fraenkel, A.S., Perl, Y., 1975. Constructions in combinatorial games with cycles. In: Infinite and Finite Sets (Colloq., Keszthely, 1973; Dedicated to P. Erdős on His 60th Birthday), Vol. II. Colloquia Mathematica Societatis Janós Bolyai, Vol. 10. North-Holland, Amsterdam, pp. 667–699.
Fraenkel, A.S., Tassa, U., 1975. Strategy for a class of games with dynamic ties. Comput. Math. Appl., 1(no. 2), 237–254.
Fraenkel, A.S., Tassa, U., Yesha, Y., 1978. Three annihilation games. Math. Mag. 51(1), 13–17.
Fraenkel, A.S., Yesha, Y., 1976. Theory of annihilation games. Bull. Amer. Math. Soc. 82(5), 775–777.
Fraenkel, A.S., Yesha, Y., 1979. Complexity of problems in games, graphs and algebraic equations. Discrete Appl. Math. 1(1–2), 15–30.
Fraenkel, A.S., Yesha, Y., 1982. Theory of annihilation games. I. J. Combin. Theory Ser. B 33(1), 60–86.
Fraenkel, A.S., Yesha, Y., 1986. The generalized Sprague-Grundy function and its invariance under certain mappings. J. Combin. Theory Ser. A 43(2), 165–177.
Gardner, M., 1964. The hypnotic fascination of sliding-block puzzles. Scientific American 210, 122–130.
Goldstein, A.S., Reingold, E.M., 1995. The complexity of pursuit on a graph. Theoret. Comput. Sci. 143(1), 93–112.
Grundy, P.M., 1939. Mathematics and games. Eureka, 2. Reprinted in *Eureka* 27 (1964), 9–11.
Hearn, R.A., 2004. The complexity of sliding block puzzles and plank puzzles. In: Tribute to a Mathemagician. A K Peters, Wellesley, MA, pp. 173–183.
Hearn, R., 2006. TipOver is NP-complete. Math. Intell. 28(3), 10–14.
Hearn, R.A., 2006. Games, Puzzles, and Computation (Ph.D. dissertation). Massachusetts Institute of Technology, Department of Electrical Engineering and Computer Science, May. http://www.swiss.ai.mit.edu/ bob/hearn-thesis-final.pdf.
Hearn, R.A., 2009. Amazons, Konane, and Cross Purposes are PSPACE-complete. In: Albert, M.H., Nowakowski, R.J., (Eds.), Games of No Chance 3, vol. 56. MSRI Publications, Cambridge University Press, Cambridge.

Hearn, R.A., Demaine, E.D., 2005. PSPACE-completeness of sliding-block puzzles and other problems through the nondeterministic constraint logic model of computation. Theor. Comput. Sci. 343(1–2), 72–96, October. Special issue "Game Theory Meets Theoretical Computer Science".

Hearn, R.A., Demaine, E.D., 2009. Games, Puzzles and Computation. A K Peters, Wellesley, MA.

Hierons, R.M., 2010. Reaching and distinguishing states of distributed systems. SIAM J. Comput. 39(8), 3480–3500.

Kalmár, L., 1928. Zur theorie der abstrakten spiele. Acta Sci. Math. Univ. Szeged, 4, 65–85.

Maxime, O., 2007. Wriggler puzzles are PSPACE-complete. Manuscript, 2007.

Morris, F.L., 1981. Playing disjunctive sums is polynomial space complete. Int. J. Game Theory, 10, 195–205.

Peterson, G.L., Reif, J.H., 1979. Multiple-person alternation. In: *FOCS*. IEEE, pp. 348–363.

Plambeck, T.E., 2005. Taming the wild in impartial combinatorial games. INTEGERS: Electr. J. Combin. Number Theory 5(1), #G05.

Plambeck, T.E., Siegel, A.N., 2008. Misère quotients for impartial games. J. Combin. Theory Ser. A, 115(4), 593–622.

Pultr, A., Morris, F.L., 1984. Prohibiting repetitions makes playing games substantially harder. Int. J. Game Theory 13(1), 27–40.

Rabin, M.O., 1957. Effective computability of winning strategies. In: Contributions to the Theory of Games, Vol. 3. Annals of Mathematics Studies. vol. 39. Princeton University Press, Princeton, NJ, pp. 147–157.

Reisch, S., 1981. Hex ist PSPACE-vollständig. Acta Inform. 15(2), 167–191.

Robson, J.M., 1983. The complexity of Go. In: Proceedings of the IFIP 9th World Computer Congress on Information Processing, pp. 413–417.

Robson, J.M., 1984. Combinatorial games with exponential space complete decision problems. In: Mathematical Foundations of Computer Science, 1984 (Prague, 1984). Lecture Notes in Computer Science, vol. 176. Springer, Berlin, pp. 498–506.

Robson, J.M., 1984. *N* by *N* Checkers is EXPTIME complete. SIAM J. Comput. 13(2), 252–267.

Schaefer, T.J., 1978. On the complexity of some two-person perfect-information games. J. Comput. Syst. Sci. 16, 185–225.

Siegel, A.N., 2013. Combinatorial Game Theory. Graduate Studies in Mathematics. American Mathematical Society, vol. 146.

Smith, C.A.B., 1966. Graphs and composite games. J. Combinatorial Theory, 1, 51–81.

Sprague, R., 1935–36. Über mathematische Kampfspiele. Tôhoku Math. J., 41, 438–444.

Stockmeyer, L.J., Chandra, A.K., 1979. Provably difficult combinatorial games. SIAM J. Comput. 8(2), 151–174.

van Leeuwen, J., 1976. Having a Grundy-numbering is NP-complete. Report No. 207, Computer Science Deparment, Pennsylvania State University, University Park, PA.

West, J., 1996. Championship-level play of Domineering. In: Nowakowski, R.J., (Ed.), Games of No Chance. MSRI Publications, vol. 29. Cambridge University Press, Cambridge, pp. 85–91.

Yedwab, L.J., 1985. On Playing Well in a Sum of Games (Master's thesis), MIT. MIT/LCS/TR-348.

CHAPTER 16

Game Theory and Distributed Control**

Jason R. Marden*, Jeff S. Shamma†

*Department of Electrical, Computer and Energy Engineering, Boulder, CO, USA
†School of Electrical and Computer Engineering, Georgia Institute of Technology, Atlanta, GA, USA

Contents

Abstract

Game theory has been employed traditionally as a modeling tool for describing and influencing behavior in societal systems. Recently, game theory has emerged as a valuable tool for controlling or prescribing behavior in distributed engineered systems. The rationale for this new perspective stems from the parallels between the underlying decision-making architectures in both societal systems and distributed engineered systems. In particular, both settings involve an interconnection

**Supported AFOSR/MURI projects #FA9550–09–1–0538 and #FA9530-12-1-0359 and ONR projects #N00014–09–1–0751 and #N0014-12-1-0643.

Handbook of game theory
http://dx.doi.org/10.1016/B978-0-12-420056-2.00016-1

of decision-making elements whose collective behavior depends on a compilation of local decisions that are based on partial information about each other and the state of the world. Accordingly, there is extensive work in game theory that is relevant to the engineering agenda. Similarities notwithstanding, there remain important differences between the constraints and objectives in societal and engineered systems that require looking at game-theoretic methods from a new perspective. This chapter provides an overview of selected recent developments of game-theoretic methods in this role as a framework for distributed control in engineered systems.

Keywords: Strategic learning, Evolutionary games, Utility design, Disequilibrium, Control systems

JEL Codes: C73 "Stochastic and dynamic games; Evolutionary games; Repeated games"

16.1. INTRODUCTION

Distributed control involves the design of decision rules for systems of interconnected components to achieve a collective objective in a dynamic or uncertain environment. One example is teams of mobile autonomous systems, such as unmanned aerial vehicles (UAVs), for uses such as search and rescue, cargo delivery, scientific data collection, and homeland security operations. Other application examples can be found in sensor and data networks, communication networks, transportation systems, and energy (Murray, 2007; Shamma, 2008). Technological advances in embedded sensing, communication, and computation all point towards an increasing potential and importance of such networks of autonomous systems.

In contrast to the traditional control paradigm, distributed control architectures do not have a central entity with access to all information or authority over all components. A lack of centralized information is possible even when components can communicate as communications can be costly, e.g., because of energy conservation, or even inadmissible, e.g., for stealthy operations. Furthermore, the latency/time-delay required to distribute information in a large-scale system may be impractical in a dynamically evolving setting. Accordingly, the collective components somehow must coordinate globally using distributed decisions based on only limited local information.

One approach to distributed control is to view the problem from the perspective of game theory. Since game theory concerns the study of interacting decision makers, the relevance of game theory to distributed control is easily recognized. Still, this perspective is a departure from the traditional study of game theory, where the focus has been the development of models and methods for applications in economic and social sciences. Following the discussion in Shoham et al. (2007) and Mannor and Shamma (2007), we will refer to the traditional role of game theory as the "descriptive" agenda, and its application to distributed control as the "engineering" agenda.

The first step in deriving a game-theoretic model is to identify the basic elements of the game, namely the players/agents and their admissible actions. In distributed control problems, there also is typically a global objective function that reflects the performance of the collective as a function of their joint actions. The following examples illustrate these elements in various applications:

- *Consensus/synchronization:* The agents are mobile platforms. The actions are agent orientations (e.g., positions or velocities). The global objective is for agents to align their orientations with each other (Olfati-Saber et al., 2007).
- *Distributed routing:* Agents are mobile vehicles. The actions are paths from sources to destinations. The global objective is to minimize network traffic congestion (Roughgarden, 2005).
- *Sensor coverage:* The agents are mobile sensors. The actions are sensor paths. The global objective is to service randomly arriving spatially distributed targets in the shortest time (Martinez et al., 2007).
- *Wind energy harvesting:* The agents are wind turbines. The actions are the blade pitch angle and rotor speed. The global objective is to maximize the overall energy generated by the turbines (Marden et al., 2012).
- *Vehicle-target assignment:* The agents are heterogeneous mobile weapons with complementary capabilities. The actions are the selection of potential targets. The global objective is to maximize the overall expected damage (Arslan et al., 2007).
- *Content distribution:* The agents are computer nodes. The actions are which files to store locally under limited storage capacity. The global objective is to service local file requests from users while minimizing peer-to-peer content requests (Dan, 2011).
- *Ad hoc networks:* The agents are mobile communication nodes. The actions are to form a network structure. Global objectives include establishing connectivity while optimizing performance specifications such as required power or communication hop lengths (Neel et al., 2005).

There is some flexibility in defining what constitutes a single player. For example in wind energy harvesting, a player could be a single turbine or a group of turbines. The determining factor is the extent to which the group can act as a single unit with shared information.

With the basic elements in place, the next step is to specify agent utility functions. Here, the difference between the descriptive and engineering agenda becomes more apparent. Whereas in the descriptive agenda, utility functions are part of the modeling process and in the engineering agenda, utility functions constitute a *design choice*.

An important consideration in specifying the utility functions is the implication on the global objective. With utility functions in place, the game is fully specified. If one takes a solution concept such as Nash equilibrium to represent the outcome of the game, then these outcomes should be desirable as measured by the global objective.

With the game now fully specified, with players, actions, and utility functions, there is another important step that again highlights a distinction between the descriptive and engineering agenda. Namely, one must specify the dynamics through which agents will arrive at an outcome or select from a set of possible outcomes.

There is extensive work in the game theory literature that explores how a solution concept such as a Nash equilibrium might emerge, i.e., the "learning in games" program (Fudenberg and Levine, 1998; Hart, 2005; Young, 2005). Quoting Arrow (1986),

"The attainment of equilibrium requires a disequilibrium process."

The work in learning in games seeks to understand how a plausible learning/adaptation disequilibrium process may (or may not) converge, and thereby reinforce the role of Nash equilibrium as a predictive outcome in the descriptive agenda. By contrast, in the engineering agenda, the role of such learning processes is to *guide* the agents towards a solution. Accordingly, the specification of the learning process also constitutes a *design choice.*

There is some coupling between the two design choices of utility functions and learning processes. In particular, it can be advantageous in designing utility functions to assure that the resulting game has an underlying structure (e.g., being a potential game or weakly acyclic game) so that one can exploit learning processes that converge for such games.

To recap, the engineering agenda requires both designing agent utility functions and learning processes. At this point, it is worthwhile highlighting various design considerations that play a more significant role in the engineering agenda:

- *Information:* One can impose various restrictions on the information available to each agent. Two natural widely considered scenarios are: (i) agents can measure the actions of other agents or (ii) agents can only measure their own action and perceived rewards. For example, in distributed routing, agents may observe the routes taken by other agents (which could be informationally intense), or, more reasonably, measure only their own experienced congestions.
- *Efficiency:* There can be several Nash equilibria, with some more desirable than others in terms of the global objective. This issue has been the subject of significant recent research in terms of the so-called "price of anarchy" (Roughgarden, 2005) in distributed routing problems.
- *Computation:* Each stage of a learning algorithm requires agent computations. Excessive computational demands per stage can render a learning algorithm impractical.
- *Dynamic constraints:* Learning algorithms can guide how agent actions evolve over time, and these actions may be restricted because of inherent agent limitations. For example, agents may have mobility limitations in that the current position restricts the possible near term positions. More generally, agent evolution may be subject to constraints in the form of physically constraining state dynamics (e.g., so-called Dubins vehicles; Bullo et al., 2011).

- *Time complexity:* A learning algorithm may exhibit several desirable features in terms of informational requirements, computational demands per stage, and efficiency, but require a excessive number of iterations to converge. One limitation is that some of the problems of distributed control, such as weapon/target assignment, have inherent computational complexity. Distributed implementation is a subclass of centralized implementation, and accordingly inherits computational complexity limitations.

Many factors contribute to the appeal of game theory for distributed control. First, in recognizing the relevance of game theory, one can benefit from the extensive existing work in game theory to build the engineering agenda. Second, the associated learning processes promise autonomous system operations in the sense of perpetual self configuration in unknown or nonstationary environments and with robustness to disruptions or component failures. Finally, the separate design of utility functions and learning processes offers a modular approach to accommodate both different global objectives and underlying physical domain-specific constraints.

The remainder of this chapter outlines selected results in the development of a engineering agenda in game theory for distributed control. Sections 16.2 and 16.3 present the design of utility functions and learning processes, respectively. Section 16.4 presents an expansion of these ideas in terms of a broader notion of game design. Finally, Section 16.5 provides some concluding remarks.

Preliminaries

A set of agents is denoted $N = \{1, 2, \ldots, n\}$. For each $i \in N$, $\mathcal{A}_i$ denotes the set of actions available to agent i. The set of joint actions is $\mathcal{A} = \mathcal{A}_1 \times \cdots \times \mathcal{A}_N$ with elements $a = (a_1, a_2, \ldots, a_n)$. The utility function of agent i is a mapping $U_i : \mathcal{A} \to \mathbb{R}$. We will often presume that there is also a global objective function $W : \mathcal{A} \to \mathbb{R}$. An action profile $a^* \in \mathcal{A}$ is a pure strategy Nash equilibrium (or just "equilibrium") if

$$U_i(a_i^*, a_{-i}^*) = \max_{a_i \in \mathcal{A}_i} U_i(a_i, a_{-i}^*)$$

for all $i \in N$.

16.2. UTILITY DESIGN

In this section, we will survey results pertaining to utility design for distributed engineered systems. Utility design for societal systems has been studied extensively in the game theory literature, e.g., cost sharing problems (Moulin, 2010; Moulin and Shenker, 2001; Moulin and Vohra, 2003; Young, 1994) and mechanism design (Fudenberg and Tirole, 1991). The underlying goal for utility design in societal systems is to augment players' utility functions in an admissible fashion to induce desirable outcomes.

Unlike mechanism design, the agents in the engineered agenda are programmable components. Accordingly, there is no concern that agents are not truthful in reporting information or obedient in executing instructions. Nonetheless, many of the contributions stemming from the cost-sharing literature is immediately applicable to utility design in distributed engineered systems.

16.2.1 Cost/Welfare-sharing games

To study formally the role of utility design in engineered systems we consider the class of welfare/cost-sharing games (Marden and Wierman, 2013). This class is particularly relevant in that many of the aforementioned applications of distributed control resemble resource allocation or sharing.

A welfare-sharing game consists of a set of agents, N, a finite set of resources, $\mathcal{R}$, and for each agent $i \in N$, an action set $\mathcal{A}_i \subseteq 2^{\mathcal{R}}$. Note that the action set represents the set of allowable resource utilization profiles. For example, if $\mathcal{R} = \{1, 2, 3\}$, the action set

$$\mathcal{A}_i = \{\{1, 2\}, \{1, 3\}, \{2, 3\}\}$$

reflects that agent i always uses two out of three resources. An example where structured actions sets emerge is in distributed routing, where resources are roads, but admissible actions are paths. Accordingly, the action set represents sets of resources induced by the underlying network structure.

We restrict our attention to the class of *separable* system level objective functions of the form

$$W(a) = \sum_{r \in \mathcal{R}} W_r(\{a\}_r)$$

where $W_r : 2^N \to \mathbb{R}^+$ is the objective function for resource r and $\{a\}_r$ is the set of agents using resource r, i.e.,

$$\{a\}_r = \{i \in N : r \in a_i\}.$$

The goal of such welfare-sharing games is to derive *admissible* agent utility functions such that the resulting game possess desirable properties. In particular, we focus on the design of *local* agent utility functions of the form

$$U_i(a_i, a_{-i}) = \sum_{r \in a_i} f_r(i, \{a\}_r) \qquad [16.1]$$

where $f_r : N \times 2^N \to \mathbb{R}$ is the *welfare-sharing protocol*, or just "protocol," at resource r. The protocol represents a mechanism for agents to evaluate the "benefit" of being at a resource given the choices of the other agents. Utility functions are "local" in the sense that the benefit of using a resource only depends on the set of other agents using that resource and not on the usage profiles of other resources.

Finally, a welfare-sharing game is now defined by the tuple $G = (N, \mathcal{R}, \{\mathcal{A}_i\}, \{W_r\}, \{f_r\})$.

One of the important design considerations associated with engineered systems is that the structure of the specific resource allocation problem, i.e., the resource set $\mathcal{R}$ or the structure of the action sets $\{\mathcal{A}_i\}_{i\in N}$, is not known to the system designer a priori. Accordingly, a challenge in welfare-sharing problems is to design a set of *scalable* protocols $\{f_r\}$ that efficiently applies to all games in the set

$$\mathcal{G} = \{G = (N, \mathcal{R}, \{\mathcal{A}_i\}, \{W_r\}, \{f_r\}) : \mathcal{A}_i \subset 2^{\mathcal{R}}\}.$$

In other words, the set $\mathcal{G}$ represents a family of welfare-sharing games with different resource availability profiles. A protocol is "scalable" in the sense that the distribution of welfare does not depend on the specific structure of resource availability. Note that the above set of games can capture both variations in the agent set and resource set. For example, setting $\mathcal{A}_i = \emptyset$ is equivalent to removing agent i from the game. Similarly, letting the action sets satisfy $\mathcal{A}_i \subseteq 2^{\mathcal{R}\backslash\{r\}}$ for each agent i is equivalent to removing resource r from the specified resource allocation problem.

The evaluation of a protocol, $\{f_r\}$, takes into account the following considerations:

Potential game structure

Deriving an efficient dynamical process that converges to an equilibrium requires additional structure on the game environment. One such structure is that of *potential games* introduced in Monderer and Shapley (1996). In a potential game there exists a potential function $\phi : \mathcal{A} \to \mathbb{R}$ such that for any action profile $a \in \mathcal{A}$, agent $i \in N$, and action choice $a_i' \in \mathcal{A}_i$,

$$U_i(a_i', a_{-i}) - U_i(a_i, a_{-i}) = \phi(a_i', a_{-i}) - \phi(a_i, a_{-i}).$$

If a game is potential game, then an equilibrium is guaranteed to exist since any action profile $a^* \in \arg\max_{a\in\mathcal{A}} \phi(a)$ is an equilibrium.[1] Furthermore, there is a wide array of distributed learning algorithms that guarantee convergence to an equilibrium (Sandholm, 2012). It is important to note that the global objective $W(\cdot)$ and the potential $\phi(\cdot)$ can be different functions.

Efficiency of equilibria

Two well-known worst case measures of efficiency of equilibria are the *price of anarchy (PoA)* and *price of stability (PoS)* (Nisan et al., 2007). The PoA provides an upper bound on the ratio between the performance of an optimal allocation versus an equilibrium. More specifically, for a game G, let $a^{\text{opt}}(G) \in \mathcal{A}$ satisfy

$$a^{\text{opt}}(G) \in \arg\max_{a\in\mathcal{A}} W(a; G);$$

[1] See Arslan et al. (2007) for examples of intuitive utility functions that do not result an equilibrium in vehicle-target assignment.

let NE(G) denote the set of equilibria for G; and define

$$\mathrm{PoA}(\mathcal{G}) = \max_{G \in \mathcal{G}} \left(\max_{a^{\mathrm{ne}} \in NE(G)} \left(\frac{W(a^{\mathrm{opt}}(G); G)}{W(a^{\mathrm{ne}}; G)} \right) \right)$$

For example, a PoA of two ensures that for any game $G \in \mathcal{G}$ any equilibrium is at least 50% as efficient as the optimal allocation. The PoS, which represents a more optimistic worst case characterization, provides a lower bound on the ratio between the performance of the optimal allocation and the best equilibrium, i.e.,

$$\mathrm{PoS}(\mathcal{G}) = \max_{G \in \mathcal{G}} \left(\min_{a^{\mathrm{ne}} \in NE(G)} \left(\frac{W(a^{\mathrm{opt}}(G); G)}{W(a^{\mathrm{ne}}; G)} \right) \right)$$

16.2.2 Achieving potential game structures

We begin by exploring the following question: is it possible to design scalable protocols and local utility functions to guarantee a potential game structure irrespective of the specific structure of the resource allocation problem? In this section, we review two constructions which originated in the traditional economic cost sharing literature (Young, 1994) that achieve this objective.

The first construction is the *marginal contribution protocol* (Wolpert and Tumor, 1999). For any resource $r \in \mathcal{R}$, player set $S \subseteq N$, and player $i \in N$,

$$f_r^{\mathrm{MC}}(i, S) = W_r(S) - W_r(S \setminus \{i\}). \tag{16.2}$$

The marginal contribution protocol provides the following guarantees.

Theorem 16.1. (Wolpert and Tumor, 1999) *Let $\mathcal{G}$ be a class of welfare-sharing games where the protocol for each resource $r \in \mathcal{R}$ is defined as the marginal contribution protocol in [16.2]. Any game $G \in \mathcal{G}$ is a potential game with potential function W.*

Irrespective of the underlying game, the marginal contribution protocol always ensures the existence of a potential game, and consequently the existence of an equilibrium. Furthermore, since the resulting potential function is $\phi = W$, the PoS is guaranteed to be 1 when using the marginal contribution protocol. In general, the marginal contribution protocol need not provide any guarantee with respect to the PoA.

The second construction is known as the *weighted Shapley value* (Haeringer, 2006; Hart and Mas-Colell, 1989; Shapley, 1953a,b). For any resource $r \in \mathcal{R}$, player set $S \subseteq N$, and player $i \in N$,

$$f_r^{\mathrm{WSV}}(i, S) = \sum_{T \subseteq S : i \in T} \frac{\omega_i}{\sum_{j \in T} \omega_j} \left(\sum_{R \subseteq T} (-1)^{|T| - |R|} W_r(R) \right). \tag{16.3}$$

where $\omega_i > 0$ is defined as the weight of player i. The Shapley value represents a special case of the weighted Shapley value when $w_i = 1$ for all agents $i \in N$. The weighted Shapley value protocol provides the following guarantees.

Theorem 16.2. (Marden and Wierman, 2013a,b) *Let $\mathcal{G}$ be a class of welfare-sharing games where the protocol for each resource $r \in \mathcal{R}$ is the weighted Shapley value protocol in [16.3]. Any game $G \in \mathcal{G}$ is a (weighted) potential game*[2] *with potential function*

$$\phi^{\mathrm{WSV}}(a) = \sum_{r \in \mathcal{R}} \phi_r^{\mathrm{WSV}}(\{a\}_r),$$

where ϕ_r^{WSV} is a resource specific potential function defined (recursively) as follows:

$$\phi_r^{\mathrm{WSV}}(\emptyset) = 0$$

$$\phi_r^{\mathrm{WSV}}(S) = \frac{1}{\sum_{i \in S} w_i}\left[W_r(S) + \sum_{i \in S} w_i \phi_r^{\mathrm{WSV}}(S \backslash \{i\})\right], \quad \forall S \subseteq N.$$

The recursion presented in the above theorem directly follows from the potential function characterization of the weighted Shapley value derived in Hart and Mas-Colell (1989). As with the marginal contribution protocol, the weighted Shapley value protocol always ensures the existence of a (weighted) potential game, and consequently the existence of an equilibrium, irrespective of the underlying structure of the resource allocation problem. However, unlike the marginal contribution protocol, the potential function is not $\phi = W$. Consequently, the PoS is not guaranteed to be 1 when using the marginal contribution protocol. In general, the weighted Shapley value protocol also does not provide any guarantees with respect to the PoA.

An important difference between the marginal contribution in [16.2] and the weighted Shapley value in [16.3] is that the weighted Shapley value protocol guarantees that the utility functions are *budget-balanced*, i.e., for any resource $r \in \mathcal{R}$ and agent set $S \subseteq N$,

$$\sum_{i \in S} f_r^{\mathrm{WSV}}(i, S) = W_r(S). \qquad [16.4]$$

The marginal contribution utility, on the other hand, does not guarantee that utility functions are budget-balanced. Budget-balanced utility functions are important for the control (or influence) of societal systems where there is a cost or revenue that needs

[2] A weighted potential game is a generalization of a potential game with the following condition on the game structure. There exist a potential function $\phi : \mathcal{A} \to \mathbb{R}$ and weights $\omega_i > 0$ for each agent $i \in N$ such that for any action profile $a \in \mathcal{A}$, agent $i \in N$, and action choice $a_i' \in \mathcal{A}_i$,

$$U_i(a_i', a_{-i}) - U_i(a_i, a_{-i}) = \omega_i\left(\phi(a_i', a_{-i}) - \phi(a_i, a_{-i})\right).$$

to be completely absorbed by the participating players, e.g., network formation (Chen et al., 2008) and content distribution (Goemans et al., 2004). Furthermore, budget-balanced (or budget-constrained) utility functions are important for engineered systems by providing desirable efficiency guarantees (Roughgarden, 2009; Vetta, 2002); see forthcoming Theorems 16.3 and 16.5. However, the design of budget-balanced utility functions is computationally prohibitive in large systems since computing a weighted Shapley value requires a summation over an exponential number of terms.

16.2.3 Efficiency of equilibria

The desirability of a potential game structure stems from the availability of various distributed learning/adaptation rules that lead to an equilibrium. Accordingly for the engineering agenda, an important consideration is the resulting PoA. This issue is related to a research thread within algorithmic game theory that focuses on *analyzing* the inefficiency of equilibria for classes of games where the agents' utility functions $\{U_i\}$ and system level objective W are specified (cf., Chapters 17–21 in Nisan et al., 2007). While these results focus on analysis and not synthesis, they can be leveraged in utility design.

The following result, expressed in the context of resource sharing and protocols, requires the notion of submodular functions. An objective function W_r is *submodular* if for any agent set $S \subseteq T \subseteq N$ and any agent $i \in N$,

$$W_r(S \cup \{i\}) - W_r(S) \geq W_r(T \cup \{i\}) - W_r(T).$$

Submodularity reflects the diminishing marginal effect of assigning agents to resources. This property is relevant in a variety of engineering applications, including the aforementioned sensor coverage and vehicle-target assignment scenarios.

Theorem 16.3. (Vetta, 2002) *Let $\mathcal{G}$ be a class of welfare-sharing games that satisfies the following conditions for each resource $r \in \mathcal{R}$:*

(i) *The objective function W_r is submodular.*

(ii) *The protocol satisfies $f_r(i, S) \geq W_r(S) - W_r(S \backslash \{i\})$ for each set of agents $S \subseteq N$ and agent $i \in S$.*

(iii) *The protocol satisfies $\sum_{i \in S} f_r(i, S) \leq W_r(S)$ for each set of agents $S \subseteq N$.*

Then for any game $G \in \mathcal{G}$, if an equilibrium exists, the PoA is 2.

Theorem 16.3 reveals two interesting properties. First, Condition (ii) parallels the aforementioned marginal contribution protocol in [16.2]. Second, Condition (iii) relates to the budget-balanced constraint associated with (weighted) Shapley value protocol in [16.4]. Since both the marginal contribution protocol and Shapley value protocol guarantee the existence of an equilibrium, we can combine Theorems 16.1–16.3 into the following corollary.

Corollary 16.1. *Let $\mathcal{G}$ be a class of welfare-sharing games with submodular resource objective functions $\{W_r\}$. Suppose one of the following two conditions is satisfied:*
(i) *The protocol for each resource $r \in \mathcal{R}$ is the marginal contribution protocol in [16.2].*
(ii) *The protocol for each resource $r \in \mathcal{R}$ is the weighted Shapley value protocol in [16.3].*
Then for any $G \in \mathcal{G}$, an equilibrium is guaranteed to exist and the PoA is 2.

Corollary 16.1 demonstrates that both the marginal contribution protocol and the Shapley value protocol guarantee desirable properties regarding the existence and efficiency of equilibria for a broad class of resource allocation problems with submodular objective functions. There are two shortcomings associated with this result. First, it does not reveal how the structure of the objective functions $\{W_r\}$ impacts the PoA guarantees beyond the factor of 2. For example, in the aforementioned vehicle-target assignment problem (cf., Arslan et al., 2007) with submodular objective functions, both the marginal contribution and weighted Shapley value protocol will ensure that all resulting equilibria are at least 50% as efficient as the optimal assignment. It is unclear whether this factor of 2 is tight or the resulting equilibria will be more efficient than this general guarantee. Second, this corollary does not differentiate between the performance associated with the marginal contribution protocol and the weighted Shapley value protocol. For example, does the marginal contribution protocol outperform the weighted Shapley value protocol with respect to PoA guarantees? The following theorem begins to address these issues.

Theorem 16.4. (Marden and Roughgarden, 2010) *Let G be a welfare-sharing game that satisfies the following conditions:*
(i) *The objective function for each resource $r \in \mathcal{R}$ is submodular and anonymous.*[3]
(ii) *The protocol for each resource $r \in \mathcal{R}$ is the Shapley value protocol as in [16.3] with $\omega_i = 1$ for all agents $i \in N$.*
(iii) *The action set for each agent $i \in N$ is $\mathcal{A}_i = \mathcal{R}$.*
Then an equilibrium is guaranteed to exist and the PoA is

$$1 + \max_{r \in \mathcal{R},\ m \leq n} \left\{ \max_{k \leq m} \left(\frac{W_r(k)}{W_r(m)} - \frac{k}{m} \right) \right\}. \qquad [16.5]$$

Theorem 16.4 demonstrates that the structure of the welfare function plays a significant role in the underlying PoA guarantees. For example, suppose that the objective function for each resource is linear in the number of agents, e.g., $W_r(S) = |S|$ for all agent sets $S \subseteq N$. For this situation, the second term in [16.5] is 0 which means that the PoA is 1.

[3] An objective function W_r is anonymous if $W_r(S) = W_r(T)$ for any agent sets $S, T \subseteq N$ such that $|S| = |T|$.

The final general efficiency result that we review in this section pertains to the efficiency of alternative classes of equilibria. In particular, we consider the class of *coarse correlated equilibria*, which represent a generalization of the class of Nash equilibria.[4] As with potential game structures, part of the interest in coarse correlated equilibria is the availability of simple adaptation rules that lead to time-averaged behavior consistent with coarse correlated equilibria (Hart, 2005; Young, 2005). A joint distribution $z \in \Delta(\mathcal{A})$ is a coarse correlated equilibrium if for any player $i \in N$ and any action $a_i' \in \mathcal{A}_i$

$$\sum_{a \in \mathcal{A}} U_i(a) z^a \geq \sum_{a \in \mathcal{A}} U_i(a_i', a_{-i}) z^a$$

where $\Delta(\mathcal{A})$ represent the simplex over the finite set $\mathcal{A}$ and z^a represents the component of the distribution z associated with the action profile a.

We extend the system level objective from allocations to a joint distribution $z \in \Delta(a)$ as

$$W(z) = \sum_{a \in \mathcal{A}} W(a) z^a.$$

Since the set of coarse correlated equilibria contains the set of Nash equilibria, the PoA associated with this more general set of equilibria can only degrade. However, the following theorem demonstrates that if the utility functions satisfy a "smoothness" condition then there is no such degradation. We will present this theorem with regards to utility functions as opposed to protocols for a more direct presentation.

Theorem 16.5. (Roughgarden, 2009) *Consider any welfare-sharing game G that satisfies the following conditions:*

(i) *There exist parameters $\lambda > 0$ and $\mu > 0$ such that for any action profiles $a, a^* \in \mathcal{A}$*

$$\sum_{i \in N} U_i(a_i^*, a_{-i}) \geq \lambda \cdot W(a^*) - \mu \cdot W(a). \qquad [16.6]$$

(ii) *For any action profile $a \in \mathcal{A}$, the agents' utility functions satisfy $\sum_{i \in N} U_i(a) \leq W(a)$.*

Then the PoA of the set of coarse correlated equilibria is

$$\inf_{\lambda > 0, \mu > 0} \left\{ \frac{1 + \mu}{\lambda} \right\}$$

where the infimum is over the set of admissible parameters that satisfy [16.6].

Many classes of games relevant to distributed engineered systems satisfy the "smoothness" condition set forth in [16.6]. For example, the class of games considered in Theorem 16.3 satisfies the conditions of Theorem 16.5 with smoothness parameters

[4] Coarse correlated equilibria are also equivalent to the set of no-regret points (Hart, 2005; Young, 2005).

$\lambda = 1$ and $\mu = 1$ (Roughgarden, 2009). Consequently, the PoA of 2 extends beyond just pure Nash equilibria to all coarse correlated equilibria.

16.3. LEARNING DESIGN

The field of learning in games concerns the analysis of distributed learning algorithms and their convergence to various solution concepts or notions of equilibrium (Fudenberg and Levine, 1998; Hart, 2005; Young, 2005). In the descriptive agenda, the motivation is that convergence of such algorithms provides some justification for a particular solution concept as a predictive model of behavior in a societal system.

This literature can be used as a starting point for the engineering agenda to offer solutions for how equilibria *should* emerge in distributed engineered systems. In this section, we will survey results pertaining to learning design and highlight their applicability to distributed control of engineered systems.

16.3.1 Preliminaries: repeated play of one-shot games

We will consider learning/adaptation algorithms in which agents repeatedly play over stages $t \in \{0, 1, 2, \ldots\}$. At each stage, an agent i chooses an action $a_i(t)$ according to the probability distribution $p_i(t) \in \Delta(\mathcal{A}_i)$. We refer to $p_i(t)$ as the *strategy* of agent i at time t. An agent's strategy at time t relies only on observations over stages $\{0, 1, 2, \ldots, t-1\}$.

Different learning algorithms are specified by the agents' information and the mechanism by which their strategies are updated as information is gathered. We categorize such learning algorithms into the following three classes of information structures.

- *Full information:* For the class of full information learning algorithms, each agent knows the structural form of his own utility function and is capable of observing the actions of all other agents at every stage but does not know other agents' utility functions. Learning rules in which agents do not know the utility functions of other agents also are referred to as *uncoupled* (Hart and Mas-Colell, 2003, 2006). Full information learning algorithms can be written as

$$p_i(t) = F_i\big(a(0), \ldots, a(t-1); U_i\big). \qquad [16.7]$$

  for an appropriately defined functions $F_i(\cdot)$.
- *Oracle-based information:* For the class of oracle-based learning algorithms, each agent is capable of evaluating the payoff associated with alternative action choices—even though these choices were not selected. More specifically, the strategy adjustment mechanism of a given agent i can be written in the form

$$p_i(t) = F_i\left(\{U_i\left(a_i, a_{-i}\left(0\right)\right)\}_{a_i \in \mathcal{A}_i}, \ldots, \{U_i\left(a_i, a_{-i}\left(t-1\right)\right)\}_{a_i \in \mathcal{A}_i}\right). \qquad [16.8]$$

- *Payoff-based information:* For the class of payoff-based learning algorithms, each agent has access to: (i) the action they played and (ii) the payoff they received. In this setting, the strategy adjustment mechanism of agent i takes the form

$$p_i(t) = F_i\left(\{a_i(0), U_i(a(0))\}, \ldots, \{a_i(t-1), U_i(a(t-1))\}\right). \qquad [16.9]$$

Payoff-based learning rules are also referred to as *completely uncoupled* (Arieli and Babichenko, 2011; Foster and Young, 2006).

The following sections review various algorithms from the literature on learning in games and highlight their relevance for the engineering agenda in terms of their limiting behavior, the resulting efficiency, and the requisite information structure.

16.3.2 Learning Nash equilibria in potential games

We begin with algorithms for the special class of potential games. The relevance of these algorithms for the engineering agenda is enhanced by the possibility of constructing utility functions, as discussed in the previous section for resource allocation problems, to ensure a potential game structure.

16.3.2.1 Fictitious play and joint strategy fictitious play

Fictitious Play (cf., Fudenberg and Levine, 1998) is representative of a full information learning algorithm. In Fictitious Play, each agent $i \in N$ tracks the empirical frequency of the actions of other players. Specifically, for any $t > 0$, let

$$q_i^{a_i}(t) = \frac{1}{t}\sum_{\tau=0}^{t-1} I\{a_i(\tau) = a_i\},$$

where $I\{\cdot\}$ denotes the indicator function.[5] The vector $q_i(t) \in \Delta(\mathcal{A}_i)$ reflects the percentage of time that agent i selected the action a_i over stages $\{0, 1, \ldots, t-1\}$. Define the empirical frequency vector for player i at time t as $q_i(t) = \{q_i^{a_i}(t)\}_{a_i \in \mathcal{A}_i}$. At each time t, each player seeks to maximize his expected utility under the presumption that all other players are playing independently accordingly the empirical frequency of their past actions. More specifically, the action of player i at time t is chosen according to

$$a_i(t) \in \arg\max_{a_i \in \mathcal{A}_i} \sum_{a_{-i} \in \mathcal{A}_{-i}} U_i(a_i, a_{-i}) \prod_{j \neq i} q_j^{a_j}(t).$$

The following theorem establishes the convergence properties of Fictitious Play for potential games.

[5] The indicator function $I\{\text{statement}\}$ equals 1 if the mathematical expression "statement" is true, and equals 0 otherwise.

Theorem 16.6. (Monderer and Shapley, 1996a,b) *Let G be a finite n-player potential game. Under Fictitious Play, the empirical distribution of the players' actions* $\{q_1(t), q_2(t), \ldots, q_n(t)\}$ *will converge to a (possibly mixed strategy) Nash equilibrium of the game G.*

One concern associated with utilizing Fictitious Play for prescribing behavior in distributed engineered systems is the informational and computational demands (Garcia et al., 2004; Lambert et al., 2005; Marden et al., 2009). Here, each agent is required to track the empirical frequency of the past actions of all other agents, which is prohibitive in large-scale systems. Furthermore, computing a best response is intractable in general since it requires computing an expectation over a joint action space whose cardinality grows exponentially in the number of agents and the cardinality of their action sets.

Inspired by the potential application of Fictitious Play for distributed control of engineered systems, several papers investigated maintaining the convergence properties associated with Fictitious Play while reducing the computational and informational demands on the agents (Arslan et al., 2007; Garcia et al., 2004; Lambert et al., 2005; Leslie and Collins, 2003, 2005, 2006; Marden et al., 2007, 2009). One such learning algorithm is *Joint Strategy Fictitious Play with inertia* introduced in Marden et al. (2009).

In Joint Strategy Fictitious Play with inertia (as with no-regret algorithms; Hart, 2005), at each time $t > 0$ each agent $i \in N$ computes the average *hypothetical utility* for each action $a_i \in \mathcal{A}_i$, defined as

$$\begin{aligned} V_i^{a_i}(t) &= \frac{1}{t}\sum_{\tau=0}^{t-1} U_i(a_i, a_{-i}(\tau)), \\ &= \left(\frac{t-1}{t}\right) V_i^{a_i}(t-1) + \frac{1}{t} U_i(a_i, a_{-i}(t-1)). \end{aligned} \qquad [16.10]$$

The average hypothetical utility for action a_i at time t is the average utility that action a_i would have received up to time t provided that all other agents did not change their action. Note that this computation only requires oracle-based information as opposed to the full information structure of Fictitious Play. Define the best response set of agent i at time t as

$$B_i(t) = \left\{ a_i \in \mathcal{A}_i : \arg\max_{a_i \in \mathcal{A}_i} V_i^{a_i}(t) \right\}.$$

The action of player i at stage t is chosen as follows:

- If $a_i(t-1) \in B_i(t)$ then $a_i(t) = a_i(t-1)$.
- If $a_i(t-1) \notin B_i(t)$ then

$$a_i(t) = \begin{cases} a_i(t-1) & \text{with probability } \epsilon \\ a_i \in B_i(t) & \text{with probability } \frac{1-\epsilon}{|B_i(t)|} \end{cases}$$

where $\epsilon > 0$ is the players' inertia. The following theorem establishes the convergence properties of Joint Strategy Fictitious Play with Inertia for *generic* potential games.[6]

Theorem 16.7. (Marden et al., 2009a,b,c) *Let G be a finite n-player generic potential game. Under Joint Strategy Fictitious Play with Inertia, the joint action profile will converge almost surely to a pure Nash equilibrium of the game G.*

As previously mentioned, Joint Strategy Fictitious Play with inertia falls under the classification of oracle-based information. Accordingly, the informational and computational demands on the agents when using Joint Strategy Fictitious Play with inertia are reasonable in large-scale systems—assuming the hypothetical utility can be measured. The availability of such measurements is application dependent. For example in distributed routing, the hypothetical utility could be estimated with some sort of "traffic report" at the end of each stage.

The name Joint Strategy Fictitious Play stems from the average hypothetical utility in [16.10] reflecting the expected utility for agent i under the presumption that all agents other than agent i select an action with a joint strategy[7] in accordance to the empirical frequency of their pasts joint decisions, i.e.,

$$V_i^{a_i}(t) = \sum_{a_{-i}\in\mathcal{A}_i} U_i(a_i, a_{-i}) z_{-i}^{a_{-i}}(t)$$

where

$$z_{-i}^{a_{-i}}(t) = \frac{1}{t}\sum_{\tau=0}^{t-1} I\{a_{-i}(\tau) = a_{-i}\}.$$

Joint strategy Fictitious Play also can be viewed as a "max-regret" variant of no-regret algorithms (Hart, 2005; Marden et al., 2007) with inertia where the regret for action $a_i \in \mathcal{A}_i$ at time t is

$$R_i^{a_i}(t) = \frac{1}{t}\sum_{\tau=0}^{t-1} \left(U_i(a_i, a_{-i}(\tau)) - U_i(a_i(\tau), a_{-i}(\tau))\right). \qquad [16.11]$$

Note that $\arg\max_{a_i\in\mathcal{A}_i} V_i^{a_i}(t) = \arg\max_{a_i\in\mathcal{A}_i} R_i^{a_i}(t)$, hence the algorithms are equivalent.

[6] Here, "generic" means that for any agent $i \in N$, action profile $a \in \mathcal{A}$, and action $a_i' \in \mathcal{A}_i \backslash a_i$, $U_i(a_i, a_{-i}) \neq U_i(a_i', a_{-i})$. Weaker versions of generosity also ensure the characterization of the limiting behavior presented in Theorem 16.7, e.g., if all equilibria are strict.

[7] That is, unlike Fictitious Play, players are not presumed to play independently according to their individual empirical frequencies.

Finally, another distinction from Fictitious Play is that Joint Strategy Fictitious Play with inertia guarantees convergence to pure equilibria almost surely.

16.3.2.2 Simple experimentation dynamics

One concern with the implementation of learning algorithms, even in the case of full information, is the need to compute utility functions and the associated utility of different action choices (as in the computation of better or best replies). Such computations presume the availability of a closed-form expression of utility functions, which may be impractical in many scenarios. A more realistic requirement is to have agents only *measure* a realized utility online, rather than compute utility values offline. Accordingly, several papers have focused on providing payoff-based dynamics with similar limiting behaviors as the preceding full information or oracle-based algorithms (Babichenko, 2010; Foster and Young, 2006; Germano and Lugosi, 2007; Marden et al., 2009; Pradelski and Young, 2012; Young, 2009).

A representative example is the learning algorithm *Simple Experimentation Dynamics*, introduced in Marden et al. (2009). Each agent $i \in N$ maintains a pair of evolving local state variables $[\bar{a}_i, \bar{u}_i]$. These variables represent

- a *benchmark action*, $\bar{a}_i \in \mathcal{A}_i$ and
- a *benchmark utility*, $\bar{u}_i$, which is in the range of $U_i(\cdot)$.

Simple Experimentation Dynamics proceed as follows:

1. **Initialization:** At stage $t = 0$, each player arbitrarily selects and plays any action, $a_i(0) \in \mathcal{A}_i$. This action will be set initially as the player's *baseline action* at stage 1, i.e., $\bar{a}_i(1) = a_i(0)$. Likewise, each player's *baseline utility* at stage 1 is initialized as $u_i(1) = U_i(a(0))$.
2. **Action selection:** At subsequent stages, each player selects his baseline action with probability $(1 - \epsilon)$ or experiments with a new random action with probability ϵ. That is,
   - $a_i(t) = \bar{a}_i(t)$ with probability $(1 - \epsilon)$ and
   - $a_i(t)$ is chosen randomly (uniformly) over $\mathcal{A}_i$ with probability ϵ

   where $\epsilon > 0$ is the player's *exploration rate*. Whenever $a_i(t) \neq \bar{a}_i(t)$, we will say that player i "experimented."
3. **Baseline action and baseline utility update:** Each player compares the utility received, $U_i(a(t))$, with his baseline utility, $\bar{u}_i(t)$, and updates his baseline action and utility as follows:
   - If player i *experimented* (i.e., $a_i(t) \neq \bar{a}_i(t)$) and if $U_i(a(t)) > \bar{u}_i(t)$, then
     $$\bar{a}_i(t+1) = a_i(t),$$
     $$\bar{u}_i(t+1) = U_i(a(t)).$$
   - If player i *experimented* and if $U_i(a(t)) \leq \bar{u}_i(t)$, then
     $$\bar{a}_i(t+1) = \bar{a}_i(t),$$
     $$\bar{u}_i(t+1) = \bar{u}_i(t).$$

- If player i *did not experiment* (i.e., $a_i(t) = \bar{a}_i(t)$), then
$$\bar{a}_i(t+1) = \bar{a}_i(t),$$
$$\bar{u}_i(t+1) = U_i(a(t)).$$

4. Return to Step 2 and repeat.

Theorem 16.8. (Marden et al., 2010) *Let G be a finite n-player potential game. Under Simple Experimentation Dynamics, given any probability $p < 1$, there exists an exploration rate $\epsilon > 0$ (sufficiently small), such that for all sufficiently large stages t, the joint action $a(t)$ is a Nash equilibrium of G with at least probability p.*

Theorem 16.8 demonstrates that one can attain convergence to equilibria even in the setting where agents have minimal knowledge regarding the underlying game. Note that for such payoff-based dynamics we attain probabilistic convergence as opposed to almost sure converges. The reasoning is that agents are unaware of whether or not they are at an equilibrium since they do not have access to oracle-based or full information. Consequently, the agents perpetually probe the system to reassess the baseline action and utility.

16.3.2.3 Equilibrium selection: log-linear learning and its variants

The previous discussion establishes how distributed learning rules under various information structures can converge to a Nash equilibrium. However, these results are silent on the issue of equilibrium *selection*, i.e., determining which equilibria may be favored or excluded. Notions such as PoA and PoS give pessimistic and optimistic bounds, respectively, on the value of a global performance measure at an equilibrium as compared to its optimal value. Equilibrium selection offers a refinement of these bounds through the specific underlying dynamics.

The topic of equilibrium selection has been widely studied within the descriptive agenda. Two standard references are Kandori et al. (1993) and Young (1993), which discuss equilibrium selection between risk dominant or payoff dominant equilibrium in symmetric 2×2 games. As would be expected, the conclusions are sensitive to the underlying dynamics (Bergin and Lipman, 1996). However, in the engineering agenda, one can exploit this dependence as an available degree of freedom (e.g., Chasparis and Shamma, 2012).

This section will review equilibrium selection in potential games for a class of dynamics, namely log-linear learning and its variants, that converge to maximizer of the underlying potential function, ϕ. The relevance for the engineering agenda stems from results such as Theorem 16.1, which illustrate how utility design can ensure the resulting interaction framework is a potential game and that the optimal allocation corresponds to the optimizer of the potential function. Hence the optimistic PoS, which equals 1 for this setting, will be achieved through the choice of dynamics.

Log-linear learning, introduced in Blume (1993), is an asynchronous oracle-based learning algorithm. At each stage $t > 0$, a single agent $i \in N$ is randomly chosen and allowed to alter his current action. All other players must repeat their actions from the previous stage, i.e. $a_{-i}(t) = a_{-i}(t-1)$. At stage t, the selected player i employs the (Boltzmann distribution) strategy $p_i(t) \in \Delta(\mathcal{A}_i)$, given by

$$p_i^{a_i}(t) = \frac{e^{\frac{1}{\tau} U_i(a_i, a_{-i}(t-1))}}{\sum_{\bar{a}_i \in \mathcal{A}_i} e^{\frac{1}{\tau} U_i(\bar{a}_i, a_{-i}(t-1))}}, \qquad [16.12]$$

for a fixed "temperature," $\tau > 0$. As is well known for the Boltzmann distribution, for large τ, player i will select any action $a_i \in \mathcal{A}_i$ with approximately equal probability, whereas for diminishing τ, player i will select a best response to the action profile $a_{-i}(t-1)$, i.e.,

$$a_i(t) \in \arg\max_{a_i \in \mathcal{A}_i} U_i(a_i, a_{-i}(t-1))$$

with increasingly high probability.

The following theorem characterizes the limiting behavior associated with log-linear learning for the class of potential games.

Theorem 16.9. (Blume, 1993) *Let G be a finite n-player potential game. Log-linear learning induces an aperiodic and irreducible process of the joint action set $\mathcal{A}$. Furthermore, the unique stationary distribution $\mu(\tau) = \{\mu^a(\tau)\}_{a \in \mathcal{A}} \in \Delta(\mathcal{A})$ is given by*

$$\mu^a(\tau) = \frac{e^{\frac{1}{\tau}\phi(a)}}{\sum_{\bar{a} \in \mathcal{A}} e^{\frac{1}{\tau}\phi(\bar{a})}}. \qquad [16.13]$$

One can interpret the stationary distribution μ as follows. For sufficiently large times $t > 0$, $\mu^a(\tau)$ equals the probability that $a(t) = a$. As one decreases the temperature, $\tau \to 0$, all the weight of the stationary distribution $\mu(\tau)$ is on the joint actions that maximize the potential function. Again, the emphasis here is that log-linear learning, coupled with suitable utility design, converges probabilistically to the maximizer of the potential function, and hence underlying global objective.

A concern with log-linear learning as a tool for the engineering agenda is whether the specific assumptions on the both the game and learning algorithm are restrictive and thereby limit the applicability of log-linear learning for distributed control. In particular, log-linear learning imposes the following assumptions:

(i) The underlying process is *asynchronous* which implies that the agents can only update their strategies one at a time, thereby requiring some sort of coordination.

(ii) The updating agent can select any action in his action set. In distributed control applications, there may be evolving constraints on the available action sets (e.g., mobile robots with limited mobility or in an environment with obstacles).
(iii) The requisite information structure is oracle based.
(iv) The agents' utility function constitute an exact potential game.

It turns out that these concerns can be alleviated through the use of similar learning rules with an alternative analysis. While Theorem 16.9 provides an explicit characterization of the resulting stationary distribution, an important consequence is that as $\tau \to 0$ the mass of the stationary distribution focuses on the joint actions that maximize the potential function. In the language of Young (1998), potential functions maximizers are *stochastically stable*.[8]

Recent work analyzes how to relax the structure of log-linear learning while ensuring that the only stochastically stable states are the potential function maximizers. Alos-Ferrer and Netzer (2010) demonstrate certain relaxations under which potential function maximizers need not be stochastically stable. Marden and Shamma (2012) demonstrate that it is possible to relax the structure carefully while maintaining the desired limiting behavior. In particular, Marden and Shamma (2012) establish a payoff-based learning algorithm, termed *payoff-based log linear learning*, that ensures that for potential games the only stochastically stable states are the potential function maximizers. We direct the readers to Marden and Shamma (2012) for details.

16.3.2.4 Near potential games

An important consideration for the engineering agenda is to understand the "robustness" of learning algorithms, i.e., how do guaranteed properties degrade as underlying modeling assumptions are violated. For example, consider the weighted Shalpey value protocol defined in [16.3]. The weighted Shapley value protocol requires a summation of an exponential number of terms, which can be computationally prohibitive in large-scale systems. While there are sampling approaches that can yield good approximations for the Shapley value Conitzer and Sandholm (2004), it is important to note that these sampled utilities will not constitute a potential game.

Accordingly, several papers have focused on analyzing dynamics in *near potential games* Candogan et al. (2011a,b) and Marden and Shamma (2012). We say that a game is $\delta > 0$ close to a potential game if there exists a potential function $\phi : \mathcal{A} \to \mathbb{R}$ such that for any player $i \in N$, actions $a_i', a_i'' \in \mathcal{A}_i$, and joint action $a_{-i} \in \mathcal{A}_{-i}$, players' utility satisfies

$$\left|\left(U_i(a_i', a_{-i}) - U_i(a_i'', a_{-i})\right) - \left(\phi(a_i', a_{-i}) - \phi(a_i'', a_{-i})\right)\right| \leq \delta.$$

A game is a near potential game for such games where δ is sufficiently small. The work in Candogan et al. (2011a,b) and Marden and Shamma (2012) proves that the

[8] An action profile $a \in \mathcal{A}$ is stochastically stable if $\lim_{\tau \to 0^+} \mu^a(\tau) > 0$.

limiting behavior associated with several classes of dynamics in near-potential games can be approximated by analyzing the dynamics on the closest potential game. Hence, the characterizations of the limiting behavior for many of the learning algorithms for potential games immediately extend to near potential games.

16.3.3 Beyond potential games and equilibria: efficient action profiles

The discussion thus far has been limited to potential games and convergence to Nash equilibrium. Nonetheless, there is an extensive body of work that discusses convergence to broader classes of games (e.g., weakly acyclic games) or alternative solution concepts (e.g., coarse and correlated equilibria). See Hart (2005) and Young (2005) for an extensive discussion. In this section, we depart from the preceding discussion on learning in games two ways. First, we do not impose a particular structure on the game.[9] Second, we focus on convergence to efficient joint actions, whether or not they may be an equilibrium of the underlying game. In doing so, we continue to exploit the prescriptive emphasis of the engineering agenda by treating the learning dynamics as a design element.

16.3.3.1 Learning efficient pure Nash equilibria

We begin by reviewing the "mood-based" learning algorithms introduced in Young (2009) and Pradelski and Young (2012). For any finite n-player "interdependent" game where a pure Nash equilibrium exists, this algorithm guarantees (probabilistic) convergence to the pure Nash equilibrium that maximizes the sum of the agents' payoffs while adhering to a payoff-based information structure. Before stating the algorithm, we introduce the following definition of interdependence.

Definition 16.1. (Interdependence; Young, 2009) *An n-person game G on the finite action space $\mathcal{A}$ is interdependent if, for every $a \in \mathcal{A}$ and every proper subset of agents $J \subset N$, there exists an agent $i \notin J$ and a choice of actions $a'_J \in \prod_{j \in J} \mathcal{A}_j$ such that $U_i(a'_J, a_{-J}) \neq U_i(a_J, a_{-J})$.*

Roughly speaking, the interdependence condition states that it is not possible to divide the agents into two distinct subsets that do not mutually interact with one another.

We will now present a version of the learning algorithm introduced in Pradelski and Young (2012), which leads to *efficient* Nash equilibria. Without loss of generality, we shall focus on the case where agent utility functions are strictly bounded between 0 and 1, i.e., for any agent $i \in N$ and action profile $a \in \mathcal{A}$ we have $1 > U_i(a) \geq 0$. As with the simple experimentation dynamics, each agent $i \in N$ maintains an evolving local state variables, now given by the triple $[\bar{a}_i, \bar{u}_i, m_i]$. These variables represent

[9] Beyond the forthcoming technical connectivity assumption of "interdependence."

- A *benchmark action* of agent i, $\bar{a}_i \in \mathcal{A}_i$.
- A *benchmark utility* of agent i, $\bar{u}_i$, which is in the range of $U_i(\cdot)$.
- A *mood* of agent i, $m_i \in \{C, D, H, W\}$. We will refer to the mood C as "content," D as "discontent," H as "hopeful," and W as "watchful."

The algorithm proceeds as follows:

1. **Initialization:** At stage $t = 0$, each player randomly selects and plays any action, $a_i(0)$. This action will be initially set as the player's *baseline action* at stage 1, i.e., $\bar{a}_i(1) = a_i(0)$. Likewise, the player's *baseline utility* at stage 1 is initialized as $u_i(1) = U_i(a(0))$. Finally, the player's *mood* at stage 1 is set as $m_i(1) = C$.
2. **Action selection:** At each subsequent stage $t > 0$, each player selects his action according to the following rules. Let $x_i(t) = [\bar{a}_i, \bar{u}_i, m_i]$ be the state of agent i at time t. If the mood of agent i is content, i.e., $m_i = C$, the agent chooses an action $a_i(t)$ according to the following probability distribution

$$p_i^{a_i}(t) = \begin{cases} \frac{\epsilon}{|\mathcal{A}_i|-1} & \text{for } a_i \neq \bar{a}_i, \\ 1 - \epsilon & \text{for } a_i = \bar{a}_i, \end{cases} \qquad [16.14]$$

where $|\mathcal{A}_i|$ represents the cardinality of the set $\mathcal{A}_i$. If the mood of agent i is discontent, i.e., $m_i = D$, the agent chooses an action a_i according to the following probability distribution

$$p_i^{a_i}(t) = \frac{1}{|\mathcal{A}_i|} \quad \text{for every } \ a_i \in \mathcal{A}_i. \qquad [16.15]$$

Note that the benchmark action and utility play no role in the agent dynamics when the agent is discontent. Lastly, if the agent is either hopeful or watchful, i.e., $m_i = H$ or $m_i = W$, the agent chooses an action $a_i(t)$ according to the following probability distribution

$$p_i^{a_i}(t) = \begin{cases} 0 & \text{for } a_i \neq \bar{a}_i, \\ 1 & \text{for } a_i = \bar{a}_i. \end{cases} \qquad [16.16]$$

3. **Baseline action, baseline utility, and mood update:** Once the agent selects an action $a_i(t) \in \mathcal{A}_i$ and receives the payoff $u_i(t) = U_i(a_i(t), a_{-i}(t))$, where $a_{-i}(t)$ is the action selected by all agents other than agent i at stage t, the state is updated according to the following rules. First, if the state of agent i at time t is $x_i(t) = [\bar{a}_i, \bar{u}_i, C]$ then the state $x_i(t+1)$ is derived from the following transition:

$$x_i(t) = [\bar{a}_i, \bar{u}_i, C] \longrightarrow x_i(t+1) = \begin{cases} [\bar{a}_i, \bar{u}_i, C] & \text{if } a_i(t) = \bar{a}_i, u_i(t) = \bar{u}_i, \\ [\bar{a}_i, u_i(t), H] & \text{if } a_i(t) = \bar{a}_i, u_i(t) > \bar{u}_i, \\ [\bar{a}_i, u_i(t), W] & \text{if } a_i(t) = \bar{a}_i, u_i(t) < \bar{u}_i, \\ [a_i(t), u_i(t), C] & \text{if } a_i(t) \neq \bar{a}_i, u_i(t) > \bar{u}_i, \\ [\bar{a}_i, \bar{u}_i, C] & \text{if } a_i(t) \neq \bar{a}_i, u_i(t) \leq \bar{u}_i. \end{cases}$$

Second, if the state of agent i at time t is $x_i(t) = [\bar{a}_i, \bar{u}_i, D]$ then the state $x_i(t+1)$ is derived from the following (probabilistic) transition:

$$x_i(t) = [\bar{a}_i, \bar{u}_i, D] \longrightarrow x_i(t+1) = \begin{cases} [a_i(t), u_i(t), C] & \text{with probability } \epsilon^{1-u_i(t)}, \\ [a_i(t), u_i(t), D] & \text{with probability } 1 - \epsilon^{1-u_i(t)}. \end{cases}$$

Third, if the state of agent i at time t is $x_i(t) = [\bar{a}_i, \bar{u}_i, H]$ then the state $x_i(t+1)$ is derived from the following transition:

$$x_i(t) = [\bar{a}_i, \bar{u}_i, H] \longrightarrow x_i(t+1) = \begin{cases} [a_i(t), u_i(t), C] & \text{if } u_i(t) \geq \bar{u}_i, \\ [a_i(t), u_i(t), W] & \text{if } u_i(t) < \bar{u}_i. \end{cases}$$

Lastly, if the state of agent i at time t is $x_i(t) = [\bar{a}_i, \bar{u}_i, W]$ then the state $x_i(t+1)$ is derived from the following transition:

$$x_i(t) = [\bar{a}_i, \bar{u}_i, W] \longrightarrow x_i(t+1) = \begin{cases} [a_i(t), u_i(t), H] & \text{if } u_i(t) > \bar{u}_i, \\ [a_i(t), u_i(t), D] & \text{if } u_i(t) \leq \bar{u}_i. \end{cases}$$

4. Return to Step 2 and repeat.

The above algorithm ensures convergence, in a stochastic stability sense, to the pure Nash equilibrium which maximizes the sum of the agents' payoffs. Before stating the theorem, we recall the notation $NE(G)$ which represents the set of action profiles that are pure Nash equilibria of the game G.

Theorem 16.10. (Pradelski and Young, 2012) *Let G be a finite n-player interdependent game where a pure Nash equilibrium exists. Under the above algorithm, given any probability $p < 1$, there exists an exploration rate $\epsilon > 0$ (sufficiently small), such that for sufficiently large times t, $a(t) \in \arg\max_{a \in NE(G)} \sum_{i \in N} U_i(a)$ of G with at least probability p.*

16.3.3.2 Learning pareto efficient action profiles

One of the main issues regarding the asymptotic guarantees associated with the learning algorithm given in Pradelski and Young (2012) is that the system performance associated with the best pure Nash equilibrium may be significantly worse than the optimal system performance, i.e., the system performance associated with the optimal action profile. Accordingly, it would be desirable if the algorithm guarantees convergence to the action profile which maximizes the sum of the agents' utilities irrespective of whether this action profile constitutes a pure Nash equilibrium. We will now present a learning algorithm, termed *Distributed Learning for Pareto Optimality*, that builds on the developments in Pradelski and Young (2012) and accomplishes such a task. As above, we shall focus on the case where agent utility functions are strictly bounded between 0 and 1. Consequently, for any action profile $a \in \mathcal{A}$ we have $n > \sum_{i \in N} U_i(a) \geq 0$. As with the dynamics presented in Pradelski and Young (2012), each agent $i \in N$ maintains an evolving local state variable given by the triple $[\bar{a}_i, \bar{u}_i, m_i]$. These variables represent

- A *benchmark action* of agent i, $\bar{a}_i \in \mathcal{A}_i$.
- A *benchmark utility* of agent i, $\bar{u}_i$, which is in the range of $U_i(\cdot)$.
- A *mood* of agent i, $m_i \in \{C, D\}$. The moods "hopeful" and "watchful" are no longer used in this setting.

Distributed Learning for Pareto Optimality proceeds as follows:

1. **Initialization:** At stage $t = 0$, each player randomly selects and plays any action, $a_i(0)$. This action will be initially set as the player's *baseline action* at stage 1, i.e., $\bar{a}_i(1) = a_i(0)$. Likewise, the player's *baseline utility* at stage 1 is initialized as $u_i(1) = U_i(a(0))$. Finally, the player's *mood* at stage 1 is set as $m_i(1) = C$.
2. **Action selection:** At each subsequent stage $t > 0$, each player selects his action according to the following rules. If the mood of agent i is content, i.e., $m_i(t) = C$, the agent chooses an action $a_i(t)$ according to the following probability distribution

$$p_i^{a_i}(t) = \begin{cases} \frac{\epsilon^c}{|\mathcal{A}_i|-1} & \text{for } a_i \neq \bar{a}_i, \\ 1 - \epsilon^c & \text{for } a_i = \bar{a}_i, \end{cases} \qquad [16.17]$$

where $|\mathcal{A}_i|$ represents the cardinality of the set $\mathcal{A}_i$ and c is a constant that satisfies $c > n$. If the mood of agent i is discontent, i.e., $m_i(t) = D$, the agent chooses an action a_i according to the following probability distribution

$$p_i^{a_i}(t) = \frac{1}{|\mathcal{A}_i|} \quad \text{for every } \; a_i \in \mathcal{A}_i. \qquad [16.18]$$

Note that the benchmark action and utility play no role in the agent dynamics when the agent is discontent.

3. **Baseline action, baseline utility, and mood update:** Once the agent selects an action $a_i(t) \in \mathcal{A}_i$ and receives the payoff $U_i(a_i(t), a_{-i}(t))$, where $a_{-i}(t)$ is the action selected by all agents other than agent i at stage t, the state is updated according to the following rules. First, the baseline action and baseline utility at stage $t+1$ are set as

$$\bar{a}_i(t+1) = a_i(t),$$
$$\bar{u}_i(t+1) = U_i(a_i(t), a_{-i}(t)).$$

The mood of agent i is updated as follows.

3a. If

$$\begin{bmatrix} \bar{a}_i(t) \\ \bar{u}_i(a(t)) \\ m_i(t) \end{bmatrix} = \begin{bmatrix} a_i(t) \\ U_i(a(t)) \\ C \end{bmatrix},$$

then $m_i(t+1) = C$.

3b. Otherwise,

$$m_i(t+1) = \begin{cases} C \text{ with probability } \epsilon^{1-U_i(a(t))}, \\ D \text{ with probability } 1-\epsilon^{1-U_i(a(t))}. \end{cases}$$

4. Return to Step 2 and repeat.

Theorem 16.11. (Marden et al., 2011) *Let G be a finite n-player interdependent game. Under Distributed Learning for Pareto Optimality, given any probability $p < 1$, there exists an exploration rate $\epsilon > 0$ (sufficiently small), such that for sufficiently large stages t, $a(t) \in \arg\max_{a \in \mathcal{A}} \sum_{i \in N} U_i(a)$ of G with at least probability p.*

Distributed Learning for Pareto Optimality guarantees probabilistic convergence to the action profile that maximizes the sum of the agents' utility functions. As stated earlier, the maximizing action profile need *not* be a Nash equilibrium. Accordingly, in games such as the classical prisoner's dilemma game, this algorithm provides convergence to the action profile where each player cooperates even though this is a strictly dominated strategy. Likewise, for the aforementioned application of wind farm optimization, where each turbine's utility function represents the power generated by that turbine, Distributed Learning for Pareto Optimality guarantees convergence to the action profile that optimizes the total power production in the wind farm. As a consequence of the payoff-based information structure, this algorithm also demonstrates that optimizing system performance in wind farms does not require a characterization of the aerodynamic interaction between the turbines nor global information available to the turbines.

Recent work (Menon and Baras, 2013) relaxes the assumption of interdependency through the introduction of simple inter-agent communications.

16.4. EXPLOITING THE ENGINEERING AGENDA: STATE-BASED GAMES

When viewed as dynamical systems, distributed learning algorithms are all described in terms of an underlying evolving state. In most cases, this state has an immediate interpretation in terms of the primitive elements of the game (e.g., empirical frequencies in Fictitious Play or immediately preceding actions in log-linear learning). In other cases, the state variable may be better interpreted as an auxiliary variable, not necessarily related to actions and payoffs. Rather, these variables are introduced into the dynamics to evoke desirable limiting behavior. One example is the "mood" in Distributed Learning for Pareto Optimality. Similarly, Shamma and Arslan (2005) illustrate how auxiliary states can overcome fundamental limitations in learning (Hart and Mas-Colell, 2003). The introduction of such states again reflects the available degrees of freedom in the engineering agenda in that these variables need not have interpretations naturally relevant to the associated game.

In this section, we continue to explore and exploit this addition degree of freedom in defining the game itself, and in particular, through a departure from utility design for normal form games. We begin this section by reviewing some of the limitations associated with the framework of strategic form games for distributed control. Next, we review the framework of *state-based games*, introduced in Marden (2012), which represents a simplification of the framework of Markov games and is better suited to address the constraints and objective inherent to engineered systems. The key distinction between strategic form games and state-based games is the introduction of an underlying state space into the game-theoretic environment. Here, the state space presents the system designer with additional design freedom to address issues pertinent to distributed engineered systems. We conclude this section by illustrating how this additional state can be exploited in distributed control.

16.4.1 Limitations of strategic form games

In this section, we review two limitations of strategic form games for distributed engineered systems. The first limitation concerns the complexity associated with utility design. The second limitation concerns on the applicability of strategic form games for distributed optimization.

16.4.1.1 Limitations of protocol design

The marginal contribution (Theorem 16.1) and the weighted Shapley value (Theorem 16.2) represent two *universal* methodologies for utility design in distributed engineering system. By universal, we mean that these methodologies will ensure that the resulting game is a (weighted) potential game irrespective of the resource set $\mathcal{R}$, the structure of the objective functions $\{W_r\}_{r\in\mathcal{R}}$, or the structure of the agents' action sets $\{\mathcal{A}_i\}_{i\in N}$. Here, universality is of fundamental importance by allowing design methodologies to be applicable to a wide array of different applications.

The natural question that emerges is whether there are other universal methodologies for utility design in distributed engineering system. To answer this question, let us first revisit the marginal contribution protocol and the weighted Shapley value protocol, defined in [16.2] and [16.3], respectively, which are derived using the true welfare functions $\{W_r\}$. Naively, a system designer could have introduced base welfare functions $\{\tilde{W}_r\}$, which may be distinct from $\{W_r\}$, as the basis for computing both the marginal contribution and the weighted Shapley value protocols and inherit the same guarantees, e.g., existence of a pure Nash equilibrium. The following theorem proves that this approach actually corresponds to the full set of universal methodologies that guarantee the existence of a pure Nash equilibrium.

Theorem 16.12. (Gopalakrishnan et al., 2014) *Let $\mathcal{G}$ be the set of welfare-sharing games. A set of protocols $\{f_r\}$ guarantees the existence of any equilibrium in any game $G \in \mathcal{G}$ if and only if the protocols can be characterized by a weighted Shapley value to some base welfare functions $\{\tilde{W}_r\}$, i.e., for any resource $r \in \mathcal{R}$, agent set $S \subseteq N$, and agent $i \in N$,*

$$f_r(i, S) = \sum_{T \subseteq S: i \in T} \frac{\omega_i}{\sum_{j \in T} \omega_j} \left(\sum_{R \subseteq T} (-1)^{|T|-|R|} \tilde{W}_r(R) \right), \quad [16.19]$$

where $\omega_i > 0$ is the (fixed) weight of agent i. Furthermore, if the set of protocols $\{f_r\}$ must also be budget-balanced, then the base welfare functions must equal the true welfare functions, i.e., $W_r = \tilde{W}_r$ for all resources $r \in \mathcal{R}$.

Theorem 16.12 shows that if a set of protocols $\{f_r(\cdot)\}$ cannot be represented by a weighted Shapley value to some base welfare functions $\{\tilde{W}_r\}$, then there exists a game $G \in \mathcal{G}$ where a pure Nash equilibrium does not exist. At first glance, it appears that Theorem 16.12 contradicts our previous analysis showing that the marginal contribution protocol, defined in [16.2], always guarantees the existence of an equilibrium. However, it turns out that the marginal contribution protocol can be expressed as a weighted Shapley value protocol where the weights and base welfare functions are chosen carefully; see Gopalakrishnan et al. (2014) for details. An alternative characterization of the space of protocols that guarantee equilibrium existence, and more directly reflects the connection to marginal contribution protocols, is as follows:

Theorem 16.13. (Gopalakrishnan et al., 2014) *Let $\mathcal{G}$ be the set of welfare-sharing games. A set of protocols $\{f_r\}$ guarantees the existence of any equilibrium in any game $G \in \mathcal{G}$ if and only if the protocols can be characterized by a weighted marginal contribution to some base welfare functions $\{\tilde{W}_r\}$, i.e., for any resource $r \in \mathcal{R}$, agent set $S \subseteq N$, and agent $i \in N$,*

$$f_r(i, S) = \omega_i \left(\tilde{W}_r(S) - \tilde{W}_r(S \backslash i) \right). \quad [16.20]$$

where $\omega_i > 0$ is the (fixed) weight of agent i.

The two characterizations presented in Theorems 16.12 and 16.13 illustrate an underlying design tradeoff between complexity and the design of budget-balanced protocols. For example, if a system-designer would like to use budget-balanced protocols, then the system-designer inherits the complexity associated with a weighted Shapley value protocol. However, if a system-designer is not partial to budget-balanced protocols, then a system-design can appeal to the far simpler weighted marginal contribution protocols.

In addition to ensuring the existence of a pure Nash equilibrium, a system-designer might also seek to optimize the efficiency of the resulting pure Nash equilibria, i.e., price of anarchy. There are currently no established methodologies for achieving this objective; however, recent results have identified limitations on achievable efficiency guarantees associated with the use of "fixed" protocols such as weighted Shapley values of weighted marginal contributions. One such limitation is as follows:

Theorem 16.14. (Marden and Wierman, 2013) *Let $\mathcal{G}$ be the set of welfare-sharing games with submodular objective functions and a fixed weighted Shapley value protocol. The PoS across the set of games $\mathcal{G}$ is* 2.

This theorem proves that in general it is impossible to guarantee that the optimal allocation is an equilibrium when using budget-balanced protocols. This conclusion is in contrast to non-budget balanced protocols, e.g., the marginal contribution protocol, which can achieve a PoS of 1 for such settings. Recall that both the marginal contribution protocol and the weighted Shapley value protocol guarantee a PoA of 2 when restricting attention to welfare-sharing games with submodular objective functions since both protocols satisfy the conditions of Theorem 16.3.

16.4.1.2 Distributed optimization: consensus

An extensively studied problem in distributed control is that of agreement and consensus (Jadbabaie et al., 2003; Olfati-Saber et al., 2007; Tsitsiklis et al., 1986). In such consensus problems, there is a group of agents, N, and each agent $i \in N$ is endowed with an initial value $v_i(0) \in \mathbb{R}$. Agents update these values over stages, $t = 0, 1, 2, \ldots$. The goal is for each agent to compute the average of the initial endowments. The challenge associated with such a task centers on the fact that each agent i is only able to communicate with neighboring agents, specified by the subset $N_i \subseteq N$. Define the *interaction graph* as the graph formed by the nodes N and edges $E = \{(i,j) \in N \times N : j \in N_i\}$. The challenge in consensus problems is then to design update rules of the form

$$v_i(t) = F_i\left(\left\{\text{information about agent } j \text{ at time } t\right\}_{j \in N_i}\right) \qquad [16.21]$$

so that $\lim_{t\to\infty} v_i(t) = v^* = \frac{1}{n}\sum_{i\in N} v_i(0)$, which represents the solution to the optimization problem

$$\begin{array}{rl} \max_{v\in\mathbb{R}^n} & -\frac{1}{2}\sum_{i\in N, j\in N_i} ||v_i - v_j||_2^2 \\ \text{s.t.} & \sum_{i\in N} v_i = \sum_{i\in N} v_i(0) \end{array} \qquad [16.22]$$

provided that the interaction graph is connected. Furthermore, the control laws $\{F_i(\cdot)\}_{i\in N}$ should achieve the desired asymptotic guarantees for any initial value profile $v(0)$ and any connected interaction graph. This implies that the underlying control design must be invariant to these parameters in addition to the specific indices assigned to the agents.

One algorithm (among many variants) that achieves asymptotic consensus is distributed averaging (Jadbabaie et al., 2003; Olfati-Saber et al., 2007; Tsitsiklis et al., 1986), given by

$$v_i(t) = v_i(t-1) + \epsilon \sum_{j \in N_i} (v_j(t-1) - v_i(t-1)),$$

where $\epsilon > 0$ is a step-size. This algorithm imposes the constraint that for all times $t \geq 0$,

$$\sum_{i \in N} v_i(t) = \sum_{i \in N} v_i(0),$$

i.e., the average value is invariant. Hence, if the agents reach consensus on a common value, this value must represent the average of the initial values.

While the above description makes no explicit reference to games, there is considerable overlap with the present discussion of games and learning. Marden et al. (2009) discuss how consensus and agreement problems can be viewed within the context of games and learning. However, the discussion is largely restricted to only asymptotic agreement, i.e., consensus but not necessarily to the original average.

Since most of the aforementioned learning rules converge to a Nash equilibrium, one could attempt to assign each agent $i \in N$ an admissible utility function such that (i) the resulting game is a potential game and (ii) all resulting equilibria solve the optimization problem in [16.22]. To ensure scalability properties, we focus on meeting these objective using "spatially invariant" utility functions of the following form

$$U_i(v) = \mathcal{U}\left(\left\{v_j, v_j(0)\right\}_{j \in N_i}\right) \qquad [16.23]$$

where the function $\mathcal{U}(\cdot)$ is invariant to specific indices assigned to agents and values take the role of the agents' actions. Note that the design of $\mathcal{U}(\cdot)$ leads to a well-defined game irrespective of the agent set, N, initial value profile, $v(0)$, or the structure of the interaction graph $\{N_i\}_{i \in N}$. The following theorem demonstrates that it is *impossible* to design $\mathcal{U}(\cdot)$ such that for any game induced by an initial value profile and an undirected and connected interaction graph all resulting Nash equilibria solve the optimization problem in [16.22].

Theorem 16.15. (Li and Marden, 2010a,b) *There does not exist a single $\mathcal{U}(\cdot)$ such that for any game induced by a connected and undirected interaction graph formed by the information sets $\{N_i\}_{i \in N}$, an initial value profile $v(0)$, and agents' utility functions of the form [16.23], the Nash equilibria of the induced game represent solutions to the optimization problem in [16.22].*

This theorem demonstrates that the framework of strategic form games is not rich enough to meet the design considerations pertinent to distributed engineered systems.

While this limitation was illustrated here on the consensus problem, one might imagine that various other system level objectives could have similar limitations.

16.4.2 State-based games

In this section, we review the framework of *state-based games*, introduced in Marden (2012), which represents an extension to the framework of strategic form games where an underlying state space to the game-theoretic framework.[10] Here, the state is introduced as a coordinating device used to improve system level behavior. A (deterministic) state-based game consists of the following elements:

(i) A set of agents N.

(ii) An underlying state space X.

(iii) An (state-invariant) action set $\mathcal{A}_i$ for each agent $i \in N$.

(iv) A state-dependent utility function $U_i : X \times \mathcal{A} \to \mathbb{R}$ for each agent $i \in N$ where $\mathcal{A} = \prod_{i \in N} \mathcal{A}_i$.

(v) A deterministic state transition function $P : X \times \mathcal{A} \to X$.

Lastly, we will restrict our attention to state-based games where agents have a null action $a^0 \in \mathcal{A}$ such that $x = P(x, a^0)$ for any $x \in X$.

Repeated play of a state-based game produces a sequence of action profiles $a(0)$, $a(1)$, $\ldots$, and a sequence of states $x(0)$, $x(1)$, $\ldots$, where $a(t) \in \mathcal{A}$ is referred to as the action profile at time t and $x(t) \in X$ is referred to as the state at time t. The sequence of actions and states is generated according to the following process: at any time $t \geq 0$, each agent $i \in N$ *myopically* selects an $a_i(t) \in \mathcal{A}_i$, optimizing only the agent's potential payoff at time t. The state $x(t)$ and the action profile $a(t) = (a_1(t), \ldots, a_n(t))$ together determine each agent's payoff $U_i(x(t), a(t))$ at time t. After all agents select their respective action, the ensuing state $x(t+1)$ is chosen according to the state transition function $x(t+1) = P(x(t), a(t))$ and the process is repeated.

We begin by introducing a class of games, termed *state based potential games* (Li and Marden, 2011), which represents an extension of potential games to the framework of state-based games.

Definition 16.2. (State-Based Potential Game; Li and Marden, 2011) *A state based game $G = \{N, X, \{\mathcal{A}_i\}, \{U_i\}, P\}$ is a state-based potential game if there exists a potential function $\phi : X \times \mathcal{A} \to \mathbb{R}$ that satisfies the following two properties for every state $x \in X$ and action profile $a \in \mathcal{A}$:*

[10] State-based games represent a simplification of the class of Markov games (Shapley, 1953a,b) where the key difference lies in the discount factor associated with future payoffs. In Markov games, an agent's utility represents a discounted sum of future payoffs. Alternatively, in state-based games, an agent's utility represents only the current payoff, i.e., the discount factor is 0. This difference greatly simplifies the analysis of such games.

1. *For any agent $i \in N$ and action $a_i' \in \mathcal{A}_i$,*

$$U_i(x, a_i', a_{-i}) - U_i(x, a) = \phi(x, a_i', a_{-i}) - \phi(x, a). \qquad [16.24]$$

2. *The potential function satisfies $\phi(\tilde{x}, a^0) = \phi(x, a)$ for the state $\tilde{x} = P(x, a)$.*

The first condition states that each agent's utility function is aligned with the potential function in the same fashion as in potential games (Monderer and Shapley, 1996). The second condition relates to the evolution on the potential function along the state trajectory. We focus on the class of state-based potential games as dynamics can be shown to converge to the following class of equilibria:

Definition 16.3. (Stationary State Nash Equilibrium; Li and Marden, 2011) *A state action pair $[x^*, a^*]$ is a stationary state Nash equilibrium if*

1. *For any agent $i \in N$ we have $a_i^* \in \arg\max_{a_i \in \mathcal{A}_i} U_i(x^*, a_i, a_{-i}^*)$.*
2. *The state is a fixed point of the state transition function, i.e., $x^* = f(x^*, a^*)$.*

It can be shown that a stationary state Nash equilibrium is guaranteed to exist in any state-based potential game (Li and Marden, 2011). Furthermore, there are several learning dynamics which will converge to such an equilibrium in state-based potential games (Li and Marden, 2011; Marden, 2012).

16.4.3 Illustrations

16.4.3.1 Protocol design

Section 16.4.1.1 highlights computational and efficiency limitations associated with designing protocols within the framework of strategic form games. Recently in Marden and Wierman (2013), the authors show that there exists a simple state-based protocol that overcomes both of these limitations. In particular, for welfare-sharing games with submodular objective functions, this state-based protocol is universal, budget-balanced, tractable, and ensures the existence of a stationary state Nash equilibrium. Furthermore, the PoS is 1 and PoA is 2 when using this state-based protocol. Hence, this protocol matches the performance of the marginal contribution protocol with respect to efficiency guarantees. We direct the readers to Marden and Wierman (2013) for the specific details regarding this protocol.

16.4.3.2 Distributed optimization

Consider the following generalization of the average consensus problem where there exists a set of agents N, a value set $\mathcal{V}_i = \mathbb{R}$ for each agent $i \in N$, a system level objective function $W : \mathbb{R}^n \to \mathbb{R}$ which is concave and continuously differentiable, and a coupled constraint on the agents' value profile which is characterized by a set of m-linear inequalities represented in matrix form as $Zv \leq C$ where $v = (v_1, v_2, \ldots, v_n) \in \mathbb{R}^n$. Here, the goal is to establish a set of local control laws of the form [16.21] such that the joint value profile converges to the solution of the following optimization problem

$$\begin{array}{ll} \max_{v \in \mathbb{R}^n, i \in N} & W(v) \\ s.t. & \sum_{i=1}^{n} Z_i^k v_i - C^k \leq 0, \;\; k \in \{1, \ldots, m\}. \end{array} \qquad [16.25]$$

Here, the interaction graph encodes the desired locality in the control laws. Note that the objective for average consensus in [16.22] is a special case of the objective presented in [16.25].

Section 16.4.1.2 demonstrates that it is impossible to design scalable agent utility functions within the framework of strategic form games which ensured that all equilibria of the resulting game represented solutions to the optimization problem in [16.25]. We will now review the methodologies developed in Li and Marden (2011, 2013) that accomplish this task using the framework of state based games. Furthermore, the forthcoming design also ensures that the resulting game is a state based potential game; hence, there are available distributed learning algorithms for reaching the stationary state Nash equilibria of the resulting game (Li and Marden, 2011; Marden, 2012). The details of the design are as follows:

Agents: The agent set is $N = \{1, 2, \ldots, n\}$.

States: The starting point of the design is an underlying state space X where each state $x \in X$ is defined as a tuple $x = (v, e, c)$ where the components are as follows:

- The term $v = (v_1, \ldots, v_n) \in \mathbb{R}^n$ is the value profile.
- The term $e = (e_1, \ldots, e_n)$ is a profile of agent-based estimation terms for the value profile v. Here, $e_i = (e_i^1, \ldots, e_i^n) \in \mathbb{R}^n$ is agent i's estimation for the joint action profile v. The term e_i^k captures agent i's estimate of agent k's value v_k.
- The term $c = (c_1, \ldots, c_n)$ is a profile of agent-based estimation terms for the constraint violations. Here, $c_i = (c_i^1, \ldots, c_i^m) \in \mathbb{R}^m$ is agent i's estimation for the constraint violation $C - Zv$. The term c_i^k captures agent i's estimate of the violation of the kth constraint, i.e., $\sum_{j \in N} Z_j^k \cdot v_j - C^k$.

Action sets: Each agent $i \in N$ is assigned an action set $\mathcal{A}_i$ that permits the agent to change their value and estimation terms by communicating with neighboring agents. Specifically, an action for agent i is defined as a tuple $a_i = (\hat{v}_i, \hat{e}_i, \hat{c}_i)$ whose components are as follows:

- The term $\hat{v}_i \in \mathbb{R}$ indicates a change in agent i's value v_i.
- The term $\hat{e}_i = (\hat{e}_i^1, \cdots, \hat{e}_i^n)$ indicates a change in agent i's estimation terms e_i. Here, $\hat{e}_i^k = \{\hat{e}_{i \to j}^k\}_{j \in N_i}$ where $\hat{e}_{i \to j}^k \in \mathbb{R}$ represents the estimation value that agent i "passes" to agent j regarding the value of agent k.
- The term $\hat{c}_i = (\hat{c}_i^1, \cdots, \hat{c}_i^m)$ indicates a change in agent i's estimation terms c_i. Here, $\hat{c}_i^k = \{\hat{c}_{i \to j}^k\}_{j \in N_i}$ where $\hat{c}_{i \to j}^k \in \mathbb{R}$ represents the estimation value that agent i "passes" to agent j regarding the violation of the kth constraint.

State dynamics: We now describe how the state evolves as a function of the joint actions $a(0)$, $a(1)$, ..., where $a(k)$ is the joint action profile at stage k. Define the initial state as $x(0) = [v(0), e(0), c(0)]$ where $v(0) = (v_1(0), \ldots, v_n(0))$ is the initial value profile, $e(0)$ is an initial estimation profile that satisfies

$$\sum_{i \in N} e_i^j(0) = n \cdot v_j(0), \ \forall j \in N, \qquad [16.26]$$

and $c(0)$ is an initial estimate of the constraint violations that satisfies

$$\sum_{i \in N} c_i^k(0) = \sum_{i \in N} Z_i^k \cdot v_i(0), \ \forall k \in M. \qquad [16.27]$$

Hence, the initial estimation terms are contingent on the initial value profile.

We represent the state transition function $x^+ = P(x, a)$ by a set of local state transition functions $\{P_i^v(x, a)\}_{i \in N}$, $\left\{P_{i,j}^e(x, a)\right\}_{i,j \in N}$, and $\left\{P_{i,k}^c(x, a)\right\}_{i \in N, k \in M}$. For any agent $i \in N$, state $x = (v, e, c)$, and action $a = (\hat{v}, \hat{e}, \hat{c})$, the state transition function pertaining to the evolution of the value profile takes on the form

$$(v_i)^+ = P_i^v(x, a) = v_i + \hat{v}_i. \qquad [16.28]$$

The state transition function pertaining to the estimate of the value profile takes on the form

$$(e_i^i)^+ = P_{i,i}^e(x, a) = e_i^i + n \cdot \hat{v}_i + \sum_{j \in N : i \in N_j} \hat{e}_{j \to i}^i - \sum_{j \in N_i} \hat{e}_{i \to j}^i, \qquad [16.29]$$

$$(e_i^j)^+ = P_{i,j}^e(x, a) = e_i^k + \sum_{j \in N : i \in N_j} \hat{e}_{j \to i}^k - \sum_{j \in N_i} \hat{e}_{i \to j}^k, \ \forall j \neq i. \qquad [16.30]$$

Lastly, the state transition function pertaining to the estimate of each constraint violations $k \in M$ takes on the form

$$(c_i^k)^+ = P_{i,k}^c(x, a) = c_i^k + Z_i^k \hat{v}_i + \sum_{j \in N : i \in N_j} \hat{c}_{j \to i}^k - \sum_{j \in N_i} \hat{c}_{i \to j}^k. \qquad [16.31]$$

It is straightforward to show that for *any* sequence of action profiles $a(0)$, $a(1)$, ..., the resulting state trajectory $x(t) = (v(t), e(t), c(t)) = P(x(t-1), a(t-1))$ satisfies

$$\sum_{i \in N} e_i^j(t) = n \cdot v_j(t), \ \forall j \in N, \qquad [16.32]$$

$$\sum_{i \in N} c_i^k(t) = \sum_{i \in N} Z_i^k v_i(t) - C^k, \ \forall k \in M, \qquad [16.33]$$

for all $t \geq 1$. Therefore, for any constraint $k \in M$ and time $t \geq 1$

$$\sum_{i \in N} c_i^k(t) \leq 0 \Leftrightarrow \left\{ \sum_{i \in N} Z_i^k v_i(t) - C^k \leq 0 \right\}. \qquad [16.34]$$

Hence, the estimation terms encode information regarding whether the constraints are violated.

Agent utility functions: The last part of our design is the agents' utility functions. For any state $x \in X$ and action profile $a \in \mathcal{A}$ the utility function of agent i is defined as

$$U_i(x, a) = \sum_{j \in N_i} W(\tilde{e}_j^1, \tilde{e}_j^2, \ldots, \tilde{e}_j^n) - \sum_{j \in N_i} \sum_{k \in N} \left[\tilde{e}_i^k - \tilde{e}_j^k\right]^2 - \mu \sum_{j \in N_i} \sum_{k=1}^{m} \left[\max\left(0, \tilde{c}_j^k\right)\right]^2$$

where $\mu > 0$ and $(\tilde{v}, \tilde{e}, \tilde{c}) = P(x, a)$ represents the ensuing state. Note that the agents' utility functions are both local and scalable.

Theorem 16.16. (Li and Marden, 2011, 2013) *Consider the state-based game depicted above. The designed game is a state based potential game with potential function*

$$\phi(x, a) = \sum_{i \in N} W(\tilde{e}_i^1, \tilde{e}_i^2, \ldots, \tilde{e}_i^n) - \frac{1}{2} \sum_{i \in N} \sum_{j \in N_i} \sum_{k \in N} \left[\tilde{e}_i^k - \tilde{e}_j^k\right]^2 - \mu \sum_{j \in N} \sum_{k=1}^{m} \left[\max\left(0, \tilde{c}_j^k\right)\right]^2$$

where $\mu > 0$ and $(\tilde{v}, \tilde{e}, \tilde{c}) = P(x, a)$ represents the ensuing state. Furthermore, if the interaction graph is connected, undirected, and nonbipartite, then a state action pair $[x, a] = \left[(v, e, c), (\hat{v}, \hat{e}, \hat{c})\right]$ is a stationary state Nash equilibrium if and only if the following conditions are satisfied:

(i) *The value profile v is an optimal point of the unconstrained optimization problem*

$$\max_{v \in \mathbb{R}^n} W(v) - \frac{\mu}{n} \left[\sum_{k \in M} \max\left(0, \sum_{i \in N} Z_i^k v_i - C^k\right)\right]^2. \qquad [16.35]$$

(ii) *The estimation of the value profile e is consistent with v, i.e., $e_i^j = v_j$ for all $i, j \in N$.*

(iii) *The estimation of the constraint violations c satisfies the following for all $i \in N$ and $k \in M$*

$$\max\left(0, c_i^k\right) = \frac{1}{n} \max\left(0, \sum_{i \in N} Z_i^k v_i - C^k\right).$$

(iv) *The change in value profile satisfies $\hat{v}_i = 0$ for all agents $i \in N$.*

(v) The net change in estimation terms for both the value profile and the constraint violation is 0, *i.e.,*

$$\sum_{j \in N: i \in N_j} \hat{e}^k_{j \to i} - \sum_{j \in N_i} \hat{e}^k_{i \to j} = 0 \;\; \forall i, k \in N,$$

$$\sum_{j \in N: i \in N_j} \hat{c}^k_{j \to i} - \sum_{j \in N_i} \hat{c}^k_{i \to j} = 0 \;\; \forall i \in N, k \in M.$$

This theorem characterizes the complete set of stationary state Nash equilibrium for the designed state-based game. There are several interesting properties regarding this characterization. First, the solutions to the unconstrained optimization problem incorporating penalty functions in [16.35] in general only represent solutions to the constrained optimization problem in [16.25] when the tradeoff parameter $\mu \to \infty$. However, in many settings, such as the consensus problem discussed in Section 16.4.1.2, any finite $\mu > 0$ will provide the equivalence between these two solution sets (Li and Marden, 2011). Second, the design methodology set forth in this section is universal and provides the desired equilibrium characteristics irrespective of the specific topological structure of the interaction graph or the agents' initial actions/values. Note that this was impossible when using the framework of strategic form games. Lastly, since the designed game represents a state-based potential game, there exists learning dynamics that guarantee convergence to a stationary state Nash equilibrium (Li and Marden, 2011).

16.5. CONCLUDING REMARKS

We conclude by mentioning some important topics not discussed in this chapter.

First, there has been work using game-theoretic formulations for engineering applications over many decades. Representative topics include cybersecurity (Alpcan and Basar, 2010), wireless networks (Srivastava et al., 2005), robust control design (Basar and Bernhard, 1995), team theory (Ho and Chu, 1972), and pursuit-evasion (Ho et al., 1965). The material in this chapter focused on more recent trends that emphasize both game design and adaptation through learning in games.

Second is the issue of convergence rates. We reviewed how various learning rules under different information structures can converge asymptotically to Nash equilibria or other solution concepts. Practical implementation for engineering applications places demands on the requisite convergence rates. Furthermore, existing computational and communication complexity results constrain the limits of achievable performance in the general case (Daskalakis et al., 2006; Hart and Mansour, 2007). Recent work has begun to address settings in which practical convergence is possible (Arieli and Young, 2011; Awerbuch et al., 2008; Kreindler and Young, 2011; Montanari and Saberi, 2009; Shah and Shin, 2010).

Finally, there is the obvious connection to distributed optimization. A theme throughout the paper is optimizing performance of a global objective function under various assumptions on available information and communication constraints. While the methods herein emphasis a game-theoretic approach, there is extensive complementary work on modifying optimization algorithms (e.g., gradient decent, Newton's method, etc.) to accommodate distributed architectures. Representative citations are Nedic and Ozdaglar (2009), Wei et al. (2013), and Wang and Elia (2010) as well as the classic reference (Bertsekas and Tsitsiklis, 1989).

REFERENCES

Alos-Ferrer, C., Netzer, N., 2010. The logit-response dynamics. Games Econ. Behav. 68, 413–427.
Alpcan, T., Basar, T., 2010. Network Security: A Decision and Game Theoretic Approach. Cambridge University Press, New York, NY.
Arieli, I., Babichenko, Y., 2011. Average testing and the efficient boundary. Discussion paper, Department of Economics, University of Oxford and Hebrew University.
Arieli, I., Young, H.P., 2011. Fast convergence in population games. University of Oxford Department of Economics Discussion Paper Series No. 570.
Arrow, K.J., 1986. Rationality of self and others in an economic system. J. Bus. 59 (4), S385–S399.
Arslan, G., Marden, J.R., Shamma, J.S., 2007. Autonomous vehicle-target assignment: a game theoretical formulation. ASME J. Dyn. Syst. Measure. Contr. 129, 584–596.
Awerbuch, B., Azar, Y., Epstein, A., Mirrokni, V.S., Skopalik, A., 2008. Fast convergence to nearly optimal solutions in potential games. In Proceedings of the ACM Conference on Electronic Commerce, pp. 264–273.
Babichenko, Y., 2010. Completely uncoupled dynamics and Nash equilibria. Working paper.
Basar, T., Bernhard, P., 1995. $\mathcal{H}^\infty$-Optimal Control and Related Minimax Design Problems: A Dynamic Game Approach. Birkhäuser.
Bergin, J., Lipman, B.L., 1996. Evolution with state-dependent mutations. Econometrica 64 (4), 943–956.
Bertsekas, D.P., Tsitsiklis, J.N., 1989. Parallel and Distributed Computation: numerical methods. Prentice Hall, Inc., Upper Saddle River, NJ.
Blume, L., 1993. The statistical mechanics of strategic interaction. Games Econ. Behav. 5, 387–424.
Bullo, F., Frazzoli, E., Pavone, M., Savla, K., Smith, S.L., 2011. Dynamic vehicle routing for robotic systems. Proc. IEEE 99 (9), 1482–1504.
Candogan, U.O., Ozdaglar, A., Parrilo, P.A., 2011a. Dynamics in near-potential games. Discussion paper, LIDS, MIT.
Candogan, U.O., Ozdaglar, A., Parrilo, P.A., 2011. Near-potential games: Geometry and dynamics. Discussion paper, LIDS, MIT.
Chasparis, G.C., Shamma, J.S., 2012. Distributed dynamic reinforcement of efficient outcomes in multiagent coordination and network formation. Dyn. Games Appl. 2 (1), 18–50.
Chen, H.-L., Roughgarden, T., Valiant, G., 2008. Designing networks with good equilibria. In Proceedings of the Nineteenth Annual ACM-SIAM Symposium on Discrete Algorithms, pp. 854–863.
Conitzer, V., Sandholm, T., 2004. Computing Shapley values, manipulating value division schemes, and checking core membership in multi-issue domains. In Proc. of AAAI.
Dan, G., 2011. Cache-to-cache: could isps cooperate to decrease peer-to-peer content distribution costs? IEEE Trans. Parallel Distrib. Syst. 22 (9), 1469–1482.
Daskalakis, C., Goldberg, P.W., Papadimitriou, C.H., 2006. The complexity of computing a Nash equilibrium. In STOC'06 Proceedings of the 38th Annual ACM Symposium on the Theory of Computing, pp. 71–78.
Foster, D.P., Young, H.P., 2006. Regret testing: learning to play Nash equilibrium without knowing you have an opponent. Theor. Econ. 1, 341–367.

Fudenberg, D., Levine, D., 1998. The Theory of Learning in Games. MIT Press, Cambridge, MA.
Fudenberg, D., Tirole, J., 1991. Game Theory. MIT Press, Cambridge, MA.
Garcia, A., Reaume, D., Smith, R., 2004. Fictitious play for finding system optimal routings in dynamic traffic networks. Transport. Res. B, Methods 34 (2), 147–156.
Germano, F., Lugosi, G., 2007. Global Nash convergence of Foster and Young's regret testing. Games Econ. Behav. 60, 135–154.
Goemans, M., Li, L., Mirrokni, V.S., Thottan, M., 2004. Market sharing games applied to content distribution in ad-hoc networks. In *Symposium on Mobile Ad Hoc Networking and Computing (MOBIHOC)*.
Gopalakrishnan, R., Marden, J.R., Wierman, A., 2014. Potential games are necessary to ensure pure Nash equilibria in cost sharing games. Math. Oper. Res.
Haeringer, G., 2006. A new weight scheme for the shapley value. Math. Soc. Sci. 52 (1), 88–98.
Hart, S., Mas-Colell, A., 1989. Potential, value, and consistency. Econometrica 57 (3), 589–614.
Hart, S., Mas-Colell, A., 2003. Uncoupled dynamics do not lead to Nash equilibrium. Am. Econ. Rev. 93 (5), 1830–1836.
Hart, S., 2005. Adaptive heuristics. Econometrica 73 (5), 1401–1430.
Hart, S., Mansour, Y., 2007. The communication complexity of uncoupled Nash equilibrium procedures. In STOC'07 Proceedings of the 39th Annual ACM Symposium on the Theory of Computing, pp. 345–353.
Hart, S., Mas-Colell, A., 2006. Stochastic uncoupled dynamics and Nash equilibrium. Games Econ. Behav. 57 (2), 286–303.
Ho, Y.-C., Chu, K.-C., 1972. Team decision theory and information structures in optimal contra problems—Part I. IEEE Trans. Autom. Contr. 17 (1), 15–22.
Ho, Y.-C., Bryson, A., Baron, S., 1965. Differential games and optimal pursuit-evasion strategies. IEEE Trans. Autom. Contr. 10 (4), 385–389.
Lambert III, T.J., Epelman, M.A., Smith, R.L., 2005. A fictitious play approach to large-scale optimization. Oper. Res. 53 (3), 477–489.
Jadbabaie, A., Lin, J., Morse, A.S., 2003. Coordination of groups of mobile autonomous agents using nearest neighbor rules. IEEE Trans. on Autom. Contr. 48 (6), 988–1001.
Kandori, M., Mailath, G., Rob, R., 1993. Learning, mutation, and long-run equilibria in games. Econometrica 61, 29–56.
Kreindler, G.H., Young, H.P., 2011. Fast convergence in evolutionary equilibrium selection. University of Oxford, Department of Economics, Discussion Paper Series No. 569.
Leslie, D., Collins, E., 2003. Convergent multiple-timescales reinforcement learning algorithms in normal form games. Ann. Appl. Probab. 13, 1231–1251.
Leslie, D., Collins, E., 2005. Individual Q-learning in normal form games. SIAM J. Contr. and Optim. 44 (2), 495–514.
Leslie, D., Collins, E., 2006. Generalised weakened fictitious play. Games Econ. Behav. 56 (2), 285–298.
Li, N., Marden, J.R., 2010. Designing games to handle coupled constraints. In Proceedings of the 48th IEEE Conference on Decision and Control.
Li, N., Marden, J.R., 2011. Decoupling coupled constraints through utility design. Discussion paper, Department of ECEE, University of Colorado, Boulder.
Li, N., Marden, J.R., 2013. Designing games for distributed optimization. IEEE J. Sel. Topics Signal Process. 7 (2), 230–242, Special Issue on Adaptation and Learning Over Complex Networks.
Mannor, S., Shamma, J.S., 2007. Multi-agent learning for engineers. Artif. Intell., pp. 417–422, Special Issue on Foundations of Multi-Agent Learning.
Marden, J.R., 2012. State based potential games. Automatica 48, 3075–3088.
Marden, J.R., Roughgarden, T., 2010. Generalized efficiency bounds for distributed resource allocation. In Proceedings of the 48th IEEE Conference on Decision and Control, December 2010.
Marden, J.R., Shamma, J.S., 2012. Revisiting log-linear learning: asynchrony, completeness and a payoff-based implementation. Games Econ. Behav. 75, 788–808.
Marden, J.R., Wierman, A., 2013a. Distributed welfare games. Oper. Res. 61, 155–168.
Marden, J.R., Wierman, A., 2013b. The limitations of utility design for multiagent systems. IEEE Trans. Autom. Contr. 58 (6), 1402–1415.

Marden, J.R., Arslan, G., Shamma, J.S., 2007. Regret based dynamics: convergence in weakly acyclic games. In Proceedings of the 2007 International Conference on Autonomous Agents and Multiagent Systems (AAMAS), Honolulu, H, May 2007.

Marden, J.R., Arslan, G., Shamma, J.S., 2009a. Connections between cooperative control and potential games. IEEE Trans. Syst. Man Cybern. B 39, 1393–1407.

Marden, J.R., Arslan, G., Shamma, J.S., 2009b. Joint strategy fictitious play with inertia for potential games. IEEE Trans. Autom. Contr. 54, 208–220.

Marden, J.R., Young, H.P., Arslan, G., Shamma, J.S., 2009. Payoff based dynamics for multi-player weakly acyclic games. SIAM J. Contr. Optim. 48, 373–396.

Marden, J.R.P., Young, H., Pao, L.Y., 2011. Achieving Pareto optimality through distributed learning. Oxford Economics Discussion Paper No. 557.

Marden, J.R., Ruben, S.D., Pao, L.Y., 2012. Surveying game theoretic approaches for wind farm optimization. In Proceedings of the AIAA Aerospace Sciences Meeting, January 2012.

Martinez, S., Cortes, J., Bullo, F., 2007. Motion coordination with distributed information. Contr. Syst. Mag. 27 (4), 75–88.

Menon, A., Baras, J.S., 2013. A distributed learning algorithm with bit-valued communications for multi-agent welfare optimization. In Proceedings of the 52nd IEEE Conference on Decision and Control, pp. 2406–2411.

Monderer, D., Shapley, L., 1996. Fictitious play property for games with identical interests. Games Econ. Theory 68, 258–265.

Monderer, D., Shapley, L., 1996. Potential games. Games Econ. Behav. 14, 124–143.

Montanari, A., Saberi, A., 2009. Convergence to equilibrium in local interaction games. In FOCS'09 Proceedings of the 2009 50th Annual IEEE Symposium on Foundations of Computer Science, pp. 303–312.

Moulin, H., Shenker, S., 2001. Strategyproof sharing of submodular costs: budget balance versus efficiency. Econ. Theory 18 (3), 511–533.

Moulin, H., 2010. An efficient and almost budget balanced cost sharing method. Games Econ. Behav. 70 (1), 107–131.

Moulin, H., Vohra, R., 2003. Characterization of additive cost sharing methods. Econ. Lett. 80 (3), 399–407.

Murray, R.M., 2007. Recent research in cooperative control of multivehicle systems. J. Dyn. Syst. Measure. Contr. 129 (5), 571–583.

Nedic, A., Ozdaglar, A., 2009. Distributed subgradient methods for multi-agent optimization. IEEE Trans. Autom Contr. 54 (1), 48–61.

Neel, J., Mackenzie, A.B., Menon, R., Dasilva, L.A., Hicks, J.E., Reed, J.H., Gilles, R.P., 2005. Using game theory to analyze wireless ad hoc networks. IEEE Commun. Surveys Tutorials 7 (4), 46–56.

Nisan, N., Roughgarden, T., Tardos, E., Vazirani, V.V., (Eds.), 2007. Algorithmic Game Theory. Cambridge University Press, New York, NY, USA.

Olfati-Saber, R., Fax, J.A., Murray, R.M., 2007. Consensus and cooperation in networked multi-agent systems. Proc. IEEE 95 (1), 215–233.

Pradelski, B.R., Young, H.P., 2012. Learning efficient Nash equilibria in distributed systems. Games Econ. Behav., 75, 882–897.

Roughgarden, T., 2005. Selfish Routing and the Price of Anarchy. MIT Press, Cambridge, MA, USA.

Roughgarden, T., 2009. Intrinsic robustness of the price of anarchy. In Proceedings of STOC.

Sandholm, W.H., 2012. Population Games and Evolutionary Dynamics. MIT Press, Cambridge, MA, USA.

Shah, D., Shin, J., 2010. Dynamics in congestion games. In ACM SIGMETRICS, pp. 107–118.

Shamma, J.S., (Ed.), 2008. Cooperative Control of Distributed Multi-Agent Systems. Wiley-Interscience, New York, NY, USA.

Shamma, J.S., Arslan, G., 2005. Dynamic fictitious play, dynamic gradient play, and distributed convergence to Nash equilibria. IEEE Trans. Autom. Contr. 50 (3), 312–327.

Shapley, L.S., 1953a. Stochastic games. Proc. Nat. Acad. Sci. U.S.A. 39 (10), 1095–1100.

Shapley, L.S., 1953b. A value for *n*-person games. In Kuhn, H.W., Tucker, A.W., (Eds.), Contributions to the Theory of Games II (Annals of Mathematics Studies 28), pp. 307–317. Princeton University Press, Princeton, NJ.

Shoham, Y., Powers, R., Grenager, T., 2007. If multi-agent learning is the answer, what is the question? Artif. Intell., 171 (7), 365–377. Special Issue on Foundations of Multi-Agent Learning.
Srivastava, V., Neel, J., Mackenzie, A.B., Menon, R., Dasilva, L.A., Hicks, J.E., Reed, J.H., Gilles, R.P., 2005. Using game theory to analyze wireless ad hoc networks. IEEE Commun. Surveys Tutorials 7 (4), 46–56.
Tsitsiklis, J.N., Bertsekas, D.P., Athans, M., 1986. Distributed asynchronous deterministic and stochastic gradient optimization algorithms. IEEE Trans. Autom. Contr. 35 (9), 803–812.
Vetta, A., 2002. Nash equilibria in competitive societies with applications to facility location, traffic routing, and auctions. In FOCS, pp. 416–425.
Wang, J., Elia, N., 2010. Control approach to distributed optimization. In Proceedings of the 2010 48th Annual Allerton Conference on Communication, Control, and Computing, pp. 557–561.
Wei, E., Ozdaglar, A., Jadbabaie, A., 2013. A distributed Newton method for network utility maximization. IEEE Trans. Autom. Contr. 58 (9), 2162–2175.
Wolpert, D., Tumor, K., 1999. An overview of collective intelligence. In Bradshaw, J.M., (Ed.), Handbook of Agent Technology. AAAI Press/MIT Press, Cambridge, MA, USA.
Young, H.P., 1993. The evolution of conventions. Econometrica 61, 57–84.
Young, H.P., 1994. Equity. Princeton University Press, Princeton, NJ.
Young, H.P., 1998. Individual Strategy and Social Structure. Princeton University Press, Princeton, NJ.
Young, H.P., 2005. Strategic Learning and its Limits. Oxford University Press, New York, NY, USA.
Young, H.P., 2009. Learning by trial and error. Games Econ. Behav. 65, 626–643.

CHAPTER 17

Ambiguity and Nonexpected Utility

Edi Karni[*,†], Fabio Maccheroni[‡], Massimo Marinacci[‡]
[*]Department of Economics, Johns Hopkins University, Baltimore, MD, USA
[†]The Warwick Business School, Warwick University, Coventry, UK
[‡]Department of Decision Sciences and IGIER, Università Bocconi, Milano, Italy

Contents

Handbook of game theory
http://dx.doi.org/10.1016/B978-0-12-420056-2.00017-3

Abstract

This chapter reviews developments in the theory of decision making under risk and uncertainty, focusing on models that, over the last 40 years, dominated the theoretical discussions. It also surveys some implications of the departures from the "linearity in the probabilities" aspect of expected utility theory to game theory. The chapter consists of two main parts: The first part reviews models of decision making under risk that depart from the independence axiom, focusing on the rank-dependent utility models and cumulative prospect theory. The second part reviews theories of decision making under uncertainty that depart from the sure thing principle and model the phenomenon of ambiguity and ambiguity aversion.

Keywords: Nonexpected utility, Rank dependent utility, Ambiguity, Ambiguity aversion, Smooth ambiguity aversion, Auction theory, Decision making under uncertainty

JEL Codes: D81, C7

17.1. INTRODUCTION

In the theory of decision making in the face of uncertainty, it is commonplace to distinguish between problems that require choice among probability distributions on the outcomes (e.g., betting on the outcome of a spin of the roulette wheel) and problems that require choice among random variables whose likely outcomes is a matter of opinion (e.g., betting on the outcome of a horse race). Following Knight (1921), problems of the former type are referred to as *decision making under risk* and problems of the latter type as *decision making under uncertainty*. In decision making under risk probabilities are a primitive aspect of the depiction of the choice set, while in decision making under uncertainty they are not.

The maximization of expected utility as a criterion for choosing among risky alternatives was first proposed by Bernoulli (1738), to resolve the St. Petersburg paradox. von Neumann and Morgenstern (1944) were the first to depict the axiomatic structure of preference relations that is equivalent to expected utility maximization, thus providing behavioral underpinnings for the evaluation of mixed strategies in games. In both instances, the use of probabilities as primitive is justifiable, since the problems at hand presume the existence of a chance mechanism that produces outcomes whose relative frequencies can be described by probabilities.

At the core of expected utility theory under risk is the independence axiom, which requires that the preferences between risky alternatives be independent of their common features.[1] This aspect of the model implies that outcomes are evaluated

[1] Variants of this axiom were formulated by Marschak (1950), Samuelson (1952), and Herstein and Milnor (1953). It was shown to be implicit in the model of von Neumann and Mogernstern (1947) by Malinvaud (1952).

separately, and is responsible for the "linearity in the probabilities" of the preference representation.

The interest in subjective probabilities that quantify the "degree of belief" in the likelihoods of events, dates back to the second half of the seventeenth century, when the idea of probability first emerged. From its inception, the idea of subjective probabilities was linked to decision making in the face of uncertainty, epitomized by Pascal's wager about the existence of God. Axiomatic treatments of the subject originated in the works of Ramsey (1926), de Finetti (1937) and culminated in the subjective expected utility theories of Savage (1954) and Anscombe and Aumann (1963). Savage's theory asserts that the preference between two random variables, or acts, is independent of the aspects on which they agree. This assertion, known as the sure thing principle, is analogous to the independence axiom in that it imposes a form of separability of the evaluation of the consequences.[2]

During the second half of the twentieth century, the expected utility model, in both its objective and subjective versions, became the paradigm of choice behavior under risk and under uncertainty in economics and game theory. At the same time it came under scrutiny, with most of the attention focus on its "core" feature, the independence axiom. Clever experiments were designed, notably by Allais (1953) and Ellsberg (1961), in which subjects displayed patterns of choice indicating systematic violations of the independence axiom and the sure thing principle. These early investigations spawned a large body of experimental studies that identify other systematic violations of the expected utility models and, in their wake, theoretical works that depart from the independence axiom and the sure thing principle. The new theories were intended to accommodate the main experimental findings while preserving, as much as possible, other aspects of the expected utility model.

In this chapter, we review some of these theoretical developments, focusing on those models that, over the last 40 years, came to dominate the theoretical discussions. We also review and discuss some implications of the departures from the "linearity in the probabilities" aspect of expected utility theory to game theory.[3] Our survey consists of two main parts: The first part reviews models of decision making under risk that depart from the independence axiom, focusing on the rank-dependent utility models and cumulative prospect theory. The second part reviews theories of decision making under uncertainty that depart from the sure thing principle and model the phenomenon of ambiguity and ambiguity aversion.

[2] The axiomatization of Anscombe and Aumann (1963) includes the independence axiom.

[3] Readers interested in a broader surveys of the literature will find them in Karni and Schmeidler (1991b), Schmidt (2004), and Sugden (2004).

PART I NONEXPECTED UTILITY THEORY UNDER RISK

17.2. NONEXPECTED UTILITY: THEORIES AND IMPLICATIONS

17.2.1 Preliminaries

Evidence of systematic violations of the independence axiom of expected utility theory served as impetus for the development of alternative theories of decision making under risk, balancing descriptive realism and mathematical tractability. These theories have in common an analytical framework consisting of a choice set, whose elements are probability measures on a set of outcomes, and a binary relation on this set having the interpretation of a preference relation. More formally, let X be a separable metric space and denote by $L(X)$ the set of all probability measures on the Borel σ-algebra on X, endowed with the topology of weak convergence.[4] Let $\succcurlyeq$ be a preference relation on $L(X)$. The strict preference relation $\succ$ and the indifference relation, $\sim$, are the asymmetric and the symmetric parts of $\succcurlyeq$, respectively. A real-valued function V on $L(X)$ is said to *represent* the preference relation $\succcurlyeq$ if, for all $\ell, \ell' \in L(X)$, $\ell \succcurlyeq \ell'$ if and only if $V(\ell) \geq V(\ell')$.

Among the properties of preference relations that are common to expected utility theory and nonexpected utility theories are weak order and continuity. Formally,

Weak order. $\succcurlyeq$ on $L(X)$ is complete (that is, for all $\ell, \ell' \in L(X)$, either $\ell \succcurlyeq \ell'$ or $\ell' \succcurlyeq \ell$) and transitive (that is, for all $\ell, \ell', \ell'' \in L(X)$, $\ell \succcurlyeq \ell'$ and $\ell' \succcurlyeq \ell''$ implies $\ell \succcurlyeq \ell''$).

The continuity property varies according to the model under consideration. For the present purpose, we assume that $\succcurlyeq$ is continuous in the sense that its upper and lower contour sets are closed in the topology of weak convergence. Formally,

Continuity. For all $\ell \in L(X)$ the sets $\{\ell' \in L(X) \mid \ell' \succcurlyeq \ell\}$ and $\{\ell' \in L \mid \ell \succcurlyeq \ell'\}$ are closed in the topology of weak convergence.

The following result is implied by Theorem 17.1 of Debreu (1954).[5]

Theorem 17.1. *A preference relation $\succcurlyeq$ on $L(X)$ is a continuous weak-order if and only if there exist continuous real-valued function, V, on $L(X)$ representing $\succcurlyeq$. Moreover, V is unique up to continuous positive monotonic transformations.*

The independence axiom asserts that the preference between two risky alternatives is determined solely by those features that make them distinct. Ignoring common features is a form of separability that distinguishes expected utility theory from nonexpected

[4] A sequence ℓ_n in $L(X)$ converges to $\ell \in L(X)$ in the topology of weak convergence if $\int_X f d\ell_n$ converges to $\int_X f d\ell$, for all bounded continuous real-valued functions on X.

[5] Note that $L(X)$ is separable topological space.

utility theories. To state the independence axiom, define a convex operation on $L(X)$ as follows: For all $\ell, \ell' \in L(X)$ and $\alpha \in [0,1]$, define $\alpha\ell + (1-\alpha)\ell' \in L(X)$ by $(\alpha\ell + (1-\alpha)\ell')(B) = \alpha\ell(B) + (1-\alpha)\ell'(B)$, for all Borel subsets of X. Under this definition, $L(X)$ is a convex set.

Independence. For all $\ell, \ell', \ell'' \in L(X)$ and $\alpha \in (0,1]$, $\ell \succcurlyeq \ell'$ if and only if $\alpha\ell + (1-\alpha)\ell'' \succcurlyeq \alpha\ell' + (1-\alpha)\ell''$.

If the preference relation satisfies the independence axiom then the representation V is a linear functional. Formally, $V(\ell) = \int_X u(x)\, d\ell(x)$, where u is a real-valued function on X, unique up to positive linear transformation.

A special case of interest is when $X = \mathbb{R}$ and $L(X)$ is the set of cumulative distribution functions on $\mathbb{R}$. To distinguish this case we denote the choice set by L. Broadly speaking, first-order stochastic dominance refers to a partial order on L according to which one element dominates another if it assigns higher probability to larger outcomes. Formally, for all $F, G \in L$, F *dominates* G *according to first-order stochastic dominance* if $F(x) \leq G(x)$, for all $x \in \mathbb{R}$, with strict inequality for some x. Monotonicity with respect to first-order stochastic dominance is a property of the preference relation on L, requiring that it ranks higher dominating distributions. If the outcomes in $\mathbb{R}$ have the interpretation of monetary payoff, monotonicity of the preference relations with respect to first-order stochastic dominance seems natural and compelling.

Monotonicity. For all $F, G \in L$, $F \succ G$ whenever F dominates G according to first-order stochastic dominance.

Another property of preference relations that received attention in the literature asserts that the preference ranking of a convex combination of two risky alternatives is ranked between them. Formally,

Betweenness. For all $\ell, \ell' \in L(X)$ and $\alpha \in [0,1]$, $\ell \succcurlyeq \alpha\ell + (1-\alpha)\ell'' \succcurlyeq \ell'$.

If a continuous weak-order satisfies betweenness then the representation V is both quasiconcave and quasiconvex. Preference relations that satisfy independence also satisfy monotonicity and betweenness but not vice versa.

17.2.2 Three approaches

The issues raised by the experimental evidence were addressed using three different approaches. The axiomatic approach, the descriptive approach, and local utility analysis.

The axiomatic approach consists of weakening or replacing the independence axiom. It includes theories whose unifying property is betweenness (e.g., Chew, 1983; Chew and MacCrimmon, 1979; Dekel, 1986; Gul, 1991), and rank-dependent utility models whose unifying characteristic is transformations of probabilities that depend on their ranks (e.g., Chew, 1989b; Quiggin, 1982; Yaari, 1987).[6]

[6] A more detailed review of the betweenness theories see Chew (1989a).

The descriptive approach addresses the violations of expected utility theory by postulating functional forms of the utility function consistent with the experimental evidence. Prominent in this approach are regret theories (e.g., Bell, 1982, 1985; Hogarth, 1980; Loomes and Sugden, 1982) whose common denominator is the notion that, facing a decision problem, decision makers anticipate their feelings upon learning the outcome of their choice. More specifically, they compare the outcome that obtains to outcomes that could have been obtained had they have chosen another feasible alternative, or to another potential outcome of their choice that did not materialized. According to these theories, decision makers try to minimize the feeling of regret or trade of regret against the value of the outcome (Bell, 1982). The main weakness of regret theories in which regret arises by comparison with alternatives not chosen, is that they necessarily violate transitivity (see Bikhchandani and Segal, 2011).

Gul's (1991) theory of disappointment aversion is, in a sense, a bridge between the axiomatic betweenness theories and regret theories. Gul's underlying concept of regret is self-referential. The anticipated sense of disappointment (and elation) is generated by comparing the outcome (of a lottery) that obtains with alternative outcomes of the same lottery that did not obtain. This model is driven by an interpretation of experimental results known as the common ratio effect. It departs from the independence axiom but preserves transitivity.

Local utility analysis, advanced by Machina (1982), was intended to salvage results developed in the context of expected utility theory, while departing from its "linearity in the probabilities" property. Machina assumes that the preference relation is a continuous weak order and that the representation functional is "smooth" in the sense of having local linear approximations.[7] Locally, therefore, a smooth preference functional can be approximated by the corresponding expected utility functionals with a local utility functions. Moreover, global comparative statics results, analogous to those obtained in expected utility theory, hold if the restrictions (e.g., monotonicity, concavity) imposed on the utility function in expected utility theory, are imposed on the set of local utility functions.

Before providing a more detailed survey of the rank-dependent expected utility models, we discuss, in the next three sections, some implications of departing from the independence axiom for the theory of games and sequential decisions.

17.2.3 The existence of Nash equilibrium

The advent of nonexpected utility theories under risk raises the issue of the robustness of results in game theory when the players do not abide by the independence axiom. Crawford (1990) broached this question in the context of 2×2 zero-sum games. He

[7] Machina's (1982) choice set is the set of cumulative distribution functions on compact interval in the real line and his "smoothness" condition is Fréchet differentiability.

showed that, since players whose preferences are quasiconvex are averse to randomized choice of pure strategies even if they are indifferent among these strategies, if a game does not have pure-strategy equilibrium then, in general, Nash equilibrium does not exists. To overcome this difficulty, Crawford introduced a new equilibrium concept, dubbed "equilibrium in beliefs," and shows that it coincides with Nash equilibrium if the players' preferences are either quasiconcave or linear, and it exists if the preferences are quasiconvex. The key to this result is that, unlike Nash equilibrium, which requires that the perception of risk associated with the employment of mixed strategies be the same for all the players, equilibrium in beliefs allows for the possibility that the risk born by a player about his own choice of pure-strategy is different form that of the opponent.

17.2.4 Atemporal dynamic consistency

Situations that involve a sequence of choices require contingent planning. For example, in ascending bid auctions, bidders must choose at each announced price whether and how much to bid. When looking for a car to buy, at each stage in the search process the decision maker must choose between stopping and buying the best available car he have seen and continuing the search. In these and similar instances decision makers choose, at the outset of the process, a strategy that determine their choices at each subsequent decision node they may find themselves in. Moreover, in view of the sequential nature of the processes under consideration, it is natural to suppose that, at each decision node, the decision maker may review and, if necessary, revise his strategies. Dynamic consistency requires that, at each decision node, the optimal substrategy as of that node agrees with the continuation, as of the same node, of the optimal strategy formulated at the outset.

If the time elapsed during the process is significant, new factors, such as time preferences and consumption, need to be introduced. Since our main concern is decision making under risk, we abstract from the issue of time and confine our attention to contexts in which the time elapsed during the process is short enough to be safely ignored.[8]

For the present purpose, we adopt the analytical framework and definitions of Karni and Schmeidler (1991a). Let X be an arbitrary set of outcomes and let $\Delta^0(X) = \{\delta_x \mid x \in X\}$, where δ_x is the degenerate probability measure that assigns x the unit probability mass. For $k \geq 1$, let $\Delta^k(X) = \Delta\left(\Delta^{k-1}(X)\right)$, where for any set, T, $\Delta(T)$ denotes the set of probability measures on the power set of T with finite supports. Elements of $\Delta^k(X)$ are *compound lotteries*. For every compound lottery $\ell \in \Delta^k(X)$, the degenerate lottery δ_ℓ is an element of $\Delta^{k+1}(X)$. Henceforth, we identify ℓ and δ_ℓ. Consequently, we have $\Delta^k(X) \supset \cup_{i=0}^{k-1}\Delta^i(X)$, for all k. Let $\Gamma(X)$ denote the set of all compound

[8] The first to address the issue of dynamic consistency in temporal context was Strotz (1955). See also Kreps and Porteus (1978, 1979).

lotteries and define $\Gamma^k(X) = \{\ell \in \Gamma(X) \mid \ell \in \Delta^k(X), \ell \notin \Delta^{k-1}(X)\}$. Then $\Gamma(X)$ is the union of the disjoint sets $\Gamma^k(X)$, $k = 0, 1, \ldots$.

Given $\ell, \bar{\ell} \in \Gamma(X)$, we write $\bar{\ell} \trianglerighteq \ell$ if $\bar{\ell} = \ell$ or if there are $k > l \geq 0$ such that $\ell \in \Gamma^k(X)$, $\bar{\ell} \in \Gamma^l(X)$ and, for some j, $k - l \geq j \geq 1$, $\ell = \ell_j, \ell_0 = \bar{\ell}$, and there are $\ell_i \in \Gamma(X)$, $j \geq i \geq 0$, such that the probability, $\mathbb{P}\{\ell_{i-1} \mid \ell_i\}$, that ℓ_i assigns to ℓ_{i-1}, is positive. If $\bar{\ell} \trianglerighteq \ell$, we refer to $\bar{\ell}$ as a *sublottery* of ℓ. Given $\ell \in \Gamma(X)$ and $\bar{\ell} \trianglerighteq \ell$, we say that ℓ' is obtained from ℓ by replacing $\bar{\ell}$ with $\bar{\ell}'$, if the only difference between ℓ and ℓ' is that $\ell_0 = \bar{\ell}'$ instead of $\ell_0 = \bar{\ell}$ in the definition of $\bar{\ell} \trianglerighteq \ell$, in one place.[9] If $\bar{\ell} \trianglerighteq \ell$, we denote by $(\bar{\ell} \mid \ell)$ the sublottery $\bar{\ell}$ given ℓ and we identify ℓ with $(\ell \mid \ell)$. Define $\Psi(X) = \{(\bar{\ell} \mid \ell) \mid \ell \in \Gamma(X), \bar{\ell} \trianglerighteq \ell\}$.

A preference relation, $\succcurlyeq$ on $\Psi(X)$, is said to exhibit *dynamic consistency* if, for all $\ell, \ell', \bar{\ell}, \bar{\ell}' \in \Gamma(X)$, such that $\bar{\ell} \trianglerighteq \ell$ and ℓ' is obtained from ℓ by replacing $\bar{\ell}$ with $\bar{\ell}'$, $(\ell \mid \ell) \succcurlyeq (\ell' \mid \ell')$ if and only if $(\bar{\ell} \mid \ell) \succcurlyeq (\bar{\ell}' \mid \ell')$. In other words, dynamic consistency requires that if a decision maker prefers the lottery ℓ over the lottery ℓ', whose only difference is that, at point "down the road," he might encounter the sublottery $\bar{\ell}$ if he chose ℓ and the sublottery $\bar{\ell}'$, had he chosen ℓ', then facing the choice between $\bar{\ell}$ and $\bar{\ell}'$ at some point in the decision tree he would prefer $\bar{\ell}$ over $\bar{\ell}'$.

Consequentialism is a property of preference relations on $\Psi(X)$ that are "forward looking" in the sense that, at each stage of the sequential process, the evaluation of alternative courses of action is history-independent. Formally, a preference relation, $\succcurlyeq$ on $\Psi(X)$, is said to exhibit *consequentialism* if, for all $\ell, \ell', \hat{\ell}, \hat{\ell}', \bar{\ell}, \bar{\ell}' \in \Gamma(X)$, such that $\bar{\ell} \trianglerighteq \ell$, $\bar{\ell} \trianglerighteq \hat{\ell}$, ℓ' is obtained from ℓ by replacing $\bar{\ell}$ with $\bar{\ell}'$, and $\hat{\ell}'$ is obtained from $\hat{\ell}$ by replacing $\bar{\ell}$ with $\bar{\ell}'$, $(\bar{\ell} \mid \ell) \succcurlyeq (\bar{\ell}' \mid \ell')$ if and only if $\left(\bar{\ell} \mid \hat{\ell}\right) \succcurlyeq \left(\bar{\ell}' \mid \hat{\ell}'\right)$.

One way of converting compound lotteries to one stage lotteries is by the application of the calculus of probabilities. Loosely speaking, a preference relation over compound lotteries satisfies reduction of compound lotteries if it treats every $\ell \in \Gamma(X)$ and its single-stage "offspring" $\ell_{(1)} \in \Delta(X)$, obtained from ℓ by the application of the calculus of probabilities as equivalent. Formally, a preference relation, $\succcurlyeq$ on $\Psi(X)$, is said to exhibit *reduction* if, for all $\ell, \ell', \hat{\ell}, \hat{\ell}', \bar{\ell}, \bar{\ell}' \in \Gamma(X)$, such that $\bar{\ell} \trianglerighteq \ell$, $\bar{\ell}' \trianglerighteq \ell'$, $\hat{\ell}$ is obtained from ℓ by replacing $\bar{\ell}$ with $\bar{\ell}_{(1)}$, and $\hat{\ell}'$ is obtained from ℓ' by replacing $\bar{\ell}'$ with $\bar{\ell}'_{(1)}$, $(\bar{\ell} \mid \ell) \succcurlyeq (\bar{\ell}' \mid \ell')$ if and only if $\left(\bar{\ell}_{(1)} \mid \hat{\ell}\right) \succcurlyeq \left(\bar{\ell}'_{(1)} \mid \hat{\ell}'\right)$.

If a preference relation on $\Psi(X)$ exhibits consequentialism and reduction, then it is completely defined by its restriction to $\Delta(X)$. We refer to this restriction as the induced preference relation on $\Delta(X)$ and, without risk of confusion, we denote it by $\succcurlyeq$. The following theorem, due to Karni and Schmeidler (1991a).

[9] If $\bar{\ell}$ appears in ℓ in one place, then the replacement operation is well defined. If $\bar{\ell}$ appears in ℓ in more than one place, we assume that there is no confusion regarding the place at which the replacement occurs.

Theorem 17.2. *If a preference relation $\succcurlyeq$ on $\Psi(X)$ exhibits consequentialism and reduction then it exhibits dynamic consistency if and only if the induced preference relation $\succcurlyeq$ on $\Delta(X)$ satisfies the independence axiom.*

To satisfy dynamic consistency, models of decision making under risk that depart from the independence axiom cannot exhibit both consequentialism and reduction. If the last two attributes are compelling, then this conclusion constitutes a normative argument in favor of the independence axiom.

Machina (1989) argues in favor of abandoning consequentialism. According to him, theories that depart from the independence axiom are intended to model preference relations that are inherently nonseparable and need not be history-independent.

Segal (1990) argues in favor of maintaining consequentialism and dynamic consistency and replacing reduction with certainty-equivalent reduction (that is, by replacing sublotteries by their certainty equivalents to obtain single-stage lotteries). Identifying compound lotteries with their certainty-equivalent reductions, the induced preference relation on the single-stage lotteries do not have to abide by the independence axiom.

Sarin and Wakker (1998) maintain consequentialism and dynamic consistency. To model dynamic choice they introduced a procedure, dubbed sequential consistency, which, as in Segal (1990), calls for evaluating strategies by replacing risky or uncertain alternatives by their certainty equivalents and than folding the decision tree back. The distinguishing characteristic of their approach is that, at each stage, the certainty equivalents are evaluated using a model from the same family (e.g., rank-dependent, betweenness).

Karni and Safra (1989b) advanced a different approach, dubbed behavioral consistency. Maintaining consequentialism and reduction, behavioral consistency requires the use of dynamic programming to choose the optimal strategy. Formally, they modeled dynamic choice of a behaviorally consistent bidder as the Nash equilibrium of the game played by a set of agents, each of which represents the bidder at a different decision node.

17.2.5 Implications for the theory of auctions

The theory of auctions, consists of equilibrium analysis of games induced by a variety of auction mechanisms. With few exceptions, the analysis is based on the presumption that the players in the induced games, seller and bidders, abide by expected utility theory. With the advent of nonexpected utility theories, the robustness of the results obtained under expected utility theory became pertinent. This issue was partially addressed by Karni and Safra (1986, 1989a) and Karni (1988). These works show that central results in the theory of independent private values auctions fail to hold when the independence axiom is relaxed.

Consider auctions in which a single object is sold to a set of bidders. Assume that the bidders' valuations of the object are independent draws from a know distribution, F, on the reals; that these valuations are private information; and that the bidders do not engage in collusive behavior. Four main auction forms have been studied extensively in the literature: (a) Ascending bid auctions, in which the price of the object being auctioned off increases as long as there are at least two bidders willing to pay the price, and stops increasing as soon as only one bidder is willing to pay the price. The remaining bidder gets the object at the price at which the penultimate bidder quit the auction. (b) Descending bid auctions, in which the price falls continuously and stops as soon as one of the bidders announces his willingness to pay the price. That bidder claims the object and pays the price at which he stopped the process. (c) First-price sealed-bid auctions, in which each bidder submits a sealed bid and the highest bidder obtains the object and pays a price equal to his bid. (d) Second-price sealed-bid auctions, in which the procedure is as in the first-price sealed-bid auction, except that the highest bidder gets the object and pays a price equal to the bid of the second highest bidder.

If the bidders are expected utility maximizers, then the following results hold[10]: First, in the games induced by ascending-bid auctions and second-price sealed bid auctions there is a unique dominant strategy equilibrium. In games induced by ascending-bid auctions, this strategy calls for staying in the auction as long as the price is below the bidder's value and quit the auction as soon as it exceeds his value. In games induced by second-price sealed-bid auctions the dominant strategy is to submit a bid equal to the value. The equilibrium outcome in both auctions forms is Pareto efficient (that is, the bidder with the highest valuation get the object), value revealing, and the payoff to the seller is the same. Second, the Bayesian-Nash equilibrium outcome of the games induced by descending-bid auctions and first-price sealed-bid auctions, are the same and they are Pareto efficient.

Neither of these results hold if the object being sold is a risky prospect (e.g., lottery ticket) and bidders do not abide by the independence axiom. The reason is that, depending on the auction form, the above conclusions require the preference relations display separability or dynamic consistency. To grasp this claim, consider a second-price sealed-bid auction with independent private values and suppose that the object being auction is a lottery, ℓ. Suppose that a bidder's initial wealth is w. The value, v, of ℓ is the maximal amount he is willing to pay to obtain the lottery. Hence, for all $\ell \in L$ and $w \in \mathbb{R}$, the *value function* $v(\ell, w)$ is defined by $(\ell - v) \sim \delta_w$, where $(\ell - v)$ is a lottery defined by $(\ell - v)(x) = \ell(x + v)$ for all x in the support of ℓ. The argument that bidding v is a dominant strategy is as follows: Consider bidding b when the value is v. If the maximal bid of the other bidders is either above $\max\{b, v\}$ or below $\min\{b, v\}$,

[10] See Milgrom and Weber (1982).

then the payoffs to the bidder under consideration are the same whether he bids b or v. However, if the maximal bid of the other bidders, β, is between b and v, then either he does not win and ends up with w instead of $(\ell - \beta)$, when $(\ell - \beta) \succ \delta_w$ or he wins and ends up with $(\ell - \beta) \prec \delta_w$. In either case he stands to lose by not bidding the value. This logic is valid provided the preference relation is separable across mutually exclusive outcomes. In other words, if the ranking of lotteries is the same independently of whether they are compared to each other directly or embedded in a larger compound lottery structure. If the preference relation is not separable in this sense, then comparing the merits of bidding strategies solely by their consequences in the events in which their consequences are distinct, while ignoring the consequences in the events on which their consequences agree is misleading.

To discuss of the implications of departing from independence (and dynamic consistency) for auction theory we introduce the following notations. For each $m > 0$ let $D_{[0,m]}$ denote the set of differentiable cumulative distribution function whose support is the interval $[0, m]$. For all $F \in D_{[0,m]}$ and $y > 0$, denote by F^y distribution function with support $[y, m]$ that is obtained from F by the application of Bayes rule. Let f and f^y be the corresponding density functions. Let $T^m = \{(y, x) \in [0, m]^2 \mid x \geq y\}$ and define the cumulative distribution function $G^a : T^m \times L \times \mathbb{R} \times L \to L$ by:

$$G^a(y, x; \ell, w, F)(t) = [1 - F^y(x)]\,\delta_w(t) + \int_y^x (\ell - z)(t)\, dF^y(z), \quad \forall t \in \mathbb{R}.$$

A bidding problem depicted by the parameters (ℓ, w, F), where w denote the decision maker's initial wealth, ℓ the object being auctioned, and F the distribution of the maximal bid of the other bidders. Then the function $G^a(y, x; \ell, w, F)$ is the payoff distribution induced by the strategy, in ascending-bid auctions with independent private values that, at the price y, calls for staying in the auction until the price x is attained and than quitting. The corresponding payoff of the strategy in descending-bid auction that at the price y calls for claiming the object, ℓ, at the price x is given by:

$$G^d(y, x; \ell, w, F)(t) = [1 - F^y(x)]\,\delta_w(t) + (\ell - x)(t)\, F^y(x), \quad \forall t \in \mathbb{R}.$$

In the case of second and first price sealed bid auctions the problem is simplified since $y = 0$. The cumulative distribution function associated with submitting the bid b is

$$G^2(b; \ell, w, F)(t) = [1 - F(b)]\,\delta_w(t) + \int_0^b (\ell - z)(t)\, dF(z), \quad \forall t \in \mathbb{R}.$$

In first-price sealed-bid auction, the cumulative distribution function associated with submitting the bid b is:

$$G^1(b; \ell, w, F)(t) = [1 - F(b)]\,\delta_w(t) + (\ell - b)(t)\, F(b), \quad \forall t \in \mathbb{R}.$$

A preference relation $\succcurlyeq$ on L is said to exhibit *dynamic consistency in ascending bid auctions* if for every given bidding problem, (ℓ, w, F), there exist $r = r(\ell, w, F) \in [0, m]$ such that $G^a(y, r; \ell, w, F) \succcurlyeq G^a(y, x; \ell, w, F)$ for all $y \in [0, r]$ and $x \in [r, m]$ and $G^a(y, r; \ell, w, F) \succ G^a(y, x; \ell, w, F)$ for all $y \in [0, r]$ such that the support of F^y is $[y, m]$. It is said to exhibit *dynamic consistency in descending bid auctions* if for every given bidding problem, (ℓ, w, F), there exist $r = r(\ell, w, F) \in [0, m]$ such that $G^d(y, r; \ell, w, F) \succcurlyeq G^d(y, x; \ell, w, F)$ for all $y \in [r, m]$ and $x \in [0, r]$ and $G^d(y, r; \ell, w, F) \succ G^d(y, y; \ell, w, F)$ for all $y \in [r, m]$.

The function $b^a : L \times \mathbb{R} \times F \to \mathbb{R}$ is said to be the *actual bid in ascending-bid auctions* if $\delta_w \succcurlyeq G^a(b^a, x; \ell, w, F)$, for all $x \in [b^a, m]$. Similarly, the function $b^d : L \times \mathbb{R} \times F \to \mathbb{R}$ is said to be the *actual bid in descending-bid auctions* if, for all $y \in [b^d, m]$, $G^d(y, x; \ell, w, F) \succ G^d(y, y; \ell, w, F)$ for some $x < y$, and $G^d(b^d, b^d; \ell, w, F) \succcurlyeq G^d(b^d, x; \ell, w, F)$, for all $x \in [0, b^d]$. We denote by $b_i^* : L \times \mathbb{R} \times F \to \mathbb{R}$, $i = 1, 2$, the optimal-bid functions in first and second price sealed-bid auctions, respectively.

In this framework Karni and Safra (1989b) and Karni (1988) prove the following results.

Theorem 17.3. *Let $\succcurlyeq$ be a continuous weak order on L satisfying monotonicity with respect to first-order stochastic dominance whose functional representation, V, is Hadamard differentiable.*[11] *Then,*

(a) *$\succcurlyeq$ on L exhibits dynamic consistency in ascending bid auctions if and only if V is linear.*
(b) *$\succcurlyeq$ on L exhibits dynamic consistency in descending bid auctions if and only if V is linear.*
(c) *For all (ℓ, w, F), $b^a(\ell, w, F) = b_2^*(\ell, w, F)$ if and only if V is linear.*
(d) *For all (ℓ, w, F), $b^d(\ell, w, F) = b_1^*(\ell, w, F)$ if and only if V is linear.*
(e) *For all (ℓ, w, F), $b_2^*(\ell, w, F) = v(\ell, w)$ if and only if V is linear.*

In addition, Karni and Safra (1989b) showed that, for all (ℓ, w, F), $b^a(\ell, w, F) = v(\ell, w)$ if and only if the bidder's preference relation satisfies betweenness.

These results imply that departure from the independence axiom while maintaining the other aspects of expected utility theory means that (a) Strategic plans in auctions are not necessarily dynamically consistent, (b) In general, the games induced by descending-bid and first-price sealed-bid auction mechanisms are not the same in strategic form, and (c) ascending-bid and second-price sealed-bid auction mechanisms are not equivalent.

[11] A path in L is a function $H_{(\cdot)} : [0, 1] \to L$. Let $\mathcal{P}$ denote the set of paths in L such that $\partial H_\alpha(x) / \partial \alpha$ exists for all $x \in \mathbb{R}$. The functional $V : L \to \mathbb{R}$ is Hadamard differentiable if there exist a bounded function $U : \mathbb{R} \times L \to \mathbb{R}$, continuous in its first argument such that, for all $H_{(\cdot)} \in \mathcal{P}$,

$$V(H_\alpha) - V(H_0) = \int U(x, H_0)\, d(H_\alpha(x) - H_0(x)) + o(\alpha).$$

Moreover, except in the case of ascending-bid auctions in which the bidder's preferences satisfy betweenness, the outcome in these auctions is not necessarily Pareto efficient.

17.3. RANK-DEPENDENT UTILITY MODELS

17.3.1 Introduction

The idea that decision makers' risk attitudes are not fully captured by the utility function and may also involve the transformation of probabilities was first suggested by Edwards (1954).[12] In its original form, a lottery $\ell = \left(x_{1,}p_1; \ldots ; x_n, p_n\right)$, where x_i denote the possible monetary payoffs and p_i their, corresponding, probabilities was evaluated according to the formula

$$\sum_{i=1}^{n} u(x_i)\, w\left(p_i\right), \qquad [17.1]$$

where the u denotes the utility function, which is taken to be monotonic increasing and continuous, and w the probability weighting function. This approach is unsatisfactory, however, because, as was pointed out by Fishburn (1978), any weighting function other than the identity function (that is, whenever the formula deviates from the expected utility criterion) implies that this evaluation is not monotonic with respect to first-order stochastic dominance. In other words, in some instances the decision maker prefers the prospect that assigns higher probability to worse outcomes. To see this, let $x < y$, and suppose, without essential loss of generality, that, $w(1) = 1$ and for some $p \in (0, 1)$, $w\left(p\right) + w\left(1 - p\right) < 1$. Then the lottery $\left(x, p; y, \left(1 - p\right)\right)$ dominates the sure outcome, x, according to first order stochastic dominance. Yet, for x and y sufficiently close and u continuous, $u(x)\, w(1) > u(x)\, w\left(p\right) + u(y)\, w\left(1 - p\right)$.

Rank-dependent utility refers to a class of models of decision making under risk whose distinguishing characteristic is that the transformed probabilities of outcomes depend on their probability ranks. Specifically, consider again the lottery $\ell = \left(x_{1,}p_1; \ldots ; x_n, p_n\right)$ and suppose that the monetary payoffs are listed in an ascending order, $x_1 < x_2, \ldots, < x_n$. Define $P_j = \sum_{i=1}^{j} p_i \delta_{x_i}, j = 1, \ldots, n$ and consider a function $g : [0, 1] \to [0, 1]$ that is continuous, strictly increasing, and onto. In the rank-dependent utility models, the decision weight of the outcome x_j, whose probability is p_j, is given by $w\left(p_j\right) = g\left(P_j\right) - g\left(P_{j-1}\right)$ and the lottery ℓ is evaluated by the formula [17.1]. Thus, rank-dependent utility models have the advantage of representing the decision-maker's risk attitudes by the shapes of his utility function and the probability transformation function, while maintaining the desired property of monotonicity with respect to first-order stochastic dominance.

[12] See also Handa (1977), Karmakar (1978), and Kahneman and Tversky (1979).

17.3.2 Representations and interpretation

Let (S, Σ, P) be a probability space and denote by $\mathcal{V}$ the set of random variables on S taking values in an interval, J, whose elements represent monetary payoffs.[13] The random variables $X, Y \in \mathcal{V}$ are said to be *comonotonic* if for all $s,t \in S$, $(X(s) - X(t))(Y(s) - Y(t)) \geq 0$. Comonotonicity implies that the two random variables involved are perfectly correlated and, hence, cannot be used as hedge against one another. In other words, if two comonotonic random variables are amalgamated, by taking their pointwise convex combination, the variability of the resulting random variable, $\alpha X + (1-\alpha) Y$, $\alpha \in (0,1)$, is not smaller than that of its components. Note also that all random variables in $\mathcal{V}$ are comonotonic with the constant functions on S taking values in J.

Let F_X denote the cumulative distribution function on J induced by $X \in \mathcal{V}$. Formally, $F_X(x) = P\{s \in S \mid X(s) \leq x\}$, for all $x \in J$. Let D_J be the set of cumulative distribution functions on J. We assume that the probability space is rich in the sense that all elements of D_J can be generated from elements of $\mathcal{V}$. Let D_J^s be the subset of D_J that consists of cumulative distribution functions with finite range.

Consider a preference relations, $\succsim$, on $\mathcal{V}$, and denote by $\succ$ and $\sim$ it asymmetric and symmetric parts, respectively. Assume that all elements of $\mathcal{V}$ that generate the same elements of D_J are equivalent (that is, for all $X, Y \in \mathcal{V}$, $F_X = F_Y$ implies $X \sim Y$). Then, every preference relation $\succsim$ on $\mathcal{V}$ induces a preference relation, $\succcurlyeq$, on D_J as follows: For all $F_X, F_Y \in D_J$, $F_X \succcurlyeq F_Y$ if and only if $X \succsim Y$.

As in expected utility theory, in rank dependent models preference relations are complete, transitive, continuous, and monotonic. One way of axiomatizing the general form of the rank-dependent utility model is by replacing independence with an axiom called comonotonic commutativity. To state this axiom, let $J_\uparrow^n$ denote the random variables in $\mathcal{V}$, whose realizations are arranged in an ascending order.[14] With slight abuse of notation, we denote the rearrangement of $X \in \mathcal{V}$ by the same symbol, $X \in J_\uparrow^n$. We denote by $\mathrm{CE}\left(\sum_{i=1} p_i \delta_{x_i}\right)$ the certainty equivalent of X, where $p_i = P\{s \in S \mid X(s) = x_i\}$, $i = 1, \ldots, n$.[15]

Comonotonic commutativity. For all $X, Y \in J_\uparrow^n$ such that $x_i \geq y_i$ for all $i = 1, \ldots, n$, and $\alpha \in (0,1)$, $\alpha \delta_{CE(\sum_{i=1}^n p_i \delta_{x_i})} + (1-\alpha)\delta_{CE(\sum_{i=1}^n p_i \delta_{y_i})} \sim \sum_{i=1}^n p_i \delta_{CE(\alpha\delta_{x_i} + (1-\alpha)\delta_{y_i})}$.

The following representation of the general rank-dependent utility model is implied by Chew (1989, Theorem 17.1).

[13] Thus, $\mathcal{V}$ is the set of Σ-measurable functions on S taking values in J.

[14] Note that any comonotonic X and Y in $\mathcal{V}$ may be rearranged in an ascending order, and, of course, $X, Y \in J_\uparrow^n$ are comonotonic.

[15] The existence of the certainty equivalents is implied by the continuity and monotonicity of the preference relation.

Theorem 17.4. *A a preference relation, $\succcurlyeq$ on D_J, satisfies weak order, continuity, monotonicity and comonotonic commutativity if and only if there exist monotonic increasing, continuous, real-valued function u on J, and a function, $g : [0,1] \to [0,1]$ continuous, strictly monotonic increasing, and onto, such that, for all $F, G \in D_J$,*

$$F \succcurlyeq G \Longleftrightarrow \int_J ud\left(g \circ F\right) \geq \int_J ud\left(g \circ G\right).$$

Moreover, u is unique up to positive linear transformation and g is unique.

For $F \in D_J^s$, the representation takes the form

$$F \mapsto \sum_{i=1}^{n} u\left(x_i\right)\left[g\left(F\left(x_i\right)\right) - g\left(F\left(x_{i-1}\right)\right)\right].$$

Expected utility theory is a special case of rank-dependent utility in which commutativity applies to all $X, Y \in D_J$, and the probability transformation function, g, is the identity function. Another special case is Quiggin's (1982) anticipated utility theory, in which comonotonic commutativity is restricted to $\alpha = 0.5$, and the representation requires that $g(0.5) = 0.5$.[16]

To grasp the intuition of the rank-dependent utility models, consider the lottery in D_J^s whose distribution function is $F = \sum_{j=1}^{n} p_j \delta_{x_j}$, where x_j and p_j, $j = 1, \ldots, n$ denote the monetary prizes and their respective probabilities. Then, the decision weight associated with the prize x_j, (that is, $\omega\left(p\left(x_j\right)\right) := g\left(F\left(x_j\right)\right) - g\left(F\left(x_{j-1}\right)\right)$) depends on its probability, p_j, and its *rank* (that is, its relative position in the rank order of the payoffs) given by $F\left(x_j\right)$.[17,18] Put differently, let x be a payoff whose probability is p, if x is ranked just above x' then its decision weight is $\omega\left(p\right) = g\left(F\left(x'\right) + p\right) - g\left(F\left(x'\right)\right)$ and if x is ranked just below x', then its decision weight is $\omega\left(p\right) = g\left(F\left(x'\right)\right) - g\left(F\left(x'\right) - p\right)$.[19] In either case the decision weight is continuous increasing function of p. Clearly, the effect of the rank on the probability weight of x depends on the nature of the probability transformation

[16] Weymark (1981) was the first to axiomatize a rank-dependent model as a measure of income inequality. A detailed discission of the rank dependent model appears in Quiggin (1993).

[17] By contrast, in models characterized by the "betweenness" property, the decision weight of an outcome depends on its probability and the equivalence class it belongs to.

[18] Note that for the lowest ranking outcome, x_1, the rank and the probability coincide (i.e., $p\left(x_1\right) = F\left(x_1\right)$). Hence, $\omega\left(p\left(x_1\right)\right)$, the decision weight of the lowest ranking outcome depends only on its probability. Because the weights add up to 1, if there are only two outcomes, $\underline{x} < \bar{x}$, whose probabilities are, respectively, p and $\left(1-p\right)$, then $\omega\left(p\right) = 1 - \omega\left(1-p\right)$.

[19] Diecidue and Wakker (2001) provide detailed and insightful discussion of the intuition of the rank-dependent utility model.

function, g. If g is the identity function then the weight is independent of the rank. More generally, by allowing the weight of an outcome to exceed (fall short of) the probability of the outcome, depending on its rank, the rank-dependent utility model accommodates optimistic as well as pessimistic attitudes. Specifically, optimistic (pessimistic) attitudes correspond to $\omega(p) = g(F(x)) - g(F(x) - p)$ increasing (resp. decreasing) in the rank, $F(x)$, of x. The preference relation displays optimistic (pessimistic) attitudes if and only if g is a convex (resp. concave) function.

A different axiomatization of the general rank-dependent utility model was advanced by Wakker (2010). We shall elaborate on Wakker's approach when we discuss cumulative prospect theory under risk below.

17.3.3 Risk attitudes and interpersonal comparisons of risk aversion

Loosely speaking, one distribution function in D_J is more risky than another if the former is a mean preserving spread of the latter. Formally, let $F, G \in D_J$ then F is *riskier than* G if they have the same mean (that is, $\int_J (F(x) - G(x))\,dx = 0$) and F is more dispersed than G (that is, $\int_{J\cap(-\infty,t)} (F(x) - G(x))\,dx \geq 0$, for all t). A preference relation $\succcurlyeq$ on D_J is said to exhibit *risk aversion* (strict risk aversion) if $G \succcurlyeq F$, $(G \succ F)$ whenever F is riskier than G.

In expected utility theory risk attitudes are completely characterized by the properties of the utility function. In particular, F is riskier than G if and only if the expected utility of F is no greater than that of G for all concave utility functions. In rank-dependent utility models, the attitudes toward risk depend on the properties of both the utility and the probability transformation functions. Chew et al. (1987) showed that individuals whose preferences are representable by rank-dependent utility exhibit risk aversion if and only if both the utility and the probability transformation functions are concave. Such individuals exhibit strict risk aversion if and only if they exhibit risk aversion and either the utility function or the probability transformation function is strictly concave. Because the probability transformation function is independent of the levels of the payoffs and variations of wealth do not affect the rank order of the payoffs, the wealth related changes of individual attitudes toward risk are completely characterized by the properties of the utility function. In other words, as in expected utility theory, a preference relation exhibits decreasing (increasing, constant) absolute risk aversion if and only if the utility function displays these properties in the sense of Pratt (1964).

Unlike in expected utility theory, in the rank-dependent model the preference relation is not smooth. Instead, it displays "kinks" at points at which the rank order of the payoffs changes, for example, at certainty.[20] At these points the preference relation

[20] In expected utility theory, risk averse attitude can only accommodate finite number of "kinks" in the utility function. By contrast, in rank-dependent utility model the utility function may have no kinks at all and yet, have "kinks" at points of where the rank order of the payoff changes.

exhibits first-order risk aversion. Formally, consider the random variable $w + t\tilde{\varepsilon}$, where $\tilde{\varepsilon}$ is a random variable whose mean is zero, w denote the decision-maker's initial wealth and $t \geq 0$. Let $CE(t; w, \tilde{\varepsilon})$ denote the certainty equivalent of this random variable as a function of t and $\pi(t; w, \tilde{\varepsilon}) = w - CE(t; w, \tilde{\varepsilon})$ the corresponding risk premium. Clearly, $\pi(0; w, \tilde{\varepsilon}) = 0$. Suppose that $\pi(t; ; w, \tilde{\varepsilon})$ is twice continuously differentiable with respect to t around $t = 0$. Following Segal and Spivak (1990), a preference relation is said to exhibit *first-order risk aversion at* w if for every nondegenerate $\tilde{\varepsilon}$, $\partial\pi(t; w, \tilde{\varepsilon}) / \partial t \mid_{t=o^+} < 0$. It is said to display *second-order risk aversion at* w if $\partial\pi(t; w, \tilde{\varepsilon}) / \partial t \mid_{t=o^+} = 0$ and $\partial^2\pi(t; w, \tilde{\varepsilon}) / \partial t^2 \mid_{t=o^+} < 0$. Unlike the expected utility model in which risk averse attitudes are generically second order, risk aversion in the rank-dependent utility model is of first order. Hence, in contrast with expected utility theory, in which risk averse decision makers take out full insurance if and only if the insurance is fair, in the rank-dependent utility theory, risk averse decision makers may take out full insurance even when insurance is slightly unfair.[21]

Given a preference relation $\succcurlyeq$ on D_J and $F, G \in D_J$, F is said to differ from G by a *simple compensating spread from the point of view of* $\succcurlyeq$ if $F \sim G$ and there exist $x^0 \in J$ such that $F(x) \geq G(x)$ for all $x < x^0$ and $F(x) \leq G(x)$ for all $x \geq x^0$. Let $\succcurlyeq$ and $\succcurlyeq^*$ be preference relations on D_J, then $\succcurlyeq$ is said to *exhibit greater risk aversion than* $\succcurlyeq^*$ if, for all $F, G \in D_J$, F differs from G by a simple compensating spread from the point of view of $\succcurlyeq$ implies that $G \succcurlyeq^* F$. If $\succcurlyeq$ and $\succcurlyeq^*$ are representable by rank-dependent functionals, with utility and probability transformation functions (u, g) and (u^*, g^*), respectively, then $\succcurlyeq$ exhibits greater risk aversion than $\succcurlyeq^*$ if and only if u^* and g^* are concave transformations of u and g, respectively.[22] The aspect of risk aversion captured by the utility function is the same as in expected utility theory. The probability transformation function translates the increase in spread of the underlying distribution function into spread of the decision weights. When the probability transformation function is concave, it reduces the weights assigned to higher ranking outcomes and increases those of lower ranking outcomes, thereby producing a pessimistic outlook that tends to lower the overall value of the representation functional.

17.3.4 The dual theory

Another interesting instance of rank-dependent utility theory is Yaari's (1987) dual theory of choice.[23] In this theory, the independence axiom is replaced with the dual-independence axiom. To introduce the dual independence axiom consider first the set of random variables $\bar{V}$ taking values in a compact interval, $\bar{J}$, in the real line. The next axiom, comonotonic independence, asserts that preference relation between two comonotonic random variables is not reversed when each of them is

[21] For a more elaborate discussion see also Machina (2001).

[22] See Chew et al. (1987, Theorem 17.1).

[23] See also Röell (1987).

amalgamated with a third random variable that is comonotonic with both of them. Without comonotonicity, the possibility of hedging may reverse the rank order, since decision makers who are not risk neutral are concerned with the variability of the payoffs. Formally,

Comonotonic independence. For all pairwise comonotonic random variables, $X, Y, Z \in \bar{\mathcal{V}}$, and all $\alpha \in [0,1]$, $X \succsim Y$ implies that $\alpha X + (1-\alpha) Z \succsim \alpha Y + (1-\alpha) Z$.

Comonotonic independence may be stated in terms of $\succcurlyeq$ on $D_{\bar{J}}$ as dual independence. Unlike the independence axiom in expected utility theory, which involves mixtures of probabilities, the mixture operation that figures in the statement of dual independence is a mixture of the comonotonic random payoffs. More concretely, the mixture operation may be portrayed as "portfolio mixture," depicting the combined yield of two assets whose yields are comonotonic. Formally, for all $F \in D_{\bar{J}}$ define the inverse function, $F^{-1} : [0,1] \to \bar{J}$ by: $F^{-1}(p) = \sup_{q \leq p}\{t \in \bar{J} \mid F(t) = q\}$. For all $F, G \in D_{\bar{J}}$ and $\alpha \in [0,1]$, let $\alpha F \oplus (1-\alpha) G \in D_{\bar{J}}$ be defined by $(\alpha F \oplus (1-\alpha) G)(r) = \left(\alpha F^{-1}(r) + (1-\alpha) G^{-1}(r)\right)^{-1}$, for all $r \in [0,1]$. For any random variables, $X, Y \in \bar{\mathcal{V}}$, and $\alpha \in [0,1]$, $\alpha F_X \oplus (1-\alpha) F_Y$ is the cumulative distribution function of the random variable $\alpha X + (1-\alpha) Y$.

A preference relation $\succsim$ satisfies comonotonic independence if and only if the induced preference relation $\succcurlyeq$ satisfies the dual independence axiom below.[24]

Dual independence. For all $F, G, H \in D_{\bar{J}}$ and $\alpha \in [0,1]$, $F \succcurlyeq G$ implies that $\alpha F \oplus (1-\alpha) H \succcurlyeq \alpha G \oplus (1-\alpha) H$.

Yaari's (1987) dual theory representation theorem may be stated as follows:

Theorem 17.5. *A preference relation* $\succcurlyeq$ *on* $D_{\bar{J}}$, *satisfies weak order, continuity, monotonicity and dual independence if and only if there exists a function,* $f : [0,1] \to [0,1]$ *continuous, non-increasing, and onto, such that, for all* $F, G \in D_{\bar{J}}$,

$$F \succcurlyeq G \Longleftrightarrow \int_{\bar{J}} f(1 - F(x))\, dx \geq \int_{\bar{J}} f(1 - G(x))\, dx.$$

Integrating by parts, it is easy to verify that $F \succcurlyeq G$ if and only if $-\int_{\bar{J}} x df(1 - F(x)) \geq -\int_{\bar{J}} x df(1 - G(x))$. Hence, the dual theory is the special case of the rank-dependent model in which the utility function is linear and $f(1-r) = 1 - g(r)$ for all $r \in [0,1]$.

Maccheroni (2004) extended the dual theory to the case of incomplete preferences, showing that the representation involves multi probability-transformation functions.

[24] See Yaari (1987, Proposition 3).

Formally, there is a set $\mathcal{F}$ of probability transformation functions such that, for all $F, G \in D_{\bar{J}}$, $F \succcurlyeq G$ if and only if $-\int_{\bar{J}} xdf(1-F(x)) \geq -\int_{\bar{J}} xdf(1-G(x))$, for all $f \in \mathcal{F}$.

The expected value of cumulative distribution functions $F \in D_J$ is the area of the epigraph of F (that is, a product of the Lebesgue measures on J and $[0,1]$ of the epigraph of F). The expected utility of F is a product measure of the epigraph of F, where the measure of an interval $[x,y] \subseteq J$ is given by $u(y) - u(x)$ instead of the Lebesgue measure, and the measure on $[0,1]$ is the Lebesgue measure. Segal (1989, 1993) axiomatized the product-measure representation of the rank-dependent utility model. In this representation the measure on J is the same as that of expected utility but the measure of $[p,q] \subseteq [0,1]$ is given by $g(q) - g(p)$.

17.4. CUMULATIVE PROSPECT THEORY

17.4.1 Introduction

Prospect theory was introduced by Kahneman and Tversky (1979) as an alternative to expected utility theory. Intended as tractable model of decision making under risk capable of accommodating large set of systematic violations of the expected utility model, the theory has several special features: First, payoffs are specified in terms of gains and losses relative to a *reference point* rather than as ultimate outcomes. Second, payoffs are evaluated by a real-valued *value* function that is concave over gains and convex over losses and steeper for losses than for gains, capturing a property dubbed *loss aversion*. Third, the probabilities of the different outcomes are transformed using a probability weighting function which overweight small probabilities and underweight moderate and large probabilities.[25]

In its original formulation the weighting function in prospect theory transformed the probabilities of the outcomes as in [17.1] and, consequently, violates monotonicity. To overcome this difficulty, cumulative prospect theory is proposed as a synthesis of the original prospect theory and the rank-dependent utility model.[26]

In what follows we present a version of cumulative prospect theory axiomatized by Chateauneuf and Wakker (1999).

17.4.2 Trade-off consistency and representation

Consider an outcomes set whose elements are monetary rewards represented by the real numbers. Denote by L_s the set of prospects (that is, simple lotteries on $\mathbb{R}$). Given

[25] See Tversky and Kahneman (1992).

[26] See Schmidt and Zank (2009) and Wakker (2010).

any $P = (x_1, p_1; \ldots; x_n, p_n) \in L_s$, assume that the outcomes are arranged in increasing order.[27] Let $\succcurlyeq$ be a preference relation on L_s. Let 0 denote the payoff representing the reference point and define the outcome x as a *gain* if $\delta_x \succ \delta_0$ and as a *loss* if $\delta_0 \succ \delta_x$.

Given a prospect, $P = (x_1, p_1, \ldots, x_n, p_n)$, denote by $P_{-j}y$ the prospect obtained by replacing the outcome x_j in P by an outcome y such that $\delta_{x_{j+1}} \succcurlyeq \delta_y \succcurlyeq \delta_{x_{j-1}}$, (i.e., $P_{-j}y = (x_1, p_1; \ldots; x_{j-1}p_{j-1}; y, p_j; x_{j+1}, p_{j+1}; \ldots; x_n, p_n)$). Following Chateauneuf and Wakker (1999), define a binary relation $\succcurlyeq^*$ on $\mathbb{R} \times \mathbb{R}$ as follows: Let $P = (x_1, p_1; \ldots; x_n, p_n)$, $Q = (y_1, p_1; \ldots; y_n, p_n)$ and j such that $p_j > 0$, then $(z, w) \succ^* (z', w')$ if and only if $P_{-j}z \succcurlyeq Q_{-j}w$ and $Q_{-j}w' \succ P_{-j}z'$ and $(z, w) \succcurlyeq^* (z', w')$ if and only if $P_{-j}z \succcurlyeq Q_{-j}w$ and $Q_{-j}w' \succcurlyeq P_{-j}z'$. The relation $\succcurlyeq^*$ represents the intensity of preference for the outcome z over w compared to that of z' over w', holding the same their ranking in the prospects in which they are embedded.[28] The following axiom requires that this intensity of preferences be independent of the outcomes and their place in the ranking in the prospects in which they are embedded.

Trade-off consistency. For no outcomes $w, x, y, z \in \mathbb{R}$, $(w, x) \succ^* (y, z)$ and $(y, z) \succcurlyeq^* (w, x)$.

The following representation theorem for cumulative prospect theory is due to Chateauneuf and Wakker (1999).

Theorem 17.6. *A preference relation, $\succcurlyeq$ on L_s, satisfies weak order, continuity, monotonicity and trade-off consistency if and only if there exist a continuous real-valued function, v on $\mathbb{R}$, satisfying $v(0) = 0$, and strictly increasing probability transformation functions, ω^+, ω^- from $[0, 1]$ to $[0, 1]$, satisfying $\omega^+(0) = \omega^-(0) = 0$ and $\omega^+(1) = \omega^-(1) = 1$, such that $\succcurlyeq$ has the following representation: For all $P \in L_s$*

$$P \mapsto \sum_{\{j|x_j \geq 0\}} v(x_j) \left[\omega^+ \left(\Sigma_{i=1}^{j} p_i\right) - \omega^+ \left(\Sigma_{i=1}^{j-1} p_i\right)\right]$$

$$+ \sum_{\{j|x_j \leq 0\}} v(x_j) \left[\omega^- \left(\Sigma_{i=1}^{j} p_i\right) - \omega^- \left(\Sigma_{i=1}^{j-1} p_i\right)\right].$$

Note that restricted to prospects whose outcomes are either all gains or all losses, the representation above is that of rank-dependent utility. Consequently, the rank-dependent

[27] More generally, the set of outcomes may be taken to be a connected topological pace and the outcomes ranked by a complete and transitive preference relation.

[28] Anticipating the next reult, let v denote the value function of cumulative prospect theory, Chateauneuf and Wakker (1999) showed that $(w, x) \succ^* (y, z)$ implies $v(w) - v(x) > v(y) - v(z)$ and $(w, x) \succcurlyeq^* (y, z)$ implies $v(w) - v(x) \geq v(y) - v(z)$.

utility model can be axiomatized using trade-off consistency without the restrictions imposed by the ranks.[29]

PART II NONEXPECTED UTILITY THEORY UNDER UNCERTAINTY

17.5. DECISION PROBLEMS UNDER UNCERTAINTY[30]

17.5.1 The decision problem

In a decision problem under uncertainty, an individual, the *decision maker*, considers a set of alternative actions whose consequences depend on uncertain factors outside his control. Formally, there is a set A of available (*pure*) *actions* a that can result in different (*deterministic*) *consequences* z, within a set Z, depending on which *state* s (*of nature* or *of the environment*) in a space S obtains. For convenience, we consider finite state spaces.

The dependence of consequences on actions and states is described by a *consequence function*

$$\rho : A \times S \rightarrow Z$$

that details the consequence $z = \rho(a, s)$ of each action a in each state s.[31] The quartet (A, S, Z, ρ) is called *decision problem (under uncertainty)*.

Example 17.1. *(i) Firms in a competitive market choose the level of production $a \in A$ being uncertain about the price $s \in S$ that will prevail. The consequence function here is their profit*

$$\rho(a, s) = sa - c(a)$$

where c is the cost of producing a units of good.

(ii) Players in a game choose their strategy $a \in A$ being uncertain about the opponents' strategy profile $s \in S$. Here the consequence function $\rho(a, s)$ determines the material consequences that result from strategy profile (a, s).

Preferences over actions are modeled by a preorder $\succsim$ on A. For each action $a \in A$, the section of ρ at a

$$\begin{aligned} \rho_a : S &\rightarrow Z \\ s &\mapsto \rho(a, s) \end{aligned}$$

[29] See Wakker and Tversky (1993) and Wakker (2010) .

[30] this and the subsequent sections are partly based on Battigalli et al. (2013). we refer to Gilboa and Marinacci (2013) for an alternative presentation.

[31] Consequences are often called *outcomes*, and accordingly ρ is called *outcome function*. Sometimes actions are called *decisions*.

associates to each $s \in S$ the consequence resulting from the choice of a if s obtains, and is called *Savage act* induced by a. In general, Savage acts are maps $f : S \to Z$ from states to consequences. The celebrated work of Savage (1954) is based on them. In a consequentialist perspective what matters about actions is not their label/name but the consequences that they determine when the different states obtain.[32] For this reason acts are, from a purely methodological standpoint, the proper concept to use, so much that most of the decision theory literature followed the Savage lead.

However, as Marschak and Radner (1972, p. 13) remark, the notions of actions, states and implied consequences "correspond more closely to the everyday connotations of the words." In particular, they are especially natural in a game theoretic context, as the previous example shows. For this reason, here we prefer to consider preferences $\succsim$ over A in a decision problem (A, S, Z, ρ). But, in line with Savage's insight, we postulate the following classic principle:

Consequentialism. Two actions that generate the same consequence in every state are indifferent. Formally,

$$\rho(a, s) = \rho(b, s) \quad \forall s \in S \Longrightarrow a \sim b$$

or, equivalently, $\rho_a = \rho_b \Longrightarrow a \sim b$.

Under this principle, the current framework and Savage's are essentially equivalent; see, again, Marschak and Radner (1972, p. 13). They also show how to enrich A and extend ρ in order to guarantee that $\{\rho_a\}_{a \in A} = Z^S$. This extension is very useful for axiomatizations and comparative statics' exercises. For example, in this case, for each $z \in Z$ there exists a *sure action* a_z such that $\rho(a_z, s) = z$ for all $s \in S$, and this allows to set

$$z \succsim_Z z' \Longleftrightarrow a_z \succsim a_{z'}.$$

By consequentialism, this definition is well posed. Although we will just write $z \succsim z'$, it is important to keep in mind how the preference between consequences has been inferred from that between actions. One might take the opposite perspective that decision makers have "basic preferences" among consequences, that is, on the material outcomes of the decision process, and these basic preferences in turn determine how they choose among actions. Be that as it may, we regard $\succsim$ as describing the decision maker's choice behavior and actions, not consequences, are the objects of choice.

Finally, observe that $\succsim_Z$ is also well defined when

$$\rho_a = \rho_b \Longrightarrow a = b. \qquad [17.2]$$

[32] In the words of Machiavelli (1532) "*si habbi nelle cose a vedere il fine e non il mezzo.*"

This assumption is not restrictive if consequentialist preferences are considered. Decision problems satisfying it essentially correspond to reduced normal forms of games in which realization equivalent pure strategies are merged.

We call *canonical decision problem* (*c.d.p.*, for short) a decision problem (A, S, Z, ρ) in which $\{\rho_a\}_{a \in A} = Z^S$ and [17.2] holds, so that A can be seen as a "suitable index set for the set of all acts".[33]

17.5.2 Mixed actions

Suppose that the decision maker can commit his actions to some random device. As a result, he can choose *mixed actions*, that is, elements α of the collection $\Delta(A)$ of all chance[34] distributions on A with finite support. Pure actions can be viewed as special mixed actions: if we denote by δ_a the Dirac measure concentrated on $a \in A$, we can embed A in $\Delta(A)$ through the mapping $a \hookrightarrow \delta_a$.

The relevance of mixed actions is well illustrated by the decision problem:

	s_1	s_2
a_1	0	1
a_2	1	0

[17.3]

with action set $A = \{a_1, a_2\}$, state space $S = \{s_1, s_2\}$, outcome space $Z = \{0, 1\}$, and consequence function

$$\rho(a_1, s_1) = \rho(a_2, s_2) = 0 \quad \text{and} \quad \rho(a_1, s_2) = \rho(a_2, s_1) = 1.$$

As Luce and Raiffa (1957, p. 279) observe, the mixed action

$$\frac{1}{2}\delta_{a_1} + \frac{1}{2}\delta_{a_2}$$

guarantees an expected value of $1/2$ regardless of which state obtains, while the minimum guaranteed by both pure actions is 0. Randomization may thus hedge uncertainty, an obviously important feature in this kind of decision problems.

Denote by $\Delta(Z)$ the collection of *random consequences*, that is, the collection of chance distributions on Z with finite support. Mixed actions (chance distributions of pure actions) induce such random consequences (chance distributions of deterministic consequences): if the decision maker takes mixed action α, the chance of obtaining consequence z in state s is

$$\alpha(\{a \in A : \rho(a, s) = z\}) = \left(\alpha \circ \rho_s^{-1}\right)(z)$$

[33] See Marschak and Radner (1972, p. 14). This assumption is obviously strong and Battigalli et al. (2013) show how it can be drastically weakened while maintaining most of the results that we present in the rest of this chapter.

[34] By chance we mean a nonepistemic, objective, probability (such as that featured by random devices).

where ρ_s is the section of ρ at s. This observation provides a natural extension of ρ given by

$$\begin{aligned} \hat{\rho} : \Delta(A) \times S &\to \Delta(Z) \\ (\alpha, s) &\mapsto \alpha \circ \rho_s^{-1} \end{aligned}$$

associating to each mixed action and each state the corresponding random consequence. Since $\hat{\rho}$ is uniquely determined by ρ, the hat is omitted. The domain of alternatives (and hence of preferences $\succsim$) has been extended to from pure actions to mixed actions, that is, from A to $\Delta(A)$ and random consequences are implied.

In this extension, each mixed action $\alpha \in \Delta(A)$ determines an *Anscombe-Aumann act*

$$\begin{aligned} \rho_\alpha : S &\to \Delta(Z) \\ s &\mapsto \alpha \circ \rho_s^{-1} \end{aligned}$$

(the section of $\hat{\rho}$ at α) that associates to each $s \in S$ the chance distribution of consequences resulting from the choice of α if s obtains. In general, Anscombe-Aumann acts are maps $f : S \to \Delta(Z)$ from states to random consequences. The analysis of Anscombe and Aumann (1963) is based on them (which they call *horse lotteries*). Consequentialism naturally extends to this mixed actions setup, with a similar motivation.

Chance consequentialism. Two mixed actions that generate the same distribution of consequences in every state are indifferent. Formally,

$$\alpha(\{a \in A : \rho(a, s) = z\}) = \beta(\{a \in A : \rho(a, s) = z\}) \quad \forall z \in Z \text{ and } s \in S \Longrightarrow \alpha \sim \beta$$

or, equivalently, $\rho_\alpha = \rho_\beta \Longrightarrow \alpha \sim \beta$.

Under this realization equivalence principle,[35] a decision problem under uncertainty in which mixed actions are available can be embedded in the Anscombe-Aumann framework, as presented in Fishburn (1970), but with a major caveat. In such setup an act $f : S \to \Delta(Z)$ is interpreted as a non-random object of choice with random consequences $f(s)$. As Fishburn (1970, p. 176) writes

> *. . .We adopt the following pseudo-operational interpretation for $f \in \Delta(Z)^S$. If f is "selected" and $s \in S$ obtains then $f(s) \in \Delta(Z)$ is used to determine a resultant consequence in Z. . .*

Randomization is interpreted to occur *ex post*: the decision maker commits to f, "observes" the realized state s, then "observes" the consequence z generated by the random mechanism $f(s)$, and receives z.[36] In contrast, the mixed actions that

[35] Remember that, in game theory, two mixed strategies of a player are called *realization equivalent* if—for any fixed pure strategy profile of the other players—both strategies induce the same chance distribution on the outcomes of the game. Analogously, here, two mixed actions can be called *realization equivalent* if—for any fixed state—both actions induce the same chance distribution on consequences.

[36] We say "interpreted" since this timeline and the timed disclosure of information are unmodeled. In principle, one could think of the decision maker committing to f and receiving the outcome of the resulting process.

we consider here are, by definition, random objects of choice. Randomization can be interpreted to occur *ex ante*: the decision maker commits to α, "observes" the realized action a, then "observes" the realized state s, and receives the consequence $z = \rho(a, s)$.[37] The next simple convexity result helps, inter alia, to understand the issue.[38]

Proposition 17.1. *For each $\alpha, \beta \in \Delta(A)$ and each $p \in [0, 1]$,*

$$\rho_{p\alpha+(1-p)\beta}(s) = p\rho_\alpha(s) + (1-p)\rho_\beta(s) \quad \forall s \in S.$$

Since each mixed action α can be written as a convex combination $\alpha = \sum_{a\in A} \alpha(a)\delta_a$ of Dirac measures, Proposition 17.1 implies the equality

$$\rho_\alpha(s) = \sum_{a\in A} \alpha(a)\rho_{\delta_a}(s) \quad \forall s \in S. \tag{17.4}$$

Hence, $\rho_\alpha(s)$ is the chance distribution on Z induced by the "*ex ante*" randomization of actions a with probabilities $\alpha(a)$ if state s obtains. Consider the act $f_\alpha : S \to \Delta(Z)$ given by

$$f_\alpha(s) = \sum_{a\in A} \alpha(a)\delta_{\rho(a,s)} \quad \forall s \in S.$$

Now $f_\alpha(s)$ is the chance distribution on Z induced by the "*ex post*" randomization of consequences $\rho(a, s)$ with probabilities $\alpha(a)$ if state s obtains. But, for all $a \in A$, $s \in S$, and $z \in Z$, denoting by $\rho_{\delta_a}(z|s)$ the chance of obtaining consequence z in state s if δ_a is chosen we have

$$\rho_{\delta_a}(z|s) = \delta_a(\{b \in A : \rho(b, s) = z\}) = \begin{cases} 1 & z = \rho(a, s) \\ 0 & \text{otherwise} \end{cases} = \delta_{\rho(a,s)}(z) \tag{17.5}$$

That is,

$$\rho_{\delta_a}(s) = \delta_{\rho(a,s)} \quad \forall (a, s) \in A \times S.$$

Therefore, $\rho_\alpha = f_\alpha$ since they assign the same chances to consequences in every state. Informally, they differ in the timing of randomization: decision makers "learn," respectively, ex ante (before "observing" the state) and ex post (after "observing" the state) the outcomes of randomization α. Whether such implementation difference matters is an empirical question that cannot be answered in our abstract setup (or in the one of Fishburn). In a richer setup, with two explicit layers of randomization, Anscombe and Aumann (1963, p. 201) are able to formalize this issue and assume that

[37] Again the timeline and the timed disclosure of information are unmodeled. In principle, one could think of the decision maker committing to α and receiving the outcome of the resulting process.

[38] See Battigalli et al. (2013).

"it is immaterial whether the wheel is spun before or after the race." Here, we do not pursue this matter any more.

If (A, S, Z, ρ) is a canonical decision problem, then $\{\rho_\alpha\}_{\alpha\in\Delta(A)} = \Delta(Z)^S$.[39] In particular, since all sure actions $\{a_z\}_{z\in Z}$ are available, all mixed actions with support in this set are available too. A special feature of these actions is that the chance distribution of consequences they generate do not depend on the realized state. For this reason, we call them *lottery actions* and denote $\Delta_\ell(A) = \{\alpha \in \Delta(A) : \mathrm{supp}\alpha \subseteq \{a_z\}_{z\in Z}\}$ the set of all of them. Setting

$$\alpha_\zeta = \sum_{z\in Z} \zeta(z)\,\delta_{a_z} \quad \forall \zeta \in \Delta(Z) \tag{17.6}$$

the correspondence $\zeta \hookrightarrow \alpha_\zeta$ is actually an affine embedding of $\Delta(Z)$ onto $\Delta_\ell(A)$ such that, for each $s \in S$,

$$\rho_{\alpha_\zeta}(s) = \sum_{z\in\mathrm{supp}\zeta} \zeta(z)\,\rho_{\delta_{a_z}}(s) = \sum_{z\in\mathrm{supp}\zeta} \zeta(z)\,\delta_{\rho(a_z,s)} = \sum_{z\in\mathrm{supp}\zeta} \zeta(z)\,\delta_z = \zeta.$$

This embedding allows to identify $\Delta(Z)$ and $\Delta_\ell(A)$, and so to set

$$\zeta \succsim_{\Delta(Z)} \zeta' \Longleftrightarrow \alpha_\zeta \succsim \alpha_{\zeta'}$$

for $\zeta, \zeta' \in \Delta(Z)$. In view of this we can write $\zeta \in \Delta_\ell(A)$.

17.5.3 Subjective expected utility

The theory of decisions under uncertainty investigates the preferences' structures underlying the choice criteria that decision makers use to rank alternatives in a decision problem. In our setting, this translates in the study of the relations between the properties of preferences $\succsim$ on $\Delta(A)$ and those of *preference functionals* which represent them, that is, functionals $V : \Delta(A) \to \mathbb{R}$ such that

$$\alpha \succsim \beta \Longleftrightarrow V(\alpha) \geq V(\beta).$$

The basic preference functional is the subjective expected utility one, axiomatized by Savage (1954); we devote this section to the version of Savage's representation due to Anscombe and Aumann (1963).

Definition 17.1. *Let (A, S, Z, ρ) be a c.d.p., a binary relation $\succsim$ on $\Delta(A)$ is a* rational preference *(under uncertainty) if it is:*

1. *reflexive: $\alpha = \beta$ implies $\alpha \sim \beta$;*
2. *transitive: $\alpha \succsim \beta$ and $\beta \succsim \gamma$ implies $\alpha \succsim \gamma$;*
3. *monotone: $\rho_\alpha(s) \succsim_{\Delta(Z)} \rho_\beta(s)$ for all $s \in S$ implies $\alpha \succsim \beta$.*

[39] But notice that the reduction assumption [17.2] does not imply that $\alpha = \beta$ if $\rho_\alpha = \rho_\beta$ when α and β are mixed actions. In the simple canonical decision problem [17.11], for example, $\alpha = \frac{1}{2}\delta_a + \frac{1}{2}\delta_b$ and $\beta = \frac{1}{2}\delta_c + \frac{1}{2}\delta_d$ have disjoint support, but $\rho_\alpha = \rho_\beta$. See also Battigalli et al. (2013).

The latter condition is a preferential version of consequentialism: if the random consequence generated by α is preferred to the random consequence generated by β irrespectively of the state, then α is preferred to β.

Proposition 17.2. *Let (A, S, Z, ρ) be a c.d.p. and $\succsim$ be a rational preference on $\Delta(A)$, then $\succsim$ satisfies chance consequentialism.*

Proof. If $\beta, \gamma \in \Delta(A)$ and $\rho_\beta(s) = \rho_\gamma(s)$ for all $s \in S$, then $\alpha_{\rho_\beta(s)} = \alpha_{\rho_\gamma(s)}$,(defined in 17.6) reflexivity implies $\alpha_{\rho_\beta(s)} \sim \alpha_{\rho_\gamma(s)}$. But then $\rho_\beta(s) \sim_{\Delta(Z)} \rho_\gamma(s)$ for all $s \in S$ by the definition of $\succsim_{\Delta(Z)}$ and monotonicity delivers $\beta \sim \gamma$.

Next we turn to a basic condition about decision makers' information.

Completeness. For all $\alpha, \beta \in \Delta(A)$, it holds $\alpha \succsim \beta$ *or* $\beta \succsim \alpha$.

Completeness thus requires that decision makers be able to compare any two alternatives. We interpret it as a statement about the quality of decision makers' information, which enables them to make such comparisons, rather than about their decisiveness. Although it greatly simplifies the analysis, it is a condition conceptually less compelling than transitivity and monotonicity (see Section 17.7.1).

Next we state a standard regularity condition, due to Herstein and Milnor (1953).

Continuity. The sets $\{p \in [0,1] : p\alpha + (1-p)\beta \succsim \gamma\}$ and $\{p \in [0,1] : \gamma \succsim p\alpha + (1-p)\beta\}$ are closed for all $\alpha, \beta, \gamma \in \Delta(A)$.

The last key ingredient is the independence axiom. Formally,

Independence. If $\alpha, \beta, \gamma \in \Delta(A)$, then

$$\alpha \sim \beta \Longrightarrow \frac{1}{2}\alpha + \frac{1}{2}\gamma \sim \frac{1}{2}\beta + \frac{1}{2}\gamma.$$

This classical axiom requires that mixing with a common alternative does not alter the original ranking of two mixed actions. When restricted to $\Delta_\ell(A) \simeq \Delta(Z)$ it reduces to the von Neumann-Morgenstern original independence axiom on lotteries. Notice that the independence axiom implies indifference to randomization, that is,

$$\alpha \sim \beta \Longrightarrow \frac{1}{2}\alpha + \frac{1}{2}\beta \sim \beta. \qquad [17.7]$$

In fact, by setting $\gamma = \beta$, the axiom yields

$$\alpha \sim \beta \Longrightarrow \frac{1}{2}\alpha + \frac{1}{2}\beta \sim \frac{1}{2}\beta + \frac{1}{2}\beta = \beta.$$

This key feature of independence will be questioned in the next section by the Ellsberg paradox.

The previous preferential assumptions lead to the Anscombe and Aumann (1963) version of Savage's result.

Theorem 17.7. *Let (A, S, Z, ρ) be a c.d.p. and $\succsim$ be a binary relation on $\Delta(A)$. The following conditions are equivalent:*

1. *$\succsim$ is a nontrivial, complete, and continuous rational preference that satisfies independence;*
2. *there exist a nonconstant $u : Z \to \mathbb{R}$ and a probability distribution $\pi \in \Delta(S)$ such that the preference functional*

$$V(\alpha) = \sum_s \pi(s) \left(\sum_a \alpha(a) u(\rho(a, s)) \right) \quad [17.8]$$

represents $\succsim$ on $\Delta(A)$.

The function u is cardinally unique[40] *and π is unique.*

To derive this version of the Anscombe-Aumann representation based on actions from the more standard one based on acts (see Schmeidler, 1989, p. 578), notice that by Proposition 17.1 the map

$$\begin{array}{rcl} F: \Delta(A) & \to & \Delta(Z)^S \\ \alpha & \mapsto & \rho_\alpha \end{array}$$

is affine and onto (since the decision problem is canonical). Chance consequentialism of $\succsim$ (implied by Proposition 17.2) allows to derive $\succsim_F$ on $\Delta(Z)^S$ by setting, for $f = F(\alpha)$ and $g = F(\beta)$ in Z^S,

$$f \succsim_F g \Longleftrightarrow \alpha \succsim \beta.$$

By affinity of F, we can transfer to $\succsim_F$ all the assumptions we made on $\succsim$. This takes us back to the standard Anscombe-Aumann setup.[41]

The embedding $a \hookrightarrow \delta_a$ of pure actions into mixed actions allows to define

$$a \succsim_A b \Leftrightarrow \delta_a \succsim \delta_b$$

Thus [17.8] implies

$$a \succsim_A b \Leftrightarrow \sum_s \pi(s) u(\rho(a, s)) \geq \sum_s \pi(s) u(\rho(b, s)) \quad [17.9]$$

[40] That is, unique up to an affine strictly increasing transformation.

[41] More on the relation between the mixed actions setup we adopt here and the standard Anscombe-Aumann setup can be found in Battigalli et al. (2013).

which is the subjective expected utility representation of Marschak and Radner (1972, p. 16). The function u is a *utility* function since [17.9] implies

$$z \succsim_Z z' \Longleftrightarrow u(z) \geq u(z').$$

Following the classic insights of Ramsey (1926) and de Finetti (1931, 1937), rankings of bets on events can be used to elicit the decision maker's subjective assessment of their likelihoods. To this end, denote by $zEz' \in A$ the bet on event $E \subseteq S$ that delivers z if E obtains and z' otherwise, with $z' \prec z$. By [17.9],

$$zEz' \succsim zE'z' \Longleftrightarrow \pi(E) \geq \pi(E')$$

and so π can be regarded as a *subjective probability*.

Each mixed action α generates, along with the subjective probability $\pi \in \Delta(S)$, a product probability distribution $\alpha \times \pi \in \Delta(A \times S)$. Since

$$\sum_s \pi(s) \left(\sum_a \alpha(a)\, u(\rho(a,s)) \right) = \sum_{(a,s) \in A \times S} u(\rho(a,s))\, \alpha(a)\, \pi(s) = \mathrm{E}_{\alpha \times \pi}(u \circ \rho)$$

we can rewrite [17.8] as follows

$$V(\alpha) = \mathrm{E}_{\alpha \times \pi}(u \circ \rho). \qquad [17.10]$$

In this perspective, $V(\alpha)$ is the expected payoff of the mixed action profile (α, π) for the decision maker in a game $(\{DM, N\}, \{A, S\}, \{u \circ \rho, v \circ \rho\})$ in which the decision maker chooses mixed strategy α and *nature* (denoted by N and assumed to determine the state of the environment) chooses mixed strategy π. But notice that, while α is a chance distribution available to the decision maker, π is a subjective belief of the decision maker on nature's behavior.

17.6. UNCERTAINTY AVERSION: DEFINITION AND REPRESENTATION

17.6.1 The Ellsberg paradox

Consider a coin that a decision maker knows to be fair, as well as an urn that he knows to contain 100 black and white balls in unknown ratio: there may be from 0 to 100 black balls.[42] To bet on heads/tails means that the decision maker wins $100 if the tossed coin lands on heads/tails (and nothing otherwise); similarly, to bet on black/white means that the decision maker wins $100 if a ball drawn from the urn is black/white (and nothing otherwise).

[42] A fair coin here is just a random device generating two outcomes with the same 1/2 chance. The original paper of Ellsberg (1961) models it through another urn that the decision maker knows to contain 50 white balls and 50 black balls.

Ellsberg's (1961) thought experiment suggests, and a number of behavioral experiments confirm, that many decision makers are indifferent between betting on either heads or tails and are also indifferent between betting on either black or white, but they strictly prefer to bet on the coin rather than on the urn. We can represent this preference pattern as

$$\text{bet on white} \sim \text{bet on black} \prec \text{bet on heads} \sim \text{bet on tails} \quad [17.11]$$

To understand the pattern, notice that the decision maker has a much better information to assess the likelihood of the outcome of the coin toss than of the urn draw. In fact, the likelihood of the coin toss is quantified by chance, it is an objective (nonepistemic) probability, here equal to 1/2; in contrast, the decision maker has to assess the likelihood of the urn draw by a subjective (epistemic) probability, based on essentially no information since the urn's composition is unknown.

The corresponding canonical decision problem under uncertainty is

	B	W
a	\$0	\$100
b	\$100	\$0
c	\$100	\$100
d	\$0	\$0

[17.12]

where the pure actions are:

- $a =$ bet on white,
- $b =$ bet on black,
- $c =$ receive \$100 irrespective of the color,
- $d =$ receive \$0 irrespective of the color.

The bet on heads (or tails) is thus represented by the mixed action $\frac{1}{2}\delta_c + \frac{1}{2}\delta_d$, and so [17.11] implies

$$a \sim b \prec \frac{1}{2}\delta_c + \frac{1}{2}\delta_d. \quad [17.13]$$

Consider now the following gamble: toss the coin, then bet on white if the coin lands on heads and bet on black otherwise. Formally, this gamble is represented by the mixed action $\frac{1}{2}\delta_a + \frac{1}{2}\delta_b$, which is easily seen to be realization equivalent to $\frac{1}{2}\delta_c + \frac{1}{2}\delta_d$.[43] By chance consequentialism, [17.13] therefore implies

$$a \sim b \prec \frac{1}{2}\delta_a + \frac{1}{2}\delta_b. \quad [17.14]$$

which violates [17.7], and so subjective expected utility maximization.

[43] Intuitively, if white is, for example, the color of the ball drawn from the urn, the probability of winning by choosing this gamble is the chance that the coin toss assigns to bet on white, and this happens with chance 1/2 since the coin is fair.

It is important to observe that such violation, and the resulting preference for randomization, is normatively compelling: randomization eliminates the dependence of the probability of winning on the unknown composition of the urn, and makes this probability a chance, thus hedging uncertainty. As observed at the beginning of Section 17.5.2, the mixed action $\frac{1}{2}\delta_a + \frac{1}{2}\delta_b$ guarantees an expected value of \$50 regardless of which state obtains, while the minimum guaranteed (expected value) by both pure actions is \$0 (although the maximum is \$100). Preference pattern [17.14] is not the result of a bounded rationality issue: the Ellsberg paradox points out a normative inadequacy of subjective expected utility, which ignores a hedging motif that decision makers may deem relevant because of the quality of their information. In what follows, we call *ambiguity* the type of information/uncertainty that makes such motif relevant.[44]

17.6.2 Uncertainty aversion

The preference for randomization that emerges from the Ellsberg paradox motivates the following assumption:

Uncertainty aversion. If $\alpha, \beta \in \Delta(A)$, then

$$\alpha \sim \beta \Longrightarrow \frac{1}{2}\alpha + \frac{1}{2}\beta \succsim \alpha.$$

This axiom, due to Schmeidler (1989), captures the idea that randomization (here in its simplest 50-50% form) hedges ambiguity by trading off epistemic uncertainty for chance. Accordingly, decision makers who dislike ambiguity should (weakly) prefer to randomize.

As Theorem 17.8 will show, rational preferences that are uncertainty averse feature a representation $V : \Delta(A) \to \mathbb{R}$ of the form

$$V(\alpha) = \inf_{\pi \in \Delta(S)} R(\alpha, \pi) \tag{17.15}$$

where $R : \Delta(A) \times \Delta(S) \to (-\infty, \infty]$ is a suitable reward function. Denoting by $\mathcal{P}$ the projection on $\Delta(S)$ of the domain of R, a mixed action α^* is optimal in $\mathcal{A} \subseteq \Delta(A)$ if and only if

$$\alpha^* \in \arg\sup_{\alpha \in \mathcal{A}} \inf_{\pi \in \mathcal{P}} R(\alpha, \pi). \tag{17.16}$$

[44] A caveat: in our setup randomization plays a major role and, accordingly, we are considering the Ellsberg paradox from this angle. However, as Gilboa and Marinacci (2013) discuss, the paradox continues to hold even without randomization as it entails a normatively compelling violation of Savage's sure thing principle.

For any fixed π, the reward function $R(\cdot,\pi)$ is an increasing transformation of the expected utility $\mathrm{E}_{\alpha\times\pi}(u\circ\rho)$. Hence, intuitively, if the decision maker knew the probability of the states—that is, if he were dealing with chances—he would maximize expected utility. Insufficient information about the environment, along with the need of taking decisions that perform well under different probabilistic scenarios $\pi\in\Delta(S)$, lead to a robust approach, that is, to min-maximization.

Some examples are:

- *Subjective expected utility*, characterized by

$$R(\alpha,\pi)=\begin{cases}\mathrm{E}_{\alpha\times\pi}(u\circ\rho) & \pi=\bar{\pi}\\ +\infty & \pi\neq\bar{\pi}\end{cases}$$

  for some $\bar{\pi}\in\Delta(S)$ (here $\mathcal{P}=\{\bar{\pi}\}$).
- *Maxmin expected utility* (Gilboa and Schmeidler, 1989), characterized by

$$R(\alpha,\pi)=\begin{cases}\mathrm{E}_{\alpha\times\pi}(u\circ\rho) & \pi\in\mathcal{C}\\ +\infty & \pi\notin\mathcal{C}\end{cases}$$

  where $\mathcal{C}\subseteq\Delta(S)$ is the set of alternative probability distributions considered by the decision maker (here $\mathcal{P}=\mathcal{C}$).
- *Variational preferences* (Maccheroni et al., 2006a), characterized by

$$R(\alpha,\pi)=\mathrm{E}_{\alpha\times\pi}(u\circ\rho)+c(\pi)\quad\forall\pi\in\Delta(S).$$

  where $c:\Delta(S)\to[0,\infty]$ is a cost function penalizing the alternative probability distributions (here $\mathcal{P}=\mathrm{dom}\,c$).

17.6.3 Uncertainty averse representations

In this section, we first characterize decision makers that, in a canonical decision problem (A,S,Z,ρ), obey Schmeidler's uncertainty aversion axiom and are expected utility maximizers on lotteries (that is, with respect to chance). We will then show how some well known classes of preferences can be derived by adding suitable independence assumptions. To state the first representation, due to Cerreia-Vioglio et al. (2011b), we need a final definition and a simplifying assumption.

A function $G:\mathbb{R}\times\Delta(S)\to(-\infty,\infty]$ is an *uncertainty aversion index* if and only if:

1. G is quasiconvex and lower semicontinuous,
2. G increasing in the first component,
3. $\inf_{\pi\in\Delta(S)}G(t,\pi)=t$ for all $t\in\mathbb{R}$,
4. for each $(\varkappa,\pi)\in\mathbb{R}^S\times\Delta(S)$ and $c\in\mathbb{R}$ such that $G(\varkappa\cdot\pi,\pi)<c$, there exist $\pi'\in\Delta(S)$ and $\delta>0$ such that $G(\varkappa\cdot\pi'+\delta,\pi')<c$.[45]

[45] Notice that the stronger assumption obtained by requiring $\pi'=\pi$ is upper semicontinuity of $G(\cdot,\pi)$ in the first component. See Kreps (2012, p. 274) for a similar condition on indirect utility functions.

The next unboundedness assumption simplifies some of the representations of this section. It implies that the utility function u is unbounded above and below, like the logarithm on $Z = \mathbb{R}^{++}$ or the "hypothetical value function" of Kahneman and Tversky on $Z = \mathbb{R}$.

Unboundedness. For every $\alpha \succ \beta$ in $\Delta_\ell(A)$, there are $\gamma, \gamma' \in \Delta_\ell(A)$ such that

$$\frac{1}{2}\gamma + \frac{1}{2}\beta \succsim \alpha \succ \beta \succsim \frac{1}{2}\alpha + \frac{1}{2}\gamma'.$$

We can now state the representation theorem.[46]

Theorem 17.8. *Let (A, S, Z, ρ) be a c.d.p. and $\succsim$ be a binary relation $\succsim$ on $\Delta(A)$. The following conditions are equivalent:*

1. *$\succsim$ is a nontrivial, complete, continuous, and unbounded rational preference that satisfies uncertainty aversion on $\Delta(A)$ and independence on $\Delta_\ell(A)$;*
2. *there exist a utility function $u : Z \to \mathbb{R}$ unbounded (above and below) and an uncertainty index $G : \mathbb{R} \times \Delta(S) \to (-\infty, \infty]$ such that the preference functional*

$$V(\alpha) = \min_{\pi \in \Delta(S)} G\left(\mathrm{E}_{\alpha \times \pi}(u \circ \rho), \pi\right) \qquad [17.17]$$

represents $\succsim$ on $\Delta(A)$.

The function u is cardinally unique and, given u, G is unique.

By setting

$$R(\alpha, \pi) = G\left(\mathrm{E}_{\alpha \times \pi}(u \circ \rho), \pi\right)$$

in [17.17] we return to [17.15]. Notice that the normalization $\inf_{\pi \in \Delta(S)} G(t, \pi) = t$ ensures that

$$V(\alpha) = \mathrm{E}_\zeta(u)$$

whenever $\rho_\alpha(s) = \zeta$ for all $s \in S$, that is, whenever there is no uncertainty on the random consequence generated by the mixed action α. In particular, the preference functional [17.17] reduces to von Neumann-Morgenstern expected utility on $\Delta_\ell(A)$.

If we strengthen independence in the first point of Theorem 17.8, we get specific forms of G in the second. First, as already observed, the independence axiom on $\Delta(A)$ rather than $\Delta_\ell(A)$, corresponds to subjective expected utility, that is,

[46] Cerreia-Vioglio et al. (2011b, p. 1283) prove a more general result without any unboundedness assumption. More important, they determine the form of G in terms of behavioral observables.

$$G(t,\pi)=\begin{cases} t & \text{if } \pi=\bar{\pi} \\ +\infty & \text{if } \pi\neq\bar{\pi} \end{cases} \quad \text{and} \quad V(\alpha)=\mathrm{E}_{\alpha\times\bar{\pi}}(u\circ\rho) \quad \forall\alpha\in\Delta(A).$$

Consider the following weakening of such axiom, due to Gilboa and Schmeidler (1989), which only requires independence with respect to lottery actions (which are not affected by state uncertainty).

C-independence. If $\alpha,\beta\in\Delta(A)$, $\gamma\in\Delta_\ell(A)$, and $p\in(0,1]$, then

$$\alpha\succsim\beta \Longleftrightarrow p\alpha+(1-p)\gamma\succsim p\beta+(1-p)\gamma.$$

This axiom corresponds to the seminal maxmin expected utility model of Gilboa and Schmeidler (1989):

$$G(t,\pi)=\begin{cases} t & \pi\in C \\ +\infty & \pi\notin C \end{cases} \quad \text{and} \quad V(\alpha)=\min_{\pi\in C}\mathrm{E}_{\alpha\times\pi}(u\circ\rho) \quad \forall\alpha\in\Delta(A) \tag{17.18}$$

where C is a compact and convex subset of $\Delta(S)$.

The following axiom, due to Maccheroni et al. (2006a), further weakens the independence axiom by requiring independence with respect to lottery actions, but only when mixing weights are kept constant.

Weak C-independence. If $\alpha,\beta\in\Delta(A)$, $\gamma,\gamma'\in\Delta_\ell(A)$, and $p\in(0,1]$, then

$$p\alpha+(1-p)\gamma\succsim p\beta+(1-p)\gamma \Longrightarrow p\alpha+(1-p)\gamma'\succsim p\beta+(1-p)\gamma'.$$

This axiom corresponds to variational preferences

$$G(t,\pi)=t+c(\pi) \quad \text{and} \quad V(\alpha)=\min_{\pi\in\Delta(S)}\{\mathrm{E}_{\alpha\times\pi}(u\circ\rho)+c(\pi)\} \quad \forall\alpha\in\Delta(A) \tag{17.19}$$

where $c:\Delta(S)\to[0,\infty]$ is a lower semicontinuous and convex function.

Strzalecki (2011) has characterized the important special case of variational preferences in which $c(\pi)$ is proportional to the *relative entropy*[47] $H(\pi||\bar{\pi})$ of π with respect to a reference probability $\bar{\pi}$. Since the mid 90s, the works of Hansen and Sargent pioneered the use of the associated preference functional

$$V(\alpha)=\min_{\pi\in\Delta(S)}\{\mathrm{E}_{\alpha\times\pi}(u\circ\rho)+\theta H(\pi||\bar{\pi})\} \quad \forall\alpha\in\Delta(A)$$

in the macro-finance literature in order to capture model uncertainty (see, e.g., Hansen and Sargent, 2001, 2008).

[47] Also known as the Kullback-Leibler divergence.

Chateauneuf and Faro (2009) require C-independence to hold only when γ is the worst lottery action (that they assume to exist). Under some mild technical conditions, they obtain

$$G(t,\pi)=\frac{t}{r(\pi)} \quad \text{and} \quad V(\alpha)=\min_{\pi\in\Delta(S)}\frac{1}{r(\pi)}\mathrm{E}_{\alpha\times\pi}(u\circ\rho) \quad \forall\alpha\in\Delta(A) \quad [17.20]$$

where $r:\Delta(S)\to[0,1]$ is upper semicontinuous and quasiconcave, and u vanishes at the worst outcome.

17.7. BEYOND UNCERTAINTY AVERSION

The uncertainty averse models that we just reviewed originate in Schmeidler (1989). For our setup is especially relevant his insight about preference for randomization as an hedge toward ambiguity. We now present two different approaches toward ambiguity, due to Bewley (1986) and Klibanoff et al. (2005), that do not rely on this assumption.

17.7.1 Incompleteness

Completeness of preferences requires that decision makers have enough information to rank all mixed actions. Under uncertainty, however, decision makers might well lack this information so that they might be unable to compare some pairs of alternatives.

This motivated Truman Bewley to present in 1986 a model of choice under uncertainty that restricts completeness to the set of lotteries alone; these alternatives induce a chance distribution that do not depend on states and so are not affected by epistemic uncertainty.

Gilboa et al. (2010) established the following form of Bewley's representation theorem.[48]

Theorem 17.9. *Let (A,S,Z,ρ) be a c.d.p. and $\succsim$ be a binary relation $\succsim$ on $\Delta(A)$. The following conditions are equivalent:*

1. *$\succsim$ is a nontrivial and continuous rational preference that is complete on $\Delta_\ell(A)$ and satisfies strong independence on $\Delta(A)$;*[49]
2. *there exists a nonconstant function $u:Z\to\mathbb{R}$ and a convex and compact set $\mathcal{S}\subseteq\Delta(S)$ of probability distributions such that, for all $\alpha,\beta\in\Delta(A)$,*

$$\alpha\succsim\beta\Longleftrightarrow\mathrm{E}_{\alpha\times\pi}(u\circ\rho)\geq\mathrm{E}_{\beta\times\pi}(u\circ\rho) \qquad \forall\pi\in\mathcal{C}. \qquad [17.21]$$

The function u is cardinally unique and $\mathcal{S}$ is unique.

[48] Relative to the original version, this version of the theorem uses a weak preference $\succsim$ rather than a strict one $\succ$ and remains true also when S is infinite.

[49] Because of incompleteness, the independence assumption have to be strengthened to $\alpha\succsim\beta\Leftrightarrow p\alpha+(1-p)\gamma\succsim p\beta+(1-p)\gamma$ for all $p\in[0,1]$ and $\gamma\in\Delta(A)$.

The set $\mathcal{C}$ of probability distributions is interpreted as the collection of subjective probabilities that are consistent with the decision maker's information, that is, the *consistent beliefs* that he entertains, or his subjective perception of ambiguity. Two actions are comparable if and only if they are unanimously ranked by all such scenarios. Any lack of sufficient information translates into a disagreement among them.

Because of incompleteness, there is no preference functional V that represents these preferences. This somehow limits the applicability of representation [17.21].[50] However, the study of these incomplete preferences turns out to be insightful even for the complete models, as shown by Ghirardato et al. (2004) and Gilboa et al. (2010). The former paper shows how some insights from Bewley's representation allow to go beyond uncertainty aversion also in the case of complete preferences, as discussed in Section 17.8.

Gilboa et al. (2010) noted that, although decisions eventually have to be made, in many decision problems preferences that can be intersubjectively justified are incomplete. For this reason they proposed a model based on two preference relations, one incomplete a la Bewley and one complete a la Gilboa and Schmeidler (1989). They showed that under certain axioms, stated on each preference relation separately as well as relating the two, the two preferences, they are jointly represented by the same set of probabilities $\mathcal{C}$: one via the unanimity rule [17.21], and the other via the maxmin rule [17.18]. This shows how maxmin expected utility behavior may result from incomplete preferences.

Further results on incompleteness in a decision problem under uncertainty appear in Ghirardato et al. (2003), Ok et al. (2012), and Galaabaatar and Karni (2013).

17.7.2 Smooth ambiguity model

The models presented in the previous section of the form

$$V(\alpha) = \inf_{\pi \in \Delta(S)} R(\alpha, \pi) \qquad [17.22]$$

are often viewed as rather extreme, almost paranoid, since they evaluate α through the worst values in the range of rewards $\{R(\alpha, \pi) \,|\, \pi \in \Delta(S)\}$. Indeed, [17.22] reminds Wald's (1950) interpretation of a decision problem under uncertainty as a zero-sum game against (a malevolent) nature.[51]

However, the reward function $R(\alpha, \pi)$ incorporates both the attitude toward ambiguity, a taste component, and its perception, an information component. To fix ideas, consider a maxmin expected utility preference that features a set $\mathcal{C}$ of subjective probabilities. A smaller set $\mathcal{C}$ may reflect both better information—that is, a lower perception of ambiguity—and/or a less averse uncertainty attitude. In other words, the size of $\mathcal{C}$ does not reflect just information, but taste as well.[52]

[50] But see Rigotti and Shannon (2005) for an economic application of Bewley's model.

[51] But remember that, as observed, subjective expected utility is a special case of this representation.

[52] See Gajdos et al. (2008) for a model that explicitly relates $\mathcal{C}$ with some underlying objective information.

The criterion [17.22] is thus not extreme; indeed, the axioms underlying it (that we have seen above) are not extreme at all.[53] That said, many economic applications that use such criterion regard the set $\mathcal{C}$—and more generally the projection on $\Delta(S)$ of the domain of R—as being essentially determined by information, that is, as the set of probabilities that are consistent with objectively available information. As such, the elements of $\mathcal{C}$ are basically viewed as chance distributions, typically data generating processes in the macro-finance applications. For example, in the Ellsberg paradox the set $\mathcal{C}$ for the unknown urn consists of all chance distributions determined by the possible compositions of the urn (hence $\mathcal{C} = \Delta(S)$ when there is no information about such compositions).

Under this "chance" interpretation of $\mathcal{C}$, criterion [17.22] is quite extreme and loses some of its appeal. Klibanoff et al. (2005) proposed a different model that tries to address this issue. They assume that decision makers have a subjective probability μ over the (Borel) set $\mathcal{C} \subseteq \Delta(S)$ of all possible chance distributions. Via reduction, the (prior) probability μ determines a (predictive) probability $\bar{\pi} = \int_{\Delta(S)} \pi \, d\mu(\pi)$ on S. The subjective expected utility criterion can be rewritten as:

$$\int_{\Delta(S)} \left(\mathrm{E}_{\alpha\times\pi}(u\circ\rho)\right) d\mu(\pi) = \mathrm{E}_{\alpha\times\bar{\pi}}(u\circ\rho). \qquad [17.23]$$

Decision makers, however, may well dislike the uncertainty about the expected payoff $\mathrm{E}_{\alpha\times\pi}(u\circ\rho)$, while criterion [17.23] just averages out these payoffs, tacitly assuming a neutral attitude toward this kind of uncertainty. If, in contrast, ambiguity aversion (or propension) prevails (as one expects to be usually the case), a nonlinear function $\phi : \mathrm{co}\,u(Z) \to \mathbb{R}$ should be introduced to capture it.[54] Criterion [17.23] then generalizes to

$$V(\alpha) = \int_{\Delta(S)} \phi\left(\mathrm{E}_{\alpha\times\pi}(u\circ\rho)\right) d\mu(\pi). \qquad [17.24]$$

Note that we have a separation between the uncertainty perception, an information feature modeled by μ and its support $\mathcal{C}$, and the uncertainty attitude, a taste trait modeled by ϕ and its shape. In particular, the concavity of ϕ corresponds to a ambiguity aversion.

Criterion [17.24], due to Klibanoff et al. (2005), has become to be known as the *smooth model* of ambiguity because, under mild assumptions, V is a smooth functional (whereas the maxmin expected utility functionals are typically not everywhere differentiable). Related models have been proposed by Segal (1987), Neilson (2010), Nau (2006), Ergin and Gul (2009), and Seo (2009).

[53] Notice that if $R(\alpha,\pi) = +\infty$ for all α then π is never considered in minimization [17.22].

[54] See also Section 17.8.2. Here $\mathrm{co}\,u(Z)$ denotes the smallest interval that contains $u(Z)$.

A main feature of the smooth model is the ability to take advantage of conventional risk theory in studying ambiguity attitudes, something that makes it especially tractable.[55] Moreover, if the concavity of ϕ becomes stronger and stronger, reflecting greater and greater uncertainty aversion, it can be shown that criterion [17.24] tends to the maxmin expected utility criterion

$$\inf_{\pi \in \operatorname{supp} \mu} \mathrm{E}_{\alpha \times \pi} (u \circ \rho)$$

which can thus be regarded as a limit case of the smooth model.

17.8. ALTERNATIVE APPROACHES

The models presented in Section 17.7 depart from uncertainty aversion in a radical way: removing completeness or taking a neo-Bayesian "prior-over-models" approach. It is natural to wonder whether the uncertainty averse models of Section 17.6.3 can be just generalized beyond uncertainty aversion thus encompassing more general attitudes.

Ghirardato et al. (2004),[56] investigated this issue by considering a complete, continuous, and nontrivial rational preference $\succsim$ on $\Delta(A)$ that satisfies the independence axiom on $\Delta_\ell(A)$, and its subrelation

$$\alpha \succsim^* \beta \Longleftrightarrow p\alpha + (1-p)\gamma \succsim p\beta + (1-p)\gamma \quad \forall p \in [0,1] \text{ and } \gamma \in \Delta(A).$$

We have $\alpha \succsim^* \beta$ when the decision maker does not find any possibility of hedging against or speculating on the ambiguity that the comparison of the two mixed actions involves. If so, Ghirardato et al. (2004) argue that such ambiguity does not affect the preference among them; for this reason they call $\succsim^*$ the "unambiguous preference."

Because of ambiguity, $\succsim^*$ is in general not complete. It is rational and satisfies the (strong) independence axiom, so that it admits a unanimity representation [17.21]. An incomplete preference a la Bewley thus emerges in studying very general preferences, with an associated utility u and consistent beliefs given by a "Bewley set" $\mathcal{C}$ of probability distributions.[57] The consistency between beliefs and behavior is confirmed by the possibility of representing the original preference $\succsim$ by a functional

$$V(\alpha) = p(\alpha) \min_{\pi \in \mathcal{C}} \mathrm{E}_{\alpha \times \pi} (u \circ \rho) + (1 - p(\alpha)) \max_{\pi \in \mathcal{C}} \mathrm{E}_{\alpha \times \pi} (u \circ \rho)$$

[55] In this regard, Maccheroni et al. (2013) established a quadratic approximation of criterion [17.24] that extends the classic expected utility results of de Finetti (1952), Pratt (1964), and Arrow (1965). On these classic results see Montesano (2009).

[56] Recently extended by Cerreia-Vioglio et al. (2011a) and Ghirardato and Siniscalchi (2012).

[57] Notably for maxmin expected utility preferences $\mathcal{C}$ is exactly the set over which the minimum is taken. We also remark that, in the original setting Ghirardato et al. (2004) assume C-independence of $\succsim$, but Cerreia-Vioglio et al. (2011a,b) show how this assumption is redundant for this part of the analysis.

where the maximum and the minimum expected utilities of α are weighted in the evaluation. This representation provides a separation between the revealed perception of ambiguity embodied by the set $\mathcal{C}$, and the decision maker reaction to it when evaluating α captured by $p(\alpha)$.

See Amarante (2009) and Eichberger et al. (2011) for further insights on this approach.

17.8.1 Choquet expected utility

The first axiomatic model of choice under ambiguity appears in the seminal paper of Schmeidler (1989) on Choquet expected utility.[58] The implied preferences are a special subclass of those considered by Ghirardato et al. (2004). The crucial axiom here is the comonotonic independence axiom, that is a stronger version of C-independence.

Extending the definition of comonotonic random variables (given in the first part of this chapter), say that mixed actions $\alpha, \beta \in \Delta(A)$ are *comonotonic* if it is never the case that both $\rho(\alpha, s) \succ \rho(\alpha, s')$ and $\rho(\beta, s) \prec \rho(\beta, s')$ for some states s and s'. As in the case of random variables, the convex combination of comonotonic mixed actions cannot provide any hedging.

Comonotonic independence. If $\alpha, \beta, \gamma \in \Delta(A)$ are pairwise comonotonic and $0 < p < 1$, then $\alpha \succ \beta$ implies $p\alpha + (1-p)\gamma \succ p\beta + (1-p)\gamma$.

Schmeidler (1989) proved a representation theorem for rational preferences that satisfy this axiom. Call *capacity* (or *nonadditive probability*) a set function $\nu : 2^S \to [0,1]$ such that

1. $\nu(\emptyset) = 0$ and $\nu(S) = 1$;
2. $E \subseteq E'$ implies $\nu(E) \leq \nu(E')$.

Expectations E_ν with respect to ν can be computed via the Choquet integral, named after Choquet (1953) and they coincide with standard expectations when ν is additive (that is a probability distribution).[59]

We can now state Schmeidler's famous result.

Theorem 17.10. *Let (A, S, Z, ρ) be a c.d.p. and $\succsim$ be a binary relation $\succsim$ on $\Delta(A)$. The following conditions are equivalent:*

1. *$\succsim$ is a nontrivial, complete, and continuous rational preference that satisfies comonotonic independence;*
2. *there exists a nonconstant $u : Z \to \mathbb{R}$ and a capacity $\nu : 2^S \to [0,1]$ such that the preference functional*

[58] For axiomatizations without randomization see Gilboa (1987) and Chew and Karni (1994).

[59] We refer to Gilboa and Marinacci (2011) for an introduction to Choquet integration, and to Schmeidler (1986), Denneberg (1994), and Marinacci and Montrucchio (2004) for more detailed studies. Ryan (2009) presents a simple pedagogical proof in the finite state case of the Schmeidler Theorem (and of other ambiguity representations).

$$V(\alpha) = \mathrm{E}_\nu [\mathrm{E}_\alpha (u \circ \rho)]$$

represents $\succsim$ *on* $\Delta(A)$.

The function u is cardinally unique and the capacity ν is unique.

In particular, in the first point, the full-fledged independence axiom holds if and only if ν is additive. In this case, we return to subjective expected utility. More interestingly, Schmeidler (1989) shows that in the previous theorem the preference $\succsim$ satisfies uncertainty aversion if and only if the capacity ν is supermodular

$$\nu(E \cup E') + \nu(E \cap E') \geq \nu(E) + \nu(E') \quad \forall E, E' \subseteq S.$$

In this case,

$$\mathrm{E}_\nu [\mathrm{E}_\alpha (u \circ \rho)] = \min_{\pi \in \mathrm{core}(\nu)} \mathrm{E}_{\alpha \times \pi} (u \circ \rho)$$

where core (ν), called the *core* of ν, is the set of probability distributions determined by ν as follows[60]:

$$\mathrm{core}(\nu) = \{\pi \in \Delta(S) : \pi(E) \geq \nu(E) \text{ for all } E \subseteq S\}.$$

Uncertainty aversion thus characterizes the overlap among Choquet expected utility and maxmin expected utility, and in this case the set of consistent beliefs (representing $\succsim^*$) is $C = \mathrm{core}(\nu)$.

Noticeably, in the special case in which $Z \subseteq \mathbb{R}^+$, u is the identity, and $\nu = f \circ \bar{\pi}$ for some $\bar{\pi} \in \Delta(S)$ and some increasing $f : [0,1] \to [0,1]$, then

$$V(\delta_a) = \int_0^\infty f\left(1 - \bar{\pi}\left(s \in S : \rho_a(s) \leq x\right)\right) dx \quad \forall a \in A$$

thus obtaining Yaari's (1987) dual theory representation. Another tractable representation, due to Chateauneuf et al. (2007), is the so called neo-additive Choquet expected utility model of the "Hurwicz" form

$$V(\alpha) = (1 - p - q)\, \mathrm{E}_{\bar{\pi}} [\mathrm{E}_\alpha (u \circ \rho)] + p \min_{s \in S} \mathrm{E}_\alpha (u \circ \rho_s) + q \max_{s \in S} \mathrm{E}_\alpha (u \circ \rho_s).$$

Through the values of the weights p and q, the preference functional V captures in a simple way different degrees of pessimism and optimism.

[60] Mathematically, capacities are a special class of transferable utility games (see Marinacci and Montrucchio, 2004, for some details on this observation). The notion of core here is adapted from the the game theoretic one.

17.8.2 Uncertainty aversion revisited

Schmeidler's uncertainty aversion axiom models ambiguity aversion through preference for randomization. As such, it crucially relies on the presence of randomization. But, ambiguity is a feature of an information state that may occur whether or not randomization is available. This led Epstein (1999) and Ghirardato and Marinacci (2002) to address ambiguity/uncertainty aversion from a different angle, inspired by the comparative analysis of risk attitudes of Yaari (1969).

Although our setup features randomization, their works still shed light on some noteworthy aspects of ambiguity aversion. Specifically, the approach of Ghirardato and Marinacci (2002), which we present here because of its sharper model implications, relies on two basic ingredients:

1. A comparative notion of ambiguity aversion that, given any two preferences $\succsim_1$ and $\succsim_2$ on $\Delta(A)$, says when $\succsim_1$ is more ambiguity averse than $\succsim_2$.
2. A benchmark for neutrality to ambiguity; that is, a class of preferences $\succsim$ on $\Delta(A)$ that are viewed as neutral to ambiguity.

In turn, the choice of these ingredients determines the absolute notion of ambiguity aversion: a preference $\succsim$ is classified as ambiguity averse provided that it is more ambiguity averse than an ambiguity neutral one.

The first ingredient, the comparative notion, is based on the comparison of mixed actions with lottery ones. Specifically, we say that $\succsim_1$ is *more ambiguity averse than* $\succsim_2$ if, for all $\alpha \in \Delta(A)$ and all $\gamma \in \Delta_\ell(A)$,

$$\alpha \succsim_1 \gamma \Longrightarrow \alpha \succsim_2 \gamma.$$

Lottery actions are unaffected by state uncertainty and so are definitely unambiguous alternatives (whatever meaning we attach to such adjective). Any other mixed action may only embody equal or more ambiguity than them. The definition above ranks ambiguity attitudes by calling decision maker 2 "less ambiguity averse" than decision maker 1 if every time 1 prefers an ambiguous alternative to an unambiguous one so does 2.

As to the second ingredient, the neutrality benchmark, consider as neutral to ambiguity all subjective expected utility preferences: it is a minimal assumption since, although more preferences can be possibly regarded as such, there is no doubt that subjective expected utility preferences are neutral to ambiguity. Fortunately, it turns out that such minimal assumption is enough to deliver some nontrivial implications.

Finally, the two ingredients lead to an absolute notion of ambiguity aversion: we say that $\succsim$ is *ambiguity averse* if it is more ambiguity averse than a subjective expected utility preference.

We can now state the main result of Ghirardato and Marinacci (2002).

Theorem 17.11. *Given any two maxmin expected utility preferences $\succsim_1$ and $\succsim_2$ on $\Delta(A)$, the following conditions are equivalent:*
1. *$\succsim_1$ is more ambiguity averse than $\succsim_2$,*
2. *u_1 is an affine strictly increasing transformation of u_2 and $\mathcal{C}_1 \supseteq \mathcal{C}_2$.*

Therefore, more ambiguity averse maxmin expected utility preferences are characterized by larger sets of consistent beliefs.

As a result, the size of the set $\mathcal{C}$ can be interpreted as a *measure of ambiguity aversion*. This provides a behavioral foundation for the comparative statics exercises in ambiguity based on the comparison of sets of priors that most economic applications of the maximin expected utility model feature. In fact, these applications are typically interested in how changes in ambiguity attitudes affect the relevant economic variables.

Inter alia, Theorem 17.11 implies that maxmin expected utility preferences are ambiguity averse, something not at all surprising since they satisfy the uncertainty aversion axiom.

Later works have extended Theorem 17.11 to other classes of preferences, also outside the domain of choice under uncertainty. In particular, for uncertainty averse preferences the condition $\mathcal{C}_1 \supseteq \mathcal{C}_2$ takes the more general form $G_1 \leq G_2$ (once $u_1 = u_2$) and for variational preferences this means $c_1 \leq c_2$. The functions G and c can thus be viewed as indices of ambiguity aversion that generalize the set $\mathcal{C}$.

17.8.3 Other models

Other important models have appeared in the literature. Segal (1987, 1990) suggested a risk-based approach to uncertainty, based on the idea that decision makers may fail to properly reduce compound lotteries. Halevy (2007) provided some experimental evidence on the link between lack of reduction of compound lotteries and ambiguity. Seo (2009) developed a full-fledged theoretical analysis of this issue. Note that in general the failure to reduce compound lotteries is regarded as a mistake; as a result, this approach to "ambiguity" has a bounded rationality flavor, while our main focus is normative (modeling rational behavior in the presence of limited information).

Stinchcombe (2003), Olszewski (2007), and Ahn (2008) model ambiguity through sets of lotteries, capturing exogenous or objective ambiguity (see also Jaffray, 1988, for related ideas).

Gajdos et al. (2008) axiomatize a model with objective information by defining preferences over pairs of acts and sets of probabilities (that represent objective information). They derive a map that connects such sets with the subjective ones that the maxmin expected utility model features.

Gul and Pesendorfer (2008) suggested *subjective expected uncertain utility theory*, according to which acts can be reduced to bilotteries, each specifying probabilities for ranges of outcome values, where these probabilities need not be allocated to sub-ranges.

Siniscalchi (2009) axiomatizes *vector expected utility*, in which acts are assessed according to an additively separable criterion whose first term is a baseline expected utility evaluation and the second term is an adjustment that reflects decision makers' perception of ambiguity and their attitudes toward it. A related approach appears in Grant and Polak (2013).

17.9. FINAL REMARKS

Another important issue in choice under uncertainty is dynamic choice. We refer the interested reader to Jaffray (1994), Epstein and Schneider (2003), Maccheroni et al. (2006), Hanany and Klibanoff (2009), Klibanoff et al. (2009), and Siniscalchi (2011).

Finally, several papers analyze notions of equilibrium in games where agents have non-neutral attitudes toward uncertainty. In situations of strategic interaction, intrapersonal consistency has to be augmented by some notion of inter-personal consistency. Dow and Werlang (1994), Klibanoff (1996), Lo (1996), Eichberger and Kelsey (2000), Marinacci (2000), and more recently Bade (2011) and Riedel and Sass (2011) extend the Nash equilibrium concept by imposing different notions of "correctness" of beliefs about other agents' strategies. Lehrer (2012) and Battigalli et al. (2014) only require that each agent's beliefs about others are "confirmed" by his imperfect information feedback. Of course, it is impossible to summarize their results here, but—in a nutshell—the general effect of ambiguity is to enlarge the set of equilibria because of the multiplicity of (intra- and inter-personal) consistent beliefs.

ACKNOWLEDGMENTS

We gratefully thank Pierpaolo Battigalli, Veronica R. Cappelli, Simone Cerreia-Vioglio, and Sujoy Mukerji for very helpful suggestions.

REFERENCES

Ahn, D., 2008. Ambiguity without a state space. Rev. Econ. Stud. 75, 3–28.
Amarante, M., 2009. Foundations of Neo-Bayesian statistics. J. Econ. Theory 144, 2146–2173.
Allais, M., 1953. Le Compotement de l'Homme Rationnel Devant le Risk: Critique des Postulates et Axiomes de l'Ecole Americaine. Econometrica 21, 503–546.
Anscombe, F.J., Aumann, R.J., 1963. A definition of subjective probability. Ann. Math. Stat. 34, 199–205.
Arrow, K., 1965. Aspects of the Theory of Risk-bearing, Yrjo Jahnsson Foundation: Helsinki.
Bade, S., 2011. Ambiguous act equilibria. Games Econ. Behav. 71, 246–260.
Battigalli, P., Cerreia-Vioglio, S., Maccheroni, F., Marinacci, M., 2013. Mixed extensions of decision problems under uncertainty, The American Economic Review, forthcoming.
Battigalli, P., Cerreia-Vioglio, S., Maccheroni, F., Marinacci, M., 2014. Selfconfirming equilibrium and model uncertainty. The American Economic Review, forthcoming.
Bell, D., 1982. Regret in decision making under uncertainty. Oper. Res. 30, 961–981.
Bell, D., 1985. Disappointment in decision making under uncertainty. Oper. Res. 33, 1–27.
Bernoulli, D., 1738. Specimen Theoriae Novae de Mensura Sortis. Commentarii Academiae Scientiatatum Imperalas Petropolitanae 5, 175–192. (Translated in Econometrica 22, 23–26, 1954).

Bewley, T., 1986. Knightian Decision Theory: Part I, Cowles Foundation DP 807. (Published in Decision in Economics and Finance 25, 79–110, 2002).

Bikhchandani, S., Segal, U., 2011. Transitive regret. Theor. Econ. 6, 95–108.

Cerreia-Vioglio, S., Ghirardato, P., Maccheroni, F., Marinacci, M., Siniscalchi, M., 2011a. Rational preferences under ambiguity. Econ. Theory 48, 341–375.

Cerreia-Vioglio, S., Maccheroni, F., Marinacci, M., Montrucchio, L., 2011b. Uncertainty averse preferences. J. Econ. Theory 146, 1275–1330.

Chateauneuf, A., Faro, J.H., 2009. Ambiguity through confidence functions. J. Math. Econ. 45, 535–558.

Chateauneuf, A., Wakker, P.P., 1999. An axiomatization of cumulative prospect theory for decision under risk. J. Risk Uncertainty 18, 137–145.

Chateauneuf, A., Eichberger, J., Grant, S., 2007. Choice under uncertainty with the best and worst in mind: neo-additive capacities. J. Econ. Theory 137, 538–567.

Chew, S.H., 1983. A generalization of the quasilinear mean with applications to the measurement of income inequality and decision theory resolving the Allais paradox. Econometrica 51, 1065–1092.

Chew, S.H., 1989a. Axiomatic utility theories with the betweenness property. Ann. Oper. Res. 19, 273–298.

Chew, S.H., 1989b. An axiomatic generalization of the quasilinear mean and Gini mean with applications to decision theory, mimeo.

Chew, S.H., Karni, E., 1994. Choquet expected utility with a finite state space: commutativity and act-independence. J. Econ. Theory 62, 469–479.

Chew, S.H., MacCrimmon, K.R., 1979. Alpha-nu Choice Theory: an Axiomatization of Expected Utility, University of British Columbia Faculty of Commerce, working paper 669.

Chew, S.H., Karni, E., Safra, Z., 1987. Risk aversion in the theory of expected utility with rank dependent probabilities. J. Econ. Theory 42, 370–381.

Choquet, G., 1953. Theory of capacities. Annales de l'Institut Fourier 5, 131–295.

Crawford, V.P., 1990. Equilibrium without independence. J. Econ. Theory 50, 127–154.

Debreu, G., 1954. Representation of a preference ordering by a numerical function. In: Thrall, R.M., Combs, C.H., Davis, R.C., (Eds.), Decision Processes, John Wiley and Sons: New York.

de Finetti, B., 1931. Sul Significato Soggettivo della Probabilità. Fundam. Math. 17, 298–329.

de Finetti, B., 1937. La Prevision: Ses Lois Logiques, ses Sources Subjectives. Annales de l'Institut Henri Poincare 7, 1–68. (Translated in Kyburg, H.E., Smokler, H.E., 1963 Studies in Subjective Probability, John Wiley and Sons: New York).

de Finetti, B., 1952. Sulla preferibilità. Giornale degli Economisti 11, 685–709.

Dekel, E., 1986. An axiomatic characterization of preferences under uncertainty: weakening the independence axiom. J. Econ. Theory 40, 304–318.

Denneberg, D., 1994. Non-additive Measure and Integral, Dordrecht. Kluwer Academic Publishers, Dordrecht.

Diecidue, E., Wakker, P.P., 2001. On the intuition of rank-dependent expected utility. J. Risk Uncertainty 23, 281–298.

Dow, J., Werlang, S.R.D.C., 1994. Nash equilibrium under Knightian uncertainty: breaking down backward induction. J. Econ. Theory 64, 305–324.

Edwards, W., 1954. The theory of decision making. Psychol. Bull. 51, 380–417.

Eichberger, J., Kelsey, D., 2000. Non-additive beliefs and strategic equilibria. Games Econ. Behav. 30, 183–215.

Eichberger, J., Grant, S., Kelsey, D., Koshevoy, G.A., 2011. The α-MEU model: a comment. J. Econ. Theory 146, 1684–1698.

Ellsberg, D., 1961. Risk, ambiguity, and the savage axioms. Q. J. Econ. 75, 643–669.

Epstein, L., 1999. A definition of uncertainty aversion. Rev. Econ. Stud. 66, 579–608.

Epstein, L.G., Schneider, M., 2003. Recursive multiple-priors. J. Econ. Theory 113, 1–31.

Ergin, H., Gul, F., 2009. A theory of subjective compound lotteries. Journal of Economic Theory 144, 899–929.

Fishburn, P.C., 1970. Utility Theory for Decision Making, John Wiley and Sons: New York.

Fishburn, P.C., 1978. On Handa's 'New Theory of Cardinal Utility' and the maximization of expected returns. J. Polit. Econ. 86, 321–324.

Gajdos, T., Hayashi, T., Tallon, J.M., Vergnaud, J.C., 2008. Attitude toward imprecise information. J. Econ. Theory 140, 27–65.

Galaabaatar, T., Karni, E., 2013. Subjective expected utility with incomplete preferences. Econometrica 81, 255–284.

Ghirardato, P., Marinacci, M., 2002. Ambiguity made precise: a comparative foundation. J. Econ. Theory 102, 251–289.

Ghirardato, P., Siniscalchi, M., 2012. Ambiguity in the small and in the large. Econometrica 80, 2827–2847.

Ghirardato, P., Maccheroni, F., Marinacci, M., Siniscalchi, M., 2003. Subjective foundations for objective randomization: a new spin on roulette wheels. Econometrica 71, 1897–1908.

Ghirardato, P., Maccheroni, F., Marinacci, M., 2004. Differentiating ambiguity and ambiguity attitude. J. Econ Theory 118, 133–173.

Gilboa, I., 1987. Expected utility with purely subjective nonadditive probabilities. J. Math. Econ. 16, 65–88.

Gilboa, I., Marinacci, M., 2013. Ambiguity and the Bayesian paradigm. In: Acemoglu, D., Arellano, M., Dekel, E., (Eds.) Advances in Economics and Econometrics: Theory and Applications. Cambridge University Press: Cambridge.

Gilboa, I., Schmeidler, D., 1989. Maxmin expected utility with a non-unique prior. J. Math. Econ. 18, 141–153.

Gilboa, I., Maccheroni, F., Marinacci, M., Schmeidler, D., 2010. Objective and subjective rationality in a multiple prior model. Econometrica 78, 755–770.

Grant, S., Polak, B., 2013. Mean-dispersion preferences and constant absolute uncertainty aversion. J. Econ. Theory 148, 1361–1398.

Gul, F., 1991. A theory of disappointment aversion. Econometrica 59, 667–686.

Gul, F., Pesendorfer, W., 2008. Measurable ambiguity, mimeo.

Halevy, Y., 2007. Ellsberg revisited: an experimental study. Econometrica 75, 503–536.

Hanany, E., Klibanoff, P., 2009. Updating ambiguity averse preferences. B.E. J. Theor. Econ. 9.

Handa, J., 1977. Risk probability and a new theory of cardinal utility. J. Polit. Econ. 85, 97–122.

Hansen, L.P., Sargent, T.J., 2001. Robust control and model uncertainty. Am. Econ. Rev. 91, 60–66.

Hansen, L.P., Sargent, T.J., 2008. Robustness, Princeton University Press: Princeton.

Herstein, I.N., Milnor, J., 1953. An axiomatic approach to measurable utility. Econometrica 21, 291–297.

Hogarth, R.M., 1980. Judgement and Choice. John Wiley and Sons, Chichester.

Jaffray, J.Y., 1988. Choice under risk and the security factor: an axiomatic model. Theory and Decision 24, 169–200.

Jaffray, J.Y., 1994. Dynamic decision making with belief functions. In: Yager, R.R., Kacprzyk, J., Fedrizzi, M., (Eds.) Advances in the Dempster-Shafer Theory of Evidence, John Wiley and Sons, New York, 331–352).

Kahneman, D., Tversky, A., 1979. Prospect theory: an analysis of decision under risk. Econometrica 47, 263–291.

Karmakar, U.S., 1978. Subjectively weighted utility: a descriptive extension of the expected utility model. Organ. Behav. Hum. Perform. 21, 61–72.

Karni, E., 1988. On the equivalence between descending bid auctions and first price sealed bid auctions. Theory Decisions 25, 211–217.

Karni, E., Safra, Z., 1986. Vickrey auctions in the theory of expected utility with rank-dependent probabilities. Econ. Lett. 20, 15–18.

Karni, E., Safra, Z., 1989a. Dynamic consistency, revelations in auctions and the structure of preferences. Rev. Econ. Stud. 56, 421–434.

Karni, E., Safra, Z., 1989b. Ascending-bid auctions with behaviorally consistent bidders. Ann. Oper. Res. 19, 435–446.

Karni, E., Schmeidler, D., 1991a. Atemporal dynamic consistency and expected utility theory. J. Econ. Theory 54, 401–408.

Karni, E., Schmeidler, D., 1991b. Utility theory with uncertainty, In: Handbook of Mathematical Economics, vol. IV, Hildenbrand, W., Sonnenschein, H. (Eds.), North Holland: Amsterdam.

Klibanoff, P., 1996. Uncertainty, decision and normal form games, mimeo.

Klibanoff, P., Marinacci, M., Mukerji, S., 2005. A smooth model of decision making under ambiguity. Econometrica 73, 1849–1892.

Klibanoff, P., Marinacci, M., Mukerji, S., 2009. Recursive smooth ambiguity preferences. J. Econ. Theory 144, 930–976.
Knight, F.H., 1921. Risk, Uncertainty and Profit, University of Chicago Press: Chicago.
Kreps, D., 2012. Microeconomic Foundations I: Choice and Competitive Markets, Princeton University Press, Princeton.
Kreps, D.M., Porteus, E.L., 1978. Temporal resolution of uncertainty and dynamic choice theory. Econometrica 46, 185–200.
Kreps, D.M., Porteus, E.L., 1979. Temporal von Neumann-Morgenstern and induced preferences. J. Econ. Theory 20, 81–109.
Lehrer, E., 2012. Partially specified probabilities: decisions and games. Am. Econ. J. Microecon. 4, 70–100.
Lo, K.C., 1996. Equilibrium in beliefs under uncertainty. J. Econ. Theory 71, 443–484.
Loomes, G., Sugden, R., 1982. Regret theory: an alternative theory of rational choice under uncertainty. Econ. J. 92, 805–824.
Luce, R.D., Raiffa, H., 1957. Games and Decisions, John Wiley and Sons: New York.
Maccheroni, F., 2004. Yaari's dual theory without the completeness axiom. Econ. Theory 23, 701–714.
Maccheroni, F., Marinacci, M., Rustichini, A., 2006a. Ambiguity aversion, robustness, and the variational representation of preferences. Econometrica 74, 1447–1498.
Maccheroni, F., Marinacci, M., Rustichini, A., 2006b. Dynamic variational preferences. J. Econ. Theory 128, 4–44.
Maccheroni, F., Marinacci, M., Ruffino, D., 2013. Alpha as ambiguity: robust mean-variance portfolio analysis. Econometrica 81, 1075–1113.
Machiavelli, N., 1532. Il Principe, Antonio Blado d'Asola: Rome.
Machina, M.J., 1982. 'Expected utility' analysis without the independence axiom. Econometrica 50, 277–323.
Machina, M.J., 1989. Dynamic consistency and non-expected utility models of choice under uncertainty. J. Econ. Lit. 27, 1622–1688.
Machina, M.J., 2001. Payoff kinks in preferences over lotteries. J. Risk Uncertainty 23, 207–260.
Malinvaud, E., 1952. Note on von Neumann-Morgenstern's strong independence axiom. Econometrica 20, 679.
Marinacci, M., 2000. Ambiguous games. Games Econ. Behav. 31, 191–219.
Marinacci, M., Montrucchio, L., 2004. Introduction to the mathematics of ambiguity. In: Gilboa, I., (Ed.) Uncertainty in Economic Theory, Routledge: New York.
Marschak, J., 1950. Rational behavior, uncertain prospects and measurable utility. Econometrica 18, 111–141.
Marschak, J., Radner, R., 1972. Economic Theory of Teams, Yale University Press: New Haven.
Milgrom P.R., Weber, R.J., 1982. A theory of auctions and competitive bidding. Econometrica 50, 1089–1122.
Montesano A., 2009. de Finetti and the Arrow-Pratt measure of risk aversion. In: Gavalotti, M.C., (Ed.) Bruno de Finetti Radical Probabilist. College Publications: London.
Nau, R.F., 2006. Uncertainty aversion with second-order utilities and probabilities. Manage. Sci. 52, 136–145.
Neilson, W., 2010. A simplified axiomatic approach to ambiguity aversion. Journal of Risk and Uncertainty 41, 113–124.
Ok, E., Ortoleva, P., Riella, G., 2012. Incomplete preferences under uncertainty: indecisiveness in beliefs vs. tastes. Econometrica 80, 1791–1808.
Olszewski, W., 2007. Preferences over sets of lotteries. The Review of Economic Studies 74, 567-595.
Pratt, J.W., 1964. Risk aversion in the small and in the large. Econometrica 32, 122–136.
Quiggin, J., 1982. A theory of anticipated utility. J. Econ. Behav. Organ. 3, 323–343.
Quiggin, J., 1993. Generalized Expected Utility Theory–The Rank-Dependent Model, Kluwer Academic Publishers: Dordrecht.
Ramsey, F.P., 1926. Truth and probability. In: Braithwaite, R.B., Plumpton, F., (Ed.) The Foundation of Mathematics and Other Logical Essays. Routledge and Kegan: London.
Riedel, F., Sass, L., 2011. The strategic use of ambiguity, mimeo.
Rigotti, L., Shannon, C., 2005. Uncertainty and risk in financial markets. Econometrica 73, 203–243.

Röell, A., 1987. Risk aversion in Quiggin and Yaari's rank-order model of choice under uncertainty. Econ. J. 87, 143–157.
Ryan, M.J., 2009. Generalizations of SEU: a geometric tour of some non-standard models. Oxford Econ. Pap. 61, 327–354.
Samuelson, P.A., 1952. Probability, utility and the independence axiom. Econometrica 20, 670–678.
Sarin, R.K., Wakker, P.P., 1998. Dynamic choice and nonexpected utility. J. Risk Uncertainty 17, 87–119.
Savage, L.J., 1954. The Foundations of Statistics, John Wiley and Sons: New York.
Schmeidler, D., 1986. Integral representation without additivity. Proc. Am. Math. Soc. 97, 255–261.
Schmeidler, D., 1989. Subjective probability and expected utility without additivity. Econometrica 57, 571–587.
Schmidt, U., Zank, H., 2009. A simple model of cumulative prospect theory. J. Math. Econ. 45, 308–319.
Seo, K., 2009. Ambiguity and second-order belief. Econometrica 77, 1575–1605.
Segal, U., 1993. The measure representation: a correction. J. Risk Uncertainty 6, 99–107.
Segal, U., 1987. The Ellsberg paradox and risk aversion: An anticipated utility approach. Int. Econ. Rev. 28, 175–202.
Segal, U., 1989. Anticipated utility: a measure representation approach. Ann. Oper. Res. 19, 359–373.
Segal, U., 1990. Two-stage lotteries without the reduction axiom. Econometrica 58, 349–377.
Segal, U., Spivak, A., 1990. First-order versus second-order risk-aversion. J. Econ. Theory 51, 111–125.
Schmidt, U., 2004. Alternatives to expected utility: formal theories. In: Barberà, S., Hammond, P.J., Seidl, C., (Eds.) Handbook of Utility Theory, Vol. II. Kluwer Academic Publishers, Dordrecht.
Schmidt, U., Zank, H., 2009. A simple model of cumulative prospect theory. J. Math. Econ. 45, 308–319.
Siniscalchi, M., 2009. Vector expected utility and attitudes toward variation. Econometrica 77, 801–855.
Siniscalchi, M., 2011. Dynamic choice under ambiguity. Theor. Econ. 6, 379–421.
Stinchcombe, M., 2003. Choice and games with ambiguity as sets of probabilities. UT Austin, mimeo.
Strotz, R.H., 1955. Myopia and inconsitency in dynamic utility maximization. Rev. Econ. Stud. 23, 165–180.
Strzalecki, T., 2011. Axiomatic foundations of multiplier preferences. Econometrica 79, 47–73.
Sugden, R., 2004. Alternatives to expected utility. In: Barberà, S., Hammond, P.J., Seidl, C., (Eds.), Handbook of Utility Theory: Vol. II. Kluwer Academic Publishers, Dordrecht.
Tversky, A., Kahneman, D., 1992. Advances in prospect theory. J. Risk Uncertainty 5, 297–323.
von Neumann, J., Morgenstern, O., 1944. Games and Economic Behavior. John Wiley and Sons. New York.
Wakker, P.P., 2010. Prospect Theory: For Risk and Ambiguity. Cambridge University Press, Cambridge.
Wakker, P.P., Tversky, A., 1993. An axiomatization of cumulative prospect theory. J. Risk Uncertainty 7, 147–176.
Wald, A., 1950. Statistical Decision Functions. John Wiley and Sons, New York.
Weymark, J.A., 1981. Generalized Gini inequality index. Math. Soc. Sci. 1, 409–430.
Yaari, M.E., 1969. Some remarks on measures of risk aversion and on their uses. Journal of Economic theory 1, 315-329.
Yaari, M.E., 1987. The dual theory of choice under risk. Econometrica 55, 95–115.

CHAPTER 18

Calibration and Expert Testing

Wojciech Olszewski[*]
[*]Department of Economics, Northwestern University, Evanston, IL, USA

Contents

Abstract

I survey and discuss the recent literature on testing experts or probabilistic forecasts, which I would describe as a literature on "strategic hypothesis testing." The starting point of this literature is some

Handbook of game theory
http://dx.doi.org/10.1016/B978-0-12-420056-2.00018-5

surprising results of the following type: suppose that a criterion for judging probabilistic forecasts (which I will call a test) has the property that if data are generated by a probabilistic model, then forecasts generated by that model pass the test. It, then, turns out an agent who knows only the test by which she is going to be judged, but knows nothing about the data-generating process, is able to pass the test by generating forecasts strategically.

The literature identifies a large number of tests that are vulnerable to strategic manipulation of uninformed forecasters, but also delivers some tests that cannot be passed without knowledge of the data-generating process. It also provides some results on philosophy of science and financial markets that are related to, and inspired by the results on testing experts.

Keywords: Probabilistic models, Calibration and other tests, Strategic forecasters

JEL Codes: C18, C70

18.1. INTRODUCTION

Probabilistic estimates of future events have long been playing a significant role in human activity. Probabilistic models are common in science, and are often used in weather forecasting. Many economic markets also rely on probabilistic forecasts, including the forecasts of financial analysts, safety assessors, earthquake locators, traffic flow managers, and sports forecasters.

Probability forecasts can be judged by several criteria (see Murphy and Epstein, 1967 for an early study on the topic). Among of most reasonable and objective ones is the criterion of calibration. This criterion has some similarity with the frequency definition of probability, but does not require a background of repeated trials under constant conditions. Dawid (1982) is one of the first theoretical studies of calibration. It shows that if data are generated by a probabilistic model, then forecasts generated by that model are (almost surely) calibrated. Murphy and Winkler (1977) argue that experienced weather forecasters are calibrated.

However, Foster and Vohra (1998) show that one need not know anything about the data-generating process, or be an experienced weather forecaster, in order to be able to produce calibrated forecasts. This is a surprising result. Foster recalls that "this paper took the longest to get published of any I have worked on. I think our first submission was about 1991. Referees simply did not believe the theorem – so they looked for amazingly tiny holes in the proof. When the proof had been compressed from its original 15-20 pages down to about 1, it was finally believed."

The follow-up literature shows that the feature of calibration observed by Foster and Vohra generalizes to a large number of other objective criteria for judging probabilistic models or forecasting. Indeed, suppose that a criterion for judging probabilistic forecasts (which I will call a *test*) has the property that if data are generated by a probabilistic model, then forecasts generated by that model pass the test. It, then, turns out an agent who knows only the test by which she is going to be judged, but knows nothing about the data-generating process, is often able to pass the test by generating forecasts strategically.

However, this follow-up literature also delivers tests that can be passed by true probabilistic models, but cannot be passed without knowledge of the data-generating process. One can compare the literature surveyed in this paper to non-Bayesian statistics.[1] More specifically, statistics is centered around hypothesis testing. It (implicitly) assumes that the hypotheses being tested were born out of thin air, and were completely unlinked to the hypothesis testing methodology. In particular the hypothesis generating entity had no incentives of its own (or at least they were ignored). The research on testing experts presented in this chapter is all about "strategic hypothesis testing." In these papers, we specifically endow the hypothesis generating entity with incentives (and strategies), which is that of passing the "test." We rebuilt the notion of hypothesis testing, alluding to criteria such as errors of types I and II.

The paper is organized as follows: I first introduce some basic terminology and notation. In Section 18.3, I present some examples that show how some simple tests can be passed without any knowledge about the data-generating process. Section 18.4 is entirely devoted to calibration. In Section 18.5, I continue the exposition of what I call negative, or impossibility results, i.e., the results which say that some tests can be passed without any knowledge about the data-generating process. Positive results, i.e., the results that provide, or prove the existence of, "good" tests are discussed in Section 18.6. The following three sections are devoted to some results that contrast with the negative results from Section 18.5, and which have been obtained in slightly different settings. Finally, Section 18.10 contains some results on philosophy of science and financial markets which are related to, and inspired by the results on testing experts.

18.2. TERMINOLOGY AND NOTATION

Each period, one out of two possible outcomes 0 or 1 is observed.[2] Define $\Omega = \{0, 1\}^\infty$ as the set of infinite sequences of outcomes. We will call each $\omega \in \Omega$ a *data set*, or simply, *data*. We will denote the outcome in period t by ω_t, and the *history of outcomes* up to period t by ω^t. That is, $\omega^t = (\omega_1, \dots, \omega_{t-1})$ for $t > 1$, and ω^1 means the empty history.

Denote by $\Delta(\Omega)$ the set of all probability measures over Ω. Measures $P \in \Delta(\Omega)$ will sometimes be called stochastic processes. We need a σ-algebra on which the probability measures are defined. A *cylinder* with base on $(\omega_1, \dots, \omega_n)$ is the set of all data sets ω with the first n elements $\omega_1, \dots, \omega_n$. We endow Ω with the Borel σ-algebra, that is, the smallest $\sigma-$algebra that contains all cylinders. We also endow Ω with the product topology, that is, the topology that comprises unions of cylinders.

[1] I thank Rann Smorodinsky for suggesting this comparison, with which I fully agree.

[2] The generalization of the model and all results to any finite set of outcomes is straightforward.

More generally, for every compact and metrizable space S, denote by $\Delta(S)$ the set of probability measures on S. We endow $\Delta(S)$ with the *weak***-topology* and with the σ-algebra of Borel sets (i.e., the smallest σ-algebra that contains all open sets in weak*-topology). The weak*-topology is defined by the condition that $P^n \to_n P$ if

$$E^{P_n}h \to_n E^P h,$$

for all real-valued and continuous functions h on S, where E is the expected-value operator. In particular, $\Delta(\Delta(\Omega))$ denotes the set of probability measures on $\Delta(\Omega)$. It is well known that $\Delta(S)$ equipped with the weak*-topology is a compact and metrizable space.

Let $\{0,1\}^t$ denote the Cartesian product of t copies of $\{0,1\}$, and let

$$\Omega^{\text{finite}} = \bigcup_{t\geq 0}\{0,1\}^t$$

be the set of all finite histories.[3] Any function

$$f : \Omega^{\text{finite}} \longrightarrow \Delta(\{0,1\})$$

that maps finite sequences of outcomes into distributions over outcomes will be called a *theory*. A theory takes finite data sets (the outcomes up to a certain period) as inputs, and returns a probabilistic forecast over outcomes for the following period as an output.

It is well known that every theory f uniquely induces a probability measure $P_f \in \Delta(\Omega)$. More precisely, given a finite history $\omega^k = (\omega_1, \ldots, \omega_{k-1})$ and an outcome ω_k, let the probability of ω_k conditional on ω^k be denoted by $f(\omega^k)[\omega_k]$. Then, the probability P_f of the cylinder C with base $(\omega_1, \ldots, \omega_n)$ is equal to the product of probabilities

$$P_f(C) = \prod_{k=1}^{n} f\left(\omega^k\right)[\omega_k].$$

We will often identify theory f with probability measure P_f.

Also, any probability measure $P \in \Delta(\Omega)$ determines a theory f by defining $f(\omega^k)[\omega_k]$ as the probability of ω_k conditional on ω^k. That is, if $P(C(\omega_1, \ldots, \omega_{k-1})) > 0$, then

$$f(\omega^k)[\omega_k] = \frac{P(C(\omega_1, \ldots, \omega_k))}{P(C(\omega_1, \ldots, \omega_{k-1}))},$$

where $C(\omega_1, \ldots, \omega_k)$ and $C(\omega_1, \ldots, \omega_{k-1})$ denote the cylinders with base $(\omega_1, \ldots, \omega_k)$ and $(\omega_1, \ldots, \omega_{k-1})$, respectively. And $f(\omega^k)[\omega_k]$ is defined in an arbitrary manner if $P(C(\omega_1, \ldots, \omega_{k-1})) = 0$.[4]

[3] By convention, $\{0,1\}^0 = \{\emptyset\}$.

[4] This last part of the definition means that there are multiple theories f determined by some probability measures P but, as will become clear shortly, this lack of uniqueness will be irrelevant for our purposes.

We consider two types of testing. The general definition requires the expert to provide a theory up front, at time 0. But an important class of tests asks for forecasts only along the sequence of observed outcomes. That is, the expert is supposed to provide at the beginning of period $t = 1, 2, \ldots$ the probability of outcome 1 in period t; the expert provides this forecast after observing the outcomes in all previous periods. The expert's forecast of outcome 1 for period t will be denoted by f_t.

Definition 18.1. *A test is a function*

$$T : \Delta(\Omega) \times \Omega \to \{\text{PASS}, \text{FAIL}\}.$$

A test is therefore an arbitrary function that takes as an input a theory (more precisely, the probability measure induced by the theory) and the observed sequence of outcomes, and returns as an output a PASS-or-FAIL verdict. In particular, we assume that the verdict is the same for any pair of theories f^1 and f^2, which induce the same probability measure $P_{f^1} = P_{f^2}$ over sequences of outcomes.

We study only measurable tests T. That is, $\{\omega \in \Omega : T(P, s) = \text{PASS}\}$ (or equivalently, set $\{\omega \in \Omega : T(P, s) = \text{FAIL}\}$) is assumed to be a measurable set for every theory P. The former set will be called the *acceptance set*, and the latter set will be called the *rejection set* for theory P.

We say that the test is *prequential* if the expert is required to give predictions only along the actual sequence of outcomes, i.e., if the verdict of test T depends only on $(f_1, f_2, \ldots)$. In such a case, we will often write $T(f_1, f_2, \ldots, \omega)$ instead of $T(P, \omega)$.

We shall now state two basic properties of empirical tests. They are versions of type I and type II errors from statistics. An important conceptual difference (compared to the classic definition of type II error) is that the second property refers to strategic behavior; instead of requiring a false theory to be rejected, we require that an ignorant but strategic expert be rejected.

Definition 18.2. *Given an* $\varepsilon \geq 0$*, a test* T *does not reject the truth with probability* $1 - \varepsilon$ *if, for any* $P \in \Delta(\Omega)$*,*

$$P(\{\omega \in \Omega : T(P, \omega) = \text{PASS}\}) > 1 - \varepsilon.$$

Suppose that there actually is a stochastic process $P \in \Delta(\Omega)$ that generates data. Definition 2 says that a test does not reject the truth if, with high probability, the actual data-generating process P, no matter what that process is, is not rejected. A theory that fails such a test can reliably be viewed as false.

Definition 18.3. *A test* T *can be ignorantly passed with probability* $1 - \varepsilon$ *if there exists a* $\xi \in \Delta\Delta(\Omega)$ *such that for every sequence of outcomes* $\omega \in \Omega$*,*

$$\xi(\{P \in \Delta(\Omega) : T(P, \omega) = \text{PASS}\}) > 1 - \epsilon.$$

We will call every $\xi \in \Delta\Delta(\Omega)$ a *random generator of theories*. The random generator of theories may depend on test T, but not on any other knowledge, such as knowledge of the actual data-generating process. If a test can be ignorantly passed, we also say that an ignorant expert can pass the test. If a test can be ignorantly passed, then an ignorant but strategic expert can randomly select theories that, with probability $1 - \varepsilon$ (according to the expert's randomization device), will not be rejected, no matter which data set is realized.

In the case of prequential tests, it will sometimes be more convenient to talk about forecasting rules, instead of random generators of theories. A *forecasting rule* specifies, for any history of outcomes ω^t and any history of forecasts $f^t = (f_1, \ldots, f_{t-1})$, a probability distribution over forecasts f_t. Then, a test can be ignorantly passed if, for every sequence of outcomes $\omega \in \Omega$, the forecasts $(f_1, f_2, \ldots)$ along ω generated by the rule are such that $T(f_1, f_2, \ldots, \omega) = \text{PASS}$ with high probability.

Suppose that for every random generator of theories $\xi \in \Delta(\Delta(\Omega))$, there exists at least one data set ω such that, with probability greater than ε, the realized theory is rejected on ω. Then, by definition, the test cannot be ignorantly passed with probability $1 - \varepsilon$. However, a stronger property may be demanded. A tester may be interested in the existence of data sets such that an ignorant expert fails the test with near certainty (as opposed to probability greater than ε), or may be interested in the existence of a larger number of data sets on which an ignorant expert fails the test.

We will sometimes call a test *good* if it does not reject the truth and cannot be ignorantly passed.

18.3. EXAMPLES

The possibility of ignorantly passing reasonable tests seems quite surprising. Therefore, before presenting more general results, I will use three simple examples to illustrate how this can be achieved.

18.3.1 Example 1

Consider the following simple test. Let

$$R(f, \omega^{m+1}) = \frac{1}{m} \sum_{t=1}^{m} (f(\omega^t) - \omega_t),$$

where $\omega^t = (\omega_1, \ldots, \omega_{t-1})$, marks the difference between the average forecast of 1 and the empirical frequency of 1. The test rejects theory f if the average forecast of 1 is not equal to the empirical frequency of 1. That is, theory f is passed on data sets ω such that

$$\lim_{m} R(f, \omega^{m+1}) = 0.$$

It readily follows from the law of large numbers that the test does not reject the truth. I omit the details of this proof. The test can, however, be ignorantly passed by using the random generator of theories that assigns probability 1 to the single theory f which predicts 1 with certainty in periods in which $R(f, \omega^{m+1}) < 0$, and predicts 0 with certainty in periods in which $R(f, \omega^{m+1}) > 0$. More precisely,

$$
\begin{aligned}
f(\omega^{m+1}) &= 1 \quad \text{if } R(f, \omega^{m+1}) < 0, \\
f(\omega^{m+1}) &= 0 \quad \text{if } R(f, \omega^{m+1}) > 0, \\
f(\omega^{m+1}) &= 0.5 \quad \text{if } R(f, \omega^{m+1}) = 0.
\end{aligned}
$$

The intuition is that when $R(f, \omega^{m+1})$ is negative, the forecast of 1 makes $R(f, \omega^{m+2})$ closer to zero, no matter whether ω_{m+1} is equal to 0 or 1. Similarly, when $R(f, \omega^{m+1})$ is positive, the forecast of 0 makes $R(f, \omega^{m+2})$ closer to zero, no matter whether ω_{m+1} is equal to 0 or 1. I omit the obvious details of the formal proof.

Alternatively, the test can be ignorantly passed by the single theory g which predicts 1 with certainty in periods m such that $\omega_{m-1} = 1$, and predicts 0 with certainty in periods m such that $\omega_{m-1} = 0$; or, more precisely, $f(\omega^m) = \omega_{m-1}$. Notice finally that the ignorant expert must know the test in order to be able pass it ignorantly. One can easily show that no random generator of theories will pass all tests at the same time.

18.3.2 Example 2

Consider for a moment the setting in which there is only one period, and consider all probability distributions from $\Delta(\{0,1\})$. Any test which does not reject the truth (with probability $1-\varepsilon$) does not reject the truth when the true P assigns equal probabilities to 0 and 1. This implies that

$$T(P, 0) = T(P, 1) = \text{PASS}$$

for any $\varepsilon < 1/2$. Thus, in this case, the test can be ignorantly passed by giving theory P.

Suppose now that there are n periods, where n is such that $1/2^n < \varepsilon$. Denote by $\{0,1\}^n$ the set of all sequences of outcomes $\omega = (\omega_1, \ldots, \omega_n)$ of length n. Consider test T to be defined as follows: let m be the lowest number such that $1/2^m < \varepsilon$. For a theory $P \in \Delta(\{0,1\}^n)$, pick any set consisting of 2^{n-m} sequences of outcomes ω such that the probability of this set is the lowest among all sets consisting of 2^{n-m} sequences of outcomes ω. Theory P fails if one of these sequences is observed. By definition, this test passes the truth with probability $1-\varepsilon$.

Since for every P, there exists an ω such that $T(P, \omega) = \text{FAIL}$, test T cannot be ignorantly passed by using a degenerated random generator of theories, i.e., by giving a single theory P. Nevertheless, the test can be ignorantly passed. For any history $\omega^{m+1} = (\omega_1, \ldots, \omega_m)$, take the theory that assigns probability 0 to history ω^{m+1}, and probability

$1/(2^m - 1)$ to any other history of the first m outcomes. Randomize uniformly over all such histories or, equivalently, over all such theories.

Given a sequence ω, a theory P that corresponds to some ω^{m+1} fails if the first m outcomes of ω coincide with ω^{m+1}. The probability that the random generator of theories selects such a P is $1/2^m < \varepsilon$.

18.3.3 Example 3

Consider now the following likelihood ratio (prequential) test. For any theory f, define the alternative theory f^A by letting

$$\begin{aligned} f^A(\omega^t) &= f(\omega^t) + 0.4 \quad \text{if } f(\omega^t) < 0.5, \\ f^A(\omega^t) &= f(\omega^t) - 0.4 \quad \text{if } f(\omega^t) > 0.5, \\ f^A(\omega^t) &= 0.6 \quad \text{if} f(\omega^t) = 0.5, \text{ and } t = 1, 3, \ldots \\ f^A(\omega^t) &= 0.4 \quad \text{if} f(\omega^t) = 0.5, \text{ and } t = 2, 4, \ldots. \end{aligned}$$

Define the likelihood of outcome ω_t according to theory f by

$$l(\omega_t) = f(\omega^t) \quad \text{if } \omega^t = 1 \quad \text{and} \quad l(\omega_t) = 1 - f(\omega^t) \text{ if } \omega^t = 0,$$

and let $l^A(\omega_t)$ be defined similarly. For any sequence of outcomes $\omega^{t+1} = (\omega_1, \ldots, \omega_t)$, let

$$L(\omega^{t+1}) = \frac{l(\omega_1) \cdot \ldots \cdot l(\omega_t)}{l^A(\omega_1) \cdot \ldots \cdot l^A(\omega_t)}$$

be the likelihood ratio of ω^{t+1} according to f compared to the alternative theory.

Finally, define test T by letting $T(f_1, f_2, \ldots, \omega) = \text{PASS}$ if

$$\lim_t L(\omega^{t+1}) = \infty, \qquad [18.1]$$

and $T(f_1, f_2, \ldots, \omega) = \text{FAIL}$ otherwise. That is, test T passes theory f if the observed sequence of outcomes is infinitely more likely according to theory f than according to theory f^A.

It readily follows from the law of large numbers that the test does not reject the truth. I omit the details of this proof. The test can be ignorantly passed by the forecasting rule that predicts $f_t = 0.4$ with probability $1/2$ and $f_t = 0.6$ with probability $1/2$, independent of the history of outcomes up to period t.

The intuition is that if the observed outcome was predicted by the expert as more likely (i.e., $\omega_t = 0$ and $f_t = 0.4$, or $\omega_t = 1$ and $f_t = 0.6$), then the ratio $l(\omega_t)/l^A(\omega_t)$ is $0.6/0.2 = 3$, while if the observed outcome was predicted by the expert as less likely (i.e., $\omega_t = 0$ and $f_t = 0.6$, or $\omega_t = 1$ and $f_t = 0.4$), then the ratio $l(\omega_t)/l^A(\omega_t)$ is only $0.4/0.8 = 1/2$. This gives the expert's theory an advantage over the alternative theory. In order to satisfy condition [18.1], it suffices that the expert predicts with frequency $1/2$ the outcome which is later observed as more likely.

The fact that the likelihood test can be ignorantly passed follows from the law of large numbers. I again omit the details of the formal proof.

18.4. CALIBRATION

18.4.1 Definition and result

The existing literature contains several similar definitions of calibration; they are not all equivalent. In this survey, calibration is defined as follows. Just before time t, after all previous outcomes have been observed, a forecast f_t is made of the probability that $\omega_t = 1$. It is assumed that this forecast takes on values that are the midpoints of one of the intervals: $[0, 1/m], [1/m, 2/m], \ldots, [(m-1)/m, 1]$. That is,

$$f_t = M_i = \frac{2i-1}{2m}, \qquad i = 1, \ldots, m.$$

Let

$$I_{f_t=M_i} = 1 \quad \text{if} \quad f_t = M_i \quad \text{and} \quad I_{f_t=M_i} = 0 \quad \text{if } f(\omega^t) \neq M_i$$

be the indicator function of the set $\{\omega^t : f_t = M_i\}$. The empirical frequency ρ_i^T of outcome 1, where $i = 1, \ldots, m$, is defined as

$$\frac{\sum_{t=1}^{T} \omega_t I_{f_t=M_i}}{\sum_{t=1}^{T} I_{f_t=M_i}},$$

if $I_{f_t=M_i} = 1$ for some t; and

$$\rho_i^T = \frac{2i-1}{2m}$$

if $I_{f_t=M_i} = 0$ for all t, where ω_t is the outcome observed in period t. The empirical frequency ρ_i^T is the frequency with which outcome 1 is observed in those periods $t < T$ for which the forecast is $f_t = M_i$.

Finally, let

$$\bar{I}_{f_t=M_i}^T = \frac{1}{T} \cdot \sum_{t=1}^{T} I_{f_t=M_i}$$

be the frequency of forecast M_i. A sequence of forecasts $(f_t)_{t=1}^{\infty}$ is $(1/m)$–*calibrated* if

$$\limsup_T \left| \rho_i^T - M_i \right| \leq \frac{1}{2m}$$

for every $i = 1, \ldots, m$ such that

$$\limsup_T \bar{I}_{f_t=M_i}^T > 0.$$

That is, if forecast $f_t = M_i$ is being made in a positive fraction of periods, then the limit empirical frequency ρ_i^T must be as close to M_i as possible, given the assumption that the forecasts must have the form $f_t = M_i$ for some $i = 1, \ldots, m$. If number m is sufficiently large, then we say that the forecasts are approximately calibrated.

Proposition 18.1. *(Foster and Vohra, 1998) For every m, there exists a forecasting rule ξ such that for every ω, the sequence of forecasts $(f_t)_{t=1}^{\infty}$ generated by ξ along the sequence of outcomes ω is almost surely $(1/m)$—calibrated.*

18.4.2 Calibrated forecasting rule

A forecasting rule with the required property is fairly easy to define. Given a theory f, and a history ω^T, let

$$\overline{d}^i = \left(\frac{i-1}{m} - \rho_i^T\right) \cdot \overline{I}^T_{f(\omega^t)=M_i}$$

and

$$\overline{e}^i = \left(\rho_i^T - \frac{i}{m}\right) \cdot \overline{I}^T_{f(\omega^t)=M_i}.$$

We define ξ by specifying a probability distribution over forecasts at each ω^T as follows:

(1) If there exists an i such that $\rho_i^T \in [(i-1)/m, i/m]$ (or, equivalently, $\overline{d}^i \leq 0$ and $\overline{e}^i \leq 0$), then $f(\omega^{T+1}) = M_i$ for any i with this property.

(2) Otherwise, there exists an i such that $\overline{d}^i > 0$ and $\overline{e}^{i-1} > 0$. Then $f(\omega^{T+1}) = M_i$ with probability

$$\frac{\overline{e}^{i-1}}{\overline{d}^i + \overline{e}^{i-1}},$$

and $f(\omega^{T+1}) = M_{i-1}$ with probability

$$\frac{\overline{d}^i}{\overline{d}^i + \overline{e}^{i-1}}.$$

A simple inductive argument shows that if there is no i such that $\rho_i^T \in [(i-1)/m, i/m]$, then there exists an i such that $\overline{d}^i > 0$ and $\overline{e}^{i-1} > 0$. Indeed, from the definition of $\overline{d}^1$, we have that $\overline{d}^1 \leq 0$. So if $\overline{e}^1 \leq 0$, condition (1) is satisfied by $i = 1$. If $\overline{e}^1 > 0$ and $\overline{d}^2 > 0$, then condition (2) is satisfied by $i = 2$. Otherwise, $\overline{d}^2 \leq 0$, and one can apply the previous argument again. Finally, if $\overline{d}^m \leq 0$, then by the definition of $\overline{e}^m$, we have that $\overline{e}^m \leq 0$, and so condition (1) is satisfied by $i = m$.

This forecasting rule, which achieves a high calibration score, can be best understood in the case of $m = 2$. If the current empirical frequency of outcome 1 over the periods in which the forecast was 1/4 happens to belong to [0, 1/2], then predict that the current-period outcome will also belong to [0, 1/2], that is, predict 1/4. Similarly, predict 3/4 if the current empirical frequency of outcome 1 over the periods in which the forecast was 3/4 happens to belong to [1/2, 1]. Choose either of the two forecasts if both empirical frequencies belong to the appropriate intervals.

Otherwise, the former empirical frequency is higher than 1/2, and the latter empirical frequency is lower than 1/2. In this case, randomize over forecasts 1/4 and 3/4. Assign to each of the two forecasts a probability that is inversely proportional to the distance of the empirical frequency to the appropriate interval, namely, $\bar{e}^1/(\bar{d}^2 + \bar{e}^1)$ and $\bar{d}^2/(\bar{d}^2 + \bar{e}^1)$, respectively.

18.4.3 Sketch of proof

Although the forecasting rule is simple, the proof that it actually achieves a high calibration score (that is, the proof of Proposition 18.1) is not that simple. I will sketch the proof which comes from Foster (1999) in the case of $m = 2$. This proof is a simplification of the general proof given by Sergiu Hart and AndreuMas-Colell for the case in which in every period only two outcomes are possible. In order to prove Proposition 18.1, we need to recall the concept of approachability and the celebrated theorem from Blackwell (1956).

Consider a two-person, zero-sum game in which each player takes actions from a finite set. Each player's payoff is an L-dimensional vector. For our purposes, it is convenient to denote the actions by $i = 1, 2, \ldots, m$ and $x \in X$, and the payoff of the first player, who will be called player I, by $c(i, x)$. This game is played repeatedly over time. Let C be a closed and convex subset of R^L. We call C *approachable* by player I if there exists a repeated-game strategy of player I which guarantees that the average payoff of player I almost surely converges to set C, regardless of the actions of player I's opponent, as the number of repetitions converges to infinity.

Theorem 18.1. *(Blackwell, 1956) A set C is approachable if and only if for all $a \in R^L$, there exists a vector $w \in \Delta(\{1, \ldots, m\})$ such that for all $x \in X$,*

$$\sum_{i=1}^{m} w_i(c(i,x) - b)^T \cdot (a - b) \leq 0, \qquad [18.2]$$

where b is the closest point to a in C.[5]

[5] By $(c(i,x) - b)^T \cdot (a - b)$ we denoted the inner product of vectors $(c(i,x) - b)$ and $(a - b)$, and the distance between two points in R^L is measured in the standard manner.

Moreover, it follows from the proof that C is approachable if condition [18.2] is satisfied for all vectors a that have the form of average payoffs of player I up to time $T = 1, 2, \ldots$. Then, w_i is the probability of taking action i in period T, when the average payoff up to period T is a. I will sketch the proof of Blackwell's theorem at the very end of this section.

For our purposes, $L = 2$ (since we are sketching the proof for $m = 2$), and

$$C = \{z = (z_1, z_2) \in R^2 : z_1, z_2 \leq 0\}. \quad [18.3]$$

Player I is the expert, action $i = 1, 2$ represents forecast M_i. One can think of player I's opponent as nature, and the set of outcomes $X = \{0, 1\}$ as nature's actions. The payoff vector $c(i, x)$ is defined as

$$c(i, x) = (d^2(i, x), e^1(i, x)), \quad [18.4]$$

where

$$d^2(i, x) = \begin{cases} 0 & \text{if } i = 1 \\ 1/2 - x & \text{if } i = 2 \end{cases}$$

and

$$e^1(i, x) = \begin{cases} x - 1/2 & \text{if } i = 1 \\ 0 & \text{if } i = 2 \end{cases}.$$

In order to prove Proposition 18.1, we need to show that condition [18.2] is satisfied. Notice that given $a = (a_1, a_2)$,

$$b = (b_1, b_2) = (\min\{0, a_1\}, \min\{0, a_2\})$$

and

$$a - b = (\max\{0, a_1\}, \max\{0, a_2\});$$

since $\max\{0, a_i\} \cdot \min\{0, a_i\} = 0$,

$$\sum_{i=1}^{2} w_i (c(i, x) - b)^T \cdot (a - b) = \sum_{i=1}^{2} w_i c(i, x)^T \cdot (a - b) =$$

$$= \sum_{i=1}^{2} w_i [d^2(i, x) \max\{0, a_1\} + e^1(i, x) \max\{0, a_2\}].$$

By the definition of $d^2(i, x)$ and $e^1(i, x)$, this is equal to

$$w_1 (x - 1/2) \max\{0, a_2\} + w_2 (1/2 - x) \max\{0, a_1\}.$$

This expression is equal to zero if we take $w_1 = 1$ and $a_2 = \overline{e}^1(1,x) \leq 0$, or if we take $w_2 = 1$ and $a_1 = \overline{d}^2(2,x) \leq 0$. Otherwise, $\overline{d}^2(2,x) > 0$ and $\overline{e}^1(1,x) > 0$. Then, since we take $w_1 = \overline{e}^1(1,x)/\left(\overline{d}^2(2,x) + \overline{e}^1(1,x)\right)$ and $w_2 = \overline{d}^2(2,x)/\left(\overline{d}^2(2,x) + \overline{e}^1(1,x)\right)$, the expression becomes

$$\frac{\overline{d}^2(2,x)}{\overline{d}^2(2,x) + \overline{e}^1(1,x)}(1-x)\overline{e}^1(1,x) + \frac{\overline{e}^1(1,x)}{\overline{d}^2(2,x) + \overline{e}^1(1,x)}(x-1)\overline{d}^2(2,x) = 0.$$

This completes the proof that condition [18.2] is satisfied.

18.4.4 Sketch of proof of Blackwell's theorem

The idea of the proof behind Blackwell's theorem can be explained as follows: denote player I's average payoff at time t by A_t, the expected average payoff at time $t+1$ contingent on A_t by $E[A_{t+1} \mid A_t]$, and the expected average payoff at time $t+1$ from the perspective of period 0 by $E[A_{t+1}]$. Then,

$$E[A_{t+1} \mid A_t] = \frac{t}{t+1}A_t + \frac{1}{t+1}E[c(i,x) \mid A_t].$$

Since the strategy of player I has the property that the inner product of $A_t - B_t$ and $E[c(i,x) \mid A_t] - B_t$ is nonpositive (B_t stands for the point in C that is closest to A_t), it follows that $E[A_{t+1} \mid A_t]$ is closer to set C than A_t is (see Figure 18.1).

Moreover, $E[A_{t+1}]$ converges to C, because the norm of the second component of $E[A_{t+1} \mid A_t]$ is at least of order $1/t$ of the norm of the first component. Together with the fact that the inner product of the two components is nonpositive, this implies that

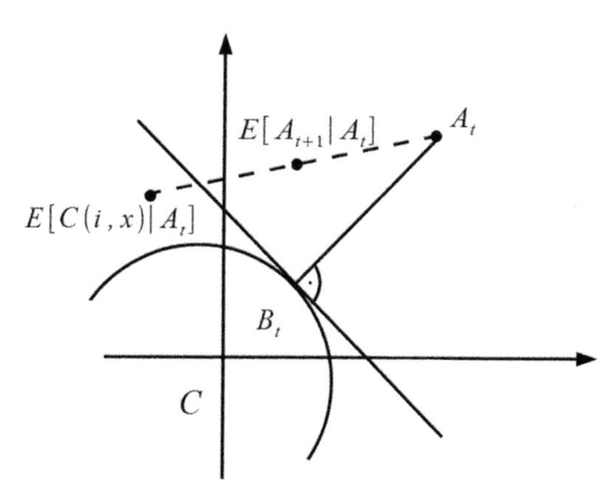

Figure 18.1 The evolution of player 1's average payoff at over time.

the distance between $E[A_{t+1}]$ and C shrinks in period t at least by order $1/t$. But if $E[A_{t+1}]$ did not converge to C, this would mean that the series

$$\sum_t \frac{1}{t}$$

was convergent.

Now, some form of the law of large numbers (more precisely, the strong law of large numbers for dependent variables) implies that A_{t+1} converges to C almost surely.

18.5. NEGATIVE RESULTS

Early follow-up papers generalize the Foster and Vohra (1998) result to "other forms" of calibration. Other papers provide simpler proofs of their result. More specifically, a *history-based checking rule* is an arbitrary function of finite sequences of outcomes (histories) to the set {active, inactive}. Given a history-based checking rule and a theory, a *forecast-based checking rule* is active if the history-based checking rule is active and the forecast takes a value from a given set $D \subset [0, 1]$. The calibration score assigned to a theory by a checking rule (either history- or forecast-based) is the difference between the frequency of an outcome and the average probability assigned to this outcome by the forecasts of the theory, where the averages are taken across the periods in which the checking rule was active. We follow here the terminology introduced in Sandroni et al. (1999).

According to this terminology, Foster and Vohra (1998) demonstrate the existence of a forecasting rule that (almost surely) calibrates the forecast-based checking rules associated with the always active history-based rule and sets $D_k = [k/m, (k+1)/m]$ (where $k = 0, \ldots, m-1$). The concept of checking rules other than calibration (i.e., other notions of calibration) was introduced in Kalai et al. (1999), which also demonstrates equivalence between notions of calibration and merging. Lehrer (2001) shows that there exists a forecasting rule that simultaneously calibrates any countable number of history-based checking rules; Sandroni et al. (1999) generalize this result to the forecast-based checking rules associated with a countable number of history-based checking rules and a countable number of sets D. Lehrer also shows that for any probability distribution over history-based checking rules, there exists a forecasting rule which simultaneously calibrates almost all these rules. All these results allow for any finite set of possible outcomes, not only 0 and 1.

Sergiu Hart first suggested a simpler proof of Foster and Vohra's result based on the minmax theorem (see the discussion in Foster and Vohra, 1998; Sandroni, 2003). A constructive version of Hart's proof has been derived independently by Fudenberg and Levine (1999). Foster (1999) contains the simple and elegant proof that has been discussed in the previous section (see also Foster and Vohra, 1997; Hart and Mas-Colell, 2000).[6] The last two authors have suggested using Blackwell's theorem. Foster and Vohra's original argument is close to the proof based on Blackwell's theorem—but uses a direct potential function instead of Blackwell's theorem. Sandroni (2003) first observed that the Foster and Vohra result generalizes well beyond calibration and scoring rules (see also Vovk and Shafer, 2005; Vovk, 2007). I discuss Sandroni's result, and the generalizations of thereof, in the following section.

18.5.1 Generalizations of Foster and Vohra's result

Theorem 18.2. *(Fan, 1953) Let X be a convex subset of a compact, Hausdorff, linear topological space, and Y be a convex subset of a linear space (not necessarily topologized). Let f be a real-valued function on $X \times Y$ such that for every $y \in Y$, $f(x, y)$ is lower semi-continuous on X. If f is convex on X and concave on Y, then*

$$\min_{x\in X} sup_{y\in Y} f(x, y) = sup_{y\in Y} \min_{x\in X} f(x, y). \qquad [18.5]$$

Theorem 18.2 is illustrated in Figure 18.2, where $X = Y = [0, 1]$, and f is a linear function of x and a linear function of y. The nontrivial part of the theorem says that the right-hand side can be as large as the left-hand side. Suppose that the left-hand side of [18.5] is "large." This means that for every x there is a y such that $f(x, y)$ is large. In Figure 18.2, $y = 1$ for $x = 0$ and $y = 0$ for $x = 1$. The linearity and continuity of $f(x, y)$ imply that there is a value of x (between 0 and 1) such that $f(x, y)$ is a constant function of y for this x, and is depicted in bold on the graph of function f. This constant must be large, since there must exist a y for this x such that $f(x, y)$ is large. However, by analogous arguments, there also exists a value of y such that $f(x, y)$ is a constant function of x for this y. This is also depicted in bold on the graph of function f. And this constant is large, because the two bold lines on the graph intersect. Thus, the right-hand side of [18.5] must also be large.

I have stated the minmax theorem in its original form. However, I will not define Hausdorff spaces or lower semi-continuity. It is enough to know that all spaces considered in this survey will be metrizable, and therefore Hausdorff, and all functions will be continuous.

We will call a test T *finite* if for every $P \in \Delta(\Omega)$, there exists an n such that for any $\omega(1), \omega(2) \in \Omega$ such that $\omega^n(1) = \omega^n(2)$, we have

$$T(P, \omega(1)) = T(P, \omega(2)).$$

[6] Foster's proof allows for only two outcomes: 0 and 1. A simple proof which allows for any finite number of outcomes is provided in Mannor and Stoltz (2010).

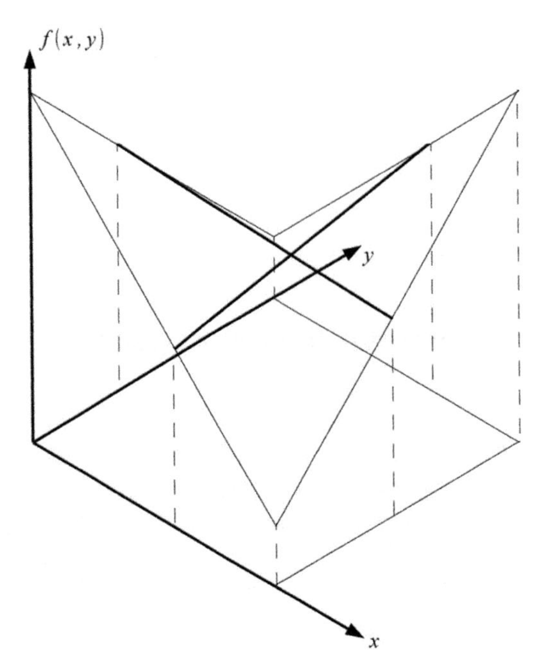

Figure 18.2 The minmax theorem for $X = Y = [0, 1]$.

That is, the verdict of a finite test T depends only on the a finite number n of observed outcomes. Notice, however, that I allow this finite number to depend on theory P.

Fix $\varepsilon \in [0, 1]$ and $\delta \in (0, 1 - \varepsilon]$.

Proposition 18.2. *(Olszewski and Sandroni, 2008; Sandroni, 2003)*[7] *Let T be a finite test that does not reject the truth with probability $1 - \varepsilon$. Then, the test T can be ignorantly passed with probability $1 - \varepsilon - \delta$.*

Proof. To provide the intuition for Proposition 18.2 consider the following zero-sum game between nature and the expert. Nature chooses a probability measure $P \in \Delta(\Omega)$. The expert chooses a random generator of theories ξ. The expert's payoff is

$$E^{\xi} E^{P} T(Q, \omega),$$

where PASS $= 1$ and FAIL $= 0$, and E^{ξ} and E^{P} are the expectation operators associated with ξ and P, respectively. By assumption, test T does not reject the truth with probability $1 - \varepsilon$. Thus, for every strategy P of nature, there is a strategy ξ_P for the expert (that assigns probability one to P) such that the expert's payoff is

[7] Sandroni (2003) proved this result for tests which are finite in a slightly stronger sense, namely, he assumes that the number n in the definition of a finite test is the same for all theories P.

$$E^{\xi_P}E^P T(Q,\omega) = P\{\omega \in \Omega : T(P,\omega) = 1\} \geq 1 - \varepsilon.$$

Hence, if the zero-sum game has value, then there is a strategy ξ_T for the expert that ensures a payoff arbitrarily close to $1 - \varepsilon$, no matter which strategy nature chooses. In particular, nature can use P_ω, selecting a single sequence of outcomes ω with certainty. Therefore, for all $\omega \in \Omega$,

$$E^{\xi_T}E^{P_\omega} T(Q,\omega) = \zeta_T\{Q \in \Delta(\Omega) : T(Q,\omega) = \text{PASS}\} \geq 1 - \epsilon - \delta.$$

Fan's minmax theorem guarantees that the zero-sum game between nature and the expert has value. More precisely, let X be $\Delta(\Omega)$, and let Y be the subset of $\Delta(\Delta(\Omega))$ consisting of all random generators of theories with finite support. That is, an element ξ of Y can be described by a finite sequence of probability measures $Q^1, \ldots, Q^n$ and positive weights $\pi^1, \ldots, \pi^n$ that add up to one such that ξ selects Q^i with probability $\pi^i, i = 1, \ldots, n$. Let function $f : X \times Y \to R$ be defined by

$$f(P,\xi) = E^{\xi}E^P T(Q,\omega) = \sum_{i=1}^{n} \pi^i \int T(Q^i,\omega)\mathrm{d}P(\omega).$$

All the conditions of the minmax theorem are satisfied. In particular, function f is continuous in P and linear, as the sum of continuous and linear functions. The continuity of functions of the form

$$E^P T(Q^i,\cdot) = \int T(Q^i,\omega)\mathrm{d}P(\omega)$$

follows immediately from the assumption that test T is finite and from the definition of weak*-topology; this guarantees that $T(Q^i,\omega)$ is a continuous function of ω.

18.5.2 Prequential principle

I conclude this section with two recent results, which show that if there is a good test, it must make use of counterfactual forecasts, which cannot be verified by any observed data.

We will say that a test rejects theories in finite time if sets $\{\omega \in \Omega : T(P,s) = \text{FAIL}\}$ are unions of cylinders.

For every theory, such a test specifies a collection of finite sequences of outcomes, which sequences contradict the theory according to the test; it therefore fails the theory if one of these sequences is observed.

Two theories f^1 and f^2 are equivalent up to period m if

$$f^1\left(\omega^{k+1}\right) = f^2\left(\omega^{k+1}\right)$$

for any $\omega^{k+1} = (\omega_1, \ldots, \omega_k)$ such that $k < m$. Two theories are therefore equivalent up to period m if they make the same predictions for periods $1, \ldots, m$.

Definition 18.4. *A test T is future-independent if, for any pair of theories f^1 and f^2 that are equivalent up to period m, and for any sequence of outcomes $\omega^{m+1} = (\omega_1, \ldots, \omega_m)$, theory f^1 is rejected at ω^{m+1} if and only if theory f^2 is rejected at ω^{m+1}.*

A test is future-independent if, whenever a theory f^1 is rejected in period m, another theory f^2, which makes exactly the same predictions as f^1 up to period m, must also be rejected in period m. In other words, if a finite sequence of outcomes contradicts a theory, then it also contradicts any theory equivalent to it.

Proposition 18.3. *(Olszewski and Sandroni, 2008) Every future-independent test which rejects theories in finite time, and which, with probability $1 - \varepsilon$, does not reject the truth, can be ignorantly passed with probability $1 - \varepsilon - \delta$.*

Recall that a test is prequential if it requires the expert to give predictions only along the actual sequence of outcomes.

Proposition 18.4. *(Shmaya, 2008) Every prequential test T that, with probability $1 - \varepsilon$, does not reject the truth, can be ignorantly passed with probability $1 - \varepsilon - \delta$.*

Propositions 18.3 and 18.4 are independent in that neither of them implies the other. There exist future-independent tests whose verdicts depend on counterfactual, "off-equilibrium" predictions. There also exist tests which require the expert to give predictions only along the actual sequence of outcomes, and which may not reject theories in finite time. Tests that reject theories in finite time have the property that sets $\{\omega \in \Omega : T(P, \omega) = \text{PASS}\}$ are closed in Ω. Shmaya's result allows for tests such that these sets are Borel, but may not be closed.

18.5.3 Interpretations

There are two interpretations of the results on testing experts. One is literal and involves informed versus ignorant experts. Informed experts know precisely the probability distribution P that generates data, and ignorant experts are completely ignorant, without even any prior over probability distributions. Although, this language is very convenient, and I am using it throughout this survey, I would argue that it should not be taken too literally.

The literal interpretation faces a number of conceptual problems. For example, what does it mean to know the probabilities of future events? But even if the concept of probability is taken as given, it is unclear whether the existence of a random generator of theories which satisfies Definition 3 really helps an ignorant but strategic expert.

To see the point, assume, as the literature often does, that an ignorant expert must make the decision whether to provide her forecasts. These forecasts will be tested. The expert receives a positive utility u from providing the forecasts, but if she fails the test, she also receives a disutility $-d$, such that $u - d < 0$. That is, the expert's utility depends on the verdict of the test, and thus on her forecasts and the observed sequence of outcomes. The ignorant expert does not know which sequence of outcomes will be observed at the time the forecasts are provided; moreover, the expert is completely ignorant, which means that she does not even have any prior over the sequences of outcomes. In other words, the expert faces Knightian uncertainty (also known as modern uncertainty or ambiguity). Suppose that the expert is most uncertainty-averse in the sense of Gilboa and Schmeidler (1989); this means that she evaluates prospects according to the worst possible scenario, i.e., the sequence of observed outcomes which gives her the lowest utility.

The existence of a random generator of theories ξ that makes the probability of passing the test high, even without any knowledge regarding the data-generating process, seemingly makes the option of providing forecasts more attractive. Put another way, if the expert forecasts according to ξ, then for every possible sequence of observed outcomes, she will end up, according to ξ, with utility $u - d$ with a low probability, and with utility u with a high probability.

However, this argument is no different from Raiffa's (1961) critique of the concept of uncertainty. Raiffa claimed that an appropriate randomization can remove uncertainty and replace it with common risk, and therefore the presence of uncertainty should have no more impact on a decision maker's utility than the presence of common risk. Subsequent studies on decision theory tend to disagree with Raiffa's critique. Intuitively, the reason for this disagreement is that the expert can randomize only before the uncertainty regarding sequence ω is resolved; as a result, once lottery ξ is resolved and a probability distribution P is selected, the decision maker faces uncertainty again. If, given the P selected, she again evaluates prospects according to the sequence of observed outcomes that gives her the lowest utility, she will typically end up with utility $-d$, because nontrivial tests fail every P on some ω.

Of course, some tests (for example, the test from Example 1) can be passed without randomizing over theories, and passing of some other tests requires "little" randomization. However, the literal interpretation of the negative results on testing experts seems to require that random generators of theories have some properties in addition to the property from Definition 18.3, and therefore future analysis is necessary.

In my opinion, this literature is about the philosophy of science, or more precisely, about probabilistic models; this is the alternative to the literal interpretation. One may disagree about the way we should interpret the concept of probability, but probabilistic models are nevertheless commonly used in scientific practice. So, if we want to test them, which is also a common scientific practice, we would do better to have tests that do not reject them when they are correct. That is, if we can generate data according to a probabilistic model, the test should not reject the model with high probability. For example, if we agree that we have a fair coin, then flipping it repeatedly should generate data sequences such that the i.i.d. fifty-fifty model will pass the test most of the time. This is the way I interpret the condition that the test passes the truth.

In my view, the major drawback in having a test that can be ignorantly passed is not that such a test cannot differentiate between informed and ignorant decision makers, but rather that the test is vulnerable to the kind of actions of malicious agents that computer scientists study. Actually, I believe that scientific testing methods should be designed so as to exclude any test that can be ignorantly passed, i.e., no test should be passable by a person with malicious intent, who is ignorant about a particular forecasting problem but who is otherwise smart.

18.6. POSITIVE RESULTS

18.6.1 Category tests

After the initial wave of negative results, described in Section 18.5, few new results appear for a short period. The revitalizing paper was Dekel and Feinberg (2006), which offered an intuitive idea of constructing a good test. Their starting point was a well-known result from measure and category theory. To get to this result, consider first the following definition:

Definition 18.5. *A subset of a compact metric space is topologically large (i.e., residual) if it contains the intersection of a countable family of open and dense sets. A subset is topologically small (i.e., first-Baire category) if it is contained in a set whose complement is topologically large.*

I refer the reader to Oxtoby (1980) for a more complete exposition of the basic concepts of category.

Theorem 18.3. *For every measure P defined on the σ-algebra of Borel subsets of a compact metric space, there exists a topologically large subset G of the metric space such that*

$$P(G) = 0.$$

Since set G contains the intersection of a countable family of open and dense sets, it follows that for every $\varepsilon > 0$, there also exists an open and dense subset U such that $P(U) \leq \varepsilon$.

Dekel and Feinberg assume that the expert must give a theory at time 0, and suggest a class of tests that they call *category tests*, which pass any theory only on a topologically small set. By Theorem 18.3, there exist tests in this class which pass the truth with probability 1, and the intuition suggests that the ignorant expert should find passing a category test difficult. For strategic reasons, the expert picks a topologically small set on which she wants to pass. So, for any data, the expert must pick with high probability a topologically small set containing that data, without knowing anything about the data-generating process.

To support this intuition, Dekel and Feinberg show that for every topologically small set S, the set of theories P such that $P(S) > 0$ is topologically small. Their intuition is, however, not entirely correct, since Olszewski and Sandroni (2011) exhibit a category test that can be ignorantly passed. In spite of that, Dekel and Feinberg provide an example of good category test. To construct such a test, they use a *Lusin set*, which is a subset of Ω with certain "exotic" properties. The existence of Lusin sets is independent of the usual axioms of set theory, and for this reason, I omit the details of Dekel and Feinberg's construction. I will provide another good category test later, but it will be useful to first provide a simpler good test with somewhat weaker properties.

18.6.2 A simple example of a good test

This simple example of a good test is provided in Olszewski and Sandroni (2011): consider any sequence of pairwise-disjoint, nonempty sets $C_n \subset \Omega$. For concreteness, let set C_n be cylinders $C(\omega^{n+1})$, $n = 1, 2, \ldots$, where ω^{n+1} is the finite history in which outcome 1 was observed in periods $1, \ldots, n-1$, and outcome 0 is observed in period n. For any theory P, consider the sets of the form

$$\overline{C}_n = \bigcup_{k=n}^{\infty} C_k.$$

There exist a number n such that $P(\overline{C}_n) < \varepsilon$, because

$$\bigcap_{n=1}^{\infty} \bigcup_{k=n}^{\infty} C_k = \emptyset.$$

Define $R(P)$ as $\overline{C}_n$ for the lowest number n such that $P(\overline{C}_n) < \varepsilon$.

The test requires the expert to provide a theory P up front (at time 0), and rejects the theory when data $\omega \in R(P)$ are observed. In other words, theory P is rejected if outcome 0 is observed after observing a sufficiently long sequence of outcomes 1.

By definition, the test rejects the truth with probability no higher than ε. It turns out that the test cannot be ignorantly passed. Indeed, for any random generator of theories $\xi \in \Delta(\Delta(\Omega))$, consider sets

$$\mathbf{Q}_n = \left\{ Q \in \Delta(\Omega) : R(Q) := \bigcup_{k=n}^{\infty} C_k \right\}.$$

Notice that the family of sets $\mathbf{Q}_n$ is a partition of $\Delta(\Omega)$. It follows that

$$\sum_{n=1}^{m} \xi(\mathbf{Q}_n) \geq 1 - \varepsilon$$

when m is sufficiently large. And thus,

$$\xi\left(\{Q \in \Delta(\Omega) : T(Q, \omega) = PASS\}\right) \leq \varepsilon$$

for any data set

$$\omega \in \bigcup_{k=m+1}^{\infty} C_k;$$

we denote this set of data sets ω by R_m. That is, by generating theories according to any ξ, the ignorant expert (like the informed expert) passes the test with probability ε or lower, if the tester observes 0 after a sufficiently long sequence of 1's. However, unlike the informed expert, the ignorant expert does not know how likely it is that the tester will observe data on which she will fail the test.

This simple test has a number of drawbacks. One of them is that although the ignorant expert cannot make sure that she will pass the test on all data sets, she has strategies enabling her to pass the test on "almost all data sets." More precisely, there exists a sequence of random generators of theories $(\xi_m)_{m=1}^{\infty}$ such that by generating theories according to ξ_m, the ignorant expert fails only on set R_m, and $(R_m)_{m=1}^{\infty}$ is a descending sequence of sets with empty intersection. Indeed, by reporting a theory $Q \in \mathbf{Q}_m$, the expert fails only on set R_m.

18.6.3 Other good tests

If we combine the simple idea for constructing a good test from the previous section, with the Dekel and Feinberg (2006) concept of category tests, we obtain tests with much stronger properties; in particular, we obtain tests that cannot be ignorantly passed on almost all data sets.

Proposition 18.5. *(Olszewski and Sandroni, 2009c) (a) For every $\varepsilon > 0$, there exists a test T that passes the truth with probability $1 - \varepsilon$, and cannot be ignorantly passed. Moreover, for every random generator of theories $\xi \in \Delta(\Delta(\Omega))$, there is an open and dense set $U \subset \Omega$ such that*

$$\xi\left(\{Q \in \Delta(\Omega) : T(Q, \omega) = \text{PASS}\}\right) \leq \varepsilon, \quad \forall\, \omega \in U.$$

(b) There also exists a test T that passes the truth with probability 1 such that for every random generator of theories $\xi \in \Delta(\Delta(\Omega))$, there is a topologically large set $G \subset \Omega$ such that

$$\xi\left(\{Q \in \Delta(\Omega) : T(Q, \omega) = \text{PASS}\}\right) = 0, \quad \forall\, \omega \in G.$$

Proof. I will prove part (a) only. The proof of part (b) is analogous, although slightly more involved. Let $D = \{\omega(1), \omega(2), \ldots\} \subset \Omega$ be any countable and dense subset of Ω. For concreteness, one can assume that D consists of all data sets such that, from some period on, only outcome 1 is observed.

For any theory P, take the lowest n such that

$$P\left(C(\omega^{n+1}(k)) - \{\omega(k)\}\right) < \frac{\varepsilon}{2^k}. \qquad [18.6]$$

That is, I consider cylinders with base on $\omega^{n+1}(k)$,[8] from which the point $\omega(k)$ has been removed, and take such a cylinder whose measure P is appropriately small. There exists an n with property [18.6], because

$$\bigcap_{n=0}^{\infty}\left[C(\omega^{n+1}(k)) - \{\omega^k\}\right] = \varnothing.$$

Denote by $C_k(P)$ the cylinder $C(\omega^{n+1}(k))$ with property [18.6].

Let

$$R(P) = \bigcup_{k=1}^{\infty}[C_k(P) - \{\omega(k)\}].$$

The test T rejects theory P when $\omega \in R(P)$ is observed.

The test rejects the truth with probability no higher than ε, because

$$P(R(P)) \leq \sum_{k=1}^{\infty} P(C_k(P) - \{\omega(k)\}) < \sum_{k=1}^{\infty} \frac{\varepsilon}{2^k} = \varepsilon.$$

Now, take any random generator of theories $\xi \in \Delta(\Delta(\Omega))$, and consider sets

$$\mathbf{Q}_{k,n} = \left\{Q \in \Delta(\Omega) : C_k(Q) = C(\omega^{n+1}(k))\right\}.$$

Notice that the family of sets $\mathbf{Q}_{k,n}$, $n = 1, 2, \ldots$, is a partition of $\Delta(\Omega)$. It follows that

$$\sum_{n=m(k)+1}^{\infty} \xi(\mathbf{Q}_{k,n}) \leq \varepsilon$$

for any sufficiently large $m(k)$.

[8] Recall that history $\omega^{n+1}(k)$ consists of the first n outcomes of $\omega(k)$.

Let

$$U = \bigcup_{k=1}^{\infty} \left[C(\omega^{m(k)+1}(k))) - \{\omega(k)\} \right].$$

Set U is open and dense in Ω. Take any $\omega \in U$. Then, $\omega \in C(\omega^{m(k)+1}(k)) - \{\omega(k)\}$ for some k, which means that

$$\{Q \in \Delta(\Omega) : T(Q, \omega) = \text{PASS}\} \subset \bigcup_{n=m(k)+1}^{\infty} \mathbf{Q}_{k,n},$$

because any $Q \in \mathbf{Q}_{k,n}$ for $n \leq m(k)$ is rejected on $C(\omega^{n+1}(k)) - \{\omega(k)\} \supset C(\omega^{m(k)+1}(k)) - \{\omega(k)\}$. It follows that

$$\{Q \in \Delta(\Omega) : T(Q, \omega) = \text{PASS}\} \leq \varepsilon.$$

18.6.4 Good "prequential" tests

Proposition 18.4 assumes that the expert must give a theory at time 0. The negative results from Section 18.5 seem to suggest the necessity of this assumption. Interestingly, and perhaps a little surprisingly, there exist good prequential tests. However, these are tests belonging to a slightly broader category than the ones studied in the previous sections.

Shmaya (2008) shows that there exists such a test T, which passes the truth with probability 1, and cannot be ignorantly passed. This test is, however, not Borel, that is, the sets

$$\left\{\omega \in \Omega : T(f_1, f_2, \ldots, \omega) = \text{PASS}\right\}$$

are not Borel. We can still talk about the probability of a theory being accepted or rejected, because these sets have the form of the union of a Borel set B and a subset of a Borel set of measure 0. Therefore, like the definition of Lebesgue measure, we can extend any probability measure defined on the Borel σ-algebra by assigning to any such set the measure of set B. Shmaya uses ideas similar to those of Dekel and Feinberg, but in his construction replaces Lusin sets with *universally null sets*. I refer the reader to Shmaya's paper for the details of his construction.

As reported in Olszewski and Sandroni (2009c),[9] Peter Grünwald has shown that there exists a *random test* $T^{\lambda}(Q, \omega)$, where λ is a random variable, whose verdict depends only on forecasts $(f_1, f_2, \ldots)$ made along the actual sequence of outcomes; this test passes the truth with probability 1, and cannot be ignorantly passed. More precisely, this random

[9] The result has been published in Olszewski and Sandroni, but was suggested to the authors by Peter Grünwald.

test takes as input the observed sequence of outcomes, the sequence of forecasts made along this sequence of outcomes, and the realization of random variable λ; it returns as output a PASS-or-FAIL verdict.

Test T passes the truth with probability 1 for all realizations of λ. And for any random generator of theories ξ, with probability 1 the realization of λ has the property that there exists a topologically large set of data sets $G \subset \Omega$ such that

$$\xi\left(\left\{Q \in \Delta(\Omega) : T^{\lambda}(Q, \omega) = \text{PASS}\right\}\right) = 0, \quad \forall\, \omega \in G.$$

Given any value of λ, test T^{λ} is a category test similar to that constructed in the proof of Proposition 18.5. I again refer the reader to the original paper for the details of this construction.

18.7. RESTRICTING THE CLASS OF ALLOWED DATA-GENERATING PROCESSES

One response to the negative results reported in Section 18.5 is that the set of allowed data-generating processes (or theories) $\Delta(\Omega)$ is too "large" and too "abstract." Stochastic processes studied in many fields of empirical research have much simpler forms, e.g., the outcomes are identically and independently distributed (i.i.d.). In other words, one may argue that the requirement that a test does not reject any true $P \in \Delta(\Omega)$ should be replaced with the requirement that the test does not reject true P's that belong to a smaller class of processes. And indeed, for many classes of processes, it is straightforward to separate informed and ignorant experts. In the deterministic world in which the informed expert knows with certainty the outcome that will occur, one can pass the expert's theory if the outcome predicted by her indeed occurs. Similarly, if one believes that the outcomes are i.i.d., a good test asks the expert for the probability distribution over outcomes, and gives the PASS verdict if the observed frequency of outcomes matches the expert's distribution. In fact, good tests exist for many classes of parametric and semi-parametric probabilistic models studied in econometrics.

However, the claim that the negative results are possible only when abstract data-generating processes are allowed does not seem to be fully justified. Olszewski and Sandroni (2008, 2009a,b) show that many of their negative results hold true when one replaces $\Delta(\Omega)$ with the class of exchangeable processes, i.e., processes which can be represented as mixtures of (i.e., probability distributions over) i.i.d. processes.

It is, however, interesting to see for what classes of data-generating processes good tests exist. For example, how large can such a class be? Consider again the one-period setting discussed in Section 18.3. In this setting, one can only allow for the distributions that assign a probability higher than $1 - \varepsilon$ to one of the outcomes. But with many periods, this bound can be much lower than $1 - \varepsilon$. Olszewski and Sandroni (2009b) provide a simple test which cannot be ignorantly passed, and which rejects the truth only when the forecasts are often close to fifty-fifty.

More interestingly, Stewart (2011) constructs a prequential test that rejects the truth only when the true probability distribution P has the property that the uniform probability distribution Q, which assigns equal probabilities to 0 and 1 contingent on any previously observed sequence of outcomes, weakly merges with P with positive probability.[10] This set of distributions P is topologically small in the space $\Delta(\Omega)$.

Al-Najjar et al. (2010) make the point that the restrictions on the theories that the expert is allowed to submit should not be guided by what seems intuitively abstract, or by what seems large in the set-theoretic sense. Instead, they should be aligned with normative standards, such as those typically expected of scientific theories and statistical models. They formalize this idea by restricting attention to theories that are learnable and predictive.

They assume that any theory is represented as a probability distribution on a set of parameters Θ, with each $\theta \in \Theta$ indexing a stochastic process. These representations are assumed to have the property that, as data accumulate, the expert is eventually able to forecast as if he knew the true parameter θ to any desired degree of precision. In addition, given a parameter θ and an integer t, the outcomes of the next t periods hardly improve predictions of outcomes in the distant future.

Al-Najjar et al. (2010) show that the class of learnable and predictive theories is "testable." Specifically, there is a finite test T such that: (1) T does not reject any (learnable and predictive) data-generating process; and (2) for any random generator of (learnable and predictive) theories, there is a (learnable and predictive) data-generating process such that the ignorant expert using this random generator of theories is rejected by T with arbitrarily high probability.

Fortnow and Vohra (2009) indicate another kind of restriction: they claim that even if there exist random generators of theories that enable ignorant experts to pass a test, the generators may not be implementable for computational reasons. For example, they construct a finite test T,[11] which can be implemented in polynomial time, such that for any $\varepsilon > 0$, for sufficiently large integer m, test T passes the truth with probability $1 - \varepsilon$,

[10] Given distributions P and Q, we say that Q weakly merges with P at ω if for every $\delta > 0$, there exists some T such that the difference in probability of ω_t contingent on ω^t according to P and Q is lower than δ for all $t > T$.
The distribution Q is said to weakly merge with P with probability π if

$$P(\omega \in \Omega : Q \text{ weakly merges with } P \text{ at } \omega) = \pi .$$

[11] This test is finite in the stronger sense of Sandroni (2003). Specifically, there exists a number n such that for all P, the test needs only n outcomes in order to give a verdict.

and any random generator of theories that can ignorantly pass test T with probability $1-\varepsilon$ can be used to factor m into prime integers.

The existence of an efficient (i.e., probabilistic polynomial-time) algorithm for factoring composite numbers is generally considered unlikely. For example, many commercially available cryptographic schemes take advantage of this fact. However, the problem of computational restrictions seems to be more complicated than it may appear from Fortnow and Vohra's analysis. In particular, Fortnow and Vohra assume that nature gives the informed expert "on a piece of paper" all the probabilities of future events, which enable her to factorize the required numbers. In practice, the informed expert may know just the method of generating correct forecasts, but may face computational restrictions similar to those faced by the ignorant expert.

Finally, it should be mentioned that Shmaya and Hu (2013) have just announced (in the paper not yet available) that if theories and tests are required to be computable (in the sense that they can be described by Turing machines), then there is a future-independent and prequential test that passes the truth with high probability, and cannot be ignorantly passed.

18.8. MULTIPLE EXPERTS

Some researchers (e.g., Al-Najjar and Weinstein, 2008; Feinberg and Stewart, 2008) argue that the negative results depend crucially on whether a single expert is tested in isolation, or multiple experts are tested at the same time. They point out, however, that some limitations on single-expert testing still have force in the multiple-expert setting.

The idea of comparative testing is very attractive. In practice, true probabilistic models may not exist, but we, nevertheless, use probabilistic models. Some models may be better than others, and with the help of data one may be able to determine which models are better (e.g., by comparing the likelihoods of observed events).

The model in which there is no true data-generating process has not yet been examined. Instead, Al-Najjar and Weinstein and Feinberg and Stewart argue that the possibility of comparative testing reverses some of the negative results.

More precisely, Al-Najjar and Weinstein consider prequential tests whose verdict depends on only a finite number n of forecasts and observed outcomes, and this number n is common for all theories P. They restrict attention to situations in which one of two experts is informed and the other is ignorant, and instead of a PASS-or-FAIL verdict their test indicates the expert which it finds to be informed. (They also allow for the verdict to be inconclusive.) They show that some likelihood tests T^n have the following property:

Proposition 18.6. *If expert i is informed and truthful, then for every* $\varepsilon > 0$, *there is an integer* K *such that for all integers* n, *data-generating processes* P, *and random generator of theories* ξ^j *of expert* $j \neq i$, *the probability of the event that*

(a) T^n *picks expert* i, *or*

(b) the probabilities assigned to outcome 1 *(or, equivalently, to outcome* 0*) by the two forecasts differ by at most* ε *in all but* K *periods,*

is no lower than $1 - \varepsilon$.[12]

Thus, the only way in which an ignorant expert can pass test T^n is for the expert to provide theories that satisfy condition (b). This seems to be difficult to achieve for all data-generating processes P. However, the test does not guarantee that an ignorant expert will fail it, even in the presence of an informed expert. Moreover, the test is unable to reveal the type of the experts when both of them are ignorant. Indeed, Al-Najjar and Weinstein show that there is no test that cannot be ignorantly passed and that can tell whether there is at least one informed expert. (An analogous result was also obtained by Feinberg and Stewart.)

Feinberg and Stewart define a cross-calibration test, under which m experts are tested simultaneously, and which reduces to the calibration test in the case of a single expert, i.e., when $m = 1$. Intuitively, just as the calibration test checks the empirical frequency of observed outcomes conditional on each forecast, the cross-calibration test checks the empirical frequency of observed outcomes conditional on each profile of forecasts.

They show the following proposition:

Proposition 18.7. ***(a)*** *For every data-generating process* P, *if an expert predicts according to* P, *the expert is guaranteed to pass the cross-calibration test with probability* 1, *no matter what strategies the other experts will use.*

(b) *In the presence of an informed expert, for every random generator of theories* ξ *of an ignorant expert, the subset of data-generating processes* P *under which the ignorant expert will pass the cross-calibration test with positive probability is topologically small in the space of data-generating processes* P.[13]

However, this test, like the test of Al-Najjar and Weinstein, may be unable to reveal the type of experts when both of them are ignorant. Feinberg and Stewart modify the cross-calibration test to obtain another test, which they call *strict cross-calibration*, and show that:

[12] The probability is measured here according to the product measure $P \times \xi^j$ on the space $\Omega \times \Delta(\Omega)$.

[13] The probability is measured here according to the product measure $P \times \xi$ on the space $\Omega \times \Delta(\Omega)$.

Proposition 18.7. *(c) For any random generator of theories* $(\xi^1, \ldots, \xi^m)$*, which are independent random variables, the set of realizations* ω *on which at least two ignorant experts simultaneously pass the strict cross-calibration test with positive probability is a topologically small set in* Ω.

Unlike cross-calibration, however, the strict cross-calibration, rejects the informed expert on some (albeit "small") set of data-generating processes P. Therefore, one may ask whether the reversion of the negative result is caused by allowing for simultaneous testing of multiple experts, or rather by allowing for some possibility of rejecting informed experts. Olszewski and Sandroni (2009a) suggest that this is due to the latter rather than the former reason.[14]

Before their result will be presented, it is important to comment on the condition that random generators of theories are independent random variables. In the interpretation, this assumption means that experts cannot collude. If we allowed for correlated generators, experts could provide identical forecasts, in which case multiple-expert tests could typically be ignorantly passed by virtue of the results for single experts.

I will now describe how Olszewski and Sandroni generalize Proposition 18.3; I will present their result for two experts, but the generalization to any number of experts is straightforward. They consider only tests that reject theories in finite time. That is, they define *comparative tests* which reject theories in finite time as functions that take as input pairs of theories and yield as output two collections of finite sequences of outcomes (one collection for each expert), and which fail the expert's theory if a sequence from her collection is observed.

A test does not reject the truth if the actual data-generating process is, with high probability, not rejected, no matter what theory is provided by the other expert. A test can be ignorantly passed if both experts can randomly select theories, independent of one another, such that the theory selected by each expert will be rejected only with small probability (no higher than $\varepsilon + \delta$), according to the experts' randomization devices, no matter how the data will unfold. Finally, a test is future-independent if the possibility that any expert's theory is rejected in period m depends only on the data observed up to period m and the predictions made by the theories of both experts up to period m.

Proposition 18.8. *Every comparative future-independent test which rejects theories in finite time, and which does not reject the truth with probability* $1 - \varepsilon$*, can be ignorantly passed with probability* $1 - \varepsilon - \delta$.

The proof of this result combines the proof of Proposition 18.3 with a fixed-point argument.

[14] Olszewski and Sandroni study only tests that reject theories in finite time. Therefore, their results have no direct implications regarding calibration tests, which give the verdict at infinity.

18.9. BAYESIAN AND DECISION-THEORETIC APPROACHES TO TESTING EXPERTS

18.9.1 Bayesian approach

Unlike most of this survey, most of economics assumes that even purely informed agents have some correct prior over future events. One may wonder whether such a prior will help a tester to separate informed from ignorant experts. Of course, a tester equipped with a prior can make forecasts herself without the help of any expert. So, the tester will find the expert's forecasts valuable only when these forecasts are more precise than her own. And only in this case would testing experts seem to make any sense. However, it seems intuitive that the likelihood-ratio test in which the tester compares her own forecasts with the expert's forecasts should reveal that the informed expert knows more than the tester. And the ignorant expert, who knows no more than the tester, should be exposed to some risk of having low likelihood ratios of her forecasts to the expert's forecasts.

Stewart (2011) formalizes this idea as follows. Suppose the informed expert knows the probability distribution P, and the tester knows only a probability distribution μ over probability distributions. Let $\overline{P}$ be the probability distribution over outcomes induced by μ. That is,

$$\overline{P}(A) = \int_{\Delta(\Omega)} P(A) \mathrm{d}\mu$$

for every Borel set $A \subset \Omega$. Given a probability distribution P, let $p(\omega^t)[\omega_t]$ be the probability of outcome ω_t conditional on history ω^t; for our purposes, it will be irrelevant how $f(\omega^t)[\omega_t]$ is defined when the probability of ω^t is zero. The conditional probabilities $\overline{p}(\omega^t)[\omega_t]$ are defined analogously. Let

$$\varepsilon = \int_{\Delta(\Omega)} P\left(\left\{\omega \in \Omega : \sum_{t=1}^{\infty} (p(\omega^t)[\omega_t] - \overline{p}(\omega^t)[\omega_t])^2 \text{ converges}\right\}\right) \mathrm{d}\mu.$$

Notice that ε is a number, and is a function of μ. This number may not be small; for example, $\varepsilon = 1$ if distribution μ is degenerated to a distribution P. However, ε is small, or even equal to zero, for many distributions μ (for example for the uniform distribution over i.i.d. processes).[15]

Let $T(P, \omega) =$ PASS if $\overline{p}(\omega^t)[\omega_t] = 0$ for some t. In addition, if $\overline{p}(\omega^t)[\omega_t] > 0$ for all t, then $T(P, \omega) =$ PASS if

$$\liminf_t \frac{p(\omega^1)[\omega_1]}{\overline{p}(\omega^1)[\omega_1]} \cdot \ldots \cdot \frac{p(\omega^t)[\omega_t]}{\overline{p}(\omega^t)[\omega_t]} > 1$$

[15] This fact is nontrivial, see Stewart (2011) for details.

and

$$\sum_{t=1}^{\infty}(p(\omega^t)[\omega_t] - \overline{p}(\omega^t)[\omega_t])^2 \text{ diverges.}$$

In all other cases $T(P, \omega) = \text{FAIL}$.

Stewart shows the following proposition:

Proposition 18.9. *(i) The test passes the truth with probability* $1 - \varepsilon$*; and (ii) for any random generator of theories* ξ*, the ignorant expert who selects a theory according to* ξ *passes the test with probability* 0*.*

The ignorant expert's probability is evaluated according to the product measure $\mu \times \xi$ on the set of all (P, Q), where P is the data-generating process and Q is the expert's theory.

18.9.2 Decision-theoretic approach

Echenique and Shmaya (2008), Olszewski and Pęski (2011), and Gradwohl and Salant (2011) make the point that in decision theory, information is only a tool for making better decisions. Of course, there is no conflict between this view and the literature on testing experts. Even if the decision problem is not explicitly modeled, one may argue that when we know the expert's type, we are able to make better decisions.

However, the impossibility of separating informed and ignorant experts may depend critically on whether the expert's forecasts play the role of advice for a specific decision problem, or whether one simply wishes to learn the expert's type. Indeed, there are two conceptual differences: (1) a tester (a decision maker) must take some default action even in the absence of any expert, and may not appreciate forecasts that suggest the same (or similar) actions; and (2) if forecasts lead to better decisions, the decision maker may appreciate them, no matter what type of expert provides them. Thus, in an analysis of forecasting in the context of a specific decision problem, it seems legitimate to relax both the requirement that a "good" test should always pass informed experts, and the requirement that it should fail ignorant ones.

The general message of these three papers is that in the decision-theoretic setting, the tester is indeed able to benefit almost fully from the possibility of obtaining the expert's advice if the expert is informed, without losing much if the expert is ignorant. The details of the model, various assumptions, and the statements of results vary across the papers. In addition, the generally positive results coexist with some negative ones (see Olszewski and Pęski, 2011). We will not discuss these papers one by one in the present survey. Nevertheless, in order to give some flavor of this kind of analysis, we will describe one result from Gradwohl and Salant (2011), which may show in the most convincing way the general message of these papers and how they contrast with the literature on testing experts.

Gradwohl and Salant study a model in which the expert observes the realizations of some stochastic process. These realization provide signals about the realizations of the data-generating process. In every period, the decision maker decides whether to bet on the outcome of the data-generating process or stay out. The decision maker has no knowledge of the data-generating process, the expert's process, or any relation between the two.

In every period, the expert provides a prediction that specifies, according to the expert's signal, the maximal expected value of betting in that period and the bet that achieves that expected value. If the decision maker decides not to follow the expert's advice, the period ends and both the decision maker and the expert get the payoff of staying out. Otherwise, the decision maker pays the expert a fixed, exogenous share of the maximal expected value of betting in that period, observes the realization of the data-generating process, and obtains the payoff from betting.

Gradwohl and Salant show that if the expert's process satisfies some condition (e.g., when that process coincides with the data-generating process), then the decision maker has a strategy that approximates the first best, that is, the payoff that the decision maker would obtain if she knew the expert's process herself. The strategy is common for all data-generating processes, and all processes of the expert that satisfy the required condition. The approximation is in terms of the average per-period expected payoff, and for any given level of approximation, the interaction is assumed to last over a sufficiently long time horizon.

In addition, the decision maker can achieve this goal with a fixed and bounded amount of money, which means that for a sufficiently long time horizon, the amount per period is sufficiently small. Therefore even if the expert is "ignorant," the decision maker will not lose much. Gradwohl and Salant show that this result holds for truthful experts, and that the any strategic behavior of the expert (whose process satisfies the required condition) that improves her own payoff over truthfulness can only increase the payoff of the decision maker.[16]

18.10. RELATED TOPICS

18.10.1 Falsifiability and philosophy of science

The literature on testing experts provides a number of insights, and stimulates a discussion on probabilistic modeling, or more generally, the philosophy of science. Olszewski and

[16] It is worth pointing out that in Gradwohl and Salant (2011), which is typical for the entire literature, the optimal strategies, or the best responses to the strategies of other agents, may not exist without imposing any conditions on the stochastic processes.

Sandroni (2011) take the relation to the philosophy of science more literally, and come up with two conclusions. First, they argue that celebrated falsifiability of Karl Popper has no power to distinguish scientific theories from worthless theories when theories can be produced by strategic experts. Second, they find formal support for the maxim that theories should never be fully accepted, and that they should be rejected when proven inconsistent with the data; this maxim clearly contrasts with an approach that accepts a theory once proven to fit the known data.

More specifically, they define a theory $P \in \Delta(\Omega)$ to be *falsifiable* if, for every history $\omega^k = (\omega_1, \ldots, \omega_{k-1})$, there is an extension $\omega^n = (\omega_1, \ldots, \omega_{n-1})$ of ω^k such that

$$P(C(\omega^n)) = 0.$$

Thus, a theory is falsifiable if, after any finite sequence of observed outcomes, there is a finite sequence of outcomes that the theory finds impossible to be observed. This *falsifiability test* rejects nonfalsifiable theories out of hand (i.e., on all data sets), while any falsifiable theory is rejected only at all finite sequences of outcomes that the theory finds impossible. Olszewski and Sandroni show that for every $\varepsilon > 0$, the falsifiability test can be ignorantly passed with probability $1 - \varepsilon$.

Olszewski and Sandroni call tests which rejects theories in finite time, *rejection tests.* Recall that such a test specifies for every theory a collection of finite sequences of outcomes, which sequences (according to the test) contradict the theory; and the test fails the theory if one of these sequences is observed. Similarly, *acceptance tests* specify for every theory a collection of finite sequences of outcomes, which sequences (according to the test) confirm the theory; and the tests pass the theory if one of these sequences is observed.

They show that every acceptance test that, with probability $1 - \varepsilon$, does not reject the truth can be ignorantly passed with probability $1 - \varepsilon - \delta$, while there exists a rejection test that, with probability $1 - \varepsilon$, does not reject the truth but cannot be ignorantly passed with probability higher than ε. Moreover, for any good test T^1, there exists a good rejection test T^2 that is harder than T^1, which means that

$$\{\omega \in \Omega : T^2(P, \omega) = \text{PASS}\} \subset \{\omega \in \Omega : T^1(P, \omega) = \text{PASS}\}, \forall_{P \in \Delta(\Omega)}.$$

18.10.2 Gaming performance fees by portfolio managers

One example of experts providing probabilistic forecasts is financial analysts. It is tempting to apply the results that ignorant experts cannot be separated from informed experts to managers of financial institutions, especially since gaming returns by portfolio managers seems to be quite common in practice. For example, it is well-known that treating gains and losses asymmetrically creates incentives for managers for taking excessive risk. Lo (2001) examines a hypothetical situation in which a manager takes short positions in S&P 500 put options that mature in one to three months, and showed that such an approach would have generated very sizable excess returns relative to the market in the 1990s.

However, the difficulty in applying our negative results to financial markets is that managers typically use their forecasts to make investment decisions. Therefore, the analysis of this application seems to be closer to that of Section 18.9.2 rather than to that of Section 18.5. Yet, Foster and Young (2010) obtain negative results similar in spirit to those from Section 18.5.

More specifically, suppose that a benchmark portfolio, such as the S&P 500, generates a sequence of stochastic returns x_t in each of T periods $t = 1, \ldots, T$, and let r_t denote a risk-free rate in period t. A fund has initial value $s_0 > 0$, which, if passively invested in the benchmark portfolio, would generate the return

$$s_0 \prod_{t=1}^{T} x_t$$

by the end of the period T.

Suppose that a skilled manager can generate a sequence of "higher" returns $m_t x_t$. A compensation contract over T periods is a sequence of functions φ_t, $t = 1, \ldots, T$, such that φ_t is a function of the realizations of m_s and x_s, $s \leq t$, that is, a function of the return of the benchmark portfolio and the excess return generated by the manager. The functions represent payments to a manager, which are made at the end of each period. The contract can also specify a payment in period 0, and the payments can be negative.

Foster and Young show that there is no compensation contract that separates skilled from ignorant managers. More precisely, for any compensation contract which attracts risk-neutral skilled managers, there is a trading strategy for risk-neutral ignorant managers, which yields them a higher expected payoff than the benchmark portfolio.

The argument can be explained as follows: suppose that $s_0 = 1$, $T = 1$ and the benchmark portfolio yields the risk-free rate r. Since the benchmark return is deterministic, the compensation scheme φ is a function of m. Suppose that the skilled manager is able to generate returns $m^* > 1 + r$. Consider the following strategy of the ignorant manager:

The ignorant manager invests s_0 entirely in the benchmark risk-free asset at the beginning of the period. Just before the end of the period, his capital will be $(1 + r)s_0$. He uses this capital as collateral to buy a lottery in the options market that is realized almost immediately, at the end of the period. The lottery is constructed so as to pay $m^*(1 + r)s_0$ with probability $1/m^*$ and to pay 0 with probability $1 - 1/m^*$.

Since the strategy of an ignorant manager generates only one of the two returns m^* or 0, we need to consider only $\varphi(m^*)$ or $\varphi(0)$, that is, the compensation of the skilled manager and the compensation in the case of bankruptcy. Consider the case in which

$\varphi(0) < 0$, that is, the manager is financially penalized in the case of bankruptcy. The case of $\varphi(0) > 0$ is simpler. To deter the ignorant manager, the expected fees earned during the period cannot be positive:

$$\left(\frac{1}{m^*}\right)\varphi(m^*) + \left(1 - \frac{1}{m^*}\right)\varphi(0) \leq 0. \qquad [18.7]$$

In order to make the penalty possible, the amount $(1+r)^{-1}|\varphi(0)|$ must be held in escrow in a safe asset which earns the risk-free rate, and is paid out to the investors if the fund goes bankrupt.

Now consider the skilled manager who can generate the return m^* with certainty. This manager must also put the amount $(1+r)^{-1}|\varphi(0)|$ in escrow, because ex ante all managers are treated alike and the investors cannot distinguish between them. However, this involves an opportunity cost for the skilled manager, because by investing $(1+r)^{-1}|\varphi(0)|$ in her own private fund, she could have generated the return $m^*(1+r)(1+r)^{-1}|\varphi(0)| = m^*|\varphi(0)|$. The resulting opportunity cost for the skilled manager is

$$m^*|\varphi(0)| - (1+r)(1+r)^{-1}|\varphi(0)| = (m^* - 1)|\varphi(0)|.$$

Therefore, she will not participate if the opportunity cost exceeds the fee, that is, if

$$(m^* - 1)|\varphi(0)| \geq \varphi(m^*). \qquad [18.8]$$

However, inequality [18.8] is equivalent to inequality [18.7].

ACKNOWLEDGMENT

I am grateful to Nabil Al-Najjar, Dean Foster, Sergiu Hart, Ehud Kalai, Shie Mannor, Alvaro Sandroni, Rann Smorodinsky, Colin Stewart, and Peyton Young for comments on an earlier draft of this survey. I thank the National Science Foundation for research support (CAREER award SES-0644930).

REFERENCES

Al-Najjar, N.I., Weinstein, J.L., 2008. Comparative testing of experts. Econometrica 76 (3), 541–559.

Al-Najjar, N.I., Smorodinsky, R., Sandroni, A., Weinstein, J.L., 2010. Testing theories with learnable and predictive representations. J. Econ. Theory 145, 2203–2217.

Blackwell, D., 1956. An analog of the minimax theorem for vector payoffs. Pacific J. Math., 6, 1–8.

Dawid, A.P., 1982. The well-calibrated bayesian. J. Am. Stat. Assoc. 77, 605–613.

Dekel, E., Feinberg, Y., 2006. Non-bayesian testing of a stochastic prediction. Rev. Econ. Stud. 73 (4), 893–906.

Echenique, F., Shmaya, E., 2008. You won't harm me if you fool me. Mimeo.

Fan, K., 1953. Minimax Theorems. Proc. Nat. Acad. Sci. USA, 39, 42–47.

Feinberg, Y., Stewart, C., 2008. Testing multiple forecasters. Econometrica, 76 (3), 541–582.

Fortnow, L.J., Vohra, R.V., 2009. The complexity of forecast testing. Econometrica, 77(1), 93–105.

Foster, D.P., 1999. A proof of calibration via Blackwell's approachability theorem. Games Econ. Behav. 29, 73–78.

Foster D.P., Vohra, R., 1997. Calibrated learning and correlated equilibrium. Games Econ. Behav. 21, 40–55.
Foster, D.P., Vohra, R.V., 1998. Asymptotic calibration. Biometrika 85 (2), 379–390.
Foster, D.P., Young, H.P., 2010. Gaming performance fees by portfolio managers. Q. J. Econ. 125 (4), 1435–1458.
Fudenberg, D., Levine, D.K., 1999. An easier way to calibrate. Games Econ. Behav. 29, 131–137.
Gilboa, I., Schmeidler, D., 1989. Maxmin expected utility with a non-unique prior. J. Math. Econ. 18 (2), 141–153.
Gradwohl, R., Salant, Y., 2011. How to buy advice. Mimeo.
Hart, S. Mas-Colell, A., 2000. A simple adaptive procedure leading to correlated equilibrium. Econometrica 68, 1127–1150.
Kalai, E., Lehrer, E., Smorodinsky, R., 1999. Calibrated forecasting and merging. Games Econ. Behav. 29, 151–169.
Lehrer, E., 2001. Any inspection rule is manipulable. Econometrica 69 (5), 1333–1347.
Lo, A.W., 2001. Risk management for hedge funds: introduction and overview. Financ. Analyst. J. 56 (6), 16–33.
Mannor, S.G., 2010. A geometric proof of calibration. Math. Oper. Res. 35 (4), 721–727.
Murphy, A.H., Winkler, R.L., 1977. Reliability of subjective probability forecasts of precipitation and temperature. J. R. Stat. Soc. Ser. C. Appl. Stat. 26, 41–47.
Murphy, A.H., Epstein, E.S., 1967. Verification of probabilistic predictions: a brief review. J. Appl. Meteorol. 6, 748–755.
Olszewski, W., Sandroni, A., 2008. Manipulability of Future-Independent Tests. Econometrica 76 (6), 1437–1466.
Olszewski, W., Sandroni, A., 2009a. Manipulability of comparative tests. Proc. Nat. Acad. Sci. USA. 106 (13), 5029–5034.
Olszewski, W., Sandroni, A., 2009b. Strategic manipulation of empirical tests. Math. Oper. Res. 34 (1) 57–70.
Olszewski, W., Sandroni, A., 2009c. A nonmanipulable test. Ann. Stat. 37 (2), 1013–1039.
Olszewski, W., Sandroni, A., 2011. Falsifiability. Am. Econ. Rev. 101, 788–818.
Olszewski, W., Pęski, M., 2011. The principal-agent approach to testing experts. Am. Econ. J. Microecon. 3(2), 89–113.
Oxtoby, J.C., 1980. Measure and Category. Graduate Texts in Mathematics. Springer Verlag, New York, Heidelberg, Berlin.
Raiffa, H., 1961. Risk, ambiguity, and the savage axioms: comment. Q. J. Econ. 75 (4) 690–694.
Sandroni, A., 2003. The reproducible properties of correct forecasts. Int. J. Game Theory 32 (1), 151–159.
Sandroni, A., Smorodinsky, R., Vohra, R.V., 1999. Calibration with many checking rules. Math. Oper. Res. 28 v(1), 141–153.
Shmaya, E., 2008. Many inspections are manipulable. Theor. Econ. 3 (3), 367–382.
Shmaya, E., Hu, T.-W., 2013. Expressible inspections. Theor. Econ. 8 (2), 263–280.
Stewart, C., 2011. Nonmanipulable Bayesian testing. J. Econ. Theory 146, 2029–2041.
Vovk, V., 2007. Predictions as statements and decisions. Mimeo.
Vovk, V., Shafer, G., 2005. Good randomized sequential probability forecasting is always possible. J. R. Stat. Soc. Ser. B, 67 (5), 747–763.

INDEX

Note: Page numbers followed by *f* indicate figures and *t* indicate tables.

C

E

N

Q

R

T

U

V

Printed and bound by CPI Group (UK) Ltd, Croydon, CR0 4YY
08/05/2025
01864967-0002